D1060987

PREFACE

The objective of the third edition is the same as the first two editions, that is, to provide the student or practitioner a meaningful introduction to the design of medium-and short-span girder bridges. However, the manner in which the material is presented has changed. Instead of the eight chapters of the second edition, the content has been spread out over twenty shorter chapters. This organization should lead to easier reading and simpler organization of classroom assignments.

To help understand how these changes have come about, it is informative to see how the process all started. It was in August 1990 that the two authors were at an International Conference on Short and Medium Span Bridges in Toronto, Canada, where both were presenting papers. They had often met at these bridge conferences and were familiar with each other's work—Puckett's on analysis and software development and Barker's fundamental application of LRFD to geotechnical materials. Both were classroom teachers in structural engineering.

At the time, a number of major changes were taking place in the design of highway bridges. Philosophically the most dramatic was the change from a deterministic (allowable stress) design approach to a probabilistic (limit state) design concepts. The other big change was a government edict that highway bridges that were built with federal dollars had to be constructed and designed in the metric system starting in 1997.

The timing was right for a comprehensive textbook on the design of highway bridges. The American Association of State Highway and Transportation Officials (AASHTO) were in the midst of a complete rewriting of their Bridge Design Specifications in a LRFD format. Finite-element analysis tools had matured, truck loads were better understood through weigh-in-motion studies, material behavior was being unified for prestressed and non-prestressed concrete by the American Concrete Institute (ACI), post-buckling strength of plate girder webs and fatigue strength of weld details were better understood.

The two professors decided that someone needed to write a textbook to present these changes to students and practicing civil engineers. So over dinner and a major league baseball game, they realized they could be the ones to do the writing. Puckett took his sabbatical with Barker at Virginia Tech in 1993, they wrote trial chapters, prepared a proposal that was accepted by John Wiley & Sons, and the first edition with ten chapters was published in 1997.

It was not long before the metric system requirement was dropped and the highway bridge designers needed a textbook written in U.S. Customary Units. Therefore, it became necessary to make revisions and to prepare a second edition of the book. Besides the units change, the LRFD specifications were in their third edition and the textbook needed to be updated. As new material was added, the number of pages was deemed too large and two chapters were dropped—Wood Bridges and Substructure Design. These two topics are found only in the metric system units of the first edition.

The remaining eight chapters of the second edition have been divided into four parts: General Aspects of Bridge Design (Chapters 1–7), Loads and Analysis (Chapters 8–12), Concrete Bridges (Chapters 13–16), and Steel Bridges (Chapters 17–20). Another change in the layout of the third edition is the addition of an insert of mainly color bridge photos. These photos have been selected to illustrate bridges of historical significance; the ones most aesthetically pleasing that are most beautiful in their surroundings, and noteworthy as the longest, tallest, or highest bridges of their type.

We suggest that a first course in bridges be based on Chapters 1–7 with Chapters 5, 6, and 7 compulsory reading. Loads and analysis should follow with required reading in Chapter 8 and selected portions of Chapter 9 and 10 depending upon the students' background and instructor's interest.

Design can be addressed with either the chapters on concrete (Chapters 13–16) or those on steel (Chapters 17–20). Instructor guidance is required to lead the student through these chapters and to address the topics of most interest. For example, concrete bridges could be addressed with nonprestressed bridges which would simplify the topic. However, teaching prestressed concrete within a bridge context could be an excellent way for students to gain broad-based knowledge in this area for both bridges and buildings. Similarly, teaching design using the steel chapter leads to a general knowledge of composite cross sections, staged construction, and plate girders. As the associated principles are common with buildings and bridges, again the bridge course can be used within a broader context.

How much of the material to present to a particular class is at the discretion of the professor, who is the best person to judge the background and maturity of the students. There is enough material in the book for more than one course in highway bridge design.

Practitioners who are entry level engineers will find the background material in Chapters 1–12 helpful in their new assignments and can use Chapters 13–16 and 17–20 for specific guidance on design of a particular bridge type. The same can be said for seasoned professionals, even though they would be familiar with the material in the loads chapter, they should find the other chapters of interest in providing background and design examples based on the AASHTO LRFD specifications.

Finally, those practitioners who just appreciate bridge history and aesthetics might find those chapters of interest from a personal enjoyment perspective. Bridges are art and so many are simply beautiful.

ACKNOWLEDGMENTS

We would like to recognize those who have made the production of the third edition possible. The first person to be acknowledged is the editorial assistant at John Wiley & Sons who prepared a twenty chapter manuscript from the contents of the eight chapters of the second edition. This reorganized manuscript became the working document that the authors could edit and assign correct numbers to equations, figures, and tables.

To accompany the description of the I-35W Bridge collapse, the new figures drafted by Philip Jennings, a structural engineering graduate student at Virginia Tech, are gratefully acknowledged. Thanks also to the following state departments of transportation who supplied photographs of their bridges: Arizona, Colorado, Washington State, and West Virginia. The authors appreciate the computer modeling and project photos provided by Julie Smith of the FIGG Engineering Group.

The patience, understanding, and support shown us by Jim Harper, Bob Argentieri, Dan Magers, and Bob Hilbert at John Wiley & Sons, especially during the time of the senior author's health issues, are greatly appreciated.

Finally, we wish to thank Marilyn Barker and Kathy Puckett for their continued patience and strong support during our time of writing.

The authors would appreciate it that if the reader should have questions or if errors are found you would contact us at puckett@uwyo.edu.

PERSONAL ACKNOWLEDGMENT TO RICHARD BARKER

I wish to recognize and thank Rich for his career of achievement in teaching, learning, research, and practice in bridge engineering, and most of all sharing it with me. Rich has made a tremendous difference to the professional lives of so many students and colleagues. I will be forever grateful for his friendship, guidance, selfless and thoughtful approach from which I have benefitted and learned so very much.

Rich was a professional in every sense of the term.

Happy trails, Rich.

Jay Puckett
Laramie, Wyoming

PREFACE TO THE SECOND EDITION

This book has the same intent as the first edition and is written for senior-level undergraduate or first-year graduate students in civil engineering. It is also written for practicing civil engineers who have an interest in the design of highway bridges. The objective is to provide the reader a meaningful introduction to the design of medium- and short-span girder bridges. This objective is achieved by providing fundamental theory and behavior, background on the development of the specifications, procedures for design, and design examples.

This book is based on the American Association of State Highway and Transportation Officials (AASHTO) LRFD Bridge Design Specifications, Third Edition, and Customary U.S. units are used throughout. The general approach is to present theory and behavior upon which a provision of the specifications is based, followed by appropriate procedures, either presented explicitly or in examples. The examples focus on the procedures involved for a particular structural material and give reference to the appropriate article in the specifications. It is, therefore, suggested that the reader have available a copy of the most recent edition of the AASHTO LRFD Bridge Design Specifications.

The scope is limited to a thorough treatment of medium- and short-span girder bridges with a maximum span length of about 250 ft. These bridge structures comprise approximately 80% of the U.S. bridge inventory and are the most common bridges designed by practitioners. Their design illustrates the basic principles used for the design of longer spans. Structure types included in this book are built of concrete and steel. Concrete cast-in-place slab, T-beam, and box-girder bridges and precast–prestressed systems are considered. Rolled steel beam and plate girder systems that are composite and noncomposite are included.

Civil engineers are identified as primary users of this book because their formal education includes topics important to a highway bridge designer. These topics include studies in transportation systems, hydrodynamics of streams and channels, geotechnical engineering, construction management, environmental engineering, structural analysis and design, life-cycle costing, material testing, quality control, professional and legal problems, and the people issues associated with public construction projects. This reference to civil engineers is not meant to exclude others from utilizing this book. However, the reader is expected to have one undergraduate course in structural design for each structural material considered. For example, if only the design of steel bridges is of interest, then the reader should have at least one course in structural analysis and one course in structural steel design.

Chapter 1 introduces the topic of bridge engineering with a brief history of bridge building and the development of bridge specifications in the United States. Added to the second edition is an expanded treatment of bridge failure case histories that brought about changes in the bridge design specifications. Chapter 2 emphasizes the need to consider aesthetics from the beginning of the design process and gives examples of successful bridge projects. Added to the second edition are a discussion of integral abutment bridges and a section on the use of computer modeling in planning and design. Chapter 3 presents the basics on load and resistance factor design (LRFD) and indicates how these factors are chosen to obtain a desirable margin of safety. Included at the end of all the chapters in the second edition are problems that can be used as student exercises or homework assignments.

Chapter 4 describes the nature, magnitude, and placement of the various loads that act on a bridge structure. Chapter 5 presents influence function techniques for determining maximum and minimum force effects due to moving vehicle loads. Chapter 6 considers the entire bridge structure as a system and how it should be analyzed to obtain a realistic distribution of forces.

Chapters 7 and 8 are the design chapters for concrete and steel bridges. Both chapters have been significantly revised to accommodate the trend toward U.S. customary units within the United States and away from SI. New to the second edition of the concrete bridge design chapter are discussions of high-performance concrete and control of flexural cracking, changes to the calculation of creep and shrinkage and its influence on prestress losses, and prediction of stress in unbonded tendons at ultimate.

Chapter 8 includes a major reorganization and rewrite of content based upon the new specifications whereby Articles 6.10 and 6.11 were completely rewritten by AASHTO. This specification rewrite is a significant simplification in the specifications from the previous editions/interims; however, the use of these articles is not simple, and hopefully Chapter 8 provides helpful guidance.

The organization of the design chapters is similar. A description of material properties is given first, followed by general design considerations. Then a discussion is given of the behavior and theory behind the member resistance expressions for the various limit states. Detailed design examples that illustrate the LRFD specification provisions conclude each chapter.

We suggest that a first course in bridges be based on Chapters 1–6, either Sections 7.1–7.6, 7.10.1, and 7.10.3 of Chapter 7 or Sections 8.1–8.4, 8.6–8.10, and 8.11.2. It is assumed that some of this material will have been addressed in prerequisite courses and can be referred to only as a reading assignment. How much of the material to present to a particular class is at the discretion of the professor, who is probably the best person to judge the background and maturity of the students. There is enough material in the book for more than one course in highway bridge design.

Practitioners who are entry-level engineers will find the background material in Chapters 1–6 helpful in their new assignments and can use Chapters 7 and 8 for specific guidance on design of a particular bridge type. The same can be said for seasoned professionals, even though they would be familiar with the material in the loads chapter, they should find the other chapters of interest in providing background and design examples based on the AASHTO LRFD specifications.

ACKNOWLEDGMENTS

In addition to the acknowledgements of those who contributed to the writing of the first edition, we would like to recognize those who have helped make this second edition possible. Since the publication of the first edition in 1997, we have received numerous emails and personal communications from students and practitioners asking questions, pointing out mistakes, making suggestions, and encouraging us to revise the book. We thank this group for their feedback and for making it clear that a revision of the book in Customary U.S. units was necessary.

We wish to acknowledge those who have contributed directly to the production of the book. The most important person in this regard was Kerri Puckett, civil engineering student at the University of Wyoming, who changed the units on all figures to Customary U.S., drafted new figures, catalogued the figures and photos, performed clerical duties, and generally kept the authors on track. Also assisting in the conversion of units was H. R. (Trey) Hamilton from the University of Florida who reworked design examples from the first edition in Customary U.S. units.

We also appreciate the contributions of friends in the bridge engineering community. Colleagues at Virginia Tech providing background material were Carin Roberts-Wollmann on unbonded tendons and Tommy Cousins on prestress losses. Thanks to John Kulicki of Modjeski & Masters for his continuing leadership in the development of the LRFD Specifications and Dennis Mertz of the University of Delaware for responding to questions on the rationale of the specifications. The authors appreciate the computer modeling and project photos provided by Linda Figg, Cheryl Maze, and Amy Kohls Buehler of Figg Engineers.

The patience and understanding shown us by Jim Harper and Bob Hilbert at John Wiley & Sons is gratefully acknowledged.

Finally we wish to thank Marilyn Barker and Kathy Puckett for their patience and strong support during our time writing.

The authors would appreciate it if the reader should have questions or if errors are found that they be contacted at marichba@aol.com or puckett@uwyo.edu.

PREFACE TO THE FIRST EDITION

This book is written for senior level undergraduate or first year graduate students in civil engineering and for practicing civil engineers who have an interest in the design of highway bridges. The object of this book is to provide the student or practitioner a meaningful introduction to the design of medium- and short-span girder bridges. This objective is achieved by providing fundamental theory and behavior, background on the development of the specifications, procedures for design, and design examples.

This book is based on the American Association of State Highway and Transportation Officials (AASHTO) LRFD Bridge Design Specifications and System International (SI) units are used throughout. The general approach is to present theory and behavior upon which a provision of the specifications is based, followed by appropriate procedures, either presented explicitly or in examples. The examples focus on the procedures involved for a particular structural material and give reference to the appropriate article in the specifications. It is, therefore, essential that the reader have available a copy of the most recent edition of the AASHTO LRFD Bridge Design Specifications in SI units. (For those who have access to the World Wide Web, addendums to the specifications can be found at http://www2.epix.net/~modjeski.)

The scope of this book is limited to a thorough treatment of medium- and short-span girder bridges with a maximum span length of about 60 m. These bridge structures comprise approximately 80% of the U.S. bridge inventory and are the most common bridges designed by practitioners, illustrating the basic principles found in bridges of longer spans. Structure types included in this book are built of concrete, steel, and wood. Concrete cast-in-place slab, *T*-beam, and box-girder bridges and precast–prestressed systems are considered. Rolled steel beam and plate girder systems that are composite and non-composite are included, as well as wood systems. This book concludes with a chapter on substructure design, which is a common component for all the bridge types.

Civil engineers are identified as primary users of this book because their formal education includes topics important to a highway bridge designer. These topics include studies in transportation systems, hydrodynamics of streams and channels, geotechnical engineering, construction management, environmental engineering, structural analysis and design, life-cycle costing, material testing, quality control, professional and legal problems, and the people issues associated with public construction projects. This reference to civil engineers is not meant to exclude others from utilizing this book. However, the reader is expected to have one undergraduate course in structural design for each structural material considered. For example, if only the design of steel bridges is of interest, then the reader should have at least one course in structural analysis and one course in structural steel design.

Chapter 1 introduces the topic of bridge engineering with a brief history of bridge building and the development of bridge specifications in the United States. Chapter 2 emphasizes the need to consider aesthetics from the beginning of the design process and gives examples of successful bridge projects. Chapter 3 presents the basics on load and resistance factor design (LRFD) and indicates how these factors are chosen to obtain a desirable margin of safety.

Chapter 4 describes the nature, magnitude, and placement of the various loads that act on a bridge structure. Chapter 5 presents influence function techniques for determining maximum and minimum force effects due to moving vehicle loads. Chapter 6 considers the entire bridge structure as a system and how it should be analyzed to obtain a realistic distribution of forces.

Chapters 7–9 are the design chapters for concrete, steel, and wood bridges. The organization of these three chapters is similar. A description of material properties is given first, followed by general design considerations. Then a discussion of the behavior and theory behind the member resistance expressions for the various limit states, and concluding with detailed design examples that illustrate the LRFD specification provisions.

Chapter 10 on substructure design completes the book. It includes general design considerations, an elastomeric bearing design example, and a stability analysis to check the geotechnical limit states for a typical abutment.

We suggest that a first course in bridges be based on Chapters 1–6, either Articles 7.1–7.6, 7.10.1, and 7.10.3 of Chapter 7 or Articles 8.1–8.4, 8.6–8.10, and 8.11.2, and conclude with Articles 10.1–10.3 of Chapter 10. It is assumed that some of this material will have been covered in prerequisite courses and can be referred to only as a reading assignment. How much of the material to present to a particular class is at the discretion of the professor, who is probably the best person to judge the background and maturity of the students. There is enough material in the book for more than one course in highway bridge design.

Practitioners who are entry level engineers will find the background material in Chapters 1–6 helpful in their new assignments and can use Chapters 7–10 for specific guidance on design of a particular bridge type. The same can be said for seasoned professionals, even though they would be familiar with the material in the loads chapter, they should find the other chapters of interest in providing background and design examples based on the AASHTO LRFD specifications.

ACKNOWLEDGMENTS

Acknowledgments to others who have contributed to the writing of this book is not an easy task because so many people have participated in the development of our engineering careers. To list them all is not possible, but we do recognize the contribution of our university professors at the University of Minnesota and Colorado State University; our engineering colleagues at Toltz, King, Duvall, Anderson & Associates, Moffatt & Nichol Engineers, and BridgeTech, Inc.; our faculty colleagues at Virginia Tech and the University of Wyoming; the government and industry sponsors of our research work; and the countless number of students who keep asking those interesting questions.

The contribution of John S. Kim, author of Chapter 10 on Substructure Design, is especially appreciated. We realize that many of the ideas and concepts presented in the book have come from reading the work of others. In each of the major design chapters, the influence of the following people is acknowledged: Concrete Bridges, Michael Collins, University of Toronto, Thomas T.C. Hsu, University of Houston, and Antoine Naaman, University of Michigan; Steel Bridges, Sam Easterling and Tom Murray, Virginia Tech, and Konrad Basler, Zurich, Switzerland; and Wood Bridges, Michael Ritter, USDA Forest Service.

We also wish to acknowledge those who have contributed directly to the production of the book. These include Elizabeth Barker who typed a majority of the manuscript, Jude Kostage who drafted most of the figures, and Brian Goodrich who made significant modifications for the conversion of many figures to SI units. Others who prepared figures, worked on example problems, handled correspondence, and checked page proofs were: Barbara Barker, Catherine Barker, Benita Calloway, Ann Crate, Scott Easter, Martin Kigudde, Amy Kohls, Kathryn Kontrim, Michelle Rambo-Roddenberry, and Cheryl Rottmann. Thanks also to the following state departments of transportation who supplied photographs of their bridges and offered encouragement: California, Minnesota, Pennsylvania, Tennessee, Washington, and West Virginia.

The patience and understanding that Charles Schmieg, Associate Editor, Minna Panfili, editorial program assistant, and Millie Torres, Associate Managing Editor at John Wiley & Sons, have shown us during the preparation and production of the manuscript are gratefully acknowledged. We also recognize the assistance provided by editors Dan Sayre and Robert Argentieri of John Wiley & Sons during the formative and final stages of this book.

Finally, on behalf of the bridge engineering community the authors wish to recognize John Kulicki of Modjeski & Masters and Dennis Mertz of the University of Delaware for their untiring leadership in the development of the LRFD Specification. The authors wish to thank these professionals for providing support and encouragement for the book and responding to many questions about the rationale and background of the specification. Others who contributed to the development of the LRFD Specification as members of the Code Coordinating Committee or as a Chair of a Task Group have also influenced the writing of this book. These include: John Ahlskog, Ralph Bishop, Ian Buckle, Robert Cassano, Paul Csagoly, J. Michael Duncan, Theodore Galambos, Andrzej Nowak, Charles Purkiss, Frank Sears, and James Witheiam. A complete listing of the members of the task groups and the NCHRP panel that directed the project is given in Appendix D.

As with any new book, in spite of numerous proofreadings, errors do creep in and the authors would appreciate it if the reader would call them to their attention. You may write to us directly or, if you prefer, use our e-mail address: barker@vt.edu or puckett@uwyo.edu.

PART I

General Aspects of Bridge Design

CHAPTER 1

Introduction to Bridge Engineering

Bridges are important to everyone. But they are not seen or understood in the same way, which is what makes their study so fascinating. A single bridge over a small river will be viewed differently because the eyes each one sees it with are unique to that individual. Someone traveling over the bridge everyday may only realize a bridge is there because the roadway now has a railing on either side. Others may remember a time before the bridge was built and how far they had to travel to visit friends or to get the children to school. Civic leaders see the bridge as a link between neighborhoods, a way to provide fire and police protection, and access to hospitals. In the business community, the bridge is seen as opening up new markets and expanding commerce. An artist may consider the bridge and its setting as a possible subject for a future painting. A theologian may see the bridge as symbolic of making a connection with God. While a boater on the river, looking up when passing underneath the bridge, will have a completely different perspective. Everyone is looking at the same bridge, but it produces different emotions and visual images in each.

Bridges affect people. People use them, and engineers design them and later build and maintain them. Bridges do not just happen. They must be planned and engineered before they can be constructed. In this book, the emphasis is on the engineering aspects of this process: selection of bridge type, analysis of load effects, resistance of cross sections, and conformance with bridge specifications. Although very important, factors of technical significance should not overshadow the *people* factor.

1.1 A BRIDGE IS THE KEY ELEMENT IN A TRANSPORTATION SYSTEM

A bridge is a key element in a transportation system for three reasons:

- ☐ It likely controls the capacity.
- ☐ It is the highest cost per mile.
- ☐ If the bridge fails, the system fails.

If the width of a bridge is insufficient to carry the number of lanes required to handle the traffic volume, the bridge will be a constriction to the traffic flow. If the strength of a bridge is deficient and unable to carry heavy trucks, load limits will be posted and truck traffic will be rerouted. The bridge controls both the volume and weight of the traffic carried.

Bridges are expensive. The typical cost per mile of a bridge is many times that of the approach roadways. This is a major investment and must be carefully planned for best use of the limited funds available for a transportation system.

When a bridge is removed from service and not replaced, the transportation system may be restricted in its function. Traffic may be detoured over routes not designed to handle the increase in volume. Users of the system experience increased travel times and fuel expenses. Normalcy does not return until the bridge is repaired or replaced.

Because a bridge is a key element in a transportation system, balance must be achieved between handling future traffic volume and loads and the cost of a heavier and wider bridge structure. Strength is always a foremost consideration but so should measures to prevent deterioration. The designer of new bridges has control over these parameters and must make wise decisions so that capacity and cost are in balance, and safety is not compromised.

1.2 BRIDGE ENGINEERING IN THE UNITED STATES

Usually a discourse on the history of bridges begins with a log across a small stream or vines suspended above a deep chasm. This preamble is followed by the development of the stone arch by the Roman engineers of the second and first centuries BC and the building of beautiful bridges across Europe during the Renaissance period of the fourteenth through seventeenth centuries. Next is the Industrial Revolution, which began in the last half of the eighteenth century and saw the emergence of cast iron, wrought iron, and finally steel for bridges. Such discourses are found in the books by Brown (1993), Gies (1963), and Kirby et al. (1956) and are not repeated here. An online search for "bridge engineering history" leads to a host of other references on this topic. Instead a few of the bridges that are typical of those found in the United States are highlighted.

1.2.1 Stone Arch Bridges

The Roman bridge builders first come to mind when discussing stone arch bridges. They utilized the semicircular arch and built elegant and handsome aqueducts and bridges, many of which are still standing today. The oldest remaining Roman stone arch structure is from the seventh century BC and is a vaulted tunnel near the Tiber River. However, the oldest surviving stone arch bridge dates from the ninth century BC and is in Smyrna, Turkey, over the Meles River. In excavations of tombs and underground temples, archaeologists found arched vaults dating to the fourth millennium BC at Ur in one of the earliest Tigris–Euphrates civilizations (Gies, 1963). The stone arch has been around a long time and how its form was first discovered is unknown. But credit is due to the Roman engineers because they are the ones who saw the potential in the stone arch, developed construction techniques, built foundations in moving rivers, and left us a heritage of engineering works that we marvel at today such as Pont du Gard (Exhibit 1 in the color insert).

Compared to these early beginnings, the stone arch bridges in the United States are relative newcomers. One of the earliest stone arch bridges is the Frankford Avenue Bridge over Pennypack Creek built in 1697 on the road between Philadelphia and New York. It is a three-span bridge, 73 ft (23 m) long and is the oldest bridge in the United States that continues to serve as part of a highway system (Jackson, 1988).

Stone arch bridges were usually small scale and built by local masons. These bridges were never as popular in the United States as they were in Europe. Part of the reason for lack of popularity is that stone arch bridges are labor intensive and expensive to build. However, with the development of the railroads in the mid- to late-nineteenth century, the stone arch bridge provided the necessary strength and stiffness for carrying heavy loads, and a number of impressive spans were built. One was the Starrucca Viaduct, Lanesboro, Pennsylvania, which was completed in 1848, and another was the James J. Hill Stone Arch Bridge, Minneapolis, Minnesota, completed in 1883.

The Starrucca Viaduct (Exhibit 2 in the color insert) is 1040 ft (317 m) in overall length and is composed of 17 arches, each with a span of 50 ft (15 m). The viaduct is located on what was known as the New York and Erie Railroad over Starrucca Creek near its junction with the Susquehanna River. Except for the interior spandrel walls being of brick masonry, the structure was of stone masonry quarried locally. The maximum height of the roadbed above the creek is 112 ft (34 m) (Jackson, 1988) and it still carries heavy railroad traffic.

The James J. Hill Stone Arch Bridge (Fig. 1.1) is 2490 ft (760 m) long and incorporated 23 arches in its original design (later, 2 arches were replaced with steel trusses to provide navigational clearance). The structure carried Hill's Great Northern Railroad (now merged into the Burlington Northern Santa Fe Railway) across the Mississippi River just below St. Anthony Falls. It played a key role in the development of the Northwest. The bridge was retired in 1982, just short of its 100th birthday, but it still stands today as a reminder of an era gone by and bridges that were built to last (Jackson, 1988).

Fig. 1.1 James J. Hill Stone Arch Bridge, Minneapolis, Minnesota. (Hibbard Photo, Minnesota Historical Society, July 1905.)

1.2.2 Wooden Bridges

Early bridge builders in the United States (Timothy Palmer, Lewis Wernwag, Theodore Burr, and Ithiel Town) began their careers as millwrights or carpenter-mechanics. They had no clear conception of truss action, and their bridges were highly indeterminate combinations of arches and trusses (Kirby and Laurson, 1932). They learned from building large mills how to increase clear spans by using the king-post system or trussed beam. They also appreciated the arch form and its ability to carry loads in compression to the abutments. This compressive action was important because wood joints can transfer compression more efficiently than tension.

The long-span wooden bridges built in the late-eighteenth and early-nineteenth centuries incorporated both the truss and the arch. Palmer and Wernwag constructed trussed arch bridges in which arches were reinforced by trusses (Fig. 1.2). Palmer built a 244-ft (74-m) trussed arch bridge over the Piscataqua in New Hampshire in the 1790s. Wernwag built his "Colossus" in 1812 with a span of 340 ft (104 m) over the Schuylkill at Fairmount, Pennsylvania (Gies, 1963).

In contrast to the trussed arch of Palmer and Wernwag, Burr utilized an arched truss in which a truss is reinforced by an arch (Fig. 1.3) and patented his design in 1817. An example of one that has survived until today is the Philippi Covered Bridge (Fig. 1.4) across the Tygant's Valley River, West Virginia. Lemuel Chenoweth completed it in 1852 as a two-span Burr arched truss with a total length of 577 ft (176 m) long. In later years, two reinforced concrete piers were added under each span to strengthen the bridge (Exhibit 3 in the color insert). As a result, it is able to carry traffic loads and is the nation's only covered bridge serving a federal highway.

One of the reasons many covered bridges have survived for well over 100 years is that the wooden arches and trusses have been protected from the weather. Palmer put a roof and siding on his "permanent bridge" (called permanent because

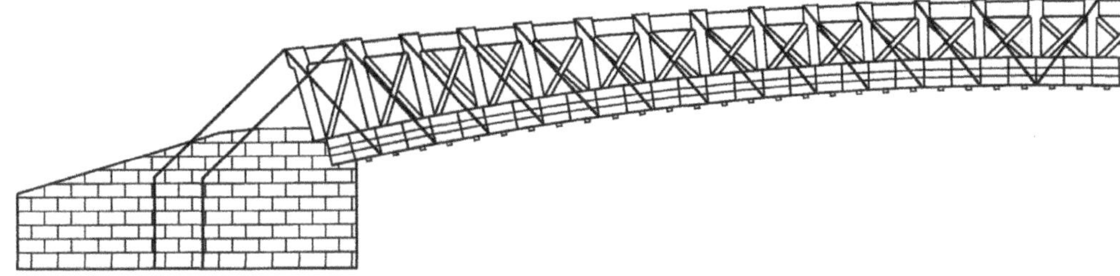

Fig. 1.2 Trussed arch—designed by Lewis Wernwag, patented 1812.

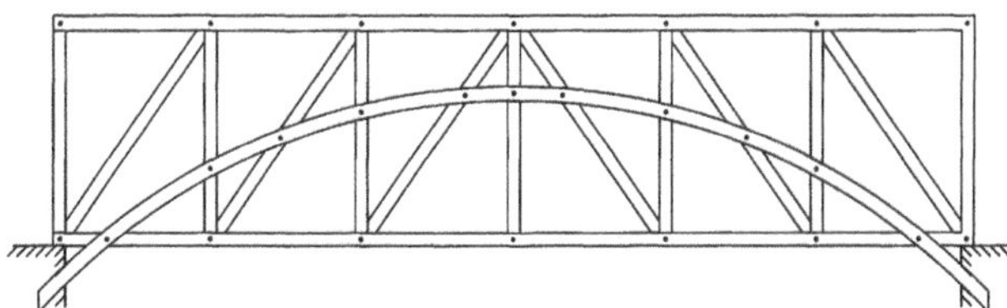

Fig. 1.3 Arched truss—designed by Theodore Burr, patented 1817. (From *Bridges and Men* by Joseph Gies. Copyright © 1963 by Joseph Gies. Used by permission of Doubleday, a division of Bantam Doubleday Dell Publishing Group, Inc.)

it replaced a pontoon bridge) over the Schuylkill at Philadelphia in 1806, and the bridge lasted nearly 70 years before it was destroyed by fire in 1875.

Besides protecting the wood from alternating cycles of wet and dry that cause rot, other advantages of the covered bridge occurred. During winter blizzards, snow did not accumulate on the bridge. However, this presented another problem; bare wooden decks had to be paved with snow because everybody used sleighs. Another advantage was that horses were not frightened by the prospect of crossing a rapidly moving stream over an open bridge because the covered bridge had a comforting barnlike appearance (so says the oral tradition). American folklore also says the covered bridges became favorite parking spots for couples in their rigs, out of sight except for the eyes of curious children who had climbed up and hid in the rafters (Gies, 1963). However, the primary purpose of covering the bridge was to prevent deterioration of the wood structure.

Another successful wooden bridge form first built in 1813 was the lattice truss, which Ithiel Town patented in 1820 (Edwards, 1959). This bridge consisted of strong top and bottom chords, sturdy end posts, and a web of lattice work (Fig. 1.5). This truss type was popular with builders because all of the web members were of the same length and could be prefabricated and sent to the job site for assembly. Another advantage is that it had sufficient stiffness by itself and did not require an arch to reduce deflections. This inherent stiffness meant that horizontal thrusts did not have to be resisted by abutments, and a true truss, with only vertical reactions, had really arrived.

The next step toward simplicity in wooden bridge truss types in the United States is credited to an army engineer named Colonel Stephen H. Long who had been assigned by the War Department to the Baltimore and Ohio Railroad

Fig. 1.4 Philippi covered bridge. (Photo by Larry Belcher, courtesy of West Virginia Department of Transportation.)

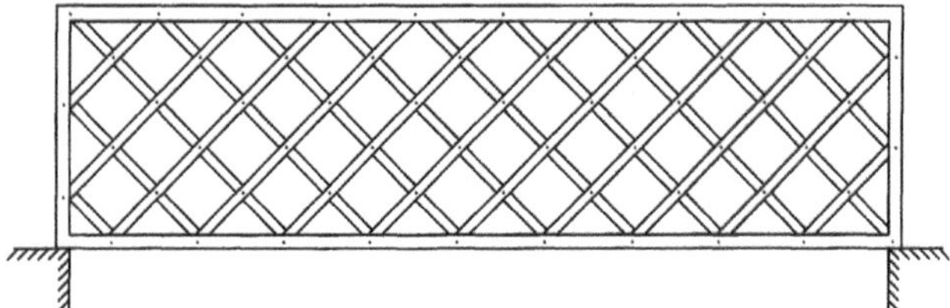

Fig. 1.5 Lattice truss—designed by Ithiel Town, patented 1820. (From *Bridges and Men* by Joseph Gies. Copyright © 1963 by Joseph Gies. Used by permission of Doubleday, a division of Bantam Doubleday Dell Publishing Group, Inc.)

Fig. 1.6 Multiple king-post truss—designed by Colonel Stephen H. Long in 1829. (From *Bridges and Men* by Joseph Gies. Copyright © 1963 by Joseph Gies. Used by permission of Doubleday, a division of Bantam Doubleday Dell Publishing Group, Inc.)

(Edwards, 1959). In 1829, Colonel Long built the first American highway–railroad grade separation project. The trusses in the superstructure had parallel chords that were subdivided into panels with counterbraced web members (Fig. 1.6). The counterbraces provided the necessary stiffness for the panels as the loading changed in the diagonal web members from tension to compression as the railroad cars moved across the bridge.

The development of the paneled bridge truss in wooden bridges enabled long-span trusses to be built with other materials. In addition, the concept of web panels is important because it is the basis for determining the shear resistance of girder bridges. These concepts are called the modified compression field theory in Chapter 14 and tension field action in Chapter 19.

1.2.3 Metal Truss Bridges

Wooden bridges were serving the public well when the loads being carried were horse-drawn wagons and carriages. Then

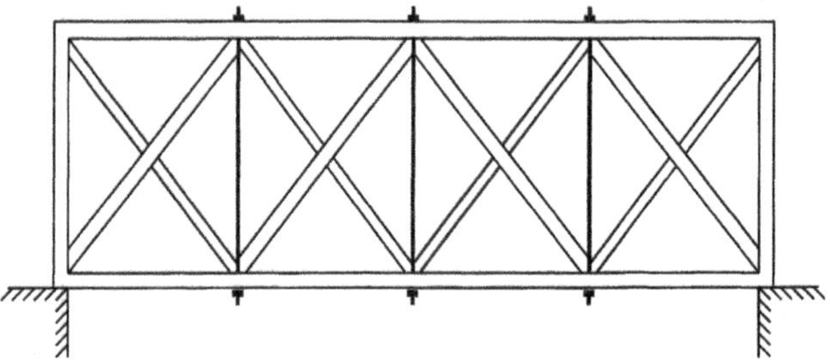

Fig. 1.7 Howe truss—designed by William Howe, patented in 1841. (From *Bridges and Men* by Joseph Gies. Copyright © 1963 by Joseph Gies. Used by permission of Doubleday, a division of Bantam Doubleday Dell Publishing Group, Inc.)

along came the railroads with their heavy loads, and the wooden bridges could not provide the necessary strength and stiffness for longer spans. As a result, wrought-iron rods replaced wooden tension members, and a hybrid truss composed of a combination of wood and metal members was developed. As bridge builders' understanding of which members were carrying tension and which were carrying compression increased, cast iron replaced wooden compression members, thus completing the transition to an all-metal truss form.

In 1841, William Howe, uncle of Elias Howe, the inventor of the sewing machine, received a patent on a truss arrangement in which he took Long's panel system and replaced the wooden vertical members with wrought-iron rods (Gies, 1963). The metal rods ran through the top and bottom chords and could be tightened by turnbuckles to hold the wooden diagonal web members in compression against cast-iron angle blocks (Fig. 1.7). Occasionally, Howe truss bridges were built entirely of metal, but in general they were composed of both wood and metal components. These bridges have the advantages of the panel system as well as those offered by counterbracing.

Thomas and Caleb Pratt (Caleb was the father of Thomas) patented a second variation on Long's panel system in 1844 with wooden vertical members to resist compression and metal diagonal members, which resist only tension (Jackson, 1988). Most of the Pratt trusses built in the United States were entirely of metal, and they became more commonly used than any other type. Simplicity, stiffness, constructability, and economy earned this recognition (Edwards, 1959). The distinctive feature of the Pratt truss (Fig. 1.8), and

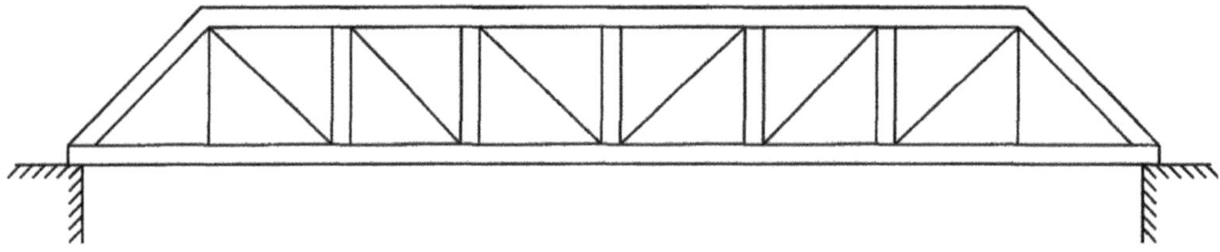

Fig. 1.8 Pratt truss—designed by Thomas and Caleb Pratt, patented in 1844. (From *Bridges and Men* by Joseph Gies. Copyright © 1963 by Joseph Gies. Used by permission of Doubleday, a division of Bantam Doubleday Dell Publishing Group, Inc.)

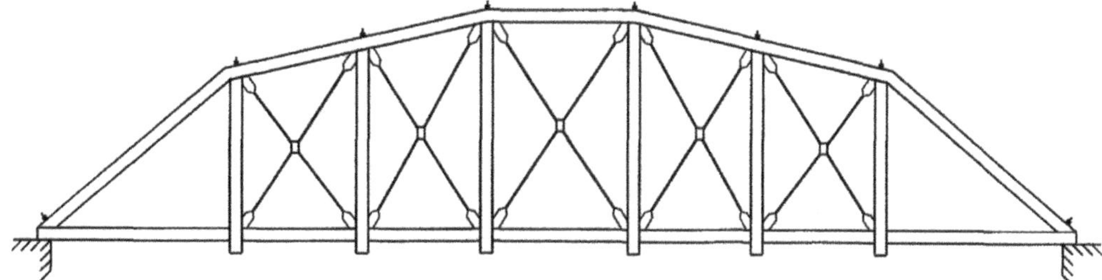

Fig. 1.9 Bowstring arch—designed by Squire Whipple, patented in 1841.

related designs, is that the main diagonal members are in tension.

In 1841, Squire Whipple patented a cast-iron arch truss bridge (Fig. 1.9), which he used to span the Erie Canal at Utica, New York (*Note:* Whipple was not a country gentleman, his first name just happened to be Squire.) Whipple utilized wrought iron for the tension members and cast iron for the compression members. This bridge form became known as a bowstring arch truss, although some engineers considered the design to be more a tied arch than a truss (Jackson, 1988). The double-intersection Pratt truss of Figure 1.10, in which the diagonal tension members extended over two panels, was also credited to Whipple because he was the first to use the design when he built railroad bridges near Troy, New York.

To implement his designs, it is implied that Squire Whipple could analyze his trusses and knew the magnitudes of the tensile and compressive forces in the various members. He was a graduate of Union College, class of 1830, and in 1847 he published the first American treatise on determining the stresses produced by bridge loads and proportioning bridge members. It was titled *A Work on Bridge Building; consisting of two Essays, the one Elementary and General, the other giving Original Plans, and Practical Details for Iron and Wooden Bridges* (Edwards, 1959). In it he showed how one could compute the tensile or compressive stress in each member of a truss that was to carry a specific load (Kirby et al., 1956).

In 1851, Herman Haupt, a graduate of the U.S. Military Academy at West Point, class of 1835, authored a book titled *General Theory of Bridge Construction*, which was published by D. Appleton and Company (Edwards, 1959). This book and the one by Squire Whipple were widely used by engineers and provided the theoretical basis for selecting cross sections to resist bridge dead loads and live loads.

One other development that was critical to the bridge design profession was the ability to verify the theoretical predictions with experimental testing. The tensile and compressive strengths of cast iron, wrought iron, and steel had to be determined and evaluated. Column load curves had to be developed by testing cross sections of various lengths. This experimental work requires large-capacity testing machines.

The first testing machine to be made in America was built in 1832 to test a wrought-iron plate for boilers by the Franklin Institute of Philadelphia (Edwards, 1959). Its capacity was about 10 tons (90 kN), not enough to test bridge components. About 1862, William Sallers and Company of Philadelphia built a testing machine that had a rated capacity of 500 tons (4500 kN) and was specially designed for the testing of full-size columns.

Two testing machines were built by the Keystone Bridge Works, Pittsburgh, Pennsylvania, in 1869–1870 for the St. Louis Bridge Company to evaluate materials for the Eads Bridge over the Mississippi River. One had a capacity of 100 tons (900 kN) while the other a capacity of 800 tons (7200 kN). At the time it was built, the capacity of the larger testing machine was greater than any other in existence (Edwards, 1959).

During the last half of the nineteenth century, the capacity of the testing machines continued to increase until in 1904 the American Bridge Company built a machine having a tension capacity of 2000 tons (18,000 kN) (Edwards, 1959) at its Ambridge, Pennsylvania, plant. These testing machines were engineering works in themselves, but they were essential to verify the strength of the materials and the resistance of components in bridges of ever increasing proportions.

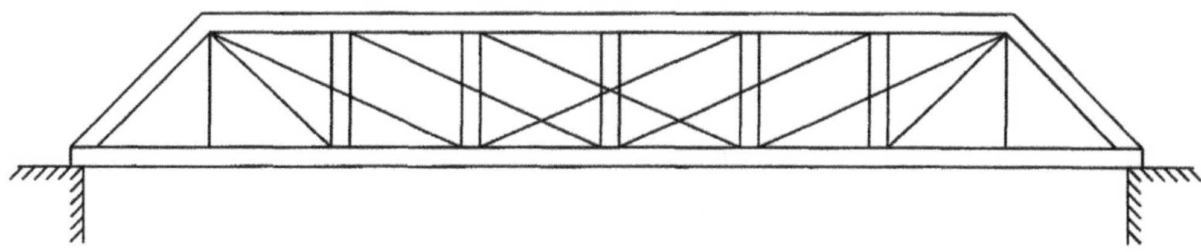

Fig. 1.10 Double-intersection Pratt—credited to Squire Whipple.

1.2.4 Suspension Bridges

Suspension bridges capture the imagination of people everywhere. With their tall towers, slender cables, and tremendous spans, they appear as ethereal giants stretching out to join together opposite shores. Sometimes they are short and stocky and seem to be guardians and protectors of their domain. Other times, they are so long and slender that they seem to be fragile and easily moved. Whatever their visual image, people react to them and remember how they felt when they first saw them.

Imagine the impression on a young child on a family outing in a state park and seeing for the first time the infamous "swinging bridge" across the raging torrent of a rock-strewn river (well, it seemed like a raging torrent). And then the child hears the jeers and challenge of the older children, daring him to cross the river as they moved side to side and purposely got the swinging bridge to swing. Well, it did not happen that first day, it felt more comfortable to stay with mother and the picnic lunch. But it did happen on the next visit, a year or two later. It was like a rite of passage. A child no longer, he was able to cross over the rock-strewn stream on the swinging bridge, not fighting it, but moving with it and feeling the exhilaration of being one with forces stronger than he was.

Suspension bridges also make strong impressions on adults and having an engineering education is not a prerequisite. People in the United States have enjoyed these structures on both coasts, where they cross bays and mouths of rivers. The most memorable are the Brooklyn Bridge (Exhibit 4 in the color insert) in the east and the Golden Gate Bridge (Exhibit 5 in the color insert) in the west. They are also in the interior of the country, where they cross the great rivers, gorges, and straits. Most people understand that the cables are the tendons from which the bridge deck is hung, but they marvel at their strength and the ingenuity it took to get them in place. When people see photographs of workers on the towers of suspension bridges, they catch their breath, and then wonder at how small the workers are compared to the towers they have built. Suspension bridges bring out the emotions: wonder, awe, fear, pleasure; but mostly they are enjoyed for their beauty and grandeur.

In 1801, James Finley erected a suspension bridge with wrought-iron chains of 70-ft (21-m) span over Jacob's Creek near Uniontown, Pennsylvania. He is credited as the inventor of the modern suspension bridge with its stiff level floors and secured a patent in 1808 (Kirby and Laurson, 1932). In previous suspension bridges, the roadway was flexible and followed the curve of the ropes or chains. By stiffening the roadway and making it level, Finley developed a suspension bridge that was suitable not only for footpaths and trails but for roads with carriages and heavy wagons.

Most engineers are familiar with the suspension bridges of John A. Roebling: the Niagara River Bridge, completed in 1855 with a clear span of 825 ft (250 m); the Cincinnati Suspension Bridge, completed in 1867 with a clear span of 1057 ft (322 m); and the Brooklyn Bridge, completed in 1883 with a clear span of 1595 ft (486 m). Of these three wire cable suspension bridges from the nineteenth century, the last two are still in service and are carrying highway traffic. However, there is one other long-span wire cable suspension bridge from this era that is noteworthy and still carrying traffic: the Wheeling Suspension Bridge completed in 1849 with a clear span of 1010 ft (308 m) (Fig. 1.11).

Fig. 1.11 Wheeling Suspension Bridge. (Photo by John Brunell, courtesy of West Virginia Department of Transportation.)

The Wheeling Suspension Bridge over the easterly channel of the Ohio River was designed and built by Charles Ellet who won a competition with John Roebling; that is, he was the low bidder. This result of a competition was also true of the Niagara River Bridge, except that Ellet walked away from it after the cables had been strung, saying that the $190,000 he bid was not enough to complete it. Roebling was then hired and he completed the project for about $400,000 (Gies, 1963).

The original Wheeling Suspension Bridge did not have the stiffening truss shown in Figure 1.11. This truss was added after a windstorm in 1854 caused the bridge to swing back and forth with increased momentum, the deck to twist and undulate in waves nearly as high as the towers, until it all came crashing down into the river (very similar to the Tacoma Narrows Bridge failure some 80 years later). A web search for "Tacoma Narrows Movie" will provide several opportunities to view movies that illustrate the failure.

The Wheeling Bridge had the strength to resist gravity loads, but it was aerodynamically unstable. Why this lesson was lost to the profession is unknown, but if it had received the attention it deserved, it would have saved a lot of trouble in the years ahead.

What happened to the Wheeling Suspension Bridge was not lost on John Roebling. He was in the midst of the Niagara River project when he heard of the failure and immediately ordered more cable to be used as stays for the double-decked bridge. An early painting of the Niagara River Bridge shows the stays running from the bottom of the deck to the shore to provide added stability.

In 1859 William McComas, a former associate of Charles Ellet, rebuilt the Wheeling Suspension Bridge. In 1872 Wilhelm Hildenbrand, an engineer with Roebling's company, modified the deck and added diagonal stay wires between the towers and the deck to increase the resistance to wind (Jackson, 1988) and to give the bridge the appearance it has today.

The completion of the Brooklyn Bridge in 1883 brought to maturity the building of suspension bridges and set the stage for the long-span suspension bridges of the twentieth century. Table 1.1 provides a summary of some of the notable long-span suspension bridges built in the United States and still standing.

Some comments are in order with regard to the suspension bridges in Table 1.1. The Williamsburg Bridge and the Brooklyn Bridge are of comparable span but with noticeable differences. The Williamsburg Bridge has steel rather than masonry towers. The deck truss is a 40-ft (12.5-m) deep lattice truss, compared to a 17-ft (5.2-m) deep stiffening truss of its predecessor. This truss gives the Williamsburg Bridge a bulky appearance, but it is very stable under traffic and wind loadings. Another big difference is that the wire in the steel cables of the Brooklyn Bridge was galvanized to protect it from corrosion in the briny atmosphere of the East River (Gies, 1963), while the wire in its successor was not. As a result, the cables of the Williamsburg Bridge have

Table 1.1 Long-Span Suspension Bridges in the United States

Bridge	Site	Designer	Clear Span, ft (m)	Date
Wheeling	West Virginia	Charles Ellet	1010 (308)	1847
Cincinnati	Ohio	John Roebling	1057 (322)	1867
Brooklyn	New York	John Roebling Washington Roebling	1595 (486)	1883
Williamsburg	New York	Leffert Lefferts Buck	1600 (488)	1903
Bear Mountain	Hudson Valley	C. Howard Baird	1632 (497)	1924
Ben Franklin	Philadelphia	Ralph Modjeski Leon Moisseiff	1750 (533)	1926
Ambassador	Detroit	Jonathon Jones Leon Moisseiff	1850 (564)	1929
George Washington	New York	Othmar Ammann Leon Moisseiff	3500 (1067)	1931
Golden Gate	San Francisco	Joseph Strauss Charles Ellis Leon Moisseiff	4200 (1280)	1937
Verrazano-Narrows	New York	Ammann and Whitney	4260 (1298)	1964

had to be rehabilitated with a new protective system that cost $73 million (Bruschi and Koglin, 1996). A web search for "Williamsburg Bridge image," or other bridge names listed in Table 1.1, provides a wealth of information and illustration.

Another observation of Table 1.1 is the tremendous increase in clear span attained by the George Washington Bridge over the Hudson River in New York. It nearly doubled the clear span of the longest suspension bridge in existence at the time it was built, a truly remarkable accomplishment.

One designer, Leon Moisseiff, is associated with most of the suspension bridges in Table 1.1 that were built in the twentieth century. He was the design engineer of the Manhattan and Ben Franklin bridges, participated in the design of the George Washington Bridge, and was a consulting engineer on the Ambassador, Golden Gate, and Oakland–Bay bridges (Gies, 1963). All of these bridges were triumphs and successes. He was a well-respected engineer who had pioneered the use of deflection theory, instead of the erroneous elastic theory, in the design of the Manhattan Bridge and those that followed. But Moisseiff will also be remembered as the designer of the Tacoma Narrows Bridge that self-destructed during a windstorm in 1940, not unlike that experienced by the Wheeling Suspension Bridge in 1854.

The use of a plate girder to stiffen the deck undoubtedly contributed to providing a surface on which the wind could act, but the overall slenderness of the bridge gave it an undulating behavior under traffic even when the wind was not blowing. Comparing the ratio of depth of truss or girder to the span length for the Williamsburg, Golden Gate, and Tacoma Narrows bridges, we have 1 : 40, 1 : 164, and 1 : 350, respectively (Gies, 1963). The design had gone one step too far in making a lighter and more economical structure. The tragedy for bridge design professionals of the Tacoma Narrows failure was a tough lesson, but one that will not be forgotten.

1.2.5 Metal Arch Bridges

Arch bridges are aesthetically pleasing and can be economically competitive with other bridge types. Sometimes the arch can be above the deck, as in a tied-arch design, or as in the bowstring arch of Whipple (Fig. 1.9). Other times, when the foundation materials can resist the thrusts, the arch is below the deck. Restraint conditions at the supports of an arch can be fixed or hinged. And if a designer chooses, a third hinge can be placed at the crown to make the arch statically determinate or nonredundant.

The first iron arch bridge in the United States was built in 1839 across Dunlap's Creek at Brownsville in southwestern Pennsylvania on the National Road (Jackson, 1988). The arch consists of five tubular cast-iron ribs that span 80 ft (24 m) between fixed supports. It was designed by Captain Richard Delafield and built by the U.S. Army Corps of Engineers (Jackson, 1988). It is still in service today.

The second cast-iron arch bridge in this country was completed in 1860 across Rock Creek between Georgetown and Washington, DC. It was built by the Army Corps of Engineers under the direction of Captain Montgomery Meigs as part of an 18.6-mile (30-km) aqueduct, which brings water from above the Great Falls on the Potomac to Washington, DC. The two arch ribs of the bridge are 4-ft (1.2-m) diameter cast-iron pipes that span 200 ft (61 m) with a rise of 20 ft (6.1 m) and carry water within its 1.5-inch (38-mm) thick walls. The arch supports a level roadway on open-spandrel posts that carried Washington's first horse-drawn street railway line (Edwards, 1959). The superstructure was removed in 1916 and replaced by a concrete arch bridge. However, the pipe arches remain in place between the concrete arches and continue to carry water to the city today.

Two examples of steel deck arch bridges from the nineteenth century that still carry highway traffic are the Washington Bridge across the Harlem River in New York and the Panther Hollow Bridge in Schenely Park, Pittsburgh (Jackson, 1988). The two-hinged arches of the Washington Bridge, completed in 1889, are riveted plate girders with a main span of 508 ft (155 m). This bridge is the first American metal arch bridge in which the arch ribs are plate girders (Edwards, 1959). The three-hinged arch of the Panther Hollow Bridge, completed in 1896, has a span of 360 ft (110 m). Due to space limitations, not all bridges noted here can be illustrated in this book; however, web searches for the bridge name and location easily takes the reader to a host of images and other resources.

One of the most significant bridges built in the United States is the steel deck arch bridge designed by James B. Eads (Exhibit 6 in the color insert) across the Mississippi River at St. Louis. It took 7 years to construct and was completed in 1874. The three-arch superstructure consisted of two 502-ft (153-m) side arches and one 520-ft (159-m) center arch that carried two decks of railroad and highway traffic (Fig. 1.12). The Eads Bridge is significant because of the very deep pneumatic caissons for the foundations, the early use of steel in the design, and the graceful beauty of its huge arches as they span across the wide river (Jackson, 1988).

Because of his previous experience as a salvage diver, Eads realized that the foundations of his bridge could not be placed on the shifting sands of the riverbed but must be set on bedrock. The west abutment was built first with the aid of a cofferdam and founded on bedrock at a depth of 47 ft (14 m). Site data indicated that bedrock sloped downward from west to east, with an unknown depth of over 100 ft (30 m) at the east abutment, presenting a real problem for cofferdams. While recuperating from an illness in France, Eads learned that European engineers had used compressed air to keep water out of closed caissons (Gies, 1963). He adapted the technique of using caissons, or wooden boxes, added a few innovations of his own, such as a sand pump, and completed the west and east piers in the river. The west pier is at a depth of 86 ft (26 m) and the east pier at a depth of 94 ft (29 m).

Fig. 1.12 Eads Bridge, St. Louis, Missouri. (Photo courtesy of Kathryn Kontrim, 1996.)

However, the construction of these piers was not without cost. Twelve workmen died in the east pier and one in the west pier from caisson's disease, or the bends. These deaths caused Eads and his physician, Dr. Jaminet, much anxiety because the east abutment had to go even deeper. Based on his own experience in going in and out of the caissons, Dr. Jaminet prescribed slow decompression and shorter working time as the depth increased. At a depth of 100 ft (30 m), a day's labor consisted of two working periods of 45 min each, separated by a rest period. As a result of the strict rules, only one death occurred in the placement of the east abutment on bedrock at a depth of 136 ft (42 m). Today's scuba diving tables suggest a 30-min stay at 100 ft (30 m) for comparison.

It is ironic that the lessons learned by Eads and Dr. Jaminet were not passed on to Washington Roebling and his physician, Dr. Andrew H. Smith, in the parallel construction of the Brooklyn Bridge. The speculation is that Eads and Roebling had a falling-out because of Eads' perception that Roebling had copied a number of caisson ideas from him. Had they remained on better terms, Roebling may not have been stricken by the bends and partially paralyzed for life (Gies, 1963).

Another significant engineering achievement of the Eads Bridge was in the use of chrome steel in the tubular arches that had to meet, for that time, stringent material specifications. Eads insisted on an elastic limit of 50 ksi (345 MPa) and an ultimate strength of 120 ksi (827 MPa) for his steel at a time when the steel producers (one of which was Andrew Carnegie) questioned the importance of an elastic limit (Kirby et al., 1956). The testing machines mentioned in Section 1.2.3 had to be built, and it took some effort before steel could be produced that would pass the tests. The material specification of Eads was unprecedented in both its scale and quality of workmanship demanded, setting a benchmark for future standards (Brown, 1993).

The cantilever construction of the arches for the Eads Bridge was also a significant engineering milestone. Falsework in the river was not possible, so Eads built falsework on top of the piers and cantilevered the arches, segment by segment in a balanced manner, until the arch halves met at midspan (Kirby et al., 1956). On May 24, 1874, the highway deck was opened for pedestrians; on June 3 it was opened for vehicles; and on July 2 some 14 locomotives, 7 on each track, crossed side by side (Gies, 1963). The biggest bridge of any type ever built anywhere up to that time had been completed. The Eads Bridge remains in service today and at the time of this writing is being rehabilitated to repair the track, ties, and rails, the deck and floor system, masonry and other structural improvements.

Since the Eads Bridge, steel arch bridges longer than its 520-ft (159-m) center span have been constructed. These include the 977-ft (298-m) clear span Hell Gate Bridge over the East River in New York, completed in 1917; the 1675-ft (508-m) clear span Bayonne Arch Bridge over the Kill van Kull between Staten Island and New Jersey, completed in 1931; and the United States' longest 1700-ft (518-m) clear span New River Gorge Bridge near Fayetteville, West Virginia, completed in 1978 and designed by Michael Baker, Jr., Inc. (Fig. 1.13). Annually the locals celebrate "New River Bridge Day" noted as the state's biggest party of the year. A web search provides a lot of detail, movies on base jumping, and so forth. This is yet another example of the importance of our bridges for social affairs perhaps not even expected by the owner or designers.

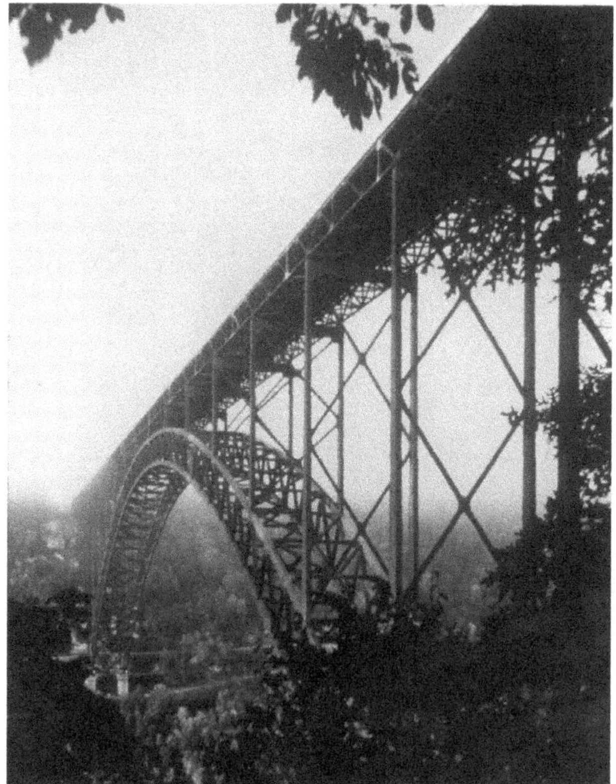

Fig. 1.13 New River Gorge Bridge. (Photo by Terry Clark Photography, courtesy of West Virginia Department of Transportation.)

1.2.6 Reinforced Concrete Bridges

In contrast to wood and metal, reinforced concrete has a relatively short history. It was in 1824 that Joseph Aspdin of England was recognized for producing Portland cement by heating ground limestone and clay in a kiln. This cement was used to line tunnels under the Thames River because it was water resistant. In the United States, D. O. Taylor produced Portland cement in Pennsylvania in 1871, and T. Millen produced it about the same time in South Bend, Indiana. It was not until the early 1880s that significant amounts were produced in the United States (MacGregor and Wight, 2008).

In 1867, a French nursery gardener, Joseph Monier, received a patent for concrete tubs reinforced with iron. In the United States, Ernest Ransome of California was experimenting with reinforced concrete, and in 1884 he received a patent for a twisted steel reinforcing bar. The first steel bar reinforced concrete bridge in the United States was built by Ransome in 1889: the Alvord Lake Bridge (Exhibit 7 in the color insert) in Golden Gate Park, San Francisco. This bridge has a modest span of 29 ft (9 m), is 64 ft (19.5 m) wide, and is still in service (Jackson, 1988).

After the success of the Alvord Lake Bridge, reinforced concrete arch bridges were built in other parks because their classic stone arch appearance fit the surroundings. One of these that remains to this day is the 137-ft (42-m) span Eden Park Bridge in Cincinnati, Ohio, built by Fritz von Emperger in 1895. This bridge is not a typical reinforced concrete arch but has a series of curved steel I-sections placed in the bottom of the arch and covered with concrete. Joseph Melan of Austria developed this design and, though it was used only for a few years, it played an important role in establishing the viability of reinforced concrete bridge construction (Jackson, 1988).

Begun in 1897, but not completed until 1907, was the high-level Taft Bridge carrying Connecticut Avenue over Rock Creek in Washington, DC. This bridge consists of five open-spandrel unreinforced concrete arches supporting a reinforced concrete deck. George Morison designed it and Edward Casey supervised its construction (Jackson, 1988). This bridge has recently been renovated and is prepared to give many more years of service. A web search for "Rock Creek Bridge DC" provides nice pictures that illustrate the rich aesthetics of this structure in an important urban and picturesque setting.

Two reinforced concrete arch bridges in Washington, DC, over the Potomac River are also significant. One is the Key Bridge (named after Francis Scott Key who lived near the Georgetown end of the bridge), completed in 1923, which connects Georgetown with Rosslyn, Virginia. It has seven open-spandrel three-ribbed arches designed by Nathan C. Wyeth and the bridge has recently been refurbished. The other is the Arlington Memorial Bridge, completed in 1932, which connects the Lincoln Memorial and Arlington National Cemetery. It has nine arches, eight are closed-spandrel reinforced concrete arches and the center arch, with a span of 216 ft (66 m), is a double-leaf steel bascule bridge that has not been opened for several years. It was designed by the architectural firm of McKim, Mead, and White (Jackson, 1988).

Other notable reinforced concrete deck arch bridges still in service include the 9-span, open-spandrel Colorado Street Bridge in Pasadena, California, near the Rose Bowl, designed by Waddell and Harrington, and completed in 1913; the 100-ft (30-m) single-span, open-spandrel Shepperd's Dell Bridge across the Young Creek near Latourell, Oregon, designed by K. R. Billner and S. C. Lancaster, and completed in 1914; the 140-ft (43-m) single-span, closed-spandrel Canyon Padre Bridge on old Route 66 near Flagstaff, Arizona, designed by Daniel Luten and completed in 1914; the 10-span, open-spandrel Tunkhannock Creek Viaduct (Exhibit 8 in the color insert) near Nicholson, Pennsylvania, designed by A. Burton Cohen and completed in 1915 (considered to be volumetrically the largest structure of its type in the world); the 13-span, open-spandrel Mendota Bridge across the Minnesota River at Mendota, Minnesota, designed by C. A. P. Turner and Walter Wheeler, and completed in 1926; the 7-span, open-spandrel Rouge River Bridge on the Oregon Coast Highway near Gold Beach, Oregon, designed by Conde B. McCullough and completed in 1932; the 5-span, open-spandrel George Westinghouse Memorial Bridge across Turtle Creek at North Versailles,

Fig. 1.14 Bixby Creek Bridge, south of Carmel, California. [From Roberts (1990). Used with permission of American Concrete Institute.]

Pennsylvania, designed by Vernon R. Covell and completed in 1931; and the 360-ft (100-m) single-span, open-spandrel Bixby Creek Bridge south of Carmel, California, on State Route 1 amid the rugged terrain of the Big Sur (Fig. 1.14), designed by F. W. Panhorst and C. H. Purcell, and completed in 1933 (Jackson, 1988).

Reinforced concrete through-arch bridges were also constructed. James B. Marsh received a patent in 1912 for the Marsh rainbow arch bridge. This bridge resembles a bowstring arch truss but uses reinforced concrete for its main members. Three examples of Marsh rainbow arch bridges still in service are the 90-ft (27-m) single-span Spring Street Bridge across Duncan Creek in Chippewa Falls, Wisconsin, completed in 1916; the eleven 90-ft (27-m) arch spans of the Fort Morgan Bridge across the South Platte River near Fort Morgan, Colorado, completed in 1923; and the 82-ft (25-m) single-span Cedar Creek Bridge near Elgin, Kansas, completed in 1927 (Jackson, 1988).

One interesting feature of the 1932 Rogue River Bridge (Exhibit 9 in the color insert), which is a precursor of things to come, is that the arches were built using the prestressing construction techniques first developed by the French engineer Ernest Freyssinet in the 1920s (Jackson, 1988). In the United States, the first prestressed concrete girder bridge was the Walnut Lane Bridge in Philadelphia, which was completed in 1950. After the success of the Walnut Lane Bridge, prestressed concrete construction of highway bridges gained in popularity and is now used throughout the United States.

1.2.7 Girder Bridges

Girder bridges are the most numerous of all highway bridges in the United States. Their contribution to the transportation system often goes unrecognized because the great suspension, steel arch, and concrete arch bridges are the ones people remember. The spans of girder bridges seldom exceed 500 ft (150 m), with a majority of them less than 170 ft (50 m), so they do not get as much attention as they perhaps should. Girder bridges are important structures because they are used so frequently.

With respect to the overall material usage, girders are not as efficient as trusses in resisting loads over long spans. However, for short and medium spans the difference in material weight is small and girder bridges are competitive. In addition, the girder bridges have greater stiffness and are less subject to vibrations. This characteristic was important to the railroads and resulted in the early application of plate girders in their bridges.

A plate girder is an I-section assembled out of flange and web plates. The earliest ones were fabricated in England with rivets connecting double angles from the flanges to the web. In the United States, a locomotive builder, the Portland Company of Portland, Maine, fabricated a number of railroad bridges around 1850 (Edwards, 1959). In early plate girders, the webs were often deeper than the maximum width of plate produced by rolling mills. As a result, the plate girders were assembled with the lengthwise dimension of the web plate in the transverse direction of the section from flange to flange. An example is a wrought-iron plate girder span of 115 ft (35 m) built by the Elmira Bridge Company, Elmira, New York, in 1890 for the New York Central Railroad with a web depth of 9 ft (2.7 m) fabricated from plates 6 ft (1.8 m) wide (Edwards, 1959).

Steel plate girders eventually replaced wrought iron in the railroad bridge. An early example is the 1500-ft (457-m) long

Fig. 1.15 Napa River Bridge. (Photo courtesy of California Department of Transportation.)

Fort Sumner Railroad Bridge on concrete piers across the Pecos River, Fort Sumner, New Mexico, completed in 1906 (Jackson, 1988). This bridge is still in service.

Other examples of steel plate girder bridges are the 5935-ft (2074-m) long Knight's Key Bridge and the 6803-ft (1809-m) long Pigeon Key Bridge, both part of the Seven Mile Bridge across the Gulf of Mexico from the mainland to Key West, Florida (Jackson, 1988). Construction on these bridges began in 1908 and was completed in 1912. Originally they carried railroad traffic but were converted to highway use in 1938.

Following the success of the Walnut Lane Bridge (Exhibit 10 in color insert) in Philadelphia in 1950, prestressed concrete girders became popular as a bridge type for highway interchanges and grade separations. In building the interstate highway system, innumerable prestressed concrete girder bridges, some with single and multiple box sections have been and continue to be built.

Some of the early girder bridges, with their multiple short spans and deep girders, were not very attractive. However, with the advent of prestressed concrete and the development of segmental construction, the spans of girder bridges have become longer and the girders more slender. The result is that the concrete girder bridge is not only functional but can also be designed to be aesthetically pleasing (Fig. 1.15).

1.2.8 Closing Remarks

Bridge engineering in the United States has come a long way since those early stone arch and wooden truss bridges. It is a rich heritage and much can be learned from the early builders in overcoming what appeared to be insurmountable difficulties. These builders had a vision of what needed to be done and, sometimes, by the sheer power of their will, completed projects that we view with awe today.

A brief exerpt from a book on the building of the Golden Gate by Kevin Starr (2010) reinforces this thought:

> *But before the bridge could be built it had to be envisioned. Imagining the bridge began as early as the 1850's and reached a crisis point by the 1920's. In this pre-design and pre-construction drama of vision, planning, and public and private organization, four figures played important roles. A Marin county businessman..., the San Francisco city engineer..., an engineering entrepreneur..., and a banker in Sonoma County..., played a crucial role in persuading the counties north of San Francisco that a bridge across the Golden Gate was in their best interest. Dreamers and doers, each of these men helped initiate a process that would after a decade of negotiations enlist hundreds of engineers, politicians, bankers, steelmakers, and, of equal importance to all of them, construction workers, in a successful effort to span the strait with a gently rising arc of suspended steel.*

The challenge for today's bridge engineer is to follow in the footsteps of these early designers and create and build bridges that other engineers will write about 100 and 200 years from now.

1.3 BRIDGE ENGINEER—PLANNER, ARCHITECT, DESIGNER, CONSTRUCTOR, AND FACILITY MANAGER

The bridge engineer is often involved with several or all aspects of bridge planning, design, and management. This situation is not typical in the building design profession where the architect usually heads a team of diverse design professionals consisting of architects and civil, structural, mechanical, and electrical engineers. In the bridge engineering profession, the bridge engineer works closely with other civil engineers who are in charge of the roadway alignment and design. After the alignment is determined, the engineer often controls the bridge type, aesthetics, and technical details. As part of the design process, the bridge engineer is often charged with reviewing shop drawing and other construction details.

Many aspects of the design affect the long-term performance of the system, which is of paramount concern to the bridge owner. The owner, who is often a department of transportation or other public agency, is charged with the management of the bridge, which includes periodic inspections, rehabilitation, and retrofits as necessary and continual prediction of the life-cycle performance or deterioration modeling. Such bridge management systems (BMS) are beginning to play a large role in suggesting the allocation of resources to best maintain an inventory of bridges. A typical BMS is designed to predict the long-term costs associated with the deterioration of the inventory and recommend maintenance items to minimize total costs for a system of bridges. Because the bridge engineer is charged with maintaining the system of bridges, or inventory, his or her role differs significantly from the building engineer where the owner is often a real estate professional controlling only one, or a few, buildings, and then perhaps for a short time.

In summary, the bridge engineer has significant control over the design, construction, and maintenance processes. With this control comes significant responsibility for public safety and resources. The decisions the engineer makes in design will affect the long-term site aesthetics, serviceability, maintainability, and ability to retrofit for changing demands. In short, the engineer is (or interfaces closely with) the planner, architect, designer, constructor, and facility manager.

Many aspects of these functions are discussed in the following chapters where we illustrate both a broad-based approach to aid in understanding the general aspects of design, and also include many technical and detailed articles to facilitate the computation/validation of design. Often engineers become specialists in one or two of the areas mentioned in this discussion and interface with others who are expert in other areas. The entire field is so involved that near-complete understanding can only be gained after years of professional practice, and then, few individual engineers will have the opportunity for such diverse experiences.

REFERENCES

Brown, D. J. (1993), *Bridges*, Macmillan, New York.

Bruschi, M. G. and T. L. Koglin (1996). "Preserving Williamsburg's Cables," *Civil Engineering*, ASCE, Vol. 66, No. 3, March, pp. 36–39.

Edwards, L. N. (1959). *A Record of History and Evolution of Early American Bridges*, University Press, Orono, ME.

Gies, J. (1963). *Bridges and Men*, Doubleday, Garden City, NY.

Jackson, D. C. (1988). *Great American Bridges and Dams*, Preservation Press, National Trust for Historic Preservation, Washington, DC.

Kirby, R. S. and P. G. Laurson (1932). *The Early Years of Modern Civil Engineering*, Yale University Press, New Haven, CT.

Kirby, R. S., S. Whithington, A. B. Darling, and F. G. Kilgour (1956). *Engineering in History*, McGraw-Hill, New York.

MacGregor, J. G. and J. K. Wight (2008). *Reinforced Concrete Mechanics and Design*, 5th ed., Prentice Hall, Englewood Cliffs, NJ.

Starr, K. (2010). *Golden Gate: The life and times of America's Greatest Bridge*, Bloomsbury Press, New York.

Roberts, J. E. (1990). "Aesthetics and Economy in Complete Concrete Bridge Design," *Esthetics in Concrete Bridge Design*, American Concrete Institute, Detroit, MI.

PROBLEMS

1.1 Explain why the *people* factor is important in bridge engineering.

1.2 In what way does a bridge control the capacity of a transportation system?

1.3 Discuss the necessity of considering life-cycle costs in the design of bridges.

1.4 How were the early U.S. wooden bridge builders able to conceive and build the long-span wooden arch and truss bridges (e.g., Wernwag's Colossus) without theoretical knowledge to analyze and proportion their structures?

1.5 What is the main reason wooden bridges were covered?

1.6 How is the bridge designer Col. Stephen H. Long linked to Long's Peak in Colorado?

1.7 Whipple in 1847 and Haupt in 1851 authored books on the analysis and design of bridge trusses. Discuss the difficulty steel truss bridge designers prior to these dates had in providing adequate safety.

1.8 Both cast-iron and wrought-iron components were used in early metal truss and arch bridges. How do they differ in manufacture? What makes the manufacture of steel different from both of them?

1.9 Explain why the development of large-capacity testing machines was important to the progress of steel bridges.

1.10 Who secured a patent, and when, for the modern suspension bridge with a stiff level floor?

1.11 The Wheeling Suspension Bridge that still carries traffic today is not the same bridge built in 1849. Explain what happened to the original.

1.12 Who was Charles Ellis and what was his contribution to the building of the Golden Gate Bridge?

1.13 List four significant engineering achievements of the Eads Bridge over the Mississippi at St. Louis.

1.14 Use the Historic American Engineering Record (HAER) digitized collection of historic bridges and obtain additional information on one of the reinforced concrete bridges mentioned in Section 1.2.6.

1.15 Explain why girder bridges are not as efficient as trusses in resisting loads (with respect to material quantities).

1.16 Comment on the significance of the Walnut Lane Bridge in Philadelphia.

CHAPTER 2

Specifications and Bridge Failures

2.1 BRIDGE SPECIFICATIONS

For most bridge engineers, it seems that bridge specifications were always there. But that is not the case. The early bridges were built under a design–build type of contract. A bridge company would agree, for some lump-sum price, to construct a bridge connecting one location to another. There were no standard bridge specifications and the contract went to the low bidder. The bridge company basically wrote its own specifications when describing the bridge it was proposing to build. As a result, depending on the integrity, education, and experience of the builder, some very good bridges were constructed and at the same time some very poor bridges were built.

Of the highway and railroad bridges built in the 1870s, one out of every four failed, a rate of 40 bridges per year (Gies, 1963). The public was losing confidence and did not feel safe when traveling across any bridge. Something had to be done to improve the standards by which bridges were designed and built.

An event took place on the night of December 29, 1876, that attracted the attention of not only the public but also the engineering profession. In a blinding snowstorm, an 11-car train with a double-header locomotive started across the Ashtabula Creek at Ashtabula, Ohio, on a 175-ft (48-m) long iron bridge, when the first tender derailed, plowed up the ties, and caused the second locomotive to smash into the abutment (Gies, 1963). The coupling broke between the lead tender and the second locomotive, and the first locomotive and tender went racing across the bridge. The bridge collapsed behind them. The second locomotive, tender, and 11 cars plunged some 70 ft (20 m) into the creek. The wooden cars burst into flames when their pot-bellied stoves were upset, and a total of 80 passengers and crew died.

In the investigation that followed, a number of shortcomings in the way bridges were designed, approved, and built were apparent. An executive of the railroad who had limited bridge design experience designed the bridge. The acceptance of the bridge was by test loading with six locomotives, which only proved that the factor of safety was at least 1.0 for that particular loading. The bridge was a Howe truss with cast-iron blocks for seating the diagonal compression members. These blocks were suspected of contributing to the failure. It is ironic that at a meeting of the American Society of Civil Engineers (ASCE), a statement was made that "the construction of the truss violated every canon of our standard practice" at a time when there were no standards of practice (Gies, 1963).

The American practice of using concentrated axle loads instead of uniformly distributed loads was introduced in 1862 by Charles Hilton of the New York Central Railroad (Edwards, 1959). It was not until 1894 that Theodore Cooper proposed his original concept of train loadings with concentrated axle loadings for the locomotives and tender followed by a uniformly distributed load representing the train. The Cooper series loading became the standard in 1903 when adopted by the American Railroad Engineering Association (AREA) and remains in use to the present day.

On December 12, 1914, the American Association of State Highway Officials (AASHO) was formed, and in 1921 its Committee on Bridges and Allied Structures was organized. The charge to this committee was the development of standard specifications for the design, materials, and construction of highway bridges. During the period of development, mimeographed copies of the different sections were circulated to state agencies for their use. The first edition of the *Standard Specifications for Highway Bridges and Incidental Structures* was published in 1931 by AASHO.

The truck train load in the standard specifications is an adaptation of the Cooper loading concept applied to highway bridges (Edwards, 1959). The "H" series loading of AASHO was designed to adjust to different weights of trucks without changing the spacing between axles and wheels. These specifications have been reissued periodically to reflect the ongoing research and development in concrete, steel, and wood structures with the final seventeenth edition of the *Standard Specifications for Highway Bridges* appearing in 2002 (AASHTO, 2002). In 1963, the AASHO became the American Association of State Highway and Transportation Officials (AASHTO). The insertion of the word *Transportation* was to recognize the officials' responsibility for all modes of transportation (air, water, light rail, subways, tunnels, and highways).

In the beginning, the design philosophy utilized in the standard specification was working stress design (also known as allowable stress design). In the 1970s, variations in the uncertainties of loads were considered and load factor design (LFD) was introduced as an alternative method. In 1986, the Subcommittee on Bridges and Structures initiated a study on incorporating the load and resistance factor design (LRFD) philosophy into the standard specification. This study recommended that LRFD be utilized in the design of highway bridges. The subcommittee authorized a comprehensive rewrite of the entire standard specification to accompany the conversion to LRFD. The result was the first edition of the AASHTO (1994) *LRFD Bridge Design Specifications*. Additional editions were published in 1998, 2004, 2007, and the fifth edition in 2010 (AASHTO, 2010). The fifth edition is used for this book.

2.2 IMPLICATION OF BRIDGE FAILURES ON PRACTICE

On the positive side of the bridge failure at Ashtabula Creek, Ohio, in 1876 was the realization by the engineering profession that standards of practice for bridge design and construction had to be codified. Good intentions and a firm handshake were not sufficient to ensure safety for the traveling public. Specifications, with legal ramifications if they were not followed, had to be developed and implemented. For railroad bridges, this task began in 1899 with the formation of the American Railway Engineering and Maintenance of Way Association and resulted in the adoption of Theodore Cooper's specification for loadings in 1903.

As automobile traffic expanded, highway bridges increased in number and size. Truck loadings were constantly increasing and legal limits had to be established. The original effort for defining loads, materials, and design procedures was made by the U.S. Department of Agriculture, Office of Public Roads in 1913 with the publication of its Circular No. 100, "Typical Specifications for the Fabrication and Erection of Steel Highway Bridges" (Edwards, 1959). In 1919, the Office of Public Roads became the Bureau of Public Roads [now the Federal Highway Administration (FHWA)] and a revised specification was prepared and issued.

The Committee on Bridges and Allied Structures of the AASHO issued the first edition of *Standard Specifications for Highway Bridges* in 1931. It is interesting to note in the Preface of the seventeenth edition of this publication the listing of the years when the standard specifications were revised: 1935, 1941, 1944, 1949, 1953, 1957, 1961, 1965, 1969, 1973, 1977, 1983, 1989, 1992, 1996, and 2002. It is obvious that this document is constantly changing and adapting to new developments in the practice of bridge engineering.

In some cases, new information on the performance of bridges was generated by a bridge failure. A number of lessons have been learned from bridge failures that have resulted in revisions to the standard specifications. For example, changes were made to the seismic provisions after the 1971 San Fernando earthquake. Other bridge failure incidents that influence the practice of bridge engineering are given in the sections that follow.

2.2.1 Silver Bridge, Point Pleasant, West Virginia, December 15, 1967

The collapse of the Silver Bridge over the Ohio River between Point Pleasant, West Virginia, and Kanauga, Ohio, on December 15, 1967, resulted in 46 deaths, 9 injuries, and 31 of the 37 vehicles on the bridge fell with the bridge (NTSB, 1970).

Description The Point Pleasant Bridge was a suspension bridge with a main span of 700 ft (213 m) and two equal side spans of 380 ft (116 m). The original design was a parallel wire cable suspension bridge but had provisions for a heat-treated steel eyebar suspension design (Fig. 2.1) that could be substituted if the bidders furnished stress sheets and specifications of the proposed materials. The eyebar suspension bridge design was accepted and built in 1927 and 1928.

Two other features of the design were also unique (Dicker, 1971): The eyebar chains were the top chord of the stiffening truss over a portion of all three spans, and the base of each tower rested on rocker bearings (Fig 2.2). As a result, redundant load paths did not exist, and the failure of a link in the eyebar chain would initiate rapid progressive failure of the entire bridge.

Cause of Collapse The National Transportation Safety Board (NTSB) found that the cause of the bridge collapse was a cleavage fracture in the eye of an eyebar of the north suspension chain in the Ohio side span (NTSB, 1970). The fracture was caused by development of a flaw due to stress corrosion and corrosion fatigue over the 40-year life of the bridge as the pin-connected joint adjusted its position with each passing vehicle.

Effect on Bridge Practice The investigation following the collapse of the Silver Bridge disclosed the lack of regular inspections to determine the condition of existing bridges. Consequently, the National Bridge Inspection Standards (NBIS) were established under the 1968 Federal Aid Highway Act. This act requires that all bridges built with federal monies be inspected at regular intervals not to exceed 2 years. As a result, the state bridge agencies were required to catalog all their bridges in a National Bridge Inventory (NBI). There are over 600,000 bridges (100,000 are culverts) with spans greater than 20 ft (6 m) in the inventory.

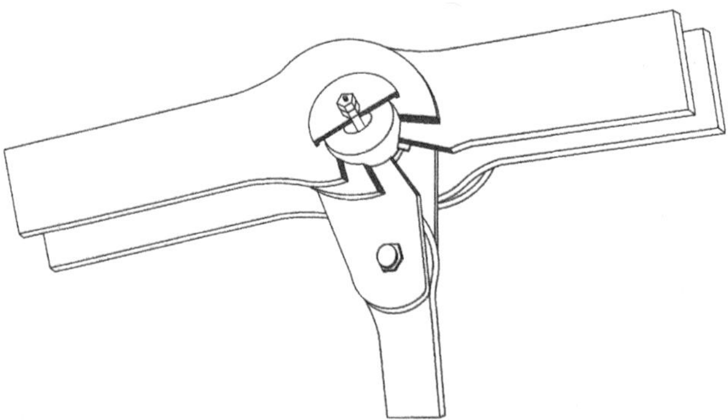

Fig. 2.1 Typical detail of eyebar chain and hanger connection (NTSB, 1970).

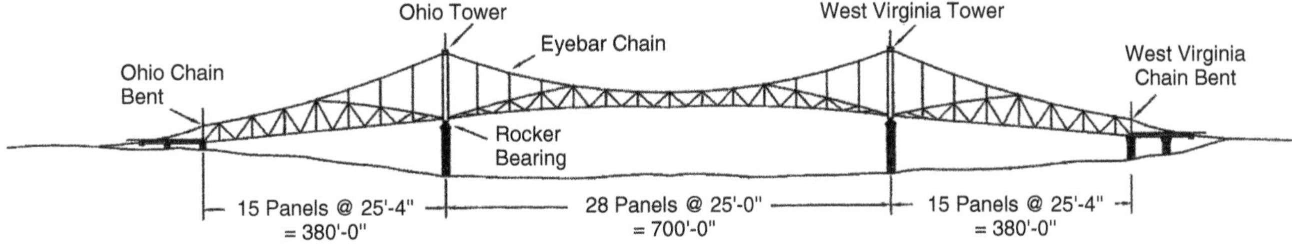

Fig. 2.2 Elevation of Silver Bridge over Ohio River, Point Pleasant, West Virginia (NTSB, 1970).

It is ironic that even if the stricter inspection requirements had been in place, the collapse of the Silver Bridge probably could not have been prevented because the flaw could not have been detected without disassembly of the eyebar joint. A visual inspection of the pin connections with binoculars from the bridge deck would not have been sufficient. The problem lies with using materials that are susceptible to stress corrosion and corrosion fatigue, and in designing structures without redundancy.

2.2.2 I-5 and I-210 Interchange, San Fernando, California, February 9, 1971

At 6:00 a.m. (Pacific Standard Time), on February 9, 1971, an earthquake with a Richter magnitude of 6.6 occurred in the north San Fernando Valley area of Los Angeles. The earthquake damaged approximately 60 bridges. Of this total, approximately 10% collapsed or were so badly damaged that they had to be removed and replaced (Lew et al., 1971). Four of the collapsed and badly damaged bridges were at the interchange of the Golden State Freeway (I-5) and Foothill Freeway (I-210). At this interchange, two men in a pickup truck lost their lives when the South Connector Overcrossing structure collapsed as they were passing underneath. These were the only fatalities associated with the collapse of bridges in the earthquake.

Description Bridge types in this interchange included composite steel girders, precast prestressed I-beam girders, and prestressed and nonprestressed cast-in-place reinforced concrete box-girder bridges. The South Connector Overcrossing structure (bridge 2, Fig. 2.3) was a seven-span, curved, nonprestressed reinforced concrete box girder, carried on single-column bents, with a maximum span of 129 ft (39 m). The North Connector Overcrossing structure (bridge 3, Fig. 2.3) was a skewed four-span, curved, nonprestressed reinforced concrete box girder, carried on multiple-column bents, with a maximum span of 180 ft (55 m). A group of parallel composite steel girder bridges (bridge group 4, Fig. 2.3) carried I-5 North and I-5 South over the Southern Pacific railroad tracks and San Fernando Road. Immediately to the east of this group, over the same tracks and road, was a two-span cast-in-place prestressed concrete box girder (bridge 5, Fig. 2.3) that was carried on a single bent, with a maximum span of 122 ft (37 m).

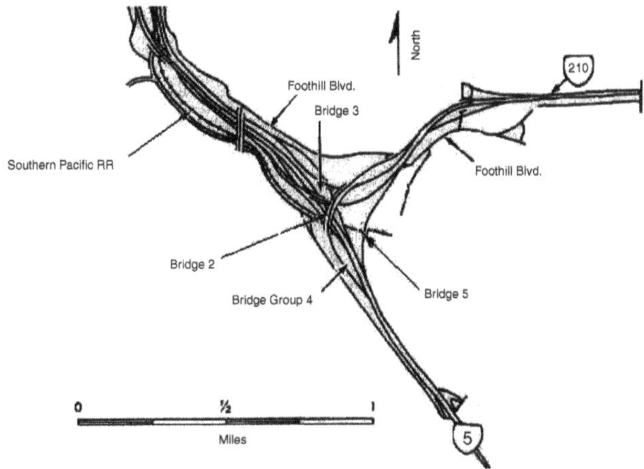

Fig. 2.3 Layout of the I-5 and I-210 Interchange (Lew et al., 1971).

When the earthquake struck, the South Connector structure (Fig. 2.4, center) collapsed on to the North Connector and I-5, killing the two men in the pickup truck. The North Connector superstructure (Fig. 2.4, top) held together, but the columns were bent double and burst their spiral reinforcement (Fig. 2.5). One of the group of parallel bridges on I-5 was also struck by the falling South Connector structure, and two others fell off their bearings (Fig. 2.4, bottom). The bridge immediately to the east suffered major column damage and was removed.

Cause of Collapse More than one cause contributed to the collapse of the bridges at the I-5 and I-210 interchange. The bridges were designed for lateral seismic forces of about 4% of the dead load, which is equivalent to an acceleration of $0.04g$, and vertical seismic forces were not considered. From field measurements made during the earthquake, the estimated ground accelerations at the interchange were from $0.33g$ to $0.50g$ laterally and from $0.17g$ to $0.25g$ vertically. The seismic forces were larger than what the structures were designed for and placed an energy demand on the structures that could not be dissipated in the column–girder and column–footing connections. The connections failed, resulting in displacements that produced large secondary effects, which led to progressive collapse. Girders fell off their supports because the seat dimensions were smaller than

Fig. 2.4 View looking north at the I-5 and I-210 interchange after the quake showing the collapsed South Connector Overcrossing structure (bridge 2) in the center, the North Connector Overcrossing structure (bridge 3) at the top, and bridge group 4 at the bottom. (Photo courtesy E. V. Leyendecker, U.S. Geological Survey.)

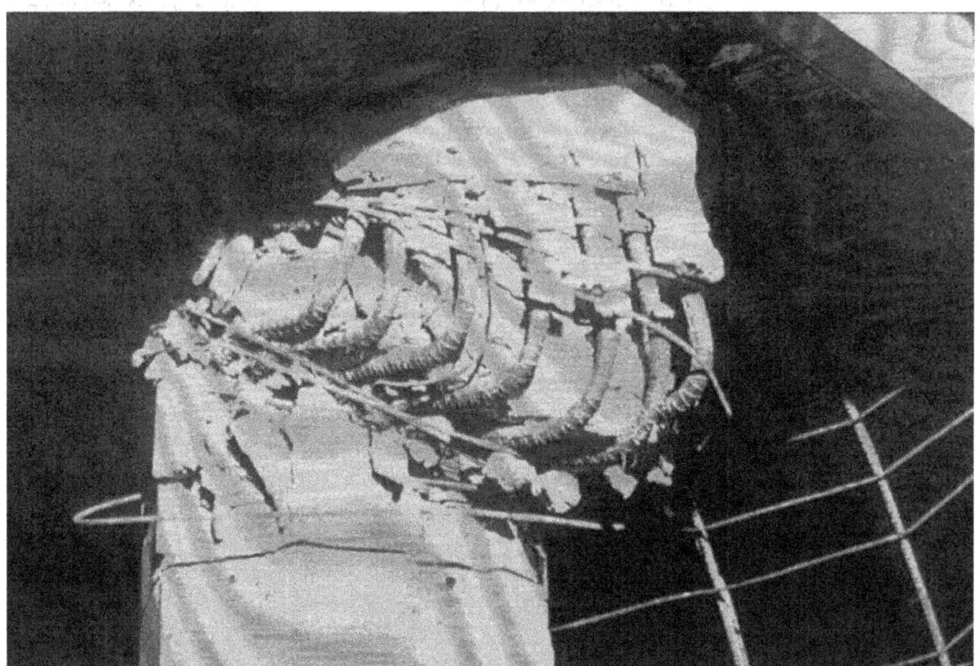

Fig. 2.5 Close up of exterior spiral column in bent 2 of bridge 3. (Photo courtesy E. V. Leyendecker, U.S. Geological Survey.)

the earthquake displacements. These displacement effects were amplified in the bridges that were curved or skewed and were greater in spread footings than in pile-supported foundations.

Effect on Bridge Practice The collapse of bridges during the 1971 San Fernando earthquake pointed out the inadequacies of the lateral force and seismic design provisions of the specifications. Modifications were made and new articles were written to cover the observed deficiencies in design and construction procedures. The issues addressed in the revisions included the following: (1) seismic design forces include a factor that expresses the probability of occurrence of a high-intensity earthquake for a particular geographic region, a factor that represents the soil conditions, a factor that reflects the importance of the structure, and a factor that considers the amount of ductility available in the design; (2) methods of analysis capable of representing horizontal curvature, skewness of span, variation of mass, and foundation conditions; (3) provision of alternative load paths through structural redundancy or seismic restrainers; (4) increased widths on abutment pads and hinge supports; and (5) dissipation of seismic energy by development of increased ductility through closely spaced hoops or spirals, increased anchorage and lap splice requirements, and restrictions on use of large-diameter reinforcing bars. Research is continuing in all of these areas, and the specifications are constantly being revised as new information on seismic safety becomes available.

2.2.3 Sunshine Skyway, Tampa Bay, Florida, May 9, 1980

The ramming of the Sunshine Skyway Bridge by the Liberian bulk carrier *Summit Venture* in Tampa Bay, Florida, on May 9, 1980, destroyed a support pier and about 1297 ft (395 m) of the superstructure fell into the bay. A Greyhound bus, a small pickup truck, and six automobiles fell 150 ft (45 m) into the bay. Thirty-five people died and one was seriously injured (NTSB, 1981).

Description The Sunshine Skyway was comprised of two parallel bridges across Lower Tampa Bay from Maximo Point on the south side of St. Petersburg to Manatee County slightly north of Palmetto, Florida. The twin bridge structures are 4.24 miles (6.82 km) long and consist of posttensioned concrete girder trestles, steel girder spans, steel deck trusses, and a steel cantilever through truss. The eastern structure was completed in 1954 and was one of the first bridges in the United States to use prestressed concrete. The western structure, which was struck by the bulk carrier, was completed in 1971. No requirements were made for structural pier protection.

The main shipping channel was spanned by the steel cantilever through truss (Fig. 2.6) with a center span of 864 ft (263 m) and two equal anchor spans of 360 ft (110 m). The through truss was flanked on either end by two steel deck trusses with spans of 289 ft (88 m). The bulk carrier rammed the second pier south of the main channel that supported the anchor span of the through truss and the first deck span. The collision demolished the reinforced concrete pier and brought down the anchor span and suspended span of the through truss and one deck truss span.

Cause of Collapse The NTSB determined that the probable cause of the accident was the failure of the pilot of the *Summit Venture* to abort the passage under the bridge when the navigational references for the channel and bridge were lost in the heavy rain and high winds of an intense thunderstorm (NTSB, 1981). The lack of a structural pier protection system, which could have redirected the vessel and reduced the amount of damage, contributed to the loss of life. The collapse of the cantilever through truss and deck truss spans of the Sunshine Skyway Bridge was due to the loss of support of the pier rammed by the *Summit Venture* and the progressive instability and twisting failure that followed.

Effect on Bridge Practice A result of the collapse of the Sunshine Skyway Bridge was the development of standards for the design, performance, and location of structural bridge

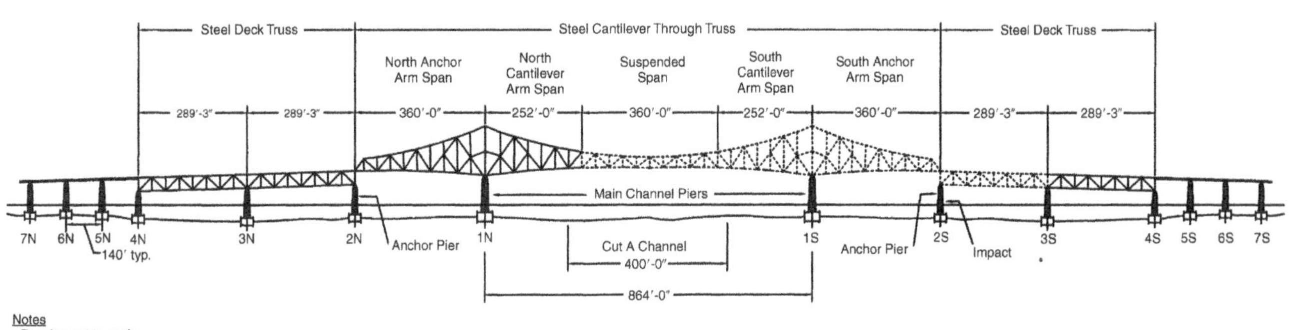

Fig. 2.6 Diagram of the damaged Sunshine Skyway Bridge (looking eastward) (NTSB, 1981).

pier protection systems. Provisions for determining vessel collision forces on piers and bridges are now incorporated in the AASHTO LRFD Bridge Specifications.

2.2.4 Mianus River Bridge, Greenwich, Connecticut, June 28, 1983

A 100-ft (30-m) suspended span of the eastbound traffic lanes of Interstate Route 95 over the Mianus River in Greenwich, Connecticut, collapsed and fell into the river on June 28, 1983. Two tractor-semitrailers and two automobiles drove off the edge of the bridge and fell 70 ft (21 m) into the river. Three people died and three received serious injuries (NTSB, 1984).

Description The Mianus River Bridge is a steel deck bridge of welded construction that has 24 spans, 19 of which are approach spans, and is 2656 ft (810 m) long. The 5 spans over water have a symmetric arrangement about a 205-ft (62.5-m) main span, flanked by a 100-ft (30-m) suspended span and a 120-ft (36.6-m) anchor span on each side (Fig. 2.7). The main span and the anchor span each cantilever 45 ft (13.7 m) beyond their piers to a pin-and-hanger assembly, which connects to the suspended span (Fig. 2.8). The highway is 6 lanes wide across the bridge, but a lengthwise expansion joint on the centerline of the bridge separates the structure into 2 parallel bridges that act independently of each other. The bridge piers in the water are skewed 53.7° to conform to the channel of the Mianus River.

The deck structure over the river consists of two parallel haunched steel girders with floor beams that frame into the girders. The continuous five-span girder has four internal hinges at the connections to the suspended spans and is, therefore, statically determinate. The inclusion of hinges raises the question of redundancy and existence of alternative load paths. During the hearing after the collapse, some engineers argued that because there were two girders, if one pin-and-hanger assembly failed, the second assembly could provide an alternative load path.

The drainage system on the bridge had been altered by covering the curb drains with steel plates when the roadway was resurfaced in 1973 with bituminous concrete. With the curb drains sealed off, rainwater on the bridge ran down the bridge deck to the transverse expansion joints between the suspended span and the cantilever arm of each anchor span. During heavy rainfall, considerable water leaked through the expansion joint where the pin-and-hanger assemblies were located.

After the 1967 collapse of the Silver Bridge, the National Bridge Inspection Standards were established, which required regular inspections of bridges at intervals not exceeding 2 years. ConnDOT's Bridge Safety and Inspection Section had inspected the Mianus River Bridge 12 times since 1967 with the last inspection in 1982. The pin-and-hanger assemblies of the inside girders were observed from a catwalk between the separated roadways, but the pin-and-hanger assemblies connecting the outside girders were visually checked from the ground using binoculars. The inspectors noted there was heavy rust on the top pins from water leaking through the expansion joints.

Cause of Collapse The eastbound suspended span that collapsed was attached to the cantilever arms of the anchor spans

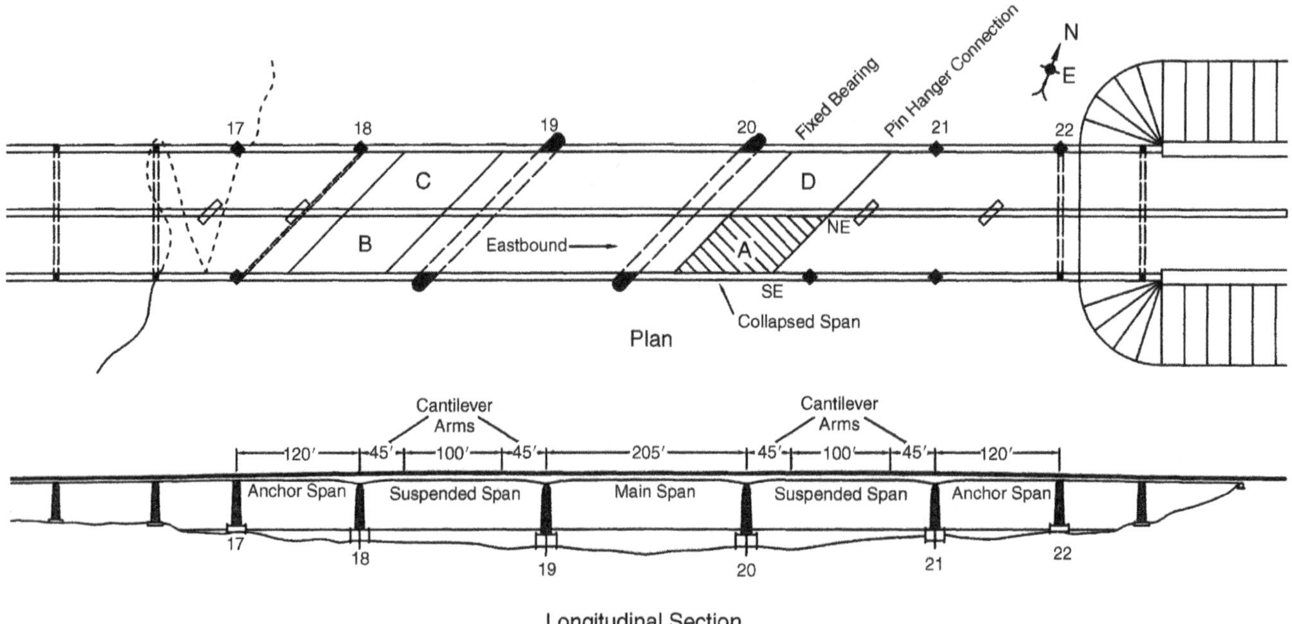

Fig. 2.7 Plan view (top) and longitudinal view (bottom) of the Mianus River Bridge. (Note that the skew of piers 17 through 22 is not depicted in the longitudinal view) (NTSB, 1984).

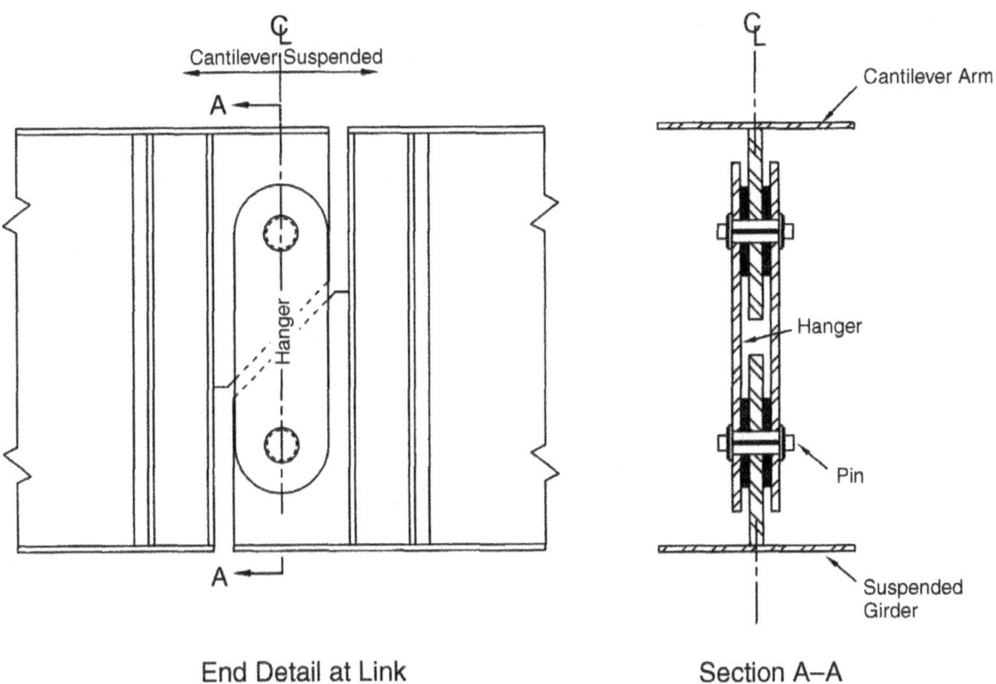

Fig. 2.8 Schematic of pin-and-hanger assembly of the Mianus River Bridge (NTSB, 1984).

at each of its four corners (Fig. 2.7). Pin-and-hanger assemblies were used to support the northeast (inside girder) and southeast (outside girder) corners of the eastern edge of the suspended span. The western edge was attached to the cantilever arms by a pin assembly without hangers. The pin-and-hanger assemblies consist of an upper pin in the cantilever arm and a lower pin in the suspended span connected by two hangers, one on either side of the web (Fig. 2.8).

Sometime before the collapse of the suspended span, the inside hanger at the southeast corner came off the lower pin, which shifted all the weight on this corner to the outside hanger. With time, the outside hanger moved laterally outward on the upper pin. Eventually, a fatigue crack developed in the end of the upper pin, its shoulder fractured, the outside hanger slipped off, and the suspended span fell into the river.

The NTSB concluded that the probable cause of the collapse of the Mianus River Bridge suspended span was the undetected lateral displacement of the hangers in the southeast corner suspension assembly by corrosion-induced forces due to deficiencies in the State of Connecticut's bridge safety inspection and bridge maintenance program (NTSB, 1984).

Effect on Bridge Practice A result of the collapse of the Mianus River Bridge was the development and enforcement of detailed and comprehensive bridge inspection procedures. The Mianus River Bridge was being inspected on a regular basis, but the inspectors had no specific directions as to what the critical elements were that could result in a catastrophic failure.

Another effect of this collapse was the flurry of activity in all the states to inspect all of their bridges with pin-and-hanger assemblies. In many cases, they found similar deterioration and were able to prevent accidents by repair or replacement of the assemblies. In designs of new bridges, pin-and-hanger assemblies have found disfavor and will probably not be used unless special provisions are made for inspectability and maintainability.

The investigation of the collapse also pointed out the importance of an adequate surface drainage system for the roadway on the bridge. Drains, scuppers, and downspouts must be designed to be self-cleaning and placed so that they discharge rainwater and melting snow with de-icing salts away from the bridge structure in a controlled manner.

Perhaps the most important result of the recommendations of the NTSB was the development of the FHWA's fracture-critical bridge inspection program. As was mentioned previously in the Silver Bridge collapse, when the eyebar failed, the whole bridge failed because there was no alternative path for the loads to be carried. In the case of the Mianus River Bridge, when the pin assembly failed, it led to the collapse of the suspended span because the structural system was *non-load-path-redundant*. Both the eyebar and the pin assembly are fracture-critical elements because their failure leads to partial or total failure of the bridge. A bridge that is non-load-path-redundant is not inherently unsafe, but it does lack redundancy in the design of its support structure. Such bridges are sometimes referred to as fracture critical and they require special attention when being inspected. According to FHWA 2007 data, of the 600,000 bridges in the National Bridge Inventory, 19,273 are considered non-load-path-redundant.

2.2.5 Schoharie Creek Bridge, Amsterdam, New York, April 5, 1987

Three spans of the Schoharie Creek Bridge on I-90 near Amsterdam, New York, fell 80 ft (24 m) into a rain-swollen creek on April 5, 1987, when two of its piers collapsed. Four automobiles and one tractor-semitrailer plunged into the creek. Ten people died (NTSB, 1988).

Description The Schoharie Creek Bridge consisted of five simply supported spans of lengths 100, 110, 120, 110, and 100 ft (30.5, 33.5, 36.6, 33.5, and 30.5 m). The roadway width was 112.5 ft (34.3 m) and carried four lanes of highway traffic (Fig. 2.9). The superstructure was composed of two main steel girders 12 ft (3.66 m) deep with transverse floor beams that spanned the 57 ft (17.4 m) between girders and cantilevered 27.75 ft (8.45 m) on either side. Stringers ran longitudinally between the floor beams and supported a noncomposite concrete deck. Members were connected with rivets.

The substructure consisted of four piers and two abutments. The reinforced concrete piers had two columns directly under the two girders and a tie beam near the top (Fig. 2.10). A spread footing on dense glacial deposits supported each pier. Piers 2 and 3 were located in the main channel of Schoharie Creek and were to be protected by riprap. Only the abutments were supported on piles. Unfortunately, in the early 1950s when this bridge was being designed, no reliable method was available to predict scour depth.

The bridge was opened to traffic on October 26, 1954, and on October 16, 1955, the Schoharie Creek experienced its flood of record (1900–1987) of 76,500 cfs (2170 m^3/s). The estimated discharge on April 5, 1987, when the bridge collapsed was 64,900 cfs (1840 m^3/s). The 1955 flood caused slight damage to the riprap, and in 1977 a consulting engineering firm recommended replacing missing riprap. This replacement was never done.

Records show that the Schoharie Creek Bridge had been inspected annually or biennially as required by the National Bridge Inspection Standards of the 1968 Federal Aid Highway Act. These inspections of the bridge were only of the above-water elements and were usually conducted by maintenance personnel, not by engineers. At no time since its completion had the bridge received an underwater inspection of its foundation.

Cause of Collapse The severe flooding of Schoharie Creek caused local scour to erode the soil beneath pier 3, which then dropped into the scour hole, and resulted in the collapse of spans 3 and 4. The bridge wreckage in the creek redirected the water flow so that the soil beneath pier 2 was eroded, and some 90 min later it fell into the scour hole and caused the collapse of span 2. Without piles, the Schoharie Creek Bridge was completely dependent on riprap to protect its foundation against scour and it was not there.

The NTSB determined that the probable cause of the collapse of the Schoharie Creek Bridge was the failure of the New York State Thruway Authority to maintain adequate riprap around the bridge piers, which led to the severe erosion of soil beneath the spread footings (NTSB, 1988). Contributing to the severity of the accident was the lack of structural redundancy in the bridge.

Effect on Bridge Practice The collapse of the Schoharie Creek Bridge resulted in an increased research effort to develop methods for estimating depth of scour in a streambed around bridge piers and for estimating size of riprap to resist a

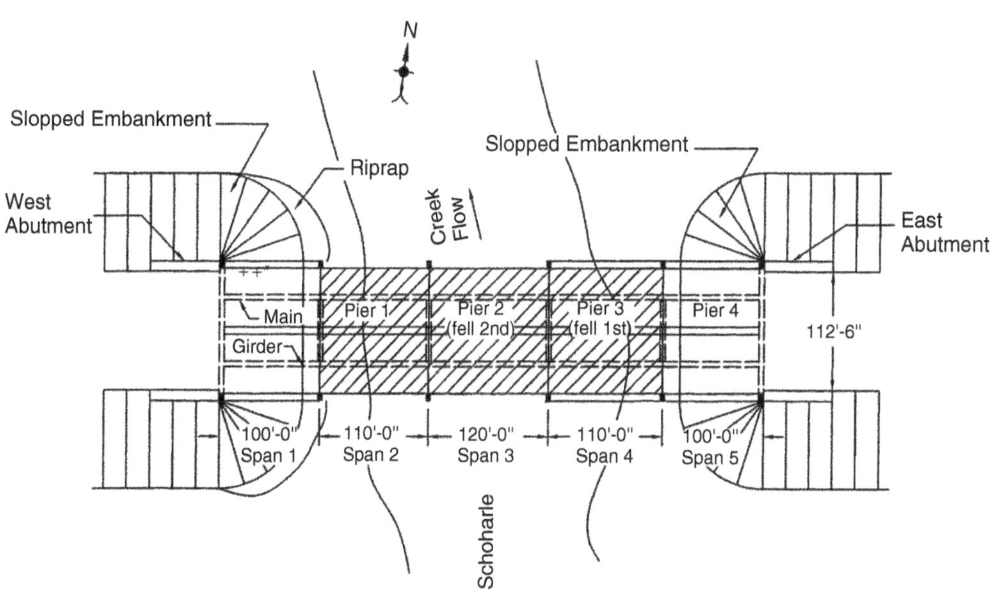

Fig. 2.9 Schematic plan of Schoharie Creek Bridge (NTSB, 1988).

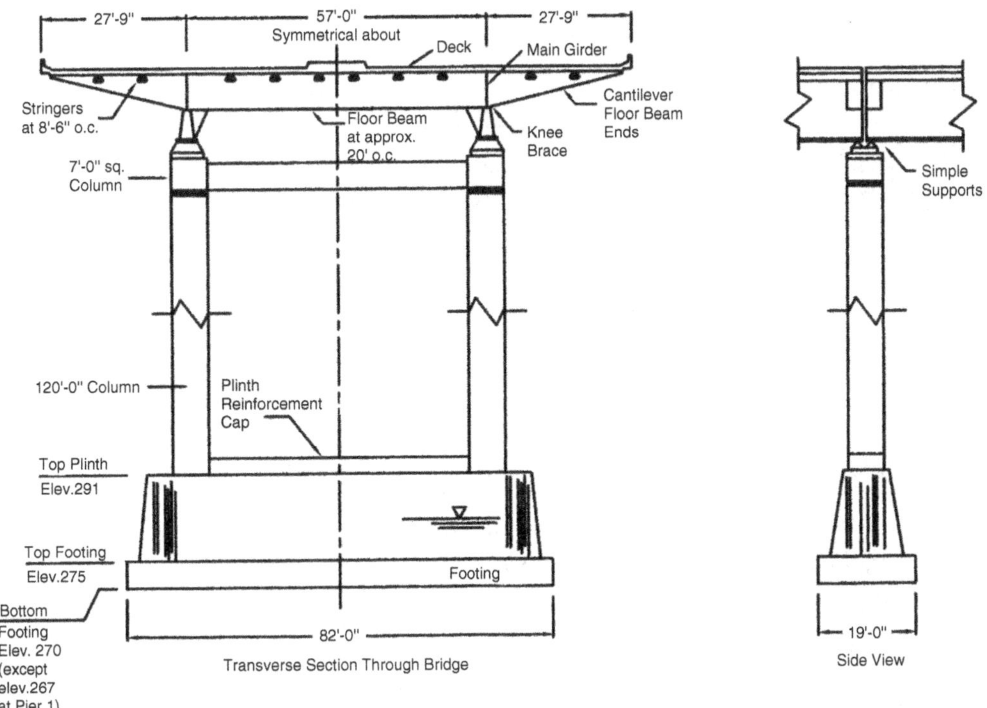

Fig. 2.10 Sections showing the Schoharie Creek Bridge pier supported on a spread footing (NTSB, 1988).

given discharge rate or velocity. Methods for predicting depth of scour are now available.

An ongoing problem that needs to be corrected is the lack of qualified bridge inspection personnel. This problem is especially true for underwater inspections of bridge foundations because there are approximately 300,000 bridges over water and 100,000 have unknown foundation conditions.

Once again the NTSB recommends that bridge structures should be redundant and have alternative load paths. Engineers should finally be getting the message and realize that continuity is one key to a successful bridge project.

2.2.6 Cypress Viaduct, Loma Prieta Earthquake, October 17, 1989

The California Department of Transportation (Caltrans) has been and is a leader in the area of seismic design and protection of bridges. Over the course of many years and numerous earthquakes, Caltrans continues to assess seismic risk, update design procedures, and evaluate existing bridges for catastrophic potential. One of the difficulties, however, is gaining the funding necessary to improve the critical design features and weakness of existing bridges within the inventory.

Description The 1989 Loma Prieta earthquake that occurred on October 17 resulted in over $8 billion in damage and loss of 62 lives. Figure 2.11 illustrates the Cypress Viaduct in Oakland. This bridge was perhaps one of the most reported-on structures by the national media as this double-deck bridge failed in shear within the columns and pancaked the bridge on traffic below.

Cause of Collapse Caltrans was aware of the critical design features that were necessary to provide the ductility and energy absorption required to prevent catastrophic failure. Unfortunately, similar details were common in other bridge substructures designed by the best practices at the time. Caltrans was working on correcting these defects, but with over 13,000 bridges in its inventory and limited resources, engineers had not been able to retrofit the Cypress Viaduct before the earthquake.

Effect on Bridge Practice With Loma Prieta the political will was generated to significantly increase the funding necessary to retrofit hundreds of bridges within the Caltrans inventory. In addition, Caltrans substantially increased its research efforts that has resulted in many of the design specification and construction details used today. From a Caltrans press release (Caltrans, 2003):

> *The Department's current Seismic Safety Retrofit Program was established following the 1989 Loma Prieta earthquake to identify and strengthen bridges that needed to be brought up to seismic safety standards.*

This reference outlines the funding and phases that California has and will use to improve thousands of bridges statewide. As illustrated in several examples in this section, sometime failures are required to provide the catalyst necessary for change either from a technical and/or political perspective.

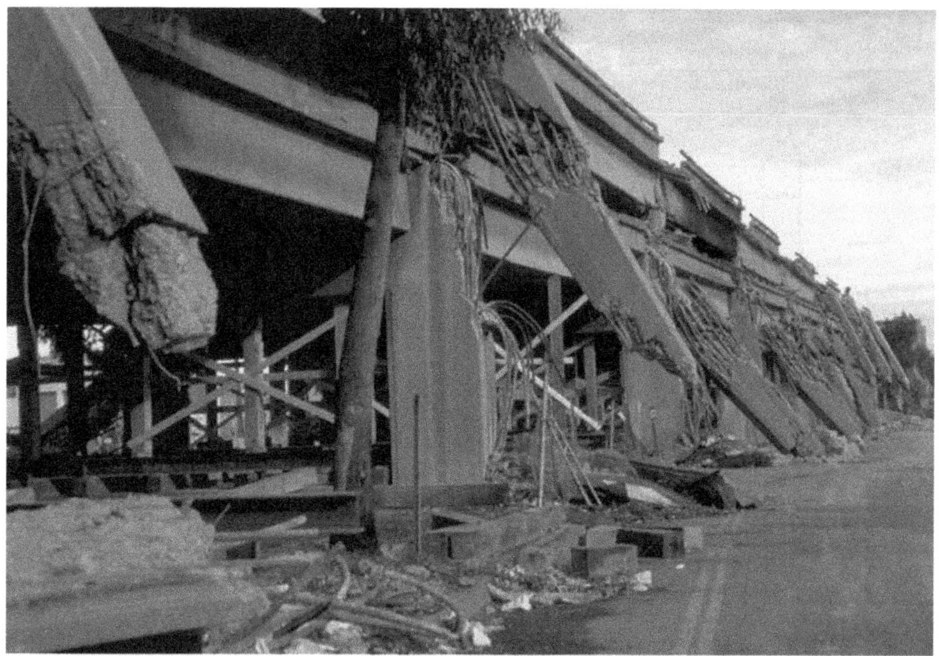

Fig. 2.11 Cypress Viaduct. (Photo courtesy H. G. Wilshire, U.S. Geological Survey.)

Fig. 2.12 Aerial view of I-35W Bridge after collapse (NTSB, 2008).

2.2.7 I-35W Bridge, Minneapolis, Minnesota, August 1, 2007

About 6:05 p.m. central daylight time on Wednesday, August 1, 2007, the eight-lane, 1907-ft (581-m) long I-35W Highway Bridge over the Mississippi River in Minneapolis, Minnesota, experienced a catastrophic failure in the main truss span of the deck truss. As a result, 1000 ft (305 m) of the deck truss collapsed, with about 456 ft (140 m) of the main span falling 108 ft (33 m) into the 15-ft (4.6-m) deep river (Fig. 2.12). A total of 111 vehicles were on the portion of the bridge that collapsed. As a result of the bridge collapse, 13 people died, and 145 were injured (NTSB, 2008).

Description The bridge elevation is shown in Figures 2.13 and 2.14. Eleven of the 14 spans were approach spans to the

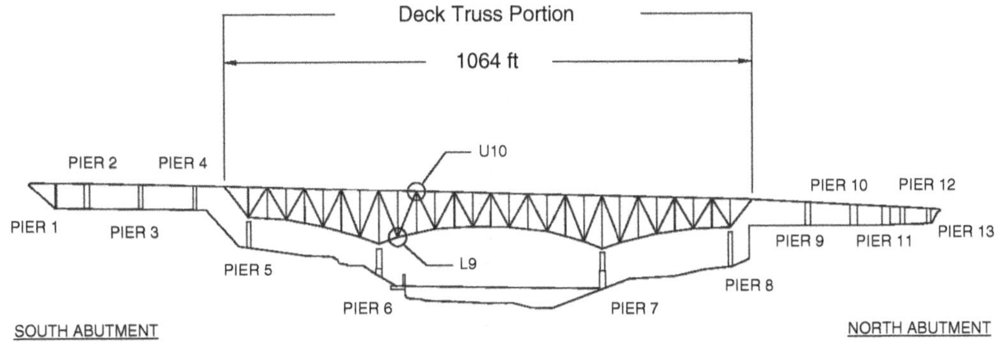

Fig. 2.13 East elevation of I-35W bridge. The deck truss portion of the bridge extends from just south of pier 5 to just north of pier 8 (NTSB, 2008).

Fig. 2.14 Center span of I-35W bridge looking northeast. The center span is supported by pier 6 on the near (south) riverbank and pier 7 on the far (north) riverbank (NTSB, 2008).

deck truss portion that failed. The original bridge design accounted for thermal expansion using a combination of fixed and expansion bearings for the bridge/pier interfaces. For the deck truss portion, a fixed bearing assembly was located at pier 7. Expansion roller bearings were used at piers 5, 6, and 8.

When opened for traffic in 1967, the cast-in-place concrete deck slab had a minimum thickness of 6.5 in. Bridge renovation projects eventually increased the average thickness by about 2 in.

The deck truss portion of the bridge was comprised of two parallel main Warren-type trusses spaced 72 ft 4 in. apart. The upper and lower chords of the main trusses were connected by straight vertical and diagonal members that made up the truss structure. The upper and lower chords were welded box members. The vertical and diagonal members were H members consisting of flanges welded to a web plate.

Riveted steel gusset plates at each of the 112 nodes (connecting points) of the two main trusses tied the ends of the truss members to one another and to the rest of the structure. The gusset plates were riveted to the side plates of the box members and to the flanges of the H members. A typical I-35W main truss node with gusset plates is shown in Figure 2.15.

Cause of Collapse On the day of the collapse, roadway work was underway on the I-35W Bridge. Four of the eight travel lanes were closed to traffic. Construction equipment and piles of sand and gravel were positioned in the deck truss portion of the bridge. The construction loads were in place by about 2:30 p.m. in preparation for a concrete pour that was to begin about 7:00 p.m.

About 6:05 p.m., a motion-activated surveillance video camera at the Lower St. Anthony Falls Lock and Dam, just southwest of the bridge center span, recorded a portion of

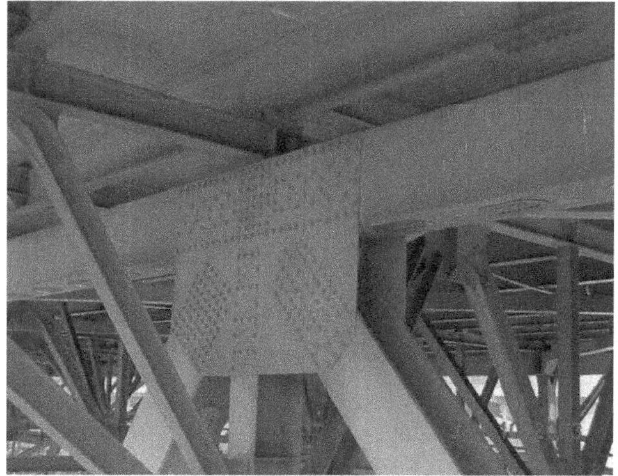

Fig. 2.15 Typical five-member node (two upper chord members, one vertical member, and two diagonal members) on I-35W Bridge (NTSB, 2008).

the collapse sequence. The video showed the bridge center span separating from the rest of the bridge with the south end dropping before the north end and falling into the river (Fig. 2.16). The center section remained relatively level east to west as it fell. Many of the vehicles remained in their lanes as the collapse occurred, indicating that the east and west main trusses at the south end fractured at about the same time (e.g., web search "minnesota dot I-35 bridge failure video").

What elements in the bridge could have fractured simultaneously in both main trusses near the south support to cause the center span to drop to the river in a flat even manner? There was no eyebar chain and hanger connection as in the Silver Bridge; nor was there a pin-and-hanger assembly as in the Mianus River Bridge.

The NTSB searched the bridge inspection reports dating from 1971 to 2006 looking for signs of a possible weak link. One detail that caught their attention was provided by a series of photographs taken in 2003 that showed visible bowing in the gusset plates at several upper truss nodes (see Fig. 2.13). At both U10 nodes, the unsupported edges of the gusset plates between the upper chords and the diagonals were bowed. At the two U10 nodes located on each side, the plate edges between the upper chords and the diagonals were bowed. The U10 gusset plates were the only plates that showed obvious evidence of bowing.

In an interview, the bridge safety inspection engineer stated that he had observed the bowing during his inspections. He said he consulted with another inspector about the bowing and concluded (NTSB, 2008, p. 63):

Our inspections are to find deterioration or findings of deterioration on maintenance. We do not note or describe construction or design problems.

As previously noted, at or near the beginning of the collapse sequence, most of the bridge center span fractured and broke away from the rest of the deck truss structure. Video and physical evidence indicated that the breaks in the span occurred just north of pier 6 and just south of pier 7. Fractures in the south and north fracture areas were at or adjacent to the U10 nodes (NTSB, 2008).

Because much of the bridge center span collapsed, its structural components did not receive detailed inspection until after their removal. The recovered truss portions were laid out relative to their original positions at a nearby park. A gusset plate fracture pattern reconstructed from the pieces at node U10W (west side) is shown in Fig. 2.17. The curved

Fig. 2.16 Collapsed bridge center section, looking southeast (NTSB, 2008).

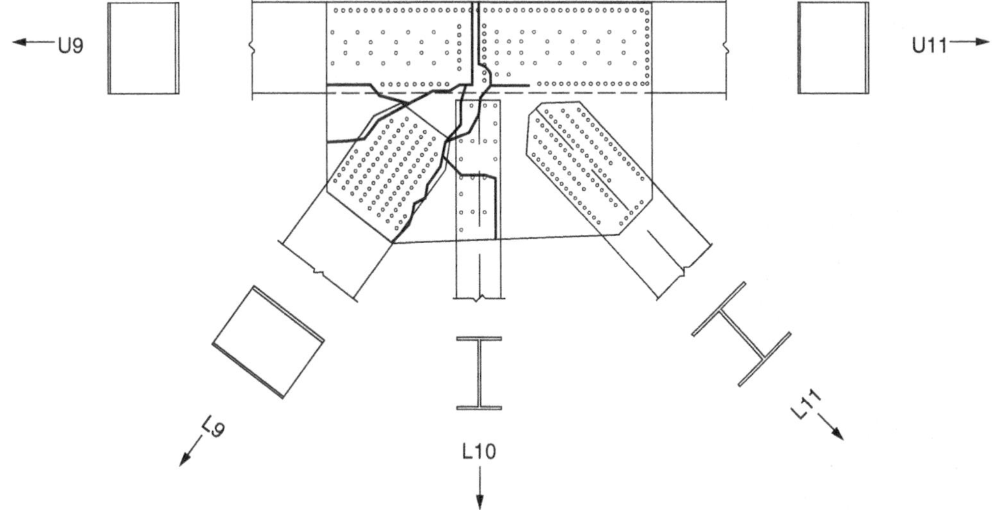

Fig. 2.17 Fracture pattern of outside (west) gusset plate at U10W (NTSB, 2008).

line indicates the fracture, which was similar for both trusses. The diagonal and chords were separated from the remaining members in the node. The remaining portion of the gusset plates kept the vertical and other diagonal connected.

During the investigation a finite-element model of the bridge deck portion with all its components was constructed. The model was able to simulate the behavior of the gusset plates when subjected to different loading conditions. This analysis showed that areas of the U10 gusset plates at the ends of the L9/U10 (see Fig. 2.17) diagonals were beyond their yield stress under the dead load of the initial bridge design. As loads on the bridge increased as a result of the added deck thickness (1977) and barriers (1998), the area of the gusset plates beyond the yield stress expanded, but large deflections were prevented by the surrounding elastic material.

With the added construction and traffic loads on the day of the accident, the areas of yielding increased further, and the finite-element analysis predicted that the failure mode under these conditions would be the unstable lateral shifting of the U10 end of the L9/U10 diagonal. The load-carrying capacity of the gusset plates would be reduced as the bending deformations and yielding increase, resulting in the tensile fracture pattern observed and the tearing away of the L9/U10 diagonal.

The finite-element analysis predicted that the lateral shifting instability of the L9/U10 diagonal would occur first at the U10W node because it was more highly stressed than the U10E node due to the placement of the construction materials. As the load-carrying capacity was reduced at the U10W node, the load would be shed to the U10E node triggering a similar fracture pattern. The failure likely proceeded rapidly, and almost simultaneously, through both the U10W and U10E nodes.

The NTSB therefore concluded that the initiating event was a lateral shifting instability of the upper end of the L9/U10W diagonal member and the subsequent failure of the U10 node gusset plates on the center portion of the deck truss.

The deck truss structure of the I-35W Bridge was non-load-path-redundant, which means that it would lose its entire load-carrying capacity if a single primary load member failed. The failure of the U10 gusset plates led to the sequential separation of the structural members connected to the plates, which placed unsupportable loads on the remainder of the structure. The total collapse followed immediately.

The NTSB determined that the probable cause of the collapse of the I-35W Bridge was due to a design error by the original design firm that resulted in an inadequate load capacity of the gusset plates at the U10 nodes. These plates failed under a combination of substantial increases in the bridge weight from previous modifications and the traffic/construction loads on the bridge on the day of the collapse (NTSB, 2008).

Effect on Bridge Practice A major effect of the collapse of the I-35W Bridge was to direct attention on the importance of proper design, quality control, and inspection of gusset plates. A number of recommendations were given by the NTSB to the FHWA and the AASHTO for implementation.

The recommendations to the FHWA included procedures to detect and correct bridge design errors before the design plans are made final, use of nondestructive evaluation technologies to assess gusset plate condition, and update of the training courses to address inspection techniques and conditions specific to gusset plates.

The recommendations to AASHTO, besides working with the FHWA on quality control, are to modify the *Manual for Bridge Evaluation* (AASHTO, 2008) to include the capacity of gusset plates as part of the load-rating calculations, develop guidelines to ensure that construction loads and stockpiled materials do not overload the structure, develop guidance for responding to potentially damaging conditions

Fig. 2.18 Bridge failure near Golden, Colorado. (Photo from Golden Fire Department Annual Report 2004, Golden, Colorado. http://ci.golden.co.us/files/2004fdreport.pdf.)

in gusset plates, such as corrosion and distortion, and revise the *AASHTO Guide for Commonly Recognized (CoRe) Structural Elements* (AASHTO, 1998) to incorporate this new information.

In addition, the NTSB issued the following safety recommendation to the FHWA on January 15, 2008: For all non-load-path-redundant steel truss bridges within the National Bridge Inventory, require that bridge owners conduct load capacity calculations to verify that the stress levels in all structural elements, including gusset plates, remain within applicable requirements whenever planned modifications or operational changes may significantly increase stresses (NTSB, 2008).

2.2.8 Failures During Construction

Most of the memorable bridge failures and the ones that most affect bridge engineering practice have occurred in structures that were in service for many years. However, in-service bridges are not the source of the most common occurrence of failures. Most failures occur during construction and are likely the most preventable kind of failure. This topic is simply too voluminous to address in this book; however, it certainly warrants discussion. Several books and many references are available; for example, in his landmark book, Feld (1996) outlines many kinds of construction failures including technical details, case studies, and litigation issues.

Discussion of one girder failure that occurred near Golden, Colorado, illustrates the importance of considering the construction process during design and construction (9News.com, 2004). An overpass bridge was being widened with the placement of a steel plate girder along the edge of the existing structure. Construction had terminated for the weekend and the girder was left with some attachments to provide lateral stability. The girder became unstable, fell, and killed three people. An aerial view is illustrated in Figure 2.18. The Web reference provided and the associated video linked on this page illustrate many aspects of this failure from a first-day perspective. Stability is the likely cause of failure and is commonly the cause—either stability of the girders supporting the deck with wet concrete or the stability of temporary formwork/shoring required to support the structure. In later chapters, construction staging is discussed related to the design. Again, see 9News.com to review what can happen when mistakes occur. This particular incident could have killed many more—the failure occurred on a Sunday morning when traffic volume was relatively light.

REFERENCES

AASHTO (1994). *LRFD Bridge Design Specifications*, 1st ed., American Association of State Highway and Transportation Officials, Washington, DC.

AASHTO (1998). *AASHTO Guide for Commonly Recognized (CoRe) Structural Elements*, American Association of State Highway and Transportation Officials, Washington, DC.

AASHTO (2002). *Standard Specification for Highway Bridges*, 17th ed., American Association of State Highway and Transportation Officials, Washington, DC.

AASHTO (2008). *The Manual for Bridge Evaluation*, 1st ed., American Association of State Highway and Transportation Officials, Washington, DC.

AASHTO (2010). *LRFD Bridge Design Specifications*, 5th ed., American Association of State Highway and Transportation Officials, Washington, DC.

Caltrans (California Department of Transportation) (2003). Public Affairs, http://www.dot.ca.gov/hq/paffairs/about/retrofit.htm.

Dicker, D. (1971). "Point Pleasant Bridge Collapse Mechanism Analyzed," *Civil Engineering*, ASCE, Vol. 41, No. 7, July, pp. 61–66.

Edwards, L. N. (1959). *A Record of History and Evolution of Early American Bridges*, University Press, Orono, ME.

Feld, J. (1996). *Construction Failure*, 2nd ed., Wiley, New York.

Gies, J. (1963). *Bridges and Men*, Doubleday, Garden City, NY.

Lew, S. L., E. V. Leyendecker, and R. D. Dikkers (1971). "Engineering Aspects of the 1971 San Fernando Earthquake," *Building Science Series 40*, National Bureau of Standards, U.S. Department of Commerce, Washington, DC.

NTSB (1970). "Collapse of U.S. 35 Highway Bridge, Point Pleasant, West Virginia, December 15, 1967," *Highway Accident Report No. NTSB-HAR-71-1*, National Transportation Safety Board, Washington, DC.

NTSB (1981). "Ramming of the Sunshine Skyway Bridge by the Liberian Bulk Carrier SUMMIT VENTURE, Tampa Bay, Florida, May 9, 1980," *Marine Accident Report No. NTSB-MAR-81-3*, National Transportation Safety Board, Washington, DC.

NTSB (1984). "Collapse of a Suspended Span of Interstate Route 95 Highway Bridge over the Mianus River, Greenwich, Connecticut, June 28, 1983," *Highway Accident Report No. NTSB-HAR-84/03*, National Transportation Safety Board, Washington, DC.

NTSB (1988). "Collapse of New York Thruway (I-90) Bridge over the Schoharie Creek, Near Amsterdam, New York, April 5, 1987," *Highway Accident Report No. NTSB/HAR-88/02*, National Transportation Safety Board, Washington, DC.

NTSB (2008). "Collapse of I-35W Highway Bridge, Minneapolis, Minnesota, August 1, 2007," *Highway Accident Report No. NTSB/HAR-08/03*, National Transportation Safety Board, Washington, DC.

9News.com (2004). "Steel Girder Collapse on to EB I-70 C-470 Crushing SUV," http://www.9news.com/acm_news.aspx?OSGNAME=KUSA&IKOBJECTID=89b09401-0abe-421a-01c8-8ca2cb4df369&TEMPLATEID=5991da4c-ac1f-02d8-0055-99a54930515e.

PROBLEMS

2.1 Before AREA and AASHO formalized the specifications for bridges, how were the requirements for design specified?

2.2 What shortcomings were evident in the collapse of the bridge over Ashtabula Creek in December 1876?

2.3 Explain how continuity is linked to redundancy and its importance in preventing progressive bridge collapse. Use one of the bridge failure examples to illustrate your point.

2.4 Discuss the difficulties often encountered in performing adequate bridge inspections.

2.5 What is a fracture-critical bridge and why does it have special inspection requirements?

2.6 The bowing of the gusset plates on the I-35W Bridge was observed by a bridge inspector prior to the collapse. Why was this observation not noted in the inspection report?

CHAPTER 3

Bridge Aesthetics

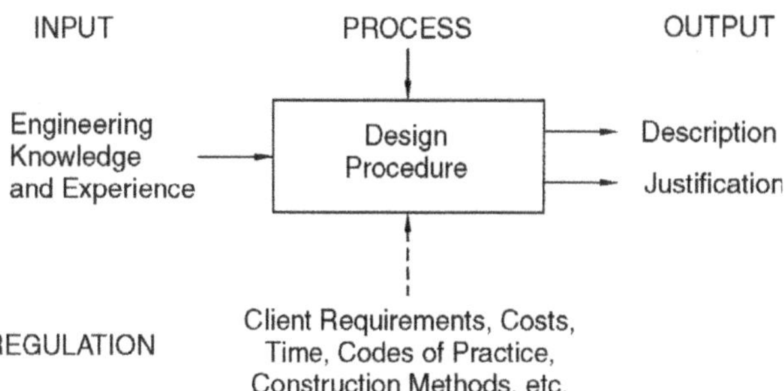

Fig. 3.1 Model of structural design process (Addis, 1990).

3.1 INTRODUCTION

Oftentimes engineers deceive themselves into believing that if they have gathered enough information about a bridge site and the traffic loads, the selection of a bridge type for that situation will be automatic. Engineers seem to subscribe to the belief that once the function of a structure is properly defined, the correct form will follow. Furthermore, that form will be efficient and aesthetically pleasing. Perhaps we believe some great differential equation exists, and, if we could only describe the relationships and the gradients between the different parameters, apply the correct boundary conditions, and set the proper limits of integration, a solution of the equation will give us the best possible bridge configuration. Unfortunately, or perhaps fortunately, no such equation exists that will define the optimal path.

If we have no equation to follow, how is a conceptual design formulated? (In this context, the word *design* is meant in its earliest and broadest sense; it is the configuration one has before any calculations are made.) Without an equation and without calculations, how does a bridge get designed? In this chapter we address this question by first examining the nature of the structural design process and then discussing aesthetics in bridge design.

3.2 NATURE OF THE STRUCTURAL DESIGN PROCESS

The structural design process itself is probably different for every engineer because it is so dependent on personal experience. However, certain characteristics about the process are common and serve as a basis for discussion. For example, we know (1) that when a design is completed in our minds, we must then be able to describe it to others; (2) that we have different backgrounds and bring different knowledge into the design process; and (3) that the design is not completely open ended, constraints exist that define an acceptable solution(s). These characteristics are part of the nature of structural design and influence how the process takes place.

A model of the design process incorporating these characteristics has been presented by Addis (1990) and includes the following components: output, input, regulation, and the design procedure. A schematic of this model is shown in Figure 3.1.

3.2.1 Description and Justification

The output component consists of description and justification. *Description* of the design will be drawings and specifications prepared by or under the direction of the engineer. Such drawings and specifications outline what is to be built and how it is to be constructed. *Justification* of the design requires the engineer to verify the structural integrity and stability of the proposed design.

In describing what is to be built, the engineer must communicate the geometry of the structure and the material from which it is made. At one time the engineer was not only the designer but also the drafter and specification writer. We would sit at our desk, do our calculations, then turn around, maybe climb up on a stool, and transfer the results onto fine linen sheets with surfaces prepared to receive ink from our pens. It seemed to be a rite of passage that all young engineers put in their time on the "board." But then the labor was divided. Drafters and spec writers became specialists, and the structural engineer began to lose the ability to communicate graphically and may have wrongly concluded that designing is mainly performing the calculations.

This trend toward separation of tasks has been somewhat reversed by the increased capabilities of personal computers. With computer-aided drafting (CAD), structural analysis software, and word processing packages all on one system, the structural engineer is again becoming drafter, analyst, and spec writer, that is, a more complete structural designer. In fact, it is becoming necessary for structural engineers to be CAD literate because the most successful structural analysis programs have CAD-like preprocessors and postprocessors.

Justification of a proposed design is where most structural engineers excel. Given the configuration of a structure, its material properties, and the loads to which it is subjected, a structural engineer has the tools and responsibility to verify that a design satisfies all applicable codes and specifications. One note of caution: A structural engineer must not fall into the trap of believing that the verification process is infallible. To provide a framework for this discussion, a few words about deductive and inductive reasoning are required.

Deductive reasoning goes from broad general principles to specific cases. Once the general principles have been established, the engineer can follow a series of logical steps based on the rules of mathematics and applied physics and arrive at an answer that can be defended convincingly. An example would be the principle of virtual work, which can be used for a number of applications such as beam deformations and element stiffness matrices. Just follow the rules, put in the numbers, and the answer has to be correct. Wrong.

Inductive reasoning goes from specific cases to general principles. An example would be going from the experimental observation that doubling the load on the end of a wire doubled its elongation to the conclusion that a linear relationship exists between stress and strain. This conclusion may be true for some materials, and then only with restrictions, but it is not true for others. If experimental observations can be put into the form of an algebraic equation, this is often convenient; however, it is also fallible.

It must be realized that deductive justification is based on quantities and concepts determined inductively. Consider, for example, a structural analysis and design program utilized to justify the adequacy of a reinforced concrete frame. Early on, screens will be displayed on the monitor asking the analyst to supply coordinates of joints, connectivity of the members, and boundary conditions. From this information, the computer program generates a mathematical model of stick members that have no depth, joints that have no thickness, and supports modeled as rollers, hinges, or are completely restrained. Often the mathematical model inductively assumes plane sections remain plane, distributed force values to be concentrated at nodes, and idealized boundary conditions at the supports.

Next, the user is asked to supply constants or parameters describing material behavior, all of which have been determined inductively from experimental observations. Finally, the values of forces at the nodes determined by the equation solvers in the program must be interpreted as to their acceptance in the real world. This acceptance is based on inductively determined safety factors, load and resistance factors, or serviceability criteria. In short, what appears to be infallible deductive justification of a proposed design is, in fact, based on inductive concepts and is subject to possible error and, therefore, is fallible.

Oftentimes engineers select designs on the basis that they are easy to justify. If an engineer feels comfortable with the analysis of a particular bridge type, that bridge configuration will be used again and again. For example, statically determinate bridge structures of alternating cantilever spans and suspended spans were popular in the 1950s before the widespread use of computers because they were easier to analyze. The same could be said of the earlier railroad truss bridges whose analysis was made simple by graphical statics. One advantage of choosing designs that are easily justified is that those responsible for checking the design have no difficulty visualizing the flow of forces from one component to another. Now, with sophisticated computer software, an engineer must understand how forces are distributed throughout the members of more complex systems to obtain a completed design. The advantage of simple analysis of statically determinate structures is easily offset by their lack of redundancy or multiple load paths. Therefore, it is better to choose continuous beams with multiple redundancies even though the justification process requires more effort to ensure that it has been done properly.

Not only is there an interrelationship between deductive and inductive reasoning, there is also an interrelationship between description and justification. The configuration described for a bridge structure determines its behavior. Triangles in trusses, continuous beams, arches, and suspension systems have distinctly different spatial characteristics and, therefore, behave differently. Description and justification are linked together, and it is important that a bridge engineer be proficient in both areas with an understanding of the interactions among them.

3.2.2 Public and Personal Knowledge

The input side of the design process shown in Figure 3.1 includes engineering knowledge and experience. An engineer brings both public and personal knowledge to the design process. Public knowledge is accumulated in books, databases, software, and libraries and can be passed on from generation to generation. Public knowledge includes handbooks of material properties, descriptions of successful designs, standard specifications, theoretical mechanics, construction techniques, computer programs, cost data, and other information too voluminous to describe here.

Personal knowledge is what has been acquired by an individual through experience and is very difficult to pass on to someone else. People with experience seem to develop an intuitive understanding of structural action and behavior. They understand how forces are distributed and how elements can be placed to gather these forces together to carry them in a simple and efficient manner. And if you were to ask them how they do it, they may not be able to explain why they know that a particular configuration will work and another will not. The link between judgment and experience has been explained this way: *Good judgment comes from experience and experience comes from bad judgment*. Sometimes experience can be a tough teacher, but it is always increasing our knowledge base.

3.2.3 Regulation

Our bridge designs are not open ended. There are many constraints that define the boundaries of an acceptable design. These constraints include client's desires, architect's design, relevant codes, accepted practice, engineer's education, available materials, contractor's capabilities, economic factors, environmental concerns, legal factors, and last, but not least, political factors. For example, if a bridge is to traverse coastal wetlands, the restrictions on how it can be built will often dictate the selection of the bridge type. If contractors in a particular region are not experienced in the construction method proposed by an engineer, then that may not be the proper design for that locality. Geometric constraints on alignment are quite different for a rural interstate overcrossing than for a densely populated urban interchange. Somehow a bridge designer must be able to satisfy all these restrictions and still have a bridge with pleasing appearance that remains personally and publicly satisfying.

3.2.4 Design Process

The process of design is what occurs within the rectangular box of Figure 3.1. An engineer knows what the output has to be and what regulations govern the design, but because each person has accumulated different knowledge and experience, it is difficult to describe a procedure for design that will work in all cases. As Addis (1990) says: "Precisely how and why a structural engineer chooses or conceives a particular structure for a particular purpose is a process so nebulous and individual that I doubt if it is possible to study it at all."

It may not be possible to definitively outline a procedure for the design process, but it is possible to identify its general stages. The first is the data gathering stage, followed by the conceptual, rhetorical, and schematic stages. In the data gathering stage, one amasses as much information as one can find about the bridge site, topography, functional requirements, soil conditions, material availability, hydrology, and temperature ranges. Above all, the designer must visit the bridge site, see the setting and its environment, and talk to local people because many of them have probably been thinking about the bridge project for a long time.

The conceptual or creative stage will vary from person to person because we all have different background, experience, and knowledge. But one thing is constant. It all begins with images in the mind. In the mind one can assimilate all of the information on the bridge site, and then mentally build the bridge, trying different forms, changing them, combining them, looking at them from different angles, driving over the bridge, walking under it, all in the mind's eye. Sometimes the configuration comes as a flash of inspiration, other times it develops slowly as a basic design is adjusted and modified in the mind of the designer.

Too often engineers associate solving problems with solving equations. So we are inclined to get out our calculation pad or get on the computer at our earliest convenience. That is not how the creative process works, in fact, putting ideas down on paper too early may restrict the process because the third spatial dimension and the feeling of spontaneity are lost.

Creative breakthroughs are not made by solving equations. Consider the words of Einstein in a letter to his friend Jacque Hadamard:

> *The words or the language, as they are written or spoken, do not seem to play any role in my mechanism of thought. The psychical entities which seem to serve as elements in thought are certain signs and more or less clear images which can be "voluntarily" reproduced and combined... this combinatory plan seems to be the essential feature in productive thought before there is any connection with logical construction in words or other kinds of signs which can be communicated to others. (Friedhoff and Benzon, 1989)*

So, if you thought the great physicist developed his theories using reams of paper and feverishly manipulating fourth-order tensors, that is wrong. You may argue Einstein was a gifted person, very abstract, and what he did would not necessarily apply to ordinary people designing bridges.

Consider then the words of Leonhardt (1982) that follow the data gathering stage:

> *The bridge must then take its initial shape in the imagination of the designer.... The designer should now find a quiet place and thoroughly think over the concept and concentrate on it with closed eyes. Has every requirement been met, will it be well built, would not this or that be better looking...?*

These are words from a successful bridge designer, one of the family so to speak, that presents what he has learned in more than 50 years of designing bridges. When in his distinguished career he realized this truth I do not know, but we should listen to him. *The design of a bridge begins in the mind*.

The rhetorical and schematic stages do not necessarily follow one another sequentially. They are simply stages that occur in the design process and may appear in any order and then reappear again. Once a design has been formulated in the mind, one may want to make some sketches to serve as a basis for discussion with one's colleagues. By talking about the design and in explaining it to others, the features of the design come into sharper focus. If there are any shortcomings, chances are they will be discovered, and improved solutions will be suggested. In addition to being willing to talk about a design, we must also be willing to have it criticized.

In the design procedure outlined by Leonhardt (1982), he encourages a designer to seek criticism by posting sketches of the proposed design on the walls around the office so others can comment on them. It is surprising what additional pairs of trained eyes can see when they look at the sketches. Well, maybe it is no surprise because behind every pair of eyes is a whole different set of experiences and knowledge, which brings to mind what de Miranda (1991) says about the three mentalities that must be brought to the design process:

> *One should be creative and aesthetic, the second analytical, and the third technical and practical, able to give a realistic*

evaluation of the possibilities of the construction technique envisaged and the costs involved. If these three mentalities do not coexist in a single mind, they must always be present on terms of absolute equality in the group or team responsible for the design.

In short, make the sketches, talk about them, make revisions, let others critique them, defend the design, be willing to make adjustments, and keep interacting until the best possible design results. It can be a stimulating, challenging, and intellectually rewarding process.

The function of the design process is to produce a bridge configuration that can be justified and described to others. Now is the time to apply the equations for justification of the design and to prepare its description on plans and in specifications. Computers can help with the analysis and the drawings, but there are still plenty of tasks to keep engineers busy. The computer software packages will do thousands of calculations, but they must be checked. Computer-driven hardware can plot full-size plan sheets, but hundreds of details must be coordinated. Model specifications may be stored in a word processor file, but every project is different and has a unique description. A lot of labor follows the selection of the bridge configuration so it must be done right. As Leonhardt (1982) says:

The phase of conceptual and aesthetic design needs a comparatively small amount of time, but is decisive for the expressive quality of the work.

In Section 3.3 we look more closely at the aesthetic design phase.

3.3 AESTHETICS IN BRIDGE DESIGN

If we recognize that the conceptual design of a bridge begins in the mind, we only need now to convince ourselves that the design we conceive in our mind is inherently beautiful. It is our nature to desire things that are lovely and appeal to our senses. We enjoy good music and soft lights. We furnish our homes with fine furniture and select paintings and colors that please our eyes. We may say that we know nothing about aesthetics, yet our actions betray us. We do know what is tasteful, delights the eye, and is in harmony with its surroundings. Perhaps we have not been willing to express it. We need to realize that it is all right to have an opinion and put confidence in what has been placed within us. We simply need to carry over the love of beauty in our daily lives to our engineering projects.

When an engineer is comparing the merits of alternative designs, some factors are more important than others. The conventional order of priorities in bridge design is safety, economy, serviceability, constructability, and so on. Somewhere down this list is aesthetics. Little doubt exists that aesthetics needs a priority boost and that it can be done without significantly infringing upon the other factors.

In recent years, engineers have come to realize that improved appearance does not necessarily increase the cost. Oftentimes the most aesthetically pleasing bridge is also the least expensive. Sometimes a modest increase in construction cost is required to improve the appearance of a bridge. Menn (1991) states that the additional construction costs are about 2% for short spans and only about 5% for long spans. Roberts (1992) seconds this conclusion in his article on case histories of California bridges.

Public expenditures on improved appearance are generally supported and appreciated. Given a choice, even with a modest increase in initial cost, the public prefers the bridge that has the nicer appearance. Unfortunately, an engineer may realize this after it is too late. Gottemoeller (1991) tells of the dedication of a pedestrian bridge over a railroad track in the heart of a community in which speaker after speaker decried the ugliness of the bridge and how it had inflicted a scar on the city. Function or costs were not primary concerns of the public, only its appearance. Needless to say, they rejected a proposal for constructing a similar bridge nearby. It is unfortunate that an engineer has to build an ugly bridge that will remain long after its cost is forgotten to learn the lesson that the public is concerned about appearance.

It is not possible in this short chapter to completely discuss the topic of bridge aesthetics. Fortunately, good references are dedicated to the subject, which summarize the thoughts and give examples of successful bridge designers throughout the world. Two of these resources are *Esthetics in Concrete Bridge Design*, edited by Watson and Hurd (1990), and *Bridge Aesthetics Around the World*, edited by Burke (1991). A third reference of note is *Bridgescape: The Art of Designing Bridges* by Gottemoeller (2004). By drawing on the expertise in these references, we will attempt to identify those qualities that most designers agree influence bridge aesthetics and to give practical guidelines for incorporating them into medium- and short-span bridges.

3.3.1 Definition of Aesthetics

The definition of the word *aesthetics* may vary according to the dictionary one uses. But usually it includes the words *beauty*, *philosophy*, and *effect* on the senses. A simple definition could be: *Aesthetics is the study of qualities of beauty of an object and of their perception through our senses*. Fernandez-Ordóñez (1991) has some wonderful quotations from the philosophers, such as:

Love of beauty is the cause of everything good that exists on earth and in heaven. (Plato)

and

Even if this particular aesthetic air be the last quality we seen in a bridge, its influence nonetheless exists and has an influence on our thoughts and actions. (Santayana)

and

> *It is impossible to discover a rule that can be used to judge what is beautiful and what is not. (Hegel)*

The last quote from Hegel seems to contradict what we propose to do in providing guidelines for aesthetically pleasing bridge designs. However, in another sense, it reinforces that some equation set or codification does not exist that will outline how to design a bridge. Lack of codification should not discourage attempts to find basic principles for aesthetic design utilized by successful bridge designers.

From the noted philosophers, it is difficult to argue against making something beautiful. Not everyone agrees about the elements that make a bridge beautiful, but it is important that designers be aware of the qualities that influence the *perception* of beauty.

3.3.2 Qualities of Aesthetic Design

In the articles compiled by Watson and Hurd (1990) and Burke (1991) and the book by Gottemoeller (2004), it becomes apparent that writers on bridge aesthetics agree on a number of qualities incorporated in successful aesthetic designs. These qualities are function, proportion, harmony, order, rhythm, contrast, texture, and use of light and shadow.

Some of these terms are familiar; others may not be, especially in the application to bridges. To explain, each term is discussed along with illustrations of its application.

Function For a bridge design to be successful, it must fulfill the purpose for which it is intended. Oftentimes the function of a bridge goes beyond the simple connection of points along a prescribed alignment with a given volume of traffic. For example, a bridge crossing a valley may have the function of safely connecting an isolated community with the schools and services of a larger community by avoiding a dangerous trip down and up steep and twisting roads. A bridge over a railroad track out on the prairie may have the function of eliminating a crossing at grade that claimed a number of lives. Sometimes a bridge has more than one function, such as the bridge across the Straits of Bosporus at Istanbul (Fig. 3.2). This bridge replaces a slow ferryboat trip,

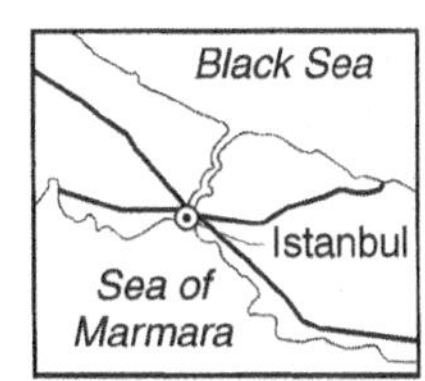

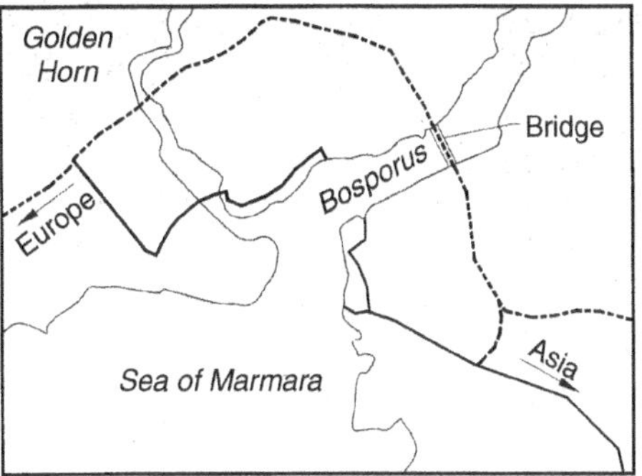

Fig. 3.2 Bosporus Straits Bridge at Istanbul (Brown et al., 1976). (Photo courtesy of Turkish Government Tourism Office, Washington, DC.)

but it also serves the function of connecting two continents (Brown et al., 1976).

Another example of a bridge with multiple functions is the Hoover Dam Bypass (Exhibit 11 in the color insert) or the Mike O'Callaghan–Pat Tillman Memorial Bridge spanning the Colorado River 1600 ft downstream from the dam. The bypass and bridge were constructed in 2005–2010 to improve safety, security, and traffic capacity. The section of U.S. 93 that approached and crossed Hoover Dam was narrow, had many dangerous curves, and poor sight distances. As a consequence of heightened security measures following the September 11, 2001, attacks, truck traffic over the dam was diverted south in an effort to safeguard the dam. Combined with sightseeing and pedestrian traffic at the dam, traffic often came to a standstill. The function of the bridge was to improve travel times, replace the dangerous roadway, and reduce the possibility of an attack or accident at the dam site (www.hooverdambypass.org/purpose_overview.htm).

The function of a bridge must be defined and understood by the designer, client, and public. How that function is satisfied can take many forms, but it must always be kept in mind as the basis for all that follows. Implied with the successful completion of a bridge that fulfills its function is the notion that it does so safely. If a bridge disappears in a flood, or other calamity, one does not take much comfort in the fact that it previously performed its function. A bridge must safely perform its function with an acceptably small probability of failure.

Proportion Artists, musicians, and mathematicians realize that for a painting, a composition, or a geometric pattern to be pleasing it must be in proper proportion. Consider the simple case of dividing a line into two segments. Dividing the line into unequal segments generates more interest than division into equal segments. Around 300 BC, Euclid proposed that a pleasing division of the line would be when the ratio of the shorter segment to the longer segment was the same as the ratio of the longer segment to the whole. Stating Euclid's proposition mathematically, if the total length of the line is x and the longer segment is unity, then the shorter segment is $x - 1$ and the equality of ratios gives $(x - 1)/1 = 1/x$. The positive root of the resulting quadratic equation is $(\sqrt{5} + 1)/2 = 1.6180339\ldots$ or simply, 1.618. This ratio of the total length of the line to the longer segment has been called the golden ratio, the golden proportion, the golden section, and the golden number.

This particular proportion between two values is not limited to mathematics but is found in biology, sculpture, painting, music, astronomy, and architecture (Livio, 2002). Throughout history, the ratio for length to width of rectangles of 1.618 has been considered the most pleasing to the eye. For example, there are golden section rectangles down to the smallest details of decoration throughout the Parthenon in Athens, Greece.

There still are advocates (Lee, 1990) of geometric controls on bridge design and an illustration of the procedure is given in Figure 3.3. The proportioning of the Mancunian Way Bridge cross section in Manchester, England, was carried out by making a layout of golden section rectangles in four columns and five rows. The three apexes of the triangles represent the eye-level position of drivers in the three lanes of traffic. The profile of the cross section was then determined by intersections of these triangles and the golden sections.

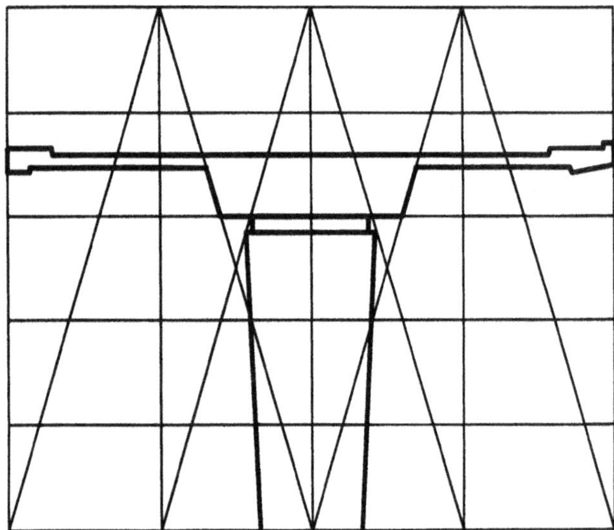

Fig. 3.3 Proportioning of Mancunian Way cross section (Lee, 1990). (Used with permission of American Concrete Institute.)

It may be that proportioning by golden sections is pleasing to the eye, but the usual procedure employed by successful designers has more freedom and arriving at a solution is often by trial and error. It is generally agreed that when a bridge is placed across a relatively shallow valley, as shown in Figure 3.4, the most pleasing appearance occurs when there are an odd number of spans with span lengths that decrease going up the side of the valley (Leonhardt, 1991). The Smart Road Bridge (Exhibit 12 in the color insert) near Blacksburg, Virginia, illustrates the application of this principle.

When artists comment on the composition of a painting, they often talk about negative space. What they mean is the space in between—the empty spaces that contrast with and help define the occupied areas. Negative space highlights what is and what is not. In Figure 3.4, the piers and girders frame the negative space, and it is this space in between that must also have proportions that are pleasing to the eye.

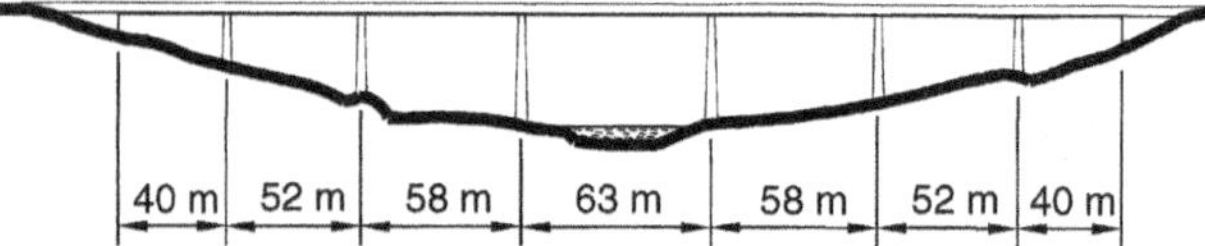

Fig. 3.4 Bridge in shallow valley: flat with varying spans; harmonious (Leonhardt, 1991). (From *Bridge Aesthetics around the World*, copyright © 1991 by the Transportation Research Board, National Research Council, Washington, DC. Reprinted with permission.)

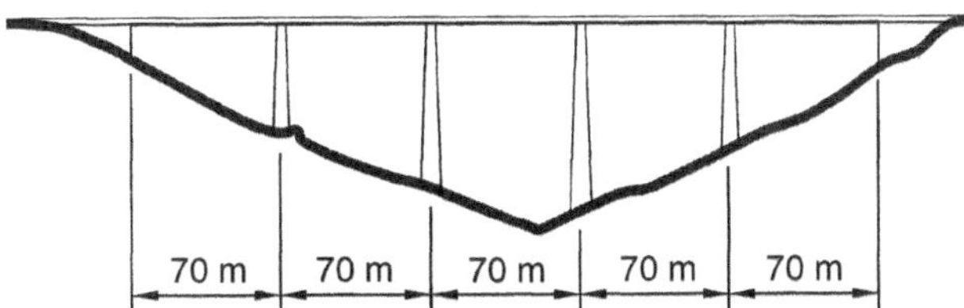

Fig. 3.5 Bridge in deep V-shaped valley: large spans and tapered piers (Leonhardt, 1991). (From *Bridge Aesthetics around the World*, copyright © 1991 by the Transportation Research Board, National Research Council, Washington, DC. Reprinted with permission.)

The bridge over a deep valley in Figure 3.5 (Leonhardt, 1991) again has an odd number of spans, but they are of equal length. In this case, the negative spaces provide a transition of pleasing rectangular shapes from vertical to horizontal. Adding to the drama of the bridge is the slender continuous girder and the tall, tapered piers. An example of such a bridge is the Magnan Viaduct, near Nizza on the French Riviera, shown in Figure 3.6 (Muller, 1991).

Another consideration is the relative proportion between piers and girders. From a strength viewpoint, the piers can be relatively thin compared to the girders. However, when a bridge has a low profile, the visual impression can be improved by having strong piers supporting slender girders. This point is illustrated in Figure 3.7 (Leonhardt, 1991).

Slender girders can be achieved if the superstructure is made continuous. In fact, Wasserman (1991) says that superstructure continuity is the most important aesthetic consideration and illustrates this with the two contrasting photographs in Figures 3.8 and 3.9. Most people would agree that the bridge in Figure 3.9 is awkward looking. It shows what can happen when the least effort by a designer drives a project. It does not have to be that way. Consider this

Fig. 3.6 Magnan Viaduct near Nizza, France (Muller, 1991). (From *Bridge Aesthetics Around the World*, copyright © 1991 by the Transportation Research Board, National Research Council, Washington, DC. Reprinted with permission.)

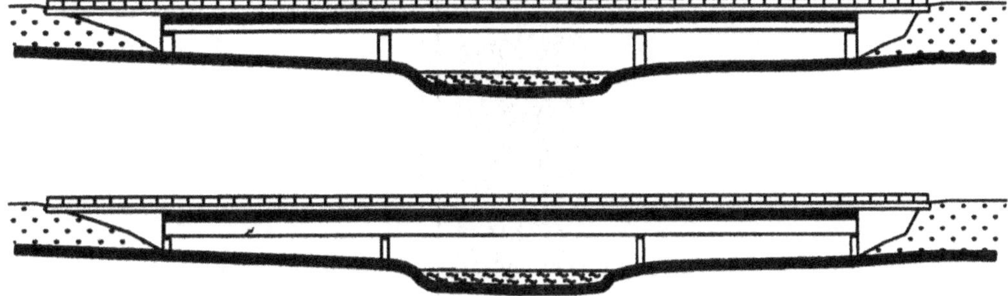

Fig. 3.7 Three-span beam: (*top*) pleasing appearance of slender beam on strong piers; (*bottom*) heavy appearance of deep beam on narrow piers (Leonhardt, 1991). (From *Bridge Aesthetics around the World*, copyright © 1991 by the Transportation Research Board, National Research Council, Washington, DC. Reprinted with permission.)

Fig. 3.8 Example of superstructure continuity on single hammerhead piers. (Photo courtesy of Tennessee DOT—Geo. Hornal, photographer.)

Fig. 3.9 Example of poor depth transitions and awkward configurations due to lack of superstructure continuity. (Photo courtesy of Tennessee DOT.)

Fig. 3.10 Single columns increase the transparency of tall bridge (Menn, 1991). (From *Bridge Aesthetics around the World*, copyright © 1991 by the Transportation Research Board, National Research Council, Washington, DC. Reprinted with permission.)

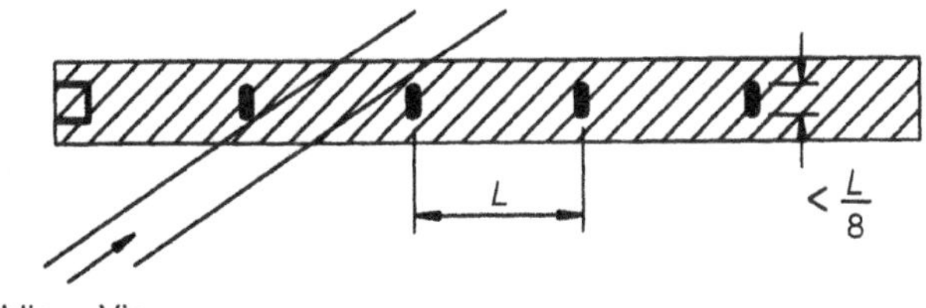

Fig. 3.11 Proportion for pier width not to exceed one-eighth of the span (Leonhardt, 1991). (From *Bridge Aesthetics around the World*, copyright © 1991 by the Transportation Research Board, National Research Council, Washington, DC. Reprinted with permission.)

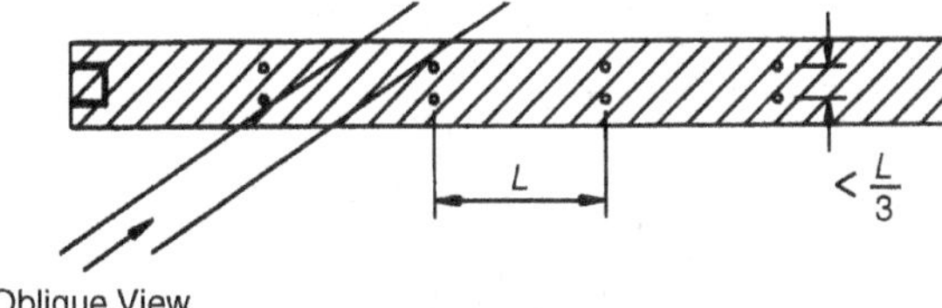

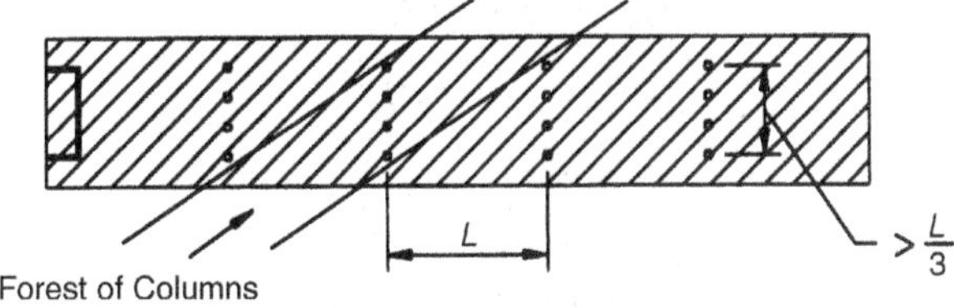

Fig. 3.12 Proportion for total width of groups of columns not to be larger than one-third of the span (Leonhardt, 1991). (From *Bridge Aesthetics around the World*, copyright © 1991 by the Transportation Research Board, National Research Council, Washington, DC. Reprinted with permission.)

quotation from Gloyd (1990): "When push comes to shove, the future generation of viewers should have preference over the present generation of penny pinchers."

A designer should also realize that the proportions of a bridge change when viewed from an oblique angle as seen in Figure 3.10 (Menn, 1991). To keep the piers from appearing as a wall blocking the valley, Leonhardt (1991) recommends limiting the width of piers to about one-eighth of the span length (Fig. 3.11). He further recommends that if groups of columns are used as piers, their total width should be limited to about one-third of the span length (Fig. 3.12). The bridge crossing the broad valley of the Mosel River (Moseltal-brücke) (Exhibit 13 in color insert) in southern Germany is a good example of tall piers with thin girders that give a pleasing appearance when viewed obliquely.

Good proportions are fundamental to achieving an aesthetically pleasing bridge structure. Words can be used to describe what has been successful for some designers, but what works in one setting may not work in another. Rules and formulas will most likely fail. It finally gets down to the responsibility of each designer on each project to make the personal choices that lead to a more beautiful structure.

Harmony In this context, harmony means getting along well with others. The parts of the structure must be in

Fig. 3.13 A graceful long bridge over a wide valley, Napa River Bridge, California. (Permission granted by California DOT.)

Fig. 3.14 Lack of harmony between adjacent bridges (Murray, 1991). (From *Bridge Aesthetics around the World*, copyright © 1991 by the Transportation Research Board, National Research Council, Washington, DC. Reprinted with permission.)

agreement with each other, and the whole structure must be in agreement with its surroundings.

Harmony between the elements of a bridge depends on the proportions between the span lengths and depth of girders, height and size of piers, and negative spaces and solid masses. The elements, spaces, and masses of the bridge in Figure 3.13 present a pleasing appearance because they are in harmony with one another. An example of lack of harmony between members and spaces is shown in Figure 3.14. This dissonance is caused by the placement of two dissimilar bridges adjacent to one another.

Harmony between the whole structure and its surroundings depends on the scale or size of the structure relative to its environment. A long bridge crossing a wide valley (Fig. 3.13) can be large because the landscape is large. But when a bridge is placed in an urban setting or used as an interstate overpass, the size must be reduced. Menn (1991) refers to this as integration of a bridge into its surroundings. Illustrations of bridges that are in harmony with their environment are the overpass in Figure 3.15 and the Blue Ridge Parkway (Linn Cove) Viaduct (Exhibit 14 in color insert) of Figure 3.16. A bridge that is in harmony derives its size and scale from its surroundings.

Fig. 3.15 Well-proportioned concrete arch, West Lilac Road overpass, I-15. (Permission granted by California DOT.)

Fig. 3.16 Blue Ridge Parkway (Linn Cove) Viaduct, Grandfather Mountain, North Carolina (Gottemoeller, 1991). (From *Bridge Aesthetics Around the World*, copyright © 1991 by the Transportation Research Board, National Research Council, Washington, DC. Reprinted with permission.)

Order and Rhythm When discussing order and rhythm in bridge structures, the same words and examples are often used to describe both. For example, the bridge in Figure 3.17 illustrates both good order and rhythm. The eye probably first sees the repeating arches flowing across the valley with the regularity of a heartbeat. But also one perceives that all of the members are tied together in an orderly manner in an uninterrupted flow of beauty with a minimum change of lines and edges. If a girder were to replace one of the arch spans, the rhythm would be lost. Rhythm can bring about order, and good order can bring about a wholeness and unity of the structure.

Fig. 3.17 Tunkhannock Viaduct near Nicholson, Pennsylvania, designed by A. Burton Cohen. (Jet Lowe, HAER Collection, Historic American Engineering Record.)

The use of the same words to describe music and bridge aesthetics is apparent. Consider these comments by Grant (1990):

> *There is beauty and order in classical music—in the harmonies of different sounds, and in their disharmonies and rhythms. There is equal beauty in geometric and arithmetic relationships, similar or equal to those of the sounds.*

An example of a bridge that exhibits beauty in geometric and arithmetic relationships is the Francis Scott Key Bridge (Exhibit 15 in color insert) over the Potomac at Georgetown, Washington, DC.

There is a downside to this analogy with music when repetition and rhythm become excessive. Repeating similar spans too many times can become boring and monotonous, just as hearing the same music with a heavy beat that is repeated over and over again can be uncomfortably similar to driving down the interstate and seeing the same standard overcrossing mile after mile. The first one or two look just fine, but after a while one has to block out appreciating the bridges to keep the mind from the monotony.

Contrast and Texture Contrast, as well as harmony, is necessary in bridge aesthetics. As often present in music and in paintings, bright sounds and bright colors are contrasted with soft and subtle tones—*all* in the same composition. Incorporation of these into our bridges keeps them from becoming boring and monotonous.

All bridges do not have to blend in with their surroundings. Fernandez-Ordóñez (1991) quotes the following from Eduardo Torroja:

> *When a bridge is built in the middle of the country, it should blend in with the countryside, but very often, because of its proportions and dynamism, the bridge stands out and dominates the landscape.*

This dominance seems to be especially true of cable-stayed and suspension bridges, such as seen in Figures 3.18 and 3.19. This dominance of the landscape does not subtract from their beauty.

Another example of a bridge that is in contrast with its surroundings yet is compatible with its urban setting is the Leonard P. Zakim Bunker Hill Memorial Bridge (Exhibit 16 in color insert) in Boston, Massachusetts.

Contrast between the elements of a bridge may emphasize the slenderness of the girders and the strength of the piers and abutments. Texture can be used to soften the hard appearance of concrete and make certain elements less dominant. Large bridges seen from a distance must develop contrast through their form and mass, but bridges with smaller spans seen up close can effectively use texture. A good example of the use of texture is the I-82 Hinzerling Road undercrossing (Exhibit 17 in color insert) near Prosser, Washington, shown in Figure 3.20. The textured surfaces on the solid concrete barrier and the abutments have visually reduced the mass of these elements and made the bridge appear to be more slender than it actually is.

The interchange between the Red Mountain Freeway (202) and U.S. 60 (Exhibit 18 in color insert) in Mesa, Arizona, is a great example of using Southwest-type motif, texture, and color to create a beautiful blend of tall piers and gracefully curved girders.

Light and Shadow To use this quality effectively, the designer must be aware of how shadows occur on the structure throughout the day. If the bridge is running north and south,

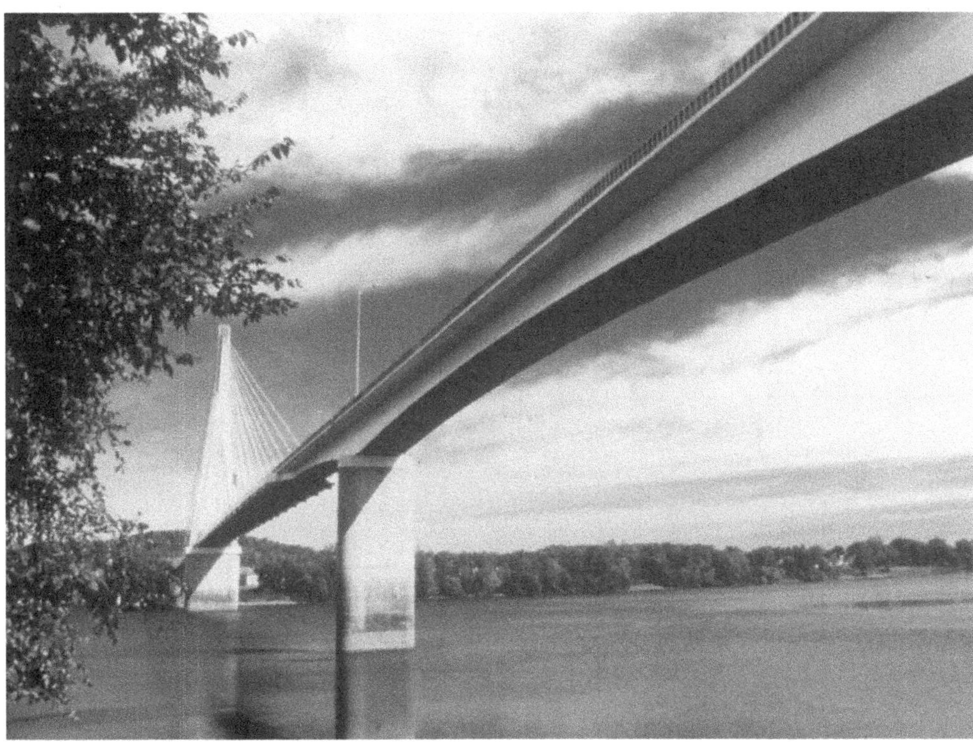

Fig. 3.18 East Huntington Bridge in Huntington, West Virginia. (Photo by David Bowen, courtesy of West Virginia DOT.)

Fig. 3.19 Brooklyn Bridge, New York City. (Jet Lowe, HAER Collection, Historic American Engineering Record.)

Fig. 3.20 Texture reduces visual mass, I-82 Hinzerling Road undercrossing, Prosser, Washington. (Photo courtesy of Washington State DOT.)

Fig. 3.21 Concrete barrier wall and short-span overpass without shadow: girders look deeper.

the shadows are quite different than if it is running east and west. When sunlight is parallel to the face of a girder or wall, small imperfections in workmanship can cast deep shadows. Construction joints in concrete may appear to be discontinuous. In steel hidden welded stiffeners may no longer be hidden due to changes in reflectivity of a web surface.

One of the most effective ways to make a bridge girder appear slender is to put it partially or completely in shadow. Creating shadow becomes especially important with the use of solid concrete safety barriers that make the girders look deeper than they actually are (Fig. 3.21). Shadows can be accomplished by cantilevering the deck beyond the exterior

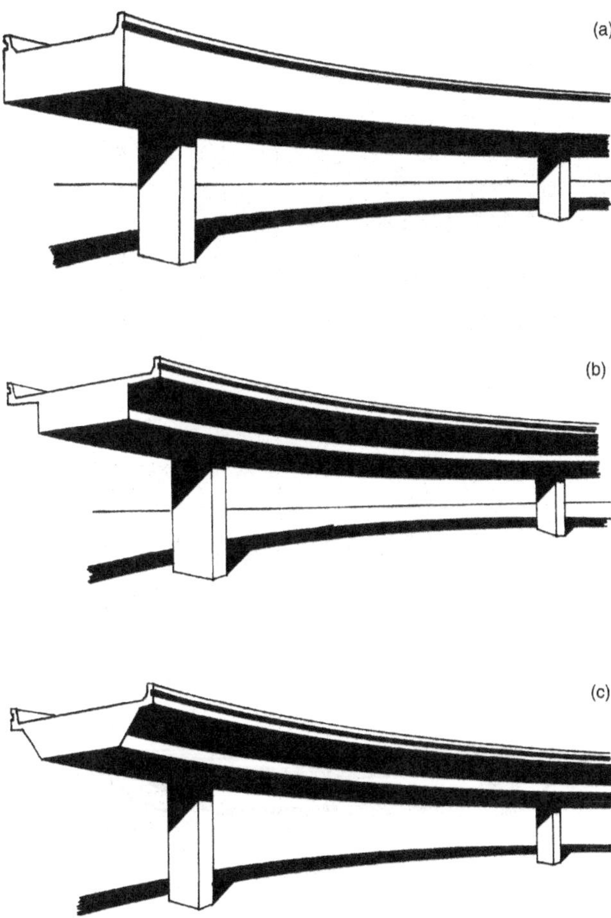

Fig. 3.22 (a) Vertical girder face without overhang presents a visual impact to the driver: Structure looks deeper. (b) Increase in overhang creates more shade on face of girder, subduing the visual impact. (c) Sloping girders recede into shadow. Brightly lit face of barrier rail contrasts with shadow and stands out as a continuous, slender band of light, accentuating the flow of the structure. Structure appears subdued, inviting flow of traffic beneath. (Permission granted by the California DOT.)

girder as shown in Figure 3.22. The effect of shadow on a box girder is further improved by giving the side of the girder an inward slope.

Shadow and light have been used effectively in the bridges shown previously. The piers in the bridge of Figure 3.6 have ribs that cast shadows and make them look thinner. The deck overhangs of the bridges in Figures 3.8, 3.10, 3.15, and 3.20 cause changes in light and shadow that improve their appearance because the girders appear more slender and the harshness of a bright fascia is reduced.

The I-35W St. Anthony Falls Bridge (Exhibit 19 in color insert) in Minneapolis, Minnesota, built to replace the collapsed I-35W Bridge (see Section 2.2.7), is another example of using light and shadow effectively.

3.3.3 Practical Guidelines for Medium- and Short-Span Bridges

The previous discussion on qualities of aesthetic design was meant to apply to all bridges in general. However, what works for a large bridge may not work for a small bridge. Medium- and short-span bridges have special problems. We address those problems and offer a few practical solutions that have worked for other designers. Most of the illustrations used are those of highway-grade separations and crossings over modest waterways.

One word of caution is in order before presenting these guidelines and that is to reemphasize that rules and formulas will likely fail. Burke (1990) provides excerpts from the literature of bridge aesthetics warning against using them exclusively. However, for the inexperienced designer, or for one who does not feel particularly gifted artistically, the guidelines may be helpful. The following quotation taken from Burke (1990) is by Munro (1956) and presents a balanced approach: "Although it is wise to report all past theories of aesthetics with some suspicion, it is equally wise to utilize them as suggestions." Therefore, let us consider the guidelines that follow as simply recommendations or suggestions.

Resolution of Duality Leonhardt (1991) makes the following statement: "An odd number of spans is always better than an even number; this is an old and approved rule in architecture." He then goes on to illustrate the balance and harmony of odd-numbered spans crossing a valley (Fig. 3.5) and a waterway (Fig. 3.7). So what is a designer to do with a grade separation over dual highways? If you are crossing two highways, the logical solution is to use a two-span layout. But this violates the principle of using odd-numbered spans and causes a split composition effect (Dorton, 1991).

This problem is often called "unresolved duality" because the observer has difficulty in finding a central focal point when viewing two large voidal spaces. He suggests increasing the visual mass of the central pier to direct attention away from the large voidal spaces. This redirection has been done successfully in the design of the 436th Avenue SE Undercrossing of I-90 (Exhibit 20 in color insert) near Olympia, Washington, shown in Figure 3.23.

Another effective way to reduce the duality effect is to reduce the emphasis on the girder by increasing its slenderness relative to the central pier. This emphasis can be accomplished by increasing the spans and moving the abutments up the slope and has the added effect of opening up the traveled way and giving the feeling of free-flowing traffic. As shown in Figure 3.24, the use of sloping lines in the abutment face and pier top provides an additional feeling of openness. Proper proportions between the girder, pier, and abutment must exist as demonstrated in Figure 3.25. The Hinzerling Road Bridge (Exhibit 17 in the color insert) of Figure 3.20 gives a fine example of applying these recommendations for resolving the duality effect.

Fig. 3.23 436th Avenue SE Undercrossing, I-90, King County, Washington. (Photo courtesy of Washington State DOT.)

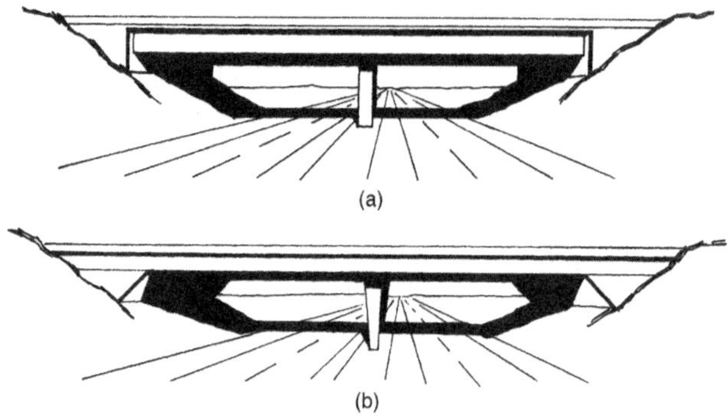

Fig. 3.24 (a) Vertical lines appear static. They provide interest and variety to the horizontal flow of the structure but do not accentuate the flow. (b) Dynamic sloping lines provide interest and variety and accentuate flow. (Permission granted by the California DOT.)

Generally speaking, the ideal bridge for a grade separation or highway interchange has long spans with the smallest possible girder depth and the smallest possible abutment size (Ritner, 1990). Continuity is the best way to minimize girder depth. In two-span applications, haunches can be used effectively, but as shown in Figure 3.26, proportions must be selected carefully. Leonhardt (1991) suggests that the haunch should follow a parabolic curve that blends in at midspan and is not deeper at the pier than twice the depth at midspan.

An elegant engineering solution to the duality problem is to eliminate the center pier and design an overpass that appears as a single span between abutments (Fig. 3.27). The low slender profile is obtained by developing end moments through anchored end spans at the abutments. Disguised externally, the superstructure is actually a three-span continuous girder system (Kowert, 1989).

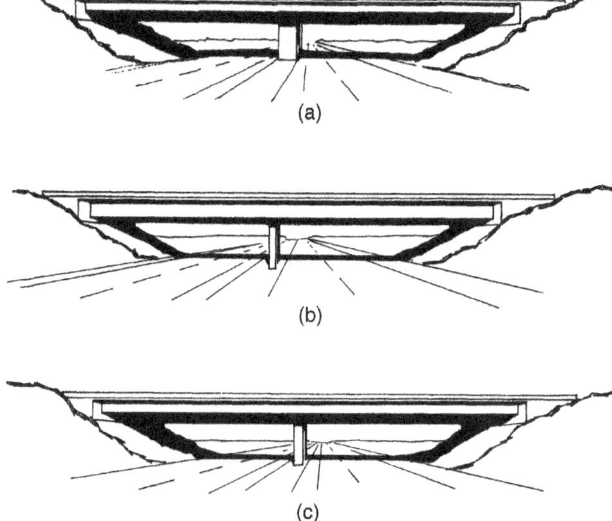

Fig. 3.25 (a) Massive columns overpower superstructure. (b) Massive superstructure overpowers spindly columns. (c) Substructure and superstructure are properly proportioned. (Permission granted by the California DOT.)

Another fine example of a single span between abutments is the Genesee Road (U.S. 40) Bridge over I-70 in Colorado. This bridge (Exhibit 21 in the color insert) provides a framework for an observer's first view of the Rocky Mountains.

By utilizing these recommendations, it is possible to overcome the duality effect and to design pleasing highway overpasses. Additional guidelines for the individual components of girders, overhangs, piers, and abutments that help the parts integrate into a unified, harmonious whole are given in the sections that follow.

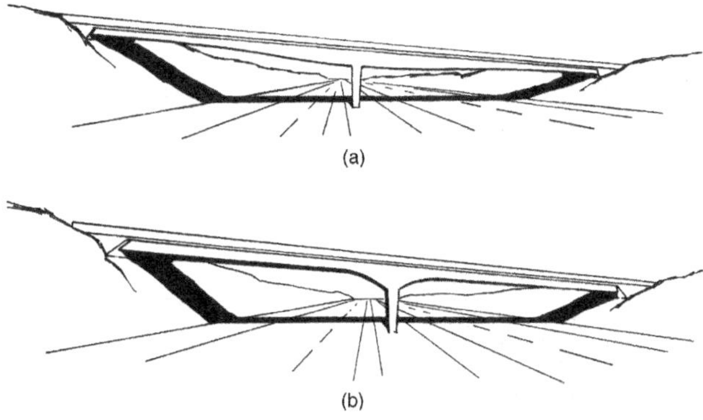

Fig. 3.26 (a) Long haunches give grace to the structure. (b) Short haunches appear awkward and abrupt, detracting from continuity of bridge. (Permission granted by the California DOT).

Girder Span/Depth Ratio According to Leonhardt (1991), the most important criterion for the appearance of a bridge is the slenderness of the beam, defined by the span length/beam depth ratio (L/d). If the height of the opening is greater than the span, he suggests L/d can be as small as 10, while for long continuous spans L/d could be up to 45. The designer has a wide range of choices in finding the L/d ratio that best fits a particular setting. In light of the general objective of using a beam with the least possible depth, the L/d ratio selected should be on the high end of the range.

Because of structural limitations, the maximum L/d ratio varies for different bridge types. Table 3.1 has been developed from recommendations given by ACI-ASCE Committee 343 (1988), and those in Table 2.5.2.6.3-1 of the AASHTO Specifications (2010). The maximum values in Table 3.1 are traditional ratios given in previous editions of the AASHTO Specifications in an attempt to ensure that vibration and deflection would not be a problem. These are not absolute maximums; they are only guidelines. They compare well with L/d ratios that are desirable for a pleasing appearance.

Table 3.1 Typical and Maximum Span/Depth Ratios

Bridge Type	Typical	Maximum
Continuous Concrete Bridges	Committee 343	AASHTO
Nonprestressed slabs	20–24	
Nonprestressed girders		
T-beams	15 ±	15
Box beams	18 ±	18
Prestressed slabs		
Cast in place	24–40	37
Precast	25–33	
Prestressed girders		
Cast-in-place box beams	25–33	25
Precast I-beams	20–28	25
Continuous Steel Bridges	Caltrans	AASHTO
Composite I-beam		
Overall		31
I-beam portion		37
Composite welded girder	22	
Structural steel box	22	

Deck Overhangs It is not possible for many of the bridge types in Table 3.1 to have L/d ratios in excess of 30. However, it is possible to increase the apparent slenderness of the superstructure by placing part or the entire girder in shadow. Cantilevering the deck slab beyond the exterior girder as shown in Figures 3.22 and 3.28 can create shadow.

When girders are spaced a distance S center to center in a multigirder bridge, a cantilevered length of the deck overhang w of about $0.4S$ helps balance the positive and negative moments in the deck slab. Another way to determine the cantilever length w is to proportion it relative to the depth of the girder h. Leonhardt (1991) suggests a ratio of w/h of 2 : 1 for single-span, low-elevation bridges and 4 : 1 for long, continuous bridges high above the ground.

Fig. 3.27 Anchored end span bridge over I-39 located in north-central Illinois.

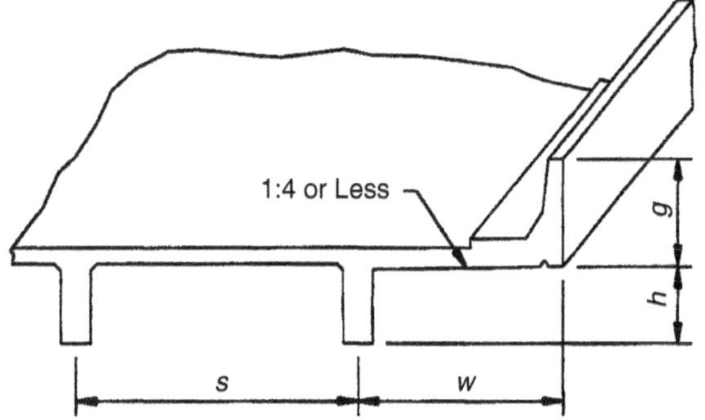

Fig. 3.28 Deck slab cantilevered over edge beam (Leonhardt, 1991). (From *Bridge Aesthetics around the World*, copyright © 1991 by the Transportation Research Board, National Research Council, Washington, DC. Reprinted with permission.)

If the slope of the underside of the overhang is less than 1 : 4, that portion of the overhang will be in deep shadow (Murray, 1991). Both Leonhardt (1991) and Murray (1991) agree that the ratio of the depth of the fascia g to the depth of the girder h should be about 1 : 3 to give a pleasing appearance. By first selecting a cantilever length w, a designer can use these additional proportions to obtain a visual effect of a more slender superstructure (Figs. 3.8 and 3.15).

When solid concrete barriers are used for safety rails, the fascia appears to have greater depth. If a box girder with a sloping side is used, it is possible for the overhang to put the entire girder in shadow (Fig. 3.23) and improve the apparent slenderness. Also, it may be advantageous to change the texture (Fig. 3.20), or to introduce an additional shadow line that breaks up the flat surface at, say, the one-third point (Fig. 3.29).

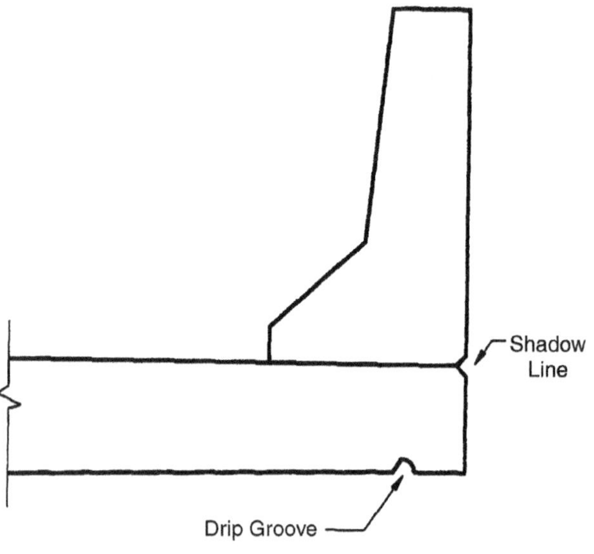

Fig. 3.29 Cantilevered overhang with drip groove (Mays, 1990). (Used with permission of the American Concrete Institute.)

Also shown in Figure 3.29 is an important and practical detail—the drip groove. This drip groove breaks the surface tension of rainwater striking the fascia and prevents it from running in sheets and staining the side of the girder. Architects have known about drip grooves for decades, but in many cases engineers have been slow to catch on and we still see many discolored beams and girders. Perhaps all that is necessary is to point it out to them one time.

Piers In addition to having proper proportions between a pier and its superstructure (Fig. 3.25), a pier has features of its own that can improve the appearance of a bridge. As shown in Figure 3.30, many styles and shapes of piers are possible. The most successful ones are those that have some flare, taper, texture, or other feature that improves the visual experience of those who pass by them. The key is that they are harmonious with the superstructure and its surroundings and that they express their structural process.

In general, tall piers should be tapered (Figs. 3.4, 3.5, and 3.31) to show their strength and stability in resisting lateral

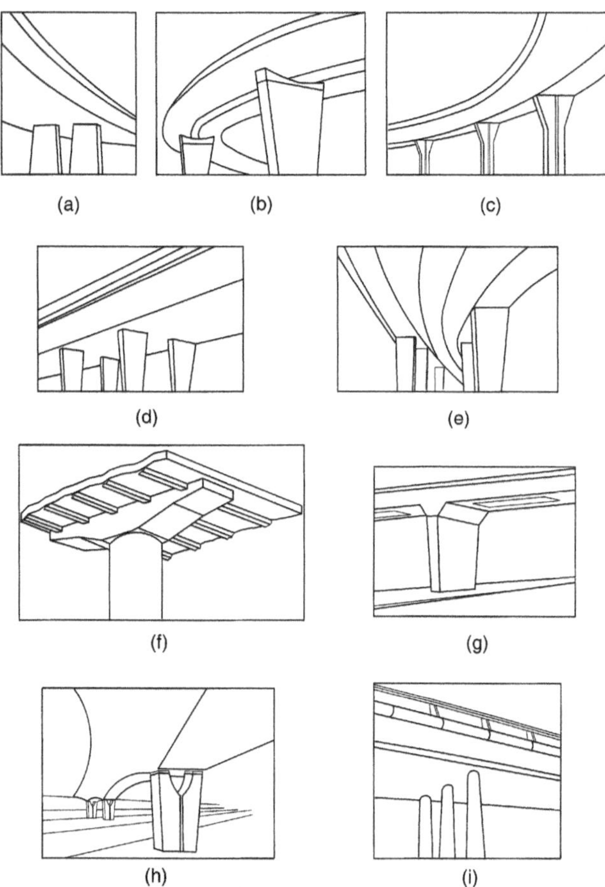

Fig. 3.30 Pier styles of contemporary bridges: wall type (a–e, g, h); T-type (f); and column type (i). (Glomb, 1991). (From *Bridge Aesthetics around the World*, copyright © 1991 by the Transportation Research Board, National Research Council, Washington, DC. Reprinted with permission.)

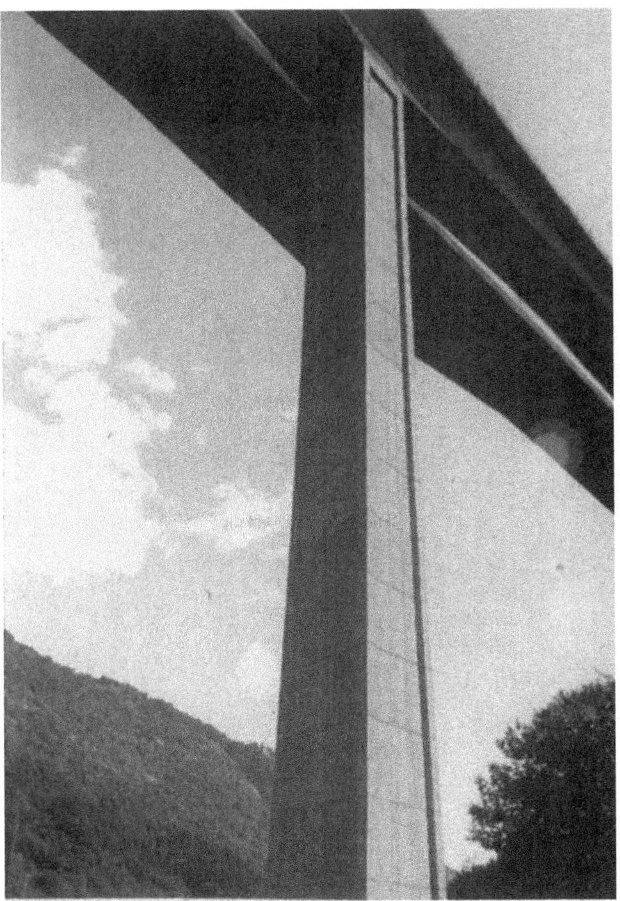

Fig. 3.31 Tall column with parabolic taper and raised edge (Menn, 1991). (From *Bridge Aesthetics around the World*, copyright © 1991 by the Transportation Research Board, National Research Council, Washington, DC. Reprinted with permission.)

loads. The tallest piers in the world are found in the spectacular Millau Viaduct (Exhibit 22 in the color insert) in France, and the slenderness of the structure gives the appearance of a floating roadway above the broad valley.

Short piers can also be tapered (Fig. 3.32) but in the opposite direction to show that less resistance is desired at the bottom than at the top. And when the piers are of intermediate height (Fig. 3.33), they can taper both ways to follow a bending moment diagram that, in this case, has a point of inflection about one-third the height from the top.

There appears to be a preference among some designers for piers that are integral with the superstructure, that is, they act together with the beams and girders to resist applied loads. Examples of integral piers are shown in Figures 3.31 and 3.34. Where nonintegral substructures are used, Wasserman (1991) recommends hammerhead or T-piers, singly (Fig. 3.8) or joined (Fig. 3.35), over the more cluttered appearance of multiple-column bents (Fig. 3.36).

When interchanges are designed, multiple columns cannot be avoided, but they should be of similar form. In the California interchange of Figure 3.34 and the Arizona interchange of (Exhibit 23 in the color insert), there are a variety of pier shapes and sizes, but they all belong to the same family. Contrast these examples with the unfortunate mixture of supports in the bridge of Figure 3.37. Harmony between the elements of the bridge has been destroyed. The wall pier is too prominent because it has not been kept in the shade and its sloping front face adds to the confusion. This mixture of supports is a good example of what not to do.

Abutments Repeating what was written earlier, to obtain a pleasing appearance for a bridge, the girder should be as slender as possible. Large abutments may be needed to anchor a suspension bridge, but they are out of place for medium- and short-span bridges.

The preferred abutment is placed near the top of the bank, well out of the way of the traffic below (Fig. 3.24), which gives the bridge a feeling of openness and invites the flow of traffic. Some designers refer to this as a stub abutment or, if it is supported on columns or piling, a spill-through abutment because the embankment material spills through the piling.

For a given length of an abutment, the flatter the slope of the embankment, the smaller the abutment appears, which can be seen in the comparisons of Figure 3.38. The preferred slope of the bank should be 1 : 2 or less.

Another feature of the abutment that improves its appearance is to slope its face back into the bank from top to bottom. Elliot (1991) explains it this way:

> *Sloping the face inward about 15 degrees, creates a magical illusion. Instead of seeming to suddenly stop against the vertical faces, the bridge now seems to flow smoothly into the supporting ground. This one feature will improve the appearance of a simple separation structure at virtually no increase in cost.*

Examples of bridges with abutments illustrating this concept are shown in Figures 3.23, 3.24, and 3.26. The mass of the vertical-faced abutments of the bridges in Figures 3.35 and 3.36 could be reduced and the appearance improved if the faces were inclined inward.

The sloping ground from the abutment to the edge of the stream or roadway beneath the bridge is usually in the shade and vegetation does not easily grow on it (see Figs. 3.20 and 3.23). Whatever materials are placed on the slope to prevent erosion should relate to the bridge or the surrounding landscape. Concrete paving blocks or cast-in-place concrete relate to the abutment while rubble stone relates to the landscape. Proper selection of slope protection materials will give the bridge a neatly defined and finished appearance.

Integral Abutments and Jointless Bridges Expansion joints in bridges have always been a maintenance problem. These mechanical devices often break loose from the deck, get bent, become a road hazard, and need to be replaced. The joints allow access of water and contaminants from the roadway that cause deterioration of the abutments, girders,

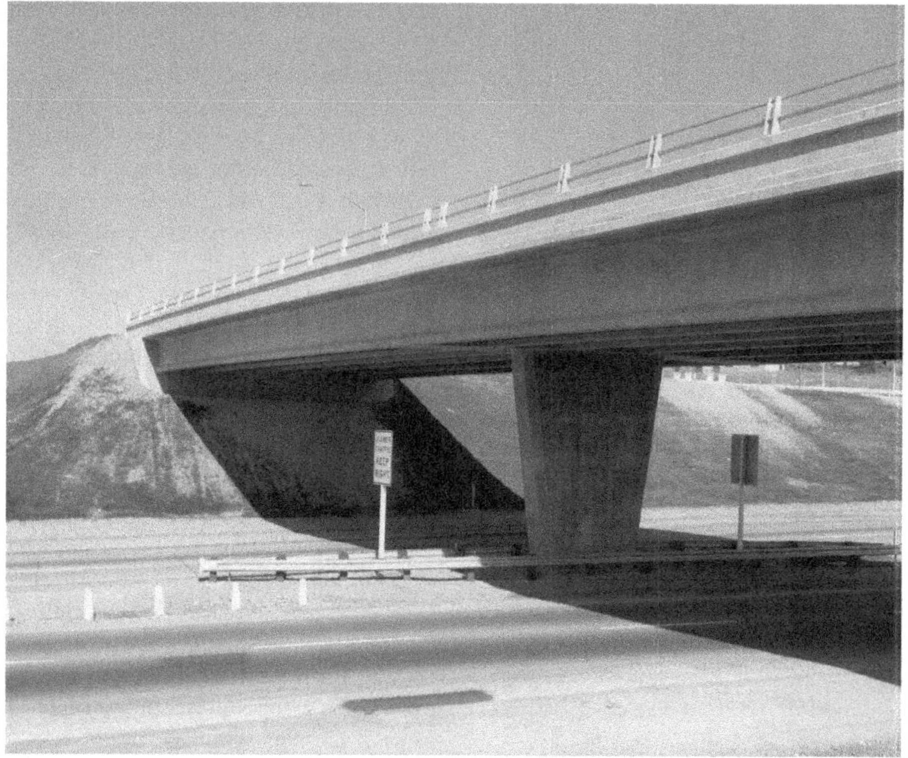

Fig. 3.32 Prestressed girders frame into the side of the supporting pier, eliminating from view the usual cap beam. (Permission granted by the California DOT.)

Fig. 3.33 Piers with double taper (Seim and Lin, 1990). (Used with permission of the American Concrete Institute.)

and piers beneath the deck. When an abutment is made integral with the girders, the deck becomes a roof that helps protect the girders and piers. When all the joints in a bridge are eliminated, the initial cost is reduced and the riding quality of the jointless roadway is improved.

An integral abutment bridge is shown in Figure 3.39. Two components make up the integral bridge: the bridge system and the approach system. The bridge system consists of a superstructure integrally connected to a stub abutment supported on a single row of piles. The superstructure may have multiple spans with intermediate piers. The jointless bridge system acts together as a single structural unit.

The approach system consists of the backfill, the approach fill, the foundation soil, and, if used, an approach slab. Some designers do not use approach slabs because they believe that remedial actions with approach slabs are more costly and inconvenient to the public than periodically regrading the settling approach (Arsoy et al., 1999). With or without approach slabs, a void between the backfill and the abutment is likely to develop as the abutments move back and forth due to temperature changes (Arsoy et al., 2004). Differential settlement between the approach system and the bridge system creates a bump at the end of the bridge. Without an approach slab, the bump is at the abutment backwall. When approach slabs are used, the bump is pushed out to the connection with the pavement at the sleeper slab.

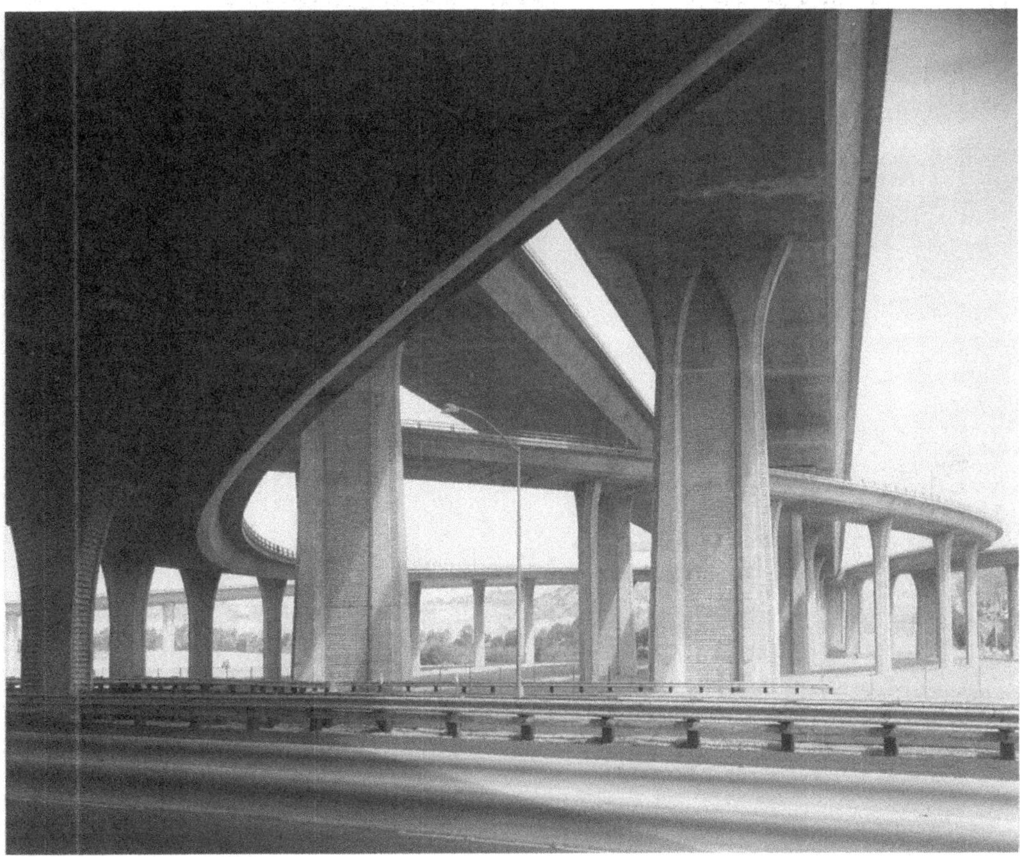

Fig. 3.34 Route 8/805 interchange, San Diego, California. (Permission granted by the California DOT.)

Fig. 3.35 Hammerhead piers. (Photo courtesy of Tennessee DOT—Geo. Hornal, photographer.)

Fig. 3.36 Multiple-column bents. (Photo courtesy of Tennessee DOT—Geo. Hornal, photographer.)

Fig. 3.37 Bridge with displeasing mixture of supports (Murray, 1991). (From *Bridge Aesthetics around the World*, copyright © 1991 by the Transportation Research Board, National Research Council, Washington, DC. Reprinted with permission.)

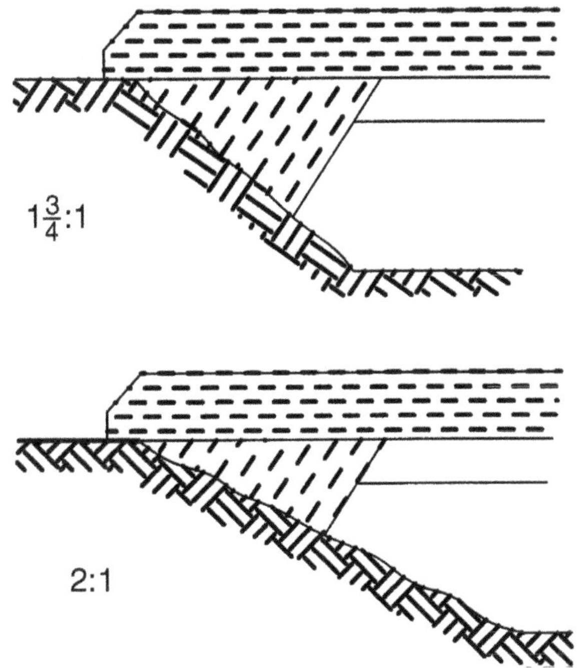

Fig. 3.38 Slope at abutment (Mays, 1990). (Used with permission of the American Concrete Institute.)

The lengths of jointless bridges continue to increase as state departments of transportation (DOTs) try to maximize the savings in maintenance costs. At one time, 500 ft (150 m) was thought to be a maximum overall length for a jointless bridge to avoid problems with the interactions between the bridge and approach systems. However, the Holston River Bridge in Tennessee has an overall jointless length of 2650 ft (800 m) and has performed well for more than 20 years (Burdette et al., 2003). Understanding the interactions of the bridge superstructure, the abutment, the approach fill, the foundation piles, and the foundation soil is important to fully utilize the advantages of jointless bridges (Arsoy et al., 2001).

3.3.4 Computer Modeling

Computer software tools are frequently used to model bridges that are large, signature spans and for bridges located in environmentally and visually sensitive areas where the bridgescape is critical. Such models are becoming integral to the planning and design process for such bridges and will likely become commonplace for more routine structures in the future. Features such as surrounding landscape, sky, and water can be added to the rendering to offer the architect, engineer, and public an accurate representation of the completed product. Additionally, detailed features such as lighting, painting, and sculpting options can be explored. View points from the drivers', waters', and aerial perspectives can be readily created from a three-dimensional (3D) model.

A few examples of modeled and completed bridges are illustrated in Figures 3.40–3.42. The Broadway Bridge, which spans the Halifax River and links the speedway at Daytona with the nearby beach, is illustrated in Figure 3.40. Figure 3.40(a) is the computer rendering of the project created during design and prior to construction; Figure 3.40(b) is the completed project photographed near the same vantage point. This type of realistic modeling helps the public to develop consensus about the bridge design and specific features. In this case, the engineer from the FIGG Engineering Group led two design charettes with the community that brought hands-on participation through consensus voting among 35 participants who voted on 40 different design features. (The term *charette* is derived from the French word for *cart* in which nineteenth-century architectural students carried their designs to the Ecole Beaux-Arts for evaluation, often finishing them en route. Today's meaning implies a work session involving all interested parties that compresses decision making into a few hours or days.) Visualization of integrated shapes, shadows, textures, color, lighting, railing, and landscaping are aided with computer modeling. The likeness of the model and actual photograph is astounding.

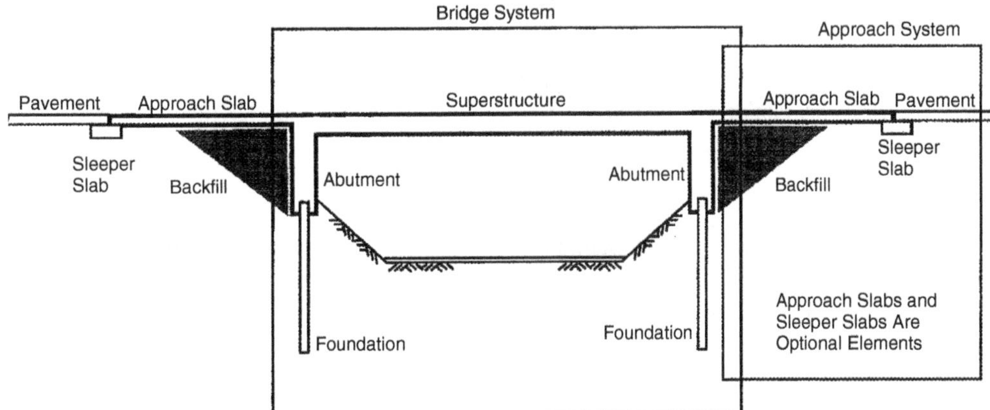

Fig. 3.39 Simplified geometry of an integral abutment bridge (Arsoy et al., 1999).

(a)

(b)

Fig. 3.40 Broadway Bridge, Daytona, Florida. (a) Computer model and (b) finished bridge in service. (Photos courtesy of FIGG Engineering Group, reprinted with permission.)

Similarly, Figure 3.41 illustrates the Lee Roy Selmon Crosstown Expressway located in Tampa, Florida, during construction. Figure 3.41(a) is a rendering of the bridge cutaway during construction, and Figure 3.41(b) is a similar photograph taken during construction.

Finally, Figure 3.42 illustrates the Smart Road Bridge located near Blacksburg, Virginia. This bridge is part of a nationally recognized smart-road research facility used to test high-tech advancements in transportation. The bridge elegantly spans the beautiful Ellett Valley.

3.3.5 Web References

The Internet offers a host of references on bridge aesthetics. Using a search engine with the keywords *bridge* and

(a)

(b)

Fig. 3.41 Lee Roy Selmon Crosstown Expressway, Tampa, Florida. (a) Computer model and (b) bridge under construction. (Photos courtesy of FIGG Engineering Group, reprinted with permission.)

(a)

(b)

Fig. 3.42 Smart Road Bridge, Blacksburg, Virginia. (a) Computer model and (b) finished bridge in service. (Photos courtesy of FIGG Engineering Group, reprinted with permission.)

aesthetics will yield many references with fine pictures and discussion. Many are related to design guidelines for specific agencies, for example, Iowa DOT (1998a,b), Australian RTA (2003), Minnesota DOT (1999), Alberta Infrastructure and Transportation (2005), while others provide specific examples and case studies for a particular crossing, for example, Federal Highway Administration (2004) and Delaware Department of Transportation (2004).

3.3.6 Closing Remarks on Aesthetics

It is important for an engineer to realize that, whether intentional or not, a completed bridge becomes an aesthetic statement. Therefore, it is necessary to understand what qualities and features of a bridge tend to make that aesthetic statement a good one. This understanding will require training and time.

Suggestions have been made regarding the improvement of the appearance of medium- and short-span bridges. Some of these suggestions include numerical values for proportions and ratios, but most of them simply point out features that require a designer's attention. No equations or design specifications can make our bridges beautiful. It is more our awareness of beauty that creates a sense of when we are in the presence of something good.

Aesthetics must be an integral part of bridge design. Beginning with the conceptual design, the engineer must consider aesthetics in the selection of spans, depths of girders, piers, abutments, and the relationship of one to another. It is an important responsibility, and we must demand it of ourselves because the public demands it of us.

REFERENCES

AASHTO (2010). *LRFD Bridge Design Specification*, 5th ed., American Association of State Highway and Transportation Officials, Washington, DC.

ACI-ASCE Committee 343 (1988). *Analysis and Design of Reinforced Concrete Bridge Structures*, American Concrete Institute, Detroit, MI.

Addis, W. (1990). *Structural Engineering: The Nature of Theory and Design*, Ellis Horwood, London.

Alberta Infrastructure and Transportation (2005). "Bridge Aesthetics Study," http://www.trans.gov.ab.ca/Content/doctype30/production/AesStudy0405.pdf.

Arsoy, S., R. M. Barker, and J. M. Duncan (1999). *The Behavior of Integral Abutment Bridges*, Report No. FHWA/VTRC 00-CR3, November, Virginia Transportation Research Council, Charlottesville, VA.

Arsoy, S., J. M. Duncan, and R. M. Barker (2001). *Experimental and Analytical Investigations of Piles and Abutments of Integral Bridges, July*, Center for Geotechnical Practice and Research, Charles E. Via Department of Civil and Environmental Engineering, Virginia Polytechnic Institute and State University, Blacksburg, VA.

Arsoy, S., J. M. Duncan, and R. M. Barker (2004). "Behavior of a Semi-Integral Abutment under Static and Temperature-Induced Cyclic Loading," *Journal of Bridge Engineering*, March, Vol. 9, No. 2, ASCE, Washington, DC, pp. 193–199.

Australian RTA (2003). "Bridge Aesthetics, Design Guideline to Improve Bridges in the NSW," http://www.rta.nsw.gov.au/publicationsstatisticsforms/downloads/bridge_aesthetics.pdf.

Brown, W. C., M. F. Parsons, and H. S. G. Knox (1976). *Bosporus Bridge: Design and Construction*, Institute of Civil Engineers, London.

Burdette, E. G., D. W. Goodpasture, and J. H. Deatherage (2003). "A Half Mile of Bridge without a Joint," *Concrete International*, Vol. 25, No. 2, February, pp. 47–51.

Burke, M. P., Jr. (1990). "Bridge Aesthetics—Rules, Formulas, and Principles—the Negative View," *Esthetics in Concrete Bridge Design*, ACI MP1-7, American Concrete Institute, Detroit, pp. 71–79.

Burke, M. P., Jr., Ed. (1991). *Bridge Aesthetics around the World*, Transportation Research Board, National Research Council, Washington, DC.

Delaware Department of Transportation (2004). "Bridge 3–156 on SR-1 over Indian River Inlet," http://www.indianriverinletbridge.com/Resources/prequalb2/2004_12_16_PreQual_Mtg_Handout.pdf.

de Miranda, F. (1991). "The Three Mentalities of Successful Bridge Design," *Bridge Aesthetics around the World*, Transportation Research Board, National Research Council, Washington, DC, pp. 89–94.

Dorton, R. A. (1991). "Aesthetics Considerations for Bridge Overpass Design," *Bridge Aesthetics around the World*, Transportation Research Board, National Research Council, Washington, DC, pp. 10–17.

Elliot, A. L. (1991). "Creating a Beautiful Bridge," *Bridge Aesthetics around the World*, Transportation Research Board, National Research Council, Washington, DC, pp. 215–229.

Federal Highway Administration (2004). "Excellence in Highway Design—Biennial Awards," http://www.fhwa.dot.gov/eihd/.

Fernandez-Ordóñez, J. A. (1991). "Spanish Bridges: Aesthetics, History, and Nature," *Bridge Aesthetics around the World*, Transportation Research Board, National Research Council, Washington, DC, pp. 205–214.

Friedhoff, R. M. and W. Benzon (1989). *Visualization*, Harry N. Abrams, New York.

Glomb, J. (1991). "Aesthetic Aspects of Contemporary Bridge Design," *Bridge Aesthetics around the World*, Transportation Research Board, National Research Council, Washington, DC, pp. 95–104.

Gloyd, C. S. (1990). "Some Thoughts on Bridge Esthetics," *Esthetics in Concrete Bridge Design*, ACI MP1-10, American Concrete Institute, Detroit, pp. 109–117.

Gottemoeller, F. (1991). "Aesthetics and Engineers: Providing for Aesthetic Quality in Bridge Design," *Bridge Aesthetics around the World*, Transportation Research Board, National Research Council, Washington, DC, pp. 80–88.

Gottemoeller, F. (2004). *Bridgescape: The Art of Designing Bridges*, 2nd ed., Wiley, Hoboken, NJ.

Grant, A. (1990). "Beauty and Bridges," *Esthetics in Concrete Bridge Design*, ACI MP1-6, American Concrete Institute, Detroit, pp. 55–65.

Iowa Department of Transportation (1998a). "Bridge Aesthetics," http://www.dot.state.ia.us/bridge/aesthetics.html.

Iowa Department of Transportation (1998b). "50th Street Bridge," http://www.dot.state.ia.us/bridge/50th_street.html.

Kowert, R. (1989). "Anchored End Span Bridges Span Interstate 39," Paper IBC-89-5, International Bridge Conference, Pittsburgh, PA.

Lee, D. J. (1990). "Bridging the Artistic Gulf," *Esthetics in Concrete Bridge Design*, ACI MP1-9, American Concrete Institute, Detroit, pp. 101–108.

Leonhardt, F. (1982). *Bridges: Aesthetics and Design*, Architectural Press, London, and MIT Press, Cambridge, MA.

Leonhardt, F. (1991). "Developing Guidelines for Aesthetic Design," *Bridge Aesthetics around the World*, Transportation Research Board, National Research Council, Washington, DC, pp. 32–57.

Livio, M. (2002). *The Golden Ratio: The Story of Phi, the World's Most Astonishing Number*, Broadway Books, New York.

Mays, R. R. (1990). "Aesthetic Rules Should Not Be Set in Concrete—A Bridge Architect's View on Bridge Design," *Esthetics in Concrete Bridge Design*, ACI MP1-16, American Concrete Institute, Detroit, pp. 203–228.

Menn, C. (1991). "Aesthetics in Bridge Design," *Bridge Aesthetics around the World*, Transportation Research Board, National Research Council, Washington, DC, pp. 177–188.

Minnesota Department of Transportation (1999). "Design Excellence through Context Sensitive Design," http://www.cts.umn.edu/education/csd/pdf-docs/Session_12.pdf.

Muller, J. M. (1991). "Aesthetics and Concrete Segmental Bridges," *Bridge Aesthetics around the World*, Transportation Research Board, National Research Council, Washington, DC, pp. 18–31.

Munro, T. (1956). *Towards Science in Aesthetics*, Liberal Arts, New York.

Murray, J. (1991). "Visual Aspects of Short- and Medium-Span Bridges," *Bridge Aesthetics around the World*, Transportation Research Board, National Research Council, Washington, DC, pp. 155–166.

Ritner, J. C. (1990). "Creating a Beautiful Concrete Bridge," *Esthetics in Concrete Bridge Design*, ACI MP1-4, American Concrete Institute, Detroit, pp. 33–45.

Roberts, J. E. (1992). "Aesthetic Design Philosophy Utilized for California State Bridges," *Journal of Urban Planning and Development*, ASCE, Vol. 118, No. 4, December, pp. 138–162.

Seim, C. and T. Y. Lin, (1990). "Aesthetics in Bridge Design: Accent on Piers," *Esthetics in Concrete Bridge Design*, ACI MP1-15, American Concrete Institute, Detroit, pp. 189–201.

Wasserman, E. P. (1991). "Aesthetics for Short- and Medium-Span Bridges," *Bridge Aesthetics around the World*, Transportation Research Board, National Research Council, Washington, DC, pp. 58–66.

Watson, S. C. and M. K. Hurd, Eds. (1990). *Esthetics in Concrete Bridge Design*, ACI MP-1, American Concrete Institute, Detroit.

PROBLEMS

3.1 Discuss the interaction between deductive and inductive reasoning in formulating the principles of structural analysis and design.

3.2 Explain the interrelationship between description and justification of a bridge design.

3.3 What makes it difficult for a person, including your professor and yourself, to pass on personal knowledge?

3.4 List the four general stages of bridge design and give a brief description of each one.

3.5 Describe how the design of a bridge begins in the mind.

3.6 Discuss the necessity of having the three "mentalities" present in the bridge design team. Imagine you are a member of the design team and indicate what abilities you think other people on the team need to have.

3.7 Explain what is meant by the following: "Whether intentional or not, every bridge structure makes an aesthetic statement."

3.8 Some of the qualities of bridge aesthetics are similar to the qualities of classical music. Choose one of these qualities common with music and describe how that quality can improve the appearance of bridges.

3.9 How can shadow be used to make a bridge appear more slender?

3.10 What is meant by "resolution of duality"? How can it be resolved in an overpass of an interstate highway?

3.11 Leonhardt (1991) states that the slenderness ratio L/d is the most important criterion for the appearance of a bridge. Explain how continuity can maximize this ratio.

3.12 In selecting abutments, what steps can be taken to give a bridge spanning traffic a feeling of openness?

3.13 In what ways do integral abutment and jointless bridge designs reduce maintenance costs? How are movements due to temperature changes accommodated in these bridges?

CHAPTER 4

Bridge Types and Selection

4.1 MAIN STRUCTURE BELOW THE DECK LINE

Arched and truss-arched bridges are included in this classification. Examples are the masonry arch, the concrete arch (Fig. 3.17), the steel truss-arch, the steel deck truss, the rigid frame, and the inclined leg frame (Fig. 3.15) bridges. Striking illustrations of this bridge type are the New River Gorge Bridge (Fig. 4.1) in West Virginia and the Salginatobel Bridge (Fig. 4.2) in Switzerland.

With the main structure below the deck line in the shape of an arch, gravity loads are transmitted to the supports primarily by axial compressive forces. At the supports, both vertical and horizontal reactions must be resisted. The arch rib can be solid or it can be a truss of various forms. Xanthakos (1994) shows how the configuration of the elements affects the structural behavior of an arch bridge and gives methods for determining the force effects.

O'Connor (1971) summarizes the distinctive features of arch-type bridges as:

- ☐ The most suitable site for this form of structure is a valley, with the arch foundations located on dry rock slopes.
- ☐ The erection problem varies with the type of structure, being easiest for the cantilever arch and possibly most difficult for the tied arch.
- ☐ The arch is predominantly a compression structure. The classic arch form tends to favor concrete as a construction material.
- ☐ Aesthetically, the arch can be the most successful of all bridge types. It appears that through experience or familiarity, the average person regards the arch form as understandable and expressive. The curved shape is almost always pleasing.

4.2 MAIN STRUCTURE ABOVE THE DECK LINE

Suspension, cable-stayed, and through-truss bridges are included in this category. Both suspension and cable-stayed bridges are tension structures whose cables are supported by towers. Examples are the Brooklyn Bridge (Fig. 3.19) and the East Huntington Bridge (Fig. 3.18).

Suspension bridges (Fig. 4.3) are constructed with two main cables from which the deck, usually a stiffened truss, is hung by secondary cables. Cable-stayed bridges (Fig. 4.4) have multiple cables that support the deck directly from the tower. Analysis of the cable forces in a suspension bridge must consider nonlinear geometry due to large deflections.

O'Connor (1971) gives the following distinctive features for suspension bridges:

- ☐ The major element of the stiffened suspension bridge is a flexible cable, shaped and supported in such a way that it can transfer the major loads to the towers and anchorages by direct tension.
- ☐ This cable is commonly constructed from high-strength wires either spun in situ or formed from component, spirally formed wire ropes. In either case the allowable stresses are high, typically of the order of 90 ksi (600 MPa) for parallel strands.
- ☐ The deck is hung from the cable by hangers constructed of high-strength wire ropes in tension.
- ☐ The main cable is stiffened either by a pair of stiffening trusses or by a system of girders at deck level.
- ☐ This stiffening system serves to (a) control aerodynamic movements and (b) limit local angle changes in the deck. It may be unnecessary in cases where the dead load is great.
- ☐ The complete structure can be erected without intermediate staging from the ground.
- ☐ The main structure is elegant and neatly expresses its function.
- ☐ It is the only alternative for spans over 2000 ft (600 m), and it is generally regarded as competitive for spans down to 1000 ft (300 m). However, even shorter spans have been built, including some very attractive pedestrian bridges.

Consider the following distinctive features for cable-stayed bridges (O'Connor, 1971):

- ☐ As compared with the stiffened suspension bridge, the cables are straight rather than curved. As a result, the stiffness is greater. It will be recalled that the nonlinearity of the stiffened suspension bridge results from changes in the cable curvature and the corresponding change in bending moment taken by the dead-load cable tension. This phenomenon cannot occur in an arrangement with straight cables.

Fig. 4.1 New River Gorge Bridge. (Photo courtesy of Michelle Rambo-Roddenberry, 1996.)

Fig. 4.2 General view of Salginatobel Bridge. [From Troitsky (1994). Reprinted with permission of John Wiley & Sons, Inc.]

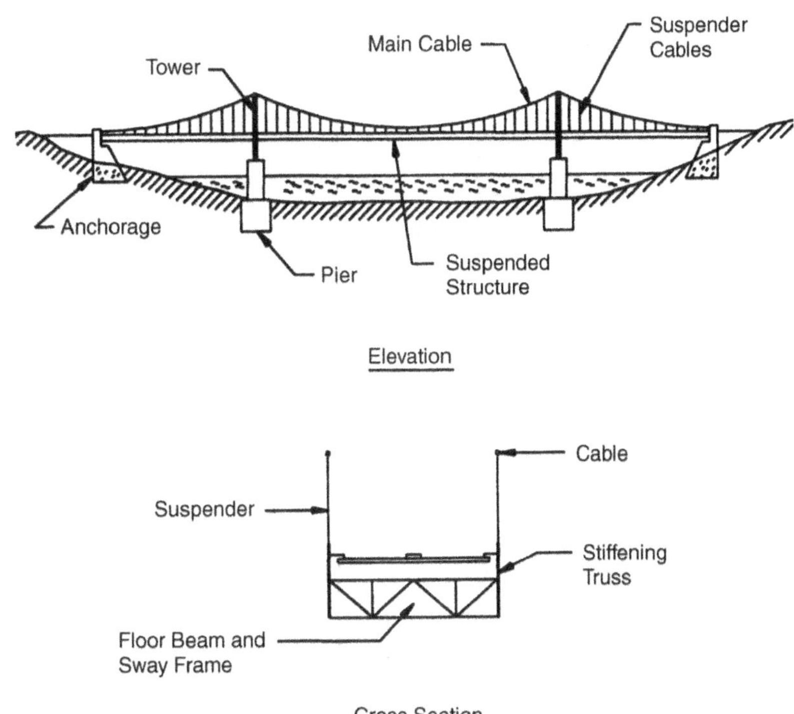

Fig. 4.3 Typical suspension bridge. [From Troitsky (1994). Reprinted with permission of John Wiley & Sons, Inc.]

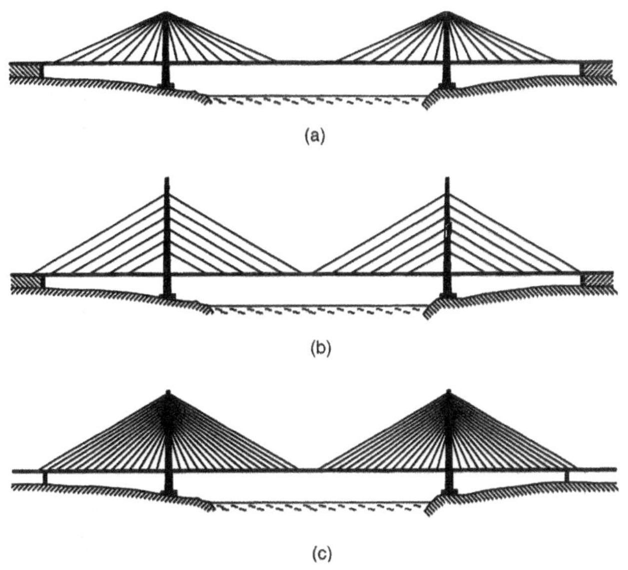

Fig. 4.4 Cable arrangements in cable-stayed bridges (Leonhardt, 1991). (From *Bridge Aesthetics around the World*, copyright © 1991 by the Transportation Research Board, National Research Council, Washington, DC. Reprinted with permission.)

- ☐ The cables are anchored to the deck and cause compressive forces in the deck. For economical design, the deck system must participate in carrying these forces. In a concrete structure, this axial force compresses the deck.
- ☐ Compared with the stiffened suspension bridge, the cable-braced girder bridge tends to be less efficient in supporting dead load but more efficient under live load. As a result, it is not likely to be economical on the longest spans. It is commonly claimed to be economical over the range of 300–1100 ft (100–350 m), but some designers would extend the upper bound as high as 2500 ft (800 m).
- ☐ The cables may be arranged in a single plane, at the longitudinal centerline of the deck. This arrangement capitalizes on the torsion capacity inherent in a tubular girder system and halves the number of shafts in the towers.
- ☐ The presence of the cables facilitates the erection of a cable-stayed girder bridge. Temporary backstays of this type have been common in the cantilever erection of girder bridges. Adjustment of the cables provides an effective control during erection.

Aerodynamic instability may be a problem with the stays in light rain and moderate winds. The water creates a small bead (or bump) that disturbs the flow of wind around the cable. This disturbance creates an oscillatory force that may create large transverse movement of the stays. This phenomenon is sometimes called "dancing in the rain."

A truss bridge (Fig. 4.5) consists of two main planar trusses tied together with cross girders and lateral bracing to form a three-dimensional truss that can resist a general system of loads. When the longitudinal stringers that support the deck

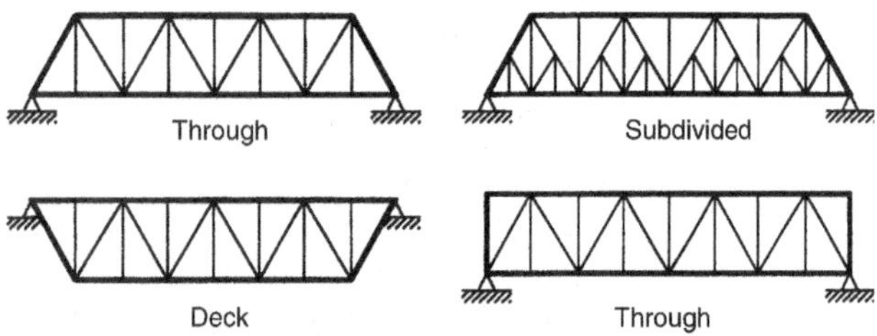

Fig. 4.5 Types of bridge trusses. [From Troitsky (1994). Reprinted with permission of John Wiley & Sons, Inc.]

slab are at the level of the bottom chord, this is a through-truss bridge as shown in Figure 4.6.

O'Connor (1971) gives the following distinctive features for truss bridges:

- ☐ A bridge truss has two major structural advantages: (1) the primary member forces are axial loads; (2) the open-web system permits the use of a greater overall depth than for an equivalent solid-web girder. Both these factors lead to economy in material and a reduced dead weight. The increased depth also leads to reduced deflections, that is, a more rigid structure.
- ☐ The conventional truss bridge is most likely to be economical for medium spans. Traditionally, it has been used for spans intermediate between the plate girder and the stiffened suspension bridge. Modern construction techniques and materials have tended to increase the economical span of both steel and concrete girders. The cable-stayed girder bridge has become a competitor to the steel truss for the intermediate spans. These factors, all of which are related to the high fabrication cost of a truss, have tended to reduce the number of truss spans built in recent years.
- ☐ The truss has become almost the standard stiffening structure for the conventional suspension bridge, largely because of its acceptable aerodynamic behavior.
- ☐ Compared with alternative solutions, the encroachment of a truss on the opening below is large if the deck is at the upper chord level but is small if the traffic runs through the bridge, with the deck at the lower chord level. For railway overpasses carrying a railway above a road or another railway, the small construction depth of a through truss bridge is a major advantage. In some structures, it is desirable to combine both arrangements to provide a through truss over the main span with a small construction depth and approaches with the deck at upper chord level.

4.3 MAIN STRUCTURE COINCIDES WITH THE DECK LINE

Girder bridges of all types are included in this category. Examples include slab (solid and voided), T-beam (cast-in-place), I-beam (precast or prestressed), wide-flange beam (composite and noncomposite), concrete box (cast-in-place and segmental, prestressed), steel box (orthotropic deck), and steel plate girder (straight and haunched) bridges.

Illustrations of concrete slab, T-beam, prestressed girder, and box-girder bridges are shown in Figure 4.7. A completed cast-in-place concrete slab bridge is shown in Figure 4.8. Numerous girder bridges are shown in the section on aesthetics.

Fig. 4.6 Greater New Orleans Through-Truss Bridge. (Photo courtesy of Amy Kohls, 1996.)

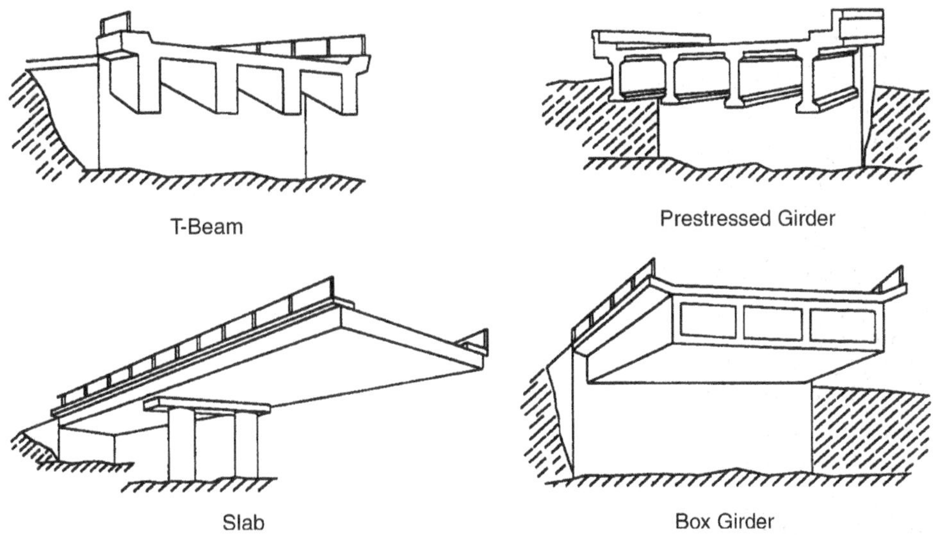

Fig. 4.7 Types of concrete bridges. (Permission granted by the California DOT.)

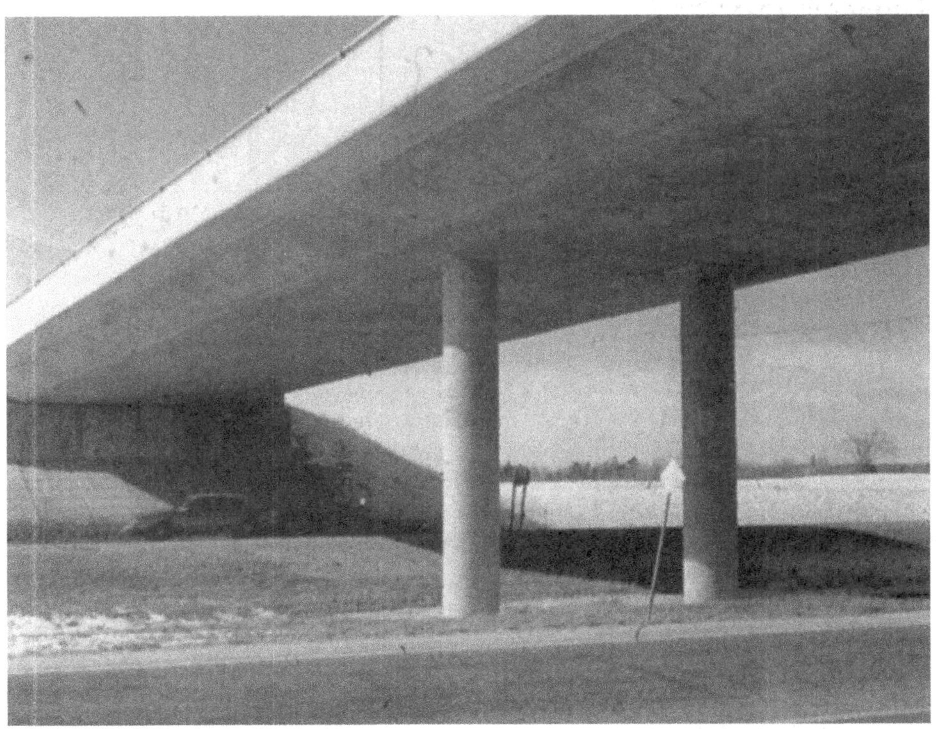

Fig. 4.8 Cast-in-place posttensioned voided slab bridge (Dorton, 1991). (From *Bridge Aesthetics around the World*, copyright © 1991 by the Transportation Research Board, National Research Council, Washington, DC. Reprinted with permission.)

Among these are prestressed girders (Fig. 3.32), concrete box girders (Figs. 3.10, 3.23, and 3.34), and steel plate girders (Figs. 3.27, 3.35, and 3.36).

Girder-type bridges carry loads primarily in shear and flexural bending. This action is relatively inefficient when compared to axial compression in arches and to tensile forces in suspension structures. A girder must develop both compressive and tensile forces within its own depth. A lever arm sufficient to provide the internal resisting moment separates these internal forces. Because the extreme fibers are the only portion of the cross section fully stressed, it is difficult to obtain an efficient distribution of material in a girder cross section. Additionally, stability concerns further limit the stresses and associated economy from a material utilization perspective. But from total economic perspective slab–girder bridges provide an economical and long-lasting solution for the vast majority of bridges. The U.S. construction industry is well tuned to provide this type of bridge. [As a result, girder bridges

Table 4.1 Common Girder Bridge Cross Sections

Supporting Components	Type of Deck	Typical Cross Section
Steel beam	Cast-in-place concrete slab, precast concrete slab, steel grid, glued/spiked panels, stressed wood	
Closed steel or precast concrete boxes	Cast-in-place concrete slab	
Open steel or precast concrete boxes	Cast-in-place concrete slab, precast concrete deck slab	
Cast-in-place concrete multicell box	Monolithic concrete	
Cast-in-place concrete T-beam	Monolithic concrete	
Precast solid, voided, or cellular concrete boxes with shear keys	Cast-in-place concrete overlay	
Precast solid, voided, or cellular concrete box with shear keys and with or without transverse posttensioning	Integral concrete	
Precast concrete channel sections with shear keys	Cast-in-place concrete overlay	
Precast concrete double T-section with shear keys and with or without transverse posttensioning	Integral concrete	
Precast concrete T-section with shear keys and with or without transverse posttensioning	Integral concrete	
Precast concrete I- or bulb T-sections	Cast-in-place concrete, precast concrete	
Wood beams	Cast-in-place concrete or plank, glued/spiked panels or stressed wood	

are typical for short- to medium-span lengths, say <250 ft (75 m).]

In highway bridges, the deck and girders usually act together to resist the applied load. Typical bridge cross sections for various types of girders are shown in Table 4.1. They include steel, concrete, and wood bridge girders with either cast-in-place or integral concrete decks. These are not the only combinations of girders and decks but represent those covered by the approximate methods of analysis in the AASHTO (2010) LRFD Specifications.

4.4 CLOSING REMARKS ON BRIDGE TYPES

For comparison purposes, typical ranges of span lengths for various bridge types are given in Table 4.2. In this book, the discussion is limited to slab and girder bridges suitable for short to medium spans. For a general discussion on other bridge types, the reader is referred to Xanthakos (1994).

4.5 SELECTION OF BRIDGE TYPE

One of the key submittals in the design process is the engineer's report to the bridge owner of the type, size, and location (TS & L) of the proposed bridge. The TS & L report includes a cost study and a set of preliminary bridge drawings. The design engineer has the main responsibility for the report, but opinions and advice will be sought from others within and without the design office. The report is then submitted to all appropriate agencies, made available for public hearings, and must be approved before starting on the final design.

4.5.1 Factors to Be Considered

Selection of a bridge type involves consideration of a number of factors. In general, these factors are related to function, economy, safety, construction experience, traffic control, soil conditions, seismicity, and aesthetics. It is difficult to prepare a list of factors without implying an order of priority, but a list is necessary even if the priority changes from bridge to bridge. The discussion herein follows the outline presented by ACI-ASCE Committee 343 (1988) for concrete bridges, but the factors should be the same, regardless of the construction material.

Geometric Conditions of the Site The type of bridge selected often depends on the horizontal and vertical alignment of the highway route and on the clearances above and below the roadway. For example, if the roadway is on a curve, continuous box girders and slabs are a good choice because they have a pleasing appearance, can readily be built on a curve, and have a relatively high torsion resistance. Relatively high

Table 4.2 Span Lengths for Various Types of Superstructure

Structural Type	Material	Range of Spans, ft (m)	Maximum Span in Service, ft (m)
Slab	Concrete	0–40 (0–12)	
Girder	Concrete	40–1000 (12–300)	988 (301), Stolmasundet, Norway, 1998
	Steel	100–1000 (30–300)	984 (300), Ponte Costa e Silva, Brazil, 1974
Cable-stayed girder	Steel	300–3500 (90–1100)	3570 (1088), Sutong, China, 2008
Truss	Steel	300–1800 (90–550)	1800 (550), Pont de Quebec, Canada, 1917 (rail) 1673 (510), Minato, Japan, 1974 (road)
Arch	Concrete	300–1380 (90–420)	1378 (420), Wanxian, China, 1997
	Steel truss	800–1800 (240–550)	1805 (550), Lupu, China, 2003
Suspension	Steel	1000–6600 (300–2000)	6530 (1991), Akashi-Kaikyo, Japan, 1998

bridges with larger spans over navigable waterways will require a different bridge type than one with medium spans crossing a floodplain. The site geometry will also dictate how traffic can be handled during construction, which is an important safety issue and must be considered early in the planning stage.

Subsurface Conditions of the Site The foundation soils at a site will determine whether abutments and piers can be founded on spread footings, driven piles, or drilled shafts. If the subsurface investigation indicates that creep settlement is going to be a problem, the bridge type selected must be one that can accommodate differential settlement over time. Drainage conditions on the surface and below ground must be understood because they influence the magnitude of earth pressures, movement of embankments, and stability of cuts or fills. All of these conditions influence the choice of substructure components that, in turn, influence the choice of superstructure. For example, an inclined leg rigid frame bridge requires strong foundation material that can resist both horizontal and vertical thrust. If this resistance is not present, then another bridge type may be more appropriate. The potential for seismic activity at a site should also be a part of the subsurface investigation. If seismicity is high, the substructure details will change, affecting the superstructure loads as well.

Functional Requirements In addition to the geometric alignment that allows a bridge to connect two points on a highway route, the bridge must also function to carry present and future traffic volumes. Decisions must be made on the number of lanes of traffic, inclusion of sidewalks and/or bike paths, whether width of the bridge deck should include medians, drainage of the surface waters, snow removal, and future wearing surface. In the case of stream and floodplain crossings, the bridge must continue to function during periods of high water and not impose a severe constriction or obstruction to the flow of water or debris. Satisfaction of these functional requirements will recommend some bridge types over others. For example, if future widening and replacement of bridge decks is a concern, multiple girder bridge types are preferred over concrete box girders.

Aesthetics Chapter 3 emphasizes the importance of designing a bridge with a pleasing appearance. It should be the goal of every bridge designer to obtain a positive aesthetic response to the bridge type selected. Details are presented earlier.

Economics and Ease of Maintenance It is difficult to separate first cost and maintenance cost over the life of the bridge when comparing the economics of different bridge types. A general rule is that the bridge with the minimum number of spans, fewest deck joints, and widest spacing of girders will be the most economical. By reducing the number of spans in a bridge layout by one span, the construction cost of one pier is eliminated. Deck joints are a high maintenance cost item, so minimizing their number reduces the life-cycle cost of the bridge.

When using the empirical design of bridge decks in the AASHTO (2010) LRFD Specifications, the same reinforcement is used for deck spans up to 13.5 ft (4100 mm). Therefore, little cost increase is incurred in the deck for wider spacing of girders, and fewer girders means less cost although at the "expense" of deeper sections.

Generally, concrete structures require less maintenance than steel structures. The cost and hazard of maintenance painting of steel structures should be considered in type selection studies (Caltrans, 1990).

One effective way to obtain the minimum construction cost is to prepare alternative designs and allow contractors to propose an alternative design. The use of alternative designs permits the economics of the construction industry at the time of bidding to determine the most economical material and bridge type. By permitting the contractor to submit an alternative design, the greatest advantage can be taken of new construction techniques to obtain less total project cost. The disadvantage of this approach is that a low initial cost may become the controlling criterion and life-cycle costs may not be effectively considered.

Construction and Erection Considerations The selection of the type of bridge to be built is often governed by construction and erection considerations. The length of time required to construct a bridge is important and varies with bridge type. In general, the larger the prefabricated or precast members, the shorter the construction time is. However, the larger the members, the more difficult they are to transport and lift into place.

Cast-in-place concrete bridges are generally economical for grade separations unless the falsework supporting the nonhardened concrete becomes a traffic problem. In that case, precast prestressed girders or welded steel plate girders would be a better choice.

The availability of skilled labor and specified materials also influences the choice of a particular bridge type. For example, if no precast plants for prestressed girders are located within easy transport, a steel fabrication plant is located nearby that could make the steel structure more economical. However, other factors in the construction industry may be at work. The primary way to determine which bridge type is more economical is to bid alternative designs. Designers are often familiar with bid histories and local economics and have significant experience regarding the lowest first cost.

Design–Build Option In the early years of bridge building in the United States, the design–build option was traditional. An owner would express an interest in having a bridge built at a particular location and solicit proposals from engineers for the design and construction of the bridge. On other occasions an engineer may see the need for a bridge and make presentations to potential owners of the merits of a particular design. Such was the case in the building of the Brooklyn Bridge (McCullough, 1972). John Roebling convinced influential people in Manhattan and Brooklyn to charter the New York Bridge Company to promote and finance his design for a great suspension bridge across the East River. The company hired his son, Washington Roebling, as chief engineer responsible for executing his father's design, preparing drawings and specifications, and supervising the construction. All services for designing and building the bridge were the responsibility of one entity.

This design–build practice of single-source responsibility faded somewhat at the end of the nineteenth century. The conventional approach became the design–bid–build model where an owner commissions an engineer to prepare drawings and specifications and separately selects a construction contractor by competitive bidding. The objective of the design–bid–build approach is to obtain the quality product defined by the drawings and specifications at a reasonable price. The approach works well with the checks and balances between the engineer and contractor when the separate parties work well together. Difficulties can occur when things go wrong on the job site or in the design office. There can be a lot of "finger pointing" that the other entity was responsible for the problem. This adversarial situation can increase the financial risk for all involved.

To alleviate some of the problems of unclear lines of responsibility, there has been a trend in recent years toward a return to the design–build option. One company is selected by the owner to prepare the engineering design and to be the construction contractor. This approach almost assures that the design group will possess the three essential mentalities: creative, analytical, and knowledge of construction techniques. If there is a question about the quality of the work or there are construction delays, only one entity is responsible. One objection to the design–build option is the absence of checks and balances because the same party that supplies a product approves it. It is important that the owner has staff people who are knowledgeable and can make independent judgments about the quality of the work provided. This knowledgeable staff is present in state DOTs, and more and more states are giving approval of the design–build option for construction of their bridges.

Legal Considerations In Figure 3.1, a model of the design process was presented. One of the components of the model was the constraint put on the design procedure by regulations. These regulations are usually beyond the control of the engineer, but they are real and must be considered.

Examples of regulations that will determine what bridge type can be built and where it can be located include: Permits over Navigable Waterways, National Environmental Policy Act, Department of Transportation Act, National Historic Preservation Act, Clean Air Act, Noise Control Act, Fish and Wildlife Coordination Act, Endangered Species Act, Water Bank Act, Wild and Scenic Rivers Act, Prime and Unique Farmlands, and Executive Orders on Floodplain Management and Protection of Wetlands. Engineers who are not conscious of the effect the design of a bridge has on the environment will soon become conscious once they begin preparing the environmental documentation required by these acts.

In addition to the environmental laws and acts defining national policy, local and regional politics are also of concern. Commitments to officials or promises made to communities often must be honored and may preclude other nonpolitical issues.

4.5.2 Bridge Types Used for Different Span Lengths

Once a preliminary span length has been chosen, comparative studies are conducted to find the bridge type best suited to the site. For each group of bridge spans (small, medium, and large), experience has shown that certain bridge types are more appropriate than others. This experience can be found in design aids prepared by associations, state agencies, and consulting firms. The comments that follow on common bridge types used for different span lengths are based on the experience of ACI-ASCE Committee 343 (1988), Caltrans (1990), and PennDOT (1993).

Small-Span Bridges [up to 50 ft (15 m)] The candidate structure types include single or multicell culverts, slab bridges, T-beam bridges, wood beam bridges, precast concrete box-beam bridges, precast concrete I-beam bridges, and composite rolled steel beam bridges.

Culvert. Culverts are used as small-span bridges to allow passage of small streams, livestock, vehicles, and pedestrians through highway embankments. These buried structures [A12.1]* are often the most economical solution for short spans. They are constructed of steel, aluminum, precast or cast-in-place reinforced concrete, and thermoplastics. Their structural form can be a pipe, pipe arch, plate arch, plate box, or rigid frame box. Either trench installations or embankment installations may be used. Minimum soil cover to avoid direct application of wheel loads is a function of the span length [Table A12.6.6.3-1] and is not less than 12 in. (300 mm). It is often cited that there are 577,000 bridges over 20 ft (6 m) long in the National Bridge Inventory. What is seldom mentioned is that 100,000 of them are structural culverts.

Slab. Slab bridges are the simplest and least expensive structure that can be built for small spans up to 40 ft (12 m). These bridges can be built on ground-supported falsework or constructed of precast elements. Construction details and formwork are the simplest of any bridge type. Their appearance is neat and simple, especially for low, short spans. Precast slab bridges constructed as simple spans require reinforcement in the topping slab to develop continuity over transverse joints at the piers, which is necessary to improve the riding quality of the deck and to avoid maintenance problems. Span lengths can be increased by use of prestressing. A design example of a simple-span solid-slab bridge is given in Chapter 14.

*References to AASHTO (2010) LRFD Specifications are enclosed in brackets and denoted by a letter A followed by the article number. A commentary is cited as the article number preceded by the letter C. Referenced figures and tables are enclosed in brackets to distinguish them from figures and tables in the text.

T-Beam. T-beam bridges, Table 4.1(e), are generally economical for spans 30–60 ft (10–20 m). These bridges usually are constructed on ground-supported falsework and require a good finish on all surfaces. Formwork may be complex, especially for skewed structures. Appearance of elevation is neat and simple, but not as desirable from below. Greatest use is for stream crossings, provided there is at least 6-ft (2-m) clearance above high water (floating debris may damage the girder stem). Usually, the T-beam superstructure is constructed in two stages: first the stems and then the slabs. To minimize cracks at the tops of the stems, longitudinal reinforcement should be placed in the stem near the construction joint. To ease concrete placement and finishing, a longitudinal joint within the structure becomes necessary for bridges wider than about 60 ft (20 m). A design example of a three-span continuous T-beam bridge is given in Chapter 16.

Wood Beam. Wood beam bridges, Table 4.1(l), may be used for low truck volume roads or in locations where a wood pile substructure can be constructed economically. Minimum width of roadway shall be 24 ft (7.2 m) curb to curb. The deck may be concrete, glued/spiked panels, or stressed wood. All wood used for permanent applications shall be impregnated with wood preservatives [A8.4.3.1]. The wood components not subject to direct pedestrian contact shall be treated with oil-borne preservatives [A8.4.3.2]. Main load-carrying members shall be precut and drilled prior to pressure treatment. For a waterway crossing, abutments and piers shall be aligned with the stream and piers shall be avoided in the stream if debris may be a problem.

Precast Concrete Box Beam. Precast prestressed concrete box-beam bridges can have spread boxes, Table 4.1(b), or butted boxes, Table 4.1(f) and (g), and can be used for spans from 30 to 150 ft (10 to 50 m). These bridges are most suitable for locations where the use of falsework is impractical or too expensive. The construction time is usually shorter than that needed for cast-in-place T-beams. Precast box beams may not provide a comfortable ride because adjacent boxes often have different camber and dead-load deflections. Unreinforced grout keys often fail between adjacent units, allowing differential live-load deflections to occur. A reinforced topping slab or transverse posttensioning can alleviate this problem. Appearance of the spread-box beam is similar to a T-beam while the butted-box beam is similar to a cast-in-place box girder. For multiple spans, continuity should be developed for live load by casting concrete between the ends of the simple-span boxes.

Precast Concrete I-Beam. Precast prestressed concrete I-beam bridges can be used for spans from 30 to 150 ft (10 to 50 m) and are competitive with steel girders. They have many of the same characteristics as precast concrete box-beam bridges including the problems with different camber and ridability. The girders are designed to carry dead load and construction loads as simple-span units. Live-load and superimposed dead-load design should use continuity and composite action with the cast-in-place deck slab. Appearance is like that of the T-beam: The elevation view is nice, but the underside looks cluttered. As in all concrete bridges, maintenance is low except at transverse deck joints, which often may be eliminated.

Rolled Steel Beam. Rolled steel wide-flange beam bridges are widely used because of their simple design and construction. See Table 4.1(a). These bridges are economical for spans up to 100 ft (30 m) when designing the deck as composite and using cover plates in maximum moment regions. The use of composite beams is strongly recommended because they make a more efficient structure. Shear connectors, usually in the form of welded studs, are designed to resist all forces tending to separate concrete and steel surfaces. The appearance of the multibeam bridge from underneath is similar to that of the T-beam, but the elevation is more slender (Table 3.1). The cost and environmental hazard of maintenance painting must be considered in any comparison with concrete bridges. Weathering steel is often used to eliminate paint.

Medium-Span Bridges [up to 250 ft (75 m)] The candidate structure types include precast concrete box-beam bridges, precast concrete I-beam bridges, composite rolled steel wide-flange beam bridges, composite steel plate girder bridges, cast-in-place concrete box-girder bridges, and steel box-girder bridges.

Precast Concrete Box Beam and Precast Concrete I-Beam. Characteristics of both of these precast prestressed concrete beams were discussed under small-span bridges. See Table 4.1(f) and (g). As span lengths increase, transportation and handling may present a problem. Most state highway departments require a permit for any load over 80 ft (24 m) long and refuse permits for loads over 115 ft (35 m) long. Girders longer than 115 ft (35 m) may have to be brought to the site in segments and then assembled. The longer girders are heavy, and firm ground is needed to store the girders and to provide support for the lifting cranes. The I-beam may be laterally unstable until incorporated into the structure and should be braced until the diaphragms are cast. A design example of a simple-span precast pretensioned concrete I-beam is given in Chapter 14.

Composite Rolled Steel Beam. Characteristics of composite rolled steel beams were discussed under small bridges. Composite construction can result in savings of 20–30% for spans over 50 ft (15 m) (Troitsky, 1994). Adding cover plates and providing continuity over several spans can increase their economic range to spans of 100 ft (30 m). A design example of a simple-span composite rolled steel beam bridge is given in Chapter 20.

Composite Steel Plate Girder. Composite steel plate girders [Table 4.1(a)] can be built to any desired size and consist of two flange plates welded to a web to form an asymmetrical I-section. These bridges are suitable for spans from 75 to 150 ft (25 to 50 m) and have been used for spans well over 300 ft (100 m). Girders must be braced against each other to provide stability against overturning and flange buckling, to resist transverse forces, and to distribute concentrated vertical loads. Construction details and formwork are simple. Transportation of prefabricated girders over 115 ft (35 m) may be a problem. Composite steel plate girder bridges can be made to look attractive and girders can be curved to follow alignment. This structure type has low dead load, which may be of value when foundation conditions are poor. A design example of a three-span continuous composite plate girder bridge is given in Chapter 20.

Cast-in-Place Reinforced Concrete Box Girder. Nonprestressed reinforced concrete box-girder bridges [Table 4.1(d)] are adaptable for use in many locations. These bridges are used for spans of 50–115 ft (15–35 m) and are often more economical than steel girders and precast concrete girders. Formwork is simpler than for a skewed T-beam, but it is still complicated. Appearance is good from all directions. Utilities, pipes, and conduits are concealed. High torsional resistance makes it desirable on curved alignment. They are an excellent choice in metropolitan areas.

Cast-in-Place Posttensioned Concrete Box Girder. Prestressed concrete box-girder bridges afford many advantages in terms of safety, appearance, maintenance, and economy. The cross section is the same as shown in Table 4.1(d) where a cast in place is shown. These bridges have been used for spans up to 600 ft (180 m). Because longer spans can be constructed economically, the number of piers can be reduced and shoulder obstacles eliminated for safer travel at overpasses. Appearance from all directions is neat and simple with greater slenderness than conventional reinforced concrete box-girder bridges. High torsional resistance makes it desirable on curved alignment. Because of the prestress, the dead-load deflections are minimized. Long-term shortening of the structure must be accommodated. Maintenance is very low, except that bearing and transverse deck joint details require attention. Addition of proper transverse and longitudinal posttensioning greatly reduces cracking. Posttensioned concrete box girders can be used in combination with conventional concrete box girders to maintain constant structure depth in long structures with varying span lengths. In areas where deck deterioration due to deicing chemicals is a consideration, deck removal and replacement is problematic.

Composite Steel Box Girder. Composite steel box-girder bridges, Table 4.1(b) and (c), are used for spans of 60–500 ft (20–150 m). These bridges are more economical in the upper range of spans and where depth may be limited. The boxes may be rectangular or trapezoidal and are effective in resisting torsion. They offer an attractive appearance and can be curved to follow alignment. Generally, multiple boxes would be used for spans up to 200 ft (60 m) and a single box for longer spans. Construction costs are often kept down by shop fabrication; therefore, designers should know the limitations placed by shipping clearances on the dimensions of the girders.

Large-Span Bridges [150–500 ft (50–150 m)] The candidate structure types include composite steel plate girder bridges, cast-in-place posttensioned concrete box-girder bridges, posttensioned concrete segmental bridges, concrete arch bridges, steel arch bridges, and steel truss bridges.

Composite Steel Plate Girder. Characteristics of composite steel plate girder bridges are presented in medium-span bridges. A design example of a medium-span bridge is given in Chapter 17.

Cast-in-Place Posttensioned Concrete Box Girder. Characteristics of cast-in-place posttensioned concrete box-girder bridges are presented in medium-span bridges.

Posttensioned Concrete Segmental Construction (ACI-ASCE Committee 343, 1988). Various bridge types may be constructed in segments and posttensioned to complete the final structure. The basic concept is to provide cost saving through standardization of details and multiple use of construction equipment. The segments may be cast-in-place or precast. If cast in place, it is common practice to use the balanced cantilever construction method with traveling forms. If the segments are precast, they may be erected by the balanced cantilever method, by progressive placement span by span, or by launching the spans from one end. Both the designer and the contractor have the opportunity to evaluate and choose the most cost-efficient method. Table 4.3 from Troitsky (1994) indicates typical span length ranges for bridge types by conventional and segmental construction methods.

The analysis and design of prestressed concrete segmental bridges is beyond the scope of this book. The reader is referred to reference books, such as Podolny and Muller (1982) and ASBI (2003), on the design and construction of segmental bridges.

Concrete Arch and Steel Arch. Characteristics of arch bridges are given in Section 4.1. Concrete arch bridges are usually below the deck, but steel arch bridges can be both above and below the deck, sometimes in the same structure. Typical and maximum span lengths for concrete and steel arch bridges are given in Table 4.2. Arch bridges are pleasing in appearance and are used largely for that reason even if a cost premium is involved. Arch bridge design is not addressed in this book, but information may be found in Xanthakos (1994) and Troitsky (1994).

Table 4.3 Range of Application of Bridge Type by Span Lengths Considering Segmental Construction

Span, ft (m)	Bridge Type
0–150 (0–45)	Precast pretensioned I-beam conventional
100–300 (30–90)	Cast-in-place posttensioned box-girder conventional
100–300 (30–90)	Precast balanced cantilever segmental, constant depth
200–600 (60–180)	Precast balanced cantilever segmental, variable depth
200–1000 (60–300)	Cast-in-place cantilever segmental
800–1500 (240–450)	Cable-stay with balanced cantilever segmental

From Troitsky (1994). *Planning and Design of Bridges*, Copyright © 1994. Reprinted with permission of John Wiley & Sons.

Steel Truss. Characteristics of steel truss bridges are given in Section 4.2. Steel truss bridges can also be below the deck and sometimes both above and below the deck in the same structure as seen in the through-truss Sydney Harbour Bridge (Fig. 4.9). Truss bridges are not addressed in this book, but they have a long history and numerous books, besides those already mentioned, can be found on truss design and construction. Few trusses are being designed and constructed now because of economic reasons. Those that are built typically are associated with maintaining a historical characteristic of a crossing where a truss bridge is being replaced.

Extra Large (long) Span Bridges [over 500 ft (150 m)] An examination of Table 4.2 shows that all of the general bridge types, except slabs, have been built with span lengths greater than 500 ft (150 m). Special bridges are designed to meet special circumstances and are not addressed in this book. Some of the bridge types in Table 4.1 were extended to their limit in attaining the long-span lengths and may not have been the most economical choice.

Two of the bridge types, cable stayed and suspension, are logical and efficient choices for long-span bridges. Characteristics of cable-stayed bridges and suspension bridges are given in Section 4.2. These tension-type structures are graceful and slender in appearance and are well suited to long water crossings. Maintenance for both is above average because of the complexity of the hanger and suspension system. Construction is actually simpler than for the conventional bridge types for long spans because falsework is usually not necessary. For additional information on the analysis, design, and construction of cable-stayed bridges, the reader is referred to Podolny and Scalzi (1986) and

Fig. 4.9 Sydney Harbour Bridge.

Troitsky (1988), while O'Connor (1971) is a good reference for stiffened suspension bridges.

4.5.3 Closing Remarks

In the selection of a bridge type, there is no unique or "correct" answer. For each span length range, more than one bridge type will satisfy the design criteria. Regional differences and preferences because of available materials, skilled workers, and knowledgeable contractors are significant. For the same set of geometric and subsurface circumstances, the bridge type selected may be different in Pennsylvania than in California. And both would be a good option for that place and time.

Because of the difficulties in predicting the cost climate of the construction industry at the time of bidding, a policy to allow the contractor the option of proposing an alternative design is prudent. This design should be made whether or not the owner has required the designer to prepare alternative designs. This policy improves the odds that the bridge type being built is the most economical.

In Section 3.2 on the design process, de Miranda (1991) was quoted as saying that for successful bridge design three "mentalities" must be present: (1) creative and aesthetic, (2) analytical, and (3) technical and practical. Oftentimes a designer possesses the first two mentalities and can select a bridge type that has a pleasing appearance and whose cross section has been well proportioned. But a designer may not be familiar with good, economical construction procedures and the third mentality is missing. By allowing the contractor to propose an alternative design, the third mentality may be restored and the original design(s) are further validated or a better design may be proposed. Either way, incorporating the three mentalities enhances the design process.

REFERENCES

AASHTO (2004). *LRFD Bridge Design Specification*, 3rd ed., American Association of State Highway and Transportation Officials, Washington, DC.

ACI-ASCE Committee 343 (1988). *Analysis and Design of Reinforced Concrete Bridge Structures*, American Concrete Institute, Detroit, MI.

ASBI (2003). *Recommended Practice for Design and Construction of Segmental Concrete Bridges*, American Segmental Bridge Institute, Phoenix, AZ.

Caltrans (1990). "Selection of Type," *Bridge Design Aids*, California Department of Transportation, Sacramento, CA, pp. 10–21 to 10–29.

Delaware Department of Transportation (2004). "Bridge 3–156 on SR-1 over Indian River Inlet," http://www.indianriverinletbridge.com/Resources/prequalb2/2004_12_16_PreQual_Mtg_Handout.pdf.

de Miranda, F. (1991). "The Three Mentalities of Successful Bridge Design," *Bridge Aesthetics around the World*, Transportation Research Board, National Research Council, Washington, DC, pp. 89–94.

Dorton, R. A. (1991). "Aesthetics Considerations for Bridge Overpass Design," *Bridge Aesthetics around the World*, Transportation Research Board, National Research Council, Washington, DC, pp. 10–17.

Leonhardt, F. (1991). "Developing Guidelines for Aesthetic Design," *Bridge Aesthetics around the World*, Transportation Research Board, National Research Council, Washington, DC, pp. 32–57.

McCullough, D. (1972). *The Great Bridge*, Simon & Schuster, New York.

O'Connor, C. (1971). *Design of Bridge Superstructures*, Wiley-Interscience, New York.
PennDOT (1993). "Selection of Bridge Types," *Design Manual—Part 4*, Vol. 1, Part A: Chapter 2, Pennsylvania Department of Transportation, Harrisburg, PA.
Podolny, W., Jr., and J. M. Muller, (1982). *Construction and Design of Prestressed Concrete Segmental Bridges*, Wiley, New York.
Podolny, W., Jr., and J. B. Scalzi (1986). *Construction and Design of Cable-Stayed Bridges*, 2nd ed., Wiley, New York.
Troitsky, M. S. (1988). *Cable-Stayed Bridges, An Approach to Modern Bridge Design*, 2nd ed., Van Nostrand Reinhold, New York.
Troitsky, M. S. (1994). *Planning and Design of Bridges*, Wiley, New York.
Xanthakos, P. (1994). *Theory and Design of Bridges*, Wiley, New York.

PROBLEMS

4.1 Discuss the interaction between deductive and inductive reasoning in formulating the principles of structural analysis and design.

4.2 Explain the interrelationship between description and justification of a bridge design.

4.3 What makes it difficult for a person, including your professor and yourself, to pass on personal knowledge?

4.4 List the four general stages of bridge design and give a brief description of each one.

4.5 Describe how the design of a bridge begins in the mind.

4.6 Discuss the necessity of having the three "mentalities" present in the bridge design team. Imagine you are a member of the design team and indicate what abilities you think other people on the team need to have.

4.7 Explain what is meant by the following: "Whether intentional or not, every bridge structure makes an aesthetic statement."

4.8 Some of the qualities of bridge aesthetics are similar to the qualities of classical music. Choose one of these qualities common with music and describe how that quality can improve the appearance of bridges.

4.9 How can shadow be used to make a bridge appear more slender?

4.10 What is meant by "resolution of duality"? How can it be resolved in an overpass of an interstate highway?

4.11 Leonhardt (1991) states that the slenderness ratio *L/d* is the most important criterion for the appearance of a bridge. Explain how continuity can maximize this ratio.

4.12 In selecting abutments, what steps can be taken to give a bridge spanning traffic a feeling of openness?

4.13 In what ways do integral abutment and jointless bridge designs reduce maintenance costs? How are movements due to temperature changes accommodated in these bridges?

4.14 What is the difference in the main load-carrying mechanism of arch bridges and suspension bridges?

4.15 If girder bridges are structurally less efficient when compared to suspension and arch bridges, why are there so many girder bridges?

4.16 List some of the factors to be considered in the selection of a bridge type. Explain why it is necessary for the factors to be kept in proper balance and considered of equal importance.

4.17 If future widening of a bridge deck is anticipated, what girder bridge type is appropriate?

4.18 Often when an engineer is comparing costs of alternative designs, the only cost considered is the initial construction cost. What other considerations affect the cost when alternatives are compared?

CHAPTER 5

Design Limit States

5.1 INTRODUCTION

The justification stage of design can begin after the selection of possible alternative bridge types that satisfy the function and aesthetic requirements of the bridge location has been completed. As discussed in the opening pages of Chapter 3, justification requires that the engineer verify the structural safety and stability of the proposed design. Justification involves calculations to demonstrate to those who have a vested interest that all applicable specifications, design, and construction requirements are satisfied.

A general statement for assuring safety in engineering design is that the resistance of the components supplied exceed the demands put on them by applied loads, that is,

$$\text{Resistance} \geq \text{effect of the loads} \tag{5.1}$$

When applying this simple principle, both sides of the inequality are evaluated for the same conditions. For example, if the effect of applied loads is to produce compressive stress on a soil, this should be compared to the bearing resistance of the soil, and not some other quantity. In other words, the evaluation of the inequality must be done for a specific loading condition that links together resistance and the effect of loads. The evaluation of both sides at the same limit state for each applicable failure mode provides this common link.

When a particular loading condition reaches its limit, failure is the assumed result, that is, the loading condition becomes a failure mode. Such a condition is referred to as a limit state:

> *A limit state is a condition beyond which a bridge system or bridge component ceases to fulfill the function for which it is designed.*

Examples of limit states for girder-type bridges include deflection, cracking, fatigue, flexure, shear, torsion, buckling, settlement, bearing, and sliding. Well-defined limit states are established so that a designer knows what is considered to be unacceptable.

An important goal of design is to prevent a limit state from being reached. However, it is not the only goal. Other goals that must be considered and balanced in the overall design are function, appearance, and economy. To design a bridge so that none of its components would ever fail is not economical. Therefore, it becomes necessary to determine what is an acceptable level of risk or probability of failure.

The determination of an acceptable margin of safety (how much greater the resistance should be compared to the effect of loads) is not based on the opinion of one individual but on the collective experience and judgment of a qualified group of engineers and officials. In the U.S. highway bridge design community, the American Association of State Highway and Transportation Officials (AASHTO) is such a group. It relies on the experience of the state department of transportation engineers, research engineers, consultants, practitioners, and engineers involved with design specifications outside the United States.

5.2 DEVELOPMENT OF DESIGN PROCEDURES

Over the years, design procedures have been developed by engineers to provide satisfactory margins of safety. These procedures were based on the engineer's confidence in the analysis of the load effects and the strength of the materials being provided. As analysis techniques improved and quality control on materials became better, the design procedures changed as well.

To understand where we are today, it is helpful to look at the design procedures of earlier AASHTO Specifications and how they have evolved as technology changed.

5.2.1 Allowable Stress Design

The earliest numerically based design procedures were developed with a primary focus on behavior of metallic structures. Structural steels were observed to behave linearly up to a relatively well-defined yield point that was safely below the ultimate strength of the material. Safety in the design was obtained by specifying that the effect of the loads should produce stresses that were a fraction of the yield stress f_y: for example, one half. This value would be equivalent to providing a safety factor F of 2; that is,

$$F = \frac{\text{resistance, } R}{\text{effect of loads, } Q} = \frac{f_y}{0.5 f_y} = 2$$

Because the specifications set limits on the stresses, this became known as allowable stress design (ASD).

When ASD methods were first used, a majority of the bridges were open-web trusses or arches. By assuming pin-connected members and using statics, the analysis indicated members that were either in tension or compression. The required net area of a tension member under uniform stress was easily selected by dividing the tension force T by an allowable tensile stress f_t:

$$\text{Required } A_{\text{net}} \geq \frac{\text{effect of load}}{\text{allowable stress}} = \frac{T}{f_t}$$

For compression members, the allowable stress f_c depended on whether the member was short (nonslender) or long (slender), but the rationale for determining the required area of the cross section remained the same; the required area was equal to the compressive force divided by an allowable stress value:

$$\text{Required } A_{\text{gross}} \geq \frac{\text{effect of load}}{\text{allowable stress}} = \frac{C}{f_c}$$

These techniques were used as early as the 1860s to design many successful statically determinate truss bridges. Similar bridges are built today, but few are statically determinate because they are no longer pin connected. As a result, the stresses in the members are no longer uniform because of the bending moments that occur due to the more rigid connections.

The ASD method is also applied to beams in bending. By assuming plane sections remain plane, and linear stress–strain response, a required section modulus S can be determined by dividing the bending moment M by an allowable bending stress f_b:

$$\text{Required } S \geq \frac{\text{effect of load}}{\text{allowable stress}} = \frac{M}{f_b}$$

Implied in the ASD method is the assumption that the stress in the member is zero before any loads are applied, that is, no residual stresses are introduced when the members are formed. This assumption is seldom accurate but is closer to being true for solid bars and rods than for thin open sections of typical rolled beams. The thin elements of rolled beams cool at different rates and residual stresses become locked into the cross section. Not only are these residual stresses highly nonuniform, they are also difficult to predict. Consequently, adjustments have to be made to the allowable bending stresses, especially in compression elements, to account for the effect of residual stresses. See Figure 17.2 for an example.

Another difficulty in applying ASD to steel beams is that bending is usually accompanied by shear, and these two stresses interact. Consequently, it is not strictly correct to use tensile coupon tests (satisfactory for pin-connected trusses) to determine the yield strength f_y for beams in bending. Another definition of yield stress that incorporates the effect of shear stress might be more logical.

What is the point in discussing ASD methods applied to steel design in a book on bridge analysis and design? Simply this:

> *ASD methods were developed for the design of statically determinate metallic structures. They do not necessarily apply in a straightforward and logical way to other materials and other levels of redundancy.*

Designers of reinforced concrete structures have realized this for some time and adopted strength design procedures many years ago. Wood designers are also moving toward strength design procedures. Both concrete and wood are nonlinear materials whose properties change with time and with changes in ambient conditions. In concrete, the initial stress state is unknown because it varies with placement method, curing method, temperature gradient, restraint to shrinkage, water content, and degree of consolidation. The only values that can be well defined are the strengths of concrete at its limit states. As described in Chapter 10, the ultimate strength is independent of prestrains and stresses associated with numerous manufacturing and construction processes, all of which are difficult to predict and are highly variable. In short, the ultimate strength is easier to determine and more reliably predicted than strengths at lower load levels. The improved reliability gives additional rationale for adoption of strength design procedures.

5.2.2 Variability of Loads

In regard to uncertainties in design, one other point concerning the ASD method needs to be emphasized. *Allowable stress design does not recognize that different loads have different levels of uncertainty*. Dead, live, and wind loads are all treated equally in ASD. The safety factor is applied to the resistance side of the design inequality of Eq. 5.1, and the load side is not factored. In ASD, safety is determined by

$$\frac{\text{Resistance, } R}{\text{Safety factor, } F} \geq \text{effect of loads, } Q \tag{5.2}$$

For ASD, fixed values of design loads are selected, usually from a specification or design code. The varying degree of predictability of the different load types is not considered.

Finally, because the safety factor chosen is based on experience and judgment, *quantitative measures of risk cannot be determined for ASD*. Only the trend is known: If the safety factor is higher, the number of failures is lower. However, if the safety factor is increased by a certain amount, it is not known by how much this increases the probability of survival. Also, it is more meaningful to decision makers to say, "This bridge has a nominal probability of 1 in 10,000 of failing in 75 years of service," than to say, "This bridge has a safety factor of 2.3."

5.2.3 Shortcomings of Allowable Stress Design

As just shown, ASD is not well suited for design of modern structures. Its major shortcomings can be summarized as follows:

1. The resistance concepts are based on elastic behavior of materials.
2. It does not embody a reasonable measure of strength, which is a more fundamental measure of resistance than is allowable stress.
3. The safety factor is applied only to resistance. Loads are considered to be deterministic (without variation).

4. Selection of a safety factor is subjective, and it does not provide a measure of reliability in terms of probability of failure.

What is needed to overcome these deficiencies is a method that is (a) based on the strength of material, (b) considers variability not only in resistance but also in the effect of loads, and (c) provides a measure of safety related to probability of failure. Such a method was first incorporated in the AASHTO LRFD Bridge Specifications in 1994 and is discussed in Section 5.2.4.

5.2.4 Load and Resistance Factor Design

To account for the variability on both sides of the inequality in Eq. 5.1, the resistance side is multiplied by a statistically based resistance factor ϕ, whose value is usually less than one, and the load side is multiplied by a statistically based load factor γ, whose value is usually greater than one. Because the load effect at a particular limit state involves a combination of different load types (Q_i) that have different degrees of predictability, the load effect is represented by a summation of $\gamma_i Q_i$ values. If the nominal resistance is given by R_n, the safety criterion is

$$\phi R_n \geq \text{effect of} \sum \gamma_i Q_i \tag{5.3}$$

Because Eq. 5.3 involves both load factors and resistance factors, the design method is called load and resistance factor design (LRFD). The resistance factor ϕ for a particular limit state must account for the uncertainties in

- ☐ Material properties
- ☐ Equations that predict strength
- ☐ Workmanship
- ☐ Quality control
- ☐ Consequence of a failure

The load factor γ_i chosen for a particular load type must consider the uncertainties in

- ☐ Magnitudes of loads
- ☐ Arrangement (positions) of loads
- ☐ Possible combinations of loads

In selecting resistance factors and load factors for bridges, probability theory has been applied to data on strength of materials and statistics on weights of materials and vehicular loads.

Some of the pros and cons of the LRFD method can be summarized as follows:

Advantages of LRFD Method

1. Accounts for variability in both resistance and load.
2. Achieves fairly uniform levels of safety for different limit states and bridge types without involving probability or statistical analysis.
3. Provides a rational and consistent method of design.
4. Provides consistency with other design specifications (e.g., ACI and AISC) that are familiar to engineers and new graduates.

Disadvantages of LRFD Method

1. Requires a change in design philosophy (from previous AASHTO methods).
2. Requires an understanding of the basic concepts of probability and statistics.
3. Requires availability of sufficient statistical data and probabilistic design algorithms to make adjustments in resistance factors.

5.3 DESIGN LIMIT STATES

5.3.1 General

The basic design expression in the AASHTO (2010) LRFD Bridge Specifications that must be satisfied for all limit states, both global and local, is given as

$$\sum \eta_i \gamma_i Q_i \leq \phi R_n \tag{5.4}$$

where Q_i is the force effect, R_n is the nominal resistance, γ_i is the statistically based load factor applied to the force effects, ϕ is the statistically based resistance factor applied to nominal resistance, and η_i is a load modification factor. For all nonstrength limit states, $\phi = 1.0$ [A1.3.2.1].*

Equation 5.4 is Eq. 5.3 with the addition of the load modifier η_i. The load modifier is a factor that takes into account the ductility, redundancy, and operational importance of the bridge. It is given for loads for which a maximum value of γ_i is appropriate by

$$\eta_i = \eta_D \eta_R \eta_I \geq 0.95 \tag{5.5a}$$

and for loads for which a minimum value of γ_i is appropriate by

$$\eta_i = 1/\eta_D \eta_R \eta_I \leq 1.0 \tag{5.5b}$$

where η_D is the ductility factor, η_R is the redundancy factor, and η_I is the operational importance factor. The first two factors refer to the strength of the bridge and the third refers to the consequence of a bridge being out of service. For all nonstrength limit states $\eta_D = \eta_R = \eta_I = 1.0$ [A1.3.2].

Ductility Factor η_D [A1.3.3] Ductility is important to the safety of a bridge. If ductility is present, overloaded portions of the structure can redistribute the load to other portions that have reserve strength. This redistribution is dependent on the ability of the overloaded component and its connections to develop inelastic deformations without failure.

*The article numbers in the AASHTO (2010) LRFD Bridge Specifications are enclosed in brackets and preceded by the letter A if a specification article and by the letter C if commentary.

If a bridge component is designed so that inelastic deformations can occur, then there will be a warning that the component is overloaded. For example, in reinforced concrete, cracking will increase and the component will show that it is in distress. For structural steel flaking of mill scale will indicate yielding and deflections will increase. The effects of inelastic behavior are elaborated in Chapter 10.

Brittle behavior is to be avoided because it implies a sudden loss of load-carrying capacity when the elastic limit is exceeded. Components and connections in reinforced concrete can be made ductile by limiting the flexural reinforcement and by providing confinement with hoops or stirrups. Steel sections can be proportioned to avoid buckling, which may permit inelastic behavior. Similar provisions are given in the specifications for other materials. In fact, if the provisions of the specifications are followed in design, experience has shown that the components will have adequate ductility [C1.3.3].

The values to be used for the strength limit state ductility factor are:

$\eta_D \geq 1.05$ for nonductile components and connections

$\eta_D = 1.00$ for conventional designs and details complying with the specifications

$\eta_D \geq 0.95$ for components and connections for which additional ductility-enhancing measures have been specified beyond those required by the specifications

For all other limit states:

$$\eta_D = 1.00$$

Redundancy Factor η_R ***[A1.3.4]*** Redundancy significantly affects the safety margin of a bridge structure. A statically indeterminate structure is redundant, that is, it has more restraints than are necessary to satisfy equilibrium. For example, a three-span continuous bridge girder may be classified as statically indeterminate to the second degree. Any combination of two supports, or two moments, or one support and one moment could be lost without immediate collapse because the applied loads could find alternative paths. The concept of multiple-load paths is the same as redundancy.

Single-load paths or nonredundant bridge systems are not encouraged. The Silver Bridge over the Ohio River between Pt. Pleasant, West Virginia, and Kanauga, Ohio, was a single-load path structure. It was constructed in 1920 as a suspension bridge with two main chains composed of eyebar links, much like large bicycle chains, strung between two towers. However, to make the structure easier to analyze, pin connections were made at the base of the towers. When one of the eyebar links failed in December 1967, there was no alternative load path, the towers were nonredundant, and the collapse was sudden and complete. Forty-six lives were lost (Section 2.2.1).

In the 1950s a popular girder bridge system was the cantilever span, suspended span, cantilever span system. These structures were statically determinate and the critical detail was the linkage or hanger that supported the suspended span from the cantilevers. The linkage was a single-load path connection, and, if it failed, the suspended span would drop to the ground or water below. This failure occurred in the bridge over the Mianus River in Greenwich, Connecticut, June 1983. Three lives were lost (Section 2.2.4).

Redundancy in a bridge system increases its margin of safety, and this is reflected in the strength limit state by redundancy factors given as

$\eta_R \geq 1.05$ for nonredundant members

$\eta_R = 1.00$ for conventional levels of redundancy

$\eta_R \geq 0.95$ for exceptional levels of redundancy

For all other limit states:

$$\eta_R = 1.00$$

Operational Importance Factor η_I ***[A1.3.5]*** Bridges can be considered of operational importance if they are on the shortest path between residential areas and a hospital or school or provide access for police, fire, and rescue vehicles to homes, businesses, and industrial plants. Bridges can also be considered essential if they prevent a long detour and save time and gasoline in getting to work and back home again. In fact, it is difficult to find a situation where a bridge would not be operationally important because a bridge must be justified on some social or security requirement to have been built in the first place. One example of a less important bridge could be on a secondary road leading to a remote recreation area that is not open year round. (But then if you were a camper or backpacker and were injured or became ill, you'd probably consider any bridge between you and the more civilized world to be operationally important!)

In the event of an earthquake, it is important that all lifelines, such as bridges, remain open. Therefore, the following requirements apply to the extreme event limit state as well as to the strength limit state:

$\eta_I \geq 1.05$ for a bridge of operational importance

$\eta_I \geq 1.00$ for typical bridges

$\eta_I \geq 0.95$ for relatively less important bridges

For all other limit states:

$$\eta_I = 1.00$$

Load Designation [A3.3.2] Permanent and transient loads and forces that must be considered in a design are designated as follows:

Permanent Loads

CR	Force effects due to creep
DD	Downdrag force
DC	Dead load of structural components and nonstructural attachments

DW Dead load of wearing surfaces and utilities
EH Horizontal earth pressure load
EL Miscellaneous locked-in force effects resulting from the construction process, including the jacking apart of cantilevers in segmental construction
ES Earth surcharge load
EV Vertical pressure from dead load of earth fill
PS Secondary forces from posttensioning
SH Force effects due to shrinkage

Transient Loads

BR Vehicular braking force
CE Vehicular centrifugal force
CT Vehicular collision force
CV Vessel collision force
EQ Earthquake load
FR Friction load
IC Ice load
IM Vehicular dynamic load allowance
LL Vehicular live load
LS Live-load surcharge
PL Pedestrian live load
SE Force effect due to settlement
TG Force effect due to temperature gradient
TU Force effect due to uniform temperature
WA Water load and stream pressure
WL Wind on live load
WS Wind load on structure

Load Combinations and Load Factors The load factors for various load combinations and permanent loads are given in Tables 5.1 and 5.2, respectively. Explanations of the different limit states are given in the sections that follow.

5.3.2 Service Limit State

The service limit state refers to restrictions on stresses, deflections, and crack widths of bridge components that occur under regular service conditions [A1.3.2.2]. For the service limit state, the resistance factors $\phi = 1.0$ and nearly all of the load factors γ_i are equal to 1.0. There are four different service limit state load combinations given in Table 5.1 to address different design situations [A3.4.1].

Service I This service limit state refers to the load combination relating to the normal operational use of the bridge with 55-mph (90-km/h) wind, and with all loads taken at their nominal values. It also relates to deflection control in buried structures, crack control in reinforced concrete structures, concrete compressive stress in prestressed concrete

Table 5.1 Load Combinations and Load Factors

Load Combination / Limit State	DC DD DW EH EL EV ES PS CR SH	LL IM CE BR PL LS	WA	WS	WL	FR	TU	TG	SE	Use One of These at a Time			
										EQ	IC	CT	CV
Strength I	γ_p	1.75	1.00	—	—	1.00	0.50/1.20	γ_{TG}	γ_{SE}				
Strength II	γ_p	1.35	1.00	—	—	1.00	0.50/1.20	γ_{TG}	γ_{SE}				
Strength III	γ_p	—	1.00	1.40	—	1.00	0.50/1.20	γ_{TG}	γ_{SE}				
Strength IV	γ_p	—	1.00	—	—	1.00	0.50/1.20	—	—				
Strength V	γ_p	1.35	1.00	0.40	1.0	1.00	0.50/1.20	γ_{TG}	γ_{SE}				
Extreme Event I	γ_p	γ_{EQ}	1.00	—	—	1.00	—	—	—	1.00			
Extreme Event II	γ_p	0.50	1.00	—	—	1.00	—	—	—	—	1.00	1.00	1.00
Service I	1.00	1.00	1.00	0.30	1.0	1.00	1.00/1.20	γ_{TG}	γ_{SE}				
Service II	1.00	1.30	1.00	—	—	1.00	1.00/1.20	—	—				
Service III	1.00	0.80	1.00	—	—	1.00	1.00/1.20	γ_{TG}	γ_{SE}				
Service IV	1.00	—	1.00	0.70	—	1.00	1.00/1.20	—	1.0				
Fatigue I—LL, IM, and CE only	—	1.50	—	—	—	—	—	—	—				
Fatigue II—LL, IM, and CE only	—	0.75	—	—	—	—	—	—	—				

AASHTO Table 3.4.1-1. From *AASHTO LRFD Bridge Design Specifications*, Copyright © 2010 by the American Association of State Highway and Transportation Officials, Washington, DC. Used by permission.

Table 5.2 Load Factors for Permanent Loads, γ_p

Type of Load, Foundation Type, and Method Used to Calculate Downdrag		Load Factor Maximum	Minimum
DC: Component and attachments		1.25	0.90
DC: Strength IV only		1.50	0.90
DD: Downdrag	Piles, α Tomlinson Method	1.4	0.25
	Piles, λ Method	1.05	0.30
	Drilled shafts, O'Neill and Reese (1999) Method	1.25	0.35
DW: Wearing surfaces and utilities		1.50	0.65
EH: Horizontal earth pressure			
• Active		1.50	0.90
• At rest		1.35	0.90
• *AEP* for anchored walls		1.35	N/A
EL: Locked-in construction stresses		1.00	1.00
EV: Vertical earth pressure			
• Overall stability		1.00	N/A
• Retaining walls and abutments		1.35	1.00
• Rigid buried structure		1.30	0.90
• Rigid frames		1.35	0.90
• Flexible buried structures			
• Metal box culverts and structural plate culverts with deep corrugations		1.5	0.9
• Thermoplastic culverts		1.3	0.9
• All others		1.95	0.9
ES: Earth surcharge		1.50	0.75

AASHTO Table 3.4.1-2. From *AASHTO LRFD Bridge Design Specifications*, Copyright © 2010 by the American Association of State Highway and Transportation Officials, Washington, DC. Used by permission.

components, and concrete tensile stress related to transverse analysis of concrete segmental girders. This load combination should also be used for the investigation of slope stability.

Service II This service limit state refers to the load combination relating only to steel structures and is intended to control yielding and slip of slip-critical connections due to vehicular live load. It corresponds to the overload provision for steel structures in past editions of the AASHTO Standard Specifications (AASHTO 2002).

Service III This service limit state refers to the load combination for longitudinal analysis relating to tension in prestressed concrete superstructures with the objective of crack control and to principal tension in the webs of segmental concrete girders. The statistical significance of the 0.80 factor on live load is that the event is expected to occur about once a year for bridges with two traffic lanes, less often for bridges with more than two traffic lanes, and about once a day for bridges with a single traffic lane. Service I is used to investigate for compressive stresses in prestressed concrete components.

Service IV This service limit state refers to the load combination relating only to tension in prestressed concrete columns with the objective of crack control. The 0.70 factor on wind represents an 84-mph (135-km/h) wind. This should result in zero tension in prestressed concrete substructures for 10-year mean reoccurrence winds.

5.3.3 Fatigue and Fracture Limit State

The fatigue and fracture limit state refers to a set of restrictions on stress range caused by a single design truck. The restrictions depend on the number of stress–range excursions expected to occur during the design life of the bridge [A1.3.2.3]. They are intended to limit crack growth under repetitive loads and to prevent fracture due to cumulative stress effects in steel elements, components, and connections. For the fatigue and fracture limit state, $\phi = 1.0$.

Because the only load effect that causes a large number of repetitive cycles is the vehicular live load, it is the only load effect that has a nonzero load factor in the fatigue limit state (see Table 5.1). A load factor of 0.75 is used for fatigue II and applied to vehicular live load, dynamic load allowance, and centrifugal force. Use of load factor less than 1.0 is justified because statistics show that trucks at slightly lower weights cause more repetitive cycles of stress than those at the weight of the design truck [C3.4.1]. Incidentally, the fatigue design truck is different than the design truck used to evaluate other force effects. It is defined as a single truck with a fixed axle spacing [A3.6.1.4.1]. The truck load models are described in detail in Chapter 8. A load factor of 1.5 is used for fatigue I and the analysis is similar to that outlined above. The resistance that is used to compare this load effect is outlined in Chapter 18.

Fracture due to fatigue occurs at stress levels below the strength measured in uniaxial tests. When passing trucks cause a number of relatively high stress excursions, cumulative damage will occur. When the accumulated damage is large enough, a crack in the material will start at a point of stress concentration. The crack will grow with repeated stress cycles, unless observed and arrested, until the member fractures. If fracture of a member may result in collapse of a bridge, the member is called *fracture critical*. The eyebar chain in the Silver Bridge (Section 2.2.1) and the hanger link in the Mianus River Bridge (Section 2.2.4) were both fracture-critical members.

5.3.4 Strength Limit State

The strength limit state refers to providing sufficient strength or resistance to satisfy the inequality of Eq. 5.4 for the statistically significant load combinations that a bridge is expected to experience in its design life [A1.3.2.4]. Strength limit states include the evaluation of resistance to bending, shear, torsion, and axial load. The statistically determined resistance factor ϕ will usually be less than 1.0 and will have different values for different materials and strength limit states.

The statistically determined load factors γ_i are given in five separate load combinations in Table 5.1 to address different design considerations. For force effects due to permanent loads, the load factors γ_p of Table 5.2 shall be selected to give the most critical load combination for a particular strength limit state. Either the maximum or minimum value of γ_p may control the extreme effect so both must be investigated. Application of two different values for γ_p could easily double the number of strength load combinations to be considered. Fortunately, not all of the strength limit states apply in every situation and some can be eliminated by inspection.

For all strength load combinations, a load factor of 0.50 is applied to TU for nondisplacement force effects to represent the reduction in this force effect with time from the value predicted by an elastic analysis. In the calculation of displacements for this load, a load factor of 1.20 is used to avoid undersized joints and bearings [C3.4.1].

Strength I This strength limit state is the basic load combination relating to normal vehicular use of the bridge without wind [A3.4.1].

Strength II This strength limit state is the load combination relating to the use of the bridge by owner-specified special design vehicles, evaluation permit vehicles, or both without wind. If a permit vehicle is traveling unescorted, or if the escorts do not provide control, the basic design vehicular live load may be assumed to occupy the other lanes on the bridge [A4.6.2.2.4].

Strength III This strength limit state is the load combination relating to the bridge exposed to wind velocity exceeding 55 mph (90 km/h). The high winds prevent the presence of significant live load on the bridge [C3.4.1].

Strength IV This strength limit state is the load combination relating to very high dead- and live-load force effect ratios. The standard calibration process used to select load factors γ_i and resistance factors ϕ for the strength limit state was carried out for bridges with spans less than 200 ft (60 m). For the primary components of large-span bridges, the ratio of dead- and live-load force effects is rather high and could result in a set of resistance factors different from those found acceptable for small- and medium-span bridges. To avoid using two sets of resistance factors with the load factors of the strength I limit state, the strength IV limit state load factors were developed for large-span bridges [C3.4.1].

Strength V This strength limit state is the load combination relating to normal vehicular use of the bridge with wind of 55-mph (90-km/h) velocity. The strength V limit state differs from the strength III limit state by the presence of live load on the bridge, wind on the live load, and reduced wind on the structure (Table 5.1).

5.3.5 Extreme Event Limit State

The extreme event limit state refers to the structural survival of a bridge during a major earthquake or flood or when collided by a vessel, vehicle, or ice floe, possibly under scoured conditions [A1.3.2.5]. The probability of these events occurring simultaneously is extremely low; therefore, they are specified to be applied separately. The recurrence interval of extreme events may be significantly greater than the design life of the bridge [C1.3.2.5]. Under these extreme conditions, the structure is expected to undergo considerable inelastic deformation by which locked-in force effects due to TU, TG, CR, SH, and SE are expected to be relieved [C3.4.1] (see Chapter 10). For the extreme event limit state, $\phi = 1.0$.

Extreme Event I This extreme event limit state is the load combination relating to earthquakes. This limit state also includes water load WA and friction FR. The probability of a major flood and an earthquake occurring at the same time is very small. Therefore, water loads and scour depths based on mean discharges may be warranted [C3.4.1].

Partial live load coincident with earthquake should be considered. The load factor for live load γ_{EQ} shall be determined on a project-specific basis [A3.4.1]. Suggested values for γ_{EQ} are 0.0, 0.5, and 1.0 [C3.4.1].

Extreme Event II This extreme event limit state is the load combination relating to ice load, collision by vessels and vehicles, and to certain hydraulic events with reduced live load. The 0.50 live-load factor signifies a low probability of the combined occurrence of the maximum vehicular live load, other than CT, and the extreme events [C3.4.1].

5.4 CLOSING REMARKS

The AASHTO LRFD Specifications outline several limit states that must be satisfied for an acceptable design. Limit states importantly consider the strength of the bridge but also consider serviceability, the chance of an extreme event, and fatigue and fracture. The LRFD format considers the variability of the material strengths and their use in various components and, also, the variability of dead, live, and other loads. Clearly, some loads are more predictable than others, and this is considered in the formulation. Chapter 6 discusses how load and resistance factors are rationally determined and how these values were set in the AASHTO LRFD Specifications.

REFERENCES

AASHTO (2002). *Standard Specifications for Highway Bridges*, 15th ed., American Association of State Highway and Transportation Officials, Washington, DC.

AASHTO (2010). *LRFD Bridge Design Specifications*, 5th ed., American Association of State Highway and Transportation Officials, Washington, DC.

PROBLEMS

5.1 What are the main reasons for choosing the probabilistic limit states philosophy of LRFD over the deterministic design philosophy of ASD?

5.2 Discuss the influence that residual stresses had on the selection of a limit states design philosophy.

5.3 The AASHTO LRFD basic design expression includes a load modifier term η. What is the purpose of this modifier? Why is it on the load side of the inequality?

5.4 How does the amount of ductility in structural members, represented by η_D, affect the reliability of bridge structures?

5.5 Why are the live-load factors in service II and service III not equal to 1.0?

5.6 Why are only live-load effects considered in the fatigue and fracture limit state? For this limit state, why is the live-load factor less than 1.0?

5.7 What is the justification for a smaller live-load factor for strength II than for strength I when the vehicles in strength II are larger than those in strength I?

5.8 Why are there no live-load factors in strength III and strength IV?

5.9 How are the maximum and minimum load factors for permanent loads γ_p to be used in various load combinations?

5.10 For the extreme event limit states, why are the load factors 1.0 for EQ, IC, CT, and CV? Why is only one used at a time?

CHAPTER 6

Principles of Probabilistic Design

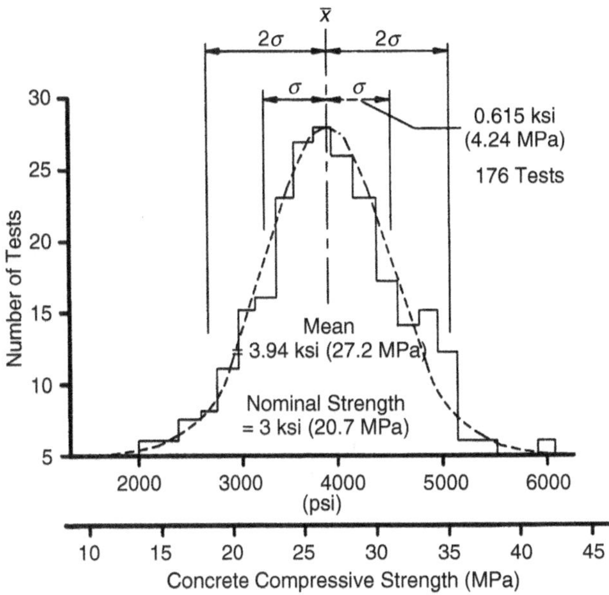

Fig. 6.1 Distribution of concrete strengths. (After J. G. MacGregor and J. K. Wight, *Reinforced Concrete: Mechanics and Design*, Copyright © 2004. Reprinted by permission of Prentice Hall, Upper Saddle River, NJ.)

6.1 INTRODUCTION

A brief primer on the basic concepts is given to facilitate the use of statistics and probability. This review provides the background for understanding how the LRFD code was developed. Probabilistic analyses are not necessary to apply the LRFD method in practice, except for rare situations that are not encompassed by the code.

There are several levels of probabilistic design. The fully probabilistic method (level III) is the most complex and requires knowledge of the probability distributions of each random variable (resistance, load, etc.) and correlation between the variables. This information is seldom available, so it is rarely practical to implement the fully probabilistic method.

Level II probabilistic methods include the first-order second-moment (FOSM) method, which uses simpler statistical characteristics of the load and resistance variables. Further, the load Q and resistance R are assumed to be statistically independent.

The load and resistance factors employed in the AASHTO (1994, 2010) LRFD Bridge Specifications were determined by using level II procedures and other simpler methods when insufficient information was available to use the level II methods. The following sections define and discuss the statistical and probabilistic terms that are involved in this level II theory.

Nowak and Collins (2000) offer an excellent treatment of probabilistic design. Several references and examples related to bridges and application to AASHTO LRFD are provided.

6.1.1 Frequency Distribution and Mean Value

Consider Figure 6.1, which is a histogram of the 28-day compressive strength distribution of 176 concrete cylinders, all intended to provide a design strength of 3 ksi (20.7 MPa). The ordinates represent the number of times a particular compressive strength (0.200 ksi = 1.38 MPa intervals) was observed.

As is well known, the mean (average) value $\bar{x}$ of the N compressive strength values x_i is calculated by

Mean

$$\bar{x} = \frac{\sum x_i}{N} \tag{6.1}$$

For the $N = 176$ tests, the mean value $\bar{x}$ is found to be 3.94 ksi (27.2 MPa).

Notice that the dashed smooth curve that approximates this histogram is the familiar bell-shaped distribution function that is typical of many natural phenomena.

6.1.2 Standard Deviation

The variance of the data from the mean is determined by summing up the square of the difference from the mean $\bar{x}$ (squared so that it is not sign dependent) and normalizing it with respect to the number of data points minus one:

$$\text{Variance} = \frac{\sum (x_i - \bar{x})^2}{N - 1} \tag{6.2}$$

The standard deviation σ is a measure of the dispersion of the data in the same units as the data x_i. It is simply the square root of the variance:

Standard deviation

$$\sigma = \sqrt{\frac{\sum (x_i - \bar{x})^2}{N - 1}} \tag{6.3}$$

For the distribution of concrete compressive strength given in Figure 6.1, the standard deviation has been calculated as 0.615 ksi (4.24 MPa).

6.1.3 Probability Density Functions

The bell-shaped curve in Figure 6.1 can also represent the probability distribution of the data if the area under the curve is set to unity (probability = 1, includes all possible concrete strengths). To make the deviation $(x - \bar{x})$ for a particular point x nondimensional, it is divided by the standard deviation σ. The result is a probability density function, which shows the range of deviations and the frequency with which they occur. If the data are typical of those encountered in natural occurrences, the normal distribution curve of Figure 6.2(a) will often result. It is given by the function (Benjamin and Cornell, 1970)

$$f_x(x) = \frac{1}{\sigma\sqrt{2\pi}}\exp\left[-\frac{1}{2}\left(\frac{x-m}{\sigma}\right)^2\right] \qquad -\infty \le x \le \infty \tag{6.4}$$

where $f_x(x)$ gives the probable frequency of occurrence of the variable x as a function of the mean $m = \bar{x}$ and the standard deviation σ of the normal distribution. The frequency distributions need not be centered at the origin. The effect of changes in m and σ is shown in Figure 6.2.

The normal probability function has been studied for many years and its properties are well documented in statistics books. An important characteristic of the areas included between ordinates erected on each side of the center of the distribution curve is that they represent probabilities at a distance of one, two, and three standard deviations. These areas are 68.26, 95.44, and 99.73%, respectively.

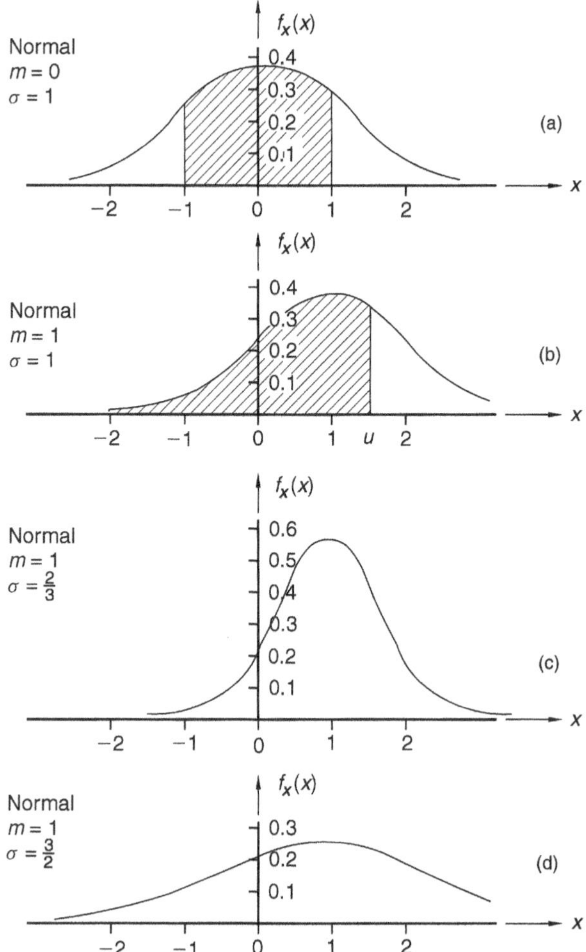

Fig. 6.2 Normal density functions. (From J. R. Benjamin and C. A. Cornell, *Probability, Statistics, and Decisions for Civil Engineers*. Copyright © 1970. Reproduced with permission of the McGraw-Hill Companies.)

Example 6.1 Statistics indicate the average height of the American male is 5 ft 9 in. (1.75 m) with a standard deviation of 3 in. (0.076 m). Table 6.1 shows the percentage of the male population in the United States in different height ranges. Basketball players greater than 7 ft (2.13 m) tall are very rare individuals indeed.

Table 6.1 U.S. Males in Different Height Ranges[a]

Standard Deviation	Height Range	Percent of Male Population
1σ	5 ft 6 in.–6 ft 0 in. (1.67–1.83 m)	68.26
2σ	5 ft 3 in.–6 ft 3 in. (1.60–1.90 m)	95.44
3σ	5 ft 0 in.–6 ft 6 in. (1.52–1.98 m)	99.73
4σ	4 ft 9 in.–6 ft 9 in. (1.45–2.05 m)	99.997
5σ	4 ft 6 in.–7 ft 0 in. (1.37–2.13 m)	99.99997

[a] $\bar{x}$ = 5 ft 9 in. (1.75 m), σ = 3 in. = 0.076 m.

Other probability density functions besides the symmetric normal function shown in Figure 6.2 are available. When the data distribution is nonsymmetrical, a logarithmic normal (or simply lognormal) probability density function is often more suitable. Stated mathematically, if $Y = \ln(x)$ is normally distributed, then x is said to be lognormal. The lognormal function was used in calibrating the AASHTO (1994) LRFD Bridge Specification because it better represented the observed distribution of resistance data, and it remains the basis in subsequent editions (AASHTO, 2010).

Lognormal probability density functions are shown in Figure 6.3 for different values of its standard deviation $\sigma = \zeta$. Notice that as the dispersion (the value of ζ) increases, the lack of symmetry becomes more pronounced.

Because $\ln(x)$ is a normal distribution, its mean λ_m and standard deviation ζ can be determined by a logarithmic transformation of the normal distribution function to give

Lognormal mean

$$\lambda_m = \ln\left(\frac{\bar{x}}{\sqrt{1+V^2}}\right) \tag{6.5}$$

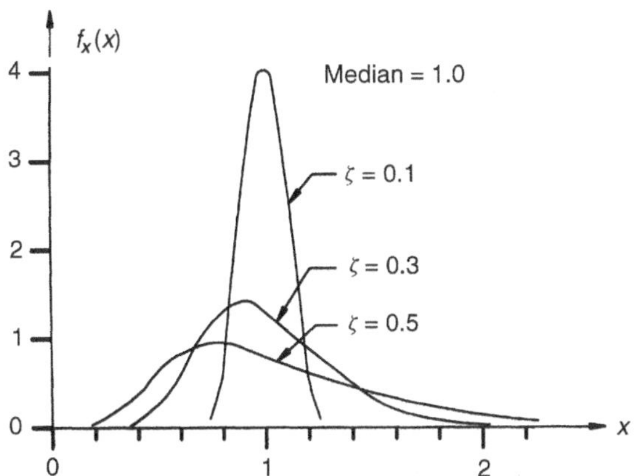

Fig. 6.3 Lognormal density functions. (From A. H-S. Ang and W. H. Tang, *Probability Concepts in Engineering Planning and Design, Volume I—Basic Principles*. Copyright © 1975. Reprinted with permission of John Wiley & Sons.)

and

Lognormal standard deviation

$$\zeta = \sqrt{\ln\left(1 + V^2\right)} \tag{6.6}$$

where $V = \sigma/\bar{x}$ is the coefficient of variation and $\bar{x}$ and σ are defined by Eqs. 6.1 and 6.3, respectively. Thus, the mean and standard deviation of the lognormal function can be calculated from the statistics obtained from the standard normal function.

6.1.4 Bias Factor

In Figure 6.1, it was observed that the mean value $\bar{x}$ for the concrete compressive strength is 3.94 ksi (27.2 MPa). The design value, or nominal value x_n, of the concrete compressive strength for this population of concrete cylinders was specified as 3 ksi (20.7 MPa). There is a clear difference between what is specified and what is delivered. This difference is referred to as the *bias*. The mean value is commonly larger than the nominal value because suppliers and manufacturers do not want their products rejected. Defining the bias factor λ as the ratio of the mean value $\bar{x}$ to the nominal value x_n, we have

Bias factor

$$\lambda = \frac{\bar{x}}{x_n} \tag{6.7}$$

For the distribution of concrete compressive strength given in Figure 6.1, the bias factor is 3.94/3.0 = 1.31.

6.1.5 Coefficient of Variation

To provide a measure of dispersion, it is convenient to define a value that is expressed as a fraction or percentage of the mean value. The most commonly used measure of dispersion is the coefficient of variation (V), which is the standard deviation (σ) divided by the mean value ($\bar{x}$):

Coefficient of variation

$$V = \frac{\sigma}{\bar{x}} \tag{6.8}$$

For the distribution of concrete compressive strength given in Figure 6.1, the coefficient of variation is 0.615/3.94 = 0.156 or 15.6%.

Table 6.2 gives typical values of the bias factor and coefficient of variation for resistance of materials collected by Siu et al. (1975). Comparing the statistics in Table 6.2 for concrete in compression to those obtained from the tests reported in Figure 6.1, the bias factor is lower but the coefficient of variation is the same.

Table 6.3 gives the same statistical parameters for highway dead and live loads taken from Nowak (1993). The largest variation is the weight of the wearing surface placed on bridge decks. Also of interest, as indicated by the bias factor, is that the observed actual loads are greater than the specified nominal values.

Table 6.2 Typical Statistics for Resistance of Materials

Limit State	Bias (λ_R)	COV (V_R)
Light-gage steel		
Tension and flexure	1.20	0.14
Hot-rolled steel		
Tension and flexure	1.10	0.13
Compression	1.20	0.15
Reinforced concrete		
Flexure	1.14	0.15
Compression	1.14	0.16
Shear	1.10	0.21
Wood		
Tension and flexure	1.31	0.16
Compression parallel to grain	1.36	0.18
Compression perpendicular to grain	1.71	0.28
Shear	1.26	0.14
Buckling	1.48	0.22

Reproduced from W. W. C. Siu, S. R. Parimi, and N. C. Lind (1975). "Practical Approach to Code Calibration," *Journal of the Structural Division*, ASCE, 101(ST7), pp. 1469–1480. With permission.

Table 6.3 Statistics for Bridge Load Components

Load Component	Bias (λ_Q)	COV (V_Q)
Dead load		
Factory made	1.03	0.08
Cast in place	1.05	0.10
Asphalt wearing surface	1.00	0.25
Live load (with dynamic load allowance)	1.10–1.20	0.18

From Nowak (1993).

6.1.6 Probability of Failure

In the context of reliability analysis, failure is defined as the realization of one of a number of predefined limit states. Load and resistance factors are selected to ensure that each possible limit state is reached only with an acceptably small probability of failure. The probability of failure can be determined if the statistics (mean and standard deviation) of the resistance and load distribution functions are known.

To illustrate the procedure, first consider the probability density functions for the random variables of load Q and resistance R shown in Figure 6.4 for a hypothetical example limit state. As long as the resistance R is greater than the effects of the load Q, a margin of safety is provided for the limit state under consideration. A quantitative measure of safety is the probability of survival given by

Probability of survival

$$p_s = P(R > Q) \tag{6.9}$$

where the right-hand side represents the probability that R is greater than Q. Because the value of both R and Q vary, there is a small probability that the load effect Q may exceed the resistance R. The shaded region in Figure 6.4 represents this situation. The complement of the probability of survival is the probability of failure, which can be expressed as

Probability of failure

$$p_f = 1 - p_s = P(R < Q) \tag{6.10}$$

where the right-hand side represents the probability that $R < Q$.

The probability density functions for R and Q in Figure 6.4 have purposely been drawn to represent different coefficients of variation, V_R and V_Q, respectively. The areas under the two curves are both equal to unity, but the resistance R is shown with greater dispersion than Q. The shaded area indicates the region of failure, but the area is not equal to the probability of failure because it is a mixture of areas coming from distributions with different ratios of standard deviation to mean value. For quantitative evaluation of probability of failure p_f, it is convenient to use a single combined probability density function $g(R, Q)$ that represents the margin of safety. From this limit state function $g(R, Q)$, with its own unique statistics, the probability of failure and the safety index can be determined in a straightforward manner.

If R and Q are normally distributed, the limit state function $g(\,)$ can be expressed as

$$g(R, Q) = R - Q \tag{6.11}$$

For lognormally distributed R and Q, the limit state function $g(\,)$ can be written as

$$g(R, Q) = \ln(R) - \ln(Q) = \ln\left(\frac{R}{Q}\right) \tag{6.12}$$

In both cases, the limit state is reached when $R = Q$ and failure occurs when $g(R, Q) < 0$. From probability theory, when two normally distributed random variables are combined, then the resulting probability density function is also normal, that is, if R and Q are normally distributed, then the function $g(R, Q)$ is also normally distributed. Similarly, if R and Q are lognormal, then the function $g(R, Q)$ is lognormal. As a result, the statistics from the individual distributions can be used to calculate the statistics (mean and standard deviation) of the combined distribution (Nowak and Collins, 2000).

Making use of these fundamental properties, the probability of failure for normally distributed R and Q can be obtained by

$$p_f = 1 - F_u\left(\frac{\bar{R} - \bar{Q}}{\sqrt{\sigma_R^2 + \sigma_Q^2}}\right) \tag{6.13}$$

and the probability of failure can be estimated for lognormally distributed R and Q by

$$p_f = 1 - F_u\left(\frac{\ln(\bar{R}/\bar{Q})}{\sqrt{V_R^2 + V_Q^2}}\right) \tag{6.14}$$

where $\bar{R}$ and $\bar{Q}$ are mean values, σ_R and σ_Q are standard deviations, V_R and V_Q are coefficients of variation of the resistance R and the load effect Q, and $F_u(\,)$ is the standard normal cumulative distribution function. The cumulative distribution function $F_u(\,)$ is the integral of $f_x(x)$ between the limits $-\infty$ to u and gives the probability that x is less than u. The shaded area in Figure 6.2(b) shows this integral. (Note that u can be interpreted as the number of standard deviations that x differs from the mean.) To determine the probability that a normal random variable lies in any interval, the difference between two values of $F_u(\,)$ gives this information. No simple expression is available for $F_u(\,)$, but it has been evaluated numerically and tabulated. Tables are available in elementary statistics textbooks.

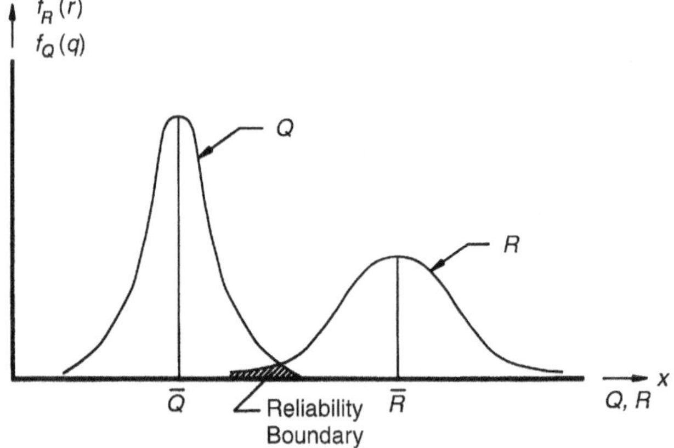

Fig. 6.4 Probability density functions for load and resistance.

6.1.7 Safety Index β

A simple alternative method for expressing the probability of failure is to use the safety index β. This procedure is illustrated using the lognormal limit state function of Eq. 6.12. As noted previously, the lognormal distribution represents actual distributions of R and Q more accurately than the normal distribution. Also, numerical calculation of the statistics for the limit state function $g(\ )$ are more stable using the ratio R/Q than for using the difference $R - Q$ because the difference $R - Q$ is subject to loss of significant figures when R and Q are nearly equal.

If the function $g(R, Q)$ as defined by Eq. 6.12 has a lognormal distribution, its frequency distribution would have the shape of the curve shown in Figure 6.5. This curve is a single-frequency distribution curve combining the uncertainties of both R and Q. The shaded area in Figure 6.5 represents the probability of attaining a limit state ($R < Q$), which is equal to the probability that $\ln(R/Q) < 0$.

The probability of failure can be reduced, and thus safety increased by either having a tighter grouping of data about the mean $\bar{g}$ (less dispersion) or by moving the mean $\bar{g}$ to the right. These two approaches can be combined by defining the position of the mean from the origin in terms of the standard deviation σ_g of $g(R, Q)$. Thus, the distance $\beta\sigma_g$ from the origin to the mean in Figure 6.5 becomes a measure of safety, and the number of standard deviations β in this measure is known as the *safety index*.

> ***Safety Index*** *β is defined as the number of standard deviations σ_g that the mean value $\bar{g}$ of the limit state function $g(\ )$ is greater than the value defining the failure condition $g(\ ) = 0$, that is, $\beta = \bar{g}/\sigma_g$.*

Normal Distributions If resistance R and load Q are both normally distributed random variables, and are statistically independent, the mean value $\bar{g}$ of $g(R, Q)$ given by Eq. 6.11 is

$$\bar{g} = \bar{R} - \bar{Q} \tag{6.15}$$

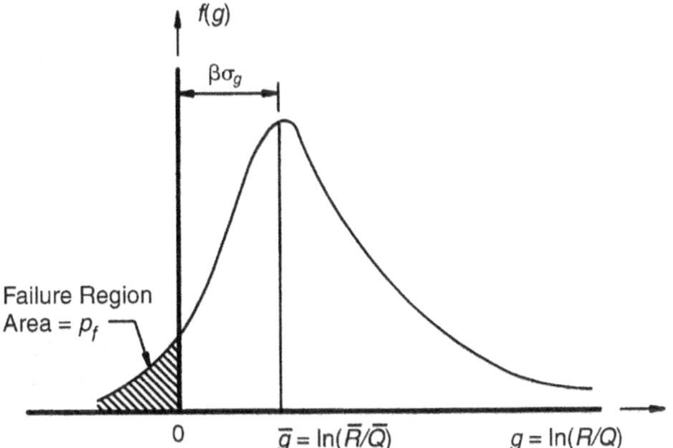

Fig. 6.5 Definition of safety index for lognormal R and Q.

and its standard deviation is

$$\sigma_g = \sqrt{\sigma_R^2 + \sigma_Q^2} \tag{6.16}$$

where $\bar{R}$ and $\bar{Q}$ are mean values and σ_R and σ_Q are standard deviations of R and Q. If the horizontal axis in Figure 6.5 represented the limit state function $g(R, Q)$ and $R - Q$, equating the distances from the origin, $\beta\sigma_g = \bar{g}$, and substituting Eqs. 6.15 and 6.16, the relationship for the safety index β for normal distributions becomes

$$\beta = \frac{\bar{R} - \bar{Q}}{\sqrt{\sigma_R^2 + \sigma_Q^2}} \tag{6.17}$$

This closed-form equation is convenient because it does not depend on the distribution of the combined function $g(R, Q)$ but only on the statistics of R and Q individually.

Comparing Eqs. 6.13 and 6.17, the probability of failure p_f, written in terms of the safety index β, is

$$p_f = 1 - F_u(\beta) \tag{6.18}$$

By relating the safety index directly to the probability of failure,

$$\beta = F_u^{-1}\left(1 - p_f\right) \tag{6.19}$$

where F_u^{-1} is the inverse standard normal cumulative distribution function.

Values for the relationship in Eqs. 6.18 and 6.19 are given in Table 6.4 based on tabulated values for F_u and F_u^{-1} found in most statistics textbooks. A change of 0.5 in β approximately results in an order of magnitude change in p_f. As mentioned earlier, no comparable relationship exists between the safety factor used in ASD and the probability of failure, which is a major disadvantage of ASD.

Lognormal Distributions If the resistance R and load Q are lognormally distributed random variables, and are statistically independent, the mean value of $g(R, Q)$ given by Eq. 6.12 is

$$\bar{g} = \ln\left(\frac{\bar{R}}{\bar{Q}}\right) \tag{6.20}$$

Table 6.4 Relationships between Probability of Failure and Safety Index for Normal Distributions

β	p_f	p_f	β
2.5	0.62×10^{-2}	1.0×10^{-2}	2.32
3.0	1.35×10^{-3}	1.0×10^{-3}	3.09
3.5	2.33×10^{-4}	1.0×10^{-4}	3.72
4.0	3.17×10^{-5}	1.0×10^{-5}	4.27
4.5	3.4×10^{-6}	1.0×10^{-6}	4.75
5.0	2.9×10^{-7}	1.0×10^{-7}	5.20
5.5	1.9×10^{-8}	1.0×10^{-8}	5.61

and its standard deviation is approximately

$$\sigma_g = \sqrt{V_R^2 + V_Q^2} \tag{6.21}$$

where $\bar{R}$ and $\bar{Q}$ are mean values and V_R and V_Q are coefficients of variation of R and Q, respectively (Nowak and Collins, 2000). If we equate the distances from the origin in Figure 6.5, the relationship for the safety index β for lognormal distributions becomes

$$\beta = \frac{\ln\left(\bar{R}/\bar{Q}\right)}{\sqrt{V_R^2 + V_Q^2}} \tag{6.22}$$

Again, this closed-form equation is convenient because it does not depend on the distribution of the combined function $g(R, Q)$ but only on the statistics of R and Q individually.

The expression of σ_g in Eq. 6.21 is an approximation that is valid if the coefficients of variation, V_R and V_Q, are relatively small, less than about 0.20. Expressing the logarithmic function in Eq. 6.6 as an infinite series illustrates the magnitude of this approximation

$$\ln\left(1 + V^2\right) = V^2 - \tfrac{1}{2}\left(V^2\right)^2 + \tfrac{1}{3}\left(V^2\right)^3 - \tfrac{1}{4}\left(V^2\right)^4 + \cdots$$

For an infinite series with alternating signs, the computational error is no more than the first neglected term. Using only the first term, a maximum relative error for $V = 0.20$ can be expressed as

$$\frac{\tfrac{1}{2}\left(V^2\right)^2}{V^2} = \frac{1}{2}V^2 = \frac{1}{2}(0.2)^2 = 0.02 \quad \text{or} \quad 2\%$$

Therefore, $\ln(1 + V^2)$ can be replaced by V^2 without large error, and Eq. 6.12 then gives $\zeta \approx V$. The typical values given in Tables 6.2 and 6.3 for coefficient of variation (COV) of resistance of materials V_R and effect of loads V_Q are generally less than about 0.20; therefore, these COVs can be used to represent the standard deviations of their respective lognormal distributions.

Rosenblueth and Esteva (1972) have developed an approximate relationship between the safety index β and the probability of failure p_f for lognormally distributed values of R and Q, given by the equation

$$p_f = 460 \exp(-4.3\beta) \qquad 2 < \beta < 6 \tag{6.23}$$

The inverse function for this relationship is

$$\beta = \frac{\ln\left(460/p_f\right)}{4.3} \qquad 10^{-1} > p_f > 10^{-9} \tag{6.24}$$

Values for both of these relationships are given in Table 6.5. Comparing Tables 6.4 and 6.5, the values for the normal and lognormal distributions are similar but not identical.

Table 6.5 Relationships between Probability of Failure and Safety Index for Lognormal Distributions

β	p_f	p_f	β
2.5	0.99×10^{-2}	1.0×10^{-2}	2.50
3.0	1.15×10^{-3}	1.0×10^{-3}	3.03
3.5	1.34×10^{-4}	1.0×10^{-4}	3.57
4.0	1.56×10^{-5}	1.0×10^{-5}	4.10
4.5	1.82×10^{-6}	1.0×10^{-6}	4.64
5.0	2.12×10^{-7}	1.0×10^{-7}	5.17
5.5	2.46×10^{-8}	1.0×10^{-8}	5.71

Example 6.2 A prestressed concrete girder bridge with a simple span of 90 ft (27 m) and girder spacing of 8 ft (2.4 m) has the following bending moment statistics for a typical girder:

Effect of loads (assumed normally distributed)

$$\bar{Q} = 3600 \text{ kip ft} \qquad \sigma_Q = 300 \text{ kip ft}$$

Resistance (assumed lognormally distributed)

$$R_n = 5200 \text{ kip ft} \qquad \lambda_R = 1.05 \qquad V_R = 0.075$$

Determine the safety index for a typical girder using Eqs. 6.17 and 6.22. To use Eq. 6.17, the mean value and standard deviation of R must be calculated. From Eqs. 6.7 and 6.18

$$\bar{R} = \lambda_R R_n = 1.05(5200) = 5460 \text{ kip ft}$$
$$\sigma_R = V_R \bar{R} = 0.075(5460) = 410 \text{ kip ft}$$

Substitution of values into Eq. 6.17 gives a safety index for normal distributions of Q and R:

$$\beta = \frac{\bar{R} - \bar{Q}}{\sqrt{\sigma_R^2 + \sigma_Q^2}} = \frac{5460 - 3600}{\sqrt{410^2 + 300^2}} = 3.66$$

To use Eq. 6.22, the coefficient of variation of Q must be calculated from Eq. 6.8:

$$V_Q = \frac{\sigma_Q}{\bar{Q}} = \frac{300}{3600} = 0.0833$$

Substitution into Eq. 6.22 yields a safety index for lognormal distributions of Q and R:

$$\beta = \frac{\ln\left(\bar{R}/\bar{Q}\right)}{\sqrt{V_R^2 + V_Q^2}} = \frac{\ln(5460/3600)}{\sqrt{0.075^2 + 0.0833^2}} = 3.72$$

The two results for β are nearly equal. The approximation used for σ_g in the development of Eq. 6.22 is reasonable. From either Table 6.4 or Table 6.5, the nominal probability of failure of one of these girders is about 1 : 10,000.

6.2 CALIBRATION OF LRFD CODE

6.2.1 Overview of the Calibration Process

Several approaches can be used in calibrating a design code. Specifications may be calibrated by use of judgment, fitting to other codes, use of reliability theory, or a combination of these approaches.

Calibration by judgment was the first approach used in arriving at specification parameters. If the performance of a specification was found to be satisfactory after many years, the parameter values were accepted as appropriate. Poor performance resulted in increasing safety margins. A fundamental disadvantage of this approach is that it results in nonuniform margins of safety because excessively conservative specification provisions will not result in problems and will therefore not be changed.

Calibration by fitting is usually done after there has been a fundamental change in either the design philosophy or the specification format. In this type of calibration, the parameters of the new specification are adjusted such that designs are obtained that are essentially the same as those achieved using the old specification. The main objective of this type of calibration is to transfer experience from the old to the new specification.

Calibration by fitting is a valuable technique for ensuring that designs obtained with the new specification do not deviate significantly from existing designs. It is also a relatively simple procedure because all that is involved is to match the parameters from the old and new specifications. The disadvantage of this type of calibration is that it does not necessarily result in more uniform safety margins or economy because the new specification essentially mimics the old specification.

A more formal process using reliability theory may also be used to calibrate a specification. The formal process for estimating suitable values of load factors and resistance factors for use in bridge design consists of the following steps (Barker et al., 1991):

Step 1. Compile the statistical database for load and resistance parameters.

Step 2. Estimate the level of reliability inherent in current design methods of predicting strengths of bridge structures.

Step 3. Observe the variation of the reliability levels with different span lengths, dead-load to live-load ratios, load combinations, types of bridges, and methods of calculating strengths.

Step 4. Select a target reliability index based on the margin of safety implied in current designs.

Step 5. Calculate load factors and resistance factors consistent with the selected target reliability index. It is also important to couple experience and judgment with the calibration results.

6.2.2 Calibration Using Reliability Theory

Calibration of the LRFD code for bridges using reliability theory followed the five steps outlined above. These steps are described in more detail in the following paragraphs.

Step 1. *Compile a Database of Load and Resistance Statistics* Calibration using reliability theory requires that statistical data on load and resistance be available. The FOSM theories require the mean value and standard deviation to represent the probability density function. For a given nominal value, these two parameters are then used to calculate the companion nondimensional bias factor and coefficient of variation for the distribution.

The statistics for bridge load components given in Table 6.3 were compiled from available data and measurements of typical bridges. The live-load statistics were obtained from surveys of truck traffic and weigh-in-motion data (Nowak, 1993).

Statistical data for resistance of materials given in Table 6.2 were obtained from material tests, component tests, and field measurements (Siu et al., 1975). Because these typical resistance statistics were not developed specifically for bridges, data from current highway bridges were utilized in the LRFD calibration process.

The highway bridges selected for evaluation numbered about 200 and were from various geographic regions in the United States (California, Colorado, Illinois, Kentucky, Maryland, Michigan, Minnesota, New York, Oklahoma, Pennsylvania, Tennessee, and Texas). When selecting bridges representative of the nation's inventory, emphasis was placed on current and anticipated future trends in materials, bridge types, and spans. Steel bridges included composite and noncomposite rolled beams and plate girders, box girders, through trusses, deck trusses, pony trusses, arches, and a tied arch. Reinforced concrete bridges included slabs, T-beams, solid frame, and a box girder. Prestressed concrete bridges included a double tee, I-beams, and box girders. Wood bridges included sawn beam, glulam beam, a truss, and decks that were either nailed or prestressed transversely. Spans ranged from 30 ft (9 m) for a reinforced concrete slab bridge to 730 ft (220 m) for a steel arch bridge.

Resistance statistics were developed for a reduced set of the selected bridges, which included only the girder-type structures. For each of the girder bridges, the load effects (moments, shears, tensions, and compressions) were calculated and compared to the resistance provided by the actual cross section. The statistical parameters of

Table 6.6 Statistical Parameters of Resistance for Selected Bridges

Type of Structure	Bias (λ_R)	COV (V_R)
Noncomposite steel girders		
Moment (compact)	1.12	0.10
Moment (noncompact)	1.12	0.10
Shear	1.14	0.105
Composite steel girders		
Moment	1.12	0.10
Shear	1.14	0.105
Reinforced concrete T-beams		
Moment	1.14	0.13
Shear w/steel	1.20	0.155
Shear w/o steel	1.40	0.17
Prestressed concrete girders		
Moment	1.05	0.075
Shear w/steel	1.15	0.14

From Nowak (1993).

resistance for steel girders, reinforced concrete T-beams, and prestressed concrete girders are shown in Table 6.6 (Nowak, 1993). Comparing the resistance statistics of Tables 6.2 and 6.6, the bias factor and coefficient of variation for the girder bridges are slightly lower than those for the general population of structures.

Step 2. *Estimate the Safety Index β in Current Design Methods* Risk levels implied in the existing specifications were determined by computing safety indexes for additional representative bridges. The additional bridges covered five span lengths from 30 to 200 ft (9 to 60 m) and five girder spacings from 4 to 12 ft (1.2 to 3.6 m). The reliability analysis was based on the FOSM methods, a normal distribution of the load, and a lognormal distribution of the resistance.

The mean value FOSM (MVFOSM) method used to derive Eqs. 6.17 and 6.22 is not the most accurate method that can be used to calculate values of the safety index β. While values of β determined from Eqs. 6.17 and 6.22 are sufficiently accurate to be useful for some purposes, it was considered worthwhile to use the more accurate advanced FOSM (AFOSM) method to derive values of β for the AASHTO (1994, 2010) LRFD Specifications.

An explicit expression for β cannot be written when the AFOSM method is used because the limit state function $g(\,)$ is linearized at a point on the failure surface, rather than at the mean values of the random variables. For the AFOSM method, an iterative procedure must be used in which an initial value of β is assumed and the process is repeated until the difference in calculated values of β on successive iterations is within a small tolerance. This iterative procedure is based on normal approximations to nonnormal distributions at the design point developed by Rackwitz and Fiessler (1978).

An estimate of the mean value of the lognormally distributed resistance R_n with bias factor λ_R and coefficient of variation V_R is given by Eq. 6.7 as

$$\bar{R} = \lambda_R R_n$$

and an assumed design point is

$$R^* = \bar{R}\left(1 - kV_R\right) = R_n\lambda_R\left(1 - kV_R\right) \quad (6.25)$$

where k is unknown. Because it is a modifier of the nominal resistance R_n, the term $\lambda_R(1 - kV_R)$ can be thought of as an estimate of a resistance factor ϕ^*, that is,

$$\phi^* = \lambda_R\left(1 - kV_R\right) \quad (6.26)$$

and the parameter k is comparable to the number of standard deviations from the mean value. As an initial guess, k is often taken as 2.

An estimate of the standard deviation of R_n is obtained from Eq. 6.8 as

$$\sigma_R = V_R\bar{R}$$

and at an assumed design point becomes

$$\sigma_R^{'} = R_n V_R \lambda_R\left(1 - kV_R\right) \quad (6.27)$$

For normally distributed R and Q, Eq. 6.17 gives the safety index for the MVFOSM method. Substituting Eqs. 6.25 and 6.27 into Eq. 6.17 and transforming the lognormally distributed R_n into a normal distribution at the design point R^*, the safety index β for normally distributed Q and lognormally distributed R can be expressed as (Nowak, 1993)

$$\beta = \frac{R_n\lambda_R\left(1 - kV_R\right)\left[1 - \ln\left(1 - kV_R\right)\right] - \bar{Q}}{\sqrt{\left[R_n V_R \lambda_R\left(1 - kV_R\right)\right]^2 + \sigma_Q^2}} \quad (6.28)$$

Example 6.3 For the prestressed girder of Example 6.1, estimate the safety index at the design point R^* using Eq. 6.28 with $k = 2$. The statistics from Example 6.2 are

$$R_n = 5200 \text{ kip ft} \quad \lambda_R = 1.05 \quad V_R = 0.075$$
$$\bar{Q} = 3600 \text{ kip ft}$$

and

$$\sigma_Q = 300 \text{ kip ft}$$
$$\phi^* = \lambda_R(1 - kV_R) = 1.05[1 - 2(0.075)] = 0.89$$
$$R^* = \phi^* R_n = 0.89(5200) = 4628 \text{ kip ft}$$
$$\sigma_R^{'} = V_R R* = 0.075(4628) = 347 \text{ kip ft}$$

Substitution into Eq. 6.28 gives

$$\begin{aligned}\beta &= \frac{R^*\left[1-\ln\left(1-kV_R\right)\right]-\bar{Q}}{\sqrt{\left(\sigma_R'\right)^2+\sigma_Q^2}} \\ &= \frac{4628\left[1-\ln\left(1-2\times 0.075\right)\right]-3600}{\sqrt{347^2+300^2}} \\ &= \frac{5380-3600}{459} = 3.88 \end{aligned} \tag{6.29}$$

This estimate of the safety index β is slightly higher than the values calculated in Example 6.2. The value of β calculated at the design point on the failure surface by the iterative AFOSM method is considered to be more accurate than the values calculated by the MVFOSM method in Example 6.2.

Step 3. *Observe the Variation of the Safety Indexes* Nowak (1993) calculated safety indexes using the iterative AFOSM method for typical girder bridges. The study covered the full range of spans and girder spacings of simple-span noncomposite steel, composite steel, reinforced concrete T-beam, and prestressed concrete I-beam bridges. For each of the bridge types five span lengths of 30, 60, 90, 120, and 200 ft (9, 18, 27, 36, and 60 m) were chosen. For each span, five girder spacing of 4, 6, 8, 10, and 12 ft (1.2, 1.8, 2.4, 3.0, and 3.6 m) were selected. For each case, cross sections were designed so that the actual resistance was equal to the required resistance of the existing code (AASHTO, 1989). In other words, the cross sections were neither overdesigned nor underdesigned. It was not possible for one cross section to satisfy this criterion for both moment and shear, so separate designs were completed for both limit states.

Calculated safety indexes for prestressed concrete girders are shown in Figure 6.6 for simple-span moment and in Figure 6.7 for shear. These results are typical of the other bridge types, that is, higher values of β for wider spacing of girders and lower values of β for shear than moment.

Observations of Figures 6.6 and 6.7 indicate for the moment a range of β from 2.0 to 4.5 with the lower value for small spans while for shear the range is 2.0–4.0 with the lower value for large spans. For these ranges of β, Tables 6.4 and 6.5 indicate that the probability of failure of designs according to AASHTO (1989) Standard Specifications varies from about 1 : 100 to 1 : 100,000. A uniform level of safety does not exist.

Step 4. *Select a Target Safety Index β_T* Relatively large ranges of safety indexes were observed for moment and shear designs using the AASHTO (1989) Standard Specifications. These safety indexes varied mostly with span length and girder spacing and to a lesser extent with bridge type. What was desired in the calibration of the AASHTO (1994) LRFD Specification was a uniform safety index for all spans, spacings, and bridge types. To achieve this objective, a desired or target safety index is chosen, and then load and resistance factors are calculated to give safety indexes as close to the target value as possible.

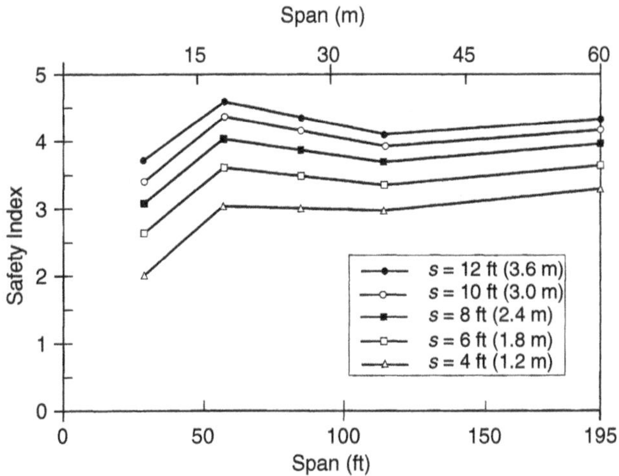

Fig. 6.6 Safety indexes for AASHTO (1989); simple-span moment in prestressed concrete girders (Nowak, 1993).

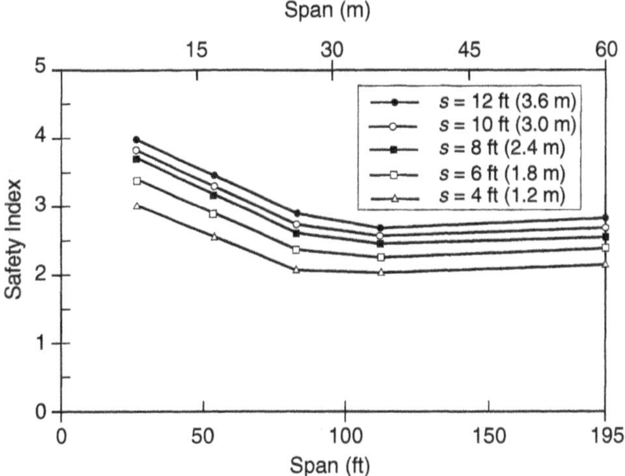

Fig. 6.7 Safety indexes for AASHTO (1989); simple-span shear in prestressed concrete girders (Nowak, 1993).

Based on the results of the parametric study by Nowak (1993), as well as calibrations of other specifications (OHBDC, 1992; AISC, 1986), a target safety index $\beta_T = 3.5$ was selected. This value of β_T corresponds to the safety index calculated for moment in a simple span of 60 ft (18 m) with a girder spacing of 6 ft (1.8 m) using the AASHTO (1989) Standard Specifications. This calibration point can be seen in Figure 6.6 for prestressed concrete girders. Similar results were obtained for the other bridge types studied.

Step 5. *Calculate Load and Resistance Factors* To achieve the desired or target safety index of $\beta_T = 3.5$, statistically based load and resistance factors must be calculated. Load factors must be common for all bridge types. The variation of β with span length is due to different ratios of dead load to live load. This effect can be minimized by proper selection of load factors for dead load and live load.

Resistance factors must account for the differences in reliability of the various limit states. For example, the safety indexes calculated for moment and shear shown in Figures 6.6 and 6.7 have different values and different trends.

It may not be possible to satisfy all of the conditions with $\beta_T = 3.5$. However, the objective of the calibration process is to select load and resistance factors that will generate safety indexes that are as close as possible to the target value. Acceptable sets of load factors and resistance factors occur when the calculated safety indexes cluster in a narrow band about the target value of $\beta_T = 3.5$.

To derive load factors γ and resistance factors ϕ from statistical considerations, assume that R and Q are normally distributed and that β is given by Eq. 6.17 so that

$$\bar{R} - \bar{Q} = \beta\sqrt{\sigma_R^2 + \sigma_Q^2} \tag{6.30}$$

It is desirable to separate the effects of R and Q, which can be done by using the approximation suggested by Lind (1971) for the value of the square-root term

$$\sqrt{\sigma_R^2 + \sigma_Q^2} \approx \alpha\left(\sigma_R + \sigma_Q\right) \tag{6.31}$$

where if $\sigma_R/\sigma_Q = 1.0$, $\alpha = \sqrt{2}/2 = 0.707$. Typical statistics for σ_R and σ_Q indicate that the maximum range for σ_R/σ_Q is between $\frac{1}{3}$ and 3.0. Taking the extreme values for σ_R/σ_Q, then $\alpha = \sqrt{10}/4 = 0.79$. The maximum error in the approximation will only be 6% if $\alpha = 0.75$.

Substitution of Eq. 6.31 into Eq. 6.30 yields

$$\bar{R} - \bar{Q} = \alpha\beta\left(\sigma_R + \sigma_Q\right)$$

which can be separated into

$$\bar{R} - \alpha\beta\sigma_R = \bar{Q} + \alpha\beta\sigma_Q \tag{6.32}$$

Recalling the definition of the bias factor (Eq. 6.7) and the coefficient of variation (Eq. 6.8), and setting $\beta = \beta_T$, we can write

$$R_n\lambda_R\left(1 - \alpha\beta_T V_R\right) = Q_n\lambda_Q\left(1 + \alpha\beta_T V_Q\right) \tag{6.33}$$

which can be written in the generic form of the basic design equation

$$\phi R_n = \gamma Q_n \tag{6.34}$$

where

$$\gamma = \lambda_Q\left(1 + \alpha\beta_T V_Q\right) \tag{6.35}$$

$$\phi = \lambda_R\left(1 - \alpha\beta_T V_R\right) \tag{6.36}$$

and the load and resistance factors are expressed only in terms of their own statistics and some fraction of the target safety index.

Establishing Load Factors Trial values for the load factors γ_i can be obtained from Eq. 6.35 using the statistics from Table 6.3 for the different load components. By taking $\alpha = 0.75$ and $\beta_T = 3.5$, Eq. 6.35 becomes

$$\gamma_i = \lambda_{Q_i}\left(1 + 2.6\ V_{Q_i}\right) \tag{6.37}$$

and the trial load factors are

Factory made	$\gamma_{\text{DC1}} = 1.03\,(1 + 2.6 \times 0.08) = 1.24$
Cast in place	$\gamma_{\text{DC2}} = 1.05\,(1 + 2.6 \times 0.10) = 1.32$
Asphalt overlay	$\gamma_{\text{DW}} = 1.00\,(1 + 2.6 \times 0.25) = 1.65$
Live load	$\gamma_{\text{LL}} = 1.10 \text{ to } 1.20\,(1 + 2.6 \times 0.18)$ $= 1.61 \text{ to } 1.76$

In the calibration conducted by Nowak (1993), the loads Q_i were considered to be normally distributed and the resistance R_n lognormally distributed. The expression used for trial load factors was

$$\gamma_i = \lambda_{Q_i}\left(1 + kV_{Q_i}\right) \tag{6.38}$$

where k was given values of 1.5, 2.0, and 2.5. The results for trial load factors were similar to those calculated using Eq. 6.37.

The final load factors selected for the strength I limit state (Table 5.1) were

$$\gamma_{\text{DC1}} = \gamma_{\text{DC2}} = 1.25 \qquad \gamma_{\text{DW}} = 1.50 \qquad \gamma_{\text{LL}} = 1.75$$

Establishing Resistance Factors Trial values for the resistance factors ϕ can be obtained from Eq. 6.36 using the statistics from Table 6.6 for the various bridge types and limit states. Because the chosen load factors represent values calculated from Eq. 6.38 with $k = 2.0$, the corresponding Eq. 6.36 becomes

$$\phi = \lambda_R\left(1 - 2.0V_R\right) \tag{6.39}$$

The calculated trial resistance factors and the final recommended values are given in Table 6.7. The recommended values were selected to give values of the safety index calculated by the iterative procedure that were close to β_T. Because of the uncertainties in calculating the resistance factors, they have been rounded to the nearest 0.05.

Table 6.7 Calculated Trial and Recommended Resistance Factors

Material	Limit State	Eq. 6.39	ϕ, Selected
Noncomposite steel	Moment	0.90	1.00
	Shear	0.90	1.00
Composite steel	Moment	0.90	1.00
	Shear	0.90	1.00
Reinforced concrete	Moment	0.85	0.90
	Shear	0.85	0.90
Prestressed concrete	Moment	0.90	1.00
	Shear	0.85	0.90

Calibration Results The test of the calibration procedure is whether or not the selected load and resistance factors develop safety indexes that are clustered around the target safety index and are uniform with span length and girder spacing. The safety indexes have been calculated and tabulated in Nowak (1993) for the representative bridges of span lengths from 30 to 200 ft (9 to 60 m).

Typical calibration results are shown in Figures 6.8 and 6.9 for moment and shear in prestressed concrete girders. Two curves for γ of 1.60 and 1.70 in the figures show the effect of changes in the load factors for live load. Figures 6.8 and 6.9 both show uniform levels of safety over the range of span lengths, which is in contrast to the variations in safety indexes shown in Figures 6.6 and 6.7 before calibration.

The selection of the higher than calculated ϕ factors of Table 6.7 is justified because Figures 6.8 and 6.9 show that they result in reduced safety indexes that are closer to the target value of 3.5. Figure 6.9 indicates that for shear in prestressed concrete girders, a ϕ factor of 0.95 could be justified. However, it was decided to keep the same value of 0.90 from the previous specification.

The selection of the live-load factor 1.75 was done after the calibration process was completed. With current highway truck traffic, this increased load factor provides a safety index greater than 3.5. This increase was undoubtedly done in anticipation of future trends in highway truck traffic.

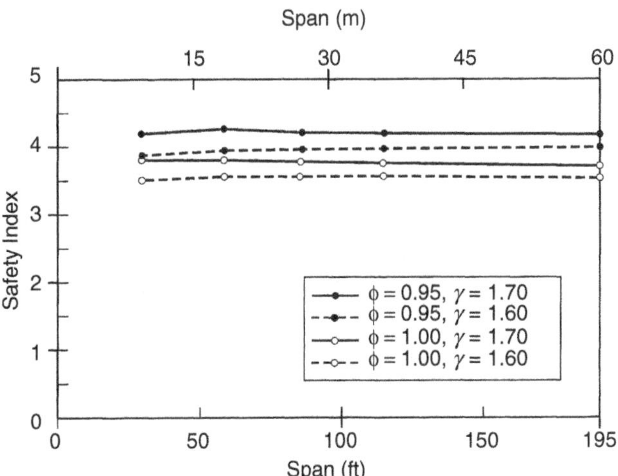

Fig. 6.8 Safety indexes for LRFD code; simple-span shears in prestressed concrete girders (Nowak, 1993).

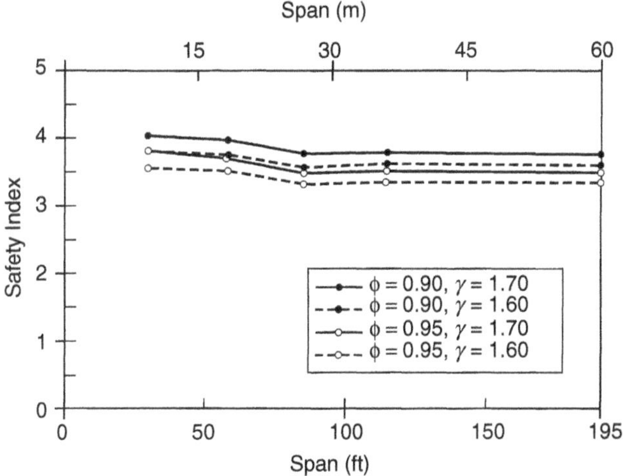

Fig. 6.9 Safety indexes for LRFD code; simple-span shears in prestressed concrete girders (Nowak, 1993).

At the time of this writing, only the strength I limit state has been formally calibrated. Other limit states were adjusted to agree with present practice. Interestingly, service II and service III often control the proportioning for steel and prestressed concrete girders, respectively.

6.2.3 Calibration of Fitting with ASD

The process of calibration with the existing ASD criteria avoids drastic deviations from existing designs. Calibration by fitting with ASD can also be used where statistical data are insufficient to calculate ϕ from an expression like Eq. 6.39.

In the ASD format, nominal loads are related to nominal resistance by the safety factor F as stated previously in Eq. 5.2:

$$\frac{R_n}{F} \geq \sum Q_i \tag{6.40}$$

Division of Eq. 5.3 by Eq. 6.40 results in

$$\phi \geq \frac{\sum \gamma_i Q_i}{F \sum Q_i} \tag{6.41}$$

If the loads consist only of dead-load Q_D and live-load Q_L, then Eq. 6.41 becomes

$$\phi = \frac{\gamma_D Q_D + \gamma_L Q_L}{F\left(Q_D + Q_L\right)} \tag{6.42}$$

Dividing both numerator and denominator by Q_L, Eq. 6.42 may be written as

$$\phi = \frac{\gamma_D\left(Q_D/Q_L\right) + \gamma_L}{F\left(Q_D/Q_L + 1\right)} \tag{6.43}$$

Example 6.4 Calculate the resistance factor ϕ for bending that is equivalent to an ASD safety factor $F = 1.6$ if the dead-load factor γ_D is 1.25, the live-load factor γ_L is 1.75, and the dead- to live-load moment ratio M_D/M_L is 1.5. Substitution of values into Eq. 6.43 gives

$$\phi = \frac{1.25\,(1.5) + 1.75}{1.6\,(1.5 + 1)} = 0.91$$

This value for ϕ is comparable to the values given for moment in Table 6.7.

6.3 CLOSING REMARKS

The background regarding the calibration of the LRFD Specifications is provided. The approach used to develop the load and resistance factors parallel those used for other specification in the United States and elsewhere. To date, only the strength I limit has been formally calibrated; however, research is underway to address the other limit states. The statistical properties associated with load and resistance are largely removed from routine structural computations by the use of codified factors. Some agencies are using weigh-in-motion methods to better understand their live loads and to recalibrate based upon more recent data. The LRFD process and format accommodates such refinements.

REFERENCES

AASHTO (1989). *Standard Specifications for Highway Bridges*, 14th ed., American Association of State Highway and Transportation Officials, Washington, DC.

AASHTO (1994). *LRFD Bridge Design Specifications* 1st ed., American Association of State Highway and Transportation Officials, Washington, DC.

AASHTO (2010). *LRFD Bridge Design Specifications*, 5th ed., American Association of State Highway and Transportation Officials, Washington, DC.

AISC (1986). *Load and Resistance Factor Design, Manual of Steel Construction*, 1st ed., American Institute of Steel Construction, New York.

Ang, A. H-S. and W. H. Tang (1975). *Probability Concepts in Engineering Planning and Design, Volume I—Basic Principles*, Wiley, New York.

Barker, R. M., J. M. Duncan, K. B. Rojiani, P. S. K. Ooi, C. K. Tan, and S. G. Kim (1991). "Load Factor Design Criteria for Highway Structure Foundations," *Final Report, NCHRP Project 24–4*, May, Virginia Polytechnic Institute and State University, Blacksburg, VA.

Benjamin, J. R. and C. A. Cornell (1970). *Probability, Statistics, and Decisions for Civil Engineers*, McGraw-Hill, New York.

Lind, N. C. (1971). "Consistent Partial Safety Factors," *Journal of the Structural Division*, ASCE, Vol. 97, No. ST6, Proceeding Paper 8166, June, pp. 1651–1669.

MacGregor, J. G. and J. K. Wight (2004). *Reinforced Concrete Design*, 4th ed., Prentice-Hall, Englewood Cliffs, NJ.

Nowak, A. S. (1993). "Calibration of LRFD Bridge Design Code," *NCHRP Project 12–33*, University of Michigan, Ann Arbor, MI.

Nowak, A. S. and K. R. Collins (2000). *Reliability of Structures*, McGraw-Hill Higher Education, New York.

OHBDC (1992). *Ontario Highway Bridge Design Code*, 3rd ed., Ministry of Transportation, Quality and Standards Division, Toronto, Ontario.

Rackwitz, R. and B. Fiessler (1978). "Structural Reliability under Combined Random Load Sequences," *Computers and Structures*, Vol. 9, No. 5, November, pp. 489–494, Pergamon Press, Great Britain.

Rosenblueth, E. and L. Esteva (1972). "Reliability Basis for Some Mexican Codes," *ACI Publication SP-31*, American Concrete Institute, Detroit, MI.

Siu, W. W. C., S. R. Parimi, and N. C. Lind (1975). "Practical Approach to Code Calibration," *Journal of the Structural Division*, ASCE, Vol. 101, No. ST7, Proceeding Paper 11404, July, pp. 1469–1480.

PROBLEMS

6.1 Discuss why this statement is true: "Every bridge structure is designed for a finite probability of failure." How would you modify the statement when making presentations to the general public?

6.2 Explain the difference between the nominal value of resistance and the mean value of resistance. Give an example.

6.3 For the bridge load statistics in Table 6.3, the bias is consistently greater than 1. What is the significance of this statistic in regard to the safety of bridges designed by ASD?

6.4 A composite steel girder bridge with a simple span of 90 ft and girder spacing of 8 ft has the following bending moment statistics: $\bar{Q} = 3028$ kip ft, $\sigma_Q = 300$ kip ft, $R_n = 4482$ kip ft, $\lambda_R = 1.12$, $V_R = 0.10$. Estimate the safety index β assuming: (a) both Q and R are normally distributed, (b) both Q and R are lognormally distributed, and (c) Q is normal and R is lognormal.

6.5 A prestressed concrete girder bridge with a simple span of 60 ft and girder spacing of 6 ft has the following bending moment statistics: $\bar{Q} = 1442$ kip ft, $\sigma_Q = 142$ kip ft, $R_n = 2084$ kip ft, $\lambda_R = 1.05$, and $V_R = 0.075$. Estimate the safety index β assuming: (a) both Q and R are normally distributed, (b) both Q and R are lognormally distributed, and (c) Q is normal and R is lognormal.

6.6 A prestressed concrete girder bridge with a simple span of 60 ft and girder spacing of 6 ft has the following shear force statistics: $\bar{Q} = 110$ kips, $\sigma_Q = 12$ kips, $R_n = 155$ kips, $\lambda_R = 1.15$, and $V_R = 0.14$. Estimate the safety index β assuming: (a) both Q and R are normally distributed, (b) both Q and R are lognormally distributed, and (c) Q is normal and R is lognormal.

6.7 Consider a reinforced concrete T-beam in bending. Using the approximate linear form for the resistance factor (Eq. 6.36), statistics from Table 6.6, and assuming $\sigma_R = \sigma_Q$, estimate a value of ϕ for a probability of failure of 1×10^{-3} and 1×10^{-4}.

6.8 Consider a composite steel girder in bending. Using the approximate linear form for the resistance factor (Eq. 6.36), statistics from Table 6.6, and assuming $\sigma_R = 1.5\sigma_Q$, estimate a value of ϕ for a probability of failure of 1×10^{-3} and 1×10^{-4}.

6.9 Calculate a resistance factor ϕ that is equivalent to an ASD safety factor F of 1.7 if the dead-load factor $\gamma_D = 1.25$, the live-load factor $\gamma_L = 1.75$, and the dead- to live-load ratio Q_D/Q_L is 2.0.

CHAPTER 7

Geometric Design Considerations

7.1 INTRODUCTION TO GEOMETRIC ROADWAY CONSIDERATIONS

In water crossings or bridges over deep ravines or across wide valleys, the bridge engineer is usually not restricted by the geometric design of the highway. However, when two highways intersect at a grade separation or interchange, the geometric design of the intersection will often determine the span lengths and selection of bridge type. In this instance, collaboration between the highway engineer and the bridge engineer during the planning stage is essential.

The bridge engineer must be aware of the design elements that the highway engineer considers to be important. Both engineers are concerned about appearance, safety, cost, and site conditions. In addition, the highway engineer is concerned about the efficient movement of traffic between the roadways on different levels, which requires an understanding of the character and composition of traffic, design speed, and degree of access control so that sight distance, horizontal and vertical curves, superelevation, cross slopes, and roadway widths can be determined.

The document that gives the geometric standards is *A Policy on the Geometric Design of Highways and Streets*, AASHTO (2004). The requirements in this publication are incorporated in the AASHTO (2010) LRFD Bridge Design Specification by reference [A2.3.2.2.3].* In the sections that follow, a few of the requirements that determine the roadway widths and clearances for bridges are given.

7.2 ROADWAY WIDTHS

Crossing a bridge should not convey a sense of restriction, which requires that the roadway width on the bridge be the same as that of the approaching highway. A typical overpass structure of a four-lane divided freeway crossing a secondary road is shown in Figure 7.1. The recommended minimum widths of shoulders and traffic lanes for the roadway on the bridge are given in Table 7.1.

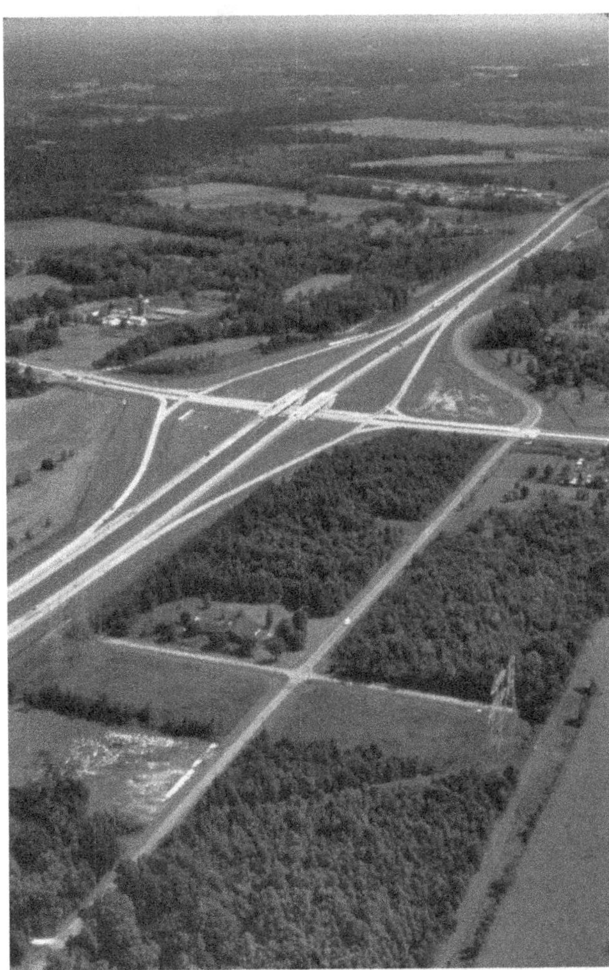

Fig. 7.1 Typical overpass structure. (Courtesy of Modjeski & Masters, Inc.)

Table 7.1 Typical Roadway Widths for Freeway Overpasses

Roadway	Width (ft)	Width (m)
Lane width	12	3.6
Right shoulder width		
Four lanes	10	3.0
Six and eight lanes	10	3.0
Left shoulder width		
Four lanes	4	1.2
Six and eight lanes	10	3.0

From *A Policy on Geometric Design of Highways and Streets*. Copyright © 2004 by the American Association of State Highway and Transportation Officials, Washington, D.C. Used by Permission.

*The article numbers in the AASHTO (2010) LRFD Bridge Specifications are enclosed in brackets and preceded by the letter A if a specification article and by the letter C if commentary.

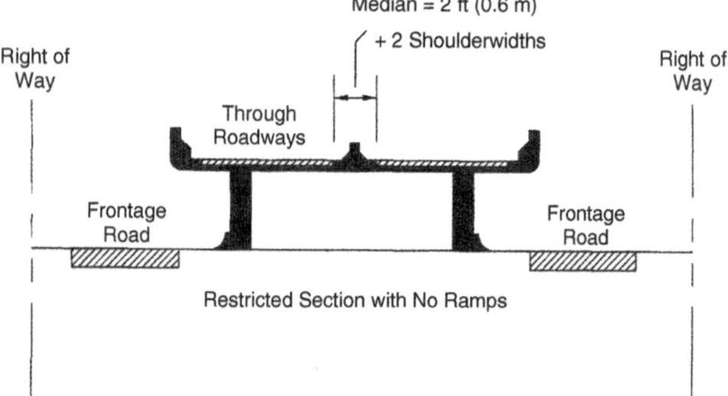

Fig. 7.2 Cross section for elevated freeways on structure with frontage roads (AASHTO Exhibit 8–10). (From *A Policy on Geometric Design of Highways, and Streets*, Copyright © 2004 by the American Association of State Highway and Transportation Officials, Washington, DC. Used by permission.)

A median barrier must separate the traffic for two-way elevated freeways in urban settings (Fig 7.2). The width of the barrier is 2 ft (0.6 m). The minimum median width is obtained by adding two left shoulder widths in Table 7.1 to give 10 ft (3.0 m) for a four-lane and 22 ft (6.6 m) for six- and eight-lane roadways.

If a highway passes under a bridge, it is difficult not to notice the structure and to get a sense of restriction. As was discussed in the aesthetics section of Chapter 3, it is possible to increase the sense of openness by placing stub abutments on top of the slopes and providing an open span beyond the right shoulder. The geometric design requirements are stated in *A Policy on Geometric Design of Highways and Streets*, AASHTO (2004) as follows:

> *Overpass structures should have liberal lateral clearances on the roadways at each level. All piers and abutment walls should be suitably offset from the traveled way. The finished underpass roadway median and off-shoulder slopes should be rounded and there should be a transition to backslopes to redirect errant vehicles away from protected or unprotected structural elements.*

In some areas it may be too costly to provide liberal lateral clearances and minimum dimensions are often used. The minimum lateral clearance from the edge of the traveled way to the face of the protective barrier should be the normal shoulder width given in Table 7.1. This clearance is illustrated in Figure 7.3 for a typical roadway underpass with a continuous wall or barrier. If the underpass has a center support, the same lateral clearance dimensions are applicable for a wall or pier on the left.

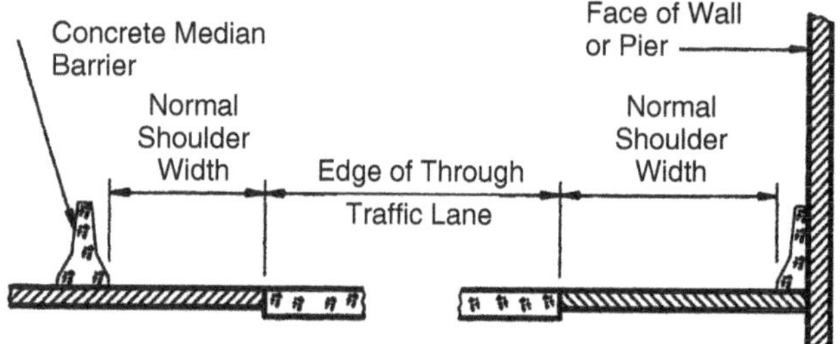

Fig. 7.3 Lateral clearances for major roadway underpasses (AASHTO Exhibit 10–6). (From *A Policy on Geometric Design of Highways and Streets*, Copyright © 2004 by the American Association of State Highway and Transportation Officials. Washington, DC. Used by permission.)

7.3 VERTICAL CLEARANCES

For bridges over navigable waterways, the U.S. Coast Guard establishes the vertical clearance [A2.3.3.1]. For bridges over highways, the vertical clearances are given by *A Policy on Geometric Design of Highways and Streets*, AASHTO (2004) [A2.3.3.2]. For freeways and arterial systems, the minimum vertical clearance is 16 ft (4.9 m) plus an allowance for several resurfacings of about 6 in. (150 mm). For other routes, a lower vertical clearance is acceptable, but in no case should it be less than 18 in. (0.5 m) greater than the vehicle height allowed by state law. In general, a desired minimum vertical clearance of all structures above the traveled way and shoulders is 16.5 ft (5.0 m).

7.4 INTERCHANGES

The geometric design of the intersection of two highways depends on the expected volumes of through and turning traffic, the topography of the site, and the need to simplify signing and driver understanding to prevent wrong-way movements. There are a number of tested interchange designs, and they vary in complexity from the simple two-level overpass with ramps shown in Figure 7.1 to the four-level directional interchange of Figure 7.4.

In comparing Figures 7.1 and 7.4, note that the bridge requirements for interchanges are dependent on the geometric design. In Figure 7.1, the bridges are simple overpasses with relatively linear ramps providing access between the two levels. In Figure 7.4, the through traffic is handled by an overpass at the lower two levels, but turning movements are handled by sweeping curved elevated ramps at levels three and four. The geometric design of the highway engineer can strongly influence the structural design of the bridge engineer. These engineers must work in concert during the

Fig. 7.4 Four-level directional interchange. (Courtesy of Modjeski & Masters, Inc.)

planning phase and share one another's needs and desires for integrating the bridge structures into the overall mission of the highway system.

REFERENCES

AASHTO (2004). *A Policy on Geometric Design of Highways and Streets*, 5th ed., American Association of State Highway and Transportation Officials, Washington, DC.

AASHTO (2010). *LRFD Bridge Design Specifications*, 5th ed., American Association of State Highway and Transportation Officials, Washington, DC.

PROBLEM

7.1 The roadway width, curb to curb, of a bridge deck is 62 ft. Using typical roadway widths given in Table 7.1, how many traffic lanes will this bridge normally carry? If vehicles are allowed to drive on the shoulders, say during an emergency, how many lanes of traffic need to be considered in the design of this bridge?

PART II

Loads and Analysis

CHAPTER 8

Loads

8.1 INTRODUCTION

The engineer must consider all the loads that are expected to be applied to the bridge during its service life. Such loads may be divided into two broad categories: permanent loads and transient loads. The permanent loads remain on the bridge for an extended period, usually for the entire service life. Such loads include the self-weight of the girders and deck, wearing surface, curbs, parapets and railings, utilities, luminaries, and pressures from earth retainments. Transient loads typically include gravity loads due to vehicular, railway, and pedestrian traffic as well as lateral loads such as those due to water and wind, ice floes, ship and vehicular collisions, and earthquakes. In addition, all bridges experience temperature fluctuations on a daily and seasonal basis and such effects must be considered. Depending on the structure type, other loads such as those from creep and shrinkage may be important, and, finally, the superstructure supports may move, inducing forces in statically indeterminate bridges.

Transient loads, as the name implies, change with time and may be applied from several directions and/or locations. Typically, such loads are highly variable. The engineer's responsibility is to anticipate which of these loads are appropriate for the bridge under consideration as well as the magnitude of the loads and how these loads are applied for the most critical effect. Finally, some loads act in combination, and such combinations must be considered for the appropriate limit state. A discussion of such considerations is presented in Chapter 5.

The loads appropriate for the design of short- and medium-span bridges are outlined in this chapter. The primary focus is on loads that are necessary for the superstructure design. Other loads are presented with only limited discussion. For example, ship impact is an important and complex load that must be considered for long-span structures over navigable waters. Similarly, seismic loads are of paramount importance in regions of high seismicity and must be considered for a bridge regardless of span length. Some bridges are an integral part of the lifeline network that must remain functional after a seismic event. An understanding of such requirements requires prerequisite knowledge of structural dynamics combined with inelastic material response due to cyclic actions and is therefore discussed only briefly. This specialized topic is considered to be beyond the scope of this book. For reference, the book edited by Chen and Duan (2004) presents additional material on seismic design of bridges.

Each type of load is presented individually with the appropriate reference to the AASHTO Specification including, where appropriate and important, a discussion regarding the development of the AASHTO provisions. The loads defined in this chapter are used in Chapter 9 to determine the load effects (shear and moment) for a girder line (single beam). In Chapter 10, the modeling of the three-dimensional system is discussed along with the reduction of the three-dimensional system to a girder line. The primary purpose of this chapter is to define and explain the rationale of the AASHTO load requirements. Detailed examples using these loads are combined with structural analysis in the subsequent chapters.

8.2 GRAVITY LOADS

Gravity loads are those caused by the weight of an object on and the self-weight of the bridge. Such loads are both permanent and transient and applied in a downward direction (toward the center of the earth).

8.2.1 Permanent Loads

Permanent loads are those that remain on the bridge for an extended period of time, perhaps for the entire service life. Such loads include:

- ☐ Dead load of structural components and nonstructural attachments (DC)
- ☐ Dead load of wearing surfaces and utilities (DW)
- ☐ Dead load of earth fill (EV)
- ☐ Earth pressure load (EH)
- ☐ Earth surcharge load (ES)
- ☐ Locked-in erection stresses (EL)
- ☐ Downdrag (DD)

The two-letter abbreviations are those used by AASHTO and are also used in subsequent discussions and examples.

The dead load of the structural components and nonstructural attachments are definitely permanent loads and must be included. Here structural components refer to those elements that are part of the load resistance system. Nonstructural attachments refer to such items as curbs, parapets, barrier rails, signs, illuminators, and guard rails. Such attachments often contribute to the stiffness and strength; however, the positive effects of such loads are traditionally and conservatively neglect in design.

Table 8.1 Unit Densities

Material	Unit Weight (kips/ft^3)
Aluminum	0.175
Bituminous wearing surfaces	0.140
Cast iron	0.450
Cinder filling	0.060
Compact sand, silt, or clay	0.120
Concrete, lightweight (includes reinforcement)	0.110
Concrete, sand—lightweight (includes reinforcement)	0.120
Concrete, normal (includes reinforcement) $f'c \leq 5$ ksi	0.145
Concrete, normal (includes reinforcement) 5 ksi $< f'c <$ 15 ksi	$0.140 + 0.001 f'c$
Loose sand, silt, or gravel	0.100
Soft clay	0.100
Rolled gravel, macadam, or ballast	0.140
Steel	0.490
Stone masonry	0.170
Hardwood	0.060
Softwood	0.050
Water, fresh	0.062
Water, salt	0.064
Transit rails, ties, and fastening per track	0.200 kip/ft

In AASHTO Table 3.5.1-1. From *AASHTO LRFD Bridge Design Specifications*. Copyright © 2010 by the American Association of State Highway and Transportation Officials, Washington, DC. Used by permission.

The weight of such items can be estimated by using the unit weight of the material combined with the geometry. For third-party attachments, for example, the guard rail, the manufacture's literature often contains weight information. In the absence of more precise information, the unit weights given in Table 8.1 may be used.

The dead load of the wearing surface (DW) is estimated by taking the unit weight times the thickness of the surface. This value is combined with the DC loads per Tables 5.1 and 5.2 [Tables 3.4.1-1 and 3.4.1-2].* Note that the load factors are different for the DC and DW loads. The maximum and minimum load factors for the DC loads are 1.25 and 0.90, respectively, and the maximum and minimum load factors for the DW loads are 1.50 and 0.65, respectively. The different factors are used because the DW loads have been determined to be more variable in load surveys than the DC loads. For example, Nowak (1993, 1995) noted the coefficients of variation (standard deviation per mean) for factory-made, cast-in-place (CIP), and asphalt surfaces are 0.08, 0.10, and 0.25, respectively. In short, it is difficult to estimate at the time of design how many layers and associated thicknesses of wearing surfaces may be applied by maintenance crews during the service life, but it is fairly easy to estimate the weight of other components.

*The article numbers in the AASHTO (2010) LRFD Bridge Specifications are enclosed in brackets and preceded by the letter A if specifications and by the letter C if commentary.

The dead load of earth fills (EV) must be considered for buried structures such as culverts. The EV load is determined by multiplying the unit weight times the depth of materials. Soil–structure interaction effects may apply. Again the load factors per Table 5.1 and 5.2 [Tables A3.4.1-1 and A3.4.1-2] apply.

The earth surcharge load (ES) is calculated like the EV loads with the only difference being in the load factors. This difference is attributed to its variability. Note that part or the entire load could be removed at some time in the future, or perhaps the surcharge material (or loads) could be changed. Thus, the ES load has a maximum load factor of 1.50, which is higher than the typical EV factors that are about 1.35. Similarly, the minimum ES and EV factors are 0.75 and 0.90 (typical), respectively.

Soil retained by a structure such as a retaining wall, wing wall, or abutment creates a lateral pressure on the structure. The lateral pressure is a function of the geotechnical characteristics of the material, the system geometry, and the anticipated structural movements. Most engineers use models that yield a fluidlike pressure against the wall. Such a procedure is outlined in AASHTO Section 3.11 and is described in more detail in AASHTO Section 10.

Locked-in erection stresses are accumulated force effects resulting from the construction process. They include secondary forces from posttensioning. Downdrag is a force exerted on a pile or drilled shaft due to soil movement around the element. Such a force is permanent and typically increases with time. The details regarding the downdrag calculations are outlined in AASHTO Section 10, Foundations.

In summary, permanent loads must always be considered in the structural analysis. Some permanent loads are easily estimated, such as component self-weight, while other loads, such as lateral earth pressures, are more difficult due to the greater variability involved. Where variabilities are greater, higher load factors are used for maximum load effects and lower factors are used for minimum load effects.

8.2.2 Transient Loads

Although the automobile is the most common vehicular live load on most bridges, the truck causes the critical load effects. In a sense, cars are "felt" very little by the bridge and come "free." More precisely, the load effects of the car traffic compared to the effect of truck traffic are negligible. Therefore, the AASHTO design loads attempt to model the truck traffic that is highly variable, dynamic, and may occur independent of, or in unison with, other truck loads.

The principal load effect is the gravity load of the truck, but other effects are significant and must be considered. Such effects include impact (dynamic effects), braking forces, centrifugal forces, and the effects of other trucks simultaneously present. Furthermore, different design limit states may require slightly different truck load models. Each of these loads is described in more detail in the following sections. Much of the research involved with the development of the live-load model and the specification calibration is presented in Nowak (1993, 1995). Readers interested in the details of this development are encouraged to obtain this reference for more background information.

Design Lanes The number of lanes a bridge may accommodate must be established and is an important design criterion. Two terms are used in the lane design of a bridge:

- ☐ Traffic lane
- ☐ Design lane

The traffic lane is the number of lanes of traffic that the traffic engineer plans to route across the bridge. A lane width is associated with a traffic lane and is typically 12 ft (3600 mm). The design lane is the lane designation used by the bridge engineer for live-load placement. The design lane width and location may or may not be the same as the traffic lane. Here AASHTO uses a 10-ft (3000 mm) design lane, and the vehicle is to be positioned within that lane for extreme effect.

The number of design lanes is defined by taking the integer part of the ratio of the clear roadway width divided by 12 ft (3600 mm) [A3.6.1.1.1]. The clear width is the distance between the curbs and/or barriers. In cases where the traffic lanes are less than 12 ft (3600 mm) wide, the number of design lanes shall be equal to the number of traffic lanes, and the width of the design lane is taken as the width of traffic lanes. For roadway widths from 20 to 24 ft (6000 to 7200 mm), two design lanes should be used, and the design lane width should be one-half the roadway width.

The direction of traffic in the present and future design scenarios should be considered and the most critical cases should be used for design. Additionally, there may be construction and/or detour plans that cause traffic patterns to be significantly restricted or altered. Such situations may control some aspects of the design loading.

Transverse positioning of trucks is automatically accounted for in the live-load distribution factors outlined in AASHTO Section 4 and Chapter 11. When positioning is required for cases where analysis is used or required, such as lever rule, rigid method, and/or rigorous analysis, the engineer must position the trucks for the critical load effect.

For exterior girders, this requires placing one wheel of a truck within 2 ft (600 mm) from the curb or barrier. The next truck, if considered, is placed within 4 ft (1200 mm) of the first. A third truck, if required, is placed within 6 ft (1800 mm) of the second so as to not infringe upon the traffic lane requirement. For an interior girder, one wheel is placed over a girder and the position of others follows a similar pattern. From a practical perspective, all trucks can be conservatively placed transversely within 4 ft (1200 mm) of each other with little loss of "accuracy" when compared to the specification intent. Patrick et al. (2006) outline this in significant detail. In several examples, they take a simple approach and place vehicles at a 4-ft (1200-mm) transverse spacing. Puckett et al. (2007) also addressed this in an appendix in NCHRP 592.

Vehicular Design Loads A study by the Transportation Research Board (TRB) was used as the basis for the AASHTO loads (TRB, 1990). The TRB panel outlined many issues regarding the development (revision of) a national policy of truck weights. This document provides an excellent summary of history and policy alternatives and associated economic trade-offs. Loads that are above the legal weight and/or length limits but are regularly allowed to operate were cataloged. Although *all states in the Northeast allow such overlegal loads . . .*, many others, from . . . *Florida to Alaska*, also routinely allow such loads. Typically, these loads are short-haul vehicles such as solid waste trucks and concrete mixers. Although above "legal" limits, these vehicles were allowed to operate routinely due to "grandfathering" provisions in state statutes. These vehicles are referred to as *exclusion vehicles*. The engineers who developed the load model felt that the exclusion trucks best represented the extremes involved in the present truck traffic (Kulicki, 1992).

Theoretically, one could use all the exclusion vehicles in each design and design for the extreme load effects (envelope of actions). As an analysis would be required for many vehicles, this is clearly a formidable task, even if automated. Hence, a simpler, more tractable model was developed called HL-93 (highway load, developed in 1993). The objective of this model is to prescribe a set of loads such that the same extreme load effects of the HL-93 model are approximately the same as the exclusion vehicles. This model consists of three distinctly different live loads:

- ☐ Design truck
- ☐ Design tandem
- ☐ Design lane

As illustrated in Figure 8.1, the design truck (the first of three separate live-load configurations) is a model load that resembles the typical semitrailer truck [A3.6.1.2]. The front axle is 8 kips (35 kN), the drive axle of 32 kips (145 kN) is located 14 ft (4300 mm) behind, and the rear trailer axle is also 32 kips (145 kN) and is positioned at a variable distance ranging between 14 and 30 ft (4300 and 9000 mm). The variable range means that the spacing used should cause a critical load effect. The long spacing typically only controls where the front and rear portions of the truck may be positioned in adjacent structurally continuous spans such as for continuous short-span bridges. The design truck is the same

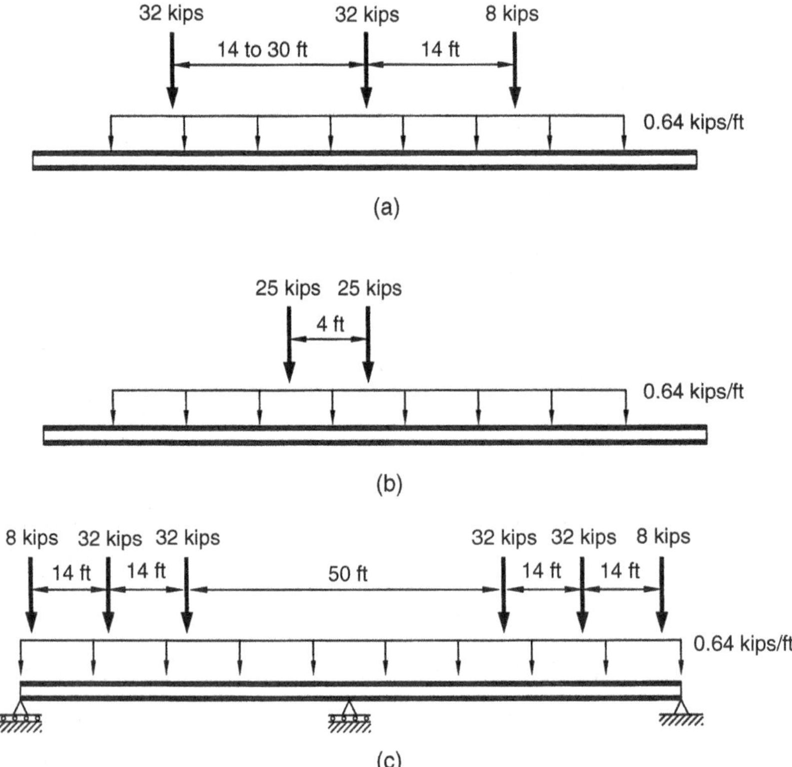

Fig. 8.1 The AASHTO HL-93 design loads. (a) Design truck plus design lane, (b) design tandem plus design lane, and (c) dual design truck plus design lane.

configuration that has been used by AASHTO (2002) Standard Specifications since 1944 and is commonly referred to as HS20. The H denotes highway, the S denotes semitrailer, and the 20 is the weight of the tractor in tons (U.S. customary units). The new vehicle combinations as described in AASHTO (2010) LRFD Bridge Specifications are designated as HL-93.

The second configuration is the design tandem and is illustrated in Figure 8.1(b). It consists of two axles weighing 25 kips (110 kN) each spaced at 4 ft (1200 mm), which is similar to the tandem axle used in previous AASHTO Standard Specifications except the load is changed from 24 to 25 kips (110 kN).

The third load is the design lane load that consists of a uniformly distributed load of 0.64 kips/ft (9.3 N/mm) and is assumed to occupy a region 10 ft (3000 mm) transversely. This load is the same as a uniform pressure of 64 lb/ft^2 (3.1 kPa) applied in a 10-ft (3000-mm) design lane. This load is similar to the lane load outlined in the AASHTO Standard Specifications for many years with the exception that the LRFD lane load does *not* require any concentrated loads.

The load effects of the design truck and the design tandem must each be *superimposed with* the load effects of the design lane load. This combination of lane and axle loads is a major deviation from the requirements of the earlier AASHTO Standard Specifications, where the loads were considered separately. These loads are *not* designed to model any one vehicle or combination of vehicles but rather the *spectra* of loads and their associated load effects. As such these types of loads are often referred to as "notional."

Although the live-load model was developed using the exclusion vehicles, it was also compared to other weigh-in-motion (WIM) studies. WIM studies obtain truck weight data by using passive weighing techniques, so the operator is unaware that the truck is being monitored. Bridges are instrumented to perform this task. Such studies include Hwang and Nowak (1991a,1991b) and Moses and Ghosen (1985). Kulicki (1992) and Nowak (1993) used these WIM studies as confirmation of the AASHTO live load.

Kulicki and Mertz (1991) compared the load effects (shear and moment) for one- and two-span continuous beams for the previous AASHTO loads and those presently prescribed. In their study, the HS20 truck and lane loads were compared to the maximum load effect of 22 trucks representative of traffic in 1991. The ratio of the maximum moments to the HS20 moment is illustrated in Figure 8.2. Similarly, the shear ratio is shown in Figure 8.3. Note that significant variation exists in the ratios and most ratios are greater than 1, indicating that the exclusion vehicle maximums are greater than the model load, a nonconservative situation. A perfect model would contain ordinates of unity for all span lengths. This model is practically not possible, but the combination of design truck with the design lane and the design tandem with

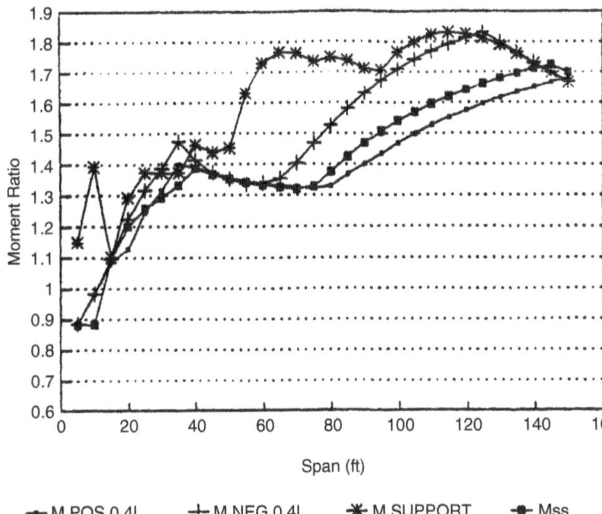

Fig. 8.2 Comparison of the exclusion vehicles to the traditional HS20 load effects—moment (AASHTO Fig. C3.6.1.2.1-1). (From *AASHTO LRFD Bridge Design Specifications*, Copyright © 2010 by the American Association of State Highway and Transportation Officials, Washington, DC. Used by permission.)

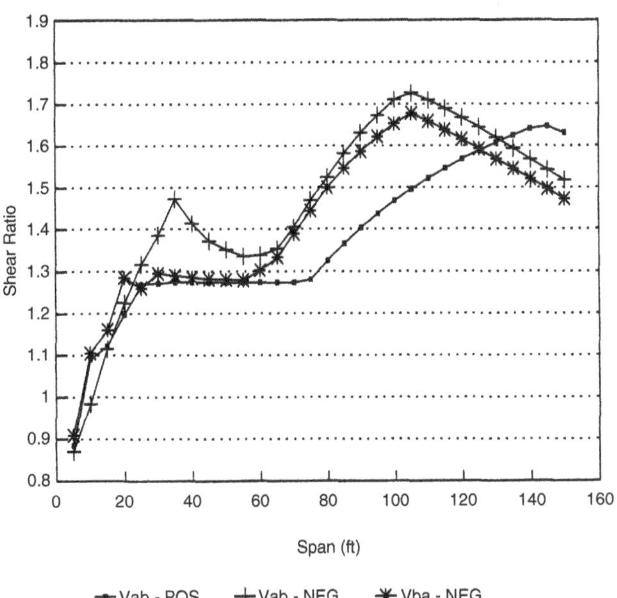

Fig. 8.3 Comparison of the exclusion vehicles to the traditional HS20 load effects—shear. (AASHTO Fig. C3.6.1.2.1-2). (From *AASHTO LRFD Bridge Design Specifications*, Copyright © 2010 by the American Association of State Highway and Transportation Officials, Washington, DC. Used by permission.)

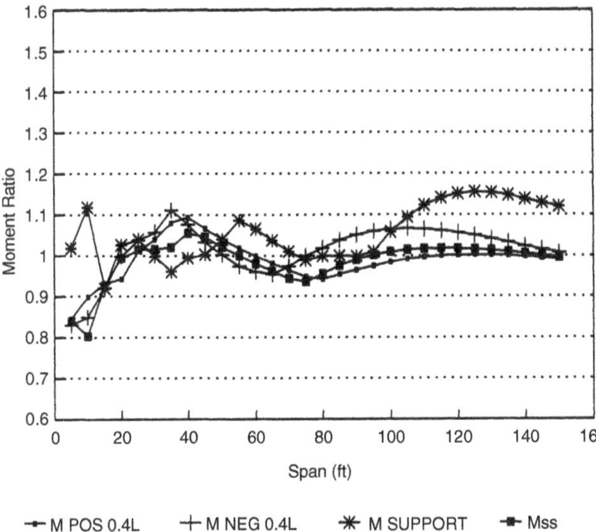

Fig. 8.4 Comparison of the design load effects with exclusion vehicle—moment (AASHTO Fig. C3.6.1.2.1-3). (From *AASHTO LRFD Bridge Design Specifications*, Copyright © 2010 by the American Association of State Highway and Transportation Officials, Washington, DC. Used by permission.)

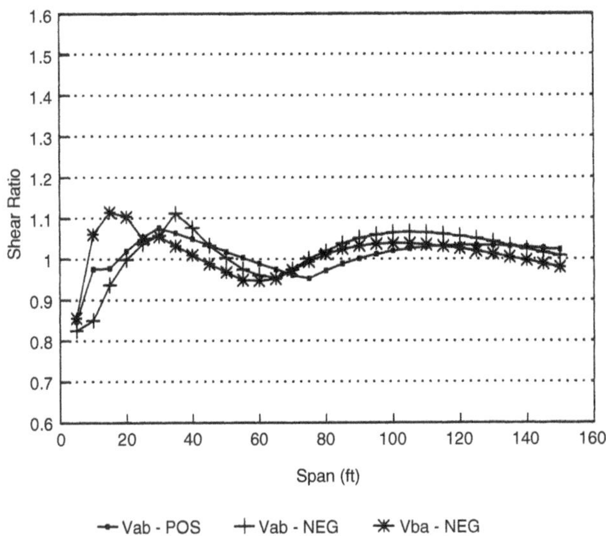

Fig. 8.5 Comparison of the design load effects with exclusion vehicle—shear (AASHTO Fig. C3.6.1.2.1-4). (From *AASHTO LRFD Bridge Design Specifications*, Copyright © 2010 by the American Association of State Highway and Transportation Officials, Washington, DC. Used by permission.)

the design lane gives improved results, which are illustrated in Figures 8.4 and 8.5.

Note that the variation is much less than in Figures 8.2 and 8.3 as the ratios are more closely grouped over the span range, for both moment and shear, and for both simple and continuous spans. The implication is that the present model adequately represents today's traffic and a single-load factor may be used for all trucks. Note that in Figure 8.4, the negative moment of the model underestimates the effect of the exclusion vehicles. This underestimation occurs because the exclusion model includes only one vehicle on the bridge at a time, likely a nonconservative assumption for the negative moments and reactions at interior supports. As it is possible that an exclusion vehicle could be closely followed by another heavily loaded truck, it was felt that a third live-load combination was required to model this case. This live-load combination is specified in AASHTO [A3.6.1.3.1]:

> *For both negative moment (tension on top) between points of contraflexure under a uniform load on all spans, and reaction at interior supports, 90 percent of the effect of two design trucks spaced a minimum of 50 ft (15,000 mm) between the lead axle of one truck and the rear axle of the other truck, combined with 90 percent of the effect of the design lane load. The distance between the 32-kip (145-kN) axles of each truck shall be taken as 14 ft (4300 mm).*

Axles that do not contribute to the extreme force effect should be neglected. Nowak (1993) compared survey vehicles with others in the same lane to the AASHTO load model, and the results are shown in Figures 8.6 and 8.7. The moments were chosen for illustration. The $M(75)$ moment represents the mean of the load effect due to the survey vehicles, HS20 is moment due to the traditional AASHTO (2002) Standard Specifications truck (same as the design truck), and LRFD is the moment due to the present AASHTO (2010) LRFD loads. Note that the present loads adequately represent the load survey with a bias of approximately 20%.

In summary, three design loads should be considered: the design truck, design tandem, and design lane. These loads are *superimposed* three ways to yield the live-load effects, which are combined with the other load effects per Tables 5.1 and 5.2. These cases are illustrated in Table 8.2 where the number in the table indicates the appropriate *multiplier* to be used prior to superposition. The term multiplier is used to avoid confusion with the load factors that are used to combine the various types of loads, for example, live and permanent loads in Tables 5.1 and 5.2.

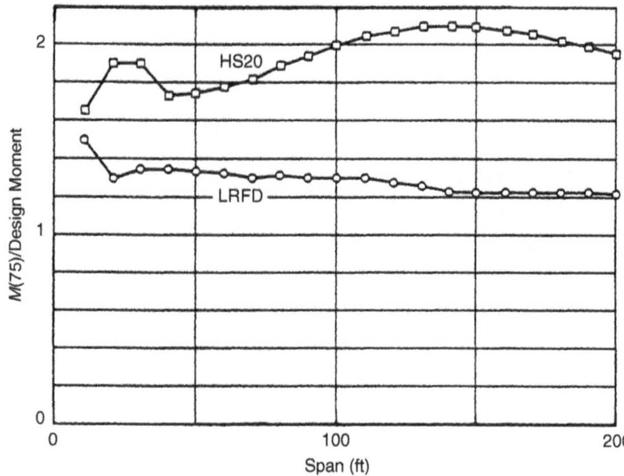

Fig. 8.6 Comparison of the design load effects with survey vehicles—simple-span moment (Nowak, 1993).

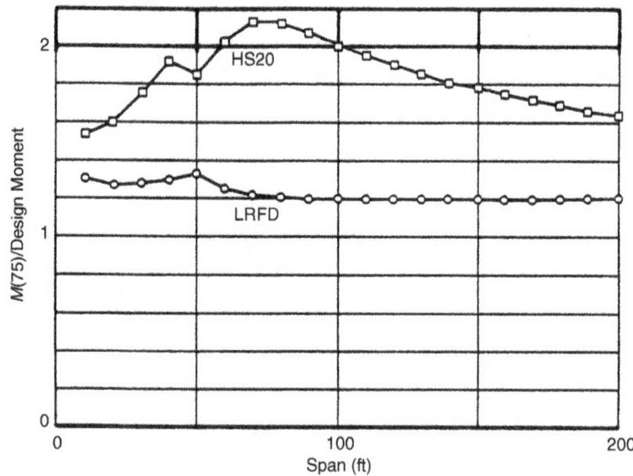

Fig. 8.7 Comparison of the design load effects with survey vehicles—negative moment (Nowak, 1993).

Fatigue Loads The strengths of various components of the bridge are sensitive to repeated stressing or fatigue. When the load is cyclic, the stress level that ultimately fractures the material can be significantly below the nominal yield strength. For example, depending on the details of the welds, steel could have a fatigue strength as low as 2.6 ksi (18 MPa) [A6.6.1.2.5]. The fatigue strength is typically related to the range of live-load stress and the number of stress cycles under service load conditions. As the majority of trucks do not exceed the legal weight limits, it would be unduly conservative to use the full live-load model, which is based on exclusion

Table 8.2 Load Multipliers for Live Loads

Live-Load Combination	Design Truck	Design Tandem	Two Design Trucks or Tandems with 50-ft (15,000-mm) Headway[a]	Design Lane
1	1.0			1.0
2		1.0		1.0
3			0.9	0.9

[a]The two design trucks and lane combination is for the negative moment and reaction at interior supports only.

vehicles to estimate this load effect. This means that a lesser load is used to estimate the live-load stress range and is accommodated by using a single design truck with the variable axle spacing set at 30 ft (9000 mm) and a load factor of 0.75 as prescribed in Table 5.1 [Table A3.4.1-1].

The dynamic load allowance (IM) [A3.6.2] must be included and the bridge is assumed to be loaded in a single lane [A3.6.1.4.3b]. The average load effect due to the survey vehicles (used to calibrate the specification) was about 75% of the moment due to the design truck (Nowak, 1993); hence a load factor of 0.75 is used.

The number of stress–range cycles is based on traffic surveys. In lieu of survey data, guidelines are provided in AASHTO [A3.6.1.4.2]. The average daily truck traffic (ADTT) in a single lane may be estimated as

$$ADTT_{SL} = p\,(ADTT)$$

where p is the fraction of traffic assumed to be in one lane as defined in Table 8.3.

Because the traffic patterns on the bridge are uncertain, the frequency of the fatigue load for a single lane is assumed to apply to all lanes.

The ADTT is usually available from the bridge owner, but in some cases only the average daily traffic (ADT) is available. In such cases, the percentage of trucks in the total traffic must be estimated. This percentage can vary widely with local conditions, and the engineer should try to estimate this with a survey. For example, it is common for interstate roadways in the rural western states to have the percentage of trucks exceed 50%. If survey data are not possible or practical, or if the fatigue limit state is not a controlling factor in the design, then AASHTO provides guidance. This guidance is illustrated in Table 8.4.

Note that the number of stress–range cycles is not used in the structural analysis directly. The fatigue truck is applied in the same manner as the other vehicles and the range of extreme stress (actions) are used. The number of stress–range cycles is used to establish the available resistance.

Table 8.3 Fraction of Truck Traffic in a Single Lane, *p*

Number of Lanes Available to Trucks	p
1	1.00
2	0.85
3 or more	0.80

In AASHTO Table 3.6.1.4.2-1. From *AASHTO LRFD Bridge Design Specifications*. Copyright © 2010 by the American Association of State Highway and Transportation Officials, Washington, DC. Used by permission.

Table 8.4 Fraction of Trucks in Traffic

Class of Highway	Fraction of Trucks in Traffic
Rural interstate	0.20
Urban interstate	0.15
Other rural	0.15
Other urban	0.10

In AASHTO Table C3.6.1.4.2-1. From *AASHTO LRFD Bridge Design Specifications*. Copyright © 2010 by the American Association of State Highway and Transportation Officials, Washington, DC. Used by permission.

Pedestrian Loads The AASHTO [A3.6.1.6] pedestrian load is 0.075 ksf (3.6×10^{-3} MPa), which is applied to sidewalks that are integral with a roadway bridge. If the load is applied to a bridge restricted to pedestrian and/or bicycle traffic, then a 0.085 ksf (4.1×10^{-3} MPa) live load is used. These loads are comparable to the building corridor load of 0.100 ksf (4.8×10^{-3} MPa) of the International Building Code (IBC, 2009).

The railing for pedestrian and/or bicycle must be designed for a load of 0.050 kip/ft (0.73 N/mm), both transversely and vertically on each longitudinal element in the railing system [A13.8.2 and A13.9.3]. In addition, as shown in Figure 8.8, railing must be designed to sustain a single concentrated load of 0.200 kip (890 N) applied to the top rail at any location and in any direction.

Deck and Railing Loads The gravity loads for the design of the deck system are outlined in AASHTO [A3.6.1.3.3]. The deck must be designed for the load effect due to the design truck or the design tandem, whichever creates the most extreme effect. The two design vehicles should not be considered together in the same load case. For example, a design

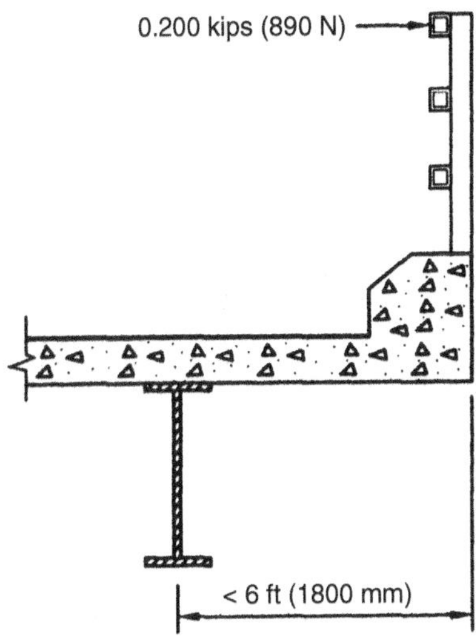

Fig. 8.8 Pedestrian rail loads.

truck in a lane adjacent to a design tandem is not considered (consider all trucks of one kind). The design lane load is not considered in the design of the deck system, except in slab bridges where the load is carried principally in the longitudinal direction (see Chapter 4 on bridge types). Several methods are available for the analysis of decks subjected to these loads. A few of the more common methods are described in Chapter 11. The vehicular gravity loads for decks may be found in AASHTO [A3.6.1.3].

The deck overhang, located outside the facia girder and commonly referred to as the cantilever, is designed for the load effect of a uniform line load of 1 kip/ft (14.6 N/mm) located 1 ft (300 mm) from the face of the curb or railing as shown in Figure 8.9. This load is derived by assuming that one-half of the 50-kip (220-kN) tandem is distributed over a length of 25 ft (7600 mm). The rationale for this rather long length is that the barrier system is structurally continuous and periodically supported by cross beams or the cantilever slab that has been strengthened. In other words, the barrier behaves as another girder located on top of the deck and distributes the load over a longer length than if the barrier was not present.

An illustration of a continuous barrier system is illustrated in Figure 8.10(a). The concrete curbs, parapets, barriers, and dividers, should be made structurally continuous with the deck [A9.4.3]. The exception requires owner approval. If the barrier is not flexurally continuous, then the load should be distributed over a lesser length, increasing the cantilever moments. An example is illustrated in Figure 8.10(b). More details regarding deck design and analysis are presented in Chapters 11 and 15.

The traffic barrier system and the deck overhang must sustain the infrequent event of a collision of a truck. The barrier is commonly referred to by many terms, such as parapet, railing, and barrier. The AASHTO uses the terms railing or railing system, and hereafter this term is used in the same manner. The deck overhang and railing design is confirmed by crash testing as outlined in AASHTO [A13.7.2]. Here the

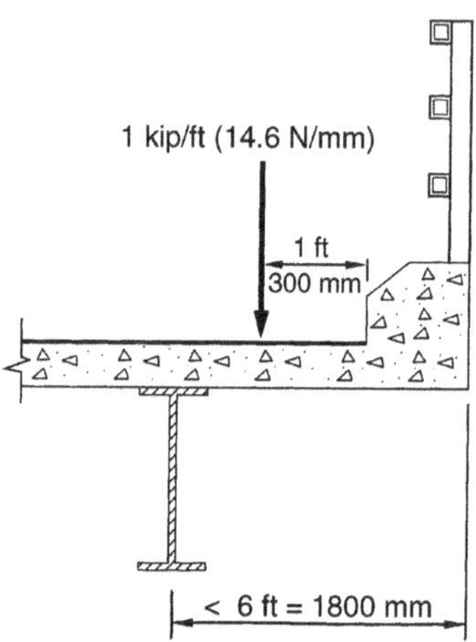

Fig. 8.9 Gravity load on cantilever.

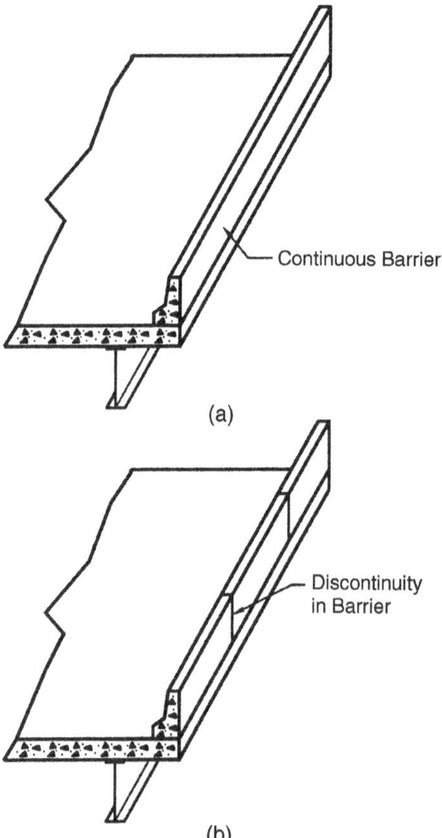

Fig. 8.10 (a) Continuous barrier and (b) discontinuous barrier.

rail/cantilever deck system is subjected to crash testing by literally moving vehicles of specified momentum (weight, velocity, and angle of attack) into the system. The momentum characteristics are specified as a function of test levels that attempt to model various traffic conditions. The design loads crash worthiness is only used in the analysis and design of the deck and barrier systems. The design forces for the rail and deck design are illustrated in Table 8.5 for six test levels (TL). The levels are described as follows [A13.7.2]:

TL-1 is used for work zones with low posted speeds and very low volume, low speed local streets.

TL-2 is used for work zones and most local and collector roads with favorable site conditions as well as where a small number of heavy vehicles is expected and posted speeds are reduced.

TL-3 is used for a wide range of high-speed arterial highways with very low mixtures of heavy vehicles and with favorable site conditions.

TL-4 is used for the majority of applications on high-speed highways, freeways, expressways, and interstate highways with a mixture of trucks and heavy vehicles.

TL-5 is used for the same applications as TL-4 and where large trucks make up a significant portion of the average daily traffic or when unfavorable site conditions justify a higher level of rail resistance.

Table 8.5 Test Configurations (Vehicle Weight in Tons)

Vehicle Characteristics	Small Automobiles		Pickup Truck	Single-Unit Van Truck	Van Type Tractor-Trailer		Tractor-Tanker Trailer
W (kips)	1.55	1.8	4.5	18.0	50.0	80.0	80.0
B (ft)	5.5	5.5	6.5	7.5	8.0	8.0	8.0
G (in.)	22	22	27	49	64	73	81
Crash angle, θ	20°	20°	25°	15°	15°	15°	15°
Test Level				Test Speeds (mph)			
TL-1	30	30	30	N/A	N/A	N/A	N/A
TL-2	45	45	45	N/A	N/A	N/A	N/A
TL-3	60	60	60	N/A	N/A	N/A	N/A
TL-4	60	60	60	50	N/A	N/A	N/A
TL-5	60	60	60	N/A	N/A	50	N/A
TL-6	60	60	60	N/A	N/A	N/A	50

In AASHTO Table A13.7.2-1. From *AASHTO LRFD Bridge Design Specifications*. Copyright © 2010 by the American Association of State Highway and Transportation Officials, Washington, DC. Used by permission.

TL-6 is used for applications where tanker-type trucks or similar high center-of-gravity vehicles are anticipated, particularly along with unfavorable site conditions.

The definitions of the TL level that previously were based upon NCHRP Report 350 (NCHRP, 1993) have been updated based upon recent research. National Cooperative Highway Research Program (NCHRP) Project 22-14(02), "Improvement of Procedures for the Safety-Performance Evaluation of Roadside Features" addressed the update. These criteria are now published in the *Manual for Assessing Safety Hardware* (MASH) AASHTO (2009a). This document changes the test matrix, installations, vehicles, evaluation criteria, documentation, and performance evaluation. FHWA has a video presentation that summarizes MASH. See http://fhwa.adobeconnect.com/mashfinal/.

Finally, videos of different barrier crash tests can be located by an Internet search for "crash bridge barrier test video" or similar keywords. These tests are used to validate the barrier and deck attachment system. Video is one of the best ways to gain an initial understanding about what a crash test is and the results.

Multiple Presence Trucks will be present in adjacent lanes on roadways with multiple design lanes, but it is unlikely that three adjacent lanes will be loaded simultaneously with the heavy loads. Therefore, some adjustments in the design loads are necessary. To account for this effect, AASHTO [A3.6.1.1.2] provides an adjustment factor for the multiple presence. Table 8.6, after AASHTO [Table A3.6.1.1.2-1], is provided.

Note that these factors should not be applied in situations where these factors have been implicitly included, such as in the load distribution factors outlined in AASHTO [A4.6.2]. If statical distribution factors are used or if the analysis is based on refined methods, then the multiple presence factors apply. The details of these analytical methods are described in Chapter 11. In addition, these factors apply in the design of bearings and abutments for the braking forces defined later. Lastly, the multiple presence factors should not be used in the case of the fatigue limit state.

Table 8.6 Multiple Presence Factors

Number of Design Lanes	Multiple Presence Factor m
1	1.20
2	1.00
3	0.85
More than 3	0.65

In AASHTO Table 3.6.1.1.2-1. From *AASHTO LRFD Bridge Design Specifications*. Copyright © 2010 by the American Association of State Highway and Transportation Officials, Washington, DC. Used by permission.

Dynamic Effects The roadway surface is not perfectly smooth, thus the vehicle suspension must react to roadway roughness by compression and extension of the suspension system. This oscillation creates axle forces that exceed the static weight during the time the acceleration is upward and is less than the static weight when the acceleration is downward. Although commonly called impact, this phenomenon is more precisely referred to as dynamic loading and the term to account for this is called dynamic load allowance (DLA).

There have been numerous experimental and analytical studies to determine the dynamic load effect. Paultre et al. (1992) provide an excellent review of analytical and experimental research regarding the effects of vehicle/bridge

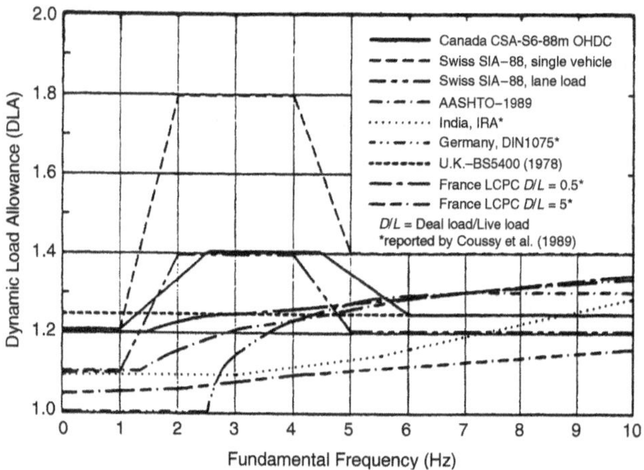

Fig. 8.11 International perspective of dynamic load allowance (Paultre et al., 1992).

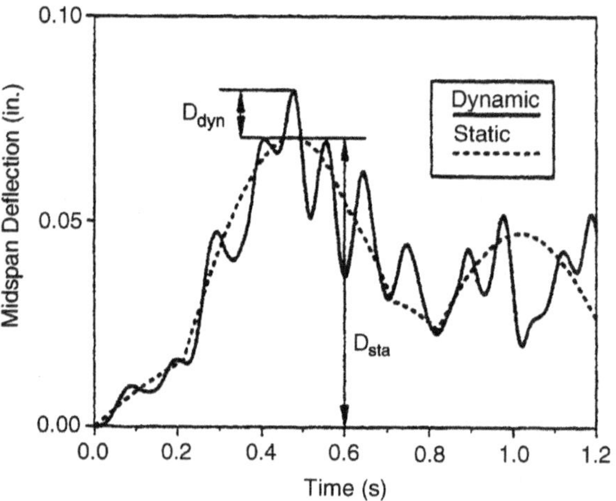

Fig. 8.12 Typical live-load response (Hwang and Nowak, 1991a).

dynamics. In this article, the writers outline the various factors used to increase the static load to account for dynamic effects. As illustrated in Figure 8.11, various bridge engineering design specifications from around the world use widely differing factors. The ordinate axis represents the load increase or DLA and the abscissa is the fundamental frequency of the structure. In cases where the specification value is a function of span length [e.g., AASHTO (2002)], the frequency is estimated using an empirically based formula. Note the wide variability for DLA. This variability indicates that the worldwide community has not reached a consensus about this issue.

One must carefully interpret and compare the results of such studies as the definitions of the dynamic effects are not consistent and are well portrayed by Bakht and Pinjarkar (1991) and Paultre et al. (1992). These writers describe the many definitions that have been used for dynamic load effects. Such definitions have a significant effect on the magnitude of the DLA reported and consequently the profession's perception of dynamic effects. It is most common to compare the static and dynamic deflections as illustrated in Figure 8.12. A typical plot of a midspan deflection is shown as a function of vehicle position. The dynamic effect is defined herein as the amplification factor applied to the static response to achieve the dynamic load effect. This effect is called by many different terms: dynamic load factor, dynamic load allowance, and impact factor. Sometimes the factor includes the static load response (>1) and other times it includes only the dynamic response (<1). The term dynamic load allowance is used by AASHTO, which is abbreviated IM (for impact). Although the terminology is inconsistent with the abbreviation, IM is traditionally used and some old habits will likely never die. When referring to Figure 8.12, the dynamic load allowance is

$$\mathrm{IM} = \frac{D_{\mathrm{dyn}}}{D_{\mathrm{sta}}}$$

where D_{sta} is the maximum static deflection and D_{dyn} is the additional deflection due to the dynamic effects.

It is important to observe that this ratio varies significantly with different vehicle positions. Thus, it is quite possible to observe impact factors that greatly exceed those at the maximum deflections (and the AASHTO value). Bakht and Pinjarkar (1991) and Paultre et al. (1992) describe this characteristic. The DLA is of concern principally because it is used for the design and evaluation of bridges for the extreme load effects. Therefore, it is reasonable to define the DLA based on the extreme values.

The principal parameters that affect the impact factor are the dynamic characteristics of the truck, the dynamic characteristics of the bridge, and the roadway roughness. These characteristics are expected as all transient structural dynamic problems involve stiffness, mass, damping, and excitation. Hwang and Nowak (1991a, 1991b) present a comprehensive analytical study involving modeling a truck as rigid bodies interconnected with nonlinear suspension springs. The simply supported bridges were modeled using the standard equation for forced beam vibration, and the excitation was derived using actual roadway roughness data.

Numerical integration was used to establish the response. The truck configurations were taken from weigh-in-motion studies of Moses and Ghosen (1985). Simply supported steel and prestressed slab girder bridges were studied. The results offer insight into vehicle bridge dynamics. The dynamic and static components of midspan deflection for the steel girder bridges are illustrated in Figures 8.13 and 8.14. Note that the dynamic component remains almost unchanged with the truck weight while the static deflection increases linearly with weight, as expected. As the ratio of the two deflections is the DLA, it follows that the DLA decreases with truck weight, which is illustrated in Figure 8.15. Note

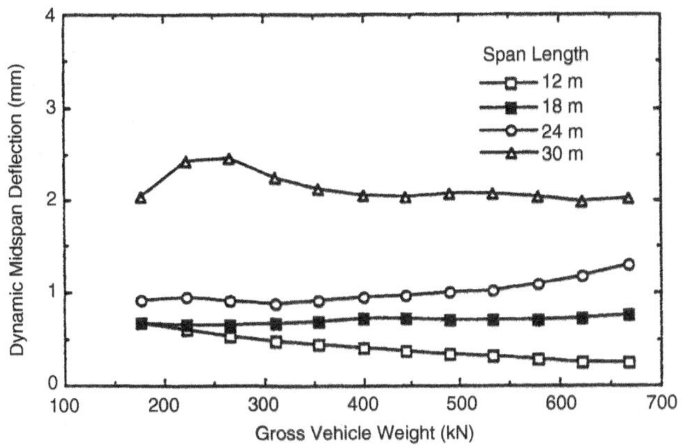

Fig. 8.13 Dynamic response (Hwang and Nowak, 1991a).

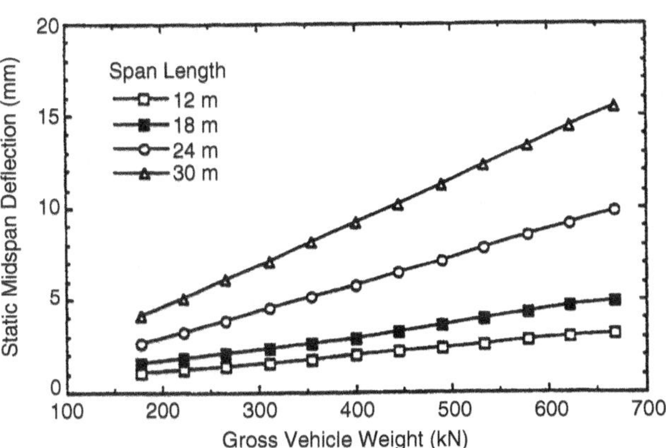

Fig. 8.14 Static response (Hwang and Nowak, 1991a).

that for the design truck weight (72 kips, 316 kN) the DLA is approximately 0.3, the AASHTO value.

Note that most of the data are below 0.3. Hwang and Nowak (1991a) also summarize their findings for various trucks and roughness profiles, where four span lengths were considered. The average impact factors ranged from a low average of 0.09 (COV = 0.43) to a maximum of 0.21 (COV = 0.72), where COV is the coefficient of variation. These results indicate that the impact load effects are typically less than 30%, but with significant variation.

The global load dynamic effects are addressed in most studies regarding impact. Global means the load effect is due to the global system response such as the deflection, moment, or shear of a main girder. Local effects are the actions that result from loads directly applied to (or in the local area of) the component being designed. These include decks and deck components. *In short, if a small variation in the live-load placement causes a large change in load effect, then this effect should be considered local.*

Impacts on such components tend to be much greater than the effects on the system as a whole and are highly dependent on roadway roughness. First, this is because the load is directly applied to these elements, and second, their stiffness is much greater than that of the system as a whole. For many years the AASHTO used an impact formula that attempted to reflect this behavior by using the span length as a parameter. The shorter spans required increased impact to an upper limit of 0.3.

Other specifications, for example, the Ontario Highway Bridge Design Code (OHBDC, 1983), an extremely progressive specification for the time, modeled this behavior as a function of the natural frequency of the system. This specification is illustrated in Figure 8.11. Although perhaps the most rational approach, it is problematic because the

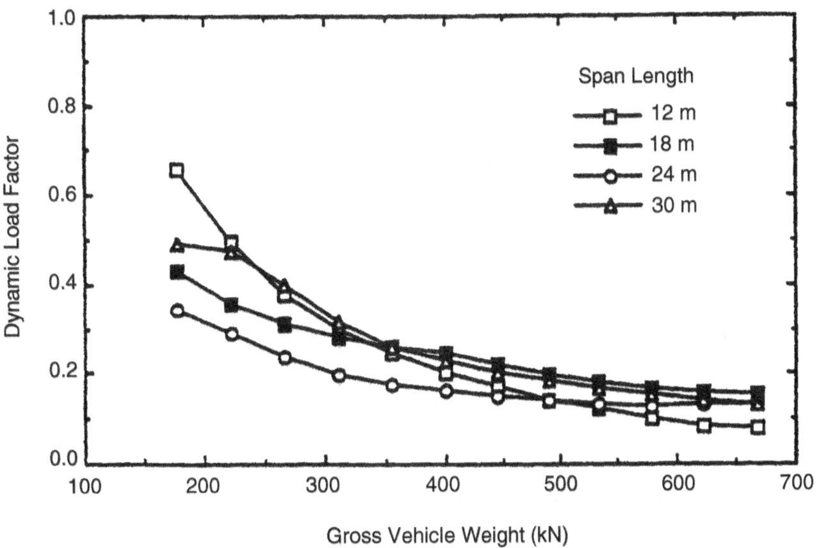

Fig. 8.15 Dynamic load allowance (Hwang and Nowak, 1991a).

frequency must be calculated (or estimated) during the design process. Obtaining a *good* estimate of the natural frequency can be difficult for an existing structure and certainly more difficult for a bridge being designed. This approach adds a level of complexity that is perhaps unwarranted. An empirical-based estimate can be obtained by a simple formula (Tilly, 1986).

The present AASHTO specification takes a very simplistic approach and defines the DLA as illustrated in Table 8.7 [A3.6.2].

These factors are to be applied to the static load as

$$U_{L+I} = U_L\,(1 + \text{IM}) \tag{8.1}$$

where U_{L+I} is the live-load effect plus allowance for dynamic loading, U_L is the live-load effect of live load, and IM is the fraction given in Table 8.7.

All other components in Table 8.7 include girders, beams, bearings (except elastomeric bearings), and columns. Clearly, the present specification does not attempt to model dynamic effects with great accuracy, but with sufficient accuracy and conservatism for design. Both experimental and analytical studies indicated that these values are reasonable estimates. Moreover, considering the variabilities involved, a flat percentage for dynamic load effect is practical, tractable for design, and reasonably based on research results. For the structural evaluation of existing bridges (rating), the engineer will likely use the criteria established in the AASHTO rating procedures that are a function of roadway roughness. At the time of design, the future roadway roughness and associated maintenance are difficult to estimate, thus more conservative values are appropriate.

Centrifugal Forces Acceleration is the time derivative of the velocity vector and as such results from either a change of magnitude or direction of velocity. A truck can increase speed, decrease speed, and/or change directions as it moves along a curvilinear path. All of these effects require an acceleration of the vehicle that causes a force between the deck and the truck. Because its mass is large compared to the power available, a truck cannot increase its speed at a rate great enough to impose a significant force on the bridge.

Table 8.7 Dynamic Load Allowance, IM

Component	IM (%)
Deck joints—all limit states	75
All other components	
Fatigue and fracture limit states	15
All other limit states	33

In AASHTO Table 3.6.2.1-1. From *AASHTO LRFD Bridge Design Specifications*.

Conversely, a decrease in speed due to braking can create a significant acceleration (deceleration) that causes large forces on the bridge in the direction of the truck movement. The braking effect is described in the next section. Finally, as a truck moves along a curvilinear path, the change in direction of the velocity causes a centrifugal acceleration in the radial direction. This acceleration is

$$a_r = \frac{V^2}{r} \tag{8.2}$$

where V is the truck speed, and r is the radius of curvature of the truck movement. The forces and accelerations involved are illustrated in Figure 8.16.

Newton's second law requires

$$F = ma \tag{8.3}$$

where m is the mass. Substitution of Eq. 8.2 into Eq. 8.3 yields

$$F_r = \frac{mV^2}{r} \tag{8.4}$$

where F_r is the force on the truck directed toward the center of the curve (outward on the bridge). The position of this force is at the center of mass, assumed to be at 6 ft (1800 mm) above the roadway surface [A3.6.3]. Note that the mass m is equal to

$$m = \frac{W}{g} \tag{8.5}$$

where W is the weight of the vehicle, and g is the gravitational acceleration: 32.2 ft/s^2 (9.807 m/s^2). Substitution of Eq. 8.5 into Eq. 8.4 yields

$$F_r = \left(\frac{V^2}{rg}\right) W \tag{8.6}$$

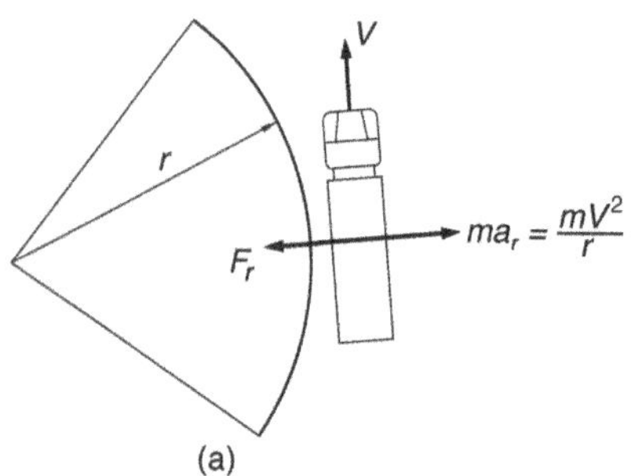

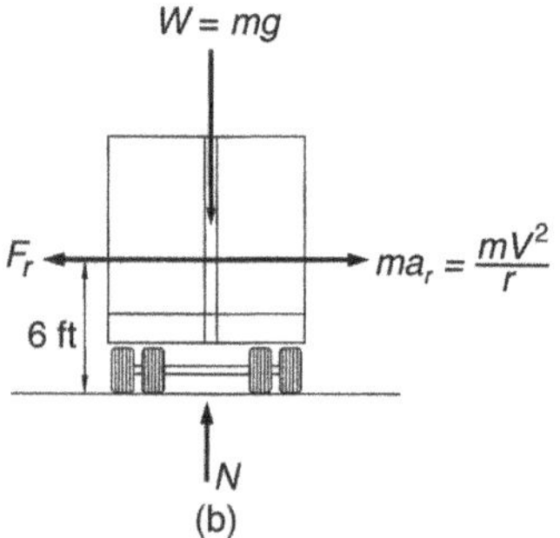

Fig. 8.16 Free-body diagrams for centrifugal force.

which is similar to the expression given in AASHTO [A3.6.3] where

$$F_r = CW \tag{8.7a}$$

where

$$C = f\left(\frac{v^2}{Rg}\right) \tag{8.7b}$$

and $f = \frac{4}{3}$ is for a combination of trucks positioned along the bridge for limit states other than fatigue; for fatigue $f = 1.0$; v is the highway design speed in feet/second (meters/second), R is radius of curvature of traffic lane in feet (meters), and F_r is applied at the assumed center of mass at a distance of 6 ft (1800 mm) above the deck surface.

Because the combination of the design truck with the design lane load gives a load approximately $\frac{4}{3}$ of the effect of the design truck considered independently, a $\frac{4}{3}$ factor is used to model the effect of a train of trucks. Equation 8.6 may be used with any system of consistent units. The multiple presence factors [A3.6.1.1.2] may be applied to this force, as it is unlikely that all lanes will be fully loaded simultaneously.

Braking Forces As described in the previous section, braking forces can be significant. Such forces are transmitted to the deck and must be taken into the substructure at the fixed bearings or supports. It is quite probable that all truck operators on a bridge will observe an event that causes the operators to apply the brakes. Thus, loading of multiple lanes should be considered in the design. Again, it is unlikely that all the trucks in all lanes will be at the maximum design level, therefore the multiple presence factors outlined previously may be applied [A3.6.1.1.2]. The forces involved are shown in Figure 8.17. The truck is initially at a velocity V, and this velocity is reduced to zero over a distance s. The braking force and the associated acceleration are assumed to be constant. The change in kinetic energy associated with the truck is assumed to be completely dissipated by the braking force. The kinetic energy is equated to the work performed by the braking force giving

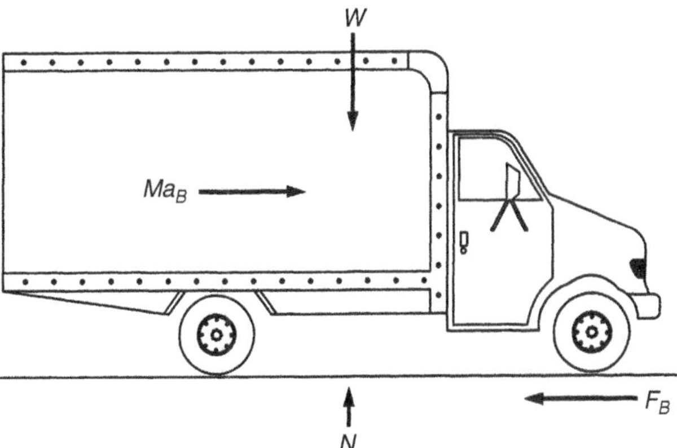

Fig. 8.17 Free-body diagram for braking force.

$$\frac{1}{2}mV^2 = \int_0^s F_B ds = F_B s \tag{8.8}$$

where F_B is the braking force transmitted into the deck and m is the truck mass. Solve for F_B and substitute the mass as defined in Eq. 8.5 to yield

$$F_B = \frac{1}{2}\left(\frac{W}{g}\right)\left(\frac{V^2}{s}\right) = \frac{1}{2}\left(\frac{V^2}{gs}\right)W = bW \tag{8.9a}$$

where

$$b = \frac{1}{2}\left(\frac{V^2}{gs}\right) \tag{8.9b}$$

where b is the fraction of the weight that is applied to model the braking force. In the development of the AASHTO braking force fraction, it was assumed that the truck is moving at a velocity of 55 mph (90 km/h) = 80 ft/s (25 m/s) and a braking distance of 400 ft (122,000 mm) is required. Substitution of these values gives the braking force fraction:

$$b = \frac{(80)^2}{2\,(32.2)\,(400)} = 0.25 = 25\%$$

The braking forces shall be taken as the larger of [A3.6.4]:

- ☐ 25% of the axle weights of the design truck or the tandem truck placed in all lanes, or
- ☐ 5% of the truck and lane, or
- ☐ 5% of the tandem and lane.

Implicit in the AASHTO value is that the coefficient of friction exceeds 0.25 for the tire–deck interface. The braking force is assumed to act horizontally at 6 ft (1800 mm) above the roadway surface in either longitudinal direction.

Permit Vehicles and Miscellaneous Considerations Transportation agencies may include other vehicle loads to model load characteristics in their particular jurisdiction. For example, the Pennsylvania Department of Transportation (PennDOT) uses a notional vehicle called P-82, which models various heavy vehicles that operate under special hauling permits (PennDOT, 2011).

For design, PennDOT incorporates such vehicles, in addition to the AASHTO LRFD live loads. See Figure 8.18a. As with HL-93, this vehicle does not represent any one vehicle but rather models an envelope. The P-82 load effect compared to other loads for a simple span is illustrated in Figure 8.18b. The use of the strength II limit state is typical for these kinds of loads. Note that the load effect is greater, but the load factor is less, 1.35. The background for the development of this is related to Koretzky et al. (1986).

Similarly, the Department of Transportation in California (Caltrans) uses a different load model for its structures (see Fig. 8.19). Caltrans' rationale is similar to PennDOT's.

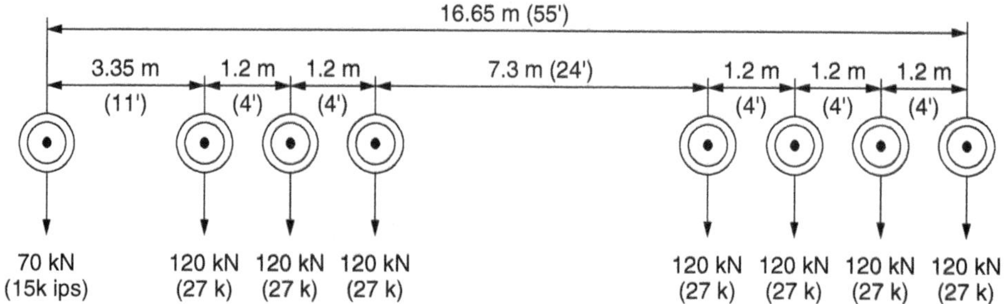

Fig. 8.18a PennDOT P-82 (PennDOT, 2007). P-82 width is the same as the design truck. Transverse wheel location is the same as the design truck.

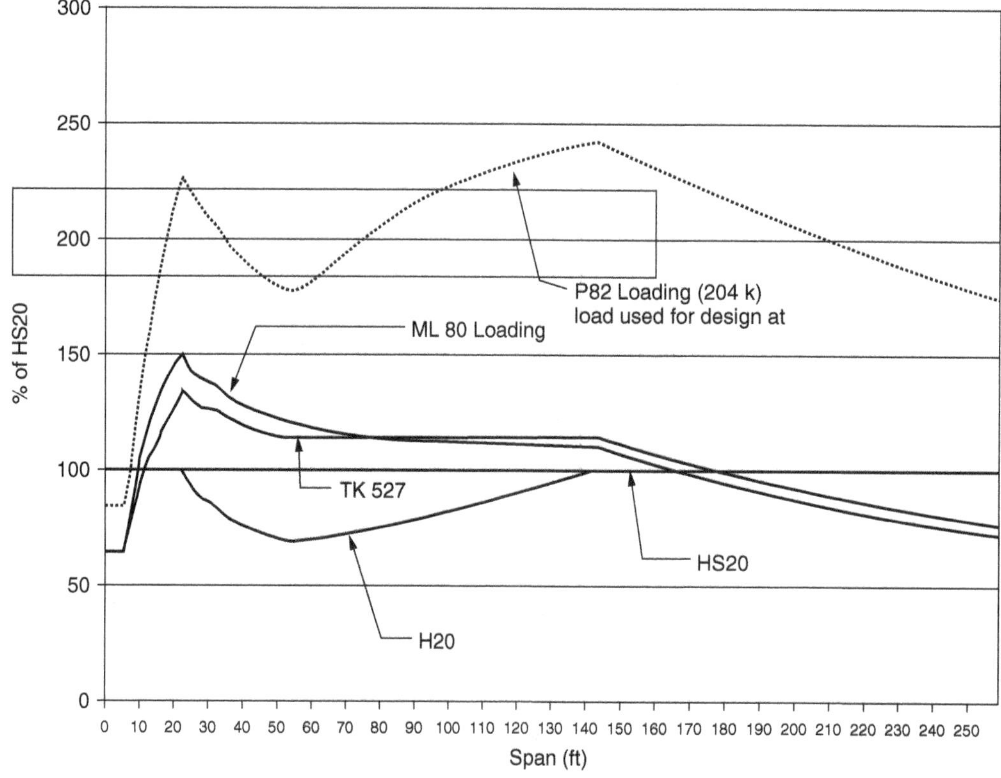

Fig. 8.18b PennDOT load effect comparison (PennDOT, 2010).

The load model should closely approximate the service conditions.

Once the bridge is in operation, the actual vehicle described by axle weights and spacing is used in the analysis to permit the heavy hauls. Handling heavy permit vehicles, and heavy truck loads in general, is a challenging part of managing the bridge inventory from both a technical and administrative perspective.

In summary, agencies use loads and procedures that exceed those found in AASHTO LRFD. Similarly, agencies have altered the AASHTO LRFD specifications to better model their specific bridge types, construction methods, loads, and environmental conditions. Most of these documents are readily available on a DOT's website.

8.3 LATERAL LOADS

8.3.1 Fluid Forces

The force on a structural component due to a fluid flow (water or air) around a component is established by Bernoulli's equation in combination with empirically established drag coefficients. Consider the object shown in an incompressible fluid in Figure 8.20. With the use of Bernoulli's equation, equating the upstream energy associated with the flow at

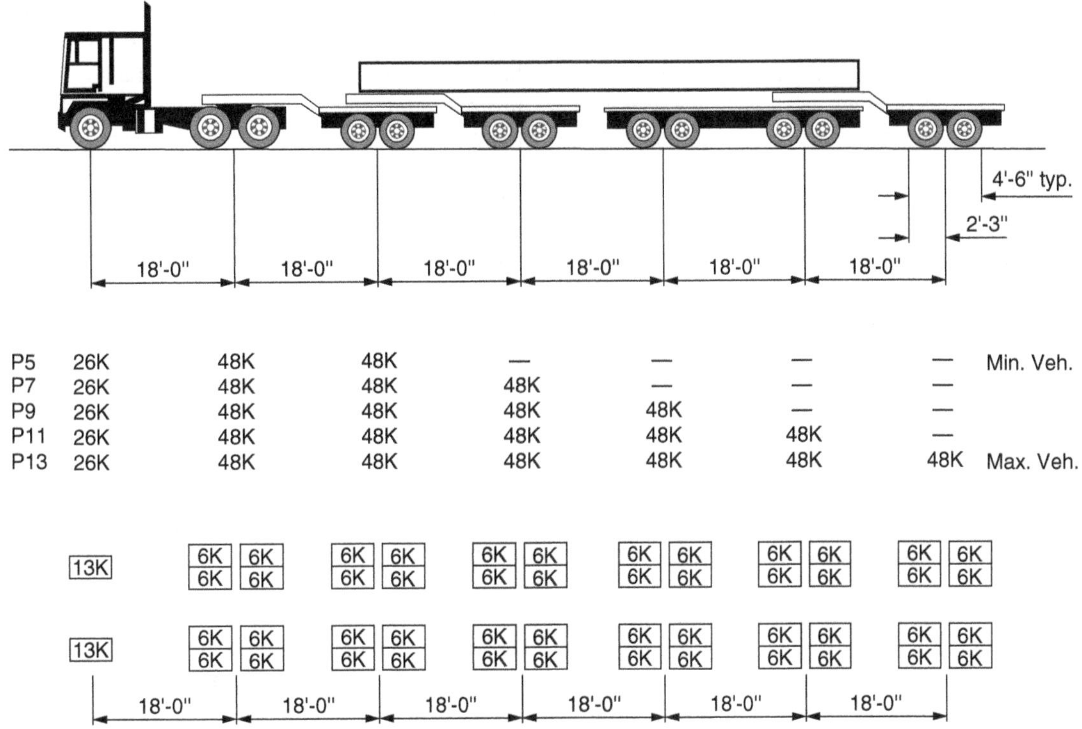

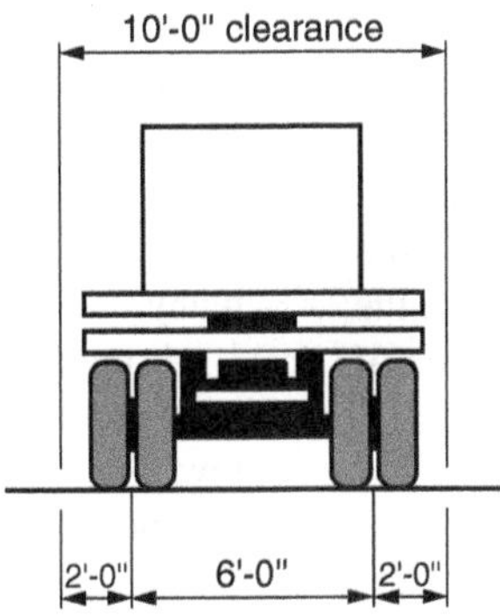

Fig. 8.19 Caltrans "purple loads".

point a with the energy associated with the stagnation point b, where the velocity is zero, yields

$$\frac{1}{2}\rho V_a^2 + p_a + \rho g h_a = \frac{1}{2}\rho V_b^2 + p_b + \rho g h_b \qquad (8.10)$$

Assuming that points a and b are at the same elevation, and that the reference upstream pressure at point a is zero, pressure at point b is

$$p_b = \frac{1}{2}\rho V_a^2 \qquad (8.11)$$

The stagnation pressure is the maximum inward pressure possible as all the upstream kinetic energy is transferred to potential energy associated with the pressure. Because every point on the surface is not at stagnation, that is, some velocity exists, the pressures at these points are less than the stagnation pressures. This effect is because the upstream energy is split between potential (pressure) and kinetic energies. The

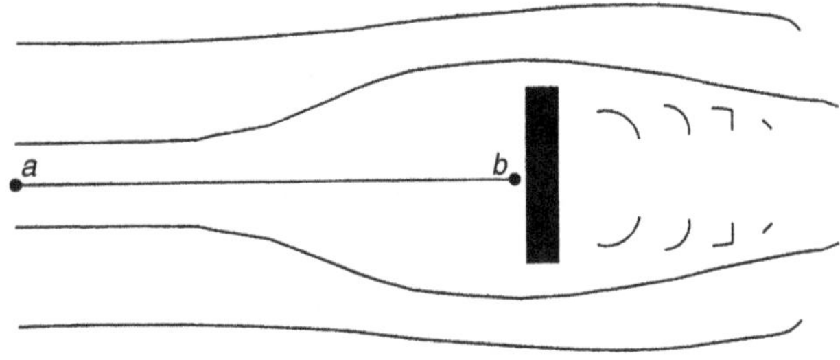

Fig. 8.20 Body in incompressible fluid.

total pressure is integrated over the surface area and is used to obtain the fluid force. It is conventional to determine the integrated effect (or force) empirically and to divide the force by the projected area. This quotient establishes the average pressure on an object, which is a fraction of the stagnation pressure. The ratio of the average pressure to the stagnation pressure is commonly called the drag coefficient C_d. The drag coefficient is a function of the object's shape and the characteristics of the fluid flow, typically Reynolds number. With the use of a known drag coefficient, the average pressure on an object may be calculated as

$$p = C_d \frac{1}{2} \rho V^2 \tag{8.12}$$

It is important to note that the fluid pressures and associated forces are proportional to the velocity squared. For example, a 25% increase in the fluid velocity creates approximately a 50% increase in fluid pressure and associated force.

Wind Forces The velocity of the wind varies with the elevation above the ground and the upstream terrain roughness, and therefore pressure on a structure is also a function of these parameters. Velocity increases with elevation, but at a decreasing rate. If the terrain is smooth, then the velocity increases more rapidly with elevation. A typical velocity profile is illustrated in Figure 8.21, where several key parameters are shown. The parameter V_g is the geotropic velocity or the velocity independent of surface (boundary) effects, δ is the boundary layer thickness, usually defined as the height where the velocity of 99% of V_g, and V_{30} is the reference velocity at 30 ft. Traditionally, this is the height at which wind velocity data is recorded. Since its introduction in 1916, the velocity profile has been modeled with a power function of the form

$$V_{\mathrm{DZ}} = C V_{30} \left(\frac{Z}{30} \right)^{\alpha}$$

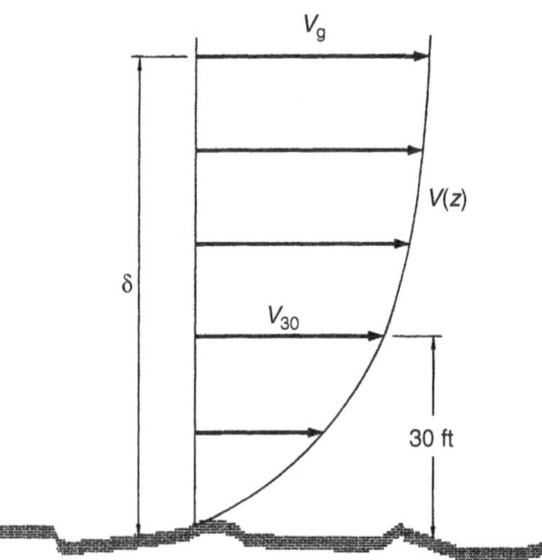

Fig. 8.21 Velocity profile.

where C and α are empirically determined constants. This model is used in many building codes. Critics of the power law point out that its exponent is not a constant for a given upstream roughness but varies with height, that the standard heights used to establish the model were somewhat subjective, and lastly, that the model is purely a best-fit function and has no theoretical basis (Simiu, 1973, 1976). More recently, meteorologists and wind engineers are modeling the wind in the boundary layer with a logarithmic function. This function is founded on boundary layer flow theory and better fits experimental results. The general form of the logarithmic velocity profile is

$$V(Z) = \frac{1}{\kappa} V_0 \ln \left(\frac{Z}{Z_0} \right) \tag{8.13}$$

where Z is the elevation above the ground, κ is von Karman's constant (~0.4), Z_0 is the friction length of the ground upstream, and

$$V_0 = \sqrt{\frac{\tau_0}{\rho}} \tag{8.14}$$

where τ_0 is the shear stress at the ground surface and ρ is the density of air. The parameter V_0 is termed the shear friction velocity because it is related to the shear force (friction), and Z_0 is related to the height of the terrain roughness upstream. As expected, these parameters are difficult to mathematically characterize, so empirical values are used. Note that for a given upstream roughness, two empirical constants Z_0 and V_0 are required in Eq. 8.13. Therefore, two measurements of velocity at different heights can be used to establish these constants. Simiu (1973, 1976) and Simiu and Scanlon (1978) report on these measurements done in the experiments of many investigators.

It is interesting that these constants are not independent and an expression can be formulated to relate them as shown below. The wind-generated shear stress at the surface of the ground is

$$\tau_0 = D_0 \rho V_{30}^2 \tag{8.15}$$

where D_0 is the surface drag coefficient and V_{30} is the wind speed at 30 ft (10 m) above the low ground or water level, miles per hour (mph).

With $Z = 30$ ft (10 m), Eq. 8.13 is used to solve for V_0:

$$V_0 = \kappa \frac{V_{30}}{\ln (30/Z_0)} \tag{8.16}$$

Equate the surface shear stress in Eqs. 8.14 and 8.15 to obtain

$$D_0 = \left(\frac{V_0}{V_{30}} \right)^2 \tag{8.17a}$$

or

$$V_0 = \sqrt{D_0} V_{30} \tag{8.17b}$$

Substitution of Eq. 8.16 into Eq. 8.17a yields

$$D_0 = \left[\frac{\kappa}{\ln (30/Z_0)} \right]^2 \tag{8.18}$$

Finally, substitute Eq. 8.18 into Eq. 8.17b to yield

$$V_0 = \left[\frac{\kappa}{\ln(30/Z_0)}\right] V_{30} \tag{8.19}$$

Equation 8.19 illustrates that for any reference velocity V_{30}, Z_0 and V_0 are related.

For unusual situations or for a more complete background, refer to Liu (1991). Liu outlines many issues in wind engineering in a format amenable to an engineer with a basic fluid mechanics background. Issues such as terrain roughness changes, local conditions, drag coefficients, and so on are discussed in a manner relevant to the bridge/structural engineer.

The equation for velocity profile used by AASHTO [A3.8.1.1] is

$$V_{DZ} = 2.5V_0 \left(\frac{V_{30}}{V_B}\right) \ln\left(\frac{Z}{Z_0}\right) \tag{8.20}$$

where V_{DZ} is the design wind speed at design elevation Z (mph) [same as $V(Z)$ in Eq. 8.13], V_B is the base wind velocity of 100 mph (160 km/h) yielding design pressures, V_0 is the "friction velocity," a meteorological wind characteristic taken as specified in Table 8.8 for upwind surface characteristics (mph), and Z_0 is the "friction length" of the upstream fetch, a meteorological wind characteristic taken as specified in Table 8.8 (ft).

The constant 2.5 is the inverse of the von Karman's constant 0.4. The ratio (V_{30}/V_B) is used to linearly proportion for a reference velocity other than 100 mph (160 km/h).

Equation 8.19 may be used to illustrate the relationship between V_0 and Z_0. For example, use the open-country exposure

$$V_0 = \frac{0.4}{\ln(10{,}000/70)}(160 \text{ km/h}) = 12.9 \text{ km/h}$$

$$V_0 = \frac{0.4}{\ln(30/0.23)}(100 \text{ mph}) = 8.2 \text{ mph}$$

which reasonably agrees with Table 8.8.

The velocity at 30 ft (V_{30}) or 10 m (V_{10}) may be established by fastest-mile-of-wind charts available in ASCE 7–02 for various recurrence intervals (ASCE, 2003), by site-specific investigations, or in lieu of a better criterion use 100 mph (160 km/h).

The wind pressure on the structure or component is established by scaling a basic wind pressure for V_B = 100 mph (160 km/h). This procedure is

$$P_D = P_B \left(\frac{V_{DZ}}{V_B}\right)^2 = P_B \frac{V_{DZ}^2}{10{,}000} \tag{8.21}$$

where the basic wind pressures are given in Table 8.9 [Table A3.8.1.2.1-1]. Table 8.9 includes the effect of gusts and the distribution of pressure on the surface (pressure coefficients). If we use Eq. 8.11 with a velocity of 100 mph (160 km/h) and a density of standard air 0.00194 slugs/ft^3 (1000 kg/m^3), we set a stagnation pressure of 25 psf (1226 Pa = 0.00123 MPa). Therefore, Table 8.9 includes a large increase of about 100% for gusts. A discussion with the code writers established that the pressures used in the previous AASHTO specifications were reasonable, seldom controlled the design of short- or medium-span bridges, and conservative values were used. So, depending on the assumed gust response and pressure coefficients, the design wind speed is likely above 100 mph (160 km/h).

Equation 8.21 uses the ratio of the design and base velocities squared because the pressure is proportional to the velocity squared. Additionally, the minimum wind loading shall not be less than 0.30 kip/ft (4.4 N/mm) in the plane of the windward chord and 0.15 kip/ft (2.2 N/mm) in the plane of the leeward chord on truss and arch components, and not less than 0.30 kip/ft (4.4 N/mm) on beam or girder spans [A3.8.1.2.1]. This wind load corresponds to the wind pressure on structure–load combination (WS) as given in Table 8.9. This wind should be considered from all directions and the extreme values are used for design. Directional adjustments are outlined in AASHTO [A3.8.1.2.2], where the pressure is separated into parallel and perpendicular component pressures as a function of the attack angle. The details are not elaborated here.

Finally, note that ASCE (2010) in its ASCE-07 document has significantly changed its approach to wind loads for the strength/extreme limit state. Here the wind speed has changed to the 3-s gust, different statistical measures are used, and the wind speed and associated pressures are much higher than the previous ASCE (2003) document. However, the wind load factor is smaller with a value of 1.0 to parallel seismic loads. AASHTO (2010) has not changed to parallel this recent work. Therefore, comparison of these approaches should be made with care.

The wind must also be considered on the vehicle (WL). This load is 0.10 kip/ft (1.46 N/mm) applied at 6 ft (1800 mm) above the roadway surface [A3.8.1.3].

For long-span structures, the possibility of aeroelastic instability exists. Here the wind causes a resonance situation with the structure, creating large deformations, actions, and possible failures. This phenomenon is best characterized by the famous Tacoma Narrows Bridge, which completely

Table 8.8 Values of V_0 and Z_0 for Various Upstream Surface Conditions

Condition	Open Country	Suburban	City
V_0, mph (km/h)	8.20 (13.2)	10.90 (17.6)	12.00 (19.3)
Z_0, ft (mm)	0.23 (70)	3.28 (1000)	8.20 (2500)

In AASHTO Table 3.8.1.1-1. From *AASHTO LRFD Bridge Design Specifications*. Copyright © 2010 by the American Association of State Highway and Transportation Officials, Washington, DC. Used by permission.

Table 8.9 Base Pressures P_B Corresponding to V_B = 100 mph (160 km/h)

Superstructure Component	Windward Load, ksf (MPa)	Leeward Load, ksf (MPa)
Trusses, columns, and arches	0.050 (0.0024)	0.025 (0.0012)
Beams	0.050 (0.0024)	N/A
Large flat surfaces	0.040 (0.0019)	N/A

In AASHTO Table 3.8.1.2.1-1. From *AASHTO Bridge Design Specifications*. Copyright © 2010 by the American Association of State Highway and Transportation Officials, Washington, DC. Used by permission.

collapsed due to aeroelastic effects. This collapse brought attention to this important design consideration that is typically a concern in the analysis of long-span bridges. Due to the complexities involved, aeroelastic instability is considered beyond the scope of the AASHTO specification and this book. Internet search for "Tacoma Narrows Bridge aeroelastic" gives several versions of video of the failure.

Water Forces Water flowing against and around the substructure creates a lateral force directly on the structure as well as debris that might accumulate under the bridge. Flood conditions are the most critical. As outlined above, the forces created are proportional to the square of velocity and to a drag coefficient. The use of Eq. 8.12 and the substitution of $\gamma = 0.062$ kip/ft^3($\rho = 1000$ kg/m^3) yields

$$p_b = \frac{1}{2}\frac{\gamma}{g}C_d V_a^2 = \frac{C_d V_a^2}{1038} \qquad (8.22\text{US})$$

$$p_b = \frac{1}{2}\rho C_d V_a^2 = 500 C_d V_a^2 \qquad (8.22\text{SI})$$

where the AASHTO equation [A3.7.3.1] is

$$p = \frac{1}{2}\frac{\gamma}{g}C_d V_d^2 = \frac{C_D V^2}{1000} \qquad (8.23\text{US})$$

$$p = 5.14 \times 10^{-4} C_D V^2 \qquad (8.23\text{SI})$$

Here C_D is the drag coefficient given in Table 8.10, and V is the design velocity of the water for the design flood in strength and service limit states, and for the check flood in the extreme event limit state [ft/s (m/s)]. Note that C_D is the specific AASHTO value and C_d is a generic term.

If the substructure is oriented at an angle to the stream flow, then adjustments must be made. These adjustments are outlined in AASHTO [A3.7.3.2]. Where debris deposition is likely, the bridge area profile should be adjusted accordingly. Some guidance on this is given in AASHTO [A3.7.3.1] and its associated references.

Although not a force, the scour of the stream bed around the foundation can result in structural failure. Scour is the movement of the stream bed from around the foundation, and this can significantly change the structural system, creating a situation that must be considered in the design [A3.7.5]. AASHTO [A2.6.4.4.2] outlines an extreme limit state for design. Because this issue is related to hydraulics, the substructure is not considered in detail here. However, it should be noted that more bridges collapse due to scour than any other cause. Assess scour with significant care and often the help of a bridge hydraulics expert.

Table 8.10 Drag Coefficient

Type	C_D
Semicircular nosed pier	0.7
Square-ended pier	1.4
Debris lodged against pier	1.4
Wedged-nosed pier with nose angle 90° or less	0.8

In AASHTO Table 3.7.3.1-1. From *AASHTO LRFD Bridge Design Specifications*. Copyright © 2010 by the American Association of State Highway and Transportation Officials, Washington, DC. Used by permission.

8.3.2 Seismic Loads

Depending on the location of the bridge site, the anticipated earthquake effects can be inconsequential or they can govern the design of the lateral load resistance system. The AASHTO Specifications have been developed to apply to all parts of the United States, so all bridges should be checked to determine if seismic loads are critical. In many cases the seismic loads are not critical and other lateral loads, such as wind, water, and/or ice govern the design.

The provisions of the AASHTO Specifications are based on the following principles [C3.10.1]:

- ☐ Small-to-moderate earthquakes should be resisted within the elastic range of the structural components without significant damage.
- ☐ Realistic seismic ground motion intensities and forces should be used in the design procedures.
- ☐ Exposure to shaking from large earthquakes should not cause collapse of all or part of the bridge. Where possible, damage should be readily detectable and accessible for inspection and repair.

The AASHTO provisions apply to bridges of conventional construction [A3.10.1]. Conventional bridges include those with slab, beam, box girder, or truss superstructures, and single- or multiple-column piers, wall-type piers, or pile-bent substructures. In addition, conventional bridges are founded on shallow or piled footings, or shafts. Nonconventional bridges include bridges with cable-stayed/cable-suspended superstructures, bridges with towers or hollow piers for substructures, and arch bridges [C3.10.1].

A discussion of the procedure used to determine when a bridge at a particular site requires a detailed seismic analysis is included in the next section. This section is followed by sections on minimum design forces and seismic load combinations.

Seismic Design Procedure The six steps in the seismic design procedure are outlined in this section. A flowchart summarizing the earthquake design provisions is presented in Appendix A to Section 3 of AASHTO (2010).

The first step is to arrive at a preliminary design describing the type of bridge, the number of spans, the height of the piers, a typical roadway cross section, horizontal alignment, type of foundations, and subsurface conditions. The nature of the connections between the spans of the superstructure, between the superstructure and the substructure, and between the substructure and the foundation are also important. For example, if a bridge superstructure has no deck joints and is integral with the abutments, its response during a seismic event is quite different from one with multiple expansion joints.

There are also innovative energy dissipating connections that can be placed below the superstructure at the abutments and pier caps to effectively isolate the superstructure from the effects of ground shaking. These devices can substantially reduce the magnitude of the inertial forces transmitted to a foundation component and can serve as a structural fuse that can be replaced or repaired if a larger earthquake occurs.

The second step is to determine the acceleration coefficients for the peak ground acceleration (PGA) (on rock), short period spectral acceleration S_s, and the one-second spectral acceleration S_1 [A3.10.2]. Contours of horizontal acceleration in rock expressed as a percent of gravity are illustrated on the maps of the United States provided in A3.10.2.1. Here there are many maps, so these are not included here. Alternatively, visit http://earthquake.usgs.gov/hazards/ where there are many resources associated with hazards including earthquakes. There is software that can be downloaded that provides the spectral acceleration values for several building codes and specifications.

At a given location, the acceleration coefficient from the map has a 93% probability of not being exceeded in 75 years. This value corresponds to a return period of about 1000 years for the design earthquake [C3.10.2.1].

Example 8.1 Determine the peak ground acceleration, short-period acceleration, and one-second accelerations for an essential bridge located in Blacksburg, Virginia (zip code 24060) Use the USGS Earthquake Ground Motion Parameters software and/or the maps in A3.10.2.1-1, -2, and -3. See discussion above. The bridge is founded on soft rock which implies site condition B.

PGA $= 8.3$, $S_s = 20.9$, $S_1 = 5.4$ as a percent of the acceleration of gravity. These values are illustrated in Figure 8.22. The horizontal axis is the natural period of the structure (or mode under study) in seconds.

The third step is to determine site coefficients F_{pga}, F_a, and F_v, which adjust the spectral accelerations for the geotechnical characteristics of the site. For example, a soft soil between bedrock and the structure can significantly amplify the accelerations and associated loads. A discussion of this topic is beyond the scope of this book. See A3.10.3 and Table 8.11.

The fourth step is to determine the operational category of a bridge [A3.10.5]. Following a seismic event, transportation routes to hospitals, police and fire departments, communication centers, temporary shelters and aid stations, power installations, water treatment plants, military installations, major airports, defense industries, refineries, and railroad and truck terminals must continue to function. Bridges on such routes should be classified as essential. In addition, a bridge that could collapse onto an essential route should also be classified as essential. Table 8.12 summarizes the characteristics of the three operational categories, one of which must be assigned to each bridge. Consideration should be given to possible future changes in the role of the bridge when assigning an operational category.

The fifth step is to determine the seismic performance zone for each bridge [A3.10.6]. These seismic zones group together regions of the United States that have similar seismic risk. The greater the acceleration coefficient, the greater is the risk. The seismic zones are given in Table 8.13, and the higher the number the greater are the seismic performance requirements for the bridge in regard to the method of analysis, the length of bridge seats, and the strength of connections.

The sixth step is to determine the response modification factors (R factors), which reduce the seismic force based on an elastic analysis of the bridge system [A3.10.7]. The force effects from an elastic analysis are to be divided by the response modification factors given in Table 8.14. The use of these R factors, generally greater than 1, recognizes that when a design seismic event (1000-year return period) occurs, energy is dissipated through inelastic deformation (hinging) in the substructure. This energy dissipation actually protects the structure from large shocks and allows it to be designed for reduced forces.

In the event a large earthquake (2500-year return period) should occur, the hinging regions may have to be repaired, but, if all of the components are properly tied together, collapse does not occur. To ensure that proper attention is given to the transfer of internal actions from one component to another, the R factors for connections given in Table 8.15 do not reduce, and in some cases amplify, the force effects from an elastic analysis.

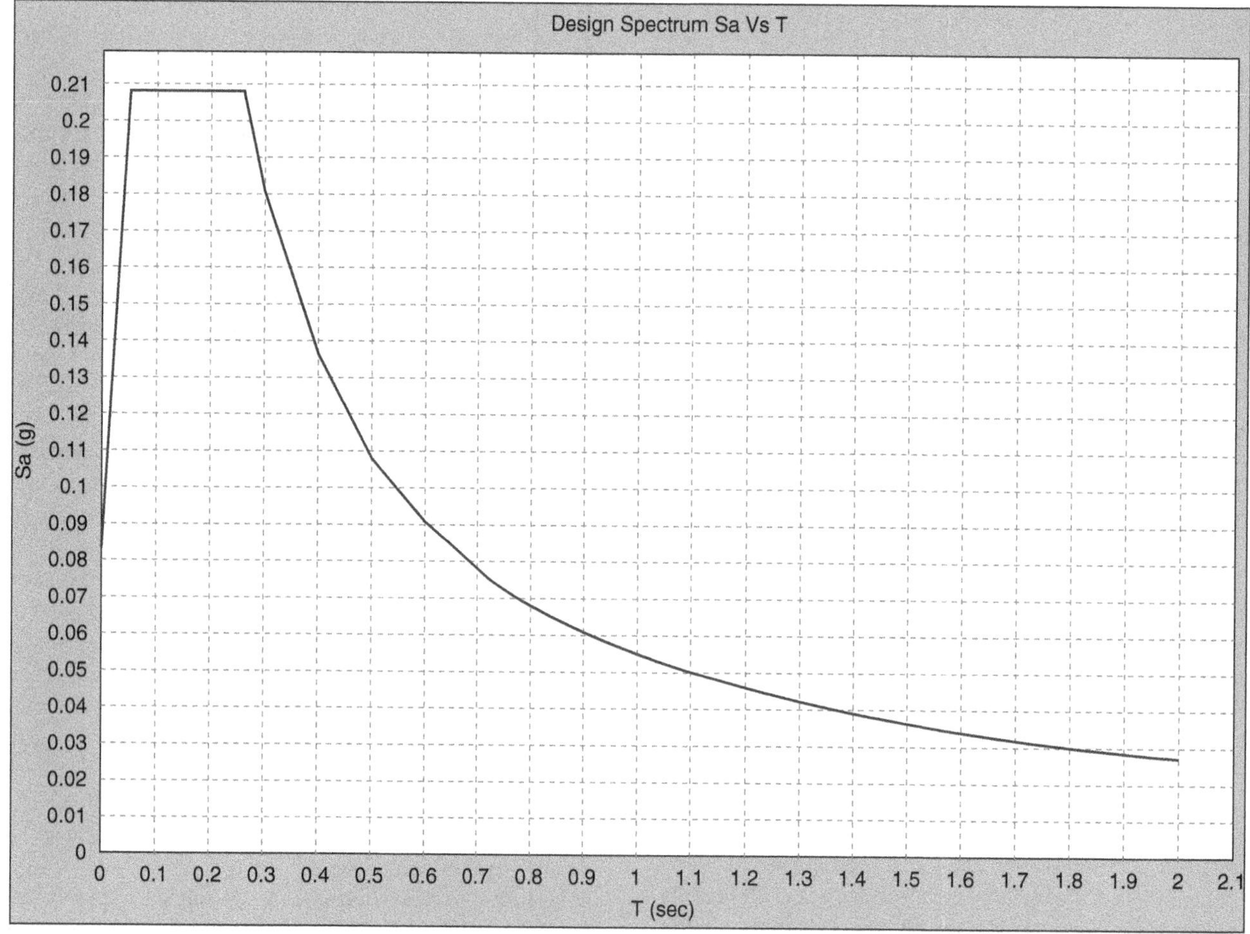

Fig. 8.22 Example plot from the USGS Ground Motion Software (http://earthquake.usgs.gov/hazards/).

Table 8.11 Soil Profiles

Type	Description
A	Hard rock (F_{pga}, F_a, *and* F_v are less than one, 0.8 is typical)
B	Rock (the base case upon which the spectral acceleration require no site adjustment; F_{pga}, F_a, *and* F_v are equal to one)
C	Dense soil and rock (F_{pga}, F_a, *and* F_v are greater than one)
D	Stiff soil
E	10 ft or more of soft clay
F	Very loose soil (peat, highly plastic, etc.) These require a detailed site investigation.

Based upon AASHTO Table 3.10.3.1-1. From *AASHTO LRFD Bridge Design Specifications*. Copyright © 2010 by the American Association of State Highway and Transportation Officials, Washington, DC. Used by permission. (The AASHTO table provides more precise descriptions.)

Based on the information obtained by completing the above steps, decisions can be made regarding the level of seismic analysis required, the design forces, and the design displacement requirements. For example, single-span bridges and bridges in seismic zone 1 do not have to be analyzed for seismic loads, while critical bridges in seismic zone 4 require a rigorous method of seismic analysis [A4.7.4].

Minimum Seismic Design Connection Forces When ground shaking due to an earthquake occurs and a bridge superstructure is set in motion, inertial forces equal to the mass times the acceleration are developed. These forces can be in any direction and must be restrained, or dissipated, at the connection between the superstructure and substructure. For a single-span bridge, the minimum design connection force in the restrained direction is to be taken as the product

Table 8.12 Operational Classification [A3.10.5, C3.10.5]

Operational Category	Description
Critical bridges	Must remain open to all traffic after the design earthquake (1000-year return period event) and open to emergency vehicles after a large earthquake (2500-year return period event).
Essential bridges	Must be open to emergency vehicles after the design earthquake.
Other bridges	May be closed for repair after a large earthquake.

Table 8.13 Seismic Performance Zones

Acceleration Coefficient	Seismic Zone
$S_{DI} \leq 0.15$	1
$0.15 < S_{DI} \leq 0.30$	2
$0.30 < S_{DI} \leq 0.50$	3
$0.50 < S_{DI}$	4

In AASHTO Table 3.10.6-1. From *AASHTO LRFD Bridge Design Specifications*. Copyright © 2010 by the American Association of State Highway and Transportation Officials, Washington, DC. Used by permission. $S_{DI} = F_v S_I$ *where* F_v *is based upon the soil as described above*.

of the acceleration coefficient and the tributary dead load associated with that connection.

Bridges in seismic zone 1 do not require a rigorous seismic analysis, and therefore nominal values are specified for the connection forces. To obtain the horizontal seismic forces in a restrained direction, the tributary dead load is multiplied by the value given in Table 8.16 [A3.10.9.2]. The tributary dead load to be used when calculating the longitudinal connection force at a fixed bearing of a continuous segment or simply supported span is the total dead load of the segment. If each bearing in a segment restrains translation in the transverse direction, the tributary dead load to be used in calculating the transverse connection force is the dead-load reaction at the bearing. If each bearing supporting a segment is an elastomeric bearing, which offers little or no restraint, the connection is to be designed to resist the seismic shear forces transmitted through the bearing, but not less than the values represented by the multipliers in Table 8.16. A_s is the ground acceleration PGA adjusted by the site factor to account for the geotechnical characteristics. Therefore, $A_s = F_{\text{pga}}$ PGA.

Example 8.2 To continue the Example of 8.1, determine the minimum longitudinal and transverse connection forces for a simply supported bridge span of 70 ft (21 m) in Blacksburg, Virginia (zone 1), with a dead load of 8 kips/ft (115 kN/m) founded on site factor B. Assume that in the longitudinal direction, one connection is free to move while the other is fixed and that in the transverse direction both connections to the abutment are restrained. Use A3.10.9.2.

SOLUTION

$A_s = F_{\text{pga}}\text{PGA} = 1.0\,(0.083) = 0.083 > 0.05$

Total dead load $W_D = 8\,(70) = 560$ kips

Connection force $F_C = ma = \left(W_D/g\right)\left(A_S \times g\right) = W_D A_S$

Longitudinal (min) $F_{\text{CL}} = (560)\,(0.083)$

$= 46$ kips does not control

Table 8.16 $F_{\text{CL}} = 560\,(0.25)$

$= 140$ kips at fixed end (controls)

Transverse $F_{\text{CT}} = \frac{1}{2} F_{\text{CL}} = 70$ kips per abutment

Note that if the nominal approach requires forces that are considered too conservative, the engineer can perform a more rigorous analysis.

Connections for bridges in seismic zone 2 are to be designed for the reaction forces determined by a single-mode elastic spectral analysis divided by the appropriate R factor of Table 8.15. Connection forces for bridges in seismic zones 3 and 4 can be determined by either a multimode elastic spectral analysis divided by R or by an inelastic step-by-step time history analysis with $R = 1.0$ for all connections. The use of $R = 1.0$ assumes that the inelastic method properly models the material hysteretic properties and the accompanying energy dissipation.

For further information for seismic analysis of bridges, the reader is referred to the *AASHTO Guide Specifications for LRFD Seismic Bridge Design*, AASHTO (2009b). This guide specification references a host of other works as well.

Combination of Seismic Forces Because of the directional uncertainty of earthquake motions, two load cases combining elastic member forces resulting from earthquakes in two perpendicular horizontal directions must be considered. The two perpendicular directions are usually the longitudinal and transverse axes of the bridge. For a curved bridge, the longitudinal axis is often taken as the line joining the two abutments. The two load cases are expressed as [A3.10.8]:

$$\text{Load case 1} \quad 1.0F_L + 0.3F_T \tag{8.24a}$$

$$\text{Load case 2} \quad 0.3F_L + 1.0F_T \tag{8.24b}$$

Table 8.14 Response Modification Factors—Substructures

Substructure	Operational Category		
	Other	Essential	Critical
Wall-type piers—larger dimension	2.0	1.5	1.5
Reinforced concrete pile bents			
(a) Vertical piles only	3.0	2.0	1.5
(b) One or more batter piles	2.0	1.5	1.5
Single columns	3.0	2.0	1.5
Steel or composite steel and concrete pile bents			
(a) Vertical piles only	5.0	3.5	1.5
(b) One or more batter piles	3.0	2.0	1.5
Multiple column bents	5.0	3.5	1.5

In AASHTO Table 3.10.7.1-1. From *AASHTO LRFD Bridge Design Specifications*. Copyright © 2010 by the American Association of State Highway and Transportation Officials, Washington, DC. Used by permission.

Table 8.15 Response Modification Factors—Connections

Connection	All Operational Categories
Superstructure to abutment	0.8
Expansion joints within a span of the superstructure	0.8
Columns, piers, or pile bents to cap beam or superstructure	1.0
Columns or piers to foundations	1.0

In AASHTO Table 3.10.7.1-2. From *AASHTO LRFD Bridge Design Specifications*. Copyright © 2010 by the American Association of State Highway and Transportation Officials, Washington, DC. Used by permission.

Table 8.16 Multiplier for Connection Force in Seismic Zone 1 [A3.10.9.2]

Acceleration Coefficient	Multiplier
$A_s \le 0.05$	0.15
$0.05 < A_s$	0.25

where

F_L = elastic member forces due to an earthquake in the direction of the longitudinal axis of the bridge

F_T = elastic member forces due to an earthquake in the direction of the transverse axis of the bridge

8.3.3 Ice Forces

Forces produced by ice must be considered when a structural component of a bridge, such as a pier or bent, is located in water and the climate is cold enough to cause the water to freeze. The usual sequence is that freeze-up occurs in late fall, the ice grows thicker in the winter, and the ice breaks up in the spring. If the bridge is crossing a lake, reservoir, harbor, or other relatively quite body of water, the ice forces are generally static. These static forces can be horizontal when caused by thermal expansion and contraction or vertical if the body of water is subject to changes in water level. If the bridge is crossing a river with flowing water, the static forces exist throughout the winter months, but when the spring breakup occurs, larger dynamic forces are produced by floating sheets of ice impacting the bridge structure.

Effective Strength of Ice Because the strength of ice is less than that of the steel and concrete used in the construction of bridge piers, the static and dynamic ice forces on bridge piers are limited by the effective strength of the ice: the static thermal forces by the crushing strength and the dynamic forces by either the crushing or the flexural strength. The strength of the ice depends on the conditions that exist at the time it is formed, at the time it is growing in thickness, and at the time it begins to melt and break up. If the ice is formed when the surface is agitated and freezes quickly, air entraps within the structure of the ice and gives it a cloudy or milky appearance. This ice is not as strong as that that is formed gradually and grows over a long period of time to be very solid and clear in appearance. The conditions during the winter months, when this ice is increasing in thickness, affects the strength of the ice. If snow cover is present and melts during a warming period and then freezes, weaker granular snow ice is formed. In fact, sections cut through ice sheets show varying layers of clear ice, cloudy ice, and snow ice. This ambiguity makes classification difficult. The conditions at the time of spring breakup also affect the strength. If the temperature throughout the thickness sheet is at the melting temperature when the ice breaks up, it has less strength than when the average ice temperature is below the melting temperature.

Table 8.17 Effective Ice Crushing Strength at Breakup

Average Ice Temperature	Condition of Ice	Effective Strength
At melting point	Substantially disintegrated	8.0 ksf (0.38 MPa)
	Somewhat disintegrated	16.0 ksf (0.77 MPa)
	Large pieces, internally sound	24.0 ksf (1.15 MPa)
Below melting point	Large pieces, internally sound	32.0 ksf (1.53 MPa)

An indication of the variation in crushing strength of ice at the time of breakup is given in AASHTO [A3.9.2.1] as shown in Table 8.17. These values are to be used in a semiempirical formula, discussed later, for determining dynamic ice forces on bridge piers.

Field Measurement of Ice Forces Haynes et al. (1991) have measured forces exerted by moving ice on a bridge pier in the St. Regis River in upstate New York. Other researchers who have measured ice forces in Canada, Alaska, and Vermont are listed in their report. The purpose of these studies is to provide data that can be used to calibrate design codes for changing local conditions.

In the Haynes et al. study, a steel panel was instrumented and placed on the upstream nose of a pier (see Fig. 8.23). The panel pivots about its base and a load cell measures a reactive force when the panel is struck by moving ice. Whenever the signal from the load cell gets above a preset threshold level, the load cell force data along with the pressure transducer reading that determines the water depth are recorded. The ice force that produced the force in the load cell is then determined by balancing moments about the pin location.

In March 1990, a major ice run took place. Ice thickness was estimated to be about 6–8 in. (152–203 mm) (nonuniform flow causes variations in ice cover thicknesses for most rivers). Plots of the ice force versus time for two of the largest ice force events during this run are shown in Figure 8.24. For the ice force record shown in Figure 8.24(a), the ice–structure interaction event lasted about 2.3 s and is believed to represent crushing failure of the ice because the force record has many oscillations without the force dropping to zero. The rapid increase and decrease of ice force shown in Figure 8.24(b) indicates an impact and possible rotation or splitting of the ice floe without much crushing. This impact event lasted only about 0.32 s and produced the maximum measured ice force of nearly 80,000 lb (356 kN). The largest ice force produced by the crushing failure of the ice was about 45,000 lb (200 kN).

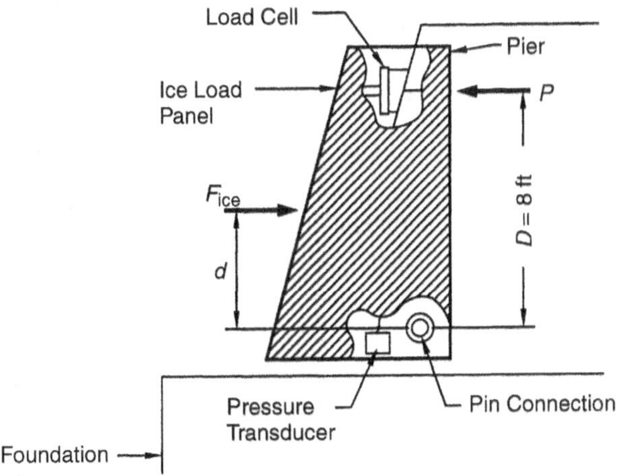

Fig. 8.23 Ice load panel on pier of the St. Regis River Bridge. (From Haynes et al., 1991.)

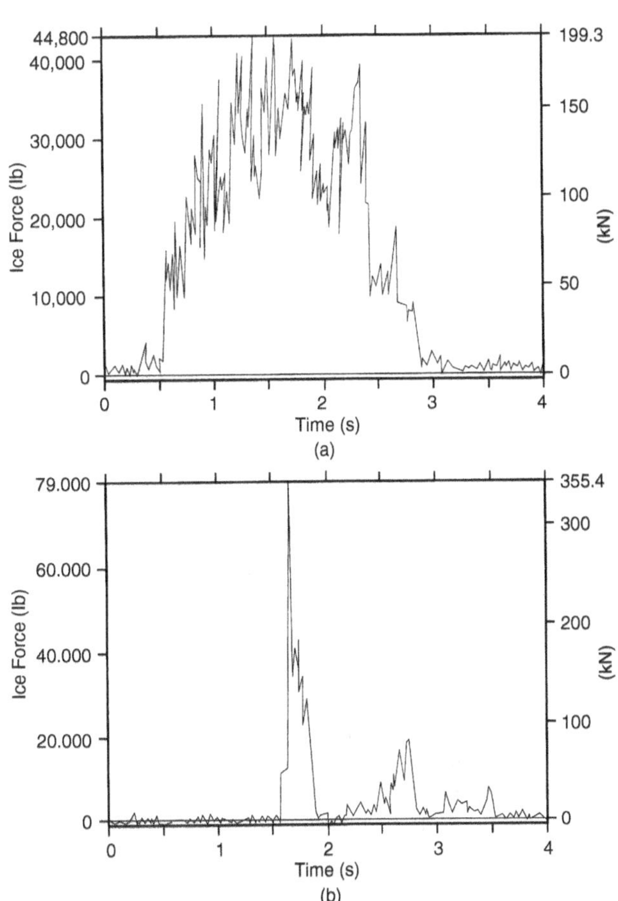

Fig. 8.24 Records of ice force versus time on March 16–17, 1990: (a) ice failure by crushing and (b) ice impact without much crushing. (From Haynes et al., 1991.)

One observation from these field measurements is the wide variation in ice forces against a pier produced in the same ice run by ice floes that were formed and broken up under similar conditions. Some of the ice sheets, probably the larger ones, were indented when they collided with the pier and failed by crushing. Other ice floes smaller in size and probably of solid competent ice banged into the pier with a larger force and then rotated and were washed past the pier. In light of this observation, it appears prudent to use only the last two categories for effective ice strength of 24 ksf (1.15 MPa) and 32 ksf (1.53 MPa) in Table 8.17, unless there is long experience with local conditions that indicate that ice forces are minimal.

Thickness of Ice The formulas used to predict horizontal ice forces are directly proportional to the effective ice strength and to the ice thickness. The thicker the ice, the larger is the ice force. Therefore, the thickness of ice selected by a designer is important, and at the same time it is the parameter with the most uncertainty. It is usually thicker at the piers where cracking, flooding, freezing, and rafting (where one ice sheet gets under another) have occurred. It is usually thinner away from the pier where the water is flowing free. Ice not only grows down into the water but also thickens on the top. Ice can thicken quite rapidly in cold weather but can also be effectively insulated by a covering of snow. On some occasions the ice can melt out in midwinter and freezeup has to begin again. And even if ice thickness has been measured over a number of years at the bridge site, this may not be the ice that strikes the bridge. It could come from as far away as 200 miles (320 km) upstream.

Probably the best way to determine ice thickness at a bridge site is to search the historical record for factual information on measured ice thickness and to talk to local people who have seen more than one spring breakup. These can be long-time residents, town or city officials, newspaper editors, state highway engineers, and representatives of government agencies. A visit to the bridge site is imperative because the locals can provide information on the thickness of ice and can also indicate what the elevation of the water level is at spring breakup.

If historical data on ice thickness is not available, a mathematical model based on how cold a region is can serve as a starting point for estimating thickness of ice. The following discussion is taken from Wortley (1984).

The measure of "coldness" is the freezing degree-day (FDD), which is defined as the departure of the daily mean temperature from the freezing temperature. For example, if the daily high was 20°F (−6.7°C) and the low was 10°F (−12°C), the daily average would be 15°F (−9.4°C), which is 17°F (9.4°C) departure from the freezing temperature. The FDD would therefore be 17°F (9.4°C). A running sum of FDDs (denoted by S_f) is a cumulative measure of winter's coldness. If this sum becomes negative due to warm weather, a new sum is started on the next freezing day.

An 80-year record of values of S_f at various sites around the Great Lakes accumulated on a daily and weekly basis

Table 8.18 Eighty-Year Mean and Extreme Freezing Degree Days (°F)

Great Lake	Station	Mean	Extreme
Lake Superior	Thunder Bay, Ontario	2500	3300
	Houghton, Michigan	1650	2400
	Duluth, Minnesota	2250	3050
Lake Michigan	Escanaba, Michigan	1400	2400
	Green Bay, Wisconsin	1350	2300
	Chicago, Illinois	500	1400
Lake Huron	Parry Sound, Ontario	1500	2550
	Alpena, Michigan	1150	2000
	Port Huron, Michigan	600	1550
Lake Erie	Detroit, Michigan	500	1350
	Buffalo, New York	500	1200
	Erie, Pennsylvania	400	1100
	Cleveland, Ohio	300	1200
Lake Ontario	Kingston, Ontario	1150	2000
	Toronto, Ontario	600	1500
	Rochester, New York	600	1300

After Assel (1980).

is given in Table 8.18. The daily basis is termed the mean S_f and weekly basis is termed the extreme S_f. The extreme sum is computed by accumulating the coldest weeks over the 80-year period.

Figure 8.25 is a map of the United States developed by Haugen (1993) from National Weather Service data covering the 30-year period from 1951 to 1980 giving contours of extreme freezing degree days in degrees Celsius (°C). For example, at Chicago, the map contour gives 700°C (1292°F). This 30-year extreme is slightly less than the 80-year extreme value of 1400°F (760°C) given in Table 8.18.

Observations have shown that the growth of ice thickness is proportional to the square root of S_f. Neill (1981) suggests the following empirical equation for estimating ice thickness:

$$t = 0.083\alpha_t\sqrt{S_f\ (^\circ F)}\ (\text{ft}) \tag{8.25US}$$

$$t = 33.9\alpha_t\sqrt{S_f\ (^\circ C)}\ (\text{mm}) \tag{8.25SI}$$

where α_t is the coefficient for local conditions from Table 8.19 [C3.9.2.2], and S_f is the sum of freezing degree days (°F or °C).

Example 8.3 Use the map of Figure 8.25 and Eq. 8.25 to estimate the maximum thickness of ice on the St. Regis River, which flows into the St. Lawrence River in northern New York, assuming it is an average river with snow. The sum of freezing degree days is 2000°F (1100°C), per Figure 8.25.

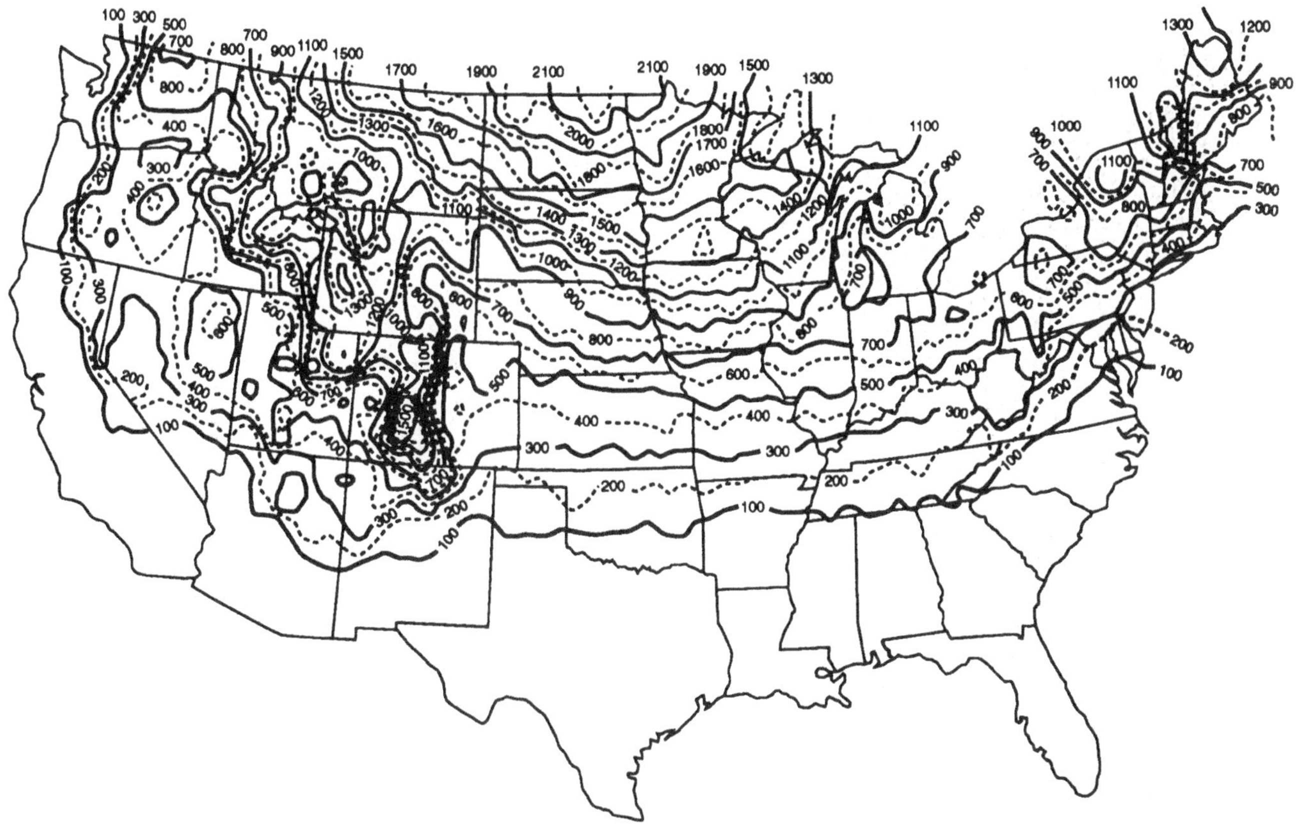

Fig. 8.25 Maximum sum of freezing degree days (FDD) in degrees Celsius (°C). (From Haugen, 1993.)

Table 8.19 Locality Factors for Estimating Ice Thickness

Local Conditions	α_t
Windy lakes with no snow	0.8
Average lake with snow	0.5–0.7
Average river with snow	0.4–0.5
Sheltered small river with snow	0.2–0.4

From Neill (1981).

Taking $\alpha_t = 0.5$, Eq. 8.25 yields

$$t = 0.083\alpha_t\sqrt{S_f} = 0.083 \times 0.5\sqrt{2000} = 1.9\,\text{ft}$$

$$t = 33.9\alpha_t\sqrt{S_f} = 33.9 \times 0.5\sqrt{1100} = 560\,\text{mm}$$

In January 1990, the ice thickness was measured to be 1.17–2.17 ft (358–660 mm) near the bridge piers (Haynes et al., 1991). The calculated value compares favorably with the measured ice thickness. As a matter of interest, the designers of the bridge at this site selected an ice thickness of 3.0 ft (914 mm).

Dynamic Horizontal Ice Forces When moving ice strikes a pier, the usual assumption is that the ice fails in crushing and the horizontal force on the pier is proportional to the width of the contact area, the ice thickness, and the effective compressive strength of the ice. During impact, the width of the contact area may increase from zero to the full width of the pier as the relative velocity of the ice floe with respect to the pier decreases. By equating the change in kinetic energy of a moving ice floe to the work done in crushing the ice, the critical velocity of the ice floe can be determined (Gershunov, 1986). The critical velocity is the velocity required to achieve full indentation of the structure into the ice. If the velocity of the ice floe is greater than the critical velocity, the ice floe continues to move and crush the ice on the full contact area.

The expressions for dynamic horizontal ice forces in AASHTO [A3.9.2.2] are independent of the velocity of the ice, which implies that the velocity of the approaching ice floe is assumed to be greater than the critical velocity. If $w/t > 6.0$, then the horizontal force F, kip (N), due to moving ice is governed by crushing over the full width of the pier and is given by

$$F = F_c = C_a p t w \tag{8.26}$$

for which

$$C_a = \left(\frac{5t}{w+1}\right)^{0.5} \tag{8.27}$$

Where

p = effective ice crushing strength from Table 8.18, ksf (MPa)

t = thickness of ice, ft (mm)

w = pier width at level of ice action, ft (mm)

When the pier nose is inclined at an angle greater than 15° from the vertical, an ice floe can ride up the inclined nose and fail in bending. If $w/t \le 6$, the horizontal ice force F, kip (N), is taken as the lesser of the crushing force F_C from Eq. 8.26 or the bending failure force F_b given by

$$F = F_b = C_n p t^2 \tag{8.28}$$

for which

$$C_n = \frac{0.5}{\tan(\alpha - 15)} \tag{8.29}$$

where α is the inclination of the pier nose from the vertical, degrees, but not less than 15°.

Example 8.4 Calculate the dynamic horizontal ice force predicted for the St. Regis River bridge pier at a water level where the pier width is 4 ft (1220 mm). The pier nose is inclined only 5.7° from the vertical, so the failure will be by crushing and Eq. 8.26 controls. Use an effective ice strength of 24 psf (1.150 MPa) and the ice thickness of 8 in. (203 mm) observed on March 16–17, 1990.

$$C_a = \left(\frac{5t}{w} + 1\right)^{0.5} = \left(\frac{5 \times 0.66}{4.0} + 1\right)^{0.5} \quad \text{or}$$

$$\times \left(\frac{5 \times 203}{1220} + 1\right)^{0.5} = 1.35$$

$$F = F_c = C_a p t w = 1.35(24\ \text{ksf})(0.66\ \text{ft})(4\ \text{ft}) = 86\ \text{kips}$$

$$F = F_c = C_a p t w = 1.35(1150\ \text{kPa})(203\ \text{mm})(1220\ \text{mm})) = 385\text{kN}$$

The maximum ice force measured during the ice run of March 16–17, 1990, was 79.9 kips (355 kN) (Haynes et al., 1991), which is comparable to the predicted value.

The above ice forces are assumed to act parallel to the longitudinal axis of the pier. When an ice floe strikes the pier at an angle, transverse forces are also developed. The magnitude of the transverse force F_t depends on the nose angle β of the pier and is given by [A3.9.2.4.1]

$$F_t = \frac{F}{2\tan[(\beta/2) + \theta_f]} \tag{8.30}$$

Where

F = horizontal ice force calculated by Eq. 8.26 or Eq. 8.28

β = angle, degrees, in a horizontal plane included between the sides of a pointed pier as shown in Figure 8.26 (for a flat nose, β is 0°;
for a round nose, β may be taken as 100°).

θ_f = friction angle between ice and pier nose, degrees

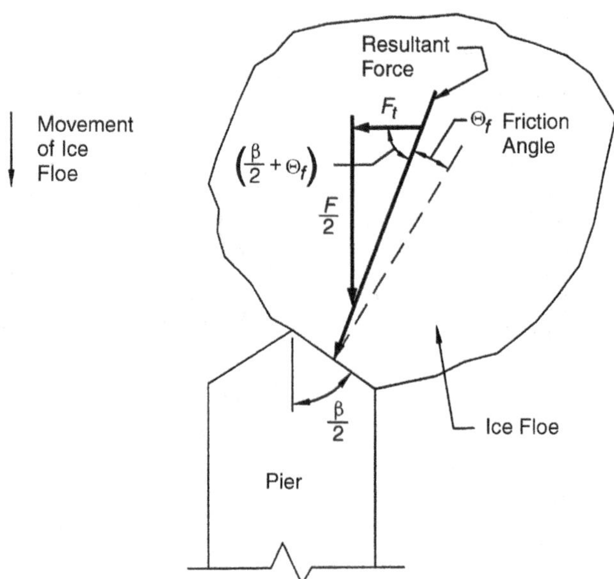

Fig. 8.26 Transverse ice force when a floe fails over a portion of a pier (AASHTO Fig. C3.9.2.4.1-1). (From *AASHTO LRFD Bridge Design Specifications*, Copyright © 2010 by the American Association of State Highway and Transportation Officials, Washington, DC. Used by permission.)

Example 8.5 Determine the transverse ice force corresponding to the dynamic horizontal ice force of Example 8.4 if the St. Regis River bridge pier has a pointed nose with an included angle of 90° and the friction angle is 10°.

$$F_t = \frac{F}{2\tan[(\beta/2) + \theta_f]} = \frac{86}{2\tan[(90/2) + 10]} = 31\ \text{kips}$$

$$= \frac{F}{2\tan\left[(\beta/2) + \theta_f\right]} = \frac{385}{2\tan[(90/2) + 10]} = 135\ \text{kN}$$

The longitudinal and transverse ice forces are assumed to act on the nose of the pier. When the ice movement is generally parallel to the longitudinal axis of the pier, two load combination cases need to be investigated [A3.9.2.4.1]:

- ☐ A longitudinal force of F shall be combined with a transverse force of $0.15F_t$.
- ☐ A longitudinal force of $0.5F$ shall be combined with a transverse force of F_t.

If the longitudinal axis of the pier is skewed with respect to the flow, the total force on the pier is calculated on the basis of the projected pier width and resolved into components.

In regions where ice forces are significant, slender and flexible piers are not recommended. Ice–structure interaction can lead to amplification of the ice forces if the piers or pier components, including piles, are flexible.

Static Horizontal Ice Forces When ice covers move slowly, the inertia can be neglected, and the ice forces can be considered static. When ice is strained slowly, it behaves in a ductile manner that tends to limit pressure. Additionally, ice creeps over time, which also decreases forces. The largest static ice forces are of thermal origin and occur when there is open water on one side of a structure and ice on the other.

Predictions of thermal ice pressures are difficult because they depend on the rate of change of temperature in the ice, the coefficient of thermal expansion ~0.000030/°F (0.000054/°C), the rheology of ice, the extent to which cracks have been filled with water, the thickness of the ice cover, and the degree of restrictions from the shores (Wortley, 1984). If thermal thrusts are calculated assuming the ice fails by crushing and using the strength values of Table 8.17, which neglect creep, the lateral loads determined will be too high.

Based on observations of ice in the Great Lakes and on stability calculations for rock-filled crib gravity structures, Wortley (1984) believes that reasonable ice thermal thrust values for this region are 5–10 kips/ft (73–146 kN/mm). If biaxial restraint conditions exist, such as in a harbor basin with a sheet piling bulkhead on most of its perimeter, the thermal thrusts can be doubled to 10–20 kips/ft (146–292 kN/mm).

Vertical Ice Forces Changes in water level cause the ice sheet to move up and down, and vertical loads result from the ice adhering to the structure. The vertical force on an embedded pile or pier is limited by the adhesive strength between the ice and the structure surface, by the shear strength of the ice, or by bending failure of the ice sheet some distance from the structure (Neill, 1981). Assuming there is no slippage at the ice–structure interface and no shear failure, a bending failure of the ice sheet will occur. If the pier is circular, this bending failure leaves a collar of ice firmly attached to the pier and a set of radial cracks in the floating ice sheet. When there is an abrupt water level fluctuation, the ice sheet will bend until the first circumferential crack occurs and a failure mechanism is formed. If the water level beneath ice sheet drops, the ice becomes a hanging dead weight [ice weighs 57 lb/ft^3 (pcf) (9.0 kN/m^3)]. If the water level rises, a lifting force is transmitted to the pier or piling that could offset the dead load of a light structure.

The AASHTO Specifications give the following expressions for the maximum vertical force F_v on a bridge pier [A3.9.5]:

☐ For a circular pier, in kips

$$F_v = 80.0t^2\left(0.35 + \frac{0.03R}{t^{0.75}}\right) \qquad (8.31)$$

☐ For a oblong pier, in kips

$$F_v = 0.2t^{1.25}L + 80.0t^2\left(0.35 + \frac{0.03R}{t^{0.75}}\right) \qquad (8.32)$$

Where

t = ice thickness, ft

R = radius of circular pier, ft

L = perimeter of pier, excluding half circles at ends of oblong pier, ft

Example 8.6 Calculate the vertical ice force on a 6-ft-diameter circular pier of a bridge crossing a reservoir that is subject to sudden changes in water level. Assume the ice is 2 ft thick:

$$F_v = 80(2)^2\left[0.35 + \frac{0.03\,(3)}{2^{0.75}}\right] = 129 \text{ kips}$$

Snow Loads on Superstructure Generally, snow loads are not considered on a bridge, except in areas of extremely heavy snowfall. In areas of significant snowfall, where snow removal is not possible, the accumulated snow loads may exceed the vehicle live loads. In some mountainous regions, snow loads up to 0.700 ksf (33.5 kPa) may be encountered. In these areas, building code roof loads, historical records, and local experience should be used to determine the magnitude of the snow loads. Finally, pedestrian bridges are sometimes used on a seasonal basis and will collect snow. In such cases, a portion of live and full snow might be appropriate.

8.4 FORCES DUE TO DEFORMATIONS

8.4.1 Temperature

Two types of temperature changes must be included in the analysis of the superstructure (see [A3.12.2] and [A3.12.3]). The first is a uniform temperature change where the entire superstructure changes temperature by a constant amount. This type of change lengthens or shortens the bridge, or if the supports are constrained it will induce reactions at the bearings and forces in the structure. This type of deformation is illustrated in Figure 8.27(a). The second type of temperature change is a gradient or nonuniform heating (or cooling) of the superstructure across its depth [see Fig. 8.27(b)]. Subjected to sunshine, the bridge deck heats more than the girders below. This nonuniform heating causes the temperature to increase more in the top portion of the system than in the bottom and the girder attempts to bow upward. If restrained by internal supports or by unintentional end restraints, compatibility actions are induced. If completely unrestrained, due to the piecewise linear nature of the imposed temperature distribution, internal stresses are introduced in the girder. In short, a statically determinate beam has internal stress due to the piecewise linear temperature gradient (even for a simply supported girder). This effect is discussed further in Chapter 12.

As expected, the temperature range is considered a function of climate. Here AASHTO defines two climatic conditions: moderate and cold. A moderate climate is when the

Table 8.20 Temperature Ranges

Climate	Steel or Aluminum °F (°C)	Concrete °F (°C)	Wood °F (°C)
Moderate	0–120 (−18–50)	10–80 (−12–27)	10–75 (−12–24)
Cold	−30–120 (−35–50)	0–80 (−18–27)	0–75 (−18–24)

In AASHTO Table 3.12.2.1.1-1. From *AASHTO LRFD Bridge Design Specifications*. Copyright © 2010 by the American Association of State Highway and Transportation Officials, Washington, DC. Used by permission.

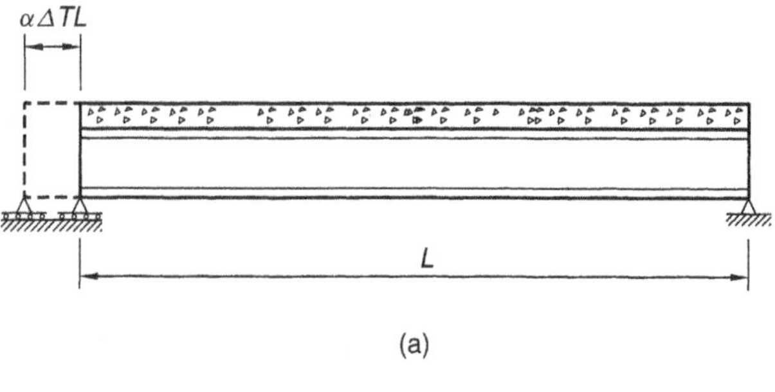

(a)

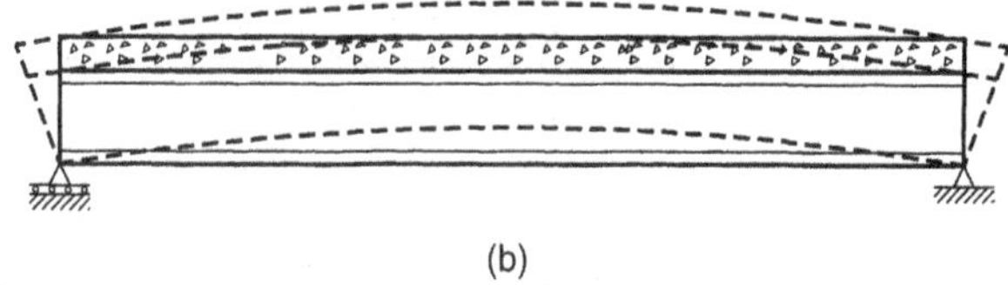

(b)

Fig. 8.27 (a) Temperature-induced elongation and (b) temperature-induced curvature.

number of freezing days per year is less than 14. A freezing day is when the average temperature is less than 32°F (0°C). Table 8.20 gives the temperature ranges. The temperature *range* is used to establish the *change* in temperature used in analysis. For example, if a concrete bridge is constructed at a temperature of 68°F (20°C), then the increase in a moderate climate for concrete is $\Delta T = 80 - 68 = 12°F$ (27 − 20 = 7°C), and the decrease in temperature is $\Delta T = 68 - (10) = 58°F$ (~41°C).

Theoretically, the range of climatic temperature is not a function of structure type, but the structure's temperature is a function of the climatic temperature record and specific heat of the material, mass, surface volume ratio, heat conductivity, wind conditions, shade, color, and so on. Because concrete bridges are more massive than steel and the specific heat of concrete is less than steel, an increase in climatic temperature causes a smaller temperature increase in the concrete structure than in the steel. Loosely stated, the concrete structure has more thermal inertia (systems with a large thermal inertia are resistant to changes in temperature) than its steel counterpart.

The temperature gradients are more sensitive to the bridge location than the uniform temperature ranges. The gradient temperature is a function of solar gain to the deck surface. In western U.S. states, where solar radiation is greater, the temperature increases are also greater. The converse is true in the eastern U.S. states. Therefore, the country is partitioned into the solar radiation zones shown in Figure 8.28 [A3.12.3]. The gradient temperatures outlined in Table 8.21 reference these radiation zones. The gradient temperature is considered in addition to the uniform temperature increase. Typically, these two effects are separated in the analysis and therefore are separated here. The AASHTO [A3.12.3] gradient temperatures are illustrated in Figure 8.29.

A temperature increase is considered positive in AASHTO. The temperature T_3 is zero unless determined from

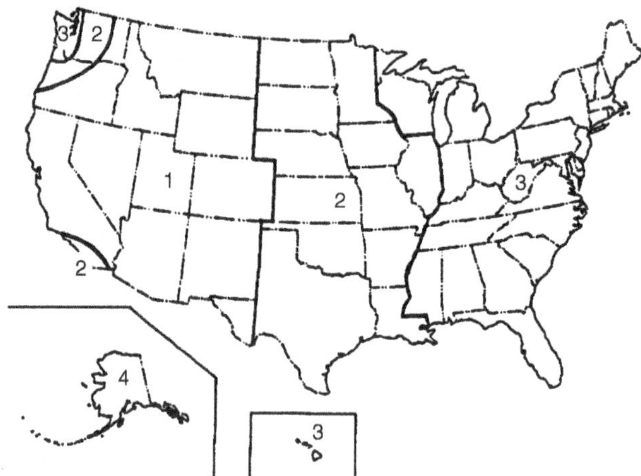

Fig. 8.28 Solar radiation zones. (AASHTO Fig. 3.12.3-1). (From *AASHTO LFRD Bridge Design Specifications*, Copyright © 2010 by the American Association of State Highway and Transportation Officials, Washington, DC. Used by permission.)

Table 8.21 Gradient Temperatures[a]

	Concrete Surface °F (°C)	
Zone	T_1	T_2
1	54 (30)	14 (7.8)
2	46 (25)	12 (6.7)
3	41 (23)	11 (6)
4	38 (21)	9 (5)

[a]To obtain negative gradients multiply by −0.3 and −0.2, for concrete and asphalt overlay decks, respectively.
In AASHTO Table 3.12.3-1. From *AASHTO LRFD Bridge Design Specifications*. Copyright © 2010 by the American Association of State Highway and Transportation Officials, Washington, DC. Used by permission.

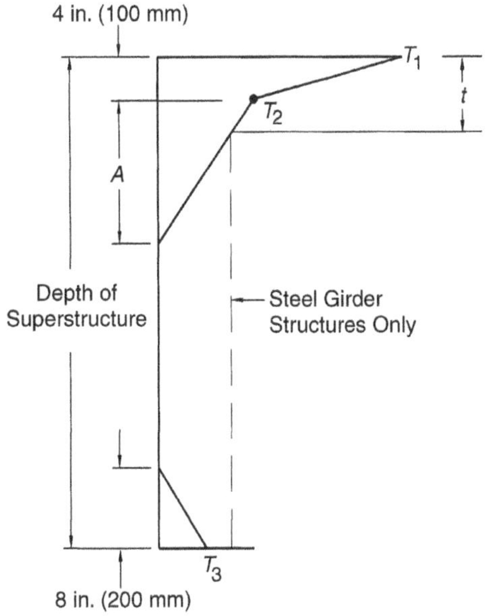

Fig. 8.29 Design temperature gradients. (AASHTO Fig. 3.12.3-2). (From *AASHTO LFRD Bridge Design Specifications*, Copyright © 2010 by the American Association of State Highway and Transportation Officials, Washington, DC. Used by permission.)

site-specific study, but in no case is T_3 to exceed 5°F (3°C). In Figure 8.29, the dimension A is determined as follows:

A = 12 in. (300 mm) for closed concrete structures that are 16 in. (400 mm) or more in depth. For shallower sections, A shall be 4 in. (100 mm) less than the actual depth.

A = 12 in. (300 mm) for steel superstructures, and the distance t shall be taken as the depth of the concrete deck.

These temperature changes are used in the structural analyses described in Chapter 12.

8.4.2 Creep and Shrinkage

The effects of creep and shrinkage can have an effect on the structural strength, fatigue, and serviceability. Traditionally, creep is considered in concrete where its effect can lead to unanticipated serviceability problems that might subsequently lead to secondary strength problems. In addition, today, however, creep is also of concern in wooden structures. Because creep and shrinkage are highly dependent on the material and the system involved, further elaboration is reserved for the chapters on design.

8.4.3 Settlement

Support movements may occur due to the elastic and inelastic deformation of the foundation. Elastic deformations include movements that affect the response of the bridge to other loads but do not lock in permanent actions. Such deformations may be modeled by approximating the stiffness of the support in the structural analysis model. This type of settlement is not a load but rather a support characteristic that should be included in the structural model. Inelastic deformations are movements that tend to be permanent and create locked-in permanent actions.

Such movements may include the settlement due to consolidation, instabilities, or foundation failures. Some such movements are the results of loads applied to the bridge, and these load effects may be included in the modeling of the structural supports. Other movements are attributed to the behavior of the foundation independent of the loads applied to the bridge. These movements must be treated as a load and hereafter are called *imposed support deformations*.

The actions due to imposed support deformations in statically indeterminate structures are proportional to the stiffness. For example, for a given imposed deformation, a stiff structure develops larger actions than a flexible one. The statically determinate structures do not develop any internal actions due to settlement, which is one of the few inherent advantages of statically determinate systems. Imposed support deformations are estimated based on the geotechnical characteristics of the site and system involved. Detailed suggestions are given in AASHTO, Section 10.

8.5 COLLISION LOADS

8.5.1 Vessel Collision

On bridges over navigable waterways, the possibility of vessel collision with the pier must be considered. Typically, this is of concern for structures that are classified as long-span bridges, which are outside the scope of this book. Vessel collision loads are defined in AASHTO [A3.14].

8.5.2 Rail Collision

If a bridge is located near a railway, the possibility of a collision with the bridge as a result of a railway derailment exists. As the possibility is remote, the bridge must be designed for collision forces using the extreme limit state given in Tables 5.1 and 5.2. The abutments and piers within 30 ft (9000 mm) of the edge of the roadway, or within a distance of 50 ft (15,000 mm) of the centerline of the track must be designed for a 400-kip (1800-kN) force positioned at a distance of 4 ft (1200 mm) above the ground [A3.6.5.1].

8.5.3 Vehicle Collision

The collision force of a vehicle with the barrier rail and parapets is described previously in the section on deck and railing loads, as well as in later chapters, and is not reiterated here.

8.6 BLAST LOADING

Unfortunately, an engineer has to consider loads from man-created blast load. The science and engineering associated with this type of load is evolving in both the bridge and building engineering professions. Design of a bridge for security with careful consideration of how the bridge functions in the overall "lifeline" of the transportation system is becoming more important for existing and new structures.

Should the owner determine that a bridge or a bridge component should be designed for intentional or unintentional blast force, the following should be considered (A3.15):

- ☐ Size of explosive charge
- ☐ Shape of explosive charge
- ☐ Stand-off distance
- ☐ Location of the charge
- ☐ Delivery methods

The literature and policy are rapidly changing in this area. For a primer on approach, issues and a host of additional references (to start), see AASHTO (2006).

8.7 SUMMARY

The various types of loads applicable to highway bridges are described with reference to the AASHTO Specification. These loads are used in the subsequent chapters to determine the load effects and to explain the use of these effects in the proportioning of the structure. For loads that are particular to bridges, background is provided on the development and use of the loads, and in other cases, the AASHTO provisions are outlined with limited explanation leaving the detailed explanation to other references.

REFERENCES

AASHTO (2002). *Standard Specifications for Highway Bridges*, 17th ed., American Association of State Highway and Transportation Officials, Washington, DC.

AASHTO (2006). *Homeland Security and State Departments of Transportation, Maintaining Strategic Direction for Protecting America's Transportation System American Association of State Highway and Transportation Officials*, Special Committee on Transportation Security, Washington,

AASHTO (2009a). *Manual for Assessing Safety Hardware*, 1st ed., American Association of State Highway and Transportation Officials, Washington, DC.

AASHTO (2009b). *AASHTO Guide Specifications for LRFD Seismic Bridge Design*, LRFDSEIS-1. American Association of State Highway and Transportation Officials, Washington, DC.

AASHTO (2010). *LRFD Bridge Design Specification*, 5th ed., American Association of State Highway and Transportation Officials, Washington, DC.

ASCE (2003). *Minimum Design Loads for Buildings and Other Structures*, ASCE 7–02 American Society of Civil Engineers, New York.

ASCE (2010) *Minimum Design Loads for Buildings and Other Structures*, (ASCE Standard ASCE/SEI 07–10), American Society of Civil Engineers, Reston, VA.

Assel, R. A. (1980). "Great Lakes Degree-Day and Temperature Summaries and Norms, 1897–1977," *NOAA Data Report ERL GLERL-15*, January, Great Lakes Environmental Research Laboratory, Ann Arbor, MI.

Bakht, B. and S. G. Pinjarkar (1991). *Dynamic Testing of Highway Bridges—A Review*, Transportation Research Record 1223, National Research Council, Washington, DC.

California Department of Transportion (2011), "*California Amendments to AASHTO LRFD Bridge Design Specifications*," Fourth Ed., Sacramento, CA.

Chen, W. F. and L. Duan (2004). *Bridge Engineering: Seismic Design*, CRC Press, Boca Raton, FL.

Gershunov, E. M. (1986). "Collision of Large Floating Ice Feature with Massive Offshore Structure," *Journal of Waterway, Port, Coastal and Ocean Engineering*, ASCE, Vol. 112, No. 3, May, pp. 390–401.

Haugen, R. K. (1993). Meteorologist, U.S. Army Corps of Engineers, Cold Regions Research and Engineering Laboratory, Hanover, NH, personal communication.

Haynes, F. D., D. S. Sodhi, L. J. Zabilansky, and C. H. Clark (1991). *Ice Force Measurements on a Bridge Pier in the St. Regis River, NY*, Special Report 91–14, October, U.S. Army Corps of Engineers, Cold Regions Research and Engineering Laboratory, Hanover, NH.

Hwang, E. S. and A. S. Nowak (1991a). "Simulation of Dynamic Load for Bridges," *Journal of Structural Engineering*, ASCE, Vol. 117, No. 5, May, pp. 1413–1434.

Hwang, E. S. and A. S. Nowak (1991b). *Dynamic Analysis of Girder Bridges*, Transportation Research Record 1223, National Research Council, Washington, DC.

IBC (International Code Council) (2009). *International Building Code*, International Code Council, Whittier, CA.

Koretzky, H. P., K. R. Patel, R. M. McClure, and D. A. VanHorn (1986). *Umbrella Loads for Bridge Design*, Transportation Research Record 1072, National Research Council, Washington, DC.

Kulicki, J. M. (1992). Personal communications.

Kulicki, J. M. and D. R. Mertz (1991). "A New Live Load Model for Bridge Design," *Proceedings of 8th Annual International Bridge Conference*, June, Pittsburgh, PA.

Liu, H. (1991). *Wind Engineering Handbook for Structural Engineers*, Prentice Hall, Englewood Cliffs, NJ.

Moses, F. and M. Ghosen (1985). *A Comprehensive Study of Bridge Loads and Reliability*, Final Report, FHWA/OH-85/005, Jan.

NCHRP (1993). NCHRP Report 350, *Recommended Procedures for the Safety Performance Evaluation of Highway Features*, National Research Council, Washington, DC.

Neill, C. R., Ed. (1981). *Ice Effects on Bridges*, Roads and Transportation Association of Canada, Ottawa, Ontario, Canada.

Nowak, A. S. (1993). "Calibration of LRFD Bridge Design Code," *NCHRP Project* 12–33, University of Michigan, Ann Arbor, MI.

Nowak, A. S. (1995). "Calibration of LRFD Bridge Code," *Journal of Structural Engineering*, *ASCE*, Vol. 121, No. 8, pp. 1245–1251.

OHBDC (1983). Ontario Ministry of Transportation and Communication, *Ontario Highway Bridge Design Code*, Toronto, Canada.

Patrick, M. D., X. S. Sharon Huo, J. A. Puckett, M. C. Jablin, and D. Mertz (2006). "Sensitivity of Live Load Distribution Factors to Vehicle Spacing," Technical Note, *Bridge Engineering Journal*, ASCE, Vol. 11, No. 1, p. 131.

Paultre, P., O. Challal, and J. Proulx (1992). "Bridge Dynamics and Dynamic Amplication Factors—A Review of Analytical and Experimental Findings," *Canadian Journal of Civil Engineering*, Vol. 19, pp. 260–278.

Pennsylvania Department of Transportation (2011), *Publication 238*, Part IP, Harrisburg, PA.

Puckett, J. A., X., Hou, M., Jablin and D. R. Mertz (2007). Simplified Live Load Distribution Factor Equations, Report 592, National Cooperative Highway Research Program, Transportation Research Board, National Research Council, Washington, DC.

Simiu, E. (1973). "Logarithmic Profiles and Design Wind Speeds," *Journal of the Engineering Mechanics Division*, ASCE, Vol. 99, No. 10, October, pp. 1073–1083.

Simiu, E. (1976). "Equivalent Static Wind Loads for Tall Building Design," *Journal of the Structures Division*, ASCE, Vol. 102, No. 4, April, pp. 719–737.

Simiu, E. and R. H. Scanlan (1978). *Wind Effects on Structures: An Introduction to Wind Engineering*, Wiley-Interscience, New York.

Tilly, G. P. (1986). "Dynamic Behavior of Concrete Structures," in *Developments in Civil Engineering*, Vol. 13, Report of the Rilem 65 MDB Committee, Elsevier, New York.

TRB Committee for the Truck Weight Study (1990). *Truck Weights and Limits—Issues and Options*, Special Report 225, Transportation Research Board, National Research Council, Washington, DC.

Wortley, C. A. (1984). Great Lakes Small-Craft Harbor and Structure Design for Ice Conditions: *An Engineering Manual, Sea Grant Advisory Report WIS-SG-84-426*, University of Wisconsin Sea Grant Institute, Madison, WI.

PROBLEMS

8.1 What is the definition of *exclusion vehicles*? Give examples of exclusion vehicles. What role did exclusion vehicles play in the development of the AASHTO live-load model HL-93?

8.2 What is the purpose in the AASHTO LRFD Specifications of the following factors? Give an example of each factor.

- Multiple presence factor, m
- Fraction of truck traffic in a single lane, p

8.3 Select a place in the United States of personal interest, such as your hometown, and determine its acceleration coefficient PGA from AASHTO Figures 3.10.2.1-1 to -21 or alternatively, use http://earthquake.usgs.gov/hazards/. To what seismic zone does this location belong?

8.4 Determine the minimum longitudinal and transverse connection forces for a simply supported concrete bridge span of 100 ft with a dead load of 12.0 kips/ft founded on soil profile C. At the location of the bridge the acceleration coefficient A_S is 0.047. State any assumptions made of the restraint provided by the connections.

8.5 The thickness of ice must be estimated to determine the static and dynamic horizontal ice forces. Using Eq. 8.25, estimate the thickness of ice for Lake Superior at Duluth, Minnesota, and for the Potomac River at Washington, DC.

CHAPTER 9

Influence Functions and Girder-Line Analysis

9.1 INTRODUCTION

As outlined in Chapter 8, bridges must carry many different types of loads, which may be present individually or in combination. The bridge engineer has the responsibility for analysis and design of the bridge subjected to these loads and for the placement of the loads in the most critical manner. For example, vehicular loads move, and hence the placement and analysis varies as the vehicle traverses the bridge. The engineer must determine the most critical load placement for all cross sections in the bridge. Frequently, this load placement is not obvious, and the engineer must rely on systematic procedures to place the loads and to analyze the structure for this placement. Structural analysis using influence functions (or influence lines) is the foundation of this procedure and is fundamental to the understanding of bridge analysis and design. The term influence function is used instead of the term influence line because it is more general, that is, the function may be one dimensional (1D) (a line) or two dimensional (2D) (a surface).

The reader may have been exposed to influence lines/functions in past coursework and/or professional practice. If so, then this chapter provides a review and perhaps a treatment unlike the previous exposure. To the novice, this chapter is intended to be comprehensive in both theory and application. The examples provide background and detailed analyses of all the structures used in the subsequent design chapters. Therefore, the reader should take careful note of the examples as they are referenced frequently throughout the remainder of this book.

Sign conventions are necessary to properly communicate the theory, procedures, and analytical results. Conventions are somewhat arbitrary and textbook writers use different conventions. Herein, the following conventions are used for shear and moment diagrams:

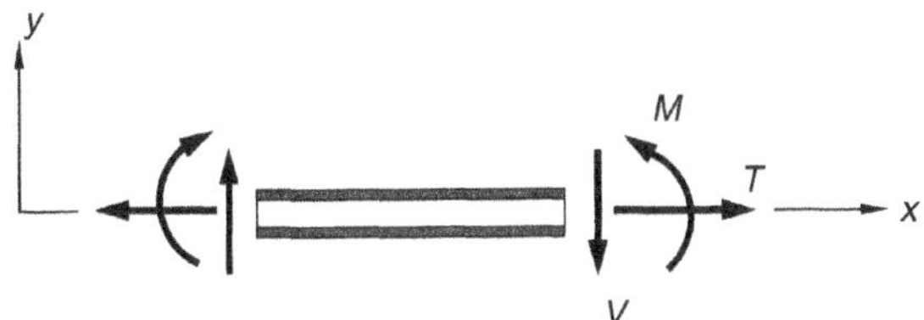

Fig. 9.1 Beam segment, with positive designer sign convention.

- ☐ For a beam, moment causing compression on the top and tension on the bottom is considered positive as shown in Figure 9.1. Moment diagrams are plotted on the compression side of the element. For frames, the distinction of positive and negative is ambiguous.
- ☐ For a beam, positive shear is upward on the left face and downward on the right face as shown in Figure 9.1. The shear diagram is plotted so that the change in shear is in the direction of the applied load. Again, for frames, the distinction of positive and negative is ambiguous.
- ☐ Axial thrust is considered positive in tension. The side of the element on which to plot this function is arbitrary but must be consistent throughout the structure and labeled appropriately to avoid misinterpretation.

These conventions are summarized in Figure 9.1 and hereafter are referred to as the *designer sign convention* because they refer to those quantities (shears, moments, and axial load) that a designer uses to select and check member resistance. Additional sign conventions such as those used in analysis procedures are given as necessary.

9.2 DEFINITION

Influence function ≡ a function that represents the load effect (force or displacement) at a point in the structure as a unit action moves along a path or over a surface.

Influence line ≡ a one-dimensional influence function (used for a beam).

Consider the two-span beam illustrated in Figure 9.2(a). The unit action is a concentrated load that traverses the structures along the beam from left to right. The dimension x represents the location of the load. For this discussion, assume that an instrument that measures the flexural bending moment is located at point n and records this action as the unit load moves across the beam. The record of the moment as a function of load position is the influence function shown in Figure 9.2(b).

The load positioned in span AB causes a positive influence (positive moment) at n. Note that the maximum value occurs directly at point n. When the load is positioned in span BC, the influence of the load on the bending moment at n is negative, that is, tension on the top (pop up). A load in span BC causes the beam to deflect upward creating tension on top of the beam or a negative bending moment.

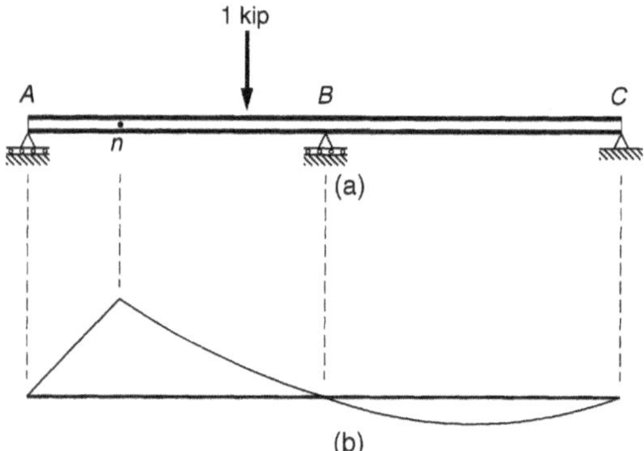

Fig. 9.2 Continuous beam influence function.

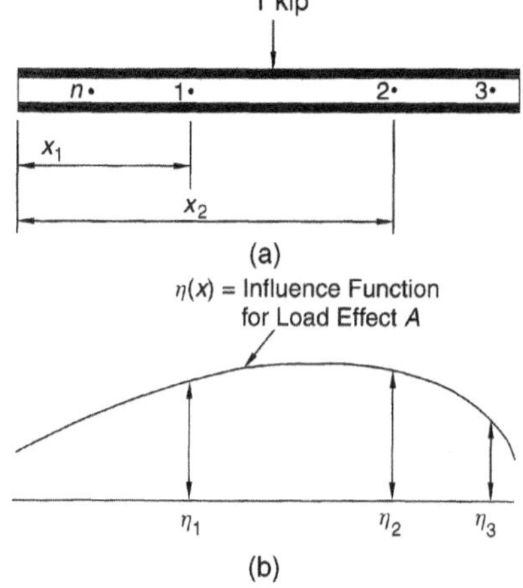

Fig. 9.3 (a) Concentrated loads on beam segment and (b) influence function for load effect A.

Consider the beam section shown in Figure 9.3(a) and the influence function for action A at point n shown in Figure 9.3(b). Assume that the influence function was created by a unit load applied downward in the same direction of the applied load shown in Figure 9.3(a). Assuming that the structure behaves linearly, the load P_1 applied at point 1 causes a load effect of P_1 times the function value $\eta(x_1) = \eta_1$. Similarly, the load P_2 applied at point 2 causes a load effect of P_2 times the function value $\eta(x_2) = \eta_2$, and so on. Superposition of all the load effects yields

$$\text{Load effect} = A = P_1\eta(x_1) + P_2\eta(x_2) + \cdots + P_n\eta(x_n)$$

$$= \sum_{i=1}^{n} P_i\eta\left(x_i\right) = \sum_{i=1}^{n} P_i\eta_i \tag{9.1}$$

Linear behavior is a necessary condition for application of Eq. 9.1, that is, the influence coefficients must be based on a linear relationship between the applied unit action and the load effect. For example, if the unit load is applied at a specific point and then doubled, the resulting load effect will also double if the response is linear. For statically determinate structures, this relationship typically holds true except for cases of large deformation where consideration of deformed geometry must be considered in the equilibrium formulation. The unit action load effect relationship in statically indeterminate structures is a function of the relative stiffness of the elements.

If stiffness changes are due to load application from either material nonlinearity and/or geometric nonlinearity (large deflections), then the application of the superposition implicit in Eq. 9.1 is incorrect. In such cases, the use of influence functions is not appropriate, and the loads must be applied sequentially as expected in the real structure. Such an analysis is beyond the scope of this book, and the reader is referred to books on advanced structural and finite-element analysis.

9.3 STATICALLY DETERMINATE BEAMS

The fundamentals of influence functions and their use are initially illustrated with statically determinate beams. Several examples are given.

9.3.1 Concentrated Loads

Example 9.1 Use the beam shown in Figure 9.4(a) to determine the influence functions for the reaction at A, and the shear and moment at B. Point B is located at midspan.

Consider the unit load at position x on the beam AC of length L. Because this system is statically determinate, the influence function may be based solely on static equilibrium. Use the free-body diagram shown in Figure 9.4(b) to balance the moments about A and to determine the reaction R_C:

$$\sum M_A = 0$$

$$1(x) - R_C(L) = 0$$

$$R_C = \frac{x}{L}$$

Similarly, balancing moments about C yields the reaction at R_A:

$$\sum M_C = 1(L - x) - R_A(L) = 0$$

$$R_A = \frac{L - x}{L}$$

Summation of the vertical forces checks the previous moment computations. If this check does not validate equilibrium, then an error exists in the calculation:

$$\sum F_V = 0$$

$$R_A + R_C - 1 = 0 \qquad \text{(OK)}$$

The influence function for R_A is shown in Figure 9.4(c). Note the function is unity when the load position is directly over A and decreases linearly to zero when the load position is at C. Linearity is characteristic of influence functions (actions and reactions) for statically determinate structures. This point is elaborated on later. Next, the influence functions for the shear

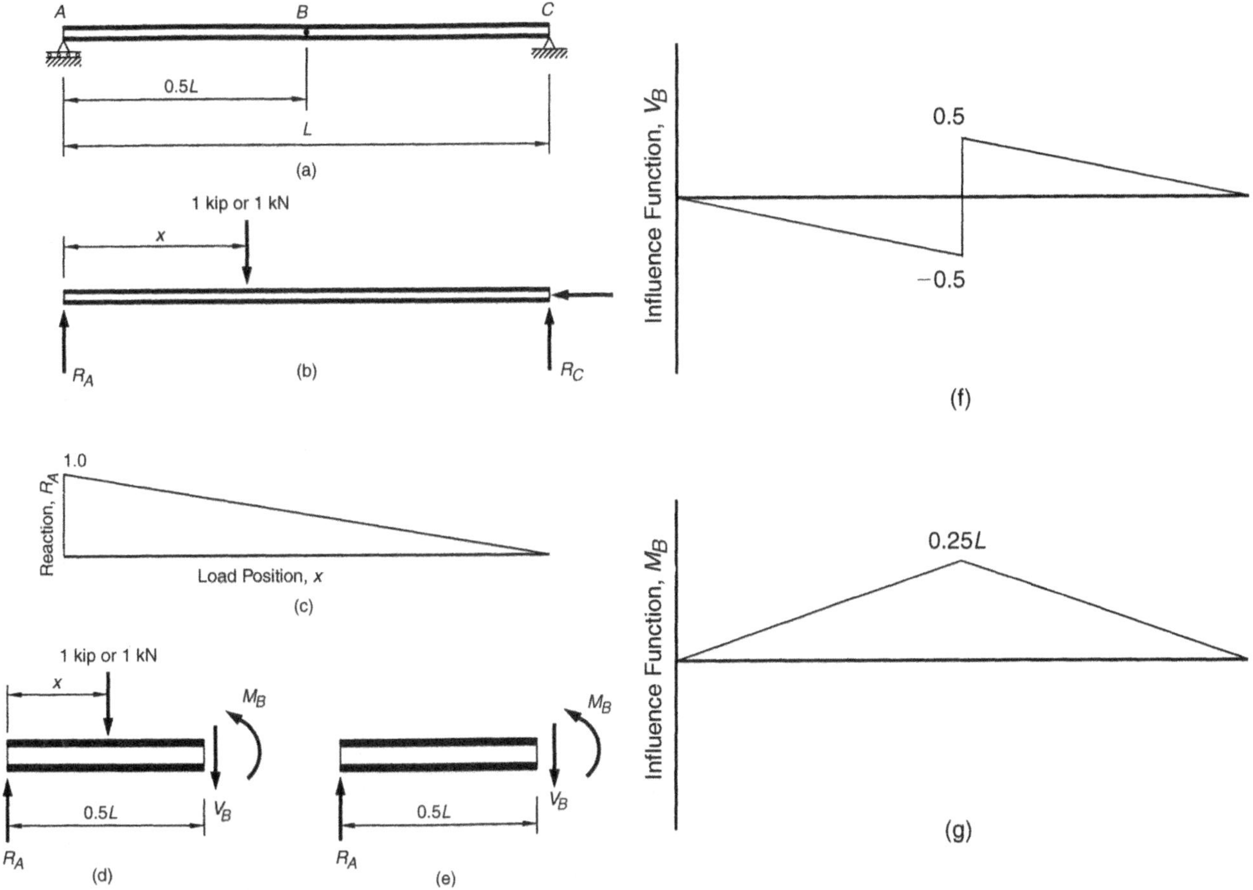

Fig. 9.4 (a) Simple beam, (b) moving unit load, (c) influence function for R_A, (d) free-body diagram AB with unit load at $x \le 0.5L$, and (e) free-body diagram AB with unit load at $x > 0.5L$. (f) Influence diagram for V_B and (g) influence diagram for M_B.

and moment at B are determined. Use the free-body diagram shown in Figure 9.4(d) to sum the vertical forces yielding V_B:

$$\sum F_V = 0$$
$$R_A - 1 - V_B = 0$$
$$V_B = \frac{L-x}{L} - 1 = -\frac{x}{L}$$

Balancing moments about B gives the internal moment at B.

$$\sum M_B = 0$$
$$R_A(0.5L) - 1(0.5L - x) - M_B = 0$$
$$M_B = \frac{x}{2} \qquad \text{when} \qquad 0 \le x \le 0.5L$$

Note that the functions for V_B and M_B are valid when $x \le L/2$. If $x > L/2$, then the unit load does not appear on the free-body diagram. The revised diagram is shown in Figure 9.4(e). Again, by balancing forces and moments, the influence functions for V_B and M_B are established:

$$V_B = R_A = \frac{L-x}{L} \qquad \text{when} \qquad 0.5L \le x \le L$$
$$M_B = \frac{L-x}{2} \qquad \text{when} \qquad 0.5L \le x \le L$$

The influence functions for V_B and M_B are illustrated in Figures 9.4(f) and 9.4(g), respectively.

Example 9.2 Use the influence functions developed in Example 9.1 to analyze the beam shown in Figure 9.5(a). Determine the reaction at A and the midspan shear and moment at B.

Use the influence function for R_A shown in Figure 9.4(c) to determine the influence ordinates at the load positions of P_1 and P_2 as shown in Figure 9.5(a). The equation developed in Example 9.1 may be used or the ordinates may be interpolated. As illustrated in Figure 9.5(b), the ordinate values are two-thirds and one-third for the positions of P_1 and P_2, respectively. Application of Eq. 9.1 yields

$$R_A = \sum_{i=1}^{2} P_i \eta_i = P_1\left(\frac{2}{3}\right) + P_2\left(\frac{1}{3}\right)$$

The parameters V_B and M_B due to the applied loads may be determined in a similar manner. With the aid of Figures 9.5(c) and 9.5(d), application of Eq. 9.1 yields

$$V_B = P_1\left(-\tfrac{1}{3}\right) + P_2\left(\tfrac{1}{3}\right)$$

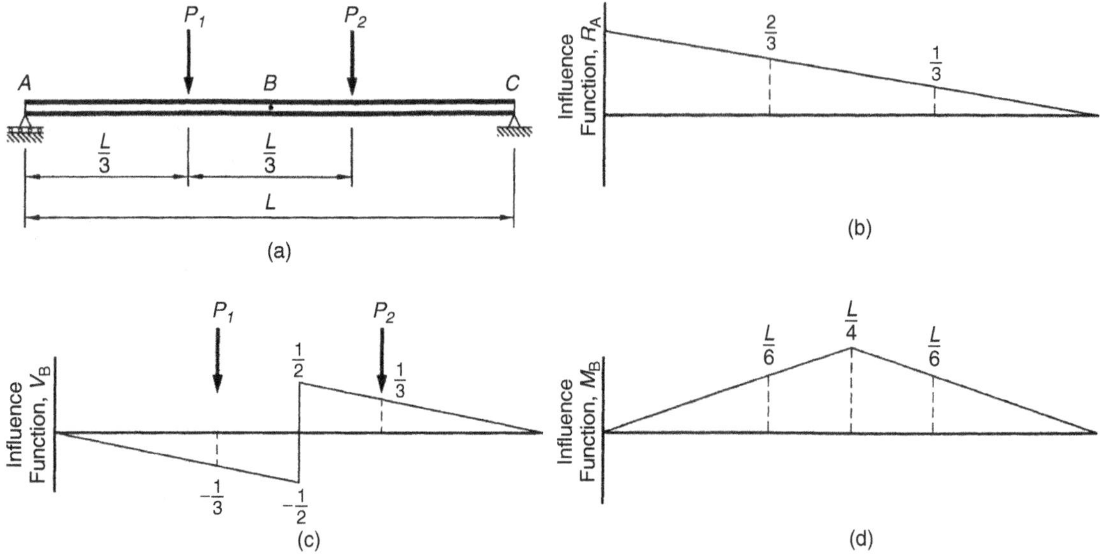

Fig. 9.5 (a) Simple beam, (b) influence function for R_A, (c) influence function for V_B, and (d) influence function for M_B.

and

$$M_B = P_1\left(\frac{L}{6}\right) + P_2\left(\frac{L}{6}\right)$$

Comparison of these results with standard statics procedures is left to the reader.

9.3.2 Uniform Loads

Distributed loads are considered in a manner similar to concentrated loads. Consider the beam segment shown in Figure 9.6(a) that is loaded with a distributed load of varying magnitude $w(x)$. The influence function $\eta(x)$ for action A is illustrated in Figure 9.6(b). The load applied over the differential element Δx is $w(x)\ \Delta x$. This load is used in Eq. 9.1. In the limit as Δx goes to zero, the summation becomes the integration:

$$\text{Load effect} = A = \sum_{i=1}^{n} P_i \eta\left(x_i\right) = \sum_{i=1}^{n} w\left(x_i\right) \eta\left(x_i\right) \Delta x$$

$$= \int_a^b w\,(x)\eta(x)\,dx \tag{9.2}$$

If the load is uniform, then the load function $w(x) = w_0$ is a constant rather than a function of x and may be placed outside of the integral. Equation 9.2 becomes

$$\text{Load effect} = A = \int_a^b w\,(x)\,\eta\,(x)\,dx = w_0 \int_a^b \eta\,(x)\,dx \tag{9.3}$$

Note that this integral is simply the area under the influence function over the range of load application.

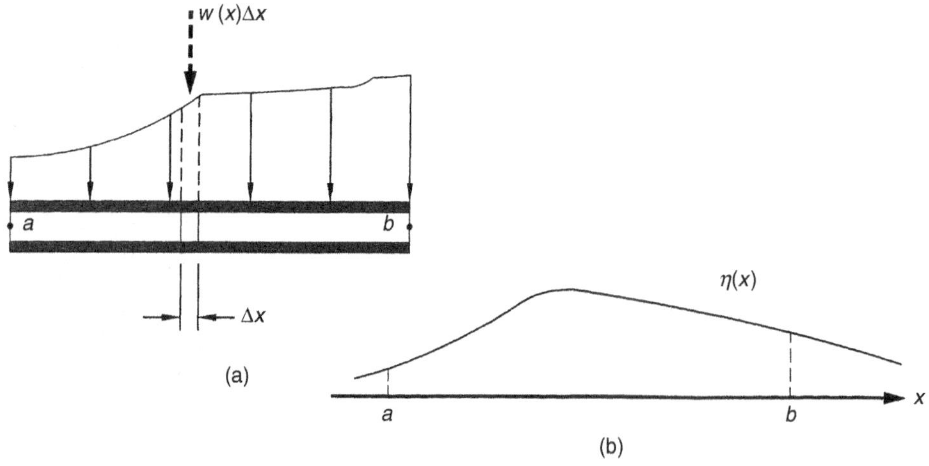

Fig. 9.6 (a) Beam segment with distributed load and (b) influence function.

Example 9.3 Determine the reaction at A and the shear and moment at midspan for the beam shown in Example 9.1 [Fig. 9.4(a)] subjected to a uniform load of w_0 over the entire span.

Application of Eq. 9.3 yields

$$R_A = \int_0^L w\,(x)\eta_{R_A}(x)\,dx = w_0\int_0^L \eta_{R_A}(x)\,dx$$

$$R_A = \frac{w_0 L}{2}$$

$$V_B = \int_0^L w\,(x)\,\eta_{V_B}(x)\,dx = w_0\int_0^L \eta_{V_B}(x)\,dx$$

$$V_B = 0$$

$$M_B = \int_0^L w\,(x)\,\eta_{M_B}(x)\,dx = w_0\int_0^L \eta_{M_B}(x)\,dx$$

$$M_B = \frac{w_0 L^2}{8}$$

Again, comparison of these results with standard equilibrium analysis is left to the reader.

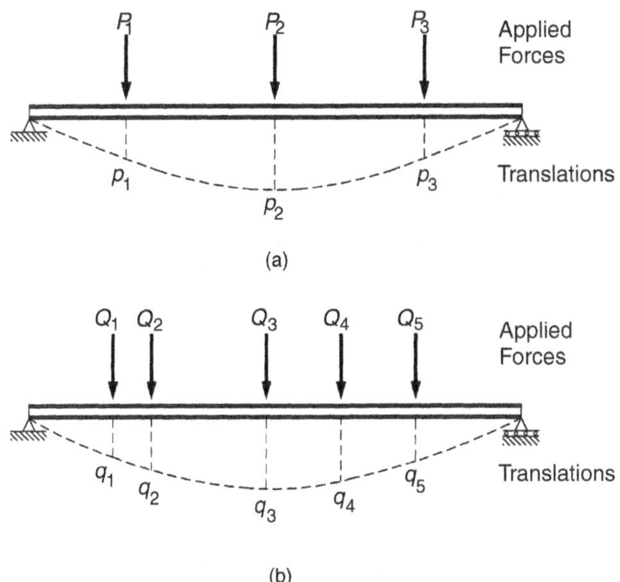

Fig. 9.7 (a) Displaced beam under system P–p and (b) displaced beam under system Q–q.

9.4 MULLER–BRESLAU PRINCIPLE

The analysis of a structure subjected to numerous load placements can be labor intensive and algebraically complex. The unit action must be considered at numerous locations requiring several analyses. The Muller–Breslau principle allows the engineer to study one load case to generate the entire influence function. Because this function has the same characteristics generated by traversing a unit action, many of the complicating features are similar. The Muller–Breslau principle has both advantages and disadvantages depending on the analytical objectives. These concerns are discussed in detail later.

The development of the Muller–Breslau principle requires application of Betti's theorem. This important energy theorem is a prerequisite and is reviewed next.

9.4.1 Betti's Theorem

Consider two force systems P and Q associated with displacements p and q applied to a structure that behaves linear elastically. These forces and displacements are shown in Figures 9.7(a) and 9.7(b). Application of the Q–q system to the structure and equating the work performed by gradually applied forces to the internal strain energy yields

$$\frac{1}{2}\sum_{i=1}^{n} Q_i q_i = U_{Qq} \qquad (9.4)$$

where U_{Qq} is the strain energy stored in the beam when the loads Q are applied quasi-statically* through displacement q.

*Because all loads are transient at some time, static load is technically a misnomer, but if the loads are applied slowly dynamic effects are small.

Now apply the forces of the second system P with the Q forces remaining in place. Note that the forces Q are now at the full value and move through displacements p due to force P. The work performed by all the forces is

$$\frac{1}{2}\sum_{i=1}^{n} Q_i q_i + \sum_{i=1}^{n} Q_i p_i + \frac{1}{2}\sum_{j=1}^{m} P_j p_j = U_{\text{final}} \qquad (9.5)$$

where U_{final} is the associated internal strain energy due to all forces applied in the order prescribed.

Use the same force systems to apply the forces in the reverse order, that is, P first and then Q. The work performed by all forces is

$$\frac{1}{2}\sum_{j=1}^{m} P_j p_j + \sum_{j=1}^{m} P_j q_j + \frac{1}{2}\sum_{i=1}^{n} Q_i q_i = U_{\text{final}} \qquad (9.6)$$

If the structure behaves linear elastically, then final displaced shape and internal strain energy are independent of the order of load application. Therefore, equivalence of the U_{final} in Eqs. 9.5 and 9.6 yields

$$\sum_{i=1}^{n} Q_i p_i = \sum_{j=1}^{m} P_j q_j \qquad (9.7)$$

In a narrative format, Eq. 9.7 states Betti's theorem:

> The product of the forces of the first system times the corresponding displacements due to the second force system is equal to the forces of the second system times the corresponding displacements of the first system.

Although derivation is performed with reference to a beam, the theorem is generally applicable to any linear elastic structural system.

9.4.2 Theory of Muller–Breslau Principle

Consider the beam shown in Figure 9.8(a), where the reaction R_A is of interest. Remove the support constraint and replace it with the reaction R_A as illustrated in Figure 9.8(b). Now replace the reaction R_A with a second force F and remove the applied forces P. Displace the released constraint a unit amount in the direction shown in Figure 9.8(c) and consistent with the remaining constraints. Application of Betti's theorem to the two systems (Eq. 9.7) yields

$$R_A\,(1) - P_1\delta_1 - P_2\delta_2 - \cdots - P_n\delta_n = F\,(0) \qquad (9.8)$$

which simplifies to

$$R_A = P_1\delta_1 + P_2\delta_2 + \cdots + P_n\delta_n = \sum_{i=1}^{n} P_i\delta_i \qquad (9.9)$$

A comparison of Eq. 9.1 to Eq. 9.9 reveals that application of Betti's theorem yields the same result as direct application of superposition combined with the definition of an influence function. Hence, the ordinates δ in Eq. 9.9 must be the same as the ordinates η in Eq. 9.1. This observation is important because ordinates δ were generated by imposing a unit displacement at the released constraint associated with the action of interest and consistent with the remaining constraints. This constitutes the Muller–Breslau principle, which is summarized below:

An influence function for an action may be established by removing the constraint associated with the action and imposing a unit displacement. The displacement at every point in the structure is the influence function. In other words *the structure's displaced shape is the influence function*.

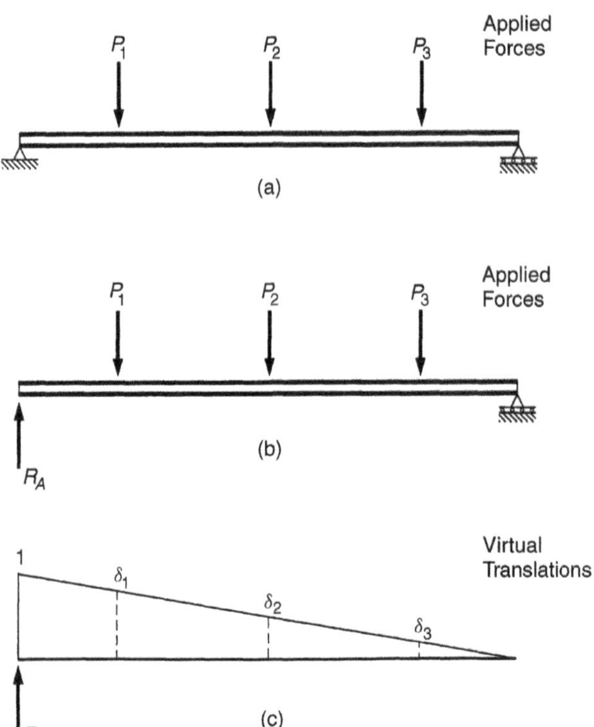

Fig. 9.8 (a) Structure with loads, (b) support A replaced with R_A, and (c) virtual displacement at A.

The sense of the displacements that define the influence function must be considered. For concentrated or distributed forces, the translation colinear with the direction of action is used as the influence ordinate or function. If the applied action is a couple, then the rotation is the associated influence function. The latter can be established by application of Betti's theorem in a similar manner.

Example 9.4 Use the Muller–Breslau principle to determine the influence function for the moment and shear at midspan of the beam shown in Example 9.1 (Fig. 9.4) and reillustrated for convenience in Figure 9.9(a). The moment is considered first. Release the moment at B by insertion of a hinge and apply a unit rotation at this hinge, that is, the relative angle between member AB and BC is one unit.

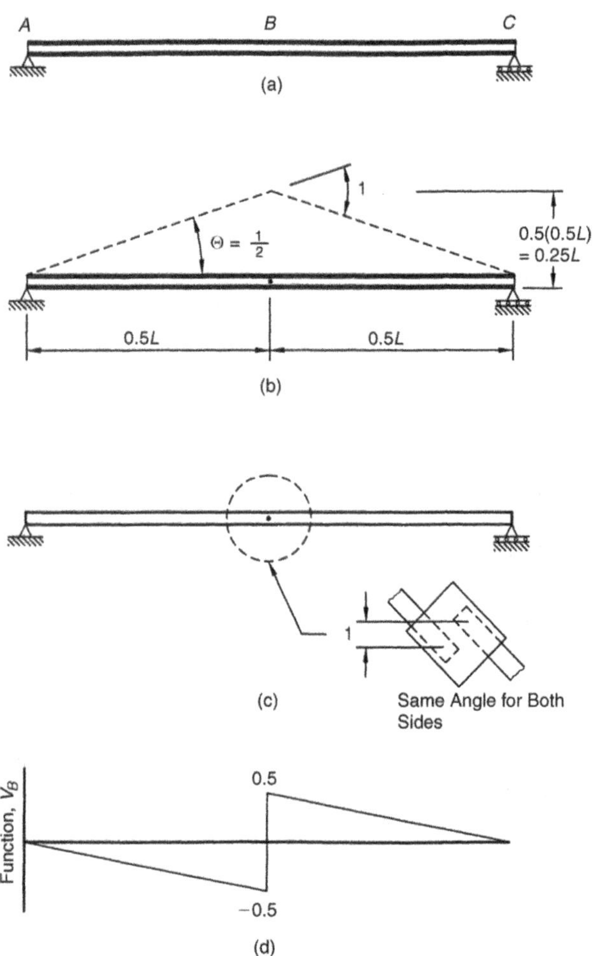

Fig. 9.9 (a) Simple beam, (b) unit rotation at B, (c) unit translation at B with mechanism for unit translation, and (d) influence function of V_B.

Geometric and symmetry considerations require that the angle at the supports be one-half unit. Assuming small displacements (i.e., $\tan\theta = \sin\theta = \theta$), the maximum ordinate at B is determined by

$$\eta_{\max} = \theta\left(\frac{L}{2}\right) = \frac{1}{2}\left(\frac{L}{2}\right) = \frac{L}{4} = 0.25L \qquad (9.10)$$

which is the same result as Example 9.1 [compare Figs. 9.4(g) and 9.9(b)].

The shear influence function is determined in a similar manner. Release the translation continuity at B and maintain the rotational continuity. Such a release device is schematically illustrated in Figure 9.9(c). Apply a relative unit translation at B and maintain the slope continuity required on both sides of the release. This displacement gives the influence function shown in Figure 9.9(d), which is the same function given in Example 9.1 [Fig. 9.4(f)], as expected.

9.4.3 Qualitative Influence Functions

The displaced shape is not always as easily established as illustrated in Example 9.4. For example, the displaced shape of a statically indeterminate structure is more involved. Related procedures are described in Section 9.5. One of the most useful applications of the Muller–Breslau principle is in the development of *qualitative influence functions*. Because most displaced shapes due to applied loads may be intuitively generated in an approximate manner, the influence functions may be determined in a similar fashion. Although exact ordinates and/or functions require more involved methods, a function can be estimated by simply releasing the appropriate restraint, inducing the unit displacement, and sketching the displaced shape. This technique is extremely useful in determining an approximate influence function that in turn aids the engineer in the placement of loads for the critical effect.

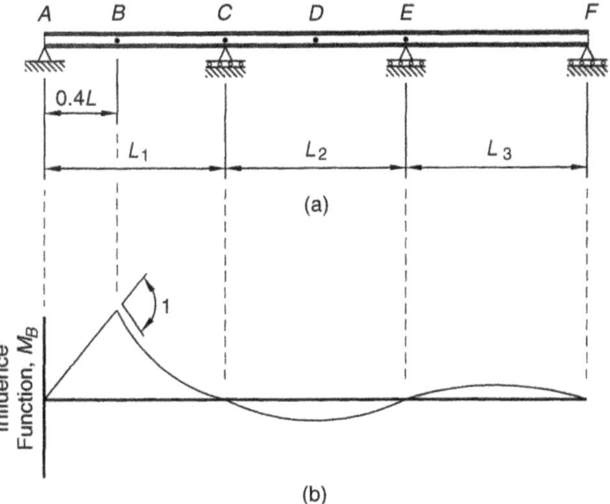

Fig. 9.10 (a) Continuous beam and (b) influence function M_B.

Example 9.5 Use the qualitative method to establish the influence function for moment at point B for the beam shown in Figure 9.10(a).

Release the moment at B and apply a relative unit rotation consistent with the remaining constraints. The resulting translation is illustrated in Figure 9.10(b). If a uniform live load is required, it is necessary to apply this load on spans AC and EF (location of positive influence) for the maximum positive moment at B and on CE for the critical negative moment at B.

9.5 STATICALLY INDETERMINATE BEAMS

Primarily, two methods exist for the determination of influence functions:

- ☐ Traverse a unit action across the structure.
- ☐ Impose a unit translation or unit rotation at the released action of interest (Muller–Breslau).

Both of these methods are viable techniques for either hand or automated analysis. The principles involved with these methods have been presented in previous sections concerning statically determinate structures. Both methods are equally applicable to indeterminate structures, but are somewhat more involved. Both methods must employ either a flexibility approach such as consistent deformations, or stiffness techniques such as slope–deflection, moment distribution, and finite-element analysis (matrix displacement analysis). Typically, stiffness methods are used in practice where slope–deflection and moment distribution are viable hand methods while the matrix approach is used in automated procedures. Both the unit load traverse and Muller–Breslau approaches are illustrated in the following sections using stiffness methods. Each method is addressed by first presenting the methodology and necessary tools required for the application, which is followed by examples.

Throughout the remainder of this book a specialized notation is used to indicate a position on the structure. This notation, termed span point notation, is convenient for bridge engineering, and is illustrated by several examples in Table 9.1. Here the span point notation is described with the points that typically control the design of a continuous girder. For example, the shear is a maximum near the supports, the positive moment is a maximum in the span, and the negative moment is the largest, often called the maximum negative moment, at the supports. Mathematically, *maximum negative moment* is poor terminology, but nevertheless it is conventional. Table 9.1 is provided for guidance in typical situations and for use in preliminary design. Final design calculations should be based on the envelope of all actions from all possible live-load placements. Actions described with span point notation are in the designer's sign convention outlined in Section 9.1 (see Fig. 9.1).

Table 9.1 Span Point Notation

Span Point Notation	Alternative Span Point Notation	Span	Percentage	Explanation	Critical Action (Typical)
100	1.00	1	0	Left end of the first span	Shear
104	1.40	1	40	Forty percent of the way across the first span	Positive moment
110	1.100	1	100	Right end of the first span immediately left of the first interior support	Shear, negative moment
200	2.00	2	0	Left end of the second span immediately right of the first interior support	Shear, negative moment
205	2.50	2	50	Middle of the second span	Positive moment

Example 9.6 For the prismatic beam shown in Figure 9.11(a), determine the influence functions for the moments at the 104, 200, and 205 points and for shear at the 100, 104, 110, 200, and 205 points. Use span lengths of 100, 120, and 100 ft (30,480, 36,576, and 30,480 mm).

A unit load is traversed across the beam and the slope–deflection method is used. This problem is repeated in Example 9.9 using the slope–deflection method combined with the Muller–Breslau principle.

With the exception of the shear at the 104 and 205 points, these points were selected because the critical actions due to vehicular loads usually occur near these locations. These influence functions are subsequently used in Example 9.12 to determine the maximum load effect due to vehicular loads. The slope–deflection relationship between the end moments and rotations for a prismatic beam on nonsettling supports is given in Eq. 9.11. The subscripts reference the locations illustrated in Figure 9.11(b).

$$\begin{aligned} M_{ij} &= \frac{4EI}{L}\theta_i + \frac{2EI}{L}\theta_j + M_{ij0} \\ M_{ji} &= \frac{2EI}{L}\theta_i + \frac{4EI}{L}\theta_j + M_{ji0} \end{aligned} \tag{9.11}$$

where

EI = flexural rigidity
L = element length
M_{ij}, M_{ji} = moments at ends i and j, respectively
M_{ij0}, M_{ji0} = fixed-end moments at ends i and j due the applied loads, respectively
θ_i, θ_j = rotations at end i and j, respectively

Counterclockwise moments and rotations are considered positive in Eq. 9.11.

A fixed–fixed beam subjected to a concentrated load located at position kL is illustrated in Figure 9.11(c). The end moments are

$$M_{ij0} = PL\,(k)\,(1-k)^2 \tag{9.12a}$$

$$M_{ji0} = -PL\,(k^2)\,(1-k) \tag{9.12b}$$

A single set of slope–deflection and equilibrium equations is desired for all locations of the unit load. Because the load must traverse all spans, the fixed-end moments must change from zero when the load is not on the span to moments based on Eq. 9.12 when the load is located on the span. To facilitate this discontinuity, a special form of MacCauley's notation (Pilkey and Pilkey, 1974) is used.

$$\begin{aligned} &\langle ij\rangle = \langle ji\rangle = 1 \quad \text{if the unit load is located between } i \text{ and } j \\ &\langle ij\rangle = \langle ji\rangle = 0 \quad \text{if the unit load is } \textit{not} \text{ located between } i \text{ and } j \end{aligned} \tag{9.13}$$

Application of Eqs. 9.11, 9.12, and 9.13 to the continuous beam shown in Figure 9.11(a) gives

$$\begin{aligned} M_{AC} &= \frac{4EI}{L_1}\theta_A + \frac{2EI}{L_1}\theta_C + k_1(1-k_1)^2 L_1\langle AC\rangle \\ M_{CA} &= \frac{4EI}{L_1}\theta_C + \frac{2EI}{L_1}\theta_A - k_1^2(1-k_1)L_1\langle AC\rangle \\ M_{CE} &= \frac{4EI}{L_2}\theta_C + \frac{2EI}{L_2}\theta_E + k_2(1-k_2)^2 L_2\langle CE\rangle \\ M_{EC} &= \frac{4EI}{L_2}\theta_E + \frac{2EI}{L_2}\theta_C - k_2^2(1-k_2)L_2\langle CE\rangle \\ M_{EF} &= \frac{4EI}{L_3}\theta_E + \frac{2EI}{L_3}\theta_F + k_3(1-k_3)^2 L_3\langle EF\rangle \\ M_{FE} &= \frac{4EI}{L_3}\theta_F + \frac{2EI}{L_3}\theta_E - k_3^2\,(1-k_3)\,L_3\,\langle EF\rangle \end{aligned}$$

where kL_i is the distance from the left end of the span i to the unit load.

The fixed-end moment terms with MacCauley's notation are zero except when the unit load is on the corresponding span. Equilibrium requires

$$\begin{aligned} M_{AC} &= 0 \\ M_{CA} + M_{CE} &= 0 \\ M_{EC} + M_{EF} &= 0 \\ M_{FE} &= 0 \end{aligned}$$

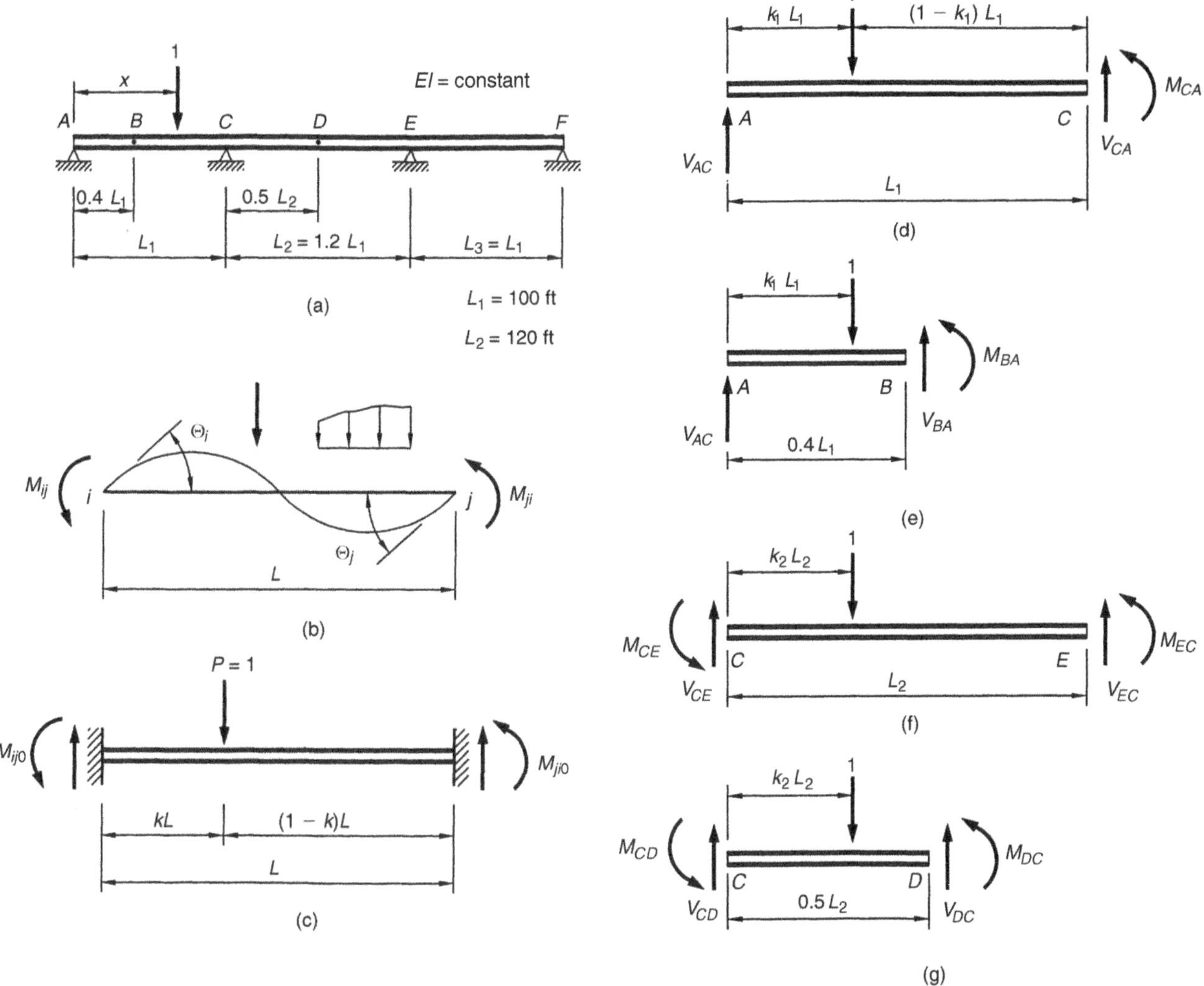

Fig. 9.11 (a) Continuous beam, (b) slope–deflection sign conventions, (c) fixed actions for concentrated loads. (d) Free-body diagram beam AC, (e) free-body diagram beam segment AB, (f) free-body diagram beam CE, and (g) free-body diagram beam segment CD.

The four rotations are determined by substitution of the slope–deflection equations into the four equilibrium equations, which can be solved for the four rotations, a system of four linear algebraic equations. The resulting rotations are back-substituted into the slope–deflection equations to obtain the end moments. Conceptually, this process is straightforward, but as a practical matter, the solution process involves significant algebraic effort. A computer-based equation solver was employed where all the required equations were entered and unknowns were automatically determined and back-substituted to achieve the end moments.

The end shears and the internal shears and moments are determined from equilibrium considerations of each element. A free-body diagram of span AC is illustrated in Figure 9.11(d). This diagram is valid if the unit load is on AC. For other cases, the diagram is valid without the unit load. Summation of moments about C [Fig. 9.11(d)] yields

$$V_{AC} = 1\left(1 - k_1\right)\langle AC\rangle + \frac{M_{CA}}{L_1}$$

Summation of moments about B [Fig. 9.11(e)] yields

$$M_{BA} = V_{AC}\left(0.4L_1\right) - (1)\left(0.4L_1 - k_1L_1\right)\langle AB\rangle$$

By using Figures 9.11(f) and 9.11(g), the end shears for span CE and the shear and moments at D are determined in a similar manner.

$$V_{CE} = (1)\left(1 - k_2\right)\langle CE\rangle + \frac{M_{EC} + M_{CE}}{L_2}$$

$$V_{EC} = 1\,\langle CE\rangle - V_{CE}$$

$$V_{DC} = 1\,\langle CD\rangle - V_{CE}$$

$$M_{DC} = V_{CD}\left(0.5L_2\right) - (1)\left(0.5L_2 - k_2L_2\right)\langle CD\rangle - M_{CE}$$

Conventional slope–deflection notation has been used [counterclockwise (CCW) positive]. The results map to the span point notation described in Table 9.1 as $M_{CA} = M_{110} = M_{200}$, $M_{BA} = M_{104}$, $M_{DC} = M_{205}$, $V_{AC} = V_{100}$, $V_{BA} = -V_{104}$, $V_{CA} = -V_{110}$, $V_{CE} = V_{200}$, and $V_{DC} = -V_{205}$. Note the designer's sign convention defined in Section 9.1 is used with all actions described with span point notation, and the

Table 9.2 Influence Ordinates and Areas (Three-Span Continuous Beam)

Location	Position	$M(104)$	$M(200)$	$M(205)$	$V(100)$	$V(104)$	$V(110)$	$V(200)$	$V(205)$
100	0	0.00	0.00	0.00	1.00	0.00	0.00	0.00	0.00
101	10	5.03	−2.43	−0.88	0.88	−0.12	−0.12	0.03	0.03
102	20	10.11	−4.71	−1.71	0.75	−0.25	−0.25	0.05	0.05
103	30	15.32	−6.70	−2.43	0.63	−0.37	−0.37	0.07	0.07
104	40	20.7	−8.25	−3.00	0.52	−0.48/0.52	−0.48	0.09	0.09
105	50	16.32	−9.21	−3.35	0.41	0.41	−0.59	0.10	0.10
106	60	12.23	−9.43	−3.43	0.31	0.31	−0.69	0.10	0.10
107	70	8.49	−8.77	−3.19	0.21	0.21	−0.79	0.09	0.09
108	80	5.17	−7.07	−2.57	0.13	0.13	−0.87	0.08	0.08
109	90	2.32	−4.20	−1.53	0.06	0.06	−0.94	0.04	0.04
110 or 200	100	0.00	0.00	0.00	0.00	0.00	−1.00	1.00	0.00
201	112	−2.04	−5.09	2.53	−0.05	−0.05	−0.05	0.93	−0.07
202	124	−3.33	−8.33	5.83	−0.08	−0.08	−0.08	0.84	−0.16
203	136	−4.00	−9.99	9.90	−0.10	−0.10	−0.10	0.73	−0.27
204	148	−4.14	−10.34	14.74	−0.10	−0.10	−0.10	0.62	−0.38
205	160	−3.89	−9.64	20.4	−0.10	−0.10	−0.10	0.50	−0.5/0.50
206	172	−3.27	−8.18	14.74	−0.08	−0.08	−0.08	0.38	0.38
207	184	−2.48	−6.21	9.90	−0.06	−0.06	−0.06	0.27	0.27
208	196	−1.60	−4.01	5.83	−0.04	−0.04	−0.04	0.16	0.16
209	208	−0.74	−1.85	2.53	−0.02	−0.02	−0.02	0.07	0.07
210 or 300	220	0.00	0.00	0.00	1.00	0.00	0.00	0.00	0.00
301	230	0.46	1.15	−1.53	0.01	0.01	0.01	−0.04	−0.04
302	240	0.77	1.93	−2.57	0.02	0.02	0.02	−0.08	−0.08
303	250	0.96	2.39	−3.19	0.02	0.02	0.02	−0.09	−0.09
304	260	1.03	2.57	−3.43	0.03	0.03	0.03	−0.10	−0.10
305	270	1.00	2.51	−3.35	0.03	0.03	0.03	−0.10	−0.10
306	280	0.90	2.25	−3.00	0.02	0.02	0.02	−0.09	−0.09
307	290	0.73	1.83	−2.44	0.02	0.02	0.02	−0.07	−0.07
308	300	0.51	1.29	−1.71	0.01	0.01	0.01	−0.05	−0.05
309	310	0.27	0.66	−0.88	0.01	0.01	0.01	−0.03	−0.03
310	320	0.00	0.00	0.00	0.00	0.00	0.00	0.00	0.00
Total positive area		1023	165.7	1036.3	45.6	15.4	1.7	66.4	20.1
Total negative area		−305.5	−1371.4	−442.0	−7.60	−17.4	−63.7	−6.4	−20.1
Net area		717.4	−1205.7	594.3	38.0	−2.0	−62.0	60.0	0.00

slope–deflection convention is used for the calculation of the end moments. The equations given are a function of load position, x. To generate the influence functions, a solution is necessary for each position considered. Typically, the load is positioned at the tenth points. This analysis is done for the present system. The results are given in Table 9.2. Each column constitutes an influence function for the associated action. These functions are illustrated in Figure 9.12(a) (moments) and 9.12(b) (shears).

9.5.1 Integration of Influence Functions

As discussed previously, the integral or area under the influence function is useful for the analysis of uniformly distributed loads. As illustrated in Figures 9.12(a) and 9.12(b), these functions are discontinuous for either value or slope. These functions could be integrated in a closed-form manner, but this is extremely tedious. An alternative approach is to numerically integrate the influence functions. A piecewise straight linear approximation to an influence function may be used and integration of this approximation results in the well-known trapezoidal rule. The integral approximation is

$$\text{Area} = b\sum_{i=1}^{n}\left(\frac{\eta_1}{2} + \eta_2 + \cdots + \eta_{n-1} + \frac{\eta_n}{2}\right) \tag{9.14}$$

where b is the regular distance between the available ordinates. Equation 9.14 integrates exactly a linear influence

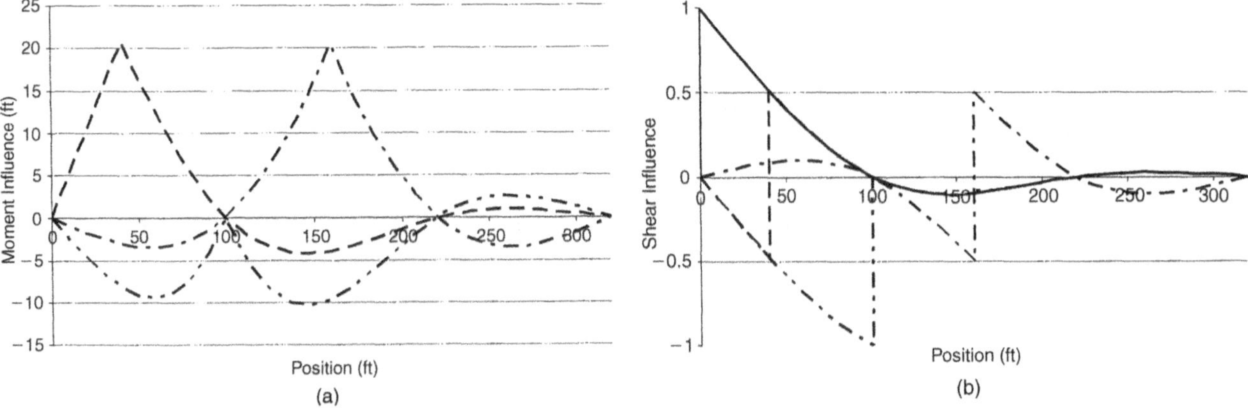

Fig. 9.12 (a) Moment influence functions and (b) shear influence functions.

function. Generally, influence functions are nonlinear and a more accurate approach is desired. Simpson's rule uses a piecewise parabolic approximation that typically approximates nonlinear functions more accurately than its linear counterpart, Eq. 9.14. Simpson's rule requires that domains have uniformly spaced ordinates and an odd number of ordinates with an even number of spaces between them. Simpson's rule is

$$\text{Area} = \frac{b}{3}\sum_{i=1}^{n}(\eta_1 + 4\eta_2 + 2\eta_3 + 4\eta_4 + \cdots + 2\eta_{n-2} + 4\eta_{n-1} + \eta_n) \tag{9.15}$$

Equation 9.15 was used to evaluate the positive, negative, and net areas for each function determined in Example 9.6. The results are given at the bottom of Table 9.2.

Example 9.7 Use the trapezoidal rule to determine the positive, negative, and net areas of the influence function for M_{104} in Table 9.2.

$$A_{\text{Span1}} = [0/2 + 5.03 + 10.11 + 15.32 + 20.7 + 16.32 + 12.23 + 8.49 + 5.17 + 2.32 + 0/2](10) = 956.9$$

$$A_{\text{Span2}} = [0/2 + (-2.04) + (-3.33) + (-4.00) + (-4.14) + (-3.89) + (-3.27) + (-2.48) + (-1.60) + (-0.74) + 0/2](10) = -305.8$$

$$A_{\text{Span3}} = [0/2 + 0.46 + 0.77 + 0.96 + 1.03 + 1.00 + 0.90 + 0.73 + 0.51 + 0.27 + 0/2](10) = 66.3$$

These areas are added to give the positive, negative, and net areas:

$$A^{+} = 956.9 + 66.3 = 1023.2 \text{ ft}^2$$
$$A^{-} = -305.8 \text{ ft}^2$$
$$A^{\text{Net}} = 717.4 \text{ ft}^2$$

9.5.2 Relationship between Influence Functions*

As illustrated in Example 9.6, the end actions, specifically end moments, are determined immediately after the displacements are established. This back-substitution process is characteristic of stiffness methods. The actions in the interior of the beam are based on static equilibrium considering the element loads and the end actions. Because the end actions and the associated influence functions are usually determined before other actions, it is convenient to establish relationships between the influence functions for the end actions and the functions for the actions in the interior portion of the span. The following discussion is rather detailed and requires techniques and associated notations that require careful study.

Consider the continuous beam shown in Figure 9.13(a) where the point of interest n is located at a distance βL from the end i. The actions at n may be determined by superposition of the actions at point n corresponding to a simple beam [Fig. 9.13(b)] and those corresponding to a simple beam with the end moment influence functions applied [Fig. 9.13(c)]. Note that the influence functions are actions that are applied on a free-body diagram and treated in a manner similar to conventional actions. To illustrate, the influence functions are shown in Figure 9.13(b)–9.13(d) instead of their corresponding actions. With the use of superposition, the influence function for an action at n is determined by

$$\eta_n = \eta_s + \eta_e \tag{9.16}$$

where η_s is the influence function for the action at n for the unit action on the simple beam [Fig. 9.13(b)] and η_e is the influence function for the action at n due to the end actions on the simple beam [Fig. 9.13(c)].

By using the free-body diagram shown in Figure 9.13(c), the shear influence function at i due to the end moments is determined from summation of moments about end j. The result is

$$\eta V_{ie} = \frac{\eta_{M_{ij}} + \eta_{M_{ji}}}{L} \tag{9.17a}$$

*Advanced material, may be skipped.

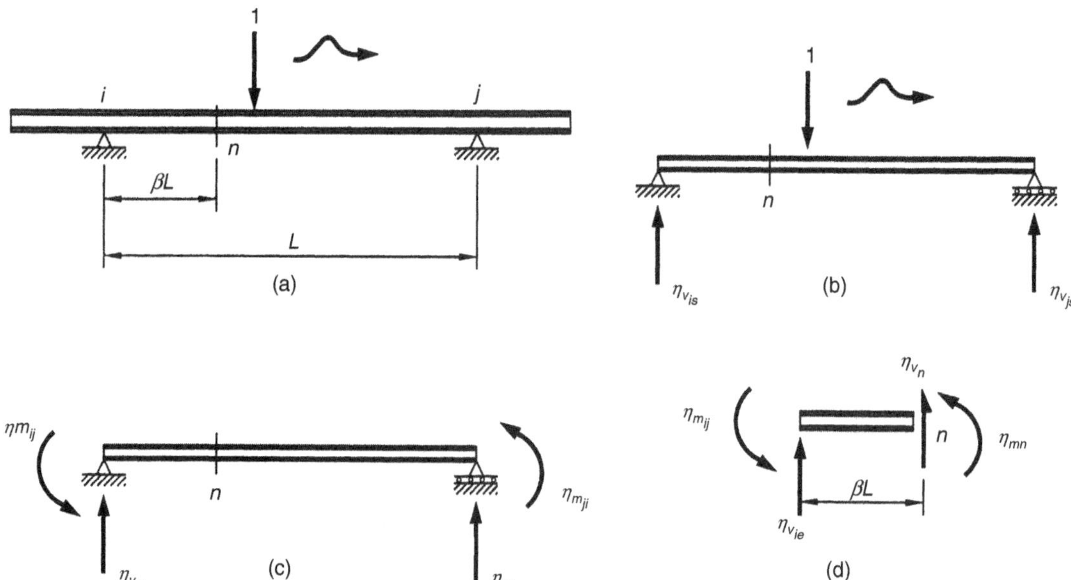

Fig. 9.13 (a) Continuous beam ij, (b) free-body diagram simple beam ij with unit load, (c) free-body diagram simple beam ij with end moments, and (d) free-body diagram beam segment in βL.

where $\eta_{M_{ij}}$ and $\eta_{M_{ji}}$ are defined in Figure 9.13(c). Equation 9.16 is used to combine the shear influence function due to the unit action on the simple beam with the shear due the end moments. The result is

$$\eta_{V_n} = \eta_{V_s} + \frac{\eta_{M_{ij}} + \eta_{M_{ji}}}{L} \tag{9.17b}$$

By using Figure 9.13(d) and summing the moments about the point of interest n, the resulting influence function (end effects only) for moment at n is

$$\eta_{M_n} = \eta V_{ie} \beta L - \eta_{M_{ij}} \tag{9.18a}$$

where βL is the distance from the left end of the span to the point of interest n.

Substitution of the left-end shear given in Eq. 9.17(a) into the moment expression given in Eq. 9.18(a) yields

$$\eta_{M_n} = (\beta - 1)\,\eta_{M_{ij}} + \beta \eta_{M_{ji}} \tag{9.18b}$$

Note the slope–deflection sign convention is used in the development of Eqs. 9.17 and 9.18. Any sign convention may be used as long as the actions are used consistently on the free-body diagram, the sense of the action is correctly considered in the equilibrium equations, and results are properly interpreted with reference to the actions on the free-body diagram.

Example 9.8*[*] Determine the influence ordinates for V_{104}, M_{104}, and M_{205} for the beam in Example 9.6 [Fig. 9.11(a)]. Use the influence functions for the end moments given in Table 9.2 and perform the calculations only for the ordinates at the 105 point.

[*] Advanced material, may be skipped.

Carefully note that the influence ordinates at 105 for actions at 104 and 205 are required. This means only *one ordinate* is established for the V_{104}, M_{104}, and M_{205} functions. The other ordinates may be determined in a similar manner.

The influence functions for the simple beam case are shown in Figure 9.14(b) and 9.14(c). The shear and moment ordinates at 105 are determined by linear interpolation. The sign convention used in Table 9.2 is the designer's sign

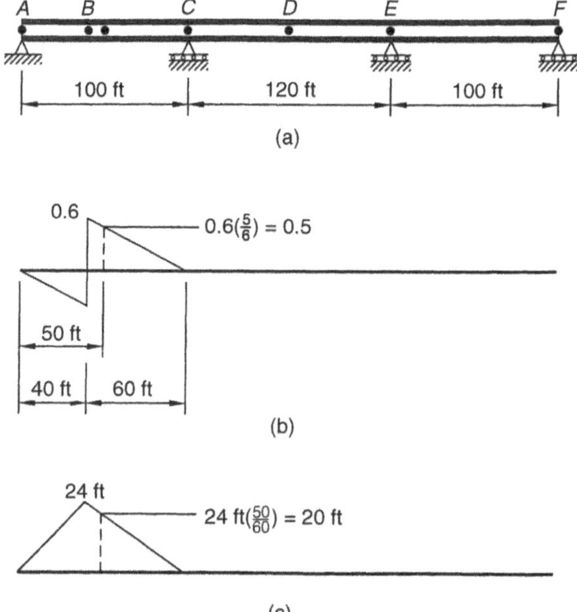

Fig. 9.14 (a) Continuous beam, (b) simple beam AC influence function for shear at 104, and (c) simple beam AC influence function for moment at 104.

convention, and the slope–deflection sign convention is used in Eqs. 9.17 and 9.18. Therefore, the appropriate transformation must be performed. This calculation tends to be a bit confusing and requires careful study. To aid the reader, symbols have been added to reference the explanatory notes given below:

$$\begin{aligned}
&\left[\eta_{V104}\right]_{\text{unit load @ 105}} \\
&= \left[\eta V_{s104} + \frac{\eta_{M100} + \eta_{M110}}{L}\right]_{\text{unit load @ 105}} \\
&= 0.6\left(\frac{5}{6}\right) + \left[\frac{0 + (-9.21)}{100}\right] \\
&= 0.408 \\
&\left[\eta_{M104}\right]_{\text{unit load @ 105}} \\
&= \left[\eta_{M_{s104}} + (\beta - 1)\,\eta_{M100} + \beta\eta_{M110}\right]_{\text{unit load @ 105}} \\
&= \left[24\left(\tfrac{5}{6}\right) + \left(0.4^{*} - 1\right)(0) + 0.4\,(-9.21)\right]_{\text{unit load @ 105}} \\
&= 16.3 \text{ ft} \\
&\left[\eta_{M205}\right]_{\text{unit load @ 105}} \\
&= \left[\eta_{M_{s205}} + (\beta - 1)\,\eta_{M200} + \beta\eta_{M300}\right]_{\text{unit load @ 105}} \\
&= 0^{\dagger} + (0.5^{\ddagger} - 1)(9.21) + (0.50)(2.51^{\P}) \\
&= -3.35 \text{ ft}
\end{aligned}$$

In summary, this method superimposes the effects of a unit load applied to the simple span with the effects of continuity (end moments). The unit load is applied only in the span containing the location of interest. Influence ordinates within this span are "affected" by the unit load and end effects. Function ordinates outside this span are affected only by the effects of continuity. Although a specific ordinate was used in this example, note that algebraic functions may be used in a similar fashion, and perhaps what is more important, general algorithms may be developed using Eqs. 9.16, 9.17, and 9.18 and subsequently coded in computer programs. In addition, this calculation is amenable to spreadsheet calculation.

9.5.3 Muller–Breslau Principle for End Moments§

The Muller–Breslau principle may be conveniently used to establish the influence functions for the end moments. Subsequently, the end moments may be used with Eqs. 9.16 and 9.17 to establish all other influence functions.

The Muller–Breslau principle requires that the displacements (in this case translation) be determined for the entire structure. The displacement of each element is solely a function of the end moments. The equation for the translation of a simple beam subjected to counterclockwise end moments M_{ij} and M_{ji} is

$$y = \frac{L^2}{6EI}\left[M_{ij}\left(2\varepsilon - 3\varepsilon^2 + \varepsilon^3\right) - M_{ji}\left(\varepsilon - \varepsilon^3\right)\right] \quad (9.19)$$

where $\varepsilon = x/L$, and y is the upward translation.

This equation can be derived many different ways, for example, direct integration of the governing equation. Verification is left to the reader. The Muller–Breslau procedure is described in Example 9.9.

Example 9.9* Use the Muller–Breslau principle to establish the influence function for the moment M_{104} for the beam of Example 9.6. Perform the calculations for the first span only.

The structure is reillustrated in Figure 9.15(a) for convenience. The influence function for the simple beam moment at the 104 point is illustrated in Figure 9.15(b). This function has been discussed previously and is not reiterated. Next determine the influence function for the end moments. Use the Muller–Breslau principle to release the moment at 110

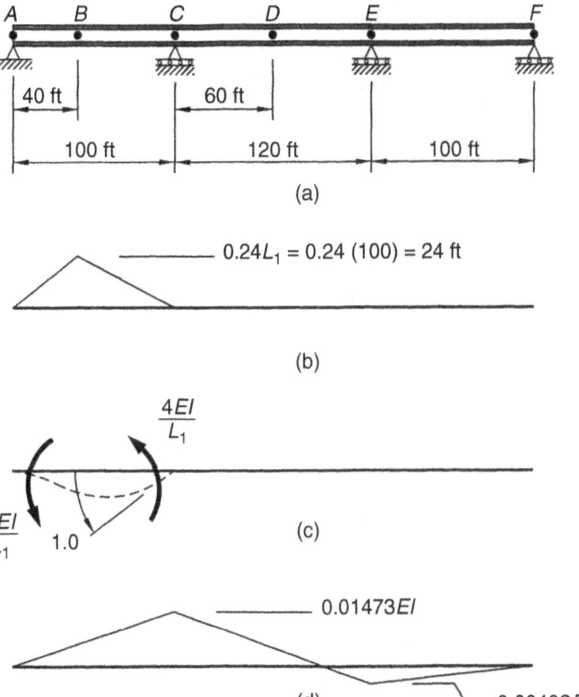

Fig. 9.15 (a) Continuous beam, (b) simple beam influence function for moment at 104, and (c) unit rotation at C member CA. (d) Moment due to unit rotation at C.

*Here β is the fraction of the span length from the left end to the point of interest, that is, 104 is 40% from the left end. Do not confuse this with the location of where the ordinate is calculated, that is, 105.

†The sign on the M_{200} and M_{300} ordinates has been changed from the table to switch to the slope–deflection convention, for example, Table 9.2 gives −9.21 and +9.21 is used here.

‡Here β is the fraction of the span length from the left end to the point of interest. In this case, 205 is located at 50% of the second span.

¶The influence function is for the moment at 205 and the ordinate is being calculated for the ordinate of this function at 105. The simple beam function is superimposed only if the location where the ordinate calculation is being performed (105) is in the same span as the location of the point of interest (205). In this case, the two locations are in different spans.

§Advanced material, may be skipped.

and impose a unit displacement. The displaced shape is shown in Figure 9.15(c). The end moments for element AB are determined using the slope–deflection equations given in Eq. 9.11. Let $\theta_j = 1.0$, $\theta_i = 0$, and the fixed-end moments due to element loads are zero. The end moments are $2EI/L_1$ and $4EI/L_1$ for the left and right ends, respectively.

These moments are the fixed-end moments used in the slope–deflection equations, that is,

$$M_{AC} = \frac{4EI}{L_1}\theta_A + \frac{2EI}{L_1}\theta_C + \frac{2EI}{L_1}$$
$$M_{CA} = \frac{4EI}{L_1}\theta_C + \frac{2EI}{L_1}\theta_A + \frac{4EI}{L_1}$$
$$M_{CE} = \frac{4EI}{L_2}\theta_C + \frac{2EI}{L_2}\theta_E$$
$$M_{EC} = \frac{4EI}{L_2}\theta_E + \frac{2EI}{L_2}\theta_C$$
$$M_{EF} = \frac{4EI}{L_3}\theta_E + \frac{2EI}{L_3}\theta_F$$
$$M_{FE} = \frac{4EI}{L_3}\theta_F + \frac{2EI}{L_3}\theta_E$$

Equilibrium requires

$$M_{AC} = 0$$
$$M_{CA} + M_{CE} = 0$$
$$M_{EC} + M_{EF} = 0$$
$$M_{FE} = 0$$

The slope–deflection equations are substituted into the equilibrium equations and the four rotations are established. The resulting rotations are $\theta_A = -0.2455$, $\theta_C = -0.5089$, $\theta_E = 0.1339$, and $\theta_F = 0.0670$. These rotations are back-substituted into the slope–deflection equations to establish the end moments given below:

$$M_{AC} = 0 \text{ ft kips}$$
$$M_{CA} = 14.73 \times 10^{-3}\ EI$$
$$M_{CE} = -14.73 \times 10^{-3} EI$$
$$M_{EC} = -4.02 \times 10^{-3} EI$$
$$M_{EF} = 4.02 \times 10^{-3} EI$$
$$M_{FE} = 0$$

The moment diagram is shown in Figure 9.15(d).

Equation 9.19 is used to determine the translation due the end moments for the first span. This equation is the influence function $\eta_{M_{110}}$ and is given in Table 9.3. A sample calculation is given for the ordinate at location 103, $\varepsilon = x/L = 0.3$:

$$\eta_{M_{110}} = \frac{100^2}{6EI}\{0.0EI\,[2(0.3) - 3(0.3)^2 + 0.3^3] - 14.73 \times 10^{-3}\,EI\,[0.3 - 0.3^3]\}$$
$$\eta_{M_{110}} = -6.70 \text{ ft}$$

A comparison of the values in Tables 9.2 and 9.3 reveals that the influence ordinate $\eta_{M_{104}}$ is the same. Use Eq. 9.18 to combine the influence function $\eta_{M_{110}}$ and the simple beam function $\eta_{S_{104}}$. The result is shown in Table 9.3 and a sample calculation for the ordinate at the 103 point is given.

$$[\eta_{M_{104}}]_{@103} = 18 + (0.4 - 1)\,(0) + (0.4)(-6.70) = 15.3 \text{ ft}$$

9.5.4 Automation by Matrix Structural Analysis

Traversing the unit action and the Muller–Breslau methods may be used in a stiffness-based matrix analysis. This unit action approach is conceptually straightforward and likely the easiest to implement in an existing stiffness-based code. Two

Table 9.3 Influence Ordinates for M_{104} for Span 1[a]

x (ft)	$\varepsilon = \frac{x}{L_1}$	$\eta_{M_{S104}}$ (Unit Load on Simple Beam) (ft)	$\eta_{M_{110}}$ (Influence Function for M_{110}) (ft) (Eq. 9.19)	$\eta_{M_{104}}$ (Influence Function for Moment at 104) (ft)
0	0	0	0	0
10	0.1	6	−2.43	5.03
20	0.2	12	−4.71	10.11
30	0.3	18	−6.70	15.32
40	0.4	24	−8.25	20.70
50	0.5	20	−9.21	16.32
60	0.6	16	−9.43	12.23
70	0.7	12	−8.77	8.49
80	0.8	8	−7.07	5.17
90	0.9	4	−4.20	2.32
100	1.0	0	0	0

[a] The parameter $L_1 = 100$ ft. = 30 480 mm. The procedure is the same for the remaining span but the simple beam contribution is zero, which is left to the reader as an exercise.

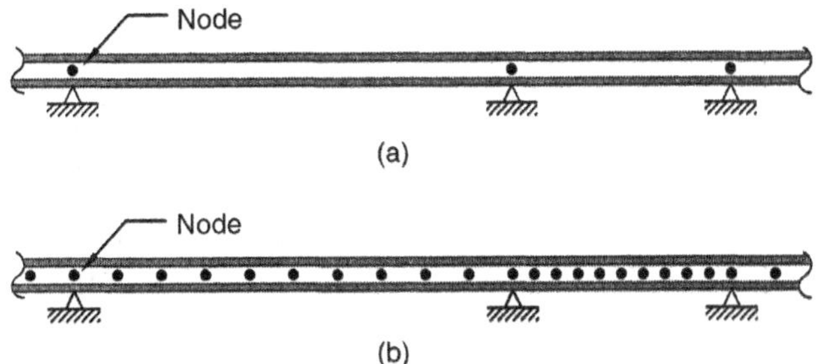

Fig. 9.16 (a) Discretized continuous beam—nodes at physical joints and (b) discretized continuous beam—nodes between physical joints.

approaches may be taken: (1) The structure is discretized with one element per span [Fig. 9.16(a)], or (2) the structure is discretized with several elements per span, often conveniently taken as 10 per span [Fig. 9.16(b)].

The use of one element per span requires special algorithms to:

- ☐ Generate the equivalent joint loads for load placement at any position within an element.
- ☐ Determine the end actions after the displacements are known. This procedure involves adding the fixed-end actions to the actions from the analysis of the released structure.
- ☐ Calculate the actions and displacements at the required locations in the interior of the element.

With these tools available, one can use the standard matrix approach to place the unit actions at regular intervals along the load path and to calculate the actions at the required locations. This involves the solution of multiple load cases on a small system of equations. The advantage is its computational efficiency. The disadvantages are its coding complexity and difficulties in including the nonprismatic effects that affect both stiffness and fixed-end action computations.

Alternatively, each span may be discretized into elements as illustrated in Figure 9.16(b). The node can be associated with the influence ordinates, eliminating the need for element load routines. The unit actions can be applied as joint loads, and each load case generates one ordinate in the influence functions. The element end actions are available at regular locations. Although this is computationally more time consuming than one element per span, it is simpler to code and the number of degrees of freedom required is relatively small by today's standards. Another advantage of this method is that the element cross-sectional properties may vary from element to element to account for the nonprismatic nature of the bridge. In addition, the displacements (e.g., influence function for translation at a point) are always available at every degree of freedom. The disadvantage is computational inefficiency, which is minor.

The Muller–Breslau method may be used with either discretation scheme. The advantage of the Muller–Breslau approach is that only one load case is required for each influence function. Because the displacements establish the function, the back-substitution process is eliminated. This process is a minor computational advantage for standard beam and frame elements. The major disadvantage is that with an automated approach, one expects to develop complete action envelopes for locations along the beam for design, usually tenth points. Therefore, one load case is required for each action considered, likely one action for each degree of freedom, which is several times (beam = 2, frame = 3, etc.) the number of load cases for the unit action traverse. The computational saving is that the displacements are the influence ordinates and action recovery is not required. If one element per span is used, then algorithms must be developed to determine the influence ordinates (displacements) in the interior of the element (e.g., Eqs. 9.16–9.19). This increases code complexity, especially if nonprismatic beams are required.

In summary, designing an automated approach for the generation of influence functions depends on the objectives and scope of the program. Typically, the combination of using multiple elements per span and applying a unit action as a joint load results in a code that is flexible, easy to maintain, and has the capability to generate influence functions for every end action in the system. Further, nonprismatic effects are naturally handled by changing the element properties. The computational efficacy for linear problems of this size becomes less important with ever increasing computational capability. With today's computers, the computation appears to be instantaneous.

9.6 NORMALIZED INFLUENCE FUNCTIONS

Influence functions may be considered a type of structural property, as they are independent of the load and dependent on the *relative* stiffness of each element. Consider the Muller–Breslau principle—the displaced shape due to an imposed displacement is dependent on the *relative*, not *absolute*, values of stiffness. For a continuous prismatic beam, the cross section and material stiffness do not vary with location; therefore, the influence functions are based on the only remaining parameter that affects stiffness, the span lengths. Note that in Figure 9.4, the influence functions for reaction and shear are independent of the span length, and the influence function for moment is proportional to span length. These relationships are similar for continuous beams, but here the shape is determined by the *relative* stiffness (in the case of a prismatic beam, the relative span lengths) and the ordinate values for moment are proportional to a characteristic span length.

For detailing and aesthetic reasons, bridges are often designed to be symmetrical about the center of the bridge; for example, the first and third span lengths of a three-span bridge are equal.

Table 9.4 Normalized Influence Functions (Three Span, Span Ratio = 1.2)[a]

Location	$M(104)$	$M(200)$	$M(205)$	$V(100)$	$V(104)$	$V(110)$	$V(200)$	$V(205)$
100	0.00000	0.00000	0.00000	1.00000	0.00000	0.00000	0.00000	0.00000
101	0.05028	−0.02431	−0.00884	0.87569	−0.12431	−0.12431	0.02578	0.02578
102	0.10114	−0.04714	−0.01714	0.75286	−0.24714	−0.24714	0.05000	0.05000
103	0.15319	−0.06703	−0.02437	0.63297	−0.36703	−0.36703	0.07109	0.07109
104	0.20700	−0.08250	−0.03000	0.51750	−0.4825/0.51750	−0.48250	0.08750	0.08750
105	0.16317	−0.09208	−0.03348	0.40792	0.40792	−0.59208	0.09766	0.09766
106	0.12229	−0.09429	−0.03429	0.30571	0.30571	−0.69429	0.10000	0.10000
107	0.08494	−0.08766	−0.03187	0.21234	0.21234	−0.78766	0.09297	0.09297
108	0.05171	−0.07071	−0.02571	0.12929	0.12929	−0.87071	0.07500	0.07500
109	0.02321	−0.04199	−0.01527	0.05801	0.05801	−0.94199	0.04453	0.04453
110 or 200	0.00000	0.00000	0.00000	0.00000	0.00000	−1.00000/0.0	0.0/1.00000	0.00000
201	−0.02037	−0.05091	0.02529	−0.05091	−0.05091	−0.05091	0.92700	−0.07300
202	−0.03333	−0.08331	0.05829	−0.08331	−0.08331	−0.08331	0.83600	−0.16400
203	−0.03996	−0.09990	0.09900	−0.09990	−0.09990	−0.09990	0.73150	−0.26850
204	−0.04135	−0.10337	0.14743	−0.10337	−0.10337	−0.10337	0.61800	−0.38200
205	−0.03857	−0.09643	0.20357	−0.09643	−0.09643	−0.09643	0.50000	−0.50/0.50
206	−0.03271	−0.08177	0.14743	−0.08177	−0.08177	−0.08177	0.38200	0.38200
207	−0.02484	−0.06210	0.09900	−0.06210	−0.06210	−0.06210	0.26850	0.26850
208	−0.01605	−0.04011	0.05829	−0.04011	−0.04011	−0.04011	0.16400	0.16400
209	−0.00741	−0.01851	0.02529	−0.01851	−0.01851	−0.01851	0.07300	0.07300
210 or 300	0.00000	0.00000	0.00000	0.00000	0.00000	0.00000	0.00000	0.00000
301	0.00458	0.01145	−0.01527	0.01145	0.01145	0.01145	−0.04453	−0.04453
302	0.00771	0.01929	−0.02571	0.01929	0.01929	0.01929	−0.07500	−0.07500
303	0.00956	0.02391	−0.03188	0.02391	0.02391	0.02391	−0.09297	−0.09297
304	0.01029	0.02571	−0.03429	0.02571	0.02571	0.02571	−0.10000	−0.10000
305	0.01004	0.02511	−0.03348	0.02511	0.02511	0.02511	−0.09766	−0.09766
306	0.00900	0.02250	−0.03000	0.02250	0.02250	0.02250	−0.08750	−0.08750
307	0.00731	0.01828	−0.02437	0.01828	0.01828	0.01828	−0.07109	−0.07109
308	0.00514	0.01286	−0.01714	0.01286	0.01286	0.01286	−0.05000	−0.05000
309	0.00265	0.00663	−0.00884	0.00663	0.00663	0.00663	−0.02578	−0.02578
310	0.00000	0.00000	0.00000	0.00000	0.00000	0.00000	0.00000	0.00000
Pos. Area span 1	0.09545	0.00000	0.00000	0.43862	0.13720	0.0000	0.06510	0.06510
Neg. Area Span 1	0.00000	−0.06138	−0.02232	0.00000	−0.09797	−0.56138	0.00000	0.00000
Pos Area Span 2	0.00000	0.00000	0.10286	0.00000	0.00000	0.00000	0.60000	0.13650
Neg. Area Span 2	−0.03086	−0.07714	0.00000	−0.07714	−0.07714	−0.07714	0.00000	−0.13650
Pos. Area Span 3	0.00670	0.01674	0.00000	0.01674	0.01674	0.01674	0.00000	0.00000
Neg. Area Span 3	0.00000	0.00000	−0.02232	0.00000	0.00000	0.00000	−0.06510	−0.06510
Total Pos. Area	0.10214	0.01674	0.10286	0.45536	0.15394	0.01674	0.66510	0.20160
Total Neg. Area	−0.03086	−0.13853	−0.04464	−0.07714	−0.17512	−0.63853	−0.06510	−0.20160
Net Area	0.07129	−0.12179	0.05821	0.37821	−0.02117	−0.62179	0.60000	0.00000

[a]Usage:
Multiply influence ordinates for moment by length of span 1.
Multiply areas for moment by length of (span 1)^2.
Multiply areas for shear by length of span 1.
Notes:
Area $M(205)+$ for span 2 is 0.1036, 0.1052, and 0.1029 for trapezoidal, Simpson's, and exact integration, respectively.
Areas $V(205)+$ and $V(205)-$ for span 2 were computed by Simpson's integration.

For economy and ease of detailing and construction, the engineer sets the span length to meet the geometric constraints and, if possible, to have similar controlling actions in all spans. In a continuous structure, making the outer spans shorter than the interior spans balances the controlling actions. Typical span ratios vary from 1.0 to 1.7. The spans for Example 9.6 are 100, 120, and 100 ft (30,480, 36,576, and 30,480 mm) that have a span ratio of 1.2. The shear influence functions from Example 9.6 can be used for a prismatic three-span continuous girder bridge with the same span ratio (i.e., 1.2). Similarly, the moment influence functions can be proportioned. For example, a bridge with spans of 35, 42, and 35 ft (10,668, 12,802, and 10,668 mm) has a span ratio of 42 : 35 = 1.2. With the use of the first span as the characteristic span, the ordinates can be proportioned by 35 : 100 = 0.35. For example, the smaller bridge has a maximum ordinate for the moment at 104 of

$$\eta_{M_{104}} = (0.35)\,(20.70) = 7.25 \text{ ft}$$

The American Institute for Steel Construction (AISC) published tables of normalized influence functions for various span configurations and span ratios (AISC, 1986). These tables were generated for a characteristic span length of 1.0. This format allows the engineer to use the tabulated values by multiplying by the actual characteristic span length. Table 9.2 (span ratio = 1.2) is normalized to a unit length for span one and the results are given in Table 9.4. This table is used in several examples that follow.

9.7 AASHTO VEHICLE LOADS

The AASHTO vehicle loads defined in Chapter 8 are used to determine the load effects for design. Because the vehicle loads are moving loads, load placement for maximum load effect may not be obvious. The influence function for a particular action is used in combination with the prescribed load to establish the load position for analysis. The engineer may place the load at one or more positions by inspection and calculate the load effect for each load placement using Eq. 9.1. The maximum and minimum values are noted and used in subsequent design calculations.

Alternatively, the load is periodically positioned along the same path used to generate the influence function. For each placement, the load effect is calculated and compared to the previous one. The maximum and minimum load effects are recorded. This approach is most appropriate for automation and is the technique most often employed in computer programs that generate load effect envelopes.

The critical load placement is sometimes obvious when the influence function is available. As illustrated in Example 9.10, this is the case for the analysis of statically determinate beams.

Critical load placement on an influence function gives the maximum or minimum load effect for the particular action at the location associated with that function. Unfortunately, this location is likely not the location that gives the absolute critical load effect in the span. For example, typical influence functions are generated at tenth points, but the critical location may be between the tenth point locations. Tenth point approximations are typically very close to the absolute maximum/minimums.

This critical location can be theoretically established for simple beams, and this formulation can be found in most elementary texts on structural analysis (e.g., Hibbeler and Hibbeler, 2004). We have chosen not to focus a great deal of attention on this aspect. From a practical perspective, the method only works for simple-span bridges. Automated approaches are written in a general way to accommodate both statically determinate and indeterminate systems with the same algorithms. Lastly, the absolute maximum or minimum load effect does not differ significantly from the tenth point approximation. The two methods are compared in the following example.

Example 9.10 Use the influence functions determined in Example 9.1 to calculate the maximum reaction R_{100}, shear V_{100}, and moment M_{105} for the AASHTO vehicle loads (AASHTO, 2010). Use a 35-ft (10,668-mm) span.

The influence lines for the actions required are shown in Figures 9.17(a)–9.17(d). The critical actions for the design truck, design tandem, and the design lane loads are determined independently and are later superimposed as necessary. The design truck is used first, followed by the design tandem, and finally, the design lane load.

Design Truck Load

The critical load placement for R_{100} is shown in Figure 9.17(e). By using Eq. 9.1, this reaction is determined as

$$R_{100} = \sum_{i=1}^{3} P_i \eta_i = 32\,(1) + 32\left(\frac{21}{35}\right) + 8\left(\frac{7}{35}\right)$$

$$= 32 + 19.2 + 1.60 = 52.8 \text{ kips}$$

Note that $R_{100} = V_{100} = 52.8$ kips.

The critical load placement for M_{105} is illustrated in Figure 9.17(f). Multiplication of the loads times the ordinates gives

$$M_{105} = 8.75[(32)(1) + 32(3.5/17.5) + 8(3.5/17.5)]$$

$$= 350 \text{ ft kips}$$

Increasing the distance between the rear axles spreads the load and decreases the load effect. Thus, the 14-ft (4300-mm) variable axle spacing is critical; this will be the case for simple spans. The variable axle spacing can become critical for short multispan beams where the truck length is approximately the same as the span lengths.

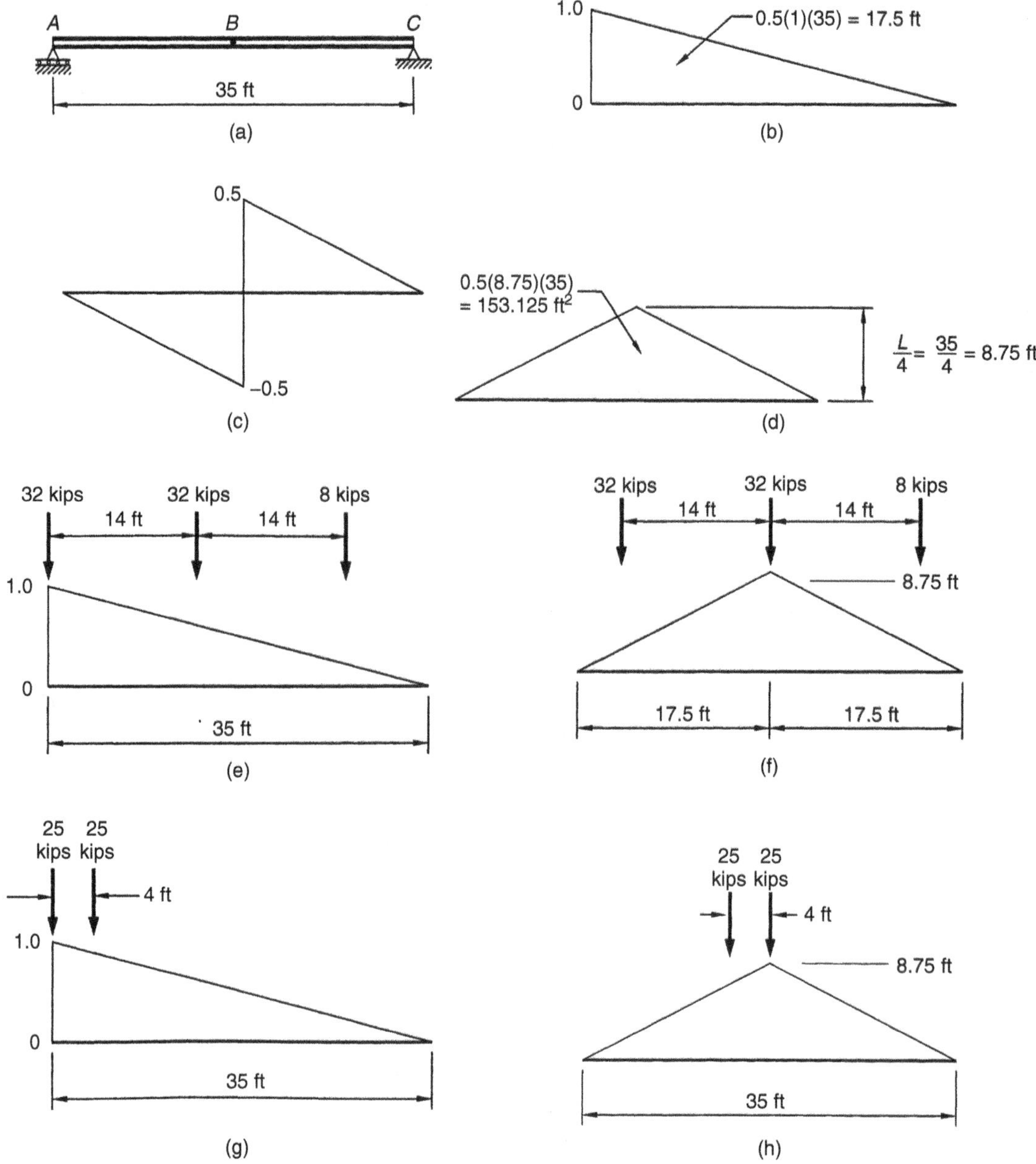

Fig. 9.17 (a) Simple beam, (b) influence function R_A and V_A, (c) influence function V_B, (d) influence function M_B, (e) design truck positioned for R_A and V_A, (f) design truck positioned for M_B, (g) tandem truck positioned for R_A and V_A, and (h) tandem truck positioned for M_B.

Design Tandem Load

To determine R_{100}, the design tandem loads are placed as illustrated in Figure 9.17(g). The reaction is

$$R_{100} = 25\,(1) + 25\left(\tfrac{31}{35}\right) = 47.1 \text{ kips}$$

Again note, $V_{100} = R_{100} = 47.1$ kips.

The maximum moment at midspan is determined by placing the design tandem as shown in Figure 9.17(h). The result is

$$\begin{aligned} M_{105} &= 8.75[25(1) + 25(13.5/17.5)] \\ &= 387.5 \text{ ft kips} \end{aligned}$$

Design Lane Load

Equation 9.3 is used to determine the shears and moments for the uniform lane load of 0.64 kip/ft (9.3 N/mm). This uniform load is multiplied by the appropriate area under the influence function. For example, the integral of the influence function for R_{100} is the area of a triangle or

$$\text{Area} = (1)\,(35)\,/2 = 17.5 \text{ ft}$$

Thus, the reaction R_{100} is calculated as

$$R_{100} = (0.64 \text{ kip/ft})(17.5 \text{ ft}) = 11.2 \text{ kips}$$

As before, $V_{100} = R_{100} = 11.2$ kips.

By using Figure 9.17(d), the moment at midspan is

$$\begin{aligned} M_{105} &= (0.64\ \text{kip/ft})\left[(8.75)(35)\left(\tfrac{1}{2}\right)\text{ft}^2\right] \\ &= (0.64\ \text{kip/ft})(153.1\ \text{ft}^2) \\ &= 98\ \text{ft kips} \end{aligned}$$

The *absolute* maximum reaction and shear are as shown above, but the *absolute* maximum moments are slightly different. The actions for *simple beams* may be established with the following rules (AASHTO, 2010):

1. The maximum shear due to moving concentrated loads occurs at one support when one of the loads is at the support. With several moving loads, the location that will produce maximum shear must be determined by trial.
2. The maximum bending moment produced by moving concentrated loads occurs under one of the loads when that load is as far from one support as the center of gravity of all the moving loads on the beam is from the other support.

Position the design truck with the rear at the short spacing of 14 ft and locate this rear axle at 6.61 ft from left support. This position results in a maximum moment under the 32-kip wheel of 360 ft kips, which is slightly greater than the value at the 105 point (350 ft kips). Position the design tandem wheel at 16.5 ft from the left, resulting in a maximum moment of 388 ft kips under the wheel. The differences between these moments and the moments at the 105 point are approximately 0.97 and 0.1% for the design truck and tandem trucks, respectively. The absolute maximum moments are also given in Table 9.5. Note that the design lane load must be added to the design truck and to the design tandem loads [A3.6.1.3.1].* The maximums for these load cases occur at different locations, that is, the uniform lane load is at a maximum at midspan. This further complicates the analysis. A rigorous approach must determine the absolute maximum for the combined factored loads, which is only reasonable for simple spans. These calculations are summarized in Table 9.5.

Example 9.11 Repeat Example 9.10 for a 100-ft (30,480-mm) span. The calculations are given below.

Design Truck Load

$$\begin{aligned} R_{100} &= 32(1) + 32\left(\tfrac{86}{100}\right) + 8\left(\tfrac{72}{100}\right) = 65.3\ \text{kips} \\ V_{100} &= R_{100} \\ M_{105} &= \left(\tfrac{100}{4}\right)\left[32(1) + 32\left(\tfrac{36}{50}\right) + 8\left(\tfrac{36}{50}\right)\right] \\ &= 1520\ \text{ft kips} \end{aligned}$$

*The article numbers in the AASHTO (2010) LRFD Bridge Specifications are enclosed in brackets and preceded by the letter A if a specification article and by the letter C if commentary.

Table 9.5 Service-Level Vehicle Design Loads[a,b,c]

Load	$R_{100} = V_{100}$ (kips)	M_{105} (ft kips)	Absolute Max (ft kips)
Design truck	52.8	350.0	360
Design tandem	47.1	387.5	389
Design lane	11.2	98.0	98.0
(1.33)Truck + lane	**81.4**	563.5	576.8
(1.33)Tandem + lane	73.8	**613.4**	**615.4**[d]

[a]Simple span = 35 ft (10 668 mm).
[b]The critical values are in boldface.
[c]The typical impact factors for the truck and tandem is 1.33 and the lane load is 1.00. These factors are discussed in Chapter 8.
[d]Used lane load moment at midspan.

Design Tandem Load

$$\begin{aligned} R_{100} &= 25(1) + 25\left(\tfrac{96}{100}\right) = 49.0\ \text{kips} \\ V_{100} &= R_{100} \\ M_{105} &= \left(\tfrac{100}{4}\right)\left[25(1) + 25\left(\tfrac{46}{50}\right)\right] \\ &= 1200\ \text{ft kips} \end{aligned}$$

Design Lane Load

$$\begin{aligned} R_{100} &= 0.64\left(\tfrac{1}{2}\right)(1)(100) = 32\ \text{kips} \\ V_{100} &= R_{100} \\ M_{105} &= 0.64\left(\tfrac{1}{2}\right)\left(\tfrac{100}{4}\right)(100) = 800\ \text{ft kips} \end{aligned}$$

The actions for the truck and tandem loads are combined with the lane load in Table 9.6.

The procedure for determining the actions in a continuous beam is similar to that illustrated for a simple beam. As illustrated previously, the influence diagrams are slightly more complicated as the functions are nonlinear with both positive and negative ordinates. To illustrate the calculation of the load effects for a continuous system, the three-span continuous beam of Example 9.6 is used. A few actions are used for illustration and the remaining actions required for design follow similar procedures.

Table 9.6 Service-Level Design Loads[a]

Load	$R_{100} = V_{100}$ (kips)	M_{104} (ft kips)
Design truck	65.3	1520
Design tandem	49.0	1200
Design lane	32.0	800
(1.33)Truck + lane	**118.8**	**2822**
(1.33)Tandem + lane	97.2	2396

[a]Span = 100 ft (30 480 mm).

Example 9.12 Determine the shear V_{100}, the moment M_{104}, and the moment $M_{110} = M_{200}$ for the beam of Example 9.6 (Fig. 9.11). Use the normalized functions given in Table 9.4. The span lengths are 100, 120, and 100 ft (30,480, 36,576, and 30,480 mm). Use the AASHTO vehicle loads.

Design Lane Load

Use the normalized areas at the bottom of Table 9.4 for the lane loads. Note that these areas require multiplication by the characteristic span length for shear and by the span length squared for moment. The positive and negative areas are used for the associated actions.

$$
\begin{aligned}
V_{100^-} &= 0.64(-0.077\,14)(100) = -4.94 \text{ kips} \\
V_{100^+} &= 0.64(0.455\,36)(100) = 29.1 \text{ kips} \\
M_{104^+} &= 0.64(0.102\,14)(100^2) = 653.6 \text{ ft kips} \\
M_{104^-} &= 0.64(-0.030\,86)(100^2) = -197.5 \text{ ft kips} \\
M_{110^-} &= 0.64(-0.138\,53)(100^2) = -886.6 \text{ ft kips} \\
M_{110^+} &= 0.64(0.016\,74)(100^2) = 107.1 \text{ ft kips}
\end{aligned}
$$

Design Tandem Load

The tandem axle is applied to the structure and the load effects are calculated with Eq. 9.1. The load placement is by inspection and noted below for each action.

For V_{100}, place the left axle at 100 and the second axle at 4 ft (1200 mm) from the left end. The influence ordinate associated with the second axle is determined by linear interpolation:

$$
\begin{aligned}
V_{100^+} &= 25(1) + 25\left[1 - \left(\tfrac{4}{10}\right)(1 - 0.875\,69)\right] \\
&= 25 + 23.75 = 48.75 \text{ kips}
\end{aligned}
$$

For the most negative reaction at 100, position the right axle at 204:

$$
\begin{aligned}
V_{100^-} &= 25(-0.103\,37) + 25\left[-0.103\,37 + \left(\tfrac{4}{12}\right)\right. \\
&\qquad \left.\times\ (0.103\,37 - 0.099\,90)\right] \\
V_{100^-} &= -2.58 - 2.56 = -5.14 \text{ kips}
\end{aligned}
$$

For the positive moment at 104, position the left axle at 104 (approximate). Again, determine the ordinate for the second axle by interpolation:

$$
\begin{aligned}
M_{104^+} &= 25(0.207\,00)(100) + 25\left[0.207\,00 - \left(\tfrac{4}{12}\right)(0.207\,00\right. \\
&\qquad \left.-0.163\,17)\right](100) \\
&= 517.5 + 481.0 = 998 \text{ ft kips}
\end{aligned}
$$

Position the right axle at 204 for the most negative moment at 104 (approximate). The result is

$$
\begin{aligned}
M_{104^-} &= 25(-0.041\,35)(100) \\
&\quad + 25\left[-0.041\,35 - \left(\tfrac{4}{10}\right)(-0.041\,35\right. \\
&\qquad \left.+0.039\,96)\right](100) \\
M_{104^-} &= -103.4 - 102.0 = -205.4 \text{ ft kips}
\end{aligned}
$$

Position the right axle at 204 for the most negative moment at 110.

$$
\begin{aligned}
M_{110^-} &= 25(-0.103\,37)(100) \\
&\quad + 25\left[-0.103\,37 - \left(\tfrac{4}{10}\right)(-0.103\,37\right. \\
&\qquad \left.+0.099\,90)\right](100) \\
&= -258.4 - 254.9 = -513.3 \text{ kips}
\end{aligned}
$$

Design Truck Load

Position the rear axle at 100 for the maximum reaction (position truck traveling to the right = forward):

$$
\begin{aligned}
R_{100^+} &= 32(1) + 32(0.8266) + 8(0.6569) \\
&= 32.0 + 26.45 + 5.26 = 63.7 \text{ kips}
\end{aligned}
$$

Position the middle axle at 104 for the positive moment at 104 (backward):

$$
\begin{aligned}
M_{104^+} &= 8\left[0.153\,19 - \left(\tfrac{4}{10}\right)(0.153\,19 - 0.101\,14)\right](100) \\
&\quad + 32(0.207\,00)(100) \\
&\quad + 32\left[0.163\,17 - \left(\tfrac{4}{10}\right)(0.163\,17 - 0.122\,29)\right](100) \\
&= 106.0 + 662.4 + 469.8 = 1238.2 \text{ ft kips}
\end{aligned}
$$

Position the middle axle at 204 for the most negative moment at 104 (forward):

$$
\begin{aligned}
M_{104^-} &= 8\left[-0.038\,57 - \left(\tfrac{2}{12}\right)(-0.038\,57 + 0.032\,71)\right](100) \\
&\quad + 32(-0.041\,35)(100) \\
&\quad + 32\left[-0.039\,96 - \left(\tfrac{2}{12}\right)(-0.039\,96 + 0.033\,33)\right] \\
&\quad \times (100) \\
&= -30.1 - 132.3 - 124.3 = -286.6 \text{ ft kips}
\end{aligned}
$$

Position the middle axle at 204 for the most negative moment at 110 (forward):

$$
\begin{aligned}
M_{110^-} &= 32\left[-0.0999 - \left(\tfrac{2}{12}\right)(-0.0999 + 0.083\,31)\right](100) \\
&\quad + 32(-0.103\,37)(100) \\
&\quad + 8\left[-0.096\,43 - \left(\tfrac{2}{12}\right)(-0.096\,43 + 0.081\,77)\right] \\
&\quad \times (100) \\
&= -310.8 - 330.8 - 75.2 = -716.8 \text{ ft kips}
\end{aligned}
$$

(A slightly different position in the automated approach gives −720 ft kips)

Position the middle axle at 304 for the maximum positive moment at 110 (backward):

$$
\begin{aligned}
M_{110^+} &= 8\left[0.0239 - \left(\tfrac{4}{10}\right)(0.0239 - 0.019\,29)\right](100) \\
&\quad + 32(0.025\,71)(100) \\
&\quad + 32\left[0.025\,11 - \left(\tfrac{4}{10}\right)(0.025\,11 - 0.022\,50)\right](100) \\
&= 17.6 + 82.2 + 77.0 = 176.8 \text{ ft kips}
\end{aligned}
$$

In the previous example, actions at selected points were determined. This procedure is generally permitted as long

Table 9.7 Action Envelopes for Three-Span Continuous Beam 100, 120, 100 ft (30 480, 36 576, and 30 480 mm)[a]

	Action Envelope Live-Load Actions (Critical Values in Bold)												
	$V+$, kips			$V-$, kips			$M+$, ft kips			$M-$, ft kips			
Location, ft	1.33 × Truck + Lane	1.33 × Tandem + Lane	Critical	1.33 × Truck + Lane	1.33 × Tandem + Lane	Critical	1.33 × Truck + Lane	1.33 × Tandem + Lane	Critical	1.33 × Truck + Lane	1.33 × Tandem + Lane	0.9 × 1.33 Train 0.9 × Lane	Critical
0	**113.9**	94.0	113.9	**−14.5**	−11.7	−14.5	**0.0**	0.0	0.0	**0.0**	0.0	0.0	0.0
10	**96.2**	79.8	96.2	**−14.9**	−12.1	−14.9	**989.9**	825.7	989.9	**−144.6**	−117.2	−137.3	−144.6
20	**79.6**	66.4	79.6	**−20.2**	−21.3	−21.3	**1687.2**	1424.8	1687.2	**−289.3**	−234.5	−274.6	−289.3
30	**64.1**	54.1	64.1	**−32.8**	−31.2	−32.8	**2107.1**	1803.8	2107.1	**−433.9**	−351.7	−411.8	−433.9
40	**49.9**	42.8	49.9	**−46.9**	−41.7	−46.9	**2301.5**	1973.0	2301.5	**−578.5**	−468.9	−549.1	−578.5
50	**37.2**	32.6	37.2	**−61.3**	−52.5	−61.3	**2273.0**	1954.7	2273.0	**−723.2**	−586.2	−686.4	−723.2
60	**26.0**	23.7	26.0	**−75.7**	−63.5	−75.7	**2051.0**	1766.6	2051.0	**−867.8**	−703.4	−823.7	−867.8
70	**16.4**	16.0	16.4	**−90.2**	−74.6	−90.2	**1618.4**	1416.4	1618.4	**−1012.4**	−820.6	−961.0	−1012.4
80	**8.8**	9.5	9.5	**−104.3**	−85.6	−104.3	**1012.1**	928.9	1012.1	**−1157.1**	−937.9	−1098.3	−1157.1
90	**3.6**	4.3	4.3	**−117.9**	−96.3	−117.9	**385.4**	436.0	436.0	−1404.8	−1158.2	**−1535.3**	−1535.3
100	**3.4**	2.8	3.4	**−130.8**	−106.5	−130.8	**341.5**	276.3	341.5	−1835.3	−1561.3	**−2428.4**	−2428.4
100	**132.4**	108.2	132.4	**−13.3**	−10.8	−13.3	**341.5**	276.3	341.5	−1835.3	−1561.3	**−2428.4**	−2428.4
112	**116.8**	95.8	116.8	**−13.6**	−11.0	−13.6	**419.8**	464.5	464.5	−1221.4	−1012.8	**−1398.8**	−1398.8
124	**100.4**	82.8	100.4	**−14.9**	−15.2	−15.2	**1101.9**	990.7	1101.9	**−934.3**	−756.2	−846.6	−934.3
136	**83.8**	69.7	83.8	**−25.3**	−23.7	−25.3	**1750.4**	1516.6	1750.4	−816.4	−668.6	**−819.4**	−819.4
148	**67.6**	56.9	67.6	**−37.8**	−33.6	−37.8	**2151.1**	1847.3	2151.1	−706.5	−589.2	**−819.4**	−819.4
160	**52.1**	44.8	52.1	**−52.1**	−44.8	−52.1	**2271.0**	1954.7	2271.0	−596.7	−509.8	**−819.4**	−819.4
172	**37.8**	33.6	37.8	**−67.6**	−56.9	−67.6	**2151.1**	1847.3	2151.1	−706.5	−589.2	**−819.4**	−819.4
184	**25.3**	23.7	25.3	**−83.8**	−69.7	−83.8	**1750.4**	1516.6	1750.4	−816.4	−668.6	**−819.5**	−819.5
196	**14.5**	15.2	15.2	**−100.4**	−82.8	−100.4	**1101.9**	990.7	1101.9	**−934.4**	−756.2	−846.6	−934.4
208	**13.6**	11.0	13.6	**−116.8**	−95.8	−116.8	**419.8**	464.5	464.5	−1221.4	−1012.9	**−1398.8**	−1398.8
220	**13.3**	10.7	13.3	**−132.4**	−108.2	−132.4	**341.4**	276.3	341.4	−1835.3	−1561.3	**−2428.4**	−2428.4
220	**130.8**	106.5	130.8	**−3.4**	−2.8	−3.4	**341.4**	276.3	341.4	−1835.3	−1561.3	**−2428.4**	−2428.4
230	**117.9**	96.3	117.9	**−3.7**	−4.3	−4.3	**358.7**	436.0	436.0	−1404.8	−1158.2	**−1535.3**	−1535.3
240	**104.3**	85.6	104.3	**−8.8**	−9.5	−9.5	**1012.1**	928.9	1012.1	**−1157.1**	−937.9	−1098.3	−1157.1
250	**90.2**	74.6	90.2	**−16.4**	−16.0	−16.4	**1618.4**	1416.4	1618.4	**−1012.5**	−820.6	−961.0	−1012.5
260	**75.7**	63.5	75.7	**−26.0**	−23.7	−26.0	**2051.1**	1766.6	2051.1	**−867.8**	−703.4	−823.7	−867.8
270	**61.3**	52.5	61.3	**−37.2**	−32.6	−37.2	**2273.1**	1954.8	2273.1	**−723.2**	−586.2	−686.4	−723.2
280	**46.9**	41.7	46.9	**−49.9**	−42.8	−49.9	**2301.5**	1973.1	2301.5	**−578.6**	−468.9	−549.2	−578.6
290	**32.8**	31.2	32.8	**−64.1**	−54.1	−64.1	**2107.1**	1803.8	2107.1	**−433.9**	−351.7	−411.9	−433.9
300	**20.2**	21.3	21.3	**−79.6**	−66.4	−79.6	**1687.2**	1424.8	1687.2	**−289.3**	−234.5	−274.6	−289.3
310	**14.9**	12.1	14.9	**−96.2**	−79.8	−96.2	**989.9**	825.7	989.9	**−144.6**	−117.2	−137.3	−144.6
320	**14.5**	11.7	14.5	**−113.9**	−94.0	−113.9	**0.0**	0.0	0.0	**0.0**	0.0	0.0	0.0

[a] The truck, tandem, and train vehicle actions are multiplied by the dynamic load allowance of 1.33 prior to combining with the lane load.

as the points and actions selected are representative of the extreme values (action envelope). These points are summarized in Table 9.1. Alternatively, all the extreme actions are determined at enough sections so that the envelope is represented. This is an extremely tedious process if performed by hand. Typically, all actions are determined at the tenth points. Therefore, this approach is most often automated.

A computer program called BT Beam—LRFD Analysis (BridgeTech, Inc., 1996) was used to develop the envelope of all actions at the tenth points. The web-based program was put online in 2010. (Search "BTBeam online" or http://www.bridgetech-laramie.com/BTBeam) This program is free. An output file name.sht provides a text file that is amenable for import into a spreadsheet.

The automated procedure performs the calculations as presented in this chapter except that it uses a matrix formulation rather than a slope–deflection analysis. For a beam analysis, these analyses are identical. The results from this analysis are given in Table 9.7. A comparison of the values in this table with those calculated previously shows minor differences. These differences are attributed to the load positioning procedures. The automated procedure moves the load along the influence diagram at relatively small intervals and the maxima/minima are stored. The hand calculations are based on a single load position estimating the maximum/minimum load effect. The critical values illustrated in Table 9.7 are plotted in Figures 9.18 and 9.19.

The AASHTO vehicle loads are also applied to a three-span continuous beam with spans of 35, 42, and 35 ft (10,668, 12,802, and 10,668 mm). The results are presented in Table 9.8 and critical values are illustrated in Figures 9.20 and 9.21. The values in Tables 9.7 and 9.8 and the associated figures are referenced in the design examples presented in the remaining chapters.

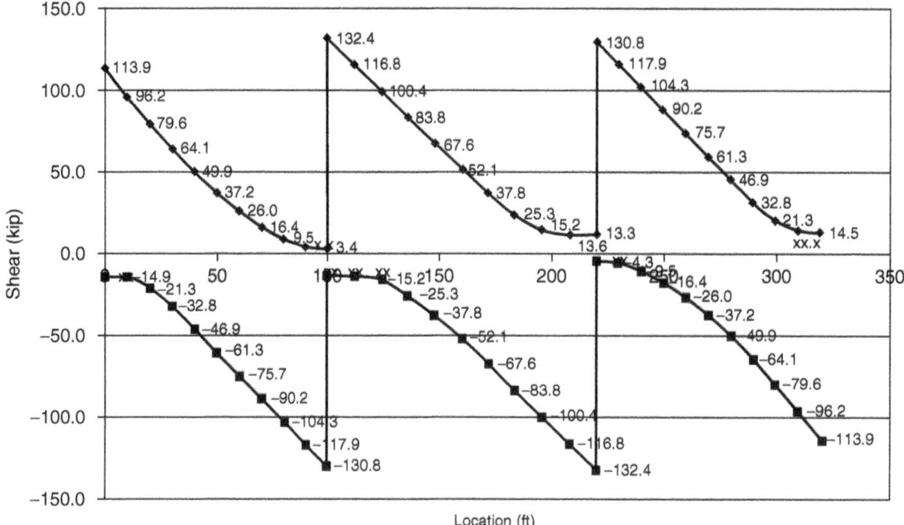

Fig. 9.18 Live-load shear envelope for HL-93 on 100–120–100 ft beam.

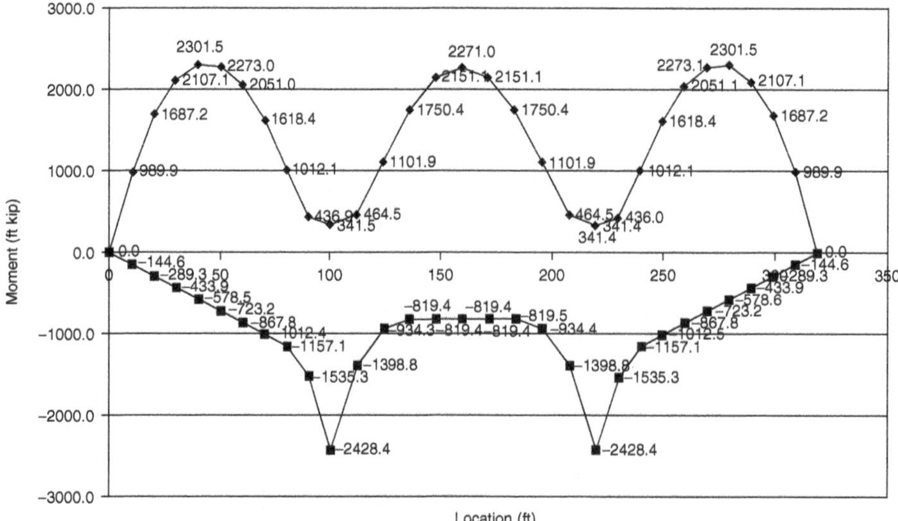

Fig. 9.19 Live-load moment envelope for HL-93 on 100–120–100 ft beam.

Table 9.8 Action Envelopes for Three-Span Continuous Beam 35, 42, 35 ft (10 668, 12 802, and 10 668 mm)[a]

	Action Envelope Live-Load Actions (Critical Values in Bold)												
	$V+$, kips			$V-$, kips			$M+$, ft kips			$M-$, ft kips			
Location, ft	1.33 × Truck + Lane	1.33 × Tandem + Lane	Critical	1.33 × Truck + Lane	1.33 × Tandem + Lane	Critical	1.33 × Truck + Lane	1.33 × Tandem + Lane	Critical	1.33 × Truck + Lane	1.33 × Tandem + Lane	0.9 × 1.33 Train 0.9 × Lane	Critical
0	**76.2**	72.0	76.2	**−9.4**	−8.5	−9.4	**0.0**	0.0	0.0	**0.0**	0.0	0.0	0.0
3.5	**63.4**	61.7	63.4	**−9.6**	−8.6	−9.6	**225.2**	219.4	252.2	−23.7	**−29.7**	−28.1	−29.7
7	**51.3**	51.8	51.8	−12.8	**−14.0**	−14.0	371.2	**374.5**	374.5	−47.3	**−59.3**	−56.2	−59.3
10.5	40.7	**42.5**	42.5	−19.0	**−22.8**	−22.8	449.7	**468.1**	468.1	−71.0	**−89.0**	−84.3	−89.0
14	31.0	**33.7**	33.7	−25.2	**−31.6**	−31.6	465.6	**504.2**	504.2	−94.7	**−118.6**	−112.5	−118.6
17.5	22.2	**25.7**	25.7	−35.6	**−40.3**	−40.3	451.7	**495.8**	495.8	−118.3	**−148.3**	−140.6	−148.3
21	14.6	**18.4**	18.4	−46.6	**−48.8**	−48.8	423.6	**449.5**	449.5	−142.0	**−178.0**	−168.7	−178.0
24.5	10.1	**12.1**	12.1	**−57.8**	−57.0	−57.8	343.6	**359.0**	359.0	−165.7	**−207.6**	−196.8	−207.6
28	6.2	**6.6**	6.6	**−68.4**	−64.8	−68.4	209.9	**232.5**	232.5	−189.4	**−237.3**	−224.9	−237.3
31.5	**2.9**	2.4	2.9	**−78.7**	−72.0	−78.7	**94.5**	92.3	94.5	−274.3	**−279.6**	−264.4	−279.6
35	**2.2**	2.0	2.2	**−89.0**	−78.5	−89.0	**75.8**	71.8	75.8	**−422.9**	−344.3	−351.0	−422.9
35	**90.2**	79.1	90.2	**−8.4**	−8.0	−8.4	**75.8**	71.8	75.8	**−422.9**	−344.3	−351.0	−422.9
39.2	**78.5**	71.0	78.5	**−8.5**	−8.1	−8.5	103.5	**106.0**	106.0	−243.5	**−245.7**	−232.6	−245.7
43.4	**66.2**	62.2	66.2	−8.8	**−9.9**	−9.9	234.5	**258.3**	258.3	−153.4	**−196.5**	−186.7	−196.5
47.6	**53.9**	52.9	53.9	−13.9	**−17.0**	−17.0	375.6	**389.9**	389.9	−132.3	−168.0	**−180.1**	−180.1
51.8	42.2	**43.4**	43.4	−21.6	**−25.1**	−25.1	457.4	**470.3**	470.3	−112.2	−140.6	**−177.8**	−177.8
56	31.5	**34.0**	34.0	−31.5	**−34.0**	−34.0	472.5	**492.8**	492.8	−92.1	−113.1	**−175.5**	−175.5
60.2	21.6	**25.1**	**25.1**	**−42.2**	**−43.4**	−43.4	457.4	**470.4**	470.4	−112.2	−140.6	**−177.8**	−177.8
64.4	13.9	**17.0**	17.0	**−53.9**	−52.9	−53.9	375.6	**389.9**	389.9	−132.3	−168.0	**−180.1**	−180.1
68.6	8.8	**9.9**	9.9	**−66.2**	−62.2	−66.2	234.5	**258.3**	258.3	−153.4	**−196.5**	−186.7	−196.5
72.8	**8.5**	8.1	8.5	**−78.5**	−71.0	−78.5	103.5	**106.0**	106.0	−243.5	**−245.7**	−232.5	−245.7
77	**8.4**	8.0	8.4	**−90.2**	−79.1	−90.2	**75.8**	71.8	75.8	**−422.9**	−344.3	−350.8	−422.9
77	**89.0**	78.5	89.0	**−2.2**	−2.0	−2.2	**75.8**	71.8	75.8	**−422.9**	−344.1	−350.8	−422.9
80.5	**78.7**	72.0	78.7	**−2.9**	−2.4	−2.9	**94.5**	92.3	94.5	−274.3	**−279.6**	−263.9	−279.6
84	**68.4**	64.8	68.4	−6.2	**−6.6**	−6.6	209.9	**232.5**	232.5	−189.4	**−237.3**	−224.5	−237.3
87.5	**57.8**	57.0	57.8	−10.1	**−12.1**	−12.1	343.7	**359.0**	359.0	−165.7	**−207.6**	−196.5	−207.6
91	46.6	**48.8**	48.8	−14.6	**−18.4**	−18.4	423.6	**449.5**	449.5	−142.0	**−178.0**	−168.4	−178.0
94.5	35.6	**40.3**	40.3	−22.2	**−25.7**	−25.7	451.7	**495.7**	495.7	−118.3	**−148.3**	−140.3	−148.3
98	25.2	**31.6**	31.6	−31.0	**−33.7**	−33.7	465.6	**504.2**	504.2	−94.7	**−118.6**	−112.3	−118.6
101.5	19.0	**22.8**	22.8	−40.7	**−42.5**	−42.5	449.7	**468.1**	468.1	−71.0	**−89.0**	−84.2	−89.0
105	12.8	**14.0**	14.0	−51.3	**−51.8**	−51.8	371.2	**374.5**	374.5	−47.3	**−59.3**	−56.1	−59.3
108.5	**9.6**	8.6	9.6	**−63.4**	−61.7	−63.4	**225.2**	219.3	225.2	−23.7	**−29.7**	−28.1	−29.7
112	**9.4**	8.5	9.4	**−76.2**	−72.0	−76.2	**0.0**	0.0	0.0	0.0	**0.0**	0.0	0.0

[a]The truck, tandem, and train vehicle actions are multiplied by the dynamic load allowance of 1.33 prior to combining with the lane load.

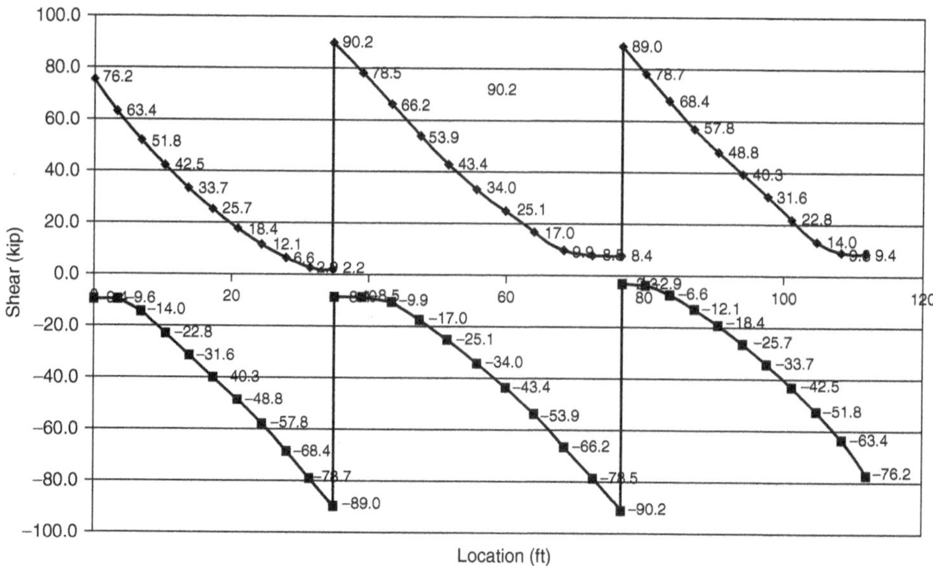

Fig. 9.20 Live-load shear envelope for HL-93 on 35–42–35 ft beam.

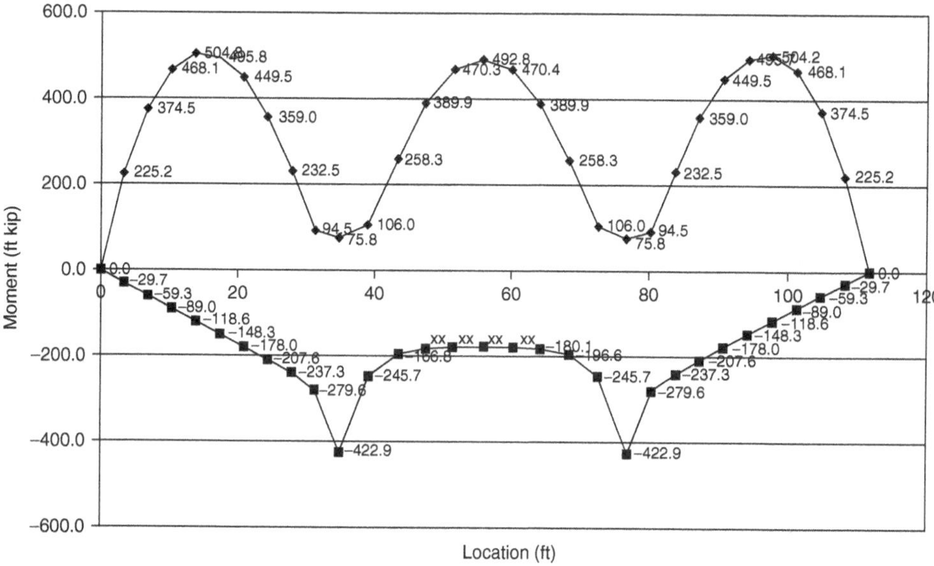

Fig. 9.21 Live-load moment envelope for HL-93 on 35–42–35 ft beam.

9.8 INFLUENCE SURFACES

Influence functions (or surfaces) can represent the load effect as a unit action moves over a surface. The concepts are similar to those presented previously. A unit load is moved over a surface, an analysis is performed for each load placement, and the response of a specific action at a fixed location is used to create a function that is two dimensional, that is, $\eta(x, y)$ where x and y are the coordinates for the load position.

This function is generated by modeling the system with the finite-element method (see Chapter 11). The function is used by employing superposition in a manner similar to Eq. 9.1. The analogous equation is

$$A = P_1\eta(x_1, y_1) + P_2\eta(x_2, y_2) + L + P_n\eta(x_n, y_n) = \sum_{i=1}^{n} P_i\eta(x_i, y_i) \tag{9.20}$$

A distributed patch load is treated in a manner similar to distributive load in Eq. 9.3. The analogous equation is

$$A = \iint_{\text{Area}} w(x, y)\,\eta(x, y)\; dA \tag{9.21}$$

where $w(x, y)$ is the distributive patch load and the integration is over the area where the load is applied. If the load

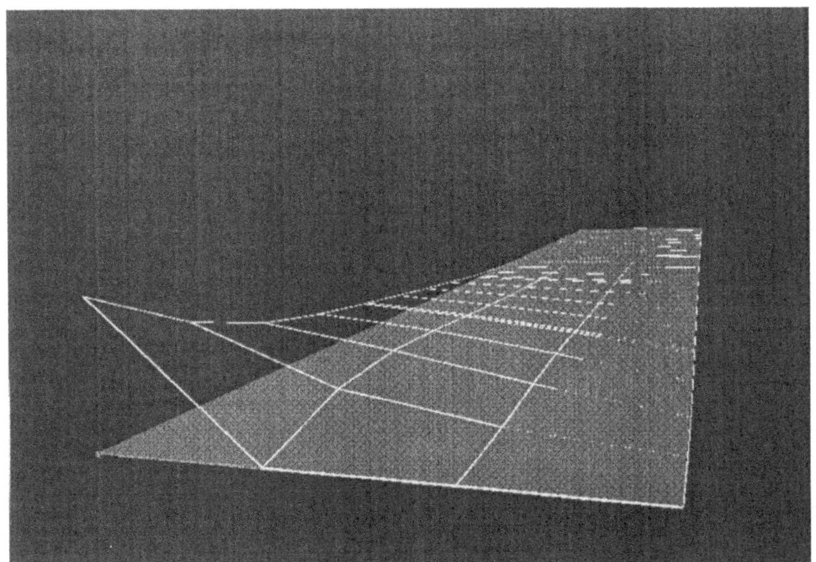

Fig. 9.22 Example of an influence surface for the corner reaction.

is uniform, then $w(x, y)$ may be removed from the integration. For example, a uniform load such as the self-weight of a bridge deck is multiplied by the volume under the influence surface to determine the load effect.

Numerical procedures similar to those described previously are used for the integration. Influence surfaces can be normalized and stored for analysis. Influence surfaces were used extensively in the development of the load distribution formulas contained in the AASHTO specification (Zokaie et al., 1991; Puckett et al., 2005). An example from work by Puckett et al. (2005) is illustrated in Figure 9.22. This work is described in detail in Chapter 11.

9.9 SUMMARY

Influence functions are important for the structural analysis of bridges. They aid the engineer in the understanding, placement, and analysis of moving loads. Such loads are required to determine the design load effects. Several methods exist to generate influence functions. All methods have advantages and disadvantages for hand and automated methods. Several methods are illustrated in this chapter. Design trucks and lane loads have been used to generate the critical actions for four bridges. These envelopes are used in design examples presented in later chapters.

REFERENCES

AASHTO (2010). *LRFD Bridge Design Specification*, 5th ed., American Association for State Highway and Transportation Officials, Washington, DC.

AISC (1986). *Moments, Shears, and Reactions for Continuous Bridges*, American Institute of Steel Construction, Chicago.

BridgeTech, Inc. (1996, 2010). *LRFD Analysis Manual*, Laramie, WY, Version 2.0.

Hibbeler, R. C. and R. C. Hibbeler (2004). *Structural Analysis*, 5th ed., Prentice Hall, New York.

Pilkey, W. D. and O. H. Pilkey (1974). *Mechanics of Solids*, Quantum, New York.

Puckett, J. A., X. S. Huo, M. D. Patrick, M. C. Jablin, D. Mertz, and M. D. Peavy (2005). "Simplified Equations for Live-Load Distribution in Highway Bridges," International Bridge Engineering Conference, TRB: 6IBECS-069.

Zokaie, T. L., T. A. Osterkamp, and R. A. Imbsen (1991). *Distribution of Wheel Loads on Highway Bridges, Final Report Project 12-26/1*, National Cooperative Highway Research Program, Transportation Research Board, National Research Council, Washington, DC.

PROBLEMS

9.1 Determine the influence lines for the shear and bending moments for the points of interest (POI) labeled. Instructor to assign structures and POIs.

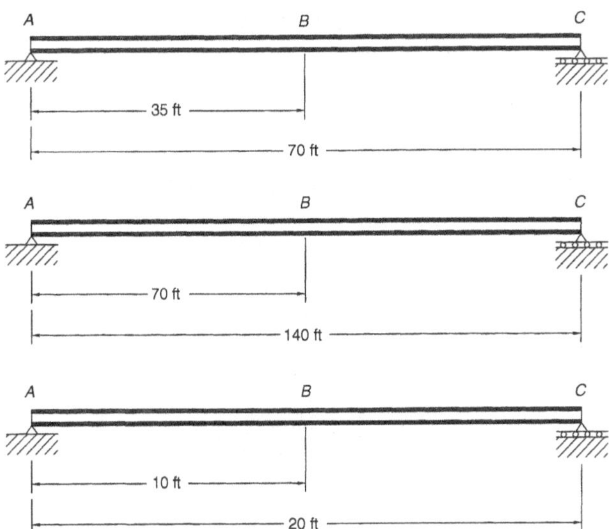

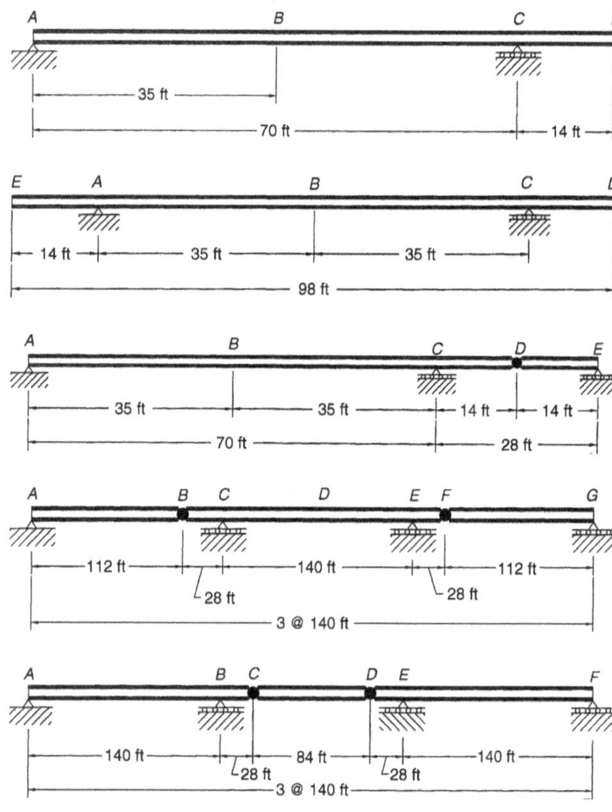

9.2 Qualitatively (without values but to scale) draw the influence lines for shear, moment, and reactions. Use the structures illustrated in Problem 9.1.

9.3 Qualitatively (without values but to scale) draw the influence lines for shear, moment, and reactions. Use the structures illustrated below.

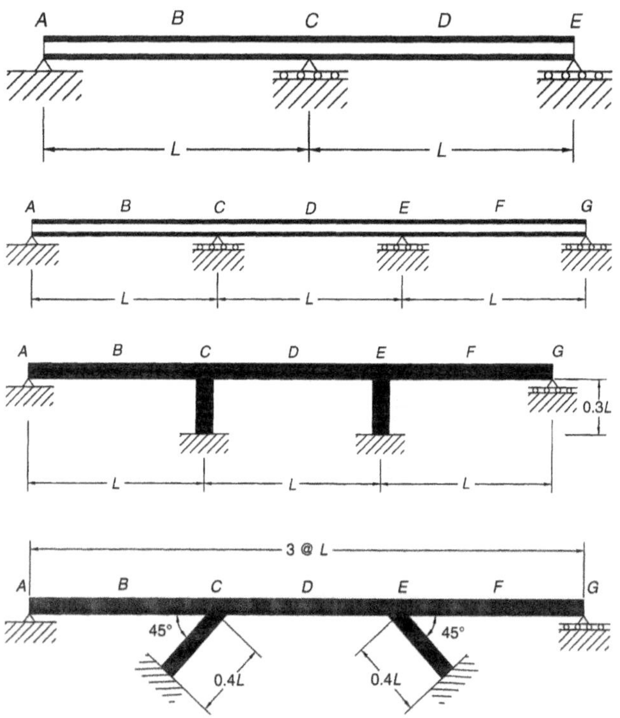

9.4 Use the design truck with a 14-ft rear-axle spacing to determine the critical (most positive and negative) shear, moment, and reactions at the points of interest for Problem 9.1. Instructor to assign structures and POIs.

9.5 Use the design tandem to determine the critical (most positive and negative) shear, moment, and reactions at the points of interest for Problem 9.1. Instructor to assign structures and POIs.

9.6 Use the design lane to determine the critical (most positive and negative) shear, moment, and reactions at the points of interest for Problem 9.1. Instructor to assign structures and POIs.

9.7 Use a dead load of $w_{DC} = 1.0$ kip/ft and $w_{DW} = 0.20$ kip/ft across the structure to determine the shear and bending moment diagrams for Problem 9.1. Instructor to assign structure(s).

9.8 Use the results from Problems 9.4–9.7 to combine these load for the strength I limit state. Use only the maximum dead-load factors, $\gamma_{DC} = 1.25$, $\gamma_{DW} = 1.50$. Assume a live-load distribution factor of $mg = 0.6$ lanes/girder.

9.9 Use the permit vehicle shown below to determine the critical live-load shears, moments, and reactions for the structures in Problem 9.1. Instructor to assign structures and POIs.

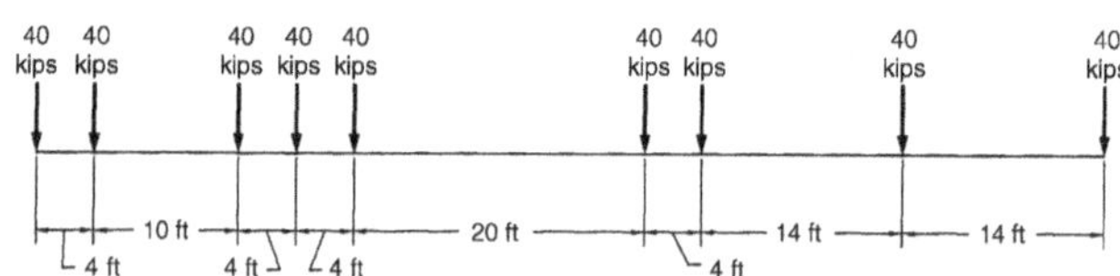

9.10 Use the design truck (with a 14-ft rear-axle spacing), design tandem, and design lane to illustrate the critical placement (show loads on influence line diagram) for the critical shear, moment, and reactions. Use the results from Problem 9.3. Instructor to assign structures and POIs.

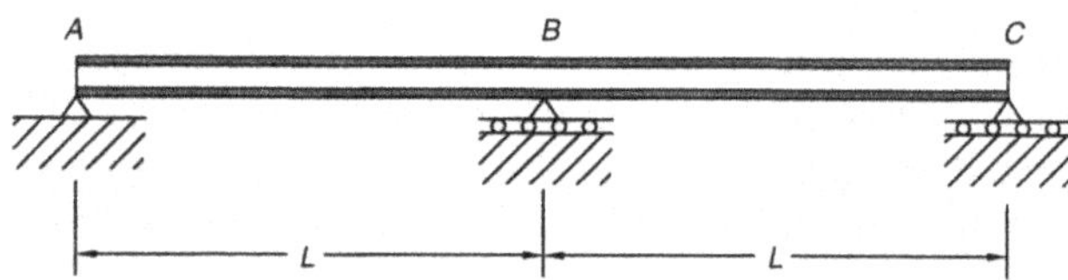

9.11 Use a structural analysis program to determine the load effects for the design truck (with a 14-ft rear-axle spacing), design tandem, and design lane. See Problem 9.10. Compute the critical shear, moment, and reactions. Instructor to assign structures and POIs.

9.12 Use a structural analysis program to determine the load effects for a dead load of $w_{DC} = 1.0$ kip/ft and $w_{DW} = 0.20$ kip/ft across the structure. Determine the shear and bending moment diagrams. See Problem 9.10. Instructor to assign structure(s).

9.13 Use the results from Problems 9.10–9.12 to combine these load for the strength I limit state. Use only the maximum dead-load factors, $\gamma_{DC} = 1.25$, $\gamma_{DW} = 1.50$. Assume a live-load distribution factor of $mg = 0.6$ lanes/girder.

9.14 Use an automated bridge analysis program, for example, QCON or BT Beam, and repeat Problem 9.8.

9.15 Use an automated bridge analysis program, for example, QCON or BT Beam, and repeat Problem 9.13.

9.16 Use a flexible rod to illustrate the influence lines for the reactions at A, B, and C. Use the rod to draw the shape of the influence line. Compare with numerical values provided in the following table. Use Problem 9.10.

Location	**A**	**B**	**C**
100	1.0000	0.0000	0.0000
104	0.5160	0.5680	−0.0840
105	0.4063	0.6875	−0.0938
110 = 200	0.0000	1.0000	0.0000
205	−0.0938	0.6875	0.4063
206	−0.0840	0.5680	0.5160
210 = 300	0.0000	0.0000	1.0000

Source: AISC (1986).

Multiply the coefficients provided by the span length for one span. *Instructor note*: Stringer–floorbeam–girder bridges show an excellent system that may be used to explain load path issues; however, few new systems are designed today. Many systems exist within the inventory and are load rated in today's engineering practice. Also, there is no discussion of this type of system in this book; the instructor will have to provide the necessary procedures and/or example.

9.17 For the stringer–floorbeam–girder system shown, determine the shear and moment diagrams in the floorbeam and the girders for the self-weight (DC) (0.150 kcf normal wt. concrete) and DW of 30 psf.

9.18 Use the design truck with a 14-ft rear-axle spacing to determine the critical (most positive and negative) shear, moment, and reactions at the points of interest for Problem 9.17. Multiple presence factors and single and multiple loaded lanes should be considered. Instructor to assign POIs.

9.19 Use the design lane to determine the critical (most positive and negative) shear and moment at the points of interest for Problem 9.17. Instructor to assign structures and POIs. Multiple presence factors and single and multiple loaded lanes should be considered.

9.20 Use the results from Problems 9.17–9.19 to combine these loads for the strength I limit state. Use only the maximum dead-load factors, $\gamma_{DC} = 1.25$, $\gamma_{DW} = 1.50$, and $\gamma_{LL} = 1.75$.

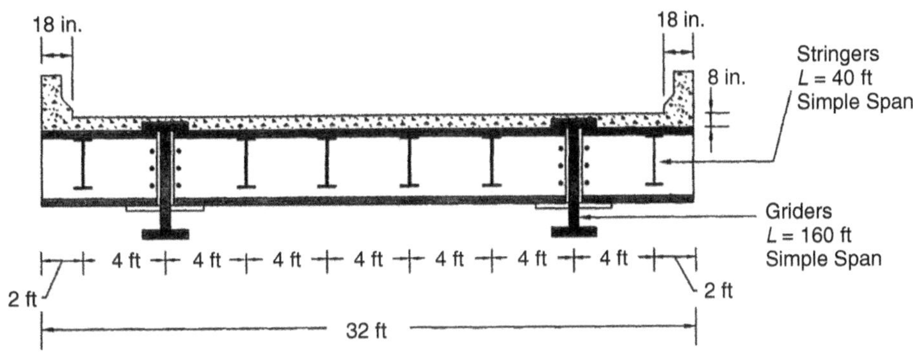

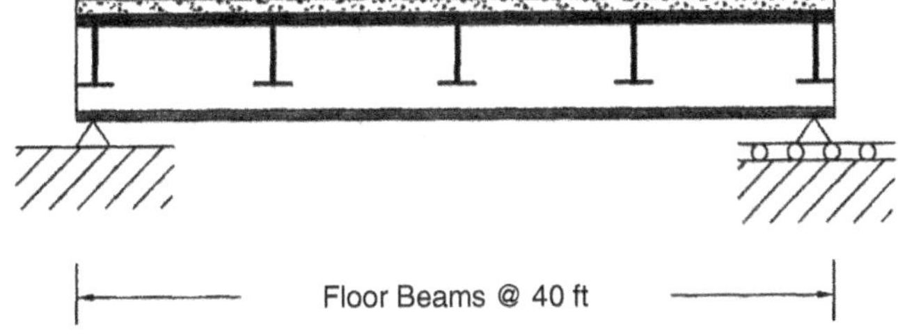

CHAPTER 10

System Analysis—Introduction

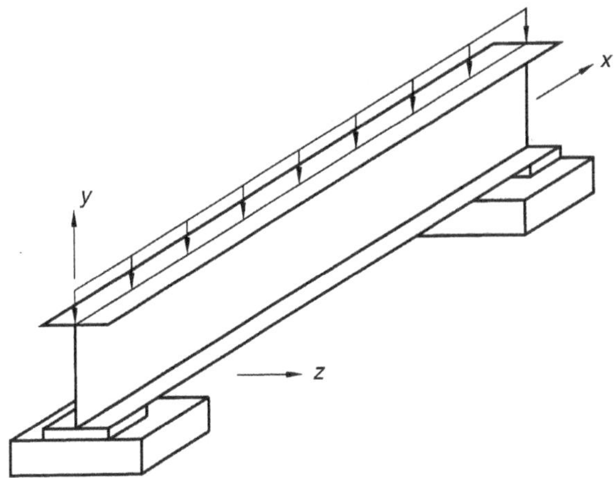

Fig. 10.1 Simple beam.

10.1 INTRODUCTION

To design a complicated system such as a bridge, it is necessary to break the system into smaller, more manageable subsystems that are comprised of components. Subsystems include the superstructure, substructure, and foundation, while the components include beams, columns, deck slab, barrier system, cross frames, diaphragms, bearings, piers, footing, piles, and caps. The forces and deformations (load effects) within the components are necessary to determine the required size and material characteristics. It is traditional and implicit in the AASHTO Specification that design be performed on a component basis. Therefore, the engineer requires procedures to determine the response of the structural system and ultimately its components.

In general, the distribution of the loads throughout the bridge requires equilibrium, compatibility, and that constitutive relationships (material properties) be maintained. These three requirements form the basis for all structural analysis, regardless of the level of complexity. Equilibrium requires that the applied forces, internal actions, and external reactions be statically in balance. Compatibility means that the deformations are internally consistent throughout the system (without gaps or discontinuities) and are consistent with the boundary conditions. Finally, the material properties, such as stiffness, must be properly characterized. Typically, the assumptions that are made regarding these three aspects of analysis determine the complexity and the applicability of the analysis model.

For example, consider the simply supported wide-flange beam subjected to uniform load shown in Figure 10.1. The beam is clearly a three-dimensional system because it has spatial dimension in all directions, but in mechanics of deformable bodies we learned that this system could be modeled by the familiar one-dimensional (1D) equation:

$$\frac{d^4y}{dx^4} = \frac{w(x)}{EI(x)} \tag{10.1}$$

Several important assumptions were used in the development of Eq. 10.1. First, the material is assumed to behave linear elastically. Second, the strain (and stress) due to flexural bending is assumed to be linear. Third, the loads are concentrically applied such that the section does not torque, and finally, the beam is proportioned and laterally braced so that instability (buckling) does not occur.

Although these assumptions are conventional and yield results comparable to laboratory results, often these conditions do not truly exist. First, for example, due to localized effects, some yielding may occur under reasonable service loads. Residual stresses from rolling result in some yielding at load levels below the predicted yield. Fortunately, these local effects do not significantly affect the global system response under service loads. Second, the bending stress profile is slightly nonlinear principally due to the load (stress) applied to the top of the beam and the reactions that create vertical normal stresses and strains that, due to Poisson's effect, also create additional horizontal stress. This effect is usually small. Third, concentric loading is difficult to achieve if the load is applied directly to the beam, but, if the beam is part of a slab system, then this assumption is perhaps more realistic. Finally, and importantly, Eq. 10.1 does not consider the local or global instability of the beam. It may be argued that other assumptions are also applicable, but a discussion of these suffices for the purpose intended here.

The purpose for discussing such a seemingly simple system is to illustrate the importance of the modeling assumptions, and their relevance to the real system. It is the engineer's responsibility to understand the assumptions and their applicability to the system under study. When the assumptions do not adequately reflect the behavior of the real system, the engineer must be confident in the bounds of the error induced and the consequences of the error. Clearly, it is impossible to exactly predict the response of any structural system, but predictions can be of acceptable accuracy. The consequences of inaccuracies are a function of the mode of failure. These phenomena are elaborated in detail later.

The application of equilibrium, compatibility, and material response, in conjunction with the assumptions, constitutes the *mathematical model* for analysis. In the case of the simple beam, Eq. 10.1, with the appropriate boundary conditions, is the mathematical model. In other cases, the mathematical model might be a governing differential equation for a beam column or perhaps a thin plate, or it may be the integral form of a differential equation expressed as an energy or variational principle. Whatever the mathematical model may be, the basis for the model and the behavior it describes must be understood.

In structural mechanics courses, numerous procedures are presented to use the *mathematical* model represented by Eq. 10.1 to predict structural response. For example, direct integration, conjugate beam analogy, moment–area, slope–deflection, and moment distribution are all well-established methods. All of these methods either directly or indirectly involve the *mathematical* model represented by Eq. 10.1.

The method used to solve the mathematical model is termed the *numerical model*. Its selection depends on many factors including availability, ease of application, accuracy, computational efficiency, and the structural response required. In theory, numerical models based on the same mathematical model should yield the same response. In practice, this is generally true for simple one-dimensional elements such as beams, columns, and trusses. Where finite elements are used to represent a continuum in two or three dimensions (2 or 3D), features such as element types, mesh characteristics, and numerical integration, complicate the comparisons. This does not mean that several solutions exist to the same problem, but rather solutions should be comparable though not *exactly* identical, even though the mathematical models are the same.

Finally, the engineer should realize that even the simplest of systems are often mathematically intractable from a rigorous closed-form approach. It is rather easy to entirely formulate a mathematical model for a particular bridge, but its solution is usually nontrivial and must be determined with approximate numerical methods such as with the finite-strip or finite-element methods. It is important to realize that approximations exist in *both* the mathematical *and* numerical models.

The process is illustrated in Figure 10.2. At the top of the diagram is the real system as either conceptualized or as built. To formulate a mathematical model, the engineer must accept some simplifying assumptions that result in a governing equation(s) or formulation. Next, the engineer must translate the characteristics of the real system into the variables of the mathematical model. This includes definition of loads, material and cross-sectional properties and boundary conditions. Likely, the engineer relies on more simplification here. The numerical model is solved. Here some numerical approximation is typically involved. The results are then interpreted, checked, and used for component design.

Should the component properties vary significantly as a result of the design, then the numerical model should be altered and the revised results should be determined. Throughout the process, the engineer must be aware of the limitations and assumptions implicit in the analysis and should take precautions to ensure that the assumptions are not violated *or* that the consequences of the violations are understood and are acceptable.

Many parameters are difficult to estimate and in such cases, the extreme conditions can be used to form an envelope of load effects to be used for design. For example, if a particular cross-sectional property is difficult to estimate due to complications such as composite action, concrete cracking, and creep effects, the engineer could model the section using the upper and lower bounds and study the sensitivity of the procedure to the unknown parameters. Such modeling provides information about the importance of the uncertainty of parameters in the structural response.

10.2 SAFETY OF METHODS

As previously stated, it is important for the engineer to understand the limitations of the mathematical and numerical models and the inaccuracies involved. As models are estimates of the actual behavior, it is important to clearly understand the design limit states and their relationship to the modes and consequences of failure. This finding is discussed in the sections that follow.

10.2.1 Equilibrium for Safe Design

An essential objective in any analysis is to establish a set of forces that satisfies equilibrium between internal actions and the applied loads at every point. The importance of equilibrium cannot be overstated and is elaborated below.

Most of the analytical models described in this book are based on linear response, that is, the load effect is proportional to the load applied. Conversely, the resistance models used by AASHTO (and most other structural design specifications) implicitly assume nonlinear material response at the strength limit state. For example, the nominal flexural capacity of a braced compact steel section is $M_n = F_y Z$, and the flexural capacity of a reinforced concrete section involves the Whitney stress block where $f_c = 0.85 f_c'$, and the steel stress is equal to the yield stress, and so on. Clearly, an inconsistency exists. The analysis is based on linear behavior and the resistance calculations are based on nonlinear behavior. The rationale for this is founded in the system behavior at and beyond yielding and is based in plasticity theory.

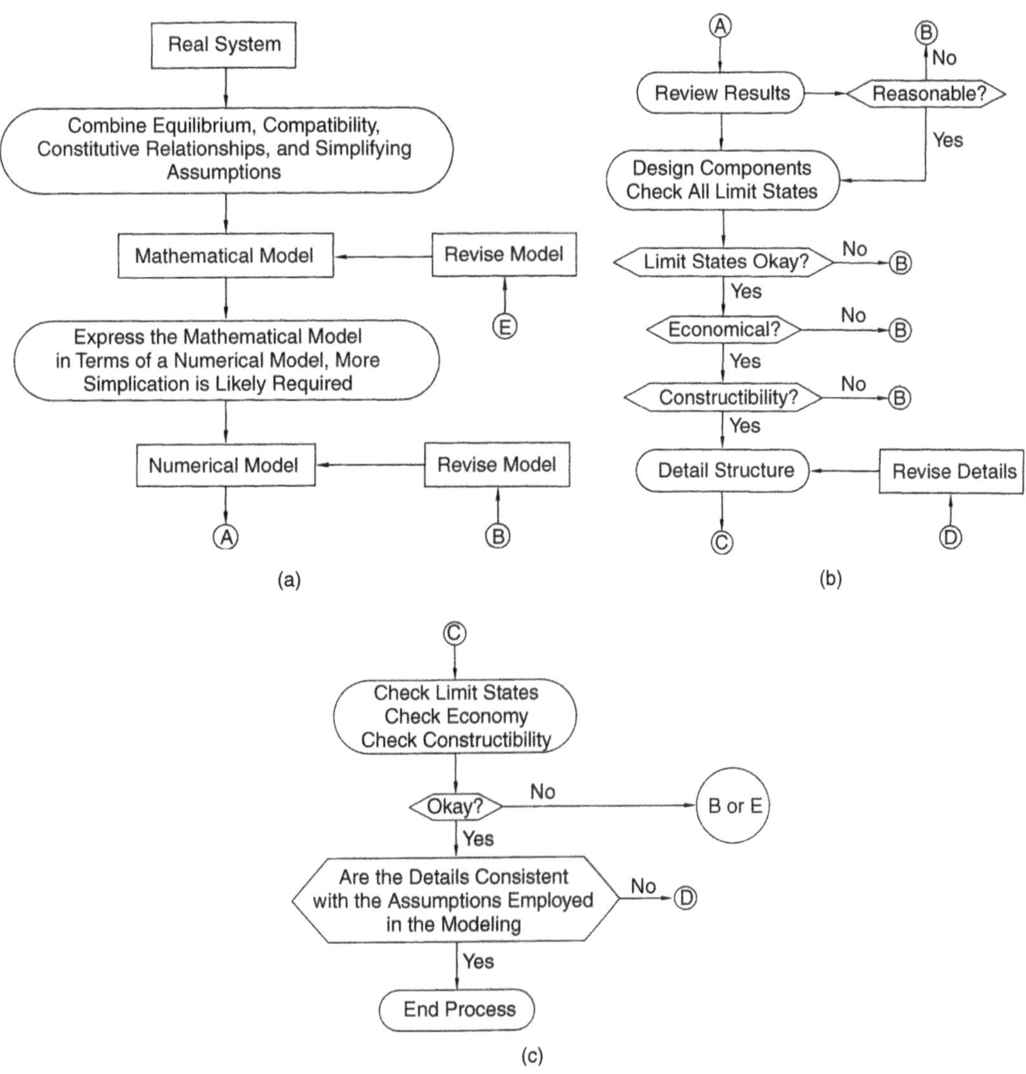

Fig. 10.2 Relationship of modeling to design.

The rationale is best explained by restating the *lower bound theorem* (Neal, 1977; Horne, 1971):

A load computed on the basis of an equilibrium moment distribution in which the moments are nowhere greater than M_p is less than or equal to the true plastic limit load.

Although stated in terms of bending moment, the theorem is valid for any type of action and/or stress. The essential requirements of the theorem are:

- ☐ Calculated internal actions and applied forces should be in equilibrium everywhere.
- ☐ Materials and the section/member behavior must be ductile; that is, the material *must* be able to yield without fracture or instability (buckling).

In simpler terms, the theorem means that if a design is based on an analysis that is in equilibrium with the applied load and the structure behaves in a ductile manner, then the ultimate failure load will meet or exceed the design load. *This is one of the most important theorems in structural mechanics and is extremely relevant to design practice*.

This theorem offers wonderful assurance! How does it work? Consider the two beams shown in Figures 10.3(a) and 10.3(c). The beams are assumed to be designed such that any section can develop its full plastic moment capacity, which is $M_p = F_y Z$, and for the sake of simplicity it is assumed that the beam behaves elastic-plastic, that is, the moment that causes first yielding is the same as the plastic capacity (these two differ by about 10–15% for steel wide-flange sections). The uniform load is applied monotonically to the simple beam of Figure 10.3(a) and the moment diagram is shown in Figure 10.3(b). When the moment at midspan reaches the capacity $M_p = w_u L^2/8$, a plastic hinge develops. This hinge creates a loss in bending rigidity that results in mechanism and collapse occurs.

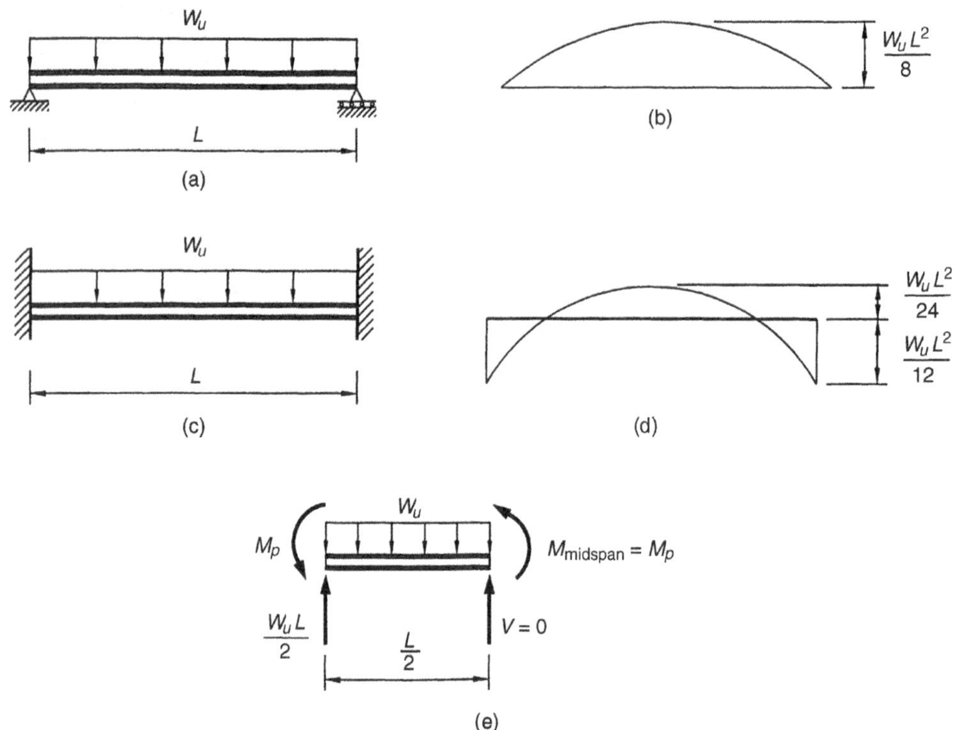

Fig. 10.3 (a) Uniformly loaded simple beam, (b) moment diagram, (c) uniformly loaded fixed–fixed beam, (d) moment diagram, and (e) free-body diagram.

Now consider the beam shown in Figure 10.3(c) and its associated linear elastic moment diagram shown in Figure 10.3(d). The load is again applied monotonically up to the level where hinges form at the supports (negative moment). A loss in bending rigidity results, and now the system becomes a simple beam with the plastic moment applied at the ends. Because the beam has not reached a mechanism, more load may be applied. A mechanism is finally reached when a hinge forms in the interior portion of the span (positive moment). This behavior illustrates redistribution of internal actions. Now assume the two beams have the same capacity M_p in positive and negative bending. For the simple beam, equate the capacities to the maximum moment at midspan yielding

$$M_p = \frac{wL^2}{8}$$
$$w\,(\text{simple beam}) = \frac{8M_p}{L^2}$$

Use a similar procedure for the fixed–fixed beam to equate the capacity to the maximum elastic moment at the support yielding

$$M_p = \frac{wL^2}{12}$$
$$w\,(\text{fixed–fixed beam}) = \frac{12M_p}{L^2}$$

A free-body diagram for the fixed–fixed beam is shown in Figure 10.3(e) for the state after the moment has reached the plastic moment capacity at the end. By equating the capacity M_p at midspan to the moment required by balancing the moment at the left support, one obtains

$$M_{\text{midspan}} = M_p + w\left(\frac{L}{2}\right)\left(\frac{L}{4}\right) = M_p$$
$$w = \frac{16M_p}{L^2}$$

Consider the relevance of this example to analysis and design. Suppose that the system is a fixed–fixed beam but the engineer designed for the simple beam moment. The design would have an additional capacity of $\frac{12}{8} = 1.5$ against initial yielding due to the neglected continuity moments and additional reserve against total collapse of $\frac{16}{8} = 2.0$ considering redistribution of internal actions.

Even though the elastic moment diagram is used, the design is safe (but quite likely unnecessarily conservative). Now consider the more likely case where the engineer designs per the fixed–fixed elastic moment diagram shown in Figure 10.3(c). Here the additional reserve against collapse is $\frac{16}{12} = 1.33$. Note these factors should not be confused with the load factors of Tables 5.1 and 5.2 as these would be included in the elastic moment diagrams for the design. The AASHTO Specification allows the engineer to consider inelastic redistribution of internal actions in various articles.

Because the amount of redistribution is related to the ductility of the component that is material and cross-section dependent, most of these provisions are outlined in the resistance articles of the specification.

The static or lower bound theorem implies that as long as equilibrium is maintained in the analytical procedures and adequate ductility is available, then the exact distribution of internal actions is *not* required. It is inevitable that the analysis and subsequent design overestimate the load effect in some locations while underestimating the effect in others. If the strength demands in the real structure are larger than the available resistance, yielding occurs and the actions redistribute to a location where the demands are less, and hence, more capacity exists to carry the redistributed actions. The requirement of ductility and equilibrium ensures that redistribution occurs and that the system has the necessary capacity to carry the redistributed actions.

In summary, it is not required that the calculated system of forces be exact predictions of the forces that exist in the real structure (this is not possible anyway). It is only necessary that the calculated system of forces satisfy equilibrium at every point. This requirement provides at least one load path. As illustrated for the fixed–fixed beam, redistribution of internal actions may also occur in statically redundant systems. As previously stated, in practice it is impossible to exactly predict the system of forces that exists in the real system, and therefore, the lower bound theorem provides a useful safety net for the strength limit state.

In the case where instability (buckling) may occur prior to reaching the plastic capacity, the static or lower bound theorem does *not* apply. If instability occurs prior to complete redistribution, then equilibrium of the redistributed actions cannot be achieved and the structure may fail in an abrupt and dangerous manner.

10.2.2 Stress Reversal and Residual Stress

In Section 10.2.1, the ultimate strength behavior was introduced, and it was assumed that the cross section achieved the full plastic moment capacity. Note we did not mention how the section reaches this state or what happens when the section is yielded and then unloaded. Both of these issues are important to understanding the behavior and design limit states for ductile materials.

Consider the cantilever beam shown in Figure 10.4(a), which has the cross section shown in Figure 10.4(b) with reference points of interest o, p, q, and r. Point o is located at the neutral axis, p is slightly above the o, r is located at the top, and q is midway between p and r. The section has residual stresses that are in self-equilibrium [see Fig. 10.4(c)]. In general, residual stresses come from the manufacturing process, construction process, temperature effects, intentional prestressing, creep, shrinkage, and so on.

The beam is *deflected* at B and the load is measured. The product of the measured load and beam length is the moment at A, which is shown in the moment–curvature diagram illustrated in Figure 10.4(d). As the tip deflection increases, the moment increases with curvature and all points remain below the yield stress up to state a. Further increase causes initial yielding in the outer fibers and the yielding progresses toward the neutral axis until the section is in a fully plastic state. Figures 10.4(e)–10.4(j) illustrate the elastic-plastic stress–strain curve for the material, the state of stress and strain of the cross-section points, and the section stress profiles. For example, at state b, points p, q, and r are all at the yield stress and the stress profile is uniform.

What happened to the residual stresses when progressing from point a to b? Because the section is initially stressed, the curvature at which yielding occurs and the rate at which the section reaches its full plastic capacity is affected, but the ultimate capacity is not. *This is an important aspect of structural design, as many residual stresses exist in a structure due to numerous reasons, some listed above. Although such stresses may affect service level behavior and/or stability, they do not affect the capacity of ductile elements.*

Upon load removal, the moment–curvature follows *b-c-d* in Figure 10.4(d). If reloaded in the initial direction, then the moment–curvature follows *d-c-b′*. The curved portion of the line is different from the *o-a-b* because the residual stresses have been removed. The curved portion must exist because the section yields incrementally with the outer fibers first and then progresses inward. Now start at state b and unload to c where the stress at r is zero. As illustrated in Figure 10.4(g), the change in stress is F_y, hence the change in moment required is M_y. Note the stress near o is still at yield because the point o is at the neutral axis and does not strain upon load removal. Points p and q have stresses that are proportionally less. Continue load removal until state d is achieved. Here all the load is removed and the structure is in self-equilibrium. The stresses are illustrated in Figure 10.4(h). Reverse the load by deflecting the beam upward at the tip until the yielding occurs at r. Because the stress at r was zero at state c, the change in moment required to produce a yield stress is again M_y [see Fig. 10.4(i)]. Further increases cause the section to reach its full plastic state at f shown in Figure 10.4(j). Note the nonlinear shape of the curve is again different because the initial stresses at state e are again different.

In summary, the initial or residual stresses do not affect the ultimate (ductile) capacity, but they do affect the load–deflection characteristics in the postyield region.

10.2.3 Repetitive Overloads

As the vehicular loading of a bridge is repetitive, the possibilities of repeated loads that are above the service level are likely and their effects should be understood. In Section 10.2.2, the lower bound theorem is introduced for a single-load application that exceeds the yield strength of the beam at localized points. Consider the uniformly loaded prismatic beam shown in Figure 10.5(a) with the moment–curvature relationship shown in Figure 10.5(c). The uniform loading

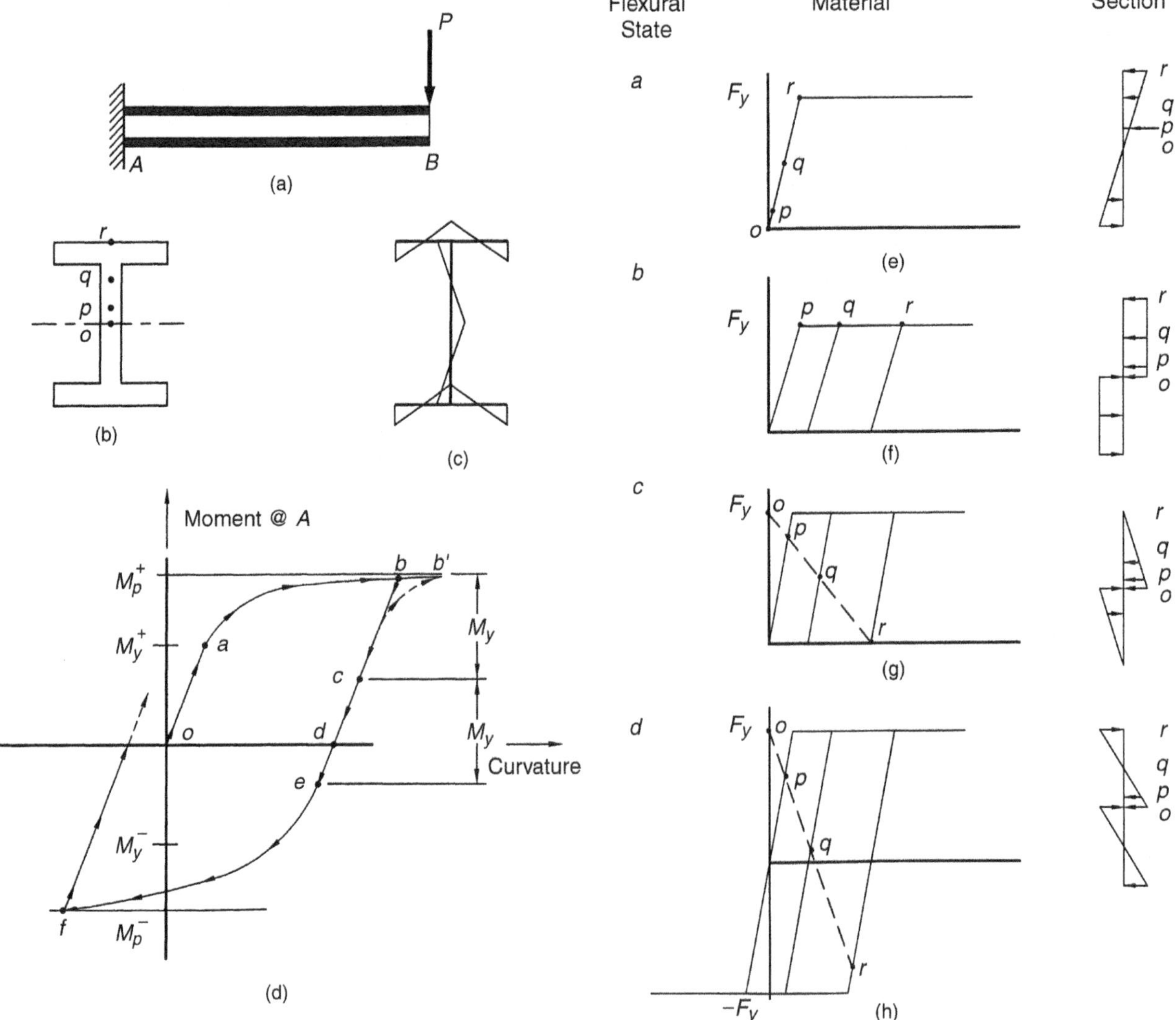

Fig. 10.4 (a) Cantilever beam, (b) cross section, (c) residual stresses, and (d) moment–curvature diagram. (e) State at a, (f) state at b, (g) state at c, and (h) state at d. (i) State at e, and (j) state at f.

shown in Figure 10.5(a) is spatially invariant and the magnitude is cycled.

Assume the load is applied slowly to a level where the moment at A and C exceeds the plastic moment, a hinge forms at A, and the section has yielded at B but just before the section is fully plastic at B [see Fig. 10.5(b)]. The resulting moment diagram is shown in Figure 10.5(d). Now the load is removed and the structure responds (unloads) elastically. The change in moment is the elastic moment diagram illustrated in Figure 10.5(e). The moment in the unloaded state is determined by superposing the inelastic (loading) and elastic (unloading) moment diagrams [Fig. 10.5(f)] and is termed the residual moment. The deflected shape of the beam is illustrated in Figure 10.5(g) where both the inelastic rotations at the beam ends and the elastic deflection due to the residual moments are shown.

To examine the effect of cyclic loads, consider the beam shown in Figure 10.6(a). The plastic collapse load for a single concentrated load is shown in Figure 10.6(b) and is used for reference. The loads W_1 and W_2 are applied independently.

First, W_1 is increased to a level such that hinges form at A and B but not to a level such that the hinge forms at D [Fig. 10.6(c)], which means that segment CD remains elastic and restrains collapse. Now remove load W_1 and the structure responses elastically, and residual moment and deflection remain. The residual deflection is illustrated in Figure 10.6(d).

Next, apply W_2 to a level such that hinges form at C and D and segment AC remains elastic. The deflected shape is also illustrated in Figure 10.6(d). It is important to note that a complete mechanism has not formed but the deflections have increased. Now if the load cycle is repeated, the deflections

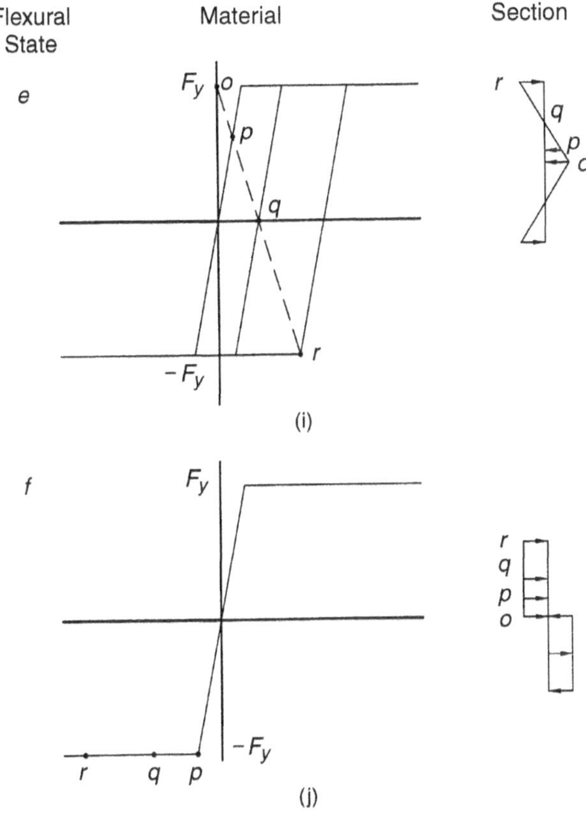

Fig. 10.4 *(Continued)*

may continue to increase, and the resultant effect is a progressive buildup of permanent deflections just as though the beam is deforming in the plastic collapse mechanism [Fig. 10.6(e)]. This limit state is termed *incremental collapse*.

The residual moments and associated deflections can be determined by incrementally applying the loads as described and performing the analysis for each load step. Although a viable method, it tends to be tedious and is limited to simple structures and loads. A bridge system is much more complex than the beam previously described. The bridge must resist moving loads and the incremental approach must include the complexity of load position and movement, which greatly complicates the analysis.

The important issue is that the repeated loads that cause incremental collapse are *less* than the static collapse loads. The load above which incremental collapse occurs is termed the *shakedown load*. If the load is below the shakedown load but above the load that causes inelastic action, then the structure experiences inelastic deformation in local areas. However, after a few load cycles, the structure behaves elastically under further loading. If the load exceeds the shakedown load (but is less than the plastic collapse load), then incremental collapse occurs. Finally, if the load exceeds the plastic collapse load, the structure collapses. This discussion is summarized in Table 10.1.

Table 10.1 Summary of Inelastic Behavior

Minimum Load	Maximum Load[a]	Result
0	Yield	Elastic behavior
Yield	Shakedown	Localized inelastic behavior initially Elastic behavior after shakedown
Shakedown	Plastic collapse	Incremental collapse
Plastic collapse	N/A	Collapse

[a]Not applicable = N/A.

Two theorems have been developed to determine the shakedown load: the lower and upper bound theorems. These theorems help to relate the elastic behavior to the inelastic behavior so complex systems can be analyzed without an incremental load analysis. The theorems are stated without proof. The interested reader is referred to the extensive references by Horne (1971) and Neal (1977).

The magnitude of variable repeated loading on a structure may be defined by a common load factor λ, where $M_{\max}$ and $M_{\min}$ represent the maximum and minimum elastic moments, $\lambda M_{\max}$ and $\lambda M_{\min}$ represent the moments for load level λ, and $\lambda_s M_{\max}$ and $\lambda_s M_{\min}$ represent shakedown moments (Horne, 1971).

Lower Bound Shakedown Theorem The lower bound shakedown limit is given by a load factor λ for which residual moment m satisfies the inequalities:

$$m + \lambda M_{\max} \leq M_p$$
$$m + \lambda M_{\min} \geq -M_p$$
$$\lambda(M_{\max} - M_{\min}) \leq 2M_y \tag{10.2}$$

The residual moment m does not necessarily have to be the exact residual moment field determined from incremental analysis but may be any self-equilibrating moment field. The third inequality is imposed to avoid an alternating plasticity failure where the material is yielded in tension and compression. Such a condition is unlikely in a bridge structure because the total change in moment at any point is far less than twice the yield moment.

The residual moment m could be set to zero, and the theorem simply implies that shakedown can be achieved if the moment is less than the plastic moment and the moment range is less than twice the yield moment. The former is similar to the ultimate strength limit state. The latter is seldom a problem with practical bridge structures (Horne, 1971).

Upper Bound Shakedown Theorem The upper bound shakedown load is determined by assuming an incremental collapse mechanism with hinges at locations j with rotations θ_j associated with the elastic moments $M_{\max}$ and $M_{\min}$.

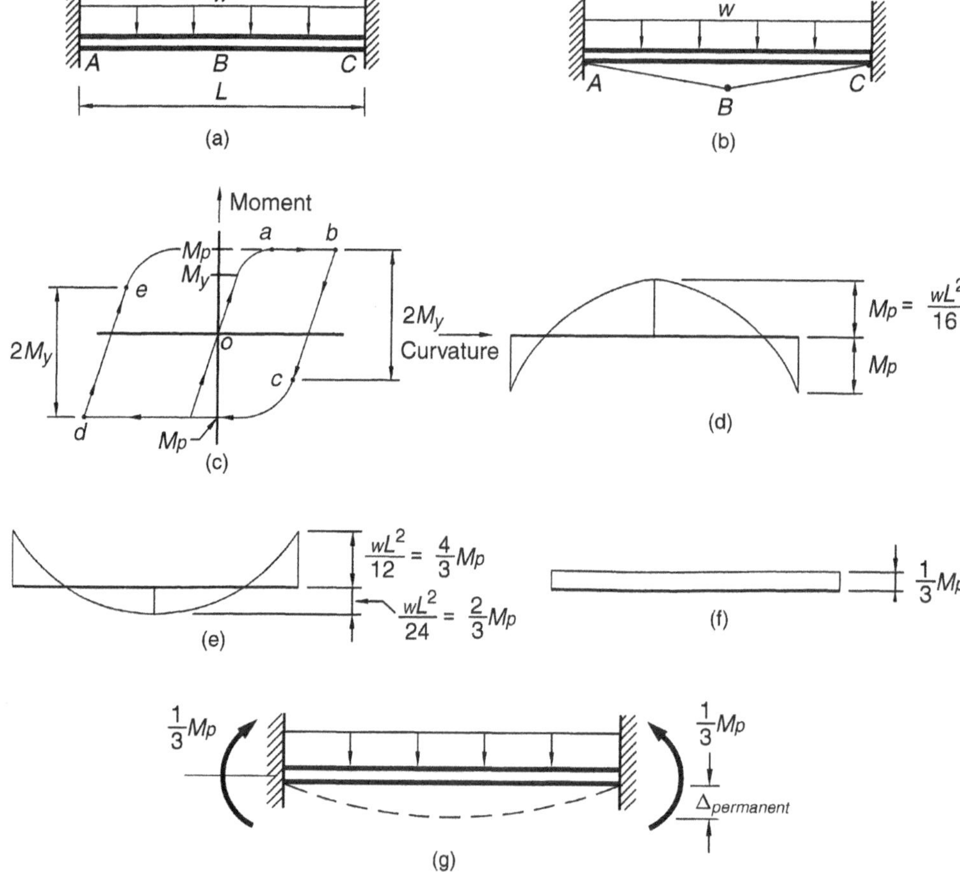

Fig. 10.5 (a) Uniformly loaded fixed–fixed beam, (b) collapse mechanism, (c) moment–curvature diagram, (d) moment diagram at collapse, (e) elastic (unloading) moment diagram, (f) residual moment diagram, and (g) residual displacement.

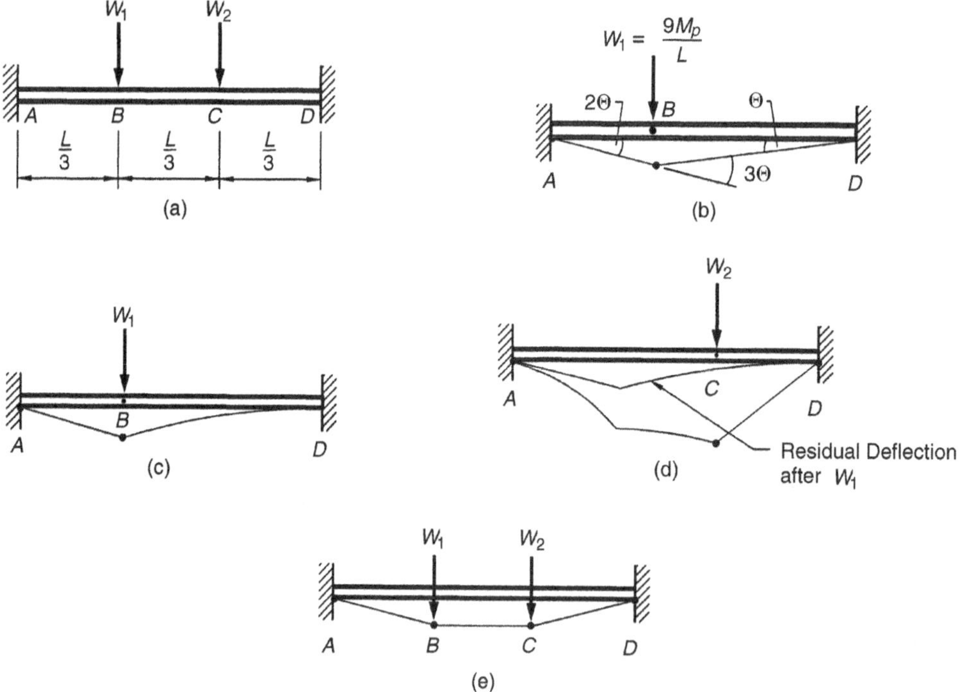

Fig. 10.6 (a) Fixed–fixed beam, (b) collapse mechanism, (c) hinges at A and B, (d) hinges at C and D, and (e) incremental collapse.

The directions of the hinge rotations are consistent with the moments (Horne, 1971):

$$\lambda_{\text{shakedown}} \sum_j \{M_{j\ \max} \text{ or } M_{j\ \min}\}\ \theta_j = \sum_j M_{pj} |\theta_j| \tag{10.3}$$

The elastic moment ($M_{j\max}$ or $M_{j\min}$) used is the one that causes curvature in the same sense as the hinge rotation.

Example 10.1 Determine the yield, shakedown, and plastic collapse load for a moving concentrated load on the prismatic beam shown in Figure 10.7(a). Compare the shakedown load with the plastic collapse load and the initial yield load for the load at midspan. Assume that $M_p = M_y$, that is, neglect the spread of plasticity.

The elastic moment envelope is illustrated in Figure 10.7(b). The elastic envelopes may be established using any method appropriate for the solution of a fixed–fixed beam subjected to a concentrated load. The envelope values represent the elastic moments $M_{\max}$ and $M_{\min}$. Use Eq. 10.3 and the mechanism shown in Figure 10.7(c) to obtain shakedown load.

$$\lambda_{\text{shakedown}} \left[\frac{4}{27} WL\,(\theta\ +\ \theta) + \frac{1}{8} WL\,(2\theta) \right] = M_p(4\theta)$$

$$\lambda_{\text{shakedown}} = \frac{432}{59} \frac{M_p}{WL} = 7.32 \frac{M_p}{WL}$$

To determine the plastic capacity, use the free-body diagram shown in Figure 10.7(d) to balance the moment about B. The result is

$$\frac{\lambda_{\text{plastic}} W}{2} \left(\frac{L}{2} \right) = 2M_p$$

$$\lambda_{\text{plastic}} = \frac{8M_p}{WL}$$

Note that the shakedown load is approximately 92% of the plastic collapse load. The initial yield load is determined by equating the maximum elastic moment to $M_y = M_p$. The result is

$$\lambda_{\text{elastic}} = \frac{27}{4} \frac{M_p}{WL} = 6.75 \frac{M_p}{WL}$$

A comparison of the results is given in Table 10.2.

In summary, it is important to understand that plastic limit collapse may not be the most critical strength limit state, but rather incremental collapse should be considered. It has been demonstrated that plastic deformation occurs at load levels below the traditional plastic collapse for repeated loads. Some procedures outlined in the AASHTO Specification implicitly permit inelastic action and assume that shakedown occurs. Such procedures are discussed in later sections.

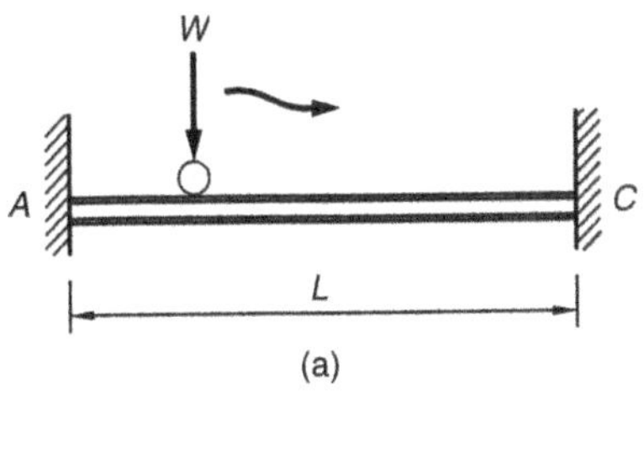

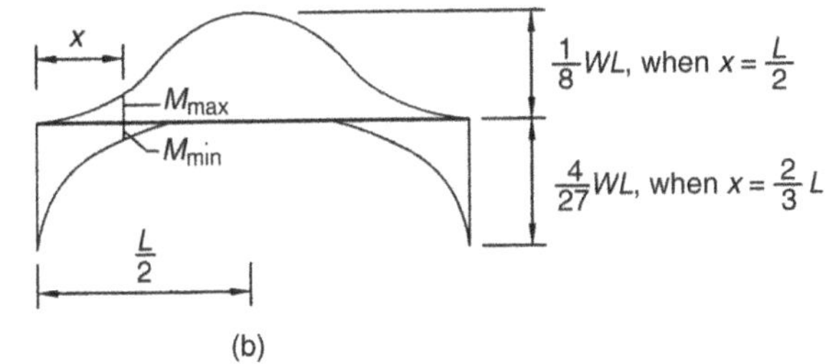

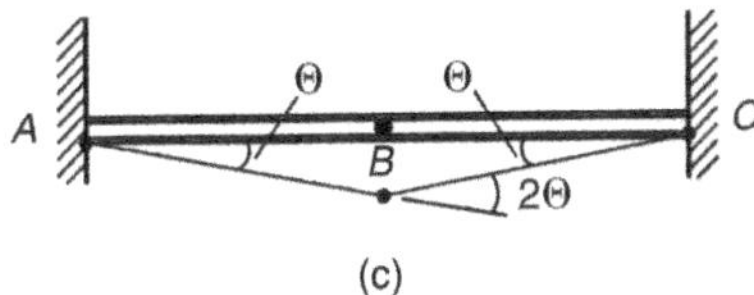

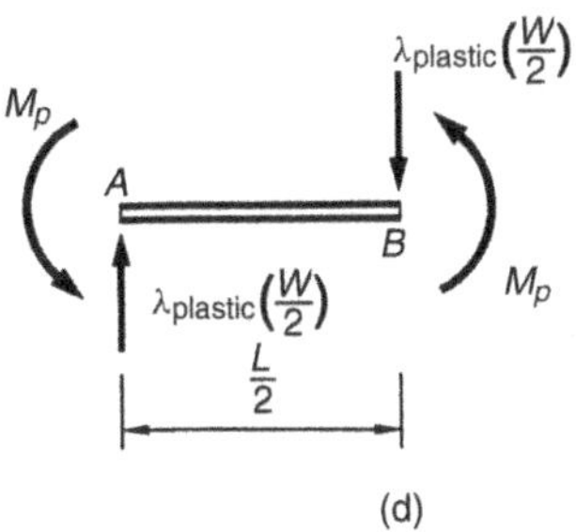

Fig. 10.7 (a) Fixed–fixed beam with moving load, (b) elastic moment envelope, (c) assumed incremental collapse mechanism, and (d) free-body diagram.

10.2.4 Fatigue and Serviceability

The static or lower bound theorem relates to the ultimate strength limit state. However, repetitive truck loads create fatigue stresses that may lead to brittle fracture under service level loads. Because the loads creating this situation are at the service level and because the failure mode is often brittle, little opportunity exists for load redistribution, hence the lower bound theorem does *not* apply. Thus, the only

Table 10.2 Example 10.1 Summary

Load Level	Load Factor (λ)	$\dfrac{\lambda}{\lambda_{\text{elastic}}}$
Elastic	$6.75\dfrac{M_p}{WL}$	1.00
Shakedown	$7.32\dfrac{M_p}{WL}$	1.08
Plastic	$8.00\dfrac{M_p}{WL}$	1.19

way to estimate the internal live-load actions accurately and safely is to properly model the relative stiffness of all components and their connections. This aspect of the analysis and subsequent design is one characteristic that differentiates bridge engineers from their architectural counterparts. The building structural engineer is typically not concerned with a large number of repetitive loads at or near service load levels.

Service or working conditions can be the most difficult to model because bridges and the ground supporting them experience long-term deformation due to creep, shrinkage, settlement, and temperature change. The long-term material properties and deformations are difficult to estimate and can cause the calculated load effects to vary widely. It is best to try to bound the model parameters involved with an analysis and design for the envelope of extreme effects. Service limit states are important and should be carefully considered (with AASHTO LRFD the service limit states often control design for steel and prestressed concrete).

A significant portion of a bridge manager's budget is spent for repair and retrofit operations. This effect is because the severe environment, including heavy loads, fluctuating temperature, and deicing chemicals cause serviceability problems that could ultimately develop into strength problems.

10.3 SUMMARY

This chapter includes numerous topics related to the structural analysis of bridge systems. It is intended to provide a broad-based perspective for analysis. It is important to understand the concepts involved with the plastic and shakedown limits and how they relate to the AASHTO design specification, which is primarily based on elastic analysis. The lower bound theorem is one of the most important theorems in structural engineering. It permits the engineer to use linear analysis methods that are clearly inconsistent with many of the specification resistance methods for the strength limit state. This inconsistency is seldom explored at the undergraduate level, but is nevertheless very important for understanding the fundamental basis for much of what is assumed in design computations.

REFERENCES

AASHTO (2010). *LRFD Bridge Design Specification*, 5th ed., American Association for State Highway and Transportation Officials, Washington, DC.

AISC (2005). *Manual of Steel Construction–LFRD*, 13th ed., American Institute for Steel Construction, Chicago.

Horne, M. R. (1971). *Plastic Theory of Structures*, Willan Clowes, London.

Neal, N. G. (1977). *The Plastic Methods of Structural Analysis*, 3rd ed., Chapman & Hall, London.

PROBLEM

10.1 Repeat Example 10.1 for a prismatic beam that is clamped at the left end and pinned at the right end.

CHAPTER 11

System Analysis—Gravity Loads

Most methods of analysis described in this chapter are based on three aspects of analysis: equilibrium, compatibility, and material properties, which are assumed to be linear elastic. The exceptions are statically determinate systems. The objective of these methods is to estimate the load effects based on the relative stiffness of the various components. The methods described vary from simplistic (beam line) to rigorous (finite strip or finite element). Equilibrium is implicit in all methods, and most methods attempt to achieve realistic estimates of the service-level behavior. Typically the materials are assumed to behave linearly; these methods will not reflect the behavior after yielding occurs. As outlined earlier, the lower bound theorem prescribes that such analyses yield a conservative distribution of actions upon which to base strength design and, hopefully, a reasonable distribution of actions upon which to base service and fatigue limit states.

In this chapter, the one method that extends beyond yield is the yield-line analysis method for slabs. Here it is instructive to revisit some of the earlier presented concepts involved with plastic hinging and the ultimate limit state. This again reinforces differences between linear and nonlinear considerations and their relationship to the lower bound theorem.

The discussion of specific analysis methods begins with the most common bridge types, the slab and slab–girder bridges. The discussion of these bridge types includes the most practical analytical procedures. As many books have been written on most of these procedures, for example, grillage, finite element, and finite strip methods, the scope must be restricted and is limited to address the basic features of each method and to address issues that are particularly relevant to the bridge engineer. Example problems are given to illustrate particular behavior or techniques.

As in the previous chapter, the examples provide guidance for the analysis and design for the bridges presented in the resistance chapters. The reader is assumed to have the prerequisite knowledge of matrix structural analysis and/or the finite-element method. If this is not the case, many topics that are based on statics and/or the AASHTO Specification provision could be read in detail. Other topics such as grillage, finite-element, and finite-strip analysis can be read with regard to observation of *behavior* rather than understanding the details of the analysis.

The discussion of slab and slab–girder bridges is followed by an abbreviated address of box systems. Many of the issues involved with the analysis of box systems are the same as slab–girder systems. Such issues are not reiterated, and the discussion is focused on the behavioral aspects that are particular to box systems.

11.1 SLAB–GIRDER BRIDGES

The slab and slab–girder bridges are the most common types of bridge in the United States. A few of these bridges are illustrated in Chapter 4. These are made of several types and combinations of materials. Several examples are listed in Table 11.1.

A schematic illustration of a slab–girder bridge is shown in Figure 11.1(a). The principal function of the slab is to provide the roadway surface and to transmit the applied loads to the girders. This load path is illustrated in Figure 11.1(b). The load causes the slab–girder system to displace as shown in Figure 11.1(c). If linear behavior is assumed, the load to each girder is related to its displacement. As expected, the girder near the location of the load application carries more load than those away from the applied load. Compare the deflection of the girders in Figure 11.1(c). Equilibrium requires that the summation of the load carried by all the girders equals the total applied load. The load carried by each girder is a function of the relative stiffness of the components that comprise the slab–girder system. The two principal components are the slab and the girders; other components include cross frames, diaphragms, and bearings. Only the slab and girder are considered here as the other components affect the behavior to a lesser extent.

The effect of relative stiffness is illustrated by considering the two slab–girder systems shown in Figures 11.1(d)

Table 11.1 Examples of Slab–Girder Bridges[a]

Girder Material	Slab Material
Steel	CIP concrete
Steel	Precast concrete
Steel	Steel
Steel	Wood
CIP concrete	CIP concrete
Precast concrete	CIP concrete
Precast concrete	Precast concrete
Wood	Wood

[a]Cast in place = CIP.

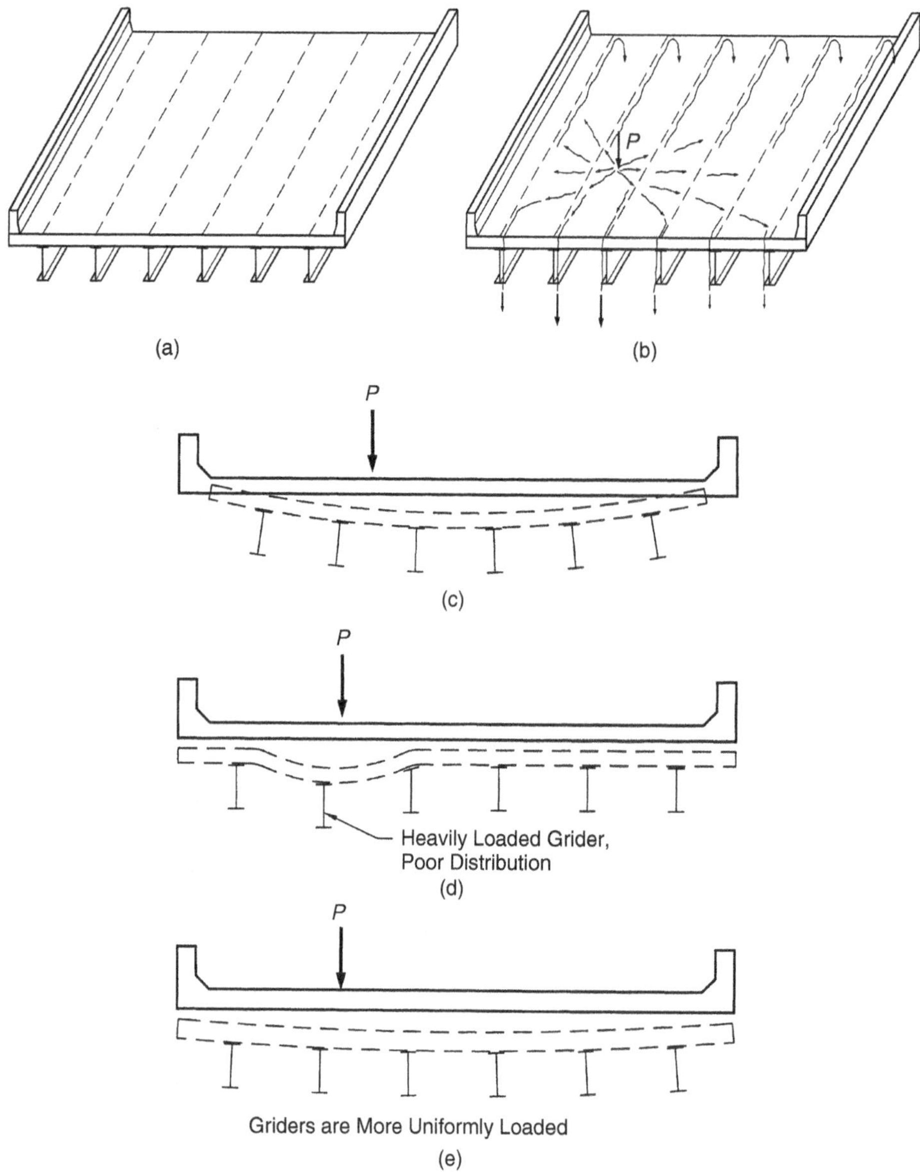

Fig. 11.1 (a) Slab–girder bridge, (b) load transfer (boldface lines indicate larger actions), (c) deflected cross section, (d) transversely flexible, and (e) transversely stiff.

and 11.1(e). The system shown in Figure 11.1(d) has a slab that is relatively flexible compared to the girder. Note the largest deflection is in the girder under the load and the other girder deflections are relatively small. Now consider the system shown in Figure 11.1(e) where the slab is stiffer than the previous case. Note the load (deflection) is distributed to the girders more evenly, therefore the load to each girder is less than shown in Figure 11.1(d).

The purpose of structural analysis is to determine the distribution of internal actions throughout the structure. Any method that is used should represent the relative stiffness of the slab and the girders. As outlined in the previous sections, the importance of accuracy of the analysis depends on the limit state considered and ductility available for the redistribution of actions after initial yielding. To illustrate, consider the simply supported slab–girder bridge shown in Figure 11.2(a). Assume the girders have adequate ductility for plastic analysis. Because of the simply supported configuration, this structure might traditionally be considered nonredundant, that is, one that does not have an alternative load path. Now assume the girder under the load yields and stiffness decreases. Any additional load is then carried by the neighboring girders. If the load continues to increase, then the neighboring girders also yield, and additional load is carried by the nonyielded girders. If the slab has the capacity to transmit the additional load, then this process continues until all girders have reached their plastic capacities and a mechanism occurs in every girder. The ultimate load is

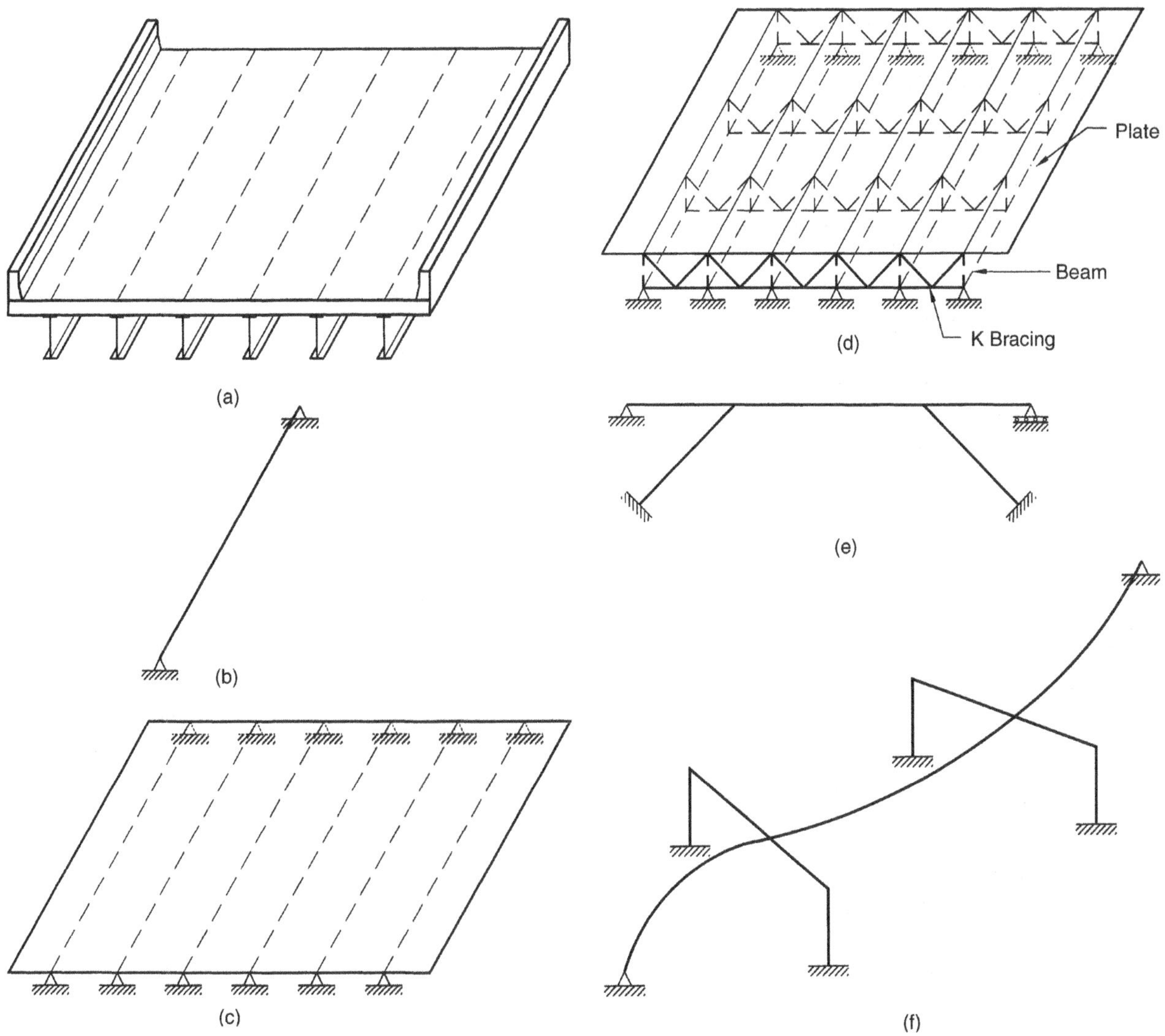

Fig. 11.2 (a) Slab–girder bridge, (b) beam-line model, and (c) flat-plate model (2D). (d) 3D model, (e) plane frame model (2D), and (f) space frame model (2.5D).

obviously greater than the load that causes first yield. Note that this is predicted by the lower bound theorem described in Section 10.2.1. The shakedown theorems also apply.

So why should the engineer perform a complicated analysis to distribute the load to the girders? There are two principal reasons: (1) The failure mode may not be ductile, such as in a fatigue-related fracture or instability, and (2) the limit state under consideration may be related to serviceability and service-level loads. Both reasons are important, and, therefore, it is traditional to model the system as linear elastic to obtain reasonable distribution of internal actions for strength, service, and fatigue limit states.

As the lower bound theorem may also apply, this approach is likely conservative and gives reasonable results for the strength limit states. In the case of the evaluation of an existing bridge where repair, retrofit, and/or posting is involved, it may be reasonable to use a linear elastic analysis for service load limit states and consider the nonlinear behavior for the strength limit states. Such a refinement could significantly influence the rehabilitation strategy or posting load.

Several methods for linear elastic analysis are described in the sections that follow, as they are used in engineering practice, and may be used for estimating the load effects for all limit states.

Behavior, Structural Idealization, and Modeling Again consider the slab–girder system shown in Figure 11.2(a). The spatial dimensionality is a primary modeling assumption. The system may be modeled as a 1-, 1.5-, 2-, 2.5-, or 3-dimensional system. The 1D system is shown in Figure 11.2(b). This system is a beam and may be modeled as such. Obviously, this is a simple model and is attractive for design. The primary issue is how the load is distributed to the girder, which is traditionally done by using empirically determined distribution factors to transform the 3D system to a 1D system. In short, the vehicle load (or load effect) from the beam analysis is multiplied by a factor that is a function of the relative stiffness of the slab–girder system. This transforms the beam load effect to the estimated load effect in the system. Herein this procedure is called the beam-line (or girder-line) method because only one girder is considered as opposed to modeling the entire bridge as a single beam.

A 2D system is shown in Figure 11.2(c). This system eliminates the vertical dimension. What results is a system that is usually modeled with thin-plate theory for the deck combined with standard beam theory for the girders. The girder is brought into the plane of the deck (or plate) and supports are considered at the slab level. The eccentricity of both may be considered and included. The in-plane effects are usually neglected. Another type of 2D system is the plane frame shown in Figure 11.2(e). Often the loads are distributed to the frame by distribution factors using the beam-line method. The analysis is performed on the plane frame.

In the 1.5D system, the distribution factors are established by a 2D system, but the girder actions are established using a 1D system. This procedure is done because several computer programs exist for beam-line analysis and designs that are 1D, but the designer wishes to use a refined procedure for the determination of the distribution factors rather than that using the empirically based methods.

A 3D system is shown in Figure 11.2(d). Here the full dimensionality is maintained. Components such as cross frames, diaphragms, and so on are often included. This model is the most refined and requires the most designer time and computer resources to perform. It is often justified for the analysis of highly skewed bridges, curved bridges, or a combination thereof.

The 2.5D system typically uses a single-girder line in combination with other components and subsystems. Such a system is shown in Figure 11.2(f) where a curved box girder and its piers are modeled with space frame elements. Such systems are often used in the western states in high seismic regions.

All of these methods are viable and have their place in engineering practice. It is not always appropriate, practical, or desirable to use the most refined method available. The complexity of the system, the load effects sought, the reason for the analysis, whether it be for design or evaluation, all are important considerations in the selection of the modeling procedures. Additionally, modeling a complex bridge for various construction stages is sometimes more critical than the bridge in its final state. The previous discussion is summarized in Table 11.2.

Beam-Line Method

Distribution Factor Method—Concepts. As previously described, the spatial dimensionality of the system can be reduced by using a distribution factor. This factor is established by analyzing the system with a refined method to establish the actions in the girders. For this discussion, bending moment is used for illustration but shear could also be used. The maximum moment at a critical location is determined

Table 11.2 Spatial Modeling

Spatial Dimensionality	Mathematical Model	Numerical Model (Examples)	Figures
1	Beam theory	Stiffness (displacement) method Flexibility (force) method Consistent deformations Slope deflection Moment distribution	11.2(b)
2	Thin-plate theory Beam theory	Grillage Finite strip Finite element Harmonic analysis Classical plate solutions	11.2(c), 11.2(e)
3	Theory of elasticity Thin-plate theory Beam theory	Grillage Finite strip Finite element Classical solutions	11.2(d)
1.5	Thin-plate theory Beam theory	Grillage Finite strip Finite element Harmonic analysis Classical plate solutions	Not shown
2.5	Beam theory	Finite element	11.2(f)

with an analytical or numerical method and is denoted as M_{refined}. Next, the same load is applied to a single girder and a 1D beam analysis is performed. The resulting maximum moment is denoted as M_{beam}. The distribution factor is defined as

$$g = \frac{M_{\text{refined}}}{M_{\text{beam}}}$$

In the case of a 1.5D analysis, this factor is used to convert the load effects established in the beam-line analysis to the estimated results of the entire system. For example, analyze the beam line for the live load and then multiply by the distribution factor g to obtain the estimated load effect in the system.

Alternatively, many analyses can be performed for numerous bridges, and the effects of the relative stiffness of the various components, geometry effects, and load configuration may be studied. The results of these analyses are then used to establish empirically based formulas that contain the system parameters as variables. These formulas can then be used by designers to estimate the distribution factors *without* performing the refined analysis. Certainly, some compromise may be made in accuracy, but this method generally gives good results. The AASHTO distribution factors are based on this concept and are presented in Table 11.3 where they are discussed in more detail.

Background. The AASHTO Specification has employed distribution factor methods for many years. In the most common case, the distribution factor was where S is the girder spacing (ft), and D is a constant depending on bridge type, the number of lanes loaded, and g may be thought of as the number of lanes carried per girder:

$$g = \frac{S}{D}$$

For example, for a concrete slab on a steel girder $D = 11.0$ was used for cases where two or more vehicles are present. Obviously, this is a simplistic formula and easy to apply, but as expected, it does not always provide good estimates of the girder load in the full system. It has been shown by Zokaie et al. (1991) and Nowak (1993) that this formulation underestimates the load effects with close girder spacing and overestimates with wider spacing. To refine this approach, research was conducted to develop formulas that are based on more parameters and provide a better estimate of the true system response. This work was performed under NCHRP Project 12–26 (Zokaie et al., 1991) and provides the basis for the distribution factors presented in AASHTO [A4.6.2.2].*

AASHTO Specification—Distribution Factors. The distribution factors may be used for bridges with fairly regular geometry. As stated in AASHTO [A4.6.2.2], the method is limited to systems with:

- ☐ Constant cross section.
- ☐ Number of beams is four or more.
- ☐ Beams are parallel and have approximately the same stiffness.
- ☐ Roadway part of the cantilever overhang does not exceed 3.0 ft (910 mm).
- ☐ Plan curvature is small [A4.6.1.2].
- ☐ Cross section is consistent with the sections shown in Table 4.1.

The provisions for load distribution factors are contained in several AASHTO articles and only a few are discussed here. These articles represent some of the most important provisions in Section 4 of the AASHTO Specification, and because of the many algebraically complex equations, these are not presented in the body of this discussion. For the sake of brevity, the most common bridge types—the slab and slab–girder bridge—are discussed here in detail. The analysis of other common types is discussed later. The distribution factors for slab–girder bridges are given in Table 11.3:

where

S = girder spacing (ft)
L = span length (ft)
t_s = slab thickness (in.)
K_g = longitudinal stiffness parameter (in.^{4})
$K_g = n\left(I_g + e_g^2 A\right)$, where
n = modular ratio ($E_{\text{girder}}/E_{\text{deck}}$)
I_g = moment of inertia of the girder (in.^{4})
e_g = girder eccentricity, which is the distance from the girder centroid to the middle centroid of the slab (in.)
A = girder area (in.^{2})
d_e = distance from the center of the exterior beam and the inside edge of the curb or barrier (in.)
θ = angle between the centerline of the support and a line normal to the roadway centerline

The *lever rule* is a method of analysis. It involves a statical distribution of load based on the assumption that each deck panel is simply supported over the girder, except at the exterior girder that is continuous with the cantilever. Because the load distribution to any girder other than one directly next to the point of load application is neglected, the lever rule is typically a conservative method of analysis.

The equations in Table 11.3 were developed by Zokaie et al. (1991). Here investigators performed hundreds of analyses on bridges of different types, geometrics, and stiffness. Many of these structures were actual bridges that were taken from the inventories nationwide. Various computer programs were used for analysis and compared to experimental results. The programs that yielded the most accurate results were selected for further analysis in developing the AASHTO formulas.

*The article number in AASHTO (2010) LRFD Bridge Specifications are enclosed in brackets and preceded by the letter A if specifications and by the letter C if commentary.

Table 11.3US Vehicles per Girder for Concrete Deck on Steel or Concrete Beams; Concrete T-Beams; T- and Double T-Sections Transversely Posttensioned Together[a]—US Units

Action/Location	AASHTO Table	Distribution Factors (mg)[b]	Skew Correction Factor[c]	Range of Applicability
A. Moment interior girder	4.6.2.2.2b-1	One design lane loaded: $mg^{SI}_{moment} = 0.06 + \left(\frac{S}{14}\right)^{0.4}\left(\frac{S}{L}\right)^{0.3}\left(\frac{K_g}{12Lt_s^3}\right)^{0.1}$		$3.5 \le S \le 16$ ft $4.5 \le t_s \le 12$ in. $20 \le L \le 240$ ft $10{,}000 \le K_g \le 7{,}000{,}000$ No. of beams ≥ 4
		Two or more (multiple) design lanes loaded: $mg^{MI}_{moment} = 0.075 + \left(\frac{S}{9.5}\right)^{0.6}\left(\frac{S}{L}\right)^{0.2}\left(\frac{K_g}{12Lt_s^3}\right)^{0.1}$	$1 - C_1(\tan\theta)^{1.5}$ $C_1 = 0.25\left(\frac{K_g}{12Lt_s^3}\right)^{0.25}\left(\frac{S}{L}\right)^{0.5}$ If $\theta < 30°$, then $C_1 = 0.0$ If $\theta > 60°$, then $\theta = 60°$	$\theta \le 30°$ then no adjustment is necessary $30° \le \theta \le 60°$
B. Moment exterior girder	4.6.2.2.2d-1	One design lane loaded: Use lever rule Two or more (multiple) design lanes loaded:[d] $mg^{ME}_{moment} = e\left(mg^{MI}_{moment}\right)$ $e = 0.77 + \frac{d_e}{9.1} \ge 1.0$ d_e is positive if girder is inside of barrier, otherwise negative	N/A	$-1.0 \le d_e \le 5.5$ ft
C. Shear interior girder	4.6.2.2.3a-1	One design lane loaded: $mg^{SI}_{shear} = 0.36 + \frac{S}{25.0}$ Two or more (multiple) design lanes loaded: $mg^{MI}_{shear} = 0.2 + \frac{S}{12} - \left(\frac{S}{35}\right)^2$	$1.0 + 0.20\left(\frac{12Lt_s^3}{K_g}\right)^{0.3}\tan\theta$	$3.5 \le S \le 16.0$ ft $4.5 \le t_s \le 12$ in. $20 \le L \le 240$ ft No. of beams ≥ 4 $\theta \le 30°$ then no adjustment is necessary $0° \le \theta \le 60°$
D. Shear exterior girder	4.6.2.2.3b-1	One design lane loaded: Use lever rule Two or more (multiple) design lanes loaded: $mg^{ME}_{shear} = e\left(mg^{MI}_{shear}\right)$ $e = 0.6 + \frac{d_e}{10}$ For $N_b = 3$, use lever rule d_e is positive if girder is inside of barrier, otherwise negative	N/A	$-1.0 \le d_e \le 5.5$ ft

[a]See Table 4.1 for applicable cross sections.
[b]Equations include multiple presence factor; for lever rule and the rigid method engineer must perform factoring by *m*.
[c]Not applicable = N/A.
[d]This value shall not be less than that computed as if the cross section is rigid (to account for diaphragms) per A4.6.2.2.2d. An example for the rigid method is provided in Appendix F.

Table 11.3SI Vehicles per Girder for Concrete Deck on Steel or Concrete Beams; Concrete T-Beams; T- and Double T-Sections Transversely Posttensioned Together[a]—SI Units

Action/Location	AASHTO Table	Distribution Factors (mg)[b]	Skew Correction Factor[c]	Range of Applicability
A. Moment interior girder	4.6.2.2.2b-1	One design lane loaded: $mg^{SI}_{moment} = 0.06 + \left(\frac{S}{4300}\right)^{0.4}\left(\frac{S}{L}\right)^{0.3}\left(\frac{K_g}{Lt_s^3}\right)^{0.1}$ Two or more (multiple) design lanes loaded: $mg^{MI}_{moment} = 0.075 + \left(\frac{S}{2900}\right)^{0.6}\left(\frac{S}{L}\right)^{0.2}\left(\frac{K_g}{Lt_s^3}\right)^{0.1}$	$1 - C_1(\tan\theta)^{1.5}$ $C_1 = 0.25\left(\frac{K_g}{Lt_s^3}\right)^{0.25}\left(\frac{S}{L}\right)^{0.5}$ If $\theta < 30°$, then $C_1 = 0.0$ If $\theta > 60°$, then $\theta = 60°$	$1100 \le S \le 4900$ mm $110 \le t_s \le 300$ mm $6000 \le L \le 73\,000$ mm $4 \times 10^9 \le K_g \le 3 \times 10^{12}$ mm^4 No. of beams ≥ 4 $30° \le \theta \le 60°$
B. Moment exterior girder	4.6.2.2.2d–1	One design lane loaded: Use lever rule Two or more (multiple) design lanes loaded:[d] $mg^{ME}_{moment} = e\,(mg^{MI}_{moment})$ $e = 0.77 + \frac{d_e}{2800} \ge 1.0$ d_e is positive if girder is inside of barrier, otherwise negative	N/A	$-300 \le d_e \le 1700$ mm
C. Shear interior girder	4.6.2.2.3a-1	One design lane loaded: $mg^{SI}_{shear} = 0.36 + \frac{S}{7600}$ Two or more (multiple) design lanes loaded: $mg^{MI}_{shear} = 0.2 + \frac{S}{3600} - \left(\frac{S}{10\,700}\right)^2$	$1.0 + 0.20\left(\frac{Lt_s^3}{K_g}\right)^{0.3}\tan\theta$	$1100 \le S \le 4900$ mm $110 \le t_s \le 300$ mm $6000 \le L \le 73{,}000$ mm $4 \times 10^9 \le K_g \le 3 \times 10^{12}$ mm^4 No. of beams ≥ 4 $0° \le \theta \le 60°$
D. Shear exterior girder	4.6.2.2.3b-1	One design lane loaded: Use lever rule Two or more (multiple) design lanes loaded: $mg^{ME}_{shear} = e\,(mg^{MI}_{shear})$ $e = 0.6 + \frac{d_e}{3000}$ d_e is positive if girder is inside of barrier, otherwise negative	N/A	$-300 \le d_e \le 1700$ mm

[a]See Table 4.1 for applicable cross sections.
[b]Equations include multiple presence factor; for lever rule and the rigid method engineer must perform factoring by *m*.
[c]Not applicable = N/A.
[d]This value shall not be less than that computed as if the cross section is rigid (to account for diaphragms) per A4.6.2.2.2d. An example for the rigid method is provided in Appendix F.

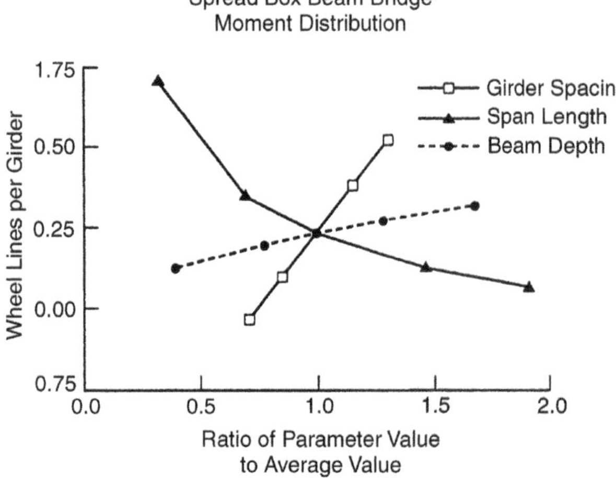

Fig. 11.3 Parametric studies (after Zokaie et al., 1991).

The database of actual bridges was used to determine "an average bridge" for each type. Within each type, the parametric studies were made to establish the distribution factor equations. Example results for the slab–girder bridge type are shown in Figure 11.3. Note that the most sensitive parameter (greatest slope) for this type of bridge is the girder spacing. This observation is consistent with the traditional AASHTO distribution factor of $S/5.5$ ft (for a wheel line or one-half lane). In fact, the division of the slope of this line, which is approximately 1.25, into the average girder spacing from the database, which is 7.5 ft yields $D = 6.0$, or approximately the value of $D = 5.5$ used by AASHTO for many years. It is important to note that the span length and girder stiffness affect the load distribution but to a lesser extent. This effect is reflected in the equations presented in Table 11.3.

Unlike the previous AASHTO equations, the parametric properties of the bridge were used to develop prediction models based on a power law. Each parameter was assumed to be independent of others in its effect on the distribution model. Although this is probably not strictly true, the resulting equations seem to work well. The results of Table 11.3 are compared to finite-element analysis (more rigorous and assumed to be more accurate) in Figure 11.4. The letters a–j reference AASHTO beam type (A4.6.2.2.2).

In Figure 11.4(a), the rigorous analyses are compared to the old AASHTO procedures [$g = (S/D)$], and in Figure 11.4(b), the rigorous analyses are compared to the equations of Table 11.3. Notice the great variability in the former and the decrease variability of the latter. Hence, the additional terms are necessary to better predict the system response. Traditionally, AASHTO has based analysis on the wheel line or half the axle weight. In the present specification, the analysis is assumed to be based on the entire vehicle weight. Thus, if one compares the distribution factors historically used by AASHTO to those presently used, then the traditional factors must be divided by 2, or the present factor must be multiplied by 2.

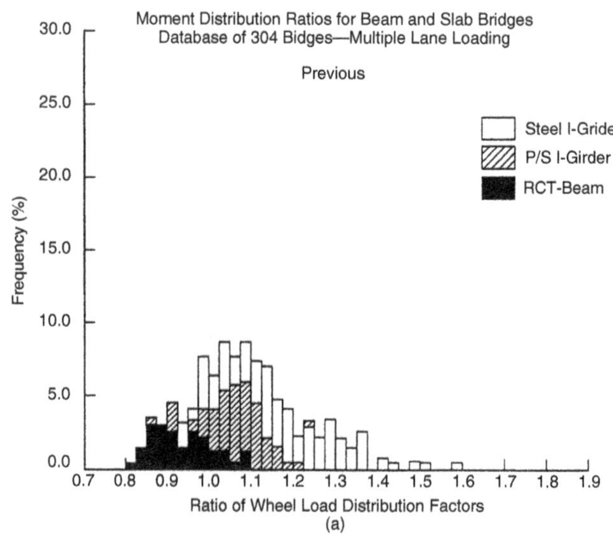

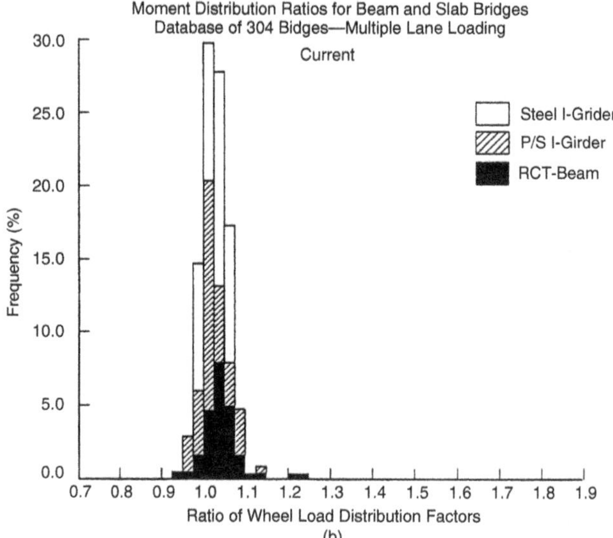

Fig. 11.4 Comparison of AASHTO distribution factor with rigorous analysis (after Zokaie et al., 1991).

The single design lane formulas were developed with a single design truck, and the multilane loaded formulas were developed with two *or* more trucks. Therefore, the most critical situation for two, three, or more vehicles was used in the development. The multiple presence factors given in Table 8.6 were included in the analytical results upon which the formulas are based. Thus, the *multiple presence factors are not to be used in conjunction with the factors given in Table 11.3, but rather the multiple presence is implicitly included in these factors.*

The development of the present AASHTO (2010) distribution factors was based on simply supported bridges. The investigators also studied systems to quantify the effect of continuity. Given the relative insensitivity of girder stiffness to the distribution factors (see Fig. 11.3), it is expected that continuity does not significantly affect the distribution factors. Zokaie et al. (1991) determined that the effect of continuity was between 1.00 and 1.10 for most systems and suggested associated adjustments. The specification writers chose to eliminate this refinement because:

- ☐ Correction factors dealing with 5% adjustments were thought to imply misleading levels of accuracy in an approximate method.
- ☐ Analysis carried out on a large number of continuous beam–slab type of bridges indicates that the distribution coefficients (factors) for negative moment exceed those obtained for positive moment by approximately 10%. On the other hand, it had been observed that stresses at or near internal bearings are reduced due to the fanning of the reaction force. This reduction is about the same magnitude as the increase in distribution factors; hence the two tend to cancel.

Note as a practical consideration the equations provided in Table 11.3 are a function of cross section. Hence, in a design scenario, the cross section must be estimated prior to determining the live-load effects. This approach is similar to dead-load estimates prior to section proportioning. To aid the engineer, estimates provided in A4.6.2.2.1 in a table are repeated in Table 11.4.

Example 11.1 The slab–girder bridge illustrated in Figure 11.5(a) with a simply supported span of 35 ft (10 688 mm) is used in this example and several others that follow. Model the entire bridge as a single beam to determine the support reactions, shears, and bending moments for one and two lanes loaded using the AASHTO design truck.

A free-body diagram is shown in Figure 11.5(b) with the design truck positioned near the critical location for flexural bending moment. Although this position does not yield the absolute maximum moment, which is 361.2 ft kips (498.7 kN m) (see Example 9.10), it is close to the critical location, and this position facilitates analysis in later examples. The resulting moment diagram is shown in Figure 11.5(c). Note the maximum moment is 358.4 ft kips (493.2 kN m) for one-lane loaded, which is within 1% of the absolute maximum moment. This value is doubled for two trucks positioned on the bridge giving a maximum of 716.8 ft kips (986.4 kN m).

These values are used repeatedly throughout several examples that follow. The critical section for design is at the location of the maximum statical moment. This location is also used in several examples that follow.

A free-body diagram is shown in Figure 11.5(d) with the design truck positioned for maximum shear/reaction force. The resulting maximum is 52.8 kips (238.3 kN) for one lane loaded and 105.6 kips (476.6 kN) for two loaded lanes. See Figure 11.5(e). These values are also used in the examples that follow.

Example 11.2 Determine the AASHTO distribution factors for bridge shown in Figure 11.5(a).

A girder section is illustrated in Figure 11.6 [see AISC (2003) for girder properties]. The system dimensions and properties are as follows:

Girder spacing, $S = 8$ ft (2438 mm)
Span length, $L = 35$ ft (10 668 mm)
Deck thickness, $t_s = 8$ in. (203 mm)
Deck modulus of elasticity, $E_c = 3600$ ksi (24.82 GPa)
Girder modulus of elasticity, $E_s = 29\,000$ ksi (200.0 GPa)
Modular ratio, $n = E_s/E_c = 29{,}000/3600 = 8.05$; use 8
Girder area, $A_g = 31.7$ in.^{2} (20,500 mm^2)
Girder moment of inertia, $I_g = 4470$ in.^{4} (1860×10^6 mm^4)
Girder eccentricity, for example, $= t_s/2 + d/2 = 8/2 + 29.83/2 = 18.92$ in. (480 mm)
Stiffness parameter, $K_g = n\left(I_g + e_g^2 A_g\right) = 8\,[4470 + \left(18.92^2\right)(31.7)] = 126{,}500$ in.^{4} (52.6×10^9 mm^4)
$d_e = 3.25$ ft (cantilever) – 1.25 ft (barrier) = 2.0 ft (610 mm)

Table 11.4 Constant Values Table 11.3 (per A4.6.2.2.2, A4.6.2.2.3)

		Simplified Value AASHTO Beam Types			
Equation Parameters	AASHTO Table Reference	a	e	k	f, g, i, j
$\left(\dfrac{K_g}{12.0Lt_s^3}\right)^{0.1}$	4.6.2.2.2b-1	1.02	1.05	1.09	
$\left(\dfrac{K_g}{12.0Lt_s^3}\right)^{0.25}$	4.6.2.2.2e-1	1.03	1.07	1.15	
$\left(\dfrac{12.0Lt_s^3}{K_g}\right)^{0.3}$	4.6.2.2.3c-1	0.97	0.93	0.85	
$\dfrac{I}{J}$	4.6.2.2.2b-1, 4.6.2.2.3a-1				$0.54\left(\dfrac{d}{b}\right) + 0.16$

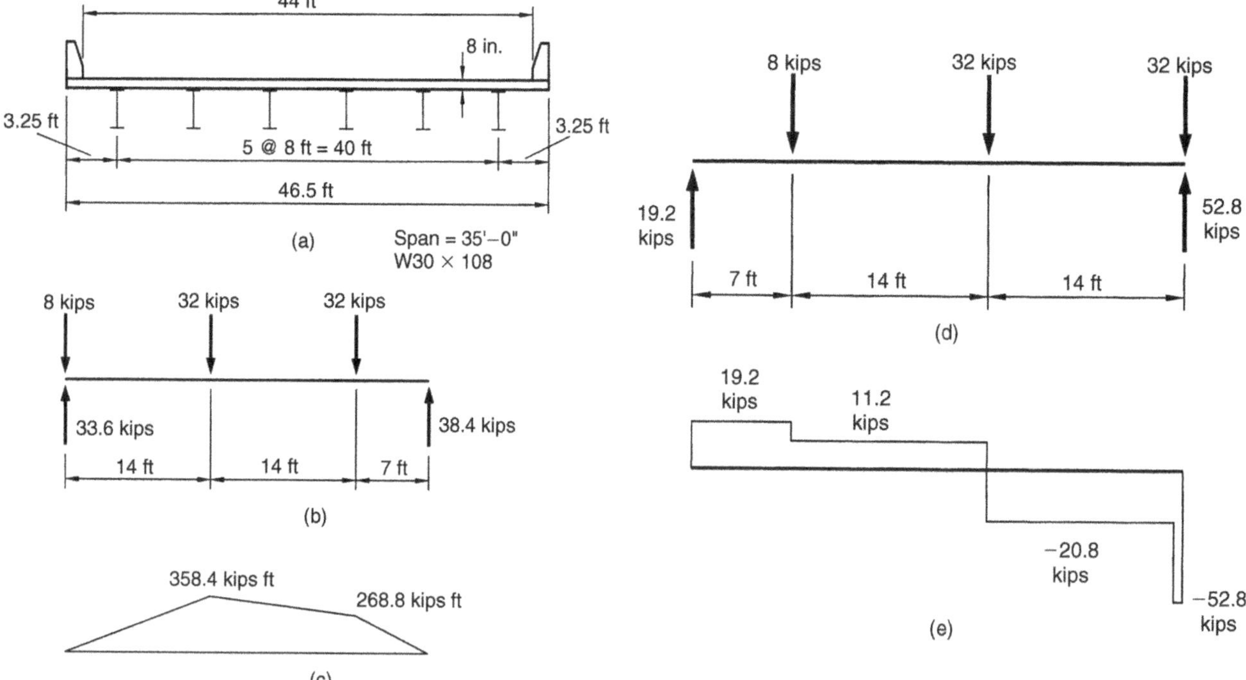

Fig. 11.5 (a) Cross section of a slab–girder bridge, (b) free-body diagram—load for near-critical flexural moment, and (c) moment diagram. (d) Free-body diagram—load for near-critical shear/reactions, and (e) shear diagram.

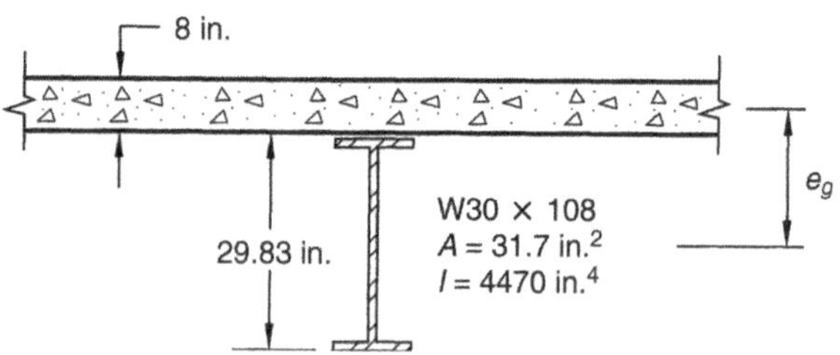

Fig. 11.6 Girder cross section.

The AASHTO distribution factors for moments are determined using rows A and B of Table 11.3.

The distribution factor for moment in the interior girder for one lane loaded is (Note the multiple presence factor m is included in the equations so this is denoted mg where m is included.)

$$mg^{\text{SI}}_{\text{moment}} = 0.06 + \left(\frac{S}{14}\right)^{0.4}\left(\frac{S}{L}\right)^{0.3}\left(\frac{K_g}{12Lt_s^3}\right)^{0.1}$$
$$= 0.06 + \left(\frac{8}{14}\right)^{0.4}\left(\frac{8}{35}\right)^{0.3}\left[\frac{126,500}{12\,(35)\,(8^3)}\right]^{0.1}$$
$$= 0.55 \text{ lane/girder}$$

The distribution factor for moment in the interior girder for multiple lanes loaded is

$$mg^{\text{MI}}_{\text{moment}} = 0.075 + \left(\frac{S}{9.5}\right)^{0.6}\left(\frac{S}{L}\right)^{0.2}\left(\frac{K_g}{12Lt_s^3}\right)^{0.1}$$
$$= 0.075 + \left(\frac{8}{9.5}\right)^{0.6}\left(\frac{8}{35}\right)^{0.2}\left[\frac{126,500}{12\,(35)\,(8^3)}\right]^{0.1}$$
$$= 0.71 \text{ lane/girder}$$

The distribution factor for moment in the exterior girder for multiple lanes loaded requires an adjustment factor:

$$e = 0.77 + \frac{d_e}{9.1} \geq 1.0$$
$$= 0.77 + \frac{2}{9.1} = 0.99 \quad \therefore \quad \text{use } e = 1.0$$

The adjustment factor for moment is multiplied by the factor for the interior girder and the result is

$$mg^{\text{ME}}_{\text{moment}} = e\left(mg^{\text{MI}}_{\text{moment}}\right) = 1.00\,(0.71)$$
$$= 0.71 \text{ lane/girder}$$

For the distribution factor for the exterior girder with one loaded lane, use the lever rule; this is done in the next example and the result is

$$mg^{\text{SE}}_{\text{moment}} = 0.75 \text{ lane/girder}$$

For the distribution factor for shear, rows C and D in Table 11.3 are used. The distribution factor for the interior girder with one lane loaded is

$$mg^{\text{SI}}_{\text{shear}} = 0.36 + \frac{S}{25} = 0.36 + \frac{8}{25} = 0.68 \text{ lane/girder}$$

Table 11.5 AASHTO Distribution Factor Method Results

Girder Location	Number of Lanes Loaded	Moment (ft kips)	Moment Distribution Factor (mg)	Girder Moment (ft kips)	Simple Beam Reaction (kips)	Shear Distribution Factor (mg)	Girder Shear (kips)
Exterior	1	358.4	$0.625 \times 1.2 = 0.75$	268.7	52.8	0.75	39.6
Exterior	2	358.4	0.71	254.3	52.8	0.65	34.3
Interior	1	358.4	0.55	197.1	52.8	0.68	35.9
Interior	2	358.4	0.71	254.3	52.8	0.81	42.8

Similarly, the factor for shear with multiple lanes loaded is

$$mg^{\mathrm{MI}}_{\mathrm{shear}} = 0.2 + \frac{S}{12} - \left(\frac{S}{35}\right)^2$$
$$= 0.2 + \frac{8}{12} - \left(\frac{8}{35}\right)^2 = 0.81 \text{ lane/girder}$$

The adjustment for shear in the exterior girder is given in row D of Table 11.3. The calculation is

$$e = 0.6 + \frac{d_e}{10} = 0.6 + \frac{2}{10} = 0.80$$

The adjustment is multiplied by the interior distribution factor, the result is

$$mg^{\mathrm{ME}}_{\mathrm{shear}} = e\left(mg^{\mathrm{MI}}_{\mathrm{shear}}\right) = 0.80\,(0.81) = 0.65 \text{ lane/girder}$$

The lever rule is used for the exterior girder loaded with one design truck. The details are addressed in the following example. The result is $mg^{\mathrm{SI}}_{\text{shear or moment}} = 0.625$ times 1.2 (multiple presence factor) = 0.75 for both shear and moment. The AASHTO results are summarized in Table 11.5.

Example 11.3 Use the lever method to determine the distribution factors for the bridge shown in Figure 11.5(a).

Exterior Girder

Consider Figure 11.7. The deck is assumed to be simply supported by each girder except over the exterior girder where the cantilever is continuous. Considering truck 1, the reaction at A (exterior girder load) is established by balancing the moment about B:

$$R_A\,(8) = \left(\frac{P}{2}\right)(8) + \left(\frac{P}{2}\right)(2)$$

which reduces to

$$R_A = \left(\frac{P}{2}\right) + \left(\frac{P}{2}\right)\left(\frac{2}{8}\right) = 0.625P$$

The fraction of the truck weight P that is carried by the exterior girder is 0.625. The multiple presence factor of 1.2 (see Table 8.6) is applicable for the one-lane loaded case. Thus, the girder distribution factors are

$$mg^{\mathrm{SE}}_{\text{shear or moment}} = (1.2)\,(0.625) = 0.75 \text{ lane/girder}$$

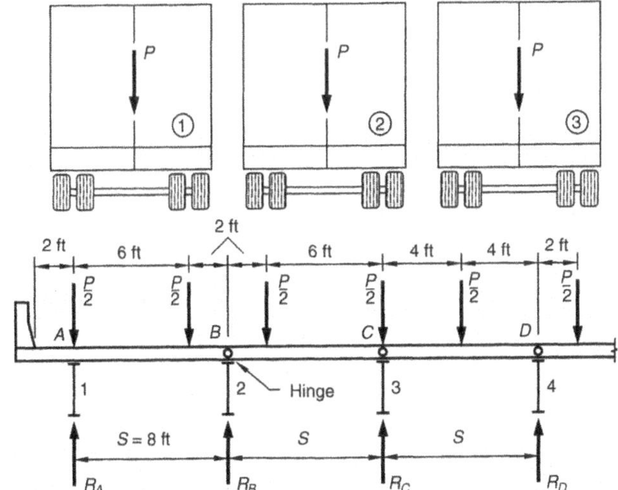

Fig. 11.7 Free-body diagram—lever rule method.

and

$$mg^{\mathrm{ME}}_{\text{shear or moment}} = (1.0)\,(0.625) = 0.625 \text{ lane/girder}$$

This factor is "statically" the same for one and two lanes loaded because the wheel loads from the adjacent truck (2) cannot be distributed to the exterior girder. Because all the wheels lie inside the first interior girder, the effect of their load cannot be transmitted across the assumed hinge. As illustrated, the difference is due to the multiple presence factor.

Interior Girder

The distribution factor for the interior girder subjected to two or more loaded lanes is established by considering trucks 2 and 3, each of weight P, positioned with axles on deck panels BC and CD, as shown in Figure 11.7. Equilibrium requires that the reaction at C is

$$R_c = \left(\frac{2}{8}\right)\left(\frac{P}{2}\right) + \left(\frac{P}{2}\right) + \left(\frac{P}{2}\right)\left(\frac{4}{8}\right) + \left(\frac{P}{2}\right)(0)$$
$$= 0.875P$$

and the distribution factor (multiple presence factor = 1.0) is

$$mg^{\mathrm{MI}}_{\text{shear or moment}} = (1.0)\,(0.875) = 0.875 \text{ lane/girder}$$

Table 11.6 Lever Rule Results

Girder Location	Number of Lanes Loaded	Moment (ft kips)	Moment Distribution Factor (*mg*)	Girder Moment (ft kips)	Simple Beam Reaction (kips)	Shear Distribution Factor (*mg*)	Girder Shear (kips)
Exterior	1	358.4	0.75	268.7	52.8	0.75	39.6
Exterior	2	358.4	0.625	223.9	52.8	0.625	33.0
Interior	1	358.4	0.75	268.7	52.8	0.75	39.6
Interior	2	358.4	0.875	313.4	52.8	0.875	46.2

Only truck 2 is considered for the case of one loaded lane on an interior girder. This truck has one wheel line directly over girder 3 and one wheel line 6 ft from the girder. By statics, the girder reaction at C is

$$R_c = \left(\frac{P}{2}\right) + \left(\frac{2}{8}\right)\left(\frac{P}{2}\right) = 0.625P$$

and the distribution factor is

$$\begin{aligned} mg^{SI}_{\text{shear or moment}} &= (1.2)(0.625) \\ &= 0.75 \text{ lane/girder} \end{aligned}$$

The distribution factors for shear and moment are the same under the pinned panel assumption. The lever rule results are summarized in Table 11.6. The format for the table is consistently used in the remaining examples in this chapter, which permits the ready comparison of results from the various methods of analysis.

Grillage Method Because the AASHTO and lever rule distribution factors are approximate, the engineer may wish to perform a more rigorous and accurate analysis. The advantages of more rigorous analysis include:

- ☐ The simplifying factors/assumptions that are made in the development of distribution factors for beam-line methods may be obviated.
- ☐ The variability of uncertain structural parameters may be studied for their effect on the system response. For example, continuity, material properties, cracking, nonprismatic effects, and support movements may be of interest.
- ☐ More rigorous models are developed in the design process and can be used in the rating of permit (overweight) vehicles and determining a more accurate overload strength.

One of the best mathematical models for the deck is the thin plate that may be modeled with the biharmonic equation (Timoshenko and Woinowsky-Kreiger, 1959; Ugural, 1981):

$$\nabla^4 w = \frac{\partial^4 w}{\partial x^4} + 2\frac{\partial^4 w}{\partial x^2 \partial y^2} + \frac{\partial^4 w}{\partial y^4} = \frac{p(x)}{D} \tag{11.1}$$

Where

w = vertical translation
x = transverse coordinate
y = longitudinal coordinate
p = vertical load
D = plate rigidity, equal to

$$D = \frac{Et^3}{12\left(1 - \nu^2\right)}$$

Where

ν = Poisson's ratio
t = plate thickness
E = modulus of elasticity

Equation 11.1 is for an isotropic (same properties in all directions) slab. Other forms are available for plates that exhibit significant orthotropy due to different reinforcement in the transverse and longitudinal directions (Timoshenko and Woinowsky-Kreiger, 1959; Ugural 1981). The development of Eq. 11.1 is based on several key assumptions: The material behaves linearly elastically, the strain profile is linear, the plate is isotropic, the vertical stresses due to the applied load are neglected, and the deformations are small relative to the dimensions of the plate.

Closed-form solutions to Eq. 11.1 are limited to cases that are based on simplified boundary conditions and loads. Even fewer solutions are available for girder-supported systems. Thus, approximate techniques or numerical models are used for the solution of Eq. 11.1; the most common methods include grillage, finite-element, and finite-strip methods.

To gain a better understanding of the development and limitations of Eq. 11.1, the reader is referred to common references on the analysis of plates (Timoshenko and Woinowsky-Kreiger, 1959; Ugural, 1981). Due to the focus and scope of this work, it suffices here to take an abbreviated and applied approach.

Consider the first term of Eq. 11.1 and neglect the transverse terms. Then Eq. 11.1 becomes

$$\frac{\partial^4 w}{\partial x^4} = \frac{p(x)}{D} \tag{11.2}$$

which is the same as Eq. 10.1, the mathematical model for a beam. Now neglect only the middle term, and Eq. 11.1 becomes

$$\nabla^4 w = \frac{\partial^4 w}{\partial x^4} + \frac{\partial^4 w}{\partial y^4} = \frac{p(x)}{D} \tag{11.3}$$

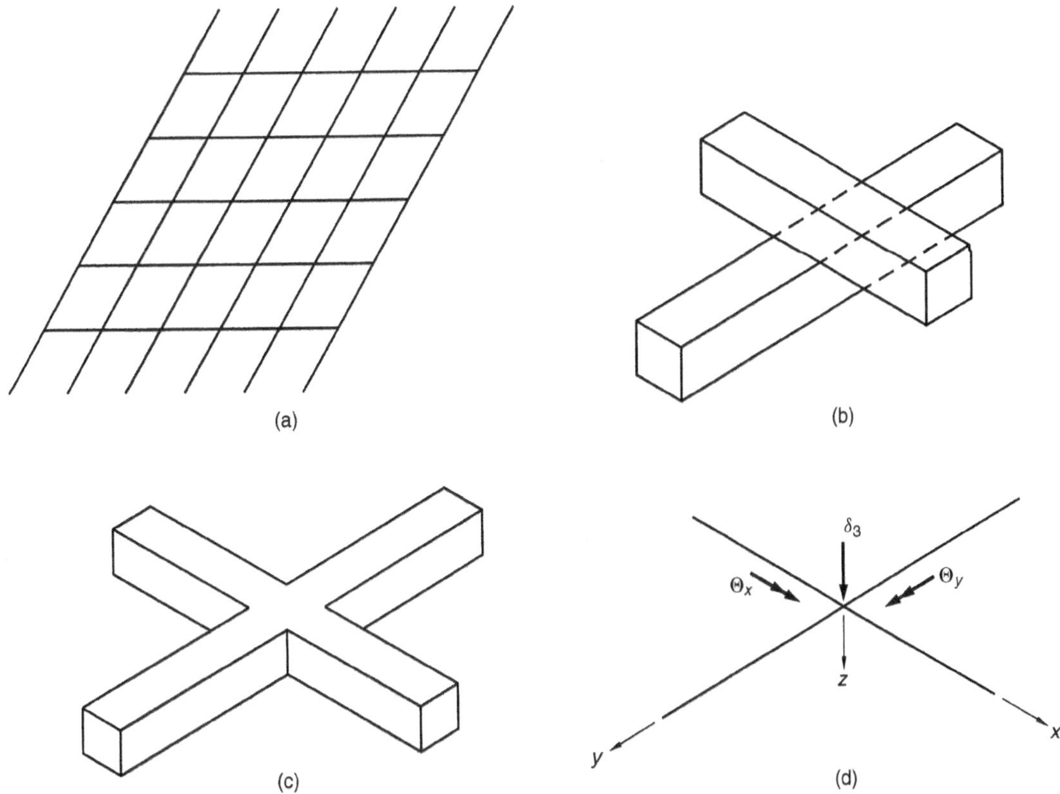

Fig. 11.8 (a) Grillage model, (b) crossing with translational continuity, (c) crossing with translational and rotational continuity, and (d) degrees of freedom in grillage (plane grid) modeling.

which is the mathematical model for a plate system that has no torsional stiffness or associated torsional actions. In a practical sense such systems do not exist and are merely mathematical models of a system where torsion exists but is neglected as well as the stiffening effect due to Poisson's effect. This type of system would be similar to modeling a plate with a series of crossing beams where one element sits on top of the other as shown in Figures 11.8(a) and 11.8(b). Note that at the intersection of the beams the only interaction force between the elements is a vertical force. This type of connection excessively simplifies the model of the deck, which is a continuum. In the continuum, a flexural rotation in one direction causes torsional rotation in an orthogonal direction. Consider the grillage joint shown in Figure 11.8(c). Here the joint is continuous for rotation in all directions, that is, the displacements of the joint is defined with the three displacements (degrees of freedom) shown in Figure 11.8(d), which includes vertical translation and two rotations. This type of joint, in combination with elements that have both flexural and torsional stiffness, is more like the continuum and, therefore, models it more accurately. This type of numerical model is called a *grillage.*

Grillage models became popular in the early 1960s with the advancement of the digital computer. As the methodologies for the stiffness analysis (or displacement method) of frames were well known, researchers looked for convenient ways to model continua with frame elements. The grillage model is such a technique.

Ideally the element stiffnesses in the grillage model would be such that when the continuum deck is subjected to a series of loads, the displacement of the continuum and the grillage are identical. In reality, the grillage can only approximate the behavior of the continuum described by Eq. 11.1. The reason for this difference is twofold: (1) The displacement in the grillage tends to be more irregular (bumpy) than the continuum, and (2) the moment in the grillage is a function of the curvature along the beam. In the plate, the moment is a function of the curvatures in two orthogonal directions due to Poisson's effect. Fortunately, these effects are small and the grillage method has been shown to be a viable method of analysis, especially for determining load effects to the girders.

Some advocates of the finite-element and strip methods are quick to discount the grillage method because it is nonrigorous. But remember that such methods are used to obtain reasonable distribution of internal actions while accounting for equilibrium (recall the lower bound theorem discussed earlier). Both advocates and critics have valid points and a few of these are listed below:

- ☐ Grillages can be used with any program that has plane grid or space frame capabilities.

- ☐ Results are easily interpreted and equilibrium is easily checked by free-body diagrams of the elements and system as a whole.
- ☐ Most all engineers are familiar with the analysis of frames.

The disadvantages are several:

- ☐ Method is nonrigorous and does not exactly converge to the exact solution of the mathematical model.
- ☐ To obtain good solutions, the method requires experience and judgment. The mesh design and refinement can be somewhat of an art form. (One could say this about any analysis method, however.)
- ☐ The assignment of the cross-sectional properties requires some discretion.

Hambly (1991) offers an excellent and comprehensive reference on modeling with grillages. The engineer interested in performing a grillage is encouraged to obtain this reference. Some of Hambly's suggestions regarding the design of meshes are paraphrased below:

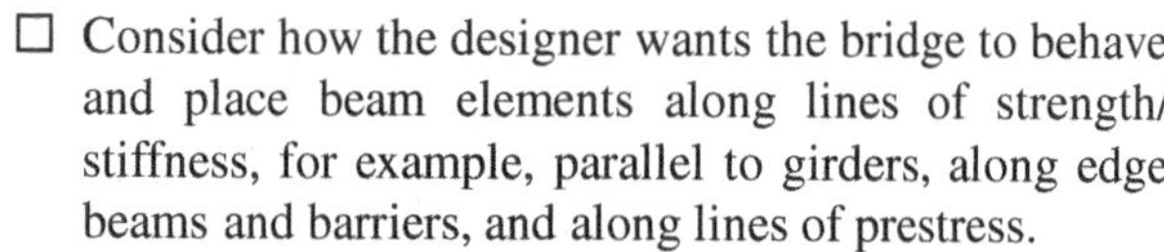

- ☐ Consider how the designer wants the bridge to behave and place beam elements along lines of strength/stiffness, for example, parallel to girders, along edge beams and barriers, and along lines of prestress.
- ☐ The total number of elements can vary widely. It can be one element in the longitudinal direction if the bridge is narrow (compared to its width) and behaves similarly to a beam, or it can be modeled with elements for the girders and other elements for the deck for wide decks where the system is dominated by the behavior of the deck. Elements need not be spaced closer than two to three times the slab thickness.
- ☐ The spacing of the transverse elements should be sufficiently small to distribute the effect of concentrated wheel loads and reactions. In the vicinity of such loads, the spacing can be decreased for improved results.

The element cross-sectional properties are usually based on the gross or uncracked section and are calculated on a per unit length basis. These properties are multiplied by the center-to-center spacing of the elements to obtain the element properties, herein called the tributary length. Two properties are required for the grillage model: flexural moment of inertia and the torsional constant. The moment of inertia is the familiar second moment of area, which is equal to

$$i_{\text{deck}} = \frac{bt^3}{12} \tag{11.4}$$

The torsional constant for a grillage element is

$$j_{\text{deck}} = \frac{bt^3}{6} = 2i_{\text{deck}} \tag{11.5}$$

The moment of inertia I_{girder} for a beam element is determined in the usual way and its eccentricity e_g (for a composite beam) is accounted by

$$I = I_{\text{girder}} + e_g^2 A_{\text{girder}} \tag{11.6}$$

For noncomposite systems, e_g is zero, and the beam is assumed to be at the middle surface of the deck.

For open sections that are comprised of thin rectangular shapes such as a wide flange or plate girder, the torsional constant is approximated by

$$J = \sum_{\text{all rectangles}} \frac{bt^3}{3} \tag{11.7}$$

where b is the long side and t is the narrower side ($b > 5t$). For open steel shapes, the torsional constant is usually small relative to the other parameters and has little affect on the response. For rectangular shapes that are not thin, the approximation is

$$J = \frac{3b^3t^3}{10\left(b^2 + t^2\right)} \tag{11.8}$$

The use of these properties is illustrated in the following example. For closed sections, such as box girders, see references on advanced mechanics for procedures to compute the torsional constant. For such sections, the torsional stiffness is significant and should be included.

Example 11.4 Use the grillage method to determine the end shear (reactions) and maximum bending moments in the girders in Figure 11.5(a), which is illustrated in Example 11.1. In addition, determine the distribution factors for moment and shear for girders for one and two lanes loaded.

The slab–girder bridge is discretized by a grillage model with the two meshes shown in Figures 11.9(a) and 11.9(b). The section properties are calculated below.

Girder Properties

$$E_s = 29\,000 \text{ ksi } (200.0 \text{ GPa})$$

$$A_g = 31.7 \text{ in.}^2 (20\,453 \text{ mm}^2)$$

$$d = 29.83 \text{ in. } (4536 \text{ mm})$$

$$e_g = \left(\frac{t_s}{2}\right) + \left(\frac{d}{2}\right) = \left(\frac{8}{2}\right) + \left(\frac{29.83}{2}\right)$$

$$= 18.92 \text{ in. } (481 \text{ mm})$$

$$I_g = 4470 \text{ in.}^4 \text{ (noncomposite girder)}$$

$$= 1.860 \times 10^9 \text{ mm}^4$$

$$J_g = 4.99 \text{ in.}^4 \text{ (noncomposite girder)}$$

$$= 2.077 \times 10^6 \text{ mm}^4$$

$$I_g = \text{(composite girder)} = I_g + e_g^2 A_g$$

$$= 4470 + 18.92^2(31.7)$$

$$= 15810 \text{ in.}^4 \text{ (steel)}$$

$$= 6.58 \times 10^9 \text{ mm}^4$$

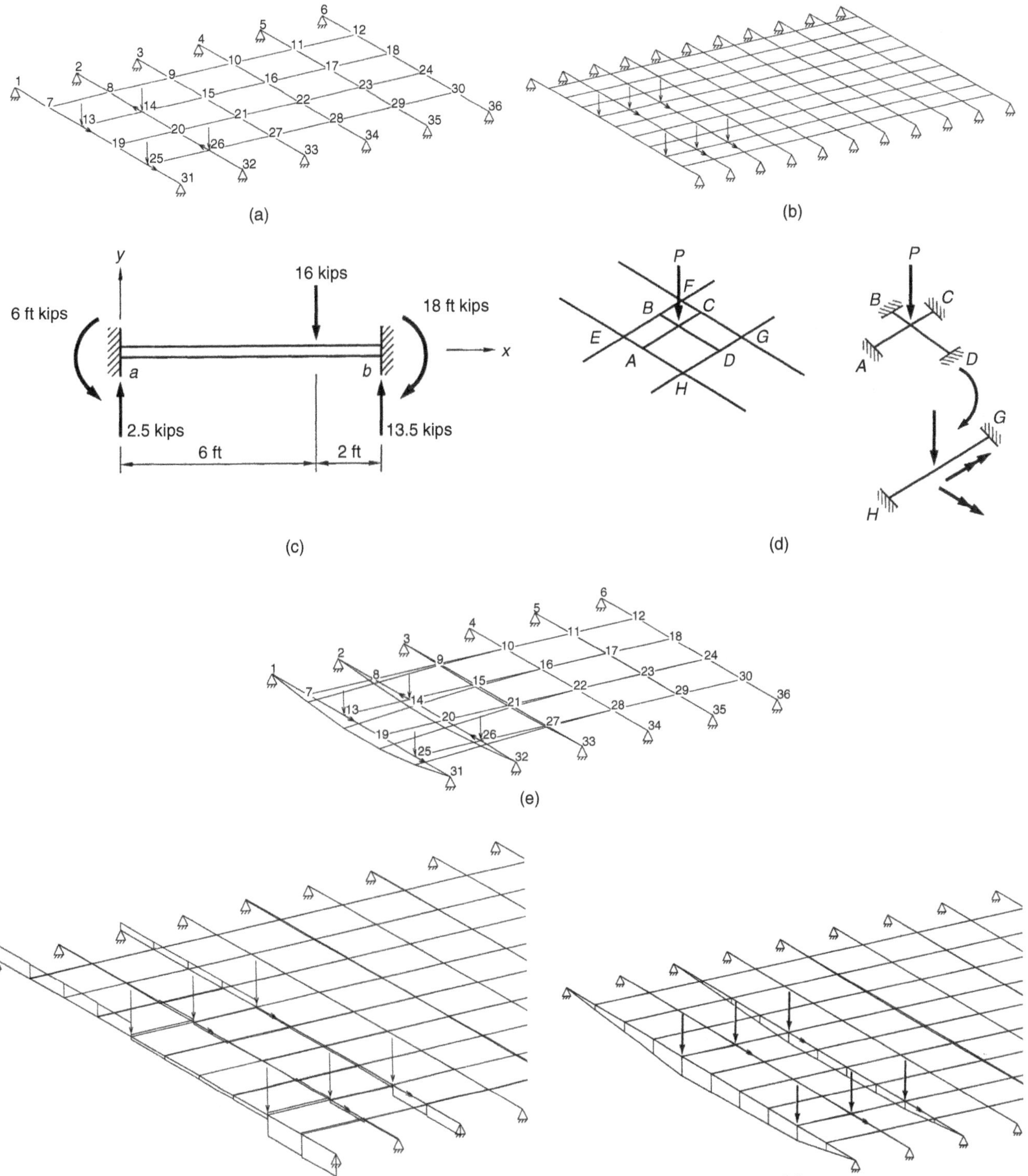

Fig. 11.9 (a) Coarse mesh—grillage and (b) fine mesh—grillage. (c) Fixed–fixed beam with wheel load (equivalent joint loads), and (d) load positioned between elements. (e) Translation of coarse mesh—load case 1. (f) Shear diagram—load case 1 and (g) moment diagram—load case 1.

Deck Properties

$$E_c = 3600 \text{ ksi } (24.82 \text{ GPa})$$
$$t_s = 8 \text{ in. } (203 \text{ mm})$$
$$v = 0.15$$
$$i_s = \tfrac{1}{12}(12)(8^3) = 512 \text{ in.}^4 (\text{per ft})$$
$$= 700000 \text{ mm}^4 \quad (\text{per mm})$$
$$j_s = \tfrac{1}{6}(12)(8^3) = 1024 \text{ in.}^4 (\text{per ft})$$
$$= 1400000 \text{ mm}^4 \quad (\text{per mm})$$

Element Properties

The elements that model the girders have the same properties as indicated above. Note that only the moment of inertia and the torsional constant are required in the grillage. The element properties for the deck are a function of the mesh size. For the coarse mesh in Figure 11.9(a), the elements oriented in the transverse (x direction) are positioned at 7 ft (2134 mm) center to center. Therefore, the properties assigned to these elements are

$$I_s = i_s \text{ (tributary length)} = 512 \text{ in.}^4/\text{ft (7 ft)}$$
$$= 3584 \text{ in.}^4 \text{ (transverse)} = 1.49 \times 10^9 \text{ mm}^4$$
$$J_s = j_s \text{ (tributary length)} = 1024 \text{ in.}^4/\text{ft (7 ft)}$$
$$= 7168 \text{ in.}^4 \text{ (transverse)} = 2.98 \times 10^9 \text{ mm}^4$$

The properties for the portion of the deck above the girders (at 8-ft centers) are

$$I_s = i_s \text{ (tributary length)} = 512 \text{ in.}^4/\text{ft (8 ft)}$$
$$= 4096 \text{ in.}^4 \text{ (longitudinal)} = 1.70 \times 10^9 \text{ mm}^4$$
$$J_s = j_s \text{ (tributary length)} = 1024 \text{ in.}^4/\text{ft (8 ft)}$$
$$= 8192 \text{ in.}^4 \text{ (longitudinal)} = 3.41 \times 10^9 \text{ mm}^4$$

For the fine mesh, the tributary width of the deck elements oriented in the transverse and longitudinal directions are 3.5 ft (1067 mm) and 4.0 ft (1219 mm), respectively. The associated element properties are

$$I_s = i_s \text{ (tributary length)} = 512(3.5)$$
$$= 1792 \text{ in.}^4 \text{ (transverse)}$$
$$= 746 \times 10^6 \text{ mm}^4$$
$$J_s = j_s \text{ (tributary length)} = 1024(3.5)$$
$$= 3584 \text{ in.}^4 \text{ (transverse)}$$
$$= 1.49 \times 10^9 \text{ mm}^4$$

and

$$I_s = i_s \text{ (tributary length)} = 512(4)$$
$$= 2048 \text{ in.}^4 \text{ (longitudinal)}$$
$$= 852 \times 10^6 \text{ mm}^4$$
$$J_s = j_s \text{ (tributary length)} = 1024(4)$$
$$= 4096 \text{ in.}^4 \text{ (longitudinal)}$$
$$= 1.70 \times 10^9 \text{ mm}^4$$

For the girder element properties, the associated properties of the beam and the slab contributions are added. The steel girder is transformed to concrete using the modular ratio of $n = 8$. The result for the fine mesh is

$$I_g = I_g \text{ (composite beam)}n + I_s$$
$$= 15,810(8) + 2048 = 128,500 \text{ in.}^4$$
$$= 52.1 \times 10^9 \text{ mm}^4$$
$$J_g = J_g \text{ (composite beam)}n + J_s$$
$$= 4.99(8) + 4096 = 4136 \text{ in.}^4$$
$$= 1.72 \times 10^9 \text{ mm}^4$$

The support boundary conditions are assumed to be restrained against translation in all directions at the girder ends. Although some torsional restraint may be present, it is difficult to estimate. By comparing the analysis of the system with both ends torsionally restrained and without, this effect was observed to be small and the torsionally unrestrained case is reported.

Eight load cases were used and are described below.

1. The design truck is positioned for near-critical maximum midspan moment and end shear in exterior girder (1) for one-lane loaded [see Figs. 11.9(a) and 11.9(b)].
2. Case 1 is repeated for two lanes loaded [see Fig. 11.9(b)].
3. The design truck is positioned for near-critical maximum midspan moment in the interior girder (3) for one lane loaded [see Fig. 11.9(b)].
4. Case 3 is repeated for two lanes loaded [see Fig. 11.9(b)].

5–8. Cases 1–4 are repeated with the design vehicles moved so that the rear 32-kip (145-kN) axle is near the support to create critical shears and reactions.

Because some of the concentrated wheel loads lie between nodes, their statical equivalence must be determined. For example, the load that lies between nodes 13 and 14 in the coarse mesh is illustrated in Figure 11.9(a). The statical equivalent actions are determined from the end actions associated with this load applied on a fixed-end beam as shown in Figure 11.9(c). The negative, or opposite, actions are applied to the grillage. The applied joint loads for the coarse mesh are illustrated in Table 11.7 for load case 1.

The nodal loads for the other load cases and for the fine mesh are established in a similar manner. It is common to neglect the joint load moments and assign the loads based on a simple beam distribution; hence the moments are not included.

Although all the loads have been assigned to a node, the distribution of the load is not correct and may lead to errors. The effect of the applied moments decreases with finer meshing. Thus, the finer mesh not only reduces the errors in

Table 11.7 Nodal Loads—Coarse Mesh

Load Case 1:	Exterior Girder—One Lane Loaded	
Node	Load, P_y (kips)	Moment, M_z (ft kips)
13	−18.5	−6
14	−13.5	18
25	−18.5	−6
26	−13.5	18
Sum	−64	

the stiffness model but also reduces the unnecessary errors due to modeling the load. If the load is applied directly to elements as member loads, then the algorithm inherent in the software should correctly determine the joint load forces and moments.

The software should correctly superimpose the fixed-end actions with the actions from the analysis of the released (joint-loaded) system to yield the correct final action accounting for the effect of the load applied directly to the member. If the load is applied within a grillage panel, then the statical equivalence becomes more difficult, as loads must be assigned to all of these nodes (this was conveniently and purposefully avoided in this example). The easiest approach in this case is to add another grillage line under the load.

If this is not viable, then the load may be assigned by using the subgrillage *A-B-C-D* shown in Figure 11.9(d). Next assign subgrillage end actions to the main grillage element *HG* and proceed as previously illustrated. The main difference is that the torque must also be considered. An alternative to this tedious approach is to refine the mesh to a point where the simple beam nodal load assignments are viable because the fixed-end torsion and bending moment are relatively small. Refinement is recommended.

Analysis Results

The translations for load case 1 for the coarse mesh are shown in Figure 11.9(e). Note that the translations are greater near the point of load application and the supports are restraining the translations as expected. The shear and moment diagrams for load case 1 for the fine mesh are shown in Figures 11.9(f) and 11.9(g).

Tables 11.8 and 11.9 summarize the maximum midspan moments and end reactions (maximum shears) for the four load cases. The simple beam actions are given for this position (see Example 11.1) and are illustrated in Figures 11.5(c) and 11.5(e). The associated actions are illustrated in Figures 11.9(f) and 11.9(g). The distribution factors are also given in the tables. The critical distribution factors are highlighted in bold in Tables 11.8 and 11.9. These distribution factors are compared with the AASHTO factors in addition to those derived from the finite-element and finite-strip methods in later examples.

The critical values for flexural moment (using the fine mesh) are highlighted in Table 11.8. The critical moment for the exterior girder with one lane loaded is 1.2 (multiple presence) × 221.2 ft kips = 265.4 ft kips with a distribution factor of $mg^{SE}_{moment} = 1.2 \times 0.62 = 0.74$, and the exterior girder moment for two lanes loaded is 1.0 × 232.8 = 232.8 ft kips with a distribution factor of $mg^{ME}_{moment} = 0.65$. The maximum interior girder moments are 1.2 × 157.4 = 188.8 ft kips $(mg^{SI}_{moment} = 0.53)$ and 1.0 × 258.8 = 258.8 ft kips $(mg^{ME}_{moment} = 0.72)$ for one and two lanes loaded, respectively. Note the coarse mesh yields approximately the same results as the fine mesh, hence convergence is deemed acceptable. The total moment at the critical section is 358.4 ft kips for one lane loaded and 716.8 ft kips for two lanes loaded. Note the summation of moments at the bottom of each load case. The differences are due to the presence of the nominal deck elements located between the girders. These elements are not shown in the table. Because of their low stiffness, they attract a small amount of load that causes the slight difference between the sum of girder moments and the statical moment. Inclusion of these elements in the summation eliminates this discrepancy. The distribution factors do not sum to 1.0 (one lane loaded) or 2.0 (two lanes loaded) for the same reason. Small differences between the reported values and these values are due to rounding.

The critical reaction/shears are highlighted in Table 11.9. The multiple presence factors (Table 8.6) are used to adjust the actions from analysis. The maximum reaction for the exterior girder with one lane loaded is 1.2 × 34.4 = 41.4 kips $(mg^{SE}_{shear} = 0.78)$ and 1.0 × 33.4 = 33.4 kips $(mg^{SE}_{shear} = 0.63)$ with two lanes loaded.

For the interior girder the reactions are 1.2 × 30.4 = 36.5 kips $(mg^{SI}_{shear} = 0.69)$ and 1.0 × 46.0 kips $(mg^{ME}_{shear} = 0.87)$ for one and two lanes loaded, respectively. The summation of the end reactions is equal (within rounding) to maximum system reaction of 52.8 (one lane) and 105.6 (two lanes). The nominal longitudinal deck elements in the fine mesh were not supported at the end; hence the total load must be distributed to the girders at the ends and the reactions check as expected.

The result of these analyzes are compared to those from other methods in a later example. The results presented in Tables 11.8 and 11.9 are summarized in Table 11.10. This tabular format is consistent with that used previously and permits ready comparison of the results from the various methods.

Finite-Element Method The finite-element method is one of the most general and powerful numerical methods. It has the capability to model many different mathematical models and to combine these models as necessary. For example, finite-element procedures are available to model Eq. 10.1 for the girders and Eq. 11.1 for the deck, and combine the two models into one that simultaneously satisfies both equations and the associated boundary conditions. Like the grillage method, the most common finite-element models are based

Table 11.8 Summary of Moments—Grillage Analysis

Load Case[a]	Girder	Beam Analysis Moment (ft kips)	Max. Moment (ft kips) (Coarse Mesh)	Distribution Factor (*mg*)	Max. Moment (ft kips) (Fine Mesh)	Distribution Factor (*mg*)
1	1	358.4	202.5	0.57	**221.2**[b]	**0.62**
1	2	358.4	123.2	0.34	116.7	0.33
1	3	358.4	37.4	0.10	18.5	0.05
1	4	358.4	0.0	0.00	−4.5	−0.01
1	5	358.4	−3.8	−0.01	−3.0	−0.01
1	6	358.4	−0.6	0.00	0.0	0.00
	Sum	Total moment = 358.4	358.7	1.00	348.9	0.98
2	1	358.4	236.2	0.66	**232.8**	**0.65**
2	2	358.4	257.5	0.72	240.6	0.67
2	3	358.4	182.3	0.51	183.2	0.51
2	4	358.4	49.2	0.14	46.4	0.13
2	5	358.4	−2.3	−0.01	−1.6	0.00
2	6	358.4	−6.3	−0.02	−5.4	−0.02
	Sum	Total moment = 2(358.4) = 716.8	716.7	2.00	695.0	1.94
3	1	358.4	36.7	0.10	19.7	0.06
3	2	358.4	148.1	0.41	123.9	0.35
3	3	358.4	132.3	0.37	**157.4**	**0.44**
3	4	358.4	45.5	0.13	48.8	0.14
3	5	358.4	1.2	0.00	1.4	0.00
3	6	358.4	−5.3	−0.01	−5.2	−0.01
	Sum	Total moment = 358.4	358.5	1.00	346.0	0.98
4	1	358.4	26.4	0.07	11.8	0.03
4	2	358.4	167.6	0.47	141.6	0.40
4	3	358.4	255.1	0.72	**258.8**	**0.72**
4	4	358.4	203.7	0.57	204.3	0.57
4	5	358.4	69.0	0.19	75.4	0.21
4	6	358.4	4.9	0.01	−6.6	0.02
	Sum	Total moment = 2(358.4) = 716.8	716.9	2.03	685.2	1.95

[a]Load cases: (1) One lane loaded for the maximum exterior girder actions (girder 1). (2) Two lanes loaded for the maximum exterior girder actions (girder 1). (3) One lane loaded for the maximum interior girder actions (girder 3). (4) Two lanes loaded for the maximum interior girder actions (girder 3).
[b]Critical values are in bold.

on a stiffness (or displacement approach), that is, a system of equilibrium equations is established and solved for the displacements at the degrees of freedom.

The scope of this method seems unending with many texts and reference books, research papers, and computer programs to address and use it. Here, only the surface is scratched and the reader is strongly encouraged to gain more information by formal and/or self-study. The method is easily used and abused. With software it is easy to generate thousands of equations and still have an inappropriate model. The discussion herein is a brief overview of the finite-element method as related to the engineering of slab–girder bridges, and it is assumed that the reader has had a course and/or experience with the method.

The finite-element formulation is commonly used in two ways: 2D and 3D models. The 2D model is the simplest and

Table 11.9 Summary of Reactions—Grillage Analysis

Load Case[a]	Girder	Beam Analysis Reaction (kips)	Max. Reactions, kips (Coarse Mesh)	Distribution Factor (*mg*)	Max. Reactions, (Fine Mesh) (kips)	Distribution Factor (*mg*)
5	1	52.8	30.3	0.57	**34.4**[b]	**0.65**
5	2	52.8	19.8	0.38	19.3	0.37
5	3	52.8	3.2	0.06	−0.4	−0.01
5	4	52.8	−0.1	−0.00	−0.4	−0.01
5	5	52.8	−0.3	−0.01	−0.2	−0.00
5	6	52.8	−0.0	−0.00	0.1	0.00
	Sum	52.8[c]	52.9	1.00	52.8	1.00
6	1	52.8	32.8	0.62	**33.4**	**0.63**
6	2	52.8	41.2	0.78	39.0	0.74
6	3	52.8	29.1	0.55	31.7	0.60
6	4	52.8	3.2	0.06	2.3	0.04
6	5	52.8	−0.3	−0.01	−0.4	−0.01
6	6	52.8	−0.4	−0.01	−0.4	−0.01
	Sum	105.6[c]	105.6	1.99	105.6	1.99
7	1	52.8	2.3	0.04	0.5	0.01
7	2	52.8	26.3	0.50	19.9	0.38
7	3	52.8	21.0	0.40	**30.4**	**0.58**
7	4	52.8	3.7	0.07	2.6	0.05
7	5	52.8	−0.1	−0.00	−0.2	−0.00
7	6	52.8	−0.4	−0.01	−0.4	−0.01
	Sum	52.8[b]	52.8	1.00	52.8	1.01
8	1	52.8	1.9	0.04	0.1	0.00
8	2	52.8	24.2	0.46	19.5	0.37
8	3	52.8	40.7	0.77	**46.0**	**0.87**
8	4	52.8	32.5	0.62	33.3	0.63
8	5	52.8	6.6	0.13	8.2	0.16
8	6	52.8	−0.3	−0.01	−1.3	−0.02
	Sum	105.6[c]	105.6	2.01	105.6	2.01

[a]Load cases: (5) One lane loaded for the maximum exterior girder actions (girder 1). (6) Two lanes loaded for the maximum exterior girder actions (girder 1). (7) One lane loaded for the maximum interior girder actions (girder 3). (8) Two lanes loaded for the maximum interior girder actions (girder 3).
[b]Critical values are in bold.
[c]Beam reaction for entire bridge.

Table 11.10 Grillage Method Summary—Fine Mesh

Girder Location	Number of Lanes Loaded	Moment (ft kips)	Distribution Factor (*mg*)	Reactions (kips)	Distribution Factor (*mg*)
Exterior	1	265.2	0.74	41.2	0.78
Exterior	2	232.8	0.65	33.4	0.63
Interior	1	190.0	0.53	37.0	0.69
Interior	2	258.8	0.72	46.0	0.87

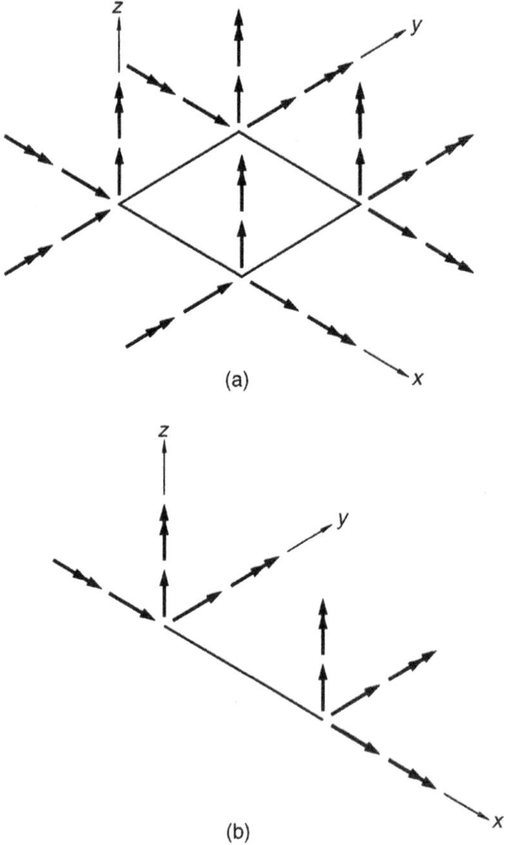

Fig. 11.10 (a) Example of shell element and (b) example of space frame element.

involves fewer degrees of freedom. Here plate elements that usually contain 3 degrees of freedom per node are used to model the deck on the basis of the mathematical model described by Eq. 11.1. The girders are modeled with grillage or plane grid elements with 3 degrees of freedom per node. Examples of these elements with 6 degrees of freedom are illustrated in Figures 11.10(a) and 11.10(b). The girder properties may be based on Eqs. 11.6–11.8. The deck properties typically include the flexural rigidities in orthogonal directions or the deck thickness and material properties upon which the rigidities can be based. The nodal loads and/or element loads are determined in the usual manner.

Because many different elements are available with differing number of degrees of freedom and response characteristics, it is difficult to provide general guidance mesh characteristics, other than those usually addressed in standard references. It is important to suggest that at least two meshes be studied to obtain some knowledge of the convergence characteristics. If the response changes significantly with refinement, a third (or fourth) mesh should be studied.

Because of the importance of maintaining equilibrium, the analytical results should be checked for global equilibrium. It is easy to mistakenly apply the loads in the wrong direction or in the wrong location. It is strongly suggested that global equilibrium be checked by hand. We have caught numerous errors in input files and in computer code by this simple check. If the program being used does not have a way to obtain reactions, then perhaps the stiff boundary spring elements can be used at the supports and the element forces are the reactions. If the program does not produce reactions, or they cannot be deduced from the element forces, then the use of another program that does is recommended. In short, no matter how complex the model, **always check statics**.

This simple check ensures that ductile elements designed on the basis of the analysis provides at least one viable load path and likely an opportunity for redistribution should yielding occur. A statics check is necessary for any method of analysis.

As an alternative to the 2D model, the bridge may be modeled as a 3D system. Here Eq. 11.1 is used to mathematically model the out-of-plane behavior of the deck, and the in-plane effects are modeled using a similar fourth-order partial differential equation (Timoshenko and Goodier, 1970). In-plane effects arise from the bending of the system, which produces compression in the deck and tension in the girder under the influence of positive bending moments. The in- and out-of-plane effects are combined into one element, commonly called a shell element. A typical shell element is shown in Figure 11.10(a) where in- and out-of-plane degrees of freedom are illustrated. Typically, the in- and out-of-plane effects are considered uncoupled, which results in a linear formulation.

The girders are usually modeled with space frame elements that have 6 degrees of freedom per node, the same as the shell element. The girder eccentricity (composite girder) is modeled by placing the elements at the centroidal axis of the girder, which creates many additional degrees of freedom. To avoid additional computational effort, the degrees of freedom at the girders may be related to the degrees of freedom of the plate by assuming that a rigid linkage exists between these two points. This linkage can be easily accommodated in the element formulation for the space frame element. This capability is typically included in commercial software and is denoted by several terms: rigid links, element offset, slave–master relationship, and element eccentricities.

An alternative approach is to use the additional degrees of freedom at the girder level but to declare these nodes to be slaves to the deck nodes directly above. A last alternative is to be lazy in the refinement of the model and just include the girder nodes, which produces a larger model, but, of course, one can complain (boast) how large the model is and how long it takes to execute. Realistically, with today's ever increasing computational power, a direct and brute force approach is acceptable. The important issue is that the engineer understands the methods used, their limitations, and their application to the problem under consideration.

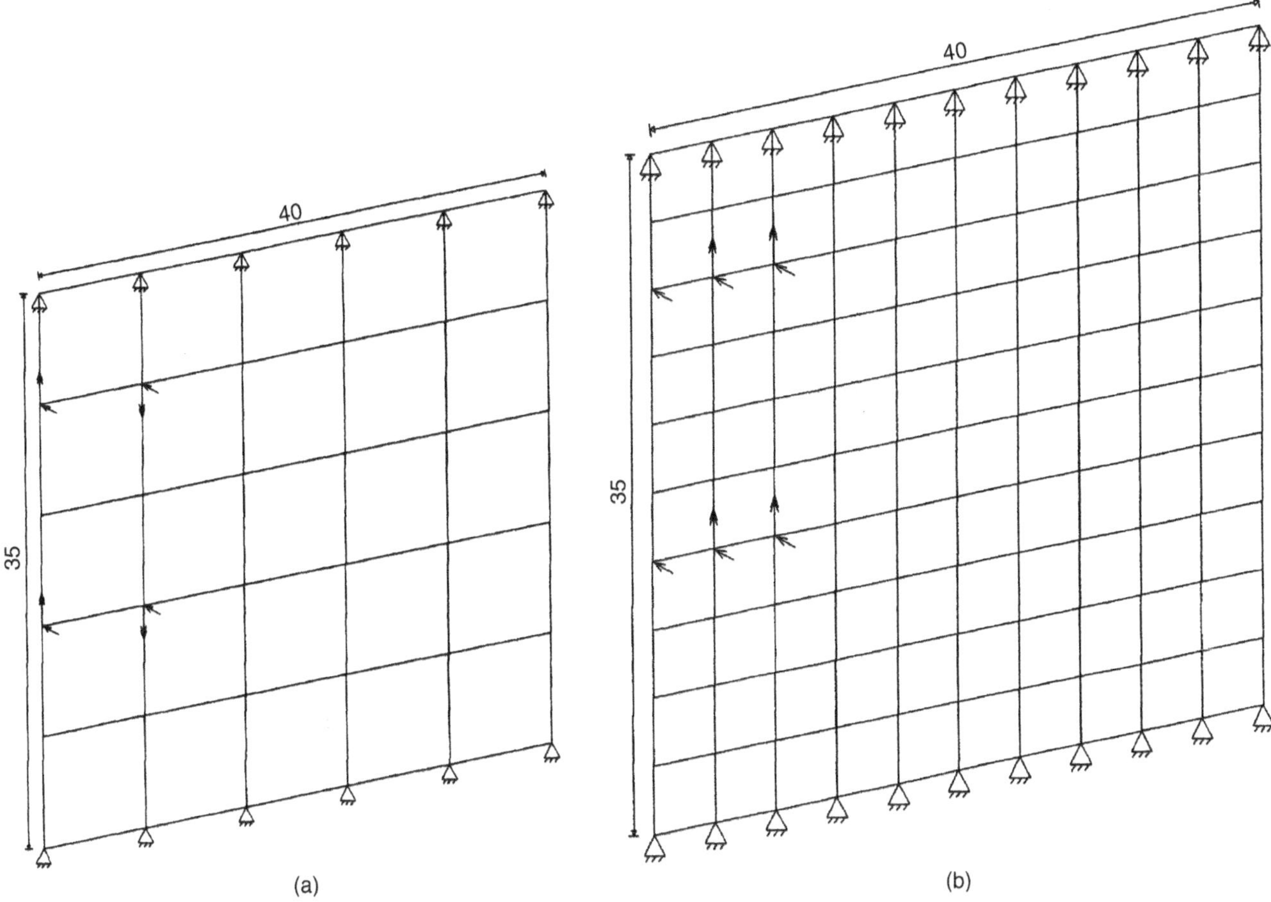

Fig. 11.11 (a) Finite-element coarse mesh and (b) finite-element fine mesh.

Example 11.5 Use the finite-element method to determine the end shear (reactions) and midspan flexural bending moments in the girders in Figure 11.5(a) as illustrated in Example 11.1. In addition, determine the distribution factors for moment and shear in girders for one and two lanes loaded.

The system is discretized with the 2D meshes shown in Figures 11.11(a) and 11.11(b). The girder properties are the same as in Example 11.4 with the exception that the deck properties are not added as before because the deck is modeled with the shell element as shown in Figure 11.10(a). Here the in-plane effects are neglected and the plate bending portion is retained. The deck rigidities are calculated internal to the finite-element program on the basis of $t = 8$ in. (203 mm) and $E = 3600$ ksi (24,800 MPa), and $\nu = 0.15$. The girder properties and nodal loads are calculated as in the previous example.

The maximum moments and reactions are summarized in Table 11.11. A table similar to Tables 11.8 and 11.9 could be developed and the results would be quite similar. For the sake of brevity, such tables are not shown and only the maximum actions are reported. The multiplication indicates the application of the multiple presence factors.

Finite-Strip Method The finite-strip method is a derivative of the finite-element method. The mathematical models described previously are the usual basis for analysis so that converged finite-element and finite-strip models should yield the same "exact" solutions. The finite-strip method employs strips to discretize the continuum as shown in Figure 11.12(a). A strip is an element that runs the entire length of the deck. With the typical polynomial shape function used in the finite-element method, this type of mesh would be unacceptable. However, the finite-strip method uses a special shape function that considers the boundary conditions at the ends to be simply supported. This condition permits the use of a Fourier sine series for the displacement in the longitudinal direction while a third-order polynomial is used in the transverse direction. A typical lower order shape function is

$$\begin{aligned} w(x, y) &= \sum_{m=1}^{r} f_m(x) Y_m \\ &= \sum_{m=1}^{r} \left(A_m + B_m x + C_m x^2 + D_m x^3\right) \sin\left(\frac{m\pi y}{L}\right) \end{aligned} \tag{11.9}$$

Table 11.11 Finite-Element Results, Critical Actions[a]

Girder Location	Number of Lanes Loaded	Moment (ft kips)	Distribution Factor (mg)	Reactions (kips)	Distribution Factor (mg)
Exterior	1	(1.2)(206.0) = 247.2	0.68	(1.2)(31.4) = 37.7	0.71
		(1.2)(196.9) = 236.3	0.66	(1.2)(29.8) = 35.8	0.68
Exterior	2	(1.0)(220.8) = 220.8	0.62	(1.0)(30.6) = 30.6	0.58
		(1.0)(219.4) = 219.4	0.61	(1.0)(30.4) = 30.4	0.58
Interior	1	(1.2)(154.9) = 186.9	0.52	(1.2)(30.2) = 36.2	0.69
		(1.2)(154.8) = 185.8	0.52	(1.2)(30.2) = 36.2	0.69
Interior	2	(1.0)(258.8) = 258.8	0.72	(1.0)(44.2) = 44.2	0.84
		(1.0)(249.0) = 249.0	0.69	(1.0)(44.9) = 44.9	0.85

[a]Coarse mesh on first line, fine mesh on the second line.

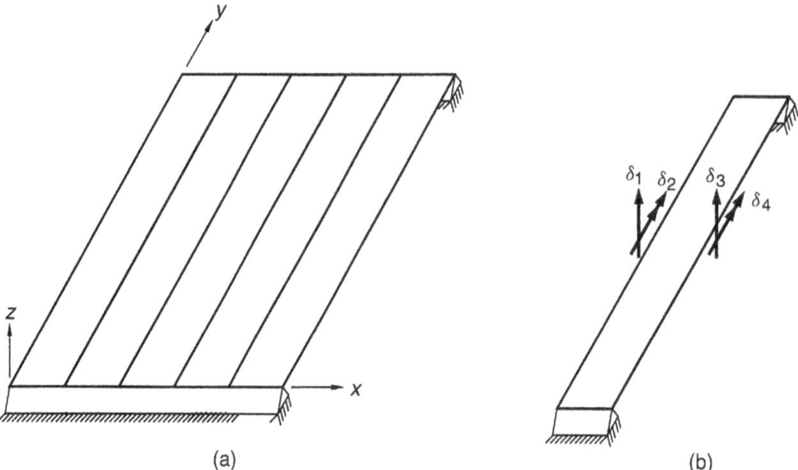

Fig. 11.12 (a) Example of a finite-strip model and (b) finite-strip element.

Where

$f_m(x)$ = third-order polynomial with coefficients A_m, B_m, C_m, and D_m
Y_m = sine function
L = span length
y = longitudinal coordinate
m = series index that has a maximum value of r

It is important to note that the polynomial function is the same one typically used in standard beam elements and may be rewritten in terms of the 4 degrees of freedom at the strip edges [see Fig. 11.12(b)]. The degrees of freedom include two translations and two rotations per harmonic considered (value of m). The total number of degrees of freedom is the number of nodal lines times 2; for example, if 50 strips are used with 50 terms, the total number of unknowns is 51(nodal lines) × 2(unknowns per nodal line) × 50(terms) = 5100(unknowns). The mathematics of the element formulation involves a procedure similar to the finite-element method. For example, the element stiffness matrix involves

$$[S] = \int_{\text{vol}} [B]^{\text{T}} [D] [B] \{\delta\} \, dV \qquad (11.10)$$

where B contains the curvatures or generalized strain, D contains the plate rigidities, and δ contains the 4 degrees of freedom. Equation 11.10 is presented to remind the reader that differentiation and integration are involved with the element formulation.

An important feature of the finite-strip method is its efficiency. When the shape function in Eq. 11.9 is twice differentiated to obtain curvatures, the polynomial function may change but the sine function remains a sine function. Upon substitution into the strain matrix B in Eq. 11.10, the summations remain. A term-by-term expansion of the series in combination with necessary matrix multiplication yields terms with the following integrals:

$$I = \int_0^L \sin\left(\frac{m\pi y}{L}\right) \sin\left(\frac{n\pi y}{L}\right) dy$$
$$I = \frac{L}{2} \qquad \text{when } m = n$$
$$I = 0 \qquad \text{when } m \neq n \qquad (11.11)$$

This integration is zero when the terms in the series are not the same (termed orthogonality). This important feature causes all terms where n is not equal to m to be zero, which permits the programmer to consider each term separately, and completely uncouples the equations to be solved. For example, if 50 strips are used with 50 terms, then the total number of degrees of freedom is 5100, as before, but this size system is never assembled or solved. Instead, the system is solved for one term at a time or 51(nodal lines) × 2 degrees of freedom per nodal line, which results in 102 degrees of freedom per mode. Thus, this system is solved repetitively for the 50 modes and the results are appropriately superimposed. Hence, a very small problem (the same as a continuous beam with 51 nodes) is solved numerous times. This approach is vastly more efficient than considering the full 5100 degrees of freedom in one solution. A typical finite-strip model runs in about 10% of the time as a finite-element model with a similar number of degrees of freedom using solvers that account for the small bandwidth and symmetry of the stiffness matrix.

A brief treatise of the finite-strip method is provided in this section, and the main objective is to introduce the reader to the rationale for its use. Complete details are presented in books by Cheung (1976) and Loo and Cusens (1978).

Although elegant in its simplicity and efficiency, the finite-strip method has lost favor to the finite-element method because the latter is less restrictive and current computers can solve tens of thousands of equation simultaneous in a short time and on the desktop at little cost. Some legacy bridge codes have finite-strip models as the basis of analysis.

Example 11.6 Use the finite-strip method to determine the end shear (reactions) and midspan flexural bending moments in the girders in Figure 11.5(a) as illustrated in Example 11.1. In addition, determine the distribution factors for moment and shear in girders for one and two lanes loaded.

The system is modeled with 20 uniform strips and 100 terms. Studies showed that this discretization is adequate for slab–girder systems (Finch and Puckett, 1992). A large number of terms is required to accurately determine the shear forces near the concentrated forces and girder ends. If only flexural effects are required near midspan, then only about 10 terms are required. The girder, deck properties, and load positioning are the same as in the previous example. The results are summarized in Table 11.12.

Example 11.7 Compare the results from the AASHTO, lever, grillage, finite-element, and finite-strip methods. Tables 11.5, 11.6, and 11.10–11.12 have been combined for comparison of the methods, and the results are given in Tables 11.13 and 11.14 for moment and shear, respectively.

Recall that the basis for the AASHTO multilanes loaded formulas includes the possibility of three or more lanes being loaded and creating a situation more critical than the two-lane case. Therefore, the AASHTO values are influenced by this and are generally, but not always, slightly higher than the two-lane numerical results. Most values compare within 10% except the lever method, which tends to be conservative for moment distribution factors; however, it is typically quite

Table 11.12 Finite-Strip Results

Girder Location	Number of Lanes Loaded	Moment (ft kips)	Distribution Factor (mg)	Reactions (kips)	Distribution Factor (mg)
Exterior	1	(1.2)(204.3) = 245.2	0.68	(1.2)(29.6) = 35.5	0.67
Exterior	2	(1.0)(218.6) = 218.6	0.61	(1.0)(30.6) = 30.6	0.58
Interior	1	(1.2)(154.1) = 184.9	0.52	(1.2)(26.4) = 31.7	0.60
Interior	2	(1.0)(250.8) = 250.8	0.70	(1.0)(41.7) = 41.7	0.79

Table 11.13 Summary of Analysis Methods—Moment (ft kips)

Girder	Number of Lanes Loaded	AASHTO	Lever	Grillage	Finite Element	Finite Strip
Exterior	1	268.8	268.8	265.2	236.3	245.2
		0.75	0.75	0.74	0.66	0.68
Exterior	2	254.5	224.0	232.3	219.4	218.6
		0.71	0.625	0.65	0.61	0.61
Interior	1	197.1	268.8	190.0	185.8	184.9
		0.55	0.75	0.53	0.52	0.52
Interior	2	254.5	313.6	258.8	249.0	250.8
		0.71	0.875	0.72	0.69	0.70

Table 11.14 Summary of Analysis Methods—Reactions (kips)

Girder	Number of Lanes Loaded	AASHTO	Lever	Grillage	Finite Element	Finite Strip
Exterior	1	39.6	39.6	41.2	35.8	35.5
		0.75	0.75	0.78	0.68	0.67
Exterior	2	34.3	33.0	33.4	30.4	30.6
		0.65	0.625	0.63	0.58	0.58
Interior	1	35.9	39.6	37.0	36.2	31.7
		0.68	0.75	0.70	0.69	0.60
Interior	2	42.8	46.2	46.0	44.9	41.7
		0.81	0.875	0.87	0.85	0.79

close to other methods for shear and reactions because of the localized behavior of loads applied near the supports.

11.2 SLAB BRIDGES

The slab bridge is another common bridge type frequently used for short spans, usually less than 50 ft (15,240 mm). The slab bridge does not have any girders, and, therefore, the load must be carried principally by flexure in the longitudinal direction. A simplistic approach (perhaps valid for the ultimate strength limit states) is to divide the total statical moment by the bridge width to achieve a moment per unit width for design. This type of analysis is valid by the lower bound theorem for consideration of the strength limit state assuming adequate transverse strength and ductility is available.

The results of this procedure are most certainly underestimates of the localized moments near the application of the load under linear elastic conditions, that is, service and fatigue limits states. Hence, it is necessary to determine the moments under service conditions. The moments are determined by establishing the width of the bridge that is assigned to carry one vehicle, or in other words the structural width per design lane. The width for one lane loaded is [A4.6.2.3]

$$E^S = 10.00 + 5.0\sqrt{L_1 W_1} \qquad (11.12\text{a-US})$$

$$E^S = 250 + 0.42\sqrt{L_1 W_1} \qquad (11.12\text{a-SI})$$

and the width for multilanes loaded is

$$E^M = 84.00 + 1.44\sqrt{L_1 W_1} \le \sqrt{\frac{W}{N_L}} \qquad (11.12\text{b-US})$$

$$E^M = 2100 + 0.12\sqrt{L_1 W_1} \le \frac{W}{N_L} \qquad (11.12\text{b-SI})$$

Where

$E^{S \text{ or } M}$ = structural width per design lane [in. (mm)], for single and multiple lanes loaded

L_1 = modified span length taken equal to the lesser of the actual span or 60.0 ft (18,000 mm)

W_1 = modified edge-to-edge width of bridge taken equal to the lesser of the actual width or 60.0 ft (18,000 mm) for multilane loading, or 30 ft (9000 mm) for single-lane loading

W = physical edge-to-edge width of the bridge [ft (mm)]

N_L = number of design lanes [A3.6.1.1.1]

The adjustment for skew is

$$r = 1.05 - 0.25 \tan \theta \le 1.00 \qquad (11.12\text{c})$$

where θ is the skew angle defined previously in Table 11.3. Note that skew *reduces* the longitudinal bending moment.

Example 11.8 Determine the slab width that is assigned to a vehicle (design lane) for the bridge described in Example 11.1 [see Fig. 11.5(a)] without the girders. Use a 20-in. (508-mm) deck thickness. Assume three design lanes are possible. By using Eq. 11.12(a) for one lane loaded, the width is

$$E^S = 10.00 + 5.0\sqrt{L_1 W_1} = 10.00 + 5.0\sqrt{(35)(30)}$$
$$= 172 \text{ in./lane} = 14.3 \text{ ft/lane}$$

and by using Eq. 11.12(b) for multiple lanes loaded, the width is

$$E^M = 84.00 + 1.44\sqrt{L_1 W_1} = 84.00 + 1.44\sqrt{(35)(44)}$$
$$= 140.5 \text{ in./lane}$$
$$= 11.7 \text{ ft/lane} \le \frac{W}{N_L} = \frac{44}{3} = 14.7 \text{ ft/lane}$$

$$\therefore \; E^M = 11.7 \text{ ft/lane}$$

The bending moment is determined for a design lane that is divided by the width E to determine the moment per unit length for design.

From the simple-beam analysis given in Example 11.1, the maximum bending moment for one lane is 358.4 ft kips. Using this moment, the moments per foot are

$$M_{LL}^S = \frac{M_{\text{beam}}}{E^S} = \frac{358.4 \text{ ft kips/lane}}{14.3 \text{ ft/lane}} = 25.0 \text{ ft kips/ft}$$

and

$$M_{LL}^{\mathrm{M}} = \frac{M_{\mathrm{beam}}}{E^{\mathrm{M}}} = \frac{358.4 \text{ ft kips/lane}}{11.7 \text{ ft/lane}} = 30.6 \text{ ft kips/ft}$$

Because the slab bridge may be properly modeled by Eq. 11.1, all the methods described earlier may be used. To illustrate, brief examples of the grillage and finite-element methods are given below. Most of the modeling details remain the same as previously presented. The girders are obviously omitted and the loading is the same as the previous examples. Shear is not typically critical with the slab bridge, and this limit state need *not* be considered [A4.6.2.3]. Zokaie et al. (1991) reexamined this long-time AASHTO provision and confirmed the validity of this approach. Only flexural bending moment is presented.

Example 11.9 Use the grillage method to model the slab bridge described in Example 11.8. Use the fine mesh used in Example 11.4 and consider two lanes loaded for bending moment. The deck may be modeled as an isotropic plate.

All the deck section properties are proportional to the thickness cubed. Hence, for the 20-in. (508-mm) slab, the properties determined in Example 11.4 are multiplied by $(20/8)^3 = 2.5^3 = 15.625$. The distribution of internal actions is not a function of the actual thickness but rather the relative rigidities in the transverse and longitudinal directions. Because isotropy is assumed in this example, any uniform thickness may be used for determining the actions. The displacements are proportional to the actual stiffness (thickness cubed as noted above).

The loads are positioned as shown in Figure 11.13(a). The moments in the grillage elements are divided by the tributary width associated with each longitudinal element, 4.0 ft. The moments M_y (beamlike) are illustrated in Figure 11.13(b) and are summarized in Table 11.15.

The critical values (in bold) are 20.56 ft kips/ft with an associated width of 358.4/20.56 = 17.4 ft. The total moment across the critical section is the summation of the grillage moments.

Equilibrium dictates that this moment be 2 lanes × 358.4 ft kips = 716.8 ft kips, which is the summation of the moments in the grillage elements at the critical section, validating equilibrium. The AASHTO value is 30.6 ft kips/ft for two lanes loaded, which is approximately 50% greater than the maximum grillage value of 20.56 ft kips/ft. This difference is discussed in more detail later with reference to one or more loaded lanes.

Example 11.10 Use the finite-element method to model the slab bridge described in Example 11.8. Use the fine mesh used in Example 11.5 and consider two and three lanes loaded for maximum bending moment. The deck may be modeled as an isotropic plate.

The deck thickness is increased to 20 in. (508 mm), and the girders are removed from the model presented in Example 11.5. The nodal loads are the same as in the previous example.

The moments that cause flexural stress in the longitudinal direction are illustrated in Table 11.16 and Figure 11.13(b). Contour plots of the flexural moments are illustrated in Figure 11.13(c). Note the values in Table 11.16 are the contour values for the bridge at the dashed line. As expected, the longitudinal moments are significantly greater than the transverse moments. These figures are provided to give the reader a sense of the distribution of internal actions in a two-way system that is traditionally modeled as a one-way system, that is, as a beam.

The maximum moment in the finite-element method is 20.54 ft kips/ft and is associated with a width of 358.4/20.54 = 17.4 ft. These values compare well with the grillage moment of 20.56 ft kips/ft and the associate width of 17.4 ft. The finite-element moments are reported at the nodes along the critical moment section for the entire system. Therefore, the total moment at the section reported is 713.1 ft kips and is slightly less than the total statical moment of 2 × 358.4 = 716.8 ft kips for two loaded lanes.

Note that the width per lane is used only if a beam-line analysis is required, that is, a 1.5D analysis where the load distribution is developed by a numerical model and the design is based on analysis of a beam. Alternatively, the entire design could be based on the mathematical/numerical model. Note: The other load cases are also required for the design.

Both the grillage and finite-element methods do not compare well with the AASHTO value of 30.6 ft kips/ft. Recall the AASHTO multilane formulas implicitly include two, three, or more lanes loaded. Because this bridge has a curb-to-curb width of 44 ft, likely three 10-ft design lanes should be considered for design and is considered below.

Why is it important to initially present the two-lane loaded case rather than the three-lane case for the refined methods? There are three reasons: (1) to highlight the assumptions included in the AASHTO distribution formulas, (2) to illustrate that the two-lane load does not always give most critical results, and (3) if the results from an analytical approach differ significantly (more than 15%) from the AASHTO value, then the differences should be understood and justified. Zokaie et al. (1991) presented numerous histograms similar to Figure 11.4 where the results of the simplified AASHTO formulas are within 15% of results based on more rigorous methods. This result suggests that significant deviation should be carefully investigated.

The two-lane loaded finite-element model is modified to include an additional vehicle placed adjacent to the others and located near the edge of the deck [see Fig. 11.13(a)]. The results are given in Table 11.17 and are plotted in Figure 11.13(b).

Note that the moment of 34.08 ft kips and the associated distribution width of 10.5 ft are critical. AASHTO [A3.6.1.1.2] provides a multiple presence factor of 0.85 for bridges with three design lanes (see Table 8.6). Hence, the

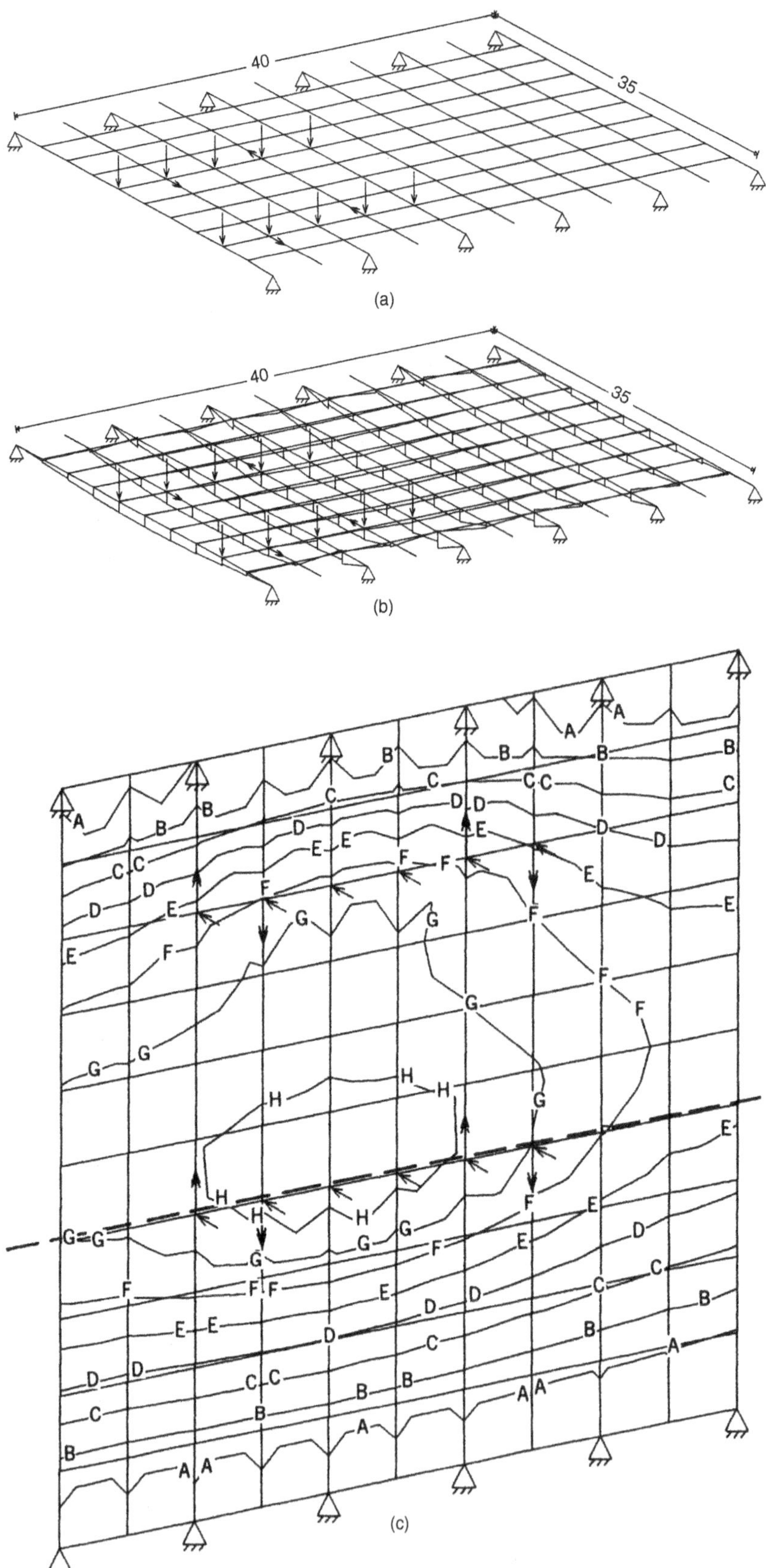

Fig. 11.13 (a) Truck positions and (b) longitudinal moment diagram. (c) Longitudinal moment contour.

Table 11.15 Analysis Results for the Grillage Model at the Critical Section (Two Lanes Loaded)

Element	1 Edge of deck	2	3	4	5	6	7	8	9	10	11 Edge of deck	Total[a]
Moment (ft kips)	58.65	62.46	69.57	75.05	**82.26**[b]	81.46	72.30	65.04	55.65	48.91	45.46	716.81
Tributary length (ft)	4.0	4.0	4.0	4.0	**4.0**	4.0	4.0	4.0	4.0	4.0	4.0	44.0
Moment (ft kips/ft)	14.66	15.62	17.39	18.76	**20.56**	20.37	18.08	16.26	13.91	12.23	11.37	N/A
Statical moment (ft kips)	358.4	358.4	358.4	358.4	**358.4**	358.4	358.4	358.4	358.4	358.4	358.4	N/A
Width per lane (ft)	24.4	22.9	20.6	19.1	**17.4**	17.6	19.8	22.0	25.8	29.3	31.5	N/A

[a]Not applicable = N/A.
[b]Critical values are in bold.

Table 11.16 Analysis Results for the Finite-Element Model near the Critical Section (Two Lanes Loaded)

Tenth Point Across Section	1 Edge of deck	2	3	4	5	6	7	8	9	10	11 Edge of deck	Total[a]
Moment (ft kips/ft) (fine mesh)	14.43	14.73	18.05	17.89	**20.54**[b]	20.24	18.29	15.46	12.78	11.37	11.06	N/A
Element width (ft)	2.2	4.4	4.4	4.4	**4.4**	4.4	4.4	4.4	4.4	4.4	2.2	44.0
Moment (ft kips per element)	31.7	64.8	79.4	78.7	**90.4**	89.1	80.5	68.0	56.2	50.0	24.3	713.1
Statical moment (ft kips)	358.4	358.4	358.4	358.4	**358.4**	358.4	358.4	358.4	358.4	358.4	358.4	358.4
Width per lane (ft)	24.8	24.3	19.9	20.0	**17.4**	17.7	19.6	23.2	28.0	31.5	32.4	N/A

[a]Not applicable = N/A.
[b]Critical values are in bold.

Table 11.17 Analysis Results for the Finite-Element Model near the Critical Section (Three Lanes Loaded)

Tenth Point Across Section	1 Edge of deck	2	3	4	5	6	7	8	9	10	11 Edge of deck	Total[a]
Moment (ft kips/ft) (fine mesh)	16.45	16.82	20.64	21.16	24.80	25.87	25.72	25.30	27.75	30.59	**34.08**[b]	N/A
Tributary width (ft)	2.2	4.4	4.4	4.4	4.4	4.4	4.4	4.4	4.4	4.4	**2.2**	44.0
Tributary moment (ft kips)	36.2	74.0	90.8	93.1	109.1	113.8	113.1	111.3	122.1	134.6	**75.0**	1073.1
Statical moment (ft kips)	358.4	358.4	358.4	358.4	358.4	358.4	358.4	358.4	358.4	358.4	**358.4**	358.4
Width per lane (ft)	21.8	21.3	17.4	16.9	14.5	13.9	13.9	14.2	12.9	11.7	**10.5**	N/A

[a]Not applicable = N/A.
[b]Critical values are in bold.

moment of 34.08 ft kips is multiplied by 0.85 yielding a critical value of (0.85)(34.08) = 29.0 ft kips, and the associated distribution width is 12.4 ft. These values compare reasonably well with the AASHTO values of 30.6 ft kips and 11.7 ft. On the basis of the preceding analysis, it is likely that the AASHTO equation for distribution width is governed by three lanes loaded for this bridge width.

11.3 SLABS IN SLAB–GIRDER BRIDGES

The slab design may be accomplished by three methods: (1) the analytical strip method approach, (2) the empirical approach, and (3) the yield-line method. The analytical method requires a linear elastic analysis upon which to proportion the slab to satisfy the strength and service limit states. The empirical approach requires that the designer satisfy a few simple rules regarding the deck thickness and reinforcement details, and limit states are assumed to be automatically satisfied without further design validations. The empirical approach is elaborated in more detail in the following chapters on design. The third method is the yield-line method and is based on inelastic yielding of the deck and, therefore, is appropriate for the strength and extreme-event limit states. All three methods may be used to proportion the slab. All three methods yield different designs that are generally viable and reasonable. In this section, the strip method is first outlined with a discussion of the AASHTO provisions and an illustrative example. A brief discussion of the yield-line method follows also reinforced with an example. Perhaps the use of the yield-line approach is for agencies that load rate decks and need to increase the load rating for the strength limit state. The empirical approach is used for design and is outlined in Chapter 16 on concrete.

Linear Elastic Method A deck slab may be considered as a one-way slab system because its aspect ratio (panel length divided by the panel width) is large. For example, a typical panel width (girder spacing) is 8–11 ft (2400–3600 mm) and a typical girder length from 30 to 200 ft (9100 to 61,000 mm). The associated aspect ratios vary from 3.75 to 10. Deck panels with an aspect ratio of 1.5 or larger may be considered one-way systems [A4.6.2.1.4]. Such systems are assumed to carry the load effects in the short-panel direction, that is, in a beamlike manner. Assuming the load is carried to the girder by one-way action, then the primary issue is the width of strip (slab width) used in the analysis and subsequent design. Guidance is provided in AASHTO [A4.6.2], Approximate Methods.

The strip width SW [in. (mm)] for a CIP section is

$$M^+: \quad SW^+ = 26.0 + 6.6S \qquad (11.13a\text{-US})$$

$$M^+: \quad SW^+ = 660 + 0.55S \qquad (11.13a\text{-SI})$$

$$M^-: \quad SW^- = 48.0 + 3.0S \qquad (11.13b\text{-US})$$

$$M^-: \quad SW^- = 1220 + 0.25S \qquad (11.13b\text{-SI})$$

$$\text{Overhang} \quad SW^{\text{Overhang}} = 45 + 10.0X \qquad (11.13c\text{-US})$$

$$\text{Overhang} \quad SW^{\text{Overhang}} = 1140 + 0.833X \qquad (11.13c\text{-SI})$$

where S is the girder spacing [ft (mm)], and X is the distance from the load point to the support [ft (mm)].

Strip widths for other deck systems are given in AASHTO [Table A4.6.2.1.3-1]. A model of the strip on top of the supporting girders is shown in Figure 11.14(a). A design truck is shown positioned for near-critical positive moment. The slab–girder system displaces as shown in Figure 11.14(b). This displacement may be considered as the superposition of the displacements associated with the local load effects [Fig. 11.14(c)] and the global load effects [Fig. 11.14(d)]. The global effects consist of bending of the strip due to the displacement of the girders. Here a small change in load position does not significantly affect these displacements; hence this is a global effect.

The local effect is principally attributed to the bending of the strip due to the application of the wheel loads on this strip. A small movement, for example, one foot transversely, significantly affects the local response. For decks, usually the local effect is significantly greater than the global effect. The global effects may be neglected and the strip may be analyzed with classical beam theory assuming that the girders provide rigid support [A4.6.2.1.5]. Because the lower bound theorem is applicable and because this distribution of internal actions accounts for equilibrium, the strip method yields adequate

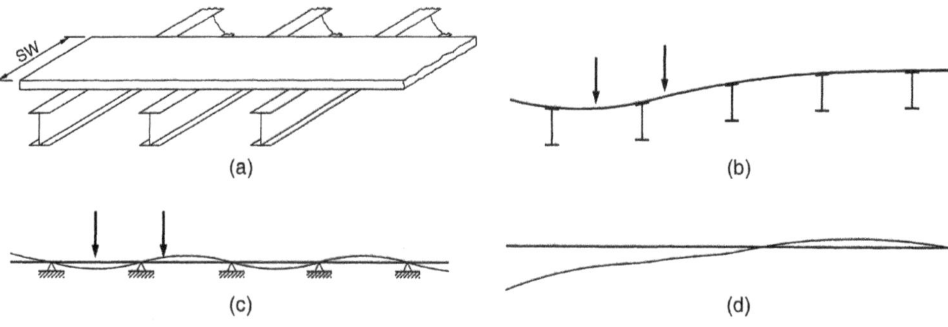

Fig. 11.14 (a) Idealized design strip, (b) transverse section under load, (c) rigid girder model, and (d) displacement due to girder translation.

strength and should, in general, yield a reasonable distribution of reinforcement. To account for the stiffening effect of the support (girder) width, the design shears and moments may be taken as critical at the face of the support for monolithic construction and at one-quarter flange width for steel girders [A4.6.2.1.5].

Sign convention for slabs: A positive slab moment creates compression on the top, and a negative moment creates compression on the bottom. Where graphed, the moment is plotted on the compression face.

Example 11.11 Determine the shear and moments required in the transverse direction for the slab shown in Figure 11.15(a). The strip widths (SW) are [A4.6.2.1.3]

$$\begin{aligned} \text{SW}^+ &= 26.0 + 6.6S = 26.0 + 6.6\,(8) \\ &= 78.8 \text{ in.} = 6.6 \text{ ft} \end{aligned}$$

and

$$\begin{aligned} \text{SW}^- &= 48.0 + 3.0S = 48.0 + 3.0\,(8) \\ &= 72.0 \text{ in.} = 6.0 \text{ ft} \end{aligned}$$

The strip model of the slab consists of the continuous beam shown in Figure 11.15(b). Here influence functions may be used to position the design truck transverse for the most critical actions. Because this approach was taken earlier, an alternative approach, moment distribution, is used here, but any beam analysis method may be employed (based on Eq. 10.1). The near-critical truck position for moment in span *BC* is shown in Figure 11.15(a). Although the beam has seven spans including the cantilevers, it may be simplified by terminating the system at joint *E* with a fixed support and neglecting the cantilever because it is not loaded and contributes no rotational stiffness as shown in Figure 11.15(b). This simplification has little affect on the response. The analysis results are shown in Table 11.18.

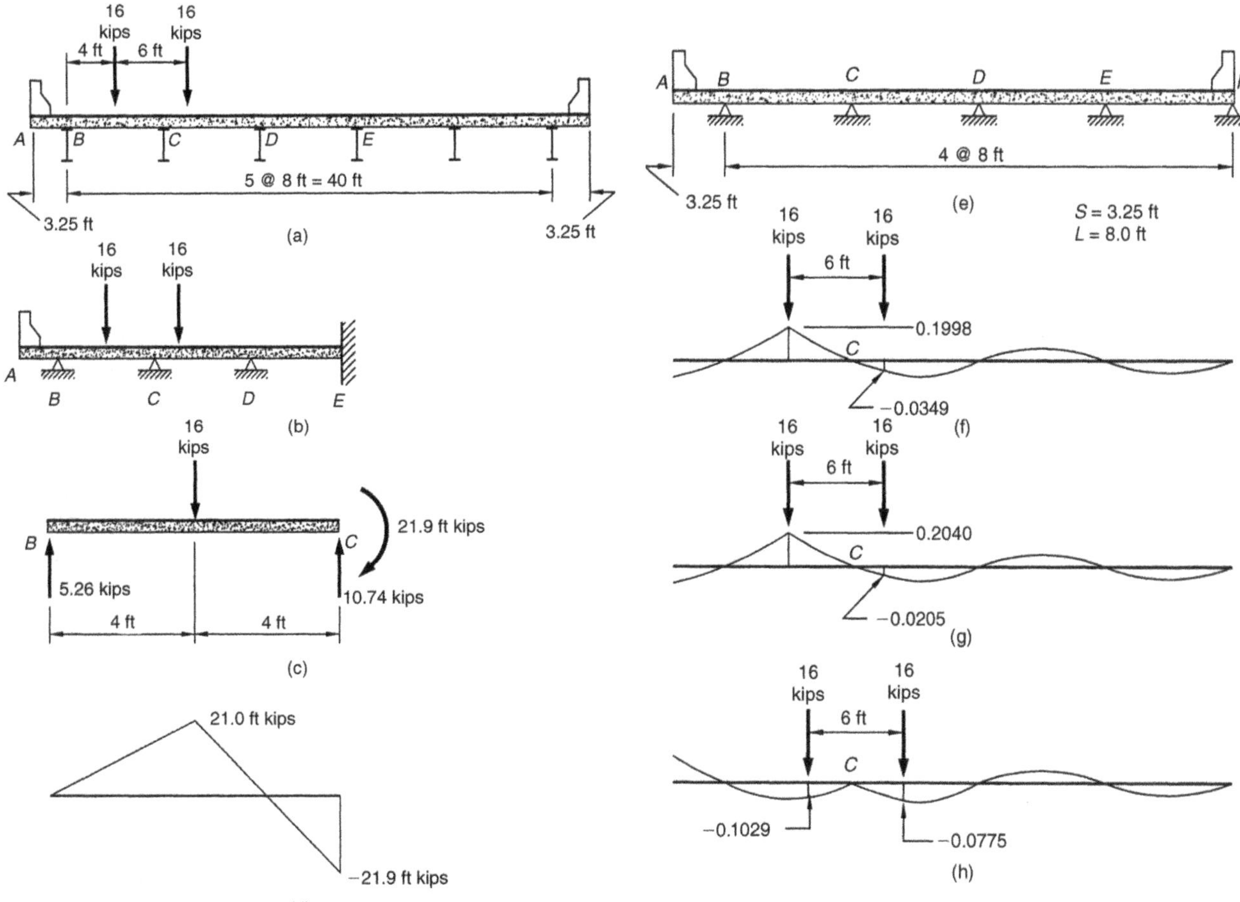

Fig. 11.15 (a) Cross section, (b) moment distribution model, (c) free-body diagram for *BC*, and (d) moment diagram for *BC*. (e) Transverse beam, (f) position for moment 205, (g) position for moment 204, and (h) position for moment 300.

Table 11.18 Moment Distribution Analysis

	BC[a]	CB	CD	DC	DE	ED
Stiffness	0	0.75	1.00	1.00	1.00	Fixed
Distribution factor	N/A	0.429	0.571	0.5	0.5	0.0
Fixed-end moment	16	−16	18	−6	0	0
Adjustment	−16					
Carryover		−8				
Fixed-end moment		−24	18	−6	0	0
Distribution		2.574	3.426	3.000	3.000	0.000
Carryover			1.500	1.713		1.500
Distribution		−0.643	−0.857	−0.857	−0.857	
Carryover			−0.429	−0.429		−0.429
Distribution		0.184	0.245	0.215	0.215	
TOTAL		−21.9	21.9	−2.36	2.36	1.07

[a]Not applicable = N/A.

The most negative moment is approximately −21.9 ft kips (nearest 0.1 ft kip). This end moment is used in the free-body diagram shown in Figure 11.15(c) to determine the end shears for element BC. The end-panel moment diagram is shown in Figure 11.15(d). The critical moments are divided by the strip width to obtain the moments per foot. The results are

$$m^+ = \frac{21.0 \text{ ft kips}}{6.6 \text{ ft}} = 3.18 \text{ ft kips/ft}$$

and

$$m^- = \frac{-21.9 \text{ ft kips}}{6 \text{ ft}} = -3.65 \text{ ft kips/ft}$$

The m^+ and m^- indicate moments at the middle of the panel and over the girder, respectively. These moments may be considered representative and used for the other panels as well.

As an alternative to the moment distribution, the beam model for transverse moments may be modeled with influence functions developed especially for slab analysis. Such functions are described in Appendix A where influence functions for four equal interior spans with length S and cantilever span with length L. This configuration is illustrated in Figure 11.15(e). The influence functions for moment at 205, 204, and 300 are shown in Figures 11.15(f)–11.15(h). The near-critical load positions are also illustrated in these figures. The calculation of the beam moments are based on Eq. 9.1 and are given below. The influence ordinates are from Table A.1 in Appendix A.

$$M_{205} = 16\,(0.1998)\,(8) + 16\left(\frac{-0.0317 - 0.0381}{2}\right)(8)$$

$$M_{205} = 16\,(0.1998)\,(8) + 16\,(-0.0349)\,(8) = 21.1 \text{ ft kips}$$

$$M^+_{205} = \frac{21.1 \text{ ft kips}}{6.6 \text{ ft}} = 3.20 \text{ ft kips/ft}$$

$$M_{204} = 16\,(0.2040)\,(8) + 16\left(\frac{-0.0155 - 0.0254}{2}\right)(8)$$

$$M_{204} = 16\,(0.2040)\,(8) + 16\,(-0.0205)\,(8) = 23.5 \text{ ft kips}$$

$$M^+_{204} = \frac{23.5 \text{ ft kips}}{6.6 \text{ ft}} = 3.56 \text{ ft kips/ft}$$

$$M_{300} = 16\,(-0.1029)\,(8) + 16\left(\frac{-0.0789 - 0.0761}{2}\right)(8)$$

$$M_{300} = 16(-0.1029)\,(8) + 16\,(-0.0775)\,(8) = -23.09 \text{ ft kips}$$

$$m^-_{300} = \frac{-23.09 \text{ ft kips}}{6 \text{ ft}} = -3.85 \text{ ft kips/ft}$$

Note that the moment at 205 is essentially the same as the moment distribution results. Repositioning the load slightly to the left, at the 204, the panel moment is increased to 23.5 ft kips (3.56 ft kips/ft). The negative moment remains essentially the same. The critical panel moments are $m^+ = 3.56$ ft kips/ft and $m^- = -3.85$ ft kips/ft. These moments are compared with the rigorous methods in the example that follows. Note that for design these moments are adjusted for multiple presence, dynamic load allowance, and load factors.

In Appendix B of AASHTO Section 4, deck moments are tabulated for these computations and thereby eliminate the need for routine transverse deck analysis. The AASHTO values for positive and negative moments are 5.69 and 6.48 ft kips/ft, respectively. These values include the multiple presence factor $m = 1.2$ and the dynamic load allowance of 0.33. Therefore, the associated unmodified values for positive and negative moments are 3.56 and 4.06 ft kips/ft, respectively. These values compare well with the computed values.

The cantilever moment is typically controlled by the crash load that must be resisted to take a load from a truck impact on a barrier or rail into the deck and superstructure. See [A13.4.1].

The grillage, finite-element, and finite-strip methods may be used to model the deck actions. The procedures outlined earlier in this chapter are generally applicable. The joint loads must be positioned transversely in the most critical

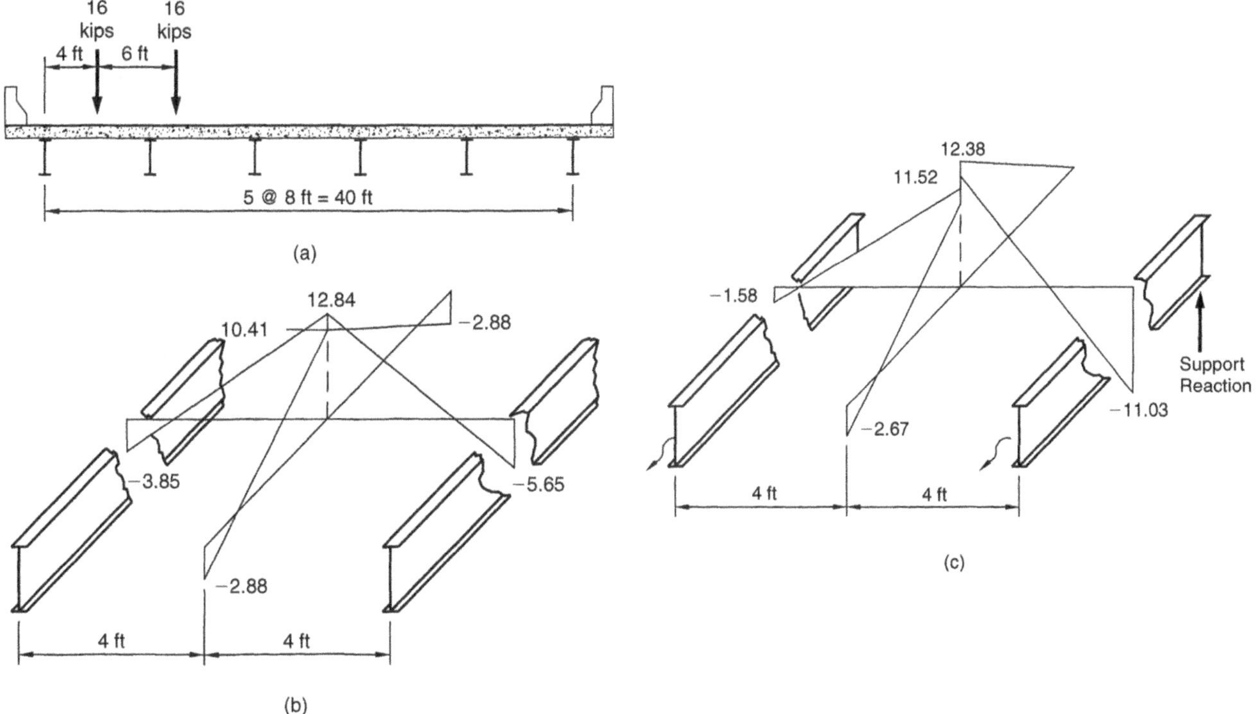

Fig. 11.16 (a) Cross section, (b) grillage moment diagram near midspan, and (c) grillage moment diagram near support.

Table 11.19 Finite-Element and AASHTO Moments (ft kips/ft) (Transverse Moments Only)

		AASHTO Strip Method	Grillage	Finite Strip	Finite Element
m^+ @ Support	Transverse	3.56	3.29	4.84	4.45
	Longitudinal		3.10	5.21	4.05
m^- @ Support	Transverse	−3.85	−3.15	−3.28	−1.26
m^+ @ Midspan	Transverse	3.56	3.67	4.38	4.26
	Longitudinal		2.60	5.78	3.63
m^- @ Midspan	Transverse	−3.85	−1.61	−2.18	−0.31

position. The longitudinal positioning affects the response of the system, and to illustrate, two positions are used in the following example. The first is near the support and the second is at midspan. The results from each are compared with the AASHTO strip method.

Example 11.12 Use the grillage, finite-element, and finite-strip methods to determine the moments in the first interior panel of the system shown in Figure 11.5(a). Position the design truck axle at 3.5 ft from the support and at midspan with the wheel positioned transversely as shown in Figure 11.16(a).

The fine meshes for the grillage, finite-element, and finite-strip methods are used as in Examples 11.4–11.6. The equivalent joint loads are determined for the truck position described, and the resulting moments are given in Table 11.19. To obtain the moment from the grillage model the element moment must be divided by the associated tributary length. The resulting flexural bending moment diagrams for the load positioned at midspan and near the support are shown in Figures 11.16(b) and 11.16(c), respectively. The transverse moment per unit length (lower case) at midpanel is

$$m^+_{\text{transverse}} = \frac{12.84 \text{ ft kips}}{3.5 \text{ ft}} = 3.67 \text{ ft kips/ft}$$

$$m^+_{\text{longitudinal}} = \frac{10.41 \text{ ft kips}}{4 \text{ ft}} = 2.60 \text{ ft kips/ft}$$

and transverse moment over the girder is

$$m^-_{\text{transverse}} = \frac{-5.65 \text{ ft kips}}{3.5 \text{ ft}} = -1.61 \text{ ft kips/ft}$$

The slab moments near the girder support are

$$m^+_{\text{transverse}} = \frac{11.52 \text{ ft kips}}{3.5 \text{ ft}} = 3.29 \text{ ft kips/ft}$$

$$m^+_{\text{longitudinal}} = \frac{12.38 \text{ ft kips}}{4 \text{ ft}} = 3.10 \text{ ft kips/ft}$$

and

$$m^-_{\text{transverse}} = \frac{-11.03 \text{ ft kips}}{3.5 \text{ ft}} = -3.15 \text{ ft kips/ft}$$

The finite-element analysis gives results directly in terms of moment per foot, and therefore no intermediate calculations are required and the results are presented in Table 11.19. The moments from the AASHTO strip method are also given for comparison.

The grillage, finite-element, and finite-strip methods include both the local and the global load effects. Note that a significant difference exists between these results that have not been exhibited in the previous examples. Also note that the grillage method gives results that are in better agreement with the AASHTO moments. It is also interesting to compare the moments in the transverse and longitudinal directions. For example, at midspan of the grillage, the ratio of the positive transverse to longitudinal moment is 3.67/2.60 = 1.4 and near the support the ratio is 3.29/3.10 = 1.06. The same ratios for the finite-strip method are 4.38/5.78 = 0.76 and 4.84/5.21 = 0.93 at the midspan and near-support values, respectively.

This result indicates that the rigorous analysis gives significant longitudinal moments, that is, moments that are considered small in the AASHTO strip method. Near the support, the behavior is affected by the boundary conditions, that is, the support assumptions at the end of the bridge. In this example, the deck is assumed to be supported across the full width (deck and girders). This boundary condition creates significant longitudinal stiffness that attracts the longitudinal moment, which is equal to or exceeds the transverse moment. If *only* the girders are supported, the transverse moments increase by about 50% and significantly exceed the AASHTO values. The longitudinal moments also decrease significantly.

The finite-strip moments are higher than all other moments because the positive moments were taken directly under the concentrated load, in an area where the curvature is increased locally due to the presence of the load (frequently called dishing). This effect decreases rapidly away from the load to values that are similar to the grillage and AASHTO values.

The finite-element values are similar to the finite-strip values when the values directly under the load are considered. The moments were significantly lower (~one-half) at the element centroids located approximately 2.5 ft (760 mm) away from the load position. Finally, it should be noted that the AASHTO strip method overestimates the negative moment over the girders in the middle of the longitudinal span because of the assumption that the girders do not translate. The results are better near the support where the girder translation is small.

The differences and difficulties that arise in modeling the deck with the various methods are particularly noteworthy. What if only one method was used with one mesh? How would the engineer know if the answers are correct? Further, is the maximum moment directly under the load the proper moment for design or should the actions be spread over a larger area? What effect does modeling the wheel load as a patch rather than a concentrated load have? How is the patch load properly modeled in the grillage, finite-strip, and finite-element models? Is a flat plate (or shell) element appropriate to represent the ultimate limit state where significant arching action has been observed in experimental research?

The answers to these questions are best established by studying the system under consideration. These questions and many others are beyond the scope of this chapter, but it suffices to note that there are many important issues that must be addressed to properly model the localized effects in structures. The best way to answer the many questions that arise in modeling is to modify the model and to observe the changes. Such modifications should likely include the use of 3D continuum elements and patch loads.

Modeling local effects takes judgment, skill, and usually significant time. Modeling for local load effects is more difficult than modeling the global response, for example, determining distribution factors. Again place analysis in perspective. The lower bound theorem requires that the one-load path be established for safe design, which makes the AASHTO method viable because the load is distributed transversely and nominal "distribution" steel is used in the longitudinal direction (see Chapter 16). The remaining limit states are associated with service loads such as cracking and fatigue, and, if these can be assured by means other than rigorous analysis, then so much the better. More information on this topic is given in Chapter 14.

In summary, the distribution of internal actions in a bridge deck is complex and not easily modeled. The AASHTO method seems to give reasonable results for this example, and, as shown in the subsequent design of this deck, the AASHTO moments result in a reasonable distribution of reinforcement.

The ultimate limit state can be modeled with the yield-line method. This method can be used to gain additional insight into the behavior of deck systems under ultimate loading conditions.

Yield-Line Analysis The yield-line method is a procedure where the slab is assumed to behave inelastically and exhibits adequate ductility to sustain the applied load until the slab reaches a plastic collapse mechanism. Because the reinforcement proportioning required by AASHTO gives underreinforced or ductile systems, this assumption is realistic. The slab is assumed to collapse at a certain ultimate load through a system of plastic hinges called yield lines. The yield lines form a pattern in the slab creating the mechanism. Two methods are available for determining the ultimate load by the yield-line method: the equilibrium approach and the energy approach. The energy approach is described here because it is perhaps the simplest to implement. The energy approach is an upper bound approach, which means that the ultimate load established with the method is either *equal to or greater* than the actual load (i.e., nonconservative). If the exact mechanism or yield-line pattern is used in the energy approach, then the solution is theoretically exact.

Practically, the yield pattern can be reasonably estimated and the solution is also reasonable for analysis. Patterns may be selected by trial or a systematic approach may be used. Frequently, the yield-line pattern can be determined in terms of a few (sometimes one) characteristic dimensions. These dimensions may be used in a general manner to establish the ultimate load, and then the load is minimized with respect to the characteristic dimensions to obtaining the lowest value. Simple differentiation is usually required.

The fundamentals and the primary assumptions of the yield-line theory are as follows (Ghali et al., 2010):

- ☐ In the mechanism, the bending moment per unit length along all yield lines is constant and equal to the moment capacity of the section.
- ☐ The slab parts (area between yield lines) rotate as rigid bodies along the supported edges.
- ☐ The elastic deformations are considered small relative to the deformation occurring in the yield lines.
- ☐ The yield lines on the sides of two adjacent slab parts pass through the point of intersection of their axes of rotation.
- ☐ The load is conservatively modeled with a concentrated load.
- ☐ Other failure modes do not occur first, for example, punching shear.

Consider the reinforcement layout shown in Figure 11.17(a) and the free-body diagram shown in Figure 11.17(b). The positive flexural capacities in the two directions are m_t and m_L. Here the axis labels t and L are introduced for the transverse and longitudinal directions, usually associated with a bridge. In general, the orthogonal directions align with the reinforcement. Assume a yield line crosses the slab at an angle α relative to the direction of reinforcement as shown in Figure 11.17(a). Equilibrium requires that

$$\begin{aligned} m_a &= m_L \cos^2 \alpha + m_t \sin^2 \alpha \\ m_{\text{twist}} &= (m_L - m_t) \sin \alpha \cos \alpha \end{aligned} \tag{11.14}$$

If the slab is isotopically reinforced, then $m_t = m_L = m$ and Eq. 11.14 simplifies to

$$\begin{aligned} m_\alpha &= m (\cos^2 \alpha + \sin^2 \alpha) = m \\ m_{\text{twist}} &= 0 \end{aligned} \tag{11.15}$$

Therefore, for isotropic reinforcement, the flexural capacity is independent of the angle of the yield line and may be uniformly assigned the value of the capacity in the direction associated with the reinforcement.

Virtual work may be used to equate the energy associated with the internal yielding along the yield lines and the external work of the applied loads. Consider the slab segment shown in Figure 11.18 where a yield line is positioned at an angle to the axis of rotation of the slab segment. By definition of work, the internal energy for yield line i is the dot product

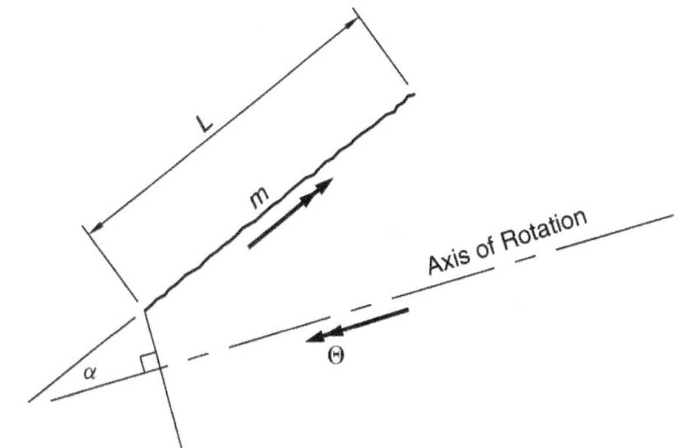

Fig. 11.18 Slab part.

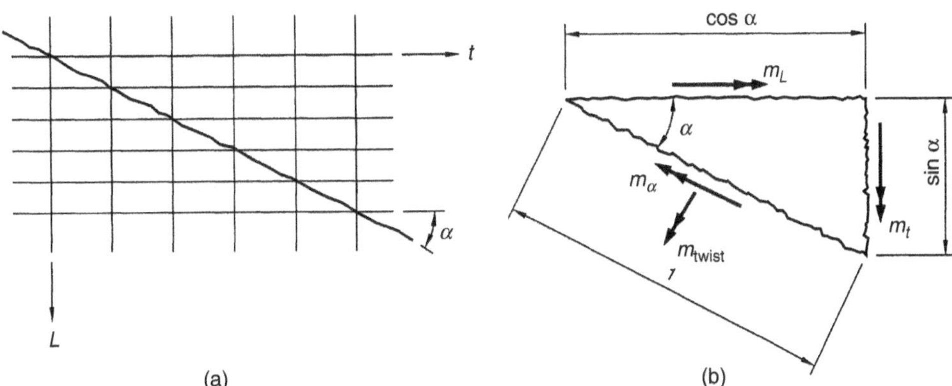

Fig. 11.17 (a) Deck reinforcement layout and (b) free-body diagram.

of the yield-line moment and the rotation, or mathematically stated

$$U_i = \int m_i \cdot \theta_i \ dL = m_i L_i \left(\cos\ \alpha_i\right) \theta_i \qquad (11.16a)$$

and the total system energy is

$$U = \sum_{i=1}^{NL} m_i L_i \ \left(\cos \alpha_i\right) \theta_i \qquad (11.16b)$$

where NL is the number of yield lines in the system and θ_i is the associated rotation.

The total system energy is the summation of the contributions from all the slab segments. It is perhaps simpler to think of the dot product as the projection of moment on the axes of rotation times the virtual rotation. To facilitate this, both moment and rotation may be divided into orthogonal components (usually associated with the system geometry), and the dot product becomes

$$U_{\text{int}} = \sum_{i=1}^{NL} \left(m_i L_i\right) \cdot \theta_i = \sum_{i=1}^{NL} M_{ti}\theta_{ti} + \sum_{i=1}^{NL} M_{Li}\theta_{Li} \qquad (11.17)$$

where M_{ti} and M_{Li} are the components of $m_i L_i$ and θ_{ti} and θ_{Li} are the components of θ.

The virtual external work for uniform and concentrated loads is

$$W_{\text{ext}} = \int pw \ dA + \sum P_i w_i \qquad (11.18)$$

where p is the distributed load, w is the virtual translation field, and w_i is the translation at concentrated load P_i.

In the examples that follow, a typical slab on a slab–girder bridge is studied with the yield-line method. This method is illustrated for a concentrated load applied in the middle portion of the bridge, near the end of the bridge, and on the cantilever. Several of the important features of the method are illustrated in this analysis. The methodology is similar for a patch load; however, the algebra becomes more complex.

Example 11.13 Determine the required moments due to the concentrated loads positioned as shown in Figure 11.19(a) in combination with uniformly distributed loads.

The assumed yield-line patterns are also illustrated in Figure 11.19(a). The girder spacing is S, the cantilever overhang is H, and G is the wheel spacing (gage), usually 6 ft (1800 mm), or the spacing between the wheels of adjacent trucks, usually 4 ft (1200 mm). First, consider the load positioned in the center of a panel near midspan as illustrated by point A in Figure 11.19(a).

Because the system is axisymmetric, the load is distributed evenly to all sectors ($d\alpha$). The analysis may be performed on the sector as shown in Figure 11.19(b), and the total energy is determined by integration around the circular path. By using Eq. 11.17, the internal work associated with yield-line rotation is

$$U_{\text{int}} = U_{\text{perimeter}} + U_{\text{radial fans}} = \int_0^{2\pi} m'\theta r \ d\alpha + \int_0^{2\pi} m\theta r \ d\alpha$$

where α is the orientation of the radial yield line, θ is the virtual rotation at the ring of the yield-line pattern, m is the positive moment capacity for the orientation α, and m' is the negative moment capacity for the orientation α. The moment capacity for a general orientation α is given in Eq. 11.14. By using Eq. 11.15, the internal strain energy becomes

$$\begin{aligned} U_{\text{int}} &= U_{\text{perimeter}} + U_{\text{radial fans}} \\ &= \int_0^{2\pi} \left(m_L' \ \cos^2\alpha + m_t' \ \sin^2 \ \alpha\right) \theta r \ d\alpha \\ &\quad + \int_0^{2\pi} \left(m_L \ \cos^2 \ \beta + m_t \ \sin^2\beta\right) \theta r \ d\beta \end{aligned}$$

where β is the compliment of α. Note that

$$\int_0^{2\pi} \sin^2 \ \alpha \ d\alpha = \int_0^{2\pi} \cos^2 \ \alpha \ d\alpha = \pi$$

U_{int} simplifies to

$$\begin{aligned} U_{\text{int}} &= U_{\text{perimeter}} + U_{\text{radial fans}} \\ &= \left(m_L' + m_t'\right) \pi r\theta + \left(m_L + m_t\right) \pi r\theta \end{aligned}$$

The energies for the perimeter and the fans are kept separate to facilitate further manipulation. The virtual rotation and translation δ at the load are related by

$$\delta = \theta r$$

The external virtual work due to the concentrated load P, in combination with a uniform loads q, is established by using Eq. 11.18, resulting in

$$W_{\text{ext}} = P\delta + q\left(\frac{\pi r^2}{3}\right) \ \delta = \text{Pr} \ \theta + q\left(\frac{\pi r^2}{3}\right) r\theta$$

Equate the external work and internal energy, which gives

$$\pi\left(m_L' + m_t'\right) + \pi\left(m_L + m_t\right) = P + \frac{q\pi r^2}{3} \qquad (11.19)$$

The moment summations may be thought of as double the average moment capacity, or

$$m' = \tfrac{1}{2}\left(m_L' + m_t'\right)$$

and

$$m = \tfrac{1}{2}\left(m_L + m_t\right)$$

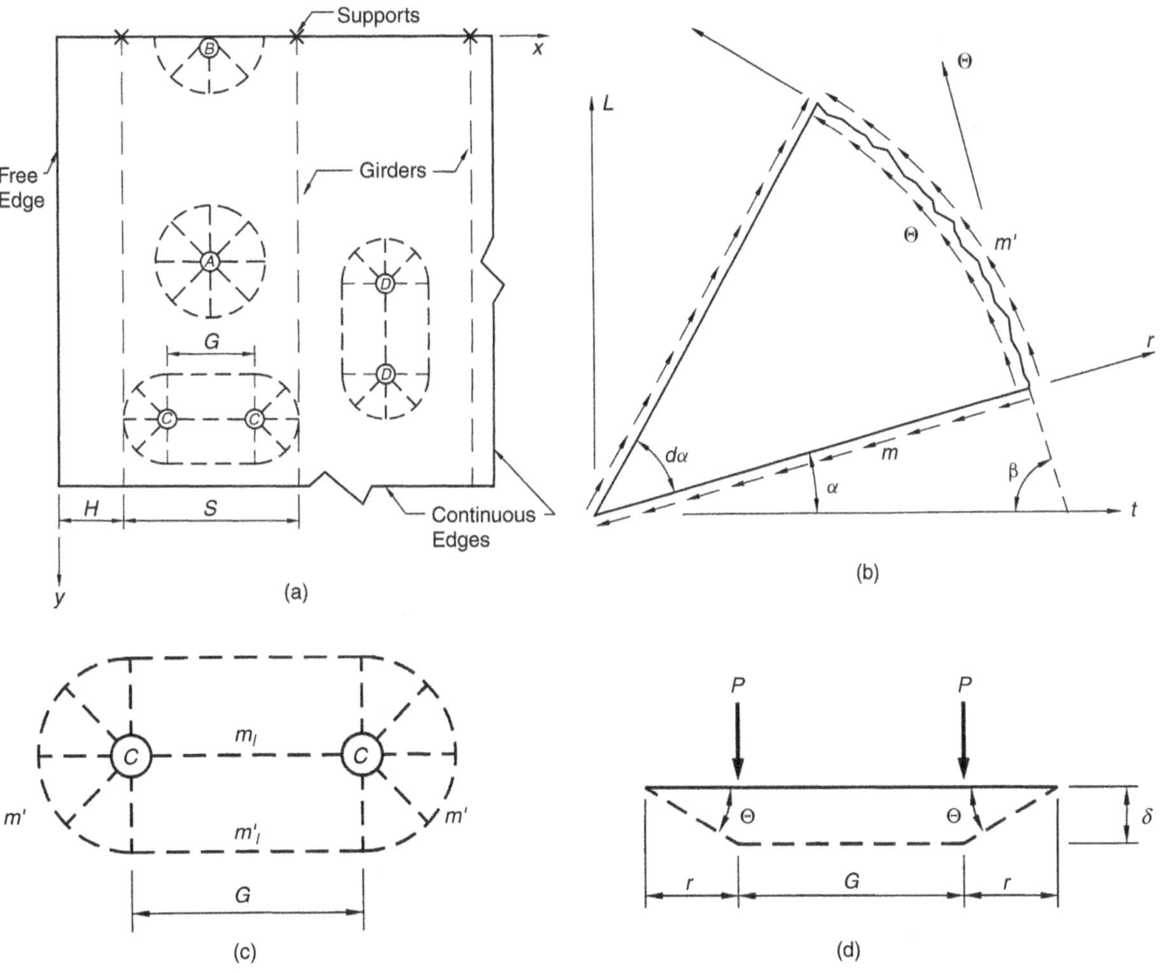

Fig. 11.19 Yield-line patterns: (a) axle positions and (b) sector (part). Yield lines for axles: (c) plan view and (d) elevation view.

Substitution of the average moments into Eq. 11.19 results in

$$m' + m = \frac{P}{2\pi} + \frac{qr^2}{6} \tag{11.20}$$

The average capacities m and m' are used for convenience in subsequent calculations. For a comparison with the elastic analysis to be presented later, neglect the uniform load and assume that the positive and negative capacities are equal, the required moment capacity is

$$m = m' = \frac{P}{4\pi} \tag{11.21}$$

Now consider the load positioned near the edge of the slab at point B as shown in Figure 11.19(a), where the yield-line pattern is also shown. Note that due to symmetry, the length of the yield lines in this system constitute one-half the length of the previous system (point A). Thus, the internal energy is one-half of that given for the yield-line pattern for point A (Eq. 11.19), the associated uniform load is also one-half of the previous value, but the concentrated load is full value, and the required moment is doubled (Eq. 11.20) giving

$$\left(m' + m\right) = \frac{P}{\pi} + \frac{qr^2}{6} \tag{11.22}$$

Next consider the two loads positioned at point C as illustrated in Figures 11.19(a) and 11.19(c). The internal energy is

$$\begin{aligned} U_{\text{int}} &= U_{\text{radial fan}} + U_{\text{straight}} \\ &= 2\pi r\theta\left(m' + m\right) + \frac{2G\delta}{r}\left(m'_L + m_L\right) \end{aligned}$$

The work due to the two concentrated loads plus the uniform load is

$$W_{\text{ext}} = 2P\delta + qr\delta\left(\frac{\pi r}{3} + G\right)$$

Again, equating the internal and external energies, one obtains

$$P + \frac{qr}{2}\left(\frac{\pi r}{3} + G\right) = \pi\left(m' + m\right) + \frac{G}{r}\left(m'_L + m_L\right) \tag{11.22}$$

Now consider the load at point D in Figure 11.19(a). The only difference between the analysis of this position and that of point C is that the moment capacity associated with the straight lines is now the capacity of the transverse reinforcement. Equation 11.22 becomes

$$P + \frac{qr}{2}\left(\frac{\pi r}{3} + G\right) = \pi\left(m' + m\right) + \frac{G}{r}\left(m_t' + m_t\right) \tag{11.23}$$

Note that as G goes to zero, Eqs. 11.22 and 11.23 reduce to Eq. 11.22, as expected. These equations may be used to estimate the ultimate strength of slabs designed with the AASHTO procedures. It is interesting to compare the moments based on the yield-line analysis to those based on the elastic methods.

Example 11.14 Determine the moments required for a wheel load of the AASHTO design truck [P = 16 kips (72.5 kN)] position in the interior panel. Compare these moments to those obtained from the AASHTO strip- and finite-element methods. By neglecting the uniform load, Eq. 11.20 can be used to determine the required moment

$$m' + m = \frac{P}{2\pi} + \frac{(q = 0)\, r^2}{6} = \frac{16}{2\pi} = 2.55 \text{ ft kips/ft}$$

If we assume that the positive and negative moment capacities are the same, the required capacity is

$$m = m' = 1.27 \text{ ft kips/ft}$$

From Table 11.19 the AASHTO strip method moments are $m^+ = 3.56$ and $m^- = -3.85$ ft kips/ft, and the finite-element transverse moments are $m^+ = 4.26$ and $m^- = -0.31$ ft kips/ft near midspan and $m^+ = 4.45$ and $m^- = -1.26$ ft kips/ft near the support. Thus, the elastic distribution is quite different from the inelastic, which is consistent with test results where slabs typically test at a minimum eight times the service-level loads (16 kips).

In summary, several methods have been described for proportioning the moment and reinforcement in a bridge deck. The AASHTO strip method is permitted by the specification and offers a simple method for all limit states. In light of the lower bound theorem, this is a conservative method. The yield-line method uses inelastic analysis techniques and is pertinent only to the strength and extreme limit states. Other methods are required in conjunction with this to ensure serviceability. Finally, the empirical design method is not an analytical approach, but rather a set of rules upon which to proportion the deck. The discussion of this method is presented in following chapters.

11.4 BOX-GIRDER BRIDGES

Behavior, Structural Idealization, and Modeling The box-girder bridge is a common structural form in both steel and concrete. The multicell box girder may be thought of as a slab–girder bridge with a bottom slab that encloses the section [see Fig. 11.20(a)]. This closure creates a "closed section" that is torsionally much stiffer than its open counterpart. This characteristic makes the box girder ideal for bridges that have significant torsion induced by horizontal curvature resulting from roadway alignments. For example, the box-girder bridge is often used for tightly spaced interchanges that require curved alignments because of its torsional resistance and fine aesthetic qualities.

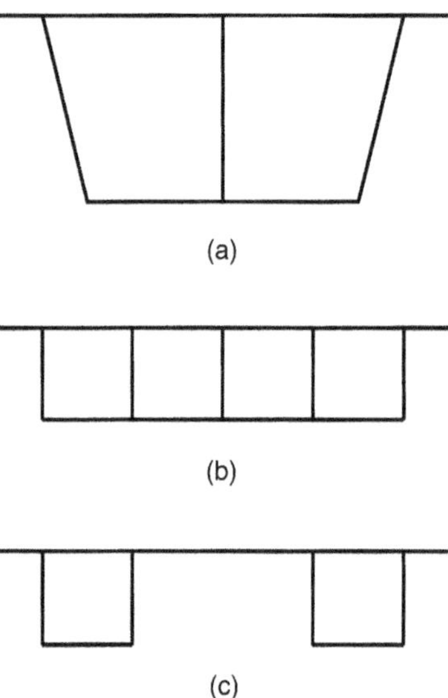

Fig. 11.20 (a) Two-cell box section, (b) multicell box section, and (c) spread box section.

Box systems are built in a wide variety of configurations, most are illustrated in Table 4.1. Examples include closed steel or precast boxes with a cast-in-place (CIP) deck (b), open steel or precast boxes with CIP deck (c), CIP multicell box (d), precast boxes with shear keys (f), and transversely posttensioned precast (g). These systems can be separated into three primary categories of box systems: single and double cell [Fig. 11.20(a)], multicell [three or more cells, Fig. 11.20(b)], and spread box systems [the boxes noncontiguous, Fig. 11.20(c)]. As expected, the behavior of these systems is distinctly different within each category and the design concern varies widely depending on construction methods. Due to the large number of systems and construction methods, selected systems are described with limited detail here.

The single- and two-cell box systems are usually narrow compared to the span and behave similar to a beam and are often modeled with space frame elements. Such systems are designed for the critical combinations of bending moment, shear, and torsion created due to global effects and the local effect of the vehicle applied directly to the deck. As stated in Chapter 8, global means the load effect is due to the global system response such as the deflection, moment, or shear of a main girder. Local effects are the actions and displacements

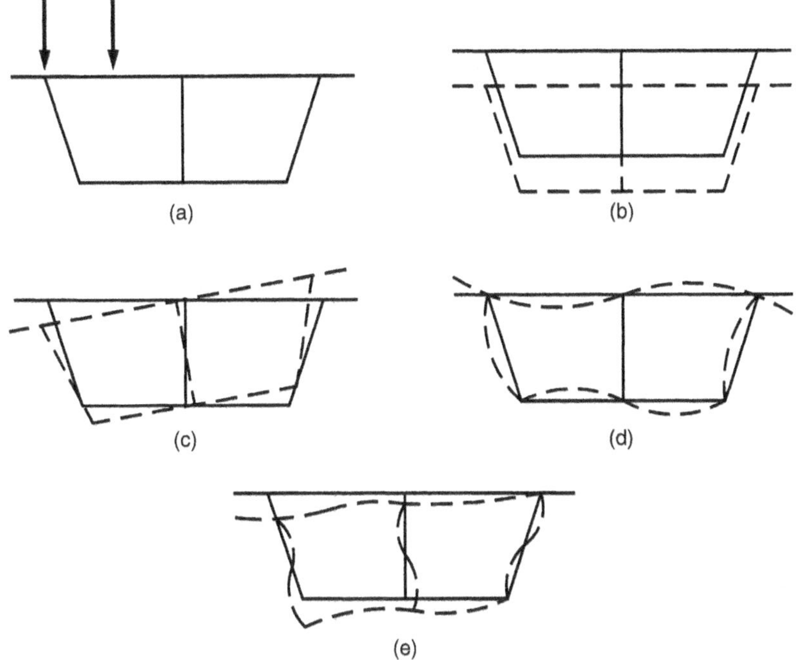

Fig. 11.21 (a) Eccentric loading, (b) global flexural deformation, (c) global torsional deformation, (d) local transverse bending deformation, and (e) distortional transverse deformation due to global displacements.

that result from loads directly applied to (or in the local area of) the component being designed. Recall that if a small spatial variation in the live-load placement causes a large change in load effect, then it is considered local.

The various displacement modes for a two-cell box-girder cross section are illustrated in Figures 11.21(a)–11.21(e). Here the total displacement is decomposed into four components: global bending, global torsion, local flexure, and local distortion due to global displacements. The global bending is due to the girder behaving as a single beam, that is, the strain profile is assumed to be linear, and there is no twisting or distortion, that is, the section shape remains unaltered [Fig. 11.21(b)]. The global torsional rotation is illustrated in Figure 11.21(c). As in the bending mode, the section shape remains unaltered by the load and the section is twisted due to eccentrically applied loads. The local bending mode is shown in Figure 11.21(d). Here the loads create out-of-plane bending in the deck. Because the girder webs are continuous with the slab, the webs and the bottom slab also bend. The intersection of elements (physical joints) rotate, but do not translate, in this mode. Finally, the distortion mode is illustrated in Figure 11.21(e). The slab and webs flex due to the translation and rotation of the physical joints, that is, the displacements shown in Figure 11.21(b) plus those shown in Figure 11.21(c). Superpose all of these modes to establish the system response.

There are numerous analytical methods available for the analysis of box-girder systems, ranging from the rigorous and complex to the simplistic and direct. The selection of the method depends on the response sought and its use. The box system may be modeled with finite elements, finite strips, and beams. All approaches are viable and the one selected depends on several factors including: the number of cells, the geometry (width/length, skew, diaphragms, cross bracing), construction method, type of box system (single, multi, or spread boxes), and, of course, the reason for and application of the results.

In general, the one- and two-cell box systems have spans that are much greater than their section widths and can be modeled as beams, usually with space frame elements. The beam is modeled so that the global flexural and torsional responses are considered. These actions are then used with the resistance provisions in the usual manner. For the case of steel boxes, the web thicknesses tend to be thin and local stiffeners are required. The local bending effects are modeled by considering the box as a frame in the transverse direction, and obtaining reasonable distribution of shear and bending moments (due to the out-of-plane deformations). This model can be based on the distribution width outlined in, for example, Eq. 11.13.

The distortional deformation is modeled by imposing the resulting beam displacements at the joints of the transverse frame (plane frame), which creates bending of the deck and web elements. The results of the local bending and distortional deformation are superposed to establish the local out-of-plane actions.

As with the single- and two-cell system, multicell (three or more) box systems can be modeled with the finite-element and finite-strip methods. Both formulations can simultaneously model the in- and out-of-plane load effects associated with global and local behavior. Finite-element method is certainly the most common rigorous methods used in engineering practice. The principle difference between the slab–girder and the multicell box bridge is that the box section has significantly more torsional stiffness.

Beam-Line Methods The AASHTO Specification has equations for distribution factors for multicell box beams. These are applied in a fashion similar to the slab–girder systems and are summarized in Table 11.20.

Other box systems, such as spread box beams and shear-keyed systems have similar formulas but due to space limitations are not presented. These formulas are given in AASHTO [A4.6.2.2]. An example is given to illustrate the use of the AASHTO formulas for a multicell system.

Example 11.15 Determine the distribution factors for one and multilanes loaded for the concrete cast-in-place box system shown in Figure 11.22. The bridge has no skew.

The AASHTO distribution factors are used as in the previous example with slab–girder bridges. The factors for one-, two-, and three-lane loadings are established for the interior and exterior girder moments and shears. Each case is considered separately. Table 11.20 is used exclusively for all calculations. Note that the multiple presence factors are included.

Interior girder moment for one lane loaded:

$$mg^{\text{SI}}_{\text{moment}} = \left(1.75 + \frac{S}{3.6}\right)\left(\frac{1}{L}\right)^{0.35}\left(\frac{1}{N_c}\right)^{0.45}$$
$$= \left(1.75 + \frac{13}{3.6}\right)\left(\frac{1}{100}\right)^{0.35}\left(\frac{1}{3}\right)^{0.45} = 0.65 \text{ lane/web}$$

Interior girder moment for multiple lanes loaded:

$$mg^{\text{MI}}_{\text{moment}} = \left(\frac{13}{N_C}\right)^{0.3}\left(\frac{S}{5.8}\right)\left(\frac{1}{L}\right)^{0.25}$$
$$= \left(\frac{13}{3}\right)^{0.3}\left(\frac{13}{5.8}\right)\left(\frac{1}{100}\right)^{0.25} = 1.10 \text{ lanes/web}$$

Exterior girder moment for one lane and multiple lanes loaded:

$$W_e = \tfrac{1}{2}S + \text{overhang} = \tfrac{1}{2}(13) + 3.75 = 10.25 \text{ ft}$$
$$mg^{(\text{S or M})\text{E}}_{\text{moment}} = \frac{W_e}{14} = \frac{10.25}{14} = 0.73 \text{ lane/web}$$

Interior girder shear for one lane loaded:

$$mg^{\text{SI}}_{\text{shear}} = \left(\frac{S}{9.5 \text{ ft}}\right)^{0.6}\left(\frac{d}{12L}\right)^{0.1}$$
$$= \left(\frac{13}{9.5 \text{ ft}}\right)^{0.6}\left[\frac{(6.66)\,(12)}{(12)\,(100)}\right]^{0.1} = 0.92 \text{ lane/web}$$

Interior girder shear for multiple lanes loaded:

$$mg^{\text{MI}}_{\text{shear}} = \left(\frac{S}{7.3 \text{ ft}}\right)^{0.9}\left(\frac{d}{12L}\right)^{0.1}$$
$$= \left(\frac{13}{7.3 \text{ ft}}\right)^{0.9}\left[\frac{(12)\,(6.66)}{(12)\,(100)}\right]^{0.1} = 1.28 \text{ lanes/web}$$

Exterior girder shear for one lane loaded: The lever rule is used for this calculation. Refer to the free-body diagram illustrated in Figure 11.22(d). Balance the moment about B to determine the reaction R_A:

$$\sum M_B = 0$$
$$\frac{P}{2}(7.5) + \frac{P}{2}(13.5) - R_A(13) = 0$$
$$R_A = 0.81P$$
$$g^{\text{SE}}_{\text{shear}} = 0.81$$
$$mg^{\text{SE}}_{\text{shear}} = 1.2(0.81) = 0.97 \text{ lane/web}$$

Exterior girder shear for two lanes loaded: The interior distribution factor is used with an adjustment factor based on the overhang length:

$$mg^{\text{ME}}_{\text{shear}} = e\left(mg^{\text{MI}}_{\text{shear}}\right) = (0.84)\,(1.28) = 1.08 \text{ lanes/web}$$

where

$$e = 0.64 + \frac{d_e}{12.5} = 0.84$$

The AASHTO distribution factors are compared with those based on a finite-element analysis in the next section. The application of these factors is similar to that of the slab–girder bridge. Details regarding the application are presented in Chapter 13.

Finite-Element Method The box-girder bridge may be modeled with the finite-element method by using shell elements. These elements must have the capability to properly model the in-plane and out-of-plane effects. One of the principal characteristics of the box girder is shear lag. This phenomenon is the decrease in stress (flange force) with increased distance away from the web. The mesh must be sufficiently fine to model this effect. A mesh that is too coarse tends to spread the flange force over a larger length, therefore decreasing the peak forces. Another important characteristic is the proper modeling of the diaphragms and support conditions.

The diaphragms are transverse walls periodically located within the span and at regions of concentrated load such as supports. The diaphragms tend to stiffen the section torsionally and reduce the distortional deformation, which produces a stiffer structure with improved load distribution characteristics. Because the supports are typically located at a significant distance below the neutral axis, the bearing stiffness is important in the modeling. For example, the bottom flange force can change significantly if the support conditions are changed from pin–roller to pin–pin.

Table 11.20US Distribution Factors for Multicell Box Beams and Box Sections Transversely Posttensioned Together—US Customary Units[a]

Action/Location	AASHTO Table	Distribution Factors (mg)[b]	Skew Correction Factor[c]	Range of Applicability
A. Moment interior girder	4.6.2.2.2b-1	Single design lane loaded: $mg^{SI}_{moment} = \left(1.75 + \frac{S}{3.6}\right)^{0.4} \left(\frac{1}{L}\right)^{0.35} \left(\frac{1}{N_c}\right)^{0.45}$		$7.0 \le S \le 13.0$ ft $60 \le L \le 240$ ft $N_c \ge 3$
		Two or more (multiple) design lanes loaded: $mg^{MI}_{moment} = \left(\frac{13}{N_c}\right)^{0.3} \left(\frac{S}{5.8}\right) \left(\frac{1}{L}\right)^{0.25}$		If $N_c > 8$, use $N_c = 8$
B. Moment exterior girder	4.6.2.2.2d-1	$mg^{ME}_{moment} = \frac{W_e}{14}$	$1.05 - 0.25 \tan\theta \le 1.0$ If $\theta > 60^\circ$, use $\theta = 60^\circ$	$W_e \le S$
C. Shear interior girder	4.6.2.2.3a-1	Single design lane loaded: $mg^{SI}_{shear} = \left(\frac{S}{9.5\text{ ft}}\right)^{0.6} \left(\frac{d}{12\ L}\right)^{0.1}$ Two or more (multiple) design lanes loaded: $mg^{SI}_{shear} = \left(\frac{S}{7.3\text{ ft}}\right)^{0.9} \left(\frac{d}{12\ L}\right)^{0.1}$		$6.0 \le S \le 13.0$ ft $20 \le L \le 240$ ft $3 \le d \le 9$ ft $N_c \ge 3$
D. Shear exterior girder	4.6.2.2.3b-1	One design lane loaded: Use lever rule	$1.0 + \left[0.25 + \frac{12L}{70d}\right] \tan\theta$	$-2.0 \le d_e \le 5.0$ ft $0 \le \theta \le 60^\circ$
		Two or more (multiple) design lanes loaded: $mg^{ME}_{shear} = e\,(mg^{MI}_{shear})$ $e = 0.64 + \frac{d_e}{12.5}$ d_e is positive if girder web is inside of barrier, otherwise negative	for shear in the obtuse corner	$6.0 \le S \le 13.0$ ft $20 \le L \le 240$ ft $35 \le d \le 110$ in. $N_c \ge 3$

[a] S = girder spacing [ft (mm)], L = span length [ft (mm)], N_c = number of cells, N_b = number of beams, W_e = half the web spacing, plus the total overhang spacing [ft (mm)], d = overall depth of a girder [ft (mm)], d_e = distance from the center of the exterior beam to the inside edge of the curb or barrier [ft (mm)], θ = angle between the centerline of the support and a line normal to the roadway centerline. The *lever rule* is in Example 11.3.

[b] Equations include multiple presence factor; for lever rule and the rigid method engineer must perform factoring by m.

[c] Not applicable = N/A.

Table 11.20SI Distribution Factors for Multicell Box Beams and Box Sections Transversely Posttensioned Together—SI Units

Action/Location	AASHTO Table	Distribution Factors (mg)	Skew Correction[a] Factor	Range of Applicability
A. Moment interior girder	4.6.2.2.2b-1	Single design lane loaded: $mg^{SI}_{moment} = \left(1.75 + \frac{S}{1100}\right)\left(\frac{300}{L}\right)^{0.35}\left(\frac{1}{N_c}\right)^{0.45}$ Two or more (multiple) design lanes loaded: $mg^{MI}_{moment} = \left(\frac{13}{N_c}\right)^{0.3}\left(\frac{S}{430}\right)\left(\frac{1}{L}\right)^{0.25}$		$2100 \le S \le 4000$ mm $18\,000 \le L \le 73\,000$ mm $N_c \ge 3$ If $N_c > 8$, use $N_c = 8$
B. Moment exterior girder	4.6.2.2.2d-1	$mg^{ME}_{moment} = \frac{W_e}{4300\text{ mm}}$	$1.05 - 0.25 \tan\theta \le 1.0$ If $\theta > 60°$, use $\theta = 60°$	$W_e \le S$
C. Shear interior girder	4.6.2.2.3a-1	Single design lane loaded: $mg^{SI}_{shear} = \left(\frac{S}{2900\text{ mm}}\right)^{0.6}\left(\frac{d}{L}\right)^{0.1}$ Two or more (multiple) design lanes loaded: $mg^{MI}_{shear} = \left(\frac{S}{2200\text{ mm}}\right)^{0.9}\left(\frac{d}{L}\right)^{0.1}$		$1800 \le S \le 4000$ mm $6000 \le L \le 73\,000$ mm $890 \le d \le 2800$ mm $N_c \ge 3$
D. Shear exterior girder	4.6.2.2.3b-1	One design lane loaded: Use lever rule Two or more (multiple) design lanes loaded: $mg^{ME}_{shear} = e\left(mg^{MI}_{shear}\right)$ $e = 0.64 - \frac{d_e}{3800\text{ mm}}$ d_e is positive if girder web is inside of barrier, otherwise negative	$1.0 + \left[0.25 + \frac{L}{70d}\right]\tan\theta$ for shear in the obtuse corner	$-600 \le d_e \le 1500$ mm $0 \le \theta \le 60°$ $1800 \le S \le 4000$ mm $6000 \le L \le 73000$ mm $900 \le d \le 2700$ mm $N_c \ge 3$

[a]Not applicable = N/A.

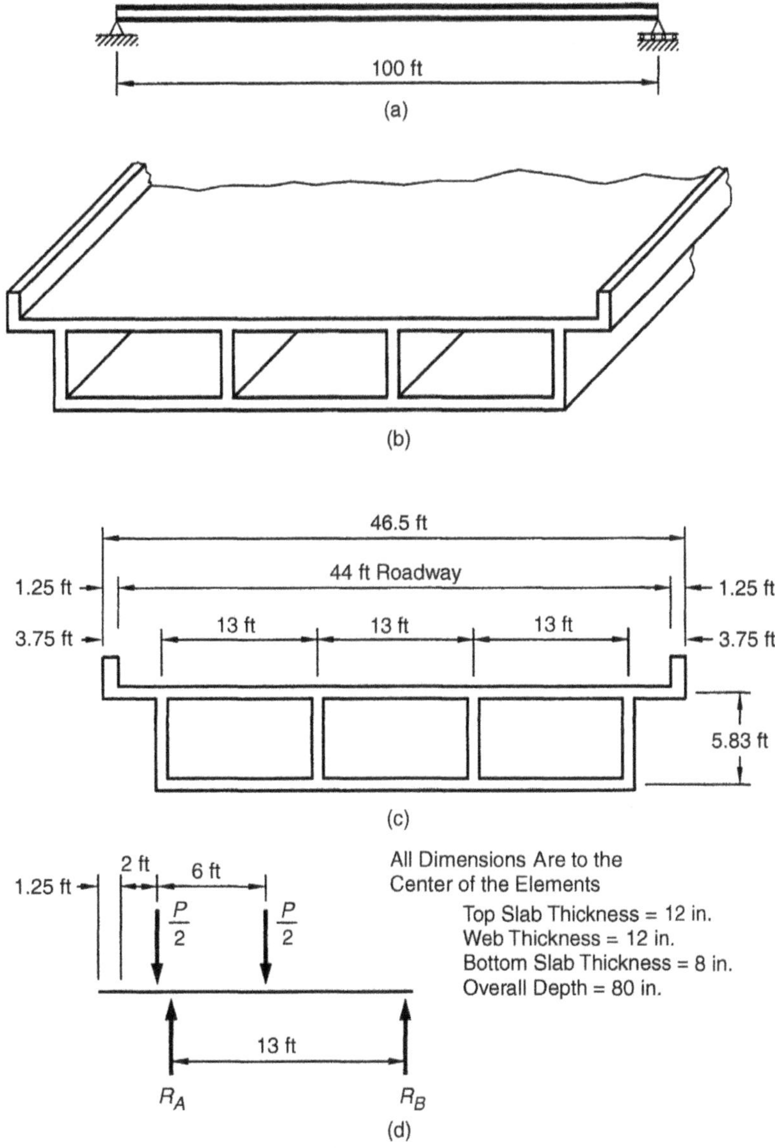

Fig. 11.22 (a) Box section, (b) dimensions, (c) span and supports, and (d) free-body diagram.

Finally, because the box section has significant torsional stiffness, the effects of skew can also be significant and must be carefully considered. For example, it is possible for an eccentrically loaded box-girder web to lift completely off its bearing seat. This effect can greatly increase the reactions and associated shear forces in the area of the support. Such forces can create cracks and damage bearings. A simply supported three-cell box girder is modeled in the example below.

Example 11.16 For the bridge illustrated in Figure 11.22, use the finite-element method to determine the distribution factors for the bending moment at midspan.

The system was modeled with the two meshes shown in Figures 11.23(a) and 11.23(b). Both meshes produced essentially the same results; hence convergence was achieved for the parameters under study. The load effect of each web was based on the longitudinal force per unit length, f, at the bottom. This quantity is readily available from the analysis. The sum of the forces for all the webs is divided by the number of lanes loaded. This ratio is then divided into the force in each girder. The distribution factor for girder i may be mathematically expressed as

$$g = \frac{f_i}{\sum_{\text{No. girders}} f_i/\text{No. lanes loaded}} \tag{11.24}$$

where f_i is the load effect in the girder web i.

The flexural bending moment for the entire web could be used as in the slab–girder bridge, but this quantity is not readily available. To determine the moment, the force per unit length must be numerically integrated over the web depth, that is, $M = \int fy\,dy$. If the end supports are not restrained and,

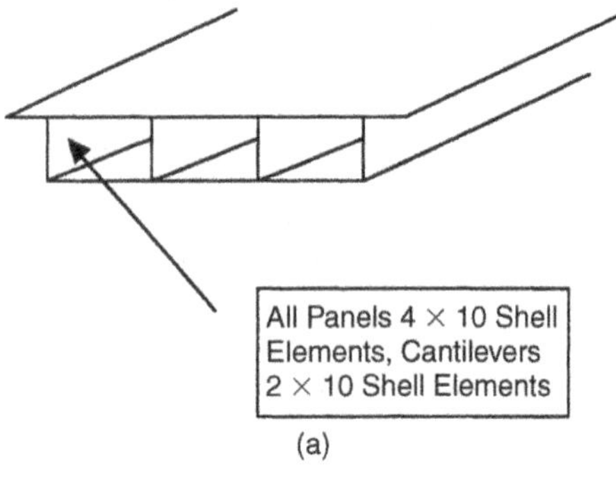

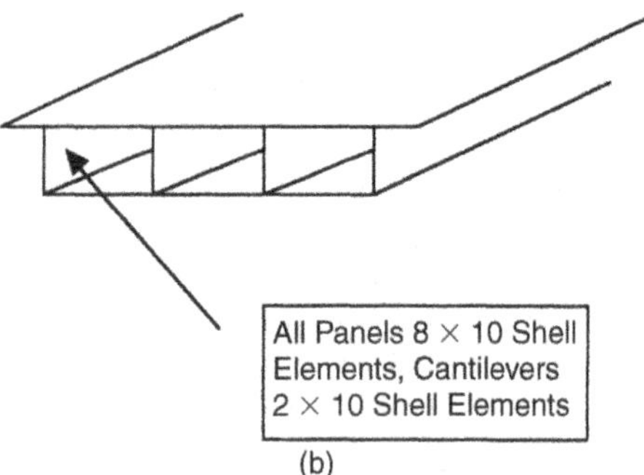

Fig. 11.23 (a) Coarse mesh and (b) fine mesh.

therefore, do not induce a net axial force in the section, then the force per length is nominally proportional to the moment. The loads are positioned for the maximum flexural effect in the exterior and interior webs for one, two, and three lanes loaded. For shear/reactions, the loads were positioned for the maximum reaction and the distribution factors were determined by Eq. 11.911.24. The distribution factors based on the finite-element methods are shown in Table 11.21. The results from the AASHTO distribution formulas are also shown for comparison.

Note that there are significant differences between the AASHTO formulas and finite-element results. In general, the finite-element model exhibits better load distribution than the AASHTO values (less conservative). Recall that the AASHTO values are empirically determined based solely on statistical observation. Therefore, it is difficult to explain the differences for this particular structure. It is interesting to note that unlike the slab–girder and slab bridges, the analytical results can vary significantly from the AASHTO method, and perhaps this highlights the need for rigorous analysis of more complex systems.

Again, note that the comparison of the distribution factors does not suggest that the only reason to perform a rigorous analysis is to establish the distribution factors. Because the finite-element-based actions are available directly, the designer may wish to proportion the bridge based on these actions. Other load cases must be included and the envelope of combined actions is used for design.

11.5 CLOSING REMARKS

This chapter includes numerous topics related to the structural analysis of bridge systems. It is intended to provide a broad-based perspective for analysis. It is important to understand the concepts and how they relate to the AASHTO design specification. It is also important to understand the development and usage of the AASHTO distribution factors, the meaning of one and multiple lanes loaded and comparisons and modeling techniques associated with advanced methods such as grillage, finite-element, and finite-strip methods. Examples were presented to address the elastic and inelastic analysis of deck systems. Here it is important to note the variation in results of these methods that are sensitive to local effects.

It is unlikely that this chapter is comprehensive on any one topic but provides a useful introduction and the background necessary for more rigorous analysis of bridges. In present U.S. bridge engineering practice, rigorous analysis is typically reserved for more complex geometries involving skew and curvature combined with the need for better analysis to model construction of these systems. Rigorous analysis is perhaps underutilized for the analysis and load rating for existing bridge where changes in the live-load distribution can yield significant economic savings associated with posting, retrofit, and satisfying operational demands for permit loads.

Additional study is recommended in NCHRP Report 592 (Puckett et al., 2007). This report outlines many aspects of live-load distribution, including a comprehensive literature

Table 11.21 Distribution Factors Based on the Finite-Element Method

Girder Location	Lanes Loaded	Finite-Element Moment Distribution Factor (mg)	AASHTO Moment Distribution Factor (mg)	Finite-Element Shear or Reaction Distribution Factor (mg)	AASHTO Shear or Reaction Distribution Factor (mg)
Exterior	1	0.53(1.2) = 0.64	0.73	0.72(1.2) = 0.86	0.97
Exterior	2	0.85(1.0) = 0.85	0.73	0.84(1.0) = 0.84	1.08
Exterior	3	1.00(0.85) = 0.85	0.73	1.34(0.85) = 1.14	1.08
Interior	1	0.38(1.2) = 0.46	0.65	0.71(1.2) = 0.85	0.92
Interior	2	0.62(1.0) = 0.62	1.10	1.21(1.0) = 1.21	1.28
Interior	3	0.80(0.85) = 0.68	1.10	1.22(0.85) = 1.04	1.28

review, comparison of various simplified to rigorous methods, international specifications, effect of skew, live-load positioning, and many other topics. The report addresses a new simplified live-load distribution method and the various appendices contain information regarding a host of topics.

A comprehensive guide is available for the analysis of steel bridges. Many of the concepts provided therein are generally applicable to other bridge types and materials (AASHTO/NSBA, 2011). A brief treatment is also available in a chapter by Puckett and Coletti (2011) in the National Steel Bridge Alliance design manual.

REFERENCES

AASHTO (2010). *LRFD Bridge Design Specification*, 5th ed., American Association for State Highway and Transportation Officials, Washington, DC.

AASHTO/NSBA (2011). *Guidelines for the Analysis of Steel Bridges*, American Association for State Highway and Transportation Officials, Washington, DC and National Steel Bridge Alliance, Chicago.

AISC (2003). *Manual of Steel Construction—LFRD*, 3rd ed., American Institute for Steel Construction, Chicago.

Cheung, Y. K. (1976). *Finite Strip Method in Structural Analysis*, Pergamon, New York.

Finch, T. R. and J. A. Puckett (1992). *Transverse Load Distribution of Slab-Girder Bridges*, Report 92–9, No. 5, Mountain Plains Consortium, Fargo, ND.

Ghali, A., A. M. Neville, and T. G. Brown (2010). *Structural Analysis—A Unified Classical and Matrix Approach*, 6th ed., Chapman & Hall, New York.

Hambly, E. C. (1991). *Bridge Deck Behavior*, 2nd ed., E & FN SPON, Chapman & Hall, London.

Loo, Y. C. and A. R. Cusens (1978). *The Finite Strip Method in Bridge Engineering*, View Point, London.

Nowak, A. S. (1993). *Calibration of LRFD Bridge Design Code—NCHRP Project 12–33*, University of Michigan, Ann Arbor, MI.

Puckett, J. A. and D. Coletti (2011). *National Steel Bridge Alliance Steel Bridge Design Handbook*, Chapter 10: Structural Analysis, in press.

Puckett, J. A., X. Hou, M. Jablin, and D. R. Mertz (2007). *Simplified Live Load Distribution Factor Equations*, Report 592, National Cooperative Highway Research Program, Transportation Research Board, National Research Council, Washington, DC.

Timoshenko, S. P. and J. N. Goodier (1970). *Theory of Elasticity*, 3rd ed., McGraw-Hill, New York.

Timoshenko, S. and S. Woinowsky-Kreiger (1959). *Theory of Plates and Shells*, 2nd ed., McGraw-Hill, New York.

Ugural, A. C. (1981). *Stresses in Plates and Shells*, McGraw-Hill, New York.

Zokaie, T., T. A. Osterkamp, and R. A. Imbsen (1991). *Distribution of Wheel Loads on Highway Bridges, Final Report Project 12-26/1, National Cooperative Highway Research Program*, Transportation Research Board, National Research Council, Washington, DC.

PROBLEMS

These problems reference the plans for the bridge over the Little Laramie River. See Wiley's website for a pdf. This bridge is a single-span, four-girder bridge that has clean details; its interpretation is straightforward for the student. These plans are used with permission.

Typically, determine the shear at the end of the bridge (100) and moment at midspan (105). Your instructor may require other points of interest.

This bridge is used again for problems in the chapters on steel bridges, so the instructor can couple assignments here with future assignments on steel.

Note: The arithmetic can be simplified by using a 100-ft span rather than the 95′–6¾″ span bearing to bearing, which is the actual analysis span length.

11.1 Determine the dead loads per length of girder for the interior girders for the girder self weigh (add 10% for cross frames and miscellaneous), the slab weight, the wearing surface weight, and the barrier weight (equally distributed to all girders).

11.2 Repeat 11.1 for the exterior girder.

11.3 Determine the live-load actions for the design truck, design tandem, and design lane for one lane (not distributed).

11.4 Use the lever rule to determine the live-load distribution factors for the interior girder for one and multiple lanes loaded. Apply the appropriate multiple presence factor to each case.

11.5 Repeat Problem 11.4 for the exterior girder.

11.6 Use the live-load distribution factor equations [A4.6.2] to determine the distribution factors for an interior girder.

11.7 Repeat Problem 11.6 for an exterior girder.

11.8 Use the results from Problems 11.3 and 11.6 to determine the distributed live-load actions for the interior girder.

11.9 Use the results from Problems 11.3 and 11.7 to determine the distributed live-load actions for the exterior girder.

11.10 Use the results from Problems 11.1 and 11.8 to determine the total factored actions for the strength I limit state for the interior girder.

11.11 Use the results from Problems 11.2 and 11.9 to determine the total factored actions for the strength I limit state for the exterior girder.

11.12 Use the results from Problems 11.10 and 11.11 to determine the critical shear and moments, that is, compare the exterior and interior cases and use the maximum.

11.13 Do a web search for "BTBEAM online" and access this program. Use BTBEAM to work any and all of Problems 11.8–11.12. Use the distribution factors computed above for data into BTBEAM. The load combination factors, eta as may be set to unity for simplicity.

11.14 Use the cross section (dead-load and live-load distribution factors) for the bridge over the Little Laramie River for a three-span girder bridge. Span lengths are 100, 120, and 100 ft.

CHAPTER 12

System Analysis—Lateral, Temperature, Shrinkage, and Prestress Loads

12.1 LATERAL LOAD ANALYSIS

As with the gravity loads, the lateral loads must also be transmitted to the ground, that is, a load path must be provided. Lateral loads may be imposed from wind, water, ice floes, and seismic events. The load due to ice floes and water is principally a concern of the substructure designer. The system analysis for wind loads is discussed in Section 12.1.1, and the analysis for seismic loads is briefly introduced in Section 12.1.2.

12.1.1 Wind Loads

The wind pressure is determined by the provisions in AASHTO [A3.8], which are described in Chapter 8 (Wind Forces). This uniform pressure is applied to the superstructure as shown in Figure 12.1(a). The load is split between the upper and lower wind-resisting systems. If the deck and girders are composite or are adequately joined to support the wind forces, then the upper system is considered to be a diaphragm where the deck behaves as a very stiff beam being bent about the y–y axis as shown in Figure 12.1(a). Note that this is a common and reasonable assumption given that the moment of inertia of the deck about the y–y axis is quite large. Wind on the upper system can be considered transmitted to the bearings at the piers and the abutments via the diaphragm acting as a deep beam. It is traditional to distribute the wind load to the supporting elements on a tributary area basis [see Fig. 12.1(c)]. If there are no piers, or if the bearing supports at the piers do not offer lateral restraint, then all the diaphragm loads must be transmitted to abutment bearings, one-half to each. If the bearings restrain lateral movement and the pier support is flexible (e.g., particularly tall), then a refined model of the system might be warranted to properly account for the relative stiffness of the piers, the bearings, and the abutments. The in-plane deformation of the deck may usually be neglected.

The wind load to the lower system is carried by the girder in weak-axis bending [y–y in Fig. 12.1(b)]. Most of the girder's strength and stiffness in this direction are associated with the flanges. Typically, the bottom flange is assumed to carry the lower system load as shown in Figure 12.1(d). The bottom flange is usually supported by intermediate bracing provided by a cross frame [see Fig. 12.1(a)], steel diaphragm element (transverse beams), or in the case of a concrete beam a concrete diaphragm (transverse). These elements provide the compression flange bracing required for lateral torsional buckling while the concrete is being placed; compression (bottom) flange bracing in the negative moment region; the transverse elements also aid in the gravity load distribution, to a minor extent; and finally, the bracing periodically supports the bottom flange, which decreases the effective span length for the wind loading.

The bracing forces are illustrated in Figure 12.1(d) where the free-body diagram is shown with the associated approximate moment diagram. In AASHTO [A4.6.2.7] an approximate analysis is permitted, where the bracing receives load on the basis of its tributary length. The moment may also be approximated with $WL_B^2/10$ [C4.6.2.7.1]. In place of the approximate analysis, a more exact beam analysis may be performed, but this refinement is seldom warranted. Once the load is distributed into the bracing elements, it is transmitted into the deck by the cross frames or by frame action in the case of bracing. Once the load is distributed into the diaphragm, it combines with the upper system loads that are transmitted to the supports. If the deck is noncomposite or the deck–girder connection is not strong enough to transmit the load, then the path is assumed to be different. This case is described later.

At the supports, the load path is designed so that the load can be transmitted from the deck level into the bearings. The cross-framing system must be designed to resist these loads. The deck diaphragm loads may be uniformly distributed to the top of each girder. Note that the end supports receive the additional load from the bottom flange for the tributary length between the first interior bracing and the support. This load is shown as P_{end} in Figure 12.1(e).

In the case of insufficient diaphragm action (or connectors to the diaphragm), the upper system load must be transmitted into the girder in weak-axis bending. The load distribution mechanism is shown in Figure 12.1(f). *This is an important strength check for the construction stage prior to deck placement*. The girders translate together because they are coupled by the transverse elements. The cross frames (or diaphragms) are very stiff axially and may be considered rigid. This system may be modeled as a plane frame, or more simply, the load may be equally shared between all girders, and the load

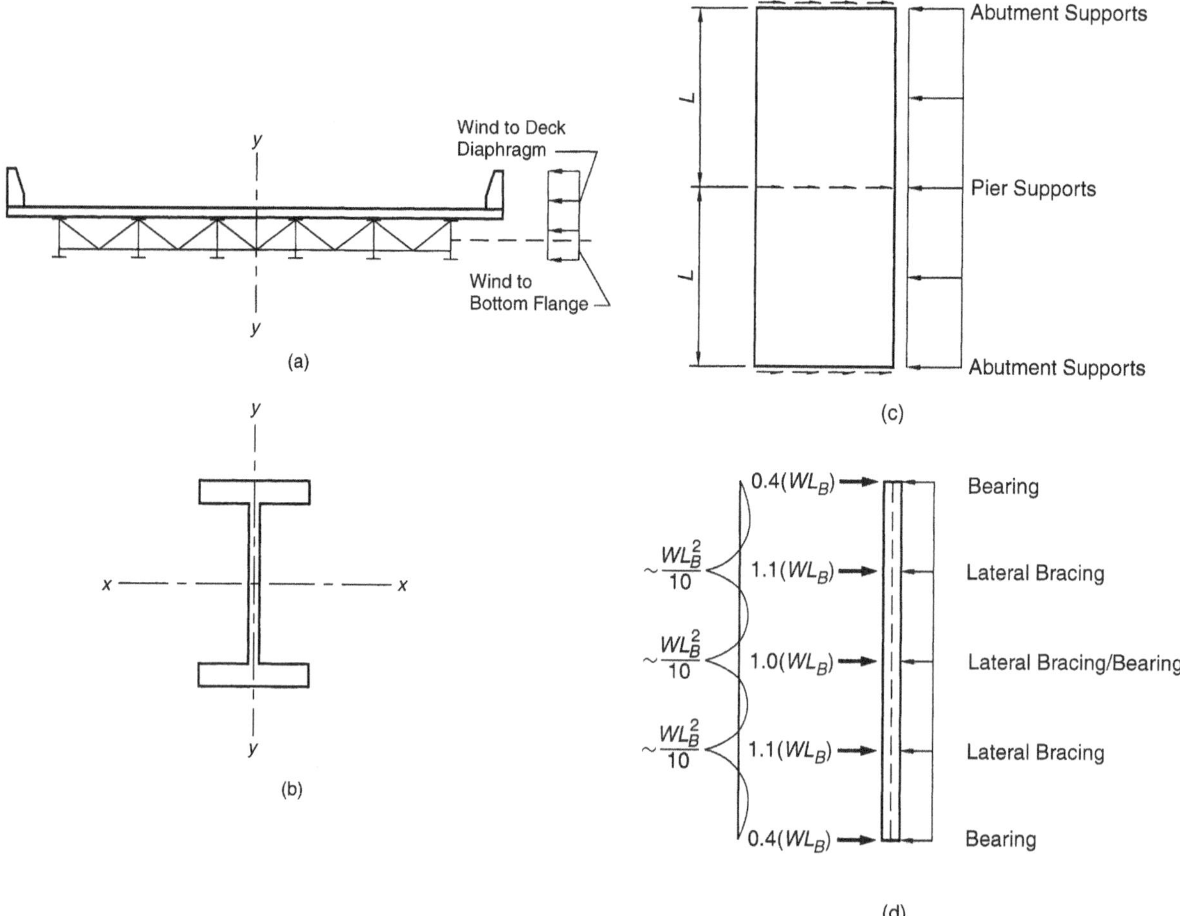

Fig. 12.1 (a) Bridge cross section with wind and (b) girder cross section. (c) Plan view and (d) load to bottom flange. (e) Load to end external bearing. (f) Girder flanges—load sharing, (g) moment diagram—exterior girder flange, (h) moment diagram—interior girder flange, and (i) moment diagram—interior girder flange, uniform load.

effect of the wind directly applied to the exterior girder may be superimposed.

The local and global effects are shown in Figures 12.1(g) and 12.1(h) and 12.1(i), respectively. For longer spans, the bracing is periodically spaced, and the global response is more like a beam subjected to a uniform load. Approximating the distribution of the load as uniform, permits the global analysis to be based on the analysis of a beam supported by bearing supports. The load is then split equally between all of the girders. Mathematically stated,

$$M_{\text{Total}} = M_{\text{Local}} + \frac{M_{\text{Global}}}{\text{No. of girders}}$$

This analysis is indeed approximate, but adequate ductility should be available and the lower bound theorem applies. The procedures described account for all the load and, therefore, equilibrium can be maintained. The procedures also parallel those outlined in the commentary in AASHTO [C4.6.2.7.1]. Similar procedures may be used for box systems.

The procedures outlined herein do not, nor do those of AASHTO, address long-span systems where the aeroelastic wind effects are expected.

Finally, the engineer should check the lateral deflection under some reasonable wind load to ensure that the system is not too flexible during construction. The wind speed in this case is left to the discretion of the engineer.

12.1.2 Seismic Load Analysis

The load path developed to resist lateral loads due to wind is the same load path followed by the seismic loads. The nature of the applied load is also similar. The wind loads acting on the superstructure are uniformly distributed along the length of the bridge and the seismic loads are proportional to the distributed mass of the superstructure along its length. What is different is the magnitude of the lateral loads and the dependence of the seismic loads on the period of vibration of various modes excited during an earthquake and the degree of inelastic deformation, which tends to limit the seismic forces.

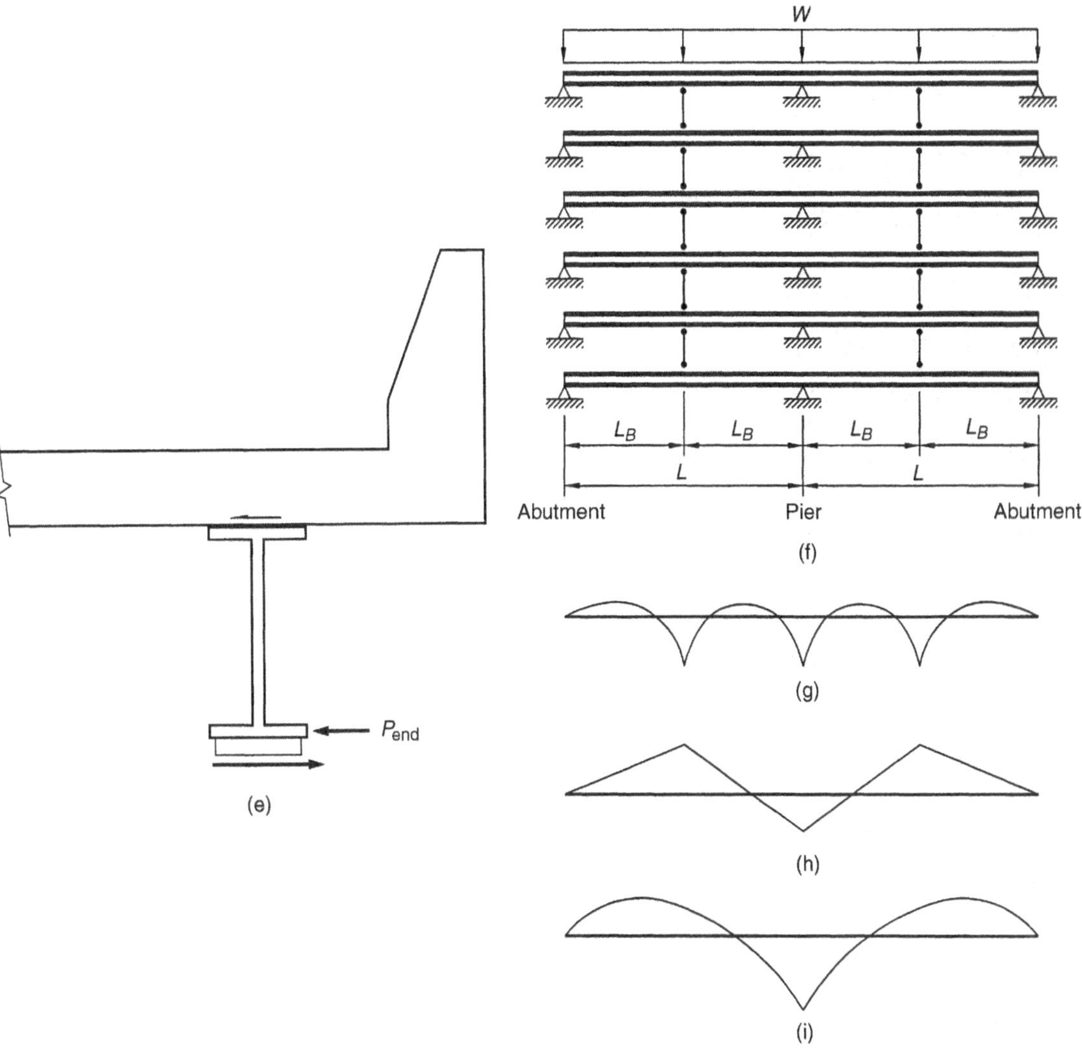

Fig. 12.1 (*Continued*).

Because of the need to resist lateral wind loads in all bridge systems, designers have already provided the components required to resist the seismic loads. In a typical superstructure cross section, the bridge deck and longitudinal girders are tied together with struts and bracing to form an integral unit acting as a horizontal diaphragm. The horizontal diaphragm action distributes the lateral loads to the restrained bearings in each of the segments of the superstructure. A segment may be a simply supported span or a portion of a multispan bridge that is continuous between deck joints.

During an earthquake, a segment is assumed to maintain its integrity, that is, the deck and girders move together as a unit. In a bridge with multiple segments, they often get out of phase with each other and may pound against one another if gaps in the joints are not large enough. In general, the deck and girders of the superstructure within a segment are not damaged during a seismic event, unless they are pulled off their supports at an abutment or internal hinge. Thus, an analysis for seismic loads must provide not only the connection force at the restrained supports, but also an estimate of the displacements at unrestrained supports. Procedures for determining these seismic forces and displacements are discussed next.

Minimum Analysis Requirements The analysis should be more rigorous for bridges with higher seismic risk and greater importance. Also, more rigorous seismic analysis is required if the geometry of a bridge is irregular. A regular bridge does not have changes in stiffness or mass that exceed 25% from one segment to another along its length. A horizontally curved bridge may be considered regular if the subtended angle at the center of curvature, from one abutment to another, is less than 60° and does not have an abrupt change in stiffness or mass. Minimum analysis requirements based on seismic zone, geometry, and importance are given in Table 12.1.

Table 12.1 Minimum Analysis Requirements for Seismic Effects

		Multispan Bridges					
		Other Bridges		Essential Bridges		Critical Bridges	
Seismic Zone	Single-Span Bridges	Regular	Irregular	Regular	Irregular	Regular	Irregular
1	None[a]	None	None	None	None	None	None
2	None	SM/UL[b]	SM	SM/UL	MM	MM	MM
3	None	SM/UL	MM[c]	MM	MM	MM	TH
4	None	SM/UL	MM	MM	MM	TH[d]	TH

[a]None = no seismic analysis is required.
[b]SM/UL = single-mode or uniform-load elastic method.
[c]MM = multimode elastic method.
[d]TH = time–history method.
In AASHTO Table 4.7.4.3.1-1. From *AASHTO LRFD Design Bridge Specification*.

Single-span bridges do not require a seismic analysis regardless of seismic zone. The minimum design connection force for a single-span bridge is the product of the acceleration coefficient and the tributary area as discussed in Chapter 8 (Minimum Seismic Design Connection Forces).

Bridges in seismic zone 1 do not require a seismic analysis. The minimum design connection force for these bridges is the tributary dead load multiplied by the coefficient given in Table 8.17.

Either a single-mode or a multimode spectral analysis is required for bridges in the other seismic zones depending on their geometry and importance classification. The single-mode method is based on the first or fundamental mode of vibration and is applied to both the longitudinal and transverse directions of the bridge. The multimode method includes the effects of all modes equal in number to at least three times the number of spans in the model [A4.7.4.3.3].

A time–history analysis is required for critical bridges in seismic zones 3 and 4. This analysis involves a step-by-step integration of the equations of motion for a bridge structure when it is subjected to ground accelerations that vary with time. Historical records of the variation in acceleration, velocity, and displacement due to ground shaking are referred to as time histories, hence the name for the analysis method. Careful attention must be paid to the modeling of the structure and the selection of the time step used. If elastic material properties are used, the R factors of Tables 8.15 and 8.16 apply to the substructures and connection forces, respectively.

If inelastic material properties are used, all of the R factors are 1.0 because the inelastic analysis accounts for the energy dissipation and redistribution of seismic forces and no further modification is needed. Oftentimes, when selecting a time–history for a specific bridge site, a historical record may not be available for the soil profile that is present. In this case, artificial time–histories are generated that include the magnitude, frequency content, and duration of the ground shaking anticipated at the bridge site. A time–history analysis requires considerable skill and judgment, and an analyst experienced in inelastic, dynamic, numerical analysis should be consulted.

Elastic Seismic Response Spectrum Both of the spectral methods of analysis require that a seismic response spectrum be given for the bridge site. The purpose of a response spectrum analysis is to change a problem in dynamics into an equivalent problem in statics. The key to this method of analysis is to construct an appropriate response spectrum for a particular soil profile. A response spectrum can be defined as a graphical representation of the maximum response of single-degree-of-freedom elastic systems to earthquake ground motions versus the periods or frequencies of the system (Imbsen, 1981). The response spectrum is actually a summary of a whole series of time–history analysis.

A response spectrum is generated by completing the steps illustrated in Figure 12.2. The single-degree-of-freedom (SDOF) system is shown as an inverted pendulum oscillator that could represent the lumped mass of a bridge superstructure supported on columns or piers. The undamped natural period of vibration(s) of the SDOF system is given by

$$T = 2\pi\sqrt{\frac{m}{k}} \tag{12.1}$$

where

m = mass of the system = W/g
W = contributing dead load for the structure (kips)
g = acceleration of gravity = 386.4 in./s^2
k = stiffness of structure supporting the mass (kips/in.)

When damping is introduced into the system, it is usually expressed as a ratio of critical damping given by

$$\xi = \frac{c}{c_c} \tag{12.2}$$

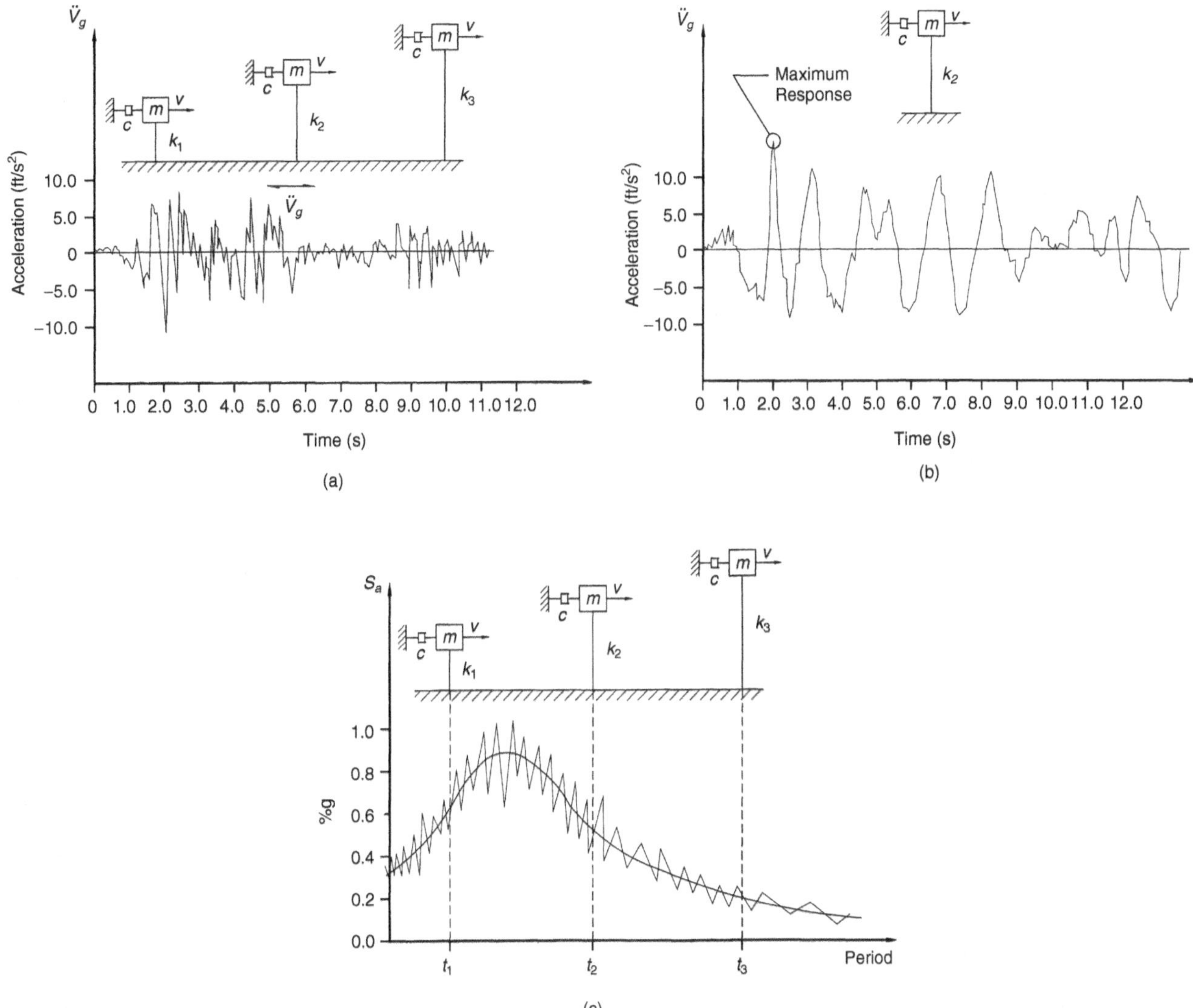

Fig. 12.2 (a) Time–history record of ground acceleration applied to a damped single-degree-of-freedom system. (b) Time–history record of structure response and (c) maximum responses of single-degree-of-freedom systems.(from Imbsen, 1981).

where

c = coefficient of viscous damping

c_c = critical damping coefficient, minimum amount of damping required to prevent a structure from oscillating = $4\pi m/T$

It can be shown that the period of vibration of a damped structure T_D is related to the undamped period T by

$$T_D = \frac{T}{\sqrt{1-\xi^2}} \tag{12.3}$$

where ξ is the ratio of actual damping to critical damping, Eq. 12.2.

In an actual structure, damping due to internal friction of the material and relative moment at connections seldom exceeds 20% of critical damping. Substitution of this value into Eq. 12.3 increases the undamped period of vibration by only 2%. In practice, this difference is neglected and the damped period of vibration is assumed to be equal to the undamped period.

In Figure 12.2(a), a time–history record of ground acceleration is shown below three SDOF systems with identical mass and damping, but with different structural stiffness to give three different periods of vibration. The acceleration response from an elastic step-by-step time–history analysis of the second SDOF system is shown in Figure 12.2(b), and the maximum response is indicated. This procedure is repeated for a large number of SDOF systems with different stiffnesses until the maximum response is determined for a whole spectrum of periods of vibration. A response spectrum for acceleration is plotted in Figure 12.2(c) and gives a graphical representation of the variation of maximum response with period of vibration. For design purposes, a curve is drawn through the average of the maximum response to give the elastic response spectrum shown by the smooth line in Figure 12.2(c). This response spectrum was developed for a single earthquake under one set of soil conditions. When a response spectrum is used for design purposes, it is usually based on more than one earthquake and includes the effects of different soil conditions.

Seismic Design Response Spectra It is well known that local geologic and soil conditions influence the intensity of ground shaking and the potential for damage during an earthquake. The 1985 earthquake in Mexico is an example of how the soils overlying a rock formation can modify the rock motions dramatically. The epicenter of the earthquake was near the west coast of Mexico, not far from Acapulco. However, most of the damage was done some distance away in Mexico City. The difference in the ground shaking at the two sites was directly attributable to their soil profile. Acapulco is quite rocky with thin overlying soil, while Mexico City is sitting on an old lake bed overlain with deep alluvial deposits. When the earthquake struck, Acapulco took a few hard shots of short duration that caused only moderate damage. In contrast, the alluvial deposits under Mexico City shook like a bowl of gelatin for some time and extensive damage occurred. The response spectra characterizing these two sites are obviously different, and these differences due to soil conditions are recognized in the response spectra developed for the seismic analysis of bridges.

To describe the characteristics of response spectra for different soil profiles statistical studies of a number of accelerometer records have been conducted. These studies defined soil profiles similar to those in Table 8.13 as a reasonable way to differentiate the characteristics of surface response. For each of the accelerometer records within a particular soil profile, an elastic response spectrum was developed as previously illustrated in Figure 12.2. An average response spectrum was obtained from the individual response system at different sites with the same soil profile but subjected to different earthquakes. This procedure was repeated for the four soil profiles that had been defined. The results of the study by Seed et al. (1976), which included the analysis of over 100 accelerometer records, are given in Figure 12.3. The elastic response spectra in this figure were developed for 5% of critical damping, and the accelerations have been normalized with respect to the maximum ground acceleration.

The shape of the average spectra in Figure 12.3 first ascends, levels off, and then decreases as the period of vibration increases. As the soil profile becomes more flexible, the period at which decay begins is delayed, so that at larger periods the softer soils have larger accelerations than the stiffer soils. These variations in acceleration with period and soil type are expressed in AASHTO [A3.10.6] by an elastic seismic response coefficient C_{sm} defined as

$$C_{sm} = \frac{1.2AS}{T_m^{2/3}} \leq 2.5A \qquad (12.4)$$

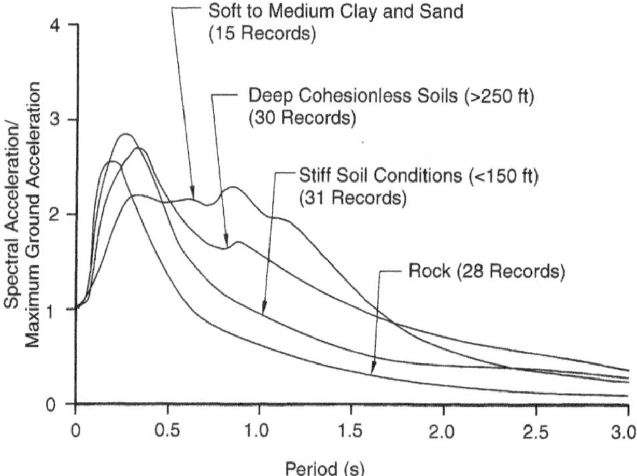

Fig. 12.3 Average acceleration spectra for different site conditions. Normalized with respect to maximum ground acceleration. (From Seed et al., 1976.)

where

T_m = period of vibration of mth mode (s)
A = acceleration coefficient from Figure 8.23
S = site coefficient from Table 8.14

The seismic response coefficient is a modified acceleration coefficient that is multiplied times the effective weight of the structure to obtain an equivalent lateral force to be applied to the structure. Because C_{sm} is based on an elastic response, the member forces resisting the equivalent lateral force used in design are divided by the appropriate R factors given in Tables 8.15 and 8.16.

The shape of the seismic response spectra defined by Eq. 12.4 does not have an ascending branch but simply levels off at $2.5A$. This characteristic can be seen in Figure 12.4 in the plot of C_{sm}, normalized with respect to the acceleration coefficient A, for different soil profiles versus the period of vibration. For soil profile types III and IV, the maximum value of $2.5A$ is overly conservative, as can be seen in Figure 12.3, so that in areas where $A \geq 0.10$, C_{sm} need not exceed $2.0A$ (Fig. 12.4). Also, for soil profiles III and IV, for short periods an ascending branch is defined as (for modes other than the first mode) [A3.10.6.2]

$$C_{sm} = A\left(0.8 + 4.0T_m\right) \leq 2.0A \qquad (12.5)$$

For intermediate periods, $0.3 \text{ s} \leq T_m \leq 4.0$ s, a characteristic of earthquake response spectra is that the average velocity spectrum for larger earthquakes of magnitudes 6.5 or greater is approximately horizontal. This characteristic implies that C_{sm} should decrease as $1/T_m$. However, because of the concern for high ductility requirements in bridges with longer periods, it was decided to reduce C_{sm} at a slower rate of $1/T_m^{2/3}$. For bridges with long periods ($T_m > 4.0$ s), the average displacement spectrum of large earthquakes becomes

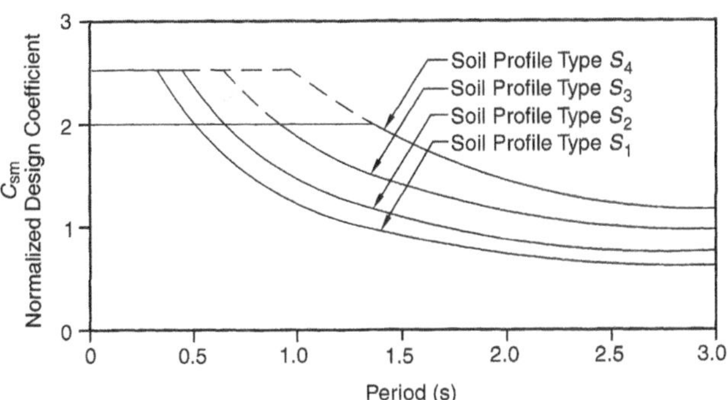

Fig. 12.4 Design response spectra for various soil profiles. Normalized with respect to acceleration coefficient A. [AASHTO Fig. C3.10.6.1-1]. (From *AASHTO LRFD Bridge Design Specifications*, Copyright © 2010 by the American Association of State Highway and Transportation Officials, Washington, DC. Used by permission.)

horizontal. This implies that C_{sm} should decay as $1/T_m^2$ and Eq. 12.4 becomes [A3.10.6.2]

$$C_{\text{sm}} = \frac{3AS}{T_m^{4/3}} \qquad T_m \geq 4.0 \text{ s} \qquad (12.6)$$

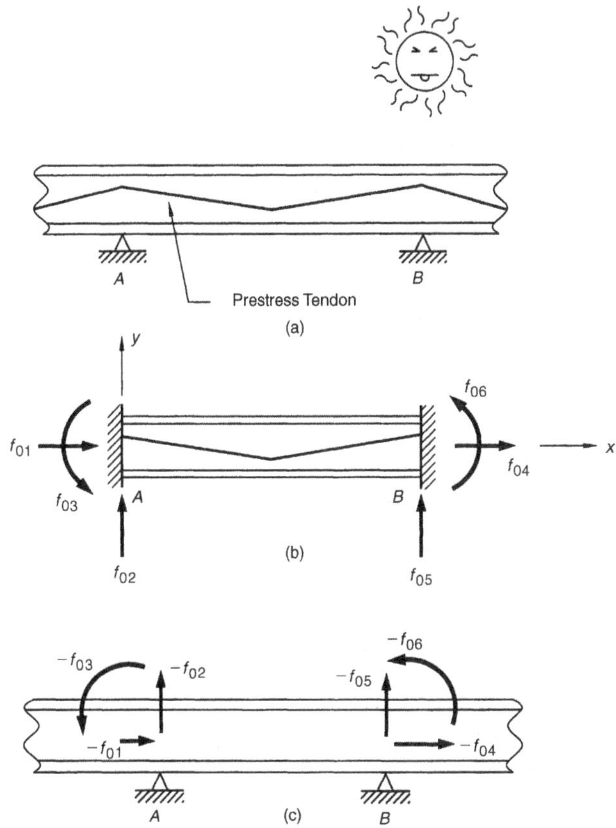

Fig. 12.5 (a) Girder subjected to prestrain, (b) restrained system subjected to prestrain, and (c) equivalent joint loads.

12.2 TEMPERATURE, SHRINKAGE, AND PRESTRESS

12.2.1 General

The effects of temperature, shrinkage, and prestress are treated in a similar manner. All of these create a state where the structure is prestrained prior to the application of gravity and/or lateral loads. The effects of these loads will likely not exceed any strength limit state, but these loads can certainly be of concern regarding serviceability. As discussed earlier in this chapter, for ductile systems, prestrains and prestress are eliminated when the ultimate limit states are reached (see Section 10.2.2). Therefore, the combination of these is typically not of concern for the ultimate limit state. They may be of concern regarding the determination of the load that creates first yield, deformation of the structure for the design of bearings and joints, and other service-level phenomena.

Because the stiffness-based methods are frequently used in the analysis of bridge systems, the discussion of prestraining is limited to these methods. The finite-element analysis of the system subjected to prestrain is similar to the stiffness method, and most finite-element textbooks address this issue. The force or flexibility method is a viable technique but is seldom used in contemporary computer codes and, therefore, is not addressed.

The general procedure for the analysis of frame elements subjected to prestraining effect is illustrated in Figure 12.5. The structure subjected to the effects of temperature and prestressing forces is shown in Figure 12.5(a). Each element in the system may be separated from the structure as illustrated in Figure 12.5(b). Here the joints are locked against rotation and translation with restraining actions located at the end of the element. The opposite (negative) of the restraining actions are applied on the joint-loaded structure, and the analysis proceeds in the same manner as with any other joint-load effect.

The displacements from the joint-loaded system [Fig. 12.5(c)] are the displacements in the entire prestrained system. However, the actions from the joint-loaded system must be superposed with the actions from the restrained system to achieve the actions in the entire system. Again, the main difference between the analysis for effects of prestrain/prestress and the analysis for directly applied load is the analysis of the restrained system. In the sections that follow, the effects of prestressing forces and temperature effects are discussed. The effects of shrinkage may be determined similar to temperature effects and is not explicitly included.

12.2.2 Prestressing

The element loads for various tendon paths are tabulated and available to aid in determining equivalent element loads that are subsequently used in establishing the restraining actions. The equivalent element loads for several commonly used tendon configurations are illustrated and are found in textbooks on advanced analysis; for example, see Ghali et al. (2010). These loads may be applied just as one would apply any other loads.

12.2.3 Temperature Effects

Most bridges experience daily and seasonal temperature variations causing material to shorten with decreased temperatures and lengthen with increased temperatures. It has been observed that these temperature fluctuations can be separated into two components: a uniform change and a gradient. The uniform change is the effect due to the entire bridge changing temperature by the same amount. The temperature gradient is created when the top portion of the bridge gains more heat due to direct radiation than the bottom. Because the strains are proportional to the temperature change, a nonuniform temperature strain is introduced. In this section, the axial strain and curvature formulas are presented for the effect of temperature. These formulas are given in discrete form. The formulas may be implemented in stiffness and flexibility formulations. An example is presented to illustrate the usage of the formulas.

Aashto Temperature Specifications Uniform temperature increases and temperature gradients are outlined in AASHTO [A3.12.2 and A3.12.3]. The uniform change is prescribed is Table 8.21 and the temperature gradient is defined in Table 8.27 and Figures 8.23 and 8.24. The temperature change creates a strain of

$$\varepsilon = \alpha\,(\Delta T)$$

where α is the coefficient of thermal expansion and ΔT is the temperature change.

This strain may be used to determine a change in length by the familiar equation

$$\Delta L = \varepsilon L = \alpha\,(\Delta T)\,L$$

where L is the length of the component, and the stress in a constrained system of

$$\sigma = \alpha\,(\Delta T)\,E$$

The response of a structure to the AASHTO multilinear temperature gradient is more complex than its uniform counterpart and can be divided into two effects: (1) gradient-induced axial strain and (2) gradient-induced curvature. The axial strain is described first.

Temperature-Gradient-Induced Axial Strain The axial strain ε due to the temperature gradient is (Ghali et al., 2010)

$$\varepsilon = \frac{\alpha}{A}\int T\,(y)\;dA \tag{12.7}$$

where

α = coefficient of thermal expansion
$T(y)$ = gradient temperature as shown in Figure 8.30
y = distance from the neutral axis
dA = differential cross-sectional area

The integration is over the entire cross section. If the coefficient of thermal expansion is the same for all cross-section materials, standard transformed section analysis may be used to establish the cross-sectional properties. For practical purposes, steel and concrete may be assumed to have the same expansion coefficients.

By discretization of the cross section into elements, Eq. 12.7 simplifies this to a discrete summation. Consider the single element shown in Figure 12.6. Although the element shown is rectangular, the element shape is arbitrary. The element's elastic centroidal axis is located at a distance $\bar{y}_i$, and y is the location of the differential area element dA. The area of the element and second moment of area are denoted by A_i and I_i, respectively. Note that $y_i = y - \bar{y}_i$. The temperature at location y is

$$T\,(y) = T_{\mathrm{ai}} + \frac{\Delta T_i}{d_i} y_i = T_{\mathrm{ai}} + \frac{\Delta T_i}{d_i}\left(y - \bar{y}_i\right) \tag{12.8}$$

where T_{ai} is the temperature at the element centroid, ΔT_i is the temperature difference from the bottom of the element to the top, and d_i is the depth of the element.

Substitution of Eq. 12.8 into Eq. 12.7 yields

$$\varepsilon = \frac{\alpha}{A}\sum\int\left[T_{\mathrm{ai}} + \frac{\Delta T_i}{d_i}\left(y - \bar{y}_i\right)\right]\,dA_i \tag{12.9}$$

where the summation is over all elements in the cross section and the integration is over the domain of the discrete element. Integration of each term in Eq. 12.9 yields

$$\varepsilon = \frac{\alpha}{A}\sum\left[T_{\mathrm{ai}}\int dA_i + \frac{\Delta T_i}{d_i}\int y\,dA_i - \frac{\Delta T_i \bar{y}_i}{d_i}\int dA_i\right] \tag{12.10}$$

Substitution of $A_i = \int dA_i$ and $\bar{y}_i A_i = \int y\,dA_i$ in Eq. 12 yields

$$\varepsilon = \frac{\alpha}{A}\sum\left[T_{\mathrm{ai}} A_i + \frac{\Delta T_i}{d_i}\bar{y}_i A_i - \frac{\Delta T_i}{d_i}\bar{y}_i A_i\right] \tag{12.11}$$

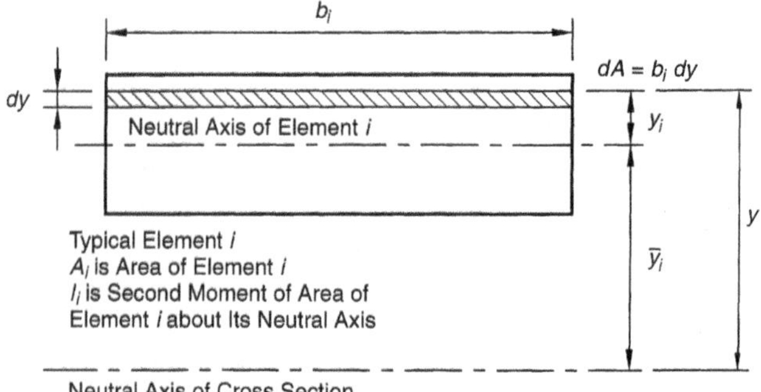

Fig. 12.6 Cross section with discrete element. Example cross section.

Note the second and third terms sum to zero, and Eq. 12.11 simplifies to

$$\varepsilon = \frac{\alpha}{A}\sum T_{\text{ai}} A_i \tag{12.12}$$

Equation 12.12 is the discrete form of Eq. 12.7, which is given in AASHTO [A4.6.6]. Note that only areas of the cross section with gradient temperature contribute to the summation.

Temperature-Gradient-Induced Curvature Temperature-induced curvature is the second deformation that must be considered. The curvature ψ due to the gradient temperature is (Ghali et al., 2010)

$$\psi = \frac{\alpha}{I}\int T(y)\, y\, dA \tag{12.13}$$

where I is the second moment of area of the entire cross section about the elastic centroidal axis.

Substitution of Eq. 12.8 into Eq. 12.13 and expansion yield

$$\psi = \frac{\alpha}{I}\sum \left[T_{\text{ai}} \int y_i\, dA_i + \frac{\Delta T_i}{d_i}\int (y^2 - \bar{y}_i y)\, dA_i \right] \tag{12.14}$$

where the summation is over all elements in the cross section and the integration is over the domain of the discrete element. Performing the required integration in Eq. 12 yields

$$\psi = \frac{\alpha}{I}\sum \left[T_{\text{ai}}\bar{y}_i A_i + \frac{\Delta T_i}{d_i} I_i - \frac{\Delta T_i}{d_i}\bar{y}_i^2 A_i \right] \tag{12.15}$$

The parallel axis theorem is used to relate the cross-section properties in Eq. 12.15, that is,

$$I_i = \bar{I}_i + \bar{y}_i^2 A_i \tag{12.16}$$

which can be rearranged as

$$\bar{I}_i = I_i - \bar{y}_i^2 A_i \tag{12.17}$$

The combination of Eq. 12.17 with Eq. 12.15 yields

$$\psi = \frac{\alpha}{I}\sum \left[T_{\text{ai}}\bar{y}_i A_i + \frac{\Delta T_i}{d_i}\bar{I}_i \right] \tag{12.18}$$

which is the discrete form of the integral equation given in Eq. 12.13 and in AASHTO [C4.6.6].

Using Strain and Curvature Formulas The axial strain and curvature may be used in both flexibility and stiffness formulations for frame elements. In the former, ε may be used in place of *P/AE*, and ψ may be used in place of *M/EI* in traditional displacement calculations. The flexibility method requires the analysis of the released statically determine and stable system. The analysis of the released system is conceptually straightforward, but this is not the case for the multilinear temperature distribution. Although the distribution does not create external reactions, it does create internal self-equilibriating stresses. These stresses must be superimposed with actions created from the redundants. The complete details are presented elsewhere (Ghali et al., 2010).

In the stiffness method, the fixed-end actions for a prismatic frame element may be calculated as

$$N = EA\varepsilon \tag{12.19}$$

$$M = EI\psi \tag{12.20}$$

where N is the axial force (constant with respect to length), M is the flexural bending moment (again constant), and E is Young's modulus.

These actions may be used to determine the equivalent joint loads in the usual manner, and the resulting displacements may be used to recover the actions in the joint-loaded systems. These actions must be superimposed with the actions (usually stresses in this case) in the restrained (fixed) system. The temperature-related stresses in the restrained system, *not the fixed-end actions* in Eqs. 12.19 and 12.20, must be used because the temperature gradient is not constant across the section. This complication forces algorithm modifications in stiffness codes because the fixed-end actions do not superimpose directly with the actions of the joint-loaded structure in the usual manner. The following example provides guidance on this issue.

Example 12.1 The transformed composite cross section shown in Figure 12.7(a) is subjected to the temperature

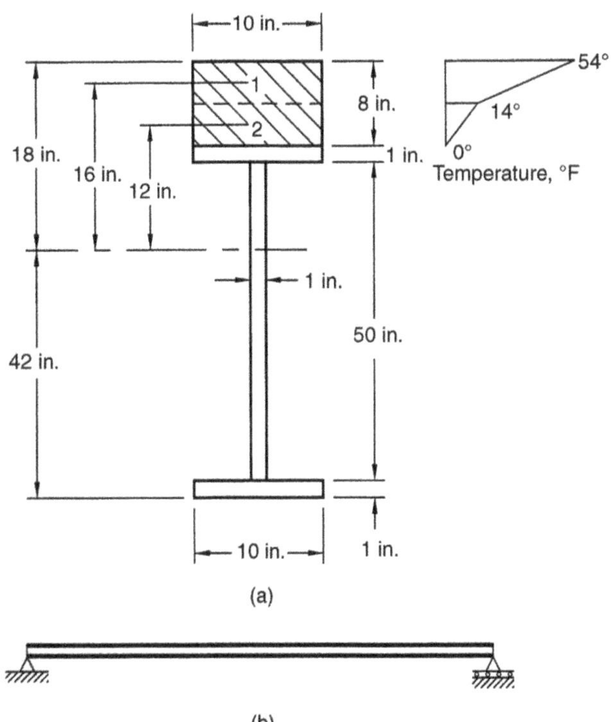

Fig. 12.7 (a) Temperature stress distribution and (b) simple span girder.

gradients associated with zone 1 (Table 8.22) for a plain concrete surface. This temperature variation is also shown in Figure 12.7(a). The dimensions of the cross section were selected for ease of computation and illustration rather than based on typical proportions. A modular ratio of 8 is used. The cross section is used in the simple beam shown in Figure 12.7(b).

The beam subjected to the temperature change has been divided into two sections labeled 1 and 2 in Figure 12.7(a).

The cross sections and material properties are listed in Table 12.2 with reference to the labeled sections. All other areas of the cross section do not have a temperature gradient and therefore are not included in the summations. These section and material properties are used to calculate the axial strain, curvature, fixed-end axial force, and flexural bending moment.

By using Eqs. 12.12 and 12.19, the gradient-induced axial strain and fixed-end axial force are

$$\varepsilon = (6.5 \times 10^{-6}/80)[34(40) + 7(40)] = 133 \times 10^{-6}$$
$$N = 29,000(80) \quad (133 \times 10^{-6}) = 309 \text{ kips}$$

Table 12.2 Cross-section Properties

Properties	Reference Section 1	Reference Section 2	Total Section[a]
A_i (in.^{2})	40	40	80
$\bar{y}_i$ (in.)	16	12	42
$\bar{I}_i$ (in.^{4})	53.3	53.3	57,450
α	6.5×10^{-6}	6.5×10^{-6}	6.5×10^{-6}
E (ksi)	29,000 ($n = 8$)	29,000 ($n = 8$)	29,000
T_{ai} (°F)	34	7	N/A
ΔT_i (°F)	40	14	N/A

[a]Not applicable = N/A.

By using Eqs. 12.18 and 12.20, the curvature and fixed-end flexural moments are calculated below.

$$\psi = (6.5 \times 10^{-6}/57,450)[34(16)(40) + (40/4)(53.3) + 7(12)(40) + (14/4)(53.3)] = 2.9 \times 10^{-6} \text{ in.}^{-1}$$
$$M = 29,000(57,450)(2.9 \times 10^{-6}) = 4853 \text{ in. kips} = 404.4 \text{ ft kips}$$

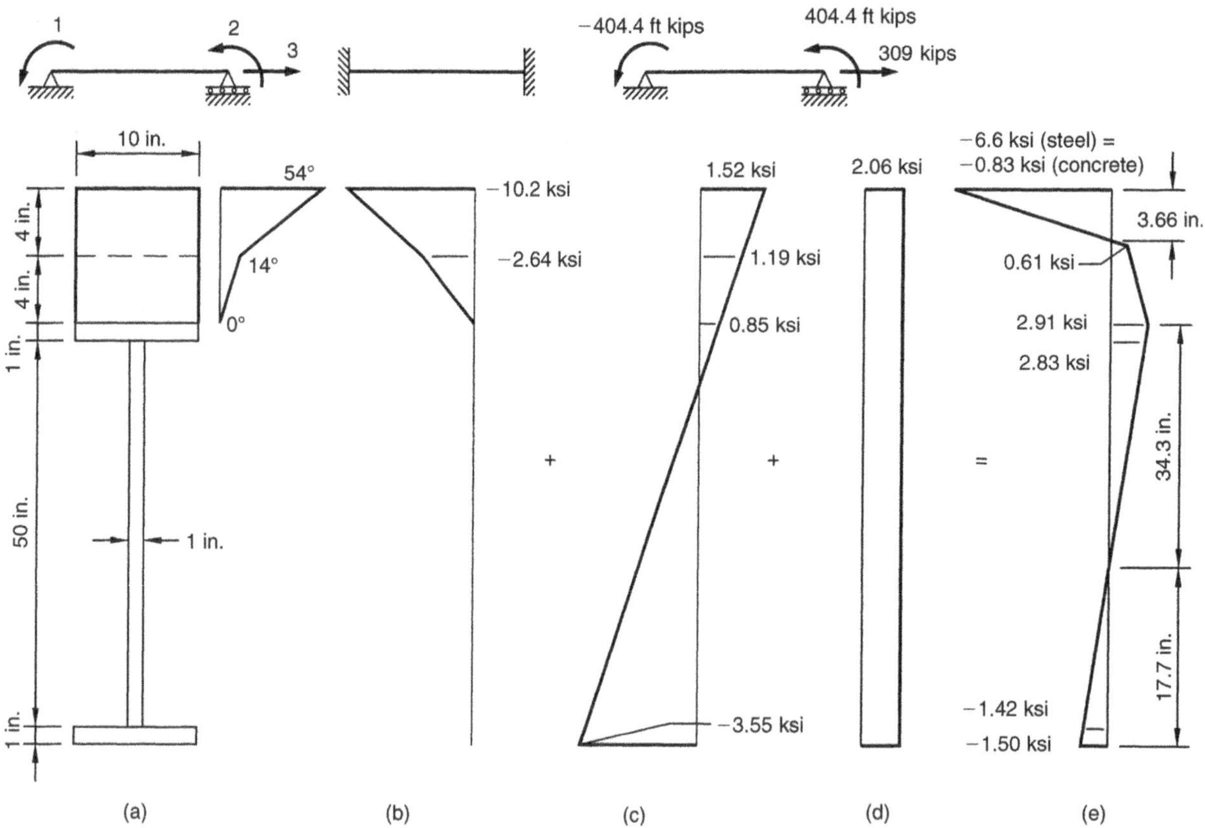

Fig. 12.8 (a) Cross section, (b) restrained system, (c) unrestrained system (bending), (d) unrestrained system (axial), and (e) total stress.

The restrained temperature stresses are determined by $\sigma = \alpha(\Delta T)E$. The restrained stress in the top of the section is

$$\sigma = 6.5 \times 10^{-6} (54) (29\ 000) = 10.2 \text{ ksi}$$

and the stress in the bottom is zero because ΔT is zero as shown in Figure 12.8(b). The moment is constant in the joint-loaded system and the associated flexural stress distribution is shown in Figure 12.8(c). The axial force in the unrestrained (joint-loaded) system is also constant and the associated axial stress is shown in Figure 12.8(d). Superimposing the stresses in the restrained and joint-loaded systems gives the stress distribution shown in Figure 12.8(e).

Note, the external support reactions are zero, but the internal stresses are not. The net internal force in each section is zero, that is, there is no axial force nor flexural moment. The equivalent joint loads shown at the top of Figure 12.8(c) are used to calculate the displacements (0.00175 rad, −0.00175 rad, 0.087 in.) referenced in Figure 12.8(a).

Summary of Temperature Effects The AASHTO (2010) LRFD Bridge Design Specification requires that a prescribed temperature gradient be used to model temperature effects in girders. The prescribed multilinear gradients are used to develop a method involving discrete summations that are used to determine the axial strain and curvature. These formulas may be implemented into stiffness and flexibility programs. An example is used to illustrate the usage of the formulas presented. The temperature load case is used for the design of joints and bearings. For strength considerations, the temperature load cases are optional and are sometimes of concern with deep box-girder systems.

12.2.4 Shrinkage and Creep

Concrete creep and shrinkage are difficult to separate as these two effects occur simultaneously in the structural system. The load effect may be estimated by an analysis similar to the procedure previously described for temperature effects. The temperature strain $\alpha(\Delta T)$ may be replaced with a shrinkage or creep strain. The strain profile is obviously different than that produced by a temperature gradient, but the appropriate strain profile may be used in a similar manner.

12.3 CLOSING REMARKS

The analysis and design for lateral loads typically requires a system analysis. For some bridges, the lateral load effects are relatively small, and most of the attention is on the substructure. Such cases are demonstrated in several examples provided in later chapters. In other cases, lateral loads dominate the design considerations, e.g., bridges in high seismic regions. Complete discussion of the analysis and design in high seismic regions must be deferred to other references and is beyond the scope of this book.

REFERENCES

AASHTO (2010). *LRFD Bridge Design Specification*, 5th ed., American Association for State Highway and Transportation Officials, Washington, DC.

Ghali, A., A. M. Neville, and T. K. Brown, (2010). *Structural Analysis—A Unified Classical and Matrix Approach*, 6th ed., Chapman & Hall, New York.

Imbsen, R. A. (1981). *Seismic Design of Highway Bridges*, Report No. FHWA-IP-81-2, Federal Highway Administration, Washington, DC.

Seed, H. B., C. Ugas, and J. Lysmer (1976). "Site Dependent Spectra for Earthquake Resistant Design," *Bulletin of the Seismological Society of America*, Vol. 66, No. 1, February, pp. 221–244.

PART III

Concrete Bridges

CHAPTER 13

Reinforced Concrete Material Response and Properties

13.1 INTRODUCTION

Concrete is a versatile building material. It can be shaped to conform to almost any alignment and profile. Bridge superstructures built of reinforced and prestressed concrete can be unique one-of-a-kind structures formed and constructed at the job site, or they can be look-alike precast girders and box beams manufactured in a nearby plant. The raw materials of concrete—cement, fine aggregate, coarse aggregate, and water—are found in most areas of the world. In many countries without a well-developed steel industry, reinforced concrete is naturally the preferred building material. However, even in North America with its highly developed steel industry, bridges built of concrete are very competitive.

Concrete bridges can be designed to satisfy almost any geometric alignment from straight to curved to doubly curved as long as the clear spans are not too large. Cast-in-place (CIP) concrete box girders are especially suited to curved alignment because of their superior torsional resistance and the ability to keep the cross section constant as it follows the curves. With the use of posttensioning, clear spans of 150 ft (45 m) are common. When the alignment is relatively straight, precast prestressed girders can be utilized for multispan bridges, especially if continuity is developed for live load. For relatively short spans, say less than 40 ft (12 m), flat slab bridges are often economical. Cast-in-place girders monolithic with the deck slab (T-beams) can be used for clear spans up to about 65 ft (20 m), longer if continuity exists. Some designers do not like the underside appearance of the multiple ribs, but if the bridge is over a small waterway rather than a traveled roadway, there is less objection.

For smaller spans, CIP and precast culverts are a mainstay. Approximately one-sixth (100,000) culverts with bridge spans greater than 20 ft (6 m) are contained within the U.S. bridge inventory. Culverts perform extremely well, exhibit few service problems, and are economical because the foundation requirements are minimal.

Cast-in-place concrete bridges may not be the first choice if speed of construction is of primary importance. Also, if formwork cannot be suitably supported, such as in a congested urban setting where traffic must be maintained, the design of special falsework to provide a construction platform may be a disadvantage.

Longer span concrete bridges have been built using match-cast and cable-supported segmental construction. These structural systems require analysis and construction techniques that are relatively sophisticated and are beyond the scope of this book. In this chapter, short- to medium-span bridges constructed of reinforced and prestressed concrete are discussed.

After a review of the behavior of the materials in concrete bridges in this chapter, the resistance of cross sections to bending and shear is presented in Chapter 14. A relatively detailed discussion of these two topics is given because of the introduction in the AASHTO (1994) LRFD Bridge Specification of a *unified* flexural theory for reinforced and prestressed concrete beams and the modified compression field theory for shear resistance. In the development of the behavior models, the sign convention adopted for strains and stresses is that tensile values are positive and compressive values are negative. This sign convention results in stress–strain curves for concrete that are drawn primarily in the third quadrant instead of the familiar first quadrant.

It is not necessary to go through each detailed step of the material response discussion. The information is given so that a reader can trace the development of the provisions in the specification. In the chapters that follow, a number of example problems are given to illustrate the application of the resistance equations that are derived. A concrete bridge deck with a barrier wall is designed followed by design examples of a solid slab, a T-beam, and a prestressed beam bridge.

13.2 REINFORCED AND PRESTRESSED CONCRETE MATERIAL RESPONSE

To predict the response of a structural element subjected to applied forces, three basic relationships must be established: (1) equilibrium of forces, (2) compatibility of strains, and (3) constitutive laws representing the stress–strain behavior of the materials in the element. For a two-dimensional (2D) element without torsion that is subjected to bending by transversely applied forces, there are three equilibrium equations between the applied external forces and the three internal resisting forces: moment, shear, and axial load. When the external forces are applied, the cross section deforms and internal longitudinal, transverse, and shear strains are developed. These internal strains must be compatible. Longitudinal strains throughout the depth of a section are related to one another through the familiar assumption that plane sections

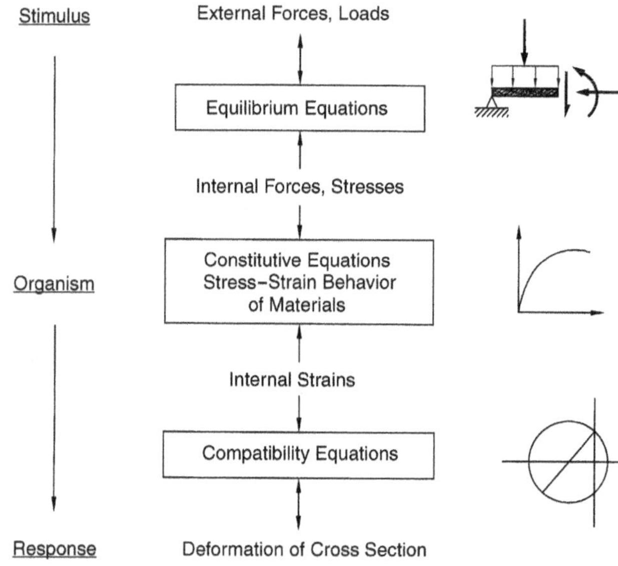

Fig. 13.1 Interrelationship between equilibrium, material behavior, and compatibility.

before bending remain plane sections after bending. The longitudinal strains are related to the transverse, shear, and principal strains through the relationships described in Mohr's circle of strain. The stress–strain relationships provide the key link between the internal forces (which are integrations over an area of the stresses) and the deformations of the cross section. These interrelationships are shown schematically in Figure 13.1 and are described in more specific terms in the sections that follow.

On the left of Figure 13.1 is a simple model used in psychology to illustrate that the manner in which individuals or groups respond to certain stimuli depends on their psychological makeup. In individuals, we often speak of one's constitution; in groups, the response depends on the constituents; in concrete, it depends on constitutive laws. The analogy to concrete elements may be imperfect, but the point is that knowledge of the behavior of the material is essential to predicting the concrete response of the element to external loads.

Another point in regard to the relationships in Figure 13.1 is that they involve both deductive and inductive reasoning. The equilibrium and compatibility equations are deductive in that they are based on general principles of physics and mechanics that are applied to specific cases. If the equations are properly written, then they lead to a set of unique correct answers. On the other hand, the constitutive equations are inductive as they are based on specific observations from which expressions are written to represent general behavior. If the trends exhibited by the data are not correctly interpreted or an important parameter is overlooked, the predicted response cannot be verified by experimental tests. As more experimental data become available, the constitutive equations change and the predicted response improves. The AASHTO (2010) Bridge Specification incorporates the current state of practice regarding material response; however, one should expect that changes may occur in the constitutive equations in the future as additional test data and/or new materials become available.

13.3 CONSTITUENTS OF FRESH CONCRETE

Concrete is a conglomerate artificial stone. It is a mixture of large and small particles held together by a cement paste that hardens and will take the shape of the formwork in which it is placed. The proportions of the coarse and fine aggregate, Portland cement, and water in the mixture influence the properties of the hardened concrete. The design of concrete mixes to meet specific requirements can be found in concrete materials textbooks (Troxell et al., 1968). In most cases a bridge engineer will select a particular class of concrete from a series of predesigned mixes, usually on the basis of the desired 28-day compressive strength, f'_c. A typical specification for different classes of concrete is shown in Table 13.1.

- ☐ Class A concrete is generally used for all elements of structures, except when another class is more appropriate, and specifically for concrete exposed to salt water.
- ☐ Class B concrete is used in footings, pedestals, massive pier shafts, and gravity walls.
- ☐ Class C concrete is used in thin sections, such as reinforced railings less than 4 in. (100 mm) thick and for filler in steel grid floors and the like.
- ☐ Class P concrete is used when strengths in excess of 4.0 ksi (28 MPa) are required. For prestressed concrete, consideration should be given to limiting the nominal aggregate size to 0.75 in. (20 mm).
- ☐ Class P (HPC), or high performance concrete, is used when strengths in excess of 10.0 ksi (70 MPa) are required.
- ☐ Class S concrete is used for concrete deposited underwater in cofferdams to seal out water.

A few brief comments on the parameters in Table 13.1 and their influence on the quality of concrete selected are in order. Air-entrained (AE) concrete improves durability when subjected to freeze–thaw cycles and exposure to deicing salts. This improvement is accomplished by adding a detergent or vinsol resin to the mixture, which produces an even distribution of finely divided air bubbles. This even distribution of pores in the concrete prevents large air voids from forming and breaks down the capillary pathways from the surface to the reinforcement.

The water–cement ratio (W/C) by weight is the single most important strength parameter in concrete. The lower the W/C ratio, the greater is the strength of the mixture. Obviously, increasing the cement content increases the strength for a given amount of water in the mixture. For each class of concrete, a minimum amount of cement in pounds per cubic yard (pcy) is

Table 13.1 Concrete Mix Characteristics by Class

Class of Concrete	Minimum Cement Content, pcy (kg/m³)	Maximum Water/Cement Ratio, lb/lb (kg/kg)	Air Content Range (%)	Coarse Aggregate per AASHTO M43 (ASTM D 448) Square Size of Openings, in. (mm)	28-Day Compressive Strength, ksi (MPa)
A	611 (362)	0.49	—	1.0 to No. 4 (25–4.75)	4.0 (28)
A(AE)	611 (362)	0.45	6.0 ± 1.5	1.0 to No. 4 (25–4.75)	4.0 (28)
B	517 (307)	0.58	—	2.0 to No. 3 and No. 3 to No. 4 (50–25)	2.4 (17)
B(AE)	517 (307)	0.55	5.0 ± 1.5	2.0 to No. 3 and No. 3 to No. 4 (25–4.75)	2.4 (17)
C	658 (390)	0.49	—	0.5 to No. 4 (12.5–4.75)	4.0 (28)
C(AE)	658 (390)	0.45	7.0 ± 1.5	0.5 to No. 4 (12.5–4.75)	4.0 (28)
P	564 (334)	0.49	As specified elsewhere	1.0 to No. 4 or 0.75 to No. 4 (25–4.75 or 19–4.75)	As specified elsewhere
P(HPC)					
S	658 (390)	0.58	—	1.0 to No. 4 (25–4.75)	—
Lightweight	564 (334)		As specified in the contract in the contract documents		

AASHTO Table C5.4.2.1-1. From *AASHTO LRFD Bridge Design Specifications*, Copyright © 2010 by the American Association of State Highway and Transportation Officials, Washington, DC. Used by permission.

specified. By increasing the cement above these minimums, it is possible to increase the water content and still obtain the same W/C ratio. This increase of water content may not be desirable because excess water, which is not needed for the chemical reaction with the cement and for wetting the surface of the aggregate, eventually evaporates and causes excessive shrinkage and less durable concrete. As a result, AASHTO [A5.4.2.1]* places an upper limit on the denominator of the W/C ratio to limit the water content of the mixture. The sum of Portland cement and other cementitious materials shall not exceed 800 pcy, except for class P (HPC) concrete where the sum of Portland cement and other cementitious materials shall not exceed 1000 pcy.

To obtain quality concrete that is durable and strong, it is necessary to limit the water content, which may produce problems in workability and placement of the mixture in the forms. To increase workability of the concrete mix without increasing the water content, chemical additives have been developed. These admixtures are called high-range water reducers (superplasticizers) and are effective in improving both wet and hardened concrete properties. They must be used with care, and the manufacturer's directions must be followed to avoid unwanted side effects such as accelerated setting times. Laboratory testing should be performed to establish both the wet and hardened concrete properties using aggregates representative of the construction mix.

*The article number in the AASHTO (2010) LRFD Bridge Specifications are enclosed in brackets and preceded by the letter A if specifications and by the letter C if commentary.

In recent years, *very* high strength concretes with compressive strengths approaching 30 ksi (200 MPa) have been developed in laboratory samples. The key to obtaining these high strengths is the same as for obtaining durable concrete and that is having an optimum graded mixture so that all of the gaps between particles are filled with extremely fine material until in the limit no voids exist. In the past, attention has been given to providing a well-graded mixture of coarse and fine aggregate so that the spaces between the maximum aggregate size would be filled with smaller particles of gravel or crushed stone, which in turn would have their spaces filled with fine aggregate or sand. Filling the spaces between the fine aggregate would be the powderlike Portland cement particles that, when reacted with water, bonded the whole conglomerate together. In very high strength concretes, a finer cementitious material is introduced to fill the gaps between the Portland cement particles. These finely divided mineral particles are typically pozzolans, fly ash, or silica fume. They can replace some of the Portland cement in satisfying the minimum cement content and must be added to the weight of the Portland cement in the denominator of the W/C ratio.

High-Performance Concrete With 28-day compressive strengths above 10 ksi (70 MPa), high-performance concretes are gaining a presence within bridge superstructures and providing span options not previously available to concrete. To fully utilize high-performance concretes, research is required so that the provisions of future AASHTO LRFD Specifications can be extended to concrete compressive strengths greater than 10 ksi. To meet that need, the National Cooperative Highway Research Program (NCHRP) has sponsored three projects to conduct research and to develop recommendations for revisions to the AASHTO LRFD Specifications. These projects are reported in Hawkins and Kuchma (2007) Ramirez and Russell (2008), and Rizkalla et al. (2007) and are briefly described below.

The objective of Project 12–56 (Hawkins and Kuchma, 2007) was to extend the shear design provisions to concrete compressive strengths greater than 10 ksi (70 MPa). Specific topics include the contribution of high-strength concrete to shear resistance, maximum and minimum transverse reinforcement limits, and bond issues related to shear.

The objective of Project 12–60 (Ramirez Russell, 2008) was to develop revisions to the specifications for normal-weight concrete having compressive strengths up to 18 ksi (125 MPa) that relate to:

- ☐ Transfer and development length of prestressing strands with diameters up to 0.62 in. (16 mm)
- ☐ Development and splice length in tension and compression of individual bars, bundled bars, and welded-wire reinforcement, and development length of standard hooks

The objective of Project 12–64 (Rizkalla et al., 2007) was to develop revisions to the specifications to extend flexural and compressive design provisions for reinforced and prestressed concrete members to concrete strengths up to 18 ksi (125 MPa). Much of this work has been included in the AASHTO LRFD Specifications (2010) that allow broader use of high-performance concrete.

13.4 PROPERTIES OF HARDENED CONCRETE

The 28-day compressive strength f_c' is the primary parameter, which affects a number of the properties of hardened concrete such as tensile strength, shear strength, and modulus of elasticity. A standard 6.0-in. diameter × 12.0-in. high (150-mm × 300-mm) cylinder is placed in a testing machine and loaded to a compressive failure to determine the value of f_c'. Note that this test is an unconfined compression test. When concrete is placed in a column or beam with lateral or transverse reinforcement, the concrete is in a state of triaxial or confined stress. The confined concrete stress state increases the peak compressive stress and the maximum strain over that of the unconfined concrete. It is necessary to include this increase in energy absorption or toughness when examining the resistance of reinforced concrete cross sections.

13.4.1 Short-Term Properties of Concrete

Concrete properties determined from a testing program represent short-term response to loads because these tests are usually completed in a matter of minutes, in contrast to a time period of months or even years over which load is applied to concrete when it is placed in a structure. These short-term properties are useful in assessing the quality of concrete and the response to short-term loads such as vehicle live loads. However, these properties must be modified when they are used to predict the response due to sustained dead loads such as self-weight of girders, deck slabs, and barrier rails.

Concrete Compressive Strength and Behavior In AASHTO [A5.4.2.1] a minimum 28-day compressive strength of 2.4 ksi (16 MPa) for all structural applications is recommended and a maximum compressive strength of 10.0 ksi (70 MPa) unless additional laboratory testing is conducted. Bridge decks should have a minimum compressive strength of 4.0 ksi (28 MPa) to provide adequate durability.

When describing the behavior of concrete in compression, a distinction has to be made between three possible stress states: uniaxial, biaxial, and triaxial. Illustrations of these three stress states are given in Figure 13.2. The uniaxial stress state of Figure 13.2(a) is typical of the unconfined standard cylinder test used to determine the 28-day compressive strength of concrete. The biaxial stress state of Figure 13.2(b) occurs in the reinforced webs of beams subjected to shear, bending, and axial load. The triaxial state of stress of Figure 13.2(c) illustrates the core of an axially load column that is confined by lateral ties or spirals.

The behavior of concrete in uniaxial compression [Fig. 13.2(a)] can be described by defining a relationship between normal stress and strain. A simple relationship for concrete strengths less than 6.0 ksi (40 MPa) is modeled with a parabola as

$$f_c = f_c' \left[2\left(\frac{\varepsilon_c}{\varepsilon_c'}\right) - \left(\frac{\varepsilon_c}{\varepsilon_c'}\right)^2 \right] \qquad (13.1)$$

where f_c is the compressive stress corresponding to the compressive strain ε_c, f_c' is the peak stress from a cylinder test,

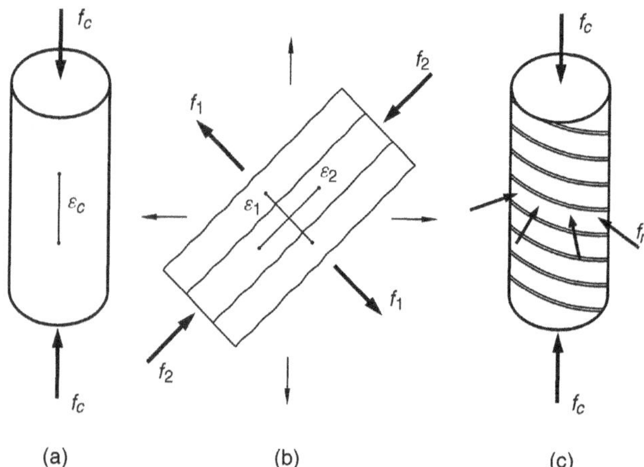

Fig. 13.2 Compressive stress states for concrete: (a) uniaxial, (b) biaxial, and (c) triaxial.

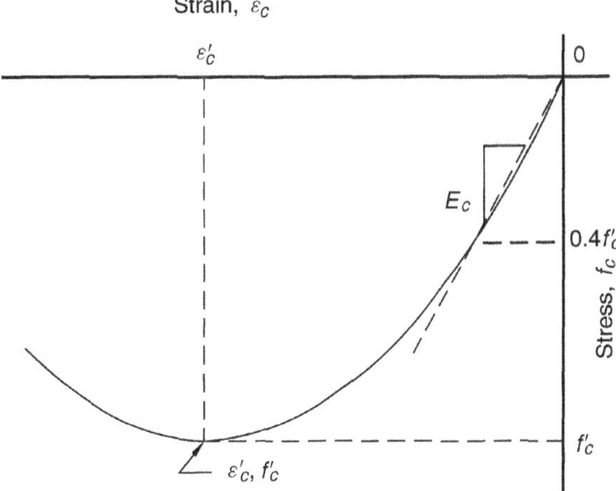

Fig. 13.3 Typical parabolic stress–strain curve for unconfined concrete in uniaxial compression.

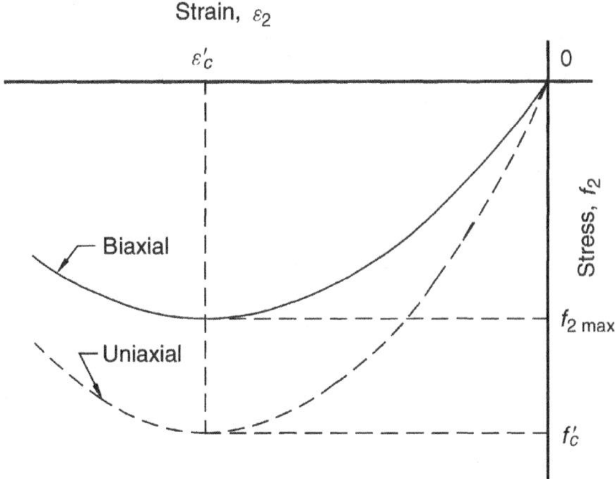

Fig. 13.4 Comparison of uniaxial and biaxial stress–strain curves for unconfined concrete in compression.

and $\varepsilon_c^{'}$ is the strain corresponding to $f_c^{'}$. This relationship is shown graphically in Figure 13.3. The sign convention adopted is that compressive stresses and compressive strains are negative values.

The modulus of elasticity given for concrete in AASHTO [A5.4.2.4] is an estimate of the slope of a line from the origin drawn through a point on the stress–strain curve at 0.4 $f_c^{'}$. This secant modulus E_c (ksi) is shown in Figure 13.3 and is given by the expression

$$E_c = 33,000 K_1 w_c^{1.5} \sqrt{f_c^{'}} \tag{13.2}$$

where K_1 is a correction factor for source of aggregate to be taken as 1.0 unless determined otherwise by a physical test, w_c is the unit weight of concrete in kips per cubic foot (kcf), and $f_c^{'}$ is the absolute value of the specified compressive strength of concrete in kips per square inch (ksi). For $K_1 = 1.0$, $w_c = 0.145$ kcf and $f_c^{'} = 4.0$ ksi:

$$\begin{aligned} E_c &= 33,000\,(1.0)\,(0.145)^{1.5}\sqrt{f_c'} \\ &= 1820\sqrt{f_c'} = 1820\sqrt{4.0} \\ &= 3640 \text{ ksi} \end{aligned}$$

When the concrete is in a state of biaxial stress, the strains in one direction affect the behavior in the other. For example, the ordinates of the stress–strain curve for principal compression f_2 in the web of a reinforced concrete beam [Fig. 13.2(b)] are reduced when the perpendicular principal stress f_1 is in tension. Vecchio and Collins (1986) quantified this phenomenon and the result is a modification of Eq. 13.1 as follows:

$$f_2 = f_{2\max}\left[2\left(\frac{\varepsilon_2}{\varepsilon_c^{'}}\right) - \left(\frac{\varepsilon_2}{\varepsilon_c^{'}}\right)^2\right] \tag{13.3}$$

where f_2 is the principal compressive stress corresponding to ε_2 and $f_{2\,\max}$ is a reduced peak stress given by

$$f_{2\,\max} = \frac{f_c^{'}}{0.8 + 170\varepsilon_1} \le f_c^{'} \tag{13.4}$$

where ε_1 is the average principal tensile strain of the cracked concrete. These relationships are illustrated in Figure 13.4. Hsu (1993) refers to this phenomenon as *compression softening* and presents mathematical expressions that are slightly different than Eqs. 13.3 and 13.4 because he includes both stress and strain softening (or reduction in peak values). When cracking becomes severe, the average strain ε_1 across the cracks can become quite large and, in the limit, causes the principal compressive stress f_2 to go to zero. A value of $\varepsilon_1 = 0.004$ results in a one-third reduction in f_2.

When the concrete within a beam or column is confined in a triaxial state of stress by lateral ties or spirals [Fig. 13.2(c)], the out-of-plane restraint provided by the reinforcement increases the peak stress and peak strain above the unconfined values. For confined concrete in compression, the limiting ultimate strain is dramatically increased beyond the 0.003 value often used for unconfined concrete. This increased strain on the descending branch of the stress–strain curve adds ductility and toughness to the element and provides a mechanism for dissipating energy without failure. As a result the confinement of concrete within closely spaced lateral ties or spirals is essential for elements located in seismic regions in order to absorb energy and allow the deformation necessary to reduce the earthquake loads.

Confined Concrete Compressive Strength and Behavior Figure 13.5 shows a comparison of typical stress–strain curves for confined and unconfined concrete in compression for an axially loaded column. The unconfined concrete is representative of the concrete in the shell outside of the lateral reinforcement, which is lost due to spalling at

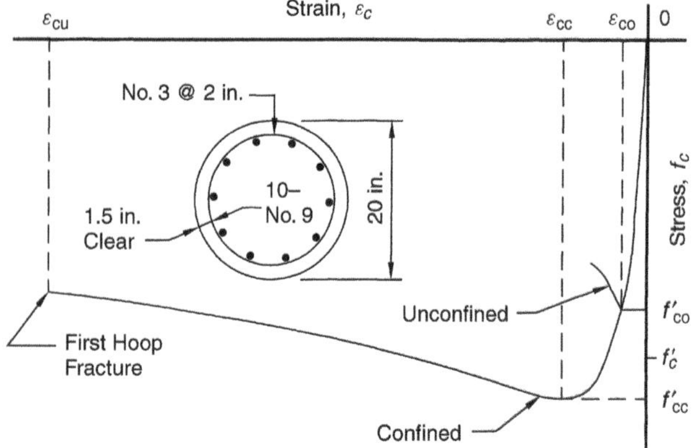

Fig. 13.5 Comparison of unconfined and confined concrete stress–strain curves in compression.

relatively low compressive strains. The confined concrete exhibits a higher peak stress f'_{cc} and a larger corresponding strain ε_{cc} than the unconfined concrete strength f'_{co} and its corresponding strain ε_{co}.

One of the earliest studies to quantify the effect of lateral confinement was by Richart et al. (1928) in which hydrostatic fluid pressure was used to simulate the lateral confining pressure f_r. The model used to represent the strength of confined concrete was similar to the Coulomb shear failure criterion used for rock (and other geomaterials):

$$f'_{cc} = f'_{co} + k_1 f_r \tag{13.5}$$

where f'_{cc} is the peak confined concrete stress, f'_{co} is the unconfined concrete strength, f_r is the lateral confining pressure, and k_1 is a coefficient that depends on the concrete mix and the lateral pressure. From these tests, Richart et al. (1928) determined that the average value of $k_1 = 4$. Setunge et al. (1993) propose that a simple lower bound value of $k_1 = 3$ be used for confined concrete of any strength below 15 ksi (120 MPa).

Richart et al. (1928) also suggested a simple relationship for the strain ε_{cc} corresponding to f'_{cc} as

$$\varepsilon_{cc} = \varepsilon_{co}\left(1 + k_2 \frac{f_r}{f'_{co}}\right) \tag{13.6}$$

where ε_{co} is the strain corresponding to f'_{co} and $k_2 = 5k_1$. Also, f'_{co} is commonly taken equal to $0.85 f'_c$ to account for the lower strength of the concrete placed in a column compared to that in the control cylinder.

The lateral confining pressure f_r in Eqs. 13.5 and 13.6, produced indirectly by lateral reinforcement, needs to be determined. Mander et al. (1988), following an approach similar to Sheikh and Uzumeri (1980), derive expressions for the effective lateral confining pressure f'_r for both circular hoop and rectangular hoop reinforcement. Variables considered are the spacing, area, and yield stress of the hoops; the dimensions of the confined concrete core, and the distribution of the longitudinal reinforcement around the core perimeter. It is convenient to use f'_{cc} on the area of the concrete core A_{cc} enclosed within centerlines of the perimeter hoops. However, not all of this area is effectively confined concrete, and f_r must be adjusted by a confinement effectiveness coefficient k_e to give an effective lateral confining pressure of

$$f'_r = k_e f_r \tag{13.7}$$

in which

$$k_e = \frac{A_e}{A_{cc}} \tag{13.8}$$

where A_e is the area of effectively confined concrete,

$$A_{cc} = A_c - A_{st} = A_c\left(1 - \rho_{cc}\right) \tag{13.9}$$

where A_c is the area of the core enclosed by the centerlines of the perimeter hoops or ties, A_{st} is the total area of the longitudinal reinforcement, and

$$\rho_{cc} = \frac{A_{st}}{A_c} \tag{13.10}$$

Example 13.1 Determine the confinement effectiveness coefficient k_e for a circular column with spiral reinforcement of diameter d_s between bar centers if the arch action between spirals with a clear vertical spacing of s' has an amplitude of $s'/4$ (see Fig. 13.6). Midway between the sprials, A_c is the smallest with a diameter of $d_s - s'/4$, that is,

$$A_e = \frac{\pi}{4}\left(d_s - \frac{s'}{4}\right)^2 = \frac{\pi}{4} d_s^2 \left(1 - \frac{s'}{4d_s}\right)^2$$
$$= \frac{\pi}{4} d_s^2 \left[1 - \frac{s'}{2d_s} + \left(\frac{s'}{4d_s}\right)^2\right]$$

Neglecting the higher order term, which is much less than one, yields

$$A_e \approx \frac{\pi}{4} d_s^2 \left(1 - \frac{s'}{2d_s}\right)$$

and with $A_c = \pi d_s^2/4$; Eqs. 13.8 and 13.9 yield

$$k_e = \frac{1 - \left(s'/2d_s\right)}{1 - \rho_{cc}} \leq 1.0 \tag{13.11}$$

(Note that the definition of d_s for this model is different than the outside diameter of the core d_c often used in selecting spiral reinforcement.)

The half-section of depth s is shown in Figure 13.7, which is confined by a spiral with hoop tension at yield exerting a uniform lateral pressure f_r (tension shown is positive) on the concrete core, the equilibrium of forces requires

$$2A_{sp} f_{yh} + f_r s d_s = 0 \tag{13.12}$$

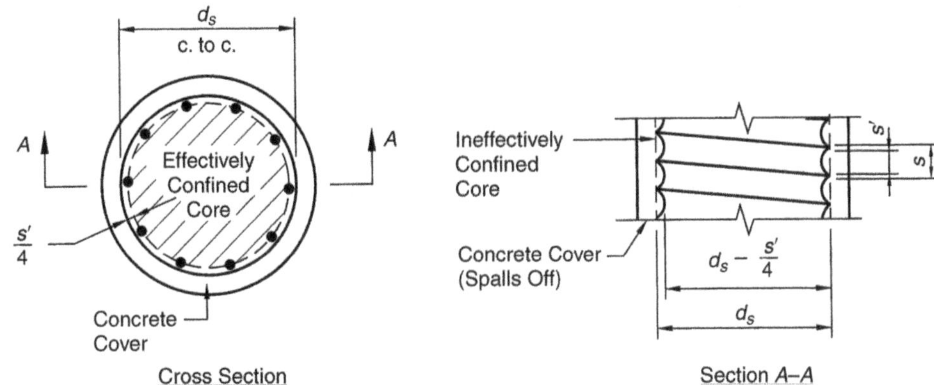

Fig. 13.6 Effectively confined core for circular spirals. [Reproduced from J. B. Mander, M. J. N. Mander and R. Park (1988). Theoretical Stress–Strain Model for Confined Concrete, *Journal of Structural Engineering*, ASCE, 14(8), pp. 1804–1826. With permission.]

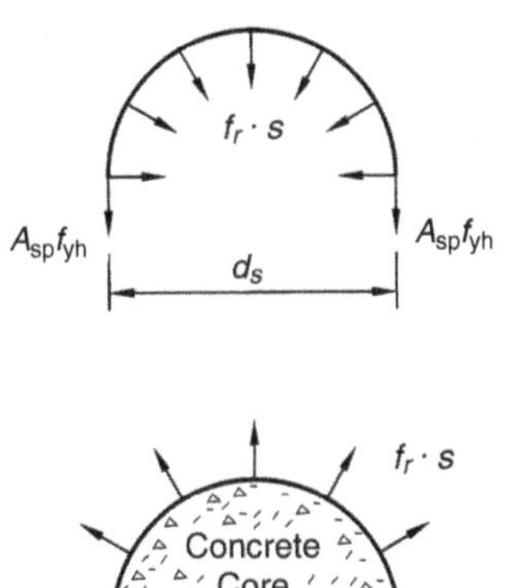

Fig. 13.7 Half-body diagrams at interface between spiral and concrete core.

where A_{sp} is the area of the spiral, f_{yh} is the yield strength of the spiral, and s is the center-to-center spacing of the spiral. Solve Eq. 13.12 for the lateral confining pressure

$$f_r = \frac{-2A_{sp}f_{yh}}{sd_s} = -\frac{1}{2}\rho_s f_{yh} \qquad (13.13)$$

where ρ_s is the ratio of the volume of transverse confining steel to the volume of confined concrete core, that is,

$$\rho_s = \frac{A_{sp}\pi d_s}{(\pi/4)\,sd_s^2} = \frac{4A_{sp}}{sd_s} \qquad (13.14)$$

Mander et al. (1988) give expressions similar to Eqs. 13.11, 13.13, and 13.14 for circular hoops and rectangular ties.

Example 13.2 Determine the peak confined concrete stress f'_{cc} and corresponding strain ε_{cc} for a 20-in.-diameter column with 10 No. 9 longitudinal bars and No. 3 round spirals at 2-in. pitch (Fig. 13.5). The material strengths are $f'_c = -4.0$ ksi and $f_{yh} = 60$ ksi. Assume that $\varepsilon_{co} = -0.002$ and that the concrete cover is 1.5 in. Use the lower bound value of $k_1 = 3$ and the corresponding value of $k_2 = 15$.

$$s = 2 \text{ in.},\ s' = 2 - 0.375 = 1.625 \text{ in.}$$

$$d_s = 20 - 2\,(1.5) - 2\left(\tfrac{1}{2}\right)(0.375) = 16.63 \text{ in.}$$

$$A_c = \frac{\pi}{4}(d_s)^2 = \frac{\pi}{4}(16.63)^2 = 217.1 \text{ in.}^2$$

$$\rho_{cc} = \frac{A_{st}}{A_c} = \frac{10\,(1.0)}{217.1} = 0.0461$$

$$\rho_s = \frac{4A_{sp}}{sd_s} = \frac{4\,(0.11)}{2\,(16.63)} = 0.0132$$

$$k_e = \frac{1 - \frac{s'}{2d_s}}{1 - \rho_{cc}} = \frac{1 - \frac{1.625}{2(16.63)}}{1 - 0.0461} = 0.997$$

$$f'_r = -\tfrac{1}{2}k_e\rho_s f_{yh} = -\tfrac{1}{2}\,(0.997)\,(0.0132)\,(60)$$
$$= -0.395 \text{ ksi} = 0.395 \text{ ksi compression}$$

$$f'_{cc} = f'_{co} + k_1 f'_r = 0.85\,(-4) + 3\,(-0.395)$$
$$= -4.59 \text{ ksi} = \underline{4.59 \text{ ksi compression}}$$

$$\varepsilon_{cc} = \varepsilon_{co}\left(1 + k_2\frac{f'_r}{f'_{co}}\right) = -0.002\left[1 + 15\left(\frac{-0.395}{-3.40}\right)\right]$$
$$= -0.0055 = \underline{0.0055 \text{ shortening}}$$

Again, note that the negative signs indicate compression.

Over the years, researchers developed stress–strain relationships for the response of confined concrete in compression that best fits their experimental data. Sheikh (1982) presented a comparison of seven models used by investigators in different research laboratories. All but one of these models use different equations for the ascending and descending branches of the stress–strain curve. The model considered to best fit the experimental data was one he and a colleague developed earlier (Sheikh and Uzumeri, 1980).

The stress–strain model proposed by Mander et al. (1988) for monotonic compression loading up to first hoop fracture is a single equation relating the longitudinal compressive stress f_c as a function of the corresponding longitudinal compressive strain ε_c:

$$f_c(x) = \frac{f'_{cc} r x}{r - 1 + x^r} \tag{13.15}$$

where

$$x = \frac{\varepsilon_c}{\varepsilon_{cc}} \tag{13.16}$$

$$r = \frac{E_c}{E_c - E_{sec}} \tag{13.17}$$

and the secant modulus of confined concrete at peak stress is

$$E_{sec} = \frac{f'_{cc}}{\varepsilon_{cc}} \tag{13.18}$$

This curve continues until the confined concrete strain reaches an ε_{cu} value large enough to cause the first hoop or spiral to fracture. Based on an energy balance approach and test results, Mander et al. (1988) present an integral equation that can be solved numerically for ε_{cu}.

Example 13.3 Determine the parameters and plot the stress–strain curves for the confined and unconfined concrete of the column section in Example 13.2 (Fig. 13.5). Assume concrete strain at the first hoop fracture $\varepsilon_{cu} = 8\varepsilon_{cc} = 8(-0.0055) = -0.044$.

$$E_c = 1820\sqrt{f'_c} = 1820\sqrt{4} = 3640 \text{ ksi}$$

$$E_{sec} = \frac{f'_{cc}}{\varepsilon_{cc}} = \frac{-4.59}{-0.0055} = 835 \text{ ksi}$$

$$r = \frac{E_c}{E_c - E_{sec}} = \frac{3640}{3640 - 835} = 1.30$$

$$f_c(\varepsilon_c) = 1.30 f'_{cc} \frac{(\varepsilon_c/\varepsilon_{cc})}{0.30 + (\varepsilon_c/\varepsilon_{cc})^{1.30}} \qquad 0 \le \varepsilon_c \le 8\varepsilon_{cc}$$

This last expression for f_c is Eq. 13.15 and has been used to plot the curve shown in Figure 13.5.

From the above discussion, it is apparent that *the behavior of concrete in compression is different when it has reinforcement within and around the concrete than when it is unreinforced*. A corollary to this concrete behavior is that *the response in tension of reinforcement embedded in concrete is different than the response of bare steel alone*. The behavior of the tension reinforcement is discussed later after a brief discussion about the tensile behavior of concrete.

Concrete Tensile Strength and Behavior Concrete tensile strength can be measured either directly or indirectly. A direct tensile test [Fig. 13.8(a)] is preferred for determining the cracking strength of concrete but requires special equipment. Consequently, indirect tests, such as the modulus of rupture test and the split cylinder test, are often used. These tests are illustrated in Figure 13.8.

The modulus of rupture test [Fig. 13.8(b)] measures the tensile strength of concrete in flexure with a plain concrete

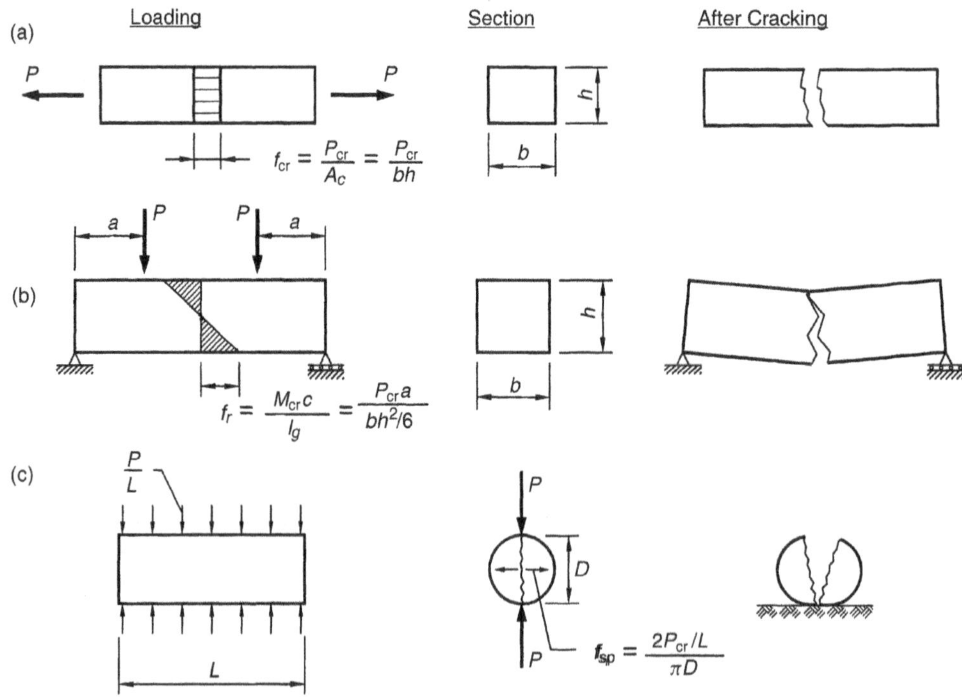

Fig. 13.8 Direct and indirect concrete tensile tests. (a) Direct tension test, (b) modulus of rupture test, and (c) split cylinder test.

beam loaded as shown. The tensile stress through the depth of the section is nonuniform and is maximum at the bottom fibers. A flexural tensile stress is calculated from elementary beam theory for the load that cracks (and fails) the beam. This flexural tensile stress is called the modulus of rupture f_r. For normal weight concrete, AASHTO [A5.4.2.6] gives a lower bound value for f_r (ksi) when considering service load cracking:

$$f_r = 0.24\sqrt{f'_c} \tag{13.19a}$$

and an upper bound value when considering minimum reinforcement

$$f_r = 0.37\sqrt{f'_c} \tag{13.19b}$$

where f'_c is the absolute value of the cylinder compressive strength of concrete (ksi).

In the split cylinder test [Fig. 13.8(c)], a standard cylinder is laid on its side and loaded with a uniformly distributed line load. Nearly uniform tensile stresses are developed perpendicular to the compressive stresses produced by opposing line loads. When the tensile stresses reach their maximum strength, the cylinder splits in two along the loaded diameter. A theory of elasticity solution (Timoshenko and Goodier, 1951) gives the splitting tensile stress f_{sp} as

$$f_{sp} = \frac{2P_{cr}/L}{\pi D} \tag{13.20}$$

where P_{cr} is the total load that splits the cylinder, L is the length of the cylinder, and D is the diameter of the cylinder.

Both the modulus of rupture (f_r) and the splitting stress (f_{sp}) overestimate the tensile cracking stress (f_{cr}) determined by a direct tension test [Fig. 13.8(a)]. If they are used, nonconservative evaluations of resistance to restrained shrinkage and splitting in anchorage zones can result. In these and other cases of direct tension, a more representative value must be used. For normal weight concrete, Collins and Mitchell (1991) and Hsu (1993) estimate the direct cracking strength of concrete, f_{cr}, as

$$\begin{aligned} f_{cr} &= 4\,(0.0316)\sqrt{f'_c} \\ &= 0.13\sqrt{f'_c} \end{aligned} \tag{13.21}$$

where f'_c is the cylinder compressive strength (ksi). Note $1/\sqrt{1000} = 0.0316$ is a unit conversion constant to place f'_c in ksi rather than the more traditional psi units.

The direct tension stress–strain curve (Fig. 13.9) is assumed to be linear up to the cracking stress f_{cr} at the same slope given by E_c in Eq. 13.2. After cracking and if reinforcement is present, the tensile stress decreases but does not go to zero. Aggregate interlock still exists and is able to transfer tension across the crack. The direct tension experiments by Gopalaratnam and Shah (1985), using a stiff testing machine, demonstrate this behavior. This response is important when predicting the tensile stress in longitudinal

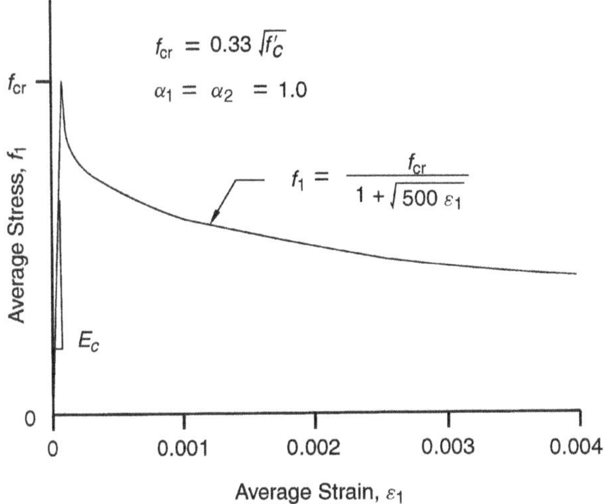

Fig. 13.9 Average stress versus average strain for concrete in tension. [From Collins and Mitchell (1991). Reprinted by permission of Prentice Hall, Upper Saddle River, NJ.]

reinforcement and the shear resistance of reinforced concrete beams. Collins and Mitchell (1991) give the following expressions for the direct tension stress–strain curve shown in Figure 13.9:

Ascending Branch $(\varepsilon_1 \le \varepsilon_{cr} = f_{cr}/E_c)$

$$f_1 = E_c\varepsilon_1 \tag{13.22}$$

where ε_1 is the average principal tensile strain in the concrete and f_1 is the average principal tensile stress.

Descending Branch $(\varepsilon_1 > \varepsilon_{cr})$

$$f_1 = \frac{\alpha_1\alpha_2 f_{cr}}{1+\sqrt{500\ \varepsilon_1}} \tag{13.23}$$

where α_1 is a factor accounting for bond characteristics of reinforcement:

$\alpha_1 = 1.0$ for deformed reinforcing bars
$\alpha_1 = 0.7$ for plain bars, wires, or bonded strands
$\alpha_1 = 0$ for unbonded reinforcement

and α_2 is a factor accounting for sustained or repeated loading:

$\alpha_2 = 1.0$ for short-term monotonic loading
$\alpha_2 = 0.7$ for sustained and/or repeated loads

If no reinforcement is present, there is no descending branch and the concrete tensile stress after cracking is zero. However, if the concrete is bonded to reinforcement, concrete tensile stresses do exist. Once again it is apparent that *the behavior of concrete with reinforcement is different than that of plain concrete*.

13.4.2 Long-Term Properties of Concrete

At times it appears that concrete is more alive than it is dead. If compressive loads are applied to concrete for a long period of time, concrete creeps to get away from them. Concrete generally gains strength with age unless a deterioration mechanism, such as that caused by the intrusion of the chloride ion occurs. Concrete typically shrinks and cracks. But even this behavior can be reversed by immersing the concrete in water and refilling the voids and closing the cracks. It appears that concrete never completely dries and there is always some gelatinous material that has not hardened and provides resiliency between the particles. These time-dependent properties of concrete are influenced by the conditions at time of placement and the environment that surrounds it throughout its service life. Prediction of the exact effect of all of the conditions is difficult, but estimates can be made of the trends and changes in behavior.

Compressive Strength of Aged Concrete If a concrete bridge has been in service for a number of years and a strength evaluation is required, the compressive strength of a core sample is a good indication of the quality and durability of the concrete in the bridge. The compressive strength can be determined by nondestructive methods by first estimating the modulus of elasticity and then back-calculating to find the compressive strength. Another device measures the rebound of a steel ball that has been calibrated against the rebound on concrete of known compressive strength.

In general, the trend is that the compressive strength of concrete increases with age. However, to determine the magnitude of the increase, field investigations are extremely useful.

Shrinkage of Concrete [A5.4.2.3.3] Shrinkage of concrete is a decrease in volume under constant temperature due to loss of moisture after concrete has hardened. This time-dependent volumetric change depends on the water content of the fresh concrete, the type of cement and aggregate used, the ambient conditions (temperature, humidity, and wind velocity) at the time of placement, the curing procedure, the amount of reinforcement, and the volume/surface area ratio. In AASHTO [A5.4.2.3.3], an empirical equation based on parametric studies by Tadros et al. (2003) is presented to evaluate the shrinkage strain ε_{sh} based on the drying time, the relative humidity, the concrete compressive strength, and the volume/surface area ratio:

$$\varepsilon_{sh} = -k_{vs}k_{hs}k_f k_{td} 0.48 \times 10^{-3} \tag{13.24}$$

in which

$$k_{vs} = 1.45 - 0.13\left(\frac{V}{S}\right) \geq 1.0$$

$$k_{hs} = 2.00 - 0.014\,H$$

$$k_f = \frac{5}{1 + f'_{ci}}$$

$$k_{td} = \frac{t}{61 - 4f'_{ci} + t}$$

Where

H = relative humidity (%). If humidity at the site is unknown, an annual average value of H depending on the geographic location may be taken from Figure 13.10 [Fig. A5.4.2.3.3-1]

k_{vs} = factor for the effect of the volume-to-surface ratio of the component

k_{hs} = humidity factor for shrinkage

k_f = factor for the effect of concrete strength

k_{td} = time development factor

t = maturity of concrete (days), defined as age of concrete between time of loading for creep calculations, or end of curing for shrinkage calculations, and time being considered for analysis of creep or shrinkage effects

V/S = volume-to-surface ratio

f'_{ci} = specified compressive strength of concrete at time of prestressing for pretensioned members and at time of initial loading for nonprestressed members. If concrete age at time of initial loading is unknown at design time, f'_{ci} may be taken as 0.80 f'_c.

Equation 13.24 is for concrete devoid of shrinkage-prone aggregates and is proposed for both precast and cast-in-place concrete components and for both accelerated curing and moist curing conditions [C5.4.2.3.2]. Equation 13.24 assumes that a reasonable estimate for the ultimate shrinkage strain is 0.00048 in./in. Correction factors are applied to this value to account for the various conditions affecting shrinkage strain.

The volume-to-surface ratio (size) correction factor k_{vs} accounts for the effect that relatively thick members do not dry as easily as thin members when exposed to ambient air. Member size affects short-term creep and shrinkage more than it does ultimate values (Tadros et al., 2003). Because ultimate values are of primary importance for most bridges, the *V/S* ratio correction factor can be simplified when ultimate prestress loss and final concrete bottom fiber stress are the primary design values. The *V/S* ratio of the member may be computed as the ratio of cross-sectional area to the perimeter exposed to the environment. Most precast concrete stemmed members have a *V/S* ratio of 3–4 in. The member size correction factor is normalized to a value of 1.0 for a *V/S* ratio of 3.5 in. and the same expression is used for both shrinkage and creep.

The relative humidity correction factor k_{hs} accounts for the effect that shrinkage is greater in dry climates than in wet climates. The value of k_{hs} is normalized to 1.0 at 70% average relative humidity and different expressions are used for shrinkage strain and for the creep coefficient.

The concrete strength correction factor k_f accounts for the effect that shrinkage strain and creep are reduced for higher strength concrete. The expression for k_f is also the same for both creep and shrinkage and is normalized to 1.0 when the initial compressive strength at prestress transfer f'_{ci} is 4.0 ksi.

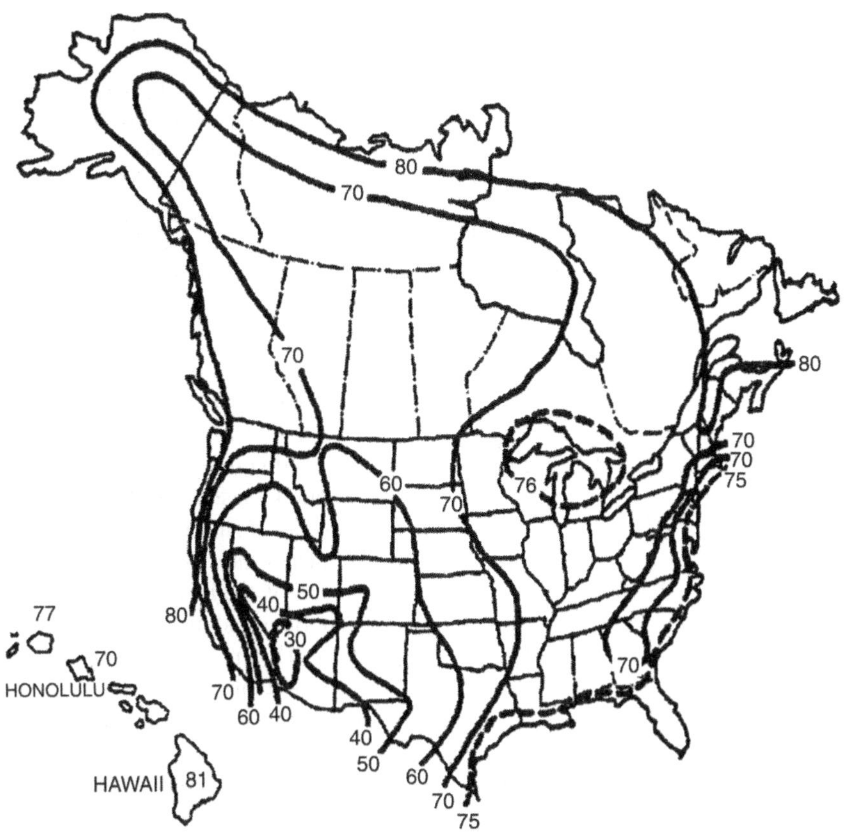

Fig. 13.10 Annual average ambient relative humidity in percent [AASHTO Fig. 5.4.2.3.3-1]. [From *AASHTO LRFD Bridge Design Specifications*. Copyright © 2010 by the American Association of State Highway and Transportation Officials, Washington, DC. Used by permission.]

The value of 4.0 ksi was taken to be 80% of an assumed final strength at service of 5.0 ksi (Tadros et al., 2003).

The time development correction factor k_{td} accounts for the effect of concrete strength on shrinkage and creep at times other than when time approaches infinity. Higher strength concretes produce accelerated shrinkage and creep at early stages of a member's life (Tadros et al., 2003). This behavior is predicted by the formula for k_{td} so it can be used for estimating camber and prestress loss at the time of girder erection. The same expression for k_{td} is used for both shrinkage and creep estimates and approaches a value of 1.0 as time approaches infinity.

Large concrete members may undergo substantially less shrinkage than that measured by laboratory testing of small specimens of the same concrete. The constraining effects of reinforcement and composite actions with other elements of the bridge tend to reduce the dimensional changes in some components [C5.4.2.3.3]. In spite of these limitations, Eq. 13.24 does indicate the trend and relative magnitude of the shrinkage strains, which are illustrated in Example 13.4.

Example 13.4 Estimate the shrinkage strain in a 8-in.-thick concrete bridge deck ($f_c' = 4.5$ ksi) whose top and bottom surfaces are exposed to drying conditions in an atmosphere with 70% relative humidity. The volume/surface area ratio for 1 in.^{2} of deck area is

$$\frac{V}{S} = \frac{\text{volume}}{\text{surface area}} = \frac{8\,(1)\,(1)}{2\,(1)\,(1)} = 4 \text{ in.}$$

For $t = 5$ years ($\approx$2000 days) and $f_{ci}' = 0.8\,f_c' = 0.8(4.5) = 3.6$ ksi

$$k_{vs} = 1.45 - 0.13\,(4) = 0.93 < 1.0, \text{ use } k_{vs} = 1.0$$

$$k_{hs} = 2.00 - 0.014\,(70) = 1.02$$

$$k_f = \frac{5}{1 + 3.6} = 1.09$$

$$k_{td} = \frac{t}{61 - 4\,(3.6) + t} = \frac{t}{46.6 + t}$$

Thus Eq. 13.24 gives

$$\begin{aligned}\varepsilon_{sh} &= -\,(1.0)\,(1.02)\,(1.09)\left(\frac{t}{46.6 + t}\right) 0.48 \times 10^{-3} \\ &= -0.00053\left(\frac{2000}{2046.6}\right) = -0.00052\end{aligned}$$

where the negative sign indicates shortening. The variation of shrinkage strain with drying time for these conditions is given in Table 13.2 and shown in Figure 13.11.

Table 13.2 Variation of Shrinkage with Time (Example 13.4)

Drying Time (day)	Shrinkage Strain ε_{sh} (in./in.)
28	−0.00020
100	−0.00036
365	−0.00047
1000	−0.00051
2000	−0.00052

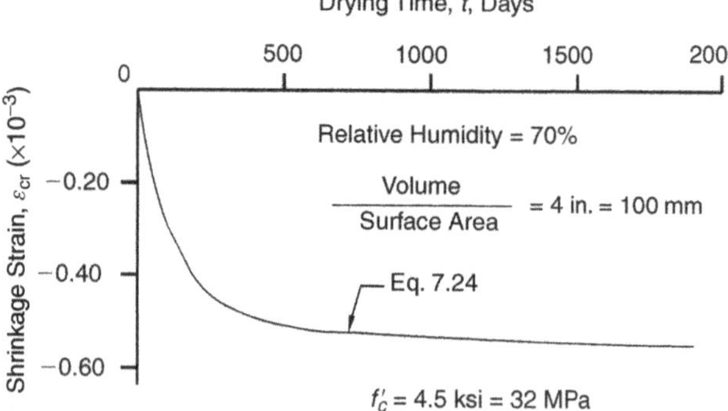

Fig. 13.11 Variation of shrinkage with time (Example 13.4).

At an early age of concrete, shrinkage strains are more sensitive to surface exposure than when t is large. For accurately estimating early deformations of such specialized structures as segmentally constructed balanced cantilever box girders, it may be necessary to resort to experimental data or use the more detailed Eq. C5.4.2.3.2-2. Because the empirical equation does not include all of the variables affecting shrinkage, the commentary in AASHTO [C5.4.2.3.1] indicates that the results may be in error by ±50% and the actual shrinkage strains could be larger than −0.0008 [C5.4.2.3.3]. Even if the values are not exact, the trend shown in Figure 13.11 of increasing shrinkage strain at a diminishing rate as drying time increases is correct. When specific information is not available on the concrete and the conditions under which it is placed, AASHTO [A5.4.2.3.1] recommends values of shrinkage strain to be taken as −0.0002 after 28 days and −0.0005 after one year of drying. These values are comparable to those in Table 13.2.

A number of measures can be taken to control the amount of shrinkage in concrete structures. One of the most effective is to reduce the water content in the concrete mixture because it is the evaporation of the excess water that causes the shrinkage. A designer can control the water content by specifying both a maximum water/cement ratio and a maximum cement content. Use of hard, dense aggregates with low absorption results in less shrinkage because they require less moisture in the concrete mixture to wet their surfaces. Another effective method is to control the temperature in the concrete before it hardens so that the starting volume for the beginning of shrinkage has not been enlarged by elevated temperatures. This temperature control can be done by using a low heat of hydration cement and by cooling the materials in the concrete mixture. High outdoor temperatures during the summer months need to be offset by shading the aggregate stockpiles from the sun and by cooling the mixing water with crushed ice. It has often been said by those in the northern climates that the best concrete (fewest shrinkage cracks) is placed during the winter months when kept sufficiently warm during cure.

Creep of Concrete Creep of concrete is an increase in deformation with time when subjected to a constant load. In a reinforced concrete beam, the deflection continues to increase due to sustained loads. In reinforced concrete beam columns, axial shortening and curvature increase under the action of constant dead loads. Prestressed concrete beams lose some of their precompression force because the concrete shortens and decreases the strand force and associated prestress. The creep phenomenon in concrete influences the selection and interaction of concrete elements and an understanding of its behavior is important.

Creep in concrete is associated with the change of strain over time in the regions of beams and columns subjected to sustained compressive stresses. This time-dependent change in strains relies on the same factors that affect shrinkage strains plus the magnitude and duration of the compressive stresses, the age of the concrete when the sustained load is applied, and the temperature of the concrete. Creep strain ε_{CR} is determined by multiplying the instantaneous elastic compressive strain due to permanent loads ε_{ci} by a creep coefficient ψ, that is,

$$\varepsilon_{CR}\left(t,\ t_i\right) = \psi\left(t,\ t_i\right)\varepsilon_{ci} \tag{13.25}$$

where t is the age of the concrete in days between time of loading and time being considered for analysis of creep effects, and t_i is the age of the concrete in days when the permanent load is applied. In AASHTO [A5.4.2.3.2], an empirical equation based on Huo et al. (2001), Al-Omaishi (2001), Tadros et al. (2003), and Collins and Mitchell (1991) is given for the creep coefficient. It is expressed as

$$\psi\left(t,\ t_i\right) = 1.9k_{vs}k_{hc}k_f k_{td}t_i^{-0.118} \tag{13.26}$$

in which

$$k_{hc} = 1.56 - 0.008H$$

where k_{hc} is the humidity factor for creep. If H is not known for the site, a value can be taken from Figure 13.10. The H factor may be higher than ambient for a water crossing due to evaporation in the vicinity of the bridge.

Equation 13.26 for estimating the creep coefficient was developed in a manner similar to the shrinkage strain prediction formula. The ultimate creep coefficient for standard conditions is assumed to be 1.90. The standard conditions are the same as defined for shrinkage: $H = 70\%$, $V/S = 3.5$ in., $f'_{ci} =$ 4 ksi, loading age = 1 day for accelerated curing and 7 days for moist curing, and loading duration = infinity. Variations from these standard conditions require correction factors to be calculated and applied to the value of 1.90 as shown in Eq. 13.26.

The loading age correction factor $t_i^{-0.118}$ can be used for both types of curing if t_i = age of concrete (days) when load is initially applied for accelerated curing and t_i = age of concrete (days) when load is initially applied minus 6 days for moist curing.

Example 13.5 Estimate the creep strain in the bridge deck of Example 13.4 after one year if the compressive stress due to sustained loads is 1.45 ksi, the 28-day compressive strength is 4.5 ksi, and $t_i = 15$ days. The modulus of elasticity from Eq. 13.2 is

$$E_c = 1820\sqrt{f'_c} = 1820\sqrt{4.5} = 3860 \text{ ksi}$$

and the initial compressive strain becomes

$$\varepsilon_{ci} = \frac{f_{cu}}{E_c} = \frac{-1.45}{3860} = -0.00038 \quad (13.27)$$

For $t - t_i = 365 - 15 = 350$ days, $V/S = 4$ in., $H = 70\%$, $f'_{ci} = 0.8$, and, $f'_c = 3.6$ ksi:

$$k_{vs} = 1.45 - 0.13\left(\frac{V}{S}\right) = 1.45 - 0.13\,(4)$$
$$= 0.93 < 1.0 \text{ use } k_{vs} = 1.0$$
$$k_{hc} = 1.56 - 0.008H = 1.56 - 0.008\,(70) = 1.0$$
$$k_f = \frac{5}{1+3.6} = 1.09$$
$$k_{td} = \frac{t}{61 - 4f'_{ci} + t} = \frac{350}{61 - 4\,(3.6) + 350} = 0.883$$

The creep coefficient is given by Eq. 13.26 as

$$\psi\,(365,\ 15) = 1.9\,(1.0)\,(1.0)\,(1.09)\,(0.883)\,15^{-0.118}$$
$$= 1.33$$

Thus, the estimated creep strain after one year is (Eq. 13.25)

$$\varepsilon_{CR}\,(365,\ 15) = 1.33\,(-0.00038) = -0.00051$$

which is of the same order of magnitude as the shrinkage strain. Again, this estimate could be in error by ±50%. For the same conditions as this example, the variation of total compressive strain with time after application of the sustained load is shown in Figure 13.12. The total compressive strain $\varepsilon_c(t, t_i)$ is the sum of the initial elastic strain plus the creep strain and the rate of increase diminishes with time. This total strain can be expressed as

$$\varepsilon_c\left(t, t_i\right) = \varepsilon_{ci} + \varepsilon_{CR}\left(t, t_i\right) = \left[1 + \psi\left(t, t_i\right)\right]\varepsilon_{ci} \quad (13.28)$$

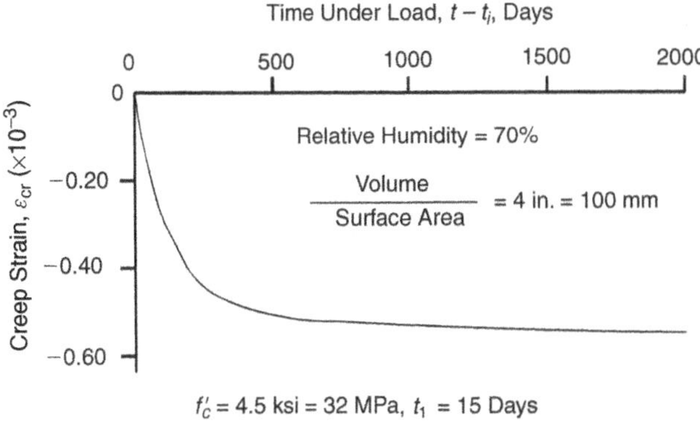

Fig. 13.12 Variation of creep strain with time (Example 13.5).

For this example, the total compressive strain after one year is

$$\varepsilon_c\,(365,\ 15) = (1 + 1.33)\,(-0.00038) = -0.00089$$

over two times the elastic value.

Creep strains can be reduced by the same measures taken to reduce shrinkage strains, that is, by using low water content in the concrete mixture and keeping the temperature relatively low. Creep strain can also be reduced by using steel reinforcement in the compression zone because the portion of the compressive force it carries is not subject to creep. By delaying the time at which permanent loads are applied, creep strains are reduced because the more mature concrete is drier and less resilient. This trend is reflected in Eq. 13.26, where larger values of t_i for a given age of concrete t result in a reduction of the creep coefficient $\psi\,(t, t_i)$.

Finally, not all effects of creep deformation are harmful. When differential settlements occur in a reinforced concrete bridge, the creep property of concrete actually decreases the stresses in the elements from those that would be predicted by an elastic analysis.

Modulus of Elasticity for Permanent Loads To account for the increase in strain due to creep under permanent loads, a reduced long-term modulus of elasticity $E_{c,LT}$ can be defined as

$$E_{c,LT} = \frac{f_{ci}}{\left[1 + \psi\left(t,\ t_i\right)\right]\varepsilon_{ci}} = \frac{E_{ci}}{1 + \psi\left(t,\ t_i\right)}$$

where E_{ci} is the modulus of elasticity at time t_i. Assuming that E_{ci} can be represented by the modulus of elasticity E_c from Eq. 13.2, then

$$E_{c,LT} = \frac{E_c}{1 + \psi\left(t,\ t_i\right)} \quad (13.29)$$

When transforming section properties of steel to equivalent properties of concrete for service limit states, the modular ratio n is used and is defined as

$$n = \frac{E_s}{E_c} \quad (13.30)$$

A long-term modular ratio n_{LT} for use with permanent loads can be similarly defined, assuming that steel does not creep,

$$n_{LT} = \frac{E_s}{E_{c,LT}} = n\left[1 + \psi\left(t,\ t_i\right)\right] \tag{13.31}$$

Example 13.6 For the conditions of Example 13.5, estimate the long-term modular ratio using $t = 5$ years.

For $t - t_i = 5(365) - 15 = 1810$ days,

$$k_{td} = \frac{1810}{61 - 4\,(3.6) + 1810} = 0.975$$

Thus,

$$\psi\,(1825,\ 15) = 1.9\,(1.0)\,(1.0)\,(1.09)\,(0.975)\,15^{-0.118} = 1.47$$

and

$$n_{LT} = 2.47n$$

In evaluating designs based on service and fatigue limit states, an effective modular ratio of $2n$ for permanent loads and prestress is assumed [A5.7.1]. In AASHTO [A5.7.3.6.2], which is applicable to the calculation of deflection and camber, the long-time deflection is estimated as the instantaneous deflection multiplied by the factor

$$3.0 - 1.2\frac{A_S'}{A_S} \geq 1.6 \tag{13.32}$$

where A_S' is the area of the compression reinforcement and A_S is the area of nonprestressed tension reinforcement. This factor is essentially $\psi\,(t, t_i)$, and, if $A_S' = 0$, Eq. 13.31 gives a value of $n_{LT} = 4n$. Based on the calculations made for the creep coefficient, it is reasonable to use the following simple expression for the modulus of elasticity for permanent loads:

$$E_{c,LT} = \frac{E_c}{3} \tag{13.33}$$

13.5 PROPERTIES OF STEEL REINFORCEMENT

Reinforced concrete is simply concrete with embedded reinforcement, usually steel bars or tendons. Reinforcement is placed in structural members at locations where it will be of the most benefit. It is usually thought of as resisting tension, but it is also used to resist compression. If shear in a beam is the limit state that is being resisted, longitudinal and transverse reinforcements are placed to resist diagonal tension forces.

The behavior of nonprestressed reinforcement is usually characterized by the stress–strain curve for bare steel bars. The behavior of prestressed steel tendons is known to be different for bonded and unbonded tendons, which suggests that we should reconsider the behavior of nonprestressed reinforcement embedded in concrete.

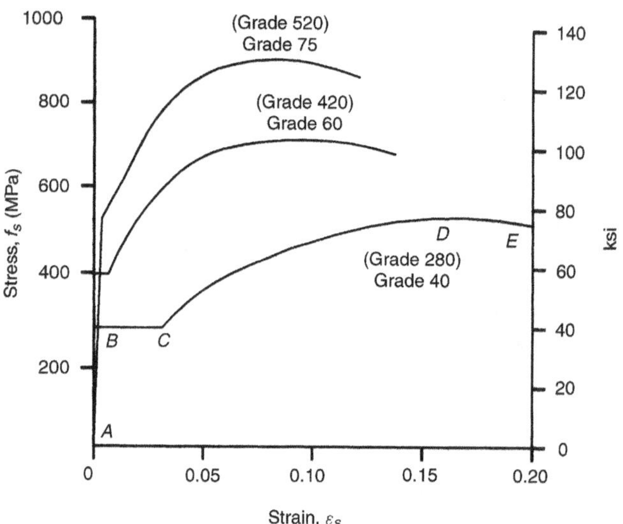

Fig. 13.13 Stress–strain curves for bare steel reinforcement. [From Holzer et al. (1975).]

13.5.1 Nonprestressed Steel Reinforcement

Typical stress–strain curves for bare steel reinforcement are shown in Figure 13.13 for steel grades 40, 60, and 75. The response of the bare steel can be broken into three parts, elastic, plastic, and strain hardening. The elastic portion *AB* of the curves respond in a similar straight-line manner with a constant modulus of elasticity E_S = 29,000 ksi (200 GPa) up to a yield strain of $\varepsilon_y = f_y/E_S$. The plastic portion *BC* is represented by a yield plateau at constant stress f_y until the onset of strain hardening. The length of the yield plateau is a measure of ductility and it varies with the grade of steel. The strain-hardening portion *CDE* begins at a strain of ε_h and reaches maximum stress f_u at a strain of ε_u before dropping off slightly at a breaking strain of ε_b. The three portions of the stress–strain curves for bare steel reinforcement can be characterized symbolically as

Elastic Portion *AB*

$$f_s = \varepsilon_s E_s \qquad 0 \leq \varepsilon_s \leq \varepsilon_y \tag{13.34}$$

Plastic Portion *BC*

$$f_s = f_y \qquad \varepsilon_y \leq \varepsilon_s \leq \varepsilon_h \tag{13.35}$$

Strain–Hardening Portion *CDE*

$$f_s = f_y\left[1 + \frac{\varepsilon_s - \varepsilon_h}{\varepsilon_u - \varepsilon_h}\left(\frac{f_u}{f_y} - 1\right)\exp\left(1 - \frac{\varepsilon_s - \varepsilon_h}{\varepsilon_u - \varepsilon_h}\right)\right]$$
$$\varepsilon_h \leq \varepsilon_s \leq \varepsilon_b \tag{13.36}$$

Equation 13.36 and the nominal limiting values for stress and strain in Table 13.3 are taken from Holzer et al. (1975). The curves shown in Figure 13.13 are calibrated to pass through the nominal yield stress values of the different steel

Table 13.3 Nominal Limiting Values for Bare Steel Stress–Strain Curves

f_y, ksi (MPa)	f_u, ksi (MPa)	ε_y	ε_h	ε_u	ε_b
40 (280)	80 (550)	0.00138	0.0230	0.140	0.200
60 (420)	106 (730)	0.00207	0.0060	0.087	0.136
75 (520)	130 (900)	0.00259	0.0027	0.073	0.115

Holzer et al. (1975).

grades. The actual values for the yield stress from tensile tests average about 15% higher. The same relationship is assumed to be valid for both tension and compression. When steel bars are embedded in concrete, the behavior is different than for the bare steel bars. The difference is due to the fact that concrete has a finite, though small, tensile strength, which was realized early in the development of the mechanics of reinforced concrete as described by Collins and Mitchell (1991). Concrete that adheres to the reinforcement and is uncracked reduces the tensile strain in the reinforcement. This phenomenon is called *tension stiffening*.

An experimental investigation by Scott and Gill (1987) confirmed the decrease in tensile strain in the reinforcement between cracks in the concrete. To measure the strains in the reinforcement without disturbing the bond characteristics on the surface of the bars, they placed the strain gages inside the bars. This internal placement of strain gages was done by splitting a bar in half, machining out a channel, placing strain gages and their lead wires, and then gluing the halves back together. The instrumented bar was then encased in concrete, except for a length at either end that could be gripped in the jaws of a testing machine. Tensile loads were then applied and the strains along the bar at 0.5-in. (12.5-mm) increments were recorded. Figure 13.14 presents the strains in one of their bars over the 39-in. (1000-mm) long section at increasing levels of tensile load.

An approximate bare bar strain for a tensile load of 9.0 kips is shown in Figure 13.14 by the horizontal dashed line. The strain is approximate because the area of the bar (0.20 in.^{2}) used in the calculation should be reduced by the area of the channel cut for placement of the gages. The actual bare bar strain would be slightly higher and more closely average out the peaks and valleys. Observations on the behavior remain the same: (a) steel tensile strains increase at locations where concrete is cracked and (b) steel tensile strains decrease between cracks because of the tensile capacity of the concrete adhering to the bar.

To represent the behavior of reinforcement embedded in concrete as shown in Figure 13.14, it is convenient to define an average stress and average strain over a length long enough to include at least one crack. The average stress–strain behavior for concrete-stiffened mild steel reinforcement is shown in Figure 13.15 and compared to the response of a bare bar (which is indicative of the bar response at a crack where the concrete contribution is lost).

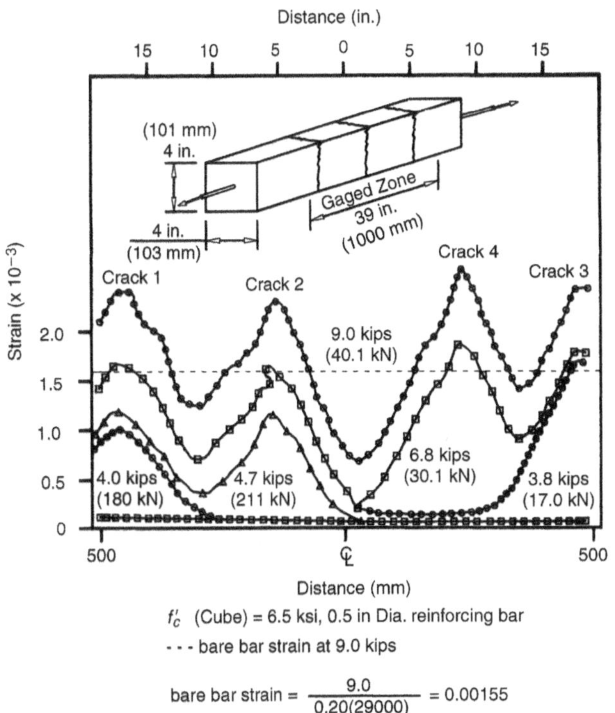

Fig. 13.14 Variation of steel strain along the length of a tension specimen tested by Scott and Gill (1987). [From Collins and Mitchell (1991). Reprinted by permission of Prentice Hall, Upper Saddle River, NJ.]

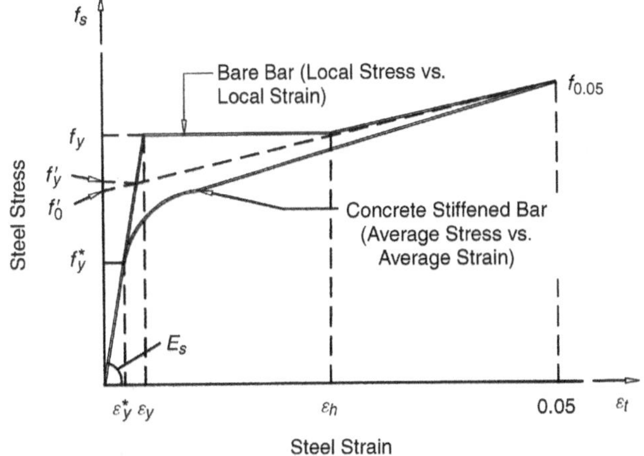

Fig. 13.15 Stress–strain curve for mild steel. [Reprinted with permission from T. T. C. Hsu (1993). *Unified Theory of Reinforced Concrete*, CRC Press, Boca Raton, FL. Copyright CRC Press, Boca Raton, FL © 1993.]

The tension stiffening effect of the concrete is greatest, as would be expected, at low strains and *tends to round off the sharp knee of the elastic perfectly plastic behavior*. This tension stiffening effect results in the average steel stress showing a reduced value of apparent yield stress f_y^* and its accompanying apparent yield strain ε_y^*. At higher strains, the concrete contribution diminishes and the embedded bar response follows the strain-hardening portion of the bare steel curve.

Also shown by the dashed line in Figure 13.15 is a linear approximation to the average stress–strain response of a mild steel bar embedded in concrete. A derivation of this approximation and a comparison with experimental data are given in Hsu (1993). The equation for these two straight lines are given by

Elastic Portion

$$f_s = E_s \varepsilon_s \qquad \text{when} \qquad f_s \le f_y' \tag{13.37}$$

Postyield Portion

$$f_s = (0.91 - 2B)\, f_y + (0.02 + 0.25B)\, E_S \varepsilon_S \qquad \text{when} \quad f_s > f_y' \tag{13.38}$$

where

$$f_y' = \text{intersection stress level} = (0.93 - 2B)\, f_y \tag{13.39}$$

$$B = \frac{1}{\rho}\left(\frac{f_{cr}}{f_y}\right)^{1.5} \tag{13.40}$$

ρ = steel reinforcement ratio based on the net concrete section $= A_s/(A_g - A_s)$

f_{cr} = tensile cracking strength of concrete, taken as $0.12\sqrt{f_c'}$ (ksi)

f_y = steel yield stress of bare bars (ksi)

Figure 13.16 compares the bilinear approximation of an average stress–strain curve ($\rho = 0.01$, $f_{cr} = 0.240$ ksi, $f_y = 60$ ksi) with a bare bar and test results by Tamai et al. (1987). This figure illustrates what was stated earlier and shows how the *response in tension of reinforcement embedded in concrete is different than the response of bare steel alone*.

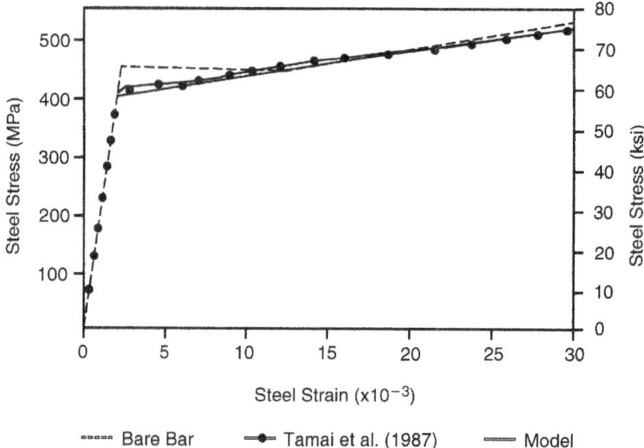

Fig. 13.16 Average stress–strain curves of mild steel: Theories and tests. [Reprinted with permission from T. T. C. Hsu (1993). *Unified Theory of Reinforced Concrete*, CRC Press, Boca Raton, FL. Copyright CRC Press, Boca Raton, FL © 1993.]

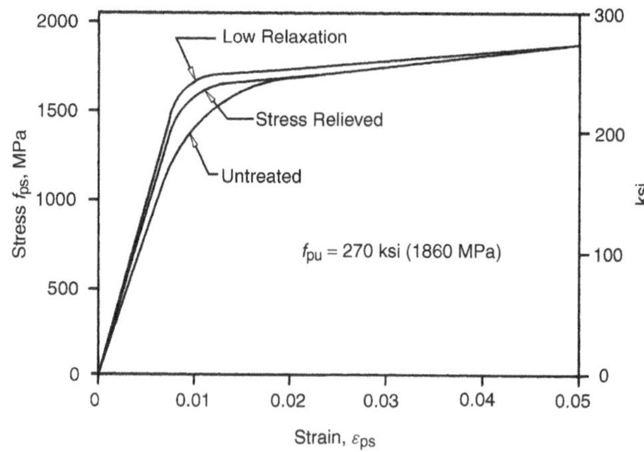

Fig. 13.17 Stress–strain response of seven-wire strand manufactured by different processes. [After Collins and Mitchell (1991). Reprinted by permission of Prentice Hall, Upper Saddle River, NJ.]

13.5.2 Prestressing Steel

The most common prestressing steel is seven-wire strand, which is available in stress-relieved strand and low-relaxation strand. During manufacture of the strands, high carbon steel rod is drawn through successively smaller diameter dies, which tends to align the molecules in one direction and increases the strength of the wire to over 250 ksi (1700 MPa). Six wires are then wrapped around one central wire in a helical manner to form a strand. The cold drawing and twisting of the wires creates locked in or residual stresses in the strands. These residual stresses cause the stress–strain response to be more rounded and to exhibit an apparently lower yield stress. The apparent yield stress can be raised by heating the strands to about 660°F (350°C) and allowing them to cool slowly. This process is called stress relieving. Further improvement in behavior by reducing the relaxation of the strands is achieved by putting the strands into tension during the heating and cooling process. This process is called strain tempering and produces the low-relaxation strands. Figure 13.17 compares the stress–strain response of seven-wire strand manufactured by the different processes. Low-relaxation strands are most commonly used and are regarded as the standard type [C5.4.4.1].

High-strength deformed bars are also used for prestressing steel. The deformations are often like raised screw threads so that devices for posttensioning and anchoring bars can be

Table 13.4 Properties of Prestressing Strand and Bar

Material	Grade or Type	Diameter (in.)	Tensile Strength, f_{pu} (ksi)	Yield Strength, f_{py} (ksi)
Strand	250 ksi	$\frac{1}{4}$ -0.6	250	85% of f_{pu} except 90% of f_{pu} for low-relaxation strand
	270 ksi	$\frac{3}{8}$ -0.6	270	
Bar	Type 1, plain	$\frac{3}{4}$ $-1\frac{3}{8}$	150	85% of f_{pu}
	Type 2, deformed	$\frac{5}{8}$ $-1\frac{3}{8}$	150	80% of f_{pu}

In AASHTO Table 5.4.4.1-1. From *AASHTO LRFD Bridge Design Specifications*, Copyright © 2010 by the American Association of State Highway and Transportation Officials, Washington, DC. Used by permission.

attached to their ends. The ultimate tensile strength of the bars is about 150 ksi (1000 MPa).

A typical specification for the properties of the prestressing strand and bar is given in Table 13.4. Recommended values for modulus of elasticity, E_p, for prestressing steels are 28,500 ksi (197 GPa) for strands and 30,000 ksi (207 GPa) for bars [A5.4.4.2].

The stress–strain curves for the bare prestressing strand shown in Figure 13.17 have been determined by a Ramberg–Osgood function to give a smooth transition between two straight lines representing elastic and plastic behavior. Constants are chosen so that the curves pass through a strain of 0.01 when the yield strengths of Table 13.4 are reached. Collins and Mitchell (1991) give the following expression for low-relaxation strands with

$$f_{pu} = 270 \text{ ksi} \,(1860 \text{ MPa})$$

$$f_{ps} = E_p \varepsilon_{ps} \left\{ 0.025 + \frac{0.975}{\left[1 + (118\varepsilon_{ps})^{10}\right]^{0.10}} \right\} \le f_{pu} \tag{13.41}$$

while for stress-relieved strands with $f_{pu} = 270$ ksi (1860 MPa)

$$f_{ps} = E_p \varepsilon_{ps} \left\{ 0.03 + \frac{0.97}{\left[1 + (121\varepsilon_{ps})^{6}\right]^{0.167}} \right\} \le f_{pu} \tag{13.42}$$

and for untreated strands with $f_{pu} = 240$ ksi (1655 MPa)

$$f_{ps} = E_p \varepsilon_{ps} \left\{ 0.03 + \frac{1}{\left[1 + (106\varepsilon_{ps})^{2}\right]^{0.5}} \right\} \le f_{pu} \tag{13.43}$$

These curves are based on the minimum specified strengths. The actual stress–strain curves of typical strands probably have higher yield strengths and can be above those shown in Figure 13.17.

A tendon can be either a single strand or bar, or it can be a group of strands or bars. When the tendons are bonded to the concrete, the *change* in strain of the prestressing steel is equal to the *change* in strain of the concrete. This condition exists in pretensioned beams where the concrete is cast around the tendons and in posttensioned beams where the tendons are pressure grouted after they are prestressed. At the time the concrete or grout is placed, the prestressing tendon has been stretched and has a difference in strain of $\Delta\varepsilon_{pe}$ when the two materials are bonded together. The strain in the prestressing tendon ε_{ps} can be determined at any stage of loading from the strain in the surrounding concrete ε_{cp} as

$$\varepsilon_{ps} = \varepsilon_{cp} + \Delta\varepsilon_{pe} \tag{13.44}$$

where ε_{cp} is the concrete strain at the same location as the prestressing tendon, and $\Delta\varepsilon_{pe}$ is

$$\Delta\varepsilon_{pe} = \varepsilon_{pe} - \varepsilon_{ce} \tag{13.45}$$

where ε_{pe} is the strain corresponding to the effective stress in the prestressing steel after losses f_{pe} expressed as

$$\varepsilon_{pe} = \frac{f_{pe}}{E_p} \tag{13.46}$$

and ε_{ce} is the strain in the concrete at the location of the prestressing tendon resulting from the effective prestress. If the tendon is located along the centrodial axis, then

$$\varepsilon_{pe} = \varepsilon_{ce} = \frac{A_{ps} f_{pe}}{E_c A_c} \tag{13.47}$$

where A_{ps} is the prestressing steel area and A_c is the concrete area. This strain is always small and is usually ignored (Loov, 1988) so that $\Delta\varepsilon_{pe}$ is approximately equal to

$$\Delta\varepsilon_{pe} \approx \frac{f_{pe}}{E_p}$$

In the case of an unbonded tendon, slip results between the tendon and the surrounding concrete and the strain in the tendon becomes uniform over the distance between anchorage points. The total change in length of the tendon must now equal the total change in length of the concrete over this distance, that is,

$$\varepsilon_{ps} = \bar{\varepsilon}_{cp} + \Delta\varepsilon_{pe} \tag{13.48}$$

where $\bar{\varepsilon}_{cp}$ is the average strain of the concrete at the location of the prestressing tendon, averaged over the distance between anchorages of the unbonded tendon.

REFERENCES

AASHTO (1994). *LRFD Bridge Design Specifications*, 1st ed., American Association of State Highway and Transportation Officials, Washington, DC.

AASHTO (2010). *LRFD Bridge Design Specifications*, 5th ed., American Association of State Highway and Transportation Officials, Washington, DC.

Al-Omaishi, N. (2001). Prestress Losses in High Strength Pretensioned Concrete Bridge Girders, Ph.D. Dissertation, University of Nebraska-Lincoln, December.

Collins, M. P. and D. Mitchell (1991). *Prestressed Concrete Structures*. Prentice-Hall, Englewood Cliffs, NJ.

Gopalaratnam, V. S. and S. P. Shah (1985). "Softening Response of Plain Concrete in Direct Tension," *ACI Journal*, Vol. 82, No. 3, May–June, pp. 310–323.

Hawkins, N. M. and D. A. Kuchma (2007). *Application of LRFD Bridge Design Specifications to High-Strength Structural Concrete: Shear Provisions*, NCHRP Report 579, National Cooperative Highway Research Program, Project 12–56, Transportation Research Board, Washington, DC.

Holzer, S. M., R. J. Melosh, R. M. Barker, and A. E. Somers (1975). "SINGER. A Computer Code for General Analysis of Two-Dimensional Reinforced Concrete Structures," *AFWL-TR-74-228 Vol*. 1, Air Force Weapons Laboratory, Kirtland AFB, New Mexico.

Hsu, T. T. C. (1993). *Unified Theory of Reinforced Concrete*, CRC Press, Boca Raton, FL.

Huo, X. S., N. Al-Omaishi, and M. K. Tadros (2001). "Creep, Shrinkage, and Modulus of Elasticity of High Performance Concrete," *ACI Materials Journal*, Vol. 98, No. 6, November–December, pp. 440–449.

Loov, R. E. (1988). "A General Equation for the Steel Stress for Bonded Prestressed Concrete Members," *PCI Journal*, Vol. 33, No. 6, Nov.–Dec., pp. 108–137.

Mander, J. B., M. J. N. Priestley, and R. Park (1988). "Theoretical Stress–Strain Model for Confined Concrete," *Journal of Structural Engineering, ASCE*, Vol. 114, No. 8, pp. 1804–1826.

Ramirez, J. A, and B. W. Russell (2008). *Transfer Length and Splice Length for Strand/Reinforcement in High-Strength Concrete*, NCHRP Report 603, National Cooperative Highway Research Program, Project 12–60, Transportation Research Board, Washington, DC.

Richart, F. E., A. Brandtzaeg, and R. L. Brown (1928). "A Study of the Failure of Concrete under Combined Compressive Stresses," *Bulletin 185*, University of Illinois Engineering Experiment Station, Champaign, IL.

Rizkalla, S., A. Z. Mirmiran, H. P. Russell, and R. Mast (2007). *Application of the LRFD Bridge Design Specifications to High-Strength Structural Concrete: Flexure and Compression Provisions*, NCHRP Report 595, National Cooperative Highway Research Program, Project 12–64, Transportation Research Board, Washington, DC.

Scott, R. H. and P. A. T. Gill (1987). "Short-Term Distribution of Strain and Bond Stress along Tension Reinforcement," *Structural Engineer*, Vol. 658, No. 2, June, pp. 39–48.

Setunge, S., M. M. Attard, and P. LeP. Darvall (1993). "Ultimate Strength of Confined Very High-Strength Concretes," *ACI Structural Journal*, Vol. 90, No. 6, Nov.–Dec., pp. 632–641.

Sheikh, S. A. (1982). "A Comparative Study of Confinement Models," *ACI Journal, Proceedings*, Vol. 79, No. 4, pp. 296–306.

Sheikh, S. A. and S. M. Uzumeri (1980). "Strength and Ductility of Tied Concrete Columns," *Journal of Structural Division*, ASCE, Vol. 106, No. ST5, May, pp. 1079–1102.

Tadros, M. K., N. Al-Omaishi, S. P. Seguirant, and J. G. Gallt (2003). *Prestress Losses in Pretensioned High-Strength Concrete Bridge Girders*, NCHRP Report 496, Transportation Research Board, Washington, DC.

Tamai, S., H. Shima, J. Izumo, and H. Okamura (1987). "Average Stress–Strain Relationship in Post Yield Range of Steel Bar in Concrete," *Concrete Library International of JSCE*, No. 11, June 1988, pp. 117–129. (Translation from *Proceedings of JSCE*, No. 378/V-6, Feb. 1987.)

Timoshenko, S. and J. N. Goodier (1951). *Theory of Elasticity*, 2nd ed., McGraw-Hill, New York.

Troxell, G. E., H. E. Davis, and J. W. Kelly (1968). *Composition and Properties of Concrete*, 2nd ed., McGraw-Hill, New York.

Vecchio, F. J. and M. P. Collins (1986). "The Modified Compression Field Theory for Reinforced Concrete Elements Subjected to Shear," *ACI Journal*, Vol. 83, No. 2, Mar–Apr, pp. 219–231.

PROBLEMS

13.1 In what ways does concrete respond differently in compression to loads when wrapped with steel than when no steel is present?

13.1 In what ways does steel respond differently in tension to loads when wrapped in concrete than when no concrete is present?

13.3 Determine the peak confined concrete stress f'_{cc} and corresponding strain ε_{cc} for a 24-in. diameter column with 10 No. 11 longitudinal bars and No. 4 round spirals at 3-in. pitch. The material strengths are $f'_c = -5.0$ ksi and $f_{yh} = 60$ ksi. Assume that $\varepsilon_{co} = -0.002$ and that the concrete cover is 1.5 in. Use the lower bound value of $k_1 = 3$ and the corresponding value of $k_2 = 15$.

13.4 Determine the parameters and plot the stress–strain curves for the unconfined and confined concrete of the column section in Problem 13.3. Assume concrete strain at first hoop fracture $\varepsilon_{cu} = 8\varepsilon_{cc}$.

13.5 Describe the *compression softening* phenomenon that occurs when concrete is subjected to a biaxial state of stress.

13.6 Why are two values given in the specifications for the modulus of rupture of concrete?

13.7 Use Eqs. 13.22 and 13.23 to plot the average stress versus average strain curve for concrete in tension. Use $f'_c = 6$ ksi and Eq. 13.21 for f_{cr}. Assume $\alpha_1 = \alpha_2 = 1.0$.

13.8 What similarities do the time-dependent responses of shrinkage and creep in concrete have? Differences?

13.9 Estimate the shrinkage strain in a 7.5-in. thick concrete bridge deck ($f_c' = 6.0$ ksi) whose top and bottom surfaces are exposed to drying conditions in an atmosphere with 60% relative humidity. Plot the variation of shrinkage strain with drying time for these conditions. Use $t = 28, 90, 365, 1000$, and 2000 days.

13.10 Estimate the creep strain in the bridge deck of Problem 13.9 after one year if the elastic compressive stress due to sustained loads is 1.55 ksi, the 28-day compressive strength is 6.0 ksi, and $t_i = 15$ days. Plot the variation of creep strain with time under load for these conditions. Use $t = 28, 90, 365, 1000$, and 2000 days.

13.11 Describe the *tension stiffening* effect of concrete on the stress–strain behavior of reinforcement embedded in concrete.

13.12 Plot the stress–strain response of seven-wire low-relaxation prestressing strands using the Ramberg–Osgood Eq. 13.41. Use $f_{pu} = 270$ ksi and $E_p = 28{,}500$ ksi.

CHAPTER 14

Behavior of Reinforced Concrete Members

14.1 LIMIT STATES

Reinforced concrete bridges must be designed so that their performance under load does not go beyond the limit states prescribed by AASHTO. These limit states are applicable at all stages in the life of a bridge and include service, fatigue, strength, and extreme event limit states. The condition that must be met for each of these limit states is that the factored resistance is greater than the effect of the factored load combinations, or simply, supply must exceed demand. The general inequality that must be satisfied for each limit state can be expressed as

$$\phi R_n \geq \sum \eta_i \gamma_i Q_i \tag{14.1}$$

where ϕ is a statistically based resistance factor for the limit state being examined; R_n is the nominal resistance; η_i is a load multiplier relating to ductility, redundancy, and operational importance; γ_i is a statistically based load factor applied to the force effects as defined for each limit state in Table 5.1; and Q_i is a force effect. The various factors in Eq. 14.1 are discussed more fully in Chapter 5 and are repeated here for convenience.

14.1.1 Service Limit State

Service limit states relate to bridge performance. Actions to be considered are cracking, deformations, and stresses for concrete and prestressing tendons under regular service conditions. Because the provisions for service limit states are not derived statistically, but rather are based on experience and engineering judgment, the resistance factors ϕ and load factors γ_i are usually taken as unity. There are some exceptions for vehicle live loads and wind loads as shown in Table 5.1.

Control of Flexural Cracking in Beams [A5.7.3.4]**** The width of flexural cracks in reinforced concrete beams, where the tensile stress is greater than 80% of the modulus of rupture, is controlled by limiting the spacings in the reinforcement under service loads over the region of maximum concrete tension:

$$s \leq \frac{700\gamma_e}{\beta_s f_s} - 2d_c \tag{14.2}$$

where

$\beta_s = 1 + \dfrac{d_c}{0.7(h - d_c)}$

γ_e = exposure factor
= 1.00 for class 1 exposure condition
= 0.75 for class 2 exposure condition

d_c = depth of concrete cover from extreme tension fiber to center of closest flexural reinforcement (in.)

f_s = tensile stress in reinforcement at the service limit state (ksi)

h = overall thickness or depth of the component (in.)

Class 1 exposure condition applies when cracks can be tolerated due to reduced concerns of appearance and/or corrosion. Class 1 exposure is calibrated to a crack width of 0.017 in. Class 2 exposure condition applies to transverse design of segmental concrete box girders for any loads prior to attaining full nominal concrete strength and when there is increased concern of appearance and/or corrosion. Deck slabs are excluded from this requirement.

The exposure factor γ_e is directly proportional to the crack width and can be adjusted as shown in Table 14.1 to obtain a desired crack width [C5.7.3.4]. The β_s factor is a geometric relationship between the crack width at the tension face and the crack width at the reinforcement level. It provides uniformity of application for flexural member depths ranging from thin slabs to deep pier caps.

An effective way to satisfy Eq. 14.2 is to use several smaller bars at moderate spacing rather than a few larger bars of equivalent area. This procedure distributes the reinforcement over the region of maximum concrete tension and provides good crack control. The minimum and maximum spacing of reinforcement shall also comply with the provisions of AASHTO articles [A5.10.3.1] and [A5.10.3.2].

To guard against excessive spacing of bars when flanges of T-beams and box girders are in tension, the flexural tension reinforcement is to be distributed over the lesser of the effective flange width or a width equal to one-tenth of the span. If the effective flange width exceeds one-tenth the span, additional longitudinal reinforcement, with area not less than 0.4% of the excess slab area, is to be provided in the outer portions of the flange.

---

*The article numbers in the AASHTO (2010) LRFD Bridge Specifications are enclosed in brackets and preceded by the letter A if a specification article and by the letter C if commentary.

Table 14.1 Exposure Factor γ_e

Exposure Condition	γ_e	Crack Width (in.)
Moderate, class 1	1.0	0.017
Severe, class 2	0.75	0.013
Aggressive	0.5	0.0085

For relatively deep flexural members, reinforcement should also be distributed in the vertical faces in the tension region to control cracking in the web. If the web depth exceeds 3.0 ft (900 mm), longitudinal skin reinforcement is to be uniformly distributed over a height of $d/2$ nearest the tensile reinforcement. The area of skin reinforcement A_{sk} in square inches per foot (in.^{2}/ft) of height required on each side face is

$$A_{sk} \geq 0.012\left(d_t - 3.0\right) \leq \frac{A_s + A_{ps}}{4} \tag{14.3}$$

where d_t is the distance from the extreme compression fiber to the centroid of the extreme tensile steel element, A_s is the area of the nonprestressed steel, and A_{ps} is the area of the prestressing tendons. The maximum spacing of the skin reinforcement is not to exceed either $d_e/6$ or 12.0 in.

Deformations Service load deformations may cause deterioration of wearing surfaces and local cracking in concrete slabs. Vertical deflections and vibrations due to moving vehicle loads can cause motorists concern. To limit these effects, optional deflection criteria are suggested [A2.5.2.6.2] as

- ☐ Vehicular load, general: span length/800
- ☐ Vehicular load on cantilever arms: span length/300

where the vehicle load includes the impact factor IM and the multiple presence factor m.

When calculating the vehicular deflection, it should be taken as the larger of that resulting from the design truck alone or that resulting from 25% of the design truck taken together with the design lane load [A3.6.1.3.2]. All of the design lanes should be loaded and *all* of the girders may be assumed to deflect equally in supporting the load. This statement is equivalent to a deflection distribution factor g equal to the number of lanes divided by the number of girders.

Calculated deflections of bridges have been difficult to verify in the field because of additional stiffness provided by railings, sidewalks, and median barriers not usually considered in the calculations. Therefore, it seems reasonable to estimate the instantaneous deflection using the elastic modulus for concrete E_c from Eq. 13.2 and the gross moment of inertia I_g [A5.7.3.6.2]. This estimate is much simpler, and probably just as reliable, as using the effective moment of inertia I_e based on a value between I_g and the cracked moment of inertia I_{cr}. It also makes the calculation of the long-term deflection more tractable because it can be taken as simply 4.0 times the instantaneous deflection [A5.7.3.6.2].

Stress Limitations for Concrete Service limit states still apply in the design of reinforced concrete members that have prestressing tendons that precompress the section so that concrete stresses f_c can be determined from elastic uncracked section properties and the familiar equation

$$f_c = -\frac{P}{A_g} \pm \frac{Pey}{I_g} \mp \frac{My}{I_g} \tag{14.4}$$

where P is the prestressing force, A_g is the cross-sectional area, e is the eccentricity of the prestressing force, M is the moment due to applied loads, y is the distance from the centroid of the section to the fiber, and I_g is the moment of inertia of the section. If the member is a composite construction, it is necessary to separate the moment M into the moment due to loads on the girder M_g and the moment due to loads on the composite section M_c, because the y and I values are different, that is,

$$f_c = -\frac{P}{A_g} \pm \frac{Pey}{I_g} \mp \frac{M_g y}{I_g} \mp \frac{M_c y_c}{I_c} \tag{14.5}$$

where the plus and minus signs for the stresses at the top and bottom fibers must be consistent with the chosen sign convention where tension is positive. (Often positive is used for compression in concrete; however, in this book a consistent approach is used for all materials.) These linear elastic concrete stress distributions are shown in Figure 14.1.

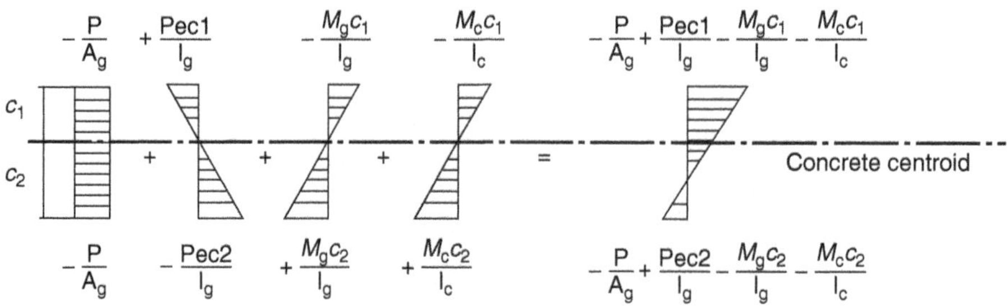

Fig. 14.1 Linear-elastic concrete stress distributions in composite prestressed beams.

Limits on the concrete stresses are given in Tables 14.2 and 14.3 for two load stages: (1) prestress transfer stage—immediately after transfer of the prestressing tendon tensile force to the concrete but prior to the time-dependent losses due to creep and shrinkage, and (2) service load stage—after allowance for all prestress losses. The concrete compressive strength at time of initial loading f'_{ci}, the 28-day concrete compressive strength f'_c, and the resulting stress limits are all in kips per square inch (ksi). A precompressed tensile zone is a region that was compressed by the prestressing tendons but has gone into tension when subjected to dead- and live-load moments. The stress limits in the tables are for members with prestressed reinforcement only and do not include those for segmentally constructed bridges. The compressive and tensile stress limits for temporary stresses before losses are prescribed in [A5.9.4.1.1] and [A5.9.4.1.2], respectively.

The reduction factor ϕ_w in Table 14.3 shall be taken as 1.0 when the web and flange slenderness ratios, calculated according to [A5.7.4.7.1], are not greater than 15. When either the web or flange slenderness ratio is greater than 15, the reduction factor ϕ_w shall be calculated according to [A5.7.4.7.2].

For the components that include both prestressed and nonprestressed reinforcement (often called partially prestressed because only a part of the reinforcement is prestressed), the compressive stress limits are those given in Tables 14.2 and 14.3, but because cracking is permitted, the tensile stress is given in [A5.7.3.4], where f_s is to be interpreted as the change in stress after decompression.

Stress Limitations for Prestressing Tendons The tendon stress, due to prestress operations or at service limit states, shall not exceed the values as specified by AASHTO in

Table 14.2 Stress Limits for Concrete for Temporary Stresses before Losses—Fully Prestressed Components

Compressive Stresses	
Pretensioned components	$0.60 f'_{ci}$ (ksi)
Posttensioned components	$0.60 f'_{ci}$ (ksi)
Tensile Stresses	
Precompressed tensile zone without bonded reinforcement	N/A[a]
Other tensile zones without bonded reinforcement	$0.0948\sqrt{f'_{ci}} \le 0.2$ (ksi)
Tensile zones with bonded reinforcement sufficient to resist the tension force in the concrete computed assuming an uncracked section, where reinforcement is proportioned using a stress of $0.5 f_y$ not to exceed 30 ksi	$0.24\sqrt{f'_{ci}}$ (ksi)
Handling stresses in prestressed piles	$0.158\sqrt{f'_{ci}}$ (ksi)

[a]Not applicable = N/A.

From AASHTO [A5.9.4.1]. From *AASHTO LRFD Bridge Design Specifications*,

Table 14.3 Stress Limits for Concrete at Service Limit State after Losses—Fully Prestressed Components

Compressive Stresses—load combination service I or fatigue I	
Due to the sum of effective prestress and permanent loads	$0.45 f'_c$ (ksi)
Due to live load (fatigue I) and one-half the sum of effective prestress and permanent loads	$0.40 f'_c$ (ksi)
Due to sum of effective prestress, permanent loads, and transient loads (service I) and during shippng and handling	$0.60\phi_w f'_c$ (ksi)
Tensile Stresses—load combination service III	
Precompressed tensile zone bridges, assuming uncracked sections	
Components with bonded prestressing tendons or reinforcement that are not subjected to worse than moderate corrosion conditions	$0.19\sqrt{f'_c}$ (ksi)
Components with bonded prestressing tendons or reinforcement that are subjected to severe corrosive conditions	$0.0948\sqrt{f'_c}$ (ksi)
Components with unbonded prestressing tendons	No tension
Other tensile zone stresses are limited by those given for the prestress transfer stage in Table 14.2.	

From AASHTO [A5.9.4.2]. From *AASHTO LRFD Bridge Design Specifications*,

Table 14.4 Stress Limits for Prestressing Tendons

Condition	Tendon Type: Stress Relieved Strand and Plain High-Strength Bars	Low-Relaxation Strand	Deformed High-Strength Bars
	Pretensioning		
Immediately prior to transfer (f_{pbt})	$0.70 f_{pu}$	$0.75 f_{pu}$	—
At service limit state after all losses (f_{pe})	$0.80 f_{py}$	$0.80 f_{py}$	$0.80 f_{py}$
	Posttensioning		
Prior to seating—short-term f_{pbt} may be allowed	$0.90 f_{py}$	$0.90 f_{py}$	$0.90 f_{py}$
At anchorages and couplers immediately after anchor set	$0.70 f_{pu}$	$0.70 f_{pu}$	$0.70 f_{pu}$
Elsewhere along length of member away from anchorages and couplers immediately after anchor set	$0.70 f_{pu}$	$0.74 f_{pu}$	$0.70 f_{pu}$
At service limit state after losses (f_{pe})	$0.80 f_{py}$	$0.80 f_{py}$	$0.80 f_{py}$

Table 14.4 or as recommended by the manufacturer of the tendons and anchorages. The tensile strength f_{pu} and yield strength f_{py} for prestressing strand and bar can be taken from Table 13.4.

14.1.2 Fatigue Limit State

Fatigue is a characteristic of a material in which damage accumulates under repeated loadings so that failure occurs at a stress level less than the static strength. In the case of highway bridges, the repeated loading that causes fatigue is the trucks that pass over them. An indicator of the fatigue damage potential is the stress range f_f of the fluctuating stresses produced by the moving trucks. A second indicator is the number of times the stress range is repeated during the expected life of the bridge. In general, the higher the ratio of the stress range to the static strength, the fewer the number of loading cycles required to cause fatigue failure.

For fatigue considerations, concrete members shall satisfy:

$$\gamma(\Delta f) \leq (\Delta F)_{TH} \qquad (14.6a)$$

where

γ = load factor specified in Table 5.1 for the fatigue I load combination

Δf = force effect, live-load stress range due to the passage of the fatigue load as specified in A3.6.1.4 (ksi)

$(\Delta F)_{TH}$ = constant-amplitude fatigue threshold, as specified in [A5.5.3.2], [A5.5.3.3], or [A5.5.3.4] as appropriate

In calculating the fatigue stress range Δf, the fatigue loading described in Chapter 8 is used. This loading consists of a special fatigue truck with constant axle spacing of 30 ft between the 32-kip axles, applied to one lane of traffic without multiple presence, and with an impact factor IM of 15% [A3.6.1.4]. The fatigue load combination (fatigue II) of Table 5.1 has a load factor of 0.75 applied to the fatigue truck, all other load factors are zero. Elastic-cracked section properties are used to calculate Δf, except when gross section properties can be used for members with prestress where the sum of the stresses due to unfactored permanent loads and prestress plus 1.5 times the unfactored fatigue load (fatigue I) does not exceed a tensile stress of $0.095\sqrt{f'_c}$ [A5.5.3.1].

For reinforced concrete components, AASHTO [A5.5.3] does not include the number of cycles of repeated loading as a parameter in determining the fatigue strength. What is implied is that the values given for the limits on the stress range are low enough so that they can be considered as representative of infinite fatigue life. Background on the development of fatigue stress limits for concrete, reinforcing bars, and prestressing strands can be found in the report by ACI Committee 215 (1992), which summarizes over 100 references on analytical, experimental, and statistical studies of fatigue in reinforced concrete. In their report, which serves as the basis for the discussion that follows, the fatigue stress limits appear to have been developed for 2–10 million cycles.

Fatigue of Plain Concrete When plain concrete beams are subjected to repetitive stresses that are less then the static strength, accumulated damage due to progressive internal microcracking eventually results in a fatigue failure. If the repetitive stress level is decreased, the number of cycles to failure N increases. This effect is shown by the *S–N* curves in Figure 14.2, where the ordinate is the ratio of the maximum stress S_{max} to the static strength and the abscissa is the number of cycles to failure N, plotted on a logarithmic scale. For the case of plain concrete beams, S_{max} is the tensile stress calculated at the extreme fiber assuming an uncracked section and the static strength is the rupture modulus stress f_r.

Exhibit 1.1 The Pont du Gard Aqueduct, Nimes, France, was built by Romans 40–60 A.D. The lower arches were widened in 1743 to accommodate a road bridge.

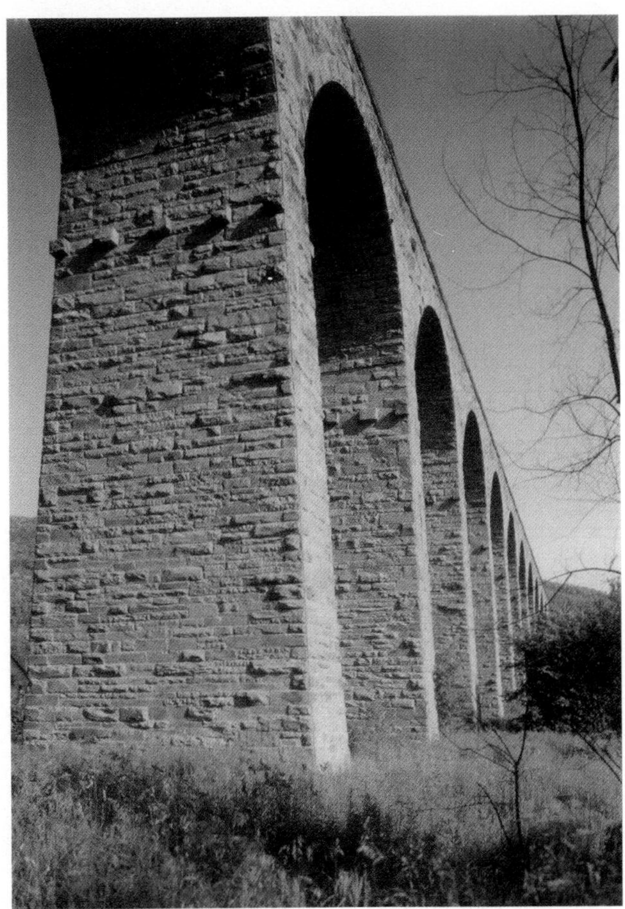

Exhibit 1.2 The Starrucca Viaduct near Lanesboro, Pennsylvania, was built in 1848 by the Erie Railway. At the time of its construction, it was the largest stone arch rail viaduct in the United States. The bridge has been in continual use for more than 160 years and still carries two tracks of the New York, Susquehanna and Western Railway. (HAER PA-6-17, photo by Jack E. Boucher, 1971.)

Exhibit 1.3 The Philippi Covered Bridge across the Tygart River Valley near Philippi, West Virginia, was built in 1852. It was used by armies of both the North and the South in the Civil War. In 1934 the bridge was strengthened and is today a part of U.S. 250. It is reportedly the only remaining two-lane "double barrel" covered bridge.

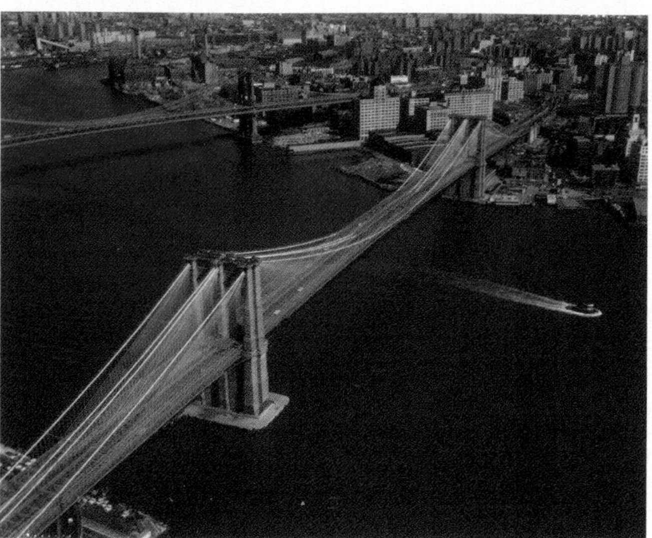

Exhibit 1.4 The Brooklyn Bridge was built 1869–1883 by John and Washington Roebling spanning the East River from Manhattan to Brooklyn, New York (photo looking east towards Brooklyn). When completed, it was the longest spanning bridge in the world and the Roebling system of suspension bridge construction became the standard throughout the world. (Jet Lowe, HAER NY-18-75.)

Exhibit 1.5 The Golden Gate Bridge was built across mouth of San Francisco Bay from 1933–1937 by design engineer Charles Ellis and chief engineer Joseph Strauss. Spanning one of the world's most spectacular channels, the bridge is internationally renowned as a superb structural and aesthetic example of suspension bridge design. (Jet Lowe, 1984, HAER CA-31-43.)

Exhibit 1.6 The Eads Bridge spanning the Mississippi River at Saint Louis, Missouri, was built 1867–1874 by James Buchanan Eads. The triple span, tubular metallic, arch construction required precise quality control and deep caissons to achieve its engineering and aesthetics success.

Exhibit 1.7 The Alvord Lake Bridge in San Francisco's Golden Gate Park was built in 1889 by Ernest Ransome. This reinforced concrete arch bridge is believed to be the oldest in the United States using steel reinforcing bars. It survived the 1906 San Francisco earthquake and several subsequent tremblers without damage and continues in service today. (sanfranciscodays.com.)

Exhibit 1.8 The Tunkhannock Creek Viaduct near Nicholson, Pennsylvania, was built in 1915 for the Lackawanna Railroad. It is 2375 feet long and 240 feet high. The viaduct is the largest concrete bridge in the United States. It has been compared to the nearly two-thousand-year old Pont du Gard in southern France because of its tall proportions and high semicircular main arches.

Exhibit 1.9 The Rogue River Bridge spans the mouth of the river on the Oregon Coast Highway near Gold Beach, Oregon, and was built 1930–1932. The bridge is the first reinforced concrete arch span built in the United States using the Freyssinet method of prestressing the arch ribs. Data collected from this bridge provided valuable insight into this technique for the engineering community. (Jet Lowe, 1990, HAER OR-38-16.)

Exhibit 1.10 The Walnut Lane Bridge spanning Lincoln Drive and Monoshone Creek, Philadelphia, Pennsylvania, was designed by Gustave Magnel and constructed in 1949–1950. This bridge was the first prestressed concrete beam bridge built in the United States. It provided the impetus for the development of methods for design and construction of this type structure in the United States. (A. Pierce Bounds, 1988, HAER PA-125-5.)

Exhibit 1.11 The Hoover Dam Bypass Bridge was completed in October, 2010, and was the first concrete-steel composite arch bridge (concrete for the arch and columns and steel for the roadway deck) built in the United States. The function of the bypass and bridge was to improve travel times, replace the dangerous approach roadway, and reduce the possibility of an attack or accident at the dam site. (www.hooverdambypass.org/Const_PhotoAlbum.htm.)

Exhibit 1.12 The Smart Road Bridge has pleasing proportions as the span lengths decrease going up the sides of the Ellett Valley near Blacksburg, Virginia. The bridge is 1985 feet long, 150 feet high, and serves the needs of researchers while protecting the scenic beauty of southwestern Virginia.

Exhibit 1.13 The bridge crossing the broad valley of the Mosel River (Moseltal-brücke) in southern Germany is a good example of tall tapered piers with thin constant-depth girders that give a pleasing appearance when viewed obliquely.

Exhibit 1.14 The Blue Ridge Parkway (Linn Cove) Viaduct, Grandfather Mountain, North Carolina, was built from the top down to protect the environment of Grandfather Mountain. This precast concrete segmental bridge was designed to blend in with the rugged environment.

Exhibit 1.15 The Francis Scott Key Bridge over the Potomac at Georgetown, Washington, DC, was built from 1917 to 1923. It has seven open-spandrel three-ribbed arches that are in an orderly and rhythmic progression.

Exhibit 1.16 The Leonard P. Zakim Bunker Hill Memorial Bridge was designed by Christian Menn, completed in 2002, and spans the Charles River at Boston, MA. The towering bridge contrasts with the skyline of the city. It has become an icon and nearly as identifiable with Boston as the Eiffel Tower is to Paris. (leonardpzakimbunkerhillbridge.org.)

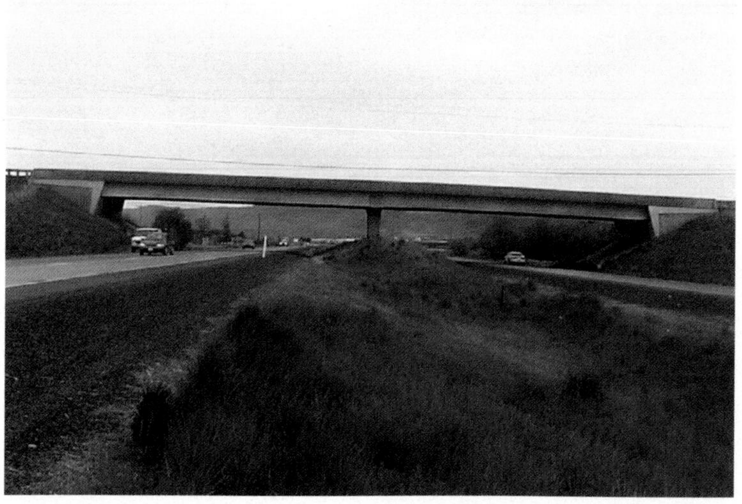

Exhibit 1.17 The I-82 Hinzerling Road undercrossing near Prosser, Washington, is a good example of the use of texture. The textured surfaces on the solid concrete barrier and the abutments have visually reduced the mass of these elements and made the bridge appear more slender than it actually is. (Photo courtesy Washington State DOT.)

Exhibit 1.18 The interchange between the Red Mountain Freeway (202) and U.S. 60 in Mesa, Arizona, is a good example of using Southwest-type texture and color to produce a beautiful blend of tall piers and gracefully curved girders. (Photo courtesy of Arizona DOT.)

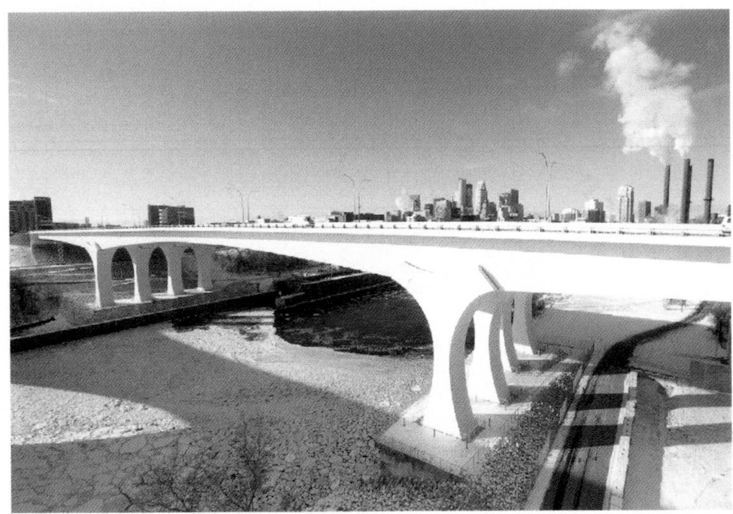

Exhibit 1.19 The I-35W St. Anthony Falls Bridge over the Mississippi River in Minneapolis, Minnesota, built in 2008 replaced the I-35W Bridge that collapsed in 2007 (see Section 2.2.7 and compare with Figure 2.14). The brightly lit girder face and sculpted piers contrast with the shadows cast by the deck overhang and the tops of the piers, accentuating the flow of the structure.

Exhibit 1.20 The 436th Avenue SE Undercrossing of I-90, King County, Washington, by increasing the mass of the central pier provides a focal point that successfully directs attention away from the split composition effect of the two-span layout and duality is resolved. (Photo courtesy of Washington State DOT.)

Exhibit 1.21 The Genesee Road (U.S. 40) Bridge over I-70 in Colorado is an elegant single span overcrossing with a slender appearance because the girder is in shadow and sloping lines on the abutment that invites the flow of traffic. It also provides a framework for an observer's first view of the Rocky Mountains. (Photo courtesy of Colorado DOT.)

Exhibit 1.22 The Millau Viaduct spans the valley of the river Tarn near Millau in southern France. Completed in 2004, it has the tallest piers of any bridge in the world. The sweeping curve of the roadway provides stability as well as breathtaking views of the broad valley.

Exhibit 1.23 The I-17/101 Interchange in Phoenix, Arizona, has tapered textured piers supporting four levels of directional roadways. The piers and girders have different dimensions, but they all belong to the same family. (Photo courtesy of Arizona DOT.)

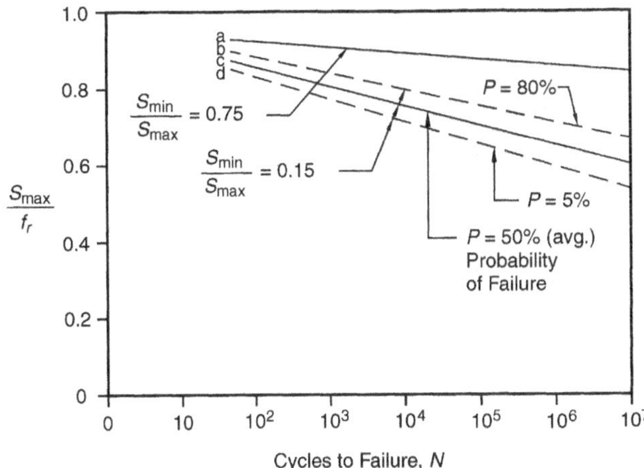

Fig. 14.2 Fatigue strength of plain concrete beams. [ACI Committee 215 (1992). Used with permission of American Concrete Institute.]

The curves a and c in Figure 14.2 were obtained from tests in which the stress range between a maximum stress and a minimum stress were equal to 75 and 15% of the maximum stress, respectively. It can be observed that an increase of the stress range results in a decreased fatigue strength for a given number of cycles. Curves b and d indicate the amount of scatter in the test data. Curve b corresponds to an 80% chance of failure while curve d represents a 5% chance of failure. Curves a and c are averages representing 50% probability of failure.

The S–N curves for concrete in Figure 14.2 are nearly linear from 100 cycles to 10 million cycles and have not flattened out at the higher number of cycles to failure. It appears that concrete does not exhibit a limiting value of stress below which the fatigue life is infinite. Thus, any statement on the fatigue strength of concrete must be given with reference to the number of cycles to failure. ACI Committee 215 (1992) concludes that the fatigue strength of concrete for the life of 10 million cycles of load and a probability of failure of 50%, regardless of whether the specimen is loaded in compression, tension, or flexure, is approximately 55% of the static strength.

In AASHTO [A5.5.3.1], the limiting tensile stress in flexure due to the sum of unfactored permanent loads and prestress, and 1.5 times the fatigue load before the section is considered as cracked is $0.095\sqrt{f_c'}$, which is 40% of the static strength $f_r = 0.24\sqrt{f_c'}$. This live load is associated with fatigue I limit state where the load is the maximum expected during the service life.

Further, because the stress range is typically the difference between a minimum stress due to permanent load and a maximum stress due to permanent load plus the transitory fatigue load, the limits on the compressive stress in Table 14.3 should keep the stress range within $0.40f_c'$. However, in this case, one-half of the permanent and prestressed load effects are used. Both of these limitations are comparable to the recommendations of ACI Committee 215 (1992) for the fatigue strength of concrete.

Fatigue of Reinforcing Bars Observations of deformed reinforcing bars subjected to repeated loads indicate that fatigue cracks start at the base of a transverse deformation where a stress concentration exists. With repeated load cycles, the crack grows until the cross-sectional area is reduced and the bar fails in tension. The higher the stress range S_r of the repeated load, the fewer the number of cycles N before the reinforcing bar fails.

Results of experimental tests on straight deformed reinforcing bars are shown by the S_r–N curves in Figure 14.3. These curves were generated by bars whose size varied from #5 to #11. The curves begin to flatten out at about 1 million cycles, indicating that reinforcing bars may have a stress endurance limit below which the fatigue life will be infinite.

The stress range S_r is the difference between the maximum stress S_{max} and the minimum stress S_{min} of the repeating load cycles. The higher the minimum stress level, the higher the average tensile stress in the reinforcing bar and the lower the fatigue strength.

The stress concentrations produced at the base of a deformation or at the intersection of deformations can also be produced by bending and welding of the reinforcing bars. Investigations reported by ACI Committee 215 (1992) indicate the fatigue strength of bars bent through an angle of 45° to be about 50% that of straight bars and the fatigue strength of bars with stirrups attached by tack welding to be about 67% that of bars with stirrups attached by tie wires.

In AASHTO [A5.5.3.2], minimum stress f_{min} is considered in setting a limit on the constant-amplitude fatigue threshold, $(\Delta F)_{TH}$, for straight reinforcement and welded wire reinforcement without a cross weld in the high-stress region, that is,

$$(\Delta F)_{TH} = 24 - 0.33 f_{min} \tag{14.6b}$$

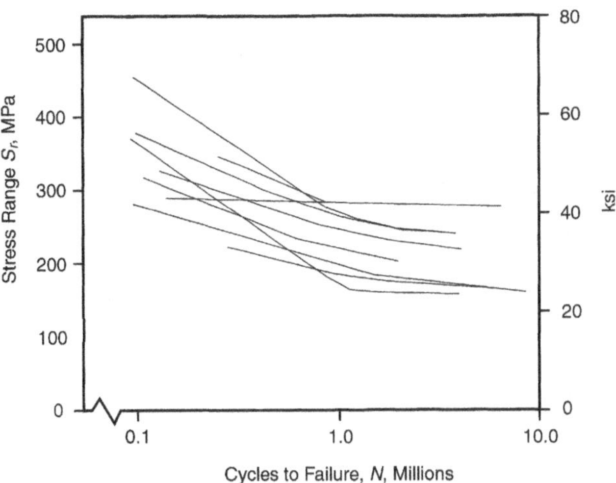

Fig. 14.3 Stress range versus fatigue life for reinforcing bars. [ACI Committee 215 (1992). Used with permission of American Concrete Institute.]

where f_{min} is the minimum live-load stress resulting from the fatigue I load combination, combined with the more severe stress from either the permanent loads or the permanent loads, shrinkage, and creep-induced external loads, and tension is positive and compression is negative (ksi) [A5.5.3.2].

The fluctuating live-load fatigue tensile stresses are in addition to any stress that exists in the reinforcement due to permanent loads. If the stress f_{min} is tensile, the average combined stress is higher and the fatigue resistance of the reinforcement is lower. If the stress f_{min} is compressive, the fatigue resistance increases.

In the case of a single-span girder, the minimum stress produced by the fatigue truck is zero. Assuming the minimum stress produced by dead loads is 15 ksi, the constant-amplitude fatigue threshold $(\Delta F)_{TH}$ is 19 ksi. This value compares well with a lower bound to the curves in Figure 14.3 for 1–10 million cycles to failure.

As recommended by ACI Committee 215 (1992), Eq. 14.6b should be reduced by 50% for bent bars or bars to which auxiliary reinforcement has been tack welded. As a practical matter, primary reinforcement should not be bent in regions of high stress range and tack welding should be avoided.

Fatigue of Prestressing Tendons If the precompression due to prestressing is sufficient so that the concrete cross section remains uncracked or cracks at or below the tensile stresses associated with service III, then fatigue of prestressing tendons is seldom a problem. In this case, AASHTO [A5.5.3.1] states that fatigue of fully prestressed sections need not be checked. Effectively this eliminates the need for this check. Similarly, the fatigue check is not required in deck slab or reinforced concrete box culverts.

However, designs are allowed that result in cracked sections beyond service loads (see Table 14.3) and it becomes necessary to consider fatigue. The AASHTO [A5.5.3.1] states that fatigue shall be considered when the compressive stress due to permanent loads and prestress is less than twice the maximum tensile live-load stress resulting from the fatigue truck. This load is associated with fatigue I limit state. A load factor of 1.5 is specified on the live-load force effect for the fatigue truck [C5.5.3.1].

Fatigue tests have been conducted on individual prestressing wires and on seven-wire strand, which are well documented in the literature cited by ACI Committee 215 (1992). However, the critical component that determines the fatigue strength of prestressing tendons is their anchorage. Even though the anchorages can develop the static strength of prestressing tendons, they develop less than 70% of the fatigue strength. Bending at an anchorage can cause high local stresses not seen by a direct tensile pull of a prestressing tendon.

The S_r–N curves shown in Figure 14.4 are for proprietary anchorages for strand and multiple wire tendons. Similar curves are also given by ACI Committee 215 (1992) for anchorages of bars. From Figure 14.4, an endurance limit for the anchorages occurs at about 2 million cycles to failure (arrows indicate specimens for which failure did not occur). A lower bound for the stress range is about $0.07f_{pu}$, which for $f_{pu} = 270$ ksi translates to $S_r = 19$ ksi.

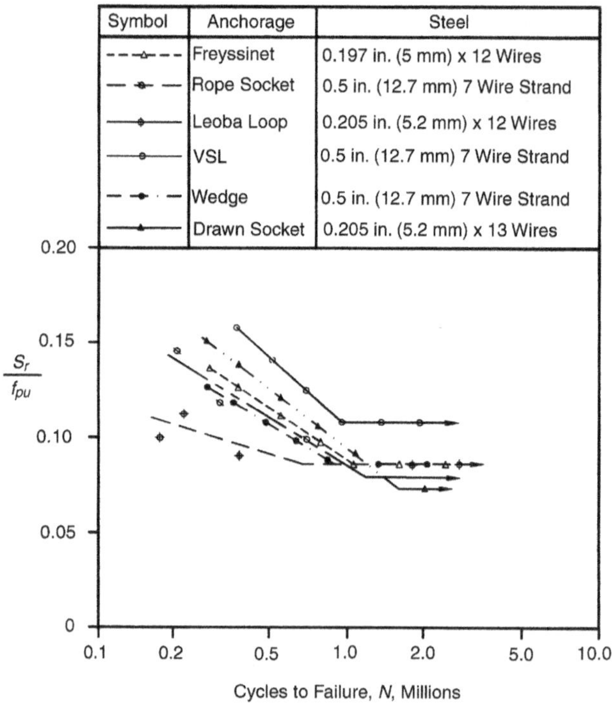 Symbol	Anchorage	Steel
	Freyssinet	0.197 in. (5 mm) x 12 Wires
	Rope Socket	0.5 in. (12.7 mm) 7 Wire Strand
	Leoba Loop	0.205 in. (5.2 mm) x 12 Wires
	VSL	0.5 in. (12.7 mm) 7 Wire Strand
	Wedge	0.5 in. (12.7 mm) 7 Wire Strand
	Drawn Socket	0.205 in. (5.2 mm) x 13 Wires

Fig. 14.4 Stress range versus fatigue life for strand and multiple wire anchorages [ACI Committee 215 (1992). Used with permission of American Concrete Institute.]

Bending of the prestressing tendons also occurs when it is held down at discrete points throughout its length. Fatigue failures can initiate when neighboring wires rub together or against plastic and metal ducts. This fretting fatigue can occur in both bonded and unbonded posttensioning systems.

The constant-amplitude fatigue threshold, $(\Delta F)_{TH}$, given for prestressing tendons [A5.5.3.3], varies with the radius of curvature of the tendon and shall not exceed

- 18.0 ksi for radii of curvature in excess of 30.0 ft
- 10.0 ksi for radii of curvature not exceeding 12.0 ft

The sharper the curvature is, the lower the fatigue strength (stress range). A linear interpolation may be used for radii between 12 and 30 ft. There is no distinction between bonded and unbonded tendons. This lack of distinction differs from ACI Committee 215 (1992), which considers the anchorages of unbonded tendons to be more vulnerable to fatigue and recommends a reduced stress range comparable to the 10 ksi for the sharper curvature.

Table 14.5 Constant-Amplitude Fatigue Threshold of Splices

Type of Splice	$(\Delta F)_{TH}$ for Greater Than 1,000,000 Cycles
Grout-filled sleeve, with or without epoxy-coated bar	18 ksi
Cold-swaged coupling sleeves without threaded ends and with or without epoxy-coated bar; intergrally forged coupler with upset NC threads; steel sleeve with a wedge; one-piece taper-threaded coupler; and single V-groove direct butt weld	12 ksi
All other types of splices	4 ksi

From AASHTO [A5.5.3.4]. From *AASHTO LRFD Bridge Design Specifications*, Copyright © 2010 by the American Association of State Highway and Transportation Officials, Washington, DC. Used by permission.

Fatigue of Welded or Mechanical Splices of Reinforcement For welded or mechanical connections that are subject to repetitive loads, the constant-amplitude fatigue threshold, $(\Delta F)_{TH}$, shall be as given in Table 14.5 [A5.5.3.4].

Review of the available fatigue and static test data indicates that any splice that develops 125% of the yield stress of the bar will sustain 1 million cycles of a 4-ksi constant-amplitude stress range [C5.5.3.4]. This lower limit agrees well with the limit of 4.5 ksi for category E from the provisions for fatigue of structural steel weldments [Table A6.6.1.2.5-3].

Where the total cycles of loading are less than one million, $(\Delta F)_{TH}$ in Table 14.5 may be increase by the quantity 24(6-log N) ksi to a total not greater than the value given by Eq. 14.6b. Higher values of $(\Delta F)_{TH}$ up to the value given by Eq. 14.6b may be used if justified by fatigue test data on splices that are the same as those that will be placed in service [A5.5.3.4].

14.1.3 Strength Limit State

A strength limit state is one that is governed by the static strength of the materials in a given cross section. There are five different strength load combinations specified in Table 5.1. For a particular component of a bridge structure, only one or maybe two of these load combinations need to be investigated. The differences in the strength load combinations are associated mainly with the load factors applied to the live load. A smaller live-load factor is used for a permit vehicle and in the presence of wind, which is logical. The load combination that produces the maximum load effect is then compared to the strength or resistance provided by the cross section of a member.

In calculating the resistance to a particular factored load effect, such as axial load, bending, shear, or torsion, the uncertainties are represented by an understrength or resistance factor ϕ. The ϕ factor is multiplied times the calculated nominal resistance R_n, and the adequacy of the design is then determined by whether or not the inequality expressed by Eq. 14.1 is satisfied.

In the case of a reinforced concrete member there are uncertainties in the quality of the materials, cross-sectional dimensions, placement of reinforcement, and equations used to calculate the resistance.

Some modes of failure can be predicted with greater accuracy than others and the consequence of their occurrence is less costly. For example, beams in flexure are usually designed as underreinforced so that failure is precipitated by gradual yielding of the tensile reinforcement while columns in compression may fail suddenly without warning. A shear failure mode is less understood and is a combination of a tension and compression failure mode. Therefore, its ϕ factor should be somewhere between that of a beam in flexure and a column in compression. The consequence of a column failure is more serious than that of a beam because when a column fails it will bring down a number of beams; therefore, its margin of safety should be greater. All of these considerations, and others, are reflected in the resistance factors specified [A5.5.4.2] and presented in Table 14.6.

Sections are compression-controlled when the net tensile strain ε_t in the extreme tension steel reaches the strain corresponding to its specific yield strength f_y just as the concrete reaches its assumed ultimate strain of −0.003 (balanced strain conditions). For Grade 60 reinforcement, and for all prestressed reinforcement, the compression controlled strain limit may be set at 0.002 [A5.7.2.1].

Table 14.6 Resistance Factors for Conventional Construction

Strength Limit State	ϕ Factor
For tension-controlled reinforced concrete sections	0.90
For tension-controlled prestressed concrete sections	1.00
For shear and torsion	
Normal weight concrete	0.90
Lightweight concrete	0.70
For compression-controlled sections with spirals or ties, except for seismic zones 2, 3, and 4	0.75
For bearing on concrete	0.70
For compression in strut-and-tie models	0.70
For compression in anchorage zones	
Normal-weight concrete	0.80
Lightweight concrete	0.65
For tension in steel in anchorage zones	1.00
For resistance during pile driving	1.00

From AASHTO [A5.5.4.2.1]. From *AASHTO LRFD Bridge Design Specifications*, Copyright © 2010 by the American Association of State Highway and Transportation Officials, Washington, DC. Used by permission.

Sections are tension controlled when the net tension strain ε_t in the extreme tension steel is equal to or greater than 0.005 just as the concrete reaches its assumed ultimate strain of −0.003. Flexural members are usually tension controlled with sufficient net tension strain to provide ample warning of failure with excessive deflection and cracking [C5.7.2.1].

The resistance factor for sections that are between compression-controlled and tension-controlled cases are determined based upon a transition between the two factors of 0.75 and 1. The basis for the interpolation is the curvature of the section at failure. Assuming that the ultimate concrete strain is 0.003, then the curvature is also proportional to the maximum tensile strain. The tensile strain and curvature are a function of the depth of the neutral axis relative to the effective section depth. This ratio (curvature) is used as a basis to transition from a compression-controlled section to a tension-controlled one as prescribed in [A5.5.4.2.1]. The associated equations are shown in Figure 14.5. Although computationally complex for the development of an axial force–moment interaction relationship, it provides a rational basis for the transition.

Note that for a typical underreinforced beam, the steel strain will be greater than 0.005 and ϕ is equal to 0.9. Alternatively, if $c/d \leq 0.375$ ($a/d \leq 0.375\beta_1$), then ϕ is equal to 0.9 or 1.0.

For beams with or without tension that are a mixture of nonprestressed and prestressed reinforcement, for tension-controlled sections, the ϕ factor depends on the partial prestressing ratio (PPR) and is given by

$$\phi = 0.90 + 0.10\text{PPR} \tag{14.7}$$

in which

$$\text{PPR} = \frac{A_{\text{ps}} f_{\text{py}}}{A_{\text{ps}} f_{\text{py}} + A_s f_y} \tag{14.8}$$

where A_{ps} is the area of prestressing steel, f_{py} is the yield strength of prestressing steel, A_s is area of nonprestressed tensile reinforcement, and f_y is the yield strength of the reinforcing bars.

No guidance is provided in AASHTO regarding partially prestressed sections with steel strains less than 0.005. However, a weighted average ϕ seems reasonable; alternatively use the nonprestressed value.

14.1.4 Extreme Event Limit State

Extreme event limit states are unique occurrences with return periods in excess of the design life of the bridge. Earthquakes, ice loads, vehicle collisions, and vessel collisions are considered to be extreme events and are to be used one at a time. However, these events may be combined with a major flood (recurrence interval >100 years but <500 years) or with the effects of scour of a major flood, as shown in Table 5.1. For example, it is possible that scour from a major flood may have reduced support for foundation components when the design earthquake occurs or when ice floes are colliding with a bridge during a major flood.

The resistance factors ϕ for an extreme event limit state are to be taken as unity. This choice of ϕ may result in excessive distress and structural damage, but the bridge structure should survive and not collapse.

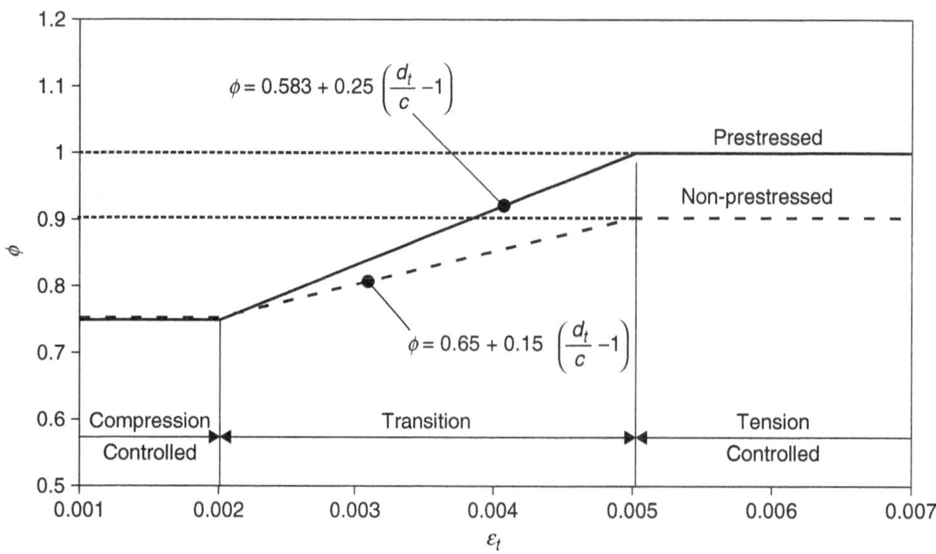

Fig. 14.5 Variation of ϕ with net tensile strain ε_t and d_t/c for Grade 60 reinforcement and for prestressing steel in AASHTO [C5.5.4.2.1]. (From *AASHTO LRFD Bridge Design Specifications*, Copyright © 2010 by the American Association of State Highway and Transportation Officials, Washington, DC. Used by permission.)

14.2 FLEXURAL STRENGTH OF REINFORCED CONCRETE MEMBERS

The AASHTO (2010) bridge specifications present unified design provisions that apply to concrete members reinforced with a combination of conventional steel bars and prestressing strands. Such members are often called partially prestressed. The expressions developed are also applicable to conventional reinforced and prestressed concrete when one reinforcement or the other is not present. Background for the development of these provisions is located in Loov (1988) and Naaman (1992).

14.2.1 Depth to Neutral Axis for Beams with Bonded Tendons

Consider the flanged cross section of a reinforced concrete beam shown in Figure 14.6 and the accompanying linear strain diagram. For bonded tendons, the compatibility condition gives the strain in the surrounding concrete as

$$\varepsilon_{cp} = -\varepsilon_{cu}\frac{d_p - c}{c} = -\varepsilon_{cu}\left(\frac{d_p}{c} - 1\right) \tag{14.9}$$

where ε_{cu} is the limiting strain at the extreme compression fiber, d_p is the distance from the extreme compression fiber to the centroid of the prestressing tendons, and c is the distance from the extreme compression fiber to the neutral axis. Again, tensile strains are considered positive and compressive strains are negative.

From Eq. 13.44, the strain in the prestressing tendon becomes

$$\varepsilon_{ps} = -\varepsilon_{cu}\left(\frac{d_p}{c} - 1\right) + \Delta\varepsilon_{pe} \tag{14.10}$$

where $\Delta\varepsilon_{pe}$ is approximately equal to f_{pe}/E_p and remains essentially constant throughout the life of the member (Collins and Mitchell, 1991). At the strength limit state, AASHTO [A5.7.2.1] defines $\varepsilon_{cu} = -0.003$ if the concrete is unconfined. For confined concrete, ε_{cu} can be an order of magnitude greater than for unconfined concrete (Mander et al., 1988). With both $\Delta\varepsilon_{pe}$ and ε_{cu} being constants depending on the prestressing operation and the lateral confining pressure, respectively, the strain in the prestressing tendon ε_{ps} and the corresponding stress f_{ps} is a function only of the ratio c/d_p.

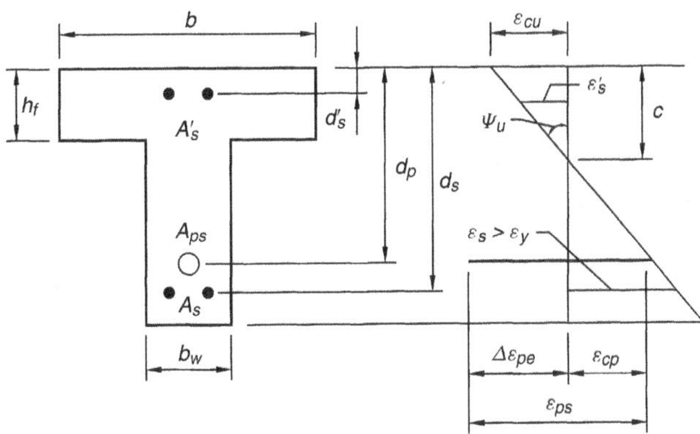

Fig. 14.6 Strains in a reinforced concrete beam. [Reproduced with permission from R. E. Loov (1988).]

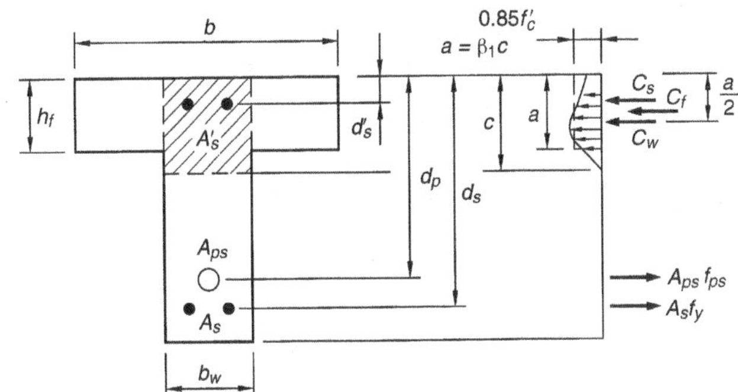

Fig. 14.7 Forces in a reinforced concrete beam.

Equilibrium of the forces in Figure 14.7 can be used to determine the depth of the neutral axis c. However, this requires the determination of f_{ps} that is a function of the ratio c/d_p. Such an equation has been proposed by Loov (1988), endorsed by Naaman (1992), and adopted by AASHTO [A5.7.3.1.1] as

$$f_{ps} = f_{pu}\left(1 - k\frac{c}{d_p}\right) \tag{14.11}$$

$$k = 2\left(1.04 - \frac{f_{py}}{f_{pu}}\right) \tag{14.12}$$

For low-relaxation strands with $f_{pu} = 270$ ksi, Table 13.4 gives $f_{py}/f_{pu} = 0.90$, which results in $k = 0.28$. By using $E_p = 28{,}500$ ksi, neglecting ε_{ce}, and assuming that $\varepsilon_{cu} = -0.003$ and $f_{pe} = 0.80(0.70)f_{pu} = 0.56f_{pu}$, Eqs. 14.10 and 14.11 become

$$\varepsilon_{ps} = 0.003\frac{d_p}{c} + 0.0023 \tag{14.10a}$$

$$f_{ps} = 270\left(1 - 0.28\frac{c}{d_p}\right) \tag{14.11a}$$

Substituting values of c/d_p from 0.05 to 0.50 into Eq. 14.10a and 14.11a, the approximate stress–strain curve has been generated and compared to the Ramberg–Osgood curve of Eq. 13.41 in Figure 14.8. Also shown on Figure 14.8 is the 0.2% offset strain often used to determine the yield point of rounded stress–strain curves and its intersection with $f_{py} = 0.9f_{pu}$. The agreement with both curves is very good.

When evaluating the compressive forces in the concrete, it is convenient to use an equivalent rectangular stress block. In AASHTO [A5.7.2.2], the following familiar provisions for the stress block factors have been adopted:

- ☐ Uniform concrete compressive stress of $0.85f'_c$
- ☐ Depth of rectangular stress block $a = \beta_1 c$

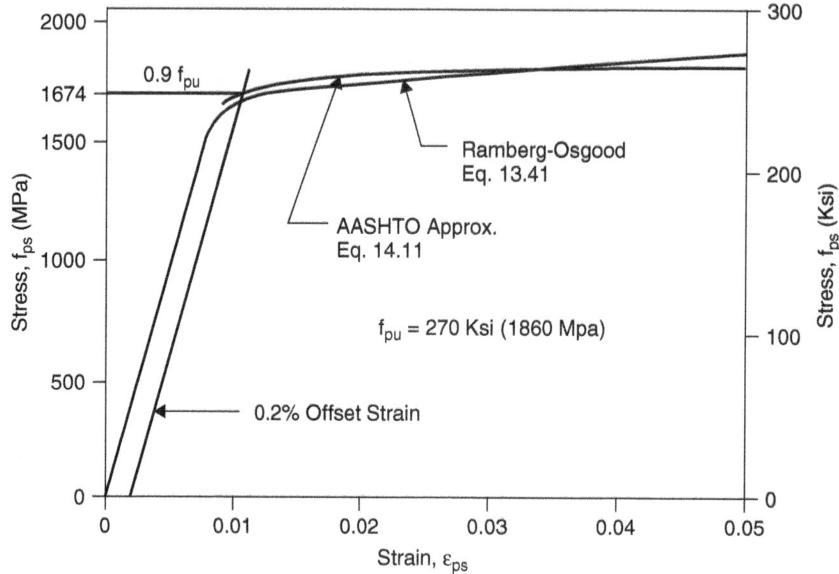

Fig. 14.8 Comparison of stress–strain curves for 270-ksi low-relaxation prestressing strands. [Reproduced with permission from R. E. Loov (1988).]

Here,

$$\beta_1 = 0.85 \text{ for } f_c' \leq 4.0 \text{ ksi}$$

$$\beta_1 = 0.65 \text{ for } f_c' \geq 8.0 \text{ ksi}$$

$$\beta_1 = 0.85 - 0.05\left(f_c' - 4.0\right) \text{ for } 4.0 \text{ ksi} \leq f_c' \leq 8.0 \text{ ksi} \tag{14.13}$$

Note that in Eq. 14.13 and in the derivations that follow, the compressive stresses f_c' and f_y' are taken as their absolute values.

Equilibrium of the forces in the beam of Figure 14.7 requires that the total nominal compressive force equals the total nominal tensile force, that is,

$$C_n = T_n \tag{14.14}$$

in which

$$C_n = C_w + C_f + C_s \tag{14.14a}$$

$$T_n = A_{ps} f_{ps} + A_s f_y \tag{14.14b}$$

where

C_w = concrete compressive force in the web
C_f = concrete compressive force in the flange
C_s = compressive force in the nonprestressed steel
A_{ps} = area of prestressing steel
f_{ps} = average stress in prestressing steel at nominal bending resistance of member as given by Eq. 14.11
A_s = area of nonprestressed tension reinforcement
f_y = specified minimum yield strength of tension reinforcement

The concrete compressive force in the web C_w is over the cross-hatched area in Figure 14.7 of width equal to the web width b_w that extends through the flange to the top fibers. It is equal to

$$C_w = 0.85 f_c' a b_w = 0.85 \beta_1 f_c' c b_w \tag{14.15a}$$

which can be thought of as an average stress in the concrete of $0.85\beta_1 f_c'$ over the area cb_w. Using stress block over the portion of the concrete in the flange,

$$C_f = 0.85 f_c' \left(b - b_w\right) h_f \tag{14.15b}$$

The compressive force in the compression steel C_s, assuming that its compressive strain ε_s' in Figure 14.6 is greater than or equal to the yield strain ε_y', is

$$C_s = A_s' f_y' \tag{14.16}$$

where A_s' is the area of the compression reinforcement and f_y' is the absolute value of specified yield strength of the compression reinforcement. The assumption of yielding of the compression steel can be checked by calculating ε_s' from similar strain triangles in Figure 14.6 and comparing to $\varepsilon_y' = f_y'/E_s$, that is,

$$\varepsilon_s' = \varepsilon_{cu} \frac{c - d_s'}{c} = \varepsilon_{cu}\left(1 - \frac{d_s'}{c}\right) \tag{14.17}$$

where d_s' is the distance from the extreme compression fiber to the centroid of the compression reinforcement.

Substitute f_{ps} from Eq. 14.11 into Eq. 14.14b and the total tensile force becomes

$$T_n = A_{ps} f_{pu}\left(1 - k\frac{c}{d_p}\right) + A_s f_y \tag{14.18}$$

By substituting the compressive forces from Eqs. 14.15 and 14.16 into Eq. 14.14a, the total compressive force becomes

$$C_n = 0.85\beta_1 f_c' c b_w + 0.85 f_c' (b - b_w) h_f + A_s' f_y' \tag{14.19}$$

Equate the total tensile and compressive forces, and solve for c to give

$$c = \frac{A_{ps} f_{pu} + A_s f_y - A_s' f_y' - 0.85 f_c' (b - b_w) h_f}{0.85\beta_1 f_c' b_w + k A_{ps} f_{pu}/d_p} \geq h_f \tag{14.20}$$

If c is less than h_f', the neutral axis is in the flange and c should be recalculated with $b_w = b$ in Eq. 14.20. This expression for c is completely general and can be used for prestressed beams without reinforcing bars ($A_s = A_s' = 0$) and for reinforced concrete beams without prestressing steel ($A_{ps} = 0$).

Equation 14.20 assumes that the compression reinforcement A_s' has yielded. If it has not yielded, the stress in the compression steel is calculated from $f_s' = \varepsilon_s' E_s$, where ε_s' is determined from Eq. 14.17. This expression for f_s' replaces the value of f_y' in Eq. 14.20 and results in a quadratic equation for determining c. As an alternative, one can simply and safely neglect the contribution of the compression steel when it has not yielded and set $A_s' = 0$ in Eq. 14.20.

14.2.2 Depth to Neutral Axis for Beams with Unbonded Tendons

When the tendons are *not* bonded, strain compatibility with the surrounding concrete cannot be used to determine the strain and the stress in the prestressing tendon. Instead the total change in length of the tendon between anchorage points must be determined by an overall structural system analysis.

Over the years, a number of researchers have proposed equations for the prediction of the stress in unbonded tendons at ultimate. The work discussed herein is based on the research of MacGregor (1989) as presented by Roberts-Wollman et al. (2005). MacGregor developed the equation for predicting the unbonded tendon stress at ultimate that is currently in AASHTO [A5.7.3.1.2].

The structural system at ultimate is modeled as a series of rigid members connected by discrete plastic hinges at various locations over supports and near midspan to form a collapse mechanism. The simplest collapse mechanism is a single span structure with a straight tendon anchored at the ends and a plastic hinge at midspan (Fig. 14.9). This model is used to illustrate the various parameters. All tendon elongation δ is assumed to occur as the hinge opens and is defined as

$$\delta = z_p \theta \tag{14.21}$$

where θ is angle of rotation at the hinge and z_p is the distance from the neutral axis to the tendon. The tendon strain increase is

$$\Delta\varepsilon_{ps} = \frac{\delta}{L} = \frac{z_p \theta}{L} = \frac{(d_{ps} - c)\,\theta}{L} \tag{14.22}$$

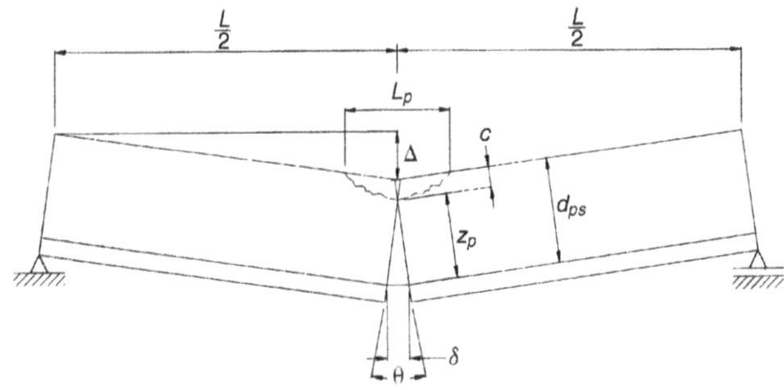

Fig. 14.9 Failure mechanism for a simple-span structure. [From Roberts-Wollmann et al. (2005). Used with permission of American Concrete Institute.]

where L is the length of the tendon between anchorages, d_{ps} is the depth from the compression face to the centroid of the prestressing tendon, and c is the depth from the compression face to the neutral axis (Fig. 14.9). Small strains are assumed so the span length L and tendon length are considered equal.

The angle θ can be defined as the integral of the curvature $\phi(x)$ over the length of the plastic hinge L_p:

$$\theta = \int_0^{L_p} \phi(x)\, dx$$

If the curvature is assumed constant, the integral can be approximated as

$$\theta = L_p \phi = L_p \frac{\varepsilon_{cu}}{c}$$

where ε_{cu} is the ultimate strain in the concrete (Fig. 14.6).

Equation 14.22 becomes

$$\Delta\varepsilon_{ps} = \frac{L_p}{c}\varepsilon_{cu}\left(\frac{d_{ps} - c}{L}\right) = \psi\varepsilon_{cu}\left(\frac{d_{ps} - c}{L}\right)$$

where $\psi = L_p/c$. Assuming the tendon remains in the elastic range,

$$\Delta f_{ps} = E_{ps}\Delta\varepsilon_{ps} = E_{ps}\psi\varepsilon_{cu}\left(\frac{d_{ps} - c}{L}\right) \tag{14.23}$$

Based on physical tests by others and observations by MacGregor (1989), a value of $\psi = 10.5$ was recommended. Using $E_{ps} = 28{,}500$ ksi and $\varepsilon_{cu} = 0.003$, Eq. 14.23 becomes

$$\Delta f_{ps} = (28{,}500)\,(10.5)\,(0.003)\left(\frac{d_{ps} - c}{L}\right)$$
$$= 900\left(\frac{d_{ps} - c}{L}\right) \text{ ksi} \tag{14.24}$$

MacGregor (1989) further developed the equation to predict tendon stress increases in structures continuous over interior supports. Consider the two collapse mechanisms for the three-span structure with anchorages at the ends shown in Figure 14.10. In Figure 14.10(a), a collapse mechanism results when one hinge forms at the interior support and a second at midspan. In Figure 14.10(b), a collapse mechanism results when hinges form at the two interior supports and a third at midspan. MacGregor recognized from his tests

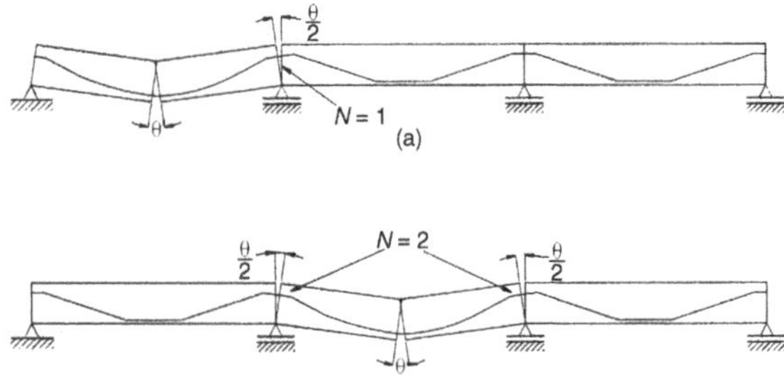

Fig. 14.10 Collapse mechanism for continuous structures. [From Roberts-Wollmann et al. (2005). Used with permission of American Concrete Institute].

that the rotation at a support hinge is only $\frac{1}{2}$ of the rotation at a midspan hinge. Equation 14.21 shows that elongation varies directly with θ so that the total tendon elongation is

$$\delta_{\text{total}} = \delta_{\text{midspan}} \left(1 + \frac{N}{2}\right)$$

where N is the number of support hinges required to form a flexural collapse mechanism that are crossed by the tendon between points of discrete bonding or anchoring, and δ_{midspan} is the elongation for a simple-span tendon.

For the simple-span case in Figure 14.9, $N = 0$. As shown in Figure 14.10, if the critical span is an end span, $N = 1$; if it is an interior span, $N = 2$. MacGregor presented his equation in the following form:

$$f_{\text{ps}} = f_{\text{pe}} + 900 \left(\frac{d_{\text{ps}} - c_y}{\ell_e}\right) \text{ksi} \qquad (14.25)$$

where f_{ps} is the stress (ksi) in the tendon at ultimate, f_{pe} is the effective prestress (ksi) in the tendon after all losses, c_y (in.) indicates that the depth to the neutral axis is calculated assuming all mild and prestressed reinforcing steel crossing the hinge opening is at yield, and

$$\ell_e = \frac{L}{(1 + N/2)} = \frac{2L}{2 + N} = \frac{2\ell_i}{2 + N}$$

where ℓ_i is the length of tendon between anchorages (in.). Equation 14.25 is the equation given for determining the average stress in unbonded prestressing steel in AASHTO [A5.7.3.1.2].

Following the same procedure as for the bonded tendon in establishing force equilibrium, the expression for the distance from the extreme compression fiber to the neutral axis for an unbonded tendon is

$$c = \frac{A_{\text{ps}} f_{\text{ps}} + A_s f_y - A_s' f_y' - 0.85 f_c' (b - b_w) h_f}{0.85 \beta_1 f_c' b_w} \geq h_f \qquad (14.26)$$

where f_{ps} is determined from Eq. 14.25. If c is less than h_f, the neutral axis is in the flange and c should be recalculated with $b_w = b$ in Eq. 14.26. If the strain in the compression reinforcement calculated by Eq. 14.17 is less than the yield strain ε_y', f_y' in Eq. 14.26 should be replaced by f_s' as described previously for Eq. 14.20.

Substitution of Eq. 14.25 into Eq. 14.26 results in a quadratic equation for c. Alternatively, an iterative method can be used starting with a first trial value for the unbonded tendon stress of [C5.7.3.1.2]

$$f_{\text{ps}} = f_{\text{pe}} + 15 \text{ ksi} \qquad (14.27)$$

in Eq. 14.26. With c known, f_{ps} is calculated from Eq. 14.25, compared with the previous trial, and a new value chosen. These steps are repeated until convergence within an acceptable tolerance is attained.

14.2.3 Nominal Flexural Strength

With c and f_{ps} known for either bonded or unbonded tendons, it is a simple matter to determine the nominal flexural strength M_n for a reinforced concrete beam section. If we refer to Figure 14.7 and balance the moments about C_w, to obtain

$$M_n = A_{\text{ps}} f_{\text{ps}} \left(d_p - \frac{a}{2}\right) + A_s f_y \left(d_s - \frac{a}{2}\right) + C_s \left(\frac{a}{2} - d_s'\right) + C_f \left(\frac{a}{2} - \frac{h_f}{2}\right)$$

where $a = \beta_1 c$ and c is not less than the compression flange thickness h_f. Substitution of Eqs. 14.15b and 14.16 for C_f and C_s results in

$$M_n = A_{\text{ps}} f_{\text{ps}} \left(d_p - \frac{a}{2}\right) + A_s f_y \left(d_s - \frac{a}{2}\right) + A_s' f_y' \left(\frac{a}{2} - d_s'\right) + 0.85 f_c' (b - b_w) h_f \left(\frac{a}{2} - \frac{h_f}{2}\right) \qquad (14.28)$$

If the depth to the neutral axis from the extreme compression fiber c is less than the compression flange thickness h_f, or if the beam has no compression flange, the nominal flexural strength M_n for the beam section is calculated from Eq. 14.28 with b_w set equal to b.

Example 14.1 For the beam cross section in Figure 14.11, determine the distance from the extreme compression fiber to the netural axis c, the average stress in the prestressing steel f_{ps}, and the nominal moment strength M_n for (a) bonded tendons and (b) unbonded tendons. Use normal weight concrete with $f_c' = 6$ ksi, Grade 60 mild steel reinforcement, and 0.5 in., 270 ksi low-relaxation prestressing tendons. The beam is uniformly loaded with a single-span length of 35 ft.

1. **Material Properties**

$$\beta_1 = 0.85 - 0.05 \left(f_c' - 4\right) = 0.85 - 0.05 (6 - 4) = 0.75$$

$$E_c = 1820\sqrt{f_c'} = 1820\sqrt{6} = 4458 \text{ ksi}$$

$$\varepsilon_{\text{cu}} = -0.003$$

$$f_y = |f_y'| = 60 \text{ ksi}$$

$$E_s = 29{,}000 \text{ ksi}$$

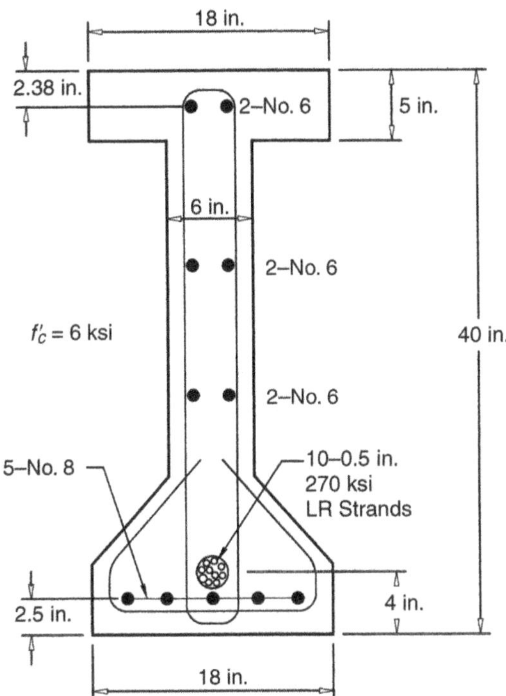

Fig. 14.11 Beam cross section used in Example 14.1.

$$\varepsilon_y = \left|\varepsilon_y'\right| = \frac{f_y}{E_s} = \frac{60}{29{,}000} = 0.00207$$

$$f_{py} = 0.9 f_{pu} = 0.9\,(270) = 243 \text{ ksi}$$

$$k = 2\left(1.04 - \frac{f_{py}}{f_{pu}}\right) = 2\,(1.04 - 0.9) = 0.28$$

Assume

$$f_{pe} = 0.6\ f_{pu} = 162 \text{ ksi}$$

$$E_p = 28{,}500 \text{ ksi}$$

2. **Section Properties**

$$b = 18 \text{ in.},\quad b_w = 6 \text{ in.},\quad h = 40 \text{ in.},\quad h_f = 5 \text{ in.}$$

$$d_s' = 2 + 0.75/2 = 2.38 \text{ in.}$$

$$d_s = h - (2 + 1.0/2) = 40 - 2.5 = 37.5 \text{ in.}$$

$$d_p = h - 4 = 36 \text{ in.}$$

$$A_s = 3.93 \text{ in.}^2,\quad A_s' = 0.88 \text{ in.}^2$$

$$A_{ps} = 10\,(0.153) = 1.53 \text{ in.}^2$$

3. **Depth to Neutral Axis and Stress in Prestressing Steel**

Bonded Case [A5.7.3.1.1]. From Eq. 14.20

$$c = \frac{A_{ps} f_{pu} + A_s f_y - A_s' f_y' - 0.85 f_c' \left(b - b_w\right) h_f}{0.85 \beta_1 f_c' b_w + \left(k A_{ps} f_{pu}/d_p\right)}$$

$$= \frac{1.53(270) + 3.93(60) - 0.88(60) - 0.85(6)(18-6)(5)}{0.85(0.75)(6)(6) + 0.28(1.53)(270)/36}$$

$$= \underline{11.1 \text{ in.}} > h_f = 5 \text{ in., neutral axis in web}$$

From Eq. 14.17

$$\varepsilon_s' = \varepsilon_{cu}\left(1 - \frac{d_s'}{c}\right) = -0.003\left(1 - \frac{2.38}{11.1}\right)$$

$$= -0.00236$$

$$|\varepsilon_s'| = 0.00236 > |\varepsilon_y'| = 0.00207$$

$\therefore$ compression steel has yielded

From Eq. 14.11

$$f_{ps} = f_{pu}\left(1 - k\frac{c}{d_p}\right)$$

$$f_{ps} = 270\left(1 - 0.28\frac{11.1}{36}\right) = \underline{247 \text{ ksi}}$$

Unbonded Case [A5.7.3.1.2]. From Eq. 14.25

$$f_{ps} = f_{pe} + 900\left(\frac{d_{ps} - c_y}{\ell_e}\right) \text{ ksi}$$

$$N = N_s = 0,\ \ell_i = 35 \text{ ft} = 420 \text{ in.},$$

$$\ell_e = \frac{2\ell_i}{2 + N_s} = \ell_i = 420 \text{ in.}$$

First Iteration: Assume $f_{ps} = f_{pe} + 15.0 = 162 + 15 = 177$ ksi [C5.7.3.1.2]

From Eq. 14.26

$$c = \frac{A_{ps} f_{ps} + A_s f_y - A_s' f_y' - 0.85 f_c' \left(b - b_w\right) h_f}{0.85 \beta_1 f_c' b_w}$$

$$\geq h_f$$

$$= \frac{1.53(177) + 3.93(60) - 0.88(60) - 0.85(6)(18-6)5}{0.85(0.75)(6)(6)}$$

$$= 6.44 \text{ in.} > h_f = 5 \text{ in., neutral axis in web.}$$

From Eq. 14.25

$$f_{ps} = 162 + 900\left(\frac{36 - 6.44}{420}\right) = 225 \text{ ksi} < f_{py}$$

$$= 243 \text{ ksi}$$

Second Iteration: Assume $f_{ps} = 225$ ksi

$$c = \frac{1.53(225) + 3.93(60) - 0.88(60) - 0.85(6)(18-6)5}{0.85(0.75)(6)(6)}$$

$$= 9.66 \text{ in.}$$

$$f_{ps} = 162 + 900\left(\frac{36 - 9.66}{420}\right) = 218 \text{ ksi}$$

Third Iteration: Assume $f_{ps} = 218$ ksi

$$c = \frac{1.53(218) + 3.93(60) - 0.88(60) - 0.85(6)(18-6)5}{0.85(0.75)(6)(6)}$$

$$= 9.17 \text{ in.}$$

$$f_{ps} = 162 + 900\left(\frac{36 - 9.17}{420}\right) = 219 \text{ ksi},$$

converged $\underline{c = 9.17 \text{ in.}}$

From Eq. 14.17

$$\varepsilon_s' = \varepsilon_{cu}\left(1 - \frac{d_s'}{c}\right) = -0.003\left(1 - \frac{2.38}{9.17}\right)$$
$$= -0.00222$$
$$|\varepsilon_s'| = 0.00222 > |\varepsilon_y'| = 0.00207$$

$\therefore$ compression steel has yielded.

4. **Nominal Flexural Strength** [A5.7.3.2.2]
Bonded Case

$$a = \beta_1 c = 0.75\,(11.1) = 8.33 \text{ in.}$$

From Eq. 14.28

$$M_n = A_{ps} f_{ps}\left(d_p - \frac{a}{2}\right) + A_s f_y\left(d_s - \frac{a}{2}\right) + A_s'\left|f_y'\right|\left(\frac{a}{2} - d_s'\right) + 0.85 f_c'\,(b - b_w)\,h_f\left(\frac{a}{2} - \frac{h_f}{2}\right)$$

$$M_n = 1.53\,(240)\left(36 - \frac{8.83}{2}\right) + 3.93\,(60)\left(37.5 - \frac{8.83}{2}\right) + 0.88\,(60)\left(\frac{8.83}{2} - 2.38\right) + 0.85\,(6)\,(18 - 6)\,5\left(\frac{8.83}{2} - \frac{5}{2}\right)$$

$$M_n = \underline{20495 \text{ k-in.} = 1708 \text{ k-ft}}$$

Unbonded Case

$$a = \beta_1 c = 0.75\,(9.17) = 6.88 \text{ in.}$$

From Eq. 14.28

$$M_n = 1.53\,(214)\left(36 - \frac{6.88}{2}\right) + 3.93\,(60)\left(37.5 - \frac{6.88}{2}\right) + 0.88\,(60)\left(\frac{6.88}{2} - 2.38\right) + 0.85\,(6)\,(18 - 6)\,5\left(\frac{6.88}{2} - \frac{5}{2}\right)$$

$$M_n = \underline{19296 \text{ k-in.} = 1608 \text{ k-ft}}$$

For the unbonded case, with the same reinforcement as the bonded case, the nominal flexural strength is approximately 6% less than that of the bonded case.

14.2.4 Ductility, Maximum Tensile Reinforcement, and Resistance Factor Adjustment

Ductility in reinforced concrete beams is an important factor in their design because it allows large deflections and rotations to occur without collapse of the beam. Ductility also allows redistribution of load and bending moments in multibeam deck systems and in continuous beams. It is also important in seismic design for dissipation of energy under hysteretic loadings.

A ductility index μ, defined as the ratio of the limit state curvature ψ_u to the yield curvature ψ_y,

$$\mu = \frac{\psi_u}{\psi_y} \tag{14.29}$$

has been used as a measure of the amount of ductility available in a beam. An idealized bilinear moment–curvature relationship for a reinforced concrete beam is shown in Figure 14.12, where the elastic and plastic flexural stiffnesses K_e and K_p can be determined from the two points (ψ_y, M_y) and (ψ_u, M_u). At the flexural limit state, the curvature ψ_u can be determined from the strain diagram in Figure 14.6 as

$$\psi_u = \frac{\varepsilon_{cu}}{c} \tag{14.30}$$

where ε_{cu} is the limit strain at the extreme compression fiber and c is the distance from the extreme compression fiber to the neutral axis. The yield curvature ψ_y is determined by dividing the yield moment M_y, often expressed as a fraction of M_u, by the flexural stiffness EI for the transformed elastic-cracked section. In design, a beam is considered to have sufficient ductility if the value of the ductility index μ is not less than a specified value. The larger the ductility index, the greater the available curvature capacity, and the larger the deformations in the member before collapse occurs.

A better measure of ductility, as explained by Skogman et al. (1988), is the rotational capacity of the member developed at a plastic hinge. A simply supported beam with a single concentrated load at midspan is shown in

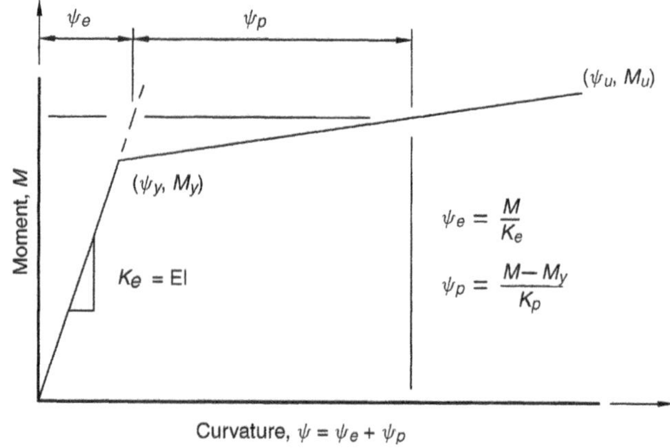

Fig. 14.12 Bilinear moment–curvature relationship.

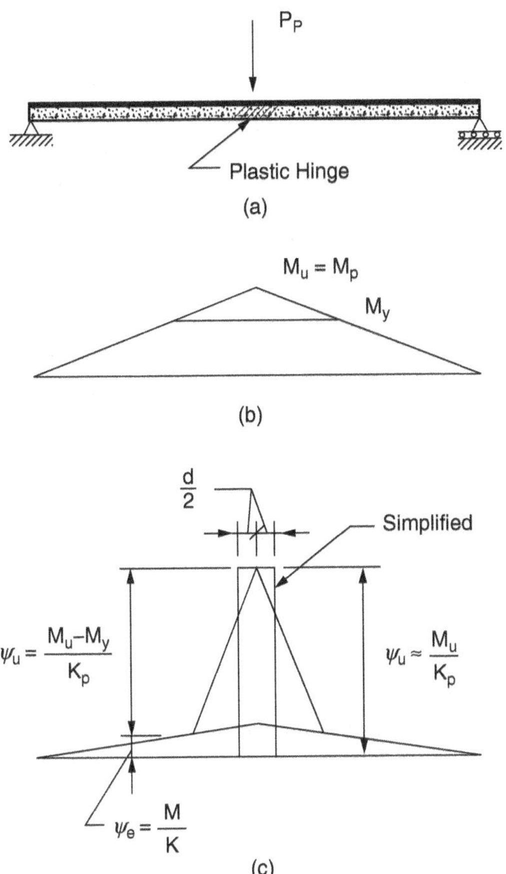

Fig. 14.13 Idealized curvature diagram at flexural limit state: (a) limit state load, (b) moment diagram, and (c) curvature diagram.

Figure 14.13. The moment diagram is a triangle and the curvature diagram at the limit state is developed from the moment–curvature relationship shown in Figure 14.12. The sharp peak in the plastic portion of the curvature diagram is not realistic because when hinging begins it spreads out due to cracking of the concrete and yielding of the steel. Sawyer (1964) recommended that the spread in plasticity extend a distance of one-half the effective depth d at each moment concentration. The elastic contribution to the curvature diagram is small compared to the plastic curvature and can be neglected. From moment–area principles, the approximate plastic rotation θ_p in the hinge is the area of the simplified curvature diagram in Figure 14.13(c). Using the relationship in Eq. 14.30, the ductility measure becomes

$$\theta_p = \psi_u d = \varepsilon_{cu}\frac{d}{c} \tag{14.31}$$

From Eq. 14.31 it is clear that ductility can be improved by increasing the limit strain ε_{cu} or by decreasing the depth to the neutral axis c. As shown in Figure 13.5, confining the concrete with sprials or lateral ties can substantially increase ε_{cu}. The neutral axis depth c depends on the total compressive force, which, in turn, must be balanced by the total tensile force. Therefore, c can be decreased by increasing the concrete compressive strength f_c', by increasing the area of the compression reinforcement A_s', or by decreasing the tensile steel areas A_{ps} and A_s. The effect of these parameters on c can also be observed in Eq. 14.20.

In previous editions of the bridge specifications, the ductility control for reinforced concrete was to limit the compressive force subject to a brittle failure by specifying a maximum tensile steel reinforcement ratio ρ_{max} as 0.75 of the balanced steel ratio ρ_b, that is,

$$\rho = \frac{A_s}{bd} \geq \rho_{max} = 0.75\rho_b \tag{14.32}$$

where ρ is the tensile reinforcement ratio and ρ_b is the reinforcement ratio that produces balanced strain conditions. The balanced strain conditions require that the concrete strain ε_c is at ε_{cu} when the steel strain ε_s reaches ε_y. By equating the balanced tensile and compressive forces in a rectangular singly reinforced concrete beam, and by using similar strain triangles, the balanced steel ratio becomes

$$\rho_b = \frac{A_{sb}}{bd} = \frac{0.85\beta_1 f_c'}{f_y}\frac{|\varepsilon_{cu}|}{|\varepsilon_{cu}| + \varepsilon_y} \tag{14.33}$$

where A_{sb} is the balanced tensile steel area. Introducing the mechanical reinforcement index ω as

$$\omega = \rho\frac{f_y}{f_c'} \tag{14.34}$$

Multiply both sides of Eq. 14.32 by f_y/f_c' and substitute Eq. 14.33 to get

$$\omega \leq 0.64\beta_1\frac{|\varepsilon_{cu}|}{|\varepsilon_{cu}| + \varepsilon_y} \tag{14.35}$$

Substituting $\varepsilon_{cu} = -0.003$, $\varepsilon_y = 0.002$ yields

$$\omega \leq 0.38\beta_1 \tag{14.36}$$

A similar limitation was also placed on the prestressed mechanical reinforcement index ω_p in previous editions of the bridge specifications as

$$\omega_p = \frac{A_{ps}f_{ps}}{bd_p f_c'} \leq 0.36\beta_1 \tag{14.37}$$

The disadvantage of using tensile reinforcement ratios to control brittle compression failures is that they must be constantly modified, sometimes in a confusing manner, to accommodate changes in the compressive force caused by the addition of flanges, compression reinforcement, and combinations of nonprestressed and prestressed tensile reinforcement. A better approach is to control the brittle concrete compressive force by setting limits on the distance c from the extreme compressive fiber to the neutral axis. Consider the left-hand side of Eq. 14.36 defined by Eq. 14.34

and substitute the compressive force in the concrete for the tensile force in the steel, so that

$$\omega = \frac{A_s}{bd_s}\frac{f_y}{f_c'} = \frac{0.85 f_c' \beta_1 cb}{bd_s f_c'} = 0.85\beta_1 \frac{c}{d_s} \qquad (14.38)$$

where d_s is the distance from the extreme compression fiber to the centroid of the nonprestressed tensile reinforcement. Similarly, for Eq. 14.37

$$\omega_p = \frac{A_{ps} f_{ps}}{bd_p f_c'} = \frac{0.85 f_c' \beta_1 cb}{bd_p f_c'} = 0.85\beta_1 \frac{c}{d_p} \qquad (14.39)$$

where d_p is the distance from the extreme compression fiber to the centroid of the prestressing tendons. Thus, putting limits on the neutral axis depth is the same as putting limits on the tensile reinforcement, only it can be much simpler. *Further, by limiting the maximum value for the ratio c/d, this assures a minimum ductility in the member as measured by the rotational capacity at the limit state of Eq.* 14.31.

All that remains is to decide on a common limiting value on the right-hand sides of Eqs. 14.36 and 14.37 and a unified definition for the effective depth to the tensile reinforcement. These topics have been presented by Skogman et al. (1988) and discussed by Naaman et al. (1990). In AASHTO [A5.7.3.3.1] (2004), the recommendations proposed by Naaman (1992) that the right-hand sides of Eqs. 14.36 and 14.37 be $0.36\beta_1$ and that the effective depth from the extreme compression fiber to the centroid of the tensile force in the tensile reinforcement be defined as

$$d_e = \frac{A_{ps} f_{ps} d_p + A_s f_y d_s}{A_{ps} f_{ps} + A_s f_y} \qquad (14.40)$$

where f_{ps} is calculated by either Eq. 14.11 or Eq. 14.25 or in a preliminary design can be assumed to be f_{py}. Finally, the ductility and maximum tensile reinforcement criterion became

$$0.85\beta_1 \frac{c}{d_e} \le 0.36\beta_1$$

or simply [C5.7.3.3.1]

$$\frac{c}{d_e} \le 0.42 \qquad (14.41)$$

The primary limitation to the approach based on $0.75\rho_b$ is that it does not address the overreinforced case typical for columns, nor does it provide a smooth transition between a ductile flexural failure mode and the less ductile column primarily in compression. As previously discussed and illustrated in Figure 14.5, the resistance factor is adjusted to provide this transition based on whether the cross section is tension controlled or compression controlled.

For the criterion represented by Eq. 14.41, the strain at the centroid of the tension steel is $\varepsilon_s = 0.003(0.58)/0.42 = 0.00414$. In the current provisions [A5.7.2.1] based on the work by Mast (1992), the net tensile strain limit at the centroid of the extreme tension reinforcement d_t was chosen to be 0.005 for tension-controlled sections. This single value applies to all types of steel (prestressed and nonprestressed) and assures sufficient ductility to provide ample warning of failure with excessive deflection and cracking [C5.7.2.1]. Henceforth, the traditional ductility check is not required in AASHTO (2010) and is replaced by adjustments in the resistance factor ϕ. A limit on tension-controlled sections that have sufficient ductility can be expressed as $c/d_t \le 0.003/(0.003 + 0.005) = 0.375$ $(a/d_t \le 0.375\beta_1)$ in which case ϕ is equal to 0.9 or 1.0.

Example 14.2 Determine the steel strain and the associated resistance factor for the beam in Figure 14.11 with the properties given in Example 14.1.

Bonded Case

$$c = 11.1 \text{ in.} \qquad f_{ps} = 247 \text{ ksi} \qquad d_t = 37.5 \text{ in.}$$

$$\varepsilon_s = \frac{0.003\,(d_t - c)}{c} = \frac{0.003\,(37.5 - 11.1)}{11.1} = 0.0071$$

$$\varepsilon_s \ge 0.005 \qquad \phi = 0.9 \quad \text{or} \quad 1.0$$

To refine the computation for the partially prestressed section, a weighted resistance factor between 0.9 and 1.0 could be computed. For tension controls, this simplifies to (Eq. 14.7)

$$\phi = 0.90 + 0.10\,(\text{PPR}) = 0.90 + 0.10\,(0.61) = 0.96$$

where (Eq. 14.8)

$$\text{PPR} = \frac{A_{ps} f_{py}}{A_{ps} f_{py} + A_s f_y} = \frac{1.53\,(243)}{1.53\,(243) + 3.93\,(60)} = 0.61$$

Unbonded Case

$$c = 9.17 \text{ in.} \qquad f_{ps} = 214 \text{ ksi} \qquad d_t = 37.5 \text{ in.}$$

$$\varepsilon_s = \frac{0.003\,(37.5 - 9.17)}{9.17} = 0.0093$$

$$\varepsilon_s \ge 0.005 \qquad \phi = 0.9 \quad \text{or} \quad 1.0$$

Refinement based upon the prestressing ratio

$$\text{PPR} = \frac{A_{ps} f_{py}}{A_{ps} f_{py} + A_s f_y} = \frac{1.53\,(243)}{1.53\,(243) + 3.93\,(60)} = 0.61$$

$$\phi = 0.90 + 0.10\,(\text{PPR}) = 0.90 + 0.10\,(0.61) = 0.96$$

14.2.5 Minimum Tensile Reinforcement

Minimum tensile reinforcement is required to guard against a possible sudden tensile failure. This sudden tensile failure could occur if the moment strength provided by the tensile reinforcement is less than the cracking moment strength of the gross concrete section. To account for the possibility that the moment resistance M_n provided by nonprestressed and prestress tensile reinforcement may be understrength while the moment resistance M_{cr} based on the concrete tensile strength may be overstrength, AASHTO [A5.7.3.3.2] gives the criteria that the amount of prestressed and nonprestressed tensile reinforcement shall be adequate to develop a factored flexural resistance $M_r = \phi M_n$ at least equal to the lesser of:

- ☐ 1.2 times the cracking moment M_{cr} determined on the basis of elastic stress distribution, that is,

$$\phi M_n \geq 1.2 M_{cr} \tag{14.42}$$

or

- ☐ 1.33 times the factored moment required by the applicable strength load combination, that is,

$$\phi M_n \geq 1.33 M_u \tag{14.43}$$

in which M_{cr} may be taken as

$$M_{cr} = S_c\left(f_r + f_{cpe}\right) - M_{dnc}\left(\frac{S_c}{S_{nc}} - 1\right) \geq S_c f_r \tag{14.44}$$

where

f_r = modulus of rupture of the concrete given by Eq. 13.19b [A5.4.2.6] (ksi)
f_{cpe} = compressive stress in concrete due to effective prestress forces only (after allowance for all prestress losses) at extreme fiber of section where tensile stress is caused by externally applied loads (ksi)
M_{dnc} = total unfactored dead-load moment acting on the monolithic or noncomposite section (k-ft)
S_c = section modulus for the extreme fiber of the composite section where tensile stress is caused by externally applied loads (in.^{3})
S_{nc} = section modulus for the extreme fiber of the monolithic or noncomposite section where tensile stress is caused by externally applied loads (in.^{3})

Where the beams are designed for the monolithic or noncomposite section to resist all loads, substitute S_{nc} for S_c in Eq. 14.44 for the calculation of M_{cr}.

Some concrete components have relatively large cross sections required by considerations other than strength, for example, pier caps that are oversized or footings with large dimensions. These large cross-section components have large cracking moments and Eq. 14.42 may require reinforcement considerably larger than what is required for strength. To avoid providing excessive reinforcement where it is not needed for strength, Eq. 14.43 provides a limit on the overstrength required.

14.2.6 Loss of Prestress

After a reinforced concrete member is precompressed by prestressing tendons, a decrease in stress occurs that reduces the effectiveness of the prestress force. In early applications of the prestressing concept using mild steel bars, the prestress losses were two-thirds of the prestress force and prestressing was not effective. It took the development of high-strength steel wire with prestress losses of about one-seventh the prestress force to make the prestressing concept work (Collins and Mitchell, 1991).

Estimating prestress losses is a complex process. The losses are affected by material factors such as mix design, curing procedure, concrete strength, and strand relaxation properties and by environmental factors such as temperature, humidity, and service exposure conditions. In spite of the difficulties, it is important to have a reasonably accurate estimate of prestress losses. If prestress losses are underestimated, the actual precompression force may be smaller than that required to prevent tensile stresses from being exceeded at the bottom fibers of the girder under full service load. If prestress losses are overestimated, a higher than necessary prestress must be provided.

To address the need of more accurate estimates of prestress losses and the impact of high-strength concrete (8 ksi $\leq f_c' \leq$ 15 ksi), research has been conducted to evaluate the AASHTO provisions and to make recommendations for estimating prestress losses. The discussion given herein is based on the work of Tadros et al. (2003), which presents the background and literature review used in developing the provisions in AASHTO [A5.9.5].

A schematic showing the changes in strand (tendon) steel stress is given in Figure 14.14. Some of the prestress losses occur almost instantaneously while others take years before they finally stabilize. Immediate prestress losses are due to slip of the tendons in the anchorages Δf_{pA} plus friction between a tendon and its conduit Δf_{pF} (AB), and elastic compression (shortening) of the concrete Δf_{pES} (CD). Long-time prestress losses Δf_{pLT} are due to the sum of shrinkage of concrete Δf_{pSR}, creep of concrete Δf_{pCR}, and relaxation of the prestressing tendon $\Delta f_{pR}(DE + FG + HK)$. Prestress loss is considered a positive quantity. There are also elastic gains shown in Figure 14.14 when the deck concrete is placed (EF) and when superimposed dead load (GH) and live load (IJ) are added. Elastic gain is considered a negative quantity in the total loss value.

Total Loss of Prestress [A5.9.5.1] The total prestress loss Δf_{pT} is the accumulation of the losses that occur at the different load stages throughout the life of the member. The total prestress losses depend on the method used to apply the prestress force.

For Pretensioned Members

$$\Delta f_{pT} = \Delta f_{pES} + \Delta f_{pLT} \tag{14.45}$$

For Posttensioned Members

$$\Delta f_{pT} = \Delta f_{pA} + \Delta f_{pF} + \Delta f_{pES} + \Delta f_{pLT} \tag{14.46}$$

The prestress losses indicated by the terms in Eqs. 14.45 and 14.46 are discussed in the sections that follow. The expressions developed to calculate the prestress losses should be considered as estimates of the magnitudes of

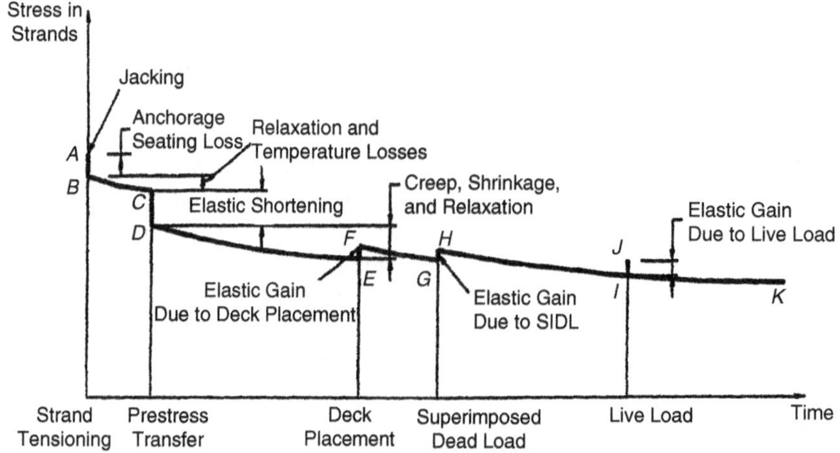

Fig. 14.14 Stress versus time in the strands in a pretensioned concrete girder. [Reproduced with permission from Tadros et al., (2003).]

the different quantities. Too many variables are associated with the prestressing operation, the placing and curing of the concrete, and the service environment to make accurate calculations. However, the expressions are sufficiently accurate for designing members with prestressing tendons and estimating their strength.

Anchorage Set Loss [A5.9.5.2.1] In posttensioned construction not all of the stress developed by the jacking force is transferred to the member because the tendons slip slightly as the wedges or other mechanical devices seat themselves in the anchorage. The anchorage slip or set Δ_A is assumed to produce an average strain over the length of a tendon L, which results in an anchorage set loss of

$$\Delta f_{\text{pA}} = \frac{\Delta_A}{L} E_p \tag{14.47}$$

where E_p is the modulus of elasticity of the prestressing tendon. The range of Δ_A varies from 0.0625 to 0.375 in. with a value of 0.25 in. often assumed. For long tendons the anchorage set loss is relatively small, but for short tendons it could become significant [C5.9.5.2.1].

The loss across stressing hardware and anchorage devices has been measured from 2 to 6% (Roberts, 1993) of the force indicated by the ram pressure times the calibrated ram area. The loss varies depending on the ram and anchor. An initial value of 3% is recommended [C5.9.5.1].

Friction Loss [A5.9.5.2.2] In posttensioned members, friction develops between the tendons and the ducts in which they are placed. If the tendon profile is curved or draped, the ducts are placed in the member to follow the profile. When the tendons are tensioned after the concrete has hardened, they tend to straighten out and develop friction along the wall of the duct. This friction loss is referred to as the *curvature* effect. Even if the tendon profile is straight, the duct placement may vary from side to side or up and down and again friction is developed between the tendon and the duct wall. This friction loss is referred to as the *wobble* effect.

Consider the posttensioned member in Figure 14.15(a) with a curved tendon having an angle change α over a length x from the jacking end. A differential element of length of the curved tendon is shown in Figure 14.15(b) with tensile forces P_1 and P_2 that differ by the friction component dP_1 developed by the normal force N, that is,

$$P_1 - P_2 = dP_1 = \mu N$$

where μ is the coefficient of friction between the tendon and the duct due to the curvature effect. Assuming P_1 and P_2 are nearly equal and that $d\alpha$ is a small angle, the normal force N can be determined from the force polygon of Figure 14.15(c) as $P_1\ d\alpha$ so that

$$dP_1 = \mu P_1\ d\alpha$$

Wobble friction losses over the tendon length dx are expressed as $KP_1\ dx$, where K is the coefficient of friction

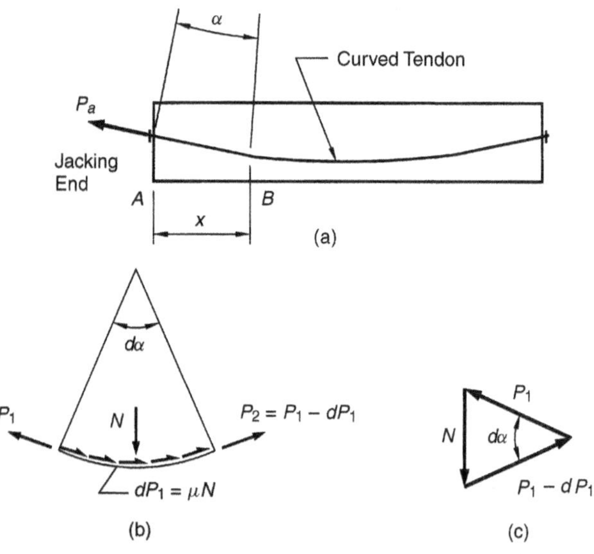

Fig. 14.15 Curvature friction loss (after Nawy, 2009): (a) tendon profile, (b) differential length element, and (c) force polygon.

between the tendon and the surrounding concrete due to the wobble effect. Thus, the total friction loss over length dx becomes

$$dP_1 = \mu P_1 \; d\alpha + KP_1 \; dx$$

or

$$\frac{dP_1}{P_1} = \mu \; d\alpha + K \; dx \tag{14.48}$$

The change in tendon force between two points A and B is given by integrating both sides of Eq. 14.48, that is,

$$\int_{P_B}^{P_A} \frac{dP_1}{P_1} = \mu \int_0^{\alpha} d\alpha + K \int_0^{x} dx$$

which results in

$$\log_e P_A - \log_e P_B = \mu\alpha + Kx$$

or

$$\log_e \frac{P_B}{P_A} = -\mu\alpha - Kx$$

By taking the antilogarithm of both sides and multiplying by P_A we get

$$P_B = P_A e^{-(\mu\alpha + Kx)}$$

By dividing both sides by the area of the prestressing tendon and subtracting from the stress at A, the change in stress between two points x distance apart can be expressed as

$$f_A - f_B = f_A - f_A e^{-(\mu\alpha + Kx)}$$

or

$$\Delta f_{\text{pF}} = f_{\text{pj}} \left[1 - e^{-(\mu\alpha + Kx)} \right] \tag{14.49}$$

where Δf_{pF} is the prestress loss due to friction and f_{pj} is the stress in the tendon at the jacking end of the member.

A conservative approximation to the friction loss is obtained if it is assumed that P_1 in Eq. 14.48 is constant over the length x, so that the integration yields

$$\Delta P_1 \approx P_1 \left(\mu\alpha + Kx \right)$$

or in terms of stresses

$$\Delta f_{\text{pF}} \approx f_{\text{pj}} \left(\mu\alpha + Kx \right) \tag{14.50}$$

This approximation is comparable to using only the first two terms of the series expansion for the exponential in Eq. 14.49. The approximation should be sufficiently accurate because the quantity in parenthesis is only a fraction of unity.

The friction coefficients μ and K depend on the type of tendons, the rigidity of the sheathing, and the form of construction. Design values for these coefficients are given in AASHTO [Table A5.9.5.2.2b-1] and are reproduced in Table 14.7. It is important to know the characteristics of the posttensioning system that is to be used to reasonably estimate friction losses.

Elastic Shortening Loss [A5.9.5.2.3] When the strands at the ends of a pretensioned member are cut, the prestress force is transferred to and produces compression in the concrete. The compressive force on the concrete causes the member to shorten. Compatibility of the strains in the concrete and in the tendon results in a reduction in the elongation of the tendon and an accompanying loss of prestress. Equating the strain in the tendon due to the change in prestress Δf_{pES} and the strain in the concrete due to the concrete stress at the centroid of the tendon f_{cgp} yields

$$\frac{\Delta f_{\text{pES}}}{E_p} = \frac{f_{\text{cgp}}}{E_{\text{ci}}}$$

Solve for the prestress loss due to elastic shortening of the concrete in a pretensioned member to give

$$\Delta f_{\text{pES}} = \frac{E_p}{E_{\text{ci}}} f_{\text{cgp}} \tag{14.51}$$

where E_{ci} is the modulus of elasticity of concrete at transfer of the prestressing force.

If the centroid of the prestressing force is below the centroid of the concrete member, the member is lifted upward at transfer and the self-weight of the member is activated. The elastic concrete stress at the centroid of the tendon is then given by the first three terms of Eq. 14.5 with $y = e_m$:

$$f_{\text{cgp}} = -\frac{P_i}{A_g} - \frac{\left(P_i e_m\right) e_m}{I_g} + \frac{M_g e_m}{I_g} \tag{14.52}$$

where P_i is the prestressing force at transfer. The third term gives the elastic gain due to the applied girder weight. These linear elastic concrete stresses are shown in Figure 14.1.

The force P_i will be slightly less than the transfer force based on the transfer stresses given in Table 14.4 because these stresses will be reduced by the elastic shortening of

Table 14.7 Friction Coefficients for Posttensioning Tendons

Type of Steel	Type of Duct	K	μ
Wire or strand	Rigid and semirigid galvanized metal sheathing	0.0002	0.15–0.25
	Polyethylene	0.0002	0.23
	Rigid steel pipe deviators for external tendons	0.0002	0.25
High-strength bars	Galvanized metal sheathing	0.0002	0.30

From AASHTO Table 5.9.5.2.2b-1. From *AASHTO LRFD Bridge Design Specifications*, Copyright © 2010 by the American Association of State Highway and Transportation Officials, Washington, DC. Used by permission.

the concrete. Thus, for low-relaxation strand, P_i can be expressed as

$$P_i = A_{ps}\left(f_{pi} - \Delta f_{pES}\right) \qquad (14.53)$$

where $f_{pi} = f_{pbt}$ = tendon stress immediately prior to transfer. Realizing that P_i is changed a relatively small amount, AASHTO [A5.9.5.2.3a] allows P_i to be based on an assumed prestressing tendon stress of $0.90f_{pbt} = 0.675f_{pu}$ for low-relaxation strands and the analysis iterated until acceptable accuracy is achieved [C5.9.5.2.3a].

To avoid iteration, Δf_{pES} can be determined by substitution of Eqs. 14.52 and 14.53 into Eq. 14.51 with elastic shortening as a positive quantity and elastic gain as negative to give

$$\Delta f_{pES} = \frac{E_p}{E_{ci}}\left[\frac{A_{ps}\left(f_{pbt} - \Delta f_{pES}\right)}{A_g} + \frac{A_{ps}\left(f_{pbt} - \Delta f_{pES}\right)e_m^2}{I_g} - \frac{M_g e_m}{I_g}\right]$$

and solving for the loss due to elastic shortening in *pretensioned members* [C5.9.5.2.3a]:

$$\Delta f_{pES} = \frac{A_{ps}f_{pbt}\left(I_g + e_m^2 A_g\right) - e_m M_g A_g}{A_{ps}\left(I_g + e_m^2 A_g\right) + \dfrac{E_{ci}}{E_p}\left(A_g I_g\right)} \qquad (14.54)$$

where

A_{ps} = area of prestressing steel (in.^{2})
A_g = gross area of concrete section (in.^{2})
E_{ci} = modulus of elasticity of concrete at transfer (ksi)
E_p = modulus of elasticity of prestressing tendons (ksi)
e_m = average prestressing steel eccentricity at midspan (ksi)
f_{pbt} = stress in prestressing steel immediately prior to transfer as specified in Table 14.4 (ksi)
I_g = moment of inertia of the gross concrete section (in.^{4})
M_g = midspan moment due to member self-weight (kip-in.)

In the case of a posttensioned member, there is no loss of prestress due to elastic shortening if all the tendons are tensioned simultaneously. No loss occurs because the posttensioning force compensates for the elastic shortening as the jacking operation progresses. If the tendons are tensioned sequentially, the first tendon anchored experiences a loss due to elastic shortening given by Eq. 14.51 for a pretensioned member.

Each subsequent tendon that is posttensioned experiences a fraction of the pretensioned loss, with the last tendon anchored without loss. The average posttensioned loss is one-half of the pretensioned loss if the last tendon also had a loss. Because the last tendon anchored does not have a loss, the loss of prestress due to elastic shortening for posttensioned members is given by [A5.9.5.2.3b]

$$\Delta f_{pES} = \frac{N-1}{2N}\frac{E_p}{E_{ci}}f_{cgp} \qquad (14.55)$$

where N is the number of identical prestressing tendons and f_{cgp} is the sum of concrete stresses at the center of gravity of prestressing tendons due to the prestressing force after jacking and the self-weight of the member at the sections of maximum moment (ksi).

Values for f_{cgp} may be calculated using a steel stress reduced below the initial value by a margin (unspecified) dependent on elastic shortening, relaxation, and friction effects. If relaxation and friction effects are neglected, the loss due to elastic shortening in *posttensioned members*, other than slab systems, may be determined by an equation developed in a manner similar to Eq. 14.54 [C5.9.5.2.3b]:

$$\Delta f_{pES} = \frac{N-1}{2N}\left[\frac{A_{ps}f_{pbt}\left(I_g + e_m^2 A_g\right) - e_m M_g A_g}{A_{ps}\left(I_g + e_m^2 A_g\right) + \frac{E_{ci}}{E_p}\left(A_g I_g\right)}\right] \qquad (14.56)$$

For posttensioned structures with bonded tendons, Δf_{pES} may be calculated at the center section of the span or, for continuous construction at the section of maximum moment. For posttensioned structures with unbonded tendons, Δf_{pES} may be calculated using the eccentricity of the prestressing steel averaged along the length of the member. For slab systems, the value of Δf_{pES} may be taken as 25% of that obtained from Eq. 14.56 [C5.9.5.2.3b].

Approximate Estimate of Time-Dependent Losses [A5.9.5.3] It is not always necessary to make detailed calculations for the time-dependent long-term prestress losses Δf_{pLT} due to creep of concrete, shrinkage of concrete, and relaxation of steel if the designs are routine and the conditions are average. The creep and shrinkage properties of concrete are discussed in Section 13.4.2. Relaxation of the prestressing tendons is a time-dependent loss of prestress that occurs when the tendon is held at constant strain.

For standard precast, pretensioned members subject to normal loading and environmental conditions, where

- ☐ Members are made from normal-weight concrete.
- ☐ The concrete is either steam or moist cured.
- ☐ Prestressing is by bars or strands with normal and low-relaxation properties.
- ☐ Average exposure conditions and temperatures characterize the site.

The long-term prestress loss, Δf_{pLT}, due to creep of concrete, shrinkage of concrete, and relaxation of steel shall be estimated using

$$\Delta f_{pLT} = 10.0\frac{f_{pi}A_{ps}}{A_g}\gamma_h\gamma_{st} + 12.0\gamma_h\gamma_{st} + \Delta f_{pR} \qquad (14.57)$$

in which

$$\gamma_h = 1.7 - 0.01H \qquad \gamma_{st} = \frac{5}{1 + f'_{ci}}$$

where

f_{pi} = prestressing steel stress immediately prior to transfer (ksi)
H = average annual ambient relative humidity (%) [A5.4.2.3.2]
γ_h = correction factor for relative humidity of the ambient air
γ_{st} = correction factor for specified concrete strength at time of prestress transfer to the concrete member
Δf_{pR} = estimate of relaxation loss taken as 2.5 ksi for low-relaxation strand, 10.0 ksi for stress-relieved strand, and in accordance with manufacturers recommendation for other types of strand (ksi)

The relative humidity correction factor γ_h is the same for both creep and shrinkage and is normalized to 1.0 when the average relative humidity is 70%. Lower values of humidity increase the long-term prestress loss due to creep and shrinkage. For example, in most of Arizona the average humidity is 40% (Fig. 13.10) and γ_h is 1.3.

The concrete strength correction factor γ_{st} is also the same for both creep and shrinkage and is normalized to 1.0 when the initial compressive strength at prestress transfer f'_{ci} is 4.0 ksi. The value of 4.0 ksi was taken to be 80% of an assumed final strength at service of 5.0 ksi (Tadros et al., 2003). The specified concrete strength is an indirect measure of the quality of the concrete. Generally, the higher strength concrete has sounder aggregate and lower water content and, therefore, has lower long-term prestress loss. For example, if the specified compressive strength at prestress transfer f'_{ci} is 6.0 ksi, then γ_{st} is 0.714.

The first term in Eq. 14.57 corresponds to creep losses, the second term to shrinkage losses, and the third to relaxation losses. The terms in Eq. 14.57 were derived as approximations of the terms in the refined method for a wide range of standard precast prestressed concrete I-beams, box beams, inverted T-beams, and voided slabs [C5.9.5.3]. They were calibrated with full-scale test results and with the results of the refined method and found to give conservative results (Al-Omaishi, 2001; Tadros et al., 2003). The approximate method should not be used for members of uncommon shapes, level of prestressing, or construction staging [C5.9.5.3].

Lump-Sum Estimate of Time-Dependent Losses For members stressed after attaining a compressive strength of 3.5 ksi, other than those made with composite slabs, previous editions of AASHTO also provided approximate lump-sum estimates of the time-dependent prestress losses, which are duplicated in Table 14.8. The losses given in Table 14.8 cover shrinkage and creep in concrete and relaxation of the prestressing tendon. The instantaneous elastic shortening Δf_{pES} must be added to these time-dependent losses to obtain the total prestress loss per Eqs. 14.45 and 14.46.

Although not part of the present specification, Table 14.8 provides a reasonable check on the value computed by Eq. 14.57 and/or the refined methods for losses provided in Appendix D. The refined methods are not presented in the body to keep the chapter focused on the larger issues, not to imply that they are not important. In fact, in many cases, the refined methods are required by AASHTO without the option for the simplified estimate provided in [A5.9.5.3] and Eq. 14.57.

Table 14.8 Time-Dependent Prestress Losses in ksi (Previous AASHTO Editions)

Type of Beam Section	Level	For Wires and Strands with f_{pu} = 235, 250, or 270 ksi	For Bars with f_{pu} = 145 or 160 ksi
Rectangular beams and solid slabs	Upper bound	$29.0 + 4.0$ PPR $(-6.0)^a$ 26.0	$19.0 + 6.0$ PPR
	Average	$+ 4.0$ PPR $(-6.0)^a$	
Box girder	Upper bound	$21.0 + 4.0$ PPR $(-4.0)^a$ 19.0	15.0
	Average	$+ 4.0$ PPR $(-4.0)^a$	
Single T, double T, hollow core, and voided slab	Upper bound	$39.0\left(1.0 - 0.15\dfrac{f'_c - 6.0}{6.0}\right)$ $+6.0$ PPR $(-8.0)^a$	$31.0\left(1.0 - 0.15\dfrac{f'_c - 6.0}{6.0}\right)$ $+6.0$ PPR
	Average	$33.0\left(1.0 - 0.15\dfrac{f'_c - 6.0}{6.0}\right)$ $+6.0$ PPR $(-8.0)^a$	

[a]Values in parentheses are subtractions for low-relaxation strands.

The values in Table 14.8 are limited to prestressed and partially prestressed, nonsegmental, posttensioned members, and pretensioned members made with normal-weight concrete under standard construction procedures and subjected to average exposure conditions. For members made with structural lightweight concrete, the values given in Table 14.8 shall be increased by 5.0 ksi.

The PPR used in Table 14.8 is defined in Eq. 14.8 and is repeated here:

$$\mathrm{PPR} = \frac{A_{\mathrm{ps}} f_{\mathrm{py}}}{A_{\mathrm{ps}} f_{\mathrm{py}} + A_s f_y}$$

In the case of wires and strands, both an upper bound estimate and an average estimate are given. A reasonable approach during preliminary design would be to use the upper bound estimate when evaluating the flexural strength and the average estimate when calculating service load effects. According to Zia et al. (1979), overestimation of prestress losses can be almost as detrimental as underestimation since the former can result in excessive camber and horizontal movement. This applies to prestressing applications based upon any loss methods.

14.3 SHEAR STRENGTH OF REINFORCED CONCRETE MEMBERS

Reinforced concrete members subjected to loads perpendicular to their axis must resist shear forces as well as flexural and axial forces. The shear force resistance mechanism is different for deep beams than for slender beams. The AASHTO Specifications [A5.8.1.1] direct a designer to use the strut-and-tie model [A5.6.3] whenever the distance from the point of zero shear to the face of a support is less than twice the effective depth of the beam, or when a load that causes at least one-half (one-third in the case of segmental box girders) of the shear at a support is within twice the effective depth. For a beam with deep-beam proportions, plane sections no longer remain plane and a better representation of the load-carrying mechanism at the ultimate strength limit state is with the concrete compression struts and steel tension ties as shown in Figure 14.16.

The proportions of typical bridge girders are slender so that plane sections before loading remain plane after loading, and engineering beam theory can be used to describe the relationships between stresses, strains, cross-sectional properties, and the applied forces. Reinforced concrete girders are usually designed for a flexural failure mode at locations of maximum moment. However, this flexural capacity cannot be developed if a premature shear failure occurs due to inadequate web dimensions and web reinforcement. To evaluate the shear resistance of typical bridge girders, the sectional design model of AASHTO [A5.8.3] is used. This model satisfies force equilibrium and strain compatibility and utilizes experimentally determined stress–strain curves for reinforcement and diagonally cracked concrete. Background and details of the sectional model are provided in Vecchio and Collins (1986, 1988) and the books by Collins and Mitchell (1991) and Hsu (1993).

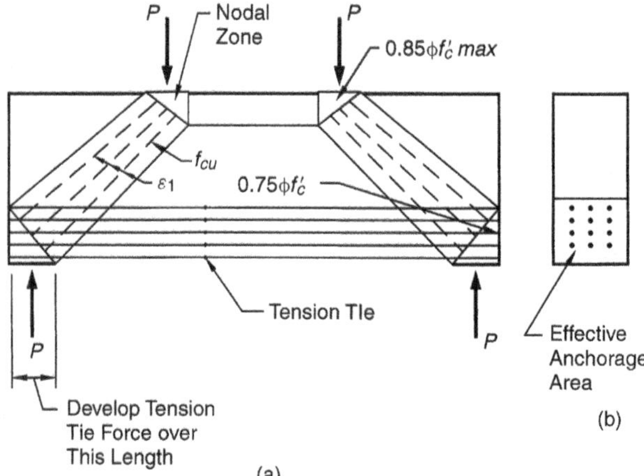

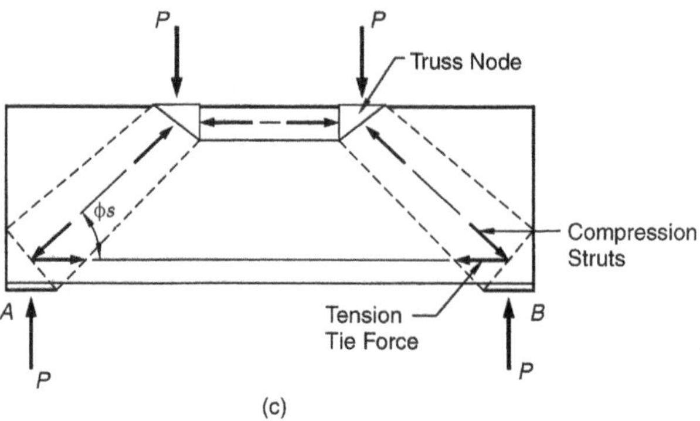

Fig. 14.16 Strut-and-tie model for a deep beam: (a) flow of forces, (b) end view, and (c) truss model [AASHTO Fig. C5.6.3.2-1]. [From *AASHTO LRFD Bridge Design Specifications*. Copyright © 2010 by the American Association of State Highway and Transportation Officials, Washington, DC. Used by permission.]

The nominal shear strength V_n for the sectional design model can be expressed as

$$V_n = V_c + V_s + V_p \tag{14.58}$$

where V_c is the nominal shear strength of the concrete, V_s is the nominal shear strength of the web reinforcement, and V_p is the nominal shear strength provided by the vertical component of any inclined prestress force. In Eq. 14.58, V_p can be determined from the geometry of the tendon profile and effective prestress while V_c and V_s are determined by satisfying equilibrium and compatibility of a diagonally cracked reinforced concrete web. The development of expressions for V_c and V_s based on a variable-angle truss model and the modified compression field theory are given in the sections that follow.

14.3.1 Variable-Angle Truss Model

The variable-angle truss model is presented to provide a connection to the past and to introduce a model that satisfies equilibrium. The truss analogy model is one of the earliest analytical explanations of shear in reinforced concrete beams. According to Collins and Mitchell (1991), it is over 100 years old since it was described by Ritter in 1899 and elaborated by Mörsch in 1902 (Mörsch, 1908).

An example of a variable-angle truss model of a uniformly loaded beam is given in Figure 14.17(a). It is similar to one in Hsu (1993). The dotted lines represent concrete compression struts for the top chord and diagonal web members of the truss. The solid lines represent steel tension ties for the bottom chord and vertical web members. The bottom chord steel area is the longitudinal reinforcement selected to resist flexure and the vertical web members are the stirrups at spacing s required to resist shear.

The top chord concrete compression zone balances the bottom chord tensile steel, and the two make up the couple that resists the moment due to the applied load. The diagonal concrete compressive struts are at an angle θ with the longitudinal axis of the beam and run from the top of a stirrup to the bottom chord. The diagonal struts fan out at the centerline and at the supports to provide a load path for the bottom and top of each stirrup. The fanning of the diagonals also results in a midspan chord force that matches the one obtained by dividing the conventional beam moment by the lever arm d_v.

In defining the lever arm d_v used in shear calculations, the location of the centroid of the tensile force is known a priori but not that of the compressive force. To assist the designer, AASHTO [A5.8.2.9] defines d_v as the effective shear depth taken as the distance, measured perpendicular to the neutral axis, between the resultants of the tensile and compressive forces due to flexure, but it need not be taken less than the greater of $0.9d_e$ or $0.72h$. The effective depth d_e from the extreme compression fiber to the centroid of the tensile force is given by Eq. 14.40 and h is the overall depth of the member.

It is not necessary in design to include every stirrup and diagonal strut when constructing a truss model for concrete beams. Stirrup forces can be grouped together in one vertical member over some tributary length of the beam to give the simplified truss design model of Figure 14.17(b). Obviously, there is more than one way to configure the design truss. For this example, the beam has been divided into six panels, each with a panel load of $wL/6$. Choosing the effective shear depth $d_v = L/9$, then $\tan\theta = \frac{2}{3}$. The bar forces in the members of the truss can then be determined using free-body diagrams such as the one in Figure 14.17(c).

The variation in the stirrup force and the tensile bar force is shown in Figure 14.17(d). Because of the discrete nature of the truss panels, these force diagrams are like stair steps. The staggered stirrup force diagram is always below the conventional shear force diagram for a uniformly loaded beam. The staggered tensile bar force diagram is always above the tensile bar force diagram derived from a conventional moment diagram divided by the lever arm d_v. If the staggered compressive bar force in the top chord had also been shown, it would be below the compressive bar force derived from the conventional moment. This variation can be explained by looking at equilibrium of joints at the top and bottom chords. The presence of compression from the diagonal strut reduces the tension required in a vertical stirrup, reduces the compression in the top chord, and increases the tension in the bottom chord.

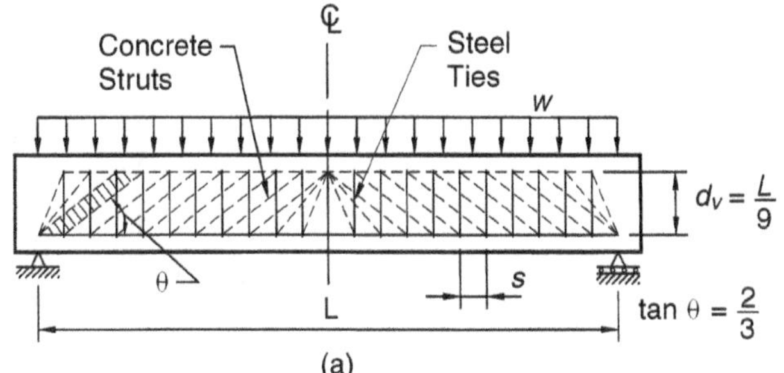

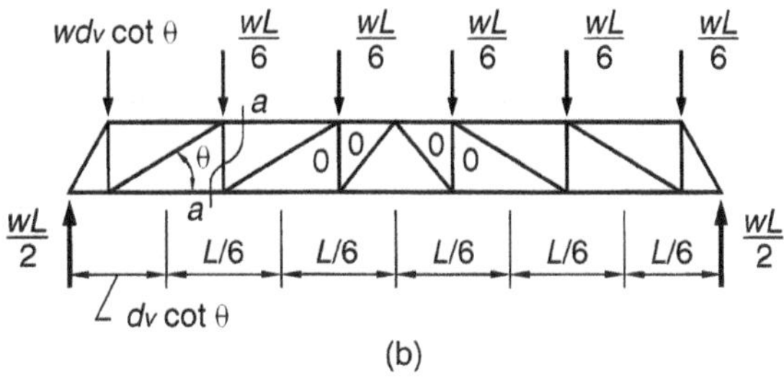

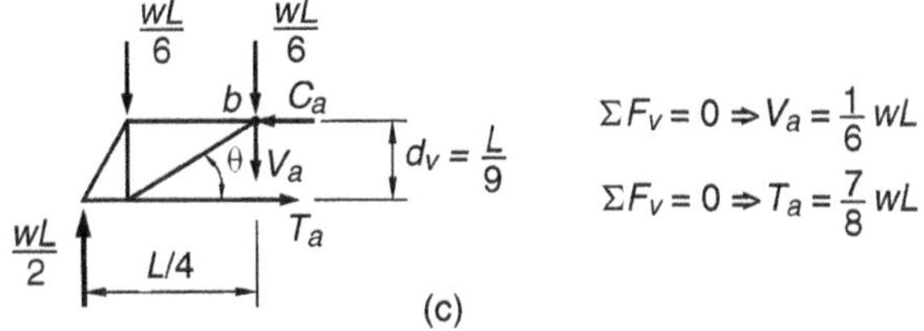

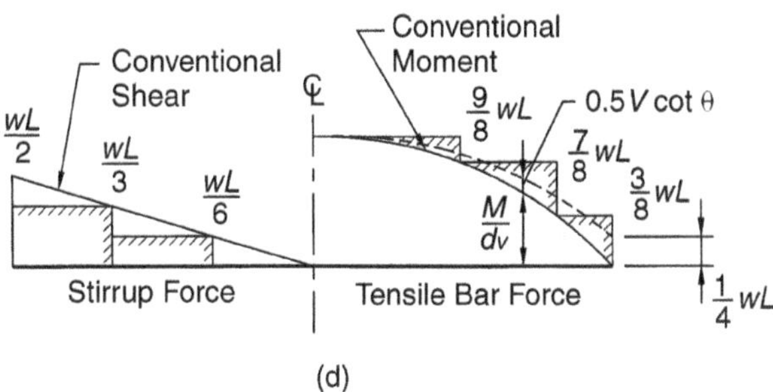

Fig. 14.17 Truss model for a uniformly loaded beam: (a) variable-angle truss model, (b) simplified strut-and-tie design model, (c) free-body diagram for section a–a, and (d) staggered diagrams for truss bar forces. [Reprinted with permission from T. T. C. Hsu (1993). *Unified Theory of Reinforced Concrete*, CRC Press, Boca Raton, FL. Copyright CRC Press, Boca Raton, FL © 1993.]

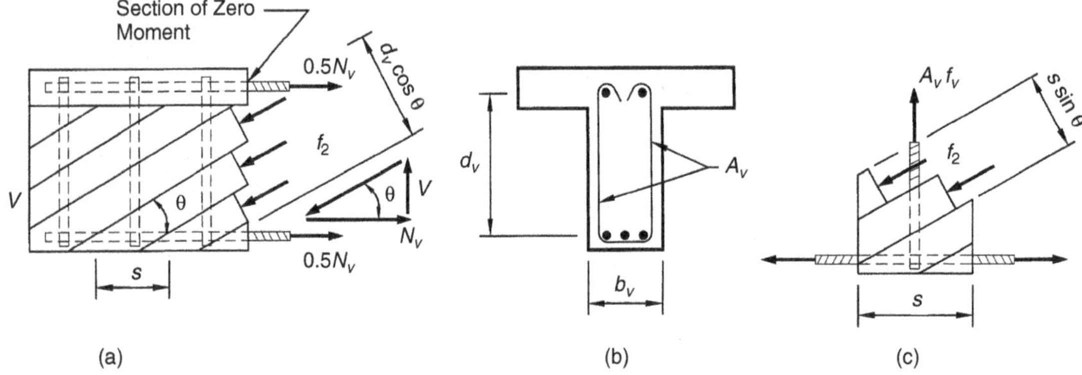

Fig. 14.18 Equilibrium conditions for variable-angle truss: (a) diagonally cracked web, (b) cross section, and (c) tension in web reinforcement. [After Collins and Mitchell (1991). Reprinted by permission of Prentice Hall, Upper Saddle River, NJ.]

To derive an expression for the shear force carried by a stirrup in the variable-angle truss, consider the equilibrium conditions in Figure 14.18 for a section of the web in pure shear ($M = 0$). The balance of vertical forces in Figure 14.18(a) results in

$$V = f_2 b_v d_v \cos\theta \sin\theta$$

or

$$f_2 = \frac{V}{b_v d_v \cos\theta \sin\theta} \tag{14.59}$$

where f_2 is the principal compressive stress in the web and b_v is the minimum web width within the depth d_v. From the force polygon, $\tan\theta = V/N_v$ and

$$N_v = V \cot\theta \tag{14.60}$$

where N_v is the tensile force in the longitudinal direction required to balance the shear force V on the section. This tensile force N_v is assumed to be divided equally between the top and bottom chords of the truss model, adding to the tension in the bottom and subtracting from the compression in the top. The additional tensile force $0.5V \cot\theta$ is shown added to the tensile force M/d_v in the right half of Figure 14.17(d). The resulting dotted line is a good approximation to a smoothed representation of the staggered tensile bar forces.

A bottom chord joint with a tributary length equal to the stirrup spacing s is shown in Figure 14.18(c). The balance of the vertical force in the stirrup with the vertical component of the diagonal compressive force applied over the stirrup spacing s results in

$$A_v f_v = f_2 s b_v \sin\theta \sin\theta$$

where A_v is the total area of the stirrup legs resisting shear and f_v is the tensile stress in the stirrup. Substitution of f_2 from Eq. 14.59 yields

$$A_v f_v = \frac{V s b_v \sin\theta \sin\theta}{b_v d_v \cos\theta \sin\theta} = \frac{Vs}{d_v} \tan\theta$$

$$V = \frac{A_v f_v d_v}{s} \cot\theta \tag{14.61}$$

It is not possible to obtain a closed-form solution for the shear capacity V from the three equilibrium equations—Eqs. 14.59–14.61—because they contain four unknowns: θ, f_v, N_v, and f_2. One design strategy is to assume $\theta = 45^\circ$ and a value for f_v, such as a fraction of f_y for strength design. In either case, Eq. 14.61 gives a shear capacity of a reinforced concrete beam that depends on the tensile stress in the stirrups and the orientation of the principal compressive stress in the concrete. *The model does not include any contribution of the tensile strength in the concrete*. In other words, by using a variable-angle truss model *only* the contribution of V_s in Eq. 14.58 is included. The contribution from the tensile strength of the concrete V_c is considered to be zero.

In summary, the variable-angle truss model clearly shows by Eq. 14.60 that a transverse shear force on a cross section results in an axial force that *increases* the tension in the longitudinal reinforcement. However, it has two shortcomings: It cannot predict the orientation of the principal stresses and it ignores the contribution of the concrete tensile strength. Both of these shortcomings are overcome by the *modified compression field theory*, where strain compatibility gives a fourth condition permitting a rational solution.

14.3.2 Modified Compression Field Theory

In the design of the relatively thin webs of steel plate girders, the web panels between transverse stiffeners subjected to shearing stresses are considered to support tensile stresses only because the compression diagonal is assumed to have buckled. The postbuckling strength of the plate girder webs depends on the orientation of the principal tensile stress, stiffener spacing, girder depth, web thickness, and yield strength of the material. A *tension field theory* has been developed to determine the relationships between these parameters and to predict the shear strength of plate girder webs. See Chapter 19 for details.

In the webs of reinforced concrete beams subjected to shearing stresses, an analogous behavior occurs, except the tension diagonal cracks, and the compression diagonal is the dominant support in the web. Instead of a tension field theory, a *compression field theory* has been developed to explain the behavior of reinforced concrete beams subjected to shear.

Originally, the compression field theory assumed that once web cracking occurred, the principal tensile stress vanished. The theory was later modified to include the principal tensile stress and to give a more realistic description of the shear failure mechanism. Hence the term "modified" compression field theory (MCFT).

Stress Considerations Figure 14.19 illustrates pure shear stress fields in the web of a reinforced concrete beam before and after cracking. A Mohr stress circle for the concrete is also shown for each of the cases. Before cracking [Fig. 14.19(a)], the reinforced concrete web is assumed to be homogeneous and Mohr's circle of stress is about the origin with radius v and $2\theta = 90°$. After cracking [Fig. 14.19(b)], the web reinforcement carries the tensile stresses and the concrete struts carry the compressive stresses. As a result, the orientation of the principal stresses changes to an angle θ less than 45°. If the concrete tensile strength is not ignored

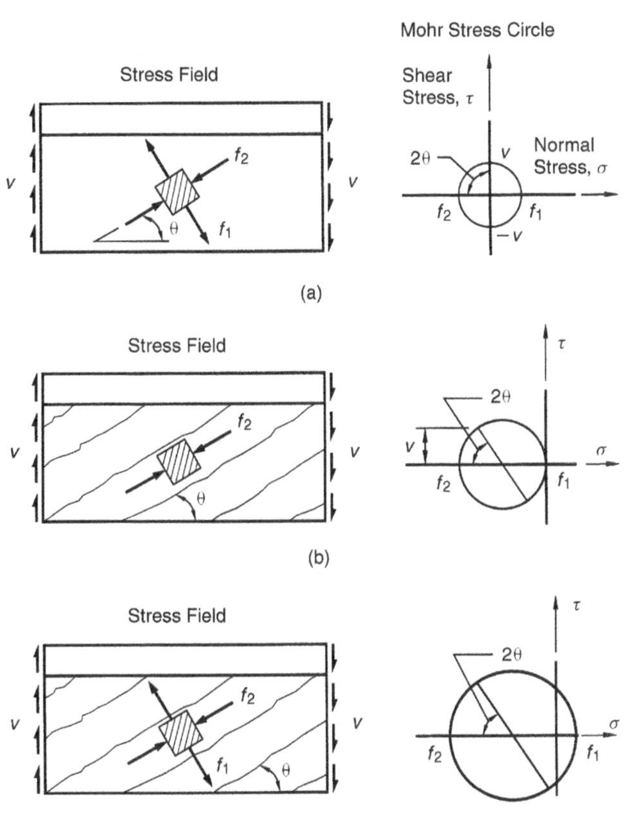

Fig. 14.19 Stress fields in web of a reinforced concrete beam subjected to pure shear: (a) Before cracking, $f_1 = f_2 = v$, $\theta = 45°$, (b) compression field theory, $f_1 = 0$, $\theta < 45°$, and (c) modified compression field theory, $f_1 \neq 0$, $\theta < 45°$. [After Collins and Mitchell (1991). Reprinted by permission of Prentice Hall, Upper Saddle River, NJ.]

and carries part of the tensile force, the stress state of the *modified compression field theory* [Fig. 14.19(c)] is used to describe the behavior.

The Mohr stress circle for the concrete compression strut of Figure 14.19(c) is more fully explained in Figure 14.20. A reinforced concrete element subjected to pure shear has a Mohr stress circle of radius v about the origin [Fig. 14.20(a)]. Interaction within the element develops compression in the concrete struts [Fig. 14.20(b)] and tension in the steel reinforcement [Fig. 14.20(c)]. The concrete portion of the element is assumed to carry all of the shear, along with the compression, which results in the Mohr stress circles of Figures 14.19(c) and 14.20(b). The angle 2θ rotates, depending on the relative values of shear and compression, even though the comparable angle of the reinforced concrete element remains fixed at 90°.

There is no stress circle for the steel reinforcement because its shear resistance (dowel action) is ignored. The tensile stresses f_s^* and f_v^* are psuedoconcrete tensile stresses, or *smeared steel tensile stresses*, that are equivalent to the tensile forces in the reinforcement. The use of superposition and diagrams in Figures 14.20(b) and 14.20(c) yields

$$f_s^* b_v s_x = f_s A_s$$

$$f_s^* = \frac{A_s}{b_v s_x} f_s = \rho_x f_s \tag{14.62}$$

and

$$f_v^* b_v s = f_v A_v$$

$$f_v^* = \frac{A_v}{b_v s} f_v = \rho_v f_v \tag{14.63}$$

where s_x is the vertical spacing of longitudinal reinforcement including skin reinforcement, and s is the horizontal spacing of stirrups:

$$\rho_x = \frac{A_s}{b_v s_x} = \text{longitudinal reinforcement ratio} \tag{14.64}$$

$$\rho_v = \frac{A_v}{b_v s} = \text{transverse reinforcement ratio} \tag{14.65}$$

The stresses between the concrete and reinforcement may be dissimilar after cracking because of different material moduli, but the strains are not. Fortunately, the condition of strain compatibility provides the additional relationships, coupled with the equilibrium equations, to uniquely determine the angle θ and the shear strength of a reinforced concrete member. This unique determination can be done by considering the web of a reinforced concrete beam to behave like a membrane element with in-plane shearing and normal stresses and strains that can be analyzed using Mohr stress and strain circles.

Strain Considerations Before writing the equilibrium equations for the modified compression field theory, the compatibility conditions based on a Mohr strain circle are developed. Consider the cracked reinforced concrete web

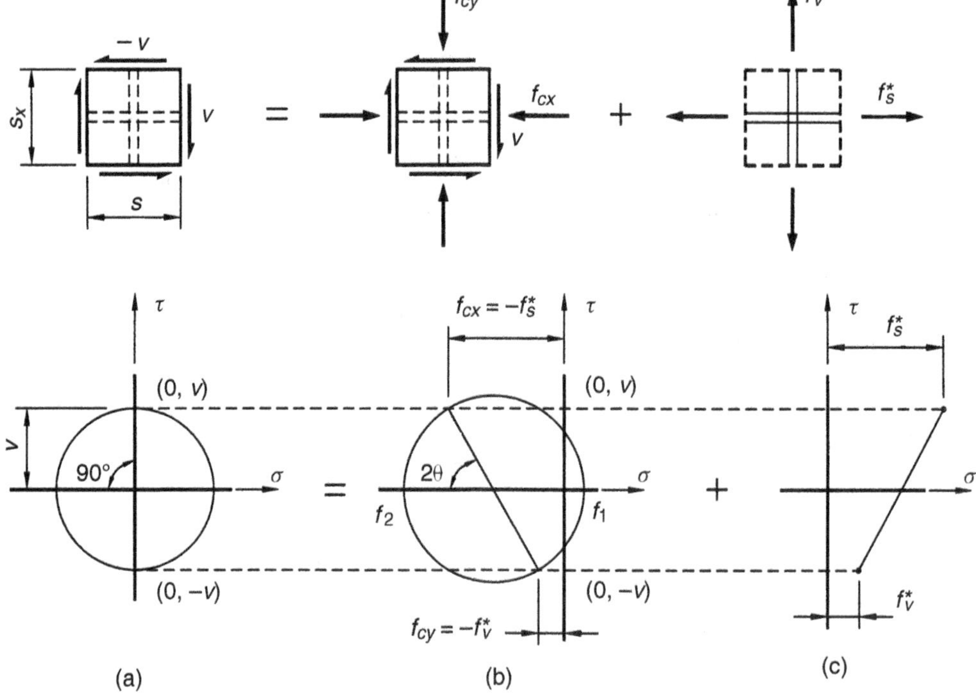

Fig. 14.20 Reinforced concrete element subjected to pure shear: (a) reinforced concrete, (b) concrete struts, and (c) reinforcement. [Reprinted with permission from T. T. C. Hsu (1993). *Unified Theory of Reinforced Concrete*, CRC Press, Boca Raton, FL. Copyright CRC Press, Boca Raton, FL © 1993.]

element in Figure 14.21(a), which is subjected to a biaxial state of stress and has strain gages placed to record average strains in the longitudinal ε_x, transverse ε_t, and 45° ε_{45} directions. The strain gages are assumed to be long enough so that the average strain is over more than one crack. The definition of normal strains [Fig. 14.21(b)] is an elongation per unit length while shearing strains [Fig. 14.21(c)] are defined as the change in angle γ from an original right angle. Because of the assumed symmetry in the material properties, this angle is split equally between the two sides originally at right angles. The direction of the shearing strains corresponds to the direction assumed for positive shearing stresses in Figure 14.20.

A Mohr strain circle [Fig. 14.21(d)] can be constructed if three strains at a point and their orientation to each other are known. The three given average strains are ε_x, ε_t, and ε_{45}. The relationships between these strains and the principal average strains ε_1 and ε_2 and the angle θ, which defines the inclination of the compression struts, are required.

In deriving the compatibility conditions, consider the top one-half of the Mohr strain circle in Figure 14.21(e) to be completely in the first quadrant, that is, all the strain quantities are positive (the $\gamma/2$ axis is to the left of the figure). [With all of the strains assumed to be positive, the derivation is straightforward and does not require intuition as to which quantities are positive and which are negative. (The positive and negative signs should now take care of themselves.)]

First, the center of the circle can be found by taking the average of ε_x and ε_t or the average of ε_1 and ε_2, that is,

$$\frac{\varepsilon_x + \varepsilon_t}{2} = \frac{\varepsilon_1 + \varepsilon_2}{2}$$

so that the principal tensile strain is

$$\varepsilon_1 = \varepsilon_x + \varepsilon_t - \varepsilon_2 \tag{14.66}$$

By using a diameter of unity in Figure 14.21(e), the radius is one-half, and the vertical line segment AE is

$$AE = \tfrac{1}{2}\sin 2\theta = \sin\theta\cos\theta$$

By recalling $\sin^2\theta + \cos^2\theta = 1$, the line segment ED is given by

$$\begin{aligned} ED &= \tfrac{1}{2}\cos 2\theta + \tfrac{1}{2} \\ &= \tfrac{1}{2}\left(\cos^2\theta - \sin^2\theta\right) + \tfrac{1}{2}\left(\cos^2\theta + \sin^2\theta\right) = \cos^2\theta \end{aligned}$$

so that the line segment BE becomes

$$BE = 1 - \cos^2\theta = \sin^2\theta$$

From these relationships and similar triangles, the following three compatibility equations can be written:

$$\varepsilon_x - \varepsilon_2 = \left(\varepsilon_1 - \varepsilon_2\right)\sin^2\theta \tag{14.67}$$

$$\varepsilon_t - \varepsilon_2 = \left(\varepsilon_1 - \varepsilon_2\right)\cos^2\theta \tag{14.68}$$

$$\gamma_{xt} = 2\left(\varepsilon_1 - \varepsilon_2\right)\sin\theta\cos\theta \tag{14.69}$$

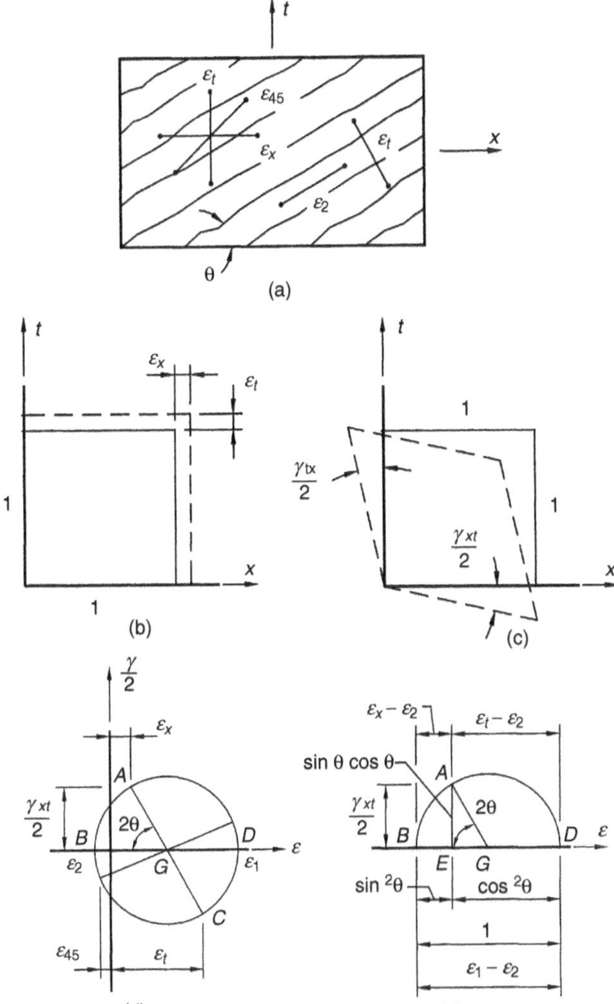

Fig. 14.21 Compatibility conditions for a cracked web element: (a) average strains in a cracked web element, (b) normal strains, (c) shearing strains, (d) Mohr strain circle, and (e) geometric relations. [Reprinted with permission from T. T. C. Hsu (1993). *Unified Theory of Reinforced Concrete*, CRC Press, Boca Raton, FL. Copyright CRC Press, Boca Raton, FL © 1993.]

Division of Eq. 14.67 by Eq. 14.68 results in an expression that does not contain ε_1, that is,

$$\tan^2\theta = \frac{\varepsilon_x - \varepsilon_2}{\varepsilon_t - \varepsilon_2} \tag{14.70}$$

The relative magnitudes of the principal strains ε_1 and ε_2 shown in Figure 14.21(d), with ε_1 being an order of magnitude greater than ε_2, are to be expected because the average tensile strain ε_1 is across cracks that offer significantly less resistance than the direct compression in the concrete struts.

Equilibrium conditions for the modified compression field theory are determined by considering the free-body diagrams in Figure 14.22. The cracked reinforced concrete web shown in Figure 14.22(a) is the same as the one in Figure 14.18(a) except for the addition of the average principal tensile stress f_1 in the concrete. The actual tensile stress distribution in the concrete struts is shown with a peak value within the strut, which then goes to zero at a crack. The constitutive laws developed for concrete in tension in cracked webs (Fig. 13.9) are based on stresses and strains measured over a *finite* length, and therefore the values for f_1 and ε_1 should be considered as *average* values over this length.

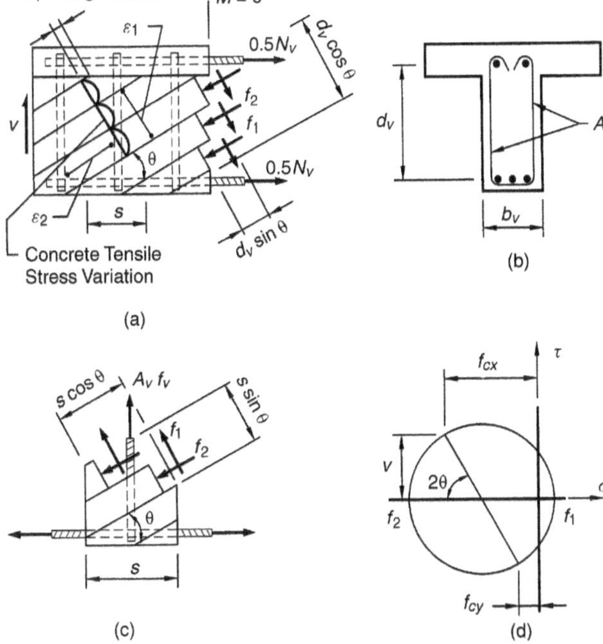

Fig. 14.22 Equilibrium conditions for modified compression field theory: (a) cracked reinforced concrete web, (b) cross section, (c) tension in web reinforcement, and (d) Mohr stress circle for concrete. [After Collins and Mitchell (1991). Reprinted by permission of Prentice Hall, Upper Saddle River, NJ.]

Equilibrium of vertical forces in Figure 14.22(a) results in

$$V = f_2 b_v d_v \cos\theta \sin\theta + f_1 b_v d_v \sin\theta \cos\theta$$

from which the principal compressive stress f_2 can be expressed as

$$f_2 = \frac{v}{\sin\theta \cos\theta} - f_1 \tag{14.71}$$

where v is the average shear stress,

$$v = \frac{V}{b_v d_v} \tag{14.72}$$

In Eq. 14.71, f_2 is assumed to be a compressive stress in the direction shown in Figures 14.22(a) and 14.22(c).

Equilibrium Considerations Equilibrium of the vertical forces in Figure 14.22(c) results in

$$A_v f_v = f_2 s b_v \sin^2\theta - f_1 s b_v \cos^2\theta$$

Substitution of Eq. 14.71 for f_2, Eq. 14.72 for v, and rearranging terms gives

$$V = f_1 b_v d_v \cot\theta + \frac{A_v f_v d_v}{s} \cot\ \theta \qquad (14.73)$$

which represents the sum of the contributions to the shear resistance from the concrete V_c and the web reinforcement tensile stresses V_s. *By comparing Eq.* 14.61 *with Eq.* 14.73, *the modified compression field theory provides the concrete tensile stress shear resistance missing from the variable-angle truss model.*

Equilibrium of the longitudinal forces in Figure 14.22(a) results in

$$N_v = f_2 b_v f_v \cos^2\theta - f_1 b_v d_v \sin^2\theta$$

Substitution for f_2 from Eq. 14.71 and combination of terms gives

$$N_v = \left(v \cot\theta - f_1\right) b_v d_v \qquad (14.74)$$

If no axial load is present on the member, N_v must be resisted by the longitudinal reinforcement, that is,

$$N_v = A_{sx} f_{sx} + A_{px} f_{px}$$

where A_{sx} is the total area of longitudinal nonprestressed reinforcement, A_{px} is the total area of longitudinal prestressing tendons, and f_{sx} and f_{px} are the "smeared stresses" averaged over the area $b_v d_v$ in the longitudinal nonprestressed reinforcement and longitudinal prestressing tendons, respectively. Equating the above two expressions for N_v and dividing by $b_v d_v$ results in

$$\rho_{sx} f_{sx} + \rho_{px} f_{px} = v \cot\theta - f_1$$

where

$$\rho_{sx} = \frac{A_{sx}}{b_v d_v} = \text{nonprestressed reinforced ratio} \qquad (14.75)$$

$$\rho_{px} = \frac{A_{px}}{b_v d_v} = \text{prestressed reinforcement ratio} \qquad (14.76)$$

Constitutive Considerations With the strain compatibility conditions and stress equilibrium requirements written, only the constitutive relations linking together the stresses and strains remain to complete the definition of the modified compression field theory. The stress–strain relations for concrete in compression (Fig. 13.4), concrete in tension (Fig. 13.9), nonprestressed reinforcement (Fig. 13.15), and prestressing reinforcement (Fig. 13.17) were presented earlier and are summarized in Figure 14.23 for convenience.

A few comments on the four stress–strain curves in Figure 14.23 are appropriate. *The importance of compression softening of concrete* [*Fig.* 14.23*(a)*] *due to tension cracking in the perpendicular direction cannot be overemphasized.* The discovery (1972) and quantification (1981) of this phenomenon was called by Hsu (1993) "the major breakthrough in understanding the shear and torsion problem in reinforced concrete."

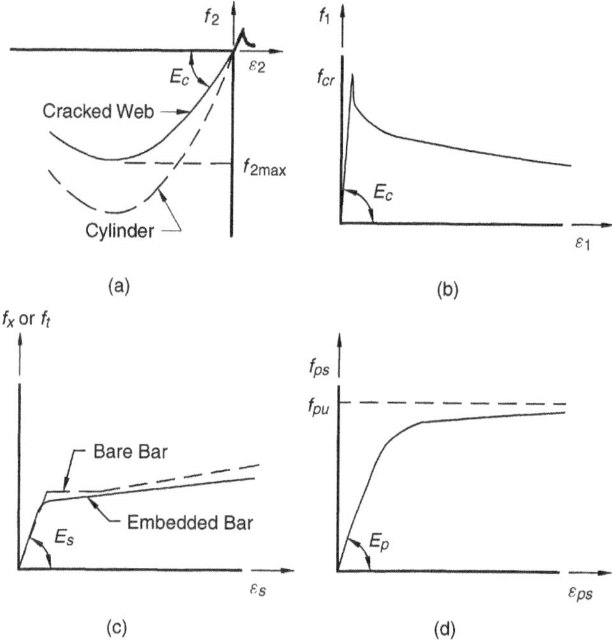

Fig. 14.23 Constitutive relations for membrane elements: (a) concrete in compression, (b) concrete in tension, (c) nonprestressed steel, and (d) prestressing tendon. [Reprinted with permission from T. T. C. Hsu (1993). *Unified Theory of Reinforced Concrete*, CRC Press, Boca Raton, FL. Copyright CRC Press, Boca Raton, FL © 1993.]

Prior to this discovery, the compression response of concrete obtained primarily from uniaxial tests on concrete cylinders and the predictions of shear strength based on the truss model consistently overestimated the tested response. The current relationships given in Eqs. 13.3 and 13.4 are based on relatively thin (3 in.) membrane elements with one layer of reinforcement (Vecchio and Collins, 1986). Additional tests (Adebar and Collins, 1994) were conducted on thicker (12 in.) elements with two layers of reinforcement. The effect of confinement provided by through-the-thickness reinforcement may change the compression relationships.

The average stress–strain response for a reinforced concrete web in tension is shown in Figure 14.23(b). The curve shown is an enlargement of the upper right-hand corner (first quadrant) of Figure 14.23(a). The concrete modulus of elasticity E_c is the same in both figures but is distorted in Figure 14.23(b) because the stress scale has been expanded while the strain has not. The maximum principal tensile strain ε_1 is of the same order of magnitude as the maximum principal compressive strain ε_2, even though the tensile stresses are not. Expressions for the ascending and descending branches are given in Eqs. 13.22 and 13.23.

As shown in Figure 14.23(c), the stress–strain response of a reinforcing bar embedded in concrete is different than that of a bare bar. The embedded bar is stiffened by the concrete surrounding it and does not exhibit a flat yield plateau. At strains beyond the yield strain of the bare bar, an embedded bar develops stresses that are lower than those in a bare bar. A bilinear approximation to the average stress–strain response of a mild steel bar embedded in concrete is given by Eqs. 13.37 and 13.38.

A typical stress–strain curve for a bare prestressing tendon is shown in Figure 14.23(d). Expression for low-relaxation and stress-relieved strands are given by Eqs. 13.41 and 13.42, respectively. For bonded and unbonded prestressing strands, approximate expressions for the prestressing stress f_{ps} are given by Eqs. 14.11 and 14.25, respectively.

In the development of the concrete tensile response shown in Figure 14.23(b), two conditions have been implied: (1) average stresses and average strains across more than one crack have been used and (2) the cracks are not so wide that shear cannot be transferred across them. The first condition has been emphasized more than once, but the second condition requires further explanation.

Behavior at the Cracks A diagonally cracked beam web is shown in Figure 14.24(a) with a diagram of the actual tensile stress variation and the average principal stress f_1 related to a principal tensile strain ε_1 taken over a finite gage length. For the cracked web the average principal tensile strain ε_1 is due mostly to the opening of the cracks because the elastic tensile strain is relatively small, that is,

$$\varepsilon_1 \approx \frac{w}{S_{m\theta}} \tag{14.77}$$

where w is the crack width and $S_{m\theta}$ is the mean spacing of the diagonal cracks. If the crack width w becomes too large, it will not be possible to transfer shear across the crack by the aggregate interlock mechanism shown in the detail at a crack. In other words, if the cracks are too wide, shear failure occurs by slipping along the crack surface.

The aggregate interlock mechanism is modeled in the definitive work by Walraven (1981). It is based on a statistical analysis of the contact areas and the wedging action that occurs between irregular crack faces. At a crack, local shear stresses v_{ci} are developed, which enable tensile forces to be transmitted across the crack. Experimental pushoff tests on externally restrained specimens were conducted and they verified the analytical model. The variables in the tests were concrete strength, maximum aggregate size, total aggregate volume per unit volume of concrete, external restraint stiffness, and initial crack width. In fitting the experimental results to the theoretical model, the best results

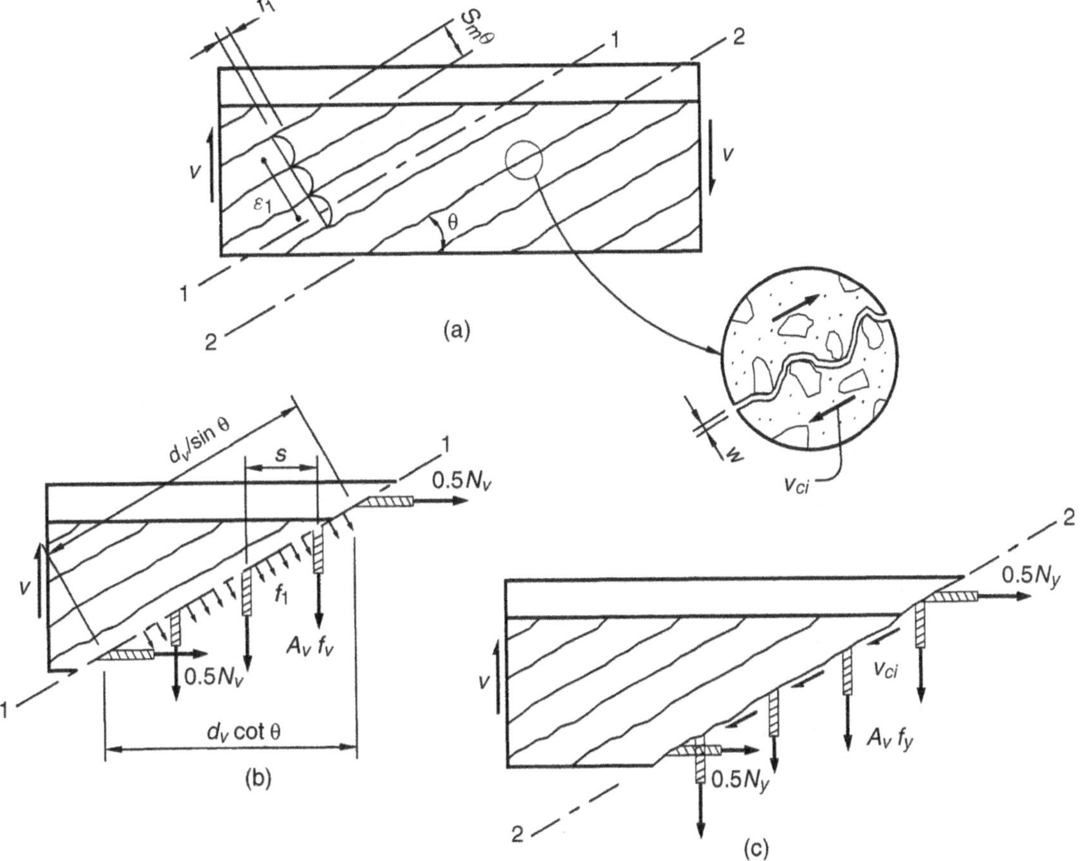

Fig. 14.24 Transmitting tensile forces across a crack: (a) beam web cracked by shear, (b) average stresses between cracks, and (c) local stresses at a crack. [After Collins and Mitchell (1991). Reprinted by permission of Prentice Hall, Upper Saddle River, NJ.]

were obtained with a coefficient of friction of 0.4 and a matrix yielding strength that is a function of the square root of the concrete compressive strength. By using Walraven's experimental data, Vecchio and Collins (1986) derived a relationship between shear transmitted across a crack and the concrete compressive strength. Their expression was further simplified by Collins and Mitchell (1991) who dropped the effect of local compressive stresses across the crack and recommended that the limiting value of v_{ci} be taken as

$$v_{ci} \leq \frac{0.0683\sqrt{f'_c}}{0.3 + 24w/\left(a_{max} + 0.63\right)} \tag{14.78}$$

where w is the crack width (in.), a_{max} is the maximum aggregate size (in.), and f'_c is concrete compressive strength (ksi). By limiting the shear stress on the crack v_{ci} to the value of Eq. 14.78, crack slipping failures should not occur.

Combined Equilibrium, Compatibility, and Constitutive Models The average stresses on section 1–1 in Figure 14.24(a) within a concrete compressive strut that were used in developing the equilibrium Eqs. 14.71, 14.73, and 14.74 are repeated in Figure 14.24(b). The stresses in the transverse and longitudinal reinforcement are also average stresses because the stiffening effect of a bar embedded in concrete shown in Figure 14.23(c) applies. At a crack along section 2–2 in Figures 14.24(a) and 14.24(c), the concrete tensile stress vanishes, the aggregate interlock mechanism is active, wedging action occurs that strains the reinforcement, and (as in Fig. 13.14) the reinforcement stress increases until it reaches its yield strength.

Both sets of stresses in Figures 14.24(b) and 14.24(c) must be in equilibrium with the same vertical shear force V. This vertical equilibrium can be stated as

$$A_v f_v \frac{d_v \cot\theta}{s} + f_1 \frac{b_v d_v}{\sin\theta} \cos\theta = A_v f_v \frac{d_v \cot\theta}{s} + v_{ci} \frac{b_v d_v}{\sin\theta} \sin\theta$$

and solving for the average principal tensile stress, we have

$$f_1 \leq v_{ci} \tan\theta + \frac{A_v}{b_v s}\left(f_y - f_v\right) \tag{14.79}$$

where f_1 is limited by the value of v_{ci} in Eq. 14.78.

The two sets of stresses in Figures 14.24(b) and 14.24(c) must also result in the same horizontal force, that is,

$$N_v + f_1 \frac{b_v d_v}{\sin\theta} \sin\theta = N_y + v_{ci} \frac{b_v d_v}{\sin\theta} \cos\theta$$

Substitution for v_{ci} from Eq. 14.79 and rearrangement of terms yields

$$N_y = N_v + f_1 b_v d_v + \left[f_1 - \frac{A_v}{b_v s}\left(f_y - f_v\right)\right] b_v d_v \cot^2\theta \tag{14.80}$$

in which

$$N_y = A_{sx} f_y + A_{px} f_{ps} \tag{14.81}$$

$$N_v = A_{sx} f_{sx} + A_{px} f_{px} \tag{14.82}$$

where A_{sx} is the total area of longitudinal nonprestressed reinforcement, A_{px} is the total area of longitudinal prestressing tendons, f_y is the yield stress of the bare nonprestressed reinforcement, f_{ps} is the stress in the prestressing tendon from Eq. 14.11, and f_{sx} and f_{px} are the *smeared* stresses averaged over the area $b_v d_v$ in the embedded longitudinal nonprestressed reinforcement and prestressing tendons, respectively. Equation 14.80 is a *second limitation* on f_1 that states that if the longitudinal reinforcement begins to yield at a crack, the maximum principal concrete tensile stress f_1 has been reached and cannot exceed

$$f_1 \leq \frac{N_y - N_v}{b_v d_v} \sin^2\theta + \frac{A_v}{b_v s}\left(f_y - f_v\right) \cos^2\theta$$

which can be written in terms of stresses as

$$f_1 \leq \left[\rho_{sx}\left(f_y - f_{sx}\right) + \rho_{px}\left(f_{ps} - f_{px}\right)\right] \sin^2\theta + \rho_v\left(f_y - f_v\right) \cos^2\theta \tag{14.83}$$

where the reinforcement ratios ρ_{sx}, ρ_{px}, and ρ_v are defined in Eqs. 14.75, 14.76, and 14.65, respectively.

The response of a reinforced concrete beam subjected to shear forces can now be determined from the relationships discussed above. In Collins and Mitchell (1991), a 17-step procedure is outlined that addresses the calculations and checks necessary to determine V of Eq. 14.73 as a function of the principal tensile strain ε_1. Similarly, Hsu (1993) presents a flowchart and an example problem illustrating the solution procedure for generating the shearing stress–strain curve.

Unfortunately, these solution procedures are cumbersome, and for practical design applications, design aids are needed to reduce the effort. These aids have been developed by Collins and Mitchell (1991) and are available in AASHTO [A5.8.3.4.2]. They are discussed in Section 14.3.3.

The computer program Response-2000 is also available as an aid to calculating the strength and ductility of a reinforced concrete cross section subject to shear, moment, and axial load. The program uses a sectional analysis and was developed at the University of Toronto by E. C. Bentz (2000) in a project supervised by M. P. Collins. Response-2000 can be downloaded at no charge, along with a manual and sample input, from the website www.ecf.utoronto.ca/~bentz/home.shtml.

14.3.3 Shear Design Using Modified Compression Field Theory

Returning to the basic expression for nominal shear resistance given by Eq. 14.58, recall

$$V_n - V_p = V_c + V_s \tag{14.84}$$

Substitution of the shear resistance from the concrete and web reinforcement determined by the modified compression field theory (Eq. 14.73) gives

$$V_n - V_p = f_1 b_v d_v \cot\theta + \frac{A_v f_v d_v}{s}\cot\theta \quad (14.85)$$

Assuming that $f_v = f_y$ when the limit state is reached, the combination of Eqs. 14.78 and 14.80 yields an upper bound for the average principal tensile stress

$$f_1 \le v_{ci}\tan\theta \le \frac{2.16\,(0.0316)\sqrt{f_c'}}{0.3 + 24w/\left(a_{max} + 0.63\right)}\tan\theta \quad (14.86)$$

and Eq. 14.85 may be written as

$$V_n - V_p = (0.0316)\,\beta\sqrt{f_c'}b_v d_v + \frac{A_v f_v d_v}{s}\cot\theta \quad (14.87)$$

where

$$\beta \le \frac{2.16}{0.3 + 24w/\left(a_{max} + 0.63\right)} \quad (14.88)$$

The nominal shear resistance expression in AASHTO [A5.8.3.3] is given by

$$V_n - V_p = (0.0316)\,\beta\sqrt{f_c'}b_v d_v + \frac{A_v f_v d_v}{s}(\cot\theta + \cot\alpha)\sin\alpha \quad (14.89)$$

where α is the angle of inclination of transverse reinforcement. If $\alpha = 90°$, Eq. 14.89 becomes Eq. 14.87. The constant $1/\sqrt{100} = 0.0316$ is necessary to keep β in familiar terms while using f_c' in ksi per the AASHTO LRFD convention.

Now the crack width w can be expressed as the product of the average principal tensile strain ε_1 and the mean spacing of the diagonal cracks $S_{m\theta}$ to yield

$$w = \varepsilon_1 S_{m\theta} \quad (14.90)$$

Simplification To simplify the calculations, Collins and Mitchell (1991) assume that the crack spacing $S_{m\theta}$ is 12 in. [C5.8.3.4.2] and that the maximum aggregate size a_{max} is 0.75 in. This results in an upper bound for β of

$$\beta \le \frac{2.16}{0.3 + 200\varepsilon_1} \quad (14.91)$$

In addition to the limitation imposed on f_1 in Eq. 14.86 by the shear stress on a diagonal crack, f_1 is also assumed to follow the constitutive relationship shown in Figure 14.23(b) and given by Eq. 13.23 with $f_{cr} = 4\,(0.0316)\sqrt{f_c'}$, that is,

$$f_1 = \frac{\alpha_1\alpha_2\,(4)\,(0.0316)\sqrt{f_c'}}{1 + \sqrt{500\varepsilon_1}} \quad (14.92)$$

Substitute this expression into Eq. 14.85 and relate it to Eq. 14.87 to give

$$\beta = \frac{\alpha_1\alpha_2\,(4)\cot\theta}{1 + \sqrt{500\varepsilon_1}}$$

assuming the tension stiffening or bond factors $\alpha_1\alpha_2$ are equal to unity, a second relationship for β that depends on the average principal tensile strain ε_1 is

$$\beta = \frac{4\cot\theta}{1 + \sqrt{500\varepsilon_1}} \quad (14.93)$$

At this point it is informative to compare the modified compression field theory Eq. 14.87 with the traditional expression for shear strength. From AASHTO (2002) standard specifications, the nominal shear strength for nonprestressed beams is (for inch-pound units)

$$V_n = 2\sqrt{f_c'}b_w d + \frac{A_v f_y d}{s} \quad (14.94)$$

By comparing this result with Eq. 14.87, and realizing that $b_w = b_v$ and d is nearly equal to d_v, the two expressions will give the same results if $\theta = 45°$ and $\beta = 2$. A simplification of Eq. 14.89 using $\theta = 45°$ and $\beta = 2$ is also allowed for nonprestressed concrete sections not subjected to axial tension and containing at least the minimum amount specified for transverse reinforcement [A5.8.3.4.1]. It is interesting to note that the associated average principal tensile strain ε_1 is 0.002 for this case; this is approximately equal to the yield strain $\varepsilon_y = 0.00207$ for Grade 60 steel.

Longitudinal Strain Thus, the improvements introduced by the modified compression field theory in Eq. 14.87 are the ability to consider a variable orientation θ and a change in magnitude β of the principal tensile stress across a cracked compression web. The orientation and magnitude are not fixed but vary according to the relative magnitude of the local shear stress and longitudinal strain.

In both Eqs. 14.91 and 14.93, an increase in tensile straining (represented by the average principal tensile strain ε_1) decreases β and the shear that can be resisted by the concrete tensile stresses. To determine this important parameter ε_1, the modified compression field theory uses the compatibility conditions of a Mohr strain circle developed in Eqs. 14.66–14.70. Substitution of Eq. 14.70 into Eq. 14.66 yields

$$\varepsilon_1 = \varepsilon_x + \left(\varepsilon_x - \varepsilon_2\right)\cot^2\theta \quad (14.95)$$

which shows that ε_1 depends on the longitudinal tensile strain ε_x, the principal compressive strain ε_2, and the orientation of the principal strains (or stresses) θ.

The principal compressive strain ε_2 can be obtained from the constitutive relationship shown in Figure 14.23(a) and given by Eq. 13.3. Set the strain ε_c' at peak compressive stress f_c' to -0.002, and solve the resulting quadratic equation to get

$$\varepsilon_2 = -0.002\left(1 - \sqrt{1 - \frac{f_2}{f_{2\,max}}}\right) \quad (14.96)$$

where $f_{2\,max}$ is the important reduced peak stress given by Eq. 13.4, that is,

$$f_{2\,max} = \frac{f_c'}{0.8 + 170\varepsilon_1} \quad (14.97)$$

which decreases as the tensile straining increases. Now the principal compressive stress f_2 is relatively large compared to the principal tensile stress f_1 as shown in Figure 14.23(a). Therefore, f_2 can reasonably and conservatively be estimated from Eq. 14.71 as

$$f_2 \approx \frac{v}{\sin\theta\,\cos\theta} \tag{14.98}$$

where the nominal shear stress on the concrete v includes the reduction provided by the vertical component V_p of an inclined prestressing tendon, that is,

$$v = \frac{V_n - V_p}{b_v d_v} \tag{14.99}$$

Substitution of $V_n = V_u/\phi$ into Eq. 14.99 and including the absolute value sign to properly consider the effects due to V_u and V_p [C5.8.3.4.2] gives the average shear stress on the concrete as [A5.8.2.9]

$$v_u = \frac{|V_u - \phi V_p|}{\phi b_v d_v} \tag{14.100}$$

Substitution of Eqs. 14.97 and 14.98 into Eq. 14.96 and then substitution of that result into Eq. 14.95 gives

$$\varepsilon_1 = \varepsilon_x + \left[\varepsilon_x + 0.002\left(1 - \sqrt{1 - \frac{v_u}{f_c'}\frac{0.8 + 170\varepsilon_1}{\sin\theta\,\cos\theta}}\right)\right]\cot^2\theta \tag{14.101}$$

which can be solved for ε_1 once θ, v_u/f_c', and ε_x are known.

Longitudinal Steel Demand Before calculating the longitudinal strain ε_x in the web on the flexural tension side of the member, the relationships between some previously defined terms need to be clarified. This clarification can be done by substituting and rearranging some previously developed equilibrium equations and seeing which terms are the same and cancel out and which terms are different and remain. First, consider Eq. 14.74, which expresses longitudinal equilibrium in Figure 14.22(a), written as

$$N_v = (v\cot\theta - f_1)\,b_v d_v = V\cot\theta - f_1 b_v d_v \tag{14.102}$$

where N_v is the total axial force due to all of the longitudinal reinforcement on the overall cross section, prestressed and nonprestressed, multiplied by smeared stresses averaged over the area $b_v d_v$. Second, consider Eq. 14.80, which expresses longitudinal equilibrium in Figure 14.24(b), written with $f_v = f_y$ as

$$N_y \geq N_v + f_1 b_v d_v + f_1 b_v d_v \cot^2\theta \tag{14.103}$$

where N_y is the total axial force due to all of the longitudinal reinforcement on the overall cross section, prestressed and nonprestressed, multiplied by the prestressing tendon stress and the yield stress of the base reinforcement, respectively. Substitution of Eq. 14.102 into Eq. 14.103 and using Eq. 14.81 to express N_y, we get

$$A_{\text{sx}} f_y + A_{\text{px}} f_{\text{ps}} \geq V\cot\theta + f_1 b_v d_v \cot^2\theta \tag{14.104}$$

Next, substitution of Eq. 14.73, which expresses vertical equilibrium in Figure 14.22(c), into Eq. 14.104 yields

$$\begin{aligned} A_{\text{sx}} f_y + A_{\text{px}} f_{\text{ps}} &\geq 2 f_1 b_v d_v \cot^2\theta + \frac{A_v f_y d_v}{s}\cot^2\theta \\ &= (2V_c + V_s)\cot\theta \end{aligned} \tag{14.105}$$

where

$$V_c = f_1 b_v d_v \cot\theta = (0.0316)\,\beta\sqrt{f_c'}\,b_v d_v \tag{14.106}$$

and

$$V_s = \frac{A_v f_y d_v}{s}\cot\theta \tag{14.107}$$

Because the majority of the longitudinal reinforcement is on the flexural tension side of a member, Eq. 14.105 can be written in terms of the more familiar tensile steel areas A_s and A_{ps} by assuming that the shear depth d_v has been divided by 2 to yield

$$A_s f_y + A_{\text{ps}} f_{\text{ps}} \geq (V_c + 0.5V_s)\cot\theta \tag{14.108}$$

However, from Eq. 14.84,

$$V_c + 0.5V_s = V_n - 0.5V_s - V_p$$

so that the longitudinal tensile force requirement caused by shear becomes

$$A_s f_y + A_{\text{ps}} f_{\text{ps}} \geq \left(\frac{V_u}{\phi_v} - 0.5V_s - V_p\right)\cot\theta \tag{14.109}$$

Thus, after all the manipulations with the previously developed equilibrium equations, it comes down to the requirement expressed in Eq. 14.109 that *additional longitudinal tensile force must be developed to resist the longitudinal force caused by shear*. This phenomenon was observed early in the study of shear resistance using truss analogies (see Fig. 14.17) where the presence of shear force was shown to add to the tensile chord force and subtract from the compressive chord force. Unfortunately, this concept was not included when the shear design procedures were originally developed. This omission can be a serious shortcoming, especially in regions of high shear force (and low moment demand).

In addition to the shear requirement given in Eq. 14.109, the longitudinal tensile reinforcement must also resist the tensile force produced by any applied moment M_u and axial load N_u as shown in Figure 14.25. This consideration leads to the following requirement for the longitudinal tensile reinforcement as given in AASHTO [A5.8.3.5]:

$$\begin{aligned} A_s f_y + A_{\text{ps}} f_{\text{ps}} \geq{}& \frac{|M_u|}{d_v \phi_f} + 0.5\frac{N_u}{\phi_a} \\ &+ \left(\left|\frac{V_u}{\phi_v} - V_p\right| - 0.5V_s\right)\cot\theta \end{aligned} \tag{14.110}$$

where ϕ_f, ϕ_a, and ϕ_v are the resistance factors from Table 14.6 for flexure, axial load, and shear, respectively.

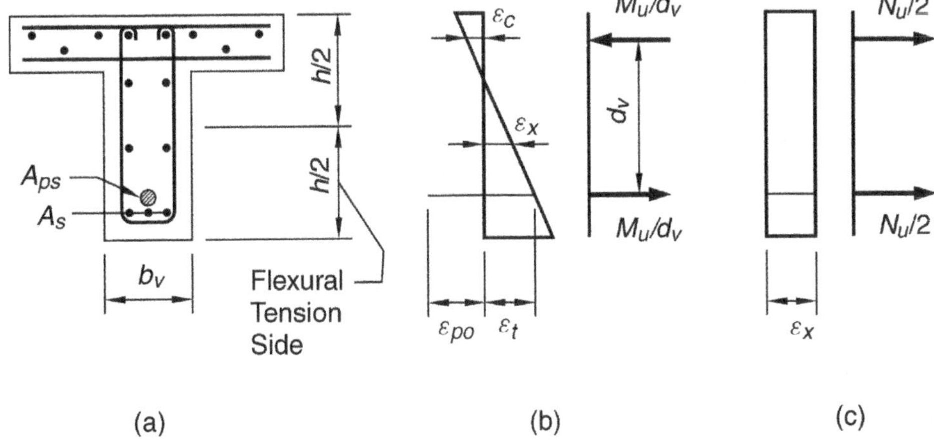

Fig. 14.25 Longitudinal strain and forces due to moment and tension: (a) cross section, (b) strains and forces due to moment M_u, and (c) strains and forces due to tension N_u. [After Collins and Mitchell (1991). Reprinted by permission of Prentice Hall, Upper Saddle River, NJ.]

Return now to the parameter ε_x, which is used to measure the stiffness of the section when it is subjected to moment, axial load, and shear. If ε_x is small, the web deformations are small and the concrete shear strength V_c is high. If ε_x is larger, the deformations are larger and V_c decreases. The strain ε_x is the average longitudinal strain in the web, and it can be reasonably estimated as one-half of the strain at the level of the flexural tensile reinforcement as shown in Figure 14.25. The longitudinal tensile force of Eq. 14.110 divided by a weighted stiffness quantity $2(E_sA_s + E_pA_{\text{ps}})$ where A_s is the area of nonprestressed steel on the flexural tension side of the member at the section as shown in Figure 14.25(a), A_{ps} is the area of prestressing steel on the flexural tension side of the member, and considering the precompression force $A_{\text{ps}}f_{\text{po}}$, results in the following expression given in AASHTO [A5.8.3.4.2]:

$$\varepsilon_x = \frac{\left(\left|M_u\right|/d_v\right) + 0.5N_u + 0.5\left|V_u - V_p\right|\cot\theta - A_{\text{ps}}f_{\text{po}}}{2\left(E_sA_s + E_pA_{\text{ps}}\right)} \tag{14.111}$$

where f_{po} is a parameter taken as the modulus of elasticity of prestressing strands multiplied by the locked-in difference in strain between the prestressing tendons and the surrounding concrete. For the usual levels of prestressing, a value for f_{po} of $0.7f_{\text{pu}}$ will be appropriate for both pretensioned and posttensioned members. Notice that the expression involving V_s and the ϕ factors are not included. The initial value for ε_x should not be taken greater than 0.001. The constant "2" assumes ε_c is small compared to ε_t so that ε_x is one half of ε_t [C5.8.3.4.2].

If Eq. 14.111 gives a negative value for ε_x because of a relatively large precompression force, then the concrete area A_c on the flexural tension side [Fig. 14.25(a)] participates and increases the longitudinal stiffness. In that case the denominator of Eq. 14.111 should be changed to $2(E_cA_c + E_sA_s + E_pA_{\text{ps}})$.

When a designer is preparing shear envelopes and moment envelopes for combined force effects, the extreme values for shear and moment at a particular location do not usually come from the same position of live load. The moment envelope values for M_u and V_u can be conservatively used when calculating ε_x. It is not necessary to calculate M_u for the same live-load position as was used in determining the maximum value of V_u (or vice versa). Some software stores and uses the coincident actions and this is certainly acceptable.

Clearly, MCFT predicts that the presence of moment affects the shear resistance; the converse is true as well. Therefore, there is an interaction curve (Fig. 14.26) associated with the section resistance. The MCFT illustrates the behavior with respect to the longitudinal strain. The load effect is represented with the small circle. If this shear–moment combination is inside the interation curve, it meets the limit state. If outside the interaction curve, it is under the capacity required. Response-2000 cited previously creates these types of plots.

Shear Resistance—Specifications MCFT was introduced into the First Edition of AASHTO LRFD Specifications (1994). Although is it an elegant and consistent theory based upon first principles, its use was perceived by some to be too complex for typical design. The AASHTO Subcommittee on Bridges and Structures in combination with several research studies simplified the method to avoid the iteration required for the determination of β and θ terms. In the Standard Specifications (2002), there are two primary methods for estimating shear resistance, and those were separate methods for prestressed concrete and conventional reinforced concrete. There were adjustments for light aggregate, compression or tension forces, and level of prestress precompression. These methods are, in part, used, as alternative methods within LRFD.

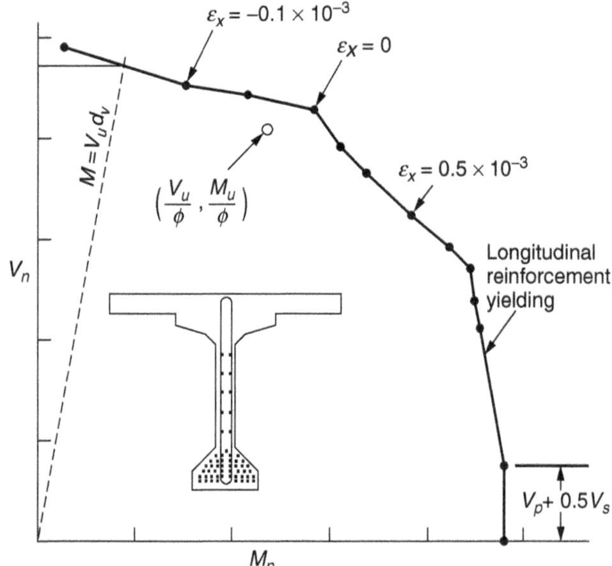

Fig. 14.26 Typical shear–moment interaction diagram [AASHTO Fig. CB5.2-6]. [From *AASHTO LRFD Bridge Design Specifications*. Copyright © 2010 by the American Association of State Highway and Transportation Officials, Washington, DC. Used by permission.]

All methods use Eq. 14.58 that combine the effects shear resisted by the concrete, steel, and longitudinal stress positioned at an angle to the section, for example, harped prestress strands. These methods vary in the approach to estimate the resistance associated with each of these three mechanisms.

In the 2010 Specifications, there are three complementary methods:

- ☐ Method 1: Simplified Procedure for Nonprestressed Sections [A5.8.3.4.1]
- ☐ Method 2: General Procedure for Shear Resistance [A5.8.3.4.2 and AASHTO Appendix B5]
- ☐ Method 3: Simplified Procedures for Prestressed and Nonprestressed Sections [A5.8.3.4.3]

Method 1 parallels the approach traditionally used by assuming that the crack angle θ is 45° and that the concrete shear term β is 2.0 [A5.8.3.4.1]. This is a fit to empirical data and details are presented in most texts on reinforced concrete design. Method 1 is demonstrated by example later. This is similar to the standard approach used for decades in the Standard Specifications (2002).

Method 2 has two forms: a simplified version that is provided in the body of the Specifications [A5.8.3.4.2] that is founded upon the MCFT presented in the previous section; however, some simplying features were added that are stated to be the "statistically equivalent" to the full theory. The MCFT is presented in its complete form above and this may be used via AASHTO and is provided in Appendix B of Section 5. This method contains the assumptions that are presented above. Clearly, many of these assumptions could be relaxed by returning to the original work.

Method 3 is based upon modifying procedures for prestressed concrete shear resistance provided in the Standard Specifications (2002) as modified by Hawkins et al. (2005) as outlined in NCHRP Report 549. This method may be used for both prestressed and nonprestressed section and is consistent with ACI Committee 318 (2011).

These methods are outlined below and the example that follows uses these methods for illustration. The most general and rigorous is the MCFT Method 2 based upon the complete theory. This is addressed first followed by the other methods.

Shear Resistance—Method 2: MCFT Appendix B Given applied forces, v_u calculated from Eq. 14.100, an estimated value of θ, and ε_x calculated from Eq. 14.111, ε_1 can be determined by Eq. 14.101. With ε_1 known, β can be determined from Eqs. 14.91 and 14.93 and a value selected. Then the concrete shear strength V_c can be calculated from Eq. 14.106, the required web reinforcement strength V_s from Eq. 14.84, and the required stirrup spacing s from Eq. 14.107. Thus, for an estimated value of θ, the required amount of web reinforcement to resist given force effects can be calculated directly.

To determine whether *the estimated θ results in the minimum amount of web reinforcement, a designer must try a series of values for θ until the optimum is found*. This possibly lengthy procedure has been shortened by the development of design aids, in the form of tables, for selecting θ and β. The tables originally appeared in Collins and Mitchell (1991) and have been expanded to include negative values of ε_x. The values of θ and β for sections with transverse reinforcement are given in Table 14.9 [Appendix B5 in AASHTO].

When developing the values in Table 14.9, Collins and Mitchell (1991) were guided by the limitations that the principal compressive stress in the concrete f_2 did not exceed $f_{2\,\max}$ and that the strain in the web reinforcement ε_v was at least 0.002, that is, $f_v = f_y$. In cases of low relative shear stress v_u/f'_c, the optimum value for θ is obtained when β is at its maximum, even though $\varepsilon_v < 0.002$. Other exceptions exist and are attributed to engineering judgment acquired through use of the proposed provisions in trial designs.

When using Table 14.9, the values of θ and β in a particular cell of the table can be applied over a range of values. For example, $\theta = 34.4°$ and $\beta = 2.26$ can be used provided ε_x is not greater than 0.75×10^{-3} and v_u/f'_c is not greater than 0.125. Linear interpolation between values given in the table may be used but is not recommended for hand calculations [C5.8.3.4.2] and is likely not warranted given the inherent level of accuracy. Additionally, a table similar to Table 14.9 for sections without transverse reinforcement is provided in Appendix 5B.

Table 14.9 Values of θ and β for Sections with Transverse Reinforcement

$\frac{v_u}{f'_c}$	$\varepsilon_x \times 1000$								
	≤ −0.20	≤ −0.10	≤ −0.05	≤ 0	≤ 0.125	≤ 0.25	≤ 0.50	≤ 0.75	≤ 1.00
≤ 0.075	22.3 6.32	20.4 4.75	21.0 4.10	21.8 3.75	24.3 3.24	26.6 2.94	30.5 2.59	33.7 2.38	36.4 2.23
≤ 0.100	18.1 3.79	20.4 3.38	21.4 3.24	22.5 3.14	24.9 2.91	27.1 2.75	30.8 2.50	34.0 2.32	36.7 2.18
≤ 0.125	19.9 3.18	21.9 2.99	22.8 2.94	23.7 2.87	25.9 2.74	27.9 2.62	31.4 2.42	34.4 2.26	37.0 2.13
≤ 0.150	21.6 2.88	23.3 2.79	24.2 2.78	25.0 2.72	26.9 2.60	28.8 2.52	32.1 2.36	34.9 2.21	37.3 2.08
≤ 0.175	23.2 2.73	24.7 2.66	25.5 2.65	26.2 2.60	28.0 2.52	29.7 2.44	32.7 2.28	35.2 2.14	36.8 1.96
≤ 0.200	24.7 2.63	26.1 2.59	26.7 2.52	27.4 2.51	29.0 2.43	30.6 2.37	32.8 2.14	34.5 1.94	36.1 1.79
≤ 0.225	26.1 2.53	27.3 2.45	27.9 2.42	28.5 2.40	30.0 2.34	30.8 2.14	32.3 1.86	34.0 1.73	35.7 1.64
≤ 0.250	27.5 2.39	28.6 2.39	29.1 2.33	29.7 2.33	30.6 2.12	31.3 1.93	32.8 1.70	34.3 1.58	35.8 1.50

From AASHTO Table B5.2-1. From *AASHTO LRFD Bridge Design Specifications*, Copyright © 2010 by the American Association of State Highway and Transportation Officials, Washington, DC. Used by permission.

The shear design of members with web reinforcement using the modified compression field theory consists of the following steps (Collins and Mitchell, 1991):

Step 1 Determine the factored shear V_u and moment M_u envelopes due to the strength I limit state. Values are usually determined at the tenth points of each span. Interpolations can easily be made for values at critical sections such as a distance d_v from the face of a support. In the derivation of the modified compression field theory, d_v is defined as the lever arm between the resultant compressive force and the resultant tensile force in flexure. The definition in AASHTO [A5.8.2.9] adds that d_v need not be less than $0.9d_e$ or $0.72h$, where d_e is the distance from the extreme compression fiber to the centroid of the tensile reinforcement and h is the overall depth of the member.

Step 2 Calculate the nominal shear stress v_u from Eq. 14.100 and divide by the concrete strength f'_c to obtain the shear stress ratio v_u/f'_c. If this ratio is higher than 0.25, a larger cross section must be chosen.

Step 3 Estimate a value of θ, say 30°, and calculate the longitudinal strain ε_x from Eq. 14.111. For a prestressed beam f_{po} is a parameter taken as the modulus of elasticity of prestressing strands multiplied by the locked-in difference in strain between the prestressing tendons and the surrounding concrete. For the usual levels of prestressing, a value for f_{po} of $0.7f_{\text{pu}}$ will be appropriate for both pretensioned and posttensioned members. If the section is within the transfer length, then an effective value of f_{po} should be determined.

Step 4 Use the calculated values of v_u/f'_c and ε_x to determine θ from Table 14.9 and compare with the value estimated in step 3. If different, recalculate ε_x and repeat step 4 until the estimated value of θ agrees with the value from Table 14.9. When it does, select β from the bottom half of the cell in Table 14.9.

Step 5 Calculate the required web reinforcement strength V_s from Eqs. 14.84 and 14.106 to give

$$V_s = \frac{V_u}{\phi_u} - V_p - 0.0316\beta\sqrt{f'_c}b_v d_v \qquad (14.112)$$

Step 6 Calculate the required spacing of stirrups from Eq. 14.107 as

$$s \le \frac{A_v f_y d_v}{V_s} \cot\theta \qquad (14.113)$$

This spacing must not exceed the value limited by the minimum transverse reinforcement of AASHTO [A5.8.2.5], that is,

$$s \le \frac{A_v f_y}{0.0316\sqrt{f'_c}b_v} \qquad (14.114)$$

Again, 0.0316 is for units. It must also satisfy the maximum spacing requirements of AASHTO [A5.8.2.7]:

☐ If $v_u < 0.125\, f'_c$, then $s \le 0.8d_v \le 24.0$ in. (14.115)

☐ If $v_u \ge 0.125\, f'_c$, then $s \le 0.4d_v \le 12.0$ in. (14.116)

Step 7 Check the adequacy of the longitudinal reinforcement using Eq. 14.111. If the inequality is not satisfied, either add more longitudinal reinforcement or increase the amount of stirrups. Appendix 5B further outlines several limitations and checks on the procedure. These are [Appendix B5.2]:

- M_u shall be taken as positive quantities and M_u shall not be taken less than $(V_u - V_p)d_v$.
- In calculating A_s and A_{ps} the area of bars or tendons, which are terminated less than their development length from the section under consideration, shall be reduced in proportion to their lack of full development.
- The value of ε_x calculated from Eqs. 14.111 (with compression term as appropriate) should not be taken as less than -0.20×10^{-3}.
- For sections closer than d_v to the face of the support, the value of ε_x calculated at d_v from the face of the support may be used in evaluating β and θ.
- If the axial tension is large enough to crack the flexural compression face of the section, the resulting increase in ε_x shall be taken into account. In lieu of more accurate calculations, the value calculated from Eq. 14.111 should be doubled.
- It is permissible to determine β and θ from Table 14.9 using a value of ε_x that is greater than that calculated from Eqs. 14.111; however, ε_x shall not be taken greater than 3.0×10^{-3}.

Example 14.3 Use Method 2 Appendix B (MCFT) to determine the required spacing of No. 3 U-shaped stirrups for the nonprestressed T-beam of Figure 14.27 at a positive moment location where $V_u = 157$ kips and $M_u = 220$ kip ft. Use $f_c' = 4.5$ ksi and $f_y = 60$ ksi.

Step 1: Given $V_u = 157$ kips and $M_u = 220$ kip ft:

$$A_s = 3.12 \text{ in.}^2 \quad b_v = 16 \text{ in.} \quad b = 80 \text{ in.}$$

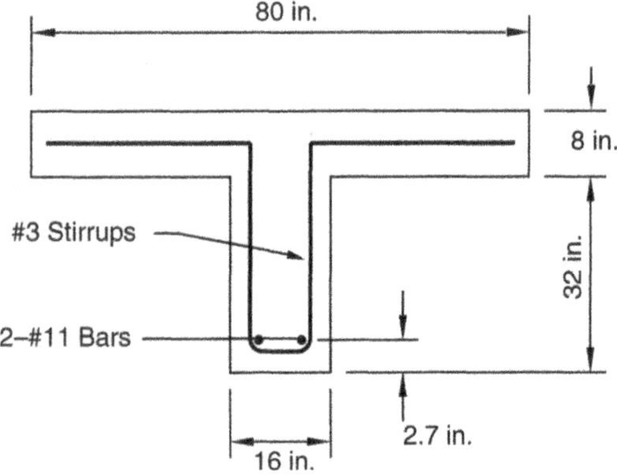

Fig. 14.27 Determination of stirrup spacing (Example 14.3).

Assume NA (i.e., neutral axis) is in flange:

$$a = \frac{A_s f_y}{0.85 f_c' b} = \frac{3.12\,(60)}{0.85\,(4.5)\,(80)}$$
$$= 0.61 \text{ in.} < h_f = 8 \text{ in.} \qquad \text{OK}$$

$$d_v = \max \begin{cases} d_e - a/2 = (40 - 2.7) - 0.61/2 \\ \quad = \underline{37.0 \text{ in.}}, \text{ governs} \\ 0.9 d_e = 0.9\,(37.3) = 33.6 \text{ in.} \\ 0.72 h = 0.72\,(40) = 28.8 \text{ in.} \end{cases}$$

Step 2: Calculate v_u / f_c':

$$V_p = 0 \qquad \phi_v = 0.9$$

Equation 14.100:

$$v_u = \frac{|V_u|}{\phi_v b_v d_v} = \frac{157}{0.9\,(16)\,(37.0)} = 0.295 \text{ ksi}$$
$$\frac{v_u}{f_c'} = \frac{0.295}{4.5} = \underline{0.066} \le 0.25 \qquad \text{OK}$$

Step 3: Calculate ε_x from Eq. 14.111. $N_u = 0$. $A_{ps} = 0$. Estimate

$$\theta = 40^\circ \qquad \cot\theta = 1.192$$
$$\varepsilon_x = \frac{|M_u|/d_v + 0.5\,|V_u| \cot\theta}{2 E_s A_s}$$
$$= \frac{220\,(12)/37.0 + 0.5\,(157)\,1.192}{2\,(29{,}000 \times 3.12)}$$
$$= 0.91 \times 10^{-3}$$

Step 4: Determine θ and β from Table 14.9 (interpolate or use higher value). $\theta = 35.4^\circ$, $\cot\theta = 1.407$:

$$\varepsilon_x = \frac{220\,(12)/37.0 + 0.5\,(157)\,1.407}{2\,(29{,}000 \times 3.12)}$$
$$= \underline{1.0 \times 10^{-3}}$$

The next iteration gives 36.4°, which converges:

$$\text{Use } \underline{\theta = 36.4^\circ} \qquad \underline{\beta = 2.23}$$

Step 5: Calculate the required V_s from Eq. 14.112:

$$V_s = \frac{|V_u|}{\phi_v} - 0.0316 \beta \sqrt{f_c'} b_v d_v$$
$$= \frac{157}{0.9} - 0.0316\,(2.23)\,\sqrt{4.5}\,(16)\,37.0$$
$$= 174.4 - 88.5 = \underline{85.9} \text{ kips}$$

Step 6: Calculate the required stirrup spacing from Eqs. 14.113–14.116 using $A_v = 0.22$ in.^{2}:

$$s \le \frac{A_v f_y d_v}{V_s} \cot\theta = \frac{0.22\,(60)\,(37.0)}{85.9}\,(1.356) = 7.7 \text{ in.}$$
$$s \le \frac{A_v f_y}{0.0316 \sqrt{f_c'} b_v} = \frac{0.22\,(60)}{0.0316 \sqrt{4.5}(16)} = 12.3 \text{ in.}$$

$$v_u < 0.125 f_c' = 0.125\,(4.5) = 0.563 \text{ ksi}$$
$$s \le 0.8 d_v = 0.8\,(37.0) = 29.6 \text{ or } 24 \text{ in.}$$

The stirrup spacing of $s = 7.7$ in. controls (use 7 in.).

Step 7: Check the additional demand on the longitudinal reinforcement caused by shear as given by Eq. 14.110:

$$V_s = \frac{7.7}{7}(85.9) = 94.5 \text{kips}$$

$$A_s f_y \ge \frac{|M_u|}{d_v \phi_f} + \left(\frac{|V_u|}{\phi_u} - 0.5 V_s\right) \cot\theta$$

$$3.12\,(60) \; ? \; \frac{220 \times 12}{37.0\,(0.9)} + \left(\frac{157}{0.9} - \frac{94.5}{2}\right) 1.356$$

$$187.2 \text{ kips} \le 79.2 + (174.4 - 47.2)\,1.356$$
$$= 252 \text{ kips}, \qquad \text{No good}$$

Stirrup strength V_s needed to satisfy the inequality

$$V_s \ge 2\left[\frac{|V_u|}{\phi_v} - \left(A_s f_y - \frac{|M_u|}{d_v \phi_f}\right)\tan\theta\right]$$
$$\ge 2\left[174.4 - (187.2 - 79.2)\tan 36.4^\circ\right]$$
$$= 189.6 \text{ kips}$$

requires the stirrup spacing to be

$$s \le \frac{0.22\,(60)\,37.0}{189.6}(1.356) = 3.5 \text{ in.}$$

which is likely not cost effective. It is better to simply increase A_s to satisfy the inequality, that is,

$$A_s \ge \frac{252}{f_y} = \frac{252}{60} = 4.2 \text{ in.}^2$$

and <u>Use 2 No. 11's plus 1 No. 10</u> $A_s = 4.39$ in.^{2}

<u>No. 3 U-stirrups at 7 in.</u>

Note that MCFT provides an alternative regarding whether to increase longitudinal and/or transverse steel. This flexibility is a benefit that can be illustrated by taking the stirrup spacing as 5 in.

Equation 14.107

$$V_s = \frac{A_v f_y d_v}{s}\cot\theta = \frac{0.22\,(60)\,(37.0)}{5.0}1.356$$
$$= 132.5 \text{ kips}$$

and As are required to satisfy the inequality from Eq. 14.110:

$$A_s = \left[\frac{|M_u|}{d_v \phi_f} + \left(\frac{|V_u|}{\phi_u} - 0.5 V_s\right)\cot\theta\right] \Big/ f_y$$
$$= \left[\frac{220\,(12)}{37\,(0.9)} + \left(\frac{157}{0.9} - \frac{132.5}{2}\right)1.356\right] \Big/ 60$$
$$= \frac{79.3 + 146.7}{60} = 3.77 \text{ in.}^2$$

Use 3 No.10's $As = 3.79$ in.^{2}

No. 3 U-stirrups at 5 in.

Shear Resistance—Method 2: MCFT [A5.8.3.4.2] The MCFT method prescribed in [A5.8.3.4.2] is essentially the same as outlined in the previous section. The primary difference is that equations are available for the coefficient β and crack angle θ and the table does not need to be used. This facilitates computation by hand or by spreadsheets or other tools. These equations are

$$\beta = \frac{4.8}{1 + 750\varepsilon_s} \tag{14.117}$$

$$\theta = 29 + 3500\varepsilon_s \tag{14.118}$$

where ε_s is the tensile strain at the centroid of the reinforcement:

$$\varepsilon_s = \frac{\left(|M_u|/d_v\right) + 0.5N_u + |V_u - V_p| - A_{ps} f_{po}}{(E_s A_s + E_p A_{ps})} \tag{14.119}$$

Note that the strain at the reinforcement level ε_s is assumed to be twice the average strain ε_x at middepth. Also, to avoid a trial-and-error iteration process, 0.5 cot θ is taken as 1.0 [C5.8.3.4.2, CB5.2]. Compare Eqs. 14.111 and 14.119.

The tension steel demand must satisfy the same requirements as outlined previously:

$$A_s f_y + A_{ps} f_{ps} \ge \frac{|M_u|}{d_v \phi_f} + 0.5\frac{N_u}{\phi_\alpha} + \left(\left|\frac{V_u}{\phi_v} - V_p\right| - 0.5 V_s\right)\cot\theta \tag{14.120}$$

Example 14.4 Use Method 2: General Procedure for Shear Resistance [A5.8.3.4.2] to rework Example 14.3. Determine the required spacing of No. 3 U-shaped stirrups for the nonprestressed T-beam of Figure 14.27 at a positive moment location where $V_u = 157$ kips and $M_u = 220$ kip ft. Use $f_c' = 4.5$ ksi and $f_y = 60$ ksi.

Step 1: Given $V_u = 157$ kips and $M_u = 220$ kip ft,

$$A_s = 3.12 \text{ in.}^2 \qquad b_v = 16 \text{ in.} \qquad b = 80 \text{ in.}$$

Assume NA is in flange:

$$a = \frac{A_s f_y}{0.85 f_c' b} = \frac{3.12\,(60)}{0.85\,(4.5)\,(80)}$$
$$= 0.61 \text{ in.} < h_f = 8 \text{ in.} \qquad \text{OK}$$

$$d_v = \max\begin{cases} d_e - a/2 = (40 - 2.7) - 0.61/2 \\ \quad = \underline{37.0 \text{ in.,}} \text{ governs} \\ 0.9 d_e = 0.9\,(37.3) = 33.6 \text{ in.} \\ 0.72h = 0.72\,(40) = 28.8 \text{ in.} \end{cases}$$

Step 2: Calculate v_u/f_c':

$$V_p = 0 \qquad \phi_v = 0.9$$

Equation 14.100

$$v_u = \frac{|V_u|}{\phi_v b_v d_v} = \frac{157}{0.9\,(16)\,(37.0)} = 0.295 \text{ ksi}$$

$$\frac{v_u}{f'_c} = \frac{0.295}{4.5} = \underline{0.066} \le 0.25 \qquad \text{OK}$$

Step 3: Calculate ε_s from Eq. 14.119. $N_u = 0$. $A_{ps} = 0$:

$$\varepsilon_s = \frac{|M_u|/d_v + |V_u|}{E_s A_s} = \frac{220 \times 12/37.0 + (157)}{(29{,}000 \times 3.12)}$$
$$= 2.52 \times 10^{-3}$$

Step 4: Determine θ and β from Eqs 14.117 and 14.118:

$$\beta = \frac{4.8}{1 + 750\varepsilon_s} = \frac{4.8}{1 + 750\,(0.00252)} = 1.66$$
$$\theta = 29 + 3500\varepsilon_s = 29 + 3500\,(0.00252) = 37.8$$

Step 5: Calculate V_s from Eq. 14.112:

$$V_s = \frac{|V_u|}{\phi_v} - 0.0316\beta\sqrt{f'_c}b_v d_v$$
$$= \frac{157}{0.9} - 0.0316\,(1.66)\,\sqrt{4.5}\,(16)\,37.0$$
$$= 174.4 - 65.9 = \underline{108.5} \text{ kips}$$

Step 6: Calculate the required stirrup spacing from Eqs. 14.113–14.116 using $A_v = 0.22$ in.^{2}:

$$s \le \frac{A_v f_y d_v}{V_s}\cot\theta = \frac{0.22(60)(37.0)}{108.5}(\cot 37.8) = 5.8 \text{ in.}$$
$$\le \frac{A_v f_y}{0.0316\sqrt{f'_c}b_v} = \frac{0.22\,(60)}{0.0316\sqrt{4.5}\,(16)} = 12.3 \text{ in.}$$
$$v_u < 0.125 f'_c = 0.125\,(4.5) = 0.563 \text{ ksi}$$
$$s \le 0.8 d_v = 0.8\,(37.0) = 29.6 \text{ or } 24 \text{ in.}$$

The stirrup spacing of $s = 5.8$ in. controls (use 5 in.).

Step 7: Check the additional demand on the longitudinal reinforcement caused by shear as given by Eq. 14.110:

$$V_s = \frac{5.8}{5.0}(108.5) = 125.9 \text{kips}$$
$$A_s f_y \ge \frac{|M_u|}{d_v \phi_f} + \left(\frac{|V_u|}{\phi_u} - 0.5 V_s\right)\cot\theta$$
$$3.12\,(60)\ ?\ \frac{220\,(12)}{37.0\,(0.9)} + \left(\frac{157}{0.9} - \frac{125.9}{2}\right)\cot 37.8$$
$$187.2 \text{ kips} \le 79.2 + (174.4 - 63.0)\,1.289$$
$$= 222.8 \text{ kips} \qquad \text{No good}$$

The spacing can be decreased or the longitudinal steel increased. See previous example for alternatives. Here, the longitudinal steel will be increased.

$$A_s \ge \frac{222.8}{f_y} = \frac{222.8}{60} = 3.71 \text{ in.}^2$$

and <u>Use 3 No. 10s</u> $A_s = 3.79$ in.^{2}

<u>No. 3 U-stirrups at 5 in.</u>

Shear Resistance—Method 1: Simplified Method [A5.8.3.4.1] This method is similar to the simple methods used in ACI Committee 318 and used in traditional reinforced concrete textbooks. It is limited to nonprestressed methods and may be used with or without transverse reinforcement. The method simply uses:

$$\beta = 2$$
$$\theta = 45^\circ$$
$$V_p = 0.0$$

Therefore,

$$V_n = V_c + V_s$$

where

$$V_c = 0.0316\beta\sqrt{f'_c}b_v d_v$$
$$V_s = \frac{A_v f_y d_v}{s}$$

Example 14.5 Use Method 1: Simplified Procedure for Nonprestressed Sections [A5.8.3.4.1] to rework Example 14.3. Determine the required spacing of No. 3 U-shaped stirrups for the nonprestressed T-beam of Figure 14.27 at a positive moment location where $V_u = 157$ kips and $M_u = 220$ kip ft. Use $f'_c = 4.5$ ksi and $f_y = 60$ ksi.

Step 1: Given $V_u = 157$ kips and $M_u = 220$ kip ft,

$$A_s = 3.12 \text{ in.}^2 \qquad b_v = 16 \text{ in.} \qquad b = 80 \text{ in.}$$

Assume NA is in flange:

$$a = \frac{A_s f_y}{0.85 f'_c b} = \frac{3.12\,(60)}{0.85\,(4.5)\,(80)}$$
$$= 0.61 \text{ in.} < h_f = 8 \text{ in.} \qquad \text{OK}$$

$$d_v = \max\begin{cases} d_e - a/2 = (40 - 2.7) - 0.61/2 \\ \quad = \underline{37.0 \text{ in.,}} \text{ governs} \\ 0.9 d_e = 0.9\,(37.3) = 33.6 \text{ in.} \\ 0.72h = 0.72\,(40) = 28.8 \text{ in.} \end{cases}$$

$$V_p = 0 \qquad \phi_v = 0.9$$

Step 2: Calculate shear strength of the concrete V_c: Equation 14.106

$$V_c = 0.0316\beta\sqrt{f'_c}b_v d_v$$
$$= 0.0316\,(2)\,\sqrt{4.5}\,(16)\,(37) = 79.4 \text{ kips}$$

Step 3: Calculate the required V_s:

$$V_s = \frac{V_u}{\phi_v} - V_c = \frac{157}{0.9} - 79.4 = 95.0 \text{ kips}$$

Step 4: Calculate the required spacing s from Eqs. 14.113–14.116 using $A_v = 0.22$ in.^{2}:

$$s \le \frac{A_v f_y d_v}{V_s} = \frac{(0.22)\,(60)\,(37.0)}{95.0} = 5.14 \text{ in.}$$

$$s \le \frac{A_v f_y}{0.0316\sqrt{f_c'}b_v} = \frac{0.22\,(60)}{0.0316\sqrt{4.5}\,(16)} = 12.3 \text{ in.}$$

Equation 14.100

$$v_u = \frac{|V_u - \phi V_p|}{\phi b_v d_v} = \frac{157 - 0}{0.9\,(16)\,(37)} = 0.295 \text{ ksi}$$
$$< 0.125 f_c' = 0.125\,(4.5) = 0.563 \text{ ksi}$$
$$s \le 0.8 d_v = 0.8\,(37.0) = 29.6 \text{ or } 24 \text{ in.}$$

The stirrup spacing of $s = 5.14$ in. controls (use 5 in.).

Step 5: Check the longitudinal reinforcement A_s:

$$A_s f_y \ge \frac{M_u}{d_v \phi_f} + \left[\frac{|V_u|}{\phi_v} - \frac{V_s}{2}\right]\cot(\theta)$$
$$V_s = \frac{A_v f_y d_v}{s} = \frac{0.22\,(60)\,(37.0)}{5} = 97.7 \text{ kips}$$
$$3.12\,(60) \ge \frac{220\,(12)}{37\,(0.9)} + \left[\frac{|157|}{0.9} - \frac{97.7}{2}\right](1)$$
$$187.2 \le 79.3 + 125.6 = 204.9 \text{ kips} \qquad \text{No good}$$

Step 6: Change stirrup spacing or increase longitudinal reinforcement, try the latter:

$$A_s \ge \frac{\dfrac{M_u}{d_v \phi_f} + \left(\dfrac{|V_u|}{\phi_v} - \dfrac{V_s}{2}\right)\cot\theta}{f_y}$$
$$\ge \frac{\dfrac{220\,(12)}{37\,(0.9)} + \left(\dfrac{|157|}{0.9} - \dfrac{97.7}{2}\right)(1.0)}{60}$$
$$\ge 3.42 \text{ in.}^2$$

Step 7: Select reinforcement:Use 2 No. 10's + 1 No. 9 ($A_s = 3.54$ in.^{2}) and No. 3 U-stirrups at 5 in.

Shear Resistance—Method 3: Simplified [A5.8.3.4.3] This method is similar to the computation of the shear resistance for prestressed concrete section in the Standard Specification (2002). Hawkins et al. (2005) provides the development. There are differences, and, importantly, this method can be applied to prestressed and nonprestressed concrete sections. Two shear resistances are computed:

V_{ci} = nominal shear resistance provided by concrete when inclined cracking results from combined shear and moment (kip), and

V_{cw} = nominal shear resistance provided by concrete when inclined cracking results from excessive principal tensions in web (kip)

The resistance assigned to the concrete is the minimum of these values:

$$V_c = \min\left[V_{ci}, V_{cw}\right]$$

where

$$V_{ci} = 0.02\sqrt{f_c'}b_v d_v + \frac{V_i M_{cre}}{M_{max}} \ge 0.06\sqrt{f_c'}b_v d_v \quad (14.121)$$
$$V_{cw} = \left(0.06\sqrt{f_c'} + 0.30 f_{pc}\right) b_v d_v + V_p$$

where

M_{cre} is the moment causing flexural cracking at section due to externally applied loads (kip-in.)

M_{max} is the maximum factored moment at section due to externally applied loads (kip-in.),

M_{cre} shall be determined as

$$M_{cre} = S_c\left[f_r + f_{cpe} - \frac{M_{dnc}}{S_{nc}}\right]$$

where

f_r = modulus of rupture (ksi) = $0.20\sqrt{f_c'}$ [A5.4.2.6]

f_{cpe} = compressive stress in concrete due to effective prestress forces only (after allowance for all prestress losses) at extreme fiber of section where tensile stress is caused by externally applied loads (ksi)

M_{dnc} = total unfactored dead-load moment acting on the monolithic or noncomposite section (kip-in.)

S_c = section modulus for the extreme fiber of the composite section where tensile stress is caused by externally applied loads (in.^{3})

S_{nc} = section modulus for the extreme fiber of the monolithic or noncomposite section where tensile stress is caused by externally applied loads (in.^{3})

In Eq. 14.121, M_{max} and V_i shall be determined from the load combination causing maximum moment at the section.

For the resistance assigned to the steel transverse reinforcement, the equations are the same as the MCFT (and standard specifications):

$$V_s = \frac{A_v f_y d_v}{s}\cot\theta$$

If $V_{ci} \le V_{cw}$, then $\cot\theta = 1.0$

$$\text{If } V_{ci} > V_{cw}, \text{ then } \cot\theta = 1.0 + 3\left(\frac{f_{pc}}{\sqrt{f_c'}}\right) \le 1.8 \quad (14.122)$$

This provides that if the section strength is controlled by web shear cracking, then the angle is assumed to be 45°. If flexural shear cracking controls, then the angle is a function of the level of prestressing at the section centriod but limited to approximately 30°.

Example 14.6 Use Method 3: Simplified Procedure for Prestressed and Nonprestressed Sections [A5.8.3.4.3] to rework Example 14.3. Determine the required spacing of

No. 3 U-shaped stirrups for the nonprestressed T-beam of Figure 14.27 at a positive moment location where $V_u = 157$ kips and $M_u = 220$ kip ft. Use $f_c' = 4.5$ ksi and $f_y = 60$ ksi. For this problem, assume that the maximum shear and moment occur at the same location. The section properties are $I = 106{,}888$ in.^{4} and $S_{\text{bottom}} = 3943$ in.^{3}.

Step 1: Given $V_u = 157$ kips and $M_u = 220$ kip ft,

$$A_s = 3.12 \text{ in.}^2 \qquad b_v = 16 \text{ in.} \qquad b = 80 \text{ in.}$$

Assume NA is in flange:

$$a = \frac{A_s f_y}{0.85 f_c' b} = \frac{3.12\,(60)}{0.85\,(4.5)\,(80)} = 0.61 \text{ in.} < h_f = 8 \text{ in.} \qquad \text{OK}$$

$$d_v = \max \begin{cases} d_e - a/2 = (40 - 2.7) - 0.61/2 = \underline{37.0 \text{ in.}}, \text{ governs} \\ 0.9 d_e = 0.9\,(37.3) = 33.6 \text{ in.} \\ 0.72 h = 0.72\,(40) = 28.8 \text{ in.} \end{cases}$$

Step 2: Determine the cracking moment M_{cre}:

$$f_r = 0.20\sqrt{f_c'} = 0.42 \text{ ksi}$$

$$M_{\text{cre}} = S_c \left[f_r + f_{\text{cpe}} - \frac{M_{\text{dnc}}}{S_{\text{nc}}} \right]$$

$$M_{\text{cre}} = (3943)\,[(0.42) + 0 + 0] = 1656 \text{ kip-in.}$$

Step 3: Determine the shear resistance assigned to the concrete:

$$V_{\text{ci}} = 0.02\sqrt{f_c'} b_v d_v + \frac{V_i M_{\text{cre}}}{M_{\text{max}}} \ge 0.06\sqrt{f_c'} b_v d_v$$

$$V_{\text{ci}} = 0.02\sqrt{4.5}\,(16)\,(37.0) + \frac{157\,(1656)}{220\,(12)}$$

$$= 123.6 \ge 0.06\sqrt{f_c'} b_v d_v = 75.3 \text{ kips}$$

$$V_{\text{cw}} = \left(0.06\sqrt{f_c'} + 0.30 f_{\text{pc}}\right) b_v d_v + V_p$$

$$V_{\text{cw}} = \left[0.06\sqrt{4.5} + 0.30\,(0)\right](16)\,(37.0) + 0 = 75.3 \text{ kips}$$

$$V_c = \min\left[V_{\text{ci}}, V_{\text{cw}}\right] = \min\,[123.6, 75.3] = 75.3 \text{ kips}$$

Step 4: Determine the shear resistance required by the transverse steel:

$$V_s = \frac{V_u}{\phi_v} - V_c = \frac{157}{0.9} - 75.3 = 99.1 \text{ kips}$$

Step 5: Determine the stirrup spacing, s, from Eqs. 14.113–14.116 using $A_v = 0.22$ in.^{2}
Equation 14.122 with $f_{\text{pc}} = 0$, then $\cot\theta = 1.0$:

$$s \le \frac{A_v f_y d_v}{V_s}\cot\theta = \frac{0.22\,(60)\,(37.0)}{99.1}(1.0) = 4.93 \text{ in.}$$

$$\le \frac{A_v f_y}{0.0316\sqrt{f_c'} b_v} = \frac{0.22\,(60)}{0.0316\sqrt{4.5}\,(16)} = 12.3 \text{ in.}$$

Equation 14.100

$$v_u = \frac{|V_u - \phi V_p|}{\phi b_v d_v} = \frac{157 - 0}{0.9\,(16)\,(37)} = 0.295 \text{ ksi}$$

$$v_u < 0.125 f_c' = 0.125\,(4.5) = 0.563 \text{ ksi}$$

$$s \le 0.8 d_v = 0.8\,(37.0) = 29.6 \text{ or } 24 \text{ in.}$$

The stirrup spacing of $s = 4.93$ in. controls (use 5 in.).

Step 6: Check the longitudinal reinforcement A_s:

$$A_s f_y \ge \frac{M_u}{d_v \phi_f} + \left[\frac{|V_u|}{\phi_v} - \frac{V_s}{2}\right]\cot\theta$$

$$V_s \le \frac{A_v f_y d_v}{s}\cot\theta = \frac{0.22\,(60)\,(37.0)}{5}(1.0) = 97.7 \text{ kips}$$

$$3.12\,(60) \ge \frac{220\,(12)}{37\,(0.9)} + \left[\frac{|157|}{0.9} - \frac{97.7}{2}\right](1.0)$$

$$187.2 \le 79.3 + 125.6 = 204.9 \text{ kips} \qquad \text{No good}$$

Step 7: Change stirrup spacing or increase longitudinal reinforcement, try the latter.

$$A_s \ge \frac{\dfrac{M_u}{d_v \phi_f} + \left(\dfrac{|V_u|}{\phi_v} - \dfrac{V_s}{2}\right)\cot\theta}{f_y}$$

$$\ge \frac{\dfrac{220\,(12)}{37\,(0.9)} + \left(\dfrac{|157|}{0.9} - \dfrac{97.7}{2}\right)(1.0)}{60}$$

$$\ge 3.42 \text{ in.}^2$$

Step 8: Select reinforcementUse 2 No. 10's + 1 No. 9 ($A_s = 3.54$ in.^{2}) and No. 3 U-stirrups at 5 in.

The shear resistance designs are compared in Table 14.10 for equal stirrup spacings of 5 in. It is interesting to note that the simplified methods 1 and 3 give the same results and so do the two MCFT methods.

Table 14.10 Comparison of Procedures for Determining Shear Resistance

Procedure	AASHTO Reference	Longitudinal Steel	Spacing of Stirrups
Method 1 Simplified	A5.8.3.4.1	2 No. 10 + 1 No. 9	$S = 5$ in.
Method 2 MCFT	A5.8.3.4.2	3 No. 10	$S = 5$ in.
Method 2 MCFT	Appendix 5B	3 No. 10	$S = 5$ in.
Method 3 Simplified	A5.8.3.4.3	2 No. 10 + 1 No. 9	$S = 5$ in.

14.4 CLOSING REMARKS

This chapter, in addition to Chapter 13, outlines some of the most fundamental behavior in reinforced and prestressed concrete beams. The behavior includes material properties, properties associated with their application in the component, that is, tension softening, compression field theory, and its modified version. There is a significant amount of information that was not presented and the interested reader may find supplementary materials in books on concrete behavior, papers, and reports. NCHRP reports on these topics are highly recommended and in many cases provide the link between the behavior, experimental work, modeling, and the specifications. Note that the specification may not end up as recommended by the researcher, but rather might be adjusted by the committee in charge of that article and adopted by AASHTO.

There are numerous prestress concrete loss models, and the primary AASHTO LRFD models are present in the body of this chapter and in an appendix for refined methods. Due to limited space and scope, this book does not address prestressed concrete beams that are made continous for live load. The design practices for these bridges vary among agencies and the reader/designer is referred to the owner's methods. Finally, the focus here is on girder elements. Substructure components such as pier caps and columns are not explicitly addressed. Similarly foundation elements such as piles, drilled shafts, and caps are not addressed. The principles presented here apply in many cases, but not all. The strut-and-tie method should be reviewed for its application in lieu of sectional analysis. The details can be found in agency design manuals and elsewhere.

In summary, the topics of concrete behavior and design (and associated literature) are voluminous. The information presented here is a start.

REFERENCES

AASHTO (1994). *LRFD Bridge Design Specifications*, 1st ed., American Association of State Highway and Transportation Officials, Washington, DC.

AASHTO (2002). *Standard Specifications for Highway Bridges*, 17th ed., American Association of State Highway and Transportation Officials, Washington, DC.

AASHTO (2004). *LRFD Bridge Design Specifications*, 3rd ed., American Association of State Highway and Transportation Officials, Washington, DC.

AASHTO (2010). *LRFD Bridge Design Specifications*, 5th ed., American Association of State Highway and Transportation Officials, Washington, DC.

ACI Committee 215 (1992). *Considerations for Design of Concrete Structures Subjected to Fatigue Loading*, ACI 215R-74 (Revised 1992), American Concrete Institute, Farmington Hills, MI.

ACI Committee 318 (2011). *Building Code Requirements for Structural Concrete and Commentary*, ACI 318–08, American Concrete Institute, Farmington Hills, MI.

Adebar, P. and M. P. Collins (1994). "Shear Design of Concrete Offshore Structures," *ACI Structural Journal*, Vol. 91, No. 3, May–June, pp. 324–335.

Al-Omaishi, N. (2001). Prestress Losses in High Strength Pretensioned Concrete Bridge Girders, Ph.D. Dissertation, University of Nebraska-Lincoln, December.

Bentz, E. C. (2000). Sectional Analysis of Reinforced Concrete Members, Ph.D. Thesis, Department of Civil Engineering, University of Toronto, ON, Canada.

Collins, M. P. and D. Mitchell (1991). *Prestressed Concrete Structures*. Prentice-Hall, Englewood Cliffs, NJ.

Hawkins, N. M, D. A. Kuchma, R. F. Mast, M. L. Marsh, and K. J. Reineck (2005). "Simplified Shear Design of Structural Concrete Members," NCHRP Report 549, National Cooperative Highway Research Program, Transportation Research Board, Washington, DC.

Hsu, T. T. C. (1993). *Unified Theory of Reinforced Concrete*, CRC Press, Boca Raton, FL.

Loov, R. E. (1988). "A General Equation for the Steel Stress for Bonded Prestressed Concrete Members," *PCI Journal*, Vol. 33, No. 6, Nov.–Dec., pp. 108–137.

MacGregor, R. J. G. (1989). Strength and Ductility of Externally Post-Tensioned Segmental Box Girders, Ph.D. Dissertation, University of Texas at Austin, Austin, TX.

Mander, J. B., M. J. N. Priestley, and R. Park (1988). "Theoretical Stress–Strain Model for Confined Concrete," *Journal of Structural Engineering*, *ASCE*, Vol. 114, No. 8, pp. 1804–1826.

Mast, R. F. (1992). "Unified Design Provisions for Reinforced and Prestressed Concrete Flexural and Compression Members," *ACI Structural Journal*, Vol. 89, No. 2, Mar.–Apr., pp. 185–199.

Mörsch, E. (1908). *Concrete-Steel Construction*, English translation by E. P. Goodrich, Engineering News, New York, 1909. [Translation from 3rd ed. (1908) of *Der Eisenbetonbau*, 1st ed., 1902.]

Naaman, A. E. (1992). "Unified Design Recommendations for Reinforced, Prestressed and Partially Prestressed Concrete Bending and Compression Members," *ACI Structural Journal*, Vol. 89, No. 2, Mar.–Apr., pp. 200–210.

Naaman, A. E., R. E. Loov, and J. F. Stanton (1990). Discussion of Skogman, B. C., M. K. Tadros, and R. Grasmick (1988), "Ductility of Reinforced and Prestressed Concrete Flexural Members," *PCI Journal*, Vol. 35, No. 2, Mar.–Apr., pp. 82–89.

Nawy, E. G. (2009). *Prestressed Concrete: A Fundamental Approach*, 5th ed. upgrade, Prentice Hall, Upper Saddle River, NJ.

Roberts, C. L. (1993). Measurements-Based Revisions for Segmental Bridge Design and Construction Criteria, Ph.D. Dissertation, The University of Texas at Austin, Austin, TX, December.

Roberts-Wollmann, C. L., M. E. Kreger, D. M. Rogowsky, and J. E. Breen (2005). "Stresses in External Tendons at Ultimate," *ACI Structural Journal*, Vol. 102, No. 2, Mar.–Apr., pp. 206–213.

Sawyer, H. A., Jr. (1964). "Design of Concrete Frames for Two Failure Stages," *Proceedings International Symposium on Flexural Mechanics of Reinforced Concrete*, Miami, FL, pp. 405–437.

Skogman, B. C., M. K. Tadros, and R. Grasmick (1988). "Ductility of Reinforced and Prestressed Concrete Flexural Members," *PCI Journal*, Vol. 33, No. 6, Nov.–Dec., pp. 94–107.

Tadros, M. K., N. Al-Omaishi, S. P. Seguirant, and J. G. Gallt (2003). *Prestress Losses in Pretensioned High-Strength Concrete Bridge Girders*, NCHRP Report 496, Transportation Research Board, Washington, DC.

Vecchio, F. J. and M. P. Collins (1986). "The Modified Compression Field Theory for Reinforced Concrete Elements Subjected to Shear," *ACI Journal*, Vol. 83, No. 2, Mar–Apr, pp. 219–231.

Vecchio, F. J. and M. P. Collins (1988). "Predicting the Response of Reinforced Concrete Beams Subjected to Shear Using Modified Compression Field Theory," *ACI Structural Journal*, Vol. 85, No. 3, May–June, pp. 258–268.

Walraven, J. C. (1981). "Fundamental Analysis of Aggregate Interlock," *Journal of the Structural Division*, ASCE, Vol. 107, No. ST11, Nov., pp. 2245–2270.

Zia, P., H. K. Preston, N. L. Scott, and E. B. Workman (1979). "Estimating Prestress Losses," *Concrete International: Design and Construction*, Vol. 1, No. 6, June, pp. 32–38.

PROBLEMS

14.1 Rework Example 14.1 with a change in the flange width b to 48 in. (instead of 18 in.) for the beam cross section in Figure 14.11.

14.2 Determine the steel strain and the associated resistance factor for the beam in Problem 14.1 with the properties given in Example 14.1 for (a) a bonded case and (b) an unbonded case.

14.3 Why are there two requirements for minimum tensile reinforcement in AASHTO [A5.7.3.3.2]? In what cases would $1.2M_{\text{cr}}$ be greater than $1.33M_u$?

14.4 Give examples of design situations where it is important to have reasonably accurate estimates of prestress loss in prestressing strands.

14.5 What are some of the consequences, both good and bad, of not estimating prestress loss with reasonable accuracy?

14.6 Give examples of loading stages where prestress gain, rather than prestress loss, can occur.

14.7 In what design situations does AASHTO [A5.6.3] recommend that a strut-and-tie model be used to represent the load-carrying mechanism in reinforced concrete members?

14.8 The variable-angle truss model has been used for years to explain shear in reinforced concrete beams. In spite of its shortcomings, one of its positive features is the prediction of increased tension in the longitudinal reinforcement produced by a transverse shear force. How has this feature been incorporated into the shear design provisions of AASHTO [A5.8.3.5]?

14.9 What is the difference between compression field theory and *modified* compression field theory for predicting shear strength of reinforced concrete members?

14.10 Apply the computer program Response-2000 to Example 14.3. The latest version of Response-2000 can be downloaded without charge from the website: www.ecf.utoronto.ca/~Bentz/home.shtml.

CHAPTER 15

Concrete Barrier Strength and Deck Design

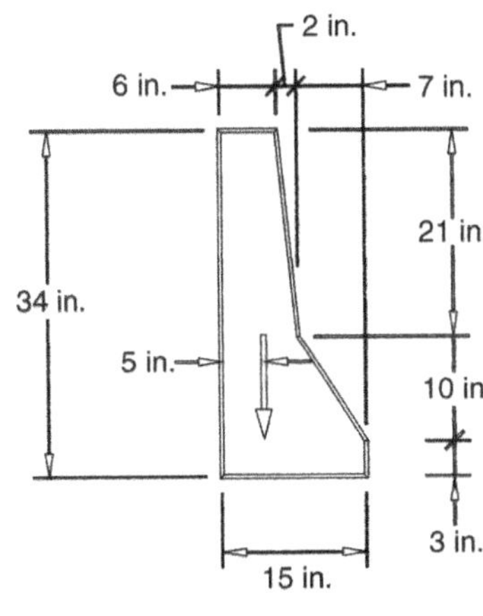

Fig. 15.1 Concrete barrier.

15.1 CONCRETE BARRIER STRENGTH

The purpose of a concrete barrier, in the event of a collision by a vehicle, is to redirect the vehicle in a controlled manner. The vehicle shall not overturn or rebound across traffic lanes. The barrier shall have sufficient strength to survive the initial impact of the collision and to remain effective in redirecting the vehicle.

To meet the design criteria, the barrier must satisfy both geometric and strength requirements. The geometric conditions will influence the redirection of the vehicle and whether it will be controlled or not. This control must be provided for the complete mix of traffic from the largest trucks to the smallest automobiles. Geometric shapes and profiles of barriers that can control collisions have been developed over the years and have been proven by crash testing. Any variation from the proven geometry may involve risk and is not recommended. A typical solid concrete barrier cross section with sloping face on the traffic side is shown in Figure 15.1, that is, the critical value.

The strength requirements for barriers depend on the truck volume and speed of the traffic anticipated for the bridge. For given traffic conditions, a performance level for the barrier can be selected and the collision forces defined [A13.7.2]*. The design forces and their location relative to the bridge deck are given for six test levels in Table 8.5. The concrete barrier in Figure 15.1 has a height sufficient for test level TL-4.

*The article numbers in the AASHTO (2010) LRFD Bridge Specifications are enclosed in breackets and preceded by the letter A if a specification article and by the letter C if commentary.

15.1.1 Strength of Uniform Thickness Barrier Wall

The lateral load-carrying capacity of a uniform thickness solid concrete barrier was analyzed by Hirsh (1978). The expressions developed for the strength of the barrier are based on the formation of yield lines at the limit state. The assumed yield line pattern caused by a truck collision that produces a force F_t that is distributed over a length L_t is shown in Figure 15.2.

The fundamentals of yield-line analysis are given in Section 11.3. Essentially, for an assumed yield-line pattern that is consistent with the geometry and boundary conditions of a wall or slab, a solution is obtained by equating the external virtual work due to the applied loads to the internal virtual work done by the resisting moments along the yield lines. The applied load determined by this method is either equal to or greater than the actual load, that is, it is nonconservative. Therefore, it is important to minimize the load for a particular yield-line pattern. In the case of the yield-line pattern shown in Figure 15.2, the angle of the inclined yield lines can be expressed in terms of the critical length L_c. The

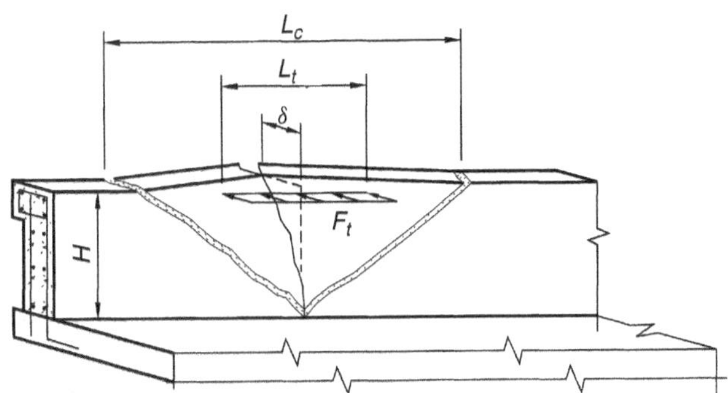

Fig. 15.2 Yield-line pattern for barrier wall. (After Hirsh, 1978.)

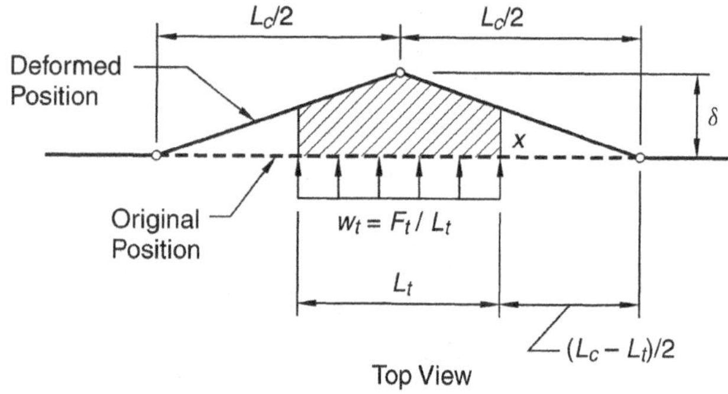

Fig. 15.3 External virtual work by distributed load. (After Calloway, 1993.)

applied force F_t is minimized with respect to L_c to get the least value of this upper bound solution.

External Virtual Work by Applied Loads The original and deformed positions of the top of the wall are shown in Figure 15.3. The shaded area represents the integral of the deformations through which the uniformly distributed load $w_t = F_t/L_t$ acts. For a virtual displacement δ, the displacement x is

$$x = \frac{L_c - L_t}{L_c}\delta \tag{15.1}$$

and the shaded area becomes

$$\begin{aligned} \text{Area} &= \frac{1}{2}(\delta + x)\,L_t = \frac{\delta}{2}\left(1 + \frac{L_c - L_t}{L_c}\right)L_t \\ &= \delta\frac{L_t}{L_c}\left(L_c - \frac{L_t}{2}\right) \end{aligned} \tag{15.2}$$

so that the external virtual work W done by w_t is

$$\begin{aligned} W = w_t\,(\text{area}) &= \frac{F_t}{L_t}\delta\frac{L_t}{L_c}\left(L_c - \frac{L_t}{2}\right) \\ &= F_t\frac{\delta}{L_c}\left(L_c - \frac{L_t}{2}\right) \end{aligned} \tag{15.3}$$

Internal Virtual Work along Yield Lines The internal virtual work along the yield lines is the sum of the products of the yield moments and the rotations through which they act. The segments of the wall are assumed to be rigid so that all of the rotation is concentrated at the yield lines. At the top of the wall (Fig. 15.4), the rotation θ of the wall segments for small deformations is

$$\theta \approx \tan\,\theta = \frac{2\delta}{L_c} \tag{15.4}$$

The barrier can be analyzed by separating it into a beam at the top and a uniform thickness wall below. At the limit state, the top beam will develop plastic moments M_b equal to its nominal bending strength M_n and form a mechanism as shown in Figure 15.4. Assuming that the negative and

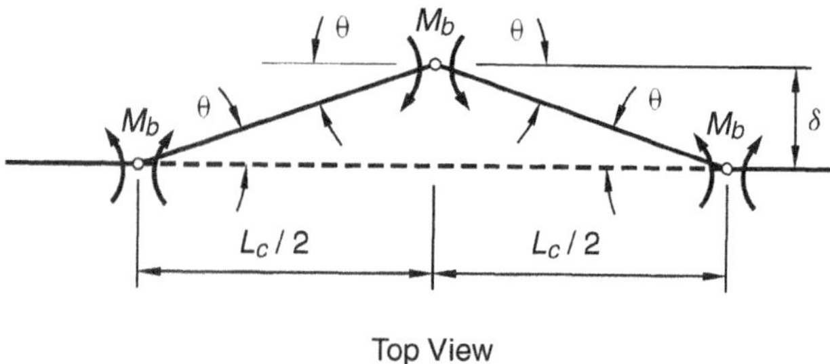

Fig. 15.4 Plastic hinge mechanism for top beam. (After Calloway, 1993.)

positive plastic moment strengths are equal, the internal virtual work U_b done by the top beam is

$$U_b = 4M_b\theta = \frac{8M_b\delta}{L_c} \tag{15.5}$$

The wall portion of the barrier will generally be reinforced with steel in both the horizontal and vertical directions. The horizontal reinforcement in the wall develops moment resistance M_w about a vertical axis. The vertical reinforcement in the wall develops a cantilever moment resistance M_c per unit length about a horizontal axis. These two components of moment will combine to develop a moment resistance M_α about the inclined yield line as shown in Figure 15.5. When determining the internal virtual work along inclined yield lines, it is simpler to use the projections of moment on and rotation about the vertical and horizontal axes.

Assume that the positive and negative bending resistance M_w about the vertical axis are equal, and use θ as the projection on the horizontal plane of the rotation about the inclined

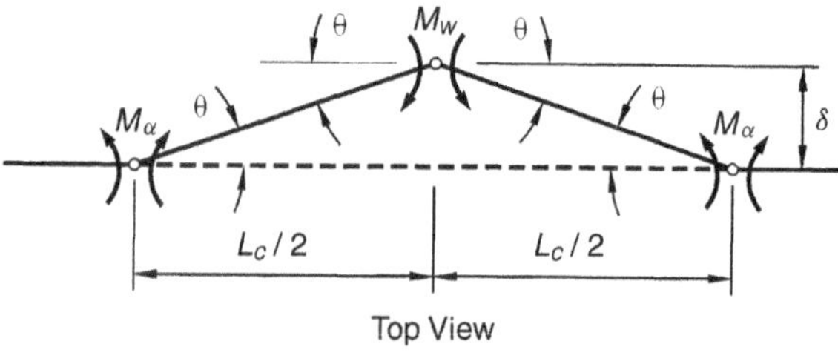

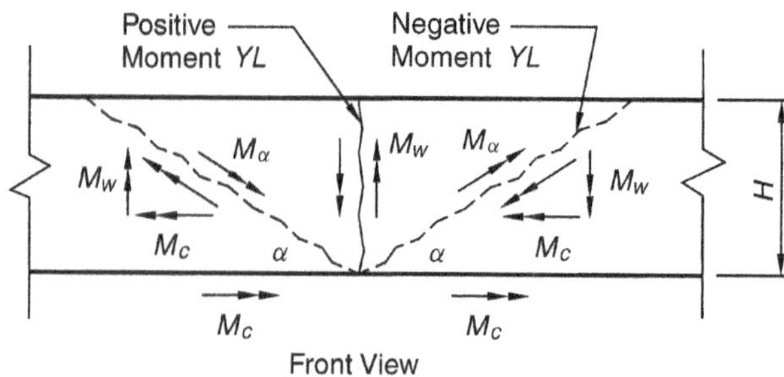

Fig. 15.5 Internal virtual work by barrier wall. (After Calloway, 1993.)

yield line. The internal virtual work U_w done by the wall moment M_w is then

$$U_w = 4M_w\theta = \frac{8M_w\delta}{L_c} \qquad (15.6)$$

The projection on the vertical plane of the rotation about the inclined yield line is δ/H, and the internal virtual work U_c done by the cantilever moment M_cL_c is

$$U_c = \frac{M_cL_c\delta}{H} \qquad (15.7)$$

Nominal Railing Resistance to Transverse Load $\mathbf{R_w}$ Equate the external virtual work W to the internal virtual work U to give

$$W = U_b + U_w + U_c$$

If we substitute Eqs. 15.3 and 15.5–15.7 to get

$$\frac{F_t}{L_c}\left(L_c - \frac{L_t}{2}\right)\delta = \frac{8M_b\delta}{L_c} + \frac{8M_w\delta}{L_c} + \frac{M_cL_c\delta}{H}$$

and solve for the transverse vehicle impact force F_t:

$$F_t = \frac{8M_b}{L_c - \frac{L_t}{2}} + \frac{8M_w}{L_c - \frac{L_t}{2}} + \frac{M_cL_c^2}{H\left(L_c - \frac{L_t}{2}\right)} \qquad (15.8)$$

This expression depends on the critical length L_c that determines the inclination of α of the negative moment yield lines in the wall. The value for L_c that minimizes F_t can be determined by differentiating Eq. 15.8 with respect to L_c and setting the result equal to zero, that is,

$$\frac{dF_t}{dL_c} = 0 \qquad (15.9)$$

This minimization results in a quadratic equation that can be solved explicitly to give

$$L_c = \frac{L_t}{2} + \sqrt{\left(\frac{L_t}{2}\right)^2 + \frac{8H\left(M_b + M_w\right)}{M_c}} \qquad (15.10)$$

When this value of L_c is used in Eq. 15.8, then the minimum value for F_t results, and the result is denoted as R_w, that is,

$$\min\ F_i = R_w \qquad (15.11)$$

where R_w is the nominal railing resistance to transverse load. By rearranging Eq. 15.8, R_w is [AA13.3.1]

$$R_w = \frac{2}{2L_c - L_t}\left(8M_b + 8M_w + \frac{M_cL_c^2}{H}\right) \qquad (15.12)$$

where

F_t = transverse force specified in Table 8.5 assumed to be acting at top of a concrete wall (kip)
H = height of wall (ft)
L_c = critical length of yield-line failure pattern given by Eq. 15.10 (ft)
L_t = longitudinal length of distribution of impact force F_t specified in Table 8.5 (ft)
R_w = total nominal transverse resistance of the railing (kip)
M_b = additional flexural resistance of beam in addition to M_w, if any, at top of wall (kip-ft)
M_c = flexural resistance of cantilevered wall about an axis parallel to the longitudinal axis of the bridge (kip-ft/ft)
M_w = flexural resistance of the wall about its vertical axis (kip-ft)

15.1.2 Strength of Variable Thickness Barrier Wall

Most of the concrete barrier walls have sloping faces, as shown in Figure 15.1, and are not of uniform thickness. Calloway (1993) investigated the yield-line approach applied to barrier walls of changing thickness. Equations were developed for R_w based on continuously varying moment resistances whose product with rotation was integrated over the height of the wall to obtain the internal virtual work. The results of this more "exact" approach were compared to those obtained from the Hirsh equations (Eqs. 15.10 and 15.12) using various methods for calculating M_w and M_c. The recommended procedure is to use the Hirsh equations with average values for M_w and M_c. In the cases examined by Calloway, the R_w calculated by using average values was 4% less (conservative) than the more "exact" approach. This procedure is illustrated in the deck overhang design of Example E15.1.

15.1.3 Crash Testing of Barriers

It should be emphasized that a railing system and its connection to the deck shall be approved only after they have been shown to be satisfactory through crash testing for the desired test level [A13.7.3.1]. If minor modifications have been made to a previously tested railing system that does not affect its strength, it can be used without further crash testing. However, any new system must be verified by full-scale crash testing. Clearly, the steel detailing of the barrier to the deck and the cantilever overhang strength is important for the transfer of the crash load into the deck diaphragm.

15.2 CONCRETE DECK DESIGN

Problem Statement Example E15.1 Use the approximate method of analysis [A4.6.2.1] to design the deck of the reinforced concrete T-beam bridge section of Figure E15.1-1 for an HL-93 live load and a TL-4 test-level concrete barrier (Fig. 15.1). The T-beams supporting the deck are 8 ft on centers and have a stem width of 14 in. The deck overhangs the exterior T-beam approximately 0.4 of the distance between T-beams. Allow for sacrificial wear of 0.5 in. of concrete

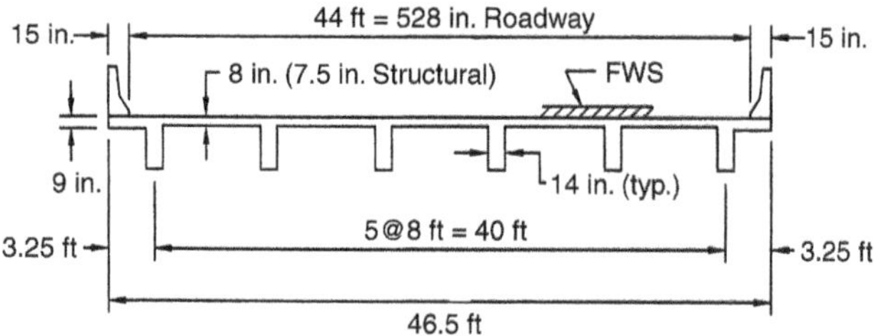

Fig. E15.1-1 Concrete deck design example.

surface and for a future wearing surface of 3.0-in.-thick bituminous overlay. The barrier concrete strength is 4.0 ksi, area is 307 in^2, the associated weight is 0.32 kip/ft, and the center of gravity is located at 5 in. from the exterior face. Class 2 exposure should be used for the service limit associated with cracking. Use $f_c' = 4.5$ ksi, $f_y = 60$ ksi, and compare the selected reinforcement with that obtained by the empirical method [A9.7.2].

A. ***Deck Thickness*** The minimum thickness for concrete deck slabs is 7 in. [A9.7.1.1]. Traditional minimum depths of slabs are based on the deck span length S (ft) to control deflection to give for continuous deck slabs with main reinforcement parallel to traffic [Table A2.5.2.6.3-1]:

$$h_{\min} = \frac{S+10}{30} = \frac{8+10}{30} = 0.6 \text{ ft} = 7.2 \text{ in.} > 7 \text{ in.}$$

Use $h_s = 7.5$ in. for the structural thickness of the deck. By adding the 0.5-in. allowance for the sacrificial surface, the dead weight of the deck slab is based on $h =$ 8.0 in. Because the portion of the deck that overhangs the exterior girder must be designed for a collision load on the barrier, its thickness has been increased to $h_o =$ 9.0 in.

B. ***Weights of Components*** [Table A3.5.1-1] Unit weight of reinforced concrete is taken as 0.150 kcf [C3.5.1]. For a 1.0-ft width of a transverse strip

Barrier

$$P_b = 0.150 \text{ kcf} \times 307 \text{ in.}^2/144 = 0.320 \text{ kip/ft}$$

Future wearing surface

$$w_{DW} = 0.140 \text{ kcf} \times 3.0 \text{ in.}/12 = 0.035 \text{ ksf}$$

Slab 8.0 in. thick

$$w_s = 0.150 \text{ kcf} \times 8.0 \text{ in.}/12 = 0.100 \text{ ksf}$$

Cantilever overhang 9.0 in. thick

$$w_o = 0.150 \text{ kcf} \times 9.0 \text{ in.}/12 = 0.113 \text{ ksf}$$

C. ***Bending Moment Force Effects—General*** An approximate analysis of strips perpendicular to girders is considered acceptable [A9.6.1]. The extreme positive moment in any deck panel between girders shall be taken to apply to all positive moment regions. Similarly, the extreme negative moment over any girder shall be taken to apply to all negative moment regions [A4.6.2.1.1]. The strips shall be treated as continuous beams with span lengths equal to the center-to-center distance between girders. The girders shall be assumed to be rigid [A4.6.2.1.6].

For ease in applying load factors, the bending moments are determined separately for the deck slab, overhang, barrier, future wearing surface, and vehicle live load.

1. *Deck Slab*

$$h = 8.0 \text{ in.} \qquad w_s = 0.100 \text{ ksf} \qquad S = 8.0 \text{ ft}$$

$$\text{FEM} = \pm\frac{w_s S^2}{12} = \pm\frac{0.100(8.0)^2}{12} = 0.533 \text{ kip-ft/ft}$$

Placement of the deck slab dead load and results of a moment distribution analysis for negative and positive moments in a 1-ft-wide strip is given in Figure E15.1-2.

A deck analysis design aid based on influence lines is given in Table A.1 of Appendix A. For a uniform load, the tabulated areas are multiplied by S for shears and by S^2 for moments.

$$R_{200} = w_s(\text{net area w/o cantilever})S = 0.100(0.3928)8 = 0.314 \text{ kip/ft}$$

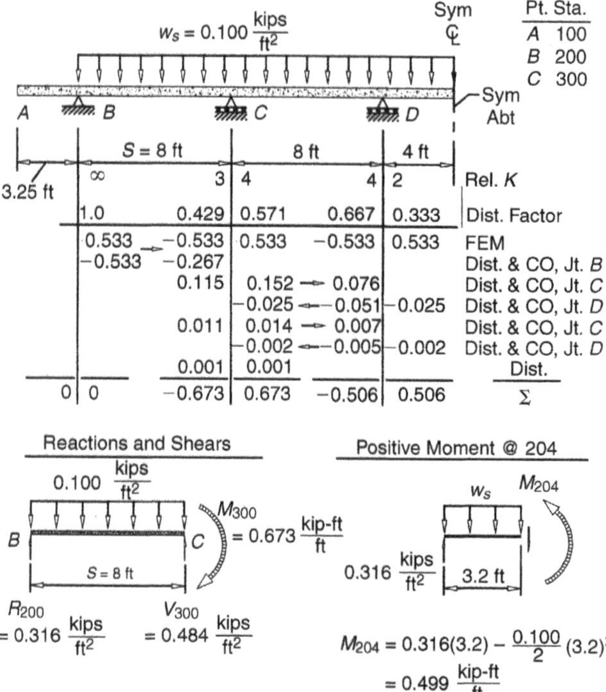

Fig. E15.1-2 Moment distribution for deck slab dead load.

$$M_{204} = w_s(\text{net area w/o cantilever})S^2$$
$$= 0.100(0.0772)8^2 = 0.494 \text{ kip-ft/ft}$$
$$M_{300} = w_s(\text{net area w/o cantilever})S^2$$
$$= 0.100(-0.1071)8^2 = -0.685 \text{ kip-ft/ft}$$

Comparing the results from the design aid with those from moment distribution shows good agreement. In determining the remainder of the bending moment force effects, the design aid of Table A.1 is used.

2. *Overhang* The parameters are $h_o = 9.0$ in., $w_o = 0.113$ ksf, and $L = 3.25$ ft. Placement of the overhang dead load is shown in Figure E15.1-3. By using the design aid of Table A.1 in Appendix A, the reaction on the exterior T-beam and the bending moments are

$$R_{200} = w_o(\text{net area cantilever})L$$
$$= 0.113\left(1.0 + 0.635\frac{3.25}{8.0}\right)3.25$$
$$= 0.462 \text{ kip/ft}$$
$$M_{200} = w_o(\text{net area cantilever})L^2$$
$$= 0.113(-0.5000)3.25^2 = -0.597 \text{ kip-ft/ft}$$
$$M_{204} = w_o(\text{net area cantilever})L^2$$
$$= 0.113(-0.2460)3.25^2 = -0.294 \text{ kip-ft/ft}$$
$$M_{300} = w_o(\text{net area cantilever})L^2$$
$$= 0.113(0.1350)3.25^2 = 0.161 \text{ kip-ft/ft}$$

3. *Barrier* The parameters are $P_b = 0.320$ kip/ft and $L = 3.25 - 0.42 = 2.83$ ft. Placement of the center of gravity of the barrier dead load is shown in Figure E15.1-4. By using the design aid of Table A.1 for the concentrated barrier load, the intensity of the load is multiplied by the influence line ordinate for shears and reactions. For bending moments, the influence line ordinate is multiplied by the cantilever length L.

$$R_{200} = P_b(\text{influence line ordinate})$$
$$= 0.320\left(1.0 + 1.270\frac{2.83}{8.0}\right)$$
$$= 0.464 \text{ kip/ft}$$
$$M_{200} = P_b(\text{influence line ordinate})L$$
$$= 0.320(-1.0000)(2.83)$$
$$= -0.906 \text{ kip-ft/ft}$$
$$M_{204} = P_b(\text{influence line ordinate})L$$
$$= 0.320(-0.4920)(2.83) = -0.446 \text{ kip-ft/ft}$$
$$M_{300} = P_b(\text{influence line ordinate})L$$
$$= 0.320(0.2700)(2.83) = 0.245 \text{ kip-ft/ft}$$

4. *Future Wearing Surface* (FWS) FWS $= w_{DW} = 0.035$ ksf. The 3 in. of bituminous overlay is placed curb to curb as shown in Figure E15.1-5. The length of the loaded cantilever is reduced by the base width of the barrier to give $L = 3.25 - 1.25 = 2.0$ ft. Using the design aid of Table A.1 gives

$$R_{200} = w_{DW}[(\text{net area cantilever})\ L + (\text{net area w/o cantilever})S]$$
$$= 0.035\left[\left(1.0 + 0.635\frac{2.0}{8.0}\right)2.0 + (0.3928)\,8.0\right]$$
$$= 0.191 \text{ kip/ft}$$
$$M_{200} = w_{DW}(\text{net area cantilever})L^2$$
$$= 0.035(-0.5000)(2.0)^2 = -0.070 \text{ kip-ft/ft}$$
$$M_{204} = w_{DW}[(\text{net area cantilever})L^2 + (\text{net area w/o cantilever})S^2]$$
$$= 0.035[(-0.2460)2.0^2 + (0.0772)8.0^2]$$
$$= 0.138 \text{ kip-ft/ft}$$
$$M_{300} = w_{DW}[(\text{net area cantilever})L^2 + (\text{net area w/o cantilever})S^2]$$
$$= 0.035[(0.135)2.0^2 + (-0.1071)8.0^2]$$
$$= -0.221 \text{ kip-ft/ft}$$

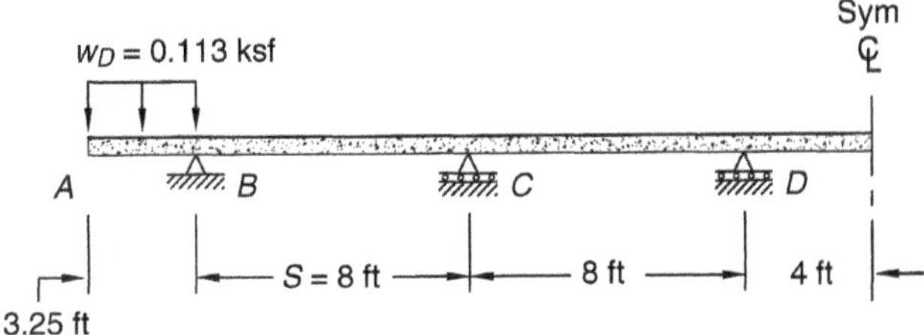

Fig. E15.1-3 Overhang dead-load placement.

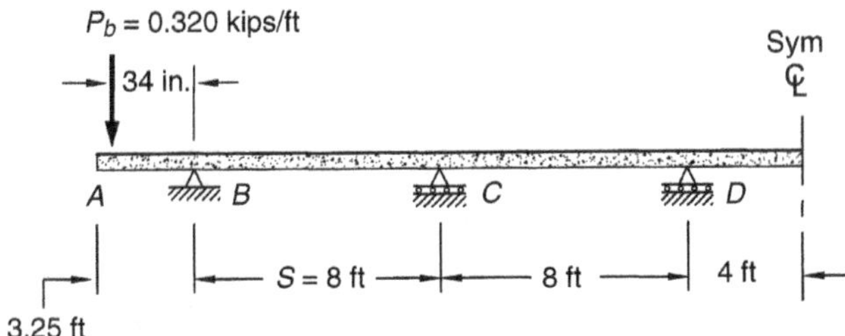

Fig. E15.1-4 Barrier dead-load placement.

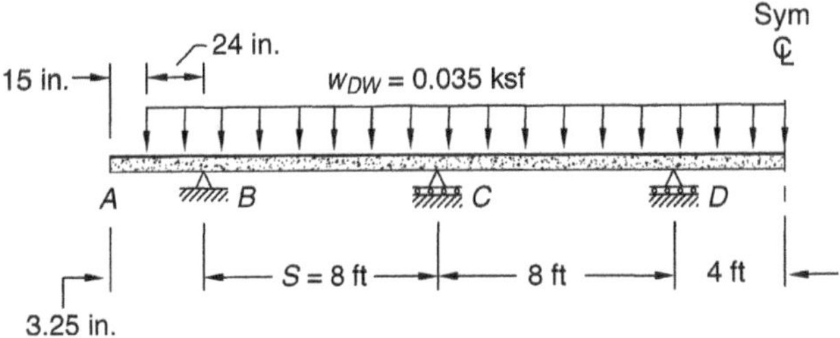

Fig. E15.1-5 Future wearing surface dead-load placement.

D. ***Vehicular Live Load—General*** Where decks are designed using the approximate strip method [A4.6.2.1], the strips are transverse and shall be designed for the 32.0-kip axle of the design truck [A3.6.1.3.3]. Wheel loads on an axle are assumed to be equal and spaced 6.0 ft apart [Fig. A3.6.1.2.2-1]. The design truck shall be positioned transversely to produce maximum force effects such that the center of any wheel load is not closer than 1.0 ft from the face of the curb for the design of the deck overhang and 2.0 ft from the edge of the 12.0-ft-wide design lane for the design of all other components [A3.6.1.3.1].

The width of equivalent interior transverse strips (in.) over which the wheel loads can be considered distributed longitudinally in CIP concrete decks is given as [Table A4.6.2.1.3-1]

- ☐ Overhang, $45.0 + 10.0X$
- ☐ Positive moment, $26.0 + 6.6S$
- ☐ Negative moment, $48.0 + 3.0S$

where X (ft) is the distance from the wheel load to the centerline of support and S (ft) is the spacing of the T-beams. For our example, X is 1.0 ft (see Fig. E15.1-6) and S is 8.0 ft.

Tire contact area [A3.6.1.2.5] shall be assumed as a rectangle with width of 20.0 in. and length of 10.0 in. with the 20.0-in. dimension in the transverse direction as shown in Figure E15.1-6.

When calculating the force effects, wheel loads may be modeled as concentrated loads or as patch loads distributed transversely over a length along the deck span of 20.0 in. plus the slab depth [A4.6.2.1.6]. This distributed model is shown in Figure E15.1-6 and represents a 1 : 1 spreading of the tire loading to mid-depth of the beam. For our example, length of patch loading = 20.0 + 7.5 = 27.5 in. If the spans are short, the calculated bending moments in the deck using the patch loading can be significantly lower than those using the concentrated load. In this design example, force effects are calculated conservatively by using concentrated wheel loads.

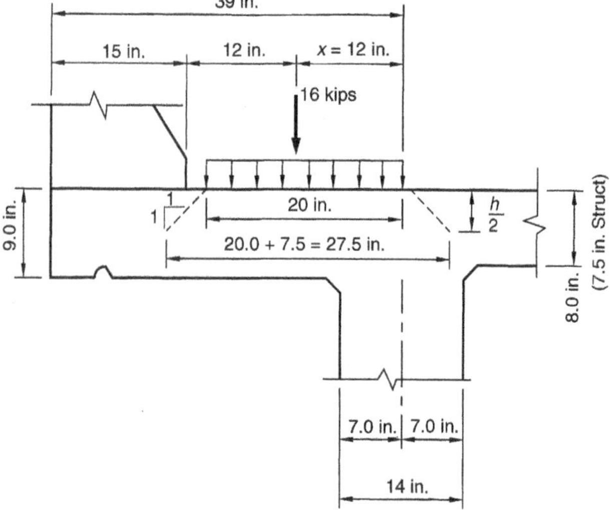

Fig. E15.1-6 Distribution of wheel load on overhang.

The number of design lanes N_L to be considered across a transverse strip is the integer value of the clear roadway width divided by 12.0 ft [A3.6.1.1.1]. For our example,

$$N_L = \text{INT}\left(\frac{44.0}{12.0}\right) = 3$$

The multiple presence factor m is 1.2 for one loaded lane, 1.0 for two loaded lanes, and 0.85 for three loaded lanes. (If only one lane is loaded, we must consider the probability that this single truck can be heavier than each of the trucks traveling in parallel lanes [A3.6.1.1.2].)

1. *Overhang Negative Live-Load Moment* The critical placement of a single wheel load is shown in Figure E15.1-6. The equivalent width of a transverse strip is $45.0 + 10.0X = 45.0 + 10.0(1.0) = 55.0$ in. $= 4.58$ ft and $m = 1.2$. Therefore,

$$M_{200} = \frac{-1.2\,(16.0)\,(1.0)}{4.58} = -4.19 \text{ kip-ft/ft}$$

The above moment can be reduced if the concrete barrier is structurally continuous and becomes effective in distributing the wheel loads in the overhang [A3.6.1.3.4]. However, as we shall see later, the overhang negative moment caused by horizontal forces from a vehicle collision [A13.7.2] is greater than the moment produced by live load.

2. *Maximum Positive Live-Load Moment* For repeating equal spans, the maximum positive bending moment occurs near the 0.4 point of the first interior span, that is, at location 204. In Figure E15.1-7, the placement of wheel loads is given for one and two loaded lanes. For both cases, the equivalent width of a transverse strip is $26.0 + 6.6S = 26.0 + 6.6(8.0) = 78.8$ in. $= 6.57$ ft. Using the influence line ordinates from Table A.1, the exterior girder reaction and positive bending moment with one loaded lane ($m = 1.2$) are

$$R_{200} = 1.2(0.5100 - 0.0510)\frac{16.0}{6.57} = 1.34 \text{ kip/ft}$$

$$M_{204} = 1.2(0.2040 - 0.0204)(8.0)\frac{16.0}{6.57} = 4.29 \text{ kip-ft/ft}$$

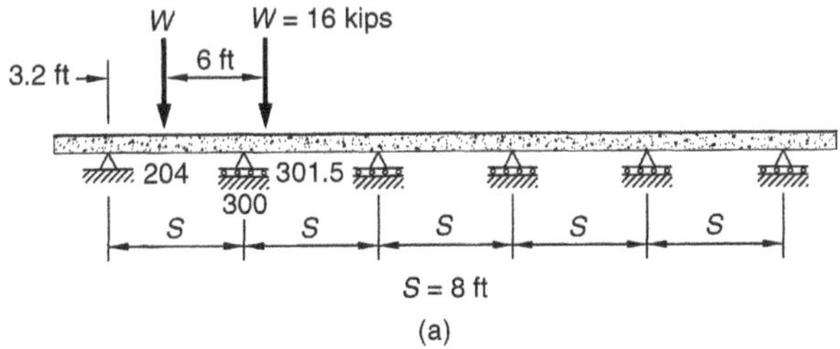

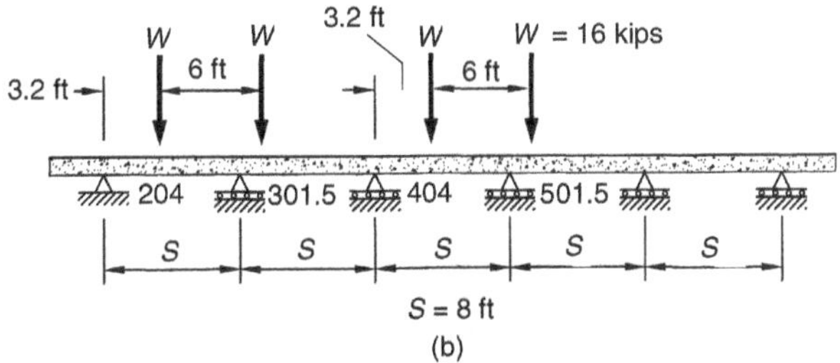

Fig. E15.1-7 Live-load placement for maximum positive moment: (a) one loaded lane, $m = 1.2$ and (b) two loaded lanes, $m = 1.0$.

and for two loaded lanes ($m = 1.0$)

$$\begin{aligned} R_{200} &= 1.0(0.5100 - 0.0510 \\ &\quad + 0.0214 - 0.0041)\frac{16.0}{6.57} \\ &= 1.16 \text{ kips/ft} \\ M_{204} &= 1.0(0.2040 - 0.0204 \\ &\quad + 0.0086 - 0.0017)(8.0)\frac{16.0}{6.57} \\ &= 3.71 \text{ kip-ft/ft} \end{aligned}$$

Thus, the one loaded lane case governs.

3. *Maximum Interior Negative Live-Load Moment* The critical placement of live load for maximum negative moment is at the first interior deck support with one loaded lane ($m = 1.2$) as shown in Figure E15.1-8. The equivalent transverse strip width is $48.0 + 3.0S = 48.0 + 3.0(8.0) = 72.0$ in. $= 6.0$ ft. Using influence line ordinates from Table A.1, the bending moment at location 300 becomes

$$\begin{aligned} M_{300} &= 1.2\,(-0.1010 - 0.0782)\,(8.0)\,\frac{16.0}{6.0} \\ &= -4.59 \text{ kip-ft/ft} \end{aligned}$$

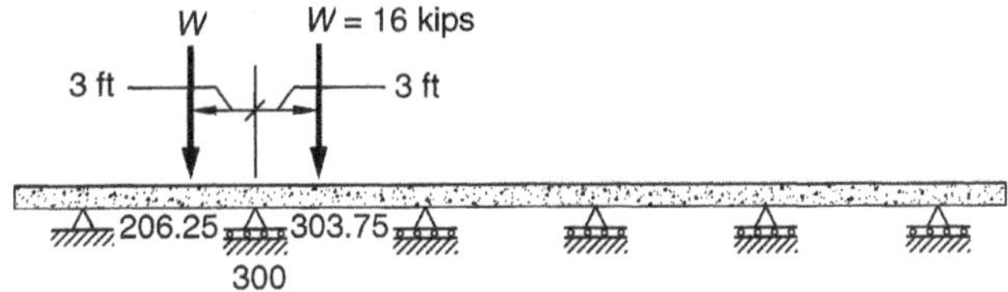

Fig. E15.1-8 Live-load placement for maximum negative moment.

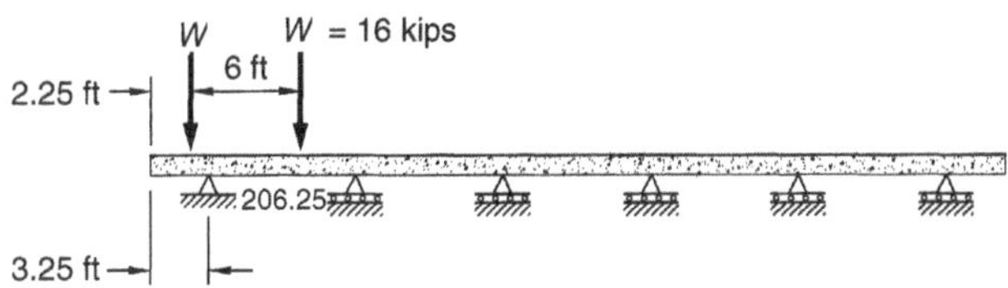

Fig. E15.1-9 Live-load placement for maximum reaction at exterior girder.

Note that the small increase due to a second truck is less than the 20% ($m = 1.0$) required to control. Therefore, only the one lane case is investigated.

4. *Maximum Live-Load Reaction on Exterior Girder* The exterior wheel load is placed 1.0 ft from the curb or 1.0 ft from the centerline of the support as shown in Figure E15.1-9. The width of the transverse strip is conservatively taken as the one for the overhang. Using influence line ordinates from Table A.1,

$$R_{200} = 1.2\,(1.1588 + 0.2739)\,\frac{16.0}{4.58} = 6.01 \text{ kips/ft}$$

E. ***Strength Limit State*** Each component and connection of the deck shall satisfy the basic design equation [A1.3.2.1]

$$\Sigma \eta_i \gamma_i Q_i \leq \phi R_n \qquad \text{(A1.3.2.1-1)}$$

in which:

For loads for which a maximum value of γ_i is appropriate

$$\eta_i = \eta_D \eta_R \eta_I \geq 0.95 \qquad \text{(A1.3.2.1-2)}$$

For loads for which a minimum value of γ_i is appropriate

$$\eta_i = \frac{1.0}{\eta_D \eta_R \eta_I} \leq 1.0 \qquad \text{(A1.3.2.1-3)}$$

For the strength limit state

$\eta_D = 1.00$ for conventional design and details complying with AASHTO (2010) [A1.3.3]
$\eta_R = 1.00$ for conventional levels of redundancy [A1.3.4]
$\eta_1 = 1.00$ for typical bridges [A1.3.5]

For these values of η_D, η_R, and η_I, the load modifier $\eta_i = 1.00(1.00)\ (1.00) = 1.00$ for all load cases and the strength I limit state can be written as [Tables 3.4.1-1]

$$\begin{aligned} \Sigma_i \eta_i \gamma_i Q_i &= 1.00\gamma_p \text{DC} + 1.00\gamma_p \text{DW} \\ &\quad + 1.00\,(1.75)\,(\text{LL} + \text{IM}) \end{aligned}$$

The factor for permanent loads γ_p is taken at its maximum value if the force effects are additive and at its minimum value if it subtracts from the dominant force effect [Table A3.4.1-2]. The dead-load DW is for the future wearing surface and DC represents all the other dead loads.

The dynamic load allowance IM [A3.6.2.1] is 33% of the live-load force effect. Factoring out the common 1.00 load modifier, the combined force effects become

$$\begin{aligned} R_{200} &= 1.00[1.25(0.314 + 0.462 + 0.464) \\ &\quad + 1.50(0.191)] + 1.75(1.33)(6.01)] \\ &= 15.83 \text{ kips/ft} \\ M_{200} &= 1.00[1.25(-0.597 - 0.906) + 1.50(-0.070) \\ &\quad + 1.75(1.33)(-4.19)] = -11.74 \text{ kip-ft/ft} \\ M_{204} &= 1.00[1.25(0.494) + 0.9(-0.294 - 0.446) \\ &\quad + 1.50(0.138) + 1.75(1.33)(4.29)] \\ &= 10.14 \text{ kip-ft/ft} \\ M_{300} &= 1.00[1.25(-0.685) + 0.9(0.161 + 0.245) \\ &\quad + 1.50(-0.221) + 1.75(1.33)(-4.59)] \\ &= -11.51 \text{ kip-ft/ft} \end{aligned}$$

The two negative bending moments are nearly equal, which confirms choosing the length of the overhang as $0.4S$. For selection of reinforcement, these moments can be reduced to their value at the face of the support [A4.6.2.1.6]. The T-beam stem width is 14.0 in., so the design sections are 7.0 in. on either side of the support centerline used in the analysis. The critical negative moment section is at the interior face of the exterior support as shown in the free-body diagram of Figure E15.1-10.

The values for the loads in Figure E15.1-10 are for a 1.0-ft-wide strip. The concentrated wheel load is for one loaded lane, that is, $W = 1.2(16.0)/4.58 = 4.19$ kips/ft. In calculating the moment effect, the loads are kept separate so that correct R_{200} values are used.

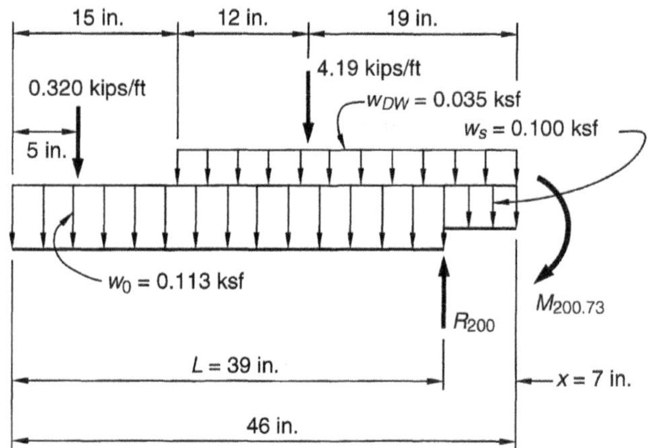

Fig. E15.1-10 Reduced negative moment at face of support.

1. *Deck Slab*

$$\begin{aligned} M_s &= -\tfrac{1}{2}w_s x^2 + R_{200}x \\ &= -\tfrac{1}{2}(0.100)\left(\tfrac{7}{12}\right)^2 + 0.314\left(\tfrac{7}{12}\right) \\ &= 0.166 \text{ kip-ft/ft} \end{aligned}$$

2. *Overhang*

$$\begin{aligned} M_o &= -w_o L\left(\frac{L}{2} + x\right) + R_{200}x \\ &= -0.113(3.25)\left(\tfrac{3.25}{2} + \tfrac{7}{12}\right) + 0.462\left(\tfrac{7}{12}\right) \\ &= -0.541 \text{ kip-ft/ft} \end{aligned}$$

3. *Barrier*

$$\begin{aligned} M_b &= -P_b\left(L + x - \tfrac{5}{12}\right) + R_{200}x \\ &= -0.320\left(\tfrac{46}{12} - \tfrac{5}{12}\right) + 0.464\left(\tfrac{7}{12}\right) \\ &= -0.823 \text{ kip-ft/ft} \end{aligned}$$

4. *Future Wearing Surface*

$$\begin{aligned} M_{\text{DW}} &= -\tfrac{1}{2}w_{\text{DW}}\left(L + x - \tfrac{15}{12}\right)^2 + R_{200}x \\ &= -\tfrac{1}{2}(0.035)\left(\tfrac{46}{12} - \tfrac{15}{12}\right)^2 + 0.191\left(\tfrac{7}{12}\right) \\ &= -0.005 \text{ kip-ft/ft} \end{aligned}$$

5. *Live Load*

$$\begin{aligned} M_{\text{LL}} &= -W\left(\tfrac{19}{12}\right) + R_{200}x \\ &= -4.19\left(\tfrac{19}{12}\right) + 6.01\left(\tfrac{7}{12}\right) \\ &= -3.128 \text{ kip-ft/ft} \end{aligned}$$

6. *Strength I Limit State*

$$\begin{aligned} M_{200.73} &= 1.00[0.9(0.166) + 1.25(-0.541 - 0.823) \\ &\quad + 1.50(-0.005) + 1.75(1.33)(-3.128)] \\ &= -8.84 \text{ kip-ft/ft} \end{aligned}$$

This negative bending design moment represents a significant reduction from the value at $M_{200} = -11.74$ kip-ft/ft. Because the extreme negative moment over any girder applies to all negative moment regions [A4.6.2.1.1], the extra effort required to calculate the reduced value is justified. Note that the moment at the outside face is smaller and can be calculated to be −5.52 kip-ft/ft.

F. ***Selection of Reinforcement—General*** The material strengths are $f_c' = 4.5$ ksi and $f_y = 60$ ksi. Use epoxy-coated reinforcement in the deck and barrier.

The effective concrete depths for positive and negative bending is different because of different cover requirements (see Fig. E15.1-11).

Concrete Cover [Table A5.12.3-1]

Deck surfaces subject to wear	2.5 in.
Bottom of CIP slabs	1.0 in.

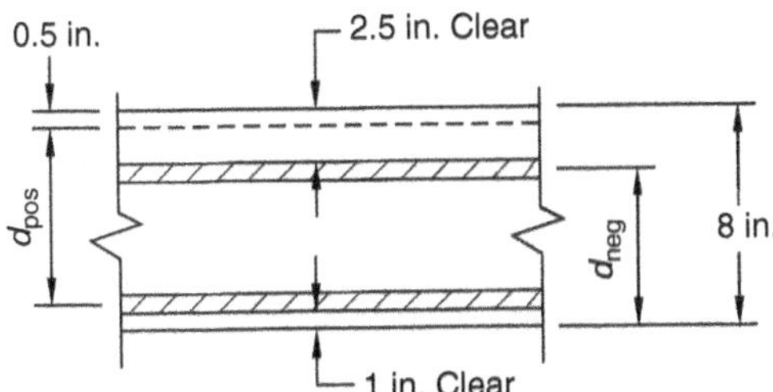

Fig. E15.1-11 Effective concrete depths for deck slabs.

Assuming a No. 5 bar, $d_b = 0.625$ in., $A_b = 0.31$ in.^{2}

$$d_{pos} = 8.0 - 0.5 - 1.0 - 0.625/2 = 6.19 \text{ in.}$$

$$d_{neg} = 8.0 - 2.5 - 0.625/2 = 5.19 \text{ in.}$$

A simplified expression for the required area of steel can be developed by neglecting the compressive reinforcement in the resisting moment to give [A5.7.3.2]

$$\phi M_n = \phi A_s f_y \left(d - \frac{a}{2}\right) \quad \text{(E15.1-1)}$$

where

$$a = \frac{A_s f_y}{0.85\ f_c' b} \quad \text{(E15.1-2)}$$

Assuming that the lever arm $(d - a/2)$ is independent of A_s, replace it by jd and solve for an approximate A_s required to resist $\phi M_n = M_u$.

$$A_s \approx \frac{M_u/\phi}{f_y\,(jd)} \quad \text{(E15.1-3)}$$

Further, substitute $f_y = 60$ ksi, $\phi = 0.9$ [A5.5.4.2.1], and assume that for lightly reinforced sections $j \approx 0.92$, a trial steel area becomes

$$\text{trial } A_s(\text{in.})^2 \approx \frac{M_u\,(\text{kip-ft})}{4d\ (\text{in.})} \quad \text{(E15.1-4)}$$

Because it is an approximate expression, it is necessary to verify the moment capacity of the selected reinforcement.

The resistance factor may be between $\phi_f = 0.75$ and $\phi_f = 0.90$ depending upon the steel strain at the failure state [A5.5.4.2.1]. This replaces the ductility requirement prescribed in previous specification. In equation format [A5.5.4.2.1] requires

$$0.75 \le \phi = 0.65 + 0.15\left(\frac{d_t}{c} - 1\right) \le 0.9 \quad \text{(E15.1-5)}$$

where c is the neutral axis depth and d_t is the depth of the tension reinforcement. When the steel strain is greater than or equal to 0.005, then ϕ is equal to 0.9. See Figure 14.5. An alternative check is $c/d_t \le 0.375$ ($a/d_t \le 0.375\beta_1$); then $\phi = 0.90$.

Minimum reinforcement [A5.7.3.3.2] for flexural components is satisfied if $\phi M_n = M_u$ is at least equal to the lesser of

- ☐ 1.2 times the cracking moment M_{cr}
- ☐ 1.33 times the factored moment required by the applicable strength load combination of [Table A3.4.1]

Where beams or slabs are designed for a noncomposite section to resist all loads

$$M_{cr} = S_{nc} f_r \quad \text{(E15.1-6)}$$

where

S_{nc} = section modulus for the extreme fiber of the noncomposite section where tensile stress is caused by external loads (in.^{3})

f_r = modulus of rupture of concrete (ksi) [A5.4.2.6]

For normal-weight concrete

$$f_r\ (\text{ksi}) = 0.37\sqrt{f_c'} \quad \text{(E15.1-7)}$$

Maximum spacing of primary reinforcement [A5.10.3.2] for slabs is 1.5 times the thickness of the member or 18.0 in. By using the structural slab thickness of 7.5 in.,

$$s_{max} = 1.5\,(7.5) = 11.25 \text{ in.}$$

1. *Positive Moment Reinforcement*

$$\text{pos } M_u = M_{204} = 10.14 \text{ kips/ft}$$

Minimum M_u depends on $M_{cr} = S_{nc} f_r$

$$S_{nc} = \tfrac{1}{6}bh^2 = \tfrac{1}{6}(12)(8.0)^2 = 128 \text{ in.}^3$$

$$f_r = 0.37\sqrt{f_c'} = 0.37\sqrt{4.5} = 0.785 \text{ ksi}$$

$$\text{min } M_u \text{ lessor of } 1.2M_{cr} = 1.2(128)(0.785)/12 = 10.05 \text{ kip-ft/ft}$$

$$\text{or } 1.33M_u = 1.33(10.14) = 13.5 \text{ kip-ft/ft}$$

therefore,

$$\text{pos } M_u = 10.14 \text{ kip-ft/ft} \qquad d_{pos} = 6.19 \text{ in.}$$

$$\text{trial } A_s \approx \frac{M_u}{4d} = \frac{10.14}{4\,(6.19)} = 0.41 \text{ in.}^2/\text{ft}$$

From Appendix B, Table B.4, try No. 5 at 9 in., provided $A_s = 0.41$ in.^{2}/ft:

$$a = \frac{A_s f_y}{0.85 f_c' b} = \frac{0.41\,(60)}{0.85\,(4.5)\,(12)} = 0.536 \text{ in.}$$

Check ductility.

$$a \le 0.375\beta_1 d = 0.375\,(0.825)\,(6.19) = 1.92 \text{ in.}, \quad \phi = 0.9$$

Check moment strength

$$\begin{aligned}\phi M_n &= \phi A_s f_y \left(d - \frac{a}{2}\right) \\ &= 0.9(0.41)(60)\left(6.19 - \frac{0.54}{2}\right)\Big/12 \\ &= 10.92 \text{ kip-ft/ft} > 10.14 \text{ kip-ft/ft} \quad \text{OK}\end{aligned}$$

For transverse bottom bars,

Use No. 5 at 9 in. $A_s = 0.41$ in.^{2}/ft

2. *Negative Moment Reinforcement*

$$\text{neg } |M_u| = |M_{200.73}| = 8.84 \text{ kip-ft/ft}$$
$$d_{\text{neg}} = 5.19 \text{ in.}$$

$$\min M_u \text{ lessor of } 1.2M_{\text{cr}} = 1.2\,(128)\,(0.785)\,/12 = 10.05 \text{ kip-ft/ft}$$

$$\text{or} \quad 1.33\,|M_u| = 1.33(8.84) = 11.6 \text{ kip-ft/ft}$$

therefore,

$$\text{neg } |M_u| = 10.05 \text{ kip-ft/ft}$$
$$\text{trial } A_s \approx \frac{10.05}{4\,(5.19)} = 0.48 \text{ in.}^2/\text{ft}$$

From Table B.4, try No. 5 at 7.5 in., provided $A_s =$ 0.49 in.^{2}/ft:

$$a = \frac{0.49\,(60)}{0.85\,(4.5)\,(12)} = 0.64 \text{ in.}$$
$$a \le 0.375\beta_1 d = 0.375\,(0.825)\,(5.19) = 1.60 \text{ in.}, \quad \phi = 0.9$$

Check moment strength

$$\begin{aligned}\phi M_n &= 0.9(0.49)(60)\left(5.19 - \frac{0.64}{2}\right)\Big/12 \\ &= 10.74 \text{ kip-ft/ft} > 10.05 \text{ kip-ft/ft} \quad \text{OK}\end{aligned}$$

For transverse top bars,

Use No. 5 at 7.5 in. $A_s = 0.49$ in.^{2}/ft

3. *Distribution Reinforcement* Secondary reinforcement is placed in the bottom of the slab to distribute wheel loads in the longitudinal direction of the bridge to the primary reinforcement in the transverse direction. The required area is a percentage of the primary positive moment reinforcement. For primary reinforcement perpendicular to traffic [A9.7.3.2]

$$\text{Percentage} = \min\left(\frac{220}{\sqrt{S_e}}, 67\%\right)$$

where S_e is the effective span length (ft) [A9.7.2.3]. For monolithic T-beams, S_e is the distance face to face of stems, that is, $S_e = 8.0 - \frac{14}{12} = 6.83$ ft, and

$$\text{Percentage} = \min\left(\frac{220}{\sqrt{6.83}} = 84\%, 67\%\right), \quad \text{Use } 67\%$$

$$\begin{aligned}\text{Distribution } A_s &= 0.67(\text{pos } A_s) = 0.67(0.41) \\ &= 0.27 \text{ in.}^2/\text{ft}\end{aligned}$$

For longitudinal bottom bars,

Use No. 4 at 8 in., $A_s = 0.29$ in.^{2}/ft

4. *Shrinkage and Temperature Reinforcement* The minimum amount of reinforcement in each direction shall be [A5.10.8]

$$0.11 \le \frac{1.3bh}{2\,(b+h)\,f_y} \le A_s\,(\text{temp}) \le 0.60$$

Substitution gives,

$$\begin{aligned}\text{temp } A_s &\ge \frac{1.3\,(12)\,(8)}{2\,(12+8)\,(60)} \\ &= 0.052 \text{ in./ft} \quad \text{Use } 0.11 \text{ in./ft}\end{aligned}$$

The primary and secondary reinforcement selected provide more than this amount; however, for members greater than 6.0 in. in thickness, the shrinkage and temperature reinforcement is to be distributed equally on both faces. The maximum spacing of this reinforcement is 3.0 times the slab thickness or 18.0 in. For the top face longitudinal bars,

$$\tfrac{1}{2}(\text{temp } A_s) = 0.09 \text{ in.}^2/\text{ft}$$

Use No. 4 at 18 in., provided $A_s = 0.13$ in.^{2}/ft

G. ***Control of Cracking—General*** Cracking is controlled by limiting the spacing in the reinforcement under service loads [A5.7.3.4]:

$$s \le \frac{700\gamma_e}{\beta_s f_s} - 2d_c$$

where

$$\beta_s = 1 + \frac{d_c}{0.7(h - d_c)}$$

γ_e = exposure factor
= 1.00 for class 1 exposure condition
= 0.75 for class 2 exposure condition

d_c = depth of concrete cover from extreme tension fiber to center of closest flexural reinforcement (in.)
f_s = tensile stress in reinforcement at the service limit state (ksi)
h = overall thickness or depth of the component (in.)

Service I limit state applies to the investigation of cracking in reinforced concrete structures [A3.4.1]. In the service I limit state, the load modifier η_i is 1.0 and the load factors for dead and live loads are 1.0. Recall IM = 1.33. Therefore, the moment used to calculate the tensile stress in the reinforcement is

$$M = M_{\text{DC}} + M_{\text{DW}} + 1.33 M_{\text{LL}}$$

The calculation of service load tensile stress in the reinforcement is based on transformed elastic, cracked section properties [A5.7.1]. The modular ratio $n = E_s/E_c$ transforms the steel reinforcement into equivalent concrete. The modulus of elasticity E_s of steel bars is 29,000 ksi [A5.4.3.2]. The modulus of elasticity E_c of concrete is given by [A5.4.2.4]

$$E_c = 33,000 K_1 w_c^{1.5} \sqrt{f_c'}$$

where

K_1 = correction factor for source of aggregate
w_c = unit weight of concrete (kcf)

For normal-weight concrete and $K_1 = 1.0$

$$E_c = 1820\sqrt{f_c'}$$

so that

$$E_c = 1820\sqrt{4.5} = 3860 \text{ ksi}$$

and

$$\eta = \frac{29,000}{3860} = 7.5 \qquad \underline{\text{Use } n = 7}$$

1. *Check of Positive Moment Reinforcement* Service I positive moment at location 204 is

$$\begin{aligned} M_{204} &= M_{\text{DC}} + M_{\text{DW}} + 1.33 M_{\text{LL}} \\ &= (0.494 - 0.294 - 0.446) + 0.138 \\ &\quad + 1.33(4.29) \\ &= 5.60 \text{ kip-ft/ft} \end{aligned}$$

The calculation of the transformed section properties is based on a 1.0-ft-wide doubly reinforced section as shown in Figure E15.1-12. Because of its relatively large cover, the top steel is assumed to be

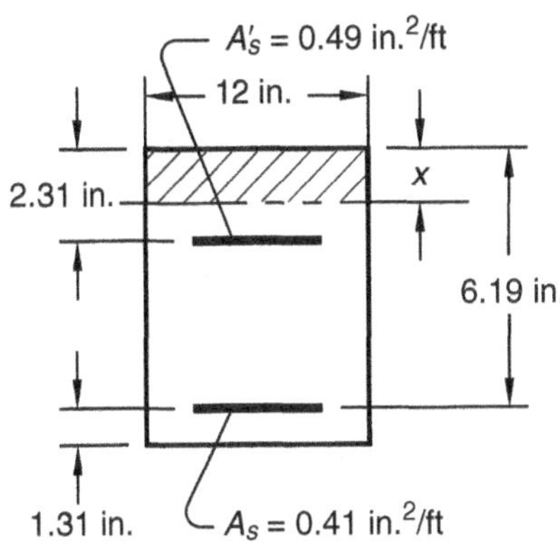

Fig. E15.1-12 Positive moment cracked section.

on the tensile side of the neutral axis. The sum of statical moments about the neutral axis yields

$$\begin{aligned} 0.5bx^2 &= nA_s'\left(d' - x\right) + nA_s(d - x) \\ 0.5(12)x^2 &= 7(0.49)(2.31 - x) \\ &\quad + 7(0.41)(6.19 - x) \\ x^2 + 1.05x - 4.28 &= 0 \end{aligned}$$

Solve $x = 1.61$ in., which is less than 2.31 in., so the assumption is correct. The moment of inertia of the transformed cracked section is

$$\begin{aligned} I_{\text{cr}} &= \frac{bx^3}{3} + nA_s'\left(d' - x\right)^2 + nA_s(d - x)^2 \\ &= \frac{12(1.61)^3}{3} + 7(0.49)(2.31 - 1.61)^2 \\ &\quad + 7(0.41)(6.19 - 1.61)^2 \\ &= 78.58 \text{ in.}^4/\text{ft} \end{aligned}$$

and the tensile stress in the bottom steel becomes

$$\begin{aligned} f_s &= n\left(\frac{My}{I_{\text{cr}}}\right) = 7\left[\frac{5.60(12)(6.19 - 1.61)}{78.58}\right] \\ &= 27.4 \text{ ksi} \end{aligned}$$

(The tensile stress was also calculated using a singly reinforced section and was found to be 28.8 ksi. The contribution of the top bars is small and can be safely neglected.)

The positive moment tensile reinforcement of No. 5 bars at 9 in. on center is located 1.31 in. from the extreme tension fiber. Therefore,

$$d_c = 1.31 \text{ in.}$$

and

$$\beta_s = 1 + \frac{1.31}{0.7(8.0 - 1.31)} = 1.28$$

For class 2 exposure conditions, $\gamma_e = 0.75$ so that

$$\begin{aligned} s_{\max} &= \frac{700(0.75)}{1.28(27.4)} - 2(1.31) \\ &= 12.3 \text{ in.} > 9.0 \text{ in.} \quad \text{OK} \quad \text{Use No. 5 at 9 in.} \end{aligned}$$

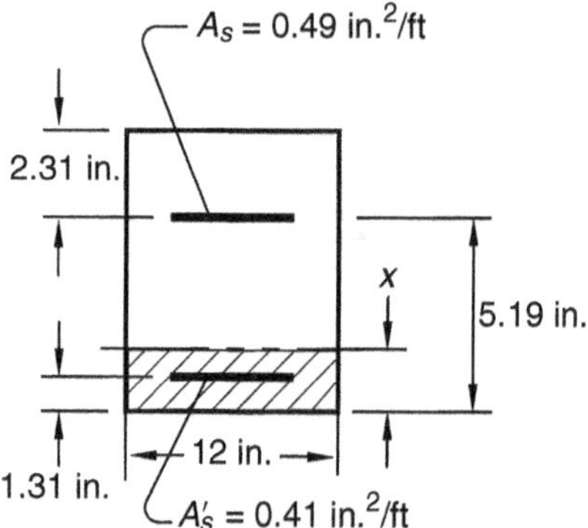

Fig. E15.1-13 Negative moment cracked section.

2. *Check of Negative Moment Reinforcement* Service I negative moment at location 200.73 is

$$
\begin{aligned}
M_{200.73} &= M_{\text{DC}} + M_{\text{DW}} + 1.33M_{\text{LL}} \\
&= (0.166 - 0.541 - 0.823) + (-0.005) \\
&\quad + 1.33(-3.128) \\
&= -5.36 \text{ kip-ft/ft}
\end{aligned}
$$

The cross section for negative moment is shown in Figure E15.1-13 with compression in the bottom. This time x is assumed greater than $d' = 1.31$ in., so that the bottom steel is in compression. Balancing statical moments about the neutral axis gives

$$
\begin{aligned}
&0.5bx^2 + (n-1)A_s'\left(x - d'\right) = nA_s(d - x) \\
&0.5(12)x^2 + (6)(0.41)(x - 1.31) \\
&\qquad = 7(0.49)(5.19 - x) \\
&x^2 + 0.982x - 3.503 = 0
\end{aligned}
$$

Solve $x = 1.44$ in., which is greater than 1.31 in., so the assumption is correct. The moment of inertia of the transformed cracked section becomes

$$
\begin{aligned}
I_{\text{cr}} &= \tfrac{1}{3}(12)(1.44)^3 + 6(0.41)(1.44 - 1.31)^2 \\
&\quad + 7(0.49)(5.19 - 1.44)^2 = 60.2 \text{ in.}^4/\text{ft}
\end{aligned}
$$

and the tensile stress in the top steel is

$$
f_s = 7\frac{(e)\,5.36(12)(5.19 - 1.44)}{60.2} = 28.0 \text{ ksi}
$$

(The tensile stress was calculated to be 27.9 ksi by using a singly reinforced section. There really is no need to do a doubly reinforced beam analysis.)

The negative moment tensile reinforcement of No. 5 bars at 7.5 in. on centers is located 2.31 in. from the tension face. Therefore, $d_c = 2.31$ in., and

$$
\beta_s = 1 + \frac{2.31}{0.7(8.0 - 2.31)} = 1.58
$$

For class 2 exposure conditions, $\gamma_e = 0.75$

$$
\begin{aligned}
s_{\max} &= \frac{700(0.75)}{1.58(28.0)} - 2(2.31) \\
&= 7.3 \text{ in.} \approx s = 7.5 \text{ in.}
\end{aligned}
$$

For class 1 exposure conditions, $\gamma_e = 1.00$

$$
\begin{aligned}
s_{\max} &= \frac{700(1.00)}{1.58(28.0)} - 2(2.31) \\
&= 11.20 \text{ in.} > s = 7.5 \text{ in.}
\end{aligned}
$$

Use No. 5 at 7.5 in.

H. ***Fatigue Limit State*** Fatigue need not be investigated for concrete decks in multigirder applications [A9.5.3].

I. ***Traditional Design for Interior Spans*** The design sketch in Figure E15.1-14 summarizes the arrangement of the transverse and longitudinal reinforcement in four layers for the interior spans of the deck. The exterior span and deck overhang have special requirements that must be dealt with separately.

J. ***Empirical Design of Concrete Deck Slabs*** Research has shown that the primary structural action of concrete decks is not flexure but internal arching. The arching creates an internal compressive dome. Only a minimum amount of isotropic reinforcement is required for local flexural resistance and global arching effects [C9.7.2.1].

1. *Design Conditions* [A9.7.2.4] Design depth subtracts the loss due to wear, h = 7.5 in. The following conditions must be satisfied:
   - ☐ Diaphragms are used at lines of support, YES
   - ☐ Supporting components are made of steel and/or concrete, YES
   - ☐ The deck is of uniform depth, YES
   - ☐ The deck is fully CIP and water cured, YES

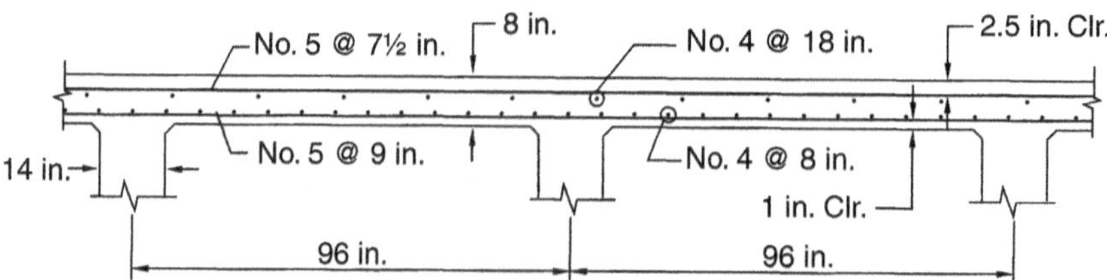

Fig. E15.1-14 Traditional design of interior deck spans.

- ☐ 6.0 < Se/h = 82/7.5 = 10.9 < 18.0, OK
- ☐ Core depth = 8.0 − 2.5 − 1.0 = 4.5 in. > 4 in., OK
- ☐ Effective length [A9.7.2.3] $= \frac{82}{12} = 6.83$ ft < 13.5 ft, OK
- ☐ Minimum slab depth = 7.0 in. < 7.5 in., OK
- ☐ Overhang = 39.0 in. > 5h = 5 × 7.5 = 37.7 in., OK
- ☐ $f_c' = 4.5$ ksi > 4.0 ksi, OK
- ☐ Deck must be made composite with girder, YES

2. *Reinforcement Requirements* [A9.7.2.5]

- ☐ Four layers of isotropic reinforcement, $f_y \geq 60$ ksi
- ☐ Outer layers placed in direction of effective length
- ☐ Bottom layers: min $A_s = 0.27$ in.^{2}/ft, No. 5 at 14 in.
- ☐ Top layers: min $A_s = 0.18$ in.^{2}/ft, No. 4 at 13 in.
- ☐ Max spacing = 18.0 in.
- ☐ Straight bars only, hooks allowed, no truss bars
- ☐ Lap splices and mechanical splices permitted
- ☐ Overhang designed for [A9.7.2.2 and A3.6.1.3.4]:
  - ☐ Wheel loads using equivalent strip method if barrier discontinuous
  - ☐ Equivalent line loads if barrier continuous
  - ☐ Collision loads using yield line failure mechanism [A.A13.2]

3. *Empirical Design Summary* With the empirical design approach analysis is not require. When the design conditions have been met, the minimum reinforcement in all four layers is predetermined. The design sketch in Figure E15.1-15 summarizes the reinforcement arrangement for the interior deck spans.

K. ***Comparison of Reinforcement Quantities*** The weight of reinforcement for the traditional and empirical design methods are compared in Table E15.1-1 for a 1.0-ft-wide by 40-ft-long transverse strip. Significant savings, in this case 67% of the traditionally designed reinforcement, can be made by adopting the empirical design method.

L. ***Deck Overhang Design*** Neither the traditional method nor the empirical method for the design of deck slabs includes the design of the deck overhang. The design loads for the deck overhang [A9.7.1.5 and A3.6.1.3.4] are applied to a free-body diagram of a cantilever that is independent of the deck spans. The resulting overhang design can then be incorporated into either the traditional or empirical design by anchoring the overhang reinforcement into the first deck span.

Two limit states must be investigated: strength I [A13.6.1] and extreme event II [A13.6.2]. The strength limit state considers vertical gravity forces and it seldom governs, unless the cantilever span is very long. The extreme event limit state considers horizontal forces caused by collision of a vehicle with the barrier. [These forces are given in Appendix A of Section 13 of the AASHTO (2010) LRFD Bridge Specifications; reference to articles here is preceded by the letters AA.] The extreme event limit state usually governs the design of the deck overhang.

1. *Strength I Limit State* The design negative bending moment is taken at the exterior face of the support shown in Figure E15.1-6 for the loads given in

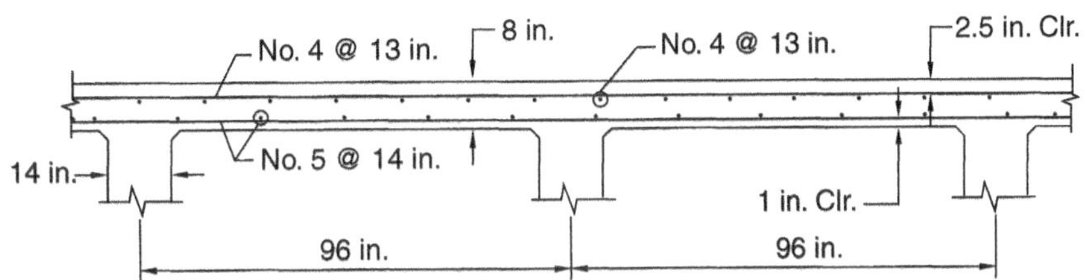

Fig. E15.1-15 Empirical design of interior deck spans.

Table E15.1-1 Comparison of Reinforcement Quantities[a]

	Transverse		Longitudinal		Totals	
Design Method	Top	Bottom	Top	Bottom	(lb)	(psf)
Traditional	No. 5 at 7.5 in.	No. 5 at 9 in.	No. 4 at 18 in.	No. 4 at 8 in.		
Weight (lb)	66.8	55.6	17.8	40.1	180.3	4.51
Empirical	No. 4 at 13 in.	No. 5 at 14 in.	No. 4 at 13 in.	No. 5 at 14 in.		
Weight (lb)	24.7	35.8	24.7	35.8	121.0	3.03

[a] Area = 1 ft × 40 ft.

Figure E15.1-10. Because the overhang has a single load path, it is a nonredundant member so that $\eta_R = 1.05$ [A1.3.4] and, for all load cases $\eta_i = \eta_D \eta_R \eta_I = (1.00)(1.05)(1.00) = 1.05$.

The individual cantilever bending moments for a 1-ft-wide design strip are

$$\begin{aligned} M_b &= -P_b\,(39.0 - 7.0 - 5.0)\,/12 \\ &= -0.320(27.0/12) \\ &= -0.720 \text{ kip-ft/ft} \\ M_o &= -w_o(39.0 - 7.0)^2/2/12^2 \\ &= -0.113(32.0)^2/2/144 \\ &= -0.402 \text{ kip-ft/ft} \\ M_{\text{DW}} &= -w_{\text{DW}}(39.0 - 7.0 - 15.0)^2/2/12^2 \\ &= -0.035(17.0)^2/2/144 \\ &= -0.035 \text{ kip-ft/ft} \\ M_{\text{LL}} &= -W(19.0 - 14.0)/12 = -4.19(5.0)/12 \\ &= -1.746 \text{ kip-ft/ft} \end{aligned}$$

The factored design moment at location 108.2 (exterior face) becomes for the common value of $\eta_i = \eta$

$$\begin{aligned} M_{108.2} &= \eta[1.25M_{\text{DC}} + 1.50M_{\text{DW}} + 1.75(1.33M_{\text{LL}})] \\ &= 1.05[1.25(-0.720 - 0.402) + 1.50(-0.035) \\ &\qquad + 1.75 \times 1.33\,(-1.746)] \\ &= -5.79 \text{ kip-ft/ft} \end{aligned}$$

When compared to the previously determined negative bending moment at the centerline of the support ($M_{200} = -11.74$ kip-ft/ft), the reduction in negative bending to the face of the support is significant. This reduced negative bending moment is *not* critical in the design of the overhang.

2. *Extreme Event II Limit State* The forces to be transmitted to the deck overhang due to a vehicular collision with the concrete barrier are determined from a strength analysis of the barrier. In this example, the loads applied to the barrier are for test level TL-4, which is suitable for [A13.7.2] *high-speed highways, freeways, expressways, and interstate highways with a mixture of trucks and heavy vehicles*.

The minimum edge thickness of the deck overhang is 8.0 in. [A13.7.3.1.2] and the minimum height of barrier for TL-4 is 32.0 in. [A13.7.3.2]. The design forces for TL-4 that must be resisted by the barrier and its connection to the deck are given in Table E15.1-2 [Table AA13.2-1]† and illustrated in Figure E15.1-16. The transverse and longitudinal forces are distributed over a length of barrier of 3.5 ft.

†Reference to articles in Appendix A of AASHTO Section 13 are preceded by the letters AA.

Table E15.1-2 Design Forces for a TL-4 Barrier

Direction	Force (kip)	Length (ft)
Transverse	54.0	3.5
Longitudinal	18.0	3.5
Vertical	18.0	18.0

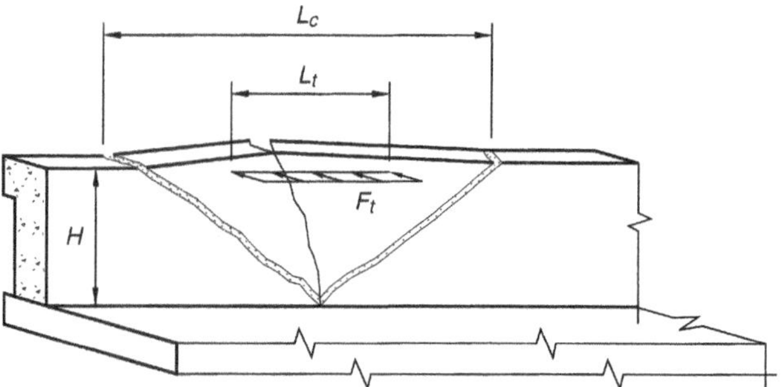

Fig. E15.1-16 Loading and yield-line pattern for concrete barrier.

This length represents the approximate diameter of a truck tire, which is in contact with the wall at time of impact. The vertical force distribution length represents the contact length of a truck lying on top of the barrier after a collision. The design philosophy is that if any failures are to occur they should be in the barrier, which can be readily repaired, rather than in the deck overhang. The procedure is to calculate the barrier strength and then to design the deck's overhang so that it is stronger. When calculating the resistance to extreme event limit states, the resistance factors ϕ are taken as 1.0 [A1.3.2.1] and the vehicle collision load factor is 1.0 [Tables A3.4.1-1 and A13.6.2].

M. ***Concrete Barrier Strength*** All traffic railing systems shall be proven satisfactory through crash testing for a desired test level [A13.7.3.1]. If a previously tested system is used with only minor modifications that do not change its performance, additional crash testing is not required [A13.7.3.1.1]. The concrete barrier and its connection to the deck overhang shown in Figure E15.1-17 is similar to the profile and reinforcement arrangement of traffic barrier type T5 analyzed by Hirsch (1978) and tested by Buth et al. (1990).

As developed by a yield-line approach in Section11.3, the following expressions [AA13.3.1] can be used to check the strength of the concrete barrier away from an end or joint and to determine the magnitude of the loads that must be transferred to the deck overhang. From Eqs. 15.12 and 15.10:

$$R_w = \left(\frac{2}{2L_c - L_t}\right)\left(8M_b + 8M_w + \frac{M_c L_c^2}{H}\right) \tag{E15.1-8}$$

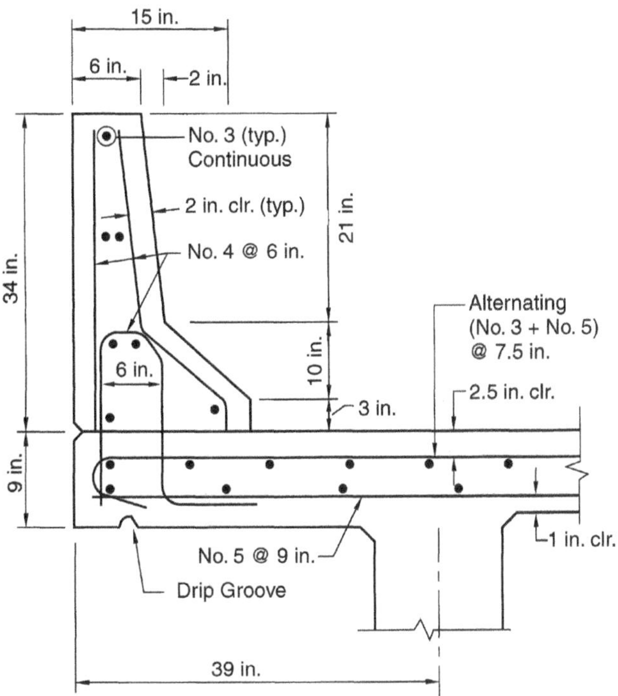

Fig. E15.1-17 Concrete barrier and connection to deck overhang.

$$L_c = \frac{L_t}{2} + \sqrt{\left(\frac{L_t}{2}\right)^2 + \frac{8H\left(M_b + M_w\right)}{M_c}} \qquad \text{(E15.1-9)}$$

where

H = height of wall (ft)

L_c = critical length of yield-line failure pattern (ft)

L_t = longitudinal distribution length of impact force (ft)

M_b = additional flexural resistance of beam, if any, at top of wall (kip-ft)

M_c = flexural resistance of wall about an axis parallel to the longitudinal axis of the bridge (kip-ft/ft)

M_w = flexural resistance of wall about vertical axis (kip-ft)

R_w = nominal railing resistance to transverse load (kips)

For the barrier wall in Figure E15.1-17, $M_b = 0$ and $H = 34.0/12 = 2.83$ ft.

1. *Flexural Resistance of Wall about Vertical Axis, M_w* The moment strength about the vertical axis is based on the horizontal reinforcement in the wall. Both the positive and negative moment strengths must be determined because the yield-line mechanism develops both types (Fig. E15.1-16). The thickness of the barrier wall varies, and it is convenient to divide it for calculation purposes into three segments as shown in Figure E15.1-18.

   Neglecting the contribution of compressive reinforcement, the positive and negative bending strengths of segment I are approximately equal and calculated as

$$\left(f_c' = 4 \text{ ksi}, f_y = 60 \text{ ksi}\right)$$

$$A_s = 2 - \text{No. } 3'\text{s} = 2\,(0.11) = 0.22 \text{ in.}^2$$

$$d_{\text{avg}} = \frac{3.0 + 2.75 + 1.375}{2} = 3.56 \text{ in.}$$

$$a = \frac{A_s f_y}{0.85 f_c' b} = \frac{0.22\,(60)}{0.85\,(4)\,(21.0)} = 0.185 \text{ in.}$$

$$\phi M_{n_1} = \phi A_s f_y \left(d - \frac{a}{2}\right)$$
$$= 1.0(0.22)(60)\,(3.56 - 0.185/2)\,/12$$
$$= 3.81 \text{ kip-ft}$$

For segment II, the moment strengths are slightly different. Considering the moment positive

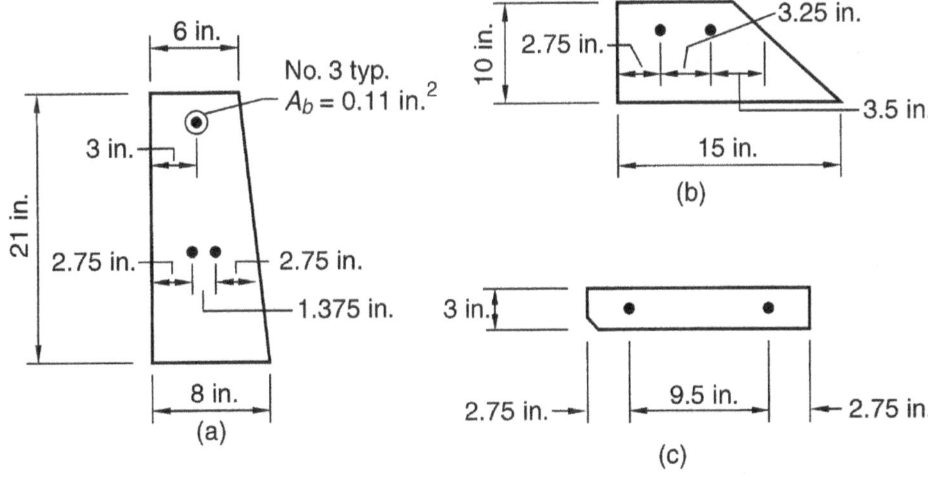

Fig. E15.1-18 Approximate location of horizontal reinforcement in barrier wall: (a) segment I, (b) segment II, and (c) segment III.

if it produces tension on the straight face, we have

$$A_s = 1 - \text{No.3} = 0.11\ \text{in.}^2$$
$$d_{\text{pos}} = 3.25 + 3.50 = 6.75\ \text{in.}$$
$$a = \frac{0.11\,(60)}{0.85\,(4)\,(10.0)} = 0.194\ \text{in.}$$
$$\phi M_{n_{\text{pos}}} = 1.0\,(0.11)\,(60)\left(6.75 - \frac{0.194}{2}\right)\Big/12$$
$$= 3.66\ \text{kip-ft}$$
$$d_{\text{neg}} = 2.75 + 3.25 = 6.0\ \text{in.}$$
$$\phi M_{n_{\text{neg}}} = 1.0\,(0.11)\,(60)\left(6.0 - \frac{0.194}{2}\right)\Big/12$$
$$= 3.25\ \text{kip-ft}$$

and the average value is

$$\phi M_{n_{\text{II}}} = \frac{\phi M_{n_{\text{pos}}} + \phi M_{n_{\text{neg}}}}{2} = 3.45\ \text{kip-ft}$$

For segment III, the positive and negative bending strengths are equal and

$$A_s = 1 - \text{No.3} = 0.11\ \text{in.}^2$$
$$d = 9.50 + 2.75 = 12.25\ \text{in.}$$
$$a = \frac{0.11\,(60)}{0.85\,(4)\,(3.0)} = 0.647\ \text{in.}$$
$$\phi M_{n_{\text{III}}} = 1.0\,(0.11)\,(60)\left(12.25 - \frac{0.647}{2}\right)\Big/12$$
$$= 6.56\ \text{kip-ft}$$

The total moment strength of the wall about the vertical axis is the sum of the strengths in the three segments:

$$M_w = \phi M_{n_{\text{I}}} + \phi M_{n_{\text{II}}} + \phi M_{n_{\text{III}}}$$
$$= 3.81 + 3.45 + 6.56 = 13.82\ \text{kip-ft}$$

It is interesting to compare this value of M_w with one determined by simply considering the wall to have uniform thickness and the same area as the actual wall, that is,

$$h_{\text{ave}} = \frac{\text{cross-sectional area}}{\text{height of wall}} = \frac{307}{34.0} = 9.03\ \text{in.}$$
$$d_{\text{ave}} = 9.03 - 2.75 = 6.28\ \text{in.}$$
$$A_s = 4 - \text{No.3's} = 4\,(0.11) = 0.44\ \text{in.}^2$$
$$a = \frac{0.44\,(60)}{0.85\,(4)\,(34.0)} = 0.228\ \text{in.}$$
$$M_w = \phi M_n = \phi A_s f_y \left(d - \frac{a}{2}\right)$$
$$= 1.0\,(0.44)\,(60)\left(6.28 - \frac{0.228}{2}\right)\Big/12$$
$$M_w = 13.56\ \text{kip-ft}$$

This value is acceptably close to that calculated previously and is calculated with a lot less effort.

2. *Flexural Resistance of Wall about an Axis Parallel to the Longitudinal Axis of the Bridge, M_c* The bending strength about the horizontal axis is determined from the vertical reinforcement in the wall. The yield lines that cross the vertical reinforcement (Fig. E15.1-16) produce only tension in the sloping face of the wall, so that only the negative bending strength need be calculated.

   The depth to the vertical reinforcement increases from bottom to top of the wall, therefore, the moment strength also increases from bottom to top. For vertical bars in the barrier, try No. 4 bars at 6 in. (A_s = 0.39 in.^{2}/ft). For segment I, the average wall thickness is 7 in. and the moment strength for a 1-ft-wide strip about the horizontal axis becomes

$$d = 7.0 - 2.0 - 0.25 = 4.75\ \text{in.}$$
$$a = \frac{A_s f_y}{0.85 f_c' b} = \frac{0.39\,(60)}{0.85\,(4)\,(12)} = 0.574\ \text{in.}$$
$$M_{c_{\text{I}}} = \phi A_s f_y \left(d - \frac{a}{2}\right)$$
$$= 1.0\,(0.39)\,(60)\left(4.75 - \frac{0.574}{2}\right)\Big/12$$
$$= 8.70\ \text{kip-ft/ft}$$

At the bottom of the wall the vertical reinforcement at the wider spread is not anchored into the deck overhang. Only the hairpin dowel at a narrower spread is anchored. The bending strength about the horizontal axis for segments II and III may increase slightly where the vertical bars overlap, but it is reasonable to assume it is constant and determined by the hairpin dowel. The effective depth for the tension leg of the hairpin dowel is (Fig. E15.1-17)

$$d = 2.0 + 0.50 + 6.0 + 0.25 = 8.75\ \text{in.}$$

and

$$M_{c_{\text{II+III}}} = 1.0\,(0.39)\,(60)\left(8.75 - \frac{0.574}{2}\right)\Big/12$$
$$= 16.50\ \text{kip-ft/ft}$$

A weighted average for the moment strength about the horizontal axis is given by

$$M_c = \frac{M_{c_{\text{I}}}\,(21.0) + M_{c_{\text{II+III}}}\,(10.0 + 3.0)}{34.0}$$
$$= \frac{8.70\,(21.0) + 16.50\,(13.0)}{34.0}$$
$$M_c = 11.68\ \text{kip-ft/ft}$$

3. *Critical Length of Yield-Line Failure Pattern, L_c* With the moment strengths determined and L_t = 3.5 ft,

Eq. E15.1-9 yields

$$L_c = \frac{L_t}{2} + \sqrt{\left(\frac{L_t}{2}\right)^2 + \frac{8H\,(M_b + M_w)}{M_c}}$$

$$= \frac{3.5}{2} + \sqrt{\left(\frac{3.5}{2}\right)^2 + \frac{8\,(2.83)\,(0 + 13.56)}{11.68}}$$

$$L_c = 7.17 \text{ ft}$$

4. *Nominal Resistance to Transverse Load, R_w* From Eq. E15.1-8, we have

$$R_w = \left(\frac{2}{2L_c - L_t}\right)\left(8M_b + 8M_w + \frac{M_c L_c^2}{H}\right)$$

$$= \frac{2}{2\,(7.17) - 3.5}\left[0 + 8\,(13.56) + \frac{11.68(7.17)^2}{2.83}\right]$$

$$= 59.1 \text{ kips} > F_t = 54.0 \text{ kips} \qquad \text{OK}$$

5. *Shear Transfer between Barrier and Deck* The nominal resistance R_w must be transferred across a cold joint by shear friction. Free-body diagrams of the forces transferred from the barrier to the deck overhang are shown in Figure E15.1-19.

Assuming that R_w spreads out at a 1 : 1 slope from L_c, the shear force at the base of the wall from the vehicle collision V_{CT}, which becomes the tensile force T per unit of length in the overhang, is given by [AA13.4.2]

$$T = V_{\text{CT}} = \frac{R_w}{L_c + 2H} \qquad \text{(E15.1-10)}$$

$$T = \frac{59.1}{7.17 + 2\,(2.83)} = 4.61 \text{ kips/ft}$$

The nominal shear resistance V_n of the interface plane is given by [A5.8.4.1]

$$V_n = \min\left\{\left[cA_{\text{cv}} + \mu\left(A_{\text{vf}} f_y + P_c\right)\right],\; K_1 f_c' A_{\text{cv}},\, K_2 A_{\text{cv}}\right\} \qquad \text{(E15.1-11)}$$

where

$$A_{\text{cv}} = \text{shear contact area} = 15(12) = 180 \text{ in.}^2/\text{ft}$$

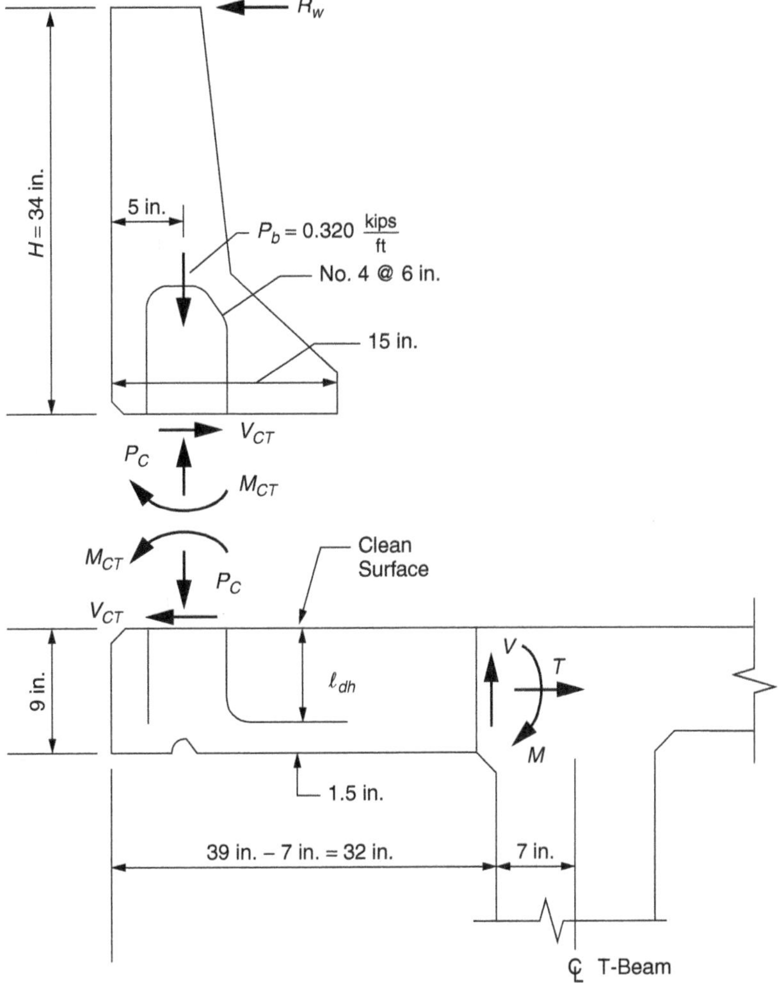

Fig. E15.1-19 Force transfer between barrier and deck.

A_{vf} = dowel area across shear plane = 0.39 in.^{2}/ft

c = cohesion factor [A5.8.4.2] = 0.075 ksi

f'_c = of weaker concrete = 4 ksi

f_y = yield strength of reinforcement = 60 ksi

P_c = permanent compressive force = P_b = 0.320 kips/ft

μ = friction factor [A5.8.4.2] = 0.6.

K_1 = fraction of concrete strength available to resist interface shear, as specified in [A5.8.4.3] = 0.2.

K_2 = limit interface shear resistance factor [A5.8.4.3] = 0.8 ksi

The factors c, μ, K_1, and K_2 are for normal-weight concrete placed against hardened concrete clean and free of laitance but not intentionally roughened. Therefore, for a 1-ft-wide design strip

$$\begin{aligned}
V_n &\le K_1 f'_c A_{\text{cv}} = 0.2(4)(180) = 144 \text{ kips/ft} \\
&\le K_2 A_{\text{cv}} = 0.8(180) = 144 \text{ kips/ft} \\
&\le c A_{\text{cv}} + \mu(A_{\text{vf}} f_y + P_c) \\
&= 0.075(180) + 0.6[0.39(60) + 0.320] \\
&= 13.5 + 14.23 = 27.73 \text{ kips/ft} \\
V_n &= 27.73 \text{ kips/ft} > V_{\text{CT}} = T \\
&= 4.61 \text{ kips/ft} \quad \text{OK}
\end{aligned}$$

where V_{CT} is the shear force produced by a truck collision.

In the above calculations, only one leg of the hairpin is considered as a dowel because only one leg is anchored in the overhang. The minimum cross-sectional area of dowels across the shear plane is [A5.8.4.4]

$$A_{\text{vf}} \ge \frac{0.05 b_v}{f_y} \tag{E15.1-12a}$$

where

$$\begin{aligned}
b_v &= \text{width of interface (in.)} \\
A_{\text{vf}} &\ge \frac{0.05\,(15.0)}{60}\,(12) \\
&= 0.15 \text{ in.}^2/\text{ft} \\
A_{\text{vf}} &\ge \text{steel necessary to carry } \frac{1.33 V_{\text{ui}}}{\phi} \\
&= \frac{1.33\,(4.65)}{0.9} = 6.9 \text{ kips}
\end{aligned} \tag{E15.1-12b}$$

The first term in Eq. E15.1-11 cA_{cv} = 13.5 kips which indicates that the minimum area may be zero by this provision. Therefore, the minimum area is satisfied by the No. 4 bars at 6 in. (A_s = 0.39 in.^{2}/ft).

The basic development length ℓ_{hb} for a hooked bar with f_y = 60 ksi is given by [A5.11.2.4.1]

$$\ell_{\text{hb}} = \frac{38 d_b}{\sqrt{f'_c}} \tag{E15.1-13}$$

and shall not be less than $8d_b$ or 6.0 in. For a No. 4 bar, d_b = 0.5 in. and

$$\ell_{\text{hb}} = \frac{38\,(0.50)}{\sqrt{4.5}} = 8.96 \text{ in.}$$

which is greater than 8(0.50) = 4 in. and 6.0 in. The modification factors of 0.8 for adequate cover and 1.2 for epoxy-coated bars [A5.11.2.4.2] apply, so that the development length ℓ_{dh} is changed to

$$\ell_{\text{dh}} = 0.8\,(1.2)\,\ell_{\text{hb}} = 0.96\,(8.96) = 8.6 \text{ in.}$$

The available development length (Fig. E15.1-19) is 9.0 − 1.5 = 7.5 in., which is not adequate, unless the required area is reduced to

$$\begin{aligned}
A_s \text{ required} &= (A_s \text{ provided})\left(\frac{7.5}{8.6}\right) \\
&= 0.39\frac{7.5}{8.6} = 0.34 \text{ in.}^2/\text{ft}
\end{aligned}$$

By using this area to recalculate M_c, L_c, and R_w, we get

$$\begin{aligned}
a &= \frac{0.34\,(60)}{0.85\,(4)\,(12)} = 0.50 \text{ in.} \\
M_{c\text{I}} &= 1.0\,(0.34)\,(60)\left(4.75 - \frac{0.50}{2}\right)\Big/12 \\
&= 7.65 \text{ kip-ft/ft} \\
M_{c\text{II+III}} &= 1.0\,(0.34)\,(60)\left(8.75 - \frac{0.50}{2}\right)\Big/12 \\
&= 14.45 \text{ kip-ft/ft} \\
M_c &= \frac{7.65\,(21.0) + 14.45\,(13.0)}{34.0} \\
&= 10.25 \text{ kip-ft/ft} \\
L_c &= \frac{3.5}{2} + \sqrt{\left(\frac{3.5}{2}\right)^2 + \frac{8\,(2.83)\,(13.56)}{10.25}} \\
&= 7.50 \text{ ft} \\
R_w &= \frac{2}{2\,(7.50) - 3.5}\left[8\,(13.56) + \frac{10.25(7.50)^2}{2.83}\right] \\
&= 54.3 \text{ kips} \\
R_w &= 54.3 \text{ kips} > 54.0 \text{ kips} \quad \text{OK}
\end{aligned}$$

The standard 90° hook with an extension of $12d_b + 4d_b$ = 16 (0.50) = 8.0 in. at the free end of the bar is adequate [C5.11.2.4.1].

6. *Top Reinforcement in Deck Overhang* The top reinforcement must resist the negative bending moment over the exterior beam due to the collision and the dead load of the overhang. Based on the strength of the 90° hooks, the collision moment M_{CT} (Fig. E15.1-19) distributed over a wall length of $(L_c + 2H)$ is

$$M_{\text{CT}} = -\frac{R_w H}{L_c + 2H} = -\frac{54.3\,(2.83)}{7.50 + 2\,(2.83)}$$
$$= -11.7 \text{ kip-ft/ft}$$

The dead-load moments were calculated previously for strength I so that for the extreme event II limit state, we have

$$M_u = \eta[1.25M_{\text{DC}} + 1.50M_{\text{DW}} + M_{\text{CT}}]$$
$$= 1.0[1.25(-0.720 - 0.402) + 1.50(-0.035) - 11.7]$$
$$= -13.2 \text{ kip-ft/ft}$$

Alternating a No. 3 bar with the No. 5 top bar at 7.5 in. on centers, the negative moment strength becomes

$$A_s = 0.18 + 0.49 = 0.67 \text{ in.}^2/\text{ft}$$
$$d = 9.0 - 2.5 - 0.625/2 = 6.19 \text{ in.}$$
$$a = \frac{0.67\,(60)}{0.85\,(4.5)\,(12)} = 0.88 \text{ in.}$$
$$\phi M_n = 1.0\,(0.67)\,(60)\left(6.19 - \frac{0.88}{2}\right)\Big/12$$
$$= 19.3 \text{ kip-ft/ft}$$

This moment strength is reduced because of the axial tension force $T = R_w/(L_c + 2H)$:

$$T = \frac{54.3}{7.50 + 2\,(2.83)} = 4.13 \text{ kips/ft}$$

By assuming the interaction curve between moment and axial tension is a straight line (Fig. E15.1-20)

$$\frac{P_u}{\phi P_n} + \frac{M_u}{\phi M_n} \le 1.0$$

and solving for M_u, we get

$$M_u \le \phi M_n \left(1.0 - \frac{P_u}{\phi P_n}\right) \qquad \text{(E15.1-14)}$$

where $P_u = T$ and $\phi P_n = \phi A_{\text{st}} f_y$. The total longitudinal reinforcement A_{st} in the overhang is the combined area of the top and bottom bars:

$$A_{\text{st}} = \text{No. 3 at 7.5 in., No. 5 at 7.5 in., No. 5 at 9 in.}$$
$$= 0.18 + 0.49 + 0.41 = 1.08 \text{ in.}^2/\text{ft}$$
$$\phi P_n = 1.0\,(1.08)\,(60) = 64.8 \text{ kips/ft}$$

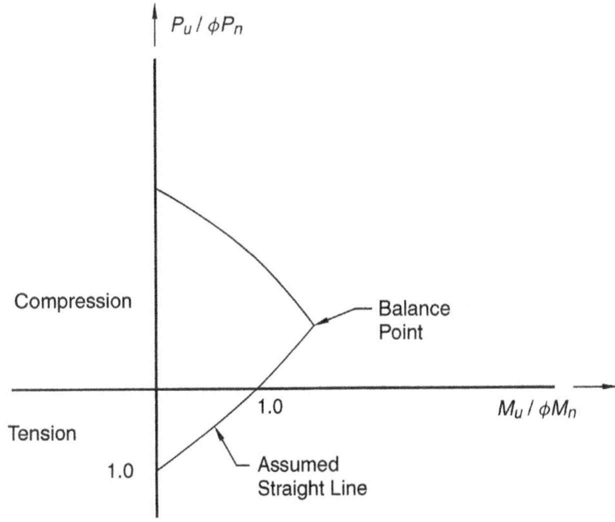

Fig. E15.1-20 Idealized interaction diagram for reinforced concrete members with combined bending and axial load.

so that

$$M_u \le 19.3\left(1.0 - \frac{4.13}{64.8}\right) = 18.1 \text{ kip-ft/ft}$$

The extreme event II design moment $M_u = 13.4$ kip-ft/ft < 18.1 kip-ft/ft, so for the top reinforcement of the overhang

Use alternating (No. 3 and No. 5) at 7.5 in.

The top reinforcement must resist $M_{\text{CT}} = 11.7$ kip-ft/ft directly below the barrier. Therefore, the free ends of the No. 3 and No. 5 bars must terminate in standard 180° hooks. The development length ℓ_{dh} for a standard hook is [A5.11.2.4.1]

$$\ell_{\text{dh}} = \ell_{\text{hb}} \cdot \text{modification factors}$$

The modification factors of 0.8 for adequate cover and 1.2 for epoxy-coated bars [A5.11.2.4.2] apply and the ratio of (A_s required)/(A_s provided) can be approximated by the ratio of (M_u required)/(ϕM_n provided). Thus, the required development length for a No. 5 bar with $\phi = 1.0$ and

$$\ell_{\text{hb}} = \frac{38d_b}{\sqrt{f'_c}} = \frac{38\,(0.625)}{\sqrt{4.5}} = 11.2 \text{ in.}$$
$$\ell_{\text{dh}} = 11.2\,(0.8)\,(1.2)\left(\frac{11.7}{18.1}\right) = 6.95 \text{ in.}$$

The development length available (Fig. E15.1-17) for the hook in the overhang before reaching the vertical leg of the hairpin dowel is

$$\text{Available } \ell_{\text{dh}} = 0.625 + 6.0 + 0.3125$$
$$= 6.94 \text{ in.} \approx 6.95 \text{ in.} \qquad \text{Close, OK}$$

and the connection between the barrier and the overhang shown in Figure E15.1-17 is satisfactory.

7. *Length of the Additional Deck Overhang Bars* The additional No. 3 bars placed in the top of the deck overhang must extend beyond the centerline of the exterior T-beam into the first interior deck span. To determine the length of this extension, it is necessary to find the distance where theoretically the No. 3 bars are no longer required. This theoretical distance occurs when the collision plus dead-load moments equal the negative moment strength of the continuing No. 5 bars at 7.5 in. This negative moment strength in the deck slab ($d = 5.19$ in.) was previously determined as -10.7 kip-ft/ft with $\phi = 0.9$. For the extreme event limit state, $\phi = 1.0$ and the negative moment strength increases to -11.9 kip-ft/ft.

Assuming a carryover factor of 0.5 and no further distribution, the collision moment diagram in the first interior deck span is shown in Figure E15.1-21. At a distance x from the centerline of the exterior T-beam, the collision moment is approximately

$$M_{\text{CT}}(x) = -11.7\left(1 - \frac{x}{5.33}\right)$$

The dead-load moments can be calculated as before from the loadings in Figure E15.1-10.

Barrier $-\ 0.320\left(\frac{34}{12} + x\right) + 0.464x$

Overhang $-\ (0.113)(3.25)(3.25/2 + x) + 0.462x$

Deck slab $-\ \left(0.100x^2/2\right) + 0.314x$

Future wearing surface is conservative to neglect

The distance x is found by equating the moment strength of -11.9 kip-ft/ft to the extreme event II load combination, that is,

$$\begin{aligned} -11.7 = M_u(x) &= \Sigma \eta_i \gamma_i Q_i \\ &= 1.0\left[1.25 M_{\text{DC}}(x) + M_{\text{CT}}(x)\right] \end{aligned}$$

Solve the resulting quadratic, $x = 0.59$ ft, say 7 in.

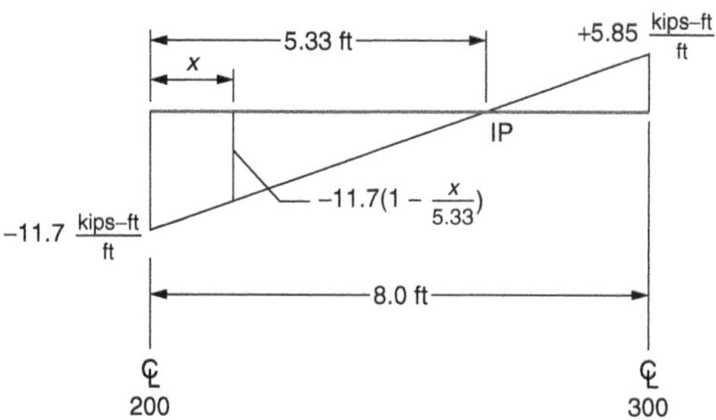

Fig. E15.1-21 Approximate moment diagram for collision forces in first interior deck span.

To account for the uncertainties in the theoretical calculation, an additional length must be added to the length x before the bar can be cut off. This length is the larger of $12d_b = 12(0.375) = 4.5$ in., 0.0625 × span length = 0.625(96)=6.0 in., and the effective depth d = 5.19 in.; 6 in. controls [A5.11.1.3]. This total length of 7 + 6 = 13 in. beyond the centerline must be compared to the development length from the face of the support and the larger length selected.

The basic tension development length ℓ_{db} for a No. 3 bar is the larger of [A5.11.2.1.1]:

$$\ell_{\text{db}} = 1.25\frac{A_b f_y}{\sqrt{f_c'}} = 1.25\frac{0.11\,(60)}{\sqrt{4.5}} = 3.9 \text{ in.}$$

but not less than

$$0.4 d_b f_y = 0.4\,(0.375)\,(60) = 9.0 \text{ in.} \qquad \text{Controls}$$

The modification factor for epoxy-coated bars [A5.11.2.1.2] = 1.2. So that the development length $\ell_d = 9.0(1.2) = 10.8$ in.

The distance from the centerline of the 14-in.-wide T-beam to the end of the development length is 10.8 + 7.0 = 17.8 in., which is greater than the 13 in. calculated from the moment requirement. The length determination of the additional No. 3 bars in the deck overhang is summarized in Figure E15.1-22. Also per [5.11.1.2.3], at least one-third of the tension steel must extend beyond the point of inflection, because the No. 5 bars extend and the No. 3 bars terminate, this is satisfied.

N. ***Closing Remarks*** This example is general in most respects for application to decks supported by different

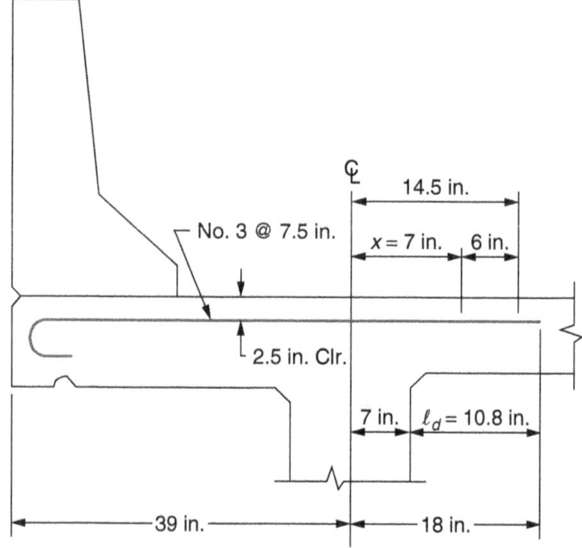

Fig. E15.1-22 Length of additional bars in deck overhang (other reinforcement not shown).

longitudinal girders. However, the effective span length must be adjusted for the different girder flange configurations. Several examples in this book use the same bridge cross section for the deck geometry, so the design is not repeated.

Designers are encouraged to use the empirical design procedure. The savings in design effort and reinforcement can be appreciable. Obviously, the details for the additional bars (Fig. E15.1-22) in the top of the deck overhang will be different for the empirical design than the traditional design. Many owners have standard details for the cantilever, barrier attachment, and deck reinforcement. This eliminates the need for detailed design.

The test level TL-4 chosen for the concrete barrier in this example may have to be increased for some traffic environments. This choice of test level is another decision that must be made when the design criteria for a project are being established.

Finally, the deck reinforcement in skewed regions is increased due to higher flexural (and torsional) actions in the deck. In the case of the empirical design method the steel is doubled [A9.7.1.2 and A9.7.2.5].

REFERENCES

AASHTO (2010). *LRFD Bridge Design Specifications*, 5th ed., American Association of State Highway and Transportation Officials, Washington, DC.

Buth, C. E., T. J. Hirsch, and C. F. McDevitt (1990). "Performance Level 2 Bridge Railings," *Transportaion Research Record 1258*, Transportation Research Board, Washington, DC.

Calloway, B. R. (1993). Yield Line Analysis of an AASHTO New Jersey Concrete Parapet Wall, M.S. Thesis, Virginia Polytechnic Institute and State University, Blacksburg, VA.

Hirsch, T. J. (1978). "Analytical Evaluation of Texas Bridge Rails to Contain Buses and Trucks," *Research Report 230–2*, August, Texas Transportation Institute, Texas A&M University, College Station, TX.

PROBLEMS

15.1 Obtain the profile dimensions and reinforcement pattern for the concrete barrier used by the DOT in your locality. Use yield-line analysis to determine if its lateral load strength is adequate to meet the requirements of test load TL-4. Calculate the lateral load strength considering (a) a uniform thickness barrier wall and (b) a variable thickness barrier wall.

15.2 In Figure E15.1-1, the girder spacing is changed to 4 at 10 ft (five girders) and the overhang changes to 4 ft. The barrier base width remains at 1.25 ft. The curb-to-curb roadway width becomes 45.5 ft. Assume that the supporting girders have a stem width of 14 in., allowance for sacrificial wear is 0.5 in., and the future wearing surface is a 3-in. thick asphalt overlay. Use $f_c' = 4.5$ ksi, $f_y = 60$ ksi to determine the deck thickness and, if the design conditions are met, the reinforcement required for the interior spans by the empirical method.

15.3 For the conditions of Problem 15.2, select the thickness and reinforcement for the deck overhang required to resist the collision force caused by test level TL-4 under extreme event II limit state. Assume the concrete barrier is the one shown in Figure 15.1.

CHAPTER 16

Concrete Design Examples

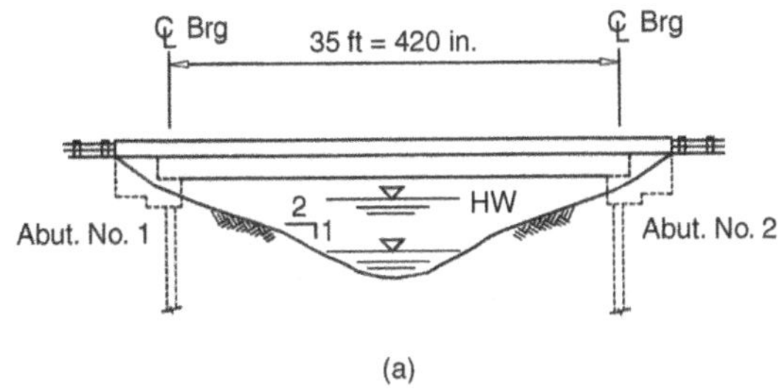

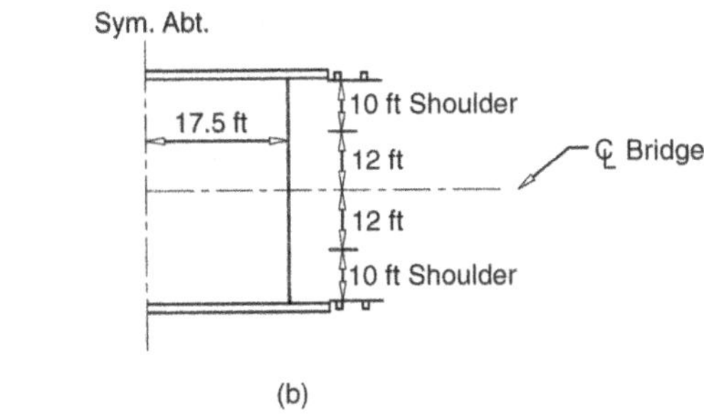

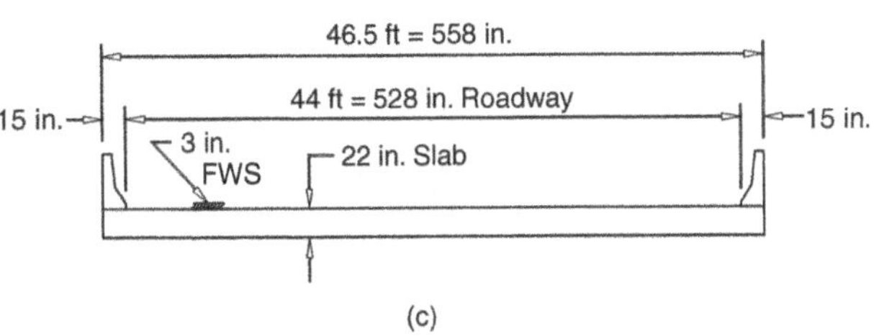

Fig. E16.1-1 Solid slab bridge design example: (a) elevation, (b) plan, and (c) section.

Three typical concrete superstructure designs are given. The first example is the design of a concrete deck followed by design examples of solid slab, T-beam, and prestressed girder bridges.

Table 5.1 describes the notation used to indicate locations of critical sections for moments and shears. This notation is used throughout the example problems.

References to the AASHTO LRFD Specifications (2010) are enclosed in brackets and denoted by the letter A followed by the article number, for example, [A4.6.2.1.3]. If a commentary is cited, the article number is preceded by the letter C. Figures and tables that are referenced are also enclosed in brackets to distinguish them from figures and tables in the text, for example, [Fig. A3.6.1.2.2-1] and [Table A4.6.2.1.3-1].

Appendix B includes tables that may be helpful to a designer when selecting bars sizes and prestressing tendons. These are referenced by the letter B followed by a number and are not enclosed in brackets. The design examples may be reviewed in any order.

The design examples generally follow the outline of Appendix A—Basic Steps for Concrete Bridges given at the end of Section 5 of the AASHTO (2010) LRFD Bridge Specifications. Care has been taken in preparing these examples, but they should not be considered as fully complete in every detail. Each designer must take the responsibility for understanding and correctly applying the provisions of the specifications. Additionally, the AASHTO LRFD Bridge Design Specifications will be altered each year by addendums that define interim versions. The computations outlined herein are based on the 2010 Specifications and may not be current with the most recent interim.

16.1 SOLID SLAB BRIDGE DESIGN

Problem Statement Example E.16.1 Design the simply supported solid slab bridge of Figure E16.1-1 with a span length of 35 ft center to center of bearings for an HL-93 live load. The roadway width is 44 ft curb to curb. Allow for a future wearing surface of 3-in.-thick bituminous overlay. A 15-in.-wide barrier weighing 0.32 k/ft is assumed to be carried by the edge strip. Use $f_c' = 4.5$ ksi and $f_y = 60$ ksi. Follow the slab bridge outline in Appendix A5.4 and the beam and girder bridge outline in Section 5, Appendix A5.3 of the AASHTO (2010) LRFD Bridge Specifications. Use exposure class 2 for crack control.

A. ***Check Minimum Recommended Depth*** [Table A2.5.2 .6.3-1]

$$h_{\min} = \frac{1.2\,(S + 10)}{30} = \frac{1.2\,(35 + 10)}{30}\,(12) = 21.6 \text{ in.}$$

Use $h = 22$ in.

B. ***Determine Live-Load Strip Width*** [A4.6.2.3] Slab bridges are typically designed using a "unit" section width. This width could be an inch, millimeter, foot, or meter. In the United State\, it is customary to use a 1-ft section. The analysis of the permanent loads is straightforward for a design strip as area loads are typically available in load per square foot. However, the live load must be distributed to width of the slab called the live-load strip width. Division of this width into the load effect associated with one lane provides the load per unit width (per foot.). In effect the inverse of this strip width is the distribution factor to the 1-ft section. It is often misunderstood that this procedure is similar to that of a beam line. Moreover, if a standard beam-line program (e.g., BTBeam) is used, then the distribution factor is $mg = 1/E$, where E is the strip width.

Similar to beam line analysis, differences exist between the load distribution on the exterior (or edge) and interior portions of the slab. Below, separate analyses are performed for exterior and interior strips, and, as with beam lines, the exterior strip should have a strength that meets or exceeds the interior strip.

Span = 35 ft, primarily in the direction parallel to traffic

Span > 15 ft, therefore the longitudinal strip method for slab-type bridges applies [A4.6.2.1.2]

1. *One Lane Loaded* Multiple presence factor included [C4.6.2.3]

$$E = \text{equivalent width (in.)}$$

$$E = 10.0 + 5.0\sqrt{L_1 W_1}$$

where

$$L_1 = \text{modified span length} = \min\begin{bmatrix} 35 \text{ ft} \\ 60 \text{ ft} \end{bmatrix} = 35 \text{ ft}$$

$$W_1 = \text{modified edge-to-edge width} = \min\begin{bmatrix} 46.5 \text{ ft} \\ 30 \text{ ft} \end{bmatrix} = 30 \text{ ft}$$

$$E = 10.0 + 5.0\sqrt{(35)(30)} = 172 \text{ in.} = 14.33 \text{ ft}$$

2. *Multiple Lanes Loaded*

$$E = 84.0 + 1.44\sqrt{L_1 W_1} \le \frac{12.0W}{N_L}$$

where $L_1 = 35$ ft.

$$W_1 = \min\begin{bmatrix} 46.5 \text{ ft} \\ 60 \text{ ft} \end{bmatrix} = 46.5 \text{ ft}$$

$$W = \text{actual edge-to-edge width} = 46.5 \text{ ft}$$

$$N_L = \text{number of design lanes [A3.6.1.1.1]} = \text{INT}\left(\frac{w}{12.0}\right)$$

where w = clear roadway width = 44.0 ft

$$N_L = \text{INT}\left(\frac{44.0}{12.0}\right) = 3$$

$$E = 84.0 + 1.44\sqrt{(35)(46.5)} = 142 \text{ in.}$$

$$\le 12.0(46.5)/3 = 186 \text{ in.}$$

$$\underline{\text{Use } E = 142 \text{ in.} = 11.83 \text{ ft}}$$

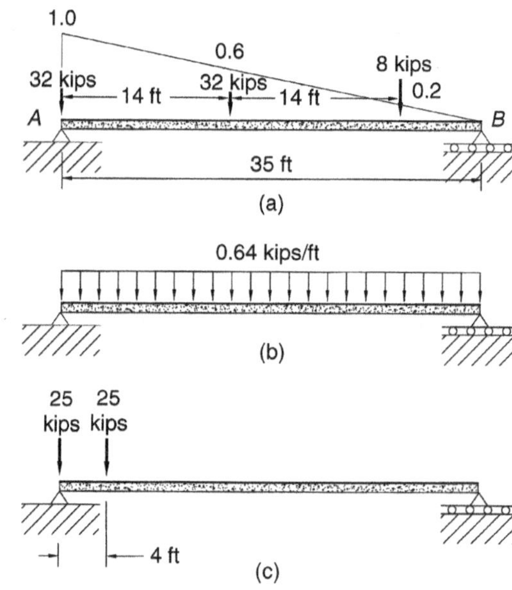

Fig. E16.1-2 Live-load placement for maximum shear force: (a) truck, (b) lane, and (c) tandem.

C. ***Applicability of Live Load for Decks and Deck Systems*** Slab-type bridges shall be designed for all of the vehicular live loads specified in AASHTO [A3.6.1.2], including the lane load [A3.6.1.3.3].

1. *Maximum Shear Force—Axle Loads* (Fig. E16.1-2)
Truck [A3.6.1.2.2]:

$$V_A^{\text{Tr}} = 32(1.0 + 0.60) + 8(0.20) = 52.8 \text{ kips}$$

Lane [A3.6.1.2.4]:

$$V_A^{\text{Ln}} = 0.64(35.0)/2 = 11.2 \text{ kips}$$

Tandem [A3.6.1.2.3]:

$$V_A^{\text{Ta}} = 25\left(1 + \frac{35 - 4}{35}\right) = 47.1 \text{ kips} \quad \text{Not critical}$$

$$\text{Impact factor} = 1 + \text{IM}/100,$$

where IM = 33% [A3.6.2.1]

Impact factor = 1.33, not applied to design lane load

$$V_{\text{LL+IM}} = 52.8(1.33) + 11.2 = 81.4 \text{ kips}$$

2. *Maximum Bending Moment at Midspan—Axle Loads* (Fig. E16.1-3)
Truck:

$$M_c^{\text{Tr}} = 32(8.75 + 1.75) + 8(1.75) = 350 \text{ kip-ft}$$

Lane:

$$M_c^{\text{Ln}} = 0.64(8.75)(35)/2 = 98.0 \text{ kip-ft}$$

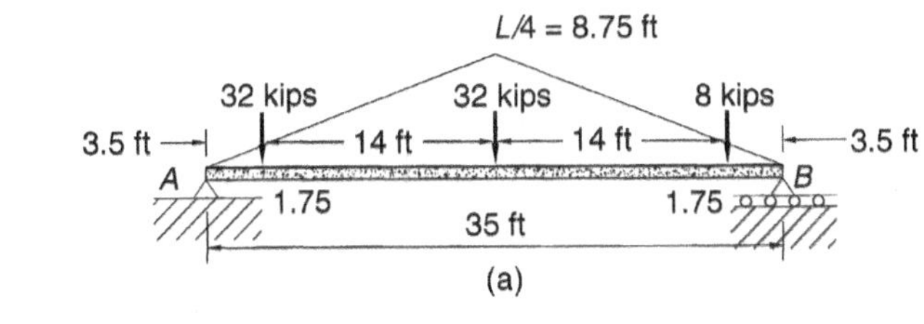

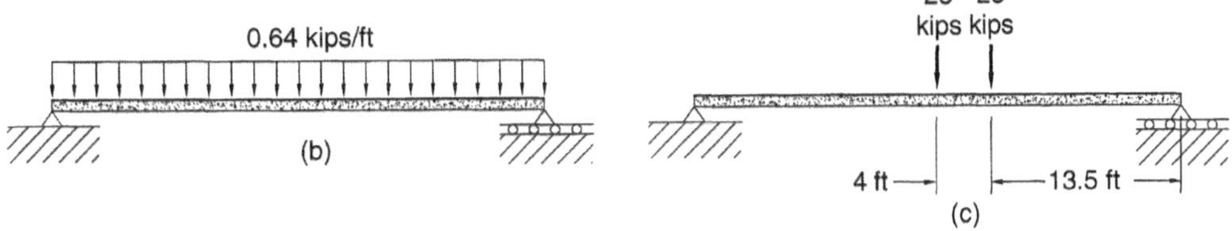

Fig. E16.1-3 Live-load placement for maximum bending moment: (a) truck, (b) lane, and (c) tandem.

Tandem:

$$M_c^{\mathrm{Ta}} = 25\,(8.75)\,(1 + 13.5/17.5)$$
$$= 387.5 \text{ kip-ft} \qquad \text{Governs}$$
$$M_{\mathrm{LL+IM}} = 387.5\,(1.33) + 98.0 = 613.4 \text{ kip-ft}$$

D. ***Select Resistance Factors*** (Table 14.6) [A5.5.4.2.1]

Strength Limit State	ϕ
Flexure and tension	0.90
Shear and torsion	0.90
Axial compression	0.75
Bearing on concrete	0.70
Compression in strut-and-tie models	0.70

E. ***Select Load Modifiers*** [A1.3.2.1]

	Strength	Service	Fatigue	
1. Ductility, η_D	1.0	1.0	1.0	[A1.3.3]
2. Redundancy, η_R	1.0	1.0	1.0	[A1.3.4]
3. Importance, η_I	1.0	N/A[a]	N/A	[A1.3.5]
$\eta_i = \eta_D \eta_R \eta_I$	1.0	1.0	1.0	

[a] N/A = not applicable.

F. ***Select Applicable Load Combinations*** (Table 5.1) [Table A3.4.1-1]
Strength I Limit State $\eta = \eta_i = 1.0$

$$U = 1.0[1.25\mathrm{DC} + 1.50\mathrm{DW} + 1.75(\mathrm{LL} + \mathrm{IM}) + 1.0\mathrm{FR} + \gamma_{\mathrm{TG}}\mathrm{TG}]$$

Service I Limit State

$$U = 1.0\,(\mathrm{DC} + \mathrm{DW}) + 1.0\,(\mathrm{LL} + \mathrm{IM}) + 0.3\,(\mathrm{WS} + \mathrm{WL}) + 1.0\mathrm{FR}$$

Fatigue I Limit State

$$U = 1.5\,(\mathrm{LL} + \mathrm{IM})$$

G. ***Calculate Live-Load Force Effects***

1. *Interior Strip* Shear and moment per lane are given in Section 16.1, Parts C.1 and C.2. Shear and moment per 1.0-ft width of strip is critical for multiple lanes loaded because one-lane live-load strip width = 14.33 ft > 11.83 ft:

$$V_{\mathrm{LL+IM}} = 81.4/11.83 = 6.88 \text{ kip/ft}$$
$$M_{\mathrm{LL+IM}} = 613.4/11.83 = 51.9 \text{ kip-ft/ft}$$

2. *Edge Strip* [A4.6.2.1.4]

Longitudinal edges strip width for a line of wheels
= distance from edge to face of barrier + 12.0 in.
+ (strip width) /4 ≤ (strip width) /2 or 72.0 in.
= 15.0 + 12.0 + 142.0/4 = 62.5 in. < 71.0 in.
Use 62.5 in.

For one line of wheels and a tributary portion of the 10-ft-wide design lane load (Fig. E16.1-4), the shear and moment per foot width of strip are

$$V_{\mathrm{LL+IM}} = [0.5\,(52.8)\,(1.33) + 11.2\,(12.0 + 35.5)/120.0]\,/\,(62.5/12)$$
$$= 7.59 \text{ kips/ft}$$
$$M_{\mathrm{LL+IM}} = [0.5\,(387.5)\,(1.33) + 98.0\,(12.0 + 35.5)/120.0]\,/\,(62.5/12)$$
$$= 56.9 \text{ kip-ft/ft}$$

For one line of wheels taken as one half the actions of the axled vehicle, the shear and moment are

$$V_{\mathrm{LL+IM}} = 0.5\,(81.4)\,/\,(62.5/12)$$
$$= 7.81 \text{ kips/ft}$$
$$M_{\mathrm{LL+IM}} = 0.5\,(613.4)\,/\,(62.5/12)$$
$$= 58.9 \text{ kip-ft/ft}$$

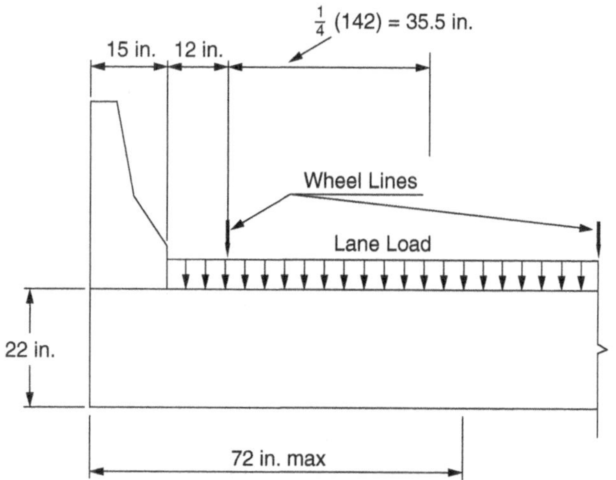

Fig. E16.1-4 Live-load placement for edge strip shear and moment.

H. ***Calculate Force Effects from Other Loads***

1. *Interior Strip, 1.0 ft Wide*

DC:

$$\rho_{\text{conc}} = 0.150 \text{ kcf}$$
$$w_{\text{DC}} = 0.150(22.0/12) = 0.275 \text{ ksf}$$
$$V_{\text{DC}} = 0.5(0.275)(35) = 4.81 \text{ kips/ft}$$
$$M_{\text{DC}} = w_{\text{DC}}L^2/8 = 0.275(35)^2/8 = 42.1 \text{ kip-ft/ft}$$

DW: Bituminous wearing surface, 3.0 in. thick

$$\rho_{\text{DW}} = 0.14 \text{ kcf} \text{ [Table A3.5.1-1]}$$
$$w_{\text{DW}} = 0.14(3.0/12) = 0.035 \text{ ksf}$$
$$V_{\text{DW}} = 0.5(0.035)(35) = 0.613 \text{ kips/ft}$$
$$M_{\text{DW}} = 0.035(35)^2/8 = 5.36 \text{ kip-ft/ft}$$

2. *Edge Strip, 1.0-ft Wide, Barrier = 0.320 kips/ft* Some owners distributed the barrier weight across the entire slab so that it is assumed to be equally shared. Others assume barrier load spread over width of live-load edge strip of 62.5 in. = 5.21 ft:

DC:

$$w_{\text{DC}} = 0.275 + 0.320/5.21 = 0.336 \text{ ksf}$$
$$V_{\text{DC}} = 0.5(0.336)(35) = 5.89 \text{ kips/ft}$$
$$M_{\text{DC}} = 0.336(35)^2/8 = 51.45 \text{ kip-ft/ft}$$

DW:

$$w_{\text{DW}} = 0.035\,(62.5 - 15.0)\,/62.5 = 0.025 \text{ ksf}$$
$$V_{\text{DW}} = 0.5(0.025)(35) = 0.438 \text{ kips/ft}$$
$$M_{\text{DW}} = 0.025(35)^2/8 = 3.83 \text{ kip-ft/ft}$$

I. ***Investigate Service Limit State***

1. *Durability* [Table A5.12.3-1]

Cover for unprotected main reinforcing steel deck surface subject to tire wear: 2.5 in.

Bottom of CIP slabs: 1.0 in.

Effective depth for No. 8 bars:

$$d = 22.0 - 1.0 - 1.0/2 = 20.5 \text{ in.}$$
$$\eta_D = \eta_R = \eta_I = 1.0, \text{ therefore } \eta_i = \eta = 1.0 \text{ [A1.3]}$$

a. Moment—Interior Strip

$$\begin{aligned} M_{\text{interior}} &= \Sigma\eta_i\gamma_i Q_i \\ &= 1.0[1.0M_{\text{DC}} + 1.0M_{\text{DW}} + 1.0M_{\text{LL+IM}}] \\ &= 1.0[42.1 + 5.36 + 51.9] \\ &= 99.36 \text{ kip-ft/ft} \end{aligned}$$

Trial reinforcement:

$$A_s \approx \frac{M}{f_s jd}$$

Assume $j = 0.875$ and $f_s = 0.6f_y = 36$ ksi

$$A_s \approx \frac{99.36 \times 12}{36\,(0.875)\,(20.5)} = 1.85 \text{ in.}^2/\text{ft}$$

Try No. 9 bars at 6 in. ($A_s = 2.00$ in.^{2}/ft) (Table B.4)

$$\begin{aligned} \text{Revised } d &= 22.0 - 1.0 - \tfrac{1}{2}\,(1.128) \\ &= 20.4 \text{ in.} \qquad \text{OK} \end{aligned}$$

b. Moment—Edge Strip

$$\begin{aligned} M_{\text{edge}} &= \Sigma\eta_i\gamma_i Q_i = 1.0(51.45 + 3.83 + 58.9) \\ &= 114.2 \text{ kip-ft/ft} \end{aligned}$$

Trial reinforcement:

$$A_s \approx \frac{M}{f_s jd} = \frac{114.2 \times 12}{36\,(0.875)\,(20.4)} = 2.13 \text{ in.}^2/\text{ft}$$

Try No. 9 bars at 5 in. ($A_s = 2.40$ in.^{2}/ft).

2. *Control of Cracking* [A5.7.3.4] Flexural cracking is controlled by limiting the bar spacing in the reinforcement closest to the tension face under service load stress f_s:

$$s \le \frac{700\gamma_e}{\beta_s f_s} - 2d_c$$

in which

$$\beta_s = 1 + \frac{d_c}{0.7\left(h - d_c\right)}$$

γ_e = exposure factor
= 1.00 for class 1 exposure condition
= 0.75 for class 2 exposure condition

d_c = concrete cover measured from extreme tension fiber to center of closest flexural reinforcement

a. Interior Strip Checking tensile stress in concrete against f_r [A5.4.2.6, A5.7.3.4]

$$M_{\text{interior}} = 99.36 \text{ kip-ft/ft}$$

$$f_c = \frac{M}{\frac{1}{6}bh^2} = \frac{99.36 \times 12}{\frac{1}{6}(12)(22)^2} = 1.23 \text{ ksi}$$

$$0.8f_r = 0.8\left(0.24\sqrt{f'_c}\right) = 0.8(0.24)\sqrt{4.5}$$
$$= 0.41 \text{ ksi}$$

$$f_c > 0.8f_r, \quad \text{section is assumed cracked}$$

Elastic-cracked section with No. 9 at 6 in. (A_s = 2.00 in.^{2}/ft) [A5.7.1] (Fig. E16.1-5)

$$n = \frac{E_s}{E_c} = 7.0, \text{ from deck design}$$

$$nA_s = 7.0(2.00) = 14.0 \text{ in.}^2/\text{ft}$$

Location of neutral axis:

$$\tfrac{1}{2}bx^2 = nA_s(d - x)$$
$$\tfrac{1}{2}(12)x^2 = (14.0)(20.4 - x)$$

solving, $x = 5.83$ in.

Moment of inertia of cracked section:

$$I_{\text{cr}} = \tfrac{1}{3}bx^3 + nA_s(d - x)^2$$
$$= \tfrac{1}{3}(12)(5.83)^3 + (14.0)(20.4 - 5.83)^2$$
$$= 3765 \text{ in.}^4/\text{ft}$$

Steel stress:

$$\frac{f_s}{n} = \frac{M(d - x)}{I_{\text{cr}}} = \frac{99.36(20.4 - 5.83)12}{3765}$$
$$= 4.61 \text{ ksi}$$
$$f_s = 7(4.61) = 32.3 \text{ ksi}$$
$$f_s \le 0.6f_y = 0.6(60) = 36 \text{ ksi}$$

For $d_c = 1.56$ in., $\gamma_e = 0.75$ (class 2 exposure)

$$\beta_s = 1 + \frac{1.56}{0.7(20.4)} = 1.11$$

$$s \le \frac{700(0.75)}{1.11(32.3)} - 2(1.56) = 11.5 \text{ in.}$$

Use No. 9 at 6 in. for interior strip for other limit state checks.

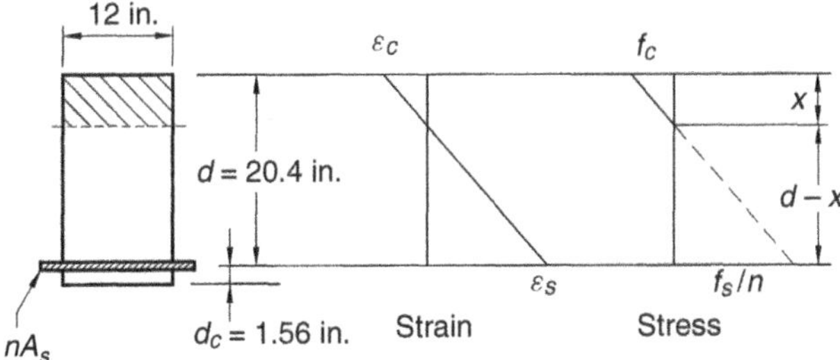

Fig. E16.1-5 Elastic-cracked section.

b. Edge Strip

$$M_{\text{edge}} = 114.2 \text{ kip-ft/ft}$$

Try No. 9 at 5 in., $A_s = 2.40$ in.^{2}/ft

$$nA_s = 7(2.40) = 16.8 \text{ in.}^2/\text{ft}$$

Location of neutral axis (Fig. E16.1-5):

$$\tfrac{1}{2}(12)\left(x^2\right) = (16.8)(20.4 - x)$$

Solving $x = 6.29$ in.

Moment of inertia of cracked section:

$$I_{\text{cr}} = \tfrac{1}{3}(12)(6.29)^3 + 16.8(20.4 - 6.29)^2$$
$$= 4338 \text{ in.}^4/\text{ft}$$

Steel stress:

$$\frac{f_s}{n} = \frac{114.2(20.4 - 6.29)12}{4338} = 4.46 \text{ ksi}$$
$$f_s = 7(4.46) = 31.2 \text{ ksi} < 36 \text{ ksi}$$

Checking spacing of No. 9 at 5 in., for d_c = 1.56 in., $\gamma_e = 0.75$, and $\beta_s = 1.11$.

$$s \le \frac{700(0.75)}{1.11(31.2)} - 2(1.56) = 12.0 \text{ in.}$$

Use No. 9 at 5 in. for edge strip.

3. *Deformations* [A5.7.3.6]

a. Dead-Load Camber [A5.7.3.6.2]:

$$w_{\text{DC}} = (0.275)(46.5) + 2(0.320)$$
$$= 13.43 \text{ kips/ft}$$
$$w_{\text{DW}} = (0.035)(44.0) = 1.54 \text{ kips/ft}$$
$$w_{\text{DL}} = w_{\text{DC}} + w_{\text{DW}} = 14.97 \text{ kips/ft}$$
$$M_{\text{DL}} = \tfrac{1}{8}w_{\text{DL}}L^2 = \frac{(14.97)(35)^2}{8} = 2292 \text{ kip-ft}$$

By using I_e:

$$\Delta_{\text{DL}} = \frac{5w_{\text{DL}}L^4}{384E_cI_e}$$
$$I_e = \left(\frac{M_{\text{cr}}}{M_a}\right)^3 I_g + \left[1 - \left(\frac{M_{\text{cr}}}{M_a}\right)^3\right] I_{\text{cr}}$$
$$M_{\text{cr}} = f_r\frac{I_g}{y_t}$$
$$f_r = 0.24\sqrt{4.5} = 0.509 \text{ ksi}$$
$$I_g = \tfrac{1}{12}(46.5 \times 12)(22)^3 = 495 \times 10^3 \text{ in.}^4$$
$$M_{\text{cr}} = 0.509\frac{495 \times 10^3}{(12)(22/2)} = 1910 \text{ kip-ft}$$
$$\left(\frac{M_{\text{cr}}}{M_a}\right)^3 = \left(\frac{1910}{2292}\right)^3 = 0.579$$

$$I_{cr} = (3765)(46.5) = 175 \times 10^3 \text{ in.}^4$$

$$I_e = (0.579)(495 \times 10^3) + (1 - 0.579)(175 \times 10^3) = 360 \times 10^3 \text{ in.}^4$$

$$\Delta_{DL} = \frac{5\,(14.97)\,(35)^4(12)^3}{384\,(3860)\,(360 \times 10^3)} = 0.36 \text{ in. instantaneous}$$

Long-time deflection factor for $A_s' = 0$ is equal to

$$3 - 1.2\left(\frac{A_s'}{A_s}\right) = 3.0$$

$$\text{Camber} = (3.0)\,(0.36) = \underline{1.08 \text{ in. upward}} \text{ (likely round to 1 in.)}$$

By using I_g [A5.7.3.6.2]:

$$\Delta_{DL} = (0.36)\left(\frac{360 \times 10^3}{495 \times 10^3}\right) = 0.26 \text{ in.}$$

Longtime deflection factor = 4.0

$$\text{Camber} = (4.0)\,(0.26) = \underline{1.05 \text{ in. upward}}$$

comparable to the value based on I_e.

b. Live-Load Deflection (Optional) [A2.5.2.6.2]:

$$\Delta_{LL+IM}^{allow} = \frac{\text{span}}{800} = \frac{35 \times 12}{800} = 0.53 \text{ in.}$$

If the owner invokes the optional live-load deflection criteria, the deflection should be the larger of that resulting from the design truck alone or design lane load plus 25% truck load [A3.6.1.3.2]. When design truck alone, it should be placed so that the distance between its resultant and the nearest wheel is bisected by the span centerline. All design lanes should be loaded [A2.5.2.6.2] (Fig. E16.1-6):

$$N_L = 3, \quad m = 0.85$$

$$\sum P_{LL+IM} = 1.33\,(32 \times 3)\,(0.85) = 108.5 \text{ kips}$$

The value of I_e changes with the magnitude of the applied moment M_a. The moment associated with the live-load deflection includes the dead-load moment plus the truck moment from Section 16.1, Part C.2:

$$M_{DC+DW+LL+IM} = 2292 + 3\,(0.85)\,(350)\,(1.33) = 3479 \text{ kip-ft}$$

so that

$$I_e = \left(\frac{1910}{3479}\right)^3 (495 \times 10^3) + \left[1 - \left(\frac{1910}{3479}\right)^3\right](175 \times 10^3) = 228 \times 10^3 \text{ in.}^4$$

$$E_c I_e = (3860)\,(228 \times 10^3) = 880 \times 10^6 \text{ kip-in.}^2$$

From case 8, AISC (2001) Manual (see Fig. E16.1-7),

$$\Delta_x\,(x < a) = \frac{Pbx}{6EIL}\left(L^2 - b^2 - x^2\right)$$

Assuming maximum deflection is under wheel load closest to the centerline, $\Delta_x = \Delta_C$.

First load: $P = 108.5$ kips, $a = 29.17$ ft, $b = 5.83$ ft, $x = 15.17$ ft (from right end):

$$\Delta_x = \frac{(108.5)\,(5.83)\,(15.17)}{6\,(880 \times 10^6)\,(35)} \times [(35)^2 - (5.83)^2 - (15.17)^2] \times 12^3 = 0.086 \text{ in.}$$

Second load: $P = 108.5$ kips, $a = x = 19.83$ ft, $b = 15.17$ ft:

$$\Delta_x = \frac{(108.5)\,(15.17)\,(19.83)}{6\,(880 \times 10^6)\,(35)} \times [(35)^2 - (15.17)^2 - (19.83)^2] \times 12^3 = 0.184 \text{ in.}$$

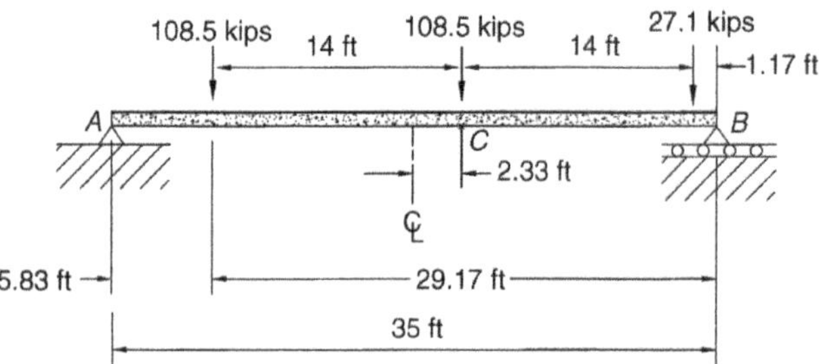

Fig. E16.1-6 Design truck placement for maximum deflection in span.

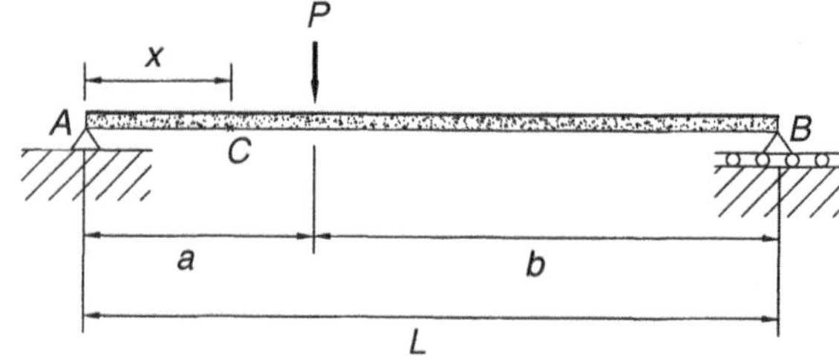

Fig. E16.1-7 Concentrated load placement for calculation of deflection.

Third load: $P = 27.1$ kips, $a = 33.83$ ft, $b = 1.17$ ft, $x = 19.83$ ft:

$$\Delta_x = \frac{(27.1)(1.17)(19.83)}{6(880 \times 10^6)(35)} \times [(35)^2 - (1.17)^2 - (19.83)^2] \times 12^3 = 0.005 \text{ in.}$$

$$\Delta_{\text{LL+IM}} = \sum \Delta_x = 0.28 \text{ in.} < 0.53 \text{ in.} \quad \text{OK}$$

Design lane load:

$$w = 1.33(0.64)(3)(0.85) = 2.17 \text{ kips/ft}$$

$$M_C \approx \frac{1}{8} wL^2 = \frac{(2.17)(35)^2}{8} = 332 \text{ kip-ft}$$

$$\Delta_C^{\text{lane}} = \frac{5}{48} \frac{M_C L^2}{E_c I_e} = \frac{5(332)(35)^2}{48(880 \times 10^6)} \times 12^3 = 0.083 \text{ in.}$$

$$25\% \text{ truck} = \tfrac{1}{4}(0.28) = 0.07 \text{ in.}$$

$$\Delta_{\text{LL+IM}} = 0.15 \text{ in., not critical}$$

The live-load deflection estimate of 0.28 in. is conservative because I_e was based on the maximum moment at midspan rather than an average I_e over the entire span. Also, the additional stiffness provided by the concrete barriers (which can be significant) has been neglected, as well as the compression reinforcement in the top of the slab. Finally, bridges typically deflect less under live load than calculations predict.

4. *Concrete Stresses* [A5.9.4.3] No prestressing, does not apply.
5. *Fatigue* [A5.5.3] The fatigue I limit state is used to check reinforcement fatigue.

$$U = 1.5(\text{LL} + \text{IM}) \quad \text{(Table 3.1)} \quad \text{[Table A3.4.1-1]}$$

$$\text{IM} = 15\% \quad \text{[A3.6.2.1]}$$

Fatigue load shall be one design truck with 30-ft axle spacing [A3.6.1.4.1]. Because of the large rear axle spacing, the maximum moment results when the two front axles are on the bridge. As shown in Figure E16.1-8, the two axle loads are placed on the bridge so that the distance between the resultant of the axle loads on the bridge and the nearest axle is divided equally by the centerline of the span (Case 42, AISC Manual, 2001). The multiple presence factor for one loaded lane includes multiple presence (m = 1.2) that must be removed from the strip width [A3.6.1.1.2]. From Figure E16.1-8,

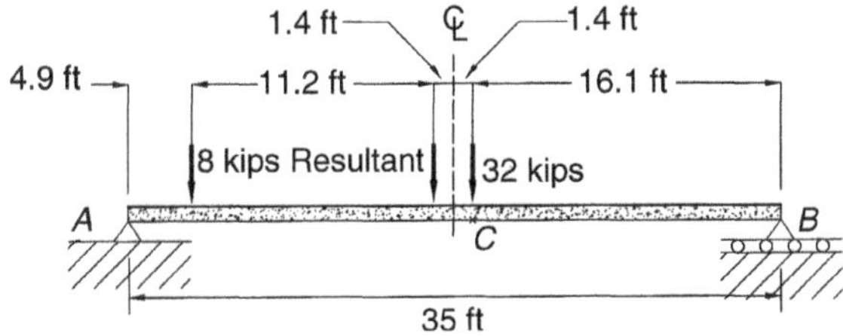

Fig. E16.1-8 Fatigue truck placement for maximum bending moment.

$$R_B = (32 + 8)\left(\frac{4.9 + 11.2}{35}\right) = 18.4 \text{ kips}$$

$$M_C = (18.4)(16.1) = 296 \text{ kip-ft}$$

$$\sum \eta_i \gamma_i Q_i = 1.0(1.5)(296)(1.15) = 511 \text{ kip-ft/lane}$$

a. Tensile Live-Load Stresses

One loaded lane, $E = 14.33$ ft,

$$M_{\text{LL+IM}} = \frac{511}{14.33(1.2)} = 29.7 \text{ kip-ft/ft}$$

$$\frac{f_s}{n} = \frac{(29.7)(12)(20.4 - 5.83)}{3765} = 1.38 \text{ ksi}$$

And the maximum steel stress due to the fatigue truck is

$$f_s = 7(1.38) = 9.66 \text{ ksi}$$

b. Reinforcing Bars [A5.5.3.2]

Maximum stress range f_f must be less than (Eq. 14.6b):

$$(\Delta F)_{\text{TH}} \le 21 - 0.33 f_{\min} + 8(r/h)$$

The dead-load moment for an interior strip is

$$M_{\text{DL}} = M_{\text{DC}} + M_{\text{DW}} = 42.1 + 5.36 = 47.46 \text{ kip-ft}$$

Using properties of a cracked section, the steel stress due to permanent loads is

$$f_{s,\text{DL}} = n \frac{M_{\text{DL}}(d - x)}{I_{\text{cr}}} = 7\left[\frac{47.46 \times 12(20.4 - 5.83)}{3765}\right] = 15.4 \text{ ksi}$$

Because the bridge is treated as a simple beam, the minimum live-load stress is zero. The minimum stress $f_{\min}$ is the minimum live-load stress combined with the stress from the permanent loads

$$f_{\min} = 0 + 15.4 = 15.4 \text{ ksi}$$

The maximum stress $f_{\max}$ is the maximum live-load stress combined with the stress from the permanent loads:

$$f_{\max} = 9.66 + 15.4 = 25.1 \text{ ksi}$$

The stress range $f_f = f_{\max} - f_{\min} = 25.1 - 15.4 = 9.7$ ksi. The limit for the stress range with $r/h = 0.3$ is

$$(\Delta F)_{\text{TH}} = 21 - 0.33(15.4) + 8(0.3) = 18.3 \text{ ksi} > f_f = 9.7 \text{ ksi} \quad \text{OK}$$

J. ***Investigate Strength Limit State***

1. *Flexure* [A5.7.3.2] Rectangular stress distribution [A5.7.2.2]

$$\beta_1 = 0.85 - 0.05\,(4.5 - 4.0) = 0.825$$

a. Interior Strip

Equation 7.73 with $A_{ps} = 0$, $b = b_w$, $A_s' = 0$. Try $A_s =$ No. 9 at 6 in. = 2.00 in.^{2}/ft from service limit state.

$$c = \frac{A_s f_y}{0.85 f_c' \beta_1 b} = \frac{(2.00)\,(60)}{0.85\,(4.5)\,(0.825)\,(12)}$$
$$= 3.17 \text{ in.}$$
$$a = \beta_1 c = (0.825)\,(3.17) = 2.61 \text{ in.}$$
$$d_s = 22 - 1.0 - \tfrac{1}{2}\,(1.128) = 20.4 \text{ in.}$$

Equation 7.76 with $A_{ps} = 0$, $b = b_w$, $A_s' = 0$, $A_s =$ 2.00 in.^{2}/ft

$$M_n = A_s f_y \left(d_s - \frac{a}{2}\right)$$
$$= 2.00\,(60)\left(20.4 - \frac{2.61}{2}\right)\Big/12$$
$$= 191 \text{ kip-ft/ft}$$

The resistance factor is determined based upon the curvature (or strain at the tension steel) per [A5.5.4.2.1]

$$\varepsilon_s = \left(\frac{d_s - c}{c}\right)(0.003)$$
$$= \left(\frac{20.4 - 3.17}{3.17}\right)(0.003) = 0.016$$
$$\varepsilon_s \geq 0.005$$
$$\phi = 0.90$$

Because the steel strain is large, the section ductility is large and the ϕ is for flexure.

$$\text{Factored resistance} = \phi M_n = 0.9\,(191)$$
$$= 172 \text{ kip-ft/ft}$$

Minimum reinforcement [A5.7.3.3.2] shall be adequate to develop $M_u = \phi M_n$ at least equal to the lessor of $1.2M_{cr}$ or $1.33M_u$ = 1.33(172) = 229 kip-ft/ft.

$$M_{cr} = S_{nc} f_r$$
$$S_{nc} = bh^2/6 = 12(22)^2/6 = 968 \text{ in.}^3/\text{ft}$$
$$f_r = 0.37\sqrt{f_c'} = 0.37\sqrt{4.5}$$
$$= 0.785 \text{ ksi} \quad [\text{A5.4.2.6}]$$
$$1.2M_{cr} = 1.2\,(968)\,(0.785)/12$$
$$= 76.0 \text{ kip-ft/ft}$$

controls minimum reinforcement

Strenth I: $\eta_i = \eta = 1.0$

$$M_u = \sum \eta_i \gamma_i Q_i$$
$$= 1.0\left(1.25M_{DC} + 1.5M_{DW} + 1.75M_{LL+IM}\right)$$
$$M_u = \eta \sum \gamma_i Q_i$$
$$= 1.0\,[1.25\,(42.1) + 1.50\,(5.36) + 1.75\,(51.9)]$$
$$M_u = 151.5 \text{ kip-ft/ft} < \phi M_n$$
$$= 172 \text{ kip-ft/ft} \quad \text{OK}$$

Service limit state governs. *Use No. 9 at 6 in. for interior strip.*

b. Edge Strip Try A_s = No. 9 at 5 in., A_s = 2.40 in.^{2}/ft from service limit state:

$$c = \frac{A_s f_y}{0.85 f_c' \beta_1 b}$$
$$= \frac{(2.40)\,(60)}{0.85\,(4.5)\,(0.825)\,(12)} = 3.80 \text{ in.}$$
$$a = \beta_1 c = (0.825)\,(3.80) = 3.14 \text{ in.}$$

The resistance factor is determined based upon the curvature (or strain at the tension steel) per [A5.5.4.2.1]:

$$\varepsilon_s = \left(\frac{d_s - c}{c}\right)(0.003)$$
$$= \left(\frac{20.4 - 3.80}{3.80}\right)(0.003) = 0.013$$
$$\varepsilon_s \geq 0.005$$
$$\phi = 0.90$$
$$\phi M_n = 0.9\,(2.40)\,(60)\left(20.4 - \frac{3.14}{2}\right)\Big/12$$
$$= 203 \text{ kip-ft/ft}$$

Minimum reinforcement [A5.7.3.3.2]:

$$M_u \geq 1.2M_{cr} = 76.0 \text{ kip-ft/ft}$$

Strength I: $\eta_i = \eta = 1.0$

$$M_u = \eta \sum \gamma_i Q_i$$
$$= 1.0\,[1.25\,(51.45) + 1.50\,(3.83) + 1.75\,(56.9)]$$
$$M_u = 169.6 \text{ kip-ft/ft} < \phi M_n$$
$$= 203 \text{ kip-ft/ft} \quad \text{OK}$$

Service limit state governs. *Use No. 9 at 5 in. for edge strip.*

2. *Shear* [A5.14.4.1] Slab bridges designed for moment in conformance with AASHTO [A4.6.2.3] may be considered satisfactory for shear. *If longitudinal tubes are placed in the slab to create voids and reduce the cross section, the shear resistance must be checked.*

K. ***Distribution Reinforcement*** [A5.14.4.1] The amount of bottom transverse reinforcement may be taken as a percentage of the main reinforcement required for positive moment as

$$\frac{100}{\sqrt{L}} \le 50\%$$

$$\frac{100}{\sqrt{35}} = 16.9\%$$

a. Interior Strip
Positive moment reinforcement = No. 9 at 6 in.,

$$A_s = 2.00 \text{ in.}^2/\text{ft}$$

$$\text{Transverse reinforcement} = 0.169\,(2.00) = 0.34 \text{in.}^2/\text{ft}$$

Try No. 5 at 10 in. transverse bottom bars, $A_s = 0.37 \text{ in.}^2/\text{ft}$.

b. Edge Strip
Positive moment reinforcement = No. 9 at 5 in.

$$A_s = 2.40 \text{ in.}^2/\text{ft}$$

$$\text{Transverse reinforcement} = 0.169\,(2.40) = 0.41 \text{ in.}^2/\text{ft}$$

Use No. 5 at 9 in., transverse bottom bars, $A_s = 0.41 \text{ in.}^2/\text{ft}$.

For ease of placement, use No. 5 at 9 in. across the entire width of the bridge.

L. ***Shrinkage and Temperature Reinforcement*** Area of reinforcement in each direction [A5.10.8.2]

$$\text{Temp } A_s \ge \frac{1.3bh}{2\,(b+h)\,f_y} = \left[\frac{1.3\,(12)\,(22)}{2\,(12+22)\,(60)}\right] = 0.084$$

$$0.11 \le \text{Temp } A_s \le 0.60$$

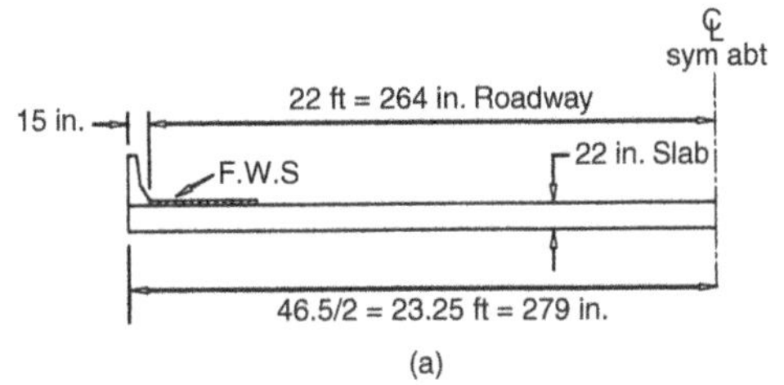

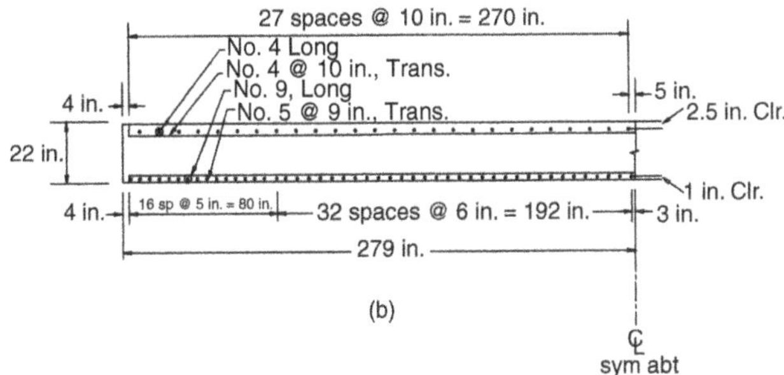

Fig. E16.1-9 Design sketch for solid slab bridge: (a) transverse half-section and (b) reinforcement half-section.

$$\text{Temp } A_s = 0.11 \tfrac{\text{in.}^2}{\text{ft}} \text{ controls}$$

$$\text{Top layer } A_s = \tfrac{1}{2}\,(0.11) = 0.055 \text{ in.}^2/\text{ft in each direction}$$

$$s_{max} \le 3h = 3\,(22) = 66 \text{ in. or } 18.0 \text{ in.}$$

Use No. 4 at 18 in., transverse and longitudinal top bars, $A_s = 0.13 \text{ in.}^2/\text{ft}$. (No. 4 bars were selected to provide more stiffness during construction for walking and concrete placement; No. 3 bar could have been used to satisfy [A5.10.8.2].

M. ***Design Sketch*** The design of the solid slab bridge is summarized in the half-section of Figure E16.1-9.

16.2 T-BEAM BRIDGE DESIGN

Problem Statement Example E16.2 Design a reinforced concrete T-beam bridge for a 44-ft-wide roadway and three-spans of 35 ft–42 ft–35 ft with a skew of 30° as shown in Figure E16.2-1 Use the concrete deck of Figures E15.1-14 and E15.1-17 previously designed for an HL-93 live load, a bituminous overlay, and a 8-ft spacing of girders in Example Problem 16.1. Use $f_c' = 4.5$ ksi, $f_y = 60$ ksi, and follow

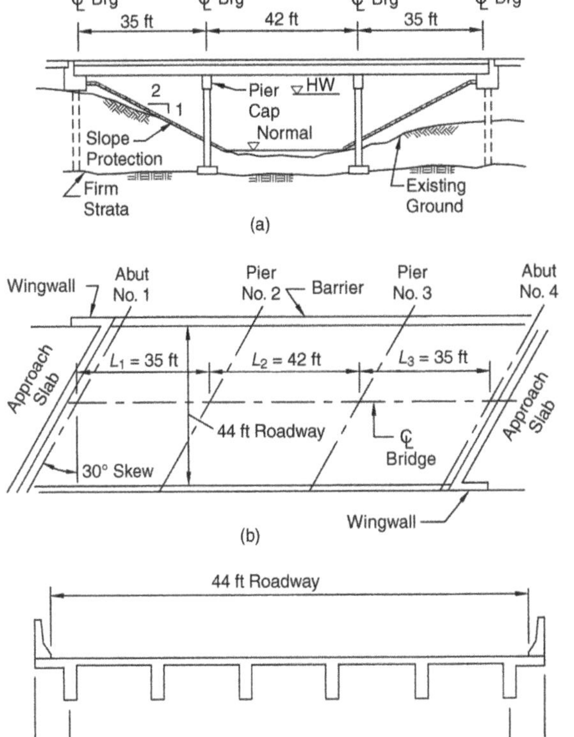

Fig. E16.2-1 T-beam bridge design example: (a) elevation, (b) plan, and (c) section.

the outline of AASHTO (2010) LRFD Bridge Specifications, Section 5, Appendix A5.3.

A. ***Develop General Section*** The bridge is to carry interstate traffic over a normally small stream that is subject to high water flows during the rainy season (Fig. E16.2-1).

B. ***Develop Typical Section and Design Basis***

1. *Top Flange Thickness* [A5.14.1.5.1a]

☐ As determined in Section 9 [A9.7.1.1]

$$\text{Minimum depth of concrete deck} = 7 \text{ in.}$$

From deck design,

$$\text{structural thickness} = \underline{7.5 \text{ in.}} \quad \text{OK}$$

☐ Maximum clear span $= 20\,(7.5/12) = 12.5\,\text{ft} >$ $8\,\text{ft} - (b_w/12)$ OK

2. *Bottom Flange Thickness* (not applicable to T-beam)
3. *Web Thickness* [A5.14.1.5.1c and C5.14.1.5.1c]

☐ Minimum of 8 in. without prestressing ducts

☐ Minimum concrete cover for main bars, exterior 2.0 in. [A5.12.3]

☐ Three No. 11 bars in one row require a beam width of [A5.10.3.1.1]

$$b_{\min} \approx 2\,(2.0) + 3d_b + 2\left(1.5d_b\right)$$
$$= 4.0 + 6\,(1.410) = 12.5 \text{ in.}$$

☐ To give a little extra room for bars, try $\underline{b_w = 14 \text{ in.}}$

4. *Structure Depth* (Table 3.1) [Table A2.5.2.6.3-1]

☐ Minimum depth continuous spans $= 0.065L$

$$h_{\min} = 0.065\,(42 \times 12)$$
$$= 33 \text{ in.,} \quad \text{try } \underline{h = 40 \text{ in.}}$$

5. *Reinforcement Limits*

☐ Deck overhang: at least $\frac{1}{3}$ of bottom layer of transverse reinforcement [A5.14.1.5.2a].

☐ Minimum reinforcement: Shall be adequate to develop the lesser of $\phi M_n > 1.2M_{\text{cr}}$ or $\phi M_n \geq$ 1.33 times the factored moment required for the strength I limit state [A5.7.3.3.2].

$$M_{\text{cr}} = S_{\text{nc}} f_r$$
$$f_r = 0.37\sqrt{f_c'} = 0.37\sqrt{4.5}$$
$$= 0.785 \text{ ksi} \quad [\text{A5.4.2.6}]$$

☐ Crack control: Cracking is controlled by limiting the spacing s in the reinforcement under service loads [A5.7.3.4]

$$s \leq \frac{700\gamma_e}{\beta_s f_s} - 2d_c$$

in which

$$\beta_s = 1 + \frac{d_c}{0.7\left(h - d_c\right)}$$

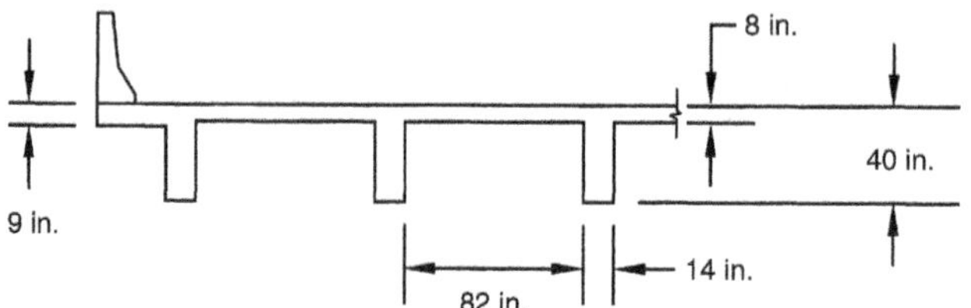

Fig. E16.2-2 Trial section for T-beam bridge.

☐ Flanges in tension at the service limit state: Tension reinforcement shall be distributed over the lesser of the effective flange width or a width equal to $\frac{1}{10}$ of the average of the adjacent spans [A4.6.2.6, A5.7.3.4].

☐ Longitudinal skin reinforcement required if web depth > 3.0 ft [A5.7.3.4].

☐ Shrinkage and temperature reinforcement [A5.10 .8.2]

$$\text{Temp } A_s \geq \frac{1.3bh}{2\,(b+h)\,f_y}$$
$$= \left[\frac{1.3\,(12)\,(22)}{2\,(12+22)\,(60)}\right]$$
$$= 0.084$$
$$0.11 \leq \text{Temp } A_s \leq 0.60$$

6. *Effective Flange Widths* [A4.6.2.6.1]

☐ Effective span length for continuous spans = distance between points of permanent load inflections.

☐ Interior beams:

$$b_i \leq \text{average spacing of adjacent beams}$$

☐ Exterior beams:

$$b_e \leq \text{width of overhang} + \tfrac{1}{2}b_i$$

7. *Identify Strut and Tie Areas, if any not applicable.* The trial section for the T-beam bridge is shown in Figure E16.2-2.

C. ***Design Conventionally Reinforced Concrete Deck*** The reinforced concrete deck for this bridge is designed in Section 16.1. The design sketches for the deck are given in Figures E15.1-14 and E12.1-17.

D. ***Select Resistance Factors*** (Table 14.6) [A5.5.4.2]

1. *Strength Limit State*	ϕ [A5.5.4.2.1]
Flexure and tension	0.90
Shear and torsion	0.90
Axial compression	0.75
Bearing	0.70
2. *Nonstrength Limit States*	1.0 [A1.3.2.1]

E. ***Select Load Modifiers*** [A1.3.2.1]

	Strength	Service	Fatigue	
Ductility, η_D	1.0	1.0	1.0	[A1.3.3]
Redundancy, η_R	1.0	1.0	1.0	[A1.3.4]
Importance, η_I	1.0	N/A	N/A	[A1.3.5]
$\eta_i = \eta_D\eta_R\eta_I$	1.0	1.0	1.0	

F. ***Select Applicable Load Combinations*** (Table 5.1) [Table A3.4.1-1]
Strength I Limit State: $\eta_i = \eta = 1.0$

$$U = \eta(1.25\text{DC} + 1.50\text{DW} + 1.75(\text{LL} + \text{IM}) + 1.0(\text{WA} + \text{FR}) + \cdots)$$

Service I Limit State:

$$U = 1.0(\text{DC} + \text{DW}) + 1.0(\text{LL} + \text{IM}) + 1.0\text{WA} + 0.3(\text{WS} + \text{WL}) + \cdots$$

Fatigue Limit State:

$$U = 0.75\,(\text{LL} + \text{IM})$$

G. ***Calculate Live-Load Force Effects***

1. ***Select Number of Lanes*** [A3.6.1.1.1]

$$N_L = \text{INT}\left(\frac{w}{12.0}\right) = \text{INT}\left(\frac{44.0}{12.0}\right) = 3$$

2. *Multiple Presence* (Table 8.6) [A3.6.1.1.2]

No. of Loaded Lanes	m
1	1.20
2	1.00
3	0.85

3. *Dynamic Load Allowance* (Table 8.7) [A3.6.2.1] Not applied to the design lane load.

Component	IM (%)
Deck joints	75
Fatigue	15
All other	33

4. *Distribution Factors for Moment* [A4.6.2.2.2] Applicability [A4.6.2.2.1]: constant deck width, at least four parallel beams of nearly same stiffness, roadway part of overhang (Fig E16.2-3), $d_e = 3.25 - 1.25 = 2.0\text{ ft} < 3.0\text{ ft}$ OK.

Cross-section type (e) (Table 4.1) [Table A4.6.2.2.1-1]

No.of beams $N_b = 6$ $\quad t_s = 7.5$ in.

$S = 8$ ft $\quad L_1 = L_3 = 35$ ft $\quad L_2 = 42$ ft

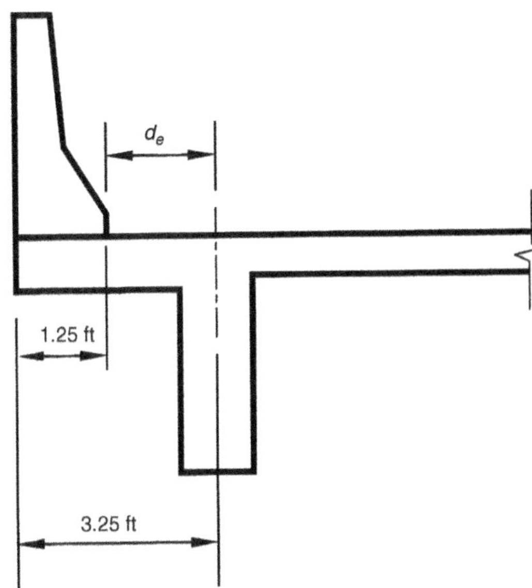

Fig. E16.2-3 Roadway part of overhang, d_e.

a. Interior Beams with Concrete Decks (Table 11.5) [A4.6.2.2.2b and Table A4.6.2.2.2b-1] (note that [Table A4.6.2.2.1-2] estimates could be used as well):

For preliminary design $\left(\dfrac{K_g}{12Lt_s^3}\right)^{0.1 \text{ or } 0.25} = 1.0$ and $\dfrac{I}{J} = 1.0$

One design lane loaded: range of applicability satisfied

$$mg_M^{\text{SI}} = 0.06 + \left(\frac{S}{14}\right)^{0.4}\left(\frac{S}{L}\right)^{0.3}\left(\frac{K_g}{12Lt_s^3}\right)^{0.1}$$

mg = girder distribution factor with multiple presence factor included

SI = single lane loaded, interior

M = moment

Two or more design lanes loaded

$$mg_M^{\text{MI}} = 0.075 + \left(\frac{S}{9.5}\right)^{0.6}\left(\frac{S}{L}\right)^{0.2}\left(\frac{K_g}{12Lt_s^3}\right)^{0.1}$$

MI = multiple lanes loaded, interior

M = moment

Distribution Factor	$L_1 = 35$ ft	$L_{\text{ave}} = 38.5$ ft	$L_2 = 42$ ft
mg_M^{SI}	0.573	0.559	0.546
mg_M^{MI}	0.746	0.734	0.722

For interior girders, distribution factors are governed by multiple lanes loaded.

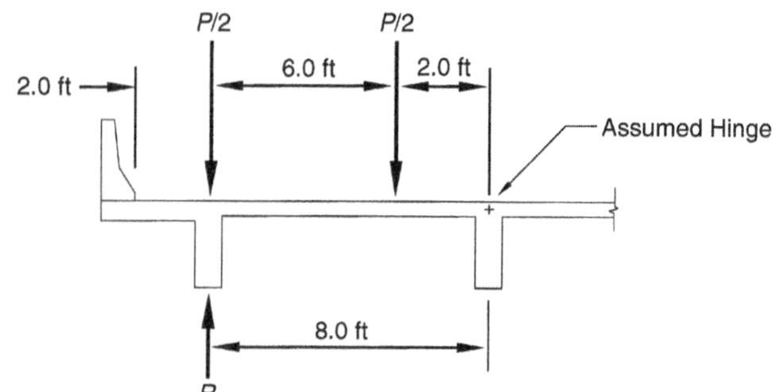

Fig. E16.2-4 Definition of level rule.

b. Exterior Beams (Table 11.5) [A4.6.2.2.2d and Table A4.6.2.2.2d-1]

One design lane loaded—lever rule, $m = 1.2$ (Fig. E16.2-4)

$$R = 0.5P\left(\frac{8.0 + 2.0}{8.0}\right) = 0.625P$$

$$g_M^{SE} = 0.625 \quad SE = \text{single lane, exterior}$$

$$mg_M^{SE} = 1.2\,(0.625) = \underline{0.750} \quad \text{governs}$$

Two or more design lanes loaded, $d_e = 2.0$ ft

$$mg_M^{ME} = emg_M^{MI}$$

ME = multiple lanes loaded, exterior

where

$$e = 0.77 + \frac{d_e}{9.1} = 0.77 + \frac{2.0}{9.1} = 0.99 < 1.0$$

Use $e = 1.0$. Therefore,

$$mg_M^{ME} = mg_M^{MI} = 0.746, \quad 0.734, \quad 0.722$$

For exterior girders, the critical distribution factor is by the lever rule with one lane loaded = 0.750.

c. Skewed Bridges (Table 11.3) [A4.6.2.2.2e] Reduction of live-load distribution factors for moment in longitudinal beam on skewed supports is permitted. $S = 8$ ft, $\theta = 30°$.

$$r_{skew} = 1 - c_1(\tan\,\theta)^{1.5} = 1 - 0.4387c_1$$

where

$$c_1 = 0.25\left(\frac{K_g}{12Lt_s^3}\right)^{0.25}\left(\frac{S}{L}\right)^{0.5} \quad \text{[Table A4.6.2.2.2e-1]}$$

Range of applicability is satisfied.

Reduction Factor	$L_1 = 35$ ft	$L_{ave} = 38.5$ ft	$L_2 = 42$ ft
c_1	0.120	0.114	0.109
r_{skew}	0.948	0.950	0.952

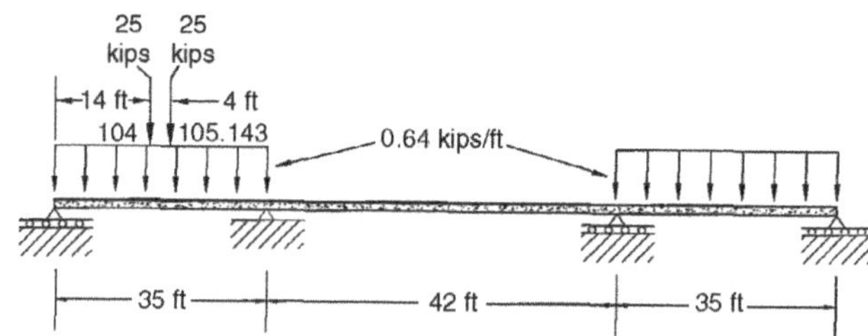

Fig. E16.2-5 Live-load placement for maximum positive moment in exterior span.

d. Distributed Live-Load Moments

$$M_{LL+IM} = mgr\left[(M_{Tr} \text{ or } M_{Ta})\left(1 + \frac{IM}{100}\right) + M_{Ln}\right]$$

Location 104 (Fig. E16.2-5) For relatively short spans, design tandem governs positive moment (see Table 9.8a). Influence line coefficients are from Table 9.4.

$$M_{Ta} = 25\,(0.20700{+}0.15732)\,35 = 318.8 \text{ kip-ft}$$

$$M_{Ln} = 0.64\,(0.10214)\,(35)^2 = 80.1 \text{ kip-ft}$$

Interior girders:

$$M_{LL+IM} = 0.746\,(0.948)\,[318.8\,(1.33) + 80.1] = 356.5 \text{ kip-ft}$$

Exterior girders:

$$M_{LL+IM} = 0.750\,(0.948)\,[318.8\,(1.33) + 80.1] = 358.4 \text{ kip-ft}$$

Location 200 (Fig. E16.2-6) For negative moment at support, a single truck governs with the second axle spacing extended to 30 ft (see Table 9.8a). The distribution factors are based on the average length of span 1 and span 2.

$$M_{Tr} = [32(-0.09429 - 0.10271) + 8(-0.05902)]35 = -237.2 \text{ kip-ft}$$

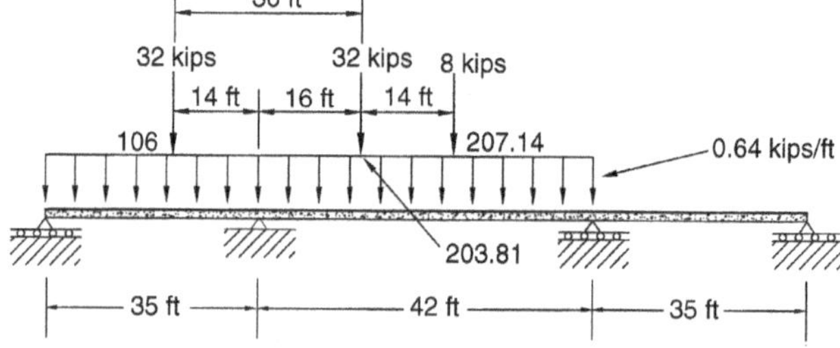

Fig. E16.2-6 Live-load placement for maximum negative moment at interior support.

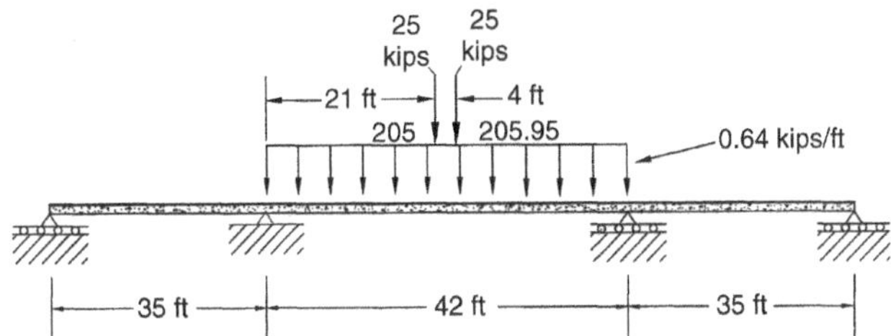

Fig. E16.2-7 Live-load placement for maximum positive moment in interior span.

$$M_{\text{Ln}} = 0.64\,(-0.13853)\,(35)^2$$
$$= -108.6 \text{ kip-ft}$$
$$1.33M_{\text{Tr}} + M_{\text{Ln}} = 1.33\,(-237.2) - 108.6$$
$$= -424.1 \text{ kip-ft}$$

Interior girders:

$$M_{\text{LL+IM}} = 0.734\,(0.950)\,(-424.1)$$
$$= -295.7 \text{ kip-ft}$$

Exterior girders:

$$M_{\text{LL+IM}} = 0.750\,(0.950)\,(-424.1)$$
$$= -302.2 \text{ kip-ft}$$

Location 205 (Fig E16.2-7) Tandem governs (see Table 9.8a)

$$M_{\text{Ta}} = 25\,(0.20357 + 0.150224)\,35$$
$$= 309.6 \text{ kip-ft}$$
$$M_{\text{Ln}} = 0.64\,(0.10286)\,(35)^2$$
$$= 80.6 \text{ kip-ft}$$
$$1.33M_{\text{Ta}} + M_{\text{Ln}} = 1.33\,(309.6) + 80.6$$
$$= 492.4 \text{ kip-ft}$$

Interior girders:

$$M_{\text{LL+IM}} = 0.722\,(0.952)\,(492.4) = 338.5 \text{ kip-ft}$$

Exterior girders:

$$M_{\text{LL+IM}} = 0.750\,(0.952)\,(492.4) = 351.6 \text{ kip-ft}$$

5. *Distribution Factors for Shear* [A4.6.2.2.3] Cross-section type (e) (Table 4.1) [Table A4.6.2.2.1-1], $S = 8$ ft, mg is independent of span length.
   a. Interior Beams (Table 11.3) [A4.6.2.2.3a and Table A4.6.2.2.3a-1]

$$mg_V^{\text{SI}} = 0.36 + \frac{S}{25} = 0.36 + \frac{8}{25} = 0.680$$
$$mg_V^{\text{MI}} = 0.2 + \frac{S}{12} - \left(\frac{S}{35}\right)^2 = 0.2 + \frac{8}{12} - \left(\frac{8}{35}\right)^2$$
$$= 0.814, \text{ governs}$$
$$V = \text{shear}$$

   b. Exterior Beams (Table 11.3) [A4.6.2.2.3b and Table A4.6.2.2.3b-1]

$$\text{Lever rule} \quad mg_V^{\text{SE}} = 0.750 \quad \text{governs}$$
$$mg_V^{\text{ME}} = emg_V^{\text{MI}}$$

where

$$e = 0.6 + \frac{d_e}{10} = 0.6 + \frac{2.0}{10} = 0.80$$
$$mg_V^{\text{ME}} = 0.80\,(0.814) = 0.651$$

   c. Skewed Bridges (Table 11.3) [A4.6.2.2.3c and Table A4.6.2.2.3c-1] All beams treated like beam at obtuse corner.

$$\theta = 30^\circ \qquad \left(\frac{12Lt_s^3}{K_g}\right) = 1.0$$
$$r_{\text{skew}} = 1.0 + 0.20\left(\frac{12Lt_s^3}{K_g}\right)^{0.3} \tan\theta$$
$$= 1.0 + 0.20(1.0)^{0.3}\,(0.577) = 1.115$$

   d. Distributed Live-Load Shears

$$V_{\text{LL+IM}} = mgr\left[\left(V_{\text{Tr}} \text{ or } V_{\text{Ta}}\right)1.33 + V_{\text{Ln}}\right]$$

Location 100 (Fig. E16.2-8) Truck governs (see Table 9.8b).

$$V_{\text{Tr}} = 32\,(1.0+0.51750)+8\,(0.12929)$$
$$= 49.6 \text{ kips}$$
$$V_{\text{Ln}} = 0.64\,(0.45536)\,35 = 10.2 \text{ kips}$$
$$1.33V_{\text{Tr}} + V_{\text{Ln}} = 1.33\,(49.6) + 10.2 = 76.2 \text{ kips}$$

Interior girders:

$$V_{\text{LL+IM}} = 0.814\,(1.115)\,(76.2)$$
$$= 69.1 \text{ kips}$$

Exterior girders:

$$V_{\text{LL+IM}} = 0.750\,(1.115)\,(76.2)$$
$$= 63.7 \text{ kips}$$

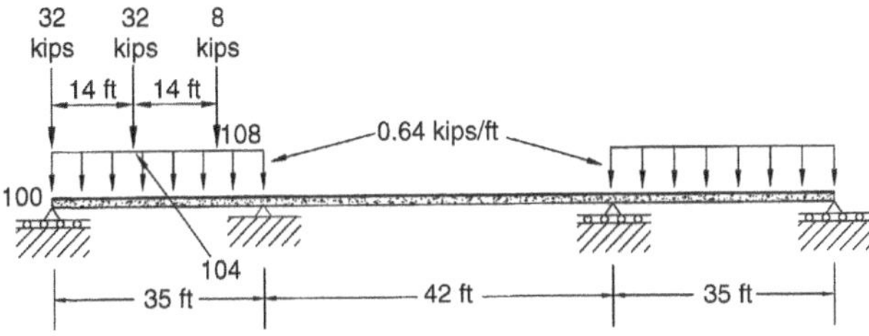

Fig. E16.2-8 Live-load placement for maximum shear at exterior support.

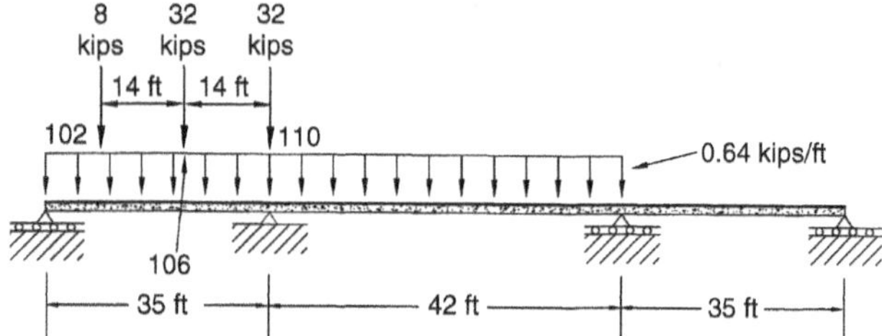

Fig. E16.2-9 Live-load placement for maximum shear to left of interior support.

Location 110 (Fig. E16.2-9) Truck governs (see Table 9.8b).

$$V_{\text{Tr}} = 32\,(-1.0 - 0.69429) + 8\,(-0.24714)$$
$$= -56.2 \text{ kips}$$
$$V_{\text{Ln}} = 0.64\,(-0.63853)\,35$$
$$= -14.3 \text{ kips}$$

$$1.33V_{\text{Tr}} + V_{\text{Ln}} = 1.33\,(-56.2) - 14.3$$
$$= -89.0 \text{ kips}$$

Interior girders:

$$V_{\text{LL+IM}} = 0.814\,(1.115)\,(-89.0) = -80.8 \text{ kips}$$

Exterior girders:

$$V_{\text{LL+IM}} = 0.750\,(1.115)\,(-89.0) = -74.5 \text{ kips}$$

Location 200 (Fig. E16.2-10) Truck governs (see Table 9.8b).

$$V_{\text{Tr}} = 32\,(1.0 + 0.69367) + 8\,(0.30633)$$
$$= 56.6 \text{ kips}$$
$$V_{\text{Ln}} = 0.64\,(0.66510)\,35 = 14.9 \text{ kips}$$

$$1.33V_{\text{Tr}} + V_{\text{Ln}} = 1.33\,(56.6) + 14.9 = 90.2 \text{ kips}$$

Interior girders:

$$V_{\text{LL+IM}} = 0.814\,(1.115)\,(90.2) = 81.8 \text{ kips}$$

Exterior girders:

$$V_{\text{LL+IM}} = 0.750\,(1.115)\,(90.2) = 75.4 \text{ kips}$$

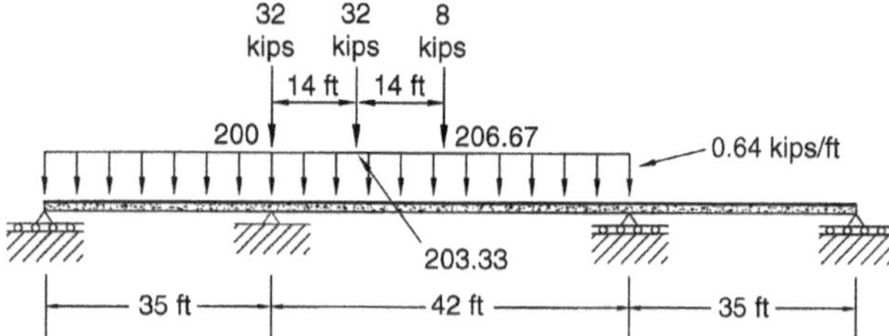

Fig. E16.2-10 Live-load placement for maximum shear to right of interior support.

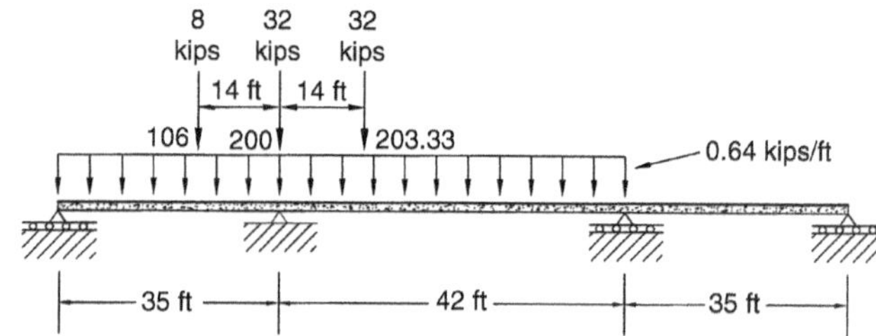

Fig. E16.2-11 Live-load placement for maximum reaction at interior support.

6. *Reactions to Substructure* [A3.6.1.3.1] The following reactions are per design lane without any distribution factors. The lanes shall be positioned transversely to produce extreme force effects.
Location 100

$$R_{100} = V_{100} = 1.33V_{\text{Tr}} + V_{\text{Ln}}$$
$$= 76.2 \text{ kips/lane}$$

Location 200 (Fig. E16.2-11)

$$R_{200} = 1.33\,[32\,(1.0 + 0.69367 + 0.10106)$$
$$+8\,(0.69429 + 0.10000)] + 14.3 + 14.9$$
$$= 114.0 \text{ kips/lane}$$

H. ***Calculate Force Effects from Other Loads*** Analysis for a uniformly distributed load w (Fig. E16.2-12). See Table 9.4 for coefficients and Fig. E16.2-13.
Moments

$$M_{104} = w\,(0.07129)\,(35)^2 = 87.33w \text{ kip-ft}$$
$$M_{200} = w\,(-0.12179)\,(35)^2 = -149.2w \text{ kip-ft}$$
$$M_{205} = w\,(0.05821)\,(35)^2 = 71.3w \text{ kip-ft}$$

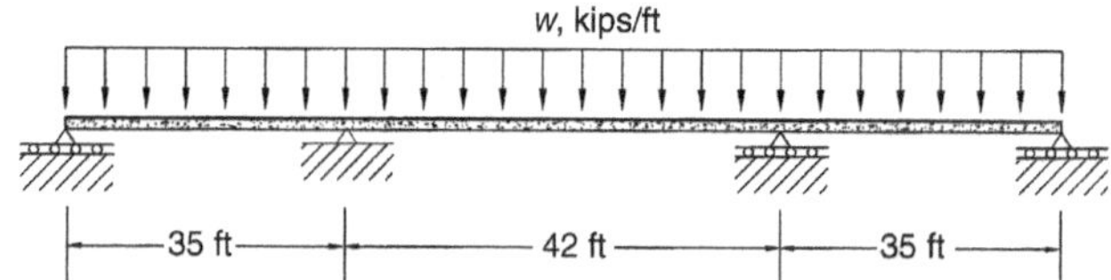

Fig. E16.2-12 Uniformly distributed dead load, w.

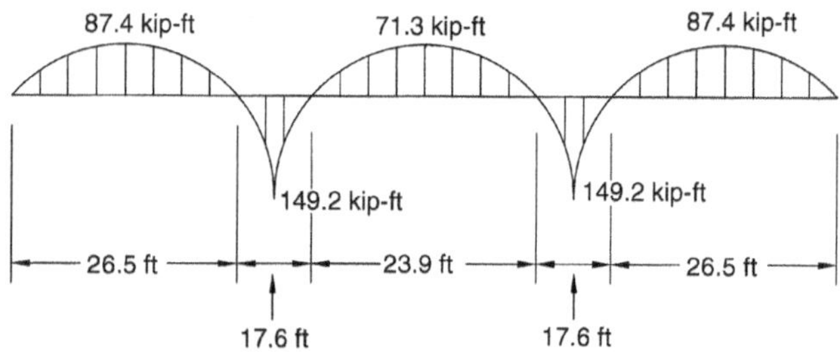

Fig. E16.2-13 Cross section with effective flange widths.

Shears

$$V_{100} = w\,(0.37821)\,(35) = 13.24w \text{ kips}$$
$$V_{110} = w\,(-0.62179)\,(35) = -21.76w \text{ kips}$$
$$V_{200} = w\,(0.60000)\,(35) = 21.0w \text{ kips}$$

1. *Interior Girders*

DC: Slab $(0.150)\,(8.0/12)\,8 = 0.800$ kips/ft

Girder stem $(0.150)\,(14)\,(40-8)\,/12^2 = \underline{0.467}$

$$w_{DC} = 1.267 \text{ kips/ft}$$

DW : FWS $w_{DW} = (0.140)\,(3.0/12)\,8 = 0.280\text{kips/ft}$

By multiplying the general expressions for uniform loads by the values of the interior girder uniform loads, the unfactored moments and shears are generated in Table E16.2-1.

2. *Exterior Girders* By using deck design results for reaction on exterior girder from Section 16.1, Part C:

DC	Deck slab	0.314 kips/ft
	Overhang	0.462
	Barrier	0.464
	Girder stem	$\underline{0.459} = 0.150 \times 7[(40-9) + (40-8)]/12^2$

$$w_{DC} = 1.699 \text{ kips/ft}$$

DW : FWS $w_{DW} = 0.191$ kips/ft

By multiplying the generic expressions for uniform loads by the values of the exterior girder uniform loads, the unfactored moments and shears in Table E16.2-2 are generated.

I. ***Investigate Service Limit State***

1–3. *Prestress Girders* Not applicable.

4. *Investigate Durability* [C5.12.1] It is assumed that concrete materials and construction procedures provide adequate concrete cover, nonreactive aggregates, thorough consolidation, adequate cement content, low water/cement ratio, thorough curing, and air-entrained concrete.

Concrete Cover for Unprotected Main Reinforcing Steel [Table 5.12.3-1]

Exposure to deicing salts	2.5 in.	cover to ties and stirrups 0.5 in. less
Exterior other than above	2.0 in.	
Bottom of CIP slabs, up to No.11	1.0 in.	

Effective Depth—assume No. 10, $d_b = 1.270$ in.

Positive Bending

$$d_{pos} = (40-0.5) - \left(2.0 + \frac{1.270}{2}\right) = 36.9 \text{ in.}$$

Negative Bending

$$d_{neg} = 40 - \left(2.5 + \frac{1.270}{2}\right) = 36.9 \text{ in.}$$

5. *Crack Control* [A5.7.3.4] Flexural cracking is controlled by limiting the spacing s in the reinforcement closest to the tension face under service load stress f_s:

$$s \le \frac{700\gamma_e}{\beta_s f_s} - 2d_c$$

in which

$$\beta_s = 1 + \frac{d_c}{0.7\,(h-d_c)}$$

Table E16.2-1 Interior Girder Unfactored Moments and Shears

		Moments (kip-ft)			Shears (kips)		
Load Type	w (k/ft)	M_{104}	M_{200}	M_{205}	V_{100}	V_{110}	V_{200}
DC	1.267	110.6	−189.0	90.3	16.8	−27.6	26.6
DW	0.280	24.5	−41.8	20.0	3.7	−6.1	5.9
LL + IM	N/A	356.5	−295.7	338.5	69.1	−80.8	81.8

Table E16.2-2 Exterior Girder Unfactored Moments and Shears[a]

		Moments (kip-ft)			Shears (kips)		
Load Type	w (k/ft)	M_{104}	M_{200}	M_{205}	V_{100}	V_{110}	V_{200}
DC	1.699	148.4	−253.5	121.1	22.5	−37.0	35.7
DW	0.191	16.7	−28.5	13.6	2.5	−4.2	4.0
LL + IM	N/A	358.4	−302.2	351.6	63.7	−74.5	75.4

[a] Interior girder has larger shears. Exterior girder has larger moments.

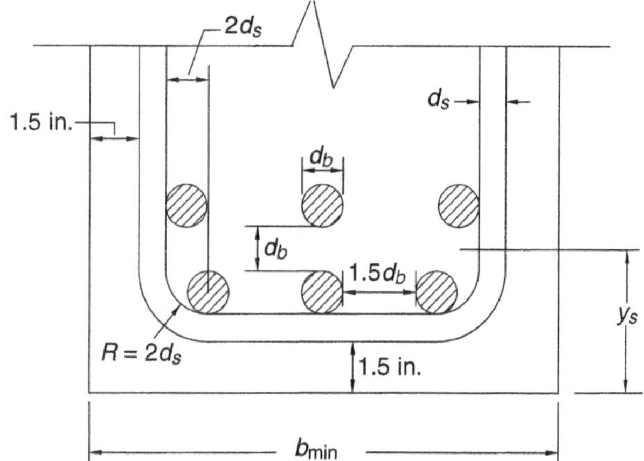

Fig. E16.2-14 Spacing of reinforcement in stem of T-beam.

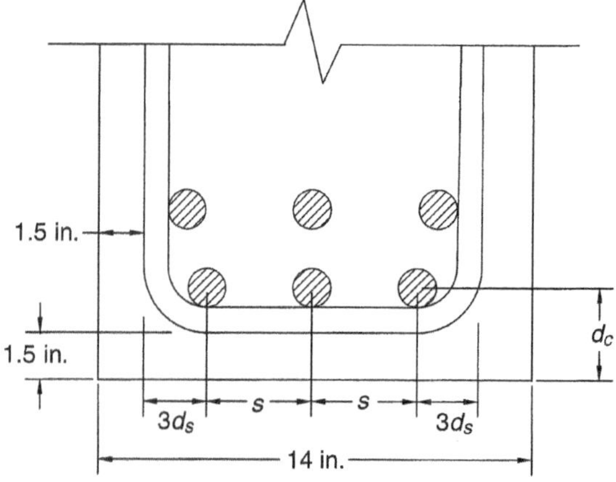

Fig. E16.2-16 Bar spacing in the reinforcement closest to the tension face.

γ_e = exposure factor

= 1.00 for class 1 exposure condition

= 0.75 for class 2 exposure condition

d_c = concrete cover measured from extreme tension fiber to center of closest flexural reinforcement (Fig. E16.2-14)

a. Effective Flange Width [A4.6.2.6.1] (Fig. E16.2-15).
Positive Bending M_{104}

$$b_i \leq S = 8 \times 12 = 96 \text{ in.}$$

$$b_e - \tfrac{1}{2}b_i \leq \text{overhang} = 39.0 \text{ in.} \quad \text{governs}$$

$$b_e = 39.0 + \tfrac{1}{2}(79.5) = 78.8 \text{ in.}$$

Use $b_i = 96$ in., $b_e = 79$ in.

b. Positive Bending Reinforcement—Exterior Girder (Table 5.1) [Table A3.4.1-1] Service I limit state, $\eta_i = 1.0$, gravity load factors = 1.0, moments from Table E16.2-2:

$$\begin{aligned} M_{104} &= \sum \eta_i \gamma_i Q_i \\ &= M_{\text{DC}} + M_{\text{DW}} + mgrM_{\text{LL+IM}} \\ &= (148.4 + 16.7 + 358.4) \\ &= 523.5 \text{ kip-ft} \end{aligned}$$

$$f_c' = 4.5 \text{ ksi} \qquad f_y = 60 \text{ ksi}$$

$$d_{\text{pos}} = 36.9 \text{ in.}$$

Assume $j = 0.875$ and $f_s = 0.6f_y = 36$ ksi:

$$A_s \approx \frac{M}{f_s jd} = \frac{523.5 \times 12}{36 \times 0.875 \times 36.9} = 5.40 \text{ in.}^2$$

Try six No. 9 bars, provided $A_s = 6.00 \text{ in.}^2$ (Table B.3).

Minimum beam width must consider bend diameter of tie [Table A5.10.2.3-1].

For No. 4 stirrup and No. 9 bar (Fig. E16.2-16)

$$\text{Inside radius} = 2d_s > \tfrac{1}{2}d_b$$

$$2(0.5) = 1.0 \text{ in.} > \tfrac{1}{2}(1.128) = 0.564 \text{ in.}$$

Center of No. 9 bar will be away from vertical leg of stirrup a distance of $2d_s = 1.0$ in.

$$\begin{aligned} b_{\min} &= 2\left(1.50 + 3d_s\right) + 2d_b + 2\left(1.5d_b\right) \\ &= 2(1.50 + 3 \times 0.5) + 5(1.128) \\ &= 11.64 \text{ in.} \end{aligned}$$

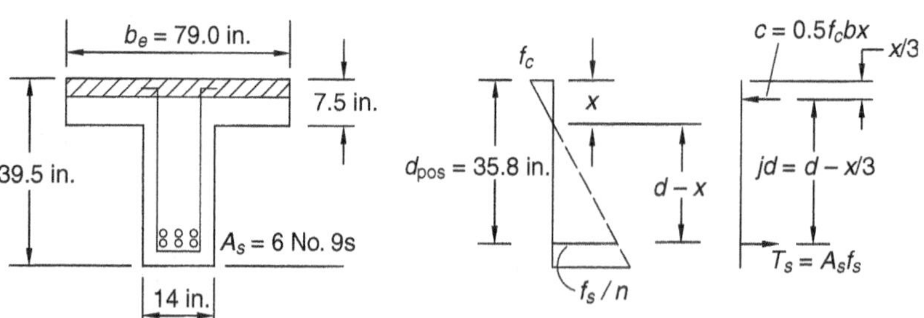

Fig. E16.2-15 Elastic-cracked transformed positive moment section at location 104.

Three No. 9 bars will fit in one layer of $b_w = 14$ in.

$$y_s = 1.5 + 0.5 + 1.128 + \tfrac{1}{2}(1.128)$$
$$= 3.69 \text{ in.}$$
$$d_{pos} = (40 - 0.5) - 3.69 = 35.8 \text{ in.}$$

Elastic-cracked transformed section analysis required to check crack control [A5.7.3.4].

$$n = \frac{E_s}{E_c} = 7 \text{ from solid-slab bridge design}$$
$$b = b_e = 79 \text{ in.}$$

Assume NA (neutral axis) in flange (Fig. E16.2-15):

$$x = -\frac{nA_s}{b} + \sqrt{\left(\frac{nA_s}{b}\right)^2 + \frac{2nA_s d}{b}}$$
$$= \frac{-7(6.0)}{79} + \sqrt{\left(\frac{7\times 6.0}{79}\right)^2 + \frac{2(7)(6.0)(35.8)}{79}}$$
$$= 5.66 \text{ in.} < h_f = 7.5 \text{ in.}$$

The neutral axis lies in flange; therefore, assumption OK.

The actual bar spacing must be compared to the maximum bar spacing allowed for crack control (Fig. E16.2-16). Actual $s = [14 - 2(1.50 + 3 \times 0.5]/2 = 4.0$ in.

$$s \leq \frac{700\gamma_e}{\beta_s f_s} - 2d_c \quad d_c = 1.5 + 0.5 + 1.128/2$$
$$= 2.56 \text{ in.}$$
$$\gamma_e = 0.75 \text{ (class 2 exposure)}$$
$$\beta_s = 1 + \frac{d_c}{0.7(h - d_c)}$$
$$= 1 + \frac{2.56}{0.7(39.5 - 2.56)} = 1.099$$
$$I_{cr} = \tfrac{1}{3}bx^3 + nA_s(d - x)^2$$
$$= \tfrac{1}{3}(79)(5.66)^3 + 7(6.0)(35.8 - 5.66)^2$$
$$= 42{,}928 \text{ in.}^4$$
$$f_s = \frac{nM(d - x)}{I_{cr}}$$
$$= \frac{7(523.5 \times 12)(35.8 - 5.66)}{42{,}928} = 30.9 \text{ ksi}$$
$$s \leq \frac{700(0.75)}{1.099(30.9)} - 2(2.56)$$
$$= 10.4 \text{ in.} > s = 4.0 \text{ in.}$$

Six No. 9 bottom bars OK for crack control.

c. Negative Bending Reinforcement—Exterior Girder Service I limit state, $\eta_i = 1.0$, gravity load factors = 1.0, moments from Table E16.2-2.

$$M_{200} = \sum \eta_i \gamma_i Q_i$$
$$= M_{DC} + M_{DW} + mgrM_{LL+IM}$$
$$= (-253.5 - 28.5 - 302.2)$$
$$= -584.2 \text{ kip-ft}$$
$$d_{neg} = 36.9 \text{ in.}$$

Assume $j = 0.875$ and $f_s = 36$ ksi

$$A_s \approx \frac{M}{f_s jd} = \frac{584.2 \times 12}{36 \times 0.875 \times 36.9} = 6.03 \text{ in.}^2$$

Try nine No. 8 bars, provided $A_s = 7.07$ in.^{2} (Table B.3).

Tension reinforcement in flange distributed over the lesser of: effective flange width or one-tenth span [A5.7.3.4].

Effective flange width b_e for an exterior girder [A4.6.2.6.1]

$$b_e = 79.0 \text{ in.}$$
$$\tfrac{1}{10} \text{ average span} = \tfrac{1}{10}(38.5 \times 12)$$
$$= 46.2 \text{ in.} \quad \text{governs}$$

Effective flange width b_e is greater than one-tenth span, additional reinforcement is required in outer portions of the flange.

$$\text{Additional } A_s > 0.004 \text{ (excess slab area)}$$
$$> 0.004\,(8.0)\,(52.8 - 46.2) = 0.21 \text{ in.}^2$$

Two No. 4 bars additional reinforcement, additional $A_s = 0.40$ in.^{2} (Fig. E16.2-17).

Spacing of nine No. 8 bars = 46.2/8 spaces = 5.8 in. Calculation of maximum allowable bar

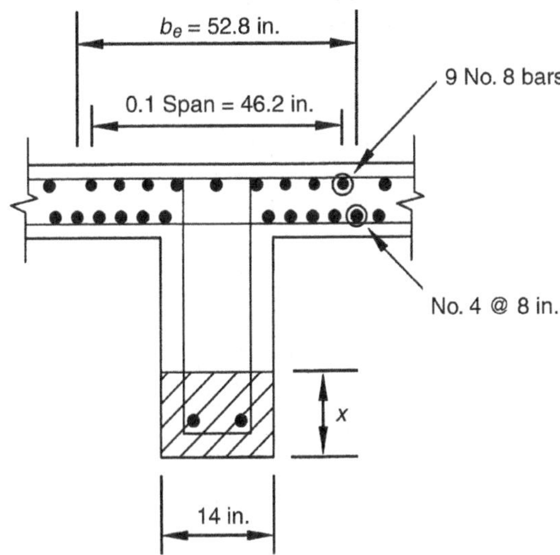

Fig. E16.2-17 Elastic-cracked transformed negative moment section at location 200.

spacing depends on service load tension stress f_s.

Revised d_{neg} for No. 8 bars below No. 4 transverse slab top bars

$$d_{\text{neg}} = 40 - 2.5 - 0.5 - \frac{1.0}{2} = 36.5 \text{ in.}$$
$$b = b_w = 14 \text{ in.}$$

Neglecting No. 4 longitudinal slab bottom bars

$$\frac{nA_s}{b} = \frac{7\,(7.07)}{14} = 3.54 \text{ in.}$$
$$\frac{2nA_s d}{b} = 2\,(3.54)\,36.5 = 258.4 \text{ in.}^2$$
$$x = -3.54 + \sqrt{3.54^2 + 258.4} = 12.9 \text{ in.}$$
$$I_{\text{cr}} = \tfrac{1}{3}(14)(12.9)^3 + 7(7.07)(36.5 - 12.9)^2 = 37{,}582 \text{ in.}^4$$
$$f_s = \frac{nM\,(d-x)}{I_{cr}} = \frac{7\,(584.2 \times 12)\,(36.5 - 12.9)}{37{,}582} = 30.8 \text{ ksi}$$
$$d_c = 2.5 + 0.5 + 1.0/2 = 3.5 \text{ in.}$$
$$h = 40 \text{ in.} \qquad \gamma_e = 0.75$$
$$\beta_s = 1 + \frac{d_c}{0.7\,(h - d_c)} = 1 + \frac{3.5}{0.7\,(40 - 3.5)} = 1.137$$
$$s \le \frac{700\gamma_e}{\beta_s f_s} - 2d_c = \frac{700\,(0.75)}{1.137 \times 30.8} - 2 \times 3.5 = 8.0 \text{ in.} > s = 5.8 \text{ in.}$$

Nine No. 8 top bars OK for crack control.

6. *Investigate Fatigue* Fatigue I Limit State (Table 5.1) [Table A3.4.1-1]

$$U_f = \sum \eta_i \gamma_i Q_i = 1.5\,(\text{LL} + \text{IM})$$

Fatigue Load

- ☐ One design truck with constant spacing of 30 ft between 32-kip axles [A3.6.1.4].
- ☐ Dynamic load allowance: IM = 15% [A3.6.2.1].
- ☐ Distribution factor for one traffic lane shall be used [A3.6.1.4.3b].
- ☐ Multiple presence factor of 1.2 shall be removed [C3.6.1.1.2].

a. Determination of Need to Consider Fatigue [A5.5.3.1] Prestressed beams may be precompressed, but for the continuous T-beam without prestress, there will be regions, sometimes in the bottom of the beam, sometimes in the top of the beam, where the permanent loads do not produce compressive stress. In these regions, such as locations 104 and 200, fatigue must be considered.

b. Allowable Fatigue Stress Range f_f in Reinforcement [A5.5.3.2]

$$f_f \le 21 - 0.33 f_{\min} + 8\left(\frac{r}{h}\right), \text{ ksi}$$

where

$f_{\min}$ = algebraic minimum stress level from fatigue load given above, positive if tension

$\frac{r}{h}$ = ratio of base radius to height of rolled-on transverse deformations if the actual value is not known, 0.3 may be used

c. Location 104 (Fig. E16.2-18) [C3.6.1.1.2] Exterior Girder—Distribution Factor

$$g_M^{\text{SE}} r = mg_M^{\text{SE}} \frac{r}{m} = \frac{0.750\,(0.948)}{1.2} = 0.593 \qquad \text{[C3.6.1.1.2]}$$

Fatigue load moment for maximum tension in reinforcement. Fatigue I limit state is used. Influence line ordinates taken from Table 9.4.

$$\text{pos } M_u = 32\,(0.20700)\,35 + 8\,(0.05171)\,35 = 246.3 \text{ kip-ft}$$
$$\text{pos } M_{104} = 1.5\left[g_M^{\text{SE}} r M_u\,(1 + \text{IM})\right] = 1.5\,[(0.593)\,(246.3)\,(1.15)] = 252.0 \text{ kip-ft}$$

Fatigue load moment for maximum compression in reinforcement (Fig. E16.2-19):

$$\text{neg } M_{\text{LL}} = [32\,(-0.04135 + 0.00574) + 8\,(0.00966)]35 = -37.2 \text{ kip-ft}$$

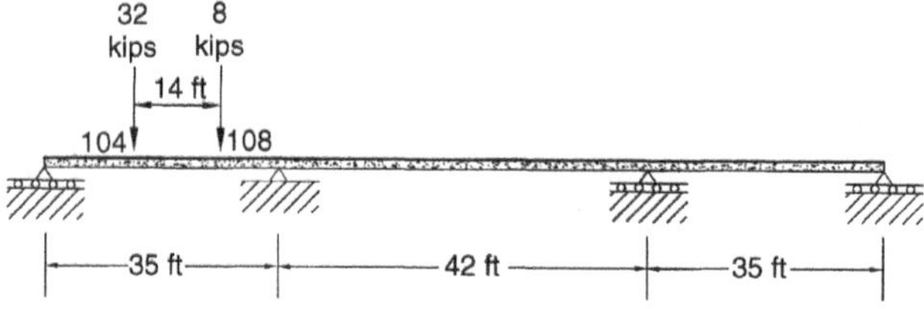

Fig. E16.2-18 Fatigue truck placement for maximum tension in positive moment reinforcement.

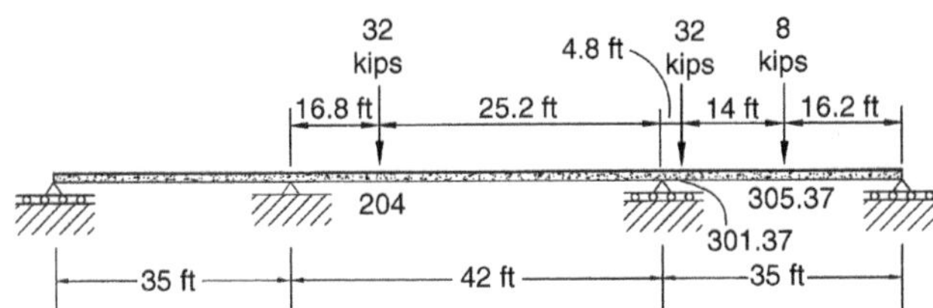

Fig. E16.2-19 Fatigue truck placement for maximum compression in positive moment reinforcement.

$$\text{neg } M_{104} = 1.5\,[0.593\,(-37.2)\,(1.15)] = -38.0 \text{ kip-ft}$$

The fatigue load moment varies from −38.0 to 252.0 kipft. The moment from dead load for an exterior girder is given in Table E16.2-2 as

$$M_{\text{DL}} = M_{\text{DC}} + M_{\text{DW}} = 148.4 + 16.7 = 165.1 \text{ kip-ft}$$

The combined moment at location 104 due to permanent loads plus the fatigue truck is always positive and never produces compression in the bottom flexural steel. Therefore, the maximum and minimum fatigue stresses are calculated using positive moment cracked section properties. The maximum fatigue stress is

$$f_{\max} = \frac{n\left(M_{\text{DL}} + M_{\text{FTr max}}\right)(d - x)}{I_{\text{cr}}} = \frac{7\,(165.1 + 252)\,(12)\,(35.8 - 5.66)}{42{,}928} = 24.6 \text{ ksi}$$

The minimum fatigue stress is

$$f_{\min} = \frac{n\left(M_{\text{DL}} + M_{\text{FTrmin}}\right)(d - x)}{I_{\text{cr}}} = \frac{7\,(165.1 - 38.0)\,12\,(35.8 - 5.66)}{42{,}928} = 7.50 \text{ ksi}$$

and the stress range f_f for fatigue at location 104 becomes

$$f_f = f_{\max} - f_{\min} = 24.6 - 7.50 = 17.1 \text{ ksi}$$

The limit for the stress range is

$$21 - 0.33 f_{\min} + 8\left(\frac{r}{h}\right) = 21 - 0.33\,(7.50) + 8\,(0.3) = 20.9 \text{ ksi} > 17.1 \text{ ksi} \quad \text{OK}$$

d. Location 200 Based on previous calculations, the moments due to LL + IM at location 200 are less than those at location 104. Therefore, by inspection, the fatigue stresses are not critical.

7. *Calculate Deflection and Camber* (Table 5.1) [Table A3.4.1-1] Service I limit state, $\eta_i = 1.0$, gravity load factors = 1.0

$$U = \sum \eta_i \gamma_i Q_i = \text{DC} + \text{DW} + (\text{LL} + \text{IM})$$

a. Live-Load Deflection Criteria (optional) [A2.5.2.6.2]

☐ Distribution factor for deflection [C2.5.2.6.2]

$$mg = m\frac{N_L}{N_B} = 0.85\frac{3}{6} = 0.425$$

N_L = No. design lanesn N_B = No. of beams [A3.6.1.1.2]:

☐ A right cross section may be used for skewed bridges.

☐ Use one design truck or lane load plus 25% design truck [A3.6.1.3.2].

☐ Live-load deflection limit, first span [A2.5 .2.6.2]:

$$\Delta_{\text{allow}} = \frac{\text{span}}{800} = \frac{35 \times 12}{800} = 0.53 \text{ in.}$$

b. Section Properties at Location 104 Transformed cracked section from Section 16.2, Part I.5b:

$$d_{\text{pos}} = 35.8 \text{ in.} \quad x = 5.66 \text{ in.}$$
$$I_{\text{cr}} = 42{,}928 \text{ in.}^4$$

Gross or uncracked section (Fig. E16.2-20)

$$A_g = 79\,(7.5) + 14\,(32) = 592.5 + 448$$
$$A_g = 1040.5 \text{ in.}^2$$
$$\bar{y} = \frac{592.5\,(32 + 3.75) + 448\,(16)}{1040.5} = 27.25 \text{ in.}$$

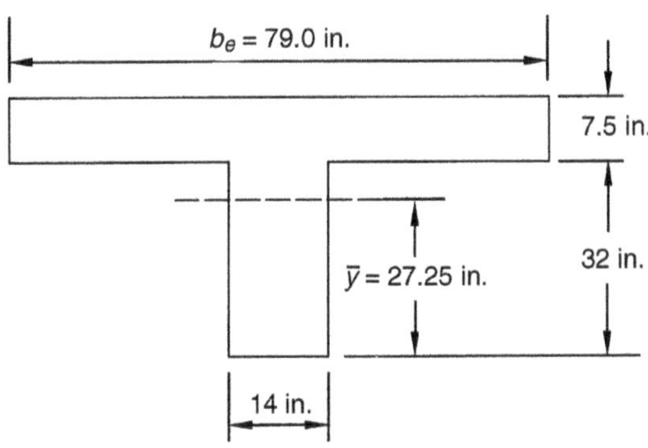

Fig. E16.2-20 Uncracked or gross section.

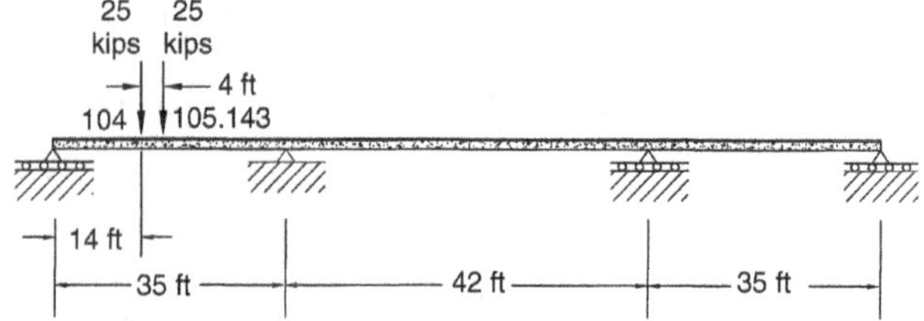

Fig. E16.2-21 Live-load placement for deflection at location 104.

$$I_g = \tfrac{1}{12}(79)(7.5)^3 + 592.5(12.25 - 3.75)^2 + \tfrac{1}{12}(14)(32)^3 + 448(27.25 - 16)^2 = 140{,}515 \text{ in.}^4$$

c. Estimated Live-Load Deflection at Location 104 Assume deflection is maximum where moment is maximum (Fig. E16.2-21):

$$M_{104} = 25\,(0.20700 + 0.15732)\,35 = 318.8 \text{ kip-ft}$$

Coefficients from Table 5.4.

$$M_{200} = 25\,(-0.08250 - 0.09240)\,35 = -153.0 \text{ kip-ft}$$

Total moment at 104,

$$M_a = M_{\text{DC}} + M_{\text{DW}} + mg M_{\text{LL}}\,(1 + \text{IM}) = 148.4 + 16.7 + 0.425\,(318.8)\,(1.33) = 345.3 \text{ kip-ft}$$

Effective moment of inertia [A5.7.3.6.2]

$$f'_c = 4.5 \text{ ksi}$$

$$E_c = 1820\sqrt{f'_c} = 1820\sqrt{4.5} = 3860 \text{ ksi} \qquad [\text{C5.4.2.4}]$$

$$f_r = 0.24\sqrt{f'_c} = 0.24\sqrt{4.5} = 0.509 \text{ ksi} \qquad [5.4.2.6]$$

$$M_{\text{cr}} = f_r \frac{I_g}{y_t} = 0.509 \frac{140{,}515}{27.25}/12 = 218.8 \text{ kip-ft}$$

$$I_e = \left(\frac{M_{\text{cr}}}{M_a}\right)^3 I_g + \left[1 - \left(\frac{M_{\text{cr}}}{M_a}\right)^3\right] I_{\text{cr}} \le I_g,$$

$$\left(\frac{M_{\text{cr}}}{M_a}\right)^3 = \left(\frac{218.8}{345.3}\right)^3 = 0.254$$

$$I_e = (0.254)(140{,}515) + (1 - 0.254)\,(42{,}928) = 67{,}715 \text{ in.}^4$$

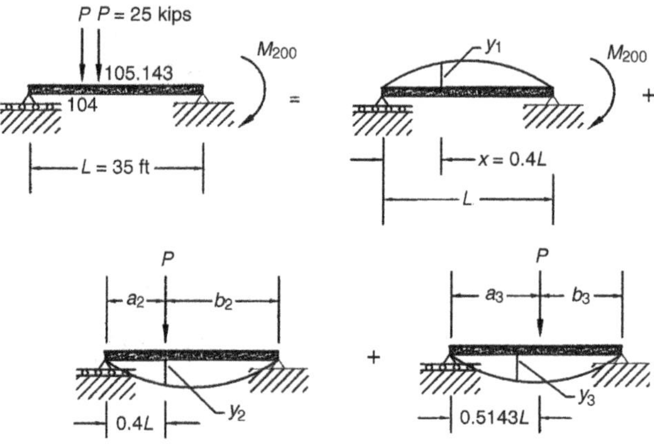

Fig. E16.2-22 Deflection estimate by superposition.

$$EI = E_c I_e = (3860)\,(67{,}715) = 261.4 \times 10^6 \text{ kip-in.}^2$$

Calculate deflection at location 104 by considering first span as a simple beam with an end moment and use superposition (Fig. E16.2-22). Deflections for a design truck are (Eq. 5.19)

$$y_1 = \frac{L^2}{6EI}\left[M_{ij}\,(2\xi - 3\xi^2 + \xi^3) - M_{ji}\,(\xi - \xi^3)\right]$$

$$\xi = \frac{x}{L}$$

$$M_{ij} = 0 \qquad M_{ji} = M_{200} = -153.0 \text{ kip-ft}$$

$$L = 35 \text{ ft} = 420 \text{ in.} \qquad \xi = 0.4$$

$$y_1 = \frac{(420)^2}{6\,(261.4 \times 10^6)} \times \left[-(-153.0 \times 12)\,(0.4 - 0.4^3)\right] = 0.069 \text{ in.} \uparrow \text{(upward)}$$

$$y_2 = \Delta_x\,(x < a) = \frac{Pbx}{6EIL}\,(L^2 - b^2 - x^2)$$

[AISC Manual (2001), Case 8]

For $P = 25$ kips, $x = 0.4L = 168$ in., $b_2 = 0.6L = 252$ in.

$$y_2 = \frac{25\,(252)\,(168)}{6\,(261.4 \times 10^6)\,(420)} \times (420^2 - 252^2 - 168^2) = 0.136 \text{ in.} \downarrow \text{(downward)}$$

For $P = 25$ kips, $x = 0.4L$, $a_3 = 0.5143(420) = 216$ in., $b_3 = L - a_3 = 204$ in.

$$y_3 = \frac{25\,(204)\,(168)}{6\,(261.4 \times 10^6)\,(420)} \times (420^2 - 204^2 - 168^2) = 0.139 \text{ in.} \downarrow \text{(downward)}$$

Estimated LL + IM Deflection at 104 With three lanes of traffic supported on six girders, each girder carries only a half-lane load. Including impact and the multiple presence factor, the estimated live-load deflection is

$$\Delta_{104}^{\text{LL+IM}} = mg\left(-y_1 + y_2 + y_3\right)(1 + \text{IM})$$
$$= 0.85\,(0.5)\,(-0.069 + 0.136 + 0.139)\,(1.33)$$
$$= 0.12 \text{ in.} < \Delta_{\text{allow}} = 0.53 \text{ in.} \quad \text{OK}$$

d. Dead-Load Camber [A5.7.3.6.2] The dead loads taken from Tables E16.2-1 and E16.2-2 are

Dead Loads	Interior Girder	Exterior Girder
w_{DC}	1.267 kips/ft	1.699 kips/ft
w_{DW}	0.280	0.191
w_{DL}	1.547 kips/ft	1.890 kips/ft

Unit Load Analysis (Fig. E16.2-23)

Deflection Equations Simple beam at distance x from left end, uniform load:

$$\Delta_x = \frac{wx}{24EI}\left(L^3 - 2Lx^2 + x^3\right)$$
$$\Delta_{\text{centerline}} = \frac{5}{384}\frac{wL^4}{EI}$$

[AISC Manual (2001), Case 1]

Simple beam at $\xi = x/L$ from i end, due to end moments:

$$y = \frac{L^2}{6EI}\left[M_{ij}\left(2\xi - 3\xi^2 + \xi^3\right) - M_{ji}\left(\xi - \xi^3\right)\right]\xi = \frac{x}{L}$$

Flexural Rigidity EI for Longtime Deflections The instantaneous deflection is multiplied by a creep factor λ to give a longtime deflection:

$$\Delta_{\text{LT}} = \lambda\Delta_i$$

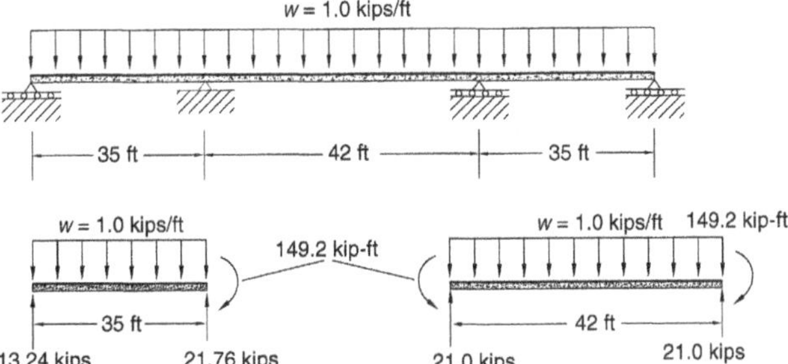

Fig. E16.2-23 Unit uniformly distributed load analysis.

so that

$$\Delta_{\text{camber}} = \Delta_i + \Delta_{\text{LT}} = (1 + \lambda)\,\Delta_{\text{i}}$$

If instantaneous deflection is based on I_g, $\lambda = 4.0$ [A 5.7.3.6.2]

If instantaneous deflection is based on I_e,

$$\lambda = 3.0 - 1.2\left(\frac{A_S'}{A_S}\right) \geq 1.6$$

Location 104, $x = 0.4L = 168$ in.

$$w = 1.0 \text{ kips/ft} \quad \text{(unit load)}$$
$$M_{ij} = 0 \qquad M_{ji} = -149.2 \text{ kip-ft}$$
$$\Delta_i = \frac{1.0\,(168/12)}{24 \times 261.4 \times 10^6} \times \left[(420)^3 - 2\,(420)\,(168)^2 + (168)^3\right] - \frac{(420)^2}{6 \times 261.4 \times 10^6} \times \left[-(-149.2 \times 12)\left(0.4 - 0.4^3\right)\right]$$
$$\Delta_i = 0.123 - 0.068$$
$$= 0.055 \text{ in.} \downarrow \quad \text{(downward)}$$

Using A_s = six No. 9 bars = 6.0 in.^{2}, A_s' = two No. 8 bars = 1.57 in.^{2}

$$\lambda = 3.0 - 1.2\frac{1.57}{6.0} = 2.69$$

Exterior girder, $w_e = 1.890$ kips/ft

$$\Delta_{\text{camber}} = 1.890\,(1 + 2.69)\,(0.055) = 0.38 \text{ in.}$$
$$\left(w_i = 1.547 \text{ kips/ft}\right) = 0.31 \text{ in.,}$$

say 0.35 in. average

Location 205 Assume same *EI* as at 104:

$$w = 1.0 \text{ kips/ft} \quad \text{(unit load)}$$
$$M_{ij} = -M_{ji} = 149.2 \text{ kip-ft}$$
$$x = 0.5L \qquad L = 42 \times 12 = 504 \text{ in.}$$
$$\Delta_i = \frac{5}{384}\frac{1.0(504)^4/12}{261.4 \times 10^6} - \frac{(504)^2}{6 \times 261.4 \times 10^6}\left[149.2 \times 12\left(1 - \tfrac{3}{4} + \tfrac{1}{8} + \tfrac{1}{2} - \tfrac{1}{8}\right)\right]$$
$$= 0.268 - 0.217$$
$$= 0.051 \text{ in.} \downarrow \quad \text{(downward)}$$

By using $\lambda = 2.69$ and $w_e = 1.890$ kips/ft

$$\Delta_{\text{camber}} = 1.890\,(1 + 2.69)\,(0.051) = 0.36 \text{ in.}$$
$$\left(w_i = 1.547 \text{ kips/ft}\right) = 0.29 \text{ in.,}$$

say 0.33 in. average

Dead-Load Deflection Diagram—All Girders (Fig. E16.2-24) Upward camber should be placed in the formwork to offset the estimated longtime dead-load deflection. The dead-load deflections are summarized in Figure E16.2-24.

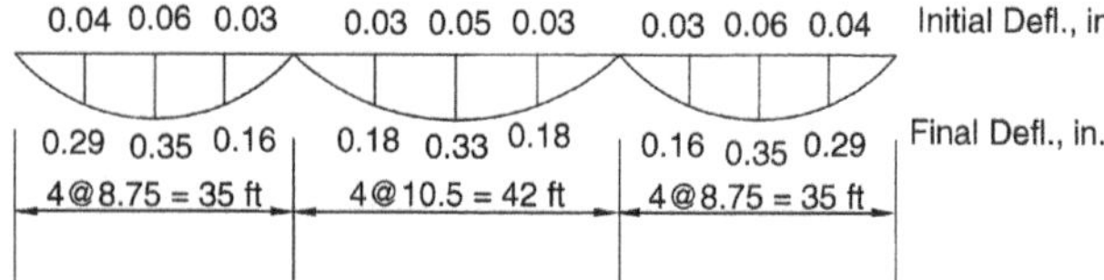

Fig. E16.2-24 Dead-load deflection diagram—all girders.

J. ***Investigate Strength Limit State*** The previous calculations for the service limit state considered only a few critical sections at locations 104, 200, and 205 to verify the adequacy of the trial section given in Figure E16.2-2. Before proceeding with the strength design of the girders, it is necessary to construct the factored moment and shear envelopes from values calculated at the tenth points of the spans. The procedure for generating the live-load values is given in Chapter 9 and summarized in Tables 9.8a and 9.8b for spans of 35, 42, and 35 ft.

The strength I limit state can be expressed as

$$\eta_i = \eta = 1.0$$

$$U = 1.0[1.25\text{DC} + 1.50\text{DW} + 1.75(mgr)\text{LL}(1 + \text{IM})] \qquad \text{(E16.2-1)}$$

With the use of permanent loads given in Tables E16.2-1 and E16.2-2, the critical live-load moments and shears from Tables 9.8a and 9.8b, and live-load distribution factors (mgr) determined earlier, the envelope values for moment and shear are generated for interior and exterior girders. Using Eq. E16.2-1, the envelope values are generated and given in Tables E16.2-3 and E16.2-4 in the columns with the factored values of moment and shear. The envelope values for moment and shear are plotted in Figure E16.2-25. Notice the closeness of the curves for the interior and exterior girders. One design will suffice for both.

1. *Flexure*
   a b. Prestressed Beams Not applicable.
   c. Factored Flexural Resistance [A5.7.3.2, Table A3.4.1-1] Exterior girder has slightly larger moment.

$$M_u = \sum \eta_i \gamma_i M_i$$
$$= 1.0\left(1.25M_{\text{DC}} + 1.50M_{\text{DW}} + 1.75M_{\text{LL+IM}}\right)$$

Location 104 Computation of the factored moment requires the unfactored values for moment from Table E16.2-2:

$$M_{104} = 1.0[1.25\,(148.4) + 1.50\,(16.7) + 1.75\,(358.4)]$$
$$= 837.8 \text{ kip-ft}$$

This number is the same as the value of 837.8 kip-ft found in Table E16.2-3.

Check resistance provided by bars selected for crack control (Fig. E16.2-26). Assume $a < t_s = 7.5$ in.

$$a = \frac{A_s f_y}{0.85 f'_c b_e} = \frac{6.0\,(60)}{0.85\,(4.5)\,(79)}$$
$$= 1.19 \text{ in.} \qquad \text{[A5.7.3.2.2]}$$

Table E16.2-3 Moment Envelope for Three-Span Continuous T-Beam 35–42–35 ft (kip-ft)

		Positive Moment			Negative Moment		
Location	Unit Uniform Load	Critical LL + IM	Factored Int. Girder	Factored Ext. Girder	Critical LL + IM	Factored Int. Girder	Factored Ext. Girder
100	0.00	0.0	0.0	0.0	0.0	0.0	0.0
101	40.21	225.2	359.3	377.1	−29.7	44.3	59.9
102	68.16	374.5	600.1	630.3	−59.3	64.2	90.3
103	83.87	468.1	747.4	784.6	−89.0	59.4	91.2
104	87.32	504.2	799.0	837.8	−118.6	30.3	62.6
105	78.53	495.8	771.0	806.2	−148.3	−23.6	4.4
106	57.49	449.5	671.5	697.8	−178.0	−102.0	−83.4
107	24.19	359.0	492.8	505.0	−207.6	−204.9	−200.5
108	−21.35	232.5	245.0	237.8	−237.3	−332.4	−347.3
109	−79.15	94.5	−41.6	−73.2	−279.6	−499.8	−539.4
110	−149.19	75.8	−205.1	−265.3	−422.9	−815.0	−886.9
200	−149.19	75.8	−205.1	−265.3	−422.9	−815.0	−886.9
201	−69.81	106.0	−8.7	−36.4	−245.7	−439.7	−474.6
202	−8.07	258.3	303.5	301.9	−196.5	−256.0	−264.5
203	36.03	389.9	554.7	572.0	−180.1	−147.6	−137.7
204	62.49	470.3	707.3	735.8	−177.8	−91.8	−71.1
205	71.31	492.8	752.8	785.0	−175.5	−71.3	−47.0

Table E16.2-4 Shear Envelope for Three-Span Continuous T-Beam 35–42–35 ft (kips)

		Positive Shear			Negative Shear		
Location	Unit Uniform Load	Critical LL + IM	Factored Int. Girder	Factored Ext. Girder	Critical LL + IM	Factored Int. Girder	Factored Ext. Girder
100	13.24	76.2	147.6	143.4	−9.4	15.1	20.2
101	9.74	63.4	120.2	116.3	−9.6	7.8	11.5
102	6.24	51.8	94.8	90.8	−14.0	−4.6	−2.4
103	2.74	42.5	73.0	68.8	−22.8	−22.3	−21.8
104	−0.76	33.7	52.0	47.5	−31.6	−40.1	−41.2
105	−4.26	25.7	32.3	27.3	−40.3	−57.7	−60.5
106	−7.76	18.4	13.7	8.2	−48.8	−75.1	−79.6
107	−11.26	12.1	−3.3	−9.4	−57.8	−93.1	−99.2
108	−14.76	6.6	−19.1	−25.9	−68.4	−113.0	−120.9
109	−18.26	2.9	−32.0	−39.8	−78.7	−132.6	−142.1
110	−21.76	2.2	−40.1	−49.2	−89.0	−152.2	−163.4
200	21.00	90.2	185.3	182.6	−8.4	31.8	40.1
201	16.80	78.5	158.3	155.4	−8.5	23.3	29.9
202	12.60	66.2	130.4	127.3	−9.9	13.2	18.0
203	8.40	53.9	102.4	99.1	−17.0	−3.9	−1.0
204	4.20	43.4	77.3	73.6	−25.1	−22.2	−21.2
205	0.00	34.0	54.0	49.8	−34.0	−41.5	−42.4

All compression is in flange.

$a \le 0.375\beta_1 \ d = 0.375\,(0.825)\,(35.8) = 11.1$ in., $\phi = 0.9$

$$\phi M_n = \phi A_s f_y \left(d - \frac{a}{2}\right)$$

$$= 0.9\,(6.0)\,(60)\left(35.8 - \frac{1.19}{2}\right)\Big/12$$

$\phi M_n = 934.5$ kip-ft $> M_u = 837.8$ kip-ft   OK

Use six No. 9 bottom bars.

Location 200 Computation of the factored moment requires the unfactored values for moment from Table E16.2-2

$$M_{200} = 1.0[1.25(-253.5) + 1.50(-28.5) + 1.75(-302.2)] = -888.5 \text{ kip-ft}$$

This number is comparable to the value of −886.9 kip-ft found in Table E16.2-3. Check resistance provided by bars selected for crack control (Fig. E16.2-27). Neglecting compression reinforcement

$$a = \frac{7.07\,(60)}{0.85\,(4.5)\,(14)} = 7.92 \text{ in.}$$

$$a \le 0.375\beta_1 d = 0.375\,(0.825)\,(36.4) = 11.3 \text{ in.}, \ \phi = 0.9$$

$$\phi M_n = 0.9\,(7.07)\,(60)\left(36.4 - \frac{7.92}{2}\right)\Big/12$$

$\phi M_n = 906.1$ kip-ft $> M_u = 888.5$ kip-ft   OK

Use nine No. 8, top bars.

d. Limits for Reinforcement

$$\beta_1 = 0.85 - 0.05\,(4.5 - 4.0) = 0.825 \quad \text{[A5.7.2.2]}$$

Minimum reinforcement such that

$$\phi M_n \ge 1.2 M_{cr} \quad \text{[A5.7.3.3.2]}$$

Gross section properties $\bar{y} = 27.25$ in., $h - \bar{y} = 12.25$ in., $I_g = 140{,}515$ in.^{4}

$$f_r = 0.37\sqrt{f'_c} = 0.37\sqrt{4.5} = 0.785 \text{ ksi}$$

Location 104

$$M_{cr} = \frac{f_r I_g}{\bar{y}} = \frac{0.785\,(140{,}515)}{27.25}\Big/12 = 337 \text{ kip-ft}$$

$$\phi M_n = 934.5 \text{ kip-ft} > 1.2 M_{cr} = 1.2\,(337) = 405 \text{ kip-ft} \quad \text{OK}$$

Location 200

$$M_{cr} = \frac{f_r I_g}{h - \bar{y}} = \frac{0.785\,(140{,}515)}{(12.25)}\Big/12 = 750 \text{ kip-ft}$$

$$\phi M_n = 906 \text{ kip-ft} > 1.2 M_{cr} = 1.2\,(750) = 900 \text{ kip-ft} \quad \text{OK}$$

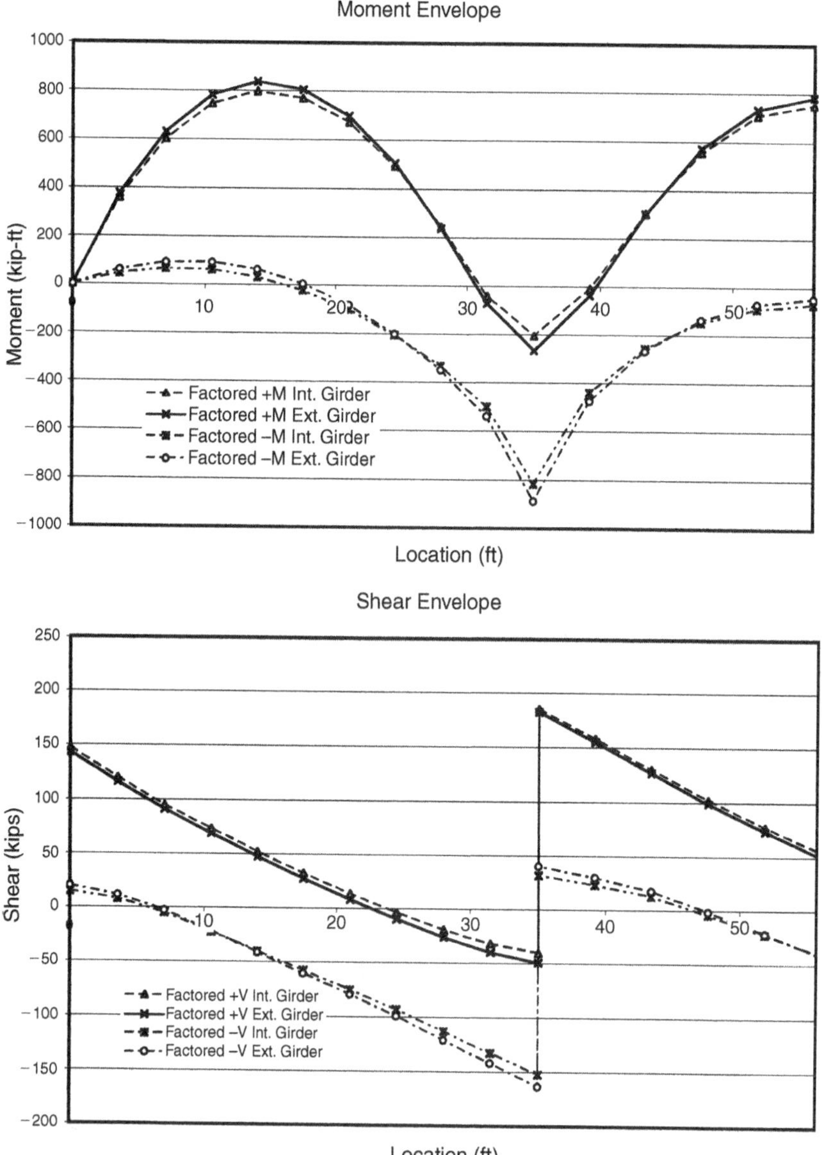

Fig. E16.2-25 Envelopes of factored moments and shears at tenth points for T-beams.

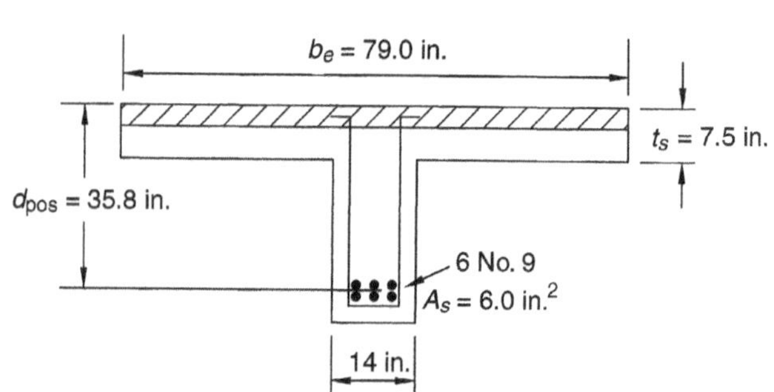

Fig. E16.2-26 Positive moment design section.

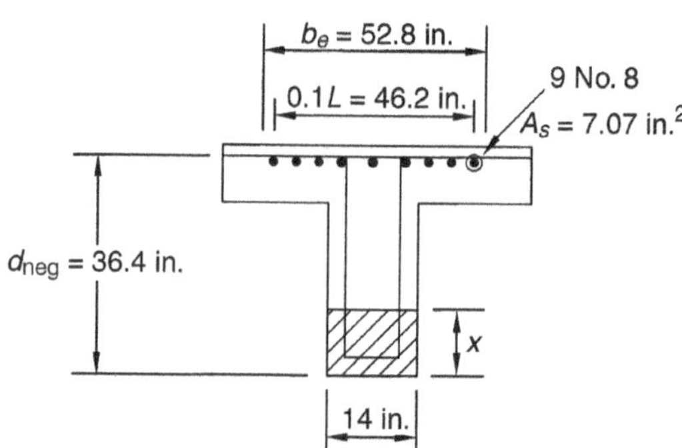

Fig. E16.2-27 Negative moment design section.

2. *Shear (Assuming No Torsional Moment)*

a. General Requirements

☐ Transverse reinforcement shall be provided where [A5.8.2.4]

$$V_u \geq 0.5\phi\left(V_c + V_p\right) \qquad \phi = \phi_v = 0.9$$

where

V_u = factored shear force (kips)
V_c = nominal shear resistance of concrete (kips)
V_p = component of prestressing force in the direction of the shear force (kips)

☐ Minimum transverse reinforcement [A5.8.2.5]

$$A_v \geq 0.0316\sqrt{f_c'}\frac{b_v s}{f_y}$$

where A_v = area of transverse reinforcement within distance s (in.^{2})
b_v = effective width of web adjusted for the presence of ducts (in.) [A5.8.2.9]
s = spacing of transverse reinforcement (in.)
f_y = yield strength of transverse reinforcement (ksi)

☐ Maximum spacing of transverse reinforcement [A5.8.2.7]

If $v_u < 0.125\ f_c'$, then $s_{\max} = 0.8d_v \leq 24$ in.
If $v_u \geq 0.125\ f_c'$, then $s_{\max} = 0.4d_v \leq 12$ in.

where

v_u = shear stress (ksi) $\frac{|V_u - \phi V_p|}{\phi b_v d_v}$ [A5.8.2.9]
b_v = minimum web width, measured parallel to the neutral axis, between the resultants of the tensile and compressive forces due to flexure, modified for the presence of ducts (in.)
d_v = effective shear depth taken as the distance, measured perpendicular to the neutral axis, between the resultants of the tensile and compressive forces due to flexure it need not be taken less than the greater of $0.9d_e$ or $0.72h$ (in.)

b. Sectional Design Model [A5.8.3]

☐ Based on equilibrium of forces and compatibility of strains (Collins and Mitchell, 1991).

☐ Where the reaction force produces compression at a support, the critical section for shear shall be taken as the larger of $0.5d_v \cot\theta$ or d_v from the internal face of the bearing (see Fig. E16.2-28) [A5.8.3.2].

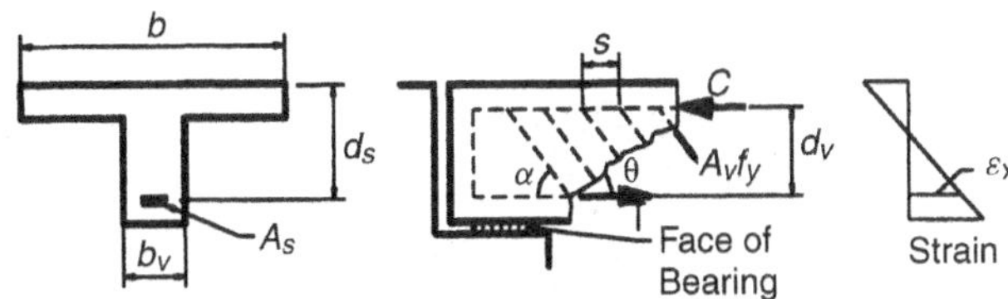

Fig. E16.2-28 Shear sectional design model.

Nominal Shear Resistance V_n [A5.8.3.3]

☐ Shall be the lesser of

$$V_n = V_c + V_s + V_p$$
$$V_n = 0.25 f_c' b_v d_v + V_p$$

☐ Nominal concrete shear resistance

$$V_c = 0.0316\beta\sqrt{f_c'}b_v d_v$$

where β is the factor indicating ability of diagonally cracked concrete to transmit tension [A5.8.3.4] (traditional value of $\beta = 2.0$) [A5.8.3.4.1].

☐ Nominal transverse reinforcement shear resistance

$$V_s = \frac{A_v f_y d_v (\cot\theta + \cot\alpha)\sin\alpha}{s}$$

for vertical stirrups $\alpha = 90^\circ$ [C5.8.3.3]

$$V_s = \frac{A_v f_y d_v \cot\theta}{s}$$

where θ is the angle of inclination of diagonal compressive stresses [A5.8.3.4] (traditional value of $\theta = 45^\circ$, $\cot\theta = 1.0$) [A5.8.3.4.1].

Determination of β and θ Use [A5.8.3.4.2] to determine β and θ. These tables depend on the following parameters for nonprestressed beams without axial load:

☐ Nominal shear stress in the concrete:

$$v_u = \frac{V_u}{\phi b_v d_v}$$

☐ Tensile strain in the longitudinal reinforcement for sections with transverse reinforcement:

$$\varepsilon_s = \frac{|M_u|/d_v + 0.5|V_u|\cot\theta}{E_s A_s}$$

Longitudinal Reinforcement [A5.8.3.5] Shear causes tension in the longitudinal reinforcement that must be added to that caused by flexure. Thus,

$$A_s f_y \geq \frac{|M_u|}{\phi_f d_v} + \left(\frac{|V_u|}{\phi_v} - 0.5V_s\right)\cot\theta$$

If this equation is not satisfied, either the tensile reinforcement A_s must be increased or the stirrups must be placed closer together to increases V_s.

The procedure outlined in Section 14.3.3 for the shear design of members with web reinforcement is illustrated for a section at a distance d_v from an interior support. The factored V_u and moment M_u envelopes for the strength I limit state are plotted in Figure E16.2-25 from the values in Tables E16.2-3 and E16.2-4.

Step 1. Determine V_u and M_u at a distance d_v from an interior support at location $200 + d_v$ [A5.8.2.7]. From Figure E16.2-27

$$A_s = \text{nine No. } 8 = 7.07 \text{ in.}^2$$

$$b_v = 14 \text{ in.} \qquad b_w = 14 \text{ in.}$$

$$a = \frac{A_s f_y}{0.85 f_c' b_w} = \frac{(7.07)(60)}{0.85(4.5)(14)} = 7.92 \text{ in.}$$

$$d = d_e = d_s = (40 - 0.5) - \left(2.5 + \frac{1.0}{2}\right)$$

$$= 36.5 \text{ in.}$$

$$d_v = \max\begin{cases} d - a/2 = 36.5 - 7.92/2 = 32.5 \text{ in.} \\ 0.9d_e = 0.9(36.5) = 32.9 \text{ in., governs} \\ 0.72h = 0.72(40) = 28.8 \text{ in.} \end{cases}$$

Distance from support as a percentage of the span

$$\frac{d_v}{L_2} = \frac{32.9}{42 \times 12} = 0.0653$$

Interpolating from Tables E16.2-3 and E16.2-4 or the factored shear and moment at location 200.653 for an interior girder:

$$V_{200.653} = 185.3 - 0.653(185.3 - 158.3)$$
$$= 167.7 \text{ kips}$$
$$M_{200.653} = -815.0 + 0.653(815 - 439.7)$$
$$= -569.9 \text{ kip-ft}$$

These values are used to calculate the strain ε_s on the flexural tension side of the member [A5.8.3.4.2]. They are both extreme values at the section and have been determined from different positions of the live load. It is conservative to take the highest value of M_u at the section, rather than a moment coincident with V_u. Moreover, if coincident actions are used, the maximum moment with coincident shear should be checked as well.

Step 2. Calculate the shear stress ratio v_u/f_c'.

$$v_u = \frac{|V_u|}{\phi b_v d_v} = \frac{167.7}{0.9(14)(32.9)} = 0.405 \text{ ksi}$$

so that

$$\frac{v_u}{f_c'} = \frac{0.405}{4.5} = 0.0899$$

Step 3. Estimate an initial value for θ and calculate ε_s from Eq. 14.130.

Try $\theta = 35°$, $\cot\theta = 1.428$, $E_s = 29{,}000$ ksi:

$$\varepsilon_s = \frac{(|M_u|/d_v) + 0.5|V_u|\cot\theta}{E_s A_s}$$
$$= \frac{\frac{569.9\times 12}{32.9} + 0.5(167.7)(1.428)}{(29{,}000)(7.07)}$$
$$= 1.60 \times 10^{-3}$$
$$\beta = \frac{4.8}{1 + 750\varepsilon_s}$$
$$= \frac{4.8}{1 + 750(0.00160)} = 2.18$$
$$\theta = 29 + 3500\varepsilon_s$$
$$= 29 + 3500(0.00160) = 34.5$$

Close enough, use 34.5°

Calculate the required web reinforcement strength V_s:

$$V_s = \frac{|V_u|}{\phi_v} - 0.0316\beta\sqrt{f_c'}b_v d_v$$
$$= \frac{167.7}{0.9} - 0.0316(2.18)\sqrt{4.5}(14)(32.9)$$
$$= 119.0 \text{ kips}$$

Step 4. Calculate the required spacing of stirrups:

No. 4 U-stirrups, $A_v = 2(0.20) = 0.40 \text{ in.}^2$

$$s \le \frac{A_v f_y d_v}{V_s}\cot\theta = \frac{0.40(60)(32.9)}{119.0}(1.455)$$
$$= 9.65 \text{ in.}$$
$$\le \frac{A_v f_y}{0.0316\sqrt{f_c'}b_v} = \frac{0.40(60)}{0.0316\sqrt{4.5}(14)}$$
$$= 25.6 \text{ in.}$$
$$v_u = 0.405 \text{ ksi} < 0.125 f_c' = 0.125(4.5)$$
$$= 0.563 \text{ ksi}$$
$$s \le 0.8d_v = 0.8(32.9) = 26.3 \text{ in.} \quad \text{or} \quad 24 \text{ in.}$$

Try $s = 9$ in.

Step 5. Check the adequacy of the longitudinal reinforcement:

$$A_s f_y \ge \frac{|M_u|}{d_v\phi_f} + \left(\frac{|V_u|}{\phi_v} - 0.5V_s\right)\cot\theta$$
$$V_s = \frac{A_v f_y d_v \cot\theta}{s}$$
$$= \frac{0.40(60)(32.9)}{9.0}(1.455)$$
$$= 127.7 \text{ kips}$$

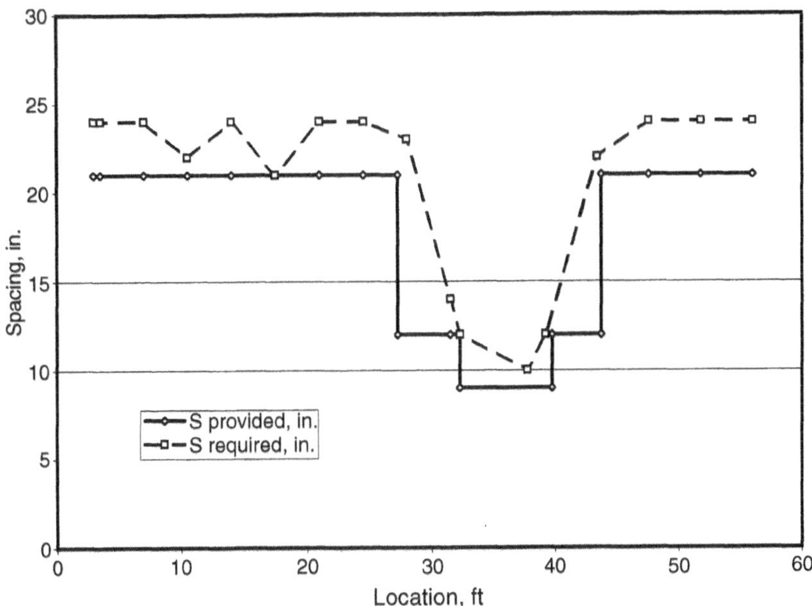

Fig. E16.2-29 Stirrup spacing for T-beam.

Table E16.2-5 Summary of Stirrup Spacing for T-Beam

Location	v_u/f'_c	θ	Strain ε_x (in./in.)	β	s Req'd (in.)	s Prov'd (in.)
$100 + d_v$	0.0624	31.6	0.000587	2.52	24	21
101	0.0602	32.1	0.000627	2.48	24	21
102	0.0475	34.1	0.000789	2.36	24	21
103	0.0366	35.1	0.000881	2.30	22	21
104	0.0261	35.2	0.000888	2.30	24	21
105	0.0303	35.4	0.000912	2.28	21	21
106	0.0399	34.8	0.000848	2.32	24	21
107	0.0497	33.2	0.000712	2.41	24	21
108	0.0649	31.2	0.000553	2.55	23	21
109	0.0763	33.6	0.000741	2.38	14	12
$110 - d_v$	0.0788	34.4	0.000809	2.34	12	12
$200 + d_v$	0.0900	34.5	0.000806	2.31	10	9
201	0.0850	33.0	0.000689	2.40	12	12
202	0.0653	31.8	0.000599	2.51	22	21
203	0.0513	33.8	0.000763	2.37	24	21
204	0.0387	34.8	0.000852	2.32	24	21
205	0.0271	34.8	0.000849	2.32	24	21

$$7.07\,(60) \geq \frac{569.9 \times 12}{32.9\,(0.9)} + \left[\frac{167.7}{0.9} - 0.5\,(127.7)\right](1.455)$$

$$424.2 \text{ kips} \geq 409.2 \text{ kips} \quad \text{OK}$$

The above procedure is repeated for each of the tenth points. The results are summarized in Table E16.2-5 and plotted in Figure E16.2-29. Stirrup spacings are then selected to have values less than the calculated spacings. Starting at the left end and proceeding to midspan of the T-beam, the spacings are 1 at 12 in., 15 at 21 in., 5 at 12 in., 10 at 9 in., 4 at 12 in., and 7 at 21 in. The selected stirrup spacings are shown by the solid line in Figure E16.2-29. This completes the design of the T-beam bridge example. Tasks remaining include the determination of cut-off points for the main flexural reinforcement, anchorage requirements for the stirrups, and side reinforcement in the beam stems.

16.3 PRESTRESSED GIRDER BRIDGE

Problem Statement Example 16.3 Design the simply supported pretensioned prestressed concrete girder bridge of Figure E16.3-1 with a span length of 100 ft center to center of bearings for an HL-93 live load. The roadway width is 44 ft curb to curb. Allow for a future wearing surface of 3-in.-thick bituminous overlay and use the concrete deck design of Example Problem 16.1 ($f_c' = 4.5$ ksi). Follow the beam and girder bridge outline in Section 5, Appendix A5.3 of the AASHTO (2010) LRFD Bridge Specifications. Use $f_{ci}' = 6$ ksi, $f_c' = 8$ ksi, $f_y = 60$ ksi, and 270 ksi, low-relaxation 0.5-in., seven-wire strands. The barrier is 15 in. wide and weighs 0.32 kips/ft. The owner requires this load to be assigned to the exterior girder.

A. ***Develop General Section*** The bridge is to carry interstate traffic in Virginia over a single-track railroad with minimum vertical clearance of 23 ft 4 in. (Fig. E16.3-1).
B. ***Develop Typical Section*** Use a precast pretensioned AASHTO-PCI bulb tee girder made composite with the deck (Fig. E16.3-2).

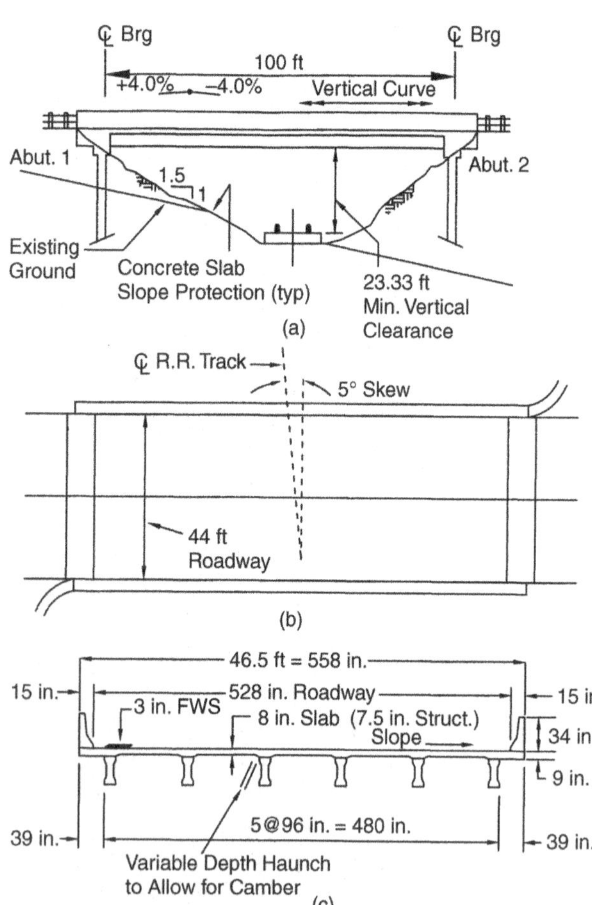

Fig. E16.3-1 Prestressed concrete girder bridge design example: (a) elevation, (b) plan, and (c) section.

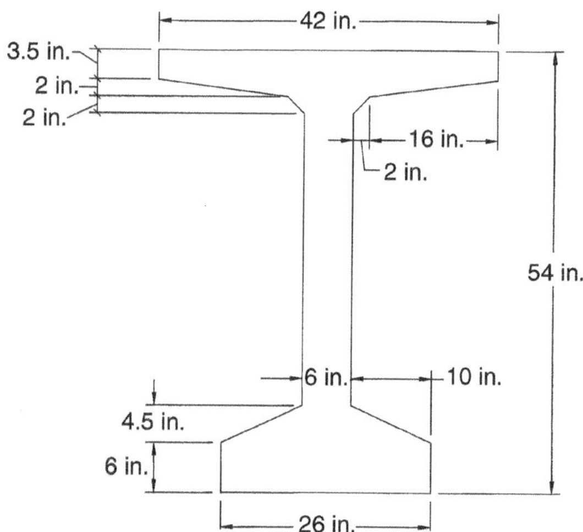

Fig. E16.3-2 Precast pretensioned AASHTO–PCI bulb tee girder BT54. $A_g = 659\,\text{in.}^2$.

1. *Minimum Thickness* [A5.14.1.2.2]

$$\text{Top flange} \geq 2.0\text{ in.}\quad \text{OK}$$
$$\text{Web} \geq 5.0\text{ in.}\quad \text{OK}$$
$$\text{Bottom flange} \geq 5.0\text{ in.}\quad \text{OK}$$

2. *Minimum Depth* (includes deck thickness) [A2.5.2.6.3]

$$\begin{aligned} h_{\min} &= 0.045L = 0.045\,(100 \times 12) \\ &= 54\text{ in.} < h = 54 + 7.5 \\ &= 61.5\text{ in.}\quad \text{OK} \end{aligned}$$

3. *Effective Flange Widths* [A4.6.2.6.1]
Effective span length $= 100 \times 12 = 1200$ in.
Interior girders

$$b_i \leq \text{center-to-center spacing of girders} = 96\text{ in.}$$

Exterior girders

$$b_e - \frac{b_i}{2} \leq \text{width of overhang} = 39\text{ in.}$$
$$b_e = \frac{96}{2} + 39 = 87\text{ in.}$$

C. ***Design Conventionally Reinforced Concrete Deck*** The design section for negative moments in the deck slab is at one-third the flange width, but not more than 15 in., from the centerline of the support for precast concrete beams [A4.6.2.1.6]. One-third of the flange width $b_f/3 = \frac{42}{3} = 14$ in. is less than 15 in.; therefore, the critical distance is 14 in. from the centerline of the support.

The deck design in Section 16.1, Part E, is for a monolithic T-beam girder and the design section is at the face of the girder or 7 in. from the centerline of the support (Fig. E16.1-10). The design negative moment for the composite deck, and resulting reinforcement, can be reduced by using the 14-in. distance rather than 7 in. By following the procedures in Section 16.1, Parts E and F.2, the top reinforcement at an interior support is reduced from No. 5 bars at 7.5 in. to No. 5 bars at 10 in. (Fig. E16.1-14).

The deck overhang design remains the same as for the T-beam (Fig. E16.1-17). It is governed by the truck collision and providing sufficient moment capacity to develop the strength of the barrier. The changes in the total design moment are small when the gravity loads are included at different distances from the centerline of the support. The dominant effect is the collision moment at the free end of the overhang and that remains the same, so the overhang design remains the same.

D. ***Select Resistance Factors*** (Table 14.6) [A5.5.4.2]

	ϕ	
1. *Strength Limit State*	ϕ	[A5.5.4.2.1]
Flexure and tension	1.00	
Shear and torsion	0.90	
Compression in anchorage zones	0.80	
2. *Nonstrength Limit States*	1.00	[A1.3.2.1]

E. ***Select Load Modifiers*** [A1.3.2.1]

	Strength	Service	Fatigue	
Ductility, η_D	1.0	1.0	1.0	[A1.3.3]
Redundancy, η_R	1.0	1.0	1.0	[A1.3.4]
Importance, η_I	1.0	N/A	N/A	[A1.3.5]
$\eta_i = \eta_D \eta_R \eta_I$	1.0	1.0	1.0	

F. ***Select Applicable Load Combinations*** (Table 5.1) [Table A3.4.1-1]

Strength I Limit State

$$\eta_i = \eta = 1.0$$

$$U = \eta\,[1.25\text{DC} + 1.50\text{DW} + 1.75\,(\text{LL} + \text{IM}) + 1.0\text{FR}]$$

Service I Limit State

$$U = 1.0(\text{DC} + \text{DW}) + 1.0(\text{LL} + \text{IM}) + 0.3(\text{WS} + \text{WL}) + 1.0\text{FR}$$

Fatigue I Limit State

$$U = 1.5\,(\text{LL} + \text{IM})$$

Service III Limit State

$$U = 1.0(\text{DC} + \text{DW}) + 0.80(\text{LL} + \text{IM}) + 1.0\text{WA} + 1.0\text{FR}$$

G. ***Calculate Live-Load Force Effects***

1. ***Select Number of Lanes*** [A3.6.1.1.1]:

$$N_L = \text{INT}\left(\frac{w}{12}\right) = \text{INT}\left(\frac{44}{12}\right) = 3$$

2. *Multiple Presence Factor* (Table 8.6) [A3.6.1.1.2]:

No. of Loaded Lanes	m
1	1.20
2	1.00
3	0.85

3. *Dynamic Load Allowance* (Table 8.7) [A3.6.2.1] *Not* applied to the design lane load.

Component	IM (%)
Deck joints	75
Fatigue	15
All other	33

4. *Distribution Factors for Moment* [A4.6.2.2.2]: Cross-Section Type (k) (Table 3.1) [Table A4.6.2.2.1-1]

$$\left.\begin{array}{ll}\text{Beam} & 8.0 - \text{ksi concrete}\\ \text{Deck} & 4.5 - \text{ksi concrete}\end{array}\right\}$$

n_c = modular ratio between beam and deck materials

$$= \sqrt{\frac{8.0}{4.5}} = 1.333$$

Stiffness factor, K_g (see Fig. E16.3-8 for additional cross section properties).

$$e_g = 26.37 + 2.0 + \frac{7.5}{2} = 32.1 \text{ in.}$$

$$K_g = n_c\left(I_g + Ae_g^2\right) = 1.333\left[268{,}077 + (659)\,(32.1)^2\right]$$

$$K_g = 1.263 \times 10^6 \text{ in.}^4$$

$$\frac{K_g}{12Lt_s^3} = \frac{1.263 \times 10^6}{12\,(100)\,(7.5)^3} = 2.494$$

$$S = 8.0 \text{ ft} \qquad L = 100 \text{ ft}$$

a. Interior Beams with Concrete Decks (Table 11.3) [A4.6.2.2.2b and Table A4.6.2.2.2b-1]

One Design Lane Loaded

$$mg_M^{\text{SI}} = 0.06 + \left(\frac{S}{14}\right)^{0.4}\left(\frac{S}{L}\right)^{0.3}\left(\frac{K_g}{12Lt_S^3}\right)^{0.1}$$

$$mg_M^{\text{SI}} = 0.06 + \left(\frac{8.0}{14}\right)^{0.4}\left(\frac{8.0}{100}\right)^{0.3}(2.494)^{0.1} = 0.47$$

Two or More Design Lanes Loaded

$$mg_M^{\text{MI}} = 0.075 + \left(\frac{S}{9.5}\right)^{0.6}\left(\frac{S}{L}\right)^{0.2}\left(\frac{K_g}{12Lt_S^3}\right)^{0.1}$$

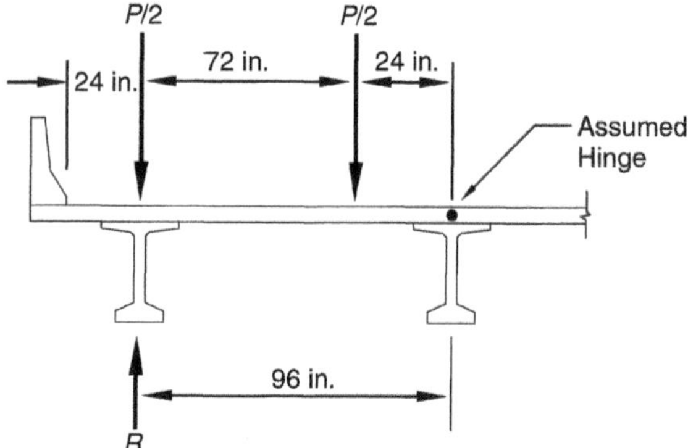

Fig. E16.3-3 Definition of lever rule for an exterior girder.

$$mg_M^{MI} = 0.075 + \left(\frac{8.0}{9.5}\right)^{0.6}\left(\frac{8.0}{100}\right)^{0.2}(2.494)^{0.1}$$
$$= 0.67 \quad \text{governs}$$

b. Exterior Beams with Concrete Decks (Table 11.3) [A4.6.2.2.2d and Table A4.6.2.2.2d-1]

One Design Lane Loaded—Lever Rule (Fig. E16.3-3)

$$R = \frac{P}{2}\left(\frac{24+96}{96}\right) = 0.625P$$
$$g_M^{SE} = 0.625$$
$$mg_M^{SE} = 1.2\,(0.625) = 0.75 \quad \text{governs}$$

Two or More Design Lanes Loaded

$$d_e = (39-15)/12 = 2.0 \text{ ft}$$
$$e = 0.77 + \frac{d_e}{9.1} = 0.77 + \frac{2.0}{9.1} = 0.990 < 1.0$$

Use $e = 1.0$

$$mg_M^{ME} = emg_M^{MI} = 0.67$$

5. *Distribution Factors for Shear* [A4.6.2.2.3] Cross-Section Type (k) (Table 3.1) [Table A4.6.2.2.1-1]

a. Interior Beams (Table 11.3) [A4.6.2.2.3a and Table A4.6.2.2.3a-1]

One Design Lane Loaded

$$mg_V^{SI} = 0.36 + \frac{S}{25} = 0.36 + \frac{8.0}{25} = 0.68$$

Two or More Design Lanes Loaded

$$mg_V^{MI} = 0.2 + \frac{S}{12} - \left(\frac{S}{35}\right)^{2.0}$$
$$mg_V^{MI} = 0.2 + \frac{8.0}{12} - \left(\frac{8.0}{35}\right)^{2.0}$$
$$= 0.81 \quad \text{governs}$$

b. Exterior Beams (Table 11.3) [A4.6.2.2.3b and Table A4.6.2.2.3b-1]

One Design Lane Loaded—Lever Rule (Fig. E16.3-3):

$$mg_V^{SE} = 0.75 \quad \text{governs}$$

Two or More Design Lanes Loaded

$$d_e = 2.0 \text{ ft}$$
$$e = 0.6 + \frac{d_e}{10} = 0.6 + \frac{2.0}{10}$$
$$= 0.800 \quad \text{Use } e = 1.0$$
$$mg_V^{ME} = emg_V^{MI} = (1.0)\,(0.81) = 0.81$$

6. *Calculation of Shears and Moments Due to Live Loads* The shears and moments at tenth points along the span are next. Calculations are shown below for locations 100, 101, and 105 only. Concentrated loads are multiplied by influence line ordinates. Uniform loads are multiplied by the area under the influence line. As discussed in Chapter 9, the influence functions are straight lines for simple spans. Shears and moments at the other locations are found in a similar manner. Results of these calculations are summarized in Tables E16.3-3 and E16.3-4.

Location 100 (Fig. E16.3-4)

Truck

$$V_{100}^{Tr} = 32\left(1 + \frac{86}{100}\right) + 8\left(\frac{72}{100}\right) = 65.28 \text{ kips}$$
$$M_{100}^{Tr} = 0$$

Lane

$$V_{100}^{Ln} = 0.64\,(0.5 \times 100) = 32 \text{ kips}$$
$$M_{100}^{Ln} = 0$$

Location 101 (Fig. E16.3-5):

Truck

$$V_{101}^{Tr} = R_A = 32\left(\frac{90+76}{100}\right) + 8\left(\frac{62}{100}\right)$$
$$= 58.08 \text{ kips}$$
$$M_{101}^{Tr} = \frac{10 \times 90}{100}\left[32\left(1 + \frac{76}{90}\right) + 8\left(\frac{62}{90}\right)\right]$$
$$= 580.8 \text{ kip-ft}$$

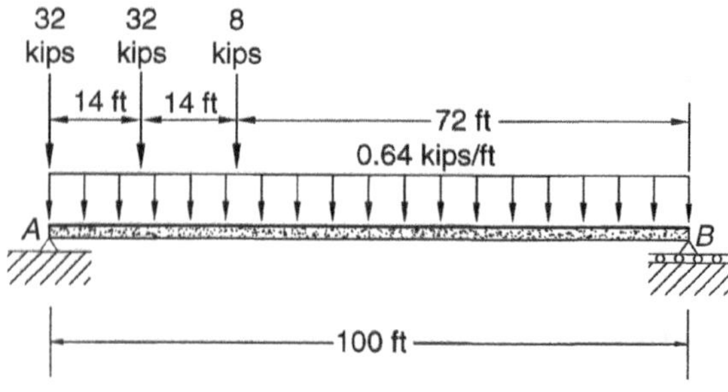

Fig. E16.3-4 Live-load placement at location 100.

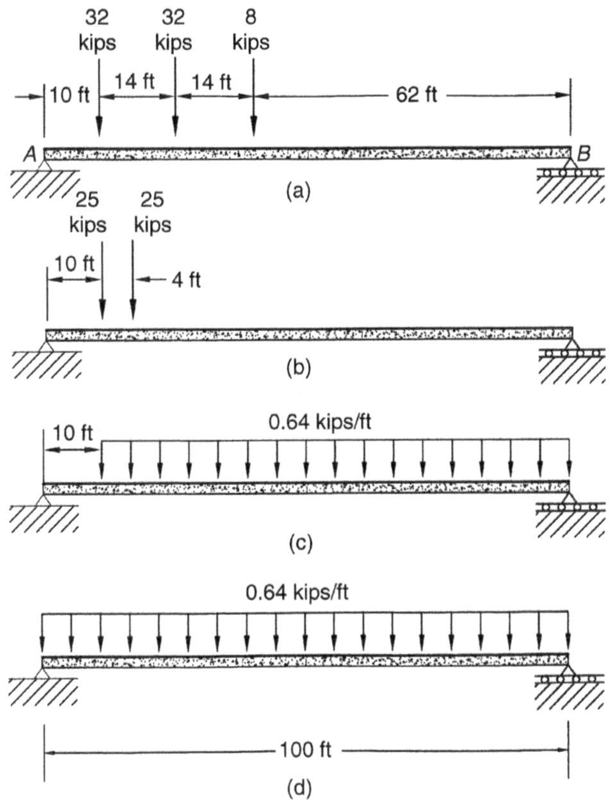

Fig. E16.3-5 Live-load placement at location 101: (a) truck, (b) tandem, (c) lane-shear, and (d) lane-moment.

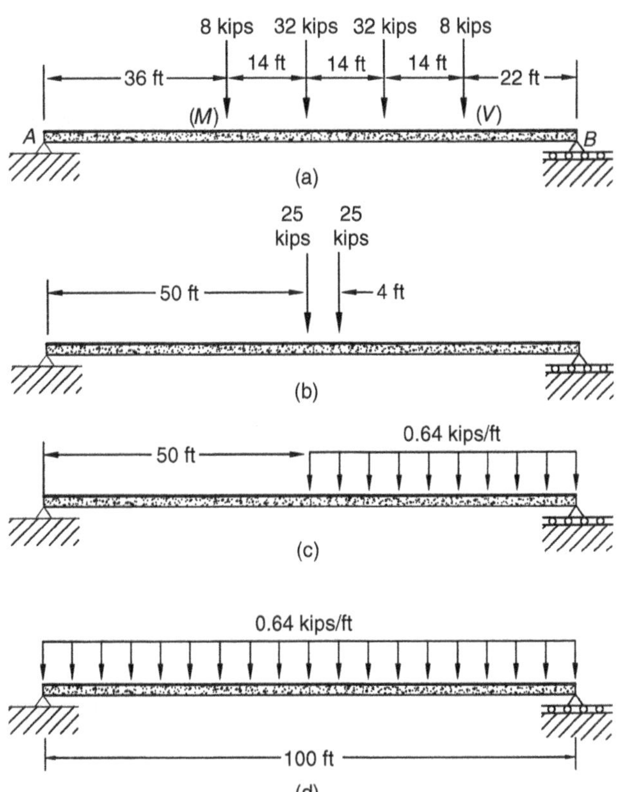

Fig. E16.3-6 Live-load placement at location 105: (a) truck-shear and moment, (b) tandem, (c) lane-shear, and (d) lane-moment.

Tandem

$$V_{101}^{\text{Ta}} = 25\left(\frac{90+86}{100}\right) = 44.0 \text{ kips}$$

$$M_{101}^{\text{Ta}} = \frac{10 \times 90}{100}(25)\left(1+\frac{86}{90}\right) = 440.0 \text{ kip-ft}$$

Lane

$$V_{101}^{\text{Ln}} = \frac{0.64\,(0.9)\,(90)}{2} = 25.92 \text{ kips}$$

$$M_{101}^{\text{Ln}} = \tfrac{1}{2}wab = \tfrac{1}{2}0.64\,(10)\,(90) = 288 \text{ kip-ft}$$

Location 105 (Fig. E16.3-6):

Truck

$$V_{105}^{\text{Tr}} = 32\left(\frac{50+36}{100}\right) + 8\left(\frac{22}{100}\right) = 29.28 \text{ kips}$$

$$M_{105}^{\text{Tr}} = \frac{(50)\,(50)}{100}\left[32\left(1+\frac{36}{50}\right) + 8\left(\frac{36}{50}\right)\right] = 1520 \text{ kip-ft}$$

Tandem

$$V_{105}^{\text{Ta}} = 25\left(\frac{50+46}{100}\right) = 24.0 \text{ kips}$$

$$M_{105}^{\text{Ta}} = 24.0\,(50) = 1200 \text{ kip-ft}$$

Lane

$$V_{105}^{\text{Ln}} = \frac{0.64\,(0.5)\,(50)}{2} = 8.0 \text{ kips}$$

$$M_{105}^{\text{Ln}} = \tfrac{1}{8}wL^2 = \tfrac{1}{8}0.64\left(100^2\right) = 800 \text{ kip-ft}$$

H. ***Calculate Force Effects from Other Loads***

1. *Interior Girders*

DC Weight of concrete = 0.150 kcf

Slab $(0.150)\left(\frac{8}{12}\right)(8) = 0.800$ kips/ft

2.0-in. haunch $(0.150)\,(2.0/12)\,(42.0/12) = 0.088$ kips/ft

Girder $(0.150)\left(659/12^2\right) = \underline{0.686}$ kips/ft

$= 1.574$ kips/ft

Estimate diaphragm size 12.0 in. thick, 36.0 in. deep

Diaphragms at $\frac{1}{3}$ points $(0.150)\,(1.0)\,(3.0)\left(8.0 - \frac{6}{12}\right)$

$= 3.38$ kips

DW 3.0-in. bituminous paving $= 0.140\,(3.0/12)\,(8)$

$= 0.280$ kips/ft

2. *Exterior Girders*

DCI Overhang $0.150(9.0/12)(39.0/12) = 0.366$ kips/ft

Slab $0.150(9.0/12)(8/2) = 0.400$ kips/ft

Girder + Haunch = $\underline{0.744}$ kips/ft

$= 1.540$ kips/ft

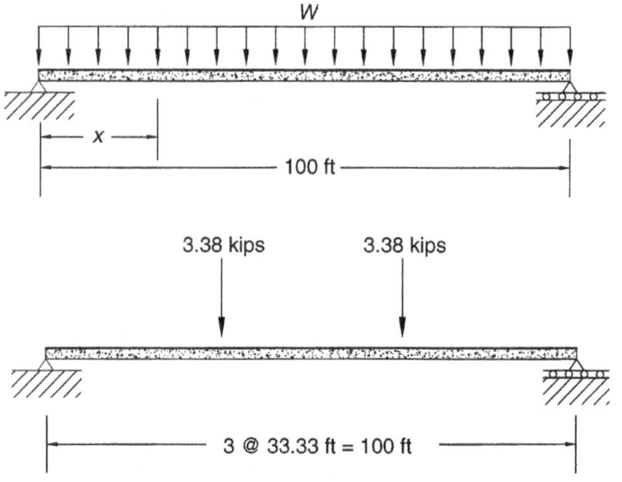

Fig. E16.3-7 Uniform dead and diaphragm loads.

Diaphragms at $\frac{1}{3}$points 3.38/2 = 1.69 kips
DC2: Barrier = 0.320kips/ft
DW 3.0-in.bituminous paving
= 0.140(3.0/12)(39 − 15 + 48)/12 = 0.210kips/ft
(DC2 and DW act on the composite section)

From Figure E16.3-7, shears and moments due to a unit uniform load are found at tenth points (Table E16.3-1), where

$$V_x = w\left(\frac{L}{2} - x\right) = wL\,(0.5 - \xi) \qquad \xi = \frac{x}{L}$$
$$M_x = \frac{w}{2}x\,(L - x) = 0.5wL^2\left(\xi - \xi^2\right)$$

From Figure E16.3-7, shears and moments due to the diaphragms for interior girders are found at tenth points (Table E16.3-2). Values for exterior girders are one-half the values for interior girders.

3. *Summary of Force Effects*
   a. Interior Girders (Table E16.3-3)

$$mg_M = 0.67 \qquad mg_V = 0.81$$
$$\text{IM}^{\text{TR}} = 33\% \qquad \text{IM}^{\text{LN}} = 0$$
$$w_g = 0.686 \text{ kips/ft}$$
$$\text{DC1} = 1.574 \text{ kips/ft} \qquad \text{Diaphragm} = 3.38 \text{ kips}$$
$$\text{DW} = 0.280 \text{ kips/ft}$$

   b. Exterior Girders (Table E16.3-4)

$$mg_M = 0.75 \quad mg_V = 0.75$$
$$\text{IM}^{\text{TR}} = 33\% \quad \text{IM}^{\text{LN}} = 0$$
$$\text{DC1} = 1.540 \text{ kips/ft} \quad \text{Diaphragm} = 1.69 \text{ kips}$$
$$\text{DC2} = 0.320 \text{ kips/ft} \quad \text{DW} = 0.210 \text{ kips/ft}$$

I. ***Investigate Service Limit State***

1. *Stress Limits for Prestressing Tendons* (Table 14.4) [A5.9.3]:

$$f_{pu} = 270 \text{ ksi, low-relaxation 0.5-in., seven-wire strands}$$
$$A = 0.153 \text{ in.}^2 \quad \text{(Table B.2)}$$
$$E_p = 28{,}500 \text{ ksi} \quad \text{[A5.4.4.2]}$$

Pretensioning [Table A5.9.3-1]
Immediately prior to transfer

$$f_{pbt} = 0.75 f_{pu} = 0.75\,(270) = 203 \text{ ksi}$$
$$f_{py} = 0.9 f_{pu} = 0.9\,(270)$$
$$= 243 \text{ ksi} \qquad \text{(Table 13.4)} \quad \text{[Table A5.4.4.1-1]}$$

At service limit state after all losses

$$f_{pe} = 0.80 f_{py} = 0.80\,(243) = 194 \text{ ksi}$$

2. *Stress Limits for Concrete* (Tables 14.2 and 14.3) [A5.9.4]:

$$f_c' = 8 \text{ ksi, 28-day compressive strength}$$
$$f_{ci}' = 0.75 f_c' = 6 \text{ ksi compressive strength at time of initial prestressing}$$

Temporary stresses before losses—fully prestressed components:

Compressive stresses $f_{ci} = 0.6\ f_{ci}' = 0.6(6) = 3.6$ ksi [A5.9.4.1.1]

Table E16.3-1 Shears and Moments for w = 1.0 kips/ft

	$\xi = 0$	$\xi = 0.1$	$\xi = 0.2$	$\xi = 0.3$	$\xi = 0.4$	$\xi = 0.5$
V_x (kips)	50	40	30	20	10	0
M_x (kip-ft)	0	450	800	1050	1200	1250

Table E16.3-2 Shears and Moments Due to Diaphragm, Interior Girders

	$\xi = 0$	$\xi = 0.1$	$\xi = 0.2$	$\xi = 0.3$	$\xi = 0.4$	$\xi = 0.5$
V_x (kips)	3.38	3.38	3.38	3.38	0	0
M_x (kip-ft)	0	33.8	67.6	101.4	112.7	112.7

Table E16.3-3 Summary of Force Effects for Interior Girder

		Distance from Support					
Force Effect	Load Type	0	0.1*L*	0.2*L*	0.3*L*	0.4*L*	0.5*L*
M_s (kip-ft)	Service I loads						
	Girder self-weight	0	309	549	720	823	858
	DC1 (incl. diaph.) on girder alone	0	742	1327	1754	2001	2080
V_s (kips)	DW on composite section	0	126	224	294	336	350
	mg_M (LL + IM)	0	712	1252	1620	1818	1893
	DC1 (incl. diaph.) on girder alone	82.1	66.3	50.6	34.9	15.7	0
M_u (kip-ft)	DW on composite section	14.0	11.2	8.4	5.6	2.8	0
	mg_V (LL + IM)	96.7	84.0	71.8	60.1	48.9	38.2
	Strength I loads						
V_u (kips)	η[1.25DC + 1.50DW + 1.75(LL + IM)]	0	2362	4185	5469	6187	6438
	η[1.25DC + 1.50DW + 1.75(LL + IM)]	292.9	246.7	201.4	157.1	109.4	66.9

Table E16.3-4 Summary of Force Effects for Exterior Girder

		Distance from Support					
Force Effect	Load Type	0	0.1*L*	0.2*L*	0.3*L*	0.4*L*	0.5*L*
M_s (kip-ft)	Service I loads						
	Girder self-weight	0	309	549	720	823	858
	DC1 (incl. diaph.) on girder alone	0	710	1266	1668	1904	1981
	DC2 (barrier) on composite section	0	144	256	336	384	400
V_s (kips)	DW on composite section	0	95	168	221	252	263
	mg_M (LL + IM)	0	795	1399	1811	2032	2116
	DC1 (incl. diaph.) on girder alone	78.7	63.3	47.9	32.5	15.4	0
	DC2 (barrier) on composite section	16.0	12.8	9.6	6.4	3.2	0
M_u (kip-ft)	DW on composite section	10.5	8.4	6.3	4.2	2.1	0
	mg_V (LL + IM)	89.1	77.4	66.1	55.3	45.0	35.2
	Strength I loads						
V_u (kips)	η[1.25DC + 1.50DW + 1.75(LL + IM)]	0	2601	4603	6005	6794	7074
	η[1.25DC + 1.50DW + 1.75(LL + IM)]	290.1	243.1	197.0	151.7	105.2	61.6

Tensile stresses [Table A5.9.4.1.2-1]

Without bonded reinforcement $f_{ti} = 0.0948\sqrt{f'_{ci}} = 0.0948\sqrt{6.0} = 0.232$ ksi > 0.2 ksi (use 0.2 ksi)

With bonded reinforcement $f_{ti} = 0.24\sqrt{f'_{ci}} = 0.24\sqrt{6.0} = 0.588$ ksi

Stresses at service limit state after losses—fully prestressed components [A5.9.4.2]:

Compressive stresses $f_c = 0.45 f'_c$

$= 0.45\,(8.0) = 3.6$ ksi Service I

Tensile stresses $f_t = 0.19\sqrt{f'_c}$

$= 0.19\sqrt{8.0} = 0.537$ ksi Service III

Modulus of Elasticity [C5.4.2.4]

$$E_{ci} = 1820\sqrt{f'_{ci}} = 1820\sqrt{6.0} = 4458 \text{ ksi}$$

$$E_c = 1820\sqrt{f'_c} = 1820\sqrt{8.0} = 5148 \text{ ksi}$$

3. *Preliminary Choices of Prestressing Tendons* Controlled either by the concrete stress limits at service

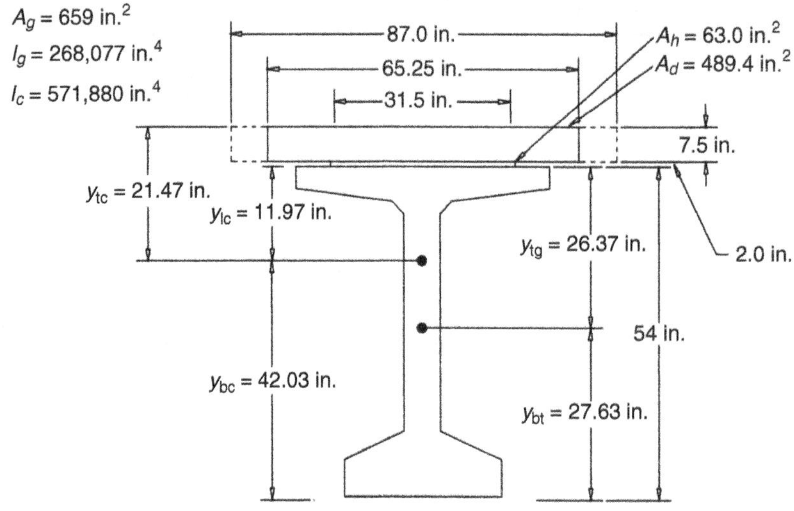

Fig. E16.3-8 Composite section properties.

loads or by the sectional strength under factored loads. For the final load condition, the composite cross-section properties are needed. To transform the CIP deck into equivalent girder concrete, the modular ratio is taken as $n_c = \sqrt{4.5/8.0} = 0.75$.

If we assume for convenience that the haunch depth is 2.0 in. and use the effective flange width of 87 in. for an exterior girder, the composite section dimensions are shown in Figure E16.3-8.

Section properties for the girder are as follows (PCI, 2003):

$$A_g = 659 \text{ in.}^2$$
$$I_g = 268{,}077 \text{ in.}^4$$
$$S_{tg} = \frac{I_g}{y_{tg}} = \frac{268{,}077}{26.37} = 10{,}166 \text{ in.}^3$$
$$S_{bg} = \frac{I_g}{y_{bg}} = \frac{268{,}077}{27.63} = 9702 \text{ in.}^3$$

Section properties for the composite girder are calculated below. The distance to the neutral axis from the top of the deck is

$$y_{tc} = \frac{(489.4)(3.75)+(63.0)(8.5)+(659)(9.5+26.37)}{489.4+63.0+659} = 21.47 \text{ in.}$$

$$I_c = (268{,}077) + (659)(26.37 - 11.97)^2 + \frac{(31.5)(2.0)^3}{12} + (63.0)(21.47 - 8.5)^2 + \frac{(65.25)(7.5)^3}{12} + (489.4)(21.47 - 3.75)^2 = 571.9 \times 10^3 \text{ in.}^4$$

$$S_{tc} = \frac{I_c}{y_{tc}} = \frac{571.9 \times 10^3}{21.47} = 26{,}636 \text{ in.}^3 \quad \text{(top of deck)}$$
$$S_{ic} = \frac{I_c}{y_{ic}} = \frac{571.9 \times 10^3}{11.97} = 47{,}776 \text{ in.}^3 \quad \text{(top of girder)}$$
$$S_{bc} = \frac{I_c}{y_{bc}} = \frac{571.9 \times 10^3}{42.03} = 13{,}606 \text{ in.}^3 \quad \text{(bottom of girder)}$$

Preliminary Analysis—Exterior Girder at Midspan The minimum value of prestress force F_f to ensure that the tension in the bottom fiber of the beam at midspan does not exceed the limit of 0.537 ksi in the composite section under *final* service conditions can be expressed as (Eq. 14.4)

$$f_{bg} = -\frac{F_f}{A_g} - \frac{F_f e_g}{S_{bg}} + \frac{M_{dg} + M_{ds}}{S_{bg}} + \frac{M_{da} + M_L}{S_{bc}} \leq 0.537 \text{ ksi}$$

where

M_{dg} = moment due to self-weight of girder = 858 kip-ft

M_{ds} = moment due to dead load of wet concrete + diaphragm = 1981 − 858 = 1123 kip-ft

M_{da} = moment due to additional dead load after concrete hardens = 663 kip-ft

M_L = moment due to live load + impact (service III) = 0.8(2116) = 1693 kip-ft

e_g = distance from center of gravity of girder to centroid of pretensioned strands = 27.63 − 5.4 = 22.23 in. (estimate $\bar{y}_{ps} = 0.1\ h_g = 5.4$ in.)

Equate the computed estimated tensile stress to the limit stress to determine the prestress force,

$$f_{bg} = -\frac{F_f}{659} - \frac{F_f\,(22.23)}{9702} + \frac{1981 \times 12}{9702} + \frac{(663 + 1693)\,12}{13{,}606}$$
$$\le 0.537 \text{ ksi}$$
$$= -\left[\left(1.517 \times 10^{-3}\right) + \left(2.291 \times 10^{-3}\right)\right] F_f + 2.450 + 2.078$$
$$\le 0.537 \quad \left(3.808 \times 10^{-3}\right) F_f \ge 3.991$$
$$F_f \ge \frac{3.991}{3.808 \times 10^{-3}} = 1048 \text{ kips}$$

Assuming stress in strands after all losses is $0.6f_{pu} = 0.6(270) = 162$ ksi,

$$A_{ps} \ge \frac{F_f}{0.6 f_{pu}} = \frac{1048}{162} = 6.47 \text{ in.}^2$$

From Collins and Mitchell (1991), in order to satisfy strength requirements (strength I), the following approximate expression can be used:

$$\phi M_n = \phi \left(0.95 f_{pu} A_{ps} + f_y A_s\right)(0.9h) \ge M_u$$

where
$\phi = 1.0$
PPR = 1.0 (prestress ratio) [A5.5.4.2.1]
h = overall depth of composite section = 63.5 in.
M_u = strength I factored moment = 7074 kip-ft

$$A_{ps} \ge \frac{M_u}{\phi 0.95 f_{pu}\,(0.9h)} = \frac{7074 \times 12}{1.0\,(0.95)\,(270)\,(0.9)\,(63.5)}$$

$A_{ps} \ge 5.79$ in.^{2} < 6.47 in.^{2}, strength limit is not likely critical.
Number of 0.5-in. strands ($A_{strand} = 0.153$ in.^{2}) = 6.47/0.153 = 42.3.
Try forty-four 0.5-in. strands; $A_{ps} = 44(0.153) = 6.73$ in.^{2} (Fig. E16.3-9).

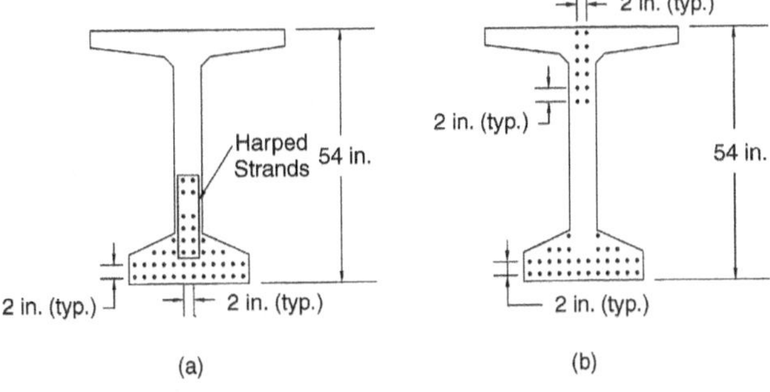

Fig. E16.3-9 Strand patterns at (a) midspan and (b) support.

(*Note*: Other strand patterns were tried. Only the final iteration is given here.)

At Midspan			At End Section		
N	Y	Ny	N	y	Ny
12	2	24	12	2	24
12	4	48	12	4	48
8	6	48	6	6	36
4	8	32	2	8	16
8	14	112	12	47	564
44		264	44		688

$$\bar{y}_m = \frac{264}{44} = 6.0 \text{ in.}$$
$$e_m = 27.63 - 6.0 = 21.63 \text{ in.}$$
$$\bar{y}_{end} = \frac{688}{44} = 15.64 \text{ in.}$$
$$e_{end} = 27.63 - 15.64 = 11.99 \text{ in.}$$

4. *Evaluate Prestress Losses* [A5.9.5]

$$\Delta f_{pT} = \Delta f_{pES} + \Delta f_{pLT} \qquad \text{[A5.9.5.1]}$$

where
Δf_{pT} = total loss (ksi)
Δf_{pES} = sum of all losses due to elastic shortening at the time of application of prestress (ksi)
Δf_{pLT} = losses due to long-term shrinkage and creep of concrete and relaxation of the steel (ksi)

a. Elastic Shortening, Δf_{pES} (Eq. 14.48) [A5.9.5.2.3a]

$$\Delta f_{pES} = \frac{E_p}{E_{ci}} f_{cgp}$$

where
E_p = 28,500 ksi
$E_{ci} = 1820\sqrt{6.0} = 4458$ ksi
f_{cgp} = sum of concrete stresses at center of gravity of A_{ps} due to F_i immediately after transfer and M_{dg} at midspan

For purposes of estimating f_{cgp}, the prestressing force immediately after transfer may be assumed to be equal to 0.9 of the force just before transfer.

$$f_{pi} = 0.9 f_{bt} = 0.9\left(0.75 f_{pu}\right)$$
$$= 0.675\,(270)$$
$$= 182.3 \text{ ksi}$$
$$F_i = f_{pi} A_{ps}$$
$$= 182.3\,(6.73)$$
$$= 1227 \text{ kips}$$

The assumed value for f_{pi} is corrected after Δf_{pES} is determined. To avoid iteration, the alternative equation [C5.9.5.2.3a-1] is used:

$$\Delta f_{pES} = \frac{A_{ps} f_{pi}\left(I_g + e_m^2 A_g\right) - e_m M_g A_g}{A_{ps}\left(I_g + e_m^2 A_g\right) + \frac{A_g I_g E_{ci}}{E_s}}$$

$$\Delta f_{pES} = \frac{\begin{array}{c}6.73(182.3)\left(268{,}077 + 21.63^2 \times 659\right)\\ -21.63\,(858 \times 12 \times 659)\end{array}}{\begin{array}{c}6.73\left(268{,}077 + 21.63^2 \times 659\right)\\ +659\,(268{,}077)\,\frac{4458}{28{,}500}\end{array}}$$

$$= 17.8 \text{ ksi}$$

b. Approximate Estimate of Time-Dependent Losses Δf_{pLT} [A5.9.5.3]

For standard precast, pretensioned members subject to normal loading and environmental conditions, where

- ☐ Members are made from normal-weight concrete.
- ☐ The concrete is either steam or moist cured.
- ☐ Prestressing is bars or strands with normal and low-relaxation properties.
- ☐ Average exposure conditions and temperatures characterize the site.

The long-term prestress loss, Δf_{pLT}, due to creep of concrete, shrinkage of concrete, and relaxation of steel may be estimated using

$$\Delta f_{pLT} = 10.0 \frac{f_{pi} A_{ps}}{A_g} \gamma_h \gamma_{st} + 12.0 \gamma_h \gamma_{st} + \Delta f_{pR}$$

in which

$$\gamma_h = 1.7 - 0.01H$$

$$\gamma_{st} = \frac{5}{1 + f'_{ci}}$$

where

f_{pi} = prestressing steel stress immediately prior to transfer (ksi)

H = average annual ambient relative humidity (%) [A5.4.2.3.2]

γ_h = correction factor for humidity

γ_{st} = correction factor for specified concrete strength at time of prestress transfer

Δf_{pR} = estimate of relaxation loss taken as 2.5 ksi for low-relaxation strand

For Virginia, $H = 70\%$ [Fig. A5.4.2.3.3], so that

$$\gamma_h = 1.7 - 0.01\,(70) = 1.0$$

$$\gamma_{st} = \frac{5}{1 + 6.0} = 0.714$$

$$f_{pi} = 0.75 f_{pu} = 0.75\,(270) = 203 \text{ ksi}$$

$$\Delta f_{pLT} = 10.0 \frac{203\,(6.73)}{659}\,(1.0)\,(0.714) + 12\,(1.0)\,(0.714) + 2.5$$
$$= 14.8 + 8.6 + 2.5 = 25.9 \text{ ksi}$$

c. Total Losses (Eq. 7.93):

$$\Delta f_{pT} = \text{(initial losses)} + \text{(long-term losses)}$$
$$= \Delta f_{pES} + \Delta f_{pLT} = 17.8 + 25.9$$
$$= 43.7 \text{ ksi}$$

5. Calculate Girder Stresses at Transfer

$$f_{pi} = 0.75 f_{pu} - \Delta f_{pES}$$
$$= 0.75\,(270) - 17.8 = 185 \text{ ksi}$$
$$F_i = f_{pi} A_{ps} = 185\,(6.73) = 1245 \text{ kips}$$
$$e_m = 21.63 \text{ in.} \quad e_{end} = 11.99 \text{ in.}$$

At midspan, the tensile stress at the top of the girder is

$$f_{ti} = -\frac{F_i}{A_g} + \frac{F_i e_m}{S_{tg}} - \frac{M_{dg}}{S_{tg}}$$
$$= -\frac{1245}{659} + \frac{(1245)\,(21.63)}{10{,}166} - \frac{858\,(12)}{10{,}166}$$
$$= -0.253 \text{ ksi} < 0.537 \text{ ksi} \quad \text{OK}$$

Again, negative denotes compression.

At midspan, the compressive stresses are checked at the bottom of the girder

$$f_{bi} = -\frac{F_i}{A_g} - \frac{F_i e_m}{S_{bg}} + \frac{M_{dg}}{S_{bg}}$$
$$= -\frac{1245}{659} - \frac{(1245)\,(21.63)}{9702} + \frac{858\,(12)}{9702}$$
$$= -3.58 \text{ ksi} > f_{ci} = -3.60 \text{ ksi} \quad \text{OK}$$

At the beam end, self-weight moments are zero and tension is possible at the top.

$$f_{ti} = -\frac{F_i}{A_g} + \frac{F_i e_{end}}{S_{tg}}$$
$$= -\frac{1245}{659} + \frac{(1245)\,(11.99)}{10{,}166}$$
$$= -0.42 \text{ ksi} < 0.537 \text{ ksi} \quad \text{OK}$$

And the compression is checked at the bottom,

$$f_{bi} = -\frac{1245}{659} - \frac{1245\,(11.99)}{9702}$$
$$= -3.43 \text{ ksi} > -3.60 \text{ ksi} \quad \text{OK}$$

In this case, the entire section remains in compression at transfer.

6. *Girder Stresses after Total Losses* Use the total loss estimates to determine the final prestress force,

$$f_{pf} = 0.75 f_{pu} - \Delta f_{pT} = 0.75\,(270) - 43.7 = 158.8 \text{ ksi}$$

$$F_f = 158.8\,(6.73) = 1069 \text{ kips}$$

At Midspan

$$f_{tf} = -\frac{F_f}{A_g} + \frac{F_f e_m}{S_{tg}} - \frac{M_{dg} - M_{ds}}{S_{tg}} - \frac{M_{da} + M_L}{S_{ic}} \quad \text{(Top of girder)}$$

$$= -\frac{1069}{659} + \frac{(1069)\,(21.63)}{10{,}166} - \frac{(1981)\,12}{10{,}166} - \frac{(663 + 2116)\,12}{47{,}776} \quad \text{Service I}$$

$$= -2.38 \text{ ksi} > -3.60 \text{ ksi} \quad \text{OK}$$

$$f_{bf} = -\frac{F_f}{A_g} - \frac{F_f e_m}{S_{bg}} + \frac{M_{dg} + M_{ds}}{S_{bg}} + \frac{M_{da} + M_L}{S_{bc}} \quad \text{(Bottom of girder)}$$

$$f_{bf} = -\frac{1069}{659} - \frac{(1069)\,(21.63)}{9702} + \frac{(1981)\,12}{9702} + \frac{(663 + 0.8 \times 2116)\,12}{13{,}606} \quad \text{Service III}$$

$$= 0.523 \text{ ksi} < 0.537 \text{ ksi} \quad \text{OK}$$

$$f_{tc} = -\frac{M_{da} + M_L}{S_{tc}} \quad \text{(Top of deck)}$$

$$f_{tc} = -\frac{(663 + 2116)\,12}{26{,}636}$$

$$= -1.25 \text{ ksi} > -0.45 f'_c = -3.60 \text{ ksi} \quad \text{OK}$$

Forty-four 0.5-in. low-relaxation strands satisfy service limit state.

7. *Check Fatigue Limit State* [A5.5.3]
   a. Live-Load Moment Due to Fatigue Truck (FTr) at Midspan (Fig. E16.3-10)

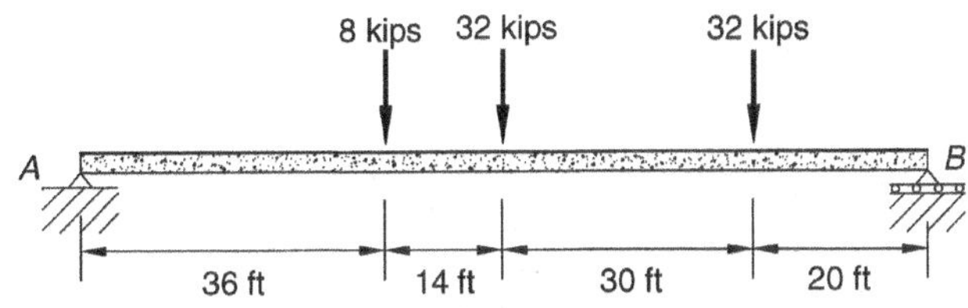

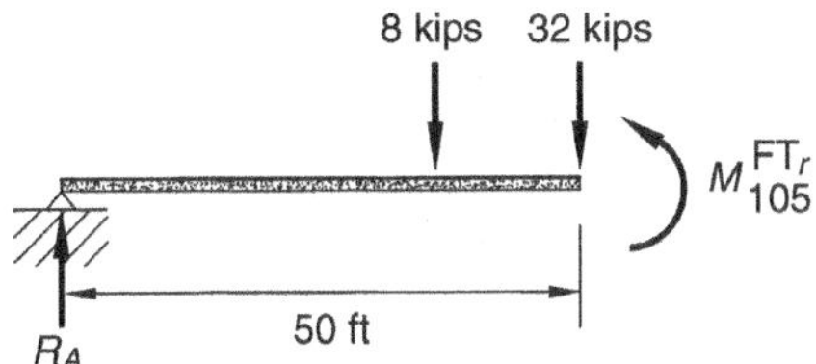

Fig. E16.3-10 Fatigue truck placement for maximum positive moment at midspan.

$$R_A = 32\left(\frac{20 + 50}{100}\right) + 8\left(\frac{64}{100}\right) = 27.52 \text{ kips}$$

$$M_{105}^{\text{FTr}} = [(27.52)\,(50) - (8)\,(14)] = 1264 \text{ kip-ft}$$

Exterior girder distribution factor for moment—remove 1.2 multiple presence for fatigue:

$$g_M^{\text{SE}} = \frac{0.75}{1.2} = 0.625$$

Distributed moment including IM = 15%:

$$M_{\text{fatigue-I}} = 1.5\,(0.625)\,(1264)\,(1.15) = 1363 \text{ kip-ft}$$

   b. Dead-Load Moments at Midspan

Exterior girder (Table E16.3-4)

Noncomposite $M_{DC1} = 1981$ kip-ft

Composite $M_{DC2} + M_{DW} = (400 + 263) = 663$ kip-ft

If section is in compression under DL and fatigue I load, fatigue does not need to be investigated [A5.5.3.1]. Concrete stress at the bottom fiber is

$$f_b = -\frac{F_f}{A_g} - \frac{F_f e_m}{S_{bg}} + \frac{M_{DC1}}{S_{bg}} + \frac{M_{DC2} + M_{DW} + M_{\text{fatigue-I}}}{S_{bc}}$$

$$= -\frac{1069}{659} - \frac{(1069)\,(21.63)}{9702} + \frac{1981\,(12)}{9702} + \frac{[663 + 1363]\,12}{13{,}606}$$

$$= 0.231 \text{ ksi, tension; therefore, fatigue shall be considered}$$

Section Properties: Cracked section properties used [A5.5.3.1] if the sum of stresses in concrete at bottom fiber due to unfactored permanent loads and prestress plus $M_{\text{fatigue-I}}$ exceeds

$$0.095\sqrt{f'_c} = 0.095\sqrt{8} = 0.269 \text{ ksi}$$

$$f_b = -\frac{1069}{659} - \frac{(1069)\,(21.63)}{9702} + \frac{1981\,(12)}{9702} + \frac{[663 + 1363]\,12}{13{,}606}$$

$$= 0.232 \text{ ksi} < 0.269 \text{ ksi; therefore, use of gross section properties is okay}$$

$$I_g = 268{,}077 \text{ in.}^4 \qquad y_{bg} = 27.63 \text{ in.}$$
$$I_c = 571{,}880 \text{ in.}^4 \qquad y_{bc} = 42.03 \text{ in.}$$
$$\bar{y}_m = 6.0 \text{ in.}$$

Eccentricity of prestress tendon in girder

$$e_{pg} = 27.63 - 6.0 = 21.63 \text{ in.}$$

Eccentricity of prestress tendon in composite section

$$e_{pc} = 42.03 - 6.0 = 36.03 \text{ in.}$$

Concrete stress at center of gravity of prestress tendons due to permanent load and prestress:

$$\begin{aligned} f_{cgp}^{DL+PS} &= -\frac{F_f}{A_g} - \frac{F_f e_m e_{pg}}{I_g} \\ &\quad + \frac{M_{DC1} e_{pg}}{I_g} + \frac{(M_{DC2} + M_{DW}) e_{pc}}{I_c} \\ &= -\frac{1069}{659} - \frac{1069\,(21.63)\,(21.63)}{268{,}077} \\ &\quad + \frac{1981 \times 12\,(21.63)}{268{,}077} \\ &\quad + \frac{(663 \times 12)\,(36.03)}{571{,}880} \\ &= -1.068 \text{ ksi} \end{aligned}$$

Concrete stress at center of gravity of prestress tendons due to fatigue moment is

$$\begin{aligned} f_{ccp}^{fatigue} &= \frac{M_{fatigue} e_{pc}}{I_c} \\ &= \frac{(1363)\,(12)\,(36.03)}{571{,}880} \\ &= 1.03 \text{ ksi} \end{aligned}$$

The maximum stress in the tendon due to permanent loads and prestress plus fatigue load is

$$\begin{aligned} f_{max} &= \frac{E_p}{E_c}\left(f_{cgp}^{DL+PS} + f_{ccp,max}^{fatigue}\right) \\ &= \frac{28{,}500}{5148}(-1.068 + 1.03) \\ &= -0.21 \text{ ksi} \end{aligned}$$

The minimum stress in the tendon due to permanent loads and prestress plus fatigue load is

$$\begin{aligned} f_{min} &= \frac{E_p}{E_c}\left(f_{cgp}^{DL+PS} + f_{ccp,min}^{fatigue}\right) \\ &= \frac{28{,}500}{5148}(-1.068 + 0.0) \\ &= -5.91 \text{ ksi} \end{aligned}$$

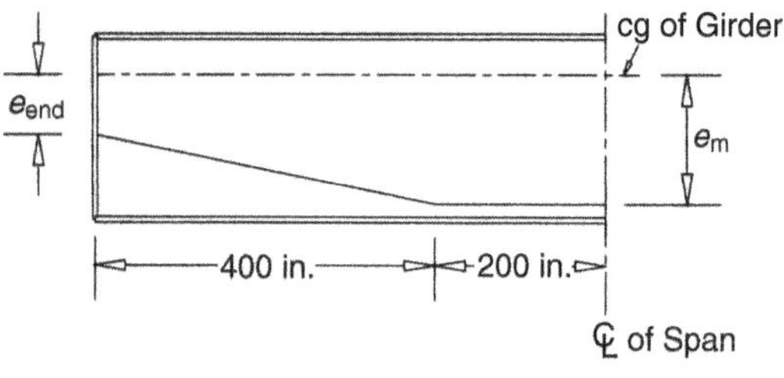

Fig. E16.3-11 Profile of center of gravity of tendons.

The fatigue stress range f_f is

$$\begin{aligned} f_f &= f_{max} - f_{min} = -0.21 - (-5.91) \\ &= 5.70 \text{ ksi} \end{aligned}$$

Stress range in prestressing tendons shall not exceed (Table 14.5) [A5.5.3.3]

□ 18 ksi for radii of curvature greater than 30 ft
□ 10 ksi for radii of curvature less than 12 ft

Harped Tendons (Fig. E16.3-11)

$$e_{end} = 11.99 \text{ in.} \qquad e_{0.33L} = e_m = 21.63 \text{ in.}$$

At hold-down point, the radius of curvature depends on the hold-down device and could be small; therefore assume $R < 12$ ft:

$$f_f = 5.70 \text{ ksi} < 10 \text{ ksi} \quad \text{OK}$$

Tendons satisfy fatigue limit state.

8. *Calculate Deflection and Camber*
   a. Immediate Deflection Due to Live Load and Impact (Fig. E16.3-12)

$$\Delta_x\,(x < a) = \frac{Pbx}{6EIL}\left(L^2 - b^2 - x^2\right)$$
$$b = L - a$$
$$\Delta_x\left(x = \frac{L}{2}\right) = \frac{PL^3}{48EI}$$

Use EI for $f_c' = 8$ ksi and composite section

$$E_c = 5148 \text{ ksi} \qquad I_c = 571{,}880 \text{ in.}^4$$
$$E_c I_c = 2.944 \times 10^9 \text{ kip-in.}^2$$

$\underline{P_1 = 8 \text{ kips}}$, $x = 50$ ft, $a = 64$ ft, $b = 36$ ft

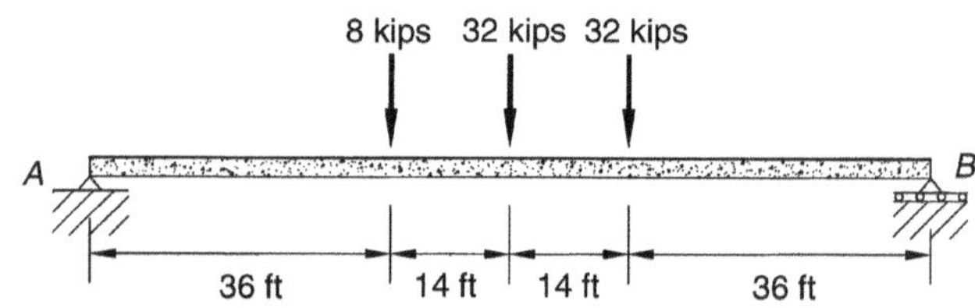

Fig. E16.3-12 Live-load placement for deflection at midspan.

$$\Delta_{x_1} = \frac{(8)(36)(50)}{6(EI)(100)}\left(100^2 - 36^2 - 50^2\right)12^3$$
$$= \frac{0.2573 \times 10^9}{EI}$$
$$= \frac{0.2573 \times 10^9}{2.944 \times 10^9} = 0.087 \text{ in.}$$

$P_2 = 32$ kips, $x = a = b = 50$ ft

$$\Delta_{x_2} = \frac{(32)(100)^3 12^3}{48EI} = \frac{1.152 \times 10^9}{EI}$$
$$= \frac{1.152 \times 10^9}{2.944 \times 10^9} = 0.391 \text{ in.}$$

$P_3 = 32$ kips, $x = 50$ ft, $a = 64$ ft, $b = 36$ ft

$$\Delta_{x_3} = \tfrac{32}{8}\Delta_{x_1} = 4\,(0.087) = 0.348 \text{ in.}$$

Total deflection due to truck:

$$\Delta_{105}^{\text{Tr}} = 0.087 + 0.391 + 0.348 = 0.826 \text{ in.}$$

$$\text{Deflection } mg = m\frac{N_L}{N_G} = 0.85\frac{3}{6}$$
$$= 0.425, \quad \text{IM} = 33\%$$
$$\Delta_{105}^{L+I} = 0.425\,(0.826)\,(1.33)$$
$$= 0.47 \text{ in.} \quad \downarrow \quad \text{(downward)}$$
$$= 0.47 \text{ in.} \le \frac{L}{800} = \frac{100 \times 12}{800}$$
$$= 1.50 \text{ in.} \quad \text{OK}$$

b. Long-Term Deflections (Collins and Mitchell, 1991) Loads on exterior girder from Section 16.3, Part H.2.

☐ Elastic deflections due to girder self-weight at release of prestress

$$E_{\text{ci}} = 4458 \text{ ksi} \qquad I_g = 268{,}077 \text{ in.}^4$$
$$E_{\text{ci}} I_g = 1.195 \times 10^9 \text{ kip-in.}^2$$
$$\Delta_{\text{gi}} = \frac{5}{384}\frac{wL^4}{EI} = \frac{5}{384}\frac{(0.686)(100)^4 12^3}{1.195 \times 10^9}$$
$$= 1.29 \text{ in.} \quad \downarrow \quad \text{(downward)}$$

☐ Elastic camber due to prestress at time of release for double harping point with $\beta L = 0.333L$ (Collins and Mitchell, 1991):

$$\Delta_{\text{pi}} = \left[\frac{e_m}{8} - \frac{\beta^2}{6}\left(e_m - e_e\right)\right]\frac{F_i L^2}{EI}$$
$$= \left[\frac{21.63}{8} - \frac{(0.333)^2}{6}(21.63 - 11.99)\right]$$
$$\times \frac{(1245)(100)^2 12^2}{1.195 \times 10^9}$$
$$= 3.79 \text{ in.} \quad \uparrow \quad \text{(upward)}$$

At release, net upward deflection:

$$3.79 - 1.29 = 2.50 \text{ in.} \quad \uparrow \quad \text{(upward)}$$

☐ Elastic deflection due to deck and diaphragms on exterior girder:

$$\text{DC1} - w_g = 1.540 - 0.686 = 0.854 \text{ kips/ft}$$
$$\text{Diaphragm} = 1.69 \text{ kips}$$

$$E_c = 5148 \text{ ksi} \qquad E_c I_g$$
$$= 1.380 \times 10^9 \text{ kip-in.}^2$$
$$b = \frac{L}{3} = 33.33 \text{ ft}$$
$$\Delta_{\text{DC}} = \frac{5}{384}\frac{wL^4}{EI} + \frac{Pb}{24EI}\left(3L^2 - 4b^2\right)$$
$$= \frac{5}{348}\frac{(0.854)(100)^4 12^3}{1.380 \times 10^9}$$
$$+ \frac{(1.69)(33.33)}{24\left(1.380 \times 10^9\right)}$$
$$\times \left[3(100)^2 - 4(33.33)^2\right]12^3$$
$$= 1.392 + 0.196$$
$$= 1.59 \text{ in.} \quad \downarrow \quad \text{(downward)}$$

☐ Elastic deflection due to additional dead load acting on composite section:

$$\text{DW} + \text{barrier} = 0.210 + 0.320$$
$$= 0.530 \text{ kips/ft}$$

$$\Delta_c = \frac{5}{384}\frac{wL^4}{EI} = \frac{5}{384}\frac{(0.530)(100)^4 12^3}{2.944 \times 10^9}$$
$$= 0.764 \text{ in.} \quad \downarrow \quad \text{(downward)}$$

Note the *full* barrier load is applied to the exterior girder. Many designers distribute this load equally to all girders.

Long-Term Deflections The calculated elastic deflections increase with time due to creep in the concrete. To approximate the creep effect, multipliers applied to the elastic deflections have been proposed. For example, AASHTO [A5.7.3.6.2] states that the long-time deflection may be taken as the instantaneous deflection multiplied by 4.0 if the instantaneous deflection is based on the gross section properties of the girder.

Additional multipliers have been developed to account for creep at different stages of loading and for changing section properties. Using the multipliers in Table E16.3-5 (PCI, 2003) to approximate the creep effect, the net upward deflection at the time the deck is placed is

$$\Delta_1 = 1.80\,(3.79) - 1.85\,(1.29)$$
$$= 4.44 \text{ in.} \quad \uparrow \quad \text{(upward)}$$

The multipliers in Table E16.3-5 were developed for precast prestressed concrete members and give

Table E16.3-5 Suggested Multipliers Used as Guide in Estimating Long-Time Cambers and Deflections for Totally Precast Concrete Members

	Without Composite Topping	With Composite Topping
At erection		
1. Deflection (downward) component—apply to the elastic deflection due to the member weight at release of prestress	1.85	1.85
2. Camber (upward) component—apply to the elastic camber due to prestress at the time of release of prestress	1.80	1.80
Final		
3. Deflection (downward) component—apply to the elastic deflection due to the member weight at release of prestress	2.70	2.40
4. Camber (upward) component—apply to the elastic camber due to prestress at the time of release of prestress	2.45	2.20
5. Deflection (downward)—apply to elastic deflection due to superimposed dead load only	3.00	3.00
6. Deflection (downward)—apply to elastic deflection caused by the composite topping		2.30

In PCI Table 4.6.2. From *PCI Design Handbook: Precast and Prestressed Concrete*, 4th ed., Copyright © 1992 by the Precast/Prestressed Concrete Institute, Chicago, IL.

reasonable estimates for camber based on the gross section properties of the precast girder prior to the placement of a cast-in-place concrete deck.

After the cast-in-place deck hardens, the stiffness of the section increases considerably and the creep strains due to prestressing, girder self-weight, and dead load of the deck are restrained. Also, differential creep and shrinkage between the precast and cast-in-place concretes can produce significant changes in member deformation. As a result, *the multipliers in Table E16.3-5 for estimating long-term final deflections should not be used for bridge beams with structurally composite cast-in-place decks* (PCI, 2003).

To estimate the final long-term deflections, it is necessary to establish the time-dependent concrete material behavior of the modulus of elasticity, the shrinkage strain (Eq.13.24), and the creep strain (Eqs. 13.25 and 13.26) as well as the time-dependent relaxation of the prestressing steel. For a series of time steps, computations are made for the section properties, the initial strains, and the changes in strains due to shrinkage, creep, and relaxation. The final long-term deflections are obtained by integrating over the time steps the change in deflections calculated for each time step. Computer programs are usually utilized to perform the calculations. A sample calculation for one time step is given in the *PCI Bridge Design Manual* (2003) based on the method presented by Dilger (1982).

A final note on long-term deflections for prestressed concrete beams is that AASHTO does not require that the final deflection be checked. The reason for a designer to compute the final deflection is to ensure that the structure does not have excessive sag or upward deflection.

J. ***Investigate Strength Limit State***

1. *Flexure*

a. Stress in Prestressing Steel-Bonded Tendons (Eq. 14.11) [A5.7.3.1.1]

$$f_{ps} = f_{pu}\left(1 - k\frac{c}{d_p}\right)$$

where (Eq. 14.12)

$$k = 2\left(1.04 - \frac{f_{py}}{f_{pu}}\right)$$
$$= 2\,(1.04 - 0.9) = 0.28$$

By using the nontransformed section for plastic behavior (Fig. E16.3-8)

$$b = 87.0 \text{ in.}$$
$$d_p = (54 + 2 + 7.5) - 6.0 = 57.5 \text{ in.}$$
$$f_c' = 8 \text{ ksi}$$
$$A_s = A_s' = 0 \qquad A_{ps} = 6.73 \text{ in.}^3$$
$$\beta_1 = 0.85 - (0.05)\,(8 - 4) = 0.65$$

Assume rectangular section behavior and check if depth of compression stress block is less than t_s:

$$c = \frac{A_{ps} f_{pu}}{0.85 f_c' \beta_1 b_w + k A_{ps}\left(f_{pu}/d_p\right)}$$

With $b_w = b = 65.25$ in.

$$c = \frac{(6.73)(270)}{0.85(8)(0.65)(87.0)+0.28(6.73)(270/57.5)}$$
$$= 4.62 \text{ in.} \quad < t_s = 7.5 \text{ in. Assumption is valid}$$

$$f_{ps} = 270\left[1 - 0.28\left(\frac{4.62}{57.5}\right)\right] = 264 \text{ ksi}$$
$$T_p = A_{ps} f_{ps} = 6.73\,(264) = 1776 \text{ kips}$$

b. Factored Flexural Resistance—Flanged Sections [A5.7.3.2.2]

$$a = \beta_1 c = 0.65\,(4.62) = 3.00 \text{ in.}$$
$$\varepsilon_s = 0.003\left(\frac{d_p - c}{c}\right)$$
$$= 0.003\left(\frac{57.5 - 4.62}{4.62}\right) = 0.034$$
$$\varepsilon_s \geq 0.005$$
$$\phi = 1.0$$

from Eq. 14.28

$$\phi M_n = \phi[A_{ps} f_{ps}\left(d_p - \frac{a}{2}\right) + A_s f_y\left(d_s - \frac{a}{2}\right)$$
$$= 1.0\left[6.73\,(262)\left(57.5 - \frac{3.00}{2}\right)\right] / 12$$
$$\phi M_n = 8228 \text{ kip-ft} > M_u$$
$$= 7074 \text{ kip-ft (Table E16.3-5)} \qquad \text{OK}$$

c. Limits for Reinforcement [A5.7.3.3]

☐ Minimum reinforcement [A5.7.3.3.2]

At any section, the amount of prestressed and nonprestressed tensile reinforcement shall be adequate to develop a factored flexural resistance M_r at least equal to the lesser of:

☐ 1.2 times the cracking moment M_{cr} determined on the basis of elastic stress distribution and the modulus of rupture f_r of concrete, or

☐ 1.33 times the factored moment required by the applicable strength load combination.

Checking at midspan: The cracking moment may be taken as [Eq. A5.7.3.3.2-1]

$$M_{cr} = S_c\left(f_r + f_{cpe}\right) - M_{dnc}\left(\frac{S_c}{S_{nc}} - 1\right)$$
$$\geq S_c f_r$$

where

f_{cpe} = compressive stress in concrete due to effective prestress forces only (after allowance for all prestress losses) at extreme fiber of section where tensile stress is caused by externally applied loads

$$= -\frac{F_f}{A_g} - \frac{F_f e_m}{S_{bg}}$$
$$= -\frac{1069}{659} - \frac{1069(21.63)}{9702} = -3.98 \text{ ksi}$$

f_r = modulus of rupture $= 0.37\sqrt{f'_c} = 0.37\sqrt{8} = 1.05$ ksi [A5.4.2.6] (use upper value)

M_{dnc} = total unfactored dead-load moment acting on the noncomposite section $= M_g + M_{DC1} = 1981$ kip-ft

S_c = section modulus for the extreme fiber of the composite section where tensile stress is caused by externally applied loads $= S_{bc} =$ 13,606 in.^{3}

S_{nc} = section modulus for the extreme fiber of the noncomposite section where tensile stress is caused by externally applied loads $= S_{bg} = 9702$ in.^{3}

$$M_{cr} = \frac{13{,}606\,(1.05 + 3.98)}{12} - 1981\left(\frac{13{,}606}{9702} - 1\right)$$
$$= 4906 \text{ kip-ft}$$
$$1.2 M_{cr} = 1.2\,(4906) = 5887 \text{ kip-ft}$$

At midspan, the factored moment required by strength I load combination is

$$M_u = 7074 \text{ kip-ft (Table E16.3-5)}, \text{ so that}$$
$$1.33 M_u = 1.33\,(7074) = 9408 \text{ kip-ft}$$

Since $1.2M_{cr} < 1.33M_u$, the $1.2M_{cr}$ requirement controls.

$$M_r = \phi M_u = 8157 \text{ kip-ft} > 1.2 M_{cr}$$
$$= 5887 \text{ kip-ft} \qquad \text{OK}$$

Forty-four 0.5-in. low-relaxation strands satisfy strength limit state.

2. *Shear* [A5.8]

a. General The nominal shear resistance V_n shall be the lesser of [A5.8.3.3]

$$V_n = V_c + V_s + V_p$$
$$V_n = 0.25 f'_c b_v d_v + V_p$$

in which the nominal concrete shear resistance is

$$V_c = 0.0316\beta\sqrt{f'_c}\, b_v d_v$$

and the nominal transverse reinforcement shear resistance is

$$V_s = \frac{A_v f_y d_v\,(\cot\theta + \cot\alpha)\sin\alpha}{s}$$

$$V_s = \frac{A_v f_y d_v \cot \theta}{s}$$

for vertical stirrups $\alpha = 90^\circ$ [C5.8.3.3]

where

b_v = minimum web width, measured parallel to the neutral axis, between the resultants of the tensile and compressive forces due to flexure, modified for the presence of ducts (in.)

d_v = effective shear depth taken as the distance, measured perpendicular to the neutral axis, between the resultants of the tensile and compressive forces due to flexure; it need not be taken less than the greater of $0.9d_e$ or $0.72h$ (in.)

s = spacing of stirrups (in.)

β = factor indicating ability of diagonally cracked concrete to transmit tension [A5.8.3.4] (traditional value of $\beta = 2.0$) [A5.8.3.4.1]

θ = angle of inclination of diagonal compressive stresses [A5.8.3.4] (traditional value of $\theta = 45^\circ$, $\cot \theta = 1.0$) [A5.8.3.4.1]

A_v = area of shear reinforcement within a distance s (in.^{2})

V_p = component in the direction of the applied shear of the effective prestressing force; positive if resisting the applied shear force (kips)

$\phi_v = 0.9$ [A5.5.4.2.1] $\eta_i = \eta = 1.0$

At midspan:

$$d_e = h - \bar{y}_m = 63.5 - 6.0 = 57.5 \text{ in.}$$

$$d_v = d_e - \frac{a}{2}$$

$$d_v \geq \max \begin{cases} 0.9d_e = 0.9\,(57.5) = 51.8 \text{ in.} \\ 0.72h = 0.72\,(63.5) = 45.7 \text{ in.} \end{cases}$$

$$a = \beta_1 c = (0.65)\,(4.62) = 3.00 \text{ in.}$$

$$d_v = 57.5 - \frac{3.00}{2} = 56.0 \text{ in.} \quad \text{[A5.8.2.7]}$$

At the end of the beam:

$$d_e = h - \bar{y}_{end} = 63.5 - 15.64 = 47.86 \text{ in.}$$

$$d_v \geq \max \begin{cases} 0.9d_e = 0.9\,(47.86) = 43.1 \text{ in.} \\ 0.72h = 0.72\,(63.5) = 45.7 \text{ in.} \\ d_e - \frac{a}{2} = 47.86 - \frac{3.00}{2} = 46.4 \text{ in., governs} \end{cases}$$

b_v = minimum web width within $d_v = 6.0$ in.

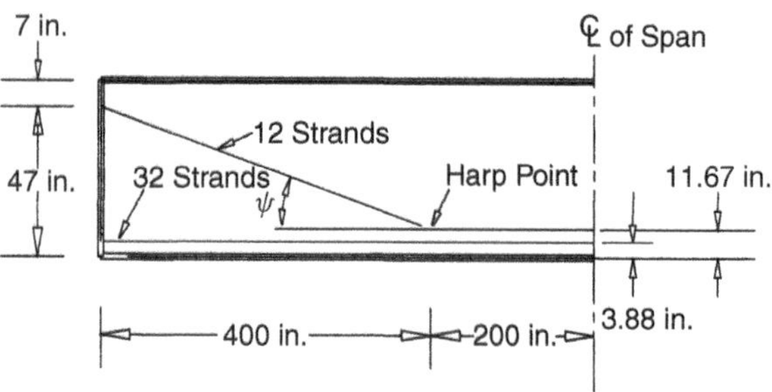

Fig. E16.3-13 Harped tendon profile.

b. Prestress Contribution to Shear Resistance

V_p = vertical component of prestressing force

The center of gravity of 12 harped strands at end of beam = 54 – 7 = 47 in. from bottom of girder

The center of gravity of 12 harped strands at midspan = 11.67 in. (Fig. E16.3-13)

$$\psi = \tan^{-1}\frac{47.0 - 11.67}{400} = 5.05^\circ$$

$$F_f = 1069 \text{ kips}$$

$$V_p = \tfrac{12}{44} F_f \sin \psi = \tfrac{12}{44}\,(1069) \sin 5.05 = 25.66 \text{ kips}$$

c. Design for Shear The location of the critical section for shear is the greater of $d_v = 46.4$ in. or $0.5d_v \cot \theta$ from the internal face of the support [A5.8.3.2]. Assuming $\theta \leq 25^\circ$,

$$0.5d_v \cot \theta \leq 0.5\,(46.4)\,(2.145) = 49.8 \text{ in.}$$

If the width of the bearing was known, the distance to the critical section from the face of the support could be increased. In this case, the critical section is conservatively taken at 48 in. from the centerline of the support.

Calculations are shown below for 48 in. from the support (Fig. E16.3-14) and location 101. The same procedure is used for the remaining tenth points with final results given in Table E16.3-6.

$$d_{critical} = 48 \text{ in.} = 4 \text{ ft}$$

$$\xi = \frac{d_{critical}}{L} = \frac{4}{100} = 0.04$$

For a unit load, $w = 1.0$ kip/ft

$$V_x = wL\,(0.5 - \xi) = 100w\,(0.5 - 0.04) = 46w \text{ kips}$$

$$M_x = 0.5wL^2\,(\xi - \xi^2) = 0.5w(100)^2\,(0.04 - 0.04^2) = 192w \text{ kip-ft}$$

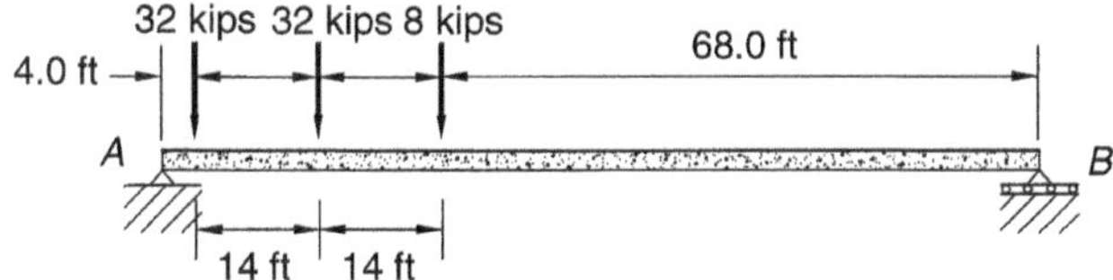

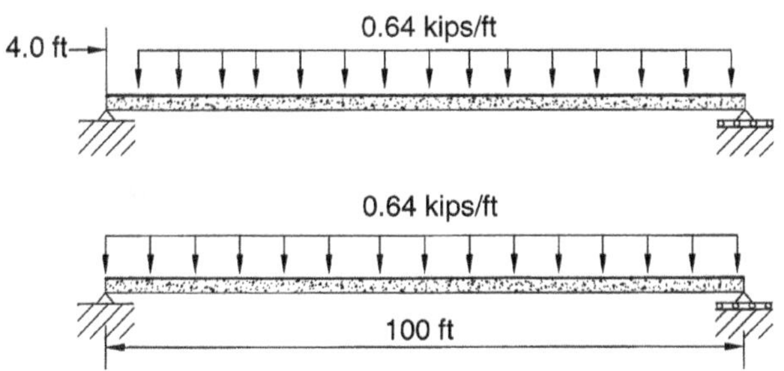

Fig. E16.3-14 Live-load placement for maximum shear and moment at location 100.4.

Exterior girders dead loads are previously presented:

$$\mathrm{DC}_1 = 1.540 \text{ kips/ft}$$
$$\mathrm{DC}_2 = 0.320 \text{ kips/ft}$$
$$\mathrm{DW} = 0.210 \text{ kips/ft}$$
$$\mathrm{DIAPH} = 1.69 \text{ kips}$$
$$\mathrm{IM} = 0.33$$

$$V_{100.4}^{\mathrm{Tr}} = \left[32\left(\frac{96+82}{100}\right) + 8\left(\frac{68}{100}\right)\right] = 62.4 \text{ kips}$$

$$M_{100.4}^{\mathrm{Tr}} = 4\,(62.4) = 249.6 \text{ kip-ft}$$

$$V_{100.4}^{\mathrm{Ln}} = \tfrac{1}{2}\,(0.64)\left(\tfrac{96}{100}\right)(96) = 29.5 \text{ kips}$$

$$M_{100.4}^{\mathrm{Ln}} = \tfrac{1}{2}\,(0.64)\,(4)\,(96) = 122.9 \text{ kip-ft}$$

$$\begin{aligned} V_u &= \eta[1.25\mathrm{DC} + 1.50\mathrm{DW} + 1.75\,(\mathrm{LL}+\mathrm{IM})] \\ &= 1.0\{1.25[(1.540 + 0.320)(46) + 1.69] \\ &\quad + 1.50[(0.210)(46)] \\ &\quad + 1.75(0.750)[(62.4)(1.33) + 29.5]\} \\ &= 271.2 \text{ kips} \end{aligned}$$

$$\begin{aligned} M_u &= 1.0\{1.25[(1.540 + 0.320)(192) \\ &\quad + 1.69(4.0)] + 1.50[(0.210)(192)] \\ &\quad + 1.75(0.750)[(249.6)(1.33) + (122.9)]\} \\ &= 1112 \text{ kip-ft} \end{aligned}$$

Determination of β and θ at critical section location 100.4:

$$d_e = d_p = 63.5 - 15.64 + \tfrac{48}{400}\,(21.62 - 11.99) = 49.0 \text{ in.}$$

$$d_v = d_e - \frac{a}{2} = 49.0 - \frac{3.00}{2} = 47.5 \text{ in.}$$

From Eq. 14.111 [A5.8.3.4.2]:

$$f_c' \text{ (girder)} = 8 \text{ ksi}$$

$$v_u = \frac{|V_u - \phi V_p|}{\phi b_v d_v} = \left[\frac{(271.2) - 0.9\,(25.66)}{0.9\,(6.0)\,(47.5)}\right] = 0.967 \text{ ksi}$$

$$\frac{v_u}{f_c'} = \frac{0.967}{8} = 0.120 < 0.125$$

Therefore,

$$s_{\max} = \min \begin{cases} 0.8d_v = 0.8\,(47.5) = 38.0 \text{ in.} \\ 24 \text{ in.} \quad \text{governs} \end{cases}$$

Table E16.3-6 Summary of Shear Design

	Location					
	100.4	101	102	103	104	105
V_u (kips)	271.2	243.1	197.0	151.7	105.2	61.6
M_u (kip-ft)	1112	2601	4603	6005	6794	7074
V_p (kips)	25.66	25.66	25.66	25.66	0	0
d_v (in.)	47.5	49.3	51.7	54.6	55.5	55.5
v_u/f_c'	0.120	0.103	0.0779	0.0546	0.0439	0.0257
θ (deg)	20.5	21	22	29	33	34
$\varepsilon_x \times 10^3$	−0.161	−0.0879	0.0243	0.424	0.724	0.789
β	3.17	3.28	3.59	2.70	2.40	2.36
V_c (kips)	80.7	86.7	99.5	79.0	71.5	70.3
Required V_s (kips)	195.0	157.8	93.8	63.9	45.4	−1.8
Required s (in.)	15.6	19.5	32.7	37.0	45.2	∞
Check $A_{ps} f_{ps} = 1763$ kips $\geq$	679	1047	1389	1490	1583	1570
Provided s (in.)	12	18	24	24	24	24

$\phi_f = 1.0$, $\phi_v = 0.9$ stirrup spacings 1 at 6 in., 10 at 12 in., 7 at 18 in., 14 at 24 in. No. 4 U-shaped stirrups each end.

For illustration, use the "long method" of Appendix B5 with iteration of the shear parameters [A5.8.2.7]

First Iteration

Assume $\theta = 25^\circ$, $f_{po} \approx 0.7 f_{pu} = 0.7(270) = 189$ ksi [A5.8.3.4.2]. From Eq. 14.122 [A5.8.3.4.2]

$$\varepsilon_x = \frac{\begin{array}{c}|M_u|/d_v + 0.5N_u \\ +0.5|V_u - V_p|\cot\theta - A_{ps}f_{po}\end{array}}{2(E_sA_s + E_pA_{ps})}$$

$$= \frac{\begin{array}{c}(1112 \times 12)/47.5 \\ +0.5(271.2 - 25.66)2.145 - (6.73)(189)\end{array}}{2(28{,}500)(6.73)}$$

$$= -0.00189 \text{ (compression)}$$

Because ε_x is negative, it shall be reduced by the factor [A5.8.3.4.2]

$$F_\varepsilon = \frac{E_sA_s + E_pA_{ps}}{E_cA_c + E_sA_s + E_pA_{ps}} = \frac{E_pA_{ps}}{E_cA_c + E_pA_{ps}}$$

where A_c is the area of concrete on flexural tension side of member defined as concrete below $h/2$ of member [Fig. A5.8.3.4.2-1]:

$$h = 54 + 2 + 7.5 = 63.5 \text{ in.}$$

$$\frac{h}{2} = \frac{63.5}{2} = 31.75 \text{ in.}$$

$$A_c = (6)(26) + 2\left(\tfrac{1}{2}\right)(4.5)(10) + (6)(31.75 - 6.0) \quad \text{(Fig. E16.3-2)}$$

$$= 355.5 \text{ in.}^2$$

$$E_c = 5148 \text{ ksi}$$

$$F_\varepsilon = \frac{(28{,}500)(6.73)}{(5148)(355.5) + (28{,}500)(6.73)} = 0.0949$$

$$\varepsilon_x = (-0.00189)(0.0949) = -0.179 \times 10^{-3}$$

Using $v_u/f'_c = 0.120$ and ε_x with [Table A5.8.3.4.2-1] $\Rightarrow \theta = 20^\circ$:

$$\cot\theta = 2.747$$

Second Iteration

$$\theta = 20^\circ$$

$$\varepsilon_x = \frac{\begin{array}{c}\frac{1112\times 12}{47.5} + 0.5(271.2 - 25.66)(2.747) \\ -(6.73)(189)\end{array}}{2(28{,}500)(6.73)}$$

$$= -0.00169$$

$$F_e\varepsilon_x = 0.0949(-0.00169) = -0.161 \times 10^{-3}$$

[Table A5.8.3.4.2-1] $\Rightarrow \theta = 20.5^\circ$ converged,

Use $\cot\theta = 2.675 \beta = 3.17$

$$V_c = 0.0316\beta\sqrt{f'_c}b_vd_v = 0.0316(3.17)\sqrt{8}(6)(47.5) = 80.7 \text{ kips}$$

Check if

$$V_u \geq 0.5\phi(V_c + V_p) = 0.5(0.9)(80.7 + 25.66) = 47.9 \text{ kips}$$

$V_u = 271.2$ kips > 47.9 kips, transverse reinforcement is required

Required

$$V_s = \frac{V_u}{\phi} - V_c - V_p = \frac{271.2}{0.9} - 80.7 - 25.66 = 195.0 \text{ kips}$$

Spacing of No. 4 U stirrups, (Eq. 14.113)

$d_s = 0.5$ in. $\qquad A_v = 2(0.20) = 0.40$ in.^{2}

$$s \leq \frac{A_vf_yd_v\cot\theta}{V_s} = \frac{(0.40)(60)(47.5)(2.675)}{195.0} = 15.6 \text{ in.}$$

$$s \leq 15.6 \text{ in.} < s_{max} = 24 \text{ in.}$$

Check Longitudinal Reinforcement (Eq. 7.14.110) [A5.8.3.5]

$$A_sf_y + A_{ps}f_{ps} \geq \frac{|M_u|}{d_v\phi_f} + 0.5\frac{N_u}{\phi_a} + \left(\left|\frac{V_u}{\phi_v} - V_p\right| - 0.5V_s\right)\cot\theta$$

Try $s = 12$ in.

$$\text{Provided } V_s = 195.0\left(\frac{15.6}{12}\right) = 253.5 \text{ kips}$$

$$(6.73)(262) \geq \frac{1112 \times 12}{(47.5)(1.0)} + \left[\frac{271.2}{0.9} - 25.66 - 0.5(253.5)\right] \times 2.675$$

1763 kips > 679 kips OK

Use s = 12-in. No. 4 U stirrups at location 100.4.

d. Location 101

$$V_u = 243.1 \text{ kips}$$

$$M_u = 2601 \text{ kip-ft} \qquad \text{(Table E16.3-5)}$$

$$d_e = 63.5 - 15.64 + \frac{120}{400}(21.63 - 11.99) = 50.75 \text{ in.}$$

$$d_v = \max \begin{cases} d_e - \frac{a}{2} = 50.75 - \frac{3.00}{2} = 49.3 \text{ in.} \quad \text{governs} \\ 0.9d_e = 0.9\,(50.75) = 45.7 \text{ in.} \\ 0.72h = 0.72\,(63.5) = 45.7 \text{ in.} \end{cases}$$

$$d_v = 49.3 \text{ in.}$$

$$v_u = \frac{|V_u - \phi V_p|}{\phi b_v d_v} = \frac{243.1 - 0.9\,(25.66)}{0.9\,(6)\,(49.3)} = 0.826 \text{ ksi}$$

$$\frac{v_u}{f_c'} = \frac{0.826}{8} = 0.103 < 0.125;$$

therefore, $s_{max} = 24$ in.

First Iteration

Assume $\theta = 21^\circ$, $\cot\theta = 2.605$, and $f_{po} = 189$ ksi.

$$\varepsilon_x = \frac{\frac{2601\times 12}{49.3} + 0.5\,(243.1 - 25.66) \times 2.605 - (6.73)\,(189)}{2\,[(5148)\,(355.5) + (28{,}500)\,(6.73)]} = -0.0879 \times 10^{-3} \qquad \text{(compression)}$$

[Table A5.8.3.4.2-1] $\Rightarrow \theta = 21^\circ$, converged

$$\beta = 3.28$$

$$V_c = 0.0316\beta\sqrt{f_c'}b_v d_v = 0.0316\,(3.28)\,\sqrt{8}\,(6)\,(49.3) = 86.7 \text{ kips}$$

Requires

$$V_s = \frac{V_u}{\phi} - V_c - V_p = \frac{243.1}{0.9} - 86.7 - 25.66 = 157.8 \text{ kips}$$

$$s \le \frac{(0.40)\,(60)\,(49.3)\,(2.605)}{157.8} = 19.5 \text{ in.} < s_{max} = 24 \text{ in.}$$

For $s = 18$ in.

$$V_s = 157.8\left(\frac{19.5}{18}\right) = 171.0 \text{ kips}$$

Check Longitudinal Reinforcement

$$(6.73)\,(262) \ge \frac{2601 \times 12}{(49.3)\,(1.0)} + \left(\frac{243.1}{0.9} - 25.66 - 0.5\,(171.0)\right) \times 2.605$$

$$1763 \text{ kips} \ge 1047 \text{ kips} \quad \text{OK}$$

Use s = 18-in. No. 4 U stirrups at location 101.

e. Summary of Shear Design (Table E16.3-6)

f. Horizontal Shear [A5.8.4] At interface between two concretes cast at different times the nominal shear resistance shall be taken as

$$V_{nh} = cA_{cv} + \mu\left(A_{vf}f_y + P_c\right) \le \min \begin{cases} \le K_1 f_c' A_{cv} \\ \le K_2 A_{cv} \end{cases}$$

where

A_{cv} = area of concrete engaged in shear transfer = (42 in.)(1 in.) = 42 in.^{2}/in.

A_{vf} = area of shear reinforcement crossing the shear plane (in.^{2}) = 2(0.20) = 0.40 in.^{2} (2 legs)

f_y = yield strength of reinforcement = 60 ksi

f_c' = compressive strength of weaker concrete = 4.5 ksi

For normal-weight concrete intentionally roughened [A5.8.4.2]

c = cohesion factor = 0.28 ksi

μ = friction factor = 1.0

$K_1 = 0.3$

$K_2 = 1.8$

P_c = permanent net compressive force normal to shear plane
= overhang + slab + haunch + barrier
= 0.366 + 0.400 + 0.088 + 0.320
= 1.174 kips/ft
= 0.098 kips/in.

Provided

$$V_{nh} = 0.28\,(42) + 1.0\left[\left(\frac{0.40}{s}\right)(60) + 0.098\right] = 11.9 + \frac{24}{s} \text{ kips/in.}$$

s = spacing of shear reinforcement, in.

$$V_{nh} \le \min \begin{cases} K_1 f_c' A_{cv} = 0.3\,(4.5)\,(42) = 56.7 \text{kips/in.} \quad \text{governs} \\ K_2 A_{cv} = 1.8\,(42) = 75.6 \text{ kips/in.} \end{cases}$$

$$\phi_v V_{nh} \ge \eta V_{uh}$$

where

V_{uh} = horizontal shear due to barrier, FWS and LL + IM = $\frac{V_u}{d_v}$ [Eq. C5.8.4.1-1]

V_u = factored shear force due to superimposed load on composite section

$V_u = 1.25\text{DC2} + 1.50\text{DW} + 1.75(\text{LL} + \text{IM})$

$d_v = = d_e - \frac{a}{2} = 49.0 - \frac{3.00}{2} = 47.5$ in.

Assume the critical section for horizontal shear is at the same location as the critical section for vertical shear.

At Location 100.4 Interpolating between locations 100 and 101 (Table E16.3-4)

$$V_u = 1.25\,(14.7) + 1.50\,(9.7) + 1.75\,(84.4)$$
$$= 180.6 \text{ kips}$$
$$V_{\text{uh}} = \frac{180.6}{47.5} = 3.80 \text{ kips/in.}$$

Required

$$V_{\text{nh}} = \frac{V_{\text{uh}}}{\phi} = \frac{3.80}{0.9}$$
$$= 4.22 \text{ kips/in.} < 33.6 \text{ kips/in.} \quad \text{OK}$$

Equating required V_{nh} to the provided V_{nh}

$$4.22 = 11.9 + \frac{24}{s}$$
$$s \le 0, \text{ no steel is required}$$

Minimum shear reinforcement

$$A_{\text{vf}} \ge \frac{0.05 b_v}{f_y} = \frac{0.05\,(42)}{60}$$
$$= 0.035 \text{ in.}^2/\text{in.} \qquad [\text{Eq. A5.8.4.4-1}]$$

Shear reinforcement provided near support to resist vertical shear are No. 4 U-shaped stirrups at 12 in. Provided

$$A_{\text{vf}} = \frac{0.40}{12}$$
$$= 0.033 \text{ in.}^2/\text{in.} < \text{required minimum } A_{vf}$$
$$= 0.035 \text{ in.}^2/\text{in.}$$

The minimum requirement of A_{vf} may be waived if V_n/A_{cv} is less than 0.210 ksi:

$$\frac{V_n}{A_{\text{cv}}} = \frac{4.22}{42} = 0.101 \text{ ksi} \quad \text{OK}$$

OK to waive minimum requirement.

Use s = 12 in. at location 100.4. By inspection, horizontal shear does not govern stirrup spacing for any of the remaining locations.

g. Check Details

Anchorage Zone [A5.10.10]

The bursting resistance provided by transverse reinforcement at the service limit state shall be taken as [A5.10.10.1]

$$P_r = f_s A_s$$

where

f_s = stress in steel not exceeding 20 ksi

A_s = total area of transverse reinforcement within $h/4$ of end of beam

h = depth of precast beam = 54 in.

The resistance P_r shall not be less than 4% of the prestressing force before transfer

$$F_{\text{pbt}} = f_{\text{pbt}} A_{\text{ps}}$$
$$= (0.75 \times 270)\,(6.73) = 1363 \text{ kips}$$
$$P_r = A_s f_s \ge 0.04 F_{\text{pbt}}$$
$$= 0.04\,(1363) = 54.5 \text{ kips}$$
$$A_s \ge 54.5/20 = 2.73 \text{ in.}^2$$
$$\text{Within} \quad \frac{h}{4} = \frac{54}{4} = 13.5 \text{ in.}$$

Number of No. 5 U stirrups required:

$$\frac{2.73}{2\,(0.31)} = 4.4$$

Use five No. 5 U stirrups, 1 at 2 in. and 4 at 3 in. from end of beam.

Confinement Reinforcement: [A5.10.10.2]

For a distance of $1.5h = 1.5(54) = 81$ in. from the end of the beam, reinforcement not less than No. 3 bars at 6 in. shall be placed to confine the prestressing steel in the bottom flange.

Use 14 No. 3 at 6 in. shaped to enclose the strands.

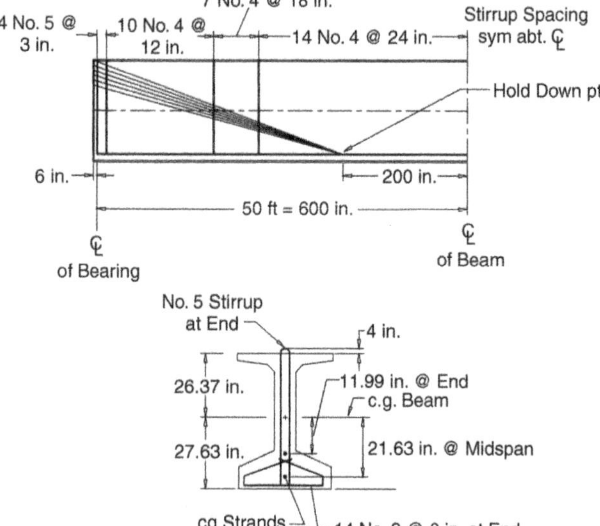

Fig. E16.3-15 Design sketch for prestressed girder.

K. ***Design Sketch*** The design of the prestressed concrete girder is summarized in Figure E16.3-15. The design utilized the AASHTO–PCI bulb tee girder, $f_c^{\prime} = 8$ ksi, and $f_{\text{ci}}^{\prime} = 6$ ksi. The prestressing steel consists of forty-four 270-ksi, low-relaxation 0.5-in. seven-wire strands.

REFERENCES

AASHTO (2004). *LRFD Bridge Design Specifications*, 3rd ed., American Association of State Highway and transportation Official, Washington, DC.

AASHTO (2010). *LRFD Bridge Design Specifications*, 5th ed., American Association of State Highway and transportation Official, Washington, DC.

AISC (2011) Load and Resistance Factor Design, *Manual of Steel Construction*, Vol.1, 13th ed., American Institute of Steel Construction, Chicago.

Collins, M.P. and D. Mitchell (1991). *Prestressed Concrete Structures*, Prentice-Hall, Englewood Cliffs, NJ.

Dilger, W.H. (1982). "Methods of Structural Creep Analysis," in Z. P. Bazant and F. H. Wittman (Eds.), *Creep and Shrinkage in Concrete Structure*, Wiley, New York.

PCI (2003). *PCI Design Handbook: Precast and Prestressed Concrete*, 2nd ed., Precast/Prestressed Concrete Institute, Chicago.

PART IV

Steel Bridges

CHAPTER 17

Steel Bridges

17.1 INTRODUCTION

Steel bridges have a long and proud history. Their role in the expansion of the railway system in the United States cannot be overestimated. The development of the long-span truss bridge was in response to the need of railroads to cross waterways and ravines without interruption. Fortunately, analysis methods for trusses (particularly graphical statics) had been developed at the same time the steel industry was producing plates and cross sections of dependable strength. The two techniques came together and resulted in a figurative explosion of steel truss bridges as the railroads pushed westward.

Steel truss bridges continue to be built today, for example, the Greater New Orleans Bridge No. 2 of Figure 4.6. However, with advances in methods of analysis and steelmaking technology, the sizes, shapes, and forms of steel bridges are almost unlimited. We now have steel bridges of many types: arches (tied and otherwise), plate girders (haunched and uniform depth), box girders (curved and straight), rolled beams (composite and noncomposite), and cable-stayed and suspension systems. More complete descriptions of these various bridge types are given in Chapter 4.

Emphasis in this book is on short- (up to 50 ft or 15 m) to medium- (up to 200 ft or 60 m) span bridges. For these span lengths, steel girder bridges are a logical choice: composite rolled beams, perhaps with cover plates, for the shorter spans and composite plate girders for the longer spans. These steel girder bridges are readily adapted to different terrain and alignment and can be erected in a relatively short time with minimum interruption of traffic.

Steel bridge material properties are described in this chapter. In the chapters that follow limit states are presented and resistance considerations are discussed. The last chapter concludes with design examples of rolled-beam and plate-girder bridges.

17.2 MATERIAL PROPERTIES

As discussed at the beginning of Chapter 13, shown in Figure 13.1, the material–stress–strain response is the essential element relating forces and deformations. At one time, there was basically a simple stress–strain curve that described the behavior of structural steel; this is no longer true because additional steels have been developed to meet specific needs such as improved strength, better toughness, corrosion resistance, and ease of fabrication.

Before presenting the stress–strain curves of the various steels, it is important to understand what causes the curves to differ from one another. The different properties are a result of a combination of chemical composition and the physical treatment of the steel (Dowling et al., 1992). In addition to knowledge of the stress–strain behavior, a steel bridge designer must also understand how fatigue and fracture resistance are affected by the selection of material, member sizes, and weld details. These topics are discussed in this section along with a brief description of the manufacturing process.

In comparing the properties of different steels, the terms strength (yield and tensile), ductility, hardness, and toughness are used. These terms are defined below:

Yield strength is the stress at which an increase in strain occurs without an increase in stress.

Tensile strength is the maximum stress reached in a tensile test.

Ductility is an index of the ability of the material to withstand inelastic deformations without fracture and can be expressed as a ratio of elongation at fracture to the elongation at first yield.

Hardness refers to the resistance to surface indentation from a standard indenter.

Toughness is the ability of a material to absorb energy without fracture.

17.2.1 Steelmaking Process: Traditional

The typical raw materials for making steel are iron ore, coke, limestone, and chemical additives. These are the basic constituents and the chemical admixtures that produce custom-designed products for specific applications, much like the process used for making concrete. However, in the case of steelmaking, it is possible to better control the process and produce a more uniformly predictable finished product.

The raw materials are placed in a ceramic-lined blast furnace and external heat is applied. The coke provides additional heat and carbon for reducing the iron ore to metallic iron. The limestone acts as a flux that combines with the impurities and accumulates on top of the liquid iron where it can be readily removed as fluid slag. The molten iron is periodically removed from the bottom of the furnace through tap holes into transfer ladles. The ladles then transfer the liquid metal to the steelmaking area.

Steel is an alloy. It is produced by combining the molten iron with other elements to give specific properties for different applications. Depending on the steel manufacturer, this can be done in a basic oxygen furnace, an open-hearth furnace, or an electric-arc furnace. At this point, the molten iron from the blast furnace is combined with steel scrap and various fluxes. Oxygen is blown into the molten metal to convert the iron into steel by oxidation. The various fluxes are often other elements added to combine with the impurities and reduce the sulfur and phosphorus contents. The steel produced flows out a tap hole and into a ladle.

The ladle is used to transport the liquid steel to either ingot molds or a continuous casting machine. While the steel is in the ladle, its chemical composition is checked and adjustments to the alloying elements are made as required. Because of the importance of these alloying elements in classifying structural steels, their effect on the behavior and characteristics of carbon and alloy steels are summarized in Table 17.1.

Aluminum and silicon are identified as deoxidizers or "killers" of molten steel. They stop the production of carbon monoxide and other gases that are expelled from the molten metal as it solidifies. Killed steel products are less porous and exhibit a higher degree of uniformity than nonkilled steel products.

Carbon is the principal strengthening element in steel. However, it has a downside as increased amounts of carbon cause a decrease in ductility, toughness, and weldability. Chromium and copper both increase the atmospheric corrosion resistance and are used in weathering steels. When exposed to the atmosphere, they build up a tight protective oxide film that tends to resist further corrosion. Sulfur is generally considered an undesirable element except where machinability is important. It adversely affects surface quality and decreases ductility, toughness, and weldability. Manganese can control the harmful effects of sulfur by combining to form manganese sulfides. It also increases the hardness and strength of steels, but to a lesser extent than does carbon.

Table 17.1 Effects of Alloying Elements

Elements	Effects
Aluminum (Al)	Deoxidizes or "kills" molten steel. Refines grain size; increases strength and toughness.
Boron (B)	Small amounts (0.0005%) increase hardenability in quenched-and-tempered steels. Used only in aluminum-killed steels. Most effective at low carbon levels.
Calcium (Ca)	Controls shape of nonmetallic inclusions.
Carbon (C)	Principal hardening element in steel. Increases strength and hardness. Decreases ductility, toughness, and weldability. Moderate tendency to segregate.
Chromium (Cr)	Increases strength and atmospheric corrosion resistance.
Copper (Cu)	Increases atmospheric corrosion resistance.
Manganese (Mn)	Increases strength. Controls harmful effects of sulfur.
Molybdenum (Mo)	Increases high-temperature tensile and creep strength.
Niobium (Nb)	Increases toughness and strength.
Nickel (Ni)	Increases strength and toughness.
Nitrogen (N)	Increases strength and hardness. Decreases ductility and toughness.
Phosphorus (P)	Increases strength and hardness. Decreases ductility and toughness. Increases atmospheric corrosion resistance. Strong tendency to segregate.
Silicon (Si)	Deoxidizes or "kills" molten steel.
Sulfur (S)	Considered undesirable except for machinability. Decreases ductility, toughness, and weldability. Adversely affects surface quality. Strong tendency to segregate.
Titanium (Ti)	Increases creep and rupture strength and hardness.
Vanadium (V) and Columbium (Nb)	Small additions increase strength.

Brockenbrough and Barsom (1992).

17.2.2 Steelmaking Process: Mini Mills

Mini mills are smaller steel mills that use recycled steel in electric-arc furnaces as the primary heat source. Because the material source is recycled scrap or other iron sources, the coke-making operation is eliminated. The downstream processing can include: casting, hot or cold rolling, wire drawing, and pickling. This is accomplished by the continuous casting process that eliminates the ingot by casting directly into the target product. Mini mills produce a smaller range of products usually for a local area. This process is expanded for larger scale production, for example, Nucor and others that use recycled steel.

17.2.3 Steelmaking Process: Environmental Considerations

As sustainable building material, steel is among the best. It is the most recycled material in the world. Approximately 96% of the beams and plates used for structural steel were producted from recycled materials. Reinforcment bar is about 60%. For each pound of steel that is recycled 5400 Btu of energy are conserved because of the elimination of the iron producing steps outlined above and the use of new technologies that are more energy efficient and environmentally friendly with respect to emission gases as well. For example, see Recycled-steel.org or Nucor.com. Also search "steel recycled" for numerous sources including stories, real-time recycling data, and new steel/iron production methods within the United States and developing countries.

Without further expansion because of space, the steel industry has made significant changes in processes and products and is now among the "greenest" industries in the construction/manufacturing area.

17.2.4 Production of Finished Products

The liquid steel from the ladle is placed in ingot molds or a continuous casting machine. Steel placed in the ingot molds solidifies as it cools. It then goes into a second process where the ingot is hot-worked into slabs (up to 9 in. thick × 60 in. wide), blooms (up to 12 in. × 12 in.), and billets (up to 5 in. × 5 in.) (230 mm × 1520 mm, 300 mm × 300 mm, 125 mm × 125 mm, respectively).

In the continuous casting process, gravity is utilized to directly form slabs, blooms, and billets from a reservoir of liquid steel as shown in Figure 17.1.

The slabs are reheated and squeezed between sets of horizontal rolls in a plate mill to reduce the thickness and produce finished plate products. The longitudinal edges are often flame-cut online to provide the desired plate width. After passing through leveling rolls, the plates are sheared to length. Heat treating can be done online or offline.

The blooms are reheated and passed sequentially through a series of roll stands in a structural mill to produce wide-flange sections, I-beams, channels, angles, tees, and zees. There are four stages of roll stands, each with multiple passes

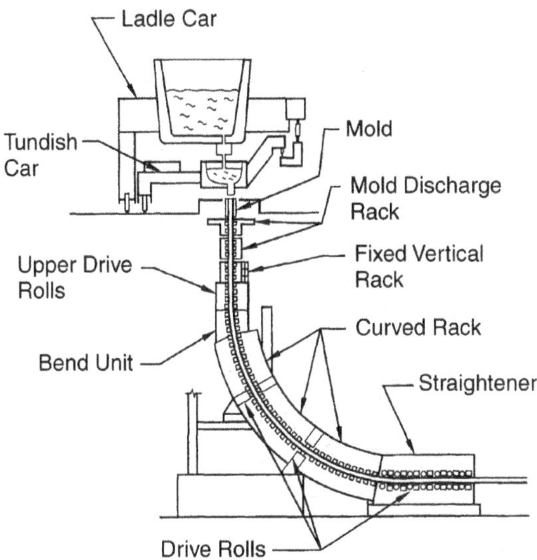

Fig. 17.1 Section schematic of a continuous caster (Brockenbrough and Barsom, 1992). [From *Constructional Steel Design: An International Guide*, P. J. Dowling, J. E. Harding, and R. Bjorhovde, eds., Copyright © 1992 by Elsevier Science Ltd (now Chapman and Hall, Andover, England), with permission.]

that are used to reduce the bloom to a finished product. They are a breakdown, a roughing, an intermediate, and finishing stands. Each stand has horizontal and vertical rolls, and in some cases edge rolls, to reduce the cross section progressively to its final shape. The structural section is cut to length, set aside to cool, and straightened by pulling or rolling.

17.2.5 Residual Stresses

Stresses that exist in a component without any applied external forces are called *residual stresses.* These forces affect the strength of members in tension, compression, and bending and can be induced by thermal, mechanical, or metallurgical processes. Thermally induced residual stresses are caused by nonuniform cooling. In general, tensile residual stresses develop in the metal that cools last. Associated compressive stresses are also introduced, and these stresses in combination create a balance of internal forces keeping the section in equilibrium.

Mechanically induced residual stresses are caused by nonuniform plastic deformations when a component is stretched or compressed under restraint. This nonuniform deformation can occur when a component is mechanically straightened after cooling or mechanically curved by a series of rollers.

Metallurgically induced residual stresses are caused by a change in the microstructure of the steel from fermite–pearlite to martensite (Brockenbrough and Barsom, 1992). This new material is stronger and harder than the original steel, but it is less ductile. The change to martensite results in a 4% increase in volume when the surface is heated to about 900°C and then cooled rapidly. If the volume change due to the transformation to marteniste is restrained, the residual stress is compressive. The tensile residual stresses induced by thermal cut edges can be partially compensated by the compressive stresses produced by the transformation.

When cross sections are fabricated by welding, complex three-dimensional (3D) residual stresses are induced by all three processes. Heating and cooling effects take place, metallurgical changes can occur, and deformation is often restrained. High tensile residual stresses of approximately 60 ksi (~400 MPa) can be developed at a weld (Bjorhovde, 1992).

In general, rolled edges of plates and shapes are under compressive residual stress while thermally cut edges are in tension. These stresses are balanced by equivalent stresses of opposite sign elsewhere in the member. For welded members, tensile residual stresses develop near the weld and equilibrating compressive stresses elsewhere. Figure 17.2 presents simplified qualitative illustrations of the global distribution of residual stresses in as-received and welded hot-rolled steel members (Brockenbrough and Barsom, 1992). Note that the stresses represented in Figure 17.2 are lengthwise or longitudinal stresses.

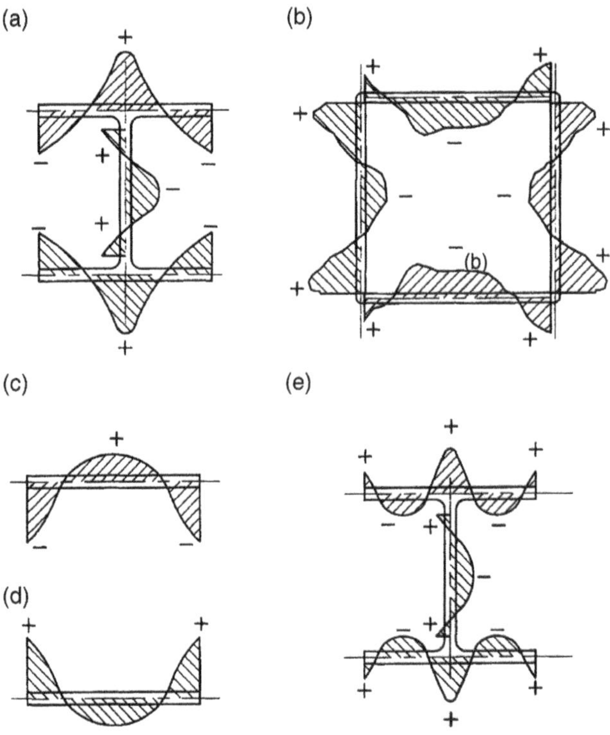

Fig. 17.2 Schematic illustration of residual stresses in as-rolled and fabricated structural components (Brockenbrough and Barsom, 1992). (a) Hot-rolled shape, (b) welded box section, (c) plate with rolled edges, (d) plate with flame-cut edges, and (e) beam fabricated from flame-cut plates. [From *Constructional Steel Design: An International Guide*, P. J. Dowling, J. E. Harding, and R. Bjorhovde, eds., Copyright © 1992 by Elsevier Science Ltd (now Chapman and Hall, Andover, England), with permission.]

17.2.6 Heat Treatments

Improved properties of steel can be obtained by various heat treatments. There are slow cooling heat treatments and rapid cooling heat treatments. Slow cooling treatments are annealing, normalizing, and stress relieving. They consist of heating the steel to a given temperature, holding for a proper time at that temperature, followed by slow cooling in air. The temperature to which the steel is heated determines the type of treatment. The slow cooling treatments improve ductility and fracture toughness, reduce hardness, and relieve residual stresses.

Rapid cooling heat treatments are indicated for the bridge steels in the AASHTO (2010) LRFD Bridge Specifications. The process is called quenching-and-tempering and consists of heating the steel to about 900°C, holding the temperature for a period of time, then rapid cooling by quenching in a bath of oil or water. After quenching, the steel is tempered by reheating to about 900°F (500°C), holding that temperature, then slowly cooling. Quenching-and-tempering changes the microstructure of the steel and increases its strength, hardness, and toughness.

17.2.7 Classification of Structural Steels

Mechanical properties of typical structural steels are depicted by the four stress–strain curves shown in Figure 17.3. Each of these curves represents a structural steel with specific composition to meet a particular need. Their behavior

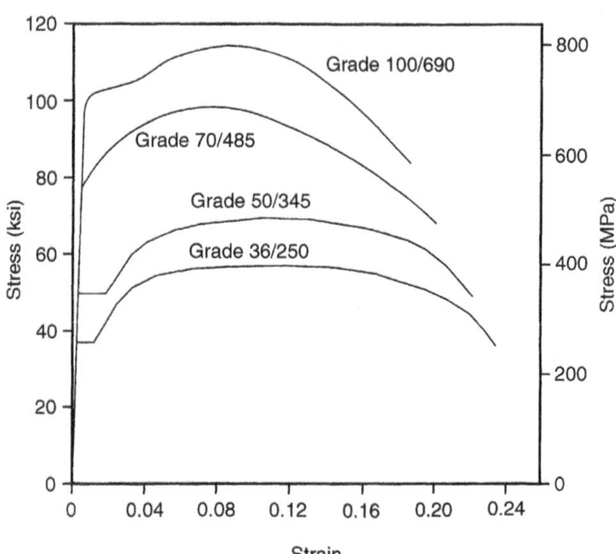

Fig. 17.3 Typical stress–strain curves for structural steels. (From R. L. Brockenbrough and B. G. Johnston, *USS Steel Design Manual*, Copyright © 1981 by R. L. Brockenbrough & Assoc, Inc., Pittsburgh, PA, with permission.)

differs from one another except for small strains near the origin. These four different steels can be identified by their chemical composition and heat treatment as (a) structural carbon steel (Grade 36/250) (ksi/MPa), (b) high-strength low-alloy steel (Grade 50/345), (c) quenched-and-tempered low-alloy steel (Grade 70/485), and (d) high-yield strength, quenched-and-tempered alloy steel (Grade 100/690). The minimum mechanical properties of these steels are given in Table 17.2.

A unified standard specification for bridge steel is given in ASTM (1995) with the designation A709/A709M-94a (M indicates metric and 94a is the year of last revision). Six grades of steel are available in four yield strength levels as shown in Table 17.2 and Figure 17.3. Steel grades with the suffix W indicate weathering steels that provide a substantially better atmospheric corrosion resistance than typical carbon steel and can be used unpainted for many applications, and the prefix HPS indicates high-performance steel.

All of the steels in Table 17.2. can be welded, but not by the same welding process. Each steel grade has specific welding requirements that must be followed.

In Figure 17.3, the number in parentheses identifying the four yield strength levels is the ASTM designation of the steel with similar tensile strength and elongation properties as the A709/A709M steel. These numbers are given because they are familiar to designers of steel buildings and other structures. The most significant difference between these steels and the A709/A709M steels is that the A709/A709M steels are specifically for bridges and must meet supplementary requirements for toughness testing. These requirements vary for nonfracture critical and fracture-critical members. This concept is discussed in Section 17.2.6.

Table 17.2-US Minimum Mechanical Properties of Structural Steel by Shape, Strength, and Thickness

AASHTO designation	M270 Grade 36	A709 Grade 50	M270 Grade 50S	M270 Grade 50W	M270 Grade HPS 50W	M270 Grade HPS 70W	M270 Grades 100/100W	
Equivalent ASTM designation	A709 Grade 36	A709 Grade 50	A709 Grade 50S	A709 Grade 50W	A709 Grade HPS 50W	A709 Grade HPS 70W	A709 Grades 100/100W	
Thickness of plates (in.)	Up to 4 incl.	Up to 4 incl.	Up to 4 incl.	Up to 4 incl.	Up to 4 incl.	Up to 4 incl.	Up to 2.5 incl.	Over 2.5–4 incl.
Shapes	All groups	All groups	All groups	All groups	N/A[a]	N/A	N/A	N/A
Minimum tensile strength, F_u, (ksi)	58	65	65	70	70	90	110	100
Minimum yield point or minimum yield strength, F_y, (ksi)	36	50	50	50	50	70	100	90

In AASHTO Table 6.4.1-1. From *AASHTO LRFD Bridge Design Specifications*, Copyright © 2010 by the American Association of State Highway and Transportation Officials. Used by permission.
[a]Not applicable = N/A.

Table 17.2-SI Minimum Mechanical Properties of Structural Steel by Shape, Strength, and Thickness

AASHTO designation	M270 Grade 250	M270 Grade 345	M270 Grade 345S	M270 Grade 345W	M270 Grade HPS 345W	M270 Grade HPS 485W	M270 Grades 690/690W	
Equivalent ASTM designation	A709M Grade 250	A709M Grade 345	A709M Grade 345S	A709M Grade 345W	A709M Grade HPS 345W	A709M Grade HPS 345W	A709M Grades 690/690W	
Thickness of plates (mm)	Up to 100 incl.	Up to 100 incl.	Up to 100 incl.	Up to 100 incl.	Up to 100 incl.	Up to 100 incl.	Up to 65 incl.	Over 65–100 incl.
Shapes	All groups	All groups	All groups	All groups	N/A[a]	N/A	N/A	N/A
Minimum tensile strength, F_u, (MPa)	400	450	450	450	485	585	760	690
Minimum yield point or minimum yield strength, F_y, (MPa)	250	345	345	345	345	485	690	620

In AASHTO Table 6.4.1-1. From *AASHTO LRFD Bridge Design Specifications*, Copyright © 2010 by the American Association of State Highway and Transportation Officials. Used by permission.
[a]Not applicable = N/A.

Table 17.3 Chemical Requirements for Bridge Steels: Heat Analysis, Percent[a]

Element	Carbon Steel Grade 36/250 Shapes	High-Strength Grade 50/345 Type 2	Low-Alloy Steel Grade 50/345W Type A	Heat-Treated Low-Alloy Steel Grade 70/485	High-Strength, Heat-Treated Alloy Steel Grade 100/690 Type C	High-Strength, Heat-Treated Alloy Steel Grade 100/690W Type F
Boron					0.001–0.005	0.0005–0.006
Carbon	0.26 max	0.23 max	0.19 max	0.19 max	0.12–0.20	0.10–0.20
Chromium			0.40–0.65	0.40–0.70		0.40–0.65
Copper	0.20 min	0.20 min	0.25–0.40	0.20–0.40		0.15–0.50
Manganese		1.35 max	0.80–1.25	0.80–1.35	1.10–1.50	0.60–1.00
Molybdenum					0.15–0.30	0.40–0.60
Nickel			0.40 max	0.50 max		0.70–1.00
Phosphorus	0.04 max	0.04 max	0.04 max	0.035 max	0.035 max	0.035 max
Silicon	0.40 max	0.40 max	0.30–0.65	0.20–0.65	0.15–0.30	0.15–0.35
Sulfur	0.05 max	0.05 max	0.05 max	0.04 max	0.035 max	0.035 max
Vanadium		0.01–0.15	0.02–0.10	0.02–0.10		0.03–0.08

From ASTM (1995).
[a] Where a blank appears in this table there is no requirement.

As discussed previously, steel is an alloy and its principal component is iron. The chemical composition of the steel grades in Table 17.2 is given in Table 17.3. One component of all the structural steels is carbon, which, as indicated in Table 17.1, increases strength and hardness but reduces ductility, toughness, and weldability. Other alloying elements are added to offset the negative effects and to custom design a structural steel for a particular application. Consequently, more than one type of steel is given in A709M for each yield strength level to cover the proprietary steels produced by different manufacturers. In general, low-alloy steel has less than 1.5% total alloy elements while alloy steels have a larger percentage.

A comparison of the chemical composition of bridge steels in Table 17.3 with the effects of the alloying elements in Table 17.1 shows the following relationships. Boron is added to the quenched-and-tempered alloy steel to increase hardenability. Carbon content decreases in the higher strength steels and manganese, molybdenum, and vanadium are added to provide the increase in strength. Chromium, copper, and nickel are found in the weathering steels and contribute to their improved atmospheric corrosion resistance. Phosphorus helps strength, hardness, and corrosion but decreases ductility and toughness so its content is limited. Sulfur is considered undesirable so its maximum percentage is severely limited. Silicon is the deoxidizing agent that kills the molten steel and produces more uniform properties.

Two properties of all grades of structural steels are assumed to be constant: the modulus of elasticity E_S of 29,000 ksi (200 GPa) and the coefficient of thermal expansion of 6.5×10^{-6} in./in./°F (11.7×10^{-6} mm/mm/°C) [A6.4.1].* A brief discussion of the properties associated with each of the four levels of yield strength is given below (Brockenbrough and Johnston, 1981). To aid in the comparison between the different steels, the initial portions of their stress–strain curves and time-dependent corrosion curves are given in Figures 17.4 and 17.5, respectively.

Structural Carbon Steel The name is somewhat misleading because all structural steels contain carbon. When reference is made to carbon steel, the technical definition is usually implied. The criteria for designation as carbon steel are (AISI, 1985): (1) No minimum content is specified for chromium, cobalt, columbium, molybdenum, nickel, titanium, tungsten, vanadium, or zirconium or any other element added to obtain a desired alloying effect; (2) the specified minimum of copper does not exceed 0.40%; or (3) the specified maximum for any of the following is not exceeded: manganese 1.65%, silicon 0.60%, and copper 0.60%. In other words, a producer can use whatever scrap steel or junked automobiles are available to put in the furnace as long as the minimum mechanical properties of Table 17.2 are met. No exotic or fancy ingredients are necessary to make it strong. As a result, engineers often refer to it as *mild* steel.

One of the main characteristics of structural carbon steel is a well-defined yield point [F_y = 36 ksi (250 MPa)] followed

*The article numbers in the AASHTO (2010) LRFD Bridge Specifications are enclosed in brackets and preceded by the letter A if specifications and by C if commentary.

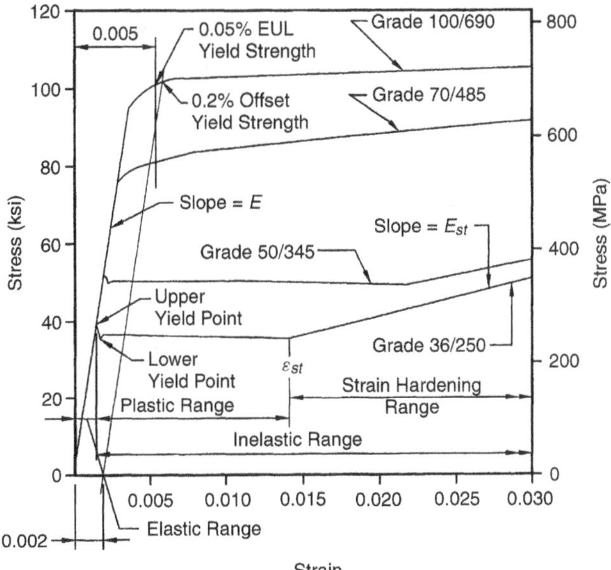

Fig. 17.4 Typical initial stress–strain curves for structural steels. (From R. L. Brockenbrough and B. G. Johnston, *USS Steel Design Manual*, Copyright © 1981 by R. L. Brockenbrough & Assoc, Inc., Pittsburgh, PA, with permission.)

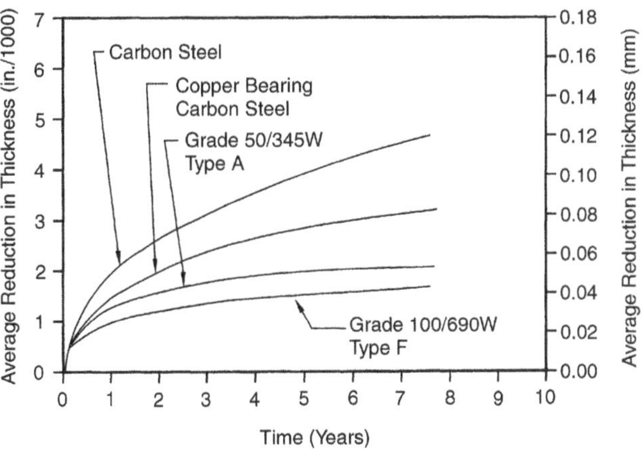

Fig. 17.5 Corrosion curves for several steels in an industrial atmosphere. (From R. L. Brockenbrough and B. G. Johnston, *USS Steel Design Manual*, Copyright © 1981 by R. L. Brockenbrough & Assoc, Inc., Pittsburgh, PA, with permission.)

by a generous yield plateau in the plastic range. This behavior is shown in Figure 17.4 and indicates significant ductility, which allows redistribution of local stresses without fracture. This property makes carbon steel especially well suited for connections.

Carbon steels are weldable and available as plates, bars, and structural shapes. They are intended for service at atmospheric temperature. The corrosion rate in Figure 17.5 for copper-bearing carbon steel (0.20% minimum, Table 17.3) is about one-half that of plain carbon steel.

High-Strength Low-Alloy Steel These steels have controlled chemical compositions to develop yield and tensile strengths greater than carbon steel, but with alloying additions smaller than those for alloy steels (Brockenbrough, 1992). The higher yield strength of $F_y = 50$ ksi (345 MPa) is achieved in the hot-rolled condition rather than through heat treatment. As a result, they exhibit a well-defined yield point and excellent ductility as shown in Figure 17.4.

High-strength low-alloy steels are weldable and available as plates, bars, and structural shapes. These alloys also have superior atmospheric corrosion resistance as shown in Figure 17.5. Because of their desirable properties, Grade 50/345 steels are often the first choice of designers of small- to medium-span bridges.

Heat-Treated Low-Alloy Steel High-strength low-alloy steels can be heat treated to obtain higher yield strengths of $F_y = 70$ ksi (485 MPa). The chemical composition for Grades 50/345W and 70/485W in Table 17.3 are nearly the same. The quenching-and-tempering heat treatment changes the microstructure of the steel and increases its strength, hardness, and toughness.

The heat treatment removes the well-defined yield point from the high-strength steels as shown in Figure 17.4. There is a more gradual transition from elastic to inelastic behavior. The yield strength for these steels is usually determined by the 0.5% extension under load (EUL) definition or the 0.2% offset definition.

The heat-treated low-alloy steels are weldable but are available only in plates. Their atmospheric corrosion resistance is similar to that of high-strength low-alloy steels.

High-Strength Heat-Treated Alloy Steel Alloy steels are those with chemical compositions that are not in the high-strength low-alloy classification (see Table 17.3). The quenching-and-tempering heat treatment is similar to that for the low-alloy steels, but the different composition of alloying elements develops higher strength of $F_y = 100$ ksi (690 MPa) and greater toughness at low temperature.

An atmospheric corrosion curve for the alloy steels (Grade 100/690) is given in Figure 17.5 and shows the best corrosion resistance of the four groups of steels.

Again the yield strength is determined by the 0.5% EUL definition or the 0.2% offset definition shown in Figure 17.4. By observing the complete stress–strain curves in Figure 17.3, note the heat-treated steels reach their peak tensile strength and decrease rapidly at lower strains than the untreated steels. This lower ductility may cause problems in some structural applications and caution must be exercised when heat-treated steels are used.

The high-strength heat-treated alloy steels are weldable but are available only in structural steel plates for bridges.

Strength, weldability, toughness, ductility, corrosion resistance, and formability are important performance characteristics. High-performance steel (HPS) has an optimized balance of these properties to give maximum performance in bridge structures. The main two differences compared to conventional Grade 50 steels are improved weldability and toughness. Corrosion resistance and ductility are nearly the same as conventional Grade 50W. The fatigue resistance is the same as well.

Lane and co-workers (1997) summarize the importance HPS and its increased weldability and the difficulties with conventional steels:

> *Conventional 485-MP (50 ksi) steels typically require preheating of plates, control of temperature between weld passes, controlled handling of welding consumables, precisely controlled energy input, and postweld heat treatment in some cases. When all of these operations are performed correctly, it is usually possible to produce high-quality welds in conventional high-strength steel. Difficulties can arise, however, when one or more of these operations deviate from prescribed procedures. Minor differences in procedure and quality control are the norm for bridge construction, where many different fabricators in different parts of the country work under different climates and conditions. The result is that conventional high-strength steels have experienced a higher percentage of weld problems compared to lower strength steels. Another disadvantage is that these controls, particularly the control of temperature, add significantly to the cost and time required for welding. The goal in developing HPS grades is to provide a steel that is forgiving enough to be welded under a variety of conditions without requiring excessive weld-process controls that increase costs.*

The minimum specifications for toughness required by a steel is set by AASHTO. For fracture-critical members in the most severe climate (zone III), AASHTO currently requires a minimum Charpy V-notch (CVN) energy of 35 ft-lb at −10°F [A6.6.2].

Experimental toughness values reported by Barsom (1996) from the first heat of HPS-70W ranged from a minimum of about 120 ft-lb to a maximum of 240 ft-lb at −10°F. This far exceeds the current AASHTO minimum requirements and provides a significant resistance to brittle fracture. This energy absorption provides added confidence to enable designers to use the full strength of this steel. Figure 17.6 illustrates the increased toughness of HPS.

Note that the brittle-ductile transition of HPS occurs at a much lower temperature than conventional Grade 50W steel. This means that HPS 70W (HPS 485W) remains fully ductile at lower temperatures where conventional Grade 50W steel begins to show brittle behavior. Although the fatigue performance is not improved with HPS, once a crack initiates its propagation in cold temperatures, it is slowed or perhaps mitigated because of the increased toughness. This increased tolerance can be the difference between catching a crack during inspection and a catastrophic brittle collapse.

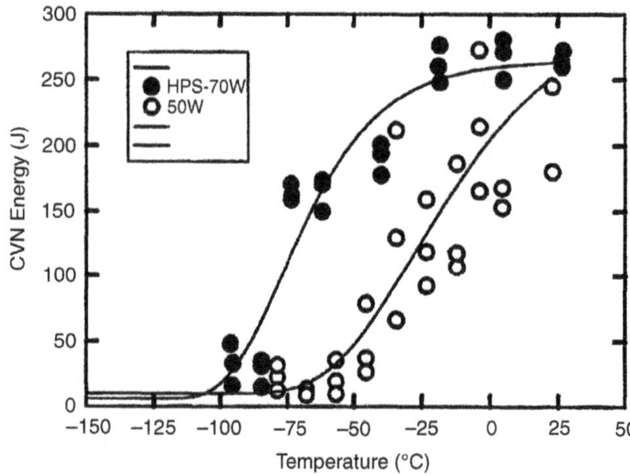

Fig. 17.6 Fracture toughness comparison. (After Hamby et al., 2002.)

Nebraska DOT was the first to use HPS 70W in the construction of the Snyder Bridge—a welded plate girder steel bridge. The bridge was opened to traffic in October 1997. The intent was to use this first HPS 70W bridge to gain experience on the HPS fabrication process. The original design utilized conventional grade 50W steel, and the designer just replaced the grade 50W steel with HPS 70W steel of equal size—not an economical design. The fabricators concluded no significant changes were needed in the HPS fabrication process.

Since that first bridge, numerous agencies have built HPS bridges and they are becoming common place in U.S. bridge engineering practice. The cost comparisons with more conventional design can be significant.

In a research project, Barker and Schrage (2000) illustrated that savings in weight and cost can be substantial; see Tables 17.4 and 17.5. Here a bridge in the Missouri DOT inventory was designed to a design ratio of nearly one in all cases. These data illustrate that the use of fewer girder lines results in lower cost, which is typical of all materials and, second, that HPS hybrid design is the most economical.

A wealth of HPS literature is available in the trade and research literature. Other examples include work by Azizinamini et al. (2004), Dexter et al. (2004), and Wasserman (2003) that might be of interest to the bridge designer.

17.2.8 Effects of Repeated Stress (Fatigue)

When designing bridge structures in steel, a designer must be aware of the effect of repeated stresses. Vehicles passing any given location are repeated time and time again. On a heavily traveled interstate highway with a typical mix of trucks in the traffic, the number of maximum stress repetitions can be millions in a year.

Table 17.4 Bridge Design Alternatives: Summary

Design Alternative	Girder Lines	Total Diaphragms	Additional Stiffeners	Steel Weight Tonnes (Tons)		
				45W (50W)	HPS	Total
9 Girder 345W (50W)	9	120	38	326.6 (360.1)	0 (0)	326.6 (360.1)
7 Girder 345W (50W)	7	90	46	310.5 (342.3)	0 (0)	310.5 (342.3)
9 Girder HPS	9	120	4	13.2 (14.6)	264.6 (291.7)	277.8 (306.3)
8 Girder HPS	8	105	2	13.0 (14.3)	259.7 (286.3)	272.7 (300.6)
7 Girder HPS	7	90	2	12.9 (14.2)	257.7 (284.1)	270.6 (298.3)
7 Girder Hybrid	7	90	28	182.7 (201.4)	94.0 (103.6)	276.7 (305.0)

After Barker and Schrage (2000).

Table 17.5 Bridge Design Alternatives: Summary of Cost Savings

	Current HPS Mat. Costs		Projected HPS Mat. Costs	
Design Alternative	% Cost Savings over 9 Girder 345W (50W)	% Cost Savings over 7 Girder 345W (50W)	% Cost Savings over 9 Girder 345W (50W)	% Cost Savings over 7 Girder 345W (50W)
9 Girder 345W (50W)	Base	−9.3%	Base	−9.3%
7 Girder 345W (50W)	8.5%	Base	8.5%	Base
9 Girder HPS	−9.7%	−20.0%	−0.5%	−9.9%
8 Girder HPS	−2.6%	−12.2%	6.4%	−2.4%
7 Girder HPS	0.4%	−8.9%	9.4%	0.9%
7 Girder Hybrid	18.6%	11.0%	21.9%	14.6%

After Barker and Schrage (2000).

These repeated stresses are produced by service loads, and the maximum stresses in the base metal of the chosen cross section are less than the strength of the material. However, if there is a stress raiser due to a discontinuity in metallurgy or geometry in the base metal, the stress at the discontinuity can easily double or triple the stress calculated from the service loads. Even though this high stress is intermittent, if it is repeated many times, damage accumulates, cracks form, and fracture of the member can result.

This failure mechanism, which consists of the formation and growth of cracks under the action of repeated stresses, each of which is insufficient by itself to cause failure, is called fatigue (Gurney, 1992). The metal just gets "tired" of being subjected to moderate-level stresses again and again. Hence, fatigue is a good word to describe this phenomenon.

Determination of Fatigue Strength Fatigue strength is not a material constant like yield strength or modulus of elasticity. It is dependent on the particular joint configuration involved and can realistically only be determined experimentally. Because most of the stress concentration problems due to discontinuities in geometry and metallurgy are associated with welded connections, most of the testing for fatigue strength has been done on welded joint configurations.

The procedure followed for each welded connection is to subject a series of identical specimens to a stress range S that is less than the yield stress of the base metal and to repeat that stress range for N cycles until the connection fails. As the stress range is reduced, the number of cycles to failure increases. The results of the tests are usually plotted as log S versus log N graphs. A typical S–N curve for a welded joint is shown in Figure 17.7. At any point on the curve, the stress

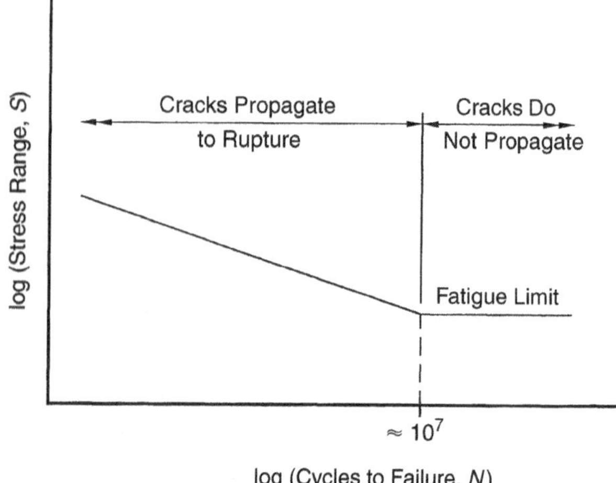

Fig. 17.7 Typical S–N curve for welded joints.

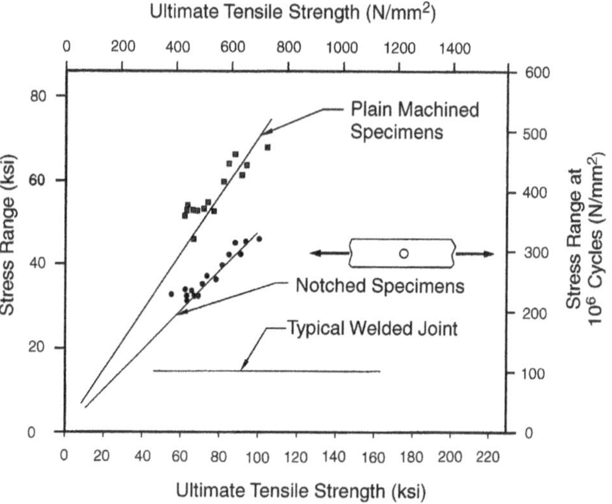

Fig. 17.8 Fatigue strength compared to static strength (Gurney, 1992). [From *Constructional Steel Design: An International Guide*, P. J. Dowling, J. E. Harding, and R. Bjorhovde, eds., Copyright © 1992 by Elsevier Science Ltd (now Chapman and Hall, Andover, England), with permission.]

value is the *fatigue strength* and the number of cycles is the *fatigue life* at that level of stress.

Notice that when the stress range is reduced to a particular value, an unlimited number of stress cycles can be applied without causing failure. This limiting stress is called the fatigue limit or endurance limit.

Influence of Strength of the Base Material The fatigue strength of unwelded components increases with the tensile strength of the base material. This fatigue strength is shown in Figure 17.8 for both solid round and notched specimens. However, if high-strength steel is used in welded components, no apparent increase in the fatigue strength is apparent.

The explanation for the difference in behavior is that in the unwelded material cracks must first be formed before they can propagate and cause failure, while in the welded joint small cracks already exist and all they need to do is propagate. Rate of crack propagation does not vary significantly with tensile strength; therefore, fatigue strength of welded joints is independent of the steel from which they are fabricated (Gurney, 1992).

Influence of Residual Stresses In general, welded joints are not stress relieved, so it is reasonable to assume that residual stresses σ_r exist somewhere in the connection. If a stress cycle with range S is applied, the actual stress range moves from σ_r to $\sigma_r \pm S$, and the nominal stress range is S. Therefore, it is possible to express the fatigue behavior of a welded joint in terms of stress range alone without knowing the actual maximum and minimum values. In the AASHTO (2010) LRFD Bridge Specifications, load-induced fatigue considerations are expressed in terms of stress range; residual stresses are not considered [A6.6.1.2].

Closing Remarks on Fatigue Fatigue is the most common cause of structural steel failure, which is caused by not considering this failure mode at the design stage. Good design requires careful assessment. Adequate attention to joint selection and detailing and knowledge of service load requirements can minimize the risk of failure, while ignorance or neglect of these factors can be catastrophic (Gurney, 1992).

17.2.9 Brittle Fracture Considerations

A bridge designer must understand the conditions that cause brittle fractures to occur in structural steel. Brittle fractures are to be avoided because they are nonductile and can occur at relatively low stresses. When certain conditions are present, cracks can propagate rapidly and sudden failure of a member can result.

One of the causes of a brittle fracture is a triaxial tension stress state that can be present at a notch in an element or a restrained discontinuity in a welded connection. When a ductile failure occurs, shear along slip planes is allowed to develop. This sliding between planes of the material can be seen in the necking down of the cross section during a standard tensile coupon test. The movement along the slip planes produces the observed yield plateau and an increase in deformation that characterizes a ductile failure. In looking at a cross section after failure, it is possible to distinguish the crystalline appearance of the brittle fracture area from the fibrous appearance of the shear plane area and its characteristic shear lip. The greater the percentage of shear area on the cross section, the greater the ductility.

In the uniaxial tension test, there is no lateral constraint to prevent the development of the shear slip planes. However, stress concentrations at a notch or stresses developed due to cooling of a restrained discontinuous weld can produce a triaxial tension state of stress in which shear cannot develop. When an impact load produces additional tensile stresses, often on the tension side caused by bending, a sudden brittle fracture may occur.

Another cause of brittle fracture is a low-temperature environment. Structural steels may exhibit ductile behavior at temperatures above 32°F (0°C) but change to brittle fracture when the temperature drops. A number of tests have been developed to measure the relative susceptibility of steel to brittle fracture with temperature. One of these is the Charpy V-notch impact test. This test consists of a simple beam specimen with a standard size V-notch at midspan that is fractured by a blow from a swinging pendulum as shown in Figure 17.9. The amount of energy required to fracture the specimen is determined by the difference in height of the pendulum before and after striking the small beam. The fracture energy can be correlated to the percent of the cross section that fails by shear. A higher energy is associated with the greater the percentage of shear failure. A typical plot of the results of a Charpy V-notch test with variation in temperature is given in Figure 17.10.

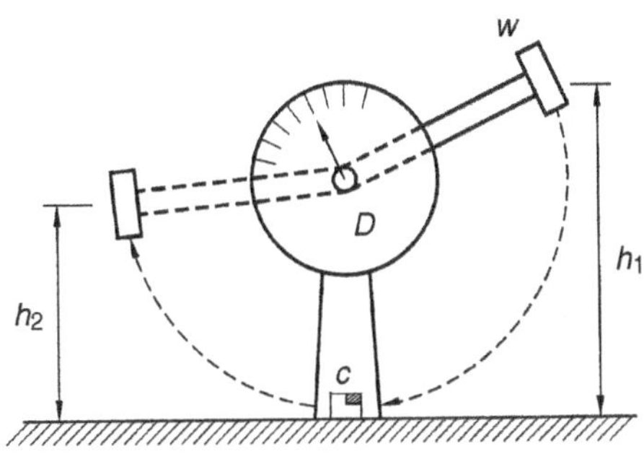

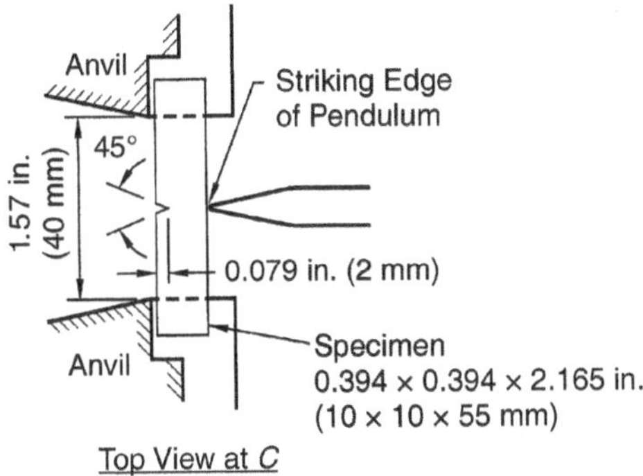

Fig. 17.9 Charpy V-notch impact test. (After Barsom, 1992.) [From *Constructional Steel Design: An International Guide*, P. J. Dowling, J. E. Harding, and R. Bjorhovde, eds., Copyright © 1992 by Elsevier Science Ltd (now Chapman and Hall, Andover, England), with permission.]

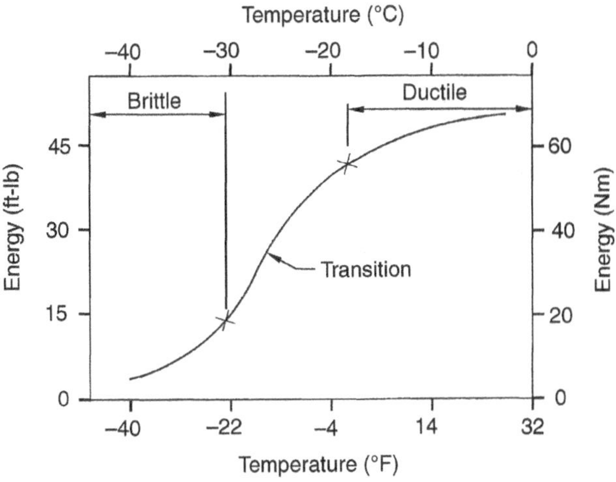

Fig. 17.10 Transition from ductile to brittle behavior for low-carbon steel.

As illustrated in Figure 17.10, the energy absorbed during fracture decreases gradually as temperature is reduced until it drops dramatically at some transition temperature. The temperature at which the specimen exhibits little ductility is called the nil ductility transformation (NDT) temperature. The NDT temperature can be determined from the Charpy V-notch test as the temperature at a specified level of absorbed energy or the temperature at which a given percentage of the cross section fails in shear. The AASHTO (2010) LRFD Bridge Specifications give minimum absorbed energy values to predict the fracture toughness of bridge steels under different temperature conditions [A6.6.2].

Welded connections must be detailed to avoid triaxial tensile stresses and the potential for brittle fracture. An example is the welded connection of intermediate stiffeners to the web of plate girders. In times past, intermediate stiffeners were full height and were often welded to both the compression and tension flanges. If the stiffener is welded to the tension flange as shown in Figure 17.11(a), restrained cooling of welds in three directions develops triaxial tensile stresses in the web. Often a notchlike stress raiser is present in the welded connection due to material flaws, cut-outs, undercuts, or arc strikes. If principal tensile stresses due to weld residual stresses, notch stress concentrations, and flexural tension in three principal directions reach the same value, then shear stresses vanish and a brittle fracture results.

If these conditions in the welded connection are accompanied by a drop in temperature, the energy required to initiate a brittle fracture drops significantly (Fig. 17.10) and the fracture can occur prematurely. Such web fractures occurred at the welded attachment of intermediate stiffeners to the tension flange of the LaFayette Street Bridge during a cold winter in St. Paul, Minnesota. After the investigation of the web

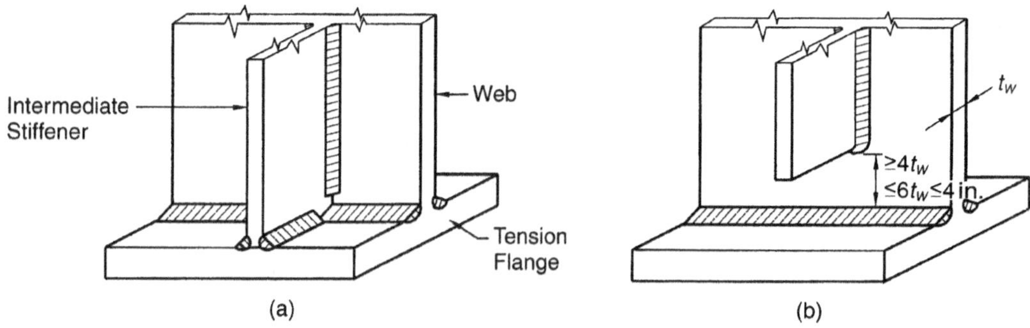

Fig. 17.11 Welded connections between an intermediate stiffener and the tension flange of a plate girder: (a) improper detail (no cope and intersecting welds) and (b) recommended detail for stiffeners *not* connected to cross frames.

fractures in the LaFayette Street Bridge, welding of intermediate stiffeners to tension flanges was no longer allowed (or permitted with a cope).

For stiffeners attached to cross frames and diaphragms, welds shall be provided on both flanges in order to avoid distortion-induced fatigue in the web. As a result, the current specifications [A6.10.11.1.1] require that the stiffener be stopped short of the tension flange [Fig. 17.11(b)] or coped, so that they cannot be inadvertently attached.

17.3 SUMMARY

The structural performance of steel bridges is strongly dependent upon material selection and connection details. Unlike steel buildings, bridges are subjected to years of repeated heavy loads that can create fatigue cracks that could lead to collapse. All welds and associated processes require careful attention.

Again, unlike buildings, steel bridges are subjected to environmental conditions that change temperature to levels that could cause brittle fracture. Environmental influences further subject steel to de-icing chemicals that can cause corrosion leading to an effective loss in cross section. Often such corrosion can be seen near deck expansion joints that leak or deck drains that are designed poorly.

Steel bridges are often used today for longer spans and curved alignment associated with "fly-over" ramps where other bridge types are more difficult to construct or lack economy.

REFERENCES

AASHTO (2010). *LRFD Bridge Design Specifications*, 5th ed., American Association of State Highway and Transportation Officials, Washington, DC.

AISI (1985). *Steel Product Manual—Plates*, American Iron and Steel Institute, Washington, DC.

ASTM (1995). "Standard Specifications for Carbon and High-Strength Low-Alloy Structural Steel Shapes, Plates, and Bars and Quenched-and-Tempered Alloy Structural Steel Plates for Bridges," *Annual Book of ASTM Standards*, Vol. 01.04, American Society for Testing and Materials, Philadelphia, PA.

Azizinamini, A, K. Barth, R. Dexter, and C. Rubeiz (2004). "High Performance Steel: Research Front—Historical Account of Research Activities," *Journal of Bridge Engineering, ASCE*, Vol. 9 No. 3, pp. 212–217.

Barker, M. G. and S. D. Schrage (2000). "High-Performance Steel Bridge Design and Cost Comparisons," *Transportation Research Record*, Issue 1740, pp. 33–39.

Barsom, J. M. (1992). "Fracture Design," in *Constructional Steel Design*, Dowling et al. (Eds.), Elsevier Science (Chapman & Hall, Andover, England), Chapter 5.5.

Barsom, J. M. (1996). "Properties of High Performance Steels and Their Use in Bridges," Proceedings, National Steel Bridge Symposium, American Institute of Steel Construction, Chicago, Ill., Oct. 15–17.

Bjorhovde, R. (1992). "Compression Members," in *Constructional Steel Design*, Dowling et al. (Eds.), Elsevier Science (Chapman & Hall, Andover, England), Chapter 2.3.

Brockenbrough, R. L. (1992). "Material Properties," in *Constructional Steel Design*, Dowling et al. (Eds.), Elsevier Science (Chapman & Hall, Andover, England), Chapter 1.2.

Brockenbrough, R. L. and J. M. Barsom (1992). "Metallurgy," in *Constructional Steel Design*, Dowling et al. (Eds.), Elsevier Science (Chapman & Hall, Andover, England), Chapter 1.1.

Brockenbrough, R. L. and B. G. Johnston (1981). *USS Steel Design Manual*, R. L. Brockenbrough & Assoc., Pittsburgh, PA.

Dexter, R. J., W. J. Wright, and J. W. Fisher (2004). "Fatigue and Fracture of Steel Girders," *Journal of Bridge Engineering, ASCE*, Vol. 9, No. 3, pp. 278–286.

Dowling, P. J., J. E. Hardin, and R. Bjorhovde (Eds.) (1992). *Constructional Steel Design—An International Guide*, Elsevier Science (Chapman & Hall, Andover, England).

Gurney, T. R. (1992). "Fatigue Design," in *Constructional Steel Design*, Dowling et al. (Eds.), Elsevier Science (Chapman & Hall, Andover, England), Chapter 5.4.

Hamby, G., G. Clinton, R. Nimis, and M. Myint Lwin (2002). *High Performance Steel Designer's Guide*, 2nd ed., Federal Highway Administration, Washington, DC.

Lane, S., E. Munley, W. Wright, M. Simon, and J. Cooper (1997). *High-Performance Materials: A Step Toward Sustainable Transportation*, Federal Highway Administration, Public Roads Online, Vol. 60, No. 4 (http://www.tfhrc.gov/pubrds/spring97/steel.htm).

Wasserman, E. P. (2003). "Optimization of HPS 70W Applications," *Journal of Bridge Engineering, ASCE*, Vol. 7, No.1, pp. 1–5.

PROBLEM

This problem references the plans for the bridge over the Little Laramie River. See Wiley's web site for a pdf. This bridge is a single-span, four-girder bridge that has clean details; its interpretation is straightforward for the student. These plans are used with permission.

17.1 For the bridge over the Little Laramie River, make a table with all the different types of steels used in this bridge and the general location, for example, bolts.

CHAPTER 18

Limit States and General Requirements

18.1 LIMIT STATES

Structural steel bridges must be designed so that their performance under load does not go beyond the limit states prescribed by the AASHTO (2010) Bridge Specifications. These limit states are applicable at all stages in the life of a bridge and include service, fatigue and fracture, strength, and extreme event limit states. The condition that must be met for each limit state is that the factored resistance is greater than the effect of the factored load combinations, which can be expressed as

$$\phi R_n \geq \sum \eta_i \gamma_i Q_i \tag{18.1}$$

where ϕ is a statistically based resistance factor for the limit state being examined; R_n is the nominal resistance; η_i is a load multiplier relating to ductility, redundancy, and operational importance; γ_i is a statistically based load factor applied to the force effects as defined for each limit state in Table 5.1; and Q_i is a load effect. The various factors in Eq. 18.1 are discussed more fully in Chapter 5.

18.1.1 Service Limit State

Service limit states relate to the performance of a bridge subjected to the forces applied when it is put into service. In steel structures, limitations are placed on deflections and inelastic deformations under service loads. By limiting deflections, adequate stiffness is provided and vibration is reduced to an acceptable level. By controlling local yielding, permanent inelastic deformations are avoided and rideability is assumed to be improved [A6.10.4].*

*The article numbers in the AASHTO (2010) LRFD Bridge Specifications are enclosed in brackets and preceded by the letter A if a specification article and by the letter C if commentary.

Because the provisions for service limit state are based on experience and engineering judgment, rather than calibrated statistically, the resistance factor ϕ, the load modifier η_i, and the load factors γ_i in Eq. 18.1 are taken as unity. One exception is the possibility of an overloaded vehicle that may produce excessive local stresses. For this case, the service II limit state in Table 5.1 with a vehicle live load factor of 1.30 is used.

Deflection Limit Deflection limitations are optional. If required by an owner, the deflection limit can be taken as the span/800 for vehicular loads or other limit specified by the owner [A2.5.2.6]. In calculating deflections, assumptions are made on load distribution to the girders, flexural stiffness of the girders in participation with the bridge deck, and stiffness contributions of attachments such as railings and concrete barriers. In general, more stiffness exists in a bridge system than is usually implied by typical engineering calculations. As a result, deflection calculations (as with all other calculations) are estimates. When this uncertainty is coupled with the subjective criteria of what constitutes an annoying vibration (or other reasons to limit deflection), the establishment of deflection limitations is not encouraged by the AASHTO Specifications. Most owners, however, require deflection limits in order to provide an acceptable stiffness that likely improves overall system performance. For example, deck durability may be indirectly improved by overall system stiffness.

Inelastic Deformation Limit Inelastic deformation limitations are mandatory. Local yielding under service II loads is not permitted [A6.10.4]. This local yielding is addressed by Eq. 18.1 for a strength limit state when the maximum force effects are determined by an elastic analysis. However, if inelastic moment redistribution follows an elastic analysis [A6.10.4.2], the concept of plastic hinging is introduced and the stresses must be checked. In this case, flange stresses in positive and negative bending shall not exceed [A6.10.4.2]:

☐ For steel top flanges of composite sections

$$f_f \leq 0.95 R_h F_{\text{yf}} \tag{18.2a}$$

☐ For steel bottom flanges of composite sections

$$f_f + \frac{f_l}{2} \leq 0.95 R_h F_{\text{yf}} \tag{18.2b}$$

☐ For both steel flanges of noncomposite sections

$$f_f + \frac{f_l}{2} \leq 0.80 R_h F_{\text{yf}} \tag{18.3}$$

where R_h is the flange-stress reduction factor for hybrid girders [A6.10.1.10.1], f_f is the elastic flange stress caused by the service II loading (ksi, MPa), f_l is the elastic flange lateral bending stress (ksi, MPa) and F_{yf} is the yield stress of the flange (ksi, MPa). For the case of a girder with the same steel in the web and flanges, $R_h = 1.0$.

Equation 18.2 (or Eq. 18.3) prevents the development of permanent deformation due to localized yielding of the flanges under an occasional service overload. The flange lateral bending (FLB) stresses are due to the bending of the bottom flange out of the plane of the girder. Such stresses arise from formwork bracing on exterior girders, forces from cross frames in skewed bridges, and forces that arise from curved aligments. The FLB computation is described in A6.10.6.1. Rigorous 3D analysis is suggested for complex bridges to determine these stresses. Herein, simplified approaches are taken for estimate the FLB. These methods are described later.

18.1.2 Fatigue and Fracture Limit State

Design for the fatigue limit state involves limiting the live-load stress range produced by the fatigue truck to a value suitable for the number of stress range repetitions expected over the life of the bridge. Design for the fracture limit state involves the selection of steel that has adequate fracture toughness (measured by Charpy V-notch test) for the expected temperature range [A6.10.5].

Load-Induced and Distortion-Induced Fatigue Load-induced fatigue can occur when live load produces a repetitive net tensile stress at a connection detail. When cross frames or diaphragms are connected to girder webs through connection plates that restrict movement, *distortion-induced* fatigue can occur. Distortion-induced fatigue is important in many cases; however, the following discussion is only for load-induced fatigue.

As discussed previously, fatigue life is determined by the tensile stress range in the connection detail. By using the stress range as the governing criteria, the values of the maximum and minimum tensile stresses need not be known. Residual stresses are implicitly considered in the determination of the fatigue resistance values. Permanent load effects need not be considered in the stress range computation [A6.6.1].

The tensile stress range is determined by considering placement of the fatigue truck live load in different spans of a bridge. If the bridge is a simple span, there is only a maximum tensile live-load stress; the minimum live-load stress is zero. In calculating these stresses, a linear elastic analysis is used.

In some regions along the span of a girder, the compressive stresses due to unfactored permanent loads (e.g., dead loads) are greater than the tensile live-load stresses due to the fatigue truck with its load factor taken from Table 5.1. However, before fatigue can be ignored in these regions, the compressive stress must be at least twice the tensile stress because the heaviest truck expected to cross the bridge is approximately twice the fatigue truck used in calculating the tensile live-load stress [A6.6.1.2.1].

Two fatigue limit states shall be checked [A3.4.1, Table 5.1]:

I—Fatigue limit state associated with infinite load-induced fatigue life, which reflects load levels expected often enough to propagate a crack.

II—Fatigue limit state associated with finite fatigue life, which reflects that load levels representative of the effective stress range of the truck population with respect to a small number of stress range cycles and to their cumulative effects.

Fatigue Design Criteria Using Eq. 18.1 in terms of fatigue load and fatigue resistance, each connection detail must satisfy

$$\phi(\Delta F)_n \geq \eta\gamma\,(\Delta f) \qquad (18.4)$$

where $(\Delta F)_n$ is the nominal fatigue resistance (ksi, MPa) and (Δf) is the live-load stress range due to the fatigue truck (ksi, MPa). For the fatigue limit state $\phi = 1.0$ and $\eta = 1.0$, so that Eq. 18.4 becomes

$$(\Delta F)_n = (\Delta F)_{\text{TH}} \geq \gamma_{\text{fatigue-I}}\,(\Delta f) \qquad (18.5a)$$

$$(\Delta F)_n = f\ (\text{number of cycles}) \geq \gamma_{\text{fatigue-II}}\,(\Delta f) \qquad (18.5b)$$

where γ are the load factors in Table 5.1 for the fatigue limit states, $(\Delta F)_{\text{TH}}$ is the infinite fatigue life resistance (threshold) (18.5a), and f(number of cycle) is the finite fatigue life resistance (18.5b).

For a bridge that is expected to experience a small number of cycles in the 75-year design life, the finite fatigue limit is appropriate and Eq. 18.5b is used. For typical bridges, the infinite fatigue limit is used as per Eq. 18.5a. The number of cycles where these two equations are equal provides the "break point" of small number of cycles. Table 18.1 provides these values.

Table 18.1 75-Year $(\text{ADTT})_{\text{SL}}$ Equivalent to Infinite Life [A6.6.1.2.3-2]

Detail Category	75-Year $(\text{ADTT})_{\text{SL}}$ Equivalent to Infinite Life (trucks per day)
A	530
B	860
B′	1035
C	1290
C′	745
D	1875
E	3530
E′	6485

From AASHTO Table 6.6.1.2.3-2. From *AASHTO LRFD Bridge Design Specifications*, Copyright © 2010 by the American Association of State Highway and Transportation Officials, Washington, DC. Used by permission.

Fatigue Load As discussed in Chapter 8, the fatigue vehicle is a single design truck with a front axle spacing of 14 ft (4300 mm), a rear axle spacing of 30 ft (9000 mm), a dynamic load allowance of 15%, and a load factor of 0.75 or 1.5.

For finite fatigue life (fatigue II), resistance depends on the number of accumulated stress-range cycles and therefore the frequency of application of the fatigue load. This frequency shall be taken as the single-lane average daily truck traffic, ADTT_{SL} [A3.6.1.4.2]. Unless a traffic survey has been conducted, the single-lane value can be estimated from the average daily truck traffic (ADTT) by

$$\text{ADTT}_{\text{SL}} = p \times \text{ADTT} \tag{18.6}$$

where p is the fraction of multiple lanes of truck traffic in a single lane taken from Table 8.3. If only the average daily traffic ADT is known, the ADTT can be determined by multiplying by the fraction of trucks in the traffic (Table 8.4). An upper bound on the total number of cars and trucks is about 20,000 vehicles per lane per day and can be used to estimate ADT.

The number N of stress-range cycles to be considered are those due to the trucks anticipated to cross the bridge in the most heavily traveled lane during its design life. For a 75-year design life, this is expressed as [A6.6.1.2.5]

$$N = (365)\,(75)\,n\,(\text{ADTT}_{\text{SL}}) \tag{18.7}$$

where n is the number of stress-range cycles per truck passage taken from Table 18.2. The values of n greater than 1.0 indicate additional cycles due to a truck passing over multiple areas of significant influence. For example, a negative moment region of a two-span bridge experiences significant stress due to loads in adjacent spans, that is, two cycles per truck crossing. Agencies have ADTT data for major routes, and the percentage of trucks can vary significantly from the specification values. For example, a cross-country interstate route in a lightly populated area can be 50% trucks or higher.

Table 18.2 Cycles per Truck Passage, *n*

	Span Length	
Longitudinal Members	> 40 ft (12 000 mm)	≤ 40 ft (12 000 mm)
Simple-span girders	1.0	2.0
Continuous girders		
1. Near interior support	1.5	2.0
2. Elsewhere	1.0	2.0
Cantilever girders		5.0
Trusses		1.0
	Spacing	
	> 20 ft (6000 mm)	≤ 20 ft (6000 mm)
Transverse members	1.0	2.0

In AASHTO Table 6.6.1.2.5-2. From *AASHTO LRFD Bridge Design Specifications*, Copyright © 2010 by the American Association of State Highway and Transportation Officials, Washington, DC. Used by permission.

For the infinite life fatigue limit state (fatigue I), the number of cycles need not be estimated, but rather the load effect is compared to the threshold value or constant amplitude fatigue limit. See Table 18.1 for limits.

Example 18.1 Estimate the number of stress-range cycles N to be considered in the fatigue design of a two-lane, 35-ft (10,670-mm) simple span bridge that carries interstate traffic in one direction. Use an ADT of 20,000 vehicles per lane per day.

Table 8.4 gives 0.20 as the fraction of trucks in interstate traffic, so that

$$\text{ADTT} = 0.20\,(2)\,(20{,}000) = 8000 \text{ trucks/day}$$

Table 8.3 gives 0.85 as the fraction of truck traffic in a single lane when two lanes are available to trucks, thus Eq. 18.6 yields

$$\text{ADTT}_{\text{SL}} = 0.85\,(8000) = 6800 \text{ trucks/day}$$

Table 18.2 gives $n = 2.0$ as the cycles per truck passage and Eq. 18.7 results in

$$N = \left(365\frac{\text{day}}{\text{yr}}\right)(75\,\text{yr})\left(2.0\frac{\text{cycle}}{\text{truck}}\right)\left(6800\frac{\text{trucks}}{\text{day}}\right)$$
$$= 372 \times 10^6 \text{ cycles}$$

Because of the large number of cycles per day, fatigue limit state II will likely be used, infinite fatigue life is the design criterion, and the number of cycles does not matter.

Detail Categories Components and details susceptible to load-induced fatigue are grouped into eight categories according to their fatigue resistance [A6.6.1.2.3]. The categories are assigned letter grades with A being the best and E′ the worst. The A and B detail categories are for plain members and well-prepared welded connections in builtup members without attachments, usually with the weld axis in the direction of the applied stress. The D and E detail categories are assigned to fillet-welded attachments and groove-welded attachments without adequate transition radius or with unequal plate thicknesses. Category C can apply to welding of attachments by providing a transition radius greater than 6 in. (150 mm) and proper grinding of the weld. The requirements for the various detail categories are summarized in Table 18.3.

Fatigue Resistance As shown in the typical S–N curve of Figure 17.7, fatigue resistance is divided into two types of behavior: one that gives infinite life and the other a finite life. If *all* the tensile stress range (load effect) is below the fatigue

Table 18.3 Detail Categories for Load-Induced Fatigue

Description	Category	Constant A (ksi^3)	Threshold $(\Delta F)_{TH}$ ksi	Potential Crack Initiation Point	Illustrative Examples
Section 1—Plain Material away from Any Welding					
1.1 Base metal, except noncoated weathering steel, with rolled or cleaned surfaces. Flame-cut edges with surface roughness value of 1,000 μ-in or less, but without re-entrant corners.	A	250×10^8	24	Away from all welds or structural connections	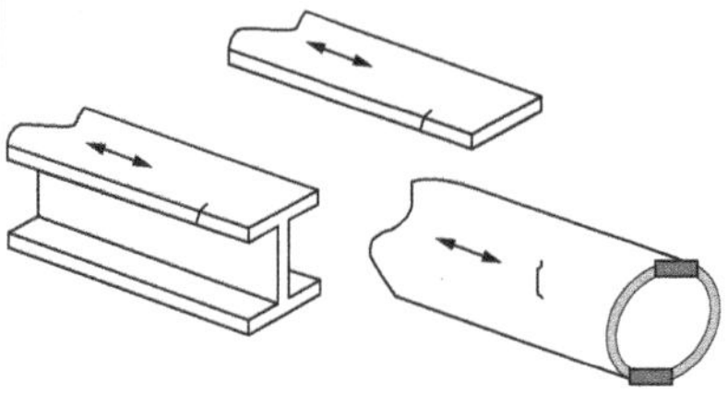
1.2 Noncoated weathering steel base metal with rolled or cleaned surfaces designed and detailed in accordance with FHWA (1989). Flame-cut edges with surface roughness value of 1,000 μ-in. or less, but without re-entrant corners.	B	120×10^8	16	Away from all welds or structural connections	
1.3 Member with re-entrant corners at copes, cuts, block-outs or other geometrical discontinuities made to the requirements of AASHTO/AWS D1.5, except weld access holes.	C	44×10^8	10	At any external edge	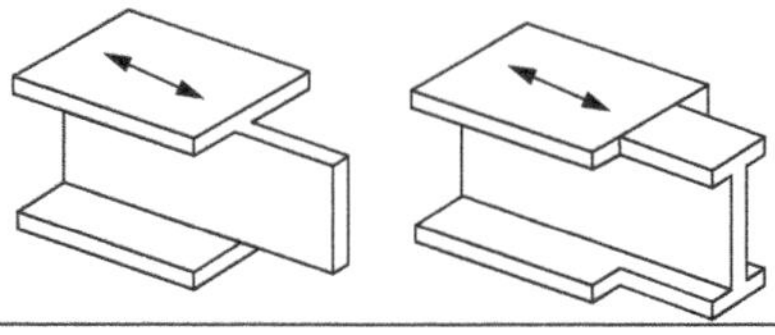
1.4 Rolled cross sections with weld access holes made to the requirements of AASHTO/AWS D1.5, Article 3.2.4.	C	44×10^8	10	In the base metal at the re-entrant corner of the weld access hole	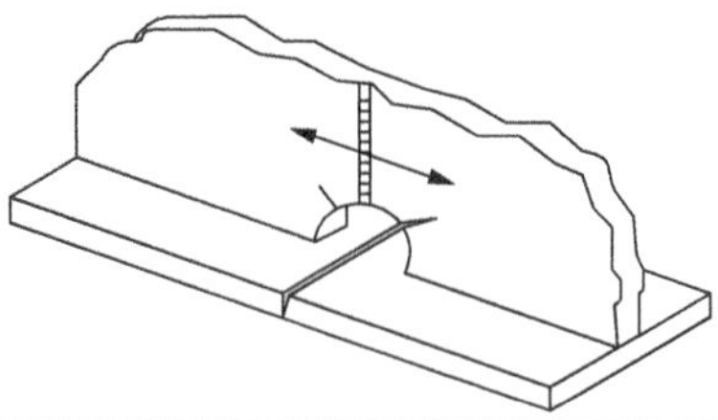
1.5 Open holes in members (Brown et al.., 2007).	D	22×10^8	7	In the net section originating at the side of the hole	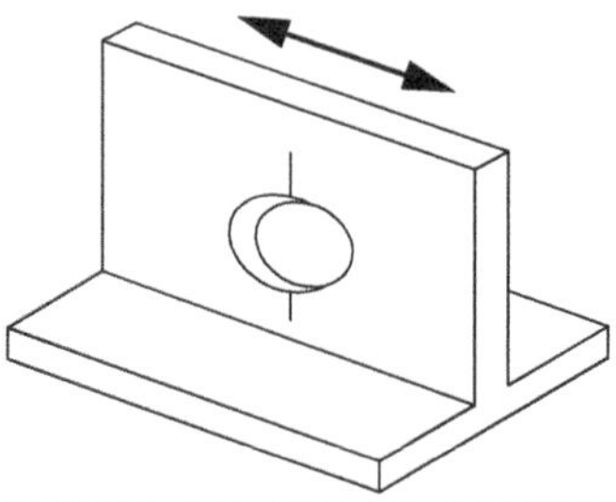
Section 2—Connected Material in Mechanically Fastened Joints					
2.1 Base metal at the gross section of high-strength bolted joints designed as slip-critical connections with pre-tensioned high-strength bolts installed in holes drilled full size or subpunched and reamed to size—e.g., bolted flange and web splices and bolted stiffeners. (Note: see Condition 2.3 for bolt holes punched full size.)	B	120×10^8	16	Through the gross section near the hole	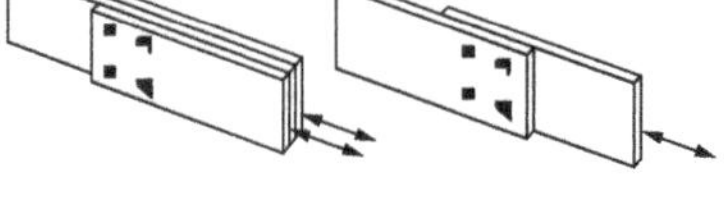

Table 18.3 *(Continued)*

Description	Category	Constant A (ksi^3)	Threshold $(\Delta F)_{TH}$ ksi	Potential Crack Initiation Point	Illustrative Examples
2.2 Base metal at the net section of high-strength bolted joints designed as bearing-type connections, but fabricated and installed to all requirements for slip-critical connections with pre-tensioned high strength bolts installed in holes drilled full size or subpunched and reamed to size. (Note: see Condition 2.3 for bolt holes punched full size.)	B	120×10^8	16	In the net section originating at the side of the hole	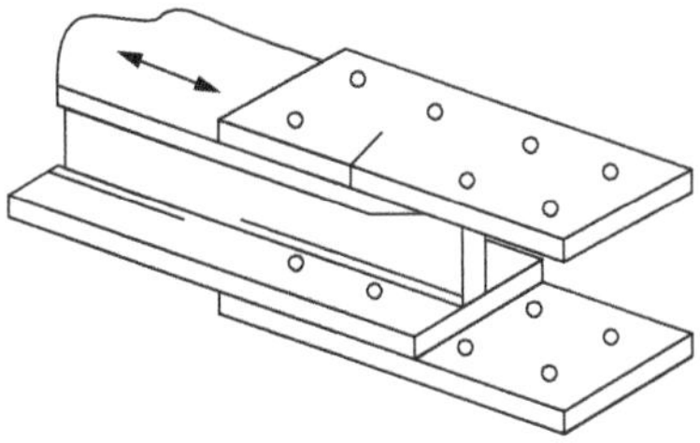
2.3 Base metal at the net section of all bolted connections in hot dipped galvanized members (Huhn and Valtinat, 2004); base metal at the appropriate section defined in Condition 2.1 or 2.2, as applicable, of high-strength bolted joints with pretensioned bolts installed in holes punched full size (Brown et al., 2007), and base metal at the net section of other mechanically fastened joints, except for eyebars and pin plates; e.g., joints using ASTM A307 bolts or non pretensioned high strength bolts.	D	22×10^8	7	In the net section originating at the side of the hole or through the gross section near the hole, as applicable	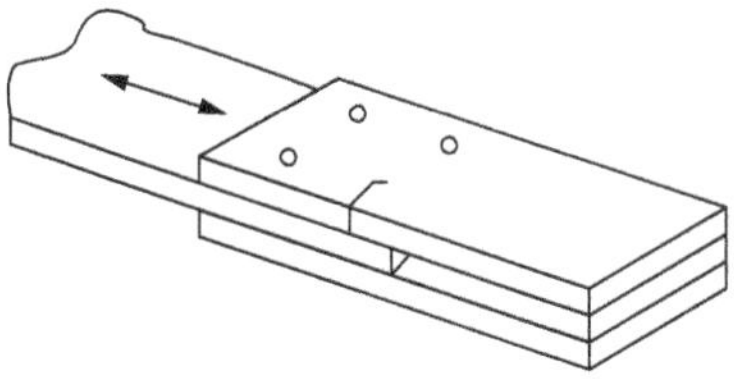
2.4 Base metal at the net section of eyebar heads or pin plates (Note: for base metal in the shank of eyebar or through the gross section of pin plates, see Condition 1.1 or 1.2, as applicable).	E	11×10^8	4.5	In the net section originating at the side of the hole	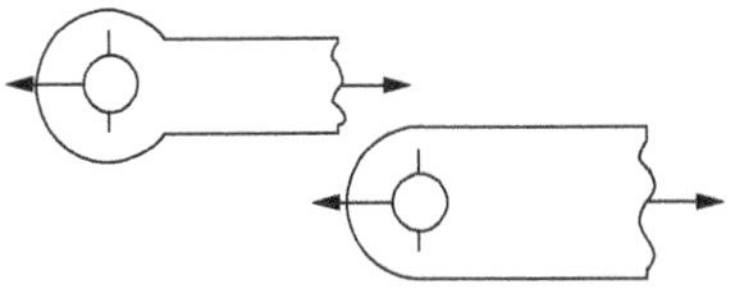
Section 3—Welded Joints Joining Components of Built-Up Members					
3.1 Base metal and weld metal in members without attachments built-up of plates or shapes connected by continuous longitudinal complete joint penetration groove welds back-gouged and welded from the second side, or by continuous fillet welds parallel to the direction of applied stress.	B	120×10^8	16	From surface or internal discontinuities in the weld away from the end of the weld	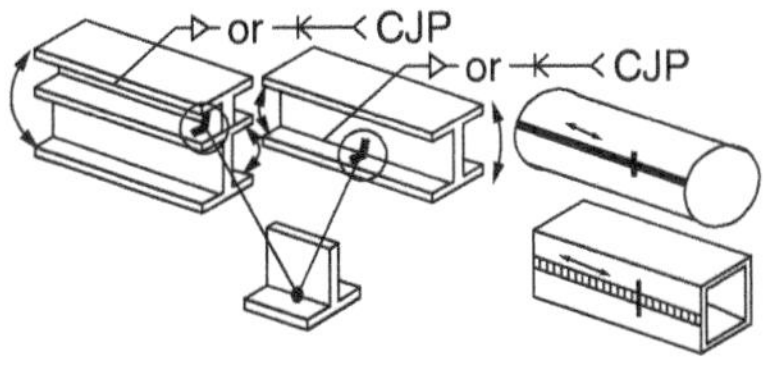 or CJP; or CJP
3.2 Base metal and weld metal in members without attachments built-up of plates or shapes connected by continuous longitudinal complete joint penetration groove welds with backing bars not removed, or by continuous partial joint penetration groove welds parallel to the direction of applied stress.	B'	61×10^8	12	From surface or internal discontinuities in the weld, including weld attaching backing bars	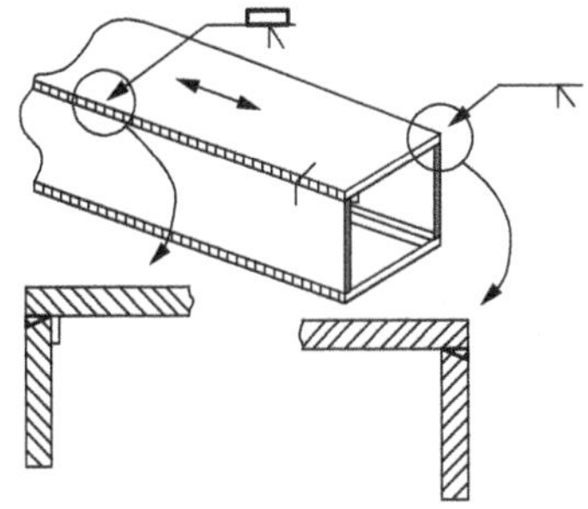

(continued overleaf)

Table 18.3 ***(Continued)***

Description	Category	Constant A (ksi^3)	Threshold $(\Delta F)_{TH}$ ksi	Potential Crack Initiation Point	Illustrative Examples
3.3 Base metal and weld metal at the termination of longitudinal welds at weld access holes made to the requirements of AASHTO/AWS D1.5, Article 3.2.4 in built-up members. (Note: does not include the flange butt splice).	D	22×10^8	7	From the weld termination into the web or flange	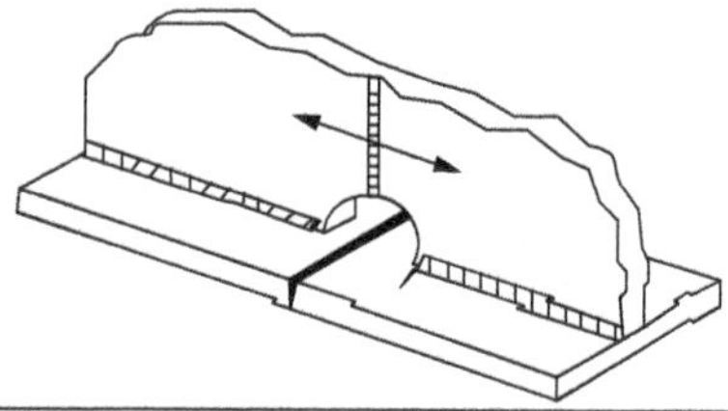
3.4 Base metal and weld metal in partial length welded cover plates connected by continuous fillet welds parallel to the direction of applied stress.	B	120×10^8	16	From surface or internal discontinuities in the weld away from the end of the weld	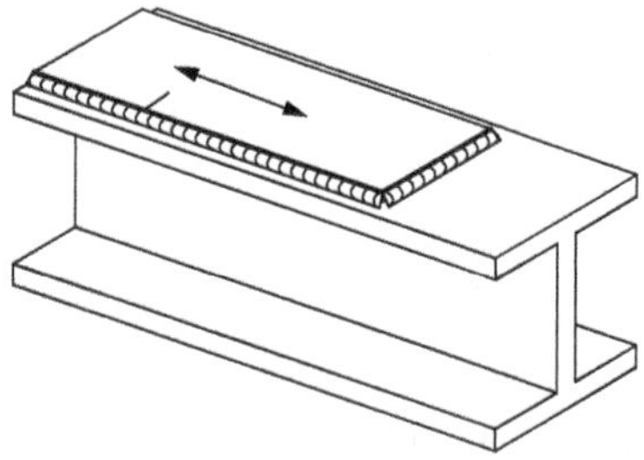
3.5 Base metal at the termination of partial length welded cover plates having square or tapered ends that are narrower than the flange, with or without welds across the ends, or cover plates that are wider than the flange with welds across the ends:				In the flange at the toe of the end weld or in the flange at the termination of the longitudinal weld or in the edge of the flange with wide cover plates	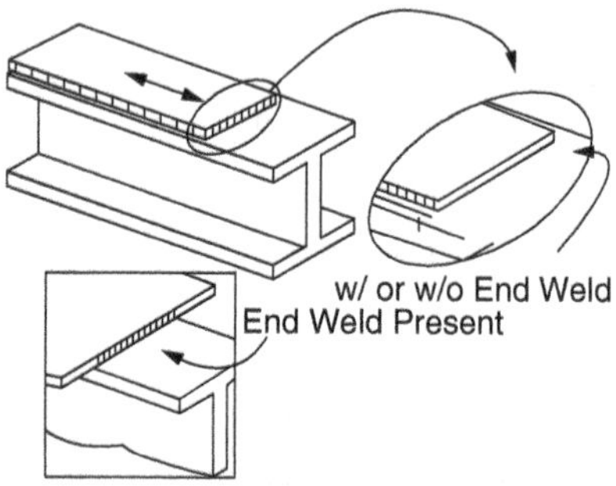 w/ or w/o End Weld End Weld Present
Flange thickness ≤ 0.8 in	E	11×10^8	4.5		
Flange thickness > 0.8 in.	E'	3.9×10^8	2.6		
3.6 Base metal at the termination of partial length welded cover plates with slip-critical bolted end connections satisfying the requirements of Article 6.10.12.2.3.	B	120×10^8	16	In the flange at the termination of the longitudinal weld	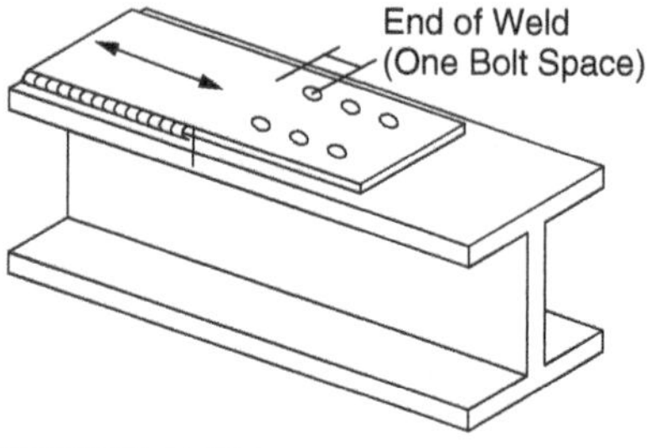 End of Weld (One Bolt Space)
3.7 Base metal at the termination of partial length welded cover plates that are wider than the flange and without welds across the ends.	E'	3.9×10^8	2.6	In the edge of the flange at the end of the cover plate weld	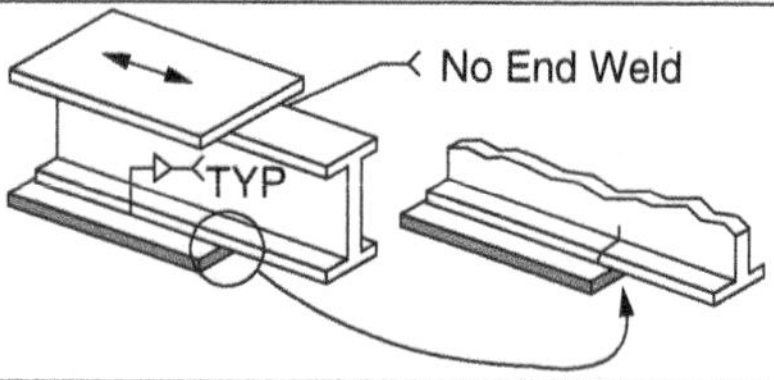 No End Weld TYP
Section 4—Welded Stiffener Connections					
4.1 Base metal at the toe of transverse stiffener-to-flange fillet welds and transverse stiffener-to-web fillet welds. (Note: includes similar welds on bearing stiffeners and connection plates).	C′	44×10^8	12	Initiating from the geometrical discontinuity at the toe of the fillet weld extending into the base metal	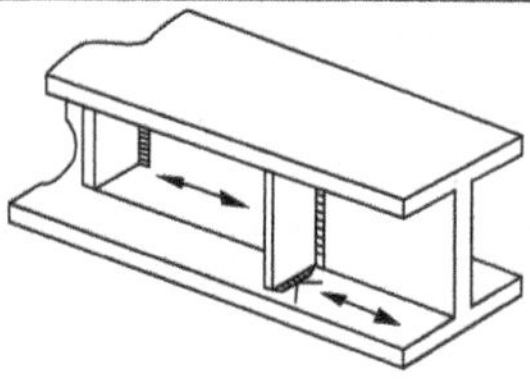

Table 18.3 (*Continued*)

Description	Category	Constant A (ksi^3)	Threshold $(\Delta F)_{TH}$ ksi	Potential Crack Initiation Point	Illustrative Examples
4.2 Base metal and weld metal in longitudinal web or longitudinal box-flange stiffeners connected by continuous fillet welds parallel to the direction of applied stress.	B	120×10^8	16	From the surface or internal discontinuities in the weld away from the end of the weld	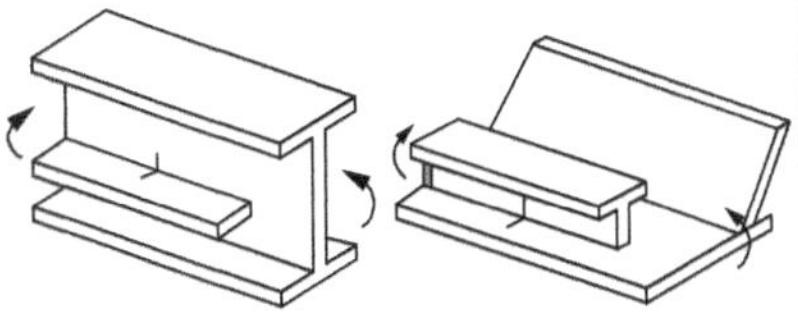
4.3 Base metal at the termination of longitudinal stiffener-to-web or longitudinal stiffener-to-box flange welds:					
With the stiffener attached by fillet welds and with no transition radius provided at the termination:				In the primary member at the end of the weld at the weld toe	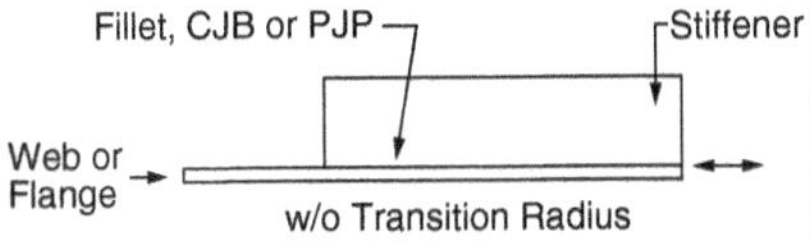 Fillet, CJB or PJP; Stiffener; Web or Flange; w/o Transition Radius
Stiffener thickness < 1.0 in.	E	11×10^8	4.5		
Stiffener thickness ≥ 1.0 in.	E′	3.9×10^8	2.6		
With the stiffener attached by welds and with a transition radius R provided at the termination with the weld termination ground smooth:					
$R \geq 24$ in.	B	120×10^8	16	In the primary member near the point of tangency of the radius	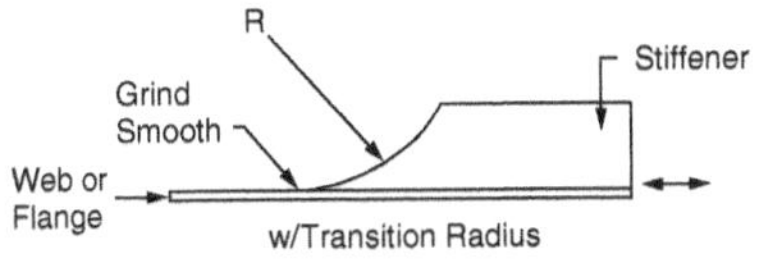 R; Stiffener; Grind Smooth; Web or Flange; w/Transition Radius
24 in. > $R \geq 6$ in.	C	44×10^8	10		
6 in. > $R \geq 2$ in.	D	22×10^8	7		
6 in. > R	E	11×10^8	4.5		
Section 5—Welded Joints Transverse to the Direction of Primary Stress					
5.1 Base metal and weld metal in or adjacent to complete joint penetration groove welded butt splices, with weld soundness established by NDT and with welds ground smooth and flush parallel to the direction of stress. Transitions in thickness or width shall be made on a slope no greater than 1:2.5 (see also Figure 6.13.6.2-1).				From internal discontinuties in the filler metal or along the fusion boundary or at the start of the transition	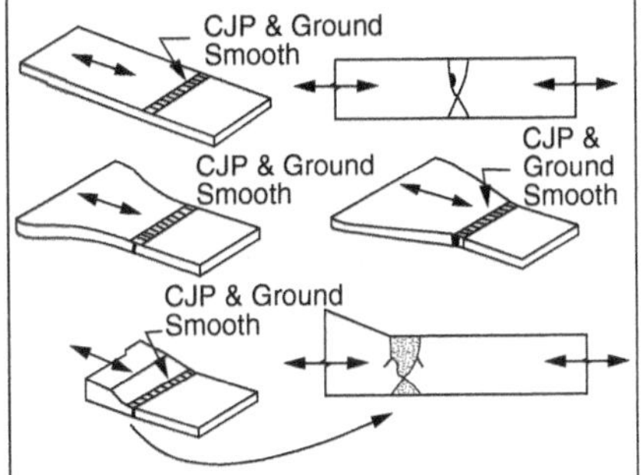 CJP & Ground Smooth; CJP & Ground Smooth; CJP & Ground Smooth; CJP & Ground Smooth
F_y < 100 ksi	B	120×10^8	16		
F_y ≥ 100 ksi	B′	61×10^8	12		

(continued overleaf)

Table 18.3 *(Continued)*

Description	Category	Constant A (ksi^3)	Threshold $(\Delta F)_{TH}$ ksi	Potential Crack Initiation Point	Illustrative Examples
5.2 Base metal and weld metal in or adjacent to complete joint penetration groove welded butt splices, with weld soundness established by NDT and with welds ground parallel to the direction of stress at transitions in width made on a radius of not less than 2 ft with the point of tangency at the end of the groove weld (see also Figure 6.13.6.2-1).	B	120×10^8	16	From internal discontinuties in the filler metal or discontinuities along the fusion boundary	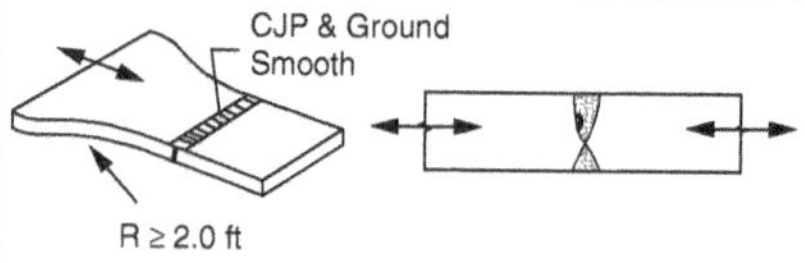 CJP & Ground Smooth; R ≥ 2.0 ft
5.3 Base metal and weld metal in or adjacent to the toe of complete joint penetration groove welded T or corner joints, or in complete joint penetration groove welded butt splices, with or without transitions in thinkness having slope no greater than 1:2.5 when weld reinforcement is not removed (Note: cracking in the flange of the 'T' may occur due to out-of-plane bending stresses induced by the stem).	C	44×10^8	10	From the surface discontinuity at the toe of the weld extending into the base metal or along the fusion boundary	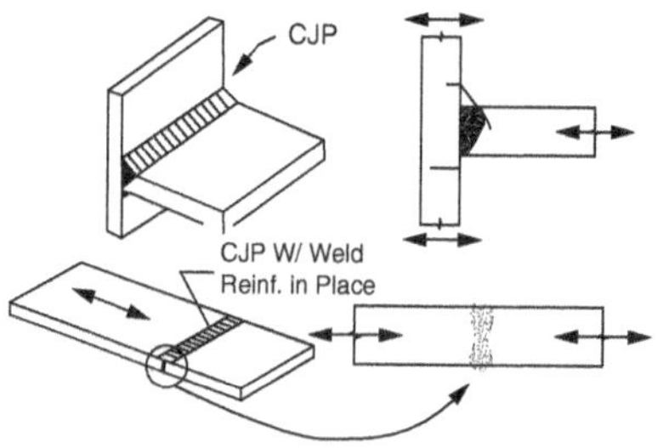 CJP; CJP W/ Weld Reinf. in Place
5.4 Base metal and weld metal at details where loaded discontinuous plate elements are connected with a pair of fillet welds or partial joint penetration groove welds on opposite sides of the plate normal to the direction of primary stress.	C as adjusted in Eq. 6.6.1.2.5-4	44×10^8	10	Initiating from the geometrical discontinuity at the toe of the weld extending into the base metal or, initiating at the weld root subject to tension extending up and then out through the weld	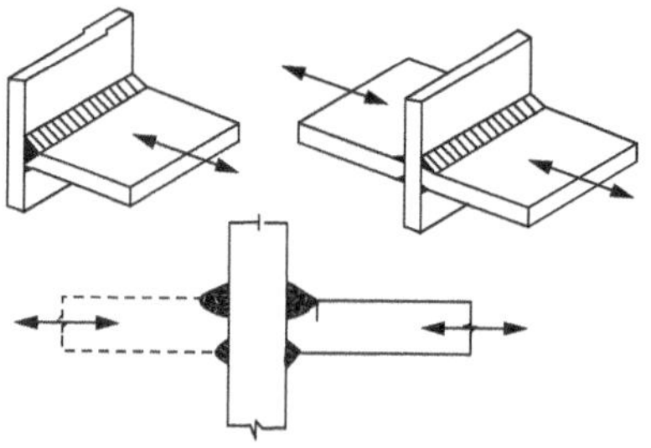
Section 6 — Transversely Loaded Welded Attachment					
6.1 Base metal in a longitudinally loaded component at a transversely loaded detail (e.g. a lateral connection plate) attached by a weld parallel to the direction of primary stress and incorporating a transition radius R with the weld termination ground smooth.				Near point of tangency of the radius at the edge of the longitudinally loaded component	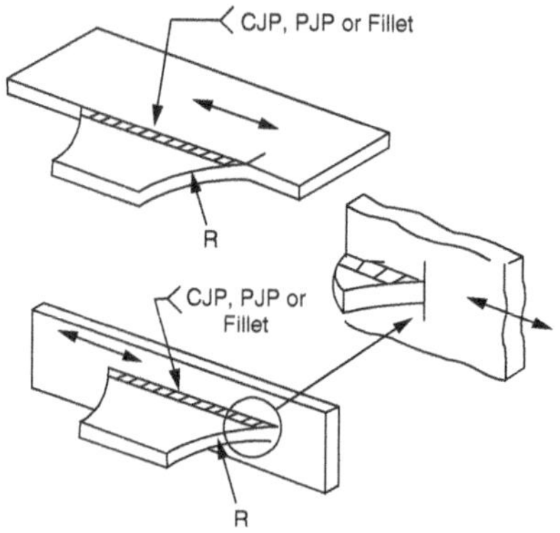 CJP, PJP or Fillet; R; CJP, PJP or Fillet; R
R ≥ 24 in.	B	120×10^8	16		
24 in. > R ≥ 6 in.	C	44×10^8	10		
6 in. > R ≥ 2 in.	D	22×10^8	7		
2 in. > R	E	11×10^8	4.5		
(Note: Condition 6.2, 6.3 or 6.4, as applicable, shall also be checked.)					

Table 18.3 (*Continued*)

Description	Category	Constant A (ksi^3)	Threshold $(\Delta F)_{TH}$ ksi	Potential Crack Initiation Point	Illustrative Examples
6.2 Base metal in a transversely loaded detail (e.g. a lateral connection plate) attached to a longitudinally loaded component of equal thickness by a complete joint penetration groove weld parallel to the direction of primary stress and incorporating a transition radius R, with weld soundness established by NDT and with the weld termination ground smooth:					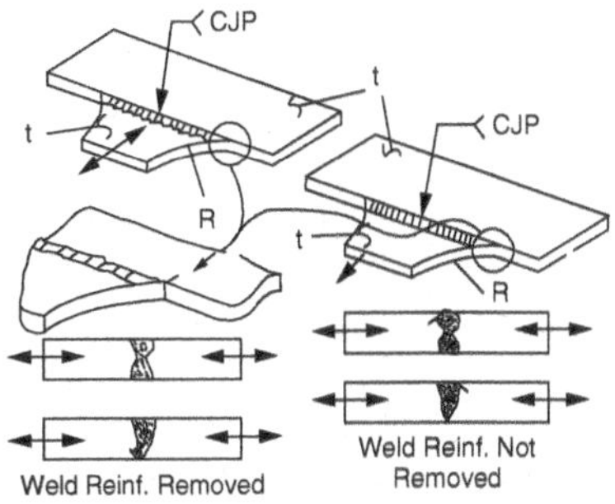 CJP; t; CJP; t; R; t; R; Weld Reinf. Removed; Weld Reinf. Not Removed
With the weld reinforcement removed:					
$R \geq 24$ in.	B	120×10^8	16	Near points of tangency of the radius or in the weld or at the fusion boundary of the longitudinally loaded component or the transversely loaded attachment	
24 in. $> R \geq 6$ in.	C	44×10^8	10		
6 in. $> R \geq 2$ in.	D	22×10^8	7		
2 in. $> R$	E	11×10^8	45		
With the weld reinforcement not removed:				At the toe of the weld either along the edge of the longitudinally loaded component or the transversely loaded attachment	
$R \geq 24$ in.	C	44×10^8	10		
24 in. $> R \geq 6$ in.	C	44×10^8	10		
6 in. $> R \geq 2$ in.	D	22×10^8	7		
2 in. $> R$	E	11×10^8	4.5		
(Note: Condition 6.1 shall also be checked.)					
6.3 Base metal in a transversely loaded detail (e.g. a lateral connection plate) attached to a longitudinally loaded component of unequal thickness by a complete joint penetration groove weld parallel to the direction of primary streess and incorporating a weld transition radius R, with weld soundness established by NDT and with the weld termination ground smooth:				At the toe of the weld along the edge of the thinner plate In the weld termination of small radius weld transitions At the toe of the weld along the edge of the thinner plate	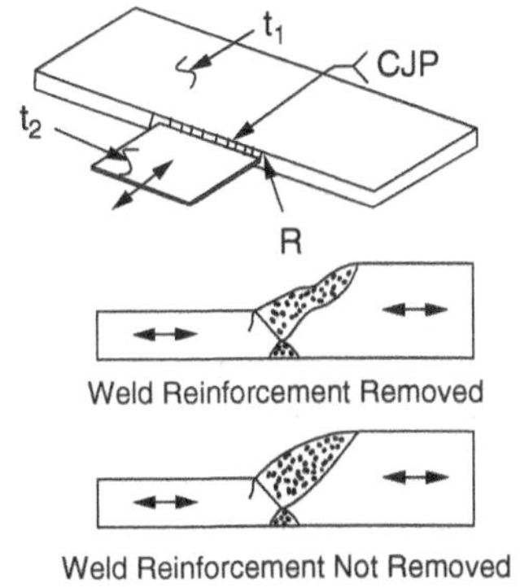 t_1; CJP; t_2; R; Weld Reinforcement Removed; Weld Reinforcement Not Removed
With the weld reinforcement removed:					
$R \geq 2$ in.	D	22×10^8	7		
$R < 2$ in.	E	11×10^8	4.5		
For any weld trasition radius with the weld reinforcement not removed:	E	11×10^8	4.5		
(Note: Condition 6.1 shall also be checked.)					

(*continued overleaf*)

Table 18.3 (*Continued*)

Description	Category	Constant A (ksi^3)	Threshold $(\Delta F)_{TH}$ ksi	Potential Crack Initiation Point	Illustrative Examples
6.4 Base metal in a transversely loaded detail (e.g. a lateral connection plate) attached to a longitudinally loaded component by a fillet weld or a partial joint penetration groove weld, with the weld parallel to the direction of primary stress (Note: Condition 6.1 shall also be checked)	See Condition 5.4				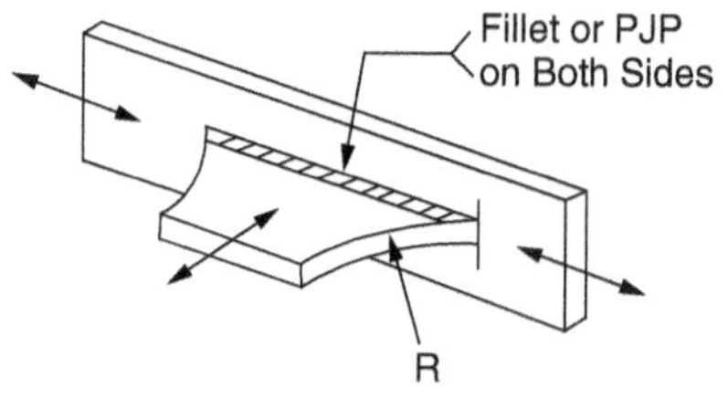 Fillet or PJP on Both Sides R
Section 7—Longitudinally Loaded Welded Attachments					
7.1 Base metal in a longitudinally loaded component at a detail with a length L in the direction of the primary stress and a thickness t attached by groove or fillet welds parallel or transverse to the direction of primary stress where the detail incorporates no transition radius:				In the primary member at the end of the weld at the weld toe	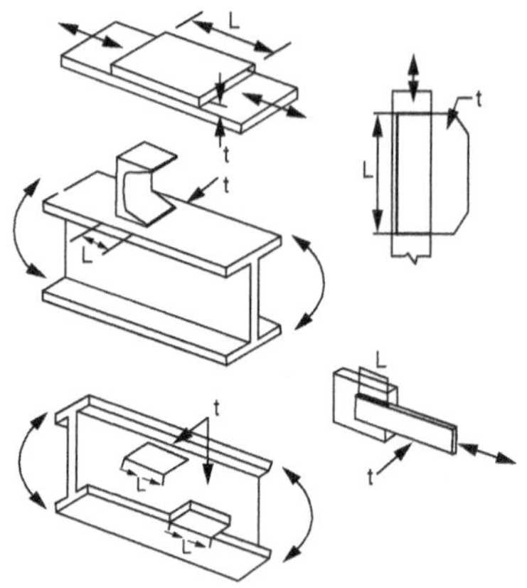 L t
$L < 2$ in.	C	44×10^8	10		
2 in. $\leq L \leq 12t$ or 4 in.	D	22×10^8	7		
$L > 12t$ or 4 in.					
$t < 1.0$ in.	E	11×10^8	4.5		
$t \geq 1.0$ in.	E'	3.9×10^8	2.6		
Section 8—Miscellaneous					
8.1 Base metal at stud-type shear connectors attached by fillet or automatic stud welding	C	44×10^8	10	At the toe of the weld in the base metal	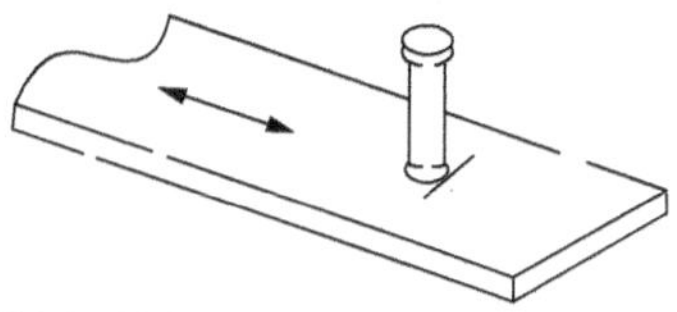
8.2 Nonpretensioned high-strength bolts, common bolts, threaded anchor rods and hanger rods with cut, ground or rolled threads. Use the stress range acting on the tensile stress area due to live load plus prying action when applicable.				At the root of the threads extending into the tensile stress area	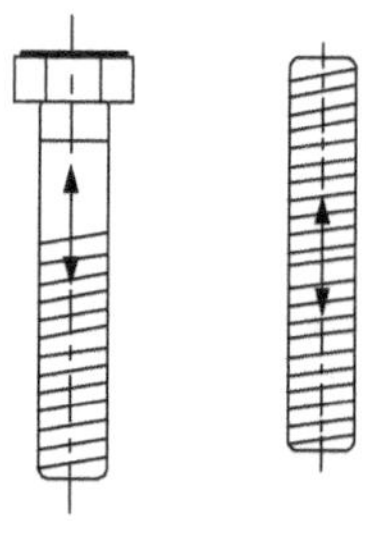
(Fatigue II) Finite Life	E'	3.9×10^8	N/A		
(Fatigue I) Infinite Life	D'	N/A	7		

From AASHTO Table 6.6.1.2.3-1. From *AASHTO LRFD Bridge Design Specifications*, Copyright © 2010 by the American Association of State Highway and Transportation Officials, Washington DC. Used by permission.

limit or threshold stress, additional loading cycles will not propagate fatigue cracks and the connection detail will have a long (infinite) life. If just a "few" trucks pass that create adequate stress ranges to exceed the threshold value, this can invalidate the threshold and the fatigue curve just continues to decrease with the number of cycles. Fatigue I attempts to avoid this situation where some heavier truck will not exceed twice the threshold value. Fischer suggested 1 : 10000 passes not exceed twice the threshold value (Fischer, 1967).

This general concept of fatigue resistance is expressed for specific conditions by the following [A6.6.1.2.5]:

For fatigue load I combination and infinite fatigue life

$$(\Delta F)_n = (\Delta F)_{\mathrm{TH}} \tag{18.8a}$$

and for fatigue load II combination and finite life

$$(\Delta F)_n = \left(\frac{A}{N}\right)^{1/3} \tag{18.8b}$$

where $(\Delta F)_n$ is the nominal fatigue resistance (ksi, MPa), A is a detail category constant taken from Table 18.4 (ksi, MPa)^3, N is the number of stress-range cycles from Eq. 18.7, and $(\Delta F)_{\mathrm{TH}}$ is the constant-amplitude fatigue threshold stress taken from Table 18.4 (ksi, MPa).

The S–N curves for all of the detail categories are represented in Eq. 18.8. These are plotted in Figure 18.1 by taking the values from Table 18.4 for A and $(\Delta F)_{\mathrm{TH}}$. In the finite life portion of the S–N curves, the effect of changes in the stress range on the number of cycles to failure can be observed by solving Eq. 18.8b for N to yield

$$N = \frac{A}{(\Delta F)_n^3} \tag{18.9}$$

Table 18.4-US Detail Category Constant, *A*, and Fatigue Thresholds

Detail Category	Constant, A Times 10^{11} (ksi)^3	Fatigue Threshold (ksi)
A	250.0	24.0
B	120.0	16.0
B′	61.0	12.0
C	44.0	10.0
C′	44.0	12.0
D	22.0	7.0
E	11.0	4.5
E′	3.9	2.6
A164 (A325M) bolts in axial tension	17.1	31.0
M253 (A490M) bolts in axial tension	31.5	38.0

From AASHTO Tables 6.6.1.2.5-1 and 6.6.1.2.5-3. From *AASHTO LRFD Bridge Design Specifications*, Copyright © 2010 by the American Association of State Highway and Transportation Officials, Washington, DC. Used by permission.

Table 18.4-SI Detail Category Constant, *A*, and Fatigue Thresholds

Detail Category	Constant, A Times 10^{11} (MPa)^3	Fatigue Threshold (MPa)
A	82.0	165
B	39.3	110
B′	20.0	82.7
C	14.4	69.0
C′	14.4	82.7
D	7.21	48.3
E	3.61	31.0
E′	1.28	17.9
A164 (A325M) bolts in axial tension	5.61	214
M253 (A490M) bolts in axial tension	10.3	262

From AASHTO Tables 6.6.1.2.5-1 and 6.6.1.2.5-3. From *AASHTO LRFD Bridge Design Specifications*, Copyright © 2010 by the American Association of State Highway and Transportation Officials, Washington, DC. Used by permission.

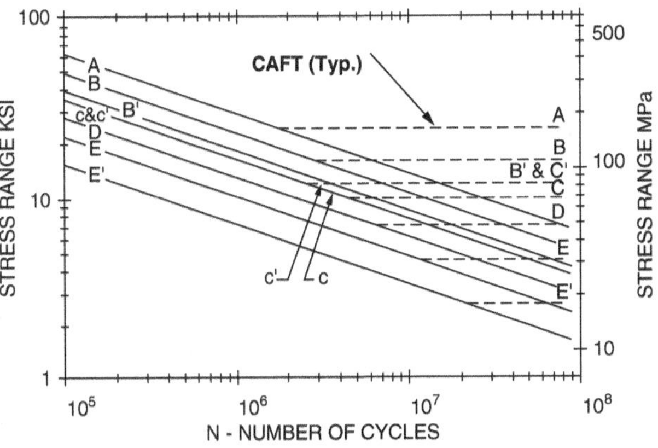

Fig. 18.1 Stress range versus number of cycles [AASHTO Fig. C6.6.1.2.5-1]. [From *AASHTO LRFD Bridge Design Specifications*, Copyright © 2010 by the American Association of State Highway and Transportation Officials, Washington, DC. Used by permission.]

Therefore, if the stress range is cut in half, the number of cycles to failure is increased by a multiple of 8. Similarly, if the stress range is doubled, the life of the detail is divided by 8.

Fracture Toughness Requirements Material in components and connections subjected to tensile stresses due to the strength I limit state of Table 5.1 must satisfy supplemental impact requirements [A6.6.2]. As discussed in Section 17.2.7 on brittle fracture considerations, these supplemental impact requirements relate to minimum energy absorbed

Table 18.5-US Temperature Zone Designations for Charpy V-Notch Requirements

Minimum Service Temperature	Temperature Zone
0°F and above	1
−1 to −30°F	2
−31 to −60°F	3

From AASHTO Table 6.6.2-1. From *AASHTO LRFD Bridge Design Specifications*, Copyright © 2010 by the American Association of State Highway and Transportation Officials, Washington, DC. Used by permission.

Table 18.5-SI Temperature Zone Designations for Charpy V-Notch Requirements

Minimum Service Temperature	Temperature Zone
−18°C and above	1
−19 to −34°C	2
−35 to −51°C	3

From AASHTO Table 6.6.2-1. From *AASHTO LRFD Bridge Design Specifications*, Copyright © 2010 by the American Association of State Highway and Transportation Officials, Washington, DC. Used by permission.

in a Charpy V-notch test at a specified temperature. The minimum service temperature at a bridge site determines the temperature zone (see Table 18.5) for the Charpy V-notch requirements.

A fracture-critical member (FCM) is defined as a member with tensile stress whose failure is expected to cause the collapse of the bridge. The material in an FCM must exhibit greater toughness and an ability to absorb more energy without fracture than a non-fracture-critical member. The Charpy V-notch fracture toughness requirements for welded components are given in Table 18.6 for different plate thicknesses and temperature zones. The FCM values for absorbed energy are approximately 50% greater than for non-FCM values at the same temperature.

Table 18.6-US Charpy V-Notch Fracture Toughness Requirements for Welded Components

Material		Min. Test	Fracture Critical		Nonfracture Critical	
Grade	Thickness (in.)	Value Energy (ft-lbs)	Zone 2 (ft-lb @ °F)	Zone 3 (ft-lb @ °F)	Zone 2 (ft-lb @ °F)	Zone 3 (ft-lb @ °F)
36	$t \le 4$	20	25 @ 40	25 @ 10	15 @ 40	15 @ 10
36/50W/50S	$t \le 2$	20	25 @ 40	25 @ 10	15 @ 40	15 @ 10
	$2 < t \le 4$	24	30 @ 40	30 @ 10	20 @ 40	20 @ 10
HPS 50W	$t \le 4$	24	30 @ 10	30 @ 10	20 @ 10	20 @ 10
HPS 70W	$t \le 4$	28	35 @ −10	35 @ −10	25 @ −10	25 @ −10
100/100W	$t \le 2.5$	28	35 @ 0	35 @ −30	25 @ 0	25 @ −30
	$2.5 < t \le 4$	36	45 @ 0	Not permitted	35 @ 0	35 @ −30

From AASHTO Table 6.6.2-2. From *AASHTO LRFD Bridge Design Specifications*, Copyright © 2010 by the American Association of State Highway and Transportation Officials, Washington, DC. Used by permission.

Table 18.6-SI Charpy V-Notch Fracture Toughness Requirements for Welded Components

Material		Min. Test	Fracture Critical		Nonfracture Critical	
Grade	Thickness (mm)	Value Energy (N m)	Zone 2 (N m @ °C)	Zone 3 (N m @ °C)	Zone 2 (N m @ °C)	Zone 3 (N m @ °C)
250	$t \le 100$	27	34 @ 4	34 @ −12	20 @ 4	20 @ −12
345/345W/345S	$t \le 50$	27	34 @ 4	34 @ −12	20 @ 4	20 @ −12
	$50 < t \le 100$	33	41 @ 4	41 @ −12	27 @ 4	27 @ −12
HPS 345W	$t \le 100$	33	41 @ −12	41 @ −12	27 @ −12	27 @ −12
HPS 485W	$t \le 100$	38	48 @ −23	48 @ −23	34 @ −23	34 @ −23
690/690W	$t \le 65$	38	48 @ −1	48 @ −18	34 @ −18	34 @ −34
	$65 < t \le 100$	49	68 @ −18	Not permitted	48 @ −18	48 @ −34

From AASHTO Table 6.6.2-2. From *AASHTO LRFD Bridge Design Specifications*, Copyright © 2010 by the American Association of State Highway and Transportation Officials, Washington, DC. Used by permission (added Min. Test Value to 2004 Table for SI).

18.1.3 Strength Limit States

A strength limit state is governed by the static strength of the materials or the stability of a given cross section. There are five different strength–load combinations specified in Table 5.1. The differences in the strength–load combinations are associated mainly with the load factors applied to the live load; for example, a smaller live-load factor is used for a permit vehicle and in the presence of wind. The load combination that produces the maximum load effect is determined and then compared to the resistance or strength provided by the member.

When calculating the resistance for a particular factored load effect such as tension, compression, bending, or shear, the uncertainties are represented by an understrength or resistance factor ϕ. The ϕ factor is multiplied times the calculated nominal resistance R_n, and the adequacy of the design is then determined by whether or not the inequality of Eq. 18.1 is satisfied. The requirements for the strength limit state are generally outlined in the AASHTO LRFD Specifications [A6.10.6] and details are outlined for positive flexure [A6.10.7], negative flexure [A6.10.8], and shear [A6.10.9]. Requirements for shear connectors, stiffeners, and cover plates are provided in the specifications [A6.10.10, A6.10.11, and A6.10.12, respectively].

In the case of structural steel members, uncertainties exist in the material properties, cross-sectional dimensions, fabrication tolerances, workmanship, and the equations used to calculate the resistance. The consequences of failure are also included in the ϕ factor. As a result, larger reductions in strength are applied to columns than beams and to connections in general. All of these considerations are reflected in the strength limit state resistance factors given in Table 18.7 [A6.5.4.2].

18.1.4 Extreme Event Limit State

Extreme event limit states are unique occurrences with return periods in excess of the design life of the bridge. Earthquakes, ice loads, vehicle, and vessel collisions are considered to be extreme events and are to be used one at a time as shown in Table 5.1. However, these events can be combined with a major flood (recurrence interval > 100 years but < 500 years) or with the effects of scour of a major flood. For example, it is possible that ice floes are colliding

Table 18.7 Resistance Factors for the Strength Limit States

Description of Mode	Resistance Factor
Flexure	$\phi_f = 1.00$
Shear	$\phi_v = 1.00$
Axial compression, steel only	$\phi_c = 0.90$
Axial compression, composite	$\phi_c = 0.90$
Tension, fracture in net section	$\phi_u = 0.80$
Tension, yielding in gross section	$\phi_y = 0.95$
Bearing on pins, in reamed, drilled, or bolted holes and milled surfaces	$\phi_b = 1.00$
Bolts bearing on material	$\phi_{bb} = 0.80$
Shear connectors	$\phi_{sc} = 0.85$
A325 and A490 bolts in tension	$\phi_t = 0.80$
A307 bolts in tension	$\phi_t = 0.80$
A307 bolts in shear	$\phi_s = 0.65$
A325 and A490 bolts in shear	$\phi_s = 0.80$
Block shear	$\phi_{bs} = 0.80$
Weld metal in complete penetration welds:	
Shear on effective area	$\phi_{e1} = 0.85$
Tension or compression normal to effective area	$\phi =$ base metal ϕ
Tension or compression parallel to axis of the weld	$\phi =$ base metal ϕ
Weld metal in partial penetration welds:	
Shear parallel to axis of weld	$\phi_{e2} = 0.80$
Tension or compression parallel to axis of weld	$\phi =$ base metal ϕ
Tension compression normal to the effective area	$\phi =$ base metal ϕ
Tension normal to the effective area	$\phi_{el} = 0.80$
Weld metal in fillet welds:	
Tension or compression parallel to axis of the weld	$\phi =$ base metal ϕ
Shear in throat of weld metal	$\phi_{e2} = 0.80$

From [A6.5.4.2]. From *AASHTO LRFD Bridge Design Specifications*, Copyright © 2010 by the American Association of State Highway and Transportation Officials, Washington, DC. Used by permission.

with a bridge during a spring flood, or that scour from a major flood has reduced support for foundation components when the design earthquake occurs.

All resistance factors ϕ for an extreme event limit state are to be taken as unity, except for bolts. For bolts, the ϕ factor at the extreme event limit state shall be taken for the bearing mode of failure in Table 18.7 [A6.5.4.2].

18.2 GENERAL DESIGN REQUIREMENTS

Basic dimension and detail requirements are given in the AASHTO (2010) LRFD Bridge Specifications. Because these requirements can influence the design as much as load effects, a brief summary of them is given in this section.

18.2.1 Effective Length of Span

The effective span length shall be taken as the center-to-center distance between bearings or supports [A6.7.1].

18.2.2 Dead-Load Camber

Steel structures should be cambered during fabrication to compensate for dead-load deflection and vertical curves associated with the alignment of the roadway [A6.7.2].

18.2.3 Minimum Thickness of Steel

In general, thickness of structural steel shall not be less than 0.3125 in. (8 mm) [A6.7.3], which includes the thickness of bracing members, cross frames, and all types of gusset plates. The exceptions are webs of rolled beams or channels and of closed ribs in orthotropic decks, which need be only 0.25 in. (7 mm) thick. If exposure to severe corrosion conditions is anticipated, unless a protective system is provided, an additional thickness of sacrificial metal shall be specified.

18.2.4 Diaphragms and Cross Frames

Diaphragms and cross frames are transverse bridge components that connect adjacent longitudinal beams or girders as shown in Figure 18.2. Diaphragms can be channels or beams and provide a flexural transverse connector. Cross frames are usually composed of angles and provide a truss framework transverse connector.

The function of these transverse connectors is threefold: (1) transfer of lateral wind loads to the deck and from the deck to the bearings, (2) provide stability of the beam or girder flanges during erection and placement of the deck, and (3) distribute the vertical dead load and live load to the longitudinal beams or girders [A6.7.4.1].

By transferring the wind loads on the superstructure up into the deck, the large stiffness of the deck in the horizontal plane carries the wind loads to the supports. At the supports,

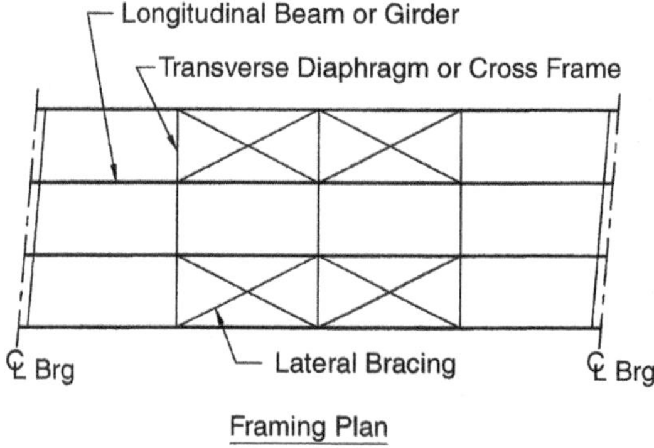

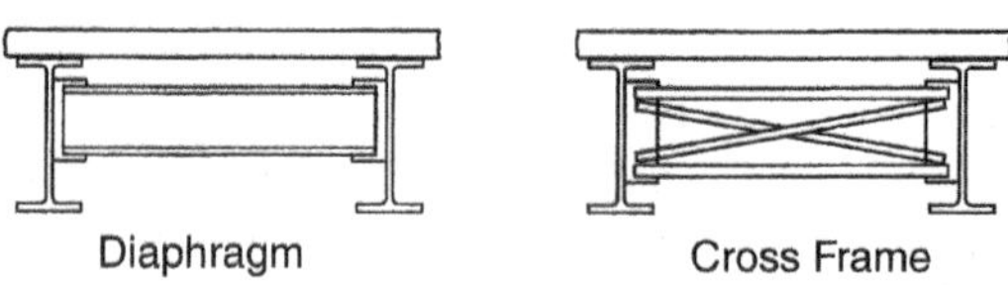

Fig. 18.2 Typical transverse diaphragm, cross frame, and lateral bracing.

diaphragms or cross frames must then transfer the wind loads down from the deck to the bearings. Typically the frames at the supports are heavier than the intermediates in order to carry the larger force.

To be effective, the diaphragms and cross frames shall be as deep as possible. They shall be at least half the beam depth for rolled sections and 0.75 times the depth for plate girders [A6.7.4.2]. Intermediate diaphragms and cross frames shall be proportioned to resist the wind forces on the tributary area between lateral connections. However, end diaphragms and cross frames shall be proportioned to transmit all the accumulated wind forces to the bearings.

A rational analysis is required to determine the lateral forces in the diaphragms or cross frames. Fewer transverse connectors are preferred because their attachment details are prone to fatigue [C6.7.4.1].

18.2.5 Lateral Bracing

The function of lateral bracing is similar to that of the diaphragms and cross frames in transferring wind loads and providing stability, except that it acts in a horizontal plane instead of a vertical plane (Fig. 18.2). All stages of construction shall be investigated [A6.7.5.1]. Where required, lateral bracing should be placed as near the horizontal plane of the flange being braced as possible. In the first stage of composite construction, the girder must support the wet concrete, associated formwork, and construction loads. It is questionable whether the formwork adequately supports the top flange in the positive moment region, and hence the unsupported length associated with the cross frames in this region must be investigated in addition to the unsupported lengths in the negative moment region for all construction stages.

Once the concrete has hardened (stage 2 and subsequent stages), the top flange is adequately braced and the unsupported region of concern is in the negative moment region near the supports. Where pattern placements of concrete are used in longer bridges, the various stages and the associated unbraced length must be considered in the analysis and resistance computations. In some cases, the girder is subjected to the most critical load effects during transportation and construction. Lastly, wind during construction should also be investigated. Inadequate bracing during construction has led to construction failures resulting in loss of life and significant financial resources.

Because of the favorable aspects of spreading the cross frames as far as possible, some of the more recent designs are requiring the constructor to place carefully designed temporary cross frames inside of the permanent ones to shorten unbraced lengths during construction. The temporary frames are then removed once the concrete has hardened and the top flange is adequately supported. This additional lateral support can eliminate fatigue-prone details associated with cross frame and transverse stiffeners.

REFERENCES

AASHTO (2010). *LRFD Bridge Design Specifications*, 5th ed., American Association of State Highway and Transportation Officials, Washington, DC.

Fischer (1967)

PROBLEMS

18.1 What is the optional live-load deflection limit for the midspan translation for a girder that spans 100 ft?

18.2 What does the service II limit attempt to address/avoid?

18.3 What is the difference between load-induced and distortion-induced fatigue?

18.4 Draw a typical S–N curve and label the finite-life and infinite-life regions.

18.5 Estimate the number of stress-range cycles N to be considered in the fatigue design of a two-lane, 30-ft simple span bridge that carries interstate traffic in one direction. Use an ADT of 10,000 vehicles per lane per day. State any assumptions.

18.6 Why are lateral bracing or diagphragms used? Provide a list.

CHAPTER 19

Steel Component Resistance

19.1 TENSILE MEMBERS

Tension members occur in the cross frames and lateral bracing of the girder bridge system shown in Figure 18.2 and are also present in truss bridges and tied-arch bridges. The cables and hangers of suspension and cable-stayed bridges are also tension members.

It is important to know how a tension member is to be used because it focuses attention on how it is to be connected to other members of the structure (Taylor, 1992). In general, it is the connection details that govern the resistance of a tension member and should be considered first.

19.1.1 Types of Connections

Two types of connections for tension members are considered: bolted and welded. A simple bolted connection between two plates is shown in Figure 19.1. Obviously, the bolt holes reduce the cross-sectional area of the member. A bolt hole also produces stress concentrations at the edge of the hole that can be three times the uniform stress at some distance from the hole (Fig. 19.1). The stress concentrations that exist while the material is elastic are reduced at higher load levels due to plasticity (Taylor, 1992).

A simple welded connection between two plates is shown in Figure 19.2. For the welded connection, the cross-sectional area of the member is not reduced. However, the stress in the plate is concentrated adjacent to the weld and is only uniform at some distance from the connection.

These stress concentrations adjacent to localized end connections are due to a phenomenon called *shear lag*. In the region near the hole or near the weld, shear stresses develop that cause the tensile stresses away from the hole or weld to lag behind the higher values at the edge.

19.1.2 Tensile Resistance—Specifications

The results of typical tensile tests on bridge steels are shown by the stress–strain curves of Figure 17.4. After the yield-point stress F_y is reached, plastic behavior begins. The stress remains relatively constant until strain hardening causes the stress to increase again before decreasing and eventually failing. The peak value of stress shown for each steel in Figure 17.3 is defined as the tensile strength F_u of the steel. Numerical values for F_y and F_u are given in Table 17.2 for the various bridge steels.

When the tensile load on an end connection increases, the highest stressed point on the critical section yields first. This point could occur at a stress concentration as shown in Figures 19.1 and 19.2 or it could occur where the tensile residual stresses (Fig. 17.2) are high. Once a portion of the critical section begins to yield and the load is increased

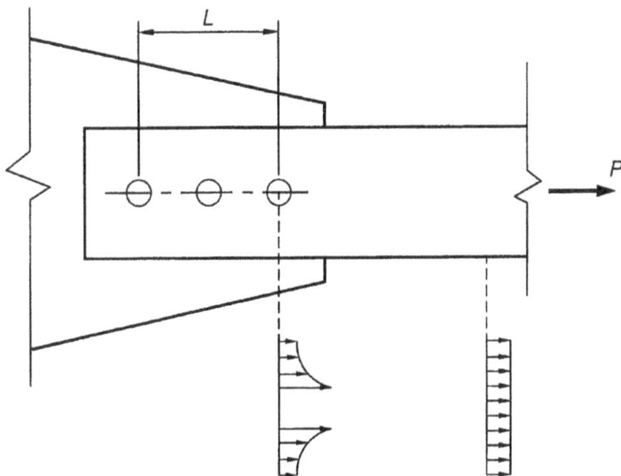

Fig. 19.1 Local stress concentration and shear lag at a bolt hole.

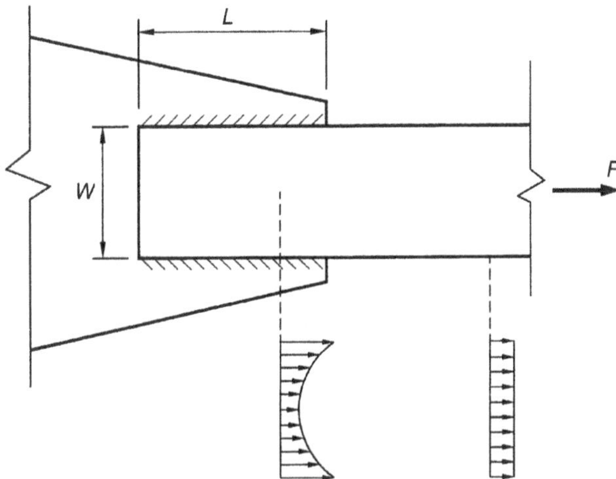

Fig. 19.2 Local stress concentration and shear lag at a welded connection.

further, a plastic redistribution of the stresses occurs. The useful tensile load-carrying limit is reached when the entire cross section becomes plastic.

The tensile resistance of an axially loaded member is governed by the lesser of [A6.8.2.1]* :

- ☐ The resistance to general yielding of the gross cross section
- ☐ The resistance to rupture on a reduced cross section at the end connection

The factored resistance to yielding is given by

$$P_r = \phi_y P_{ny} = \phi_y F_y A_g \quad (19.1)$$

where ϕ_y is the resistance factor for yielding of tension members taken from Table 18.7, P_{ny} is the nominal tensile resistance for yielding in the gross section (kip, N), F_y is the yield strength (ksi, MPa), and A_g is the gross cross-sectional area of the member (in.^{2}, mm^2).

The factored resistance to rupture is given by

$$P_r = \phi_u P_{nu} = \phi_u F_u A_e \quad (19.2)$$

where ϕ_u is the resistance factor for fracture of tension members taken from Table 18.7, P_{nu} is the nominal tensile resistance for fracture in the net section (kip, N), F_u is the tensile strength (ksi, MPa), and A_e is the effective net area of the member (in.^{2}, mm^2). For bolted connections, the effective net area is

$$A_e = UA_n \quad (19.3)$$

where A_n is the net area of the member (in.^{2}, mm^2) and U is the reduction factor to account for shear lag. For welded connections, the effective net area is

$$A_e = UA_g \quad (19.4)$$

The reduction factor U does not apply when checking yielding on the gross section because yielding tends to equalize the nonuniform tensile stresses caused over the cross section by shear lag [C6.8.2.1]. The resistance factor for fracture ϕ_u is smaller than the resistance factor for yielding ϕ_y because of the possibility of a brittle fracture in the strain-hardening range of the stress–strain curve.

Reduction Factor* U *[A6.8.2.2] When all the elements of a component (flanges, web, legs, and stem) are connected by splice or gusset plates so that force effects are transmitted uniformly, $U = 1.0$. If only a portion of the elements are connected (e.g., one leg of a single angle), the connected elements are stressed more than the unconnected ones. In the case of a partial connection, stresses are nonuniform, shear lag occurs, and $U < 1.0$.

*The article numbers in the AASHTO (2010) LRFD Bridge Specifications are enclosed in brackets and preceded by the letter A if a specification article and by the letter C if commentary.

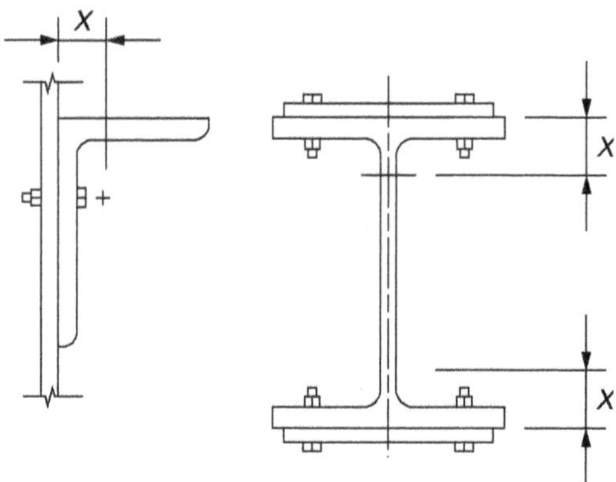

Fig. 19.3 Determination of x. (From William T. Segui, *LRFD Steel Design*, Copyright © 2003 by PWS Publishing Company, Boston, MA, with permission.)

For partial bolted connections, Munse and Chesson (1963) observed that a decrease in joint length L (Fig. 19.1) increases the shear lag effect. They proposed the following approximate expression for the reduction factor:

$$U = 1 - \left(\frac{x}{L}\right) \quad (19.5)$$

$$A_{ne} = UA_{gn} \quad (19.6)$$

where x is the distance from the centroid of the connected area of the component to the shear plane of the connection, A_{ne} is the net area of the connected elements, A_{gn} is the net area of the rolled shape outside the connected length, and U is the shear lag reduction factor. If a member has two symmetrically located planes of connection, x is measured from the centroid of the nearest one-half of the area. Illustrations of the distance x are given in Figure 19.3.

The reduction factor formula, values, and lower bounds are provided in AASHTO Table A6.8.2.2-1 [A6.8.2.2] for different types of cross sections and connection types.

For welded connections with longitudinal welds along some but not all of the connected elements (Fig. 19.2), the strength is controlled by the weld strength.

Example 19.1 Determine the net effective area and the factored tensile resistance of a single angle $L\ 6 \times 4 \times \frac{1}{2}$ tension member welded to a gusset plate as shown in Figure 19.4. Use Grade 36 structural steel.

Solution

Because only one leg of the angle is connected, the net area must be reduced by the factor U. Using Table A6.8.2.2-1, Case 4, with $L = 8$ in. and $W = 6$ in.

$$L = \tfrac{8}{6}W = 1.33W \qquad U = 0.75$$

and from Eq. 19.4 with $A_g = 4.75$ in.^{2}

$$A_e = UA_g = 0.75\,(4.75) = 3.56\ \text{in.}^2$$

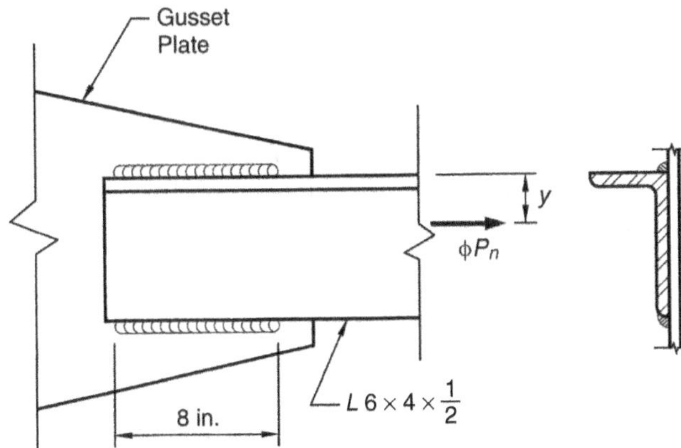

Fig. 19.4 Single-angle tension member welded to a gusset plate.

The factored resistance to yielding is calculated from Eq. 19.1 with $\phi_y = 0.95$ (Table 18.7) and $F_y = 36$ ksi (Table 17.2) to give

$$\phi_y P_{nu} = \phi_y F_y A_g = 0.95\,(36)\,(4.75) = 162.5 \text{ kips}$$

The factored resistance to rupture is calculated from Eq. 19.2 with $\phi_u = 0.80$ (Table 18.7) and $F_u = 58$ ksi (Table 17.2) to give

$$\phi_u P_{nu} = \phi_u F_u A_e = 0.80\,(58)\,(3.56) = 165.2 \text{ kips}$$

Answer The factored tensile resistance is governed by yielding of the gross section away from the connection and is equal to 162.5 kips.

Net Area [A6.8.3] The net area A_n of a tension member is the sum of the products of thickness t and the smallest net width w_n of each element. If the connection is made with bolts, the maximum net area is with all of the bolts in a single line (Fig. 19.1). Sometimes space limitations require that more than one line be used. The reduction in cross-sectional area is minimized if a staggered bolt pattern is used (Fig. 19.5). The net width is determined for each chain of holes extending across the member along any transverse, diagonal, or zigzag line. All conceivable failure paths should be considered and the one corresponding to the smallest w_n should be used. The net width for a chain of holes is computed by subtracting from the gross width of the element the sum of the widths of all holes and

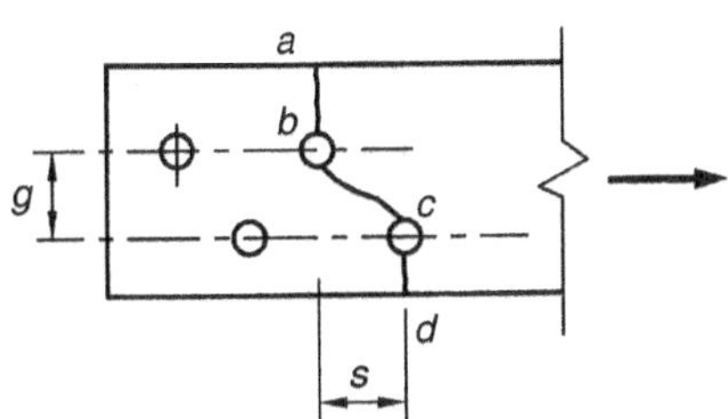

Fig. 19.5 Staggered bolt pattern.

adding the quantity $s^2/4g$ for each inclined line in the chain, that is,

$$w_n = w_g - \sum d + \sum \frac{s^2}{4g} \tag{19.7}$$

where w_g is the gross width of the element (in., mm), d is the nominal diameter of the bolt (in., mm) plus 0.125 in. (3.2 mm), s is the pitch of any two consecutive holes (in., mm), and g is the gage of the same two holes (Fig. 19.5).

Example 19.2 Determine the net effective area and the factored tensile resistance of a single angle $L\ 6 \times 4 \times \frac{1}{2}$ tension member bolted to a gusset plate as shown in Figure 19.6. The holes are for $\frac{7}{8}$-in.-diameter bolts. Use Grade 36 structural steel.

Solution

The gross width of the cross section is the sum of the legs minus the thickness [A6.8.3]:

$$w_g = 6 + 4 - \tfrac{1}{2} = 9.5 \text{ in.}$$

The effective hole diameter is $d = \frac{7}{8} + \frac{1}{8} = 1$ in.

Using Eq. 19.7, the net width on line *abcd* is

$$w_n = 9.5 - 2\,(1) + \frac{(1.375)^2}{4\,(2.375)} = 7.70 \text{ in.}$$

and on line *abe*

$$w_n = 9.5 - 1\,(1) = 8.5 \text{ in.}$$

The first case controls, so that

$$A_n = tw_n = 0.5\,(7.70) = 3.85 \text{ in.}^2$$

Because only one leg of the cross section is connected, the net area must be reduced by the factor U. From the properties table in AISC (2005), the distance from the centroid to the outside face of the leg of the angle is $x = 0.987$ in. Using Eq. 19.5 with $L = 3(2.75) = 8.25$ in.

$$U = 1 - \frac{x}{L} = 1 - \frac{0.987}{8.25} = 0.88 > 0.85$$

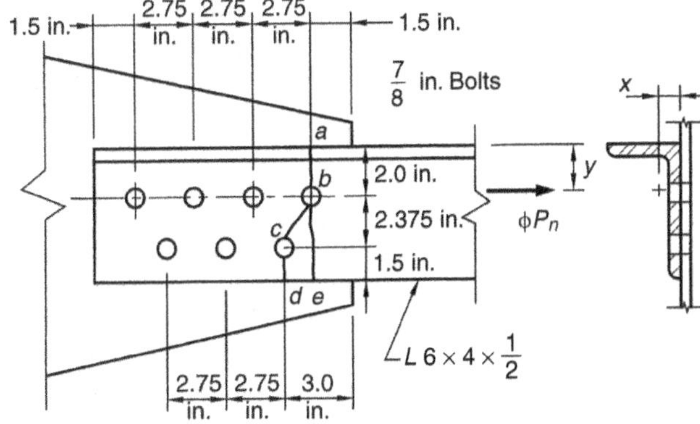

Fig. 19.6 Single-angle tension member bolted to a gusset plate.

Table 19.1 Maximum Slenderness Ratios for Tension Members

Tension Member	max (L/r)
Main members	
Subject to stress reversals	140
Not subject to stress reversals	200
Bracing members	240

and from Eq. 19.3

$$A_e = UA_n = 0.88\,(3.85) = 3.39 \text{ in.}^2$$

The factored resistance to yielding is the same as in Example 19.1:

$$\phi_y P_{ny} = \phi_y F_y A_g = 0.95\,(36)\,(4.75) = 162.5 \text{ kips}$$

The factored resistance to rupture is calculated from Eq. 19.2 to give

$$\phi_u P_{uy} = \phi_u F_u A_e = 0.80\,(58)\,(3.39) = 157 \text{ kips}$$

Answer The factored tensile resistance is governed by rupture on the net section and is equal to 157 kips.

Limiting Slendernes Ratio [A6.8.4] Slenderness requirements are usually associated with compression members. However, it is good practice also to limit the slenderness of tension members. If the axial load in a tension member is removed and small transverse loads are applied, undesirable vibrations or deflections might occur (Segui, 2003). The slenderness requirements are given in terms of L/r, where L is the member length and r is the least radius of gyration of the cross-sectional area.

Slenderness requirements for tension members other than rods, eyebars, cables, and plates are given in Table 19.1 [A6.8.4].

19.1.3 Strength of Connections for Tension Members

Strength calculations for welded and bolted connections are not given in this book. The reader is referred to standard steel design textbooks and manuals that cover this topic in depth. Examples of textbooks are Gaylord et al. (1992) and Segui (2003). Also see *Detailing for Steel Construction* (AISC, 2002).

19.2 COMPRESSION MEMBERS

Compression members are structural elements that are subjected only to axial compressive forces that are applied along the longitudinal axis of the member and produce uniform stress over the cross section. This uniform stress is an idealized condition as there is always some eccentricity between the centroid of the section and the applied load. The resulting bending moments are usually small and of secondary importance. The most common type of compression member is a *column*. If calculated bending moments exist, due to continuity or transverse loads, they cannot be ignored and the member must be considered as a *beam column*. Compression members exist in trusses, cross frames, and lateral bracing systems where the eccentricity is small and the secondary bending can be reasonably ignored.

Compression members made of open shapes such as wide flange beam, plate girders, and angles are susceptible to three types of buckling:

- ☐ Global member instability (Euler buckling)
- ☐ Local buckling of cross-section element (flange or web)
- ☐ Torsonal buckling

The focus in this section will be primarily on global buckling

19.2.1 Column Stability—Behavior

In structural steel, column cross sections are often slender and other limit states are reached before the material yields. These other limit states are associated with inelastic and slender member buckling. They include lateral buckling, local buckling, and lateral-torsional buckling of the compression member. Each of the limit states must be incorporated in the design rules developed to select compression members.

The starting point for studying the buckling phenomenon is an idealized perfectly straight elastic column with pin ends. As the axial compressive load on the column increases, the column remains straight and shortens elastically until the critical load P_{cr} is reached. The critical load is defined as the lowest axial compressive load for which a small lateral displacement causes the column to bow laterally and seek a new equilibrium position. This definition of critical load is depicted schematically in the load–deflection curves of Figure 19.7.

In Figure 19.7, the point at which the behavior changes is the *bifurcation point*. The load–deflection curve is vertical until this point is reached, and then the midheight of the column moves right or left depending on the direction of the lateral disturbance. Once the lateral deflection becomes nonzero, a buckling failure occurs and small deflection theory predicts that no further increase in the axial load is possible. If large deflection theory is used, additional stress resultants are developed and the load–deflection response follows the dashed line in Figure 19.7.

The small deflection theory solution to the buckling problem was published by Euler in 1759. He showed that the critical buckling load P_{cr} is given as

$$P_{cr} = \frac{\pi^2 EI}{L^2} \tag{19.8}$$

where E is the modulus of elasticity of the material, I is the moment of inertia of the column cross section about the

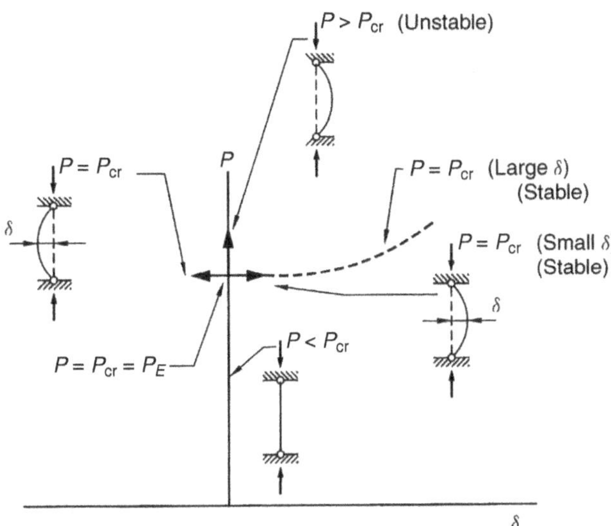

Fig. 19.7 Load–deflections curves for elastic columns (Bjorhovde, 1992). [From *Constructional Steel Design: An International Guide*, P. J. Dowling, J. E. Harding, and R. Bjorhovde, eds., Copyright © 1992 by Elsevier Science Ltd (now Chapman and Hall, Andover, England), with permission.]

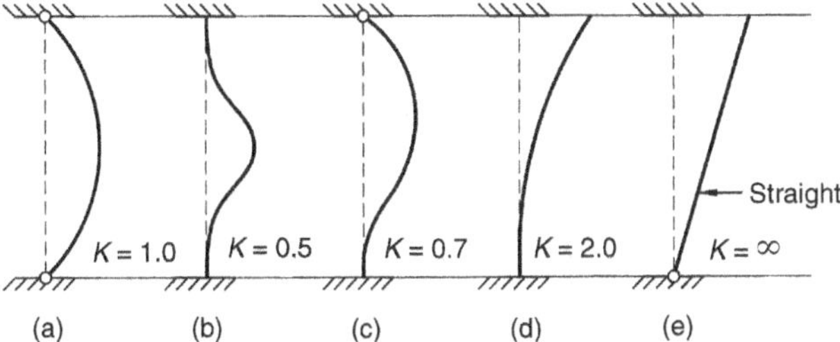

Fig. 19.8 End restraint and effective length of columns: (a) pinned-pinned, (b) fixed-fixed, (c) fixed-pinned, (d) fixed-free, and (e) pinned-free (Bjorhovde, 1992). [From *Constructional Steel Design: An International Guide*, P. J. Dowling, J. E. Harding, and R. Bjorhovde, eds., Copyright © 1992 by Elsevier Science Ltd (now Chapman and Hall, Andover, England), with permission.]

centroidal axis perpendicular to the plane of buckling, and L is the pin-ended column length. This expression is well known in mechanics and its derivation is not repeated here.

Equation 19.8 can also be expressed as a critical buckling stress σ_{cr} by dividing both sides by the gross area of the cross section A_g to give

$$\sigma_{cr} = \frac{P_{cr}}{A} = \frac{\pi^2 (EI/A_g)}{L^2}$$

By using the definition of the radius of gyration r of the section as $I = Ar^2$ and rewriting the above equation, we get

$$\sigma_{cr} = \frac{\pi^2 E}{(L/r)^2} \qquad (19.9)$$

where L/r is commonly referred to as the *slenderness ratio* of the column. Given axisymmetric boundary conditions, buckling occurs about the centroidal axis with the least moment of inertia I (Eq. 19.8) or the least radius of gyration r (Eq. 19.9). Sometimes the critical centroidal axis is inclined, as in the case of a single-angle compression member. In any event, the maximum slenderness ratio must be found because it governs the critical stress.

The idealized critical buckling stress given in Eq. 19.9 is influenced by three major parameters: end restraint, residual stresses, and initial crookedness. The first depends on how the member is connected and the last two on how it was manufactured. These parameters are discussed in the following sections.

Effective Length of Columns The buckling problem solved by Euler was for an idealized column without any moment restraint at its ends. For a column of length L whose ends do not move laterally (nor sidesway), end restraint provided by connections to other members causes the location of the points of zero moment to move away from the ends of the column. The distance between the points of zero moment is the effective pinned-pinned column length KL, where in this case $K < 1$. If the end restraint is either pinned or fixed, typical values of K for the no-sidesway case are shown in the first three deformed shapes of Figure 19.8.

If the ends of a column move laterally with respect to one another, the effective column length KL can be large with K considerably greater than 1. This behavior is shown in the last two deformed shapes of Figure 19.8 with one end free and the other end either fixed or pinned.

In general, the critical buckling stress for a column with effective length KL can be obtained by rewriting Eq. 19.9 as

$$\sigma_{cr} = \frac{\pi^2 E}{(KL/r)^2} \qquad (19.10)$$

where K is the effective length factor.

Actual column end conditions are going to be somewhere between pinned and fixed depending on the stiffness provided by the end connections. For bolted or welded connections at both ends of a compression member in which sidesway is prevented, K may be taken as 0.75 [A4.6.2.5]. Therefore, the effective length of the compression members in cross frames and lateral bracing can be taken as $0.75L$, where L is the laterally unsupported length of the member.

Residual Stresses Residual stresses are discussed previously. In general, they are caused by nonuniform cooling of the elements in a component during the manufacturing or fabrication process. The basic principle of residual stress can be summarized as follows: The fibers that cool first end up in residual compression; those that cool last have residual tension (Bjorhovde, 1992).

The magnitude of the residual stresses can be almost equal to the yield stress of the material. Additional applied axial compressive stress can cause considerable yielding in the cross section at load levels below that predicted by $F_y A_g$.

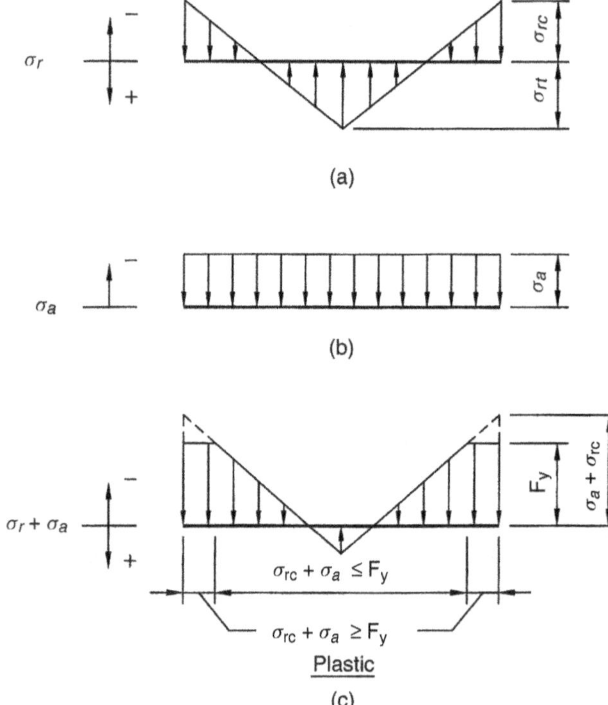

Fig. 19.9 (a) Residual stress, (b) applied compressive stress, and (c) combined residual and applied stress (Bjorhovde, 1992). [From *Constructional Steel Design: An International Guide*, P. J. Dowling, J. E. Harding, and R. Bjorhovde, eds., Copyright © 1992 by Elsevier Science Ltd (now Chapman and Hall, Andover, England), with permission.]

This combined stress is shown schematically in Figure 19.9, where σ_{rc} is the compressive residual stress, σ_{rt} is the tensile residual stress, and σ_a is the additional applied axial compressive stress. The outer portions of the element have gone plastic while the inner portion remains elastic.

Initial Crookedness Residual stresses develop in an element along its length, and each cross section is assumed to have a stress distribution similar to that shown in Figure 19.9. This uniform distribution of stress along the length of the element occurs only if the cooling process is uniform. What usually happens is that a member coming off the rolling line in a steel mill is cut to length and then set aside to cool. Other members are placed along side it on the cooling bed and will influence the rate of cooling.

If a hot member is on one side and a warm member is on the other side, the cooling is nonuniform across the section. Further, the cut ends cool faster than the sections at midlength and the cooling is nonuniform along the length of the member. After the member cools, the nonuniform residual stress distribution causes the member to bow, bend, and even twist. If the member is used as a column, it can no longer be assumed to be perfectly straight, but must be considered to have initial crookedness.

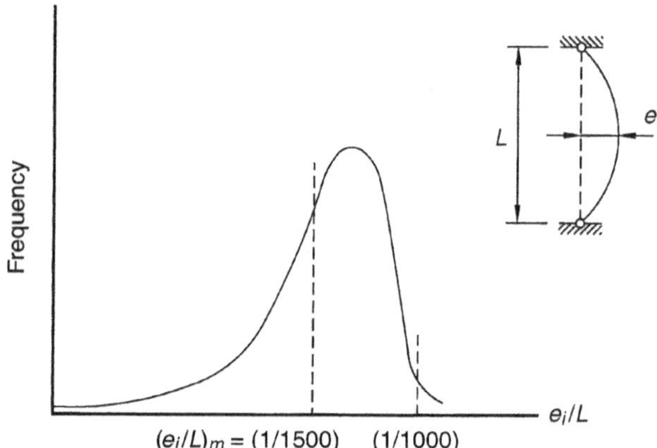

Fig. 19.10 Statistical variation of initial crookedness (Bjorhovde, 1992). [From *Constructional Steel Design: An International Guide*, P. J. Dowling, J. E. Harding, and R. Bjorhovde, eds., Copyright © 1992 by Elsevier Science Ltd (now Chapman and Hall, Andover, England), with permission.]

A column with initial crookedness introduces bending moments when axial loads are applied. Part of the resistance of the column is used to carry these bending moments and a reduced resistance is available to support the axial load. Therefore, the imperfect column exhibits a load-carrying capacity that is less than that of the ideal column.

The amount of initial crookedness in wide-flange shapes is shown in Figure 19.10 as a fraction of the member length. The mean value of the random eccentricity e_1 is $L/1500$ with a maximum value of about $L/1000$ (Bjorhovde, 1992).

19.2.2 Inelastic Buckling—Behavior

The Euler buckling load of Eq. 19.8 was derived assuming elastic material behavior. For long, slender columns this assumption is reasonable because buckling occurs at a relatively low level and the stresses produced are below the yield strength of the material. However, for short, stubby columns buckling loads are higher and yielding of portions of the cross section occurs.

For short columns, not all portions of the cross section reach yield simultaneously because the locations with compressive residual stresses yield first as illustrated in Figure 19.9. Therefore, as the axial compressive load increases the portion of the cross section that remains, elastic decreases until the entire cross section becomes plastic. The transition from elastic to plastic behavior is gradual as demonstrated by the stress–strain curve in Figure 19.11 for a stub column. This stress–strain behavior is different from the relatively abrupt change from elastic to plastic usually observed in a bar or coupon test of structural steel (Fig. 17.4).

The stub column stress–strain curve of Figure 19.11 deviates from elastic behavior at the proportional limit σ_{prop} and gradually changes to plastic behavior when F_y is reached. The modulus of elasticity E represents elastic behavior until

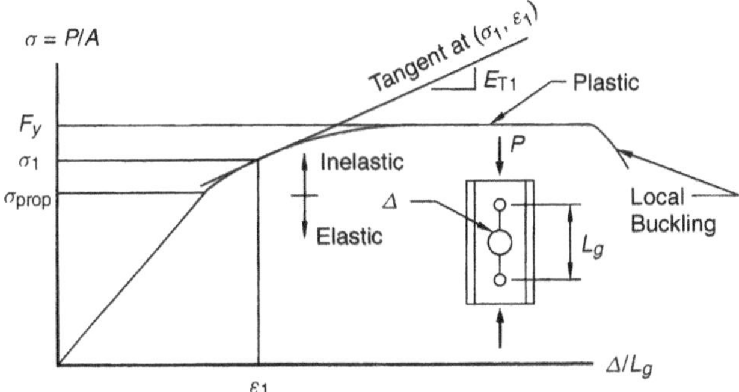

Fig. 19.11 Stub column stress–strain curve (Bjorhovde, 1992). [From *Constructional Steel Design: An International Guide*, P. J. Dowling, J. E. Harding, and R. Bjorhovde, eds., Copyright © 1992 by Elsevier Science Ltd (now Chapman and Hall, Andover, England), with permission.]

the sum of the compressive applied and maximum residual stress in Figure 19.9 equals the yield stress, that is,

$$\sigma_a + \sigma_{rc} = F_y$$

or

$$\sigma_{prop} = F_y - \sigma_{rc} \tag{19.11}$$

In the transition between elastic and plastic behavior, the rate of change of stress over strain is represented by the tangent modulus E_T as shown in Figure 19.11. This region of the curve where the cross section is a mixture of elastic and plastic stresses is called *inelastic*. The inelastic or tangent modulus column buckling load is defined by substituting E_T for E in Eq. 19.10 to yield

$$\sigma_T = \frac{\pi^2 E_T}{(KL/r)^2} \tag{19.12}$$

A combined Euler (elastic) and tangent modulus (inelastic) column buckling curve is shown in Figure 19.12. The transition point that defines the change from elastic to inelastic behavior is the proportional limit stress σ_{prop} of Eq. 19.11 and the corresponding slenderness ratio $(KL/r)_{prop}$.

19.2.3 Compressive Resistance—Specifications

The short or stub column resistance to axial load is at maximum when no buckling occurs and the entire gross cross-sectional area A_g is at the yield stress F_y. The fully plastic yield load P_y is the maximum axial load the column can support and can be used to normalize the column curves so that they are independent of structural steel grade. The axial yield load is

$$P_y = A_g F_y \tag{19.13}$$

$$P_o = QP_y = QA_g F_y \tag{19.14}$$

where Q is a factor that adjusts the resistance downward if the cross sections contain slender elements subject to local buckling. See [A6.9.4.1] for guidance on the application of Q. Here, the focus will be global (member) buckling of shapes with compact elements and $Q = 1.0$.

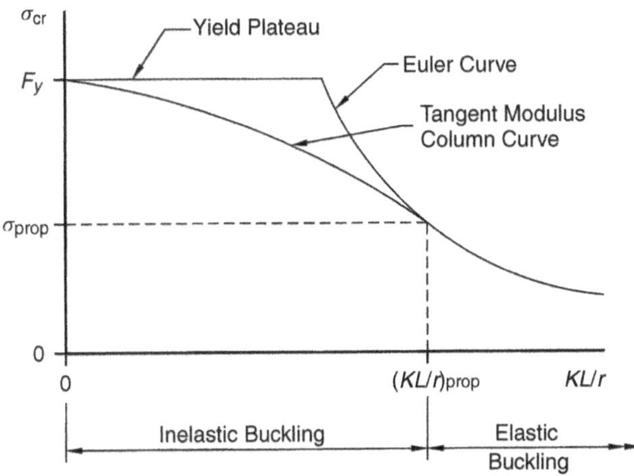

Fig. 19.12 Combined tangent modulus and Euler column curves (Bjorhovde, 1992). [From *Constructional Steel Design: An International Guide*, P. J. Dowling, J. E. Harding, and R. Bjorhovde, eds., Copyright © 1992 by Elsevier Science Ltd (now Chapman and Hall, Andover, England), with permission.]

For long columns, the critical elastic buckling load $P_{cr} = P_e$ is obtained by multiplying Eq. 19.10 by A_g to give [A6.9.4.1.2]

$$P_e = \frac{\pi^2 E A_g}{(KL/r)^2} \tag{19.15}$$

Dividing Eq. 19.15 by Eq. 19.14, the normalized elastic column equation with $Q = 1.0$ is

$$\frac{P_e}{P_o} = \left(\frac{\pi}{KL/r}\right)^2 \left(\frac{E}{F_y}\right) \tag{19.16}$$

It is convenient to define a slenderness term λ_c where

$$\lambda_c^2 = \left(\frac{P_e}{P_o}\right)^{-1} = \left(\frac{KL}{\pi r}\right)^2 \frac{F_y}{E} \tag{19.17}$$

The normalized plateau and Euler column curve are shown as the top curves in Figure 19.13. The inelastic transition curve due to residual stresses is also shown. The column curve that includes the additional reduction in buckling load caused by initial crookedness is the bottom curve in Figure 19.13. This bottom curve is the column strength curve given in the specifications.

The column strength curve represents a combination of inelastic and elastic behavior. Inelastic buckling occurs for intermediate length columns from $\lambda_c = 0$ to $\lambda_c = \lambda_{prop}$, where λ_{prop} is the slenderness term for an Euler critical stress equal to the proportional limit σ_{prop} (Eq. 19.18). Elastic buckling occurs for long columns with λ_c greater than λ_{prop}. Substitution of Eq. 19.11 and these definitions into Eq. 19.16 results in

$$\frac{F_y - \sigma_{rc}}{F_y} \frac{A_g}{A_g} = \frac{1}{\lambda_{prop}^2}$$

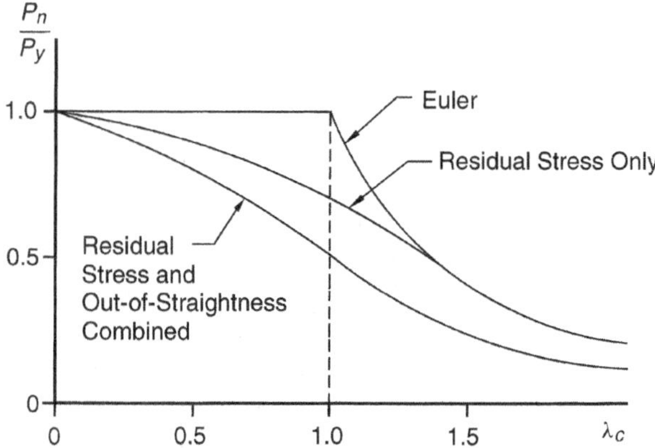

Fig. 19.13 Normalized column curves with imperfection effects (Bjorhovde, 1992). [From *Constructional Steel Design: An International Guide*, P. J. Dowling, J. E. Harding, and R. Bjorhovde, eds., Copyright © 1992 by Elsevier Science Ltd (now Chapman and Hall, Andover, England), with permission.]

or

$$\lambda_{\text{prop}}^2 = \frac{1}{1 - \dfrac{\sigma_{\text{rc}}}{F_y}} \tag{19.18}$$

The value for λ_{prop} depends on how large the residual compressive stress σ_{rc} is relative to the yield stress F_y. For example, if $F_y = 50$ ksi (345 MPa) and $\sigma_{\text{rc}} = 28$ ksi (190 MPa), then Eq. 19.17 gives

$$\lambda_{\text{prop}}^2 = \frac{1}{1 - \frac{28}{50}} = 2.27 \approx 2.25$$

and $\lambda_{\text{prop}} = 1.5$. The larger the residual stress the larger the slenderness term at which the transition to elastic buckling occurs. Nearly all of the columns designed in practice behave as inelastic intermediate-length columns. Seldom are columns slender enough to behave as elastic long columns that buckle at the Euler critical load.

The transition point between inelastic buckling and elastic buckling or between intermediate-length columns and long columns is specified as $\lambda = 2.25$. For long columns ($\lambda \geq 2.25$), the nominal column strength P_n is given by

$$P_n = \frac{0.88 F_y A_g}{\lambda_c} \tag{19.19}$$

which is the Euler critical buckling load of Eq. 19.15 reduced by a factor of 0.88 to account for initial crookedness of $L/1500$ [C6.9.4.1].

For intermediate-length columns ($\lambda < 2.25$), the nominal column strength P_n is determined from a tangent modulus curve that provides a smooth transition between $P_n = P_y$ and the Euler buckling curve. The formula for the transition curve is

$$P_n = 0.66^{\lambda_c} F_y A_g \tag{19.20}$$

Nominal Compressive Resistance [A6.9.4.1] The curves representing Eqs. 19.19, 19.20, and 19.21 are plotted in Figure 19.14.

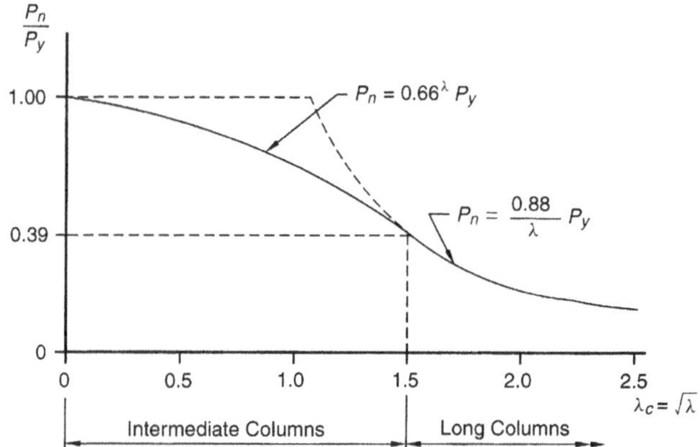

Fig. 19.14 Column design curves.

In lieu of the column slenderness parameter λ_c, AASHTO uses the ratio P_e/P_o (and its inverse) directly. So the column resistance equations become

If $P_e/P_o \geq 0.44$, then

$$P_n = 0.658^{(P_o/P_e)} P_o \tag{19.21a}$$

If $P_e/P_o < 0.44$, then

$$P_n = 0.877 P_e \tag{19.21b}$$

The final step in determining the compressive resistance of P_r of columns is to multiply the nominal resistance P_n by the resistance factor for compression ϕ_c taken from Table 18.7, that is,

$$P_r = \phi_c P_n \tag{19.21c}$$

Limiting Width/Thickness Ratios [A6.9.4.2] Compressive strength of columns of intermediate length is based on the tangent modulus curve obtained from tests of stub columns. A typical stress–strain curve for a stub column is given in Figure 19.11. Because the stub column is relatively short, it does not exhibit flexural buckling. However, it could experience local buckling with a subsequent decrease in load if the width/thickness ratio of the column elements is too high. Therefore, the slenderness is limited so that the yield stress can be achieved without buckling:

$$\frac{b}{t} \leq k\sqrt{\frac{E}{F_y}} \tag{19.22}$$

where k is the plate buckling coefficient taken from Table 19.2, b is the width of the plate described in Table 19.2 (in., mm), and t is the plate thickness (in., mm). The requirements given in Table 19.2 for plates supported along one edge and plates supported along two edges are illustrated in Figure 19.15.

Elements that do not satisfy Eq. 19.22 have an associated strength that is estimated by the factor Q outlined in A6.9.4.1.1. Here, compactness is assumed and $Q = 1.0$.

Limiting Slenderness Ratio [A6.9.3] If a column is too slender, it has little strength and is not economical and is susceptible to buckling and perhaps sudden collapse

Table 19.2 Limiting Width–Thickness Ratios

Plates Supported along One Edge	k	b
Flanges and projecting legs of plates	0.56	• Half-flange width of I-sections • Full-flange width of channels • Distance between free edge and first line of bolts or welds in plates • Full width of an outstanding leg for pairs of angles in continuous contact
Stems of rolled tees	0.75	• Full depth of tee
Other projecting elements	0.45	• Full width of outstanding leg for single-angle strut or double-angle strut with separator • Full projecting width for others

Plates Supported along Two Edge	k	b
Box flanges and cover plates	1.40	• Clear distance between webs minus inside corner radius on each side for box flanges • Distance between lines of welds or bolts for flange cover plates
Webs and other plate elements	1.49	• Clear distance between flanges minus fillet radii for webs of rolled beams • Clear distance between edge supports for all others
Perforated cover plates	1.86	• Clear distance between edge supports

AASHTO Table 6.9.4.2-1. From *AASHTO LRFD Bridge Design Specifications*, Copyright © 2010 by the American Association of State Highway and Transportation Officials, Washington, DC. Used by permission.

depending on the importance of the member and the system. The recommended limit for main members is $(KL/r) \leq 120$ and for bracing member it is $(KL/r) \leq 140$. From a practical perspective such slender members are not typically used in today's designs. An exception is the use of "counters" or tension-only members found in older trusses that are designed to buckle and shed their load to adjacent tension members. Such members were used in small- to medium-span trusses built in the 1900s.

Example 19.3 Calculate the design compressive strength $P_r = \phi_c P_n$ of a $W14 \times 74$ column with a length of 240 in. and pinned ends. Use Grade 50 structural steel.

Properties

From AISC (2005), $A_g = 21.8\,\text{in.}^2$, $d = 14.17\,\text{in.}$, $t_w = 0.45\,\text{in.}$, $b_f = 10.07\,\text{in.}$, $t_f = 0.785\,\text{in.}$, $h_c/t_w = 25.3$, $r_x = 60.4\,\text{in.}$, and $r_y = 2.48\,\text{in.}$

Solution

Slenderness Ratio

$$\max\frac{KL}{r} = \frac{1.0\,(240)}{2.48} = 96.8 < 120 \qquad \text{OK}$$

$$\frac{\text{Width}}{\text{Thickness}} : \frac{b_f}{2t_f} = \frac{10.07}{2\,(0.785)} = 6.4 < k\sqrt{\frac{E}{F_y}}$$

$$= 0.56\sqrt{\frac{29{,}000}{50}} = 13.5 \qquad \text{OK}$$

$$\frac{h_c}{t_w} = 25.3 < k\sqrt{\frac{E}{F_y}} = 1.49\sqrt{\frac{29{,}000}{50}} = 35.9 \qquad \text{OK}$$

Therefore, the section is compact and $Q = 1.0$.
Column Slenderness Term

$$\frac{P_e}{P_o} = \pi^2\left(\frac{1}{KL/r}\right)^2\left(\frac{E}{F_y}\right) = \pi^2\left(\frac{1}{96.8}\right)^2\left(\frac{29\,000}{50}\right)$$

$$= 0.611 \geq 0.44$$

Intermediate Length Column

$$P_o = QF_yA_g = (1.0)\,(50)\,(21.8) = 1090\text{ kips}$$

$$P_n = 0.66^{(P_o/P_e)}P_o = (0.658)^{1/0.611}\,(1090) = 549\text{ kips}$$

Note that P_e/P_o is for the region check but P_o/P_e (inverse) is used in the strength equation.

Answer Design compressive strength $= \phi_c P_n = 0.90(549) = 495$ kips.

19.2.4 Connections for Compression Members

Strength calculations for welded and bolted connections are not given in this book. The reader is referred to standard steel design textbooks and manuals that cover this topic in depth. Examples of textbooks are Gaylord et al. (1992) and Segui (2003). *Detailing for Steel Construction* (AISC, 2002) also provides guidance.

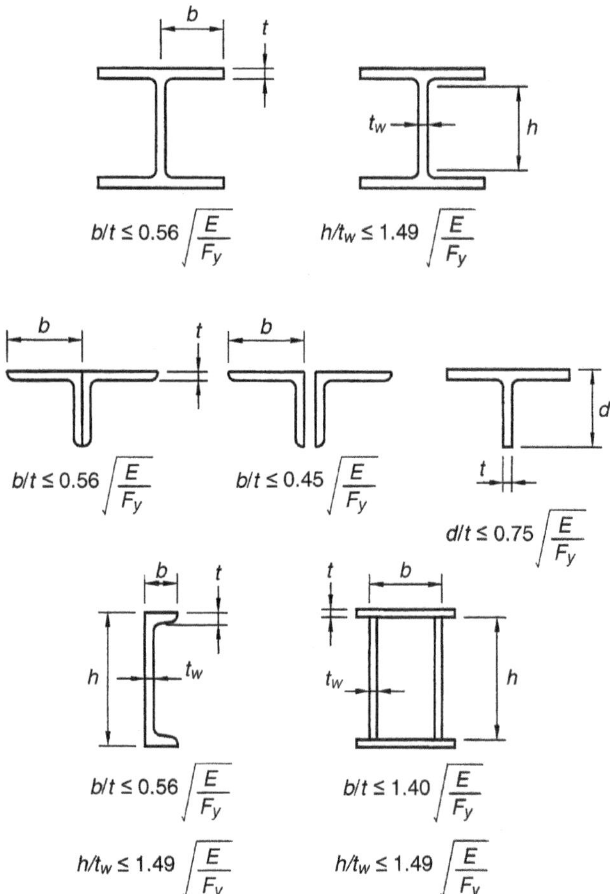

Fig. 19.15 Limiting width–thickness ratios (after Segui, 2003). (From William T. Segui, *LRFD Steel Design*, Copyright © 2003 by PWS Publishing Company, Boston, MA, with permission.)

19.3 I-SECTIONS IN FLEXURE

I-sections in flexure are structural members that carry transverse loads perpendicular to their longitudinal axis primarily in a combination of bending and shear. Axial loads are usually small in most bridge girder applications and are often neglected. If axial loads are significant, then the cross section should be considered as a *beam column*. If the transverse load is eccentric to the shear center of the cross section, then combined bending and torsion must be considered. The discussion that follows is limited to the basic behavior and design of rolled or fabricated straight steel I-sections that are symmetrical about a vertical axis in the plane of the web and are primarily in flexure and shear. The approach is unified for curved girders; see White and Grubb (2005).

All types of I-section flexural members must generally satisfy [A6.10.1]:

- ☐ Cross-section proportions to avoid local buckling and to provide ease in handling [A6.10.2]
- ☐ Constructibility requirements [A6.10.3]
- ☐ Service limit states [A6.10.4]
- ☐ Fatigue and fracture limit states [A6.10.5]
- ☐ Strength limit states [A6.10.6]

These items are listed in the preferred order for computational checks. With the AASHTO (2010) LRFD Bridge Specifications for steel bridges, construction and service limits often control the design. Fatigue and strength limit states are typically satisfied and do not control the design.

19.3.1 General

The resistance of I-sections in flexure is largely dependent on the degree of stability provided, either locally or in a global manner. If the section is stable at high loads, then the I-section can develop a bending resistance beyond the first yield moment M_y to the full plastic moment resistance M_p. If stability is limited by either local or global buckling, then the bending resistance is less than M_p and if the buckling is significant, less than M_y.

***Plastic Moment* $\mathbf{M_p}$** Consider the doubly symmetric I-section of Figure 19.16(a) that is subjected to pure bending

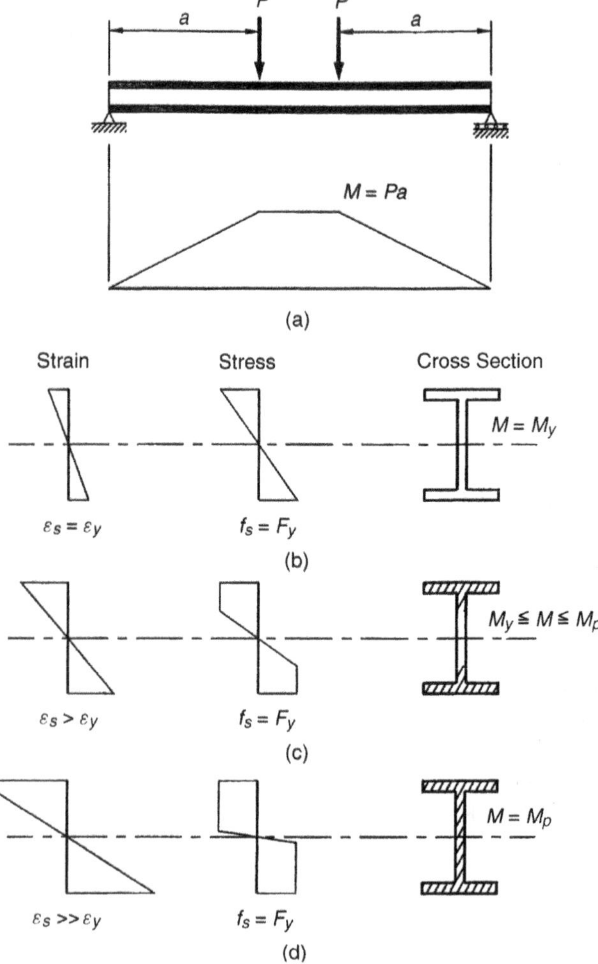

Fig. 19.16 Progressive yielding in flexure: (a) simple beam with twin concentrated loads, (b) first yield at extreme fibers, (c) partially plastic and partially elastic, and (d) fully plastic.

at midspan by two equal concentrated loads. Assume stability is provided and the steel stress–strain curve is elastic perfectly plastic. As the loads increase, plane sections remain plane, and the strains increase until the extreme fibers of the section reach $\varepsilon_y = F_y/E$ [Fig. 19.16(b)]. The bending moment at which the first fibers reach yield is defined as the yield moment M_y.

Further increase of the loads causes the strains and rotations to increase and more of the fibers in the cross section to yield [Fig. 19.16(c)]. The limiting case is when the strains caused by the loads are so large that the entire cross section can be considered at the yield stress F_y [Fig. 19.16(d)]. When this occurs, the section is fully plastic and the corresponding bending moment is defined as the plastic moment M_p.

Any attempt to further increase the loads only results in increased deformations without any increase in moment resistance. This limit of moment can be seen in the idealized moment–curvature curve in Figure 19.17. Curvature ψ is defined as the rate of change of strain or simply the slope of the strain diagram, that is,

$$\psi = \frac{\varepsilon_c}{c} \tag{19.23}$$

where ε_c is the strain at a distance c from the neutral axis.

The moment–curvature relation of Figure 19.17 has three parts: elastic, inelastic, and plastic. The inelastic part provides a smooth transition between elastic to plastic behavior as more of the fibers in the cross section yield. The length of the plastic response ψ_p relative to the elastic curvature ψ_y is a measure of ductility of the section.

Moment Redistribution When the plastic moment M_p is reached at a cross section, additional rotation occurs at the section and a hinge resisting constant moment M_p forms. When this plastic hinge forms in a statically determinate structure, such as the simple beam of Figure 19.16, a collapse mechanism is formed.

However, if a plastic hinge forms in a statically indeterminate structure, collapse does not occur and additional load-carrying capacity remains. This increase in load is illustrated with the propped cantilever beam of Figure 19.18(a) that is subjected to a gradually increasing concentrated load at midspan. The limit of elastic behavior is when the load causes the moment at the fixed end of the beam to reach M_y. This limiting load P_y produces moments that are consistent with an elastic analysis as shown in Figure 19.18(b).

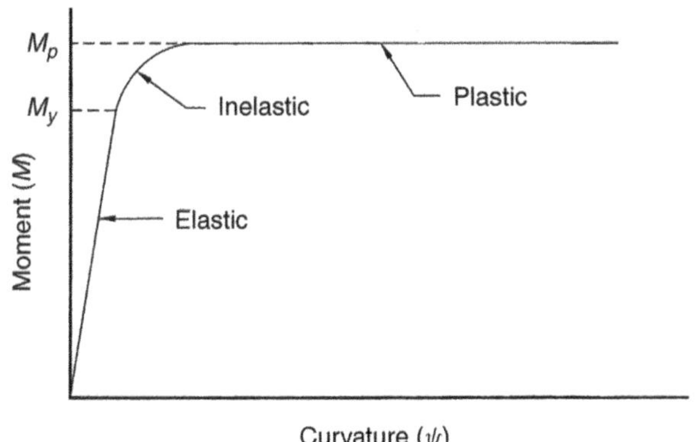

Fig. 19.17 Idealized moment–curvature response.

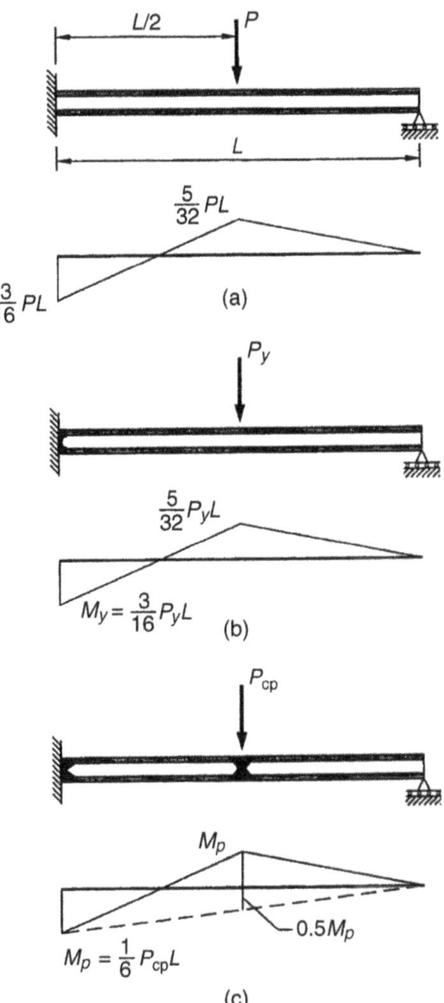

Fig. 19.18 Moment redistribution in a propped cantilever: (a) elastic moments, (b) first yield moments, and (c) collapse mechanism moments.

Further increase in the load causes a plastic hinge to form at the fixed end. However, the structure will not collapse because a mechanism has not been formed. The beam with one fixed end has now become a simple beam with a known moment M_p at one end. A mechanism does not form until a second plastic hinge develops at the second highest moment location under the concentrated load. This condition is shown in Figure 19.18(c). This behavior for moving loads is described in detail in Chapter 10.

By assuming that $M_y = 0.9M_p$, the ratio of the collapse load P_{cp} to the yield load P_y is

$$\frac{P_{cp}}{P_y} = \frac{6M_p/L}{\frac{16}{3}\left(0.9M_p\right)/L} = 1.25$$

For this example, there is an approximate 25% increase in resistance to load beyond the load calculated by elastic analysis. However, for this to take place, rotation capacity had to exist in the plastic hinge at the fixed end so that moment redistribution could occur.

Another way to show that moment redistribution has taken place when plastic hinges form is to compare the ratio of positive moment to negative moment. For the elastic moment diagram in Figure 19.18(b), the ratio is

$$\left(\frac{M_{\text{pos}}}{M_{\text{neg}}}\right)_e = \frac{\frac{5}{32}PL}{\frac{3}{16}PL} = 0.833$$

while for the moment diagram at collapse [Fig. 19.18(c)]

$$\left(\frac{M_{\text{pos}}}{M_{\text{neg}}}\right)_{\text{cp}} = \frac{M_p}{M_p} = 1.0$$

Obviously, the moments have been redistributed.

An extensive procedure is outlined in the AASHTO (2010) LRFD Bridge Design Specifications in Section 6, Appendix B. Here optional simplified and rigorous procedures are outlined and explained in the commentary. More details are not presented here.

Composite Considerations Sections are classified as *composite* or *noncomposite*. A *composite section* is one where a properly designed shear connection exists between the concrete deck and the steel beam (Fig. 19.19). A section where the concrete deck is not connected to the steel beam is considered as a *noncomposite section*. During construction prior to and during concrete hardening, the steel section is noncomposite and must be checked against the deck dead load and construction loads such as equipment concrete screed and associated rails.

When the shear connection exists, the deck and beam act together to provide resistance to bending moment. In regions of positive moment, the concrete deck is in compression and the increase in flexural resistance can be significant. In regions of negative moment, the concrete deck is in tension and its tensile reinforcement adds to the flexural resistance of the steel beam. Additionally, well-distributed reinforcement enhances the effective stiffness of the concrete (tension stiffening, see Section 14.3.2). The flexural resistance of the composite section is further increased because the connection of the concrete deck to the steel beam provides continuous lateral support for its compression flange and prevents lateral-torsional buckling for positive moment. However, for negative moment (bottom flange in compression), the section is susceptible to lateral movement and buckling. In this case, the bracing is provided by the bearings and cross frames.

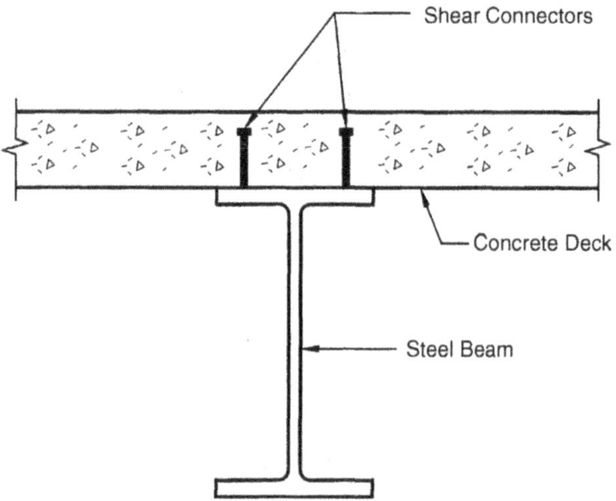

Fig. 19.19 Composite section.

Because of these advantages, the AASHTO (2010) LRFD Bridge Specification recommends that, wherever technically feasible, structures should be made composite for the entire length of the bridge [A6.10.1.1]. "Noncomposite sections are not recommended, but are permitted" [C6.10.1.2].

Stiffness Properties [A6.10.1.3] In the analysis of flexural members for loads applied to a noncomposite section, only the stiffness properties of the steel beam should be used. In the analysis of flexural members for loads applied to a composite section, the transformed area of concrete used in calculating the stiffness properties shall be based on a modular ratio of n (Table 19.3) [A6.10.1.1b] for transient loads and $3n$ for permanent loads. The modular ratio of $3n$ is to account for the larger increase in strain due to the creep of concrete under permanent loads. The concrete creep tends to transfer long-term stresses from the concrete to the steel, effectively increasing the relative stiffness of the steel. The multiplier on n accounts for this increase. The stiffness of the full composite section may be used over the entire bridge length, including regions of negative bending. This constant stiffness is reasonable, as well as convenient; field tests of continuous composite bridges have shown there is considerable composite action in the negative bending regions [C6.10.1.1.1].

Table 19.3 Ratio of Modulus of Elasticity of Steel to That of Concrete, Normal Weight Concrete

f'_c (ksi)	n
$2.4 \le f'_c < 2.9$	10
$2.9 \le f'_c < 3.6$	9
$3.6 \le f'_c < 4.6$	8
$4.6 \le f'_c < 6.0$	7
$6.0 \le f'_c$	6

From [C6.10.1.1.1b]. From *AASHTO LRFD Bridge Design Specifications*,

19.3.2 Yield Moment and Plastic Moment

The bending moment capacity of I-sections depends primarily on the compressive force capacity of the compression flange. If the compression flange is continuously laterally supported and the web has stocky proportions, no buckling of the compression flange occurs and the cross section develops its full plastic moment, that is, $M_n = M_p$. Cross sections that satisfy the restrictions for lateral support and width/thickness ratios for flanges and web are called *compact sections*. These sections exhibit fully plastic behavior and their moment–curvature response is similar to the top curve in Figure 19.20.

If the compressed flange is laterally supported at intervals large enough to permit the compression flange to buckle locally, but not globally, then the compression flange behaves like an inelastic column. The section of the inelastic column is T-shaped and part of it reaches the yield stress and part of it does not. These cross sections are intermediate between plastic and elastic behavior and are called *noncompact sections*. They can develop the yield moment M_y but have limited plastic response as shown in the middle curve of Figure 19.20.

If the compression flange is laterally unsupported at intervals large enough to permit lateral-torsional buckling, then the compression flange behaves as an elastic column whose capacity is an Euler-like critical buckling load reduced by the effect of torsion. The buckling of these sections with relatively high-compression flange slenderness ratio occurs before the yield moment M_y can be reached and are called *slender sections*. The slender sections behavior is shown by the bottom curve in Figure 19.20. The slender sections do not use materials effectively and most designers avoid them by providing sufficient lateral support.

Yield Moment of a Composite Section The yield moment M_y is the moment that causes first yielding in either flange of the steel section. Because the cross section behaves elastically until first yielding, superposition of moments is valid.

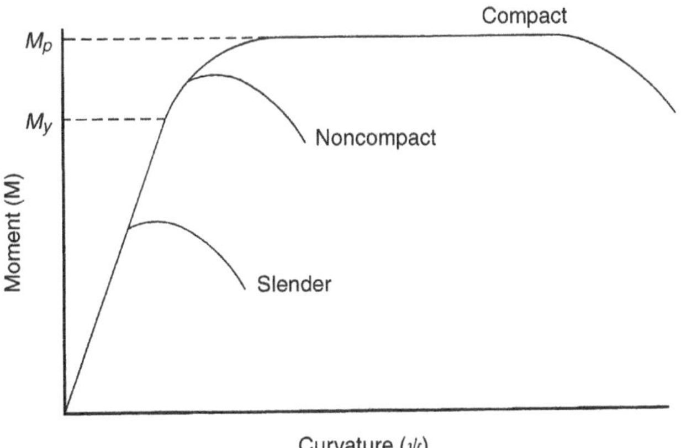

Fig. 19.20 Response of three beam classes.

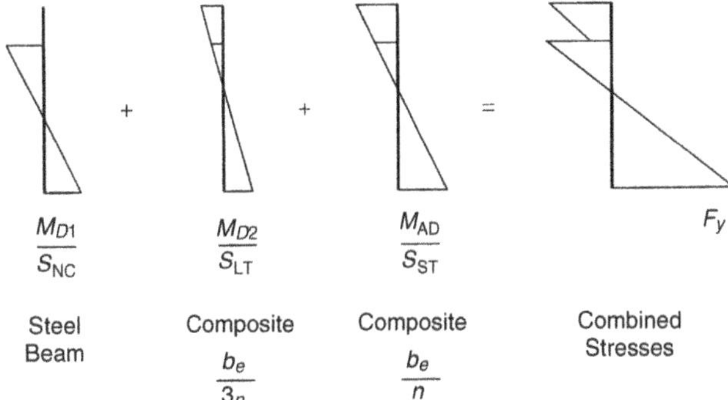

Fig. 19.21 Flexural stresses at first yield.

Therefore, M_y is the sum of the moment applied separately on the steel section, the short-term composite section, and the long-term composite section [A6.10.1.1 and A6.10.4.2].

The three stages of loading on a composite section are shown for a positive bending moment region in Figure 19.21. The moment due to factored permanent loads on the steel section before the concrete reaches 75% of its 28-day compressive strength is M_{D1}, and it is resisted by the noncomposite section modulus S_{NC}. The moment due to the remainder of the factored permanent loads (wearing surface, concrete barrier) is M_{D2}, and it is resisted by the long-term composite section modulus S_{LT}. The additional moment required to cause yielding in one of the steel flanges is M_{AD}. This moment is due to factored live load and is resisted by the short-term composite section modulus S_{ST}. The moment M_{AD} can be solved from the equation

$$F_y = \frac{M_{D1}}{S_{NC}} + \frac{M_{D2}}{S_{LT}} + \frac{M_{AD}}{S_{ST}} \tag{19.24}$$

and the yield moment M_y calculated from

$$M_y = M_{D1} + M_{D2} + M_{AD} \tag{19.25}$$

Again, these moments are factored. Details are presented in Section 6, Appendix D6.2.2, and an example is presented next.

Example 19.4 Determine the yield moment M_y for the composite girder cross section in Figure 19.22 subjected to factored positive moments $M_{D1} = 900$ kip-ft and $M_{D2} = 300$ kip-ft. Use $f_c' = 4$ ksi for the concrete deck slab and Grade 50 structural steel for the girder.

Properties

The noncomposite, short-term, and long-term section properties are calculated in Tables 19.4–19.6. The modular ratio of $n = 8$ is taken from Table 19.3 for $f_c' = 4$ ksi. The transformed effective width of the slab is b_e divided by n for short-term properties and by $3n$, to account for creep, for long-term properties. The centroid of the section at each stage is calculated from the top of the steel beam, and then the parallel axis

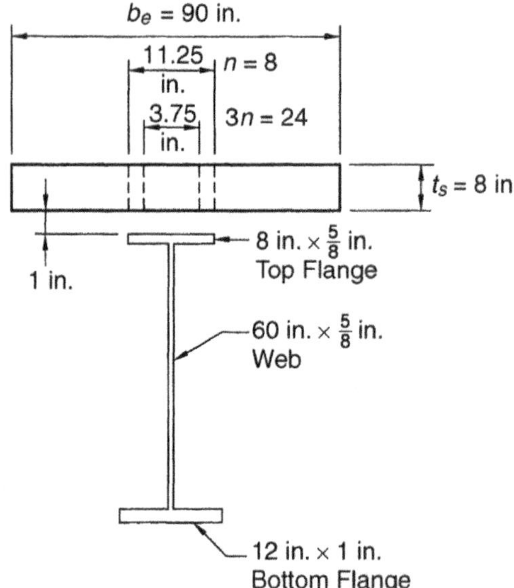

Fig. 19.22 Example 19.4. Composite positive moment section.

Table 19.4 Noncomposite Section Properties

Component (in.)	A (in.^{2})	y (in.)	Ay (in.^{3})	$A(y-\bar{y})^2$ (in.^{4})	I_0 (in.^{4})
Top flange 0.625 × 8	5.0	61.313	306.6	5864	~0
Web 0.625 × 60	37.5	31.0	1162.5	581	11,250
Bottom flange 1 × 12	12.0	0.5	6.0	8469	1.00
Sum	54.5		1475.1	14,914	11,251

$I = 14{,}914 + 11{,}251$
$= 26{,}165\ \text{in.}^4$

Table 19.5 Short-Term Section Properties, n = 8

Component (in.)	A (in.^{2})	y (in.)	Ay (in.^{3})	$A(y-\bar{y})^2$ (in.^{4})	I_0 (in.^{4})
Slab [90/(n = 8) × 8]	90.0	66.625	5,996.25	20,036	480
Top flange 0.625 × 8	5.0	61.313	306.6	462	~0
Web 0.625 × 60	37.5	31.0	1,162.5	16,075	11,250
Bottom flange 1 × 12	12.0	0.50	6.0	31,463	1.00
Sum	144.5		7,471.6	68,036	11,731

$I = 68{,}036 + 11{,}731$
$= 79{,}767\ \text{in.}^4$

Table 19.6 Long-Term Section Properties, 3n = 24

Component (in.)	A (in.^{2})	y (in.)	Ay (in.^{3})	$A(y-\bar{y})^2$ (in.^{4})	I_0 (in.^{4})
Slab [90/(n = 24) × 8]	30.0	66.625	1,998.75	19,530	160
Top flange 0.625 × 8	5.0	62.313	306.6	2,041	~0
Web 0.625 × 60	37.5	31.0	1,162.50	3,833	11,250
Bottom flange 1 × 12	12.0	0.50	6.0	19,790	1.00
Sum	84.5		3,473.8	45,194	11,411

$I = 45{,}194 + 11{,}411$
$= 56{,}605\ \text{in.}^4$

theorem is used to get the moment of inertia of the components about this centroid:

$$\bar{y}_{NC} = \frac{1475.1}{54.5} = 27.1\ \text{in.} \quad \text{from bottom}$$

$$S^t_{NC} = \frac{26{,}165}{61.625 - 27.1} = 757\ \text{in.}^3 \quad \text{top}$$

$$S^b_{NC} = \frac{26{,}165}{27.1} = 967\ \text{in.}^3 \quad \text{bottom}$$

$$\bar{y}_{ST} = \frac{7471.6}{144.5} = 51.7\ \text{in.} \quad \text{bottom}$$

$$S^t_{ST} = \frac{79{,}767}{70.625 - 51.7} = 4216\ \text{in.}^3 \quad \text{top of deck}$$

$$S^t_{ST} = \frac{79{,}767}{61.625 - 51.7} = 8041\ \text{in.}^3 \quad \text{top of steel}$$

$$\bar{y}_{LT} = \frac{3473.8}{84.5} = 41.1\ \text{in.} \quad \text{from bottom}$$

$$S^t_{LT} = \frac{56{,}605}{70.625 - 41.1} = 1918\ \text{in.}^3 \quad \text{top of deck}$$

$$S^t_{LT} = \frac{56{,}605}{61.625 - 41.1} = 2758\ \text{in.}^3 \quad \text{top of steel}$$

$$S^b_{LT} = \frac{56{,}605}{41.1} = 1377\ \text{in.}^3 \quad \text{to bottom}$$

Solution

The stress at the bottom of the girder reaches yield first. From Eq. 19.24,

$$F_y = \frac{M_{D1}}{S_{NC}} + \frac{M_{D2}}{S_{LT}} + \frac{M_{AD}}{S_{ST}}$$

$$50 = \frac{900\,(12)}{967} + \frac{300\,(12)}{1377} + \frac{M_{AD}}{1543}$$

$$M_{AD} = 1543\,(50 - 11.2 - 2.6) = 55{,}883\ \text{in. kips}$$

$$M_{AD} = 4657\ \text{ft kips}$$

Answer From Eqs. 19.24 and 19.25, the yield moment is

$$M_y = M_{D1} + M_{D2} + M_{AD}$$
$$M_y = 900 + 300 + 4657 = 5857 \text{ ft kips}$$

Yield Moment of a Noncomposite Section For a noncomposite section, the section moduli in Eq. 19.24 are all equal to S_{NC} and the yield moment M_y is

$$M_y = F_y S_{NC} \quad (19.26)$$

Plastic Neutral Axis of a Composite Section The first step in determining the plastic moment strength of a composite section is to locate the neutral axis of the plastic forces. The plastic forces in the steel portions of the cross section is the product of the area of the flanges, web, and reinforcement times their appropriate yield strengths. The plastic forces in the concrete portions of the cross section, which are in compression, are based on the equivalent rectangular stress block with uniform stress of 0.85 f_c'. Concrete in tension is neglected.

The location of the plastic neutral axis (PNA) is obtained by equating the plastic forces in compression to the plastic forces in tension. If it is not obvious, it may be necessary to assume a location of the PNA and then to prove or disprove the assumption by summing plastic forces. If the assumed location does not satisfy equilibrium, then a revised expression is solved to determine the correct location of the PNA.

Example 19.5 Determine the location of the plastic neutral axis for the composite cross section of Example 19.4 subjected to positive moment bending. Use $f_c' = 4$ ksi for the concrete and $F_y = 50$ ksi for the steel. Neglect the plastic forces in the longitudinal reinforcement of the deck slab.

Plastic Forces

The general dimensions and plastic forces are shown in Figure 19.23.

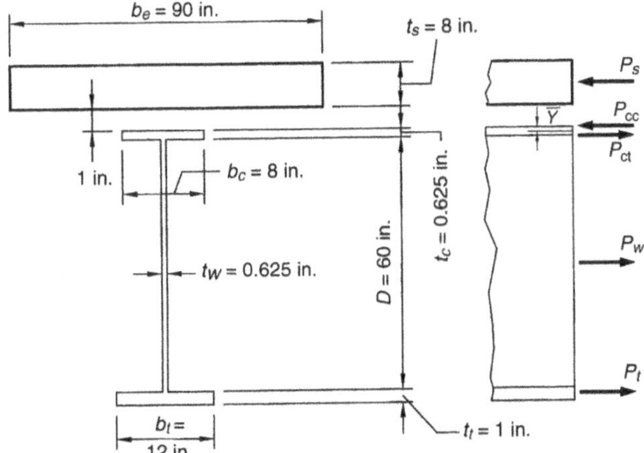

Fig. 19.23 Example 19.5. Plastic forces for composite positive moment section.

☐ Slab

$$P_s = 0.85 f_c' b_e t_s = 0.85\,(4)\,(90)\,(8) = 2448 \text{ kips}$$

☐ Top flange

$$P_c = F_y b_c t_c = 50\,(8)\,(0.625) = 250 \text{ kips}$$

☐ Web

$$P_w = F_y D t_w = 50\,(60)\,(0.625) = 1875 \text{ kips}$$

☐ Bottom flange

$$P_t = F_y b_t t_t = 50\,(12)\,(1) = 600 \text{ kips}$$

Solution

The PNA lies in the top flange because

$$P_s + P_c > P_w + P_t$$
$$2448 + 250 > 1875 + 600$$
$$2698 > 2475$$

Only a portion of the top flange is required to balance the plastic forces in the steel beam. Balancing compression and tensile forces yields

$$2448 + 50\,(8)\,(\bar{Y}) = 2475 + 400\,(0.625 - \bar{Y})$$

so that the PNA is located a distance $\bar{Y}$ from the top of the top flange:

$$\bar{Y} = 0.346 \text{ in.}$$

Answer By substituting the values from above, the tension and compression force equal about 2586 kips and the plastic neutral axis is

$$\text{PNA depth} = 8 + 1 + 0.346 = 9.346 \text{ in. from top of deck}$$

In a region of negative bending moment where shear connectors develop composite action, the reinforcement in a concrete deck slab can be considered effective in resisting bending moments. In contrast to the positive moment region where their lever arms are small and their contribution is dominated by the concrete deck, the contribution of the reinforcement in the negative moment region can make a difference.

Example 19.6 Determine the location of the plastic neutral axis for the composite cross section of Figure 19.24 when subjected to negative bending moment. Use $f_c' = 4$ ksi and $F_y = 50$ ksi. Consider the plastic forces in the longitudinal reinforcement of the deck slab to be provided by two layers with nine No. 4 bars ($A_s = 0.20$ in.^{2}/bar) in the top layer and seven No. 5 bars ($A_s = 0.31$ in.^{2}/bar) in the bottom layer. Use $f_y = 60$ ksi.

Plastic Forces

The general dimensions and plastic forces are shown in Figure 19.24. The concrete slab is in tension and is considered to be noneffective, that is, $P_s = 0$.

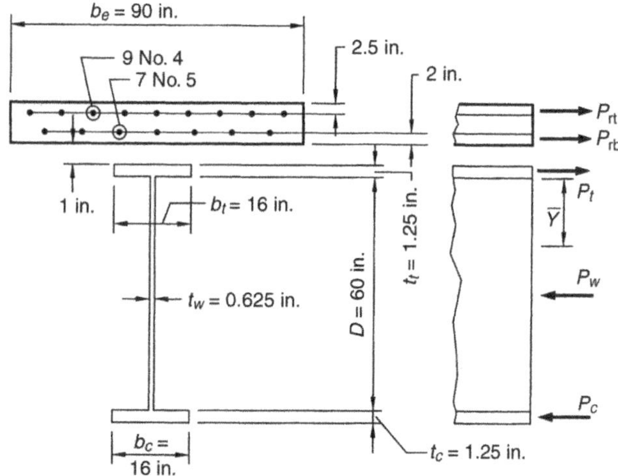

Fig. 19.24 Example 19.6. Plastic forces for composite negative moment section.

☐ Top reinforcement

$$P_{rt} = A_{rt} f_y = 9\,(0.20)\,(60) = 108 \text{ kips}$$

☐ Bottom reinforcement

$$P_{rb} = A_{rb} f_y = 7\,(0.31)\,(60) = 130 \text{ kips}$$

☐ Top flange

$$P_t = F_y b_t t_t = 50\,(16)\,(1.25) = 1000 \text{ kips}$$

☐ Web

$$P_w = F_y D t_w = 50\,(60)\,(0.625) = 1875 \text{ kips}$$

☐ Bottom flange

$$P_c = F_y b_c t_c = 50\,(16)\,(1.25) = 1000 \text{ kips}$$

Solution
By inspection, the PNA lies in the web because

$$\begin{aligned} P_c + P_w &> P_t + P_{rb} + P_{rt} \\ 1000 + 1875 &> 1000 + 130 + 108 \\ 2875 &> 1238 \end{aligned}$$

The plastic force in the web must be divided into tension and compression plastic forces to obtain equilibrium, that is,

$$P_c + P_w \left(1 - \frac{\bar{Y}}{D}\right) = P_w \left(\frac{\bar{Y}}{D}\right) + P_t + P_{rb} + P_{rt}$$

where $\bar{Y}$ is the distance from the top of the web to the PNA. Solving for $\bar{Y}$, we get

$$\bar{Y} = \frac{D}{2}\left(\frac{P_c + P_w - P_t - P_{rb} - P_{rt}}{P_w}\right)$$

Answer By substituting the values from above

$$\begin{aligned} \bar{Y} &= \frac{60}{2}\frac{(1000 + 1875 - 1000 - 108 - 130)}{1875} \\ &= 26.2 \text{ in.} \quad \text{(of web in tension)} \end{aligned}$$

Plastic Neutral Axis of a Noncomposite Section For a noncomposite section, there is no contribution from the deck slab and the PNA is determined above with $P_{rb} = P_{rt} = 0$. If the steel beam section is symmetric with equal top and bottom flanges, then $P_c = P_t$ and $\bar{Y} = D/2$.

Plastic Moment of a Composite Section The plastic moment M_p is the sum of the moments of the plastic forces about the PNA. It can best be described by examples. Global and local buckling is assumed to be prevented so that plastic forces can be developed. Details on the plastic moment computations are located in AASHTO [Section 6, Appendix D6.1].

Example 19.7 Determine the positive plastic moment for the composite cross section of Example 19.5 shown in Figure 19.23. The plastic forces were calculated in Example 19.5 and $\bar{Y}$ was determined to be 9.346 in. from the top of the slab.

Moment Arms
The moment arms about the PNA for each of the plastic forces can be found from the dimensions given in Figure 19.23.

☐ Slab in compression

$$d_s = 9.346 - \tfrac{8}{2} = 5.346 \text{ in.}$$

☐ Top flange in compression

$$d_{\text{top flange comp}} = \frac{0.346}{2} = 0.173 \text{ in.}$$

☐ Top flange in tension

$$d_{\text{top flange tension}} = \frac{0.625 - 0.346}{2} = \frac{0.279}{2} = 0.140 \text{ in.}$$

☐ Web in tension

$$d_{\text{web in tension}} = 30 + 0.279 = 30.279 \text{ in.}$$

☐ Tension flange

$$d_{\text{bottom flange in tension}} = 0.279 + 60 + 0.5 = 60.779 \text{ in.}$$

Solution
The sum of the moments of the plastic forces about the PNA is the plastic moment:

$$M_p = \sum_{\text{elements}} \left| F_{\text{element}} \left(d_{\text{elements}}\right) \right| \qquad (19.27)$$

Answer By substituting the values from above:

Element	Force, kips	Lever arm, in.	Element Contribution, in. kips
Deck (slab)	0.85(4)(90)(8) = 2448 (c)	5.346	13,087
Top flange in compression	50(8)(0.346) = 138.4 (c)	0.173	23.9
Top flange in tension	50(8)(0.279) = 111.6 (t)	0.140	15.6
Web in tension	50(60)(0.625) = 1875 (t)	30.279	56,773
Bottom flange in tension	50(12)(1) = 600 (t)	60.779	36,467
Total	~0		100,367 = 8864 ft kips

(c) indicates compression, and (t) indicates tension.

Note that the yield moment for this section is 5867 ft kips and the plastic moment is 8864 ft kips. This illustrates the significant capacity of the composite section after the first yield. The ratio of these moments is 1.51, which is also termed a shape factor.

Example 19.8 Determine the negative plastic moment for the composite cross section of Example 19.6 shown in Figure 19.24. The plastic forces were calculated in Example 19.6; $\bar{Y}$ was determined to be 26.2 in. from the top of the web.

Moment Arms

The moment arms about the PNA for each of the plastic forces can be found from the dimensions given in Figure 19.24.

☐ Top reinforcement (in tension)

$$\begin{aligned} d_{\text{rt}} &= \text{web depth} + \text{top flange thickness} + \text{haunch} \\ &\quad + \text{deck thickness} - \text{top cover} - \text{one-half bar} \\ &= 26.2 + 1.25 + 1 + 8 - 2.5 - 0.25 = 33.7 \text{ in.} \end{aligned}$$

☐ Bottom reinforcement (in tension)

$$\begin{aligned} d_{\text{rb}} &= \text{web depth} + \text{top flange thickness} + \text{haunch} \\ &\quad + \text{bottom cover} + \text{one-half bar} \\ &= 26.2 + 1.25 + 1 + 2.0 + 0.313 = 30.76 \text{ in.} \end{aligned}$$

☐ Top flange (in tension)

$$\begin{aligned} d_t &= \text{PNA} + \tfrac{1}{2}\text{flange} \\ &= 26.2 + \tfrac{1.25}{2} = 26.83 \text{ in.} \end{aligned}$$

☐ Web (in tension)

$$d_{\text{wt}} = \frac{\text{PNA}}{2} = \frac{26.2}{2} = 13.1 \text{ in.}$$

☐ Web (in compression)

$$\begin{aligned} d_{\text{wc}} &= \tfrac{1}{2}(\text{web} - \text{PNA}) \\ &= \tfrac{1}{2}(60 - 26.2) = 16.9 \text{ in.} \end{aligned}$$

☐ Bottom flange (in compression)

$$\begin{aligned} d_c &= \text{web} - \text{PNA} + \tfrac{1}{2}\text{ flange} \\ &= 60 - 26.2 + \tfrac{1.25}{2} = 34.43 \text{ in.} \end{aligned}$$

Solution

The plastic moment is the sum of the moments of the plastic forces about the PNA:

$$M_p = \sum_{\text{elements}} \left| F_{\text{element}} \left(d_{\text{elements}} \right) \right| \tag{19.28}$$

Answer By substituting the values from above:

Element	Force, kips	Lever arm, in.	Element Contribution, in. kips
Top rebar in tension	60(9)(0.2) = 108 (t)	33.70	3640
Bottom rebar in tension	60(7)(0.31) = 130 (t)	30.76	3999
Top flange in tension	50(16)(1.25) = 1000 (t)	26.83	26,830
Web in tension	50(26.2)(0.625) = 818.8 (t)	13.10	10,726
Web in compression	50(60–26.2)(0.625) = 1056.3 (c)	16.90	17,846
Bottom flange in compression	50(16)(1.25) = 1000 (c)	34.42	34,420
Total	~0		97,461 = 8122 ft kips

(c) indicates compression, and (t) indicates tension.

$$M_p = 97{,}461 \text{ in. kips} = 8122 \text{ ft kips}$$

Plastic Moment of a Noncomposite Section If no shear connectors exist between the concrete deck and the steel cross section, the concrete slab and its reinforcement do not contribute to the section properties for the computation of resistance (stresses or forces). However, it should be considered in the modeling of the stiffness of the beam in the structural analysis of continuous structures.

Consider the cross section of Figure 19.24 to be noncomposite. Then $P_{\text{rt}} = P_{\text{rb}} = 0$ and $\bar{Y} = D/2$, and the plastic moment is

$$M_p = P_t\left(\frac{D}{2} + \frac{t_t}{2}\right) + P_w\left(\frac{D}{4}\right) + P_c\left(\frac{D}{2} + \frac{t_c}{2}\right)$$

Element	Force, kips	Lever arm, in.	Element Contribution, in. kips
Top flange in tension	50(16)(1.25) = 1000 (t)	30 + 1.25/2 = 30.625	30,625
Web in tension	50(30)(0.625) = 937.5 (t)	15	14,062.5
Web in compression	50(30)(0.625) = 937.5 (c)	15	14,062.5
Bottom flange in compression	50(16)(1.25) = 1000 (c)	30 + 1.25/2 = 30.625	30,625
Total	0		89,375 = 7448 ft kips

(c) indicates compression, and (t) indicates tension.

Depth of Web in Compression When evaluating the slenderness of a web as a measure of its stability, the depth of the web in compression is important. In a noncomposite cross section with a doubly symmetric steel beam, one-half of the web depth D is in compression. For unsymmetric noncomposite cross sections and composite cross sections, the depth of web in compression will not be $D/2$ and varies with the direction of bending in continuous girders.

When stresses due to unfactored loads remain in the elastic range, the depth of the web in compression D_c shall be the depth over which the algebraic sum of stresses due to the dead-load D_{c1} on the steel section plus the dead-load D_{c2} and live-load LL + IM on the short-term composite section are compressive [Section 6, Appendix D6.3.1].

Example 19.9 Determine the depth of web in compression D_c for the cross section of Figure 19.22 whose elastic properties were calculated in Example 19.4. The cross section is subjected to *factored* positive moments $M_{D1} = 900$ ft kips, $M_{D2} = 300$ ft kips, and $M_{\text{LL+IM}} = 1200$ ft kips.

Solution

The stress at the top of the steel for the given moments and section properties is (see Fig. 19.22)

$$f_t = \frac{M_{D1}}{S^t_{\text{NC}}} + \frac{M_{D2}}{S^t_{\text{LT}}} + \frac{M_{\text{LL+IM}}}{S^t_{\text{ST}}}$$

$$= \frac{900\,(12)}{736} + \frac{300\,(12)}{2758} + \frac{1200\,(12)}{8037}$$

$$= 14.7 + 1.3 + 1.8 = 17.8 \text{ ksi} \qquad \text{(compression)}$$

$$f_b = \frac{M_{D1}}{S^b_{\text{NC}}} + \frac{M_{D2}}{S^b_{\text{LT}}} + \frac{M_{\text{LL+IM}}}{S^b_{\text{ST}}}$$

$$= \frac{900\,(12)}{967} + \frac{300\,(12)}{1377} + \frac{1200\,(12)}{1543}$$

$$= 11.2 + 2.6 + 9.3 = 23.1 \text{ ksi} \qquad \text{(tension)}$$

Answer Using the proportion of the section in compression and subtracting the thickness of the compression flange with $d = 60 + 0.625 + 1.00 = 61.625$ in.

$$D_c = d\frac{f_t}{f_t + f_b} - t_c = 61.625\frac{17.8}{17.8 + 23.1} - 0.625$$

$$= 26.2 \text{ in.}$$

The depth of web in compression at plastic moment D_{cp} is usually determined once the PNA is located. In Example 19.5, positive bending moment is applied and the PNA is located in the top flange. The entire web is in tension and $D_{\text{cp}} = 0$.

In Example 19.6, the cross section is subjected to negative bending moment and the PNA is located 26.2 in. from the top of the web. The bottom portion of the web is in compression, so that

$$D_{\text{cp}} = D - \bar{Y} = 60 - 26.2 = 33.8 \text{ in.}$$

AASHTO Eq. D6.3.1-1 takes the same approach.

Hybrid Strength Reduction—Behavior A hybrid section has different strength steel in the flanges and/or the web. Typically, higher strength materials are used for the tension flanges where buckling is of no concern and lower strength materials are used for the web and compression flange. Also, note that the compression flange is functionally replaced by the composite deck after the concrete hardens. Therefore, a common hybrid section contains Grade 70 steel in the tension flange and Grade 50 for the web and compression flanges.

Consider the location where the Grade 50 steel in the web is welded to the Grade 70 steel of a tension flange; here the flexural strain is the same and the modulus of elasticity of the two materials is the same as well. Therefore prior to yield, the stress at this location is the same for the web and flange. Under increased load, the web yields prior to the flange and exhibits a constant (yield) stress with depth. This behavior is simply quantified by performing a strain compatibility analysis at ultimate considering the web yielding (rationale analysis). However, the AASHTO specifications simplifies this analysis by requiring that the strength of the hybrid section be computed, initially, by neglecting this effect, that is, using the elastic section properties, S, and the flange yield stress.

The solid line in Figure 19.25 represents the "true" stress with the materials in a portion of the web yielded and the dotted line illustrates the assumed stress neglecting the yielding.

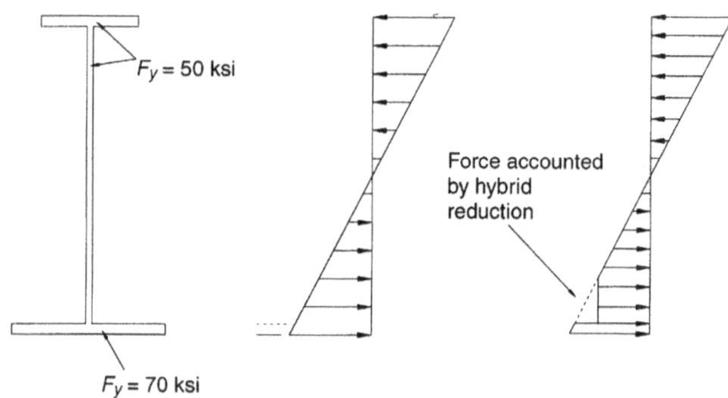

Fig. 19.25 Stress profile after web yield in a hybrid section.

This computation, without some adjustment, overestimates the section resistance; therefore, a factor is used to scale the resistance downward. This factor is referred to as the hybrid reduction factor.

Hybrid Strength Reduction—Specifications The hybrid reduction is only applicable for nonhomogeneous sections. The reduction factor is [A6.10.1.10.1]

$$R_h = \frac{12 + \beta\left(3\rho - \rho^3\right)}{12 + 2\beta} \tag{19.29}$$

where

$$\beta = \frac{2D_w t_w}{A_{\text{fn}}} \tag{19.30}$$

where ρ is the smaller of F_{yw}/f_n and 1.0; A_{fn} is the sum of the flange area and the area of any cover plates on the side of the neutral axis corresponding to D_n. For composite section in negative flexure, the area of the longitudinal reinforcement may be included in calculating A_{fn} for the top flange; D_n is the larger of the distance from the elastic neutral axis of the cross section to the inside of the face of the flange where yielding occurs (in.). For sections where the neutral axis is at the middepth of the web, D_n is the distance from the neutral axis to the inside of the neutral axis where yielding occurs first. For sections where yielding occurs first in the flange, a cover plate or longitudinal reinforcement on the side of the neutral axis corresponding to D_n,f_n is the largest of the specified minimum yield strengths of each component included in the calculation of A_{fn} (ksi). Otherwise, f_n is the largest of the elastic stresses in the flange, cover plate, or longitudinal reinforcement on the side of the neutral axis corresponding to D_n at first yield on the opposite side of the neutral axis.

This "simplification" makes a direct rationale method of assuming a linear strain profile and equating compressive and tensile forces to obtain neutral axis, and finally, summing moments to obtain the section flexural resistance appear not only logical but straightforward as well. Simply put, Figure 19.25 may be used in the usual manner similar to computation of plastic moment capacity.

19.3.3 Stability Related to Flexural Resistance

For the development of the plastic moment resistance M_p adequate stability must be provided. If global or local buckling occurs, M_p cannot be reached.

Global buckling can occur if the compression flange of a section in flexure is not laterally supported. A laterally unsupported compression flange behaves similar to a column and tends to buckle out-of-plane between points of lateral support. However, because the compression flange is part of a beam cross section with a tension zone that keeps the opposite flange in line, the cross section twists when it moves laterally. This behavior is shown in Figure 19.26 and is referred to as *lateral-torsional buckling (LTB)*.

Local buckling can occur if the width–thickness ratio (slenderness) of elements in compression becomes too

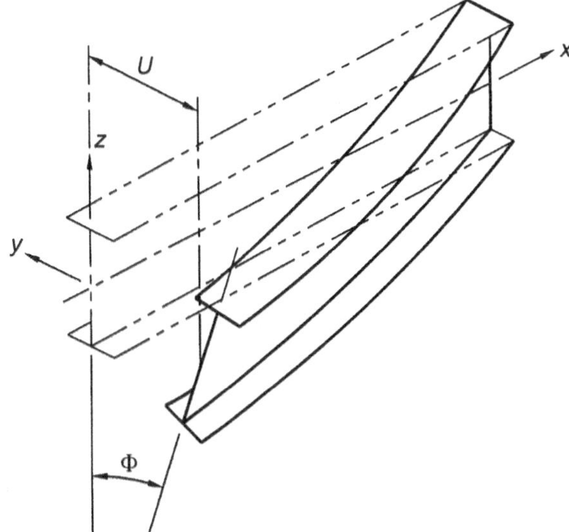

Fig. 19.26 Isometric of lateral torsional buckling (Nethercot, 1992). [From *Constructional Steel Design: An International Guide*, P. J. Dowling, J. E. Harding, and R. Bjorhovde, eds., Copyright © 1992 by Elsevier Science Ltd (now Chapman and Hall, Andover, England), with permission.]

large. Limitations on these ratios are similar to those given for columns in Figure 19.15. If the buckling occurs in the compression flange, it is called *flange local buckling (FLB)*. If it occurs in the compression portion of the web, it is called *web local buckling (WLB)*. Illustrations of local buckling are shown in the photographs of Figure 19.27 of a full-scale test to failure of a roof beam. Flange local buckling can be seen in the top flange of the overall view (Fig. 19.27). A closeup of the compression region of the beam (Fig. 19.28) shows the buckled flange and measurement of the out-of-plane web deformation indicating web local buckling has also occurred.

Classification of Sections and Elements within Cross Sections (Flanges and Web) Cross-sectional shapes are classified as *compact, noncompact*, or *slender* depending on the width–thickness ratios of their compression elements and bracing requirements. A *compact section* is one that can develop a fully plastic moment M_p before lateral torsional buckling or local buckling of its flange or web occurs. A *noncompact section* is one that can develop a moment equal to or greater than M_y, but less than M_p, before local buckling of any of its compression elements occurs. A *slender section* is one whose compression elements are so slender that buckling occurs locally before the moment reaches M_y. A comparison of the moment–curvature response of these shapes in Figure 19.20 illustrates the differences in their behavior.

As discussed previously, another classification is hybrid where a girder is comprised of two steel strengths. This is also termed nonhomogeneous. Hybrid girders pose special

Fig. 19.27 Local buckling of flange. (Courtesy of Structures/Materials Laboratory, Virginia Tech.)

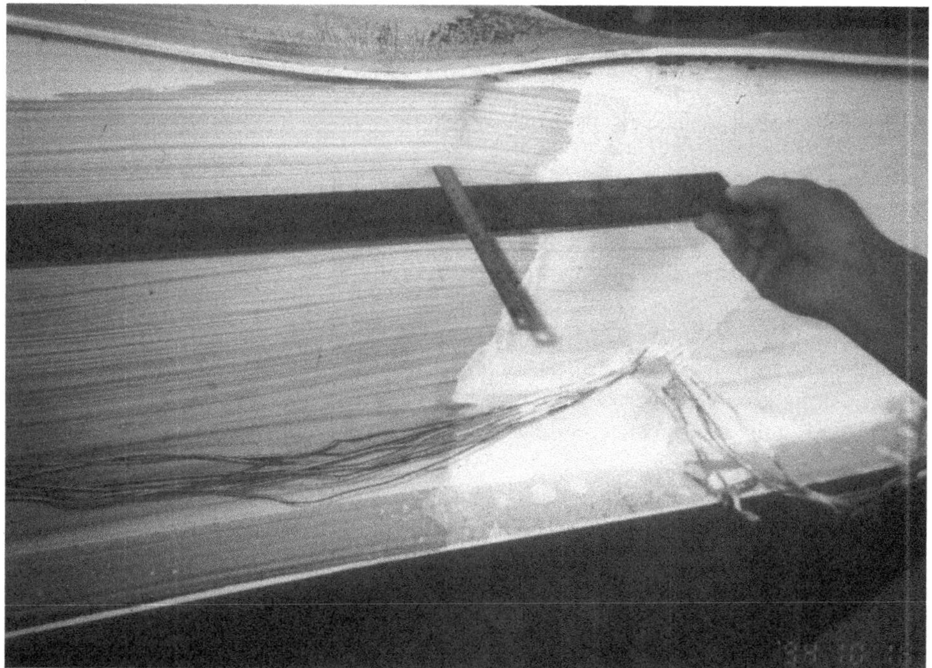

Fig. 19.28 Local buckling of web. (Courtesy of Structures/Materials Laboratory, Virginia Tech.)

problems as two materials of different strengths are located next to the other within the cross section; here one yields at a different curvature than the other. This is explained in more detail next.

General Stability Treatment in the AASHTO Specification The stability of I-sections related to flexural and shear behavior and performance is related to local and global buckling, yielding, and the relationship between the two modes of failure. Figure 19.29 illustrates modeling of nearly all compression behavior in the AASHTO LRFD Specification [A6.10.8.2]. A slenderness ratio of a flange, web, or beam is expressed in terms of a dimensionless ratio that is a characteristic width divided by length. For example,

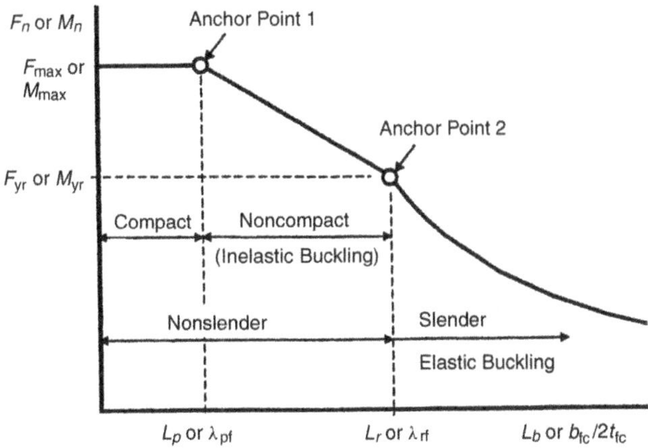

Fig. 19.29 Typical behavior for slenderness effects. [Figure C6.10.8.2.1-1]. (From *AASHTO LRFD Bridge Design Specifications*, Copyright © 2010 by the American Association of State Highway and Transportation Officials, Washington, DC. Used by permission.)

the familiar (kl/r) for columns is one ratio. Another is the width-to-thickness ratio $(b_f/2t_f)$ and the web slenderness ratio (D/t_w) is yet another. Depending upon the slenderness, the component may yield if it is "stocky" or buckle if it is very slender. Sections between experience inelastic behavior or some of both. As discussed in the section on columns, the specification-based behavioral models include a combination of theoretical formulation with some empirical results included to address issues such as out-of-plumbness and residual stresses.

Referring to Figure 19.29, in the region where the slenderness ratio λ is less than λ_p (anchor point 1), the component is assumed to be able to support the yield stress and is considered compact. In the region where the slenderness ratio λ is greater than λ_r (anchor point 2), the componenet is considered slender and elastic buckling controls strength. In the region between these two regions, that is, $\lambda_p < \lambda < \lambda_r$, the component behaves inelastically and the behavior is modeled with a linear interpolation between the two anchor points. The AASHTO LRFD specifications [C6.10.8.2.1] follow this model for most compression behavior in steel.

Local Buckling In addition to resisting shear forces, the web has the function of supporting the flanges far enough apart so that bending is resisted effectively. When an I-section is subjected to bending, two failure mechanisms or limit states can occur in the web. The web can buckle as a vertical column that carries the compressive force that keeps the flanges apart or the web can buckle as a plate due to horizontal in-plane bending stresses. Both of these failure mechanisms require limitations on the slenderness of the web. Shear failure mechanisms are considered separately and are discussed later. Similarly, the web can buckle due to bending stresses as well as the flange. All behavior is similar yet requires separate discussion because of unique characteristics associated with the boundary conditions and/or the stress fields present for each element.

Web Vertical Buckling—Behavior When bending occurs in an I-section, curvature produces compressive stresses between the flanges and the web of the cross section. These compressive stresses are a result of the vertical component of the flange force as shown schematically for a doubly symmetric I-section in Figure 19.30. To develop the yield moment of the cross section requires that the compression flange reach its yield stress F_{yc} before the web buckles. If the web is too slender, it buckles as a column, which causes the compression flange to lose its lateral support, and it buckles vertically into the web before the yield moment is reached.

Vertical buckling of the flange into the web can be shown by considering the elemental length of web dx along the axis of the beam in Figure 19.31. It is subjected to an axial compressive stress f_{wc} from the vertical component of the compression flange force P_c. From Figure 19.30, the vertical component is $P_c\, d\phi$, which for a doubly symmetrical I-section

$$d\phi = \frac{2\varepsilon_{fc}}{D} dx \tag{19.31}$$

where ε_{fc} is the strain in the compression flange and D is the web depth. The axial compressive stress in the web then becomes

$$f_{wc} = \frac{P_c\, d\phi}{t_w\, dx} = \frac{2A_{fc} f_c \varepsilon_{fc}}{Dt_w} \tag{19.32}$$

where A_{fc} is the area of the compression flange and f_c is the stress in the compression flange. Equation 19.32 can be written in terms of the cross-sectional area of the web $A_w = Dt_w$ as

$$f_{wc} = \frac{2A_{fc}}{A_w} f_c \varepsilon_{fc} \tag{19.33}$$

Thus, the vertical compressive stress in the web is proportional to the ratio of flange area to web area in the cross

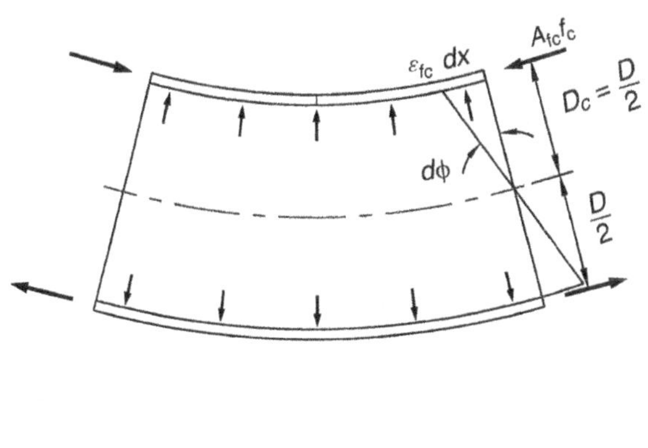

Fig. 19.30 Web compression due to curvature. (After Basler and Thürlimann, 1961.)

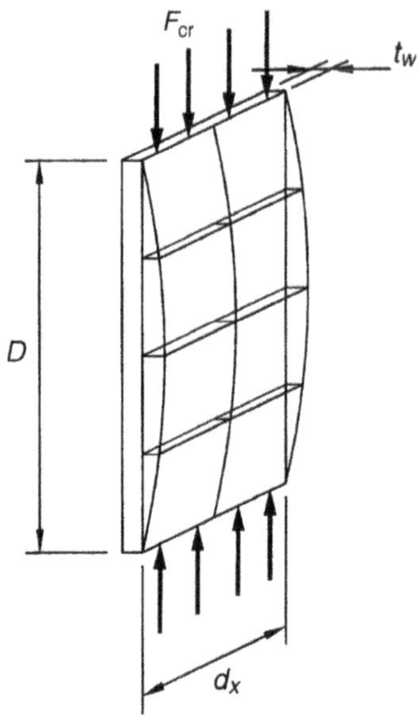

Fig. 19.31 Vertical buckling of the web.

section, the compressive stress in the flange, and the compressive strain in the flange. The strain ε_{fc} is not simply f_c/E, but must also include the effect of residual stress f_r in the flange (Fig. 19.9), that is,

$$\varepsilon_{fc} = \frac{f_c + f_r}{E}$$

so that Eq. 19.33 becomes

$$f_{wc} = \frac{2A_{fc}}{EA_w} f_c \left(f_c + f_r\right) \quad (19.34)$$

and a relationship between the compressive stress in the web and the compressive stress in the flange is determined.

By assuming the element in Figure 19.31 is from a long plate that is simply supported along the top and bottom edges, the critical elastic buckling or Euler load is

$$P_{cr} = \frac{\pi^2 EI}{D^2} \quad (19.35)$$

for which the moment of inertia I for the element plate length dx is

$$I = \frac{t_w^2 dx}{12\left(1 - \mu^2\right)} \quad (19.36)$$

where Poisson's ratio μ takes into account the stiffening effect of the two-dimensional (2D) action of the web plate. The critical buckling stress F_{cr} is obtained by dividing Eq. 19.35 by the elemental area $t_w\, dx$ to yield

$$F_{cr} = \frac{\pi^2 E t_w^3 dx}{12\left(1 - \mu^2\right) D^2 t_w\, dx} = \frac{\pi^2 E}{12\left(1 - \mu^2\right)} \left(\frac{t_w}{D}\right)^2 \quad (19.37)$$

To prevent vertical buckling of the web, the stress in the web must be less than the critical buckling stress, that is,

$$f_{wc} < F_{cr} \quad (19.38)$$

Substitution of Eqs. 19.34 and 19.37 into Eq. 19.38 gives

$$\frac{2A_{fc}}{EA_w} f_c \left(f_c + f_r\right) < \frac{\pi^2 E}{12\left(1 - \mu^2\right)} \left(\frac{t_w}{D}\right)^2$$

Solving for the web slenderness ratio D/t_w results in

$$\left(\frac{D}{t_w}\right)^2 < \frac{A_w}{A_{fc}} \frac{\pi^2 E^2}{24\left(1 - \mu^2\right)} \frac{1}{f_c\left(f_c + f_r\right)} \quad (19.39)$$

To develop the yield moment M_y in the symmetric I-section, it is required that the compressive stress in the flange f_c reach the yield stress f_{yc} before the web buckles vertically. Assume a minimum value of 0.5 for A_w/A_{fc} and a maximum value of $0.5F_{yc}$ for f_r; then a minimum upper limit on the web slenderness ratio can be estimated from Eq. 19.39:

$$\frac{D}{t_w} < \sqrt{\frac{0.5\pi^2 E^2}{24\left(1 - 0.3^2\right) F_{yc}^2 (1.5)}} = 0.388 \frac{E}{F_{yc}} \quad (19.40)$$

where Poisson's ratio for steel has been taken as 0.3. Equation 19.40 is not rigorous in its derivation because of the assumptions about A_w/A_{fc} and f_r, but it can be useful as an approximate measure of web slenderness to avoid vertical buckling of the flange into the web. For example, if $E = 29{,}000$ ksi and $F_{yc} = 36$ ksi, then Eq. 19.40 requires that D/t_w be less than 310. For $F_{yc} = 50$ ksi, D/t_w should be less than 225, and for $F_{yc} = 70$ ksi, D/t_w should be less than 160. Often, 70-ksi steels are used for tension flanges. Additional failure modes and handling requirements restrict the web slenderness to be less than 150 [A6.10.2.1]; therefore vertical web buckling should not be critical. This value is used in the Specifications.

Web Vertical Buckling—Specifications Vertical web buckling is not addressed directly in the AASHTO LRFD Specifications. The general limits on web slenderness (transversely stiffened sections) of

$$\frac{D}{t_w} \le 150 \quad (19.41)$$

And for longitudinally stiffened sections

$$\frac{D}{t_w} \le 300 \quad (19.42)$$

address this for F_{cy} less than 85 ksi. Research has indicated that the effect is small on the overall strength [see C10.2.1.1].

Web Bend Buckling—Behavior Because bending produces compressive stresses over a part of the web, buckling out of the plane of the web can occur as shown in Figure 19.32. The elastic critical buckling stress is given by a generalization of Eq. 19.37, that is,

$$F_{cr} = \frac{k\pi^2 E}{12\left(1 - \mu^2\right)} \left(\frac{t_w}{D}\right)^2 \quad (19.43)$$

where k is the buckling coefficient that depends on the boundary conditions of the four edges, the aspect ratio

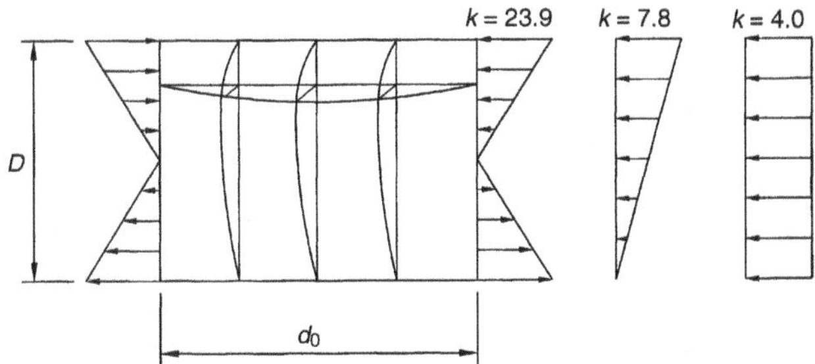

Fig. 19.32 Bending buckling of the web.

(Eq. 19.35) of the plate, and the distribution of the in-plane stresses. For all four edges simply supported and an aspect ratio greater than 1, Timoshenko and Gere (1969) give the values of k for the different stress distributions shown in Figure 19.32.

Using a Poisson ratio of 0.3, Eq. 19.43 becomes

$$F_{cr} = \frac{0.9kE}{(D/t_w)^2} \qquad (19.44)$$

where

$$k = \frac{\pi^2}{(D_c/D)^2} \approx \frac{9}{(D_c/D)^2}$$

The solution for the web slenderness ratio yields in Eq. 19.43

$$\left(\frac{D}{t_w}\right)^2 = \frac{k\pi^2}{12(1-\mu^2)}\frac{E}{F_{cr}}$$

For the I-section to reach the yield moment before the web buckles, the critical buckling stress F_{cr} must be greater than F_{yc}. Therefore, setting $\mu = 0.3$, the web slenderness requirement for developing the yield moment becomes

$$\frac{D}{t_w} \leq \sqrt{\frac{k\,(0.904)\,E}{F_{yc}}} = 0.95\sqrt{k}\sqrt{\frac{E}{F_{yc}}} \qquad (19.45)$$

For the pure bending case of Figure 19.32, $k = 23.9$.

$$\frac{D}{t_w} \leq 0.95\sqrt{23.9}\sqrt{\frac{E}{F_{yc}}} = 4.64\sqrt{\frac{E}{F_{yc}}} \qquad (19.46)$$

Comparisons with experimental tests indicate that Eq. 19.46 is too conservative because it neglects the postbuckling strength of the web.

Web Bend Buckling—Specifications The AASHTO (2010) LRFD Bridge Specifications give slightly different expressions for defining the web slenderness ratio that separates elastic and inelastic buckling. To generalize the left side of Eq. 19.45 for unsymmetric I-sections, the depth of the web in compression D_c, defined in Figure 19.22 and calculated in Example 19.9, replaces $D/2$ for the symmetric case to yield

$$\frac{D}{t_w} = \frac{2D_c}{t_w} \qquad (19.47)$$

The right side of Eq. 19.45 for unsymmetric I-sections may be modified for the case of a stress in the compression flange f_c less than the yield stress F_{yc}. Further, to approximate the postbuckling strength and the effect of longitudinal stiffeners, the value for k is effectively taken as 50 and 150 for webs without and with longitudinal stiffeners, respectively. The AASHTO expressions are [Table A6.10.5.3.1-1 (2010)]:

- ☐ Without longitudinal stiffeners

$$\frac{2D_c}{t_w} \leq 6.77\sqrt{\frac{E}{f_c}} \qquad (19.48\text{-}2004)$$

$$\frac{2D}{t_w} \leq 150 \qquad (19.48\text{-}2010)$$

- ☐ With longitudinal stiffeners

$$\frac{2D_c}{t_w} \leq 11.63\sqrt{\frac{E}{f_c}} \qquad (19.49\text{-}2004)$$

$$\frac{2D}{t_w} \leq 300 \qquad (19.49\text{-}2010)$$

The typical compression flange is Grade 50 and the 2010 AASHTO LRFD specifications simplify the f_c term (a load effect) to the max f_c, which is F_{cy} (material property). This obviates the need for strength to be a function of the load effects and significantly simplies computations. This is one of many simplifications achived with the 2005 Interims.

Web Buckling Load Shedding—Behavior When an I-section is noncompact, the nominal flexural resistance based on the nominal flexural stress F_n is given by [A6.10.7.2.2]

$$F_n = R_b R_h F_{yf} \qquad (19.50)$$

where R_b is the load shedding factor, R_h is the hybrid factor, and F_{yf} is the yield strength of the flange. When the flange and web have the same yield strength, $R_h = 1.0$. A hybrid girder has a lower strength material in the web than the flange.

Load shedding occurs when the web buckles prior to yielding of the compression flange. Part of the web transfers its load to the flange, which creates an apparent decrease in strength as the computation of the section properties does not include this effect. So, the elastic properties are used with the yield strength, but the computation is modified to decrease the strength by a factor (load shedding) to account for this web buckling.

The load shedding factor R_b provides a transition for inelastic sections with web slenderness properties between λ_p and λ_r (Fig. 19.33). From analytical and experimental studies conducted by Basler and Thürlimann (1961), the transition was given by

$$\frac{M_u}{M_y} = 1 - C(\lambda - \lambda_0) \qquad (19.51)$$

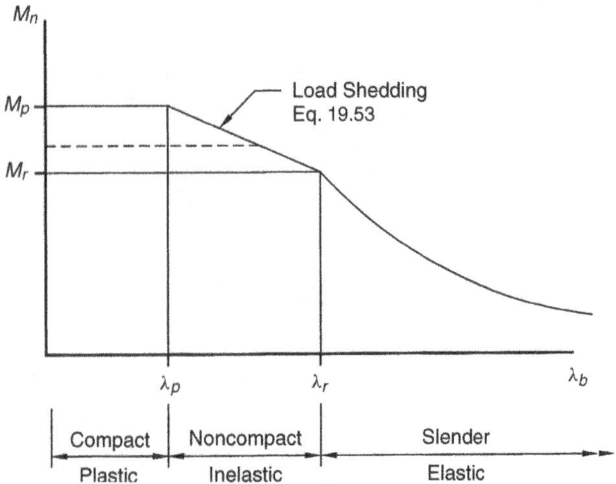

Fig. 19.33 Flexural resistance of I-sections versus slenderness ratio.

in which C is the slope of the line between λ_p and λ_r, and λ_0 is the value of λ when $M_u/M_y = 1$. The constant C was expressed as

$$C = \frac{A_w/A_f}{1200 + 300A_w/A_f} \tag{19.52}$$

Web Buckling Load Shedding—Specifications The AASHTO (2010) LRFD Bridge Specifications use the same form as Eqs. 19.51 and 19.52 for R_b [A6.10.1.10.2], that is,

$$R_b = 1 - \left(\frac{a_{\text{wc}}}{1200 + 300a_{\text{wc}}}\right)\left(\frac{2D_c}{t_w} - \lambda_{\text{rw}}\right) \leq 1.0 \tag{19.53}$$

in which

$$a_{\text{wc}} = \frac{2D_c t_w}{t_{\text{fc}} b_{\text{fc}}}$$

and

$$\lambda_{\text{rw}} = 5.7\sqrt{\frac{E}{F_{\text{cy}}}}$$

If the web meets

$$\frac{D}{t_w} \leq \lambda_{\text{rw}}$$

then load shedding of the web is assumed to not occur and

$$R_b = 1.0$$

Additional requirements are outlined in AASHTO [A6.10.1.10.2] for web stiffened with longitudinal stiffeners. Due to space, these are not discussed here.

Compression Flange Local Buckling—Behavior Because of the postbuckling strength due to increased strain capacity of the web, an I-section does not fail in flexure when the web-buckling load is reached. However, it fails in flexure when one of the framing members on the edges of a web panel fails. If one of the flanges or transverse stiffeners should fail, then the web displacements are unrestrained, the web could no longer resist its portion of the bending moment, and the I-section then fails.

In a doubly symmetric I-section subjected to bending, the compression flange fails first in local or global buckling. Therefore, the bracing and proportioning of the compression flange are important in determining the flexural resistance of I-sections. To evaluate the buckling strength of the compression flange, it is considered as an isolated column.

Consider the connection between the web and the flange: One-half of the compression flange can be modeled as a long uniformly compressed plate (Fig. 19.34) with one longitudinal edge free and the other simply supported. Usually, the plate is long compared to its width, and the boundary conditions on the loaded edges are not significant and the buckling coefficient is $k = 0.425$ for uniform compression (Maquoi, 1992).

To develop the plastic moment M_p resistance in the I-section, the critical buckling stress F_{cr} must exceed the yield stress F_{yc} of the compression flange. In a similar manner to the development of Eq. 19.40, the limit for the compression flange slenderness becomes

$$\frac{b_f}{2t_f} \leq 0.95\sqrt{\frac{Ek}{F_{\text{yc}}}} \tag{19.54}$$

For an ideally perfect plate, $k = 0.425$ and the slenderness limit can be written as

$$\frac{b_f}{2t_f} \leq 0.62\beta\sqrt{\frac{E}{F_{\text{yc}}}} \tag{19.55}$$

where β is a factor that accounts for both geometrical imperfections and residual stresses in the compression flange (Maquoi, 1992).

If the compression flange is too slender, elastic local buckling occurs prior to yielding. To ensure that some inelastic behavior takes place in the flange, the AASHTO (2010) LRFD Bridge Specifications provide a method to compute

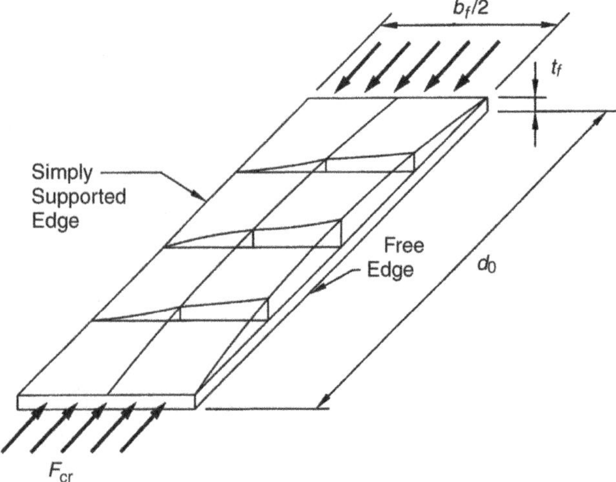

Fig. 19.34 Model of half a compression flange.

the moment resistance for noncompact and slender flanges [Section 6, Appendix A6.3.2]:

$$\frac{b_f}{2t_f} \le 1.38\sqrt{\frac{E}{f_c\sqrt{2D_c/t_w}}} \tag{19.56}$$

where f_c is the stress in the compression flange due to factored loading. Equation 19.56 is dependent on the web slenderness ratio $2D_c/t_w$ because it can vary between the values given by Eqs. 19.48 and 19.49 for noncompact sections.

As the web slenderness increases, the simply supported longitudinal edge in Figure 19.34 loses some of its vertical and transverse restraint. The effect of web slenderness on buckling of the compression flange can be shown by rewriting Eq. 19.56 as

$$\frac{b_f}{2t_f} \le C_f\sqrt{\frac{E}{f_c}} \tag{19.57}$$

in which

$$C_f = \frac{1.38}{\sqrt[4]{\dfrac{2D_c}{t_w}}} \tag{19.58}$$

where C_f is a compression flange slenderness factor that varies with $2D_c/t_w$ as shown in Figure 19.35. The value of C_f is comparable to the constant in Eq. 19.56 for compact sections. In fact, if $2D_c/t_w = 170$, they are the same. For values of $2D_c/t_w > 170$, the upper limit on $b_f/2t_f$ decreases until at $2D_c/t_w = 300$, and

$$\left(\frac{b_f}{2t_f}\right)_{300} = 0.332\sqrt{\frac{E}{f_c}} \tag{19.59}$$

Compression Flange Local Buckling Specifications The AASHTO (2010) LRFD Bridge Specifications take $\beta \approx 0.61$, and the compact section compression flange slenderness requirement becomes [A6.10.8.2.2]

$$\lambda_f = \frac{b_f}{2t_f} \le \lambda_{\text{pf}} = 0.38\sqrt{\frac{E}{F_{\text{yc}}}} \tag{19.60}$$

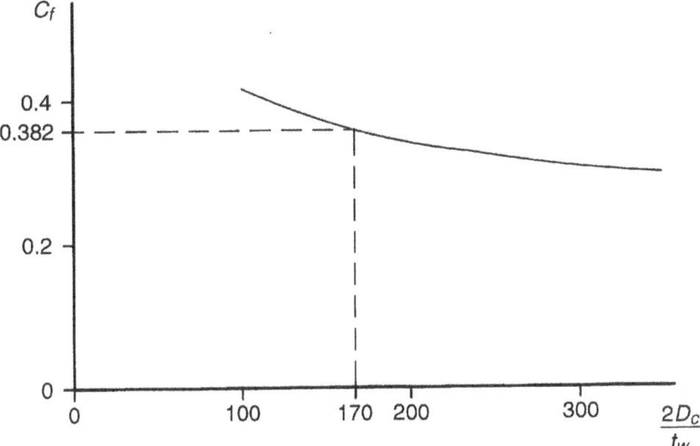

Fig. 19.35 Compression flange slenderness factor as a function of web slenderness.

If the steel I-section is composite with a concrete deck in a region of positive bending moment, the compression flange is fully supported throughout its length and the slenderness requirement does *not* apply.

Reference again to Figure 19.29 and the now familiar plot showing three types of behavior, the slenderness parameter λ for the compression flange is

$$\lambda_f = \frac{b_f}{2t_f} \tag{19.61}$$

and the values at the transition anchor points are

$$\lambda_{\text{pf}} = 0.38\sqrt{\frac{E}{F_{\text{yc}}}} \tag{19.62}$$

and

$$\lambda_{\text{rf}} = 0.56\sqrt{\frac{E}{F_{\text{yr}}}}$$

$$F_{\text{yr}} = \min\left(0.7F_{\text{yc}},\ F_{\text{yw}}\right) \ge 0.5F_{\text{yc}} \tag{19.63}$$

If $\lambda_f \le \lambda_{\text{pf}}$, then the compression flange is compact and

$$F_{\text{nc}} = R_b R_h F_{\text{yc}} \tag{19.64}$$

and the plastic moment resistance M_p is based on F_{yc} and plastic section properties. If $\lambda \ge \lambda_p$, then the compression flange is noncompact and the two anchor points are used to establish the compression flange strength:

$$F_{\text{nc}} = \left[1 - \left(1 - \frac{F_{\text{yr}}}{R_h F_{\text{yc}}}\right)\left(\frac{\lambda_f - \lambda_{\text{pf}}}{\lambda_{\text{rf}} - \lambda_{\text{pf}}}\right)\right] R_b R_h F_{\text{yc}} \tag{19.65}$$

Lateral Torsion Buckling Behavior Previous sections on web slenderness and compression flange slenderness were concerned with local buckling of the compression region in I-sections subjected to bending. The problem of global buckling of the compression region as a column between brace points must also be addressed. As described by the stability limit state and illustrated in Figure 19.26, an unbraced compression flange moves laterally and twists in a mode known as lateral torsional buckling (LTB).

If the compression flange is braced at sufficiently close intervals (less than L_p), the compression flange material can yield before it buckles and the plastic moment M_p may be reached if other compactness requirements are also met. If the distance between bracing points is greater than the inelastic buckling limit L_r, the compression flange buckles elastically at a reduced moment capacity. This behavior can once again be shown by the generic resistance–slenderness relationship of Figure 19.29 with the slenderness parameter given by

$$\lambda = \frac{L_b}{r_t} \tag{19.66}$$

where L_b is the distance between lateral brace points and r_t is the minimum radius of gyration of the compression flange

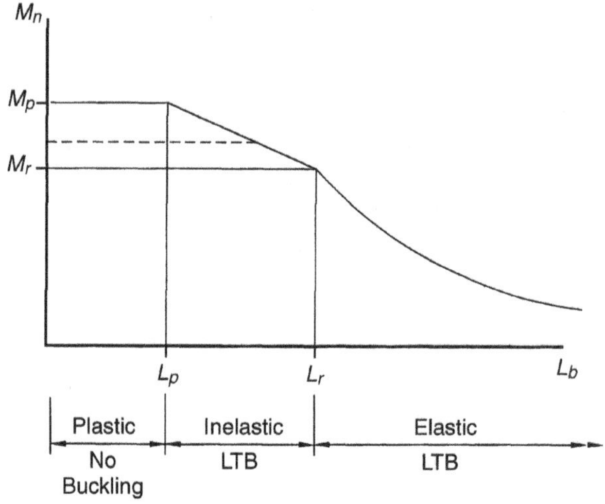

Fig. 19.36 Flexural resistance of I-sections versus unbraced length of compression flange.

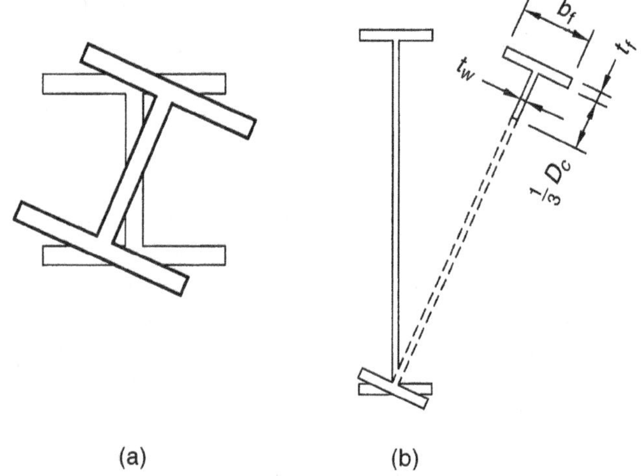

Fig. 19.37 (a) St. Venant torsion and (b) warping torsion in lateral buckling.

plus one-third of the web in compression taken about the vertical axis in the plane of the web.

Because the unbraced length L_b is the primary concern in the design of I-sections for flexure, it is taken as the independent parameter rather than the slenderness ratio L_b/r_t in determining the moment resistance. Figure 19.29 is, therefore, redrawn as Figure 19.36 with L_b replacing λ. The same three characteristic regions remain: plastic (no buckling), inelastic lateral-torsional buckling, and elastic lateral-torsional buckling.

For L_b less than L_p in Figure 19.36, the compression flange is considered laterally supported and the moment resistance M_n is constant. The value of M_n depends on the classification of the cross section. If the cross section is classified as *compact*, the value of M_n is M_p. If the cross section is *noncompact* or *slender*, then the value of M_n is less than M_p. The dashed horizontal line on Figure 19.36 indicates a typical value of M_n for a section that is not compact.

For $L_b > L_r$, the compression flange fails by elastic LTB. This failure mode has a classical stability solution (Timoshenko and Gere, 1969) in which the moment resistance is the square root of the sum of the squares of two contributions: torsional buckling (St. Venant torsion) and lateral buckling (warping torsion), that is,

$$M_n^2 = M_{n,v}^2 + M_{n,w}^2 \tag{19.67}$$

where $M_{n,v}$ is the St. Venant torsional resistance and $M_{n,w}$ is the warping contribution. For the case of constant bending between brace points, Gaylord et al. (1992) derive the following expressions:

$$M_{n,v}^2 = \frac{\pi^2}{L_b^2} EI_y GJ \tag{19.68}$$

$$M_{n,w}^2 = \frac{\pi^4}{L_b^4} EI_y EC_w \tag{19.69}$$

where I_y is the moment of inertia of the steel section about the vertical axis in the plane of the web, G is the shear modulus of elasticity, J is the St. Venant torsional stiffness constant, and C_w is the warping constant. When an I-section is short and stocky [Fig. 19.37(a)], pure torsional strength (St. Venant's torsion) dominates. When the section is tall and thin [Fig. 19.37(b)], warping torsional strength dominates.

Substitution of Eqs. 19.68 and 19.69 into Eq. 19.67 and along with approximations for I_y and C_w (see below) results in

$$\begin{aligned} M_n &= \frac{\pi E C_b}{L_b}\sqrt{(2I_{yc})(0.385)J + \frac{\pi^2}{L_b^2}(2I_{yc})\frac{d^2}{2}(I_{yc})} \\ &= \pi E C_b \frac{I_{yc}}{L_b}\sqrt{0.77\left(\frac{J}{I_{yc}}\right) + \pi^2\left(\frac{d}{L_b}\right)^2} \le M_y \end{aligned} \tag{19.70}$$

For L_b between L_p and L_r, the compression flange fails by inelastic LTB. Because of its complexity, the inelastic behavior is usually approximated from observations of experimental results. A straight-line estimate of the inelastic lateral-torsional buckling resistance is often used between the values at L_p and L_r.

The bending moment field within the unbraced region affects the LTB. A uniform bending moment is most critical and is the basis for the elastic buckling equation. A spacially variant moment field (presence of shear) is less critical and a region that changes between positive and negative moment (contains the point of contraflexure) is even stronger. The moment gradient factor C_b accounts for the nonuniform moment field. This factor is outlined in a later section.

Finally, a requirement of the elastic buckling equation is that the section proportions meet [A6.10.2.2]:

$$0.1 \le \frac{I_{yc}}{I_{yt}} \le 10 \tag{19.71}$$

where I_{yc} and I_{yt} are the moment of inertia of the compression and tension flanges of the steel section about the vertical axis in the plane of the web. If the member proportions are not within the limits given, the formulas for lateral-torsional buckling used in AASHTO (2010) LRFD Bridge Specifications are not valid.

If the unbraced length is greater than the L_r

$$L_b > L_r = \pi r_t \sqrt{\frac{E}{F_{yc}}} \tag{19.72}$$

then the cross section behaves elastically and has a nominal resisting moment (horizontal dashed line in Fig. 19.36) less than or equal to M_y [A6.10.8.2.3].

If the web is relatively stocky, or if a longitudinal stiffener is provided, bend buckling of the web cannot occur and both the pure torsion and warping torsion resistances in Eq. 19.70 can be included in calculating M_n. Some simplification to Eq. 19.70 occurs if it is assumed that the I-section is doubly symmetric and the moment of inertia of the steel section about the weak axis I_y, neglecting the contribution of the web, is

$$I_y \approx I_{yc} + I_{yt} = 2I_{yc} \tag{19.73}$$

Also, the shear modulus G can be written for Poisson's ratio $\mu = 0.3$ as

$$G = \frac{E}{2(1+\mu)} = \frac{E}{2(1+0.3)} = 0.385E \tag{19.74}$$

and the warping constant C_w for a webless I-section becomes (Kitipornchai and Trahair, 1980)

$$C_{tw} \approx I_{yc}\left(\frac{d}{2}\right)^2 + I_{yt}\left(\frac{d}{2}\right)^2 = \frac{d^2}{2} I_{yc} \tag{19.75}$$

where d is the depth of the steel section. Substitution of Eqs. 19.73–19.75 gives Eq. 19.70 and factoring out the common terms results in

$$\begin{aligned} M_n &= \frac{\pi E C_b}{L_b}\sqrt{(2I_{yc})(0.385)J + \frac{\pi^2}{L_b^2}(2I_{yc})\frac{d^2}{2}(I_{yc})} \\ &= \pi E C_b \frac{I_{yc}}{L_b}\sqrt{0.77\left(\frac{J}{I_{yc}}\right) + \pi^2\left(\frac{d}{L_b}\right)^2} \le M_y \end{aligned} \tag{19.76}$$

which is valid as long as

$$\frac{2D_c}{t_w} \le 5.7\sqrt{\frac{E}{F_{yc}}} \tag{19.77}$$

and

$$L_b > L_p = 1.0 r_t \sqrt{\frac{E}{F_{yc}}} \tag{19.78}$$

Even though Eq. 19.76 was derived for a doubly symmetric I-section ($I_{yc}/I_{yt} = 1.0$), it can be used for a singly symmetric I-section that satisfies Eq. 19.77. For I-sections composed of narrow rectangular elements, the St. Venant torsional stiffness constant J can be approximated by

$$J = \frac{Dt_w^3}{3} + \sum \frac{b_f t_f^3}{3} \tag{19.79}$$

In the development of Eq. 19.76, the hybrid factor R_h was taken as 1.0, that is, the material in the flanges and web have the same yield strength.

For I-sections with webs more slender than the limit of Eq. 19.77 or without longitudinal stiffeners, cross-sectional distortion is possible and the St. Venant torsional stiffness can be neglected [C6.10.8.2.3]. Setting $J = 0$ in Eq. 19.76, the elastic LTB moment for $L_b > L_r$ becomes

$$M_n = \pi^2 E C_b \frac{I_{yc} d}{L_b^2} \le M_y \tag{19.80a}$$

$$F_{cn} = \frac{C_b R_b \pi^2 E}{\left(\frac{L_b}{r_t}\right)^2} \tag{19.80b}$$

Reintroducing the load shedding factor R_b of Eq. 19.53 and defining L_r as the unbraced length at which $M_n = 0.5M_y$ (anchor point), then Eq. 19.80 becomes

$$M_n = C_b R_b (0.5M_y)(L_r/L_b)^2 \le R_b M_y \tag{19.81}$$

for which

$$M_y = F_{yc} S_{xc} \tag{19.82}$$

where F_{yc} is the yield strength of the compression flange and S_{xc} is the section modulus about the horizontal axis of the I-section at the compression flange. Inserting Eq. 19.82 into Eq. 19.81, multiplying by R_b, equating to the modified Eq. 19.81, and solving for L_r gives

$$L_r = \sqrt{\frac{2\pi^2 I_{yc} d}{S_{xc}}\frac{E}{F_{yc}}} \tag{19.83a}$$

which may be conservatively approximated by [C6.10.8.2.3]

$$L_r = \pi r_t \sqrt{\frac{E}{F_{yc}}} \tag{19.83b}$$

For values of L_b between L_p and L_r a straight-line anchor points between $M_n = M_y$ and $M_n = 0.5M_y$ is given by

$$M_n = C_b R_b M_y \left(1 - 0.5\frac{L_b - L_p}{L_r - L_p}\right) \le R_b M_y \tag{19.84a}$$

which in terms of stress is

$$F_{cn} = \left(1 - 0.5\frac{L_b - L_p}{L_r - L_p}\right) C_b R_b R_h F_{cy} \le R_b R_h F_{yc} \tag{19.84b}$$

Because the moment gradient factor C_b can be greater than 1.0 (Eq. 19.92), the elastic upper limit of M_n is given on the right side of Eq. 19.84 as $R_b R_h M_y$.

Lateral Torsion Buckling—Specifications [A6.10.8.2.3] For unbraced lengths less than L_p the section is compact and the LTB compression strength is

$$F_{nc} = R_b R_h F_{yc} \tag{19.85}$$

where

$$L_b < L_p = 1.0 r_t \sqrt{\frac{E}{F_{yc}}} \tag{19.86}$$

If the unbraced length is greater than that required for compactness but not elastic, then, if $L_r \geq L_b \geq L_p$, where

$$L_r = \pi r_t \sqrt{\frac{E}{F_{yc}}} \tag{19.87}$$

then the compression flange is noncompact and the anchor-point interpolation is used to establish the compress flange strength,

$$F_{nc} = C_b \left[1 - \left(1 - \frac{F_{yr}}{R_h F_{yc}} \right) \left(\frac{L - L_p}{L_r - L_p} \right) \right] \times R_b R_h F_{yc} \leq R_b R_h F_{yc} \tag{19.88}$$

where

$$F_{yr} = \min\left[0.7 F_{yc}, \; F_{yw}\right] \geq 0.5 F_{yc} \tag{19.89}$$

If the unbraced length is greater than that required for inelastic buckling and the buckling is elastic and if $L_b \geq L_r$, then

$$F_{nc} = F_{cr} \leq R_b R_h F_{yc} \tag{19.90}$$

where

$$F_{cr} = \frac{C_b R_b \pi^2 E}{\left(L_b / r_t\right)^2} \tag{19.91}$$

***Moment Gradient Correction Factor* C_b** Equations 19.68 and 19.69 were derived for constant (uniform) moment between brace points. This worst-case scenario is overly conservative for the general case of varying applied moment over the unbraced length. To account for I-sections with both variable depth and variable moment (anywhere shear is present), the force in the compression flange at the brace points is used to measure the effect of the moment gradient. The expression for the correction factor is given as [A6.10.8.2.3]

$$C_b = 1.75 - 1.05 \left(\frac{f_1}{f_2} \right) + 0.3 \left(\frac{f_1}{f_2} \right)^2 \leq 2.3 \tag{19.92}$$

where f_1 is the stress in the compression flange at the brace point with the smaller force due to factored loading and f_2 is the stress in the compression flange at the brace point with the larger force due to factored loading.

An I-section with moments M_1 and M_2 with associated flange stresses f_1 and f_2 at the brace points is shown in Figure 19.38. The moment diagram (stress variation) between the brace points is given in Figure 19.38(a) and the compression flange corresponding to f_1 and f_2 in Figure 19.38(b). If $f_1 = f_2$, Eq. 19.92 gives $C_b = 1.0$. As the compression flange stress f_1 decreases, the LTB strength increases. If $f_1 = 0$ [Fig. 19.38(c)], then $C_b = 1.75$. If f_1 goes into tension, C_b continues to increase until it reaches its maximum value of 2.3 at $f_1 = -0.46 f_2$ [Fig. 19.38(d)].

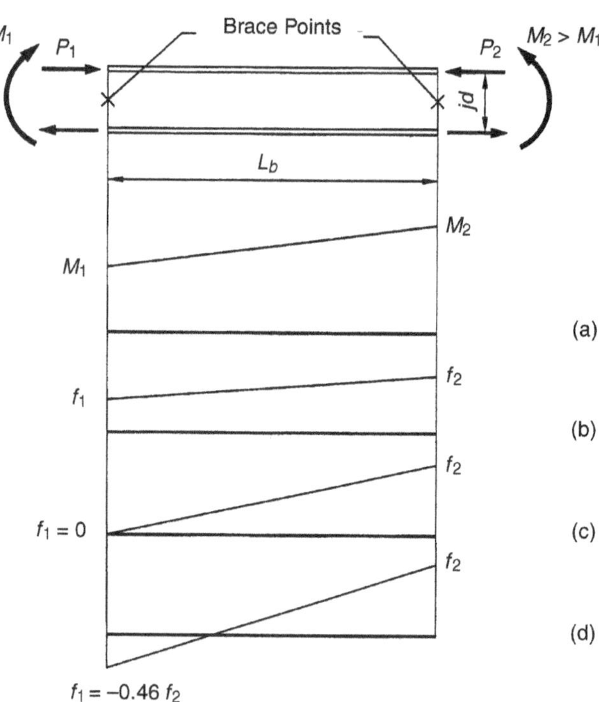

Fig. 19.38 (a) Moment gradient between brace points, (b) compression flange forces corresponding to f_1 and f_2, (c) compression flange forces when $f_1 = 0$, and (d) compression flange forces when $f_1 = -0.46 f_2$.

Many of the articles in the AASHTO LRFD Specifications were taken directly or indirectly from specifications that address primarily stationary loads (e.g., AISC), and these articles are sometimes difficult to use for bridges. Theoretically, the values for the compression flanges forces should be those forces that are coincident with the forces that cause the critical load effect of the section of interest. The coincident actions are actions at other sections when the cross section of interest is loaded for critical effect. Such actions or the load effects such as flange forces are not easily computed. The AASHTO LRFD Specifications address this issue by permitting the use of the moment envelope to estimate coincident actions in many such cases.

Another complicating feature of the moment gradient effect is where a point of contraflexure occurs in a composite section in the "unbraced" length, for example, with a negative moment transitioning to a positive before a brace point is encountered. See the extensive commentary where moment gradient factor issues are discussed. The commentary continues for several pages and is recommended regarding details of unbraced length and LTB issues [C6.10.8.2.3].

Noncomposite Elastic I-Section Specifications [A6.10.8] For noncomposite I-sections, the compactness requirements are the same as for composite sections in negative flexure [A6.10.6.3 and A6.10.8], when the unbraced length L_b exceeds the noncompact (inelastic) section requirement [A6.10.8.2.3]

Composite Noncompact Sections For composite I-sections in negative flexure with L_b greater than the value of Eq. 19.86 but less than the value of Eq. 19.87, then the nominal flexural resistance is based on the nominal flexural stress of the compression flange and Eq. 19.88:

$$f_c \leq F_{\mathrm{nc(LTB)}} \tag{19.93}$$

Limited Redistribution—Behavior If the positive moment region sustains plastic hinging, moments attempt to redistribute to other areas of the girder, most likely the negative moment region. If the negative moment does not have sufficient strength and ductility to sustain the increased moment, then the negative moment region could fail. Typically, economical design gives a negative moment section that is noncompact and discretely braced. Hence if the section is noncompact, then full plastic moment is limited to a lesser value to avoid this situation.

Limited Redistribution—Specification [A6.10.7.2] For composite sections in negative flexure with L_b less than or equal to the value given by Eq. 19.86, the nominal flexural resistance is equal to the plastic moment, that is,

$$M_n = M_p \tag{19.94}$$

For continuous spans with compact positive bending sections and *non*compact interior negative moment sections, the nominal positive flexural resistance is limited to [A6.10.7.2]

$$M_n = 1.3R_h M_y \tag{19.95}$$

This limits the shape factor for the compact positive bending section to 1.3. This is necessary in continuous spans because excessive yielding in the positive moment region can redistribute moments to the negative moment region that are greater than those predicted by an elastic analysis [C6.10.7.1.2]. The negative moment section may not be able to carry the redistributed moment in a ductile, fully plastic manner.

Section A6.10.7.1.2 further outlines equations that interpolate between M_y and M_p depending upon the neutral axis depth. These equations, along with those in the Appendix 6B likely better represent that ultimate behavior. A detailed description of these is beyond the scope of this book and the reader is referred to extensive commentary of AASHTO and the cited references therein.

Philosphically, sharping the "design pencil" to this degree is likely unnecessary. However, for the engineer evaluating the strength of an existing structure for load rating, the expanded approach could be helpful.

Ductility of Composite Compact Sections Behavior For compact composite sections in positive flexure, a limitation is imposed on the depth of the composite section in compression to ensure that the tension flange of a steel section reaches strain hardening before the concrete slab crushes [C6.10.7.3]. The higher the neutral axis is the greater the curvature at failure and the more ductility exhibited by the section. Figure 19.39 illustrates the strain profile for crushing concrete (0.003) and yield in a Grade 36 bottom flange. Consistent with concrete, the level of the neutral axis is to be higher than 0.42 times the section depth.

Ductility of Composite Compact Sections Specifications [A6.10.7.3] The ductility requirements for steel composite sections are outlined in [A6.10.7.3]. In summary,

$$D_{\mathrm{cp}} \leq 0.42 D_{\mathrm{total}} \tag{19.96}$$

19.3.4 Limit States

I-sections in flexure must be designed to resist the load combinations for the strength, service, and fatigue limit states of Table 5.1. Often the most critical state of the bridge is during construction when the girders are braced only with cross frames prior to deck construction and hardening.

Constructibility Checks [A6.10.3] During construction, adequate strength shall be provided under factored loads of [A3.4.2], nominal yielding should be prevented as well as postbuckling behavior. The yielding associated with the hybrid reduction factor is permitted. The flexural strength is checked considering the unbraced lengths encountered during construction, assuming that the deck does not provide restraint to the top flange. Cross-frame locations are critical in establishing the LTB strength of the girder. The AASHTO Specifications for Section 6.10 are summarized in several flowcharts for each limit state. The constructibility limit state [6.10.3] is illustrated in Figure 19.40.

Service Limit State [A6.10.4] The service II load combination of Table 5.1 shall apply. This load combination is

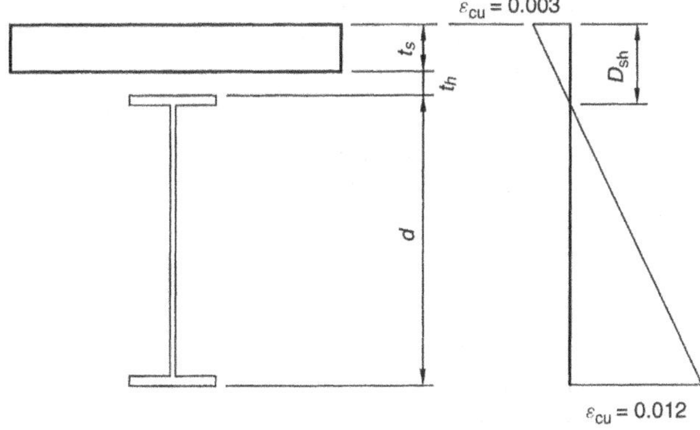

Fig. 19.39 Strain-hardening depth to neutral axis.

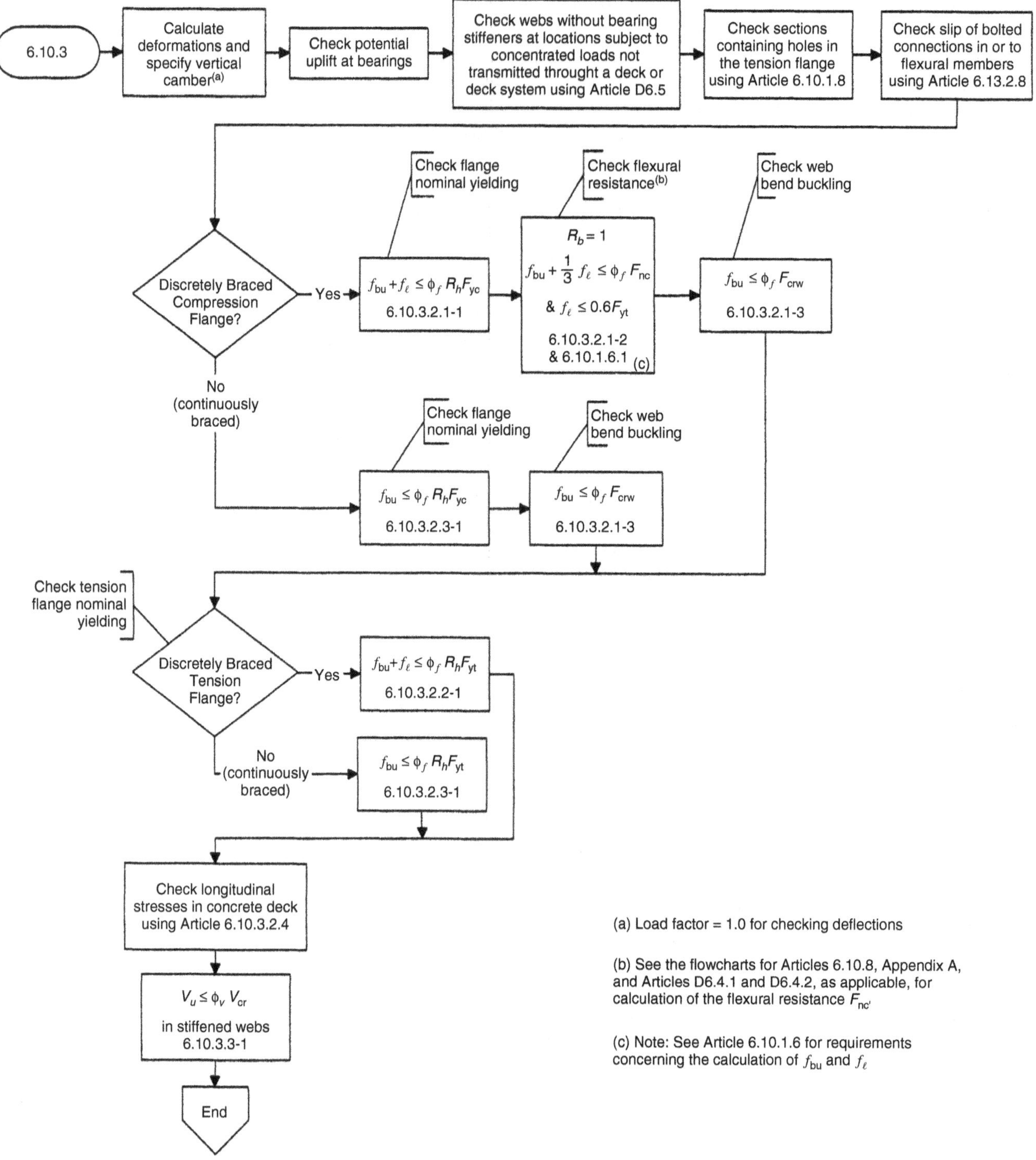

Fig. 19.40 Flowchart for LRFD 6.10.3—Constructibility. (After AASHTO, 2010.)

intended to control yielding of steel structures and to prevent objectionable permanent deflections that would impair rideability [C6.10.4.2]. When checking the flange stresses, moment redistribution may be considered if the section in the negative moment region is compact. Flange stresses in positive and negative bending for composite sections shall not exceed

$$f_f + \frac{f_l}{2} \le 0.95 R_h F_{yf} \tag{19.97}$$

where the lateral bending stress f_l is zero in the top flange and is considered in the bottom flange.

For noncomposite sections

$$f_f + \frac{f_l}{2} \leq 0.80 R_h F_{yf} \qquad (19.98)$$

where f_f is the elastic flange stress caused by the factored loading, f_l is the flange lateral bending stress at the section under consideration due to service II loads, R_h is the hybrid flange stress reduction factor [A6.10.1.10] (for a homogeneous section, $R_h = 1.0$), and F_{yf} is the yield stress of the flange.

Lateral flange bending is due to a bending moment in the flange [about beam's minor (weak) axis]. This load effect is due to the 3D system effects in skewed, curved, and skew-curved bridges for gravity loads. Lateral flange bending is also due to lateral load applied directly to the girders from wind, construction shoring, and the like. The specification addresses the resistance of the combined major- and minor-axis bending by considering the compression flange as a beam column. The load effects are addressed with a refined analysis and/or with simplified procedures outlined in [A4.6.1.2.4b] (gravity loads) and [A6.10.3] for construction loads.

AASHTO does not have shear checks for the service II limit state. The details for the service II limit state are provided in Figure 19.41.

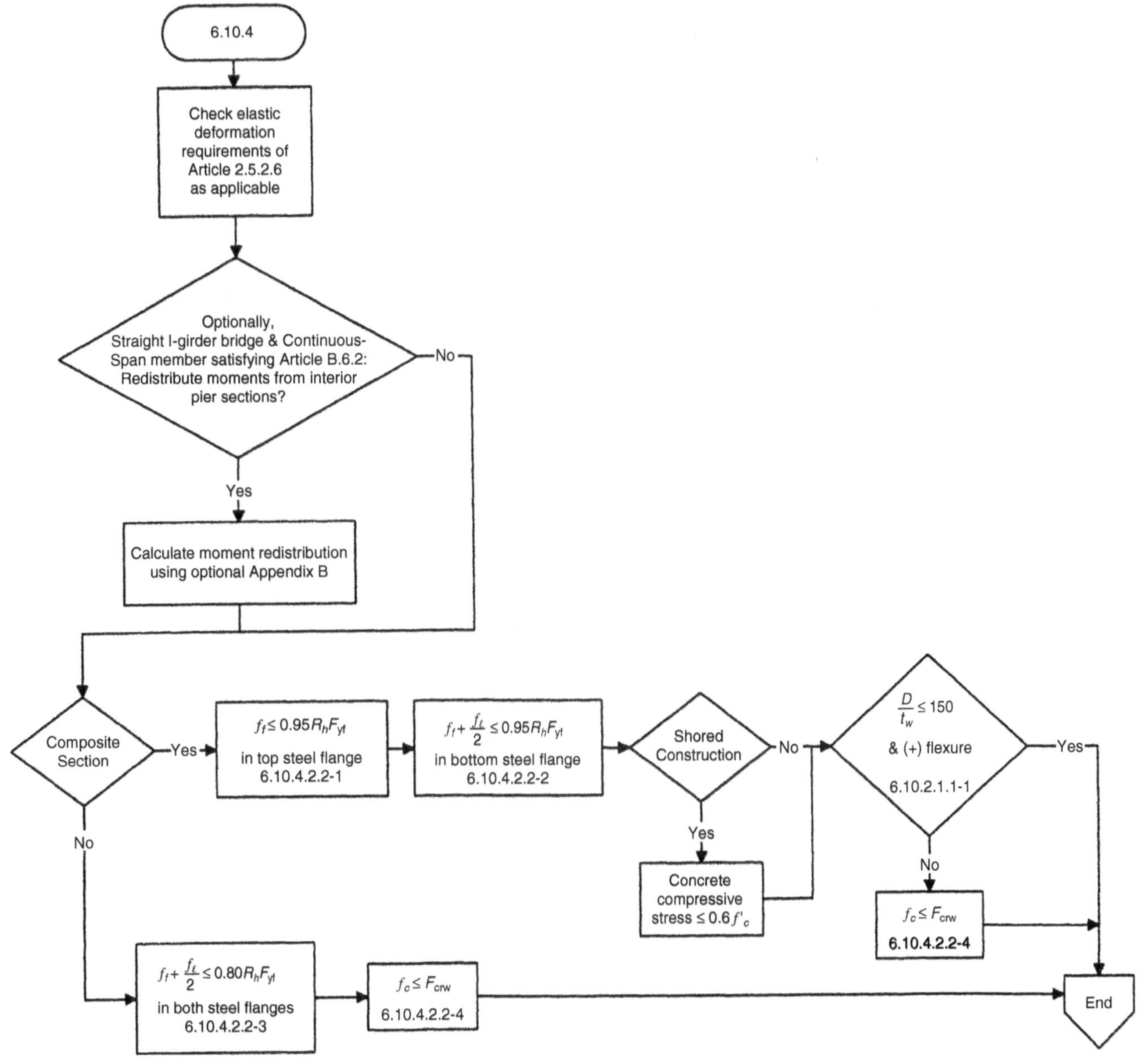

Fig. 19.41 Flowchart for LRFD 6.10.4—Service II. (After AASHTO, 2010.)

Fatigue Limit State [A6.10.5] The fatigue limit state for I-section is addressed by four failure modes. First, the details associated with welds must be checked. Reference the earlier discussion on fatigue details, fatigue categories, and fatigue life required. Shear connectors require welds to the top flange and therefore are susceptible to fatigue. The shear live-load range is used for this check. This check is best illustrated by example and two are provided in the composite girder design examples in the next chapter (E20.2 and E20.3) [A6.6.1]. Fracture toughness requirements are checked per temperature requirement and steels specified [A6.6.2]. Finally, the web should not buckle under the unfactored service permanent loads and the factored load for fatigue I.

This eliminates out-of-plane distortions due to service-level traffic [A6.10.5.3]:

$$V_u = V_{\text{cr}} \tag{19.99}$$

where V_u is the shear in the web due to unfactored permanent loads plus factored at fatigue limit state I, and V_{cr} is the shear-buckling resistance determined by shear provisions, which are discussed later [A6.10.9.3.3].

For shear connectors, the live-load shear range is used. The details for the fatigue and fracture states are provided in Figure 19.42.

Strength Limit State [A6.10.6] The strength limit state is addressed by checking flexural resistance and shear resistance. Because of the many permutations of compactness of various local elements (web and flange), LTB, hybrid factors, and load shedding, the AASHTO specifications are a complex web of checks and paths. The discussion above outlines the behavioral consideration and references the specification equations/articles. A short summary follows:

For compact sections, the factored flexural resistance is in terms of moments

$$M_r = \phi_f M_n \tag{19.100}$$

where ϕ_f is the resistance factor for flexure, and $M_n = M_p$ is the nominal resistance specified for a compact section.

For noncompact sections, the factored flexural resistance is defined in terms of stress:

$$F_r = \phi_f F_n \tag{19.101}$$

where F_n is the nominal resistance specified for a noncompact section.

The factored shear resistance V_r shall be taken as

$$V_r = \phi_v V_n \tag{19.102}$$

where ϕ_v is the resistance factor for shear and V_n is the nominal shear resistance specified for unstiffened and stiffened webs. Other elements such as transverse and bearing stiffeners must also be checked. The details for the strength limit state are provided in Figures 19.43–19.46.

19.3.5 Summary of I-Sections in Flexure

The behavior of I-sections in flexure is complex in details and yet simple in concept. The details are complex because of the many different conditions for which requirements must be established. Both composite and noncomposite sections subjected to positive and negative flexure must be considered for the three classes of shapes: compact, noncompact, and slender.

The concept is straightforward because all of the limit states follow the same pattern—web slenderness, compression flange slenderness, or compression flange bracing—these three failure modes are identified: no buckling, inelastic buckling, and elastic buckling. The numerous formulas describe the behavior and define the anchor points for the three segments that represent the design requirements.

19.3.6 Closing Remarks on I-Sections in Flexure

When rolled steel shapes are used as beams, the web and flange slenderness requirements do not have to be checked because all of the webs satisfy the compact section criterion. Further, if Grade 36 steel is used, all but the $W150 \times 22$ satisfy the flange slenderness criterion for a compact section. If Grade 50 steel is used, six of the 253 W-shapes listed in AISC (2005) do not satisfy the flange slenderness criterion for a compact section. Therefore, local buckling is seldom a problem with rolled steel shapes; and, when they are used, the emphasis is on providing adequate lateral support for the compression flange to prevent global buckling.

Plate girders are seldom economically proportioned as compact in the negative moment region. Recent trends using Grade 70 HPS has led to a renewal of hybrid girders. Additionally, more agencies try to use unstiffened sections with a thicker web in order to save labor costs. The next major topic is the design of girders for shear and bearing where stiffener considerations are elaborated.

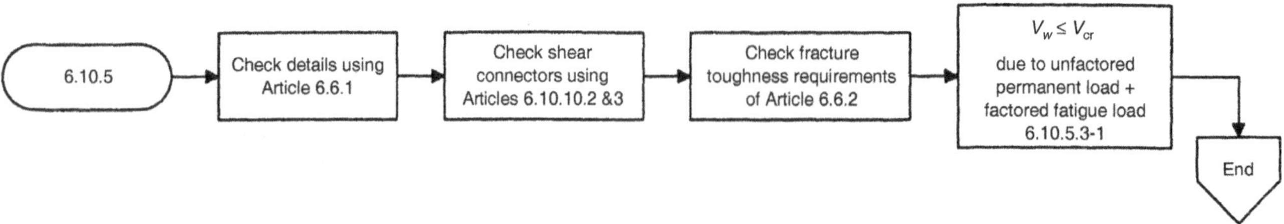

Fig. 19.42 Flowchart for LRFD 6.10.5—fatigue and fracture. (After AASHTO, 2010.)

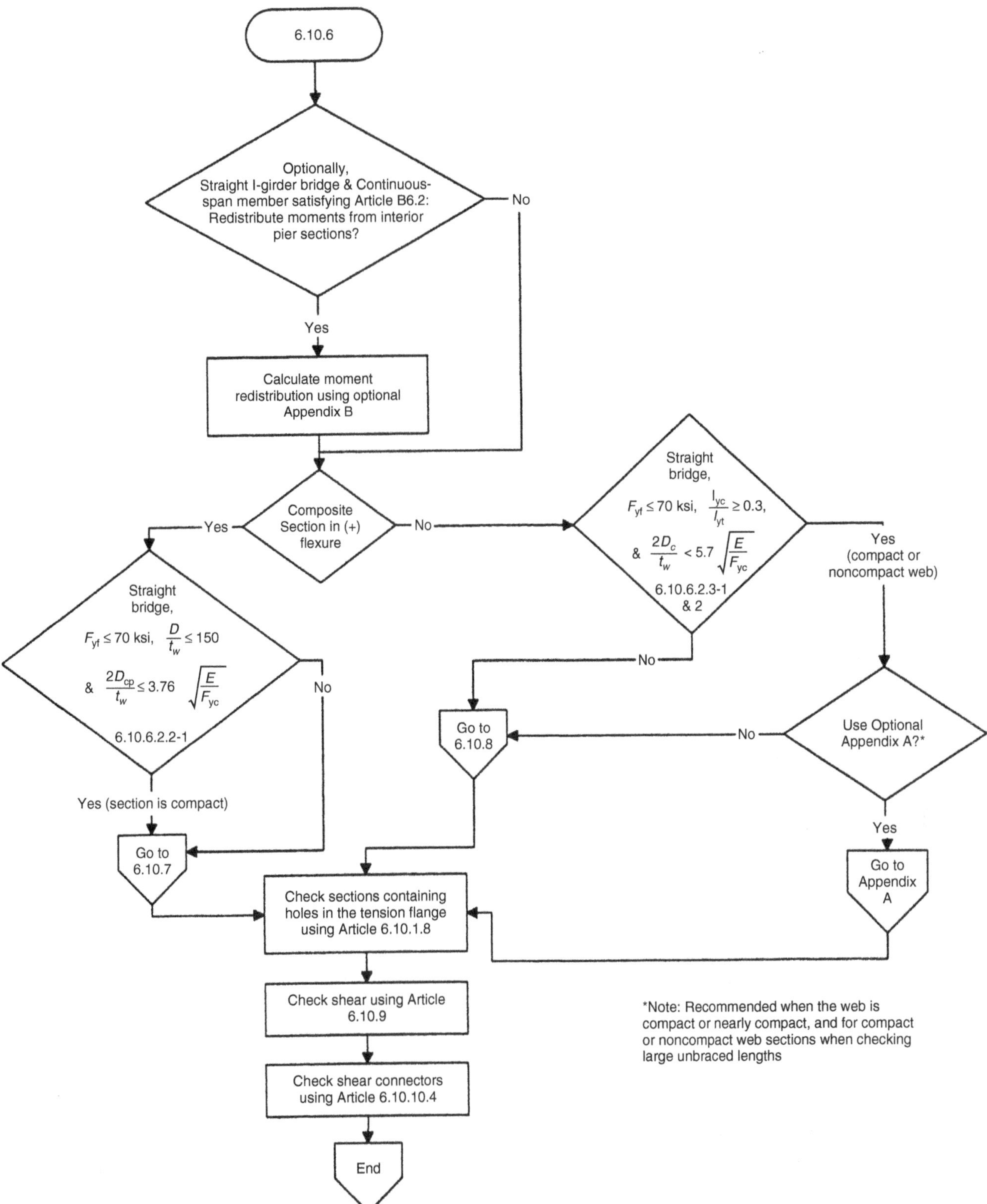

Fig. 19.43 Flowchart for LRFD 6.10.6—strength limit state. (After AASHTO, 2010.)

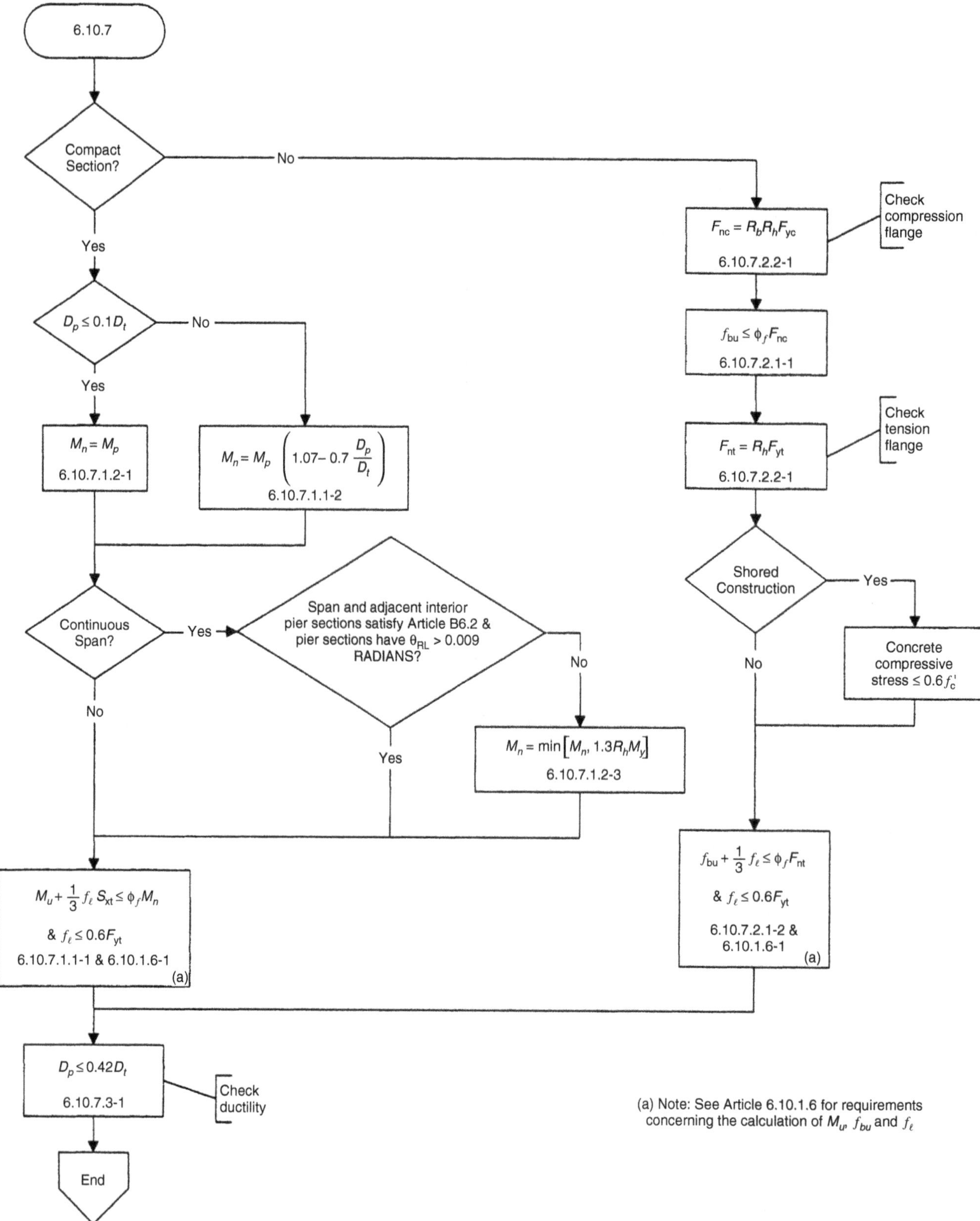

Fig. 19.44 Flowchart for LRFD 6.10.6—composite sections in positive flexure. (After AASHTO, 2010.)

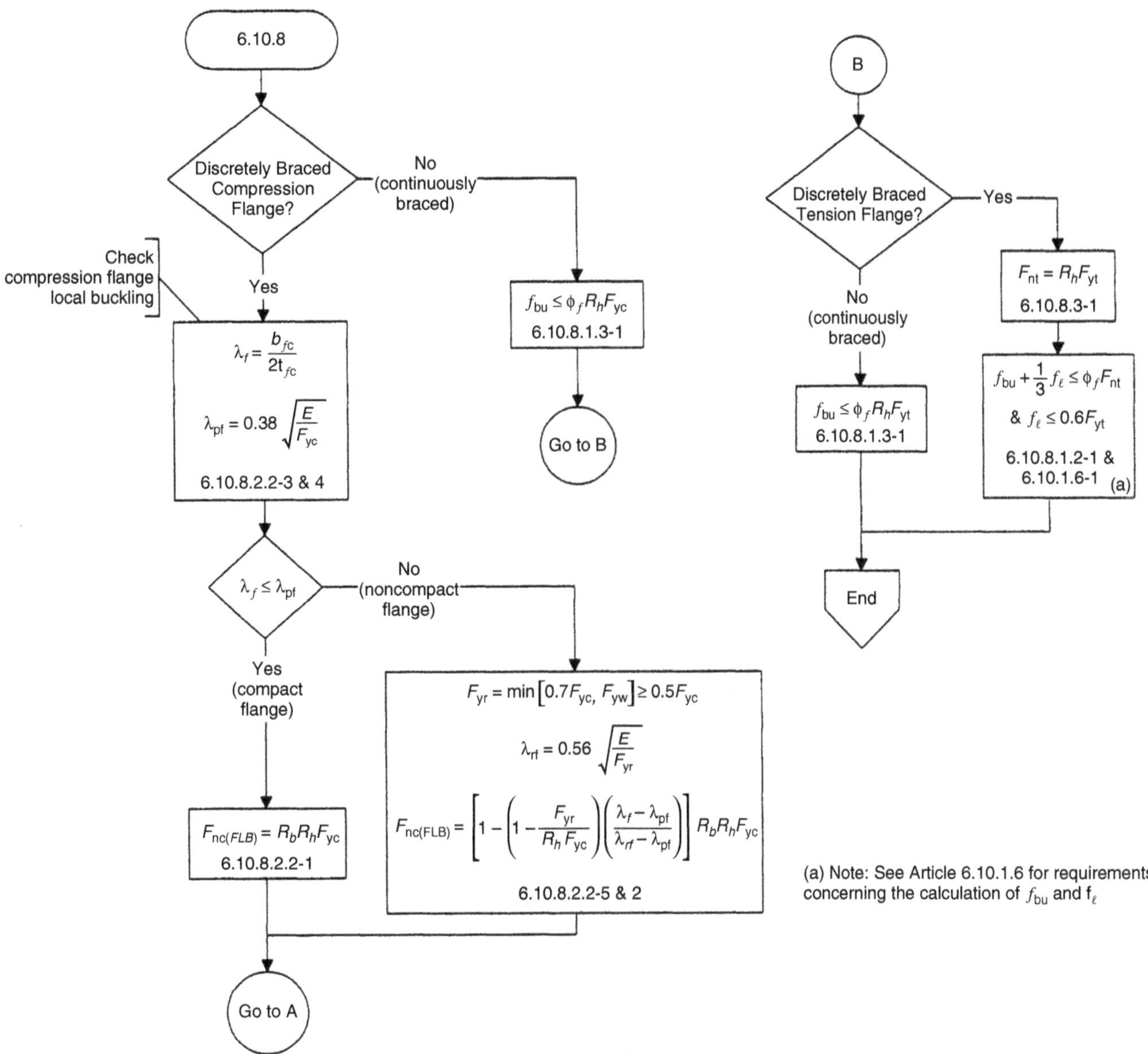

Fig. 19.45 Flowchart for LRFD 6.10.6—Composite sections in negative flexure and noncomposite sections. (After AASHTO, 2010.)

19.4 SHEAR RESISTANCE OF I-SECTIONS

When the web of an I-section is subjected to in-plane shear forces that are progressively increasing, small deflection beam theory can be used to predict the shear strength until the critical buckling load is reached. If the web is stiffened, additional postbuckling shear strength due to tension field action is present until web yielding occurs. Using the notation in Basler (1961), the nominal shear resistance V_n can be expressed as

$$V_n = V_\tau + V_\sigma \tag{19.103}$$

where V_τ is the beam action shear resistance and V_σ is the tension field action shear resistance.

19.4.1 Beam Action Shear Resistance

A stress block at the neutral axis of a web of an I-section is shown in Figure 19.47(a). Because the flexural stresses are zero, the stress block is in a state of pure shear. A Mohr circle of stress [Fig. 19.47(b)] indicates principal stresses σ_1 and σ_2 that are equal to the shearing stress τ. These principal stresses are oriented at 45° from the horizontal. When using beam theory, it is usually assumed that the shear force V is resisted by the area of the web of an I-section shape, and maximum shear stress is close to the average,

$$\tau = \frac{V}{Dt_w} \tag{19.104}$$

where D is the web depth and t_w is the web thickness.

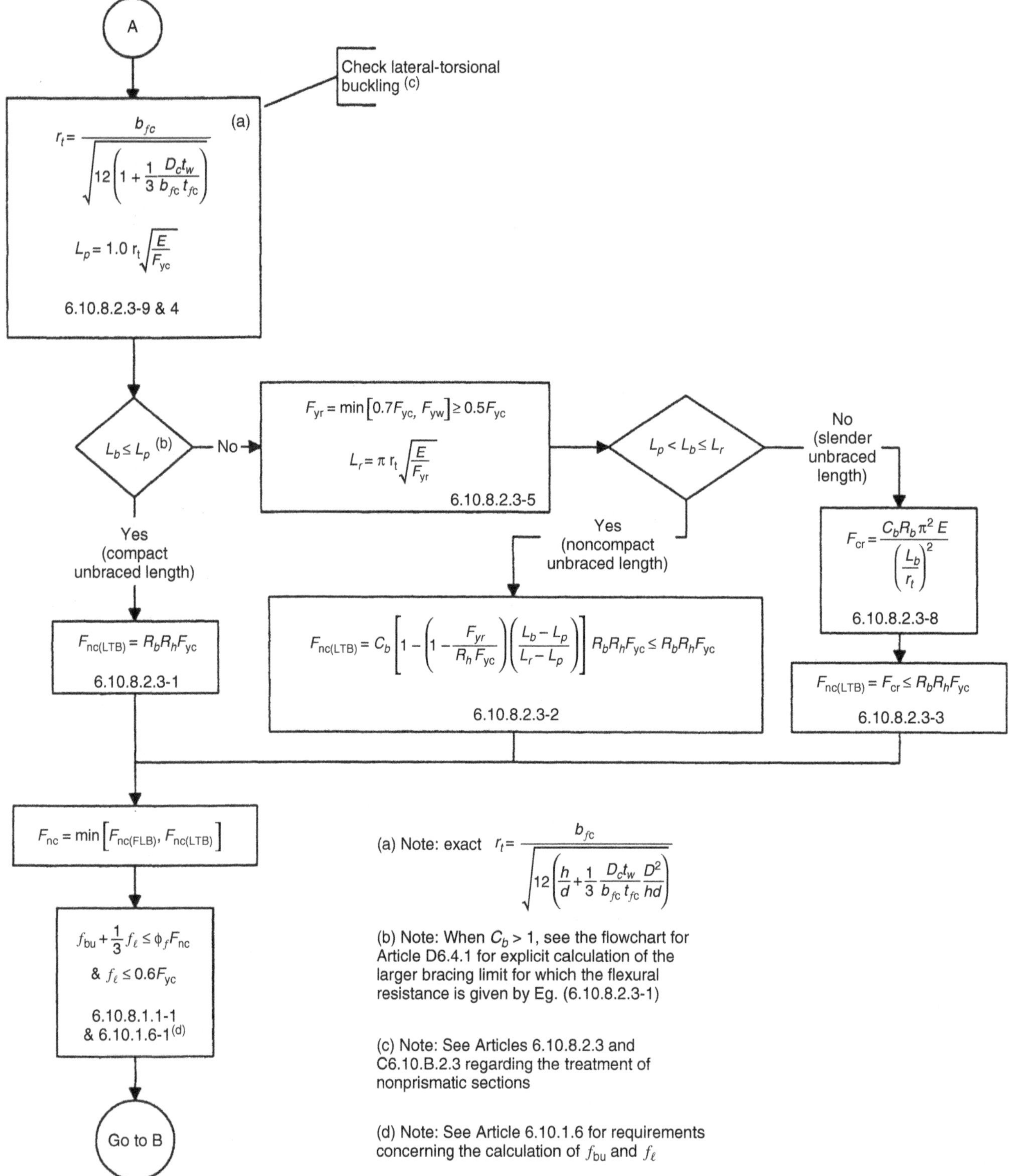

Fig. 19.46 Flowchart for LRFD 6.10.6 (continued)—Composite sections in negative flexure and noncomposite sections. (After AASHTO, 2010.)

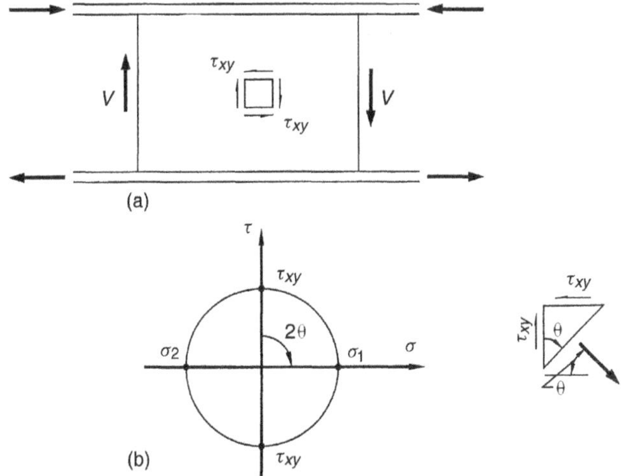

Fig. 19.47 Beam action states of stress: (a) stress block at neutral axis and (b) Mohr circle of stress.

If no buckling occurs, the shear stress can reach its yield strength τ_y and the full plastic shear force V_p can be developed. Substitution of these values into Eq. 19.104 and rearranging,

$$V_p = \tau_y D t_w \tag{19.105}$$

The shear yield strength τ_y cannot be determined by itself but is dependent on the shear failure criteria assumed. By using the Mises shear failure criterion, the shear yield strength is related to the tensile yield strength of the web σ_y by

$$\tau_y = \frac{\sigma_y}{\sqrt{3}} \approx 0.58\sigma_y \tag{19.106}$$

If buckling occurs, the critical shear buckling stress τ_{cr} for a rectangular panel (Fig. 19.48) is given by

$$\tau_{cr} = k\frac{\pi^2 E}{12\left(1-\mu^2\right)}\left(\frac{t_w}{D}\right)^2 \tag{19.107}$$

in which

$$k = 5.0 + \frac{5.0}{\left(d_0/D\right)^2} \tag{19.108}$$

where d_0 is the distance between transverse stiffeners.

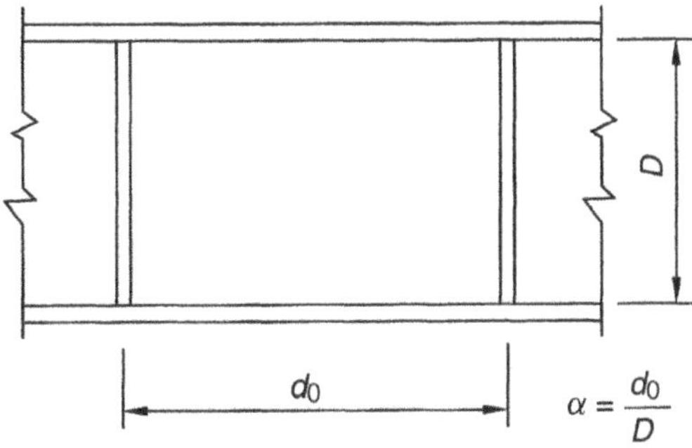

Fig. 19.48 Definition of aspect ratio α.

By assuming that shear is carried in a beamlike manner up to τ_{cr} and then remains constant, we can express V_τ as a linear fraction of V_p, that is,

$$V_\tau = \frac{\tau_{cr}}{\tau_y} V_p \tag{19.109}$$

19.4.2 Tension Field Action Shear Resistance

Tension Field Action Behavior When a rectangular web panel subjected to shear is supported on four edges, *tension field action* (TFA) on the diagonal can develop. The web panel of an I-section (Fig. 19.48) has two edges that are at flanges and two edges that are at transverse stiffeners. These two pairs of boundaries are very different. The flanges are relatively flexible in the vertical direction and cannot resist stresses from a tension field in the web. On the other hand, the transverse stiffeners can serve as compression struts to balance the tension stress field. As a result, the web area adjacent to the junction with the flanges is not effective and the trusslike load-carrying mechanism of Figure 19.49 can be assumed. In this truss analogy, the flanges are the chords, the transverse stiffeners are compression struts, and the web is a tension diagonal.

The edges of the effective tension field in Figure 19.49 are assumed to run through the corners of the panel. The tension field width s depends on the inclination from the horizontal θ of the tensile stresses σ_t and is equal to

$$s = D\cos\theta - d_0 \sin\theta \tag{19.110}$$

The development of this tension field has been observed in numerous laboratory tests. An example of one from Lehigh University is shown in Figure 19.50. At early stages of loading the shear in the web is carried by beam action until the compressive principal stress σ_2 of Figure 19.47(b) reaches its critical stress and the compression diagonal of the panel buckles. At this point, no additional compressive stress can be carried, but the tensile stresses σ_t in the tension diagonal continue to increase until they reach the yield

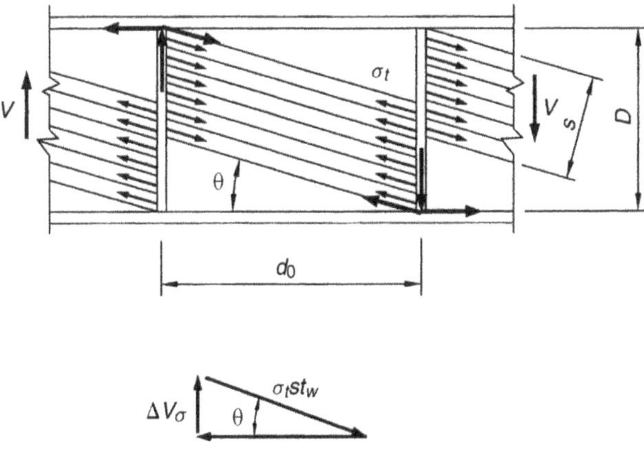

Fig. 19.49 Tension field action.

Fig. 19.50 Thin-web girder after testing. (Photo courtesy of John Fisher, ATLSS Engineering Research Center, Lehigh University.)

stress $\sigma_y = F_{yw}$ of the web material. The stiffened I-section in Figure 19.50 shows the buckled web, the postbuckling behavior of the tension field, and the trusslike appearance of the failure mechanism.

The contribution to the shear force V_σ from the tension field action ΔV_σ is the vertical component of the diagonal tensile force (Fig. 19.49), that is,

$$\Delta V_\sigma = \sigma_t s t_w \sin\theta \tag{19.111}$$

To determine the inclination θ of the tension field, assume that when $\sigma_t = \sigma_y$ the orientation of the tension field is such that ΔV_σ is a maximum. This condition can be expressed as

$$\frac{d}{d\theta}\left(\Delta V_\sigma\right) = \frac{d}{d\theta}\left(\sigma_y s t_w \sin\theta\right) = 0$$

Substitute Eq. 19.110 for s:

$$\sigma_y t_w \left[\frac{d}{d\theta}\left(D\cos\theta\sin\theta - d_0 \sin^2\theta\right)\right] = 0$$

which reduces to

$$D\tan^2\theta + 2d_0\tan\theta - D = 0$$

Solve for $\tan\theta$:

$$\tan\theta = \frac{-2d_0 + \sqrt{4d_0^2 + 4D^2}}{2D} = \sqrt{1+\alpha^2} - \alpha \tag{19.112}$$

where α is the aspect ratio of the web panel d_0/D. Use trigonometric identities to obtain

$$\cos\theta = \left(\tan^2\theta + 1\right)^{-1/2} = \left[2\sqrt{1+\alpha^2}\left(\sqrt{1+\alpha^2} - \alpha\right)\right]^{-1/2} \tag{19.113}$$

and

$$\sin\theta = \left(\cot^2\theta + 1\right)^{-1/2} = \left(\frac{1}{2} - \frac{\alpha}{2\sqrt{1+\alpha^2}}\right)^{1/2} \tag{19.114}$$

Consider equilibrium of the free-body $ABDC$ in Figure 19.51 taken below the neutral axis of the web and between the middle of the web panels on either side of a transverse stiffener. By assuming a doubly symmetric I-section, the components of the partial tension field force on the vertical sections AC and BD are $V_\sigma/2$ vertically and F_w horizontally in the directions shown in Figure 19.51. On the horizontal

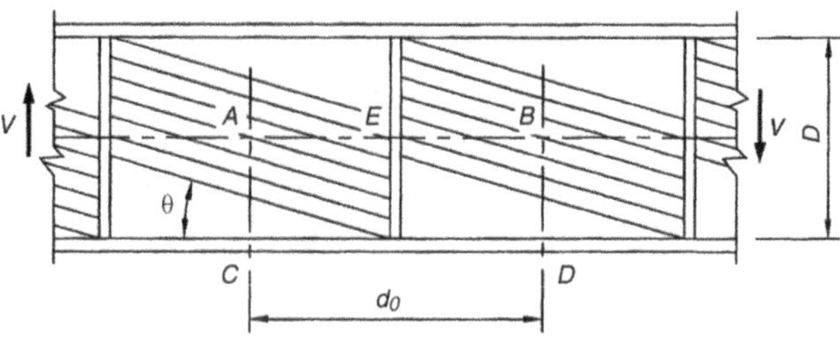

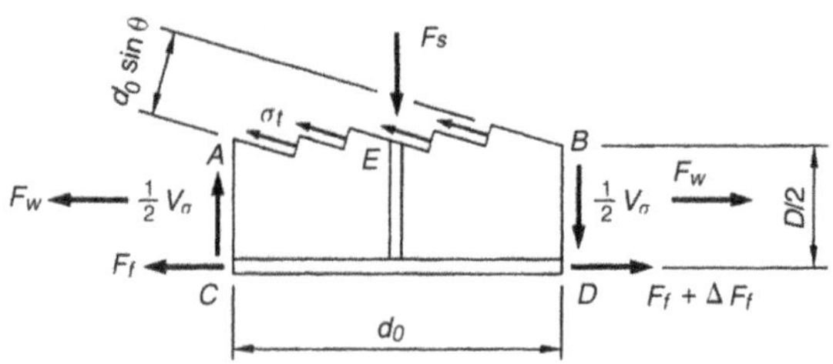

Fig. 19.51 Free-body diagram of tension field action.

section AB, the tension field stresses σ_t are inclined at an angle θ and act on a projected area $t_w d_0 \sin\theta$. Equilibrium in the vertical direction gives the axial load in the stiffener F_s as

$$F_s = \sigma_t t_w d_0 \sin\theta \sin\theta = \sigma_t t_w (\alpha D) \sin^2\theta$$

Substitution of Eq. 19.114 gives

$$F_s = \sigma_t t_w D\left(\frac{\alpha}{2} - \frac{\alpha^2}{2\sqrt{1+\alpha^2}}\right) \quad (19.115)$$

Equilibrium in the horizontal direction gives the change in the flange force ΔF_f as

$$\Delta F_f = \sigma_t t_w (\alpha \ D) \sin\theta \cos\theta$$

Substitution of Eqs. 19.113 and 19.114 into the above expression for ΔF_f gives

$$\Delta F_f = \sigma_t t_w D \frac{\alpha}{2\sqrt{1+\alpha^2}} \quad (19.116)$$

Balance of the moments about point E results in

$$\frac{1}{2}V_\sigma (d_0) - \Delta F_f\left(\frac{D}{2}\right) = 0$$

$$V_\sigma = \Delta F_f \frac{D}{d_0} = \frac{\Delta F_f}{\alpha}$$

So that the shear force contribution of TFA V_σ becomes

$$V_\sigma = \sigma_t t_w D \frac{1}{2\sqrt{1+\alpha^2}} \quad (19.117)$$

With the use of Eqs. 19.105 and 19.106, V_σ can be written in terms of V_p as

$$V_\sigma = \frac{\sqrt{3}}{2}\frac{\sigma_t}{\sigma_y}\frac{1}{\sqrt{1+\alpha^2}}V_p \quad (19.118)$$

19.4.3 Combined Shear Resistance

Substituting Eqs. 19.109 and 19.118 into Eq. 19.103, the expression for the combined nominal shear resistance of the web of an I-section is

$$V_n = V_p\left[\frac{\tau_{cr}}{\tau_y} + \frac{\sqrt{3}}{2}\frac{\sigma_t}{\sigma_y}\frac{1}{\sqrt{1+\alpha^2}}\right] \quad (19.119)$$

where the first term in brackets is due to beam action and the second term is due to TFA. These two actions should *not* be thought of as two separately occurring phenomena where first one is observed and later the other becomes dominant. Instead they occur together and interact to give the combined shear resistance of Eq. 19.119.

Basler (1961) develops a simple relation for the ratio σ_t/σ_y in Eq. 19.119 based on two assumptions. The first assumption is that the state of stress anywhere between pure shear and pure tension can be approximated by a straight line when using the Mises yield criterion. The second assumption is that θ is equal to the limiting case of 45°. By using these two assumptions, substitution into the stress equation representing the Mises yield criterion results in

$$\frac{\sigma_t}{\sigma_y} = 1 - \frac{\tau_{cr}}{\tau_y} \quad (19.120)$$

Basler (1961) conducted a numerical work comparing the nominal shear resistance of Eq. 19.119 with that using the approximation of Eq. 19.120 where the difference was demonstrated to be less than 10% for values of α between zero and infinity. Substituting Eq. 19.120 into Eq. 19.119, the combined nominal shear resistance of the web becomes

$$V_n = V_p\left[\frac{\tau_{cr}}{\tau_y} + \frac{\sqrt{3}}{2}\frac{1-(\tau_{cr}/\tau_y)}{\sqrt{1+\alpha^2}}\right] \quad (19.121)$$

Combined Shear Specifications [A6.10.9] In the AASHTO (2010) LRFD Bridge Specifications, Eq. 19.121 appears as [A6.10.9.3]

$$V_n = V_p\left[C + \frac{0.87(1-C)}{\sqrt{1+(d_0/D)^2}}\right] \quad (19.122)$$

for which

$$C = \frac{\tau_{cr}}{\tau_y} \quad (19.123)$$

$$\alpha = \frac{d_0}{D} \quad (19.124)$$

$$V_p = 0.58F_{yw}Dt_w \quad (19.125)$$

Because τ_{cr} is a function of panel slenderness, so is C. Table 19.7 provides the ratio C for plastic, inelastic, and elastic behavior.

Table 19.7 Ratio of Shear Buckling Stress to Shear Yield Strength [A6.10.9.3]

	No Buckling	Inelastic Buckling	Elastic Buckling
Web slenderness	$\frac{D}{t_w} \le 1.12\sqrt{\frac{Ek}{F_{yw}}}$	$\frac{D}{t_w} \le 1.40\sqrt{\frac{Ek}{F_{yw}}}$	$\frac{D}{t_w} > 1.40\sqrt{\frac{Ek}{F_{yw}}}$
$C = \frac{\tau_{cr}}{\tau_y}$	$C = 1.0$	$C = \frac{1.12}{D/t_w}\sqrt{\frac{Ek}{F_{yw}}}$	$C = \frac{1.57}{(D/t_w)^2}\frac{Ek}{F_{yw}}$

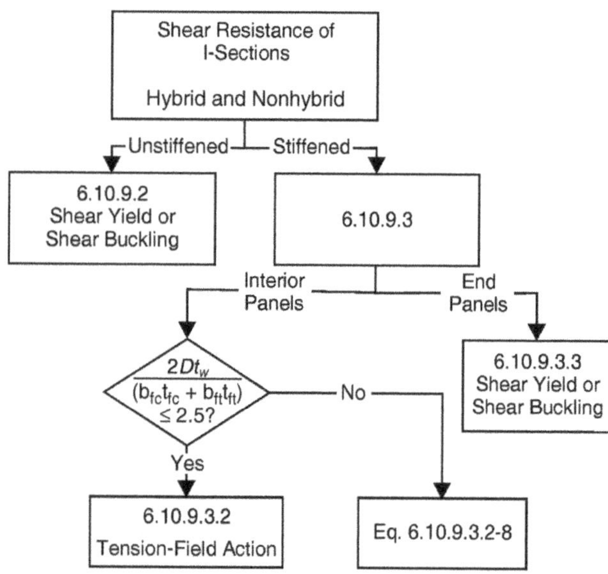

Fig. 19.52 Flowchart for shear design of I-sections. (After AASHTO, 2010.)

The exception to the above is if the section along the entire panel does not satisfy

$$\frac{2Dt_w}{b_{fc}t_{fc} + b_{ft}t_{ft}} \leq 2.5$$

Then

$$V_n = V_p \left[C + \frac{0.87\,(1 - C)}{\sqrt{1 + (d_0/D)^2} + d_0/D} \right] \qquad (19.126)$$

The flowchart for the design of I-sections for shear is provided in Fig. 19.52.

19.4.4 Shear Resistance of Unstiffened Webs

Unstiffened Web Behavior and Specifications The nominal shear resistance of unstiffened webs of I-sections can be determined from Eq. 19.122 by setting d_0 equal to infinity, that is, only the beam action resistance remains and the shear-buckling coefficient is $k = 5$:

$$V_n = V_{cr} = CV_p = 0.58CF_{yw}Dt_w \qquad (19.127)$$

where no additional strength is available due to postbuckling behavior.

Note that with $D/t_w = 150$, d_0 is $3D$, which is the maximum stiffener spacing for an interior panel. In order to avoid postbuckling behavior in the end panel where the tension field is not anchored, the end panel stiffener should be spaced at less than $d_0 \leq 1.5D$ [A6.10.9.3.3].

Example 19.10 Determine the web shear strength of the I-section of Example 19.6 shown in Figure 19.24 if the spacing of transverse stiffeners is 80 in. for an interior web panel. The yield strength of the web F_{yw} is 50 ksi.

Solution

$$V_n = V_p \left[C + \frac{0.87\,(1 - C)}{\sqrt{1 + (d_0/D)^2}} \right]$$

$$\alpha = \frac{d_0}{D} = \frac{80}{60} = 1.33$$

and

$$V_p = 0.58F_{yw}Dt_w$$
$$= 0.58\,(50)\,(60)\,(0.625) = 1087.5 \text{ kips}$$

The calculation of k from Eq. 19.108 is

$$k = 5.0 + \frac{5.0}{\alpha^2} = 5.0 + \frac{5.0}{(1.33)^2} = 7.81$$

so that

$$1.40\sqrt{\frac{Ek}{F_{yw}}} = 1.40\sqrt{\frac{(29{,}000)\,(7.81)}{50}} = 94$$

and

$$\frac{D}{t_w} = \frac{60}{0.625} = 96 > 1.40\sqrt{\frac{Ek}{F_{yw}}} = 94$$

Thus,

$$C = \frac{1.57}{(D/t_w)^2}\frac{Ek}{F_{yw}} = \frac{1.57}{(96)^2}\frac{(29{,}000)\,(7.81)}{50} = 0.77$$

and

$$V_n = 1087.5 \left[0.77 + \frac{0.87\,(1 - 0.77)}{\sqrt{1 + 1.33^3}} \right]$$
$$= 1087.5\,(0.89) = 968 \text{ kips}$$

Answer The factored web shear strength is

$$V_r = \phi_v V_n = 1.0\,(968) = 968 \text{ kips}$$

where ϕ_v is taken from Table 18.7.

19.5 SHEAR CONNECTORS

To develop the full flexural strength of a composite member, horizontal shear must be resisted at the interface between the steel section and the concrete deck slab. To resist the horizontal shear at the interface, connectors are welded to the top flange of the steel section that are embedded in the deck slab when the concrete is placed. These shear connectors come in various types: headed studs, channels, spirals, inclined stirrups, and bent bars. Only the welded headed studs (Fig. 19.53) are discussed in this section.

In simple-span composite bridges, shear connectors shall be provided throughout the length of the span [A6.10.10.1].

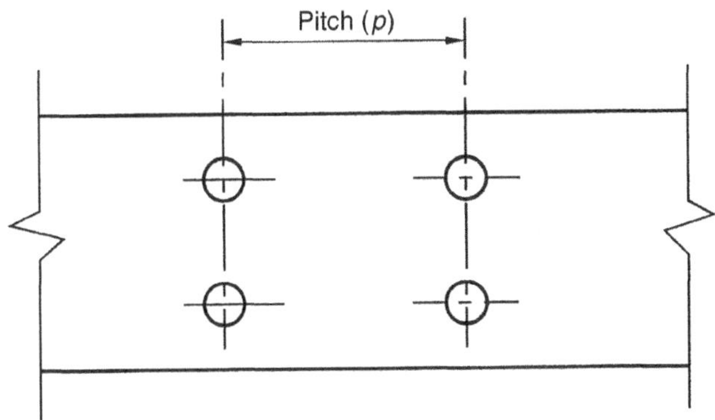

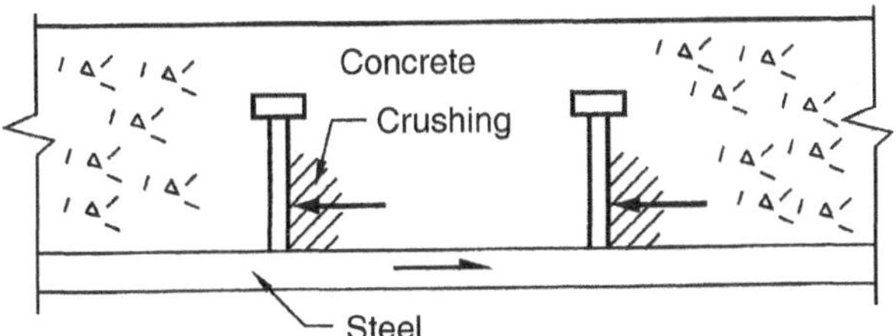

Fig. 19.53 Forces acting on a shear connector in a solid slab.

In continuous composite bridges, shear connectors are often provided throughout the length of the bridge. Placing shear connectors in the negative moment regions prevents the sudden transition from composite to noncomposite section and assists in maintaining flexural compatibility throughout the length of the bridge (Slutter and Fisher, 1967).

The larger diameter head of the stud shear connector enables it to resist uplift as well as horizontal slip. Calculations are not made to check the uplift resistance. Experimental tests (Ollgaard et al., 1971) indicate failure modes associated with shearing of the stud or failure of the concrete (Fig. 19.53). The headed studs did not pull out of the concrete and can be considered adequate to resist uplift.

Data from experimental tests are used to develop empirical formulas for resistance of welded headed studs. Tests have shown that to develop the full capacity of the connector, the height of the stud must be at least four times the diameter of its shank. Therefore, this condition becomes a design requirement [A6.10.10.1.1].

Two limit states must be considered when determining the resistance of stud shear connectors: fatigue and strength. The fatigue limit state is examined at stress levels in the elastic range. The strength limit state depends on plastic behavior and the redistribution of horizontal shear forces among connectors.

19.5.1 Fatigue Limit State for Stud Connectors

Studs in Fatigue—Behavior In the experimental tests conducted by Slutter and Fisher (1967), the shear stress range was found to be the governing factor affecting the fatigue life of shear connectors. Concrete strength, concrete age, orientation of connectors, size effect, and minimum stress did not significantly influence the fatigue strength. As a result, the fatigue resistance of stud connectors can be expressed by the relationship between allowable shear stress range S_r and the number of load cycles to failure N. The log–log plot of the S–N data for both $\frac{3}{4}$-in. and $\frac{7}{8}$-in. diameter studs is given in Figure 19.54. The shear stress was calculated as the average stress on the nominal diameter of the stud. The mean curve resulting from a regression analysis is given by (Slutter and Fisher, 1967)

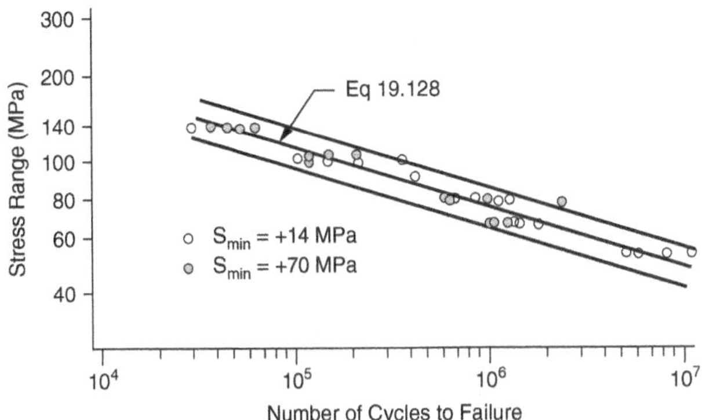

Fig. 19.54 Comparison of regression curve with test data for stud shear connectors (Slutter and Fisher, 1967).

$$S_r = 1065N^{-0.19} \qquad (19.128\text{-SI})$$

$$S_r = 153N^{-0.19} \qquad (19.128\text{-US})$$

where S_r is the shear stress range in ksi (MPa) and N is the number of loading cycles given by Eq. 18.7.

The data fits nicely within the 90% confidence limits shown in Figure 19.54. No endurance limit was found within 10 million cycles of loading.

Fatigue Resistence of Studs—Specifications In AASHTO (2010) LRFD Bridge Specifications, the shear stress range S_r (ksi) becomes an allowable shear force Z_r (kips) for a specific life of N loading cycles by multiplying S_r by the cross-sectional area of the stud, that is, the finite-life fatigue resistance is

$$Z_r = \frac{\pi}{4}d^2 S_r = \left(120N^{-0.19}\right) d^2 \qquad (19.129)$$

where d is the nominal diameter of the stud connector in inches.

The AASHTO (2010) LRFD Bridge Specifications represent Eq. 19.129 in a different format, which approximates Eq. 19.129 as [A6.10.10.2]

$$Z_r \text{ (finite life)} = \alpha d^2 \qquad (19.130\text{a})$$

for which

$$\alpha = 34.5 - 4.28 \log N \qquad (19.130\text{b})$$

however, where the projected 75-year single-lane average daily truck traffic $(\text{ADTT})_{SL}$ is greater than or equal to 960 trucks per day, the fatigue I limit state ($\gamma_{\text{fatigue I}} = 1.50$) shall

Table 19.8 Comparison of α with Regression Equation

N	34.5–4.28 log N, ksi	153 $N^{-0.19}$, ksi
2×10^4	16.1	18.3
1×10^5	13.1	13.5
5×10^5	10.1	9.9
2×10^6	7.5	7.6
6×10^6	5.5	6.2

be used, the resistance shall be based upon the Z_r (kips) (infinite life) represented the resistance plateau commonly observed in the test data:

$$Z_r \text{ (infinite life)} = 5.5d^2 \tag{19.131}$$

The finite-life resistance Eq. 19.130 is used to estimate the resistance and the limit state is fatigue II ($\gamma_{\text{fatigue II}} = 0.75$). However, the use of Eq. 19.131 is typical because of the usual large number of truck crossings per day.

Values for α are compared in Table 19.8 with those for the quantity in parenthesis in Eq. 19.128 over the test data range of N. (This variable α is *not* the same or related to the shear panel aspect ratio used in the previous section.) The expression for α in Eq. 19.130b is a reasonable approximation.

Equations 19.130 and 19.131 can be used to determine the fatigue shear resistance of a single stud connector with diameter d for a specified life N or infinite life (independent of number of cycles, i.e., a large number). The spacing or pitch of these connectors along the length of the bridge depends on how many connectors n are at a transverse section and how large the shear force range V_{sr} (kips) due to the fatigue truck is at the section of interest.

Because fatigue is critical under repetitions of working loads, the design criteria are based on elastic conditions. If complete composite interaction is assumed, the horizontal shear per unit of length v_h (kip/in.) can be obtained from the familiar elastic relationship

$$v_h = \frac{V_{\text{sr}} Q}{I} \tag{19.132}$$

where Q (in.^{3}) is the first moment of the transformed deck area about the neutral axis of the short-term composite section and I (in.^{4}) is the moment of inertia of the short-term composite section. The shear force per unit length that can be resisted by n connectors at a cross section with a distance p (in.) between groups (Fig. 19.53) is

$$v_h = \frac{nZ_r}{p} \tag{19.133}$$

The combination of Eqs. 19.132 and 19.133 yields the pitch p in inches as

$$p = \frac{nZ_r I}{V_{\text{sr}} Q} \tag{19.134}$$

The center-to-center pitch of shear connectors shall not exceed 24 in. and shall not be less than six stud diameters [A6.10.10.1.2].

Stud shear connectors shall not be closer than four stud diameters center-to-center transverse to the longitudinal axis of the supporting member. The clear distance between the edge of the top flange of the steel section and the edge of the nearest shear connector shall not be less than 1 in. [A6.10.10.1.3].

The clear depth of cover over the tops of the shear connectors should not be less than 2 in. In regions where the haunch between the top of the steel section and the bottom of the deck is large, the shear connectors should penetrate at least 2 in. into the deck [A6.10.10.1.4].

19.5.2 Strength Limit State for Stud Connectors

Stud Connector Strength—Behavior Experimental tests were conducted by Ollgaard et al. (1971) to determine the shear strength of stud connectors embedded in solid concrete slabs. Variables considered in the experiments were the stud diameter, number of stud connectors per slab, type of aggregate in the concrete (lightweight and normal weight), and the concrete properties. Four concrete properties were evaluated: compressive strength, split cylinder tensile strength, modulus of elasticity, and density.

Two failure modes were observed. Either the studs sheared off the steel beam and remained embedded in the concrete slab or the concrete failed and the connectors were pulled out of the slab together with a wedge of concrete. Sometimes both of these failure modes were observed in the same test.

An examination of the data indicated that the nominal shear strength of a stud connector Q_n is proportional to its cross-sectional area A_{sc}. Multiple regression analyses of the concrete variables indicate that the concrete compressive strength f_c' and modulus of elasticity E_c are the dominant properties in determining connector shear strength. The empirical expression for the concrete modulus of elasticity (Eq. 13.2) includes the concrete density w_c and, therefore, the effect of the aggregate normal type, that is, for $w_c =$ 0.145 ksi

$$E_c = 1820\sqrt{f_c'} \tag{13.2}$$

where f_c' is the concrete compressive strength (ksi). Including the split cylinder tensile strength in the regression analyses did not significantly improve the correlation with the test results and it was dropped from the final prediction equation.

Stud Connector—Specifications After rounding off the exponents from the regression analysis to convenient design values, the prediction equation for the nominal shear resistance Q_n (kips) for a single shear stud connector embedded in a solid concrete slab is [A6.10.10.4.3]

$$Q_n = 0.5A_{\text{sc}}\sqrt{f_c' E_c} \le A_{\text{sc}} F_u \tag{19.135}$$

where A_{sc} is the cross-sectional area of a stud shear connector (in.^{2}), f_c' is the specified 28-day concrete-compressive strength (ksi), E_c is the concrete modulus of elasticity (ksi),

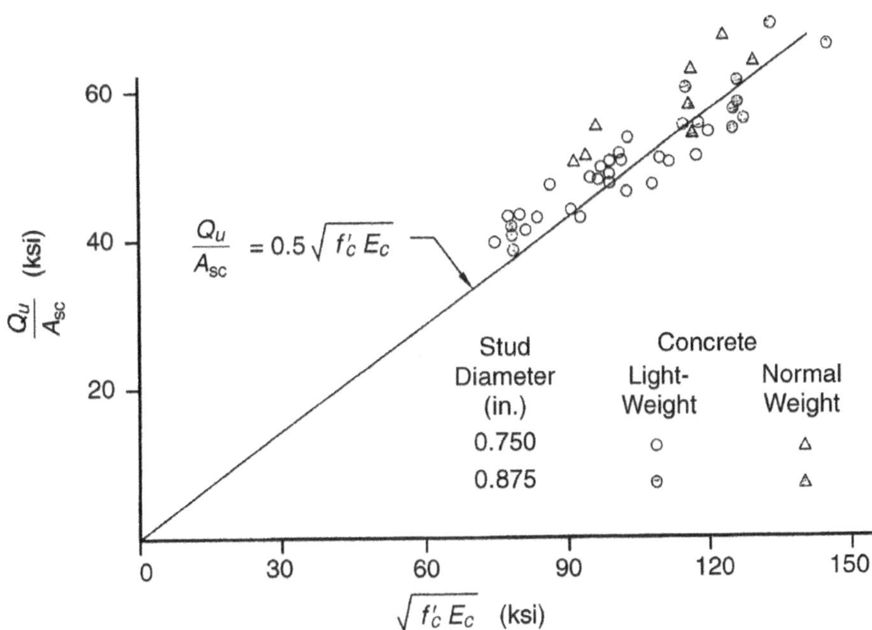

Fig. 19.55 Comparison of connector strength with concrete strength and modulus of elasticity (Ollgaard et al., 1971).

and F_u is the specified minimum tensile strength of a stud shear connector (ksi). The upper bound on the nominal stud shear strength is taken as its ultimate tensile force.

When Eq. 19.135 is compared with the test data from which it was derived (Fig. 19.55), it provides a reasonable estimate to the nominal strength of a stud shear connector. The factored resistance of one shear connector Q_r must take into account the uncertainty in the ability of Eq. 19.135 to predict the resistance at the strength limit state, that is [A6.10.10.4.1],

$$Q_r = \phi_{sc} Q_n \tag{19.136}$$

where ϕ_{sc} is the resistance factor for shear connectors taken from Table 18.7 as 0.85.

Number of Shear Connectors Required If sufficient shear connectors are provided, the maximum possible flexural strength of a composite section can be developed. The shear connectors placed between a point of zero moment and a point of maximum positive moment must resist the compression force in the slab at the location of maximum moment. This resistance is illustrated by the free-body diagrams at the bottom of Figure 19.56 for two different loading conditions. From either of these free-body diagrams, equilibrium requires that

$$n_s Q_r = V_h$$

or

$$n_s = \frac{V_h}{Q_r} \tag{19.137}$$

where n_s is the total number of shear connectors between the points of zero and maximum positive moment, V_h is the nominal horizontal shear force at the interface that must be resisted, and Q_r is the factored resistance of a single shear connector as given by Eqs. 19.135 and 19.136.

Spacing of the Shear Connectors Spacing of the shear connectors along the length L_s needs to be examined. For the concentrated loading of Figure 19.56(a), the vertical shear force is constant. Therefore, the horizontal shear per unit of length calculated from the elastic relationship of Eq. 19.132 is constant and spacing becomes uniform. For the uniformly distributed loading of Figure 19.56(b), the elastic

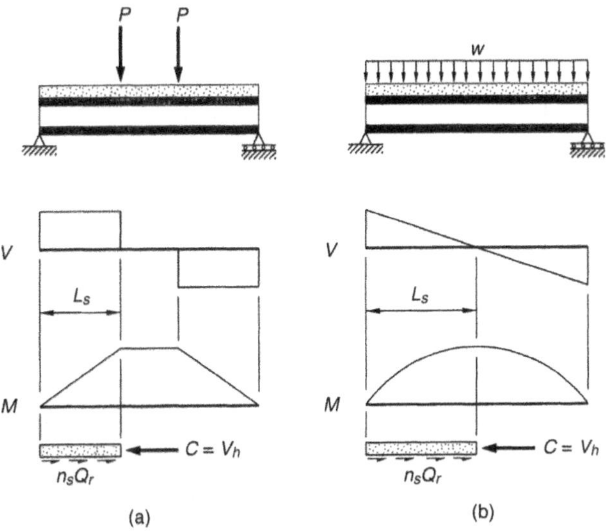

Fig. 19.56 Total number of shear connectors required: (a) concentrated loading and (b) uniformly distributed loading.

horizontal shear per unit of length is variable and indicates that the connectors be closer together near the support than near midspan. These are the conditions predicted by elastic theory. At the strength limit state, conditions are different if ductile behavior permits redistribution of the horizontal shear forces.

To test the hypothesis that stud shear connectors have sufficient ductility to redistribute horizontal shear forces at the strength limit state, Slutter and Driscoll (1965) tested three uniformly loaded simple composite beams with different connector spacings. They designed the beams with about 90% of the connectors required by Eq. 19.137 so that the connectors would control the flexural resistance. The normalized moment versus deflection response for the three beams is shown in Figure 19.57. Considerable ductility is observed and for all practical purposes the response is the same for the three beams. The conclusion is that spacing of the shear connectors along the length of the beam is not critical and can be taken as uniform [C6.10.10.4.2].

Nominal Horizontal Shear Force $\mathbf{V_h}$ At the flexural strength limit state of a composite, the two stress distributions in Figure 19.58 are possible. A haunch is shown to indicate a gap where the shear connectors must transfer the horizontal shear from the concrete slab to the steel section.

For the first case, the plastic neutral axis is in the slab and the compressive force C is less than the full strength of the slab. However, equilibrium requires that C equal the tensile force in the steel section, so that

$$C = V_h = F_{yw}Dt_w + F_{yt}b_tt_t + F_{yc}b_ct_c \qquad (19.138)$$

where V_h is the nominal horizontal shear force shown in Figure 19.56; F_{yw}, F_{yt}, and F_{yc} are the yield strengths of the web, tension flange, and compression flange; D and t_w are the depth and thickness of the web; b_t, t_t and b_c, t_c are the

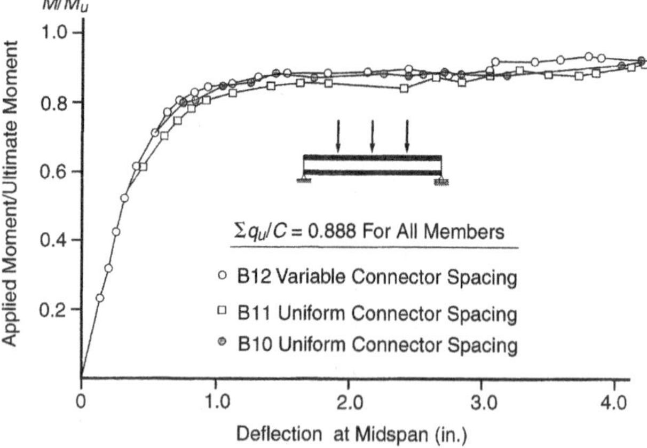

Fig. 19.57 Experimental moment–deflection curves. [Reproduced from R. G. Slutter and G. C. Driscoll (1965). "Flexural Strength of Steel-Concrete Composite Beams," *Journal of Structural Division*, ASCE, 91(ST2), pp. 71–99. With permission.]

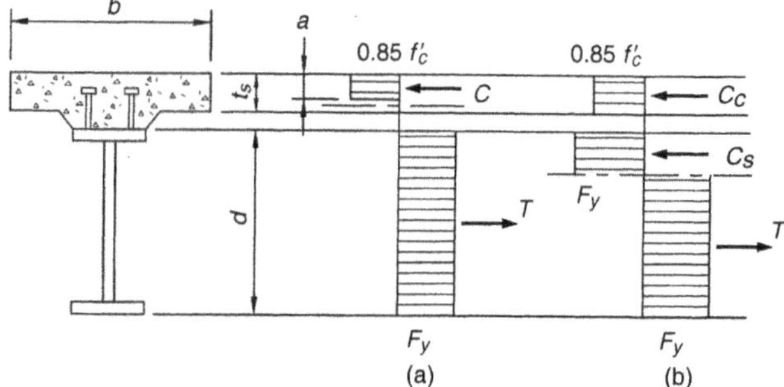

Fig. 19.58 Nominal horizontal shear force: (a) PNA in slab and (b) PNA in steel.

width and thickness of the tension and compression flanges. For the homogeneous steel sections, this simplifies to

$$V_h = F_yA_s \qquad (19.139)$$

where F_y is the yield strength (ksi) and A_s is the total area (in.^{2}) of the steel section.

For the second case, the plastic neutral axis is in the steel section and the compressive force $C = V_h$ is the full strength of the slab given by

$$V_h = 0.85f'_cbt_s \qquad (19.140)$$

where f'_c is the 28-day compressive strength of the concrete (ksi), b is the effective width of the slab (in.), and t_s is the slab thickness (in.).

Techniques for locating the plastic neutral axis in positive moment regions were illustrated in Example 19.5 and Figure 19.23. In calculating V_h, this procedure can be bypassed by simply selecting the *smaller* value of V_h obtained from Eqs. 19.139 and 19.140.

Continuous Composite Sections When negative moment regions in continuous beams are made composite, the nominal horizontal shear force V_h to be transferred between the point of zero moment and maximum moment at an interior support shall be

$$V_h = A_rF_{yr} \qquad (19.141)$$

where A_r is the total area of longitudinal reinforcement (in.^{2}) over the interior support within the effective slab width and F_{yr} is the yield strength (ksi) of the longitudinal reinforcement. Figure 19.24 shows the forces acting on a composite section in a negative moment region. The number of shear connectors required for this region is determined by Eq. 19.141.

Example 19.11 Design stud shear connectors for the positive moment composite section of Example 19.4 shown in Figure 19.22. Assume that the shear range V_{sr} for the fatigue loading is nearly constant and equal to 60 kips in the positive moment region. Use $\frac{3}{4}$-in.-diameter studs 4 in. high, $F_u = 58$ ksi for the studs, $f'_c = 4$ ksi for the concrete deck, and Grade 50 for the steel beam.

General

The haunch depth is 1 in., so the connectors project 4 − 1 = 3 in. into the concrete deck. This projection is greater than the minimum of 2 in. The ratio of stud height to stud diameter is [A6.10.10.1.1]

$$\frac{h}{d} = \frac{4}{0.75} = 5.33 > 4 \qquad \text{OK}$$

The minimum center-to-center transverse spacing of studs is four stud diameters and the minimum clear edge distance is 1 in. The minimum top flange width for three $\frac{3}{4}$-in. studs side by side is

$$b_{f,\min} = 2\,(1) + 3\left(\tfrac{3}{4}\right) + 2\,(4)\left(\tfrac{3}{4}\right) = 10.25 \text{ in.}$$

which is less than the 12 in. provided. Therefore, use three $\frac{3}{4}$-in. stud connectors at each transverse section. [A6.10.10.1.3]

Fatigue Limit State

The center-to-center pitch of shear connectors in the longitudinal direction shall not exceed 24 in. and shall not be less than six stud diameters (6 × 0.75 = 4.5 in. [A6.10.10.1.2].

The pitch is controlled by the fatigue strength of the studs as given by Eq. 19.134:

$$p = \frac{nZ_r I}{V_{sr} Q}$$

where I and Q are elastic properties of the short-term composite section and from Eq. 19.130a [A6.10.10.1.2]

$$Z_r = \alpha d^2$$

for which Eq. 19.130b gives

$$\alpha = 34.5 - 4.28 \log N$$

In Example 18.1, the number of cycles N was estimated for a 75-year life of a rural interstate bridge as 372×10^6 cycles. This value for N is definitely larger than that associated with the limit of 960 trucks per day. Therefore, the design criterion is for fatigue I limit state and infinite fatigue life.

The resistance per stud is (Eq. 19.131)

$$Z_r = 5.5d^2 = 5.5(0.75)^2 = 3.09 \text{ kips}$$

The values of I and Q for the short-term composite section are taken from Table 19.5 as

$$I = 79{,}767 \text{ in.}^4$$

$$Q = Ay = (90)\left(66.625 - \frac{7471.6}{144.5}\right) = 1343 \text{ in.}^3$$

For three stud connectors at a transverse section and $\gamma_{\text{fatigue I}} V_{sr} = 1.50(60 \text{ kips}) = 90$ kips, the pitch is calculated as

$$p = \frac{nZ_r I}{V_{sr} Q} = \frac{3\,(3.09)\,79{,}767}{90 \times 1343} = 6.12 \text{ in.}$$

This pitch is between the limits of 4.5 and 24 in. given earlier. By assuming that the distance from the maximum positive moment to the point of zero moment is 40 ft = 480 in. and that V_{sr} is relatively unchanged, the total number of $\frac{3}{4}$-in. stud connectors over this distance is

$$n = 3\left(\frac{480}{6}\right) = 240 \text{ connectors}$$

Strength Limit State

The total number of shear connectors required to satisfy the strength limit state between the maximum positive moment and the point of zero moment is given by substituting Eq. 19.136 into Eq. 19.137:

$$n_s = \frac{V_h}{Q_r} = \frac{V_h}{\phi_{sc} Q_n}$$

where $\phi_{sc} = 0.85$, Q_n is given by Eq. 19.135, and V_h is given by either Eq. 19.139 or Eq. 19.140. From Eq. 19.135 [A6.10.10.4]

$$Q_n = 0.5A_{sc}\sqrt{f_c' E_c} \le A_{sc} F_u$$

For $\frac{3}{4}$-in. stud connectors

$$A_{sc} = \frac{\pi}{4}(0.75)^2 = 0.44 \text{ in.}^2$$

and for $f_c' = 4$ ksi, Eq. 13.2 yields

$$E_c = 1820\sqrt{f_c'} = 1820\sqrt{4} = 3640 \text{ ksi}$$

so that

$$Q_n = 0.5\,(0.44)\,\sqrt{4\,(3640)} = 26.5 \text{ kips}$$

which is greater than the upper bound of

$$A_{sc} F_u = 0.44\,(58) = 25.5 \text{ kips}$$

Therefore, $Q_n = 25.5$ kips.

The nominal horizontal shear force is the lesser of the values given by Eq. 19.139 or Eq. 19.140. From Eq. 19.139 with A_s taken from Table 19.5

$$V_h = F_y A_s = 50\,(144.5) = 7225 \text{ kips}$$

From Eq. 19.140 with $b = 90$ in. and $t_s = 8$ in. taken from Figure 19.22

$$V_h = 0.85 f_c' b t_s = 0.85\,(4)\,(90)\,(8) = 2448 \text{ kips}$$

Therefore, $V_h = 2448$ kips and the total number of connectors required in the distance from maximum moment to zero moment is

$$n_s = \frac{V_h}{\phi_{sc} Q_n} = \frac{2448}{0.85\,(25.5)} = 113 \text{ connectors}$$
$$= 113\,(2) = 226 \quad \text{(both sides)}$$

Answer The required number of shear connectors is governed by the fatigue I limit state (as it often is). For the assumptions made in this example, the $\frac{3}{4}$-in. diameter stud connectors placed in groups of three are spaced at a pitch of 6 in. throughout the positive moment region.

19.6 STIFFENERS

Webs of standard rolled sections have proportions such that they can reach the bending yield stress and the shear yield stress without buckling. These proportions are not the case with many built-up plate girder and box sections and to prevent buckling their webs must be stiffened. Both transverse and longitudinal stiffeners can be used to improve the strength of webs. In general, transverse stiffeners increase the resistance to shear while longitudinal stiffeners increase the resistance to flexural buckling of the web. The requirements for selecting the sizes of these stiffeners are discussed in the following sections.

19.6.1 Transverse Intermediate Stiffeners

Transverse intermediate stiffeners do not prevent shear buckling of web panels, but they do define the boundaries of the web panels within which the buckling occurs. These stiffeners serve as anchors for the tension field forces so that postbuckling shear resistance can develop (Fig. 19.50). The design of transverse intermediate stiffeners includes consideration of slenderness, stiffness, and strength.

Slenderness Behavior When selecting the thickness and width of a transverse intermediate stiffener (Fig. 19.59), the slenderness of projecting elements must be limited to prevent local buckling. For projecting elements in compression, Eq. 19.22 yields

$$\frac{b_t}{t_p} \le k\sqrt{\frac{E}{F_{ys}}} \tag{19.142}$$

where b_t is the width of the projecting stiffener element, t_p is the thickness of the projecting element, k is the plate buckling coefficient taken from Table 19.2, and F_{ys} is the yield strength of the stiffener. For plates supported along one edge, Table 19.2 gives $k = 0.45$ for projecting elements not a part of rolled shapes.

Other design rules are more empirical, but are nevertheless important for the satisfactory performance of transverse intermediate stiffeners. These are the width of the stiffener b_t must not be less than 2 in. plus one-thirtieth of the depth d of the steel section and not less than one-fourth of the full-width b_f of the steel flange. Further, the slenderness ratio b_t/t_p must be less than 16 [A6.10.11.1.2].

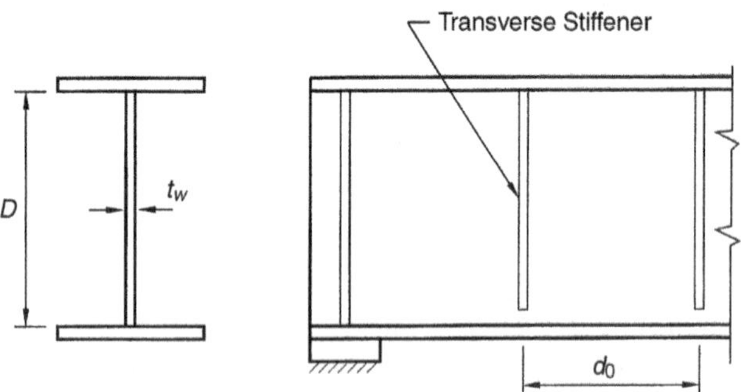

Fig. 19.59 Transverse intermediate stiffener.

Specification Requirements [A6.10.11.1] All of these slenderness requirements for transverse intermediate stiffeners are summarized by two expressions in the AASHTO (2010) LRFD Bridge Specifications as limits on the width b_t of each projecting stiffener element [A6.10.11.1.2]:

$$2 + \frac{d}{30} \le b_t \tag{19.143}$$

and

$$0.25b_f \le b_t \le 16t_p \tag{19.144}$$

Transverse Intermediate Stiffener Behavior (Stiffness) Transverse intermediate stiffeners define the vertical boundaries of the web panel. They must have sufficient stiffness so that they remain relatively straight and permit the web to develop its postbuckling strength.

A theoretical relationship can be developed by considering the relative stiffness between a transverse intermediate stiffener and a web plate. This relationship employs the nondimensional parameter (Bleich, 1952)

$$\gamma_t = \frac{(EI)_{\text{stiffener}}}{(EI)_{\text{web}}}$$

for which

$$(EI)_{\text{web}} = \frac{ED\,t_w^3}{12\left(1-\mu^2\right)}$$

so that

$$\gamma_t = \frac{12\left(1-\mu^2\right)I_t}{Dt_w^3} \tag{19.145}$$

where μ is Poisson's ratio, D is the web depth, t_w is the web thickness, and I_t is the moment of inertia of the transverse intermediate stiffener taken about the edge in contact with the web for single stiffeners and about the midthickness of the web for stiffener pairs. With $\mu = 0.3$, Eq. 19.145 can be rearranged to give

$$I_t = \frac{Dt_w^3}{10.92}\gamma_t \tag{19.146}$$

For a web without longitudinal stiffeners, the value of γ_t to ensure that the critical shear buckling stress τ_{cr} is sustained is approximately (Maquoi, 1992)

$$\gamma_t = m_t\left(\frac{21}{\alpha} - 15\alpha\right) \ge 6 \tag{19.147}$$

where α is the aspect ratio d_0/D and m_t is a magnification factor that allows for postbuckling behavior and the detrimental effect of imperfections. Taking $m_t = 1.3$ and then substituting Eq. 19.147 into Eq. 19.146,

$$I_t = 2.5Dt_w^3\left(\frac{1}{\alpha} - 0.7\alpha\right) \ge 0.55\,Dt_w^3 \tag{19.148}$$

Transverse Intermediate Stiffener Specifications (Stiffness) [A6.10.11.1] The AASHTO (2010) LRFD Bridge Specifications give the requirement for the moment of inertia of any transverse stiffener by two equations [A6.10.11.1.3]:

$$I_t \geq I_{t1} = \min\left[d_o, D\right] t_w^3 J \quad (19.149)$$

where

$$J = 2.5\left(\frac{D}{d_0}\right)^2 - 2.0 \geq 0.5 \quad (19.150)$$

where d_0 is the spacing of transverse intermediate stiffeners and D is the web depth (Fig. 19.59). The transverse stiffener moment of inertia must also be greater than

$$I_t \geq I_{t1} = \frac{D^4}{40}\rho_t^{1.3}\left[\frac{F_{yw}}{E}\right]^{1.5} \quad (19.151a)$$

where

$$\rho_t = \min\left[\frac{F_{yw}}{F_{crs}}, 1.0\right] \quad (19.151b)$$

and F_{crs} is the local buckling stress for the stiffeners,

$$F_{crs} = \frac{0.31E}{\left(b_t/t_p\right)^2} \leq F_{ys} \quad (19.151c)$$

Equation 19.151 was developed by Kim et al. (2007). They determined that the transverse stiffness is critical to develop the full tension field action and restrain the lateral translation of the web. The axial strength requirements (Eq. 19.115) do not control given practical dimensions. Strength is however necessary, but the area requirement for the transverse stiffener is no longer in the specifications. C6.10.11 explains this in more detail.

Example 19.12 Select a one-sided transverse intermediate stiffener for the I-section used in Example 19.11 and shown in Figure 19.60. Use Grade 36 structural steel for the stiffener. The steel in the web is Grade 50. Assume $V_u = 440$ kips at the section.

Slenderness [A6.10.11.1.2]

The size of the stiffener is selected to meet slenderness requirements and then checked for stiffness (and strength, which is implied in the Specifications). From Eqs. 19.143 and 19.144, the width of the projecting element of the stiffener must satisfy

$$b_t \geq 0.25b_f = 0.25\,(8) = 2 \text{ in.}$$

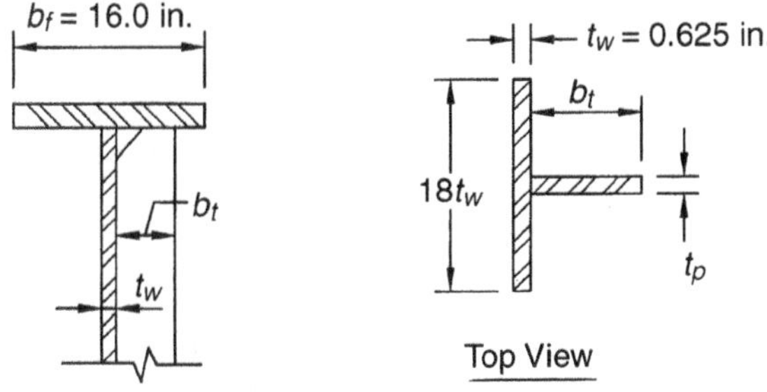

Fig. 19.60 One-sided transverse stiffener. Example 19.12.

and the thickness of the projecting element must satisfy

$$t_p \geq \frac{b_t}{16} = \frac{4}{16} = 0.25 \text{ in.}$$

The minimum thickness of steel elements is $\frac{5}{16}$ in. [A6.7.3], so try a $\frac{7}{8} \times 4.5$ in. transverse intermediate stiffener (Fig. 19.60).

From Eq. 19.143, the width b_t of the stiffener must also satisfy

$$b_t \leq 0.48t_p\sqrt{\frac{E}{F_{ys}}} = 0.48\,(0.875)\sqrt{\frac{29{,}000}{36}} = 11.9 \text{ in.} \quad \text{OK}$$

and

$$b_t \geq 2 + \frac{d}{30} = 2 + \frac{60 + 1.00 + 0.625}{30} = 4.08 \text{ in.} \quad \text{OK}$$

Stiffness [A6.10.11.1.3]

The moment of inertia of the one-sided stiffener is to be taken about the edge in contact with the web. For a rectangular plate, the moment of inertia taken about its base is

$$I_t = \tfrac{1}{3}t_p b_t^3 = \tfrac{1}{3}(0.875)\,(4.5)^3 = 26.6 \text{ in.}^4$$

From Eqs. 19.149 and 19.150, the moment of inertia must satisfy

$$I_t \geq I_{t1} = \min\left[d_0, D\right] t_w^3 J$$

where

$$J = 2.5\left(\frac{D}{d_0}\right)^2 - 2.0 \geq 0.5$$

There are no longitudinal stiffeners, so that $D = 60$ in. From Example 19.11, $d_0 = 80$ in. and $t_w = 0.3125$ in. Hence,

$$J = 2.5\left(\frac{60}{80}\right)^2 - 2.0 = -0.59 \geq 0.5 \qquad \text{Use } J = 0.5$$

Therefore,

$$I_{t1} = \min\left[d_0, D\right] t_w^3 J = \min\,[80{,}60]\,(0.625)^3\,(0.5) = 7.32 \text{ in.}^4$$

which is satisfied by the $\frac{7}{8}$-in. × 4.5-in. stiffener. The transverse stiffener moment of inertia must also satisfy

$$I_t \geq I_{t2} = \frac{D^4}{40}\rho_t^{1.3}\left[\frac{F_{yw}}{E}\right]^{1.5}$$

where

$$\rho_t = \min\left[\frac{F_{yw}}{F_{crs}}, 1.0\right] = \min\left[\frac{50}{36}, 1.0\right] = 1.0$$

Therefore,

$$F_{crs} = \frac{0.31E}{\left(b_t/t_p\right)^2} - \frac{0.31\,(29{,}000)}{(4.5/0.875)^2} = 340 \leq F_{ys} = 36 \text{ ksi}$$

and

$$I_{t2} = \frac{60^4}{40}(1)^{1.3}\left[\frac{50}{29000}\right]^{1.5} = 23.2 \text{ in.}^4$$

$$I_t \geq I_{t2}$$

Because the actual value is greater than required, the requirement is satisfied.

Answer Use a one-sided transverse intermediate stiffener with a thickness of $t_p = \frac{7}{8}$ in. and a width $b_t = 4.5$ in.

19.6.2 Bearing Stiffeners

Bearing stiffeners are transverse stiffeners placed at locations of support reactions and other concentrated loads. The concentrated loads are transferred through the flanges and supported by bearing on the ends of the stiffeners. The bearing stiffeners are connected to the web and provide a vertical boundary for anchoring shear forces from tension field action.

Rolled Beam Shapes Bearing stiffeners are required on webs of rolled beams at points of concentrated forces whenever the factored shear force V_u exceeds [A6.10.9.2.1]

$$V_u > 0.75\phi_b \ V_n \tag{19.152}$$

where ϕ_b is the resistance factor for bearing taken from Table 18.7 and V_n is the nominal shear resistance determined in Section 19.4.

Slenderness Bearing stiffeners are designed as compression members to resist the vertical concentrated forces. They are usually comprised of one or more pairs of rectangular plates placed symmetrically on either side of the web (Fig. 19.61). They extend the full depth of the web and are as close as practical to the outer edges of the flanges. The projecting elements of the bearing stiffener must satisfy the slenderness requirements of [A6.10.11.2.2]

$$\frac{b_t}{t_p} \leq 0.48\sqrt{\frac{E}{F_{ys}}} \tag{19.153}$$

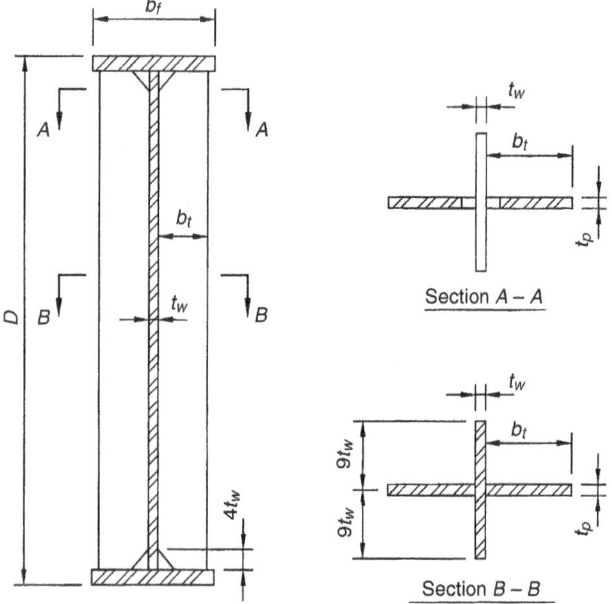

Fig. 19.61 Bearing stiffener cross sections.

where b_t is the width of the projecting stiffener element, t_p is the thickness of the projecting element, and F_{ys} is the yield strength of the stiffener.

Bearing Resistance The ends of bearing stiffeners are to be milled for a tight fit against the flange from which it receives its reaction, the bottom flange at supports and the top flange for interior concentrated loads. If they are not milled, they are to be attached to the loaded flange by a full-penetration groove weld [A6.10.11.2.1].

The effective bearing area is less than the gross area of the stiffener because the end of the stiffener must be notched to clear the fillet weld between the flange and the web (Section A–A, Fig. 19.61). The bearing resistance is based on this reduced bearing area and the yield strength F_{ys} of the stiffener to give [A6.10.11.2.3]

$$\left(R_{sb}\right)_r = \phi_b\left(R_{sb}\right)_n \tag{19.154a}$$

$$\left(R_{sb}\right)_n = 1.4A_{pn}F_{ys} \tag{19.154b}$$

where $(R_{sb})_r$ is the factored bearing resistance, ϕ_b is the bearing resistance factor taken from Table 18.7, A_{pn} is the net area of the projecting elements of the stiffener, and 1.4 is an empirical adjustment. AISC (2005) uses 1.8 here, but they have a difference resistance factor. The 1.4 reflects this difference so that result is the same as AISC [C6.10.11.2.3].

Axial Resistance The bearing stiffeners plus a portion of the web combine to act as a column to resist an axial compressive force (Section B–B, Fig. 19.61). The effective area of the column section is taken as the area of all stiffener elements, plus a centrally located strip of web extending not more than $9t_w$ on each side of the outer projecting elements of the stiffener group [A6.10.11.2.4b].

Because the bearing stiffeners fit tightly against the flanges, rotational restraint is provided at the ends and the effective pin-ended column length KL can be taken as $0.75D$, where D is the web depth [A6.10.11.2.4a]. The moment of inertia of the column section used in the calculation of the radius of gyration is taken about the centerline of the web. Designers often conservatively ignore the contribution of the web when calculating the moment of inertia and simply take the sum of the moments of inertia of the stiffeners about their edge in contact with the web.

The factored axial resistance P_r is calculated from

$$P_r = \phi_c P_n \tag{19.155}$$

where ϕ_c is the resistance factor for compression taken from Table 18.7 and P_n is the nominal compressive resistance determined in Section 19.2.

Example 19.13 Select bearing stiffeners for the I-section used in Example 19.12 and shown in Figure 19.62 to support a factored concentrated reaction $R_u = 900$ kips. Use Grade 36 structural steel for the stiffener.

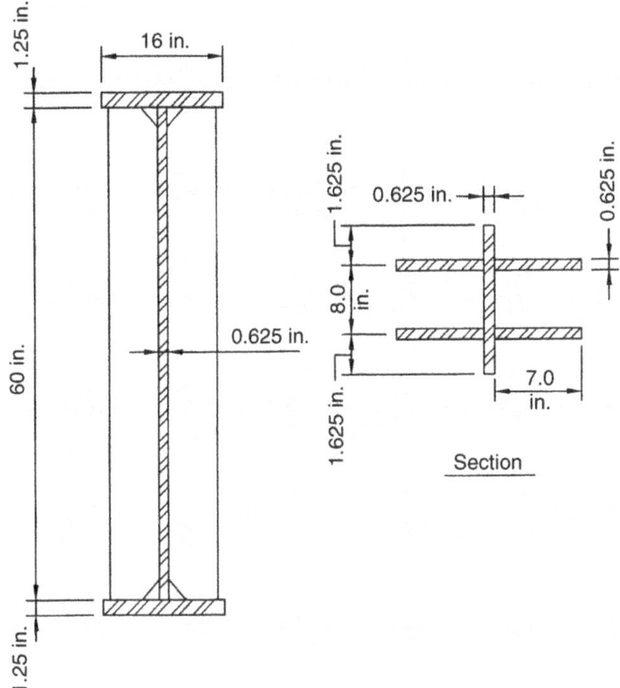

Fig. 19.62 Bearing stiffener Example 19.13.

Slenderness

Selecting the width b_t of the bearing stiffener as 7 in. to support as much of the 16-in. flange width as practical, the minimum thickness for t_p is obtained from Eq. 19.153:

$$\frac{b_t}{t_p} \leq 0.48\sqrt{\frac{E}{F_{ys}}} = 0.48\sqrt{\frac{29{,}000}{36}} = 13.6$$

$$t_p \geq \frac{b_t}{13.6} = \frac{7}{13.6} = 0.51 \text{ in.}$$

Try a $\frac{5}{8}$-in. × 7-in. bearing stiffener element.

Bearing Resistance

The required area of all the bearing stiffener elements can be calculated from Eq. 19.154 for $(R_{sb})_r = 900$ kips, $\phi_b = 1.0$ (milled surface), and $F_{ys} = 36$ ksi:

$$(R_{sb})_r = \phi_b 1.4 A_{pn} F_{ys} = (1.0)(1.4) A_{pn} (36)$$

$$A_{pn} = \frac{900}{1.4(36)} = 17.9 \text{ in.}^2$$

By using two pairs of $\frac{5}{8}$-in. × 7-in. stiffener elements on either side of the web (Fig. 19.62), and allowing 2.5 in. to clear the web to flange fillet weld, the provided bearing area is

$$4(0.625)(7-2.5) = 11.25 \text{ in.}^2 + \text{web contribution} > 17.9 \text{ in.}^2 \qquad \text{OK}$$

Try a bearing stiffener composed of four $\frac{5}{8}$-in. × 7-in. elements placed in pairs on either side of the web. (Note that the 45° notch with $4t_w$ sides prevents the development of the unwanted triaxial tensile stress in the welds at the junction of the web, stiffener, and flange.)

Axial Resistance

By spacing the pairs of stiffeners 8 in. apart as shown in Figure 19.62, the effective area of the column cross section is

$$A = 4A_s + t_w(18t_w + 8)$$

$$A = 4(0.625)(7) + 0.625(11.25 + 8) = 29.5 \text{ in.}^2$$

and the moment of inertia of the stiffener elements about the centerline of the web is

$$I = 4I_0 + 4A_g y^2$$

$$= 4\left[\frac{1}{12}(0.625)(7)^3\right] + 4(0.625)(7)\left(\frac{7}{2} + \frac{0.625}{2}\right)^2$$

$$= 326 \text{ in.}^4$$

so that the radius of gyration for the column cross section becomes

$$r = \sqrt{\frac{I}{A}} = \sqrt{\frac{326}{29.5}} = 3.3 \text{ in.}$$

Therefore,

$$\frac{KL}{r} = \frac{0.75D}{r} = \frac{0.75(60)}{3.3} = 13.5 < 120 \qquad \text{OK}$$

and Eq. 19.16 gives

$$\frac{P_e}{P_o} = \left(\frac{\pi}{KL/r}\right)^2 \left(\frac{E}{F_y}\right) = \left(\frac{\pi}{13.5}\right)^2 \left(\frac{29000}{36}\right)$$

$$= 43.6 \approx 0.44$$

so that the nominal column strength is given by Eq. 19.21a:

$$P_n = 0.658^{(P_o/P_e)} F_y A_g = (0.658)^{1/43.6}(36)(29.5)$$

$$= 1052 \text{ kips}$$

that is essentially 100% of yield strength. The factored axial resistance is calculated from Eq. 19.21a with $\phi_c = 0.90$:

$$P_r = \phi_c P_n = 0.90(1052) = 947 \text{ kips} > 900 \text{ kips} \qquad \text{OK}$$

Answer Use a bearing stiffener composed of two pairs of $\frac{5}{8}$-in. × 7-in. stiffener elements arranged as shown in Figure 19.62.

REFERENCES

AASHTO (2010). *LRFD Bridge Design Specifications*, 5th ed., American Association of State Highway and Transportation Officials, Washington, DC.

AISC (2002). *Detailing for Steel Construction*, 2nd ed., American Institute of Steel Construction, Chicago.

AISC (2005). *LRFD Manual of Steel Construction*, 13th ed., American Institute of Steel Construction, Chicago.

Basler, K. (1961). "Strength of Plate Girders in Shear," *Journal of Structural Division*, ASCE, Vol. 87, No. ST7, October, pp. 151–180.

Basler, K. and B. Thürlimann (1961). "Strength of Plate Girders in Bending," *Journal of Structural Divisoin*, ASCE, Vol. 87, No. ST6, August, pp. 153–181.

Bjorhovde, R. (1992). "Compression Members," in *Constructional Steel Design*, Dowling et al. (Eds.), Elsevier Science (Chapman & Hall, Andover, England), Chapter 2.3.

Bleich, F. (1952). *Buckling Strength of Metal Structures*, McGraw-Hill, New York.

Dowling, P. J., J. E. Hardin, and R. Bjorhovde (Eds.) (1992). *Constructional Steel Design—An International Guide*, Elsevier Science (Chapman & Hall, Andover, England).

Gaylord, E. H., Jr., C. N. Gaylord, and J. E. Stallmeyer (1992). *Design of Steel Structures*, 3rd ed., McGraw-Hill, New York.

Kim, Y. D, S. K. Jung, and D. W. White (2007). Transverse Stiffener Requirements in Straight and Horizontally Curved Steel I-Girders, *Journal of Bridge Engineering*, ASCE, Vol. 12, No. 2, March.

Kitipornchai, S. and N. S. Trahair (1980). "Buckling Properties of Monosymmetric I-Beams," *Journal of Structural Division*, ASCE, Vol. 106, No. ST5, May, pp. 941–957.

Maquoi, R. (1992). "Plate Girders," in *Constructional Steel Design*, Dowling et al. (Eds.), Elsevier Science (Chapman & Hall, Andover, England), Chapter 2.6.

Munse, W. H. and E. Chesson, Jr. (1963). "Riveted and Bolted Joints: Net Section Design," *Journal of Structural Division*, ASCE, Vol. 89, No. ST1, February, pp. 107–126.

Nethercot, D. A. (1992). "Beams," in *Constructional Steel Design*, Dowling, et al. (Eds.), *Elsevier Science* (Chapman & Hall, Andover, England), Chapter 2.2.

Ollgaard, J. G., R. G. Slutter, and J. W. Fisher (1971). "Shear Strength of Stud Connectors in Lightweight and Normal-Weight Concrete," *AISC Engineering Journal*, Vol. 8, No. 2, April, pp. 55–64.

Segui, W. T. (2003). *LRFD Steel Design*, 3rd ed., Brooks-Cole, Boston.

Slutter, R. G. and G. C. Driscoll, Jr. (1965). "Flexural Strength of Steel-Concrete Composite Beams," *Journal of Structural Division*, ASCE, Vol. 91, No. ST2, April, pp. 71–99.

Slutter, R. G. and J. W. Fisher (1967). "Fatigue Strength of Shear Connectors," *Steel Research for Construction*, Bulletin No. 5, American Iron and Steel Institute, Washington, DC.

Taylor, J. C. (1992). "Tension Members," in *Constructional Steel Design*, Dowling et al. (Eds.), Elsevier Science (Chapman & Hall, Andover, England), Chapter 2.1.

Timoshenko, S. and J. M. Gere (1969). *Theory of Elastic Stability*, 3rd ed., McGraw-Hill, New York.

White, D. W. and M. A. Grubb (2005). "Unified Resistance Equations for Design of Curved and Tangent Steel Bridge I-Girders," Proceedings of the 2005 TRB Bridge Engineering Conference, Transportation Research Board, Washington, D.C., July.

PROBLEMS

These problems reference the plans for the bridge over the Little Laramie River. See Wiley's website for a pdf. This bridge is a single-span, four-girder bridge that has clean details; its interpretation is straightforward for the student. These plans are used with permission.

Note: The arithmetic can be simplified by using a 100-ft span rather than the 95′–6 3/4″ span (bearing-to-bearing span, which is the actual analysis span length .

19.1 The intermediate cross frame contains diagonal steel angles $L3 \times 3 \times \frac{5}{16}$. Determine the tensile resistance of this member. Note that bolts are used on the center-of-bay gusset (see half-section—intermediate cross frame).

19.2 The intermediate cross frame contains diagonal steel angles $L3 \times 3 \times \frac{5}{16}$. Determine the compression resistance of this member. Note that bolts are used on the center-of-bay gusset (see half-section—intermediate cross frame).

19.3 The intermediate cross frame contains diagonal steel angles $L3\,^1\!/_2 \times 3\,^1\!/_2 \times \frac{5}{16}$. Determine the tensile resistance of this member. Note that bolts are used on the center-of-bay gusset (see half-section—intermediate cross frame).

19.4 The intermediate cross frame contains a bottom steel angle $L\,^1\!/_2 \times 3\,^1\!/_2 \times \frac{5}{16}$. Determine the compression resistance of this member. Note that bolts are used on the center-of-bay gusset (see half-section—intermediate cross frame).

19.5 For the midspan section and interior girder, determine:

a. The noncomposite section properties
b. The composite section properties for live loads
c. The composite section properties for sustained loads
d. The plastic moment resistance for the noncomposite section
e. The plastic moment resistance for the composite section

Use an effective width of 92 in.

19.6 Near the end of the bridge, determine the shear resistance. Although there are stiffeners at the bearing and cross frames, this is considered an unstiffened girder, that is, d_0 is about 25 ft.

19.7 How many shear groups of connectors (each side of midspan) are required to develop the full plastic moment capacity at midspan.

19.8 This bridge has an "integral" abutment. Under what load conditions are the bearing stiffeners important?

19.9 Repeat 19.5 for the exterior girder.

CHAPTER 20

Steel Design Examples

20.1 NONCOMPOSITE ROLLED STEEL BEAM BRIDGE

Problem Statement Example 20.1 Design the simple-span noncomposite rolled steel beam bridge of Figure E20.1-1 with 35-ft span for an HL-93 live load. Roadway width is 44 ft curb to curb. Allow for a future wearing surface of 3-in.-thick bituminous overlay. Use $f_c' = 4$ ksi and M270 Grade 50W steel. The steel is assumed not to be coated, therefore, the fatigue detail at midspan is category B. The barrier is 15 in. wide and weighs 0.5 k/ft and may be assigned to the exterior girders. Consider the outline of AASHTO (2010) LRFD Bridge Specifications, Section 6, Appendix C.

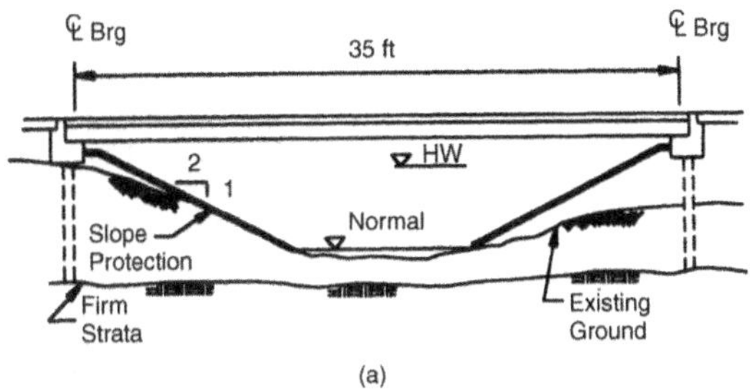

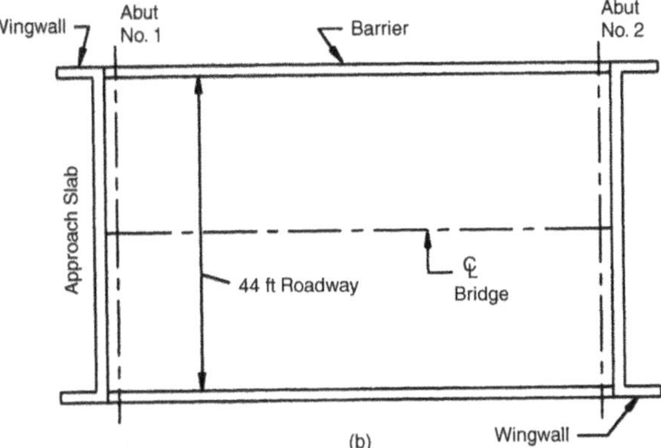

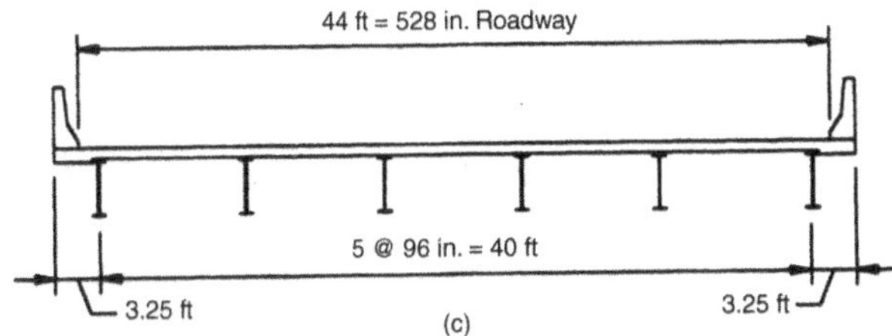

Fig. E20.1-1 Noncomposite rolled steel beam bridge design example: (a) general elevation, (b) plan view, and (c) cross section.

A. Develop General Section The bridge is to carry interstate traffic over a normally small stream that is subject to high water flows during the rainy season (Fig. E20.1-1).

1. *Roadway Width (Highway Specified)* Roadway width is 44-ft curb to curb.
2. *Span Arrangements* [A2.3.2]* [A2.5.4] [A2.5.5] [A2.6] Simple span, 35 ft.
3. *Select Bridge Type* A noncomposite steel plate I-girder is selected for this bridge.

***B.* Develop Typical Section**

1. *I-Girder*
   a. Composite or Noncomposite Section [A6.10.1.1] This bridge is noncomposite, does not have shear connectors, and the shear strength should follow [A.6.10.10]. Noncomposite design is discouraged by the AASHTO LRFD Bridge Design Specifications, however, this example is provided in order to begin a comprehensive example with a simple bridge. This same span configuration is repeated for a composite bridge in the next example.
   b. Nonhybrid [A6.10.1.3] This cross section is a rolled beam and the same material properties are used throughout the cross section. The section is nonhybrid.
   c. Variable Web Depth [A6.10.1.4] The section depth is prismatic and variable-depth provisions are not applicable.

C. Design Conventionally Reinforced Concrete Deck The deck was designed in Example Problem 16.1.

D. Select Resistance Factor

	ϕ	
1. *Strength Limit State*		[A6.5.4.2]
Flexure	1.00	
Shear	1.00	
2. *Nonstrength Limit States*	1.00	[A1.3.2.1]

E. Select Load Modifiers For simplicity in this example, these factors are set to unity, and $\eta_i = \eta$.

		Strength	Service	Fatigue
1. Ductility, η_D	[A1.3.3]	1.0	1.0	1.0
2. Redundancy, η_R	[A1.3.4]	1.0	1.0	1.0
3. Importance, η_I	[A1.3.5]	1.0	N/A	N/A
$\eta = \eta_D\eta_R\eta_I$	[A1.3.2.1]	1.0	1.0	1.0

*The article numbers in the AASHTO (2010) LRFD Bridge Specifications are enclosed in brackets and preceded by the letter A if a specification article and by the letter C if commentary.

F. Select Load Combination and Load Factors

1. *Strength I Limit State*

$$U = \eta\,[1.25\text{DC} + 1.50\text{DW} + 1.75\,(\text{LL} + \text{IM}) + 1.0\text{FR} + \gamma_{\text{TG}}\text{TG}]$$

2. *Service I Limit State*

$$U = \eta\,[1.0\,(\text{DC} + \text{DW}) + 1.0\,(\text{LL} + \text{IM}) + 0.3\,(\text{WS} + \text{WL}) + 1.0\text{FR}]$$

3. *Service II Limit State*

$$U = \eta\,[1.0\,(\text{DC} + \text{DW}) + 1.3\,(\text{LL} + \text{IM})]$$

4. *Fatigue I and II, and Fracture Limit State*

$$U_{\text{fatigue I}} = \eta\,[1.5\,(\text{LL} + \text{IM})]$$
$$U_{\text{fatigue II}} = \eta\,[0.75\,(\text{LL} + \text{IM})]$$

5. *Construction State Strength I*

$$U = \eta[1.25\,(\text{DC}) + 1.75\,(\text{Construction live loads plus 1.5 IM})]$$

G. Calculate Live-Load Force Effects

1. *Select Live Loads* [A3.6.1] *and Number of Lanes* [A3.6.1.1.1] *Select Number of Lanes* [A3.6.1.1.1]:

$$N_L = \text{INT}\left(\frac{w}{12}\right) = \text{INT}\left(\frac{44}{12}\right) = 3$$

2. *Multiple Presence* [A3.6.1.1.2] (Table 8.6)

No. of Loaded Lanes	M
1	1.20
2	1.00
3	0.85

3. *Dynamic Load Allowance* [A3.6.2] (Table 8.7)

Component	IM (%)
Deck joints	75
Fatigue	15
All other	33

Not applied to the design lane load.

4. *Distribution Factor for Moment* [A4.6.2.2.2] Assume for preliminary design, $K_g/12Lt_s^3 = 1.0$. This should be conservative for a noncomposite beam and this value is checked later.

a. Interior Beams [A4.6.2.2.2b] (Table 6.5) One design lane loaded:

$$mg_M^{\text{SI}} = 0.06 + \left(\frac{S}{14}\right)^{0.4}\left(\frac{S}{L}\right)^{0.3}\left(\frac{K_g}{12Lt_s^3}\right)^{0.1}$$
$$mg_M^{\text{SI}} = 0.06 + \left(\frac{8}{14}\right)^{0.4}\left(\frac{8}{35}\right)^{0.3}(1.0)^{0.1} = 0.573$$

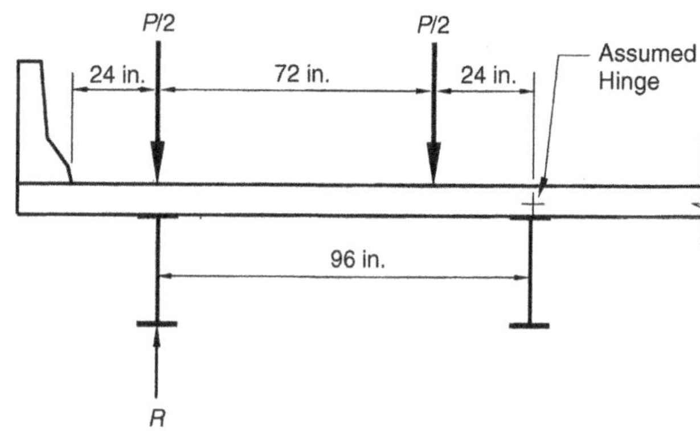

Fig. E20.1-2 Lever rule for the determination of distribution factor for moment in exterior bean, one lane loaded.

Two or more design lanes loaded:

$$mg_M^{\text{MI}} = 0.075 + \left(\frac{S}{9.5}\right)^{0.6}\left(\frac{S}{L}\right)^{0.2}\left(\frac{K_g}{12Lt_s^3}\right)^{0.1}$$
$$mg_M^{\text{MI}} = 0.075 + \left(\frac{8}{9.5}\right)^{0.6}\left(\frac{8}{35}\right)^{0.2}(1.0)^{0.1}$$
$$= 0.746 \quad \text{governs}$$

b. Exterior Beams [A4.6.2.2.2d] (Table 11.3) [Table A4.62.2.2d-1] One design lane loaded—lever rule (Fig. E20.1-2):

$$R = \frac{P}{2}\left(\frac{2+8}{8}\right) = 0.625P$$
$$g_M^{\text{SE}} = 0.625$$
$$mg_M^{\text{SE}} = 1.2\,(0.625) = 0.75 \quad \text{governs}$$

Two or more design lanes loaded:

$$d_e = 3.25 - 1.25 = 2 \text{ ft}$$
$$e = 0.77 + \frac{d_e}{9.1} = 0.77 + \frac{2}{9.1} = 0.99$$
$$mg_M^{\text{ME}} = e \cdot mg_M^{\text{MI}} = 0.743$$

The rigid method of [A4.6.2.2.2] requires stiff diagraphms or cross frame that affects the transverse stiffness. Here we assume end diaphragms and others at one-quarter point. This is not significant for this case and [A4.6.2.2.2] rigid method is neglected. If computed, it yields a slightly higher distribution factor for the exterior girder.

c. Skewed Bridge [A4.6.2.2.2e] This is a straight bridge and no adjustment is required for skew.

Live-Load Moments (See Figs. E20.1-3 and E20.1-4)

$$M_{\text{LL+IM}} = mg\left[(M_{\text{Truck}} \text{ or } M_{\text{Tanderm}})\left(1 + \frac{\text{IM}}{100}\right) + M_{\text{Lane}}\right]$$

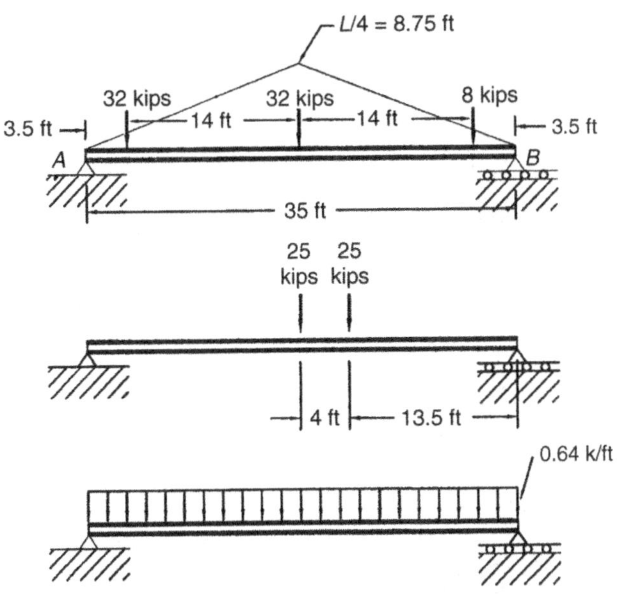

Fig. E20.1-3 Truck, tandem, and lane load placement for maximum moment at location 105.

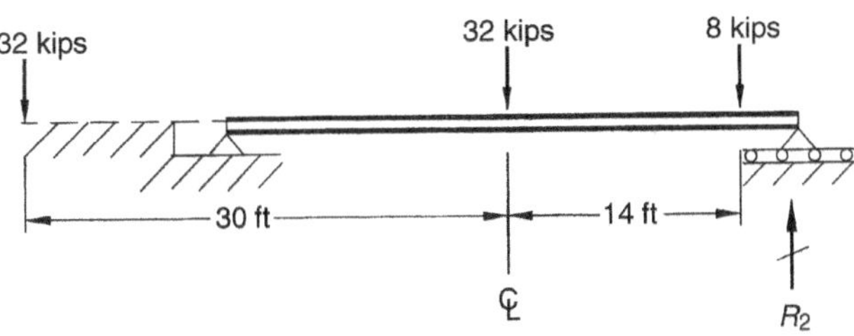

Fig. E20.1-4 Fatigue truck placement for maximum moment.

$$M_{\text{Truck}} = 32\,(8.75) + (32 + 8)\,(1.75) = 350 \text{ k ft}$$
$$M_{\text{Tandem}} = 25\,(8.75 + 6.75) = 387.5 \text{ k ft} \quad \text{governs}$$
$$M_{\text{Fatigue}} = 32\,(8.75) + 8\,(1.75) = 294 \text{ k ft} \quad \text{(used later)}$$

The absolute moment due to the tandem actually occurs under the wheel closest to the resultant when the center of gravity of the wheels on the span and the critical wheel are equidistant from the centerline of the span. For this span, the absolute maximum moment is 388 k ft. However, the value of 387.5 k ft is used because the moments due to other loads are maximum at the centerline and thus can be added to the tandem load moment:

$$M_{\text{Lane}} = \frac{0.64(35)^2}{8} = 98.0 \text{ k ft}$$

Interior Beams

$$M_{\text{LL+IM}} = 0.743\,[387.5\,(1.33) + 98.0] = 455.7 \text{ k ft}$$

$$M_{\text{Fatigue+IM}} = (0.573/1.2)\,[294\,(1.15)] = 161.4 \text{ k ft} \quad \text{(used later)}$$

Exterior Beams

$$m_{\text{LL+IM}} = 0.75\,[387.5\,(1.33) + 98.0] = 460.0 \text{ k ft}$$

$$M_{\text{Fatigue+IM}} = (0.75/1.2)\,[294\,(1.15)] = 211.3 \text{ k ft} \quad \text{(used later)}$$

5. *Distribution Factor for Shear* [A4.6.2.2.3] Use cross-section type (a) (Table 4.1).

a. Interior Beams [A4.6.2.2.2a] One design lane loaded (Table 11.3) [Table 4.6.2.2.3a-1]:

$$mg_V^{\text{SI}} = 0.36 + \frac{S}{25} = 0.36 + \frac{8}{25} = 0.68$$

Two design lanes loaded:

$$mg_V^{\text{MI}} = 0.2 + \frac{S}{12} - \left(\frac{S}{L}\right)^{2.0}$$
$$mg_V^{\text{MI}} = 0.2 + \frac{8}{12} - \left(\frac{8}{35}\right)^{2.0} = 0.81 \quad \text{governs}$$

b. Exterior Beams [A4.6.2.2.2b] One design lane loaded—lever rule (Table 11.3) [Table A4.6.2.2.3b-1] (Fig. E20.1-2):

$$mg_V^{\text{SE}} = 0.75 \quad \text{governs}$$

Two or more design lanes loaded:

$$d_e = 2 \text{ ft}$$
$$e = 0.6 + \frac{d_e}{10} = 0.6 + \frac{2}{10} = 0.80$$
$$mg_V^{\text{ME}} = e \cdot mg_V^{\text{MI}} = (0.80)\,(0.81) = 0.65$$

Again, the rigid method is not used.
Distributed live-load shears (Fig. E20.1-5):

$$V_{\text{LL+IM}} = mg\left[\left(V_{\text{Truck}} \text{ or } V_{\text{Tandem}}\right)\left(1 + \frac{\text{IM}}{100}\right) + V_{\text{Lane}}\right]$$
$$V_{\text{Truck}} = 32\,(1 + 0.60) + 8\,(0.20) = 52.8 \text{ kips} \quad \text{governs}$$
$$V_{\text{Tandem}} = 25\,(1 + 0.886) = 47.1 \text{ kips}$$
$$V_{\text{Lane}} = \frac{0.64\,(35)}{2} = 11.2 \text{ kips}$$
$$V_{\text{Fatigue}} = 32\,(1) + 8\,(0.6) = 36.8 \text{ kips} \quad \text{(used later)}$$

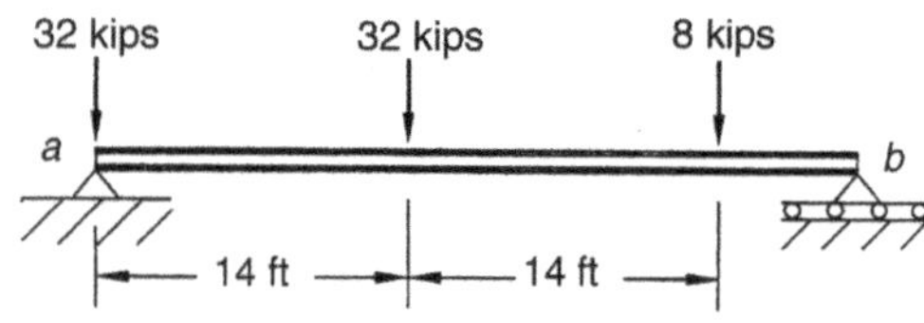

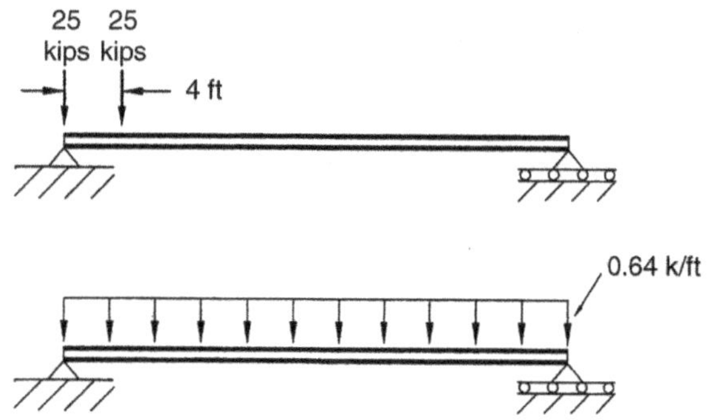

Fig. E20.1-5 Truck, tandem, and lane load placement for maximum shear at location 100.

Interior Beams

$$V_{LL+IM} = 0.81\,[52.8\,(1.33) + 11.2] = 66.0 \text{ kips}$$
$$V_{Fatigue+IM} = (0.68/1.2)\,[36.8\,(1.15)]$$
$$= 24.0 \text{ kips} \quad \text{(used later)}$$

Exterior Beams

$$V_{LL+IM} = 0.75\,[52.8\,(1.33) + 11.2] = 61.1 \text{ kips}$$
$$V_{Fatigue+IM} = (0.75/1.2)\,[36.8\,(1.15)] = 26.5 \text{ kips} \quad \text{(used later)}$$

c. Skewed Bridge [A4.6.2.2.2c] Again, this is a straight bridge and no adjustment is necessary for skew.

6. *Stiffness* [A6.10.1.5] Loads are applied to the bare steel noncomposite section.
7. *Wind Effects* [A4.6.2.7] The wind pressure on superstructure is 50 psf = 0.050 ksf. This load is applied to girders, deck, and barriers. The diaphragm design uses these loads.
8. *Reactions to Substructure* [A3.6] The following reactions are per design lane without any distribution factors:

$$R_{100} = V_{100} = 1.33 V_{Truck} + V_{Lane}$$
$$= 1.33\,(52.8) + 11.2 = 81.4 \text{ kips/lane}$$

H. Calculate Force Effects from Other Loads Analysis for a uniformly distributed load w (Fig. E20.1-6):

$$M_{max} = M_{105} = \frac{wL^2}{8} = \frac{w(35)^2}{8} = 153.1 \times w \text{ kip ft}$$

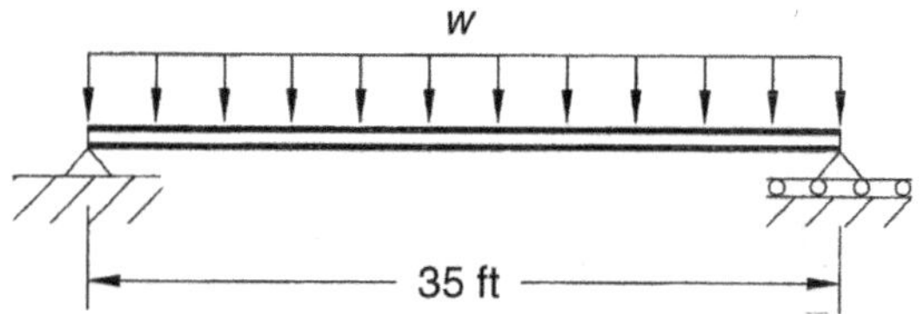

Fig. E20.1-6 Uniform distributed load.

Table E20.1-1 Interior Girder Unfactored Moments and Shears

Load Type	w (k/ft)	Moment (k ft) M_{105}	Shear (kips) V_{100}
DC	0.90	137.8	15.75
DW	0.28	42.9	4.90
LL + IM (distributed)	N/A	455.7	66.0
(Fatigue + IM) (distributed)	N/A	161.4	24.0

$$V_{max} = V_{100} = \frac{wL}{2}$$
$$= \frac{w\,(35)}{2} = 17.5 \times w \text{ kips}$$

Assume a beam weight of 0.10 k/ft:

1. *Interior Girders*

$$\text{DC} \quad \text{Deck slab} \quad (0.15)\left(\tfrac{8}{12}\right)(8) = 0.80 \text{ k/ft}$$
$$\text{Girder} = 0.10 \text{ k/ft}$$
$$w_{DC} = 0.90 \text{ k/ft}$$
$$\text{DW} \quad \text{75-mm bituminous paving} = (0.140)(3/12)(8)$$
$$w_{DW} = 0.28 \text{ k/ft}$$

Unfactored moments and shears for an interior girder are summarized in Table E20.1-1.

2. *Exterior Girders* The barrier weight is assigned to the exterior girder for this example:

$$\text{DC} \quad \text{Deck slab} = (0.15)\left(\tfrac{8}{12}\right)\left(3.25 + \tfrac{8}{2}\right)$$
$$= 0.72 \text{ k/ft}$$
$$\text{Barrier} = 0.50 \text{ k/ft}$$
$$\text{Girder} = 0.10 \text{ k/ft}$$
$$w_{DC} = 1.33 \text{ k/ft}$$

$$\text{DW} \quad \text{3-in. bituminous paving}$$
$$w_{DW} = (0.14)\left(\tfrac{3}{12}\right)\left(2 + \tfrac{8}{2}\right)$$
$$= 0.21 \text{ k/ft}$$

Unfactored moments and shears for an exterior girder are summarized in Table E20.1-2.

Table E20.1-2 Exterior Girder Unfactored Moments and Shears

Load Type	w (k/ft)	Moment (k ft) M_{105}	Shear (kips) V_{100}
DC	1.33	203.7	23.3
DW	0.21	32.20	3.68
(LL + IM) (distributed)	N/A	460.0	61.1
(Fatigue + IM) (distributed)	N/A	211.3	26.5

I. Design Required Sections Flexural design.

1. Factored Loads

a. Interior Beam

Factored Shear and Moment

$$U_{\text{Strength I}} = \eta\,[1.25\text{DC} + 1.50\text{DW} + 1.75\,(\text{LL} + \text{IM})]$$

$$V_u = 1.0\,[1.25\,(15.75) + 1.50\,(4.9) + 1.75\,(66)] = \mathbf{142.5\ kips} \quad \textbf{(strength I)}$$

$$M_u = 1.0\,[1.25\,(137.8) + 1.50\,(42.9) + 1.75\,(455.7)] = 1034.1 \text{ k ft} \quad \text{(strength I)}$$

$$U_{\text{Service II}} = \eta\,[1.0\text{DC} + 1.0\text{DW} + 1.30\,(\text{LL} + \text{IM})]$$

$$V_u = 1.0\,[1.0\,(15.75) + 1.0\,(4.9) + 1.3\,(66)] = \mathbf{106.5\ kips} \quad \textbf{(service II)}$$

$$M_u = 1.0\,[1.0\,(137.8) + 1.0\,(42.9) + 1.3\,(455.7)] = 773.1 \text{ k ft} \quad \text{(service II)}$$

$$U_{\text{Fatigue I}} = \eta\left[1.5\,(\text{Range of }(\text{LL} + \text{IM}))\right]$$

$$U_{\text{Fatigue II}} = \eta\left[0.75\,(\text{Range of }(\text{LL} + \text{IM}))\right]$$

Dead loads are considered in some fatigue computations and not in others. Details are provided later. Critical values are in boldface.

$$V_u = 1.0\,[1.5\,(24)] = \mathbf{36\ kips} \quad \textbf{(fatigue I)}$$

$$V_u = 1.0\,[0.75\,(24)] = \mathbf{18\ kips} \quad \textbf{(fatigue II)}$$

$$M_u = 1.0\,[1.5\,(161.4)] = 242.1 \text{ k ft} \quad \text{(fatigue I)}$$

$$= 1.0\,[0.75\,(161.4)] = 121.1 \text{ k ft} \quad \text{(fatigue II)}$$

$$U_{\text{Construction}} = \eta\,[1.25\text{DC}]$$

$$V_u = 1.0\,[1.25\,(15.75)] = 19.7 \text{ kips} \quad \text{(construction)}$$

$$M_u = 1.0\,[1.25\,(137.8)] = 172.3 \text{ k ft} \quad \text{(construction)}$$

b. Exterior Beam

Factored Shear and Moment

$$U_{\text{Strength I}} = \eta\,[1.25\text{DC} + 1.50\text{DW} + 1.75\,(\text{LL} + \text{IM})]$$

$$V_u = 1.0\,[1.25\,(23.3) + 1.50\,(3.68) + 1.75\,(61.1)] = 141.6 \text{ kips} \quad \text{(strength I)}$$

$$M_u = 1.0\,[1.25\,(203.7) + 1.50\,(32.2) + 1.75\,(460.0)] = \mathbf{1107.9\ k\ ft} \quad \textbf{(strength I)}$$

$$U_{\text{Service II}} = \eta\,[1.0\text{DC} + 1.0\text{DW} + 1.30\,(\text{LL} + \text{IM})]$$

$$V_u = 1.0\,[1.0\,(23.3) + 1.0(3.68) + 1.3\,(61.1)] = 106.4 \text{ kips} \quad \text{(service II)}$$

$$M_u = 1.0\,[1.0\,(203.7) + 1.0\,(32.2) + 1.3\,(460.0)] = \mathbf{833.9\ k\ ft} \quad \textbf{(strength II)}$$

$$U_{\text{Fatigue I}} = \eta\,[1.5\,(\text{LL} + \text{IM})]$$

$$U_{\text{Fatigue II}} = \eta\,[0.75\,(\text{LL} + \text{IM})]$$

Dead loads are considered in some fatigue computations and not in others. Details are provided later.

$$V_u = 1.0\,[1.5\,(26.5)] = 39.8 \text{ kips} \quad \text{(fatigue I)}$$

$$V_u = 1.0\,[0.75\,(26.5)] = 19.9 \text{ kips} \quad \text{(fatigue II)}$$

$$M_u = 1.0\,[1.5\,(211.3)] = \mathbf{317.0\ k\ ft} \quad \textbf{(fatigue I)}$$

$$M_u = 1.0\,[0.75\,(211.3)] = \mathbf{158.5\ k\ ft} \quad \textbf{(fatigue II)}$$

$$U_{\text{Construction}} = \eta\,[1.25\text{DC}]$$

$$V_u = 1.0\,[1.25\,(23.3)] = \mathbf{29.1\ kips} \quad \textbf{(construction)}$$

$$M_u = 1.0\,[1.25\,(203.7)] = \mathbf{254.6\ k\ ft} \quad \textbf{(construction)}$$

2. Trial Section

$$\phi_f M_n \ge M_u \qquad \phi_f = 1.0 \qquad M_n = M_p = ZF_y$$

$$ZF_y \ge M_u$$

Assume that the compression flange is fully braced and section is compact:

$$\text{Req'd } Z \ge \frac{M_u}{F_y} = \frac{1107.9\,(12)}{50} = 265.9 \text{ in.}^3$$

Try W30 × 90, $Z = 283\text{ in.}^3$, $S = 245\text{ in.}^3$,

$$I = 3620\text{ in.}^4 \quad b_f = 10.400\text{ in.} \quad t_f = 0.610\text{ in.}$$
$$t_w = 0.470 \quad d = 29.53\text{ in.} \quad w_g = 0.090\text{ k/ft}$$

Lateral bracing of the compression flange is later addressed.

Cross-Section Proportion Limits [A6.10.2] For most rolled sections, this one included, only bracing is an issue regarding the proportions and compactness. All checks are illustrated later for completeness. Plate girders should be completely checked for proportion limits.

$$\frac{D}{t_w} \leq 150$$
$$\frac{29.53 - 2\,(0.61)}{0.47} = 60.3 \leq 150 \qquad \text{OK}$$

This is conservative for the wide-flange section, as expected. For flange stability,

$$\frac{b_f}{2t_f} \leq 12$$
$$\frac{10.4}{2\,(0.61)} = 8.52 \leq 12 \qquad \text{OK}$$

and

$$b_f \geq \frac{D}{6}$$
$$10.4 \geq \frac{29.53}{6} = 4.92 \qquad \text{OK}$$
$$t_f \geq 1.1\ t_w$$
$$0.61 \geq 1.1\,(0.47) = 0.52 \qquad \text{OK}$$

And for handling

$$0.1 \leq \frac{I_{\text{compression flange}}}{I_{\text{tension flange}}} \leq 1$$
$$0.1 \leq 1.0 \leq 1 \qquad \text{OK}$$

a. Composite Section Stresses [A6.10.1.1.1] Composite stresses and stage construction is not of concern for this noncomposite bridge.
b. Flange Stresses and Member Bending Moments [A6.10.1.6] Lateral torsional buckling is considered below. The lateral flange bending is considered small for this example.
c. Fundamental Section Properties [AASHTO Appendices D6.1, D6.2, D6.3] The fundamental section properties are shown above.
d. Constructibility [A6.10.3]

(1) General [A2.5.3] [A6.10.3.1] The resistance of the girders during construction is checked. Note that the unbraced length is important when checking lateral torsion buckling under the load of wet concrete.

(2) Flexure [A6.10.3.2] [A6.10.1.8] [A6.10.1.9] [A6.10.1.10.1] [A6.10.8.2] [A6.3.3—optional] Lateral support for compression flange is not available when fresh concrete is being placed [A6.10.3.2.1 and A6.10.8.2]:

$$L_p = 1.0 r_t \sqrt{\frac{E}{F_{\text{yc}}}}$$
$$= 1.0\,(2.56)\sqrt{\frac{29{,}000}{50}} = 61.7\text{ in.}$$
$$L_r = \pi r_t \sqrt{\frac{E}{F_{\text{yc}}}} = \pi\,(61.7) = 193.7\text{ in.}$$

Try bracing at one-quarter points, $L_b = 8.75$ ft $= 105$ in. As L_b is between L_p and L_r, the resistance is [A10.8.2.3], the interpolation is between these two achor points. C_b is conservatively considered 1.0 and could be refined as necessary. No significant construction live load is anticipated.

$$F_{\text{nc}} = C_b\left[1 - \left(1 - \frac{F_{\text{yr}}}{R_h F_{\text{yc}}}\right)\left(\frac{L_b - L_p}{L_r - L_p}\right)\right] \times R_b R_h F_{\text{yc}} \leq R_b R_h F_{\text{yc}}$$
$$= 1.0\left[1 - \left(1 - \frac{0.7F_{\text{yc}}}{1.0F_{\text{yc}}}\right)\left(\frac{105 - 61.7}{193.7 - 61.7}\right)\right] \times (1.0)\,(1.0)\,(50) \leq 50$$
$$= 45.1\text{ ksi}$$
$$M = 1.25\,(\text{DC}) = 1.25\,(203.7) = 254.6\text{ k ft}$$
$$f_c = \frac{M}{S} = \frac{254.6\,(12)}{245}$$
$$= 12.5\text{ ksi} \leq 45\text{ ksi} \quad \text{OK}$$

The quarter-point cross framing is considered in the wind bracing design later.

(3) Shear [A6.10.3.3] The shear resistance is computed and then used for constructability, strength I, and fatigue limit states.

$$V_u \leq \phi V_{\text{cr}} = \phi C V_p = \phi C\,(0.58)\,F_y D t_w$$

where

V_u = maximum shear force due to unfactored permanent load and twice the fatigue loading [A6.10.5.3]
V_{cr} = critical buckling resistance
C = shear buckling coefficient
V_p = plastic shear resistance

For a wide-flange section this should not be an issue but the computations are provided for completeness.

Shear resistance of unstiffened web is applicable [C6.10.9.2]. For the wide flange, the shear resistance should be the plastic shear resistance. The computations illustrate that this resistance is slightly within the inelastic range:

$$D = d - 2t_f = 29.54 - 2\,(0.61)$$
$$= 28.32 \text{ in.}$$
$$\frac{D}{t_w} = \frac{28.32}{0.47} = 60.2$$
$$1.12\sqrt{\frac{Ek}{F_{yw}}} = 1.12\sqrt{\frac{29000\,(5)}{50}} = 60.3$$
$$\frac{D}{t_w} \le 60.3$$
$$C = 10$$
$$V_p = 1.0\,(0.58)\,(50)\,(29.53)\,(0.47)$$
$$= 402 \text{ kips}$$
$$V_u \le \phi V_{cr} = \phi C V_p = (1.0)\,(1.0)\,(402)$$
$$= 402 \text{ kips}$$
$$= 29.1 \le 402 \text{ kips}$$

Normally a rolled wide flange has the full plastic shear resistance.

(4) Deck Placement [A6.10.3.4] This deck is a noncomposite section and because the span is short, placement is at the same time. This article is not applicable.

(5) Dead-Load Placement [A6.10.3.5] Deck placement is not patterned or staged. This article is not applicable.

e. Service Limit State [A6.5.2] [A6.10.4]

(1) Elastic Deformations [A6.10.4.1]

(a) Optional Live-Load Deflection [A2.5.2.6.2] *Optional Deflection Control* [A2.5.2.6.2]

$$\text{Allowable service load deflection} \le \tfrac{1}{800}\text{ span} = \frac{35\,(12)}{800} = 0.53 \text{ in.}$$

From [A3.6.1.3.2], deflection is taken as the larger of that:

☐ Resulting from the design truck alone

☐ Resulting from 25% of the design truck taken together with the design lane load

The distribution factor for deflection may be taken as the number of lanes divided by the number of beams [C2.5.2.6.2] because all design lanes should be loaded, and all supporting components should be assumed to deflect equally:

$$mg_{\text{deflection}} = m\left(\frac{\text{No.lanes}}{\text{No.beams}}\right) = 0.85\left(\frac{3}{6}\right) = 0.43$$

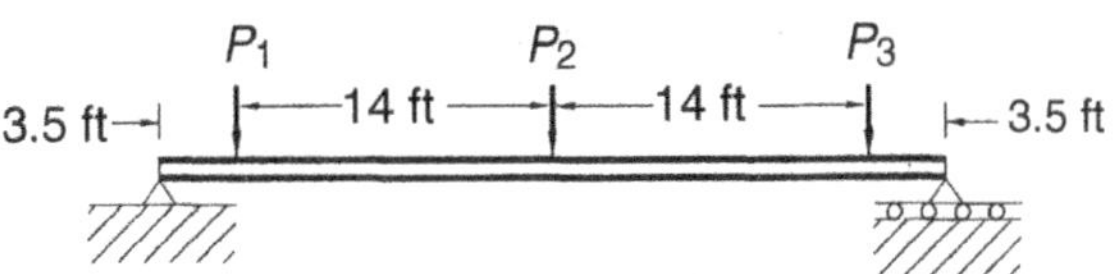

Fig. E20.1-7 Truck placement for maximum deflection.

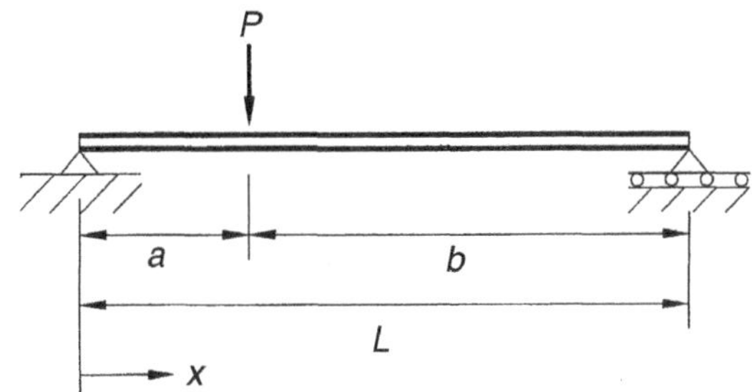

Fig. E20.1-8 General placement of point load P.

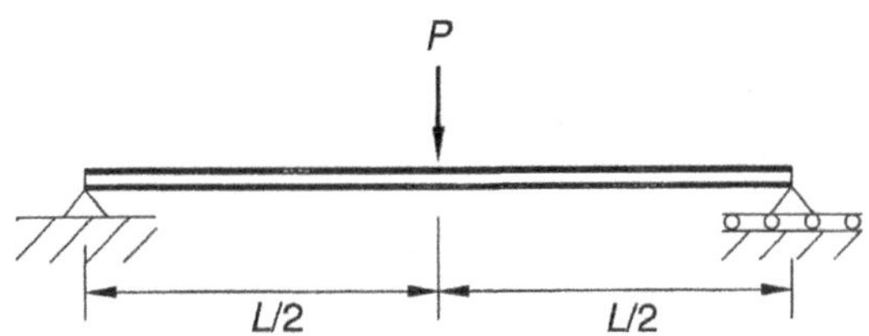

Fig. E20.1-9 Point load P at center of the span.

(i) Deflection resulting from design truck alone (Fig. E20.1-7):

$$P_1 = P_2 = 0.43\,(32)\left(1 + \frac{\text{IM}}{100}\right)$$
$$= 0.43\,(32)\,(1.33)$$
$$= 18.3 \text{ kips}$$
$$P_3 = 0.43\,(8)\,(1.33)$$
$$= 4.58 \text{ kips}$$

The deflection at any point, Δ_x, due to a point load P can be found from AISC Manual (2005) (Fig. E20.1-8) for $x \le a$:

$$\Delta_x = \frac{Pbx}{6EIL}\left(L^2 - b^2 - x^2\right)$$

The maximum deflection (located at the center) of a simply supported span, due to a concentrated load at the center of the span, can be found from AISC Manual (2010) (Fig. E20.1-9):

$$\Delta_{\text{CL}} = \frac{PL^3}{48EI}$$

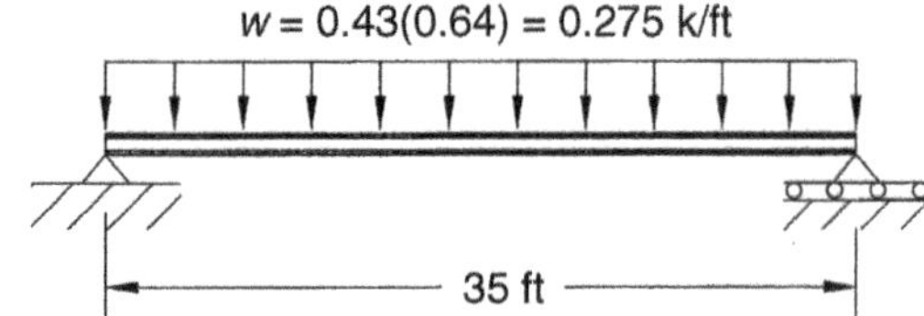

Fig. E20.1-10 Uniform lane load on the span.

$$
\begin{aligned}
\Delta_{\text{CLTruck}} &= \left(\Delta_{P_1} + \Delta_{P_2}\right) + \Delta_{P_3} \\
&= \frac{(18.3 + 45.8)\,(42)\,(420/2)}{6\,(29{,}000)\,(3620)\,(420)} \\
&\quad \times \left[(420)^2 - (42)^2 - (420/2)^2\right] \\
&\quad + \frac{18.3(420)^3}{48\,(29{,}000)\,(3620)} \\
&= 0.0995 + 0.269 = 0.37 \text{ in.}
\end{aligned}
$$

(ii) Deflection resulting from 25% of design truck together with the design lane load:

$$\Delta_{\text{CL25\%Truck}} = 0.25\,(0.37) = 0.092 \text{ in.}$$

The deflection due to lane load can be found from AISC Manual (2005) (Fig. E20.1-10):

$$
\begin{aligned}
\Delta_{\max} &= \frac{5wL^4}{384EI} \\
\Delta_{\text{CL Lane}} &= \frac{5\,(0.43)\,(0.64/12)\,(420)^4}{384\,(29{,}000)\,(3620)} \\
&= 0.089 \text{ in.} \\
\Delta_{\text{CL}} &= \Delta_{\text{CL25\%Truck}} + \Delta_{\text{CL Lane}} \\
&= 0.0922 + 0.089 = 0.18 \text{ in.} \\
\therefore\ \Delta_{\text{CL Truck}} &= 0.37 \text{ in.} \quad \text{controls} \\
\Delta_{\text{CL}} &= 0.37 \text{ in.} < \Delta_{\text{all}} \\
&= 0.5 \text{ in.} \quad \text{OK}
\end{aligned}
$$

(b) Optional Criteria for Span-to-Depth Ratio [A6.10.4.2.1] From [A2.5.2.6.3], the owner may choose to invoke a requirement for minimum section depth based upon a span-to-depth ratio. Per AASHTO [Table E20.2-5.2.6.3-1], the depth of the steel portion of a composite steel beam is $0.033L$ ($L/30$) for simple spans, or

$$\text{Min depth} = (12)\,(35)\,/30 = 14 \text{ in.}$$

The trial design section easily meets this optional requirement.

(2) Permanent Deformations [A6.10.4.2] This limit state is checked to prevent permanent deflection that would impair rideability. Lateral bending stresses are considered small, that is, $f_l = 0.0$. For both flanges of noncomposite sections:

$$f_f + \frac{f_l}{2} \le 0.80 R_h F_{\text{yf}}$$

$$R_h = 1.0 \text{ for homogeneous sections} \quad \text{[A6.10.5.4.1a]}$$

$$F_{\text{yf}} = 50 \text{ ksi}$$

The maximum service II moment, which occurs at location 105 in the exterior beam is

$$
\begin{aligned}
M &= 833.9 \text{ k ft} \\
f_f &= \frac{M}{S_x} = \frac{833.9\,(12)}{245} = 40.8 \text{ ksi} \\
&= 40.8 \text{ ksi} \approx 0.8\,(1.0)\,(50) = 40 \text{ ksi} \quad \text{close}
\end{aligned}
$$

At this juncture, the engineer makes a decision that this is close enough, increases the cross section size, or adds a cover plate. The latter is likely the most expensive solution because of the additional welding. Also note that the live-load distribution factor assumption (K_g term) is conservative and the self-weight of the beam is slightly conservative as well. We continue with this section and discuss this near the end of this example.

(a) General [A6.10.4.2.1]

(b) Flexure [A6.10.4.2.2] [Appendix B—optional] [A6.10.1.9] [A6.10.1.10.1]

f. Fatigue and Fracture Limit State [A6.6.1.2] [A6.10.5]

(1) Fatigue [A6.10.5.1] [A6.6.1] Allowable fatigue stress range depends on load cycles and connection details. Fracture depends on material grade and temperature.

(a) Stress Cycles Assuming a rural interstate highway with 20,000 vehicles per lane per day,

Fraction of trucks in traffic = 0.20 (Table 4.4) [Table C3.6.1.4.2-1]

$$
\begin{aligned}
\text{ADTT} &= 0.20 \times \text{ADT} \\
&= 0.20\,(20000)\,(2 \text{ lanes}) \\
&= 8000 \text{ trucks/day} \\
p &= 0.85 \text{ (Table 4.3)} \quad \text{[Table A3.6.1.4.2-1]} \\
\text{ADTT}_{\text{SL}} &= p \times \text{ADTT} = 0.85\,(8000) \\
&= 6800 \text{ trucks/day}
\end{aligned}
$$

(b) Allowable Fatigue Stress Range—Category A Per A6.10.6.1.2.3, as the number of trucks is more than 530 per day, the fatigue I limit state ($\gamma_{\text{fatigue I}} = 1.5$) shall be used and the resistance shall be based upon the infinite fatigue life. Assuming that the girder is weathering steel that is not coated, the fatigue category is B (Table A6.6.1.2.3-1 Case 1.2 or Table 18.3):

$$(\Delta F)_n = (\Delta F)_{\text{TH}} = 16 \text{ ksi} \quad (\text{fatigue I})$$

(c) Dynamic load allowance for fatigue is IM = 15%.

$M_{\text{LL+IM}}$ is maximum in the exterior girder, no multiple presence (live-load range only):

$$M_{\text{fatigue I}} = 1.5\,(211.3) = 317 \text{ k ft}$$

$$f = \frac{M}{S} = \frac{317\,(12)}{245} = 15.5 \text{ ksi} \leq 16 \text{ ksi}$$

Infinite fatigue life is OK

Because infinite fatigue life is met, then the finite fatigue life check is not applicable.

(2) Fracture [A6.10.5.2] [A6.6.2] The steel specified meets fracture requirements for this non-fracture-critical system.

(3) Special Fatigue Requirements for Webs [A6.10.5.3] The shear force due the fatigue truck is determined with the use of the fatigue truck (exterior girder governs and no multiple presence) [A3.6.1.1.2]). Here dead load is considered with 1.5 × fatigue the load.

$$V_u = mg_{\text{fatigue+IM}}\left(1.5\,V_{\text{fatigue+IM}}\right) + V_{\text{DC}} + V_{\text{DW}} = 66.7 \text{ kips}$$

$$66.7 \text{ kips} \leq 402 \text{ kips} \quad \text{OK}$$

g. Strength Limit State [A6.5.4] [A6.10.6]

(1) Composite Sections in Positive Flexure [A6.10.6.2.2] [A6.10.7] This article is not applicable.

(2) Composite Sections in Negative Flexure [A6.10.6.2.3] [A6.10.8] [Appendix A—optional] [Appendix B—optional] [D6.4—optional] This article is not applicable.

(3) Net Section [A6.10.1.8] This article is not applicable. No splices are required.

(4) Flange-Strength Reduction Factors [A6.10.1.10] [A6.10.2.1] is satisfied and therefore there are not reductions per [A6.10.1.10] required.

3. *Shear Design*

a. General [A6.10.9.1] [A6.10.9.2] The section is a wide flange, and shear resistance should be at the plastic shear capacity. No transverse stiffeners are required; the computation is for an unstiffened section. The shear resistance was previously computed as

$$V_u \leq \phi V_{\text{cr}} = \phi C V_p = (1.0)\,(1.0)\,(402) = 402 \text{ kips}$$

$$V_u = 142.5 \text{ kips} \leq V_r = 402 \text{ kips} \quad \text{OK}$$

J. Dimension and Detail Requirements

1. *Material Thickness* Material Thickness [A6.7.3] Bracing and cross frames shall not be less than 0.3125 in. thickness. Web thickness of rolled beams shall not be less than 0.25 in.

$$t_w = 0.47 \text{ in.} > 0.25 \text{ in.} \quad \text{OK}$$

2. *Diaphragms and Cross Frames* [A6.7.4] See computation below.
3. *Lateral Support of Compression Flange Prior to Curing of Deck*

- ☐ Transfer of wind load on exterior girder to all girders
- ☐ Distribution of vertical dead and live loads applied to the structure
- ☐ Stability of the bottom flange for all loads when it is in compression

For straight I-sections, cross frames shall be at least half the beam depth.

a. Intermediate Diaphragms (Fig. E20.1-11) Try C15 × 33.9 intermediate diaphragms at one-quarter points, for $A_s = 9.96 \text{ in.}^2$, $r_y = 0.904$ in., and $L_b = 35/4 = 8.75 \text{ ft} = 105$ in.

The wind load acting on the bottom half of the beam goes to the bottom flange and is

$$w_{\text{Bot}} = \frac{\gamma P_D d}{2} = \frac{1.4\,(0.050)\,(30/12)}{2} = 0.0875 \text{ k/ft}$$

$$P_{w\text{Bot}} = w_{\text{Bot}} L_b = (0.0875)\,(8.75)\,(1/2) = 0.38 \text{ kips}$$

The remaining wind load is transmitted to the abutment region by the deck diaphragm. The end reaction must be transferred to the bearings equally by all six girders. The resultant force is F_{uD}.

$$P_{w\text{Top}} = \left[1.4\,(0.050)\,(30 + 8 + 34)\left(\frac{1}{12}\right)\right] \times \left[\frac{(32 - 87.5)/2}{6 \text{ girders}}\right] = 0.81 \text{ kips}$$

$$F_{uD} = P_{w\text{Bot}} + P_{w\text{Top}} = 0.38 + 0.81 = 1.19 \text{ kips}$$

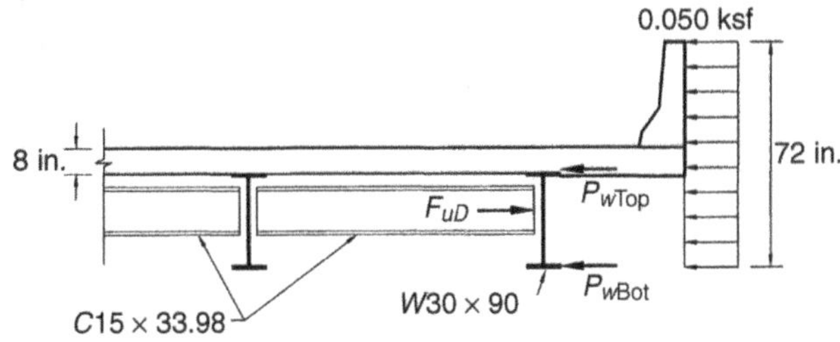

Fig. E20.1-11 Wind load acting on exterior elevation.

The axial resistance is [A6.9.3, A6.9.4]

$$\frac{kL}{r_y} = \frac{1.0\,(96)}{0.904} = 106 < 140$$

$$P_e = \frac{\pi^2 E A_g}{\left(\dfrac{KL}{r_y}\right)^2} = \frac{\pi^2\,(29000)\,(9.96)}{(106)^2} = 253.7 \text{ kips}$$

$$P_o = Q F_y A_g = 1.0\,(50)\,(9.96) = 498.0 \text{ kips}$$

$$P_n = 0.658^{P_o/P_e} F_y A_s = 0.658^{1.96}\,(498.0) = 219 \text{ kips}$$

$$P_r = \phi_c P_n = 0.9\,(220) = 198 \text{ kips} \gg P_{w\text{Bot}} = 0.38 \text{ kips} \quad \text{OK}$$

b. End Diaphragms Must adequately transmit all the forces to the bearings:

$$P_r = 198 \text{ kips} \gg F_{uD} = 1.19 \text{ kips} \quad \text{OK}$$

Use the same section as intermediate diaphragm. This component is overdesigned, however, the 15-in. deep section facilitates simple connection to the girders.

Use C15 × 33.9, M270 Grade 50W, for all diaphragms.

Lateral Bracing [A6.7.5] Lateral bracing shall be provided at quarter points.

Use same section as diaphragms.

Use C15 × 33.9, M270 Grade 50W, for all lateral braces.

K. Dead-Load Camber

Exterior Beam

$$w_D = w_{\text{DC}} + w_{\text{DW}} = 1.33 + 0.21 = 1.54 \text{ k/ft}$$

Interior Beam

$$w_D = 0.9 + 0.28 = 1.18 \text{ k/ft}$$

$$\Delta_{\text{CL}} = \frac{5}{384}\frac{(w_D L^4)}{EI} = \frac{5}{384}\frac{(1.54/12)(420)^4}{(29{,}000)(3620)} = 0.50 \text{ in.}$$

Use 0.5-in. camber on all beams. Alternatively, some agencies thicken the CIP deck in the middle by 0.5 in. to account for the dead-load deflection rather than cambering rolled sections.

L. Check Assumptions Made in Design Nearly all the requirements are satisfied, using a W30 × 90. This beam has a self-weight of 0.090 k/ft; thus, our assumed beam weight of 0.10 k/ft is conservative. Also, for preliminary design, the value for $\left[K_g/12Lt_s^3\right]^{0.1}$ was taken as 1.0 in calculating the distribution factors for moment. The actual value is calculated below.

$$K_g = n\left(I + Ae_g^2\right)$$

$$n = \frac{E_s}{E_c} = \frac{29{,}000}{1820\sqrt{4}} = 7.98 \quad \text{Use } 8$$

$$I = 3620 \text{ in.}^4$$

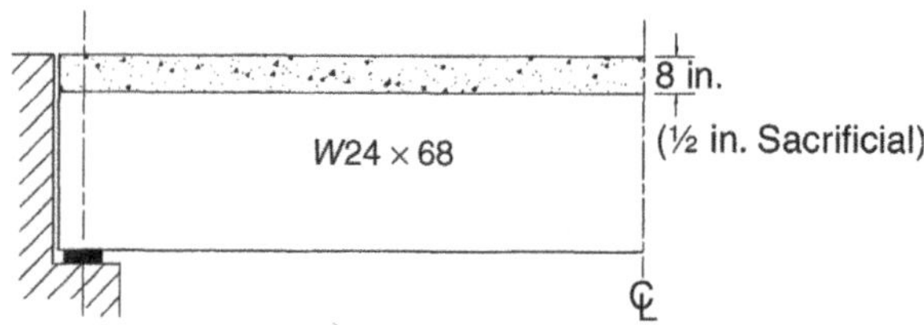

Fig. E20.1-12 Design sketch of noncomposite rolled steel girder.

Because section is noncomposite $e_g = 0$:

$$K_g = 8\,(3620) = 28{,}960 \text{ in.}^4$$

$$\frac{K_g}{12Lt_s^3} = \frac{28{,}960}{(12)\,(35)\,(8)^3} = 0.13$$

$$\left(\frac{K_g}{12Lt_s^3}\right)^{0.1} = 0.82$$

Recompute the distribution factors:

$$mg_M^{\text{SI}} = 0.06 + \left(\frac{S}{14}\right)^{0.4}\left(\frac{S}{L}\right)^{0.3}\left(\frac{K_g}{12Lt_s^3}\right)^{0.1}$$

$$mg_M^{\text{SI}} = 0.06 + \left(\frac{8}{14}\right)^{0.4}\left(\frac{8}{35}\right)^{0.3}(0.82) = 0.48$$

Two design lanes loaded:

$$mg_M^{\text{MI}} = 0.075 + \left(\frac{S}{9.5}\right)^{0.6}\left(\frac{S}{L}\right)^{0.2}\left(\frac{K_g}{12Lt_s^3}\right)^{0.1}$$

$$mg_M^{\text{MI}} = 0.075 + \left(\frac{8}{9.5}\right)^{0.6}\left(\frac{8}{35}\right)^{0.2}(0.82) = 0.626 \quad \text{governs}$$

This demonstrates that the live loads calculated in the preliminary design are about 18% higher than actual, which is conservative (interior girder only). However, the distribution factor is not applied to the dead load so that when the live- and dead-load effects are combined, the preliminary design loads are less conservative, which is more acceptable. Also, the exterior girder does not have this factor, so the distribution factors are unchanged with the better estimate of the longitudinal to transverse stiffness (so-called K_g term). The K_g term equal to 1.0 is closer for composite beams as will be illustrated below.

M. Design Sketch The design of the noncomposite, simple span, rolled steel beam bridge is summarized in Figure E20.1-12.

20.2 COMPOSITE ROLLED STEEL BEAM BRIDGE

Design the simple-span composite rolled steel beam bridge of Figure E20.2-1 with 35-ft span for an HL-93 live load. Roadway width is 44 ft curb to curb. Allow for a future wearing surface of 3-in.-thick bituminous overlay. Use $f_c' = 4$ ksi

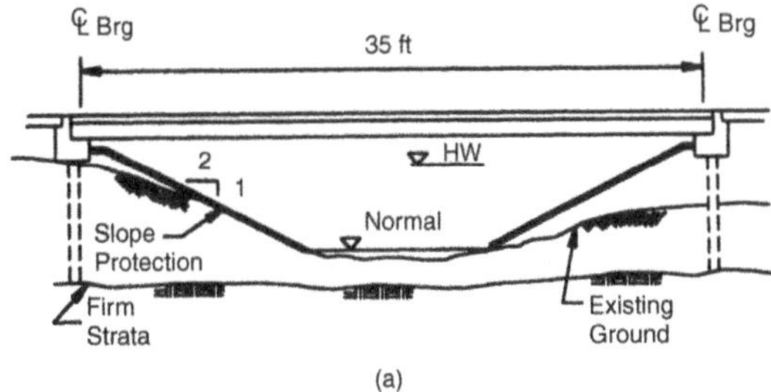

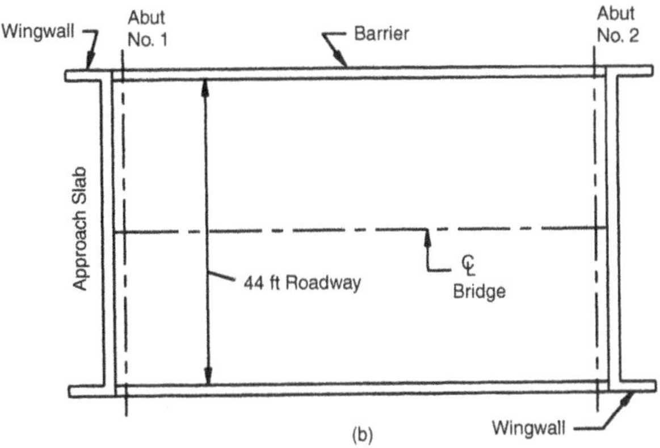

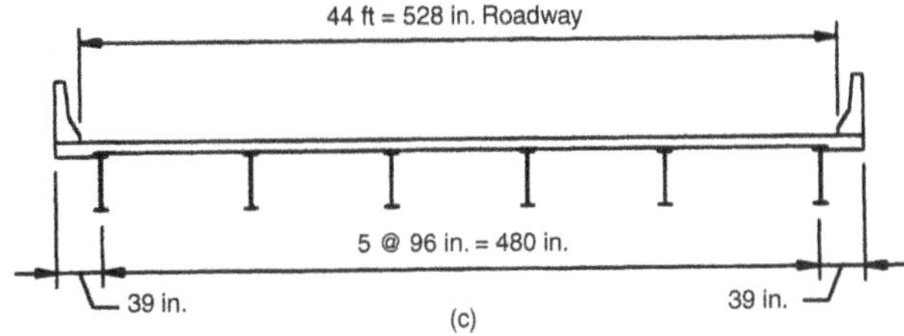

Fig. E20.2-1 Composite rolled steel beam bridge design example: (a) general elevation, (b) plan view, and (c) cross section.

and M270 Grade 50W steel. For this example, the barrier weight will be assigned equally to each girder.

A-G. Same as Example Problem 20.1

H. Calculate Force Effects from Other Loads

$D1$ = dead load of structural components and their attachments, acting on the noncomposite section (DC)

$D2$ = future wearing surface (DW)

$D3$ = barriers that have a cross sectional area of $300\text{ in.}^2 = 2.08\text{ ft}^2$, and weight of 0.32 k/ft (DC).

A 2-in. × 12-in. average concrete haunch at each girder is used to account for camber and unshored construction. A 1-in. depth is assumed for resistance computations due to variabilities and flange embedment. Assume a beam weight of 0.10 k/ft.

For a uniformly distributed load w on the simple span,

$$M_{105} = \left(\tfrac{1}{8}\right)(35)^2 w = 153.1w$$
$$V_{100} = 17.5w$$

1. Interior Girders

$$\begin{aligned} &D1 \quad \text{Deck slab} \quad (0.15)\left(\tfrac{8}{12}\right)(8) = 0.80\text{ k/ft} \\ &\quad\ \text{Girder} \qquad\qquad\qquad = 0.10\text{ k/ft} \\ &\quad\ \text{Haunch} \quad (2)(12)(0.150)/144 = \underline{0.025\text{ k/ft}} \\ &\qquad\qquad\qquad\qquad w_{D1}^{I} = 0.93\text{ k/ft} \end{aligned}$$

$$D2 \quad \text{3-in. bituminous paving} \quad w_{D2}^{I} = \left(\tfrac{3}{12}\right)(0.14)(8) = 0.28\text{ k/ft}$$

$$D3 \quad \text{Barriers, one-sixth share} \quad w_{D3}^{I} = \frac{2(0.32\text{ k/ft})}{6} = 0.11\text{ k/ft}$$

Table E20.2-1 summarizes the unfactored moments and shears at critical sections for interior girders. The values for LL + IM were determined in the previous example.

2. Exterior Girders

$$\begin{aligned} &D1 \quad \text{Deck slab} \quad (0.15)\left(\tfrac{8}{12}\right)\left(3.25 + \tfrac{8}{2}\right) = 0.73\text{ k/ft} \\ &\quad\ \text{Girder} \qquad\qquad = 0.10\text{ k/ft} \\ &\quad\ \text{Haunch} \qquad\qquad = \underline{0.025\text{ k/ft}} \\ &\qquad\qquad\qquad w_{D1}^{E} = 0.86\text{ k/ft} \end{aligned}$$

$$D2 \quad \text{3-in. bituminous paving} \quad w_{D2}^{E} = (0.14)\left(\tfrac{3}{12}\right) \times \left(3.25 + \tfrac{8}{2}\right) = 0.25\text{ k/ft}$$

$$D3 \quad \text{Barriers, one-sixth share} \quad w_{D3}^{E} = 0.11\text{ k/ft}$$

Table E20.2-2 summarizes the unfactored moments and shears at critical sections for exterior girders. The values for LL + IM were determined in the previous example.

Table E20.2-1 Interior Girder Unfactored Moments and Shears[a]

Load Type	w (k/ft)	Moment (k ft) M_{105}	Shear (kips) V_{100}
$D1$ (DC)	0.93	142.4	16.3
$D2$ (DW)	0.28	42.9	4.9
$D3$ (DC)	0.11	16.8	1.9
LL + IM (distributed)	N/A	458.8	66.1
Fatigue + IM (distributed)	N/A	161.4	24.0

[a]Critical values are in boldface.

Table E20.2-2 Exterior Girder Unfactored Moments and Shears[a]

Load Type	w (k/ft)	Moment (k ft) M_{105}	Shear (kips) V_{100}
$D1$ (DC)	0.86	131.7	15.1
$D2$ (DW)	0.25	38.3	4.4
$D3$ (DC)	0.11	16.8	1.9
LL + IM	N/A	467.4	61.0
Fatigue + IM (distributed)	N/A	211.3	26.5

[a]Critical values are in boldface.

Factored Load Effects

a. Interior Beam—Factored Shear and Moment

Strength I $U = \eta[1.25D1 + 1.50D2 + 1.25D3 + 1.75\,(\text{LL} + \text{IM})]$

$$V_{\text{Strength I}} = 1.0[1.25\,(16.3) + 1.50\,(4.9) + 1.25\,(1.9) + 1.75\,(66.1)]$$
$$= \mathbf{145.8\ kips}\quad \textbf{(strength I)}$$

$$M_{\text{Strength I}} = 1.0[1.25\,(142.4) + 1.50\,(42.9) + 1.25\,(16.8) + 1.75\,(458.8)]$$
$$= \mathbf{1066.3\ k\ ft}\quad \textbf{(strength I)}$$

Fatigue I $U = \eta\,[1.5\,(\text{LL} + \text{IM})]$

Fatigue II $U = \eta\,[0.75\,(\text{LL} + \text{IM})]$

$$V_{\text{Fatigue I}} = 1.0\,[1.5\,(24.0)] = 36.0 \text{ kips (fatigue I)}$$

$$V_{\text{Fatigue II}} = 1.0\,[0.75\,(24.0)] = 18.0 \text{ kips (fatigue II)}$$

$$M_{\text{Fatigue I}} = 1.0\,[1.5\,(161.4)] = 242.1 \text{ k ft (fatigue I)}$$

$$M_{\text{Fatigue II}} = 1.0\,[0.75\,(161.4)] = 121.1 \text{ k ft (fatigue II)}$$

Service II $U = \eta[1.0D1 + 1.0D2 + 1.0D3 + 1.30\,(\text{LL} + \text{IM})]$

$$V_{\text{Service II}} = 1.0[1.0\,(16.3) + 1.0\,(4.9) + 1.0\,(1.9) + 1.3\,(66.0)]$$
$$= \mathbf{108.9\ kips\ (service\ II)}$$

$$M_{\text{Service II}} = 1.0[1.0\,(142.4) + 1.0\,(42.6) + 1.0\,(16.8) + 1.3\,(458.8)]$$
$$= \mathbf{798.2\ k\ ft\ (service\ II)}$$

Construction $U = \eta\,[1.25D1]$

$$V_{\text{Construction}} = 1.0\,[1.25\,(16.3)] = \mathbf{20.4\ kips\ (construction)}$$

$$M_{\text{Construction}} = 1.0\,[1.25\,(142.4)] = \mathbf{178.0\ k\ ft\ (construction)}$$

b. Exterior Beam—Factored Shear and Moment

$$V_{\text{Strength I}} = 1.0[1.25\,(15.1) + 1.50\,(4.4) + 1.25\,(1.9) + 1.75\,(61)]$$
$$= 134.6 \text{ kips (strength I)}$$

$$M_{\text{Strength I}} = 1.0[1.25\,(131.7) + 1.50\,(38.3) + 1.25\,(16.8) + 1.75\,(467.4)]$$
$$= 1061.0 \text{ k ft (strength I)}$$

$$V_{\text{Fatigue I}} = 1.0\,[1.5\,(26.5)] = 39.8 \text{ kip} \quad \text{(fatigue I)}$$

$$V_{\text{Fatigue II}} = 1.0\,[0.75\,(26.5)] = 19.9 \text{ kip} \quad \text{(fatigue I)}$$

$$M_{\text{Fatigue I}} = 1.0\,[1.5\,(211.3)] = 317.0 \text{ k ft} \quad \text{(fatigue I)}$$

$$M_{\text{Fatigue II}} = 1.0\,[0.75\,(211.3)] = 158.5 \text{ k ft} \quad \text{(fatigue II)}$$

$$V_{\text{Service II}} = 1.0[1.0\,(15.1) + 1.0\,(4.4) + 1.0\,(1.9) + 1.3\,(61)]$$
$$= 100.7 \text{ kip} \quad \text{(service II)}$$

$$M_{\text{Service II}} = 1.0[1.0\,(131.7) + 1.0\,(38.3) + 1.0\,(16.8) + 1.3\,(467.4)]$$
$$= 794.4 \text{ k ft} \quad \text{(service II)}$$

$$V_{\text{Construction}} = 1.0\,[1.25\,(15.1)] = 18.9 \text{ kips} \quad \text{(construction)}$$

$$M_{\text{Construction}} = 1.0\,[1.25\,(131.7)] = 164.6 \text{ k ft} \quad \text{(construction)}$$

Critical values are in boldface.

I. Design Required Sections

1. Flexural Design

a. Composite Section Stresses [A6.10.1.1.1] The composite cross-section properties computed below include the bare steel, composite deck for long-term loading, and composite deck for short-term loading.

b. Flange Stresses and Member Bending Moments [A6.10.1.6] Because this is a straight (nonskewed) bridge, the lateral bending effects are considered to be minimal and the lateral bending stress f_l is considered here as zero:

$$f_l = 0.0$$

c. Fundamental Section Properties

(1) Consider Loading and Concrete Placement Sequence [A6.10.5.1.1a]

Case 1 Weight of girder and slab (D1). Supported by steel girder alone.

Case 2 Superimposed dead load (FWS, curbs, and railings) (D2 and D3). Supported by long-term composite section.

Case 3 Live load plus impact (LL + IM). Supported by short-term composite section.

(2) Determine Effective Flange Width [A4.6.2.6] For interior girders the effective flange width is the tributary width:

$$b_i = 8 \text{ ft} = 96 \text{ in.}$$

For exterior girders the effective flange width is

$$b_e = \frac{b_i}{2} + 39 = \frac{96}{2} + 39 = 88 \text{ in.}$$

(3) Modular Ratio [A6.10.5.1.1b] For

$$f_c' = 4 \text{ ksi} \qquad n = 8$$

(4) Trial Section Properties At this point in the design, analysis indicates that strength I moment and shear are critical in the interior girder and fatigue moment and shear are critical in the exterior girder. The effective widths are nearly the same. For the trial check, use $M_u = 1066.3$ k ft, $V_u = 145.8$ kips, for fatigue limit I including dead loads, $M_{\text{fatigue}} = 345.3$ k ft, and $V_{\text{fatigue}} = 41.3$ kips. For fatigue limit I for live-load range only, $M_{\text{fatigue}} = 317.0$ k ft, and $V_{\text{fatigue}} = 39.8$ kips. The trial section properties are based upon the interior girder effective slab width of 96 in.

(a) Steel Section at Midspan Try W24 × 68. Properties of W24 × 68 are taken from AISC (2010). The calculations for the steel section properties are summarized below and are shown in Figure E20.2-2

$$I_x = 1830 \text{ in.}^4 \qquad I_y = 70.4 \text{ in.}^4$$
$$A = 20.1 \text{ in.}^2$$
$$Z_x = 177 \text{ in.}^3 \qquad S_x = 154 \text{ in.}^3$$
$$b_f = 8.965 \text{ in.} \qquad t_f = 0.585 \text{ in.}$$
$$t_w = 0.415 \text{ in.}$$
$$d = 23.73 \text{ in.}$$

(b) Composite Section, $n = 8$, at Midspan Figure E20.2-3 shows the composite section with a haunch of 1 in., a net slab thickness (without 0.5 in. sacrificial wearing surface) of 7.5 in., and an effective width of 96 in. The composite section properties calculations are summarized in Table E20.2-3. See Fig. E20.2-3.

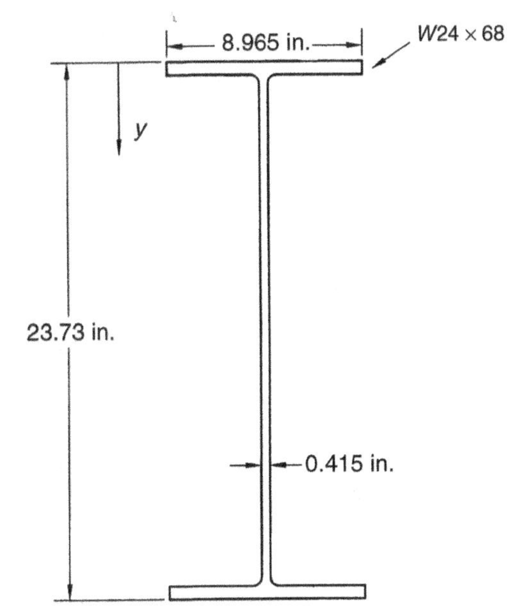

Fig. E20.2-2 Noncomposite steel section at midspan.

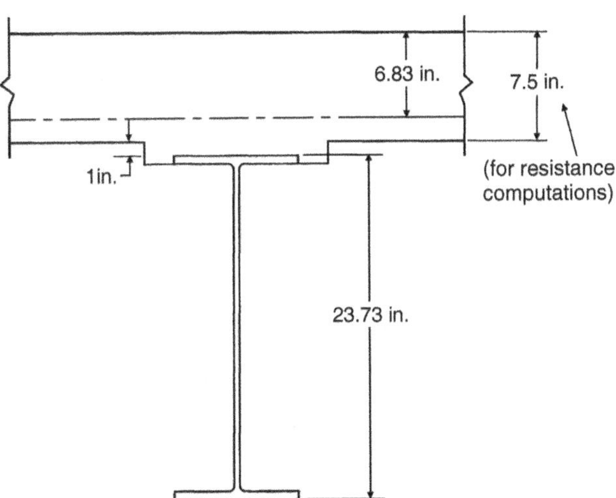

Fig. E20.2-3 Composite steel section at midspan.

$$\bar{y} = \frac{\sum Ay}{\sum A} = \frac{2802}{110.1} = 25.5 \text{ in.}$$

$$y_t = 23.73 + 1 + 7.5 - 25.4 = 6.83 \text{ in.}$$

$$I_x = 4533 + 2252 = 6785 \text{ in.}^4$$

$$S_t = \frac{6785}{6.78} = 1001 \text{ in.}^3$$

$$S_b = \frac{6785}{25.5} = 266 \text{ in.}^3$$

Table E20.2-3 Short-Term Composite Section Properties, $n = 8, b_i = 94$ in

Component	A	y	Ay	$\lvert y - \bar{y}\rvert$	$A(y - \bar{y})^2$	I_0
Concrete $(b_i \times t_s/n)^a$	7.5(96)/8 = 90.0	$1 + \frac{7.5}{2} + 23.73 = 28.48$	2563	2.98	799	422
Steel	20.1	11.87	239	13.6	3734	1830
$\sum$	110.1		2802		4533	2252

[a]The parameter b_i is used because interior girders control the moment design. As an aside, computations are not very sensitive to this width.

Table E20.2-4 Long-Term Composite Section Properties, $3n = 24, b_i = 94$ in

Component	A	y	Ay	$\lvert y - \bar{y}\rvert$	$A(y - \bar{y})^2$	I_0
Concrete$(b_i \times t_s/3n)^a$	7.5(96)/24 = 30.0	$1 + \frac{7.5}{2} + 23.73 = 28.48$	854.4	6.68	1339	141
Steel	20.1	11.87	239	9.93	1982	1830
$\sum$	50.1		1093		3293	1971

[a]The parameter b_i is used because interior girders control the moment design.

(c) Composite Section, $3n = 24$, at Midspan The composite section properties calculations, reduced for the effect of creep in the concrete slab, are summarized in Table E20.2-4.

$$\bar{y} = \frac{\sum Ay}{\sum A} = \frac{1093}{50.1} = 21.8 \text{ in.}$$

$$y_t = 10.4 \text{ in.}$$

$$I_x = 3321 + 1971 = 5292 \text{ in.}^4$$

$$S_t = \frac{5292}{10.4} = 509 \text{ in.}^3$$

$$S_b = \frac{5292}{21.8} = 243 \text{ in.}^3$$

(5) Member Proportions [A6.10.1.1]

$$\frac{D}{t_w} \le 150$$

$$\frac{23.73 - 2\,(0.585)}{0.415} = \frac{22.56}{0.415} = 54 \le 150 \quad \text{OK}$$

This is conservative for the wide-flange section, as expected. For flange stability,

$$\frac{b_f}{2t_f} \le 12$$

$$\frac{8.965}{2\,(0.415)} = 10.8 \le 12 \quad \text{OK}$$

And

$$b_f \ge \frac{D}{6}$$

$$8.965 \ge \frac{22.56}{6} = 3.76 \quad \text{OK}$$

$$t_f \ge 1.1\, t_w$$

$$0.585 \ge 1.1\,(0.415) = 0.46 \quad \text{OK}$$

And for handling

$$0.1 \le \frac{I_{\text{compression flange}}}{I_{\text{tension flange}}} \le 1$$

$$0.1 \le 1.0 \le 1 \quad \text{OK}$$

All the general proportions are met for this wide flange as expected.

d. Constructibility [A6.10.3]

(1) General [A2.5.3] [A6.10.3.1] The resistance of the girders during construction is checked. Note that the unbraced length is important for lateral torsion buckling under the load of wet concrete. Nominal yielding or postbuckling behavior is not permitted during construction.

(a) Local Buckling [A6.10.3.2] The wide flange trial section will not have local buckling issues.

(b) Flexure [A6.10.3.2] [A6.10.8.2] Lateral support for compression flange is not available when fresh concrete is being placed and should be checked to ensure that bracing is adequate.

Compression flange bracing [A6.10.8]

$$f_{\text{bu}} + \tfrac{1}{3} f_l \le \phi_f F_{\text{nc}}$$

where

$$F_{\text{nc}} = R_b R_h F_{\text{yc}}$$

and is computed per [A6.10.8.2].

For the rolled beam [A.6.10.8.2.2], local buckling resistance is satisfied. The lateral torsional buckling resistance [A6.10.8.2.3] is dependent upon the unbraced length. The two anchor points associated with the inelastic buckling L_p and elastic buckling L_r, are

$$L_p \leq 1.0\, r_t \sqrt{\frac{E}{F_{yc}}}$$

and

$$L_r \leq \pi r_t \sqrt{\frac{E}{F_{yc}}}$$

where

r_t = minimum radius of gyration of the compression flange of the steel section (without one-third of the web in compression) taken about the vertical axis

$$r_t = \frac{b_{fc}}{\sqrt{12\left(1 + \frac{1}{3}\frac{D_c}{b_{fc}}\frac{t_w}{t_{fc}}\right)}}$$

$$r_t = \frac{8.965}{\sqrt{12\left(1 + \frac{1}{3}\left(\frac{22.56/2}{8.965}\right)\left(\frac{0.415}{0.585}\right)\right)}} = 2.58 \text{ in.}$$

Therefore,

$$L_p \leq 1.0\,(2.58)\sqrt{\frac{29{,}000}{50}} = 62 \text{ in.}$$

and

$$L_r \leq \pi\,(2.58)\sqrt{\frac{29{,}000}{50}} = 195 \text{ in.}$$

Provide braces for the compression flange at quarter points so that

$$L_b = \frac{(35)\,(12)}{4} = 105 \text{ in.}$$

These braces need not be permanent because the composite slab provides compression flange bracing once it is cured.

Nominal flexural resistance:

$$F_{nc} = (\text{LTB factor})\, R_b R_h F_{yc}$$

$$R_h = 1.0 \text{ for homogeneous sections [A6.10.1.10.1]}$$

Therefore, for $R_b = 1.0$:

$$F_{nc} = C_b\left[1 - \left(1 - \frac{F_{yr}}{R_h F_{yc}}\right) \times \left(\frac{L_b - L_p}{L_r - L_p}\right)\right] R_b R_h F_{yc}$$

$$= 1.0\left[1 - \left(1 - \frac{0.7F_y}{1.0F_{yc}}\right) \times \left(\frac{105 - 62}{195 - 62}\right)\right](1.0)\,(1.0)\,F_{yc}$$

$$= 0.90F_{yc} = 45.1 \text{ ksi} \leq 50 \text{ ksi} \quad \text{OK}$$

Compare the resistance to the load effect under construction:

$$M_{105} = 1.0\,(1.25)\,(142.5) = 178 \text{ k ft}$$

$$f_{bu} = \frac{178(12)}{154} 13.9 \text{ ksi} \leq 45.1 \text{ ksi} \quad \text{OK}$$

(2) Shear [A6.10.10.3] This article does not apply to sections with unstiffened webs because the shear force is limited to the shear yield or shear buckling force at the strength limit state. Per [A6.10.3.3], the shear resistance is

$$V_u \leq \phi_v V_{cr} = CV_p$$

$$C = 1.0$$

$$\phi_v V_{cr} = 1.0\,(0.58)\,(50)\,(23.73)\,(0.415) = 286 \text{ kips}$$

$$V_u = 1.25\,(20.4) = 25.5 \text{ kips} \leq 286 \text{ kips} \quad \text{OK}$$

(3) Deck Placement [A6.10.3.4] This deck is a composite section and because the span is short, placement is at the same time and pattern dead load need not be considered.

(4) Dead-Load Placement [A6.10.3.5] Deck placement is considered in the computation of the cross-section properties.

e. Service Limit State [A6.5.2] [A6.10.4]

(1) Elastic Deformations [A6.10.4.1]

(a) Optional Live-Load Deflection [A2.5.2.6.2] Optional Deflection Control [A2.5.2.6.2]

This requirement was met in Example Problem 20.1. The only difference between this example and Example 20.1 is the moment of inertia, I, of the section for which I is equal to 3620 in.^{4}. In this example, I is equal to 6785 in.[4], determined previously. Because I is greater than I in Example Problem 20.1, the deflections are less, and the optional deflection control requirement is met.

(b) Optional Criteria for Span-to-Depth Ratio [A2.5.2.6.3] The optional span-to-depth ratio of $0.033L$ for the bare steel and $0.040L$ for the total section is computed as

$$0.033L = 0.033\,(35)\,(12) = 13.9 \text{ in.} \leq 23.73 \text{ in.} \quad \text{OK}$$

$$0.040L = 0.040\,(35)\,(12) = 16.8 \text{ in.} \leq 32.23 \text{ in.} \quad \text{OK}$$

These minimum depths are easily met, which is consistent with the deflection computation.

(2) Permanent Deformations [A6.10.4.2] For tension flanges of composite sections

$$f_f + \frac{f_l}{2} \leq 0.95 R_h F_{\text{yf}} = 0.95\,(1.0)\,(50) = 47.5 \text{ ksi}$$

where

f_f = elastic flange stress caused by the factored loading

f_l = elastic flange stress caused by lateral bending that is assumed to be zero here

The maximum service II moment, which occurs near location 105 in the interior beam, is due to unfactored dead loads $D1$, $D2$, and $D3$, and factored live load, 1.3(LL + IM) taken from Table E20.2-1. The stresses calculated from these moments are given in Table E20.2-5.

$$\max\ f_f = 40.9 \text{ ksi} < 47.5 \text{ ksi} \quad \text{OK}$$

f. Fatigue and Fracture Limit State [A6.5.3] [A6.10.5]

(1) Fatigue [A6.10.5.1] [A6.6.1] From Example Problem 20.1,

$$(\Delta F)_n = (\Delta F)_{\text{TH}} = 16 \text{ ksi}$$

From Example Problem 20.1,

$$M_{\text{fatigue}} = 317.0 \text{ k ft}$$

$$f = \frac{M}{S_b} = \frac{317.0\,(12)}{266} = 14.3 \text{ ksi} < 16 \text{ ksi} \quad \text{OK}$$

where S_b is the section modulus for the short-term composite section, calculated previously.

(2) Fracture [A6.10.5.2] [A6.6.2] The steel specified meets fracture requirements for this non-fracture-critical system.

(3) Special Fatigue Requirements for Webs [A6.10.5.3] The shear force due the fatigue truck is determined with the use of Figure E20.1-6 (interior girder governs and no multiple presence) [A3.6.1.1.2])

$$V_u = mg_{\text{fatigue}}\ 1.5\, V_{\text{fatigue}} + V_{\text{DC}} + V_{\text{DW}} = 1.5\,(26.5) + 15.1 + 1.9 + 4.4 = 61.1 \text{ kips}$$

$$61.1 \text{ kips} \leq 286 \text{ kips} \quad \text{OK}$$

g. Strength Limit State [A6.5.4] [A6.10.6]

Composite Sections in Positive Flexure [A6.10.6.2.2] [A6.10.7] The positive moment sections may be considered compact composite if the following are satisfied:

- ☐ The specified minimum yield strength of the flange does not exceed 70 ksi.
- ☐ The web satisfies [A6.10.2.1.1]; see previous computation.

$$\frac{2D_{\text{cp}}}{t_w} \leq 3.76\sqrt{\frac{E}{F_y}} = 90.6$$

Table E20.2-5 Stresses in Bottom Flange of Steel Beam Due to Service II Moments

Load	M_{D1} (k ft)	M_{D2} (k ft)	M_{D3} (k ft)	$1.3M_{\text{LL+IM}}$	S_b Steel (in.^{3})	S_b Composite (in.^{3})	Stress (ksi)
$D1$	142.4				154		11.1
$D2$		42.9				243	2.1
$D3$			16.8			243	0.8
LL + IM				596.4		266	26.9
Total							40.9

Table E20.2-6 Shear Range for Fatigue Loading and Required Shear Connector Spacing

Location	Unfactored Maximum Positive Shear (kips)[a]	Unfactored Maximum Negative Shear (kips)	Factored Shear Range (kips)	Pitch (in.)
100	26.5	0	39.6	6
101	23.6	−2.4	39.05	5.9
102	20.7	−4.6	38.0	6.1
103	17.9	−7.0	37.5	6.4
104	15.0	−9.2	36.2	6.4
105	12.1	−12.1	36.2	6.4

[a]Distributed with IM included.

As shown below, the depth of the plastic neutral axis is in the deck, therefore, no portion of the web is in compression and the last provision is satisfied.

The $A_{steel} = 20.1$ in.^{2} and the yield stress is 50 ksi, therefore, the steel tensile capacity is 1005 kips. If the deck is completely in compression, then the force would be

$$\begin{aligned} C &= 0.85\left(f_c'\right) b_e t_s \\ &= 0.85\,(4)\,(96)\,(7.5) \\ &= 2448 \text{ kips} \geq 1005 \text{ kips} \end{aligned}$$

Therefore, the neutral axis lies within the deck:

$$\begin{aligned} C &= 0.85\left(f_c'\right) b_e a \\ &= 0.85\,(4)\,(96)\,(a) \\ &= 326.4a = 1005 \text{ kips} \\ a &= 3.08 \text{ in.} \end{aligned}$$

The lever arm between the compression and tension force is

$$\begin{aligned} \text{Lever} &= (23.73 + 1 + 7.5) - \frac{23.73}{2} - \frac{3.08}{2} \\ &= 18.8 \text{ in.} \end{aligned}$$

and the flexural capacity is

$$\begin{aligned} \phi_m M_n &= 1.0\,(1005)\,(18.8) = 18894 \text{ k in.} \\ &= 1575 \text{ k ft} \end{aligned}$$

$$M_u \leq \phi_m M_n$$

$$M_u = 1066.3 \text{ k ft} \leq 1575 \text{ k ft} \quad \text{OK}$$

The ductility requirement of [A6.10.7.3] is

$$D_p \leq 0.42D$$

$$3.15 \leq 0.42\,(32.23) = 13.5 \quad \text{OK}$$

2. *Shear Design*
   a. General [A6.10.9.1] The section is a wide flange and shear resistance should be at the plastic shear capacity. No transverse stiffeners are required; the computations are for an unstiffened section. The shear resistance was previously computed as

$$\begin{aligned} V_u &\leq \phi V_{cr} = \phi C V_p \\ &= (1.0)\,(1.0)\,(286) \\ &= 286 \text{ kips} \end{aligned}$$

$$V_u = 145.8 \text{ kips} \leq V_r = 286 \text{ kips} \quad \text{OK}$$

   The details of the resistance computation are illustrated above.

3. *Shear Connectors* [A6.10.10] Shear connectors must be provided throughout the length of the span for simple-span composite bridges.
   a. Use $\frac{3}{4}$-in. diameter studs, 4 in. high. The ratio of height to diameter is

$$\frac{4}{0.75} = 5.33 > 4 \quad \text{OK [A6.10.10.1.1]}$$

      1. Transverse spacing [A6.10.10.1.3]: The center-to-center spacing of the connectors cannot be closer than 4 stud diameters, or 3 in. The clear distance between the edge of the top flange and the edge of the nearest connector must be at least 1 in.
      2. Cover and Penetration [A6.10.10.1.4]: Penetration into the deck should be at least 2 in. Clear cover should be at least 2.5 in.
   b. General [A6.10.10.1] No computations are necessary for this article.
   c. Fatigue Resistance [A6.10.10.2] Fatigue resistance for the shear connectors is controlled by the infinite life criterion as the number of cycles is larger than 960 trucks per day; therefore the resistance per stud is

$$Z_r = 5.5d^2 = 5.5\left(0.75^2\right) = 3.1 \text{ kips}$$

$$p = \frac{nZ_r I}{V_{sr} Q}$$

I = moment of inertia of short-term composite section = 6785 in.^{4}

n = 3 shear connectors in a cross section

Q = first moment of the transformed area about the neural axis of the short-term composite section

$$= \left(y_b - \frac{d}{2}\right) A_{steel}$$

$$= \left(25.5 - \frac{23.73}{2}\right)(20.1)$$
$$= 274 \text{ in.}^3$$

V_{sr} = shear force range under LL + IM determined for the fatigue limit state

Shear ranges at tenth points, with required pitches, are located in Table E20.2-6. The shear range is computed by finding the difference in the positive and negative shears at that point due to the fatigue truck, multiplied by the dynamic load allowance for fatigue (1.15), the maximum distribution factor for one design lane loaded without multiple presence (0.765/1.2) for the exterior beam, and by the load factor for the fatigue limit state I (1.5). Values are symmetric about the center of the bridge, location 105.

An example calculation of the pitch is performed below, for the shear range at location 101:

$$V_{sr} = [23.6 - (-2.4)](1.5) = 39.0 \text{ kips}$$
$$p = \frac{(3)(3.1)(6785)}{39.0(272)} = 5.9 \text{ in.}$$
$$6d_s = 4.5 \text{ in.} < p < 24 \text{ in.} \quad \text{OK}$$

Because the required pitch from Table E20.2-6 does not vary much between tenth points, use a pitch of 6 in. along the entire span. The minor difference between 5.9 in. and 6 in. is neglected.

d. Special Requirements for Point of Permanent Load Contraflexure [A6.10.10.3] This article is not applicable for a simple-span beam.

e. Strength Limit State [A6.10.10.4]
Strength Limit State [A6.10.10.10.4.3]

$$Q_r = \phi_{sc} Q_n$$
$$\phi_{sc} = 0.85$$
$$Q_n = 0.5 A_{sc}\sqrt{f'_c E_c} \le A_{sc} F_u$$
$$A_{sc} = \frac{\pi}{4}(0.75)^2 = 0.44 \text{ in.}^2$$
$$E_c = 1820\sqrt{f'_c} = 1820\sqrt{4} = 3640 \text{ ksi}$$
$$Q_n = 0.5(0.44)\sqrt{4(3640)} = 26.5 \text{ kips}$$
$$A_{sc} F_u = (0.44)(60) = 26.4 \text{ kips use } 26.5 \text{ kips}$$
$$Q_n = 26.5 \text{ kips}$$
$$Q_r = 0.85(26.5) = 22.5 \text{ kips}$$

Between sections of maximum positive moment and points of zero moment, the number of shear connectors required is

$$n = \frac{V_h}{Q_r}$$

for which

$$V_h = \min \begin{cases} 0.85 f'_c b t_s = 0.85(4)(96)(7.5) = 2448 \text{ kips} \\ A_s F_y = (20.1)(50) = 1005 \text{ kips} \end{cases}$$

Therefore use a nominal horizontal shear force, V_h, of 1005 kips:

$$n = \frac{V_h}{Q_f} = \frac{1005}{22.5} = 44.7$$

Therefore a minimum of 45 shear connectors are required at the strength limit state in half the span (or 15 groups of 3). This requirement is more than satisfied by the 6-in. pitch of the three shear connector group required for fatigue resistance.

J. Dimension and Detail Requirements

1. *Material Thickness* [A6.7.3] Bracing and cross frames shall not be less than 0.3125 in. thickness. Web thickness of rolled beams shall not be less than 0.25 in.

$$t_w = 0.47 \text{ in.} > 0.25 \text{ in.} \quad \text{OK}$$

2. *Bolted Connections* [A6.13.2] Bolts are not addressed in this example.
3. *Diaphragms and Cross Frames* [A6.7.4] Diaphragms were designed for the noncomposite bridge of Example Problem 20.1. The same diaphragms are adequate for this bridge. Use C15 × 33.9, M270 Grade 50W for all diaphragms.
4. *Lateral Support of Compression Flange Prior to Curing of the Deck* Lateral bracing shall be provided at quarter points, as determined in the previous example. Use the same section as diaphragms, C15 × 33.9, M270 Grade 50W. The two braces other than the brace at midspan may be removed after the concrete cures.

K. Dead-Load Camber The centerline deflection due to a uniform load on a simply support span is

$$\Delta_{CL} = \frac{5}{384}\frac{(w_D/12)L^4}{EI}$$
$$= \frac{5}{384}\frac{(w_D/12)(420)^4}{29{,}000I}$$
$$= 1164\frac{w_D}{I}$$

By substituting the dead loads from Tables E20.2-1 and E20.2-2, and using the I values determined previously for long-term loads, the centerline deflections are calculated in Tables E20.2-7 and E20.2-8. Use a $\frac{3}{4}$-in. camber on all beams.

L. Check Assumptions Made in Design Nearly all the requirements are satisfied, using a W24 × 68. This beam has a self-weight of 0.068 k/ft; thus,

Table E20.2-7 Exterior Beam Deflection Due to Dead Loads

Load Type	Load, w (k/ft)	I (in.^{4})	Δ_{CL} (in.)
$D1$	0.86	1830	0.55
$D2$	0.25	5292	0.06
$D3$	0.11	6785	0.02
Total			0.63

Table E20.2-8 Interior Beam Deflection Due to Dead Loads

Load Type	Load, w (k/ft)	I (in.^{4})	Δ_{CL} (in.)
$D1$	0.93	1830	0.59
$D2$	0.28	5292	0.06
$D3$	0.11	6785	0.02
Total			0.67

our assumed beam weight of 0.10 k/ft is conservative. Also, for preliminary design, the value for $\left[K_g/12Lt_s^3\right]^{0.1}$ was taken as 1.0 in calculating the distribution factors for moment. The actual value is calculated below.

$$K_g = n\left(I + Ae_g^2\right)$$

$$n = \frac{E_s}{E_c} = \frac{29{,}000}{1820\sqrt{4}} = 7.98 \quad \text{use } 8$$

$$I = 1830 \text{ in.}^4$$

$$e_g = \left(\frac{d}{2} + t_h + \frac{t_s}{2}\right)$$

$$e_g = \frac{23.73}{2} + 1 + \frac{7.5}{2} = 16.6 \text{ in.}$$

$$K_g = 8\left[(1830) + (20.1)\left(16.6^2\right)\right] = 58{,}950 \text{ in.}^4$$

$$\frac{K_g}{12Lt_s^3} = \frac{58{,}950}{12\,(35)\left(7.5^3\right)} = 0.33$$

$$\left(\frac{K_g}{12Lt_s^3}\right)^{0.1} = 0.90$$

Recompute the distribution factors:

$$mg_M^{SI} = 0.06 + \left(\frac{S}{14}\right)^{0.4}\left(\frac{S}{L}\right)^{0.3}\left(\frac{K_g}{12Lt_s^3}\right)^{0.1} = 0.06 + \left(\frac{8}{14}\right)^{0.4}\left(\frac{8}{35}\right)^{0.3}(0.90) = 0.52$$

Two design lanes loaded:

$$mg_M^{MI} = 0.075 + \left(\frac{S}{9.5}\right)^{0.6}\left(\frac{S}{L}\right)^{0.2}\left(\frac{K_g}{12Lt_s^3}\right)^{0.1} = 0.075 + \left(\frac{8}{9.5}\right)^{0.6}\left(\frac{8}{35}\right)^{0.2}(0.90) = 0.68 \quad \text{governs}$$

This demonstrates that the live loads calculated in the preliminary design are about 12% higher than actual, which is conservative (interior girder only). However, the distribution factor is not applied to the dead load so that when the live- and dead-load effects are combined, the preliminary design loads are less conservative, which is more acceptable. Also the exterior girder does not have this factor, so the distribution factors are unchanged with the better estimate of the longitudinal to transverse stiffness (so-called K_g term).

M. ***Design Sketch*** The design of the composite, simple-span, rolled steel beam bridge is summarized in Figure E20.2-4.

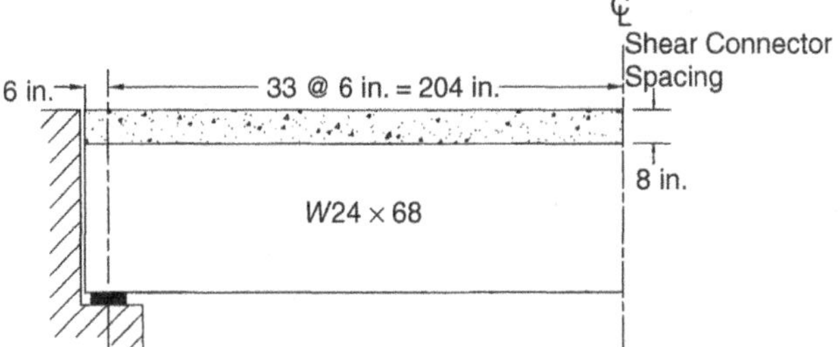

Fig. E20.2-4 Design sketch of composited rolled steel girder.

20.3 MULTIPLE-SPAN COMPOSITE STEEL PLATE GIRDER BEAM BRIDGE

Problem Statement Example 20.3 Design the continuous steel plate girder bridge of Figure E20.3-1 with 100, 120, and 100-ft spans for an HL-93 live load. Roadway width is 44 ft curb to curb and carries an interstate highway. Allow for a future wearing surface of 3-in.-thick bituminous overlay. Use $f_c' = 4$ ksi and M270 Grade 50W steel (50 ksi).

Note that the computer program *BT-Beam* was used to generate the actions. The sample computations are presented to illustrate the hand and computer computations. The computer results are slightly different due to a refined live-load positioning as compared to the hand-based critical position estimates.

A. Develop General Section

1. *Roadway Width (Highway Specified)* The general elevation and plan of the three-span continuous steel plate girder bridge is shown in Figure E20.3-1. The

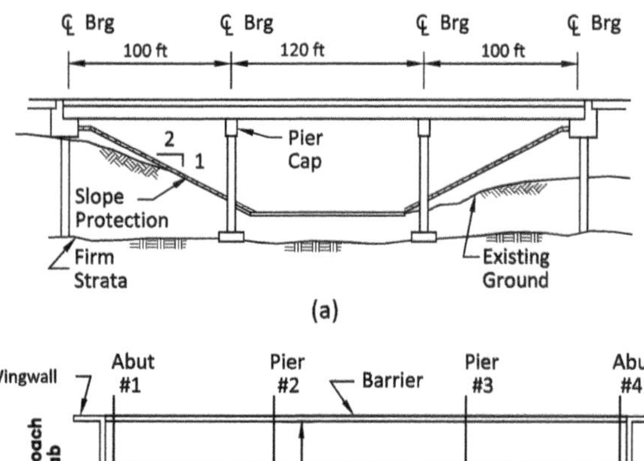

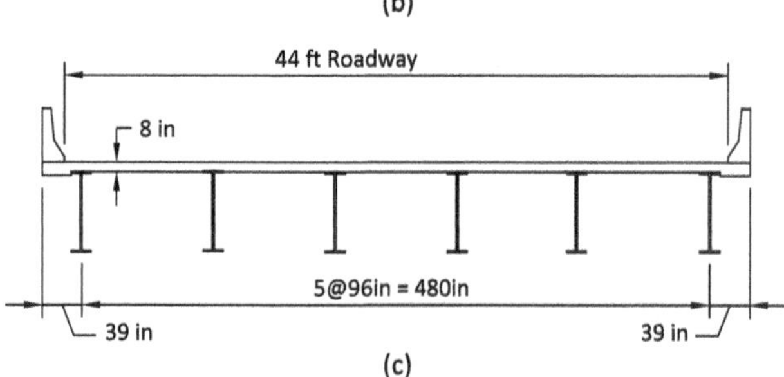

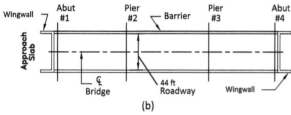

Fig. E20.3-1 Steel plate girder bridge design example: (a) general elevation, (b) plan view, and (c) cross section.

bridge will carry two lanes of urban interstate traffic over a secondary road.

2. *Span Arrangements* [A2.3.2] [A2.5.4] [A2.5.5] [A2.6]
3. *Select Bridge Type* A composite steel plate girder is selected. The concrete acts compositely in the positive moment region and the reinforcement acts compositely in the negative moment region.

B. *Develop Typical Section* A section of the bridge is shown in Figure E20.3-1(c). Six equally spaced girders are composite with the 8-in.-thick concrete deck. The flanges and web of the plate girder are of the same material, so that $R_h = 1.0$.

C. *Design Reinforced Concrete Deck* Use same design as in Example Problem 20.1.

D. *Select Resistance Factors* [A6.5.4.2]

1. For flexure $\phi_f = 1.00$
2. For shear $\phi_v = 1.00$
3. For axial compression $\phi_c = 0.90$
4. For shear connectors $\phi_{sc} = 0.85$

E. *Select Load Modifiers* The welded plate girder is considered to be ductile. The multiple girders and continuity of the bridge provide redundancy. The bridge cross section is ductile and redundant. In this example, we consider these adjustments. Additionally, because the bridge supports an interstate highway, we consider it "important" as well. This combination yields a net load modifier of 0.95, which demonstrates its application. (In previous examples, the load modifer of unity was used.)

		Strength	**Service, Fatigue**	
1. Ductility	η_D	0.95	1.0	[A1.3.3]
2. Redundancy	η_R	0.95	1.0	[A1.3.4]
3. Importance	η_I	1.05	1.0	[A1.3.5]
	$\eta = \eta_D\eta_R\eta_I$	0.95	1.0	

F. *Determine Combinations and Factors* Load factors are outlined as used throughout this example.

G. *Calculate Live-Load Force Effects*

1. *Select Live Loads* [A3.6.1] *and Number of Lanes* [A3.6.1.1.1]

$$N_L = \text{INT}\left(\frac{w}{12}\right)$$
$$= \text{INT}\left(\frac{44}{12}\right) = 3 \text{ lanes}$$

2. *Multiple Presence* [A3.6.1.1.2] (Table 8.6)

No. Loaded Lanes	Multiple Presence Factor
1	1.20
2	1.00
3	0.85

3. *Dynamic Load Allowance* [A3.6.2] (Table 8.7) [Table A3.6.2.1-1]

$$\text{Impact} = 33\%$$
$$\text{Fatigue and fracture} = 15\%$$

4. *Distribution Factor for Moment* [A4.6.2.2.2]
   a. Interior Beams [A4.6.2.2.2b] (Table 11.3) Check that design parameters are within the range of applicability.

$$3.5 \text{ ft} \le (S = 8 \text{ ft}) \le 240 \text{ ft}$$
$$20 \text{ ft} \le (L = 120 \text{ ft or } 100 \text{ ft}) \le 240 \text{ ft}$$
$$4.5 \text{ int} \le \left(t_s = 8\right) \le 12 \text{ in.}$$
$$N_b = 6 > 4$$

Therefore, the design parameters are within range, and the approximate method is applicable for concrete deck on steel beams cross section. (Note: The following properties are based on $t_s = 8$ in. A more conservative approach is to use a deck thickness reduced by the sacrificial wear thickness to give $t_s = 8 - 0.5 = 7.5$ in.) This approach was illustrated in the last example.

After initially designing the superstructure this value was determined to be 1.02 and 1.03 for the positive and negative moment regions. Per A4.6.2.2.1-2 the K_g is estimated to be 1.02, which is used throughout this example.

$$mg_M^{SI} = 0.06 + \left(\frac{S}{14}\right)^{0.4}\left(\frac{S}{L}\right)^{0.3}\left(\frac{K_g}{12Lt_s^3}\right)^{0.1}$$

$$mg_M^{MI} = 0.075 + \left(\frac{S}{9}\right)^{0.6}\left(\frac{S}{L}\right)^{0.2}\left(\frac{K_g}{12Lt_s^3}\right)^{0.1}$$

$L = 100$ ft, positive flexure

$$mg_M^{SI} = 0.06 + \left(\frac{8}{14}\right)^{0.4}\left(\frac{8}{100}\right)^{0.3}(1.02) = 0.45$$

$$mg_M^{MI} = 0.075 + \left(\frac{8}{9}\right)^{0.6}\left(\frac{8}{100}\right)^{0.2}(1.02) = 0.65$$

$L_{ave} = 110$ ft, negative flexure

$$mg_M^{SI} = 0.06 + \left(\frac{8}{9}\right)^{0.4}\left(\frac{8}{110}\right)^{0.3}(1.02) = 0.44$$

$$mg_M^{MI} = 0.075 + \left(\frac{8}{9}\right)^{0.6}\left(\frac{8}{110}\right)^{0.2}(1.02) = 0.64$$

$L = 120$ ft, positive flexure

$$mg_M^{SI} = 0.06 + \left(\frac{8}{9}\right)^{0.4}\left(\frac{8}{120}\right)^{0.3}(1.02) = 0.42$$

$$mg_M^{MI} = 0.075 + \left(\frac{8}{9}\right)^{0.6}\left(\frac{8}{120}\right)^{0.2}(1.02) = 0.64$$

Two or more lanes loaded controls. Because there is little difference between the maximum values, use a distribution factor of 0.65 for moment for all interior girders. Similarly, use 0.45 for all interior girders with one loaded lane and $0.45/1.2 = 0.38$ for interior girders flexural stress check due to fatigue.

b. Exterior Beams [A4.6.2.2.2d] (Table 11.3) [Table A4.62.2.2d-1] One design lane loaded. Use the lever rule to determine mg_M^{SE}, where $m = 1.2$, from Figure E20.3-2.

$$\sum M_{hinge} = 0$$

$$R = 0.5P\left(\frac{8+2}{8}\right) = 0.625P$$

$$mg_M^{SE} = 1.2\,(0.625) = 0.75$$

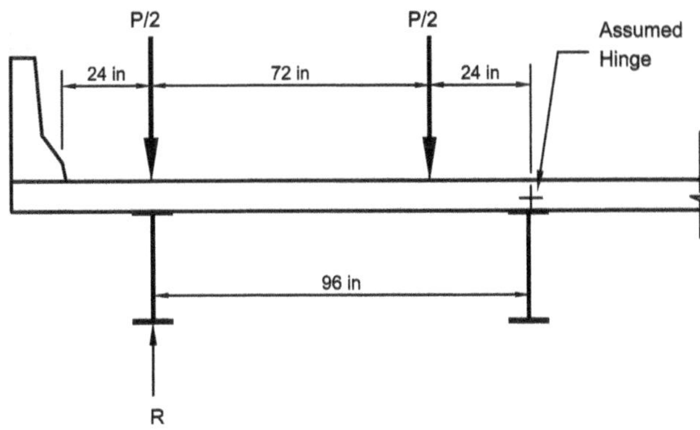

Fig. E20.3-2 Lever rule for determination of distribution factor for moment in exterior beam, one lane loaded.

Two or more design lanes loaded:

$$mg_M^{ME} = e \cdot mg_M^{MI}$$

$$e = 0.77 + \frac{d_e}{9.1} \geq 1.0$$

$$d_e = 39 - 15 = 24 \text{ in.} = 2 \text{ ft}$$

$$e = 0.77 + \frac{2}{9.1} = 0.99 < 1.0 \quad \therefore \text{ use } 1.0$$

$$mg_M^{ME} = (1.0)\,(0.625) = 0.625$$

Use distribution factor of 0.762 for moment for all exterior girders. The rigid method of [A4.6.2.2.2] requires stiff diagraphms or cross frame that affects the transverse stiffness. Here the rigid method is not used.

c. Skewed Bridge [A4.6.2.2.2e] This is a straight bridge and no adjustment is required for skew.

5. *Distribution Factor for Shear* [A4.6.2.2.3]

a. Interior Beams [A4.6.2.2.2a] Check that design parameters are within the range of applicability:

$$3.5 \text{ ft} \leq (S = 8 \text{ ft}) \leq 16 \text{ ft}$$

$$20 \text{ ft} \leq (L = 120 \text{ ft and } 120 \text{ ft}) \leq 240 \text{ ft}$$

$$4.5 \text{ in.} \leq (t_s = 8 \text{ in.}) \leq 12 \text{ in.}$$

$$N_b = 6 > 4$$

Therefore, the approximate method is applicable for concrete deck on steel beam cross sections.

$$mg_V^{SI} = 0.36 + \frac{S}{25} = 0.36 + \frac{8}{25} = 0.68$$

$$mg_V^{MI} = 0.2 + \frac{S}{12} - \left(\frac{S}{35}\right)^{2.0} = 0.2 + \frac{8}{12} - \left(\frac{8}{35}\right)^{2.0} = 0.81$$

Use distribution factor of 0.81 for shear for interior girders.

b. Exterior Beams [A4.6.2.2.2b] For one design lane loaded, use the lever rule as before, therefore,

$$mg_V^{SE} = 0.75$$

For two or more design lanes loaded,

$$mg_V^{ME} = e \cdot mg_V^{MI}$$
$$e = 0.6 + \frac{d_e}{10} = 0.6 + \frac{2}{10} = 0.80$$
$$mg_V^{ME} = 0.800\,(0.81) = 0.65$$

Use a distribution factor of 0.75 for shear for exterior girders and 0.68/1.2 = 0.57 for interior girders fatigue-related shear checks.

c. Skewed Bridge [A4.6.2.2.2c] This is a straight bridge and no adjustment is necessary for skew.

Example Live-Load Computations

$$M_{\text{LL+IM}} = mg\left[\left(M_{\text{Truck}} \text{ or } M_{\text{Tandem}}\right)\left(1 + \frac{\text{IM}}{100}\right) + M_{\text{Lane}}\right]$$

$$V_{\text{LL+IM}} = mg\left[\left(V_{\text{Truck}} \text{ or } V_{\text{Tandem}}\right)\left(1 + \frac{\text{IM}}{100}\right) + V_{\text{Lane}}\right]$$

a. Location 205 (Maximum moment at midspan in exterior girder) Influence line has the general shape shown in Figure E20.3-3 and ordinates are taken from Table 9.4. The placement of truck, tandem, and lane live loads are shown in Figures E20.3-4–E20.3-6.

$$M_{\text{Truck}} = [32\,(0.13823 + 0.20357) + 8\,(0.13823)]\,(100) = 1204.3 \text{ k ft}$$

$$M_{\text{Tandem}} = [25\,(0.20357 + 0.18504)]\,(100) = 971.5 \text{ k ft}$$

$$M_{\text{Lane}} = 0.64\,(0.10286)\,(100)^2 = 658.3 \text{ k ft}$$

$$M_{\text{LL+IM}} = (0.75)\,[1204.3\,(1.33) + 658.3] = 1695 \text{ k ft}$$

which compares well with the BT-Beam result, 1703 k ft.

b. Location 205 (Shear at midspan in interior girder) Influence line has the general shape shown in Figure E20.3-7 and ordinates are taken from Table 9.4. The placement of truck, tandem, and lane live loads are shown in Figures E20.3-8–E20.3-10.

$$V_{\text{Truck}} = [32\,(0.5 + 0.36044) + 8\,(0.2275)] = 29.4 \text{ kips}$$

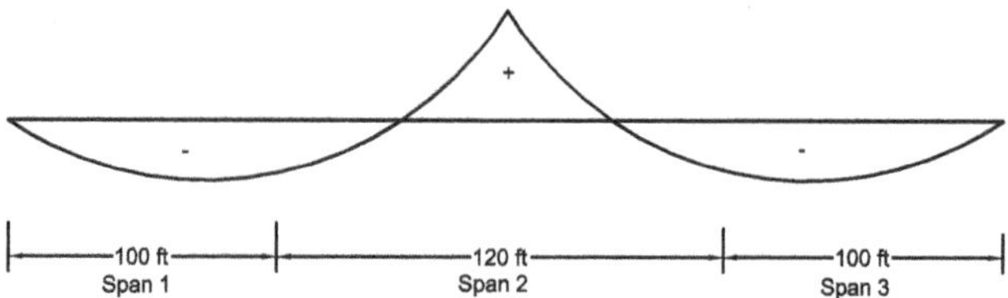

Fig. E20.3-3 Influence line for maximum moment at location 205.

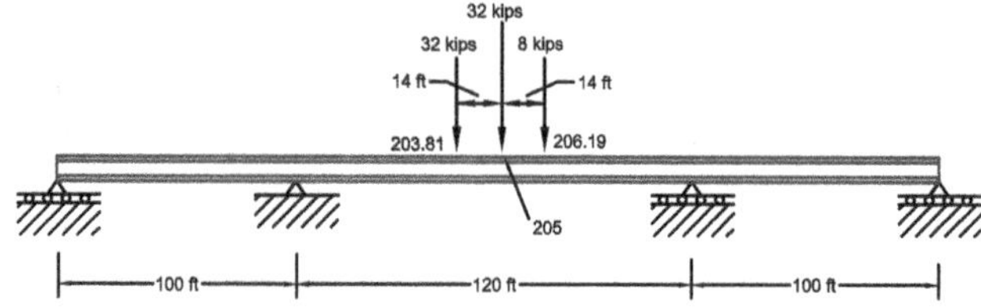

Fig. E20.3-4 Truck placement for maximum moment at location 205.

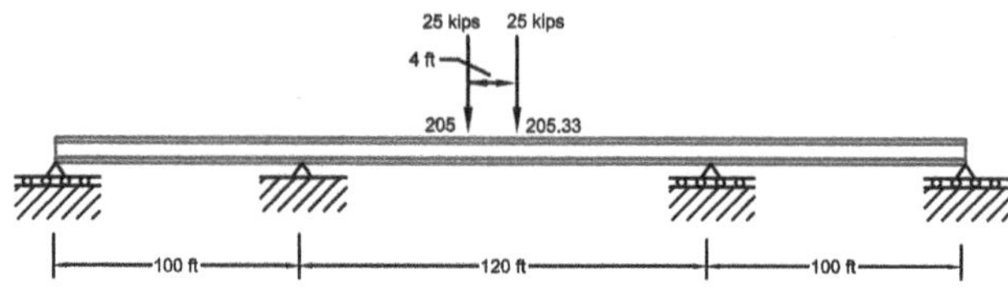

Fig. E20.3-5 Tandem placement for maximum moment at location 205.

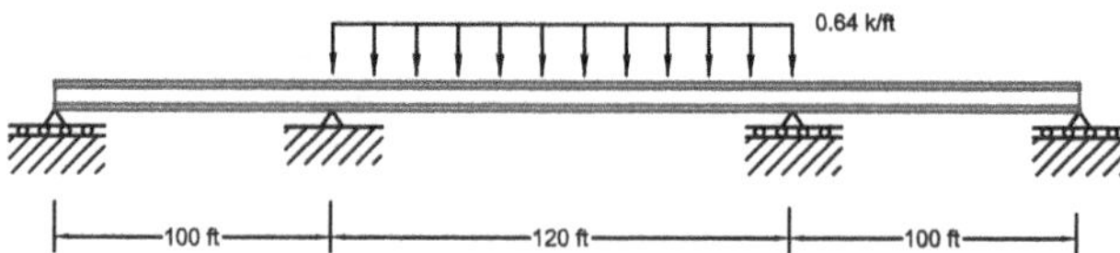

Fig. E20.3-6 Lane load placement for maximum moment at location 205.

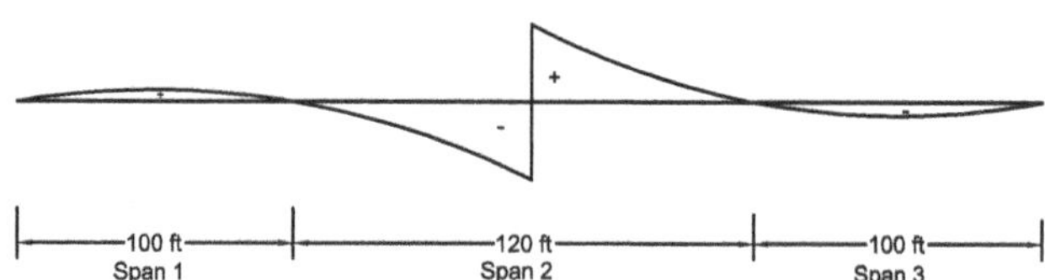

Fig. E20.3-7 Influence line for shear at location 205.

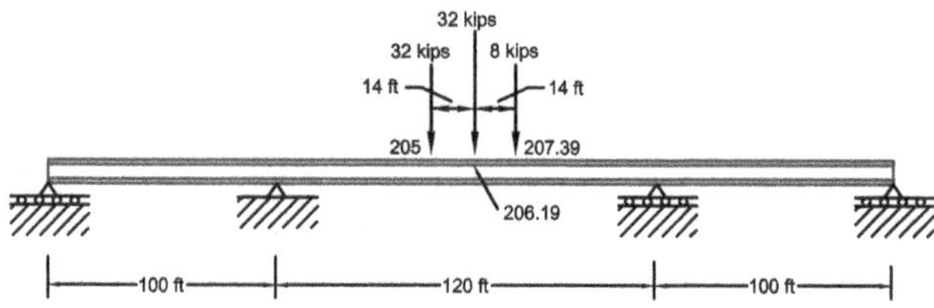

Fig. E20.3-8 Truck placement for maximum shear at location 205.

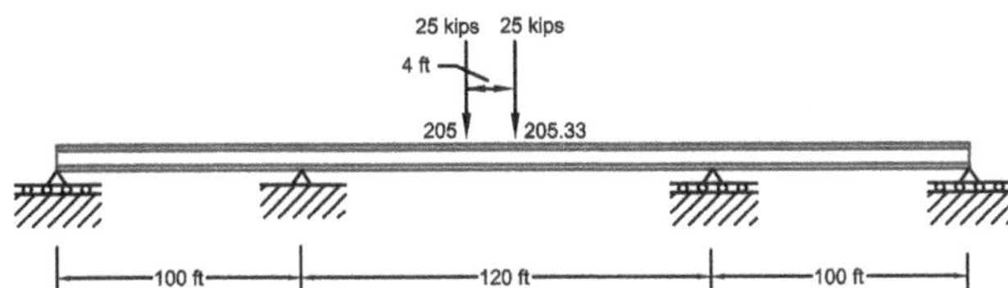

Fig. E20.3-9 Tandem placement for maximum shear at location 205.

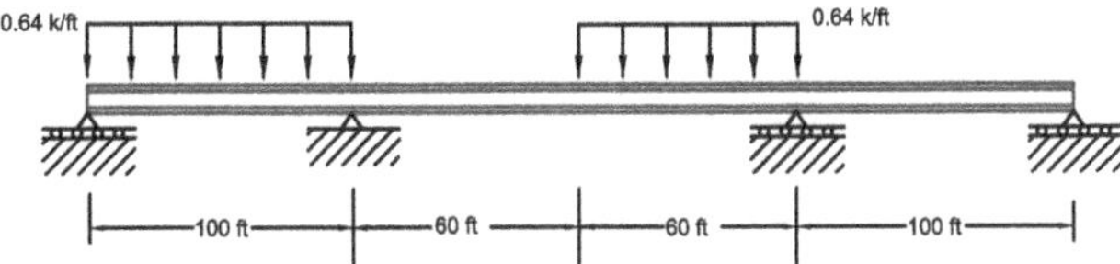

Fig. E20.3-10 Lane load placement for maximum shear at location 205.

$$
\begin{aligned}
V_{\text{Tandem}} &= [25\,(0.5 + 0.46106)] \\
&= 24.0 \text{ kips} \\
V_{\text{Lane}} &= 0.64\,(0.06510 + 0.13650)\,(100) \\
&= 12.9 \text{ kips} \\
V_{\text{LL+IM}} &= (0.81)\,[29.4\,(1.33) + 12.9] \\
&= 42.1 \text{ kips}
\end{aligned}
$$

c. Location 200 (Truck Train in negative moment region) [A3.6.1.3.1] Influence line ordinates and areas are taken from Table 9.4.

The truck train is applicable in the negative moment regions and for the reactions at the interior supports of continuous superstructures. The truck train (Fig. E20.3-11) is composed of 90% of the effect of two design trucks spaced a minimum of 50 ft between the rear axle of one and the front axle of the other, combined with 90% of the design lane loading. The spacing between the 32-kip axles on each truck is taken as 14 ft. The lane load is placed on spans 1 and 2 for maximum negative moment at location 200 (Fig. E20.3-12). Note that the impact factor of 33% is only applied to the combined truck load.

$$
\begin{aligned}
\text{Area span 1} &= (0.06138)\,(100)^2 = 613.8 \text{ft}^2 \\
\text{Area span 2} &= (0.07714)\,(100)^2 = 771.4 \text{ft}^2 \\
M_{\text{Lane}} &= (0.64)\,(613.8 + 771.4) \\
&= 886.5 \text{ k ft} \\
M_{\text{train}} &= 0.9\left(1.33 M_{\text{Tr}}\right) + 0.9 M_{\text{Ln}} \\
&= 0.9\,(1.33)\,(1367) \\
&\quad + 0.9\,(886.5) = 2435 \text{ k ft}
\end{aligned}
$$

which compares well with the BT-Beam result, 2420 k ft.

6. *Stiffness* [A6.10.1.5] The beam stiffness is modeled as prismatic and the same moment of inertia is used for the entire cross section. The specification recommends this approach; however, traditionally, some designers change the moment of interia in the negative moment region to model concrete cracking and include only the bare steel and reinforcement. Whatever stiffness model is used, the designer must be consistent throughout the design. See the discussion of the lower-bound theorem in Chapter 10.
7. *Wind Effects* [A4.6.2.7] Wind effects are considered later in this example.
8. *Reactions to Substructure* [A3.6] Reactions to the substructure do not include the distribution factor. Because the substructure is not considered here, the undistributed reactions are not provided.

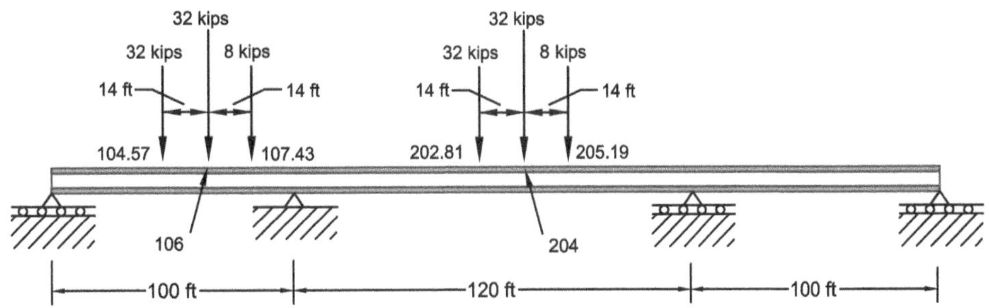

Fig. E20.3-11 Truck train placement for maximum moment at location 200.

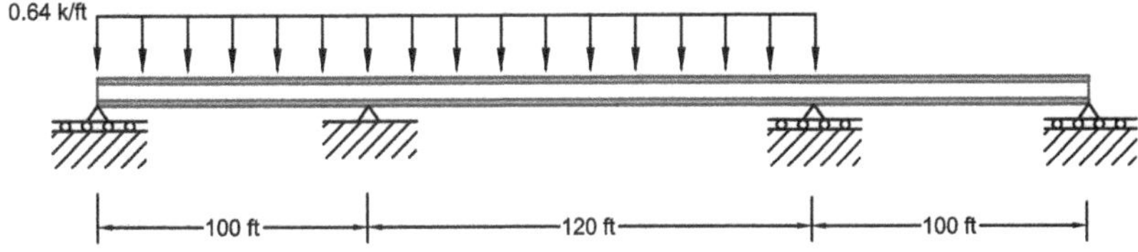

Fig. E20.3-12 Lane load placement for maximum moment at location 200.

H. Calculate Force Effects from Other Loads

1. *Interior Girders* Three separate dead loads must be calculated. The first is the dead load of the structural components and their attachments, $D1$, acting on the noncomposite section. The second type of dead load is $D2$, which represents the future wearing surface. The third load, $D3$, is caused by the barriers, where each has a cross-sectional area of 306 in^2. For this design it was assumed that the barrier loads were distributed equally among the interior and exterior girders. The initial cross section consists of a $\frac{7}{16}$-in. × 60-in. web and $1\frac{1}{4}$ × 16-in. $1\frac{1}{4}$ = 1.25 in. top and bottom flanges. The decimal equivalents are used in the computations: $1\frac{1}{4}$ = 1.25 in. and $\frac{7}{16}$ = 0.4375 in. The girder spacing is 96 in., and a 2-in. × 12-in. concrete haunch at each girder is used to accommodate camber and unshored construction. The density of the concrete and steel are taken as 0.15 and 0.49 k/ft^3, respectively. The density of the 3-in. bituminous future wearing surface (FWS) is taken as 0.14 k/ft^3.

$D1$

Slab $\quad 0.15\,(8)\,(96)\,/144 = 0.80$ k/ft

Haunch $\quad 0.15\,(12)\,(2)\,/144 = 0.025$ k/ft

Web $\quad 0.49\,(60)\,(0.4275)\,/144 = 0.089$ k/ft

Flanges $\quad 2\,(0.49)\,(12)\,(1.25)\,/144 = \underline{0.136 \text{ k/ft}}$

$$w_{D1}^{I} = 1.05 \text{ k/ft}$$

$D2$ 3-in. bituminous overlay

$$w_{D2}^{\mathrm{I}} = 0.14\,(3)\,(93)\,/144 = 0.28 \text{ k/ft}$$

$D3$ Barriers, one-sixth share

$$W_{D3}^{\mathrm{I}} = \frac{2\,(306)\,(0.15)}{6\,(144)} = 0.106 \text{ k/ft}$$

2. *Exterior Girders* The loads for the exterior girders are based on tributary areas. This approach gives smaller loads on an exterior girder than from a consideration of the deck as a continuous beam with an overhang and finding the reaction at the exterior support.

$D1$

Slab $\quad 0.15\,[(9 \times 40) + (8 \times 48)]\,/144 = 0.78$ k/ft

Haunch $\quad 0.15\,(12)\,(2)\,/144 = 0.025$ k/ft

Web $\quad 0.49\,(60)\,(0.4375)\,/144 = 0.089$ k/ft

Flanges $\quad 2\,(0.49)\,(16)\,(1.25)\,/144 = \underline{0.136 \text{ k/ft}}$

$$w_{D1}^{\mathrm{E}} = 1.03 \text{ k/ft}$$

$D2$ 3-in.bituminous overlay

$$w_{D2}^{\mathrm{E}} = 0.14\,(3)\,[48 + 40 - 15] = 0.071 \text{ k/ft}$$

$D3$ Barriers, one-sixth share

$$w_{D3}^{\mathrm{E}} = \frac{2\,(306)\,(0.15)}{6\,(144)} = 0.106 \text{ k/ft}$$

3. *Analysis of Uniformly Distributed Load* w (Fig. E20.3-13)

a. Moments

$$M_{104} = 0.07129w(100)^2 = -712.9w \text{ kft (BT-Beam: } 717.7w \text{ k ft)}$$

$$M_{200} = -0.12179w(100)^2 = -1217.9w \text{ kft (BT-Beam: } -1205.7w \text{ k ft)}$$

$$M_{205} = 0.05821w(100)^2 = 582.1w \text{ kft (BT-Beam: } 593.3w \text{ k ft)}$$

b. Shears

$$V_{100} = 0.37821w\,(100) = 37.821w \text{ kips (BT-Beam: } 37.94w \text{ kips)}$$

$$V_{110} = -0.62179w\,(100) = -62.179w \text{ kips (BT-Beam: } -62.06w \text{ kips)}$$

$$V_{200} = 0.6000w\,(100) = 60.0w \text{ kips (BT-Beam: } 60.0w \text{ kips)}$$

By substituting the values determined for dead load into the BT-Beam equations for moments and shears, the values at critical locations are generated in Table E20.3-1. The LL + IM values listed in Table E20.3-1 include the girder distribution factors as previously illustrated.

c. Effective Span Length The effective span length is defined as the distance between points of permanent load inflection for continuous spans:

Span 1 (Fig. E20.3-14)

$$M = 0 = 37.821x - \tfrac{1}{2}\,(1.0)\,x^2$$

$$x = L_{\text{eff}} = 75.64 \text{ ft}$$

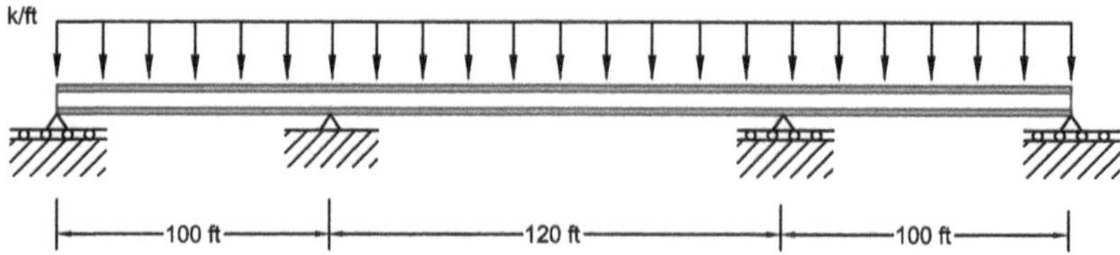

Fig. E20.3-13 Uniformly distributed load.

Table E20.3-1 Moments and Shears at Typical Critical Locations

Load Type	Value (k/ft)	Moments (k ft)[a] M_{104}	M_{200}	M_{205}	Shears (kips)[b] V_{100}	V_{110}	V_{200}
Uniform	1.0	717.7	−1205.7	594.3	37.9	−62.1	60.0
Truck/tandem/train		1238.1	−1369	1208.9	−63.7	−67.7	67.6
Lane		654.8	−877.7	663.2	−29.2	−40.8	42.52
			INTERIOR GIRDER				
$D1^I$	1.05	753.6	−1266.0	624.0	39.8	−65.2	63.0
$D2^I$	0.28	201.0	−337.6	166.9	10.6	−17.4	16.8
$D3^I$	0.106	76.1	−129.8	63.0	4.02	−6.58	6.36
$mg^I_{M\ or\ V}$		0.65	0.65	0.65	0.81	0.81	0.81
Live Load with distribution and IM		1496	1579(0.9) = 1421	1476	92.2	−106.0	107.3
Strength I Interior		3749	5042	3499	**220**	**−286**	**284**
			EXTERIOR GIRDER				
$D1^E$	1.03	739.2	−1241.9	612.1	39.0	−63.9	63.6
$D2^E$	0.071	50.96	−85.6	42.4	2.69	−12.6	12.4
$D3^E$	0.106	76.1	−127.8	63.0	4.02	−6.58	6.36
$mg^E_{M\ or\ V}$		0.75	0.75	0.75	0.65	0.65	0.65
Live Load with distribution and IM		1726	2024(0.9) = 1822	1703	74.0	85.1	86.1
Strength I Exterior		**3900**[d]	**4772**	**3684**	187	242	228
Strength I Controlling		**3900**	**4772**	**3684**	**220**	**286**	**284**
BT Beam[c]		**3910**	−4776	**3694**	220	−286	284

[a]Exterior girders govern for moments.
[b]Interior girders govern for shears.
[c]*See Table* E20.3-2 *and Table* E20.3-3.
[d]*Bold values are controlling*.

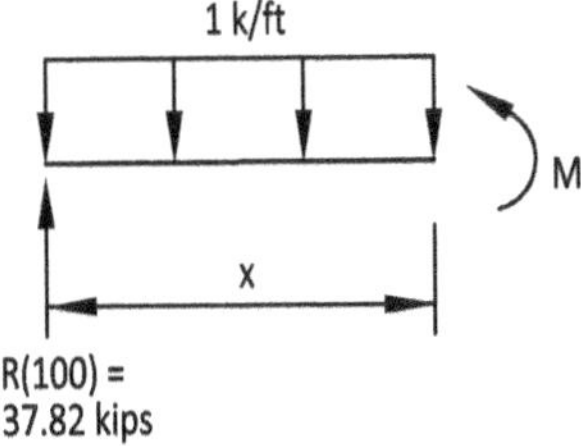

Fig. E20.3-14 Uniform load inflection point for span 1.

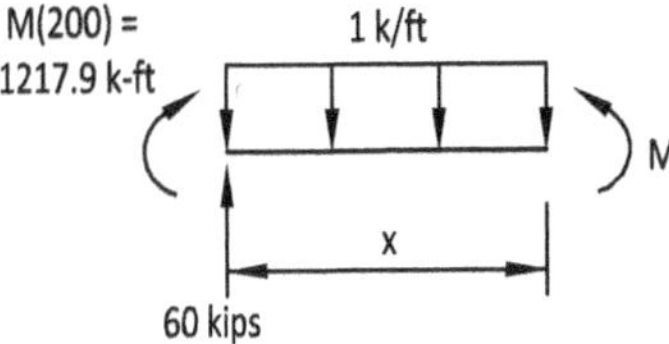

Fig. E20.3-15 Uniform load inflection point for span 2.

Span 2 (Fig. E20.3-15)

$$M = 0 = 60x - \tfrac{1}{2}(1.0)\,x^2 - 1217.9$$
$$x = 25.88, 94.12 \text{ ft}$$
$$L_{\text{eff}} = 94.12 - 25.88 = 68.24$$

Points of inflection are points where zero moment occurs. The points of inflection due to dead load are important because at these locations the flange plate transitions are used.

d. Maximum Dead-Load Moment The maximum moment occurs where the shear is equal to zero (Fig. E20.3-16):

$$M = (37.82)^2 - \frac{(37.82)^2}{2}(1.0) = 715.2 \text{ k ft}$$

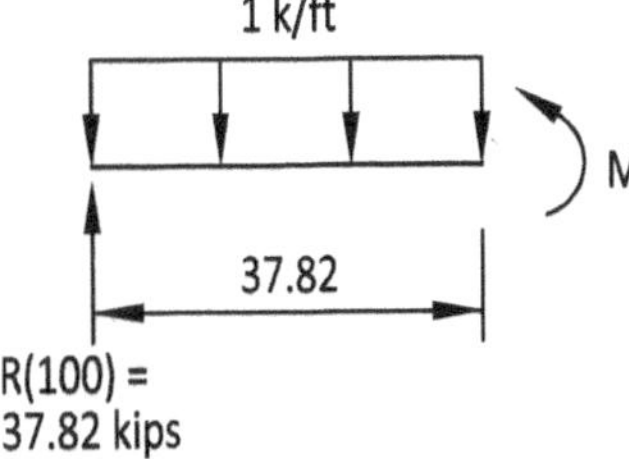

Fig. E20.3-16 Maximum moment due to uniform load in span 1.

The shears and moments, due to dead and live loads, at the tenth points are calculated. The procedures are the same as those illustrated for the critical locations, only with different placement of the live load. The results are summarized in Tables E20.3-2 and E20.3-3 for the strength I limit state. The second

Table E20.3-2 Moment Envelope for 100-, 120-, and 100-ft Plate Girder (k ft)[a]

Location	Unit Dead Load	Positive Moment Truck or Tandem	Positive Moment Lane	Positive Moment Strength I Int. Gir.	Positive Moment Strength I Ext. Gir.	Negative Moment Truck or Tandem	Negative Moment Lane	Negative Moment Strength I Int. Gir.	Negative Moment Strength I Ext. Gir.
100	0	0	0	0.00	0.00	0	0	0.00	0.00
101	329.43	549.03	259.71	1653.39	1712.03	−72	−48.88	0.00	0.00
102	558.85	926.13	455.42	2813.34	2914.12	−143.99	−97.75	0.00	0.00
103	688.28	1142.82	587.12	3496.42	3625.39	−215.99	−146.62	0.00	0.00
104	717.71	1238.07	654.83	**3758.62**	**3910.44**	−287.98	−195.5	0.00	0.00
105	647.14	1213.89	658.55	3602.85	3772.63	−359.98	−244.38	−29.95	−244.76
106	476.57	1092.28	598.26	3060.73	3248.45	−431.98	−293.26	−384.33	−598.25
107	206.01	860.44	473.98	2113.86	2316.67	−503.99	−342.14	−854.83	−1053.26
108	−164.55	546.17	285.71	902.63	1094.94	−623.56[b]	−391.02	−1541.93	−1681.38
109	−635.11	225.21	136.5	0.00	0.00	−874.39[b]	−542.97	−2784.36	−2835.37
110	−1205.67	176.98	106.07	0.00	0.00	−1368.79[b]	−877.7	−4760.31	−4776.31
200	−1205.67	176.98	106.07	0.00	0.00	−1368.79[b]	−877.7	**−4760.31**	**−4776.31**
201	−557.67	265.53	111.31	0.00	13.02	−816.56[b]	−468.22	−2499.65	−2552.87
202	−53.67	635.55	256.63	1128.42	1319.46	−488.46[b]	−290.97	−1104.75	−1242.83
203	306.33	956	478.9	2434.24	2626.73	−401.14	−282.85	−526.45	−706.94
204	522.33	1153.37	617.14	3249.98	3439.64	−318.55	−282.85	−156.92	−350.72
205	594.33	1208.86	663.22	**3507.10**	**3693.53**	−235.97	−282.85	0.00	−140.68

[a]Value computed with BT-Beam (web search "btbeam online" to find this application).
[b]Truck train with trucks spaced 58 ft apart governs.

Table E20.3-3 Shear Envelope for 100-, 120-, and 100-ft Plate Girder (kips)

Location	Unit Dead Load	Positive Shear Truck or Tandem	Positive Shear Lane	Positive Shear Strength I Int. Gir.	Positive Shear Strength I Ext. Gir.	Negative Shear Truck or Tandem	Negative Shear Lane	Negative Shear Strength I Int. Gir.	Negative Shear Strength I Ext. Gir.
100	37.94	63.71	29.17	**220.60**	**197.04**	−7.2	−4.89	0.00	0.00
101	27.94	54.9	23.17	179.04	160.46	−7.2	−5.29	0.00	0.00
102	17.94	46.31	17.96	138.91	125.20	−11.13	−6.47	−7.81	−8.31
103	7.94	38.02	13.52	100.38	91.43	−18.28	−8.44	−34.88	−32.78
104	−2.06	30.15	9.84	64.87	60.19	−26.84	−11.16	−66.74	−61.41
105	−12.06	22.81	6.88	36.11	34.16	−35.08	−14.6	−103.84	−93.85
106	−22.06	16.09	4.6	9.40	10.03	−42.88	−18.71	−141.08	126.43
107	−32.06	10.11	2.94	0.00	0.00	−50.15	−23.46	−178.20	158.90
108	−42.06	5.75	1.85	0.00	0.00	−56.77	−28.76	−214.92	191.00
109	−52.06	2.32	1.25	0.00	0.00	−62.65	−34.56	−250.97	222.47
110	−62.06	1.77	1.06	0.00	0.00	−67.67	−40.78	**−286.05**	**253.05**
200	60	67.59	42.52	**284.62**	**252.12**	−6.88	−4.13	0.00	
201	48	61.41	35.13	242.33	215.24	−6.88	−4.41	0.00	0.00
202	36	54.18	28.36	199.01	177.42	−7.44	−5.32	0.00	0.00
203	24	46.24	22.34	155.41	139.33	−13.74	−6.98	−6.14	0.00
204	12	37.9	17.15	112.24	101.64	−21.33	−9.47	−37.02	−7.13
205	0	29.49	12.86	70.13	64.94	−29.49	−12.86	−70.13	−35.00

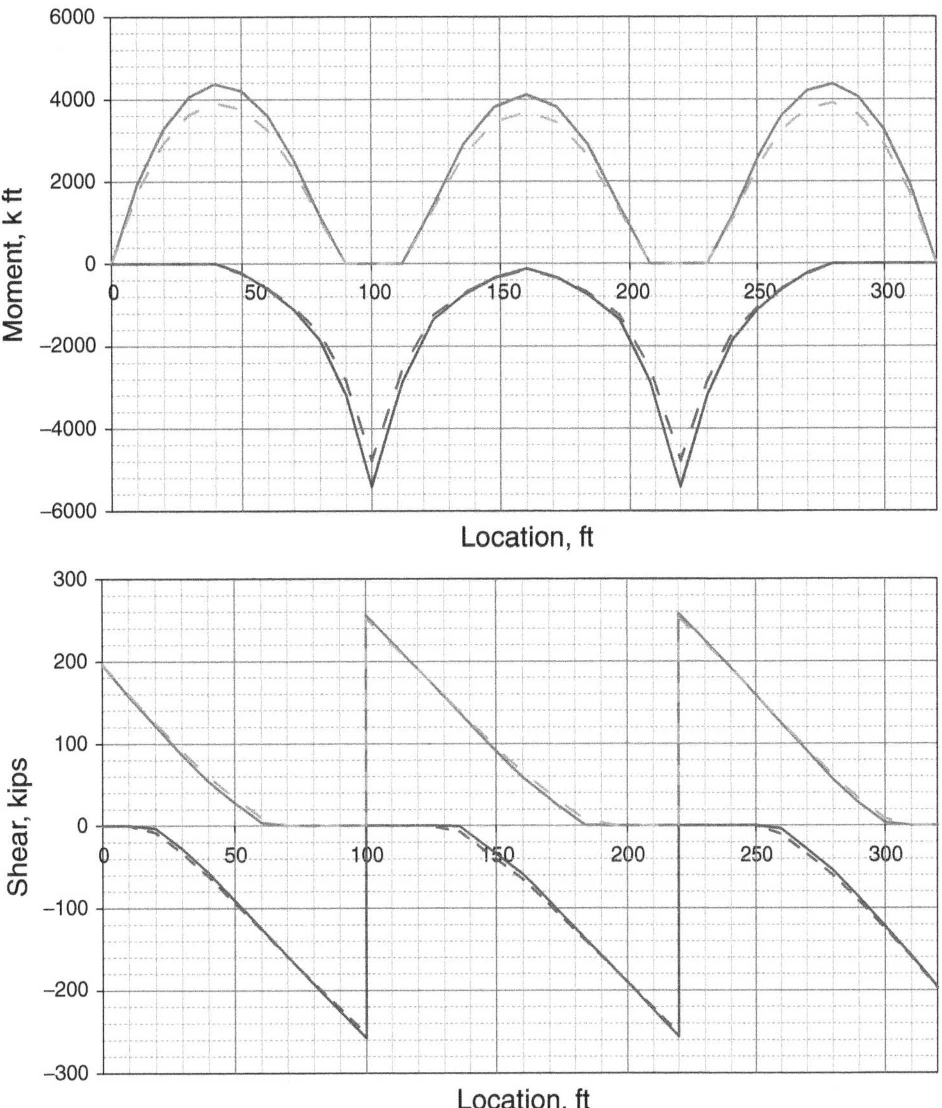

Fig. E20.3-17 Moment and shear envelopes for three-span plate girder.

column in these tables provides either the moment or the shear at each tenth point due to a unit distributed load. These are multiplied by the actual distributed load given in Table E20.3-1, combined with appropriate load factors, and added to the product of the distribution factor times the factored live load plus impact. This sum is then multiplied by the load modifier η and tabulated as η [sum]. The shear and moment envelopes are plotted in Figure E20.3-17.

Fatigue Load Example Computations Positive shear @ 100 (Fig. E20.3-18)

$$
\begin{aligned}
V_{100} &= 32\,(1.0 + 0.63297) + 8\,(0.47038) \\
&= 56.0 \text{ kips}
\end{aligned}
$$

Negative shear @ 100 (Fig. E20.3-19)

$$
\begin{aligned}
V_{100} &= 8\,(-0.09675) + 32\,(-0.10337 - 0.07194) \\
&= -6.38 \text{ kips}
\end{aligned}
$$

Positive shear @ 104 (Fig. E20.3-20)

$$
\begin{aligned}
V_{104} &= 32\,(0.51750 + 0.21234) + 8\,(0.09864) \\
&= 24.1 \text{ kips}
\end{aligned}
$$

Negative shear @ 104 (Fig. E20.3-21)

$$
\begin{aligned}
V_{104} &= 32\,[-0.12431 - (1 - 0.51750)] \\
&= -19.4 \text{ kips}
\end{aligned}
$$

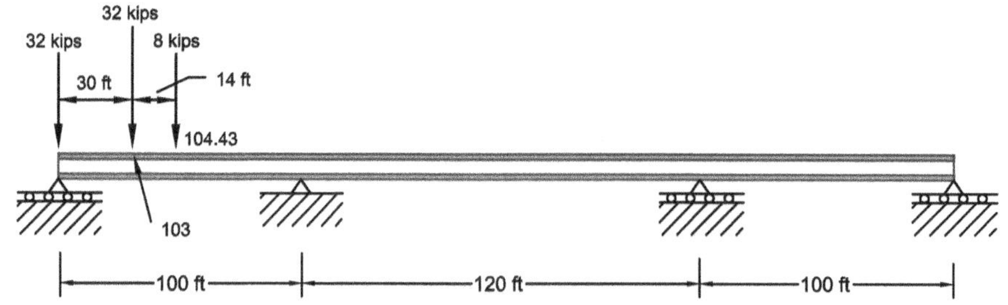

Fig. E20.3-18 Fatigue truck placement for maximum positive shear at location 100.

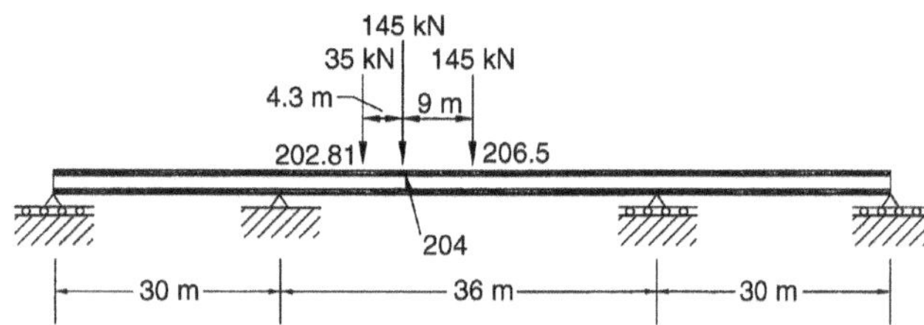

Fig. E20.3-19 Fatigue truck placement for maximum negative shear at location 100.

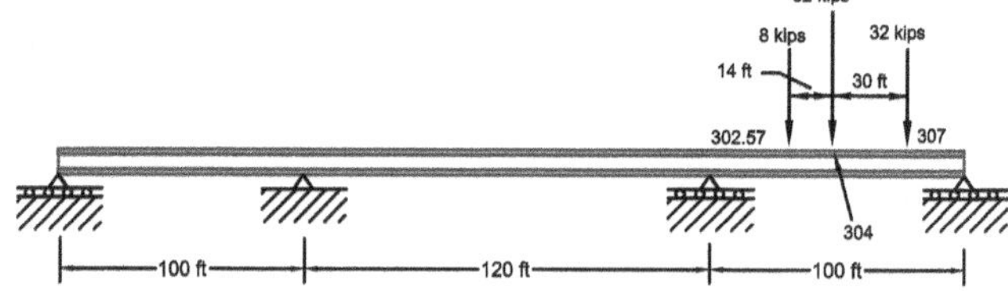

Fig. E20.3-22 Fatigue truck placement for maximum positive shear at location 110.

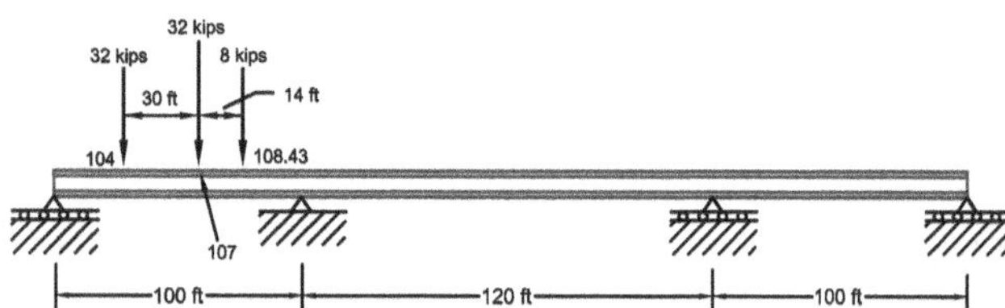

Fig. E20.3-20 Fatigue truck placement for maximum positive shear at location 104.

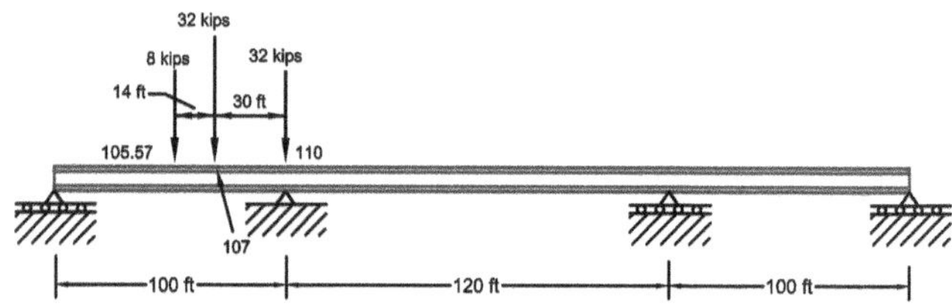

Fig. E20.3-23 Fatigue truck placement for maximum negative shear at location 110.

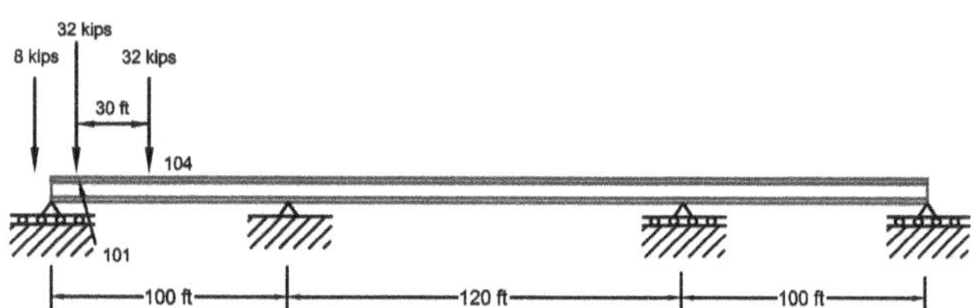

Fig. E20.3-21 Fatigue truck placement for maximum negative shear at location 104.

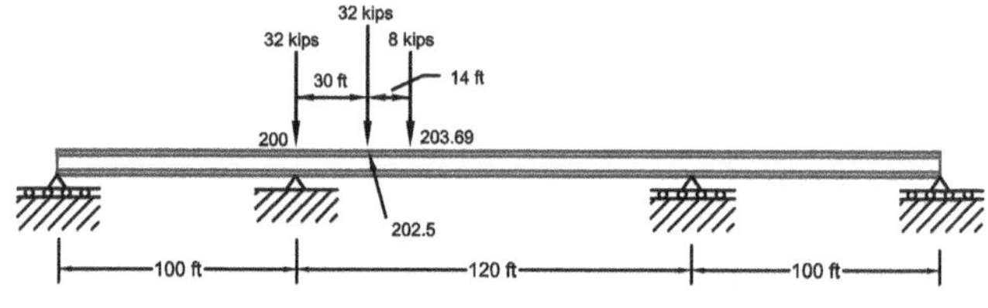

Fig. E20.3-24 Fatigue truck placement for maximum positive shear at location 200.

Positive shear @ 110 (Fig. E20.3-22)

$$\begin{aligned} V_{110} &= 8\,(0.02192) + 32\,(0.02571 + 0.01828) \\ &= 1.58 \text{ kips} \end{aligned}$$

Negative shear @ 110 (Fig. E20.3-23)

$$\begin{aligned} V_{110} &= 8\,(-0.65034) + 32\,(-0.78766 - 1.0) \\ &= -62.4 \text{ kips} \end{aligned}$$

Positive shear @ 200 (Fig. E20.3-24)

$$\begin{aligned} V_{200} &= 32\,(1.0 + 0.78375) + 8\,(0.653185) \\ &= 62.3 \text{ kips} \end{aligned}$$

Negative shear @ 200 (Fig. E20.3-25)

$$\begin{aligned} V_{200} &= 32\,(-0.10 - 0.07109) + 8\,(-0.039585) \\ &= -5.79 \text{ kips} \end{aligned}$$

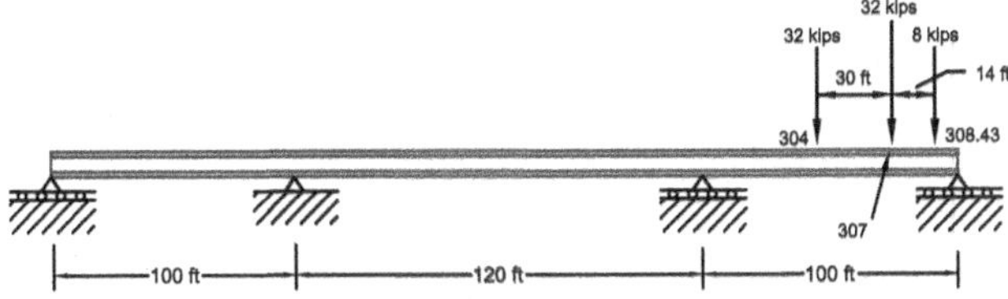

Fig. E20.3-25 Fatigue truck placement for maximum negative shear at location 200.

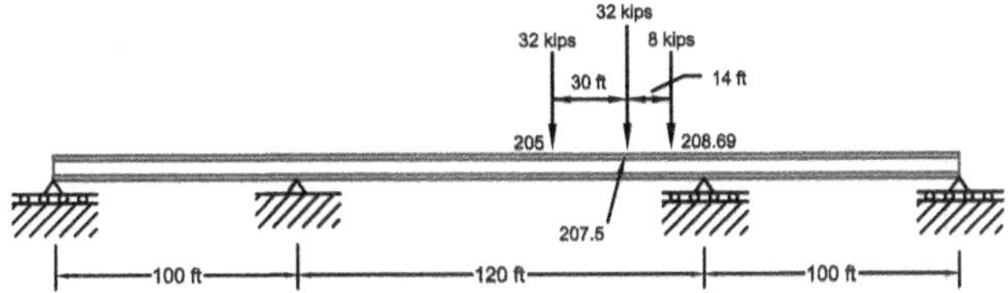

Fig. E20.3-26 Fatigue truck placement for maximum positive shear at location 205.

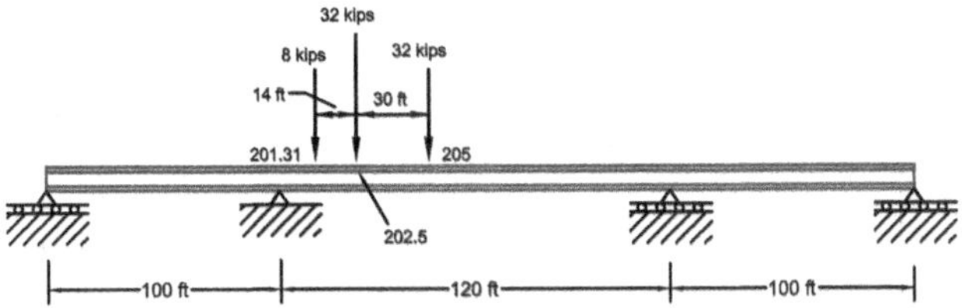

Fig. E20.3-27 Fatigue truck placement for maximum negative shear at location 205.

Positive shear @ 205 (Fig. E20.3-26)

$$V_{205} = 32\,(0.5 + 0.21625) + 8\,(0.10121)$$
$$= 23.7 \text{ kips}$$

Negative shear @ 205 (Fig. E20.3-27)

$$V_{205} = 8\,(-0.10121) + 32\,(-0.21625 - 0.5)$$
$$= -23.7 \text{ kips}$$

Positive moment @ 100

$$M_{100} = 0 \text{ k ft}$$

Negative moment @ 100

$$M_{100} = 0 \text{ k ft}$$

Positive moment @ 104 (Fig. E20.3-28)

$$M_{104} = [32\,(0.207 + 0.08494) + 8\,(0.130652)]\,100$$
$$= 1039 \text{ k ft}$$

Negative moment @ 104 (Fig. E20.3-29)

$$M_{104} = [32\,(-0.037105 - 0.0383025)$$
$$+8\,(-0.029213)]\,100$$
$$= -264 \text{ k ft}$$

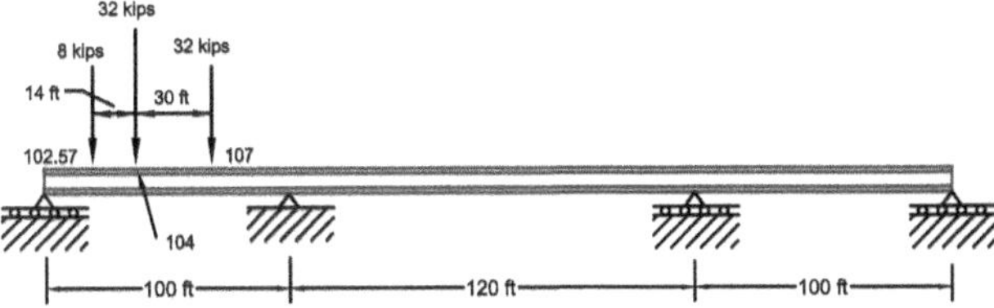

Fig. E20.3-28 Fatigue truck placement for maximum positive moment at location 104.

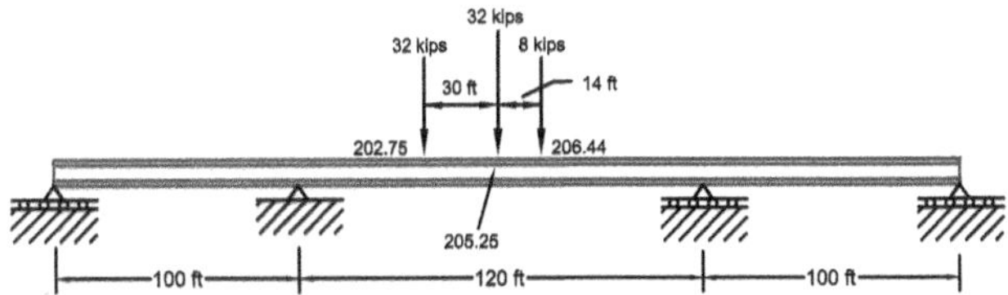

Fig. E20.3-29 Fatigue truck placement for maximum negative moment at location 104.

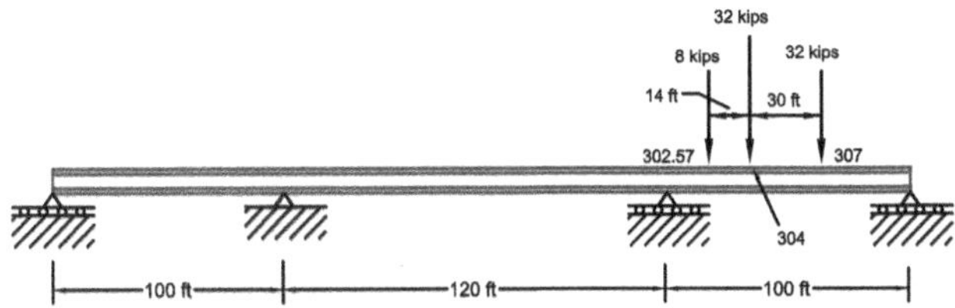

Fig. E20.3-30 Fatigue truck placement for maximum positive moment at location 110.

Positive moment @ 110 (Fig. E20.3-30)

$$M_{110} = [32\,(0.02571 + 0.01828)$$
$$+8\,(0.02191)]\,100$$
$$= 158 \text{ k ft}$$

Negative moment @ 110 (Fig. E20.3-31)

$$M_{110} = [32\,(-0.0957525 - 0.092765)$$
$$+8\,(-0.073028)]\,100$$
$$= -667 \text{ k ft}$$

Positive moment @ 200 same as M_{110}
Negative moment @ 200 same as M_{110}

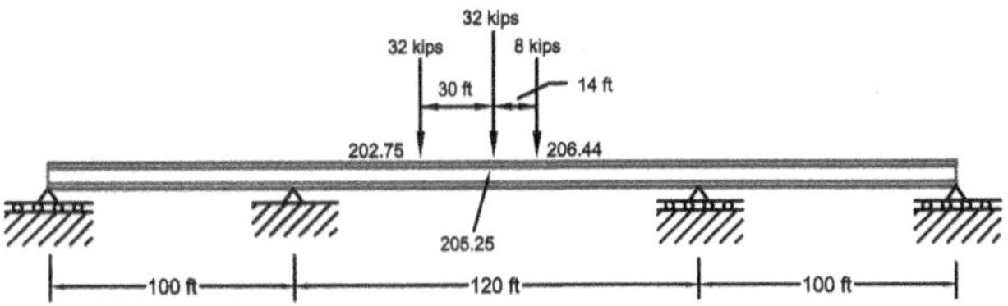

Fig. E20.3-31 Fatigue truck placement for maximum negative moment at location 110.

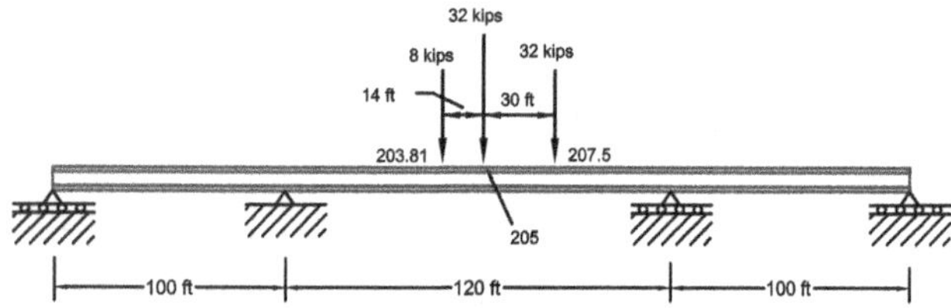

Fig. E20.3-32 Fatigue truck placement for maximum positive moment at location 205.

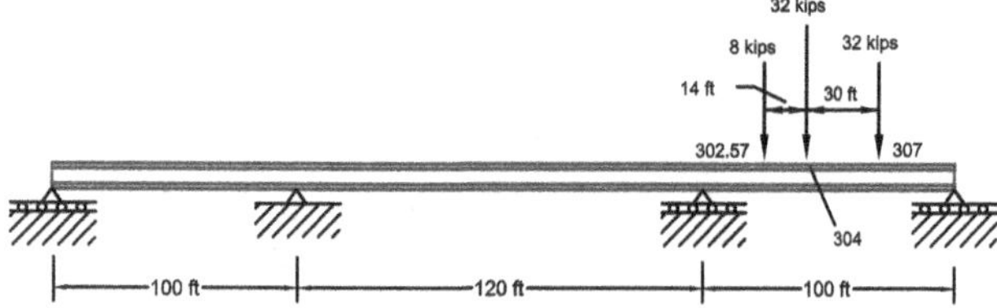

Fig. E20.3-33 Fatigue truck placement for maximum negative moment at location 205.

Positive moment @ 205 (Fig. E20.3-32)

$$\begin{aligned} M_{205} &= [32\,(0.20357 + 0.078645) \\ &\quad +8\,(0.138035)]\,100 \\ &= 1014\ \text{k ft} \end{aligned}$$

Negative moment @ 205 (Fig. E20.3-33)

$$\begin{aligned} M_{205} &= [8\,(-0.029208) \\ &\quad +32\,(-0.03429 - 0.02437)]\,100 \\ &= -211\ \text{k ft} \end{aligned}$$

The maximum flexural fatigue stress in the web at location 200 is calculated in Table E20.3-8. Maximum negative moment is −667 k ft from Table E20.3-4. The factored fatigue moments are given in Table E20.3-5.

$$\begin{aligned} \text{Critical LL + IM (IM} = 0.15) &= -767\ \text{k ft} \\ \text{LL + IM (with DF} = 0.65) &= -498\ \text{k ft} \\ \text{LL + IM (with LF} = 0.75) &= -374\ \text{k ft} \end{aligned}$$

Table E20.3-4 Unfactored Shear and Moment Due to Fatigue Loads at Critical Points[a]

Shear (kips)	V_{100}	V_{104}	V_{110}	V_{200}	V_{205}
Positive	56.0	24.1	1.58	62.3	23.7
Negative	−6.38	−19.4	−62.4	−5.79	−23.7
Moment (k-ft)	M_{100}	M_{104}	M_{110}	M_{200}	M_{205}
Positive	0	1039	158	158	1014
Negative	0	−264	−667	−667	−211

[a]IM = 1.15, Fatigue II $\gamma = 0.75$ and Fatigue I $\gamma = 1.5$ are applied later.

For checking fatigue in web, the moment for fatigue limit I is used [A6.10.4.2],

$$-1241.8 - 85.6 - 127.8 - 242 = 1697\ \text{k ft}$$

M_{D1}, M_{D2}, and M_{D3} in Table E20.3-8 are the moments at location 200 due to unfactored dead loads on the exterior girder, from Table E20.3-1. The maximum calculated stress of 18.0 ksi is less than the allowable flexural fatigue stress of 226.1 MPa calculated earlier; therefore, the section is adequate.

I. ***Design Required Sections*** The bridge is composite in both the positive and negative moment regions and continuous throughout. Homogeneous sections are used and the depth of the web is constant. Only one flange plate transition is used. Longitudinal stiffeners are not used. Minimum thickness of steel is 0.3175 in. [A6.7.3]. The optional minimum depth requirement of [A2.5.2.6.3] is

$$\begin{aligned} \text{Min depth steel} &= 0.027L = 0.027\,(120)\,(12) \\ &= 38.9\ \text{in.} \end{aligned}$$

and

$$\begin{aligned} \text{Min depth total composite section} &= 0.032L\,(120)\,(12) \\ &= 46.1\ \text{in.} \end{aligned}$$

The design girder depth of 62.5 in. exceeds the minimums provided. The plate girder is initially designed for flexural requirements. The lateral bending stress F_l is considered to be small and is neglected.

1. Flexural Section Properties for Negative Moment The negative moment region is designed first to set the overall controlling proportions for the girder section. Following this step, the section is designed for maximum positive moment. An initial section is chosen based on similar designs. The final section for both the negative and positive moment regions is determined by iterations. Although a number of sections are investigated, only the final design of the section is illustrated herein. As stated before, the cross section for the maximum negative moment region consists of a 0.4375-in. × 60-in. web and 1.25-in. × 16-in. top and bottom flanges. Cross-sectional properties are computed for the steel girder alone and for the composite section. In the negative moment regions of continuous spans, the composite section is composed of the steel girder and the longitudinal reinforcement within an effective width of the slab. The concrete is neglected because it is considered cracked under tensile stress. At the interior support, stresses are checked at the top and bottom of the steel girder and in the reinforcing bars using factored moments. The steel girder alone resists moment due to $D1$. The composite section resists the moments due to $D2$, $D3$, and LL + IM.

Table E20.3-5 Factored Moments for Fatigue Limit State for Exterior Girder

	Positive Moment (k-ft)				Negative Moment (k-ft)			
Location	LL + IM w/IM = 0.15	LL + IM w/DF = 0.65	LL + IM w/LF = 0.75	LL + IM w/LF = 1.5	LL + IM w/IM = 0.15	LL + IM w/DF = 0.65	LL + IM w/LF = 0.75	LL + IM w/LF = 1.50
100	0	0	0	0	0	0	0	0
104	1195	777	551	1102	−230	−150	−113	−226
110	182	118	90	180	−767	−499	−374	−748
205	1166	759	569	1138	−247	−161	−121	−242

a. Sequence of Loading Consider the sequence of loading as specified in AASHTO [A6.10.1.1.1]. This article states that at any location on the composite section the elastic stress due to the applied loads shall be the sum of the stresses caused by the loads applied separately to:
  (1) Steel girder
  (2) Short-term composite section (use $n = E_s/E_c$)
  (3) Long-term composite section (use $3n = 3E_s/E_c$ to account for concrete creep)
  (4) For computation of flexural concrete deck stresses in the negative moment region, that is, tensile stresses, use the short-term section (use $n = E_s/E_c$)

  Permanent load that is applied before the slab reaches 75% of f_c' shall be carried by the steel girder alone. Any permanent load and live load applied after the slab reaches 75% of f_c' shall be carried by the composite section.

b. Effective Flange Width Determine the effective flange width specified in AASHTO [A4.6.2.6]. For interior girders the effective flange width is the tributary width.

  Therefore $b_i = 96$ in.

  For exterior girders the effective flange width may be taken as one half the effective width of the adjacent interior girder, plus the overhang:

$$b_e = b_i/2 + 39 = 96/2 + 39 = 135 \text{ in}$$

c. Section Properties Calculate the section properties for the steel girder alone and the composite section. Figure E20.3-34 illustrates the dimensions of the section. From Table E20.3-6 the following section properties are calculated for the steel section alone:

$$y_c = \frac{\sum A_y}{\sum A} = 0$$

$$I_{NA} = I - \left(y_c \times \sum Ay\right) = 45397 \text{ in.}^4$$

$$y_{\text{top of steel}} = \frac{D}{2} + t_f - y_c = \frac{60}{2} + 1.25 - 0 = 31.25 \text{ in.}$$

$$y_{\text{bottom of steel}} = \frac{D}{2} + t_f + y_c = \frac{60}{2} + 1.25 + 0 = 31.25 \text{ in.}$$

$$D_c = \frac{D}{2} = \frac{60}{2} = 30 \text{ in.}$$

$$S_{\text{top of steel}} = \frac{I_{NA}}{y_t} = \frac{45397}{31.25} = 1453 \text{ in.}^3$$

$$S_{\text{bottom of steel}} = \frac{I_{NA}}{y_b} = \frac{45397}{31.25} = 1453 \text{ in.}^3$$

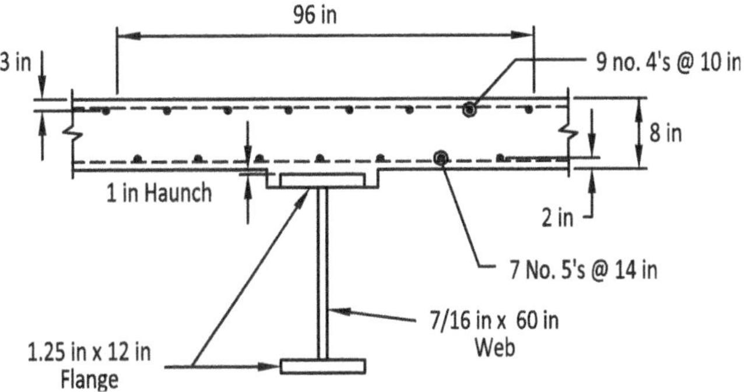

Fig. E20.3-34 Negative moment composite section.

Table E20.3-6 Steel Section Properties (Negative Flexure)[a]

Component	A (in.^{2})	y (in.)	Ay	Ay^2	I_0 (in.^{4})	I (in.^{4})
Top flange 1.25 in. × 16 in.	20.00	30.625	612.5	18 758	2.60	18 761
Web 0.4375 in. × 60 in.	26.25	0	0	0	7875	7875
Bottom flange 1.25 in. × 16 in.	20.00	−30.625	−612.5	18 758	2.60	18 761
Total	**66.25**		**0**			**45 397**

[a] y is the distance from the neutral axis of the web to the neutral axis of the component.

Table E20.3-7 Composite Section Properties (Negative Flexure)

Component	A (in.^{2})	y (in.)	Ay	Ay^2	I_0 (in.^{4})	I (in.^{4})
Top flange 1.25 in. × 16 in.	20.00	30.625	612.5	18 758	2.60	18 761
Web 0.4375 in. × 60 in.	26.25	0	0	0	7875	7875
Bottom flange 1.25 in. × 16 in.	20.00	−30.625	−612.5	18 758	2.60	18 761
Top reinforcement (9 No. 4's)	1.8	37.25	67.05	2498		2498
Bottom reinforcement (7 No. 5's)	2.17	34.25	74.32	2546		2546
Total	**70.22**		**141.4**			**50 441**

From Table E20.3-7 the following section properties were calculated for the composite section in negative flexure:

$$y_c = \frac{\sum Ay}{\sum A} = \frac{141.4}{70.22} = 2.00 \text{ in.}$$

$$I_{NA} = I - \left(y_c \times \sum Ay\right)$$
$$= 50441 - 2.0\,(141.4)$$
$$= 50\,153 \text{ in.}^4$$

$$y_{\text{top reinf.}} = \frac{D}{2} + t_f + \text{haunch} + \text{cover} - y_c$$
$$y_{\text{top reinf.}} = \tfrac{60}{2} + 1.25 + 1.0 + 5.0 - 2.0$$
$$= 35.25 \text{ in.}$$
$$y_{\text{bottom reinf.}} = \tfrac{60}{2} + 1.25 + 1.0 + 1.31 - 2.0$$
$$= 31.56 \text{ in.}$$
$$S_{\text{top reinf.}} = \frac{I_{NA}}{y_{rt}} = \frac{50153}{35.25} = 1423 \text{ in.}^3$$
$$S_{\text{bottom reinf.}} = \frac{I_{NA}}{y_{rb}} = \frac{50\,153}{31.56} = 1589 \text{ in.}^3$$
$$y_{\text{top of steel}} = \frac{D}{2} + t_f - y_c = \frac{60}{2} + 1.25 - 2.0$$
$$= 29.25 \text{ in.}$$
$$y_{\text{bottom of steel}} = \frac{D}{2} + t_f + y_c$$
$$= \frac{60}{2} + 1.25 + 2.0$$
$$= 33.35 \text{ in.}$$
$$S_{\text{top of steel}} = \frac{I_{NA}}{y_t} = \frac{1161861}{29.25} = 1715 \text{ in.}^3$$
$$S_{\text{bottom of steel}} = \frac{I_{NA}}{y_b} = \frac{50\,153}{33.25} = 1508 \text{ in.}^3$$

Cross-Section Proportion Limits [A6.10.2] Check the member proportions [A6.10.2.1]. This article states that the web shall be proportioned to meet the following requirement:

$$\frac{D}{t} \le 150$$
$$\frac{60}{0.4375} = 137 \qquad \text{OK}$$

And [A6.10.2.2] states that the flanges shall meet

$$\frac{b_f}{2t_f} \le 12$$
$$\frac{16}{2\,(1.25)} = 6.4 \le 12 \qquad \text{OK}$$

$$b_f \geq \frac{D}{6}$$
$$16 \geq \tfrac{60}{6} = 10 \quad \text{OK}$$
$$t_f \geq 1.1 t_w$$
$$1.25 \geq 1.1\,(0.4375) = 0.48 \quad \text{OK}$$
$$0.1 \leq \frac{I_{yc}}{I_{yt}} \leq 10$$
$$0.1 \leq 1 \leq 10 \quad \text{OK}$$

where

I_{yt} = moment of inertia of the tension flange of the steel section about the vertical axis in the plane of the web (in.)

I_{yc} = moment of inertia of the compression flange of the steel section about the vertical axis in the plane of the web (in.^{4})

$I_{yt} = \frac{1}{12}\,(1.25)\,(16)^3 = 426.7$ in.^{4}

$I_{yc} = \frac{1}{12}\,(1.25)\,(16)^3 = 426.7$ in.^{4}

$I_{yt} = I_{yc}$

$\dfrac{I_{yc}}{I_{yt}} = 1.0$; therefore the section is adequate

Compactness requirements use the depth on web in compression for the plastic case. For sections in negative flexure, where the plastic neutral axis is in the web:

$$D_{cp} = \frac{D}{2A_w F_{yw}}\left(F_{yt}A_t + F_{yw}A_w + F_{yr}A_r - F_{yc}A_c\right)$$

where

A_w = area of web (in.^{2})

A_t = area of tension flange (in.^{2})

A_r = area of longitudinal reinforcement in the section (in.^{2})

A_c = area of compression flange (in.^{2})

F_{yt} = minimum yield strength of tension flange (ksi)

F_{yw} = minimum yield strength of web (ksi)

F_{yr} = minimum yield strength of longitudinal reinforcement (ksi)

F_{yc} = minimum yield strength of compression flange (ksi)

For all other sections in negative flexure, D_{cp} shall be taken as equal to D. Find the location of the plastic neutral axis using AASHTO (2010) LRFD Bridge Specifications, Appendix D of Section 6. The diagrams in Figure E20.3-35 illustrate the dimensions of the section and the plastic forces. The diagrams are taken from Section 6, Appendix D of AASHTO (2010) LRFD Bridge Specifications.

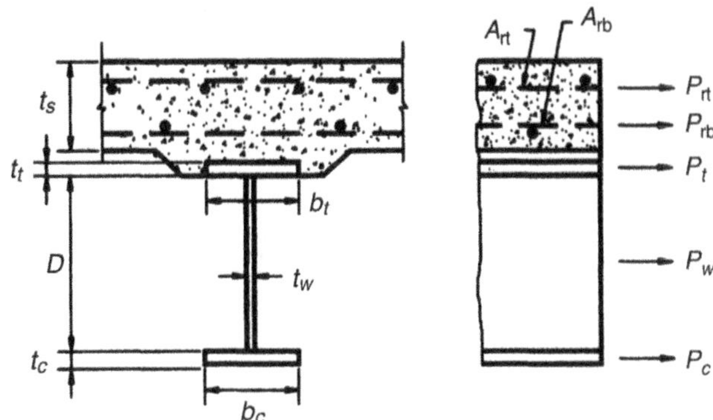

Fig. E20.3-35 Plastic neutral axis for negative moment section.

Plastic Forces

Top reinforcement $P_{rt} = F_{yr}A_{rt} = (60)\,(1.80) = 108.0$ kips

Bottom reinforcement $P_{rb} = F_{yr}A_{rb} = (60)\,(2.17) = 130.2$ kips

Tension flange $= P_t = F_{yt}b_t t_t = (50)\,(16)\,(1.25) = 1000$ kips

Compression flange $= P_c = F_{yc}b_t t_c = (50)\,(16)\,(1.25) = 1000$ kips

Web $= P_w = F_{yw}Dt_w = (50)\,(60)\,(0.4375) = 1312.5$ kips

Plastic Neutal Axis: $C = T$

Check if PNA is in the web:

$$P_c + P_w \geq P_t + P_{rb} + P_{rt}$$
$$1000 + 1312.5 = 2313 \geq 1000 + 130.2 + 108.0 = 1238$$

Therefore the PNA is in the web.

$$D_{cp} = \frac{60}{2\,(1312.5)}\,(1000 + 1312.5 + 108.0 + 130.2 - 1000) = 35.4 \text{ in}$$

Compactness is checked with [A6.10.6.2.3], Composite Sections in Negative Flexure:

$$\frac{2D_{cp}}{t_w} = \frac{2\,(35.4)}{0.4375} = 162$$
$$5.7\sqrt{\frac{E}{F_{yc}}} = 5.7\sqrt{\frac{29{,}000}{50}} = 137$$

Since 162 > 137, the web is noncompact as expected and [A6.10.8] for noncompact sections should be used. The flange local buckling criteria per [A6.10.8.2.2–3&4]:

$$\lambda_f = \frac{b_f}{2t_f} = \frac{16}{2\,(1.25)} = 6.4$$

$$\lambda_{\text{pf}} = 0.38\sqrt{\frac{E}{F_{\text{yc}}}} = 0.38\sqrt{\frac{29000}{50}} = 9.2$$

$$\lambda_f \le \lambda_{\text{pf}}$$

Therefore, the flange is compact and the flexural resistance is

$$F_{\text{nc(FLB)}} = R_b R_h F_{\text{yc}}$$

which is a function of the unbraced lengths.

The two anchor points associated with the inelastic lateral torsional buckling, L_p, and elastic buckling L_r, are [A6.10.8.2.3]

$$L_p \le 1.0 r_t \sqrt{\frac{E}{F_{\text{yc}}}} \quad \text{and} \quad L_r \le \pi r_t \sqrt{\frac{E}{F_{\text{yc}}}}$$

where

r_t = minimum radius of gyration of the compression flange of the steel section (without one-third of the web in compression) taken about the vertical axis:

$$r_t = \frac{b_{\text{fc}}}{\sqrt{12\left(1 + \dfrac{1}{3}\dfrac{D_c\,t_w}{b_{\text{fc}}\,t_{\text{fc}}}\right)}} = \frac{16}{\sqrt{12\left[1 + \dfrac{1}{3}\left(\dfrac{30+2}{16}\right)\left(\dfrac{0.4375}{1.25}\right)\right]}} = 4.16 \text{ in.}$$

Therefore,

$$L_p \le 1.0\,(4.16)\sqrt{\frac{29000}{50}} = 100 \text{ in.}$$

and

$$L_r \le \pi\,(4.16)\sqrt{\frac{29000}{50}} = 315 \text{ in.}$$

Preliminarily, check brace spacing for the compression flange assumed to be L_r or less:

$$L_b = 315 \text{ in.}$$

This is a fairly long spacing, however, the actual will likely be less to align with intermediate stiffeners.

Nominal flexural resistance [A6.10.8.2.3]:

$$F_{\text{nc(LTB)}} = R_b R_h F_{\text{yc}}$$
$$R_h = 1.0$$

for homogeneous sections [A6.10.5.4.1a] and C_b is conservatively assumed to be 1.0 and may be refined if necessary:

$$F_{\text{cr}} = \frac{C_b R_b \pi^2 E}{\left(L_b/r_t\right)^2} = \frac{1.0\,(1.0)\,(\pi^2)\,(29000)}{(315/4.16)^2} = 49.9 \text{ ksi}, \quad \text{say } 50 \text{ ksi}$$

$$F_{\text{nc(LTB)}} = F_{\text{cr}} \le R_b R_h F_{\text{yc}} = 1.0\,(1.0)\,(50) = 50 \text{ ksi}$$

This stress is used for the compression flange in the negative moment region for constructibility and for strength I.

Various other stresses are computed in Tables E20.3-8–E20.3-10. These stresses are used in subsequent computations.

2. *Flexural Section Properties for Positive Flexure* For the positive moment region, a steel section consisting of a 0.625-in. × 12-in. top flange, 0.4375-in. × 60-in. web, and 1-in. × 16-in. bottom flange is used. As stated earlier, the cross section of the web remained constant. The top flange is smaller than the bottom flange due to the additional strength provided by the concrete. Section properties are computed for the steel section alone, the short-term composite section

Table E20.3-8 Maximum Flexural Fatigue Stress in the Web for Negative Flexure at Location 200

Load	M_{D1} (k ft)	M_{D2} (k ft)	M_{D3} (k ft)	$M_{\text{LL+IM}}$ (k ft)	S_b Steel (in.^{3})	S_b Composite (in.^{3})	Stress (ksi)
$D1$	1241.9				1453		10.25
$D2$		85.6				1508	0.68
$D3$			127.8			1508	1.02
LL + IM				748		1508	5.95
Total							17.9
						$\eta = 1.0$	17.9

Table E20.3-9 Stress in Top of Steel Girder (Tension) for Negative Flexure Due to Factored Loading

Load	M_{D1} (k ft)	M_{D2} (k ft)	M_{D3} (k ft)	M_{MLL+IM} (k ft)	S_t Steel (in.^{3})	S_t Composite (in.^{3})	Stress (ksi)
$D1$	1.25(1241.9) = 1552				1453		12.82
$D2$		1.5(85.6) = 128.4				1715	0.90
$D3$			1.25(127.8) = 159.8			1715	1.12
LL + IM				1.75(1822) = 3189		1715	22.3
Total							37.2
						$\eta = 0.95$	35.3

Table E20.3-10 Stress in Bottom of Steel Girder (Compression) for Negative Flexure Due to Factored Loading (Strength I) Interior Girder

Load	M_{D1} (k ft)	M_{D2} (k ft)	M_{D3} (k ft)	M_{LL+IM} (k ft)	S_b Steel (in.^{3})	S_b Composite (in.^{3})	Stress (ksi)
$D1$	1.25(1241.9) = 1552				1453		12.82
$D2$		1.5(85.6) = 128,4				1508	1.02
$D3$			1.25(127,8) = 159.8			1508	1.27
LL + IM				1.75(1822) = 3189		1508	25.37
Total							40.5
						$\eta = 0.95$	38.5

with n equal to 8, and the long-term composite section with $3n$ equal to 24, where n is the modular ratio. The composite section in positive flexure consists of the steel section and a transformed area of an effective width of concrete slab [A6.10.1.1.1.1b]. For normal weight concrete, the modular ratio n for 25 MPa $\leq f_c' \leq$ 32 MPa is taken as 8, where f_c' is the 28-day compressive strength of the concrete [A6.10.5.1.1b]. Stresses are computed at the top and bottom of the steel girder and in the concrete using factored moments. The steel girder alone resists moments due to $D1$. The short-term composite section resists moments due to LL + IM, and the long-term composite section resists moments due to $D2$ and $D3$. The sequence of loading and the effective flange width are identical to that determined for the negative moment region previously, respectively. The moments for the exterior girders control the positive moment region also. Therefore, the effective width used is b_e equal to 96 in.

a. Section Properties Calculate the section properties for the steel girder alone and the short-term and long-term composite sections. Figure E20.3-36 illustrates the dimensions of the section. From Table E20.3-11 the following section properties are calculated for the steel section alone:

$$y_c = \frac{\sum Ay}{\sum A} = \frac{-260.2}{49.75} = -5.24 \text{ in.}$$

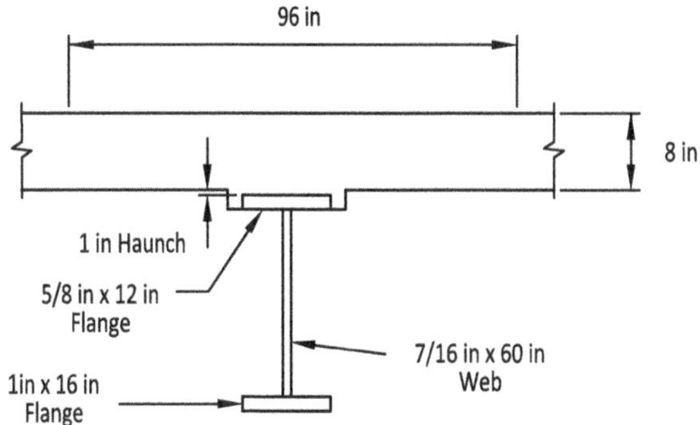

Fig. E20.3-36 Composite section for positive moment.

$$\begin{aligned} I_{NA} &= I - \left(y_c \times \sum Ay\right) \\ &= 29650 - (5.24)\,(260.7) \\ &= 28{,}284 \text{ in.}^4 \end{aligned}$$

$$\begin{aligned} y_{\text{top of steel}} &= \frac{D}{2} + t_f - y_c \\ &= \frac{60}{2} + 0.625 + 5.24 \\ &= 35.87 \text{ in.} \end{aligned}$$

$$\begin{aligned} y_{\text{bottom of steel}} &= \frac{D}{2} + t_f + y_c = \frac{60}{2} + 1 - 5.24 \\ &= 25.88 \text{ in.} \end{aligned}$$

$$S_{\text{top of steel}} = \frac{I_{NA}}{y_t} = \frac{28284}{35.81} = 789 \text{ in.}^3$$

$$S_{\text{bottom of steel}} = \frac{I_{NA}}{y_b} = \frac{28284}{25.82} = 1095 \text{ in.}^3$$

Table E20.3-11 Steel Section Properties (Positive Flexure)

Component	A (in.^{2})	y (in.)	Ay	Ay^2	I_0 (in.^{4})	I (in.^{4})
Top flange 0.625 in. × 12 in.	7.5	30.31	227.3	6,890	0.24	6,890
Web 0.4375 in. × 60 in.	7.5	0	0	0	7,875	7,875
Bottom flange	26.25	−30.5	−488.0	14,884	520 833	14,885
1 in. × 16 in.	16.00				1.33	
Total	**49.75**		**−260.2**			**29,650**

Table E20.3-12 Short-Term Composite Section Properties, n = 8 (Positive Flexure)

Component	A (in.)	y (in.)	Ay	Ay^2	I_0 (in.^{4})	I (in.^{4})
Top flange 0.625 in. × 12 in.	7.50	30.31	227.3	6,890	0.24	6,890
Web 0.4375 in. × 60 in.	26.25	0	0	0	7,875	7,875
Bottom flange 1 in. × 16 in.	16.00	−30.5	−488.0	14,884	1.33	14,885
Concrete $b_e \times t_s/n$	96.00	35.63	3,420	12,1871	512	122,383
Total	**145.8**		**3160**			**152,033**

From Table E20.3-12, the following section properties are calculated for the short-term composite section, where $n = 8$, $b_e = 96$ in., and $t_s = 8$ in.:

$$y_c = \frac{\sum Ay}{\sum A} = \frac{3160}{145.8} = 21.7 \text{ in.}$$

$$I_{NA} = I - \left(y_c \times \sum Ay\right)$$
$$= 152{,}033 - (21.7)(3160)$$
$$= 83{,}461 \text{ in.}^4$$

$$y_{\text{top of steel}} = \frac{D}{2} + t_f - y_c = \frac{60}{2} + 0.625 - 21.7$$
$$= 8.93 \text{ in.}$$

$$y_{\text{bottom of steel}} = \frac{D}{2} + t_f + y_c = \frac{60}{2} + 1 + 21.7$$
$$= 52.7 \text{ in.}$$

$$S_{\text{top of steel}} = \frac{I_{NA}}{y_t} = \frac{83{,}461}{8.93} = 9346 \text{ in.}^3$$

$$D_c = \frac{60}{2} + 5.24 = 35.5 \text{ in}$$

$$S_{\text{bottom of steel}} = \frac{I_{NA}}{y_b} = \frac{83{,}461}{52.7} = 1584 \text{ in.}^3$$

$$y_{\text{top of concrete}} = \frac{D}{2} + t_f + \text{haunch} + t_s - y_c$$

$$y_{\text{top of concrete}} = \tfrac{60}{2} + 0.625 + 1 + 8 - 21.7$$
$$= 17.93 \text{ in.}$$

$$S_{\text{top of concrete}} = \frac{I_{NA}}{y_{tc}} = \frac{83461}{17.93} = 4653 \text{ in.}^3$$

From Table E20.3-13 the following section properties are calculated for the long-term composite section, where $3n = 24$:

$$y_c = \frac{\sum Ay}{\sum A} = \frac{879.3}{81.75} = 10.75 \text{ in.}$$

$$I_{NA} = I - \left(y_c \times \sum Ay\right)$$
$$= 70{,}445 - (10.75)(879.3)$$
$$= 60{,}992 \text{ in.}^4$$

$$y_{\text{top of steel}} = \frac{D}{2} + t_f - y_c$$
$$= \frac{60}{2} + 0.625 - 10.75$$
$$= 19.88 \text{ in.}$$

$$y_{\text{bottom of steel}} = \frac{D}{2} + t_f + y_c$$
$$= \frac{60}{2} + 1.0 + 10.75$$
$$= 41.75 \text{ in.}$$

$$S_{\text{top of steel}} = \frac{I_{NA}}{y_t} = \frac{60{,}992}{19.88} = 3068 \text{ in.}^3$$

$$S_{\text{bottom of steel}} = \frac{I_{NA}}{y_b} = \frac{60{,}992}{41.75} = 1461 \text{ in.}^3$$

Table E20.3-13 Long-Term Composite Section Properties, 3n = 24 (Positive Flexure)

Component	A (in.^{2})	y (in.)	Ay	Ay^2	I_0 (in.^{4})	I (in.^{4})
Top flange 0.625 in. × 12 in.	7.50	31.31	227.3	6,890	0.24	6,890
Web 0.4375 in. × 60 in.	26.25	0	0	0	7,875	7,875
Bottom flange 1 in. × 16 in.	16.00	−30.50	−488.0	14,884	1.33	14,885
Concrete $b_e \times t_s/3n$	32.00	35.63	1,140	40,624	170.7	40,795
Total	**81.75**		**879.3**			**70,445**

b. Member Proportions Check the member proportions [A6.10.2.1]. This article states that the web shall be proportioned to meet the following requirement:

$$\frac{D}{t} \leq 150$$

$$\frac{60}{0.4375} = 137 \qquad \text{OK}$$

And [A6.10.2.2] states that the flanges shall meet

$$\frac{b_f}{2t_f} \leq 12$$

$$\frac{12}{2\,(0.625)} = 9.6 \leq 12 \qquad \text{OK}$$

$$b_f \geq \frac{D}{6}$$

$$12 \geq \tfrac{60}{6} = 10 \qquad \text{OK}$$

$$t_f \geq 1.1\ t_w$$

$$0.625 \geq 1.1\,(.4375) = 0.48 \qquad \text{OK}$$

$$0.1 \leq \frac{I_{\text{yc}}}{I_{\text{yt}}} \leq 10$$

where

I_{yt} = moment of inertia of tension flange of steel section about vertical axis in plane of the web (in.)

I_{yc} = moment of inertia of compression flange of steel section about vertical axis in plane of the web (in.^{4})

$$I_{\text{yc}} = \tfrac{1}{12}\,(0.625)\,(12)^3 = 90\ \text{in.}^4$$

$$I_{\text{yt}} = \tfrac{1}{12}\,(1)\,(16)^3 = 341\ \text{in.}^4$$

$$\frac{I_{\text{yc}}}{I_{\text{yt}}} = 0.26$$

$$0.1 \leq 0.26 \leq 10 \qquad \text{OK}$$

c. Composite Section Stresses for Positive Moment [A6.10.1.1.1] The factored stresses are computed for the fatigue and strength I limit states in Tables E20.3-14–E20.3-16.

From AASHTO [A6.10.5.1.4b] for sections in positive flexure, where the plastic neutral axis is in the web:

$$D_{\text{cp}} = \frac{D}{2}\left(\frac{F_{\text{yt}}A_t - F_{\text{yc}}A_c - 0.85 f_c' A_s - F_{\text{yr}}A_r}{F_{\text{yw}}A_w} + 1\right)$$

For all other sections in positive flexure, D_{cp} shall be taken equal to 0 and the web slenderness requirement is considered satisfied.

Find the location of the plastic neutral axis using AASHTO (2010) LRFD Appendix D of Section 6. Figure E20.3-37 illustrates the dimensions of the

Table E20.3-14 Maximum Flexural Fatigue Stress in the Web for Positive Flexure, Interior Girder

Load	M_{D1} (k ft)	M_{D2} (k ft)	M_{D3} (k ft)	$M_{\text{LL+IM}}$ (k ft)	S_t Steel (in.^{3})	S_t Composite (in.^{3})	Stress (ksi)
$D1$	739.2				789		11.24
$D2$		50.96				3068	0.20
$D3$			76.1			3068	0.30
LL + IM				1102		9346	1.41
Total							13.2
						$\eta = 1.0$	13.2

Table E20.3-15 Stress in Top of Exterior Steel Girder (Compression) for Positive Flexure Due to Factored Loading, Strength I

Load	M_{D1} (k ft)	M_{D2} (k ft)	M_{D3} (k ft)	M_{LL+IM} (k ft)	S_t Steel (in.^{3})	S_t Composite (in.^{3})	Stress (ksi)
$D1$	924				789		14.05
$D2$		76.44				3068	0.30
$D3$			95.1			3068	0.37
LL + IM				3021		9346	3.88
Total							18.6
						$\eta = 0.95$	17.7

Table E20.3-16 Stress in Bottom of Exterior Steel Girder (Tension) for Positive Flexure Due to Factored Loading, Strength I

Load	M_{D1} (k ft)	M_{D2} (k ft)	M_{D3} (k ft)	M_{LL+IM} (k ft)	S_b Steel (in.^{3})	S_b Composite (in.^{3})	Stress (ksi)
$D1$	924				1095		10.13
$D2$		76.44294				1461	0.63
$D3$			95.1			1461	0.78
LL + IM				3021		1584	22.89
Total							34.4
						$\eta = 0.95$	32.7

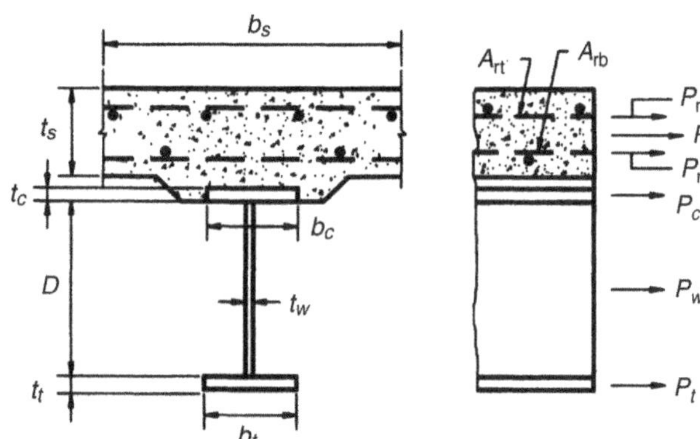

Fig. E20.3-37 Plastic neutral axis for positive moment section.

section and the plastic forces. The diagrams are taken from Section 6 Appendix D of Section 6 of AASHTO (2010).

Plastic Forces

Top reinforcement $P_{rt} = F_{yr}A_{rt} = (60)(9)(0.20) = 108$ kips

Concrete slab $P_s = 0.85 f_c' a b_e = 0.85(4)(96)a = 326.4a$ (kips)

Bottom reinforcement $P_{rb} = F_{yr}A_{rb} = (60)(7)(0.31) = 130.2$ kips

Tension flange $= P_t = F_{yt}b_t t_t = (50)(16)(1) = 800$ kips

Compression flange $= P_c = F_{yc}b_c t_c = (50)(12)(0.625) = 375.0$ kips

Web $= P_w = F_{yw}Dt_w = (50)(60)(0.4375) = 1312.5$ kips

Plastic Neutral Axis (PNA): $C = T$

Assume plastic neutral axis is in the slab below bottom reinforcement:

$$108.0 + 326.4a + 130.2 = 800 + 1312.5 + 375$$

Therefore $a = 6.89$ in.:

$$c = \frac{a}{\beta_1} = \frac{6.89}{0.85} = 8.11 \text{ in.} > 6 \text{ in.} \quad \text{OK}$$

$$P_s = (0.85)(4)(96)(6.89) = 2249 \text{kips}$$

The plastic neutral axis is in the slab below the bottom reinforcement; therefore D_{cp} is equal to zero. The web slenderness requirement is satisfied.

Calculate M_p:

$$\bar{y} = c = 8.11 \text{ in.}$$

$$M_p = \left(\frac{\bar{y}P_s}{t_s}\right)\left(\frac{\bar{y}}{2}\right) + (P_{rt}d_{rt} + P_c d_c + P_w d_w + P_t d_t + P_{rb}d_{rb})$$

where

d_{rt} = distance from PNA to centroid of top reinforcement

$d_{rt} = 8.11 - 3 = 5.11$ in.

d_{rb} = distance from PNA to centroid of bottom reinforcement

$d_{rb} = 8.11 - 6 = 2.11$ in.

d_w = distance from PNA to centroid of web

$$d_w = 60/2 + 0.625 + 1.0 + 8 - 8.11 = 31.5\text{ in.}$$

d_t = distance from PNA to centroid of tension flange

$$d_t = 1.0/2 + 60 + 0.625 + 1.0 + 8.0 - 8.11 = 62.0\text{ in.}$$

d_c = distance from PNA to centroid of compression flange

$$d_c = 0.625/2 + 1.0 + 8.0 - 8.11 = 1.20\text{ in.}$$

$$M_p = \left[\frac{(8.11)(2249)}{8}\right]\left(\frac{8.11}{2}\right) = 9245\text{ k ft}$$

$$+ \left[\begin{array}{l}(108)(5.11) + (375)(1.2) \\ + (1213)(31.5) + (800)(62.0) \\ + (130.2)(2.11)\end{array}\right]$$

M_p = 89090 k in. = 7420 k ft > 3910 k ft m from Table E20.3-2 (at kips location 104)

Refer to Section 6 Appendix D in AASHTO (2005), the yield moment of a composite section is calculated below.

First calculate the additional stress required to cause yielding in the tension flange:

$$f_{AD} = F_y - (f_{D1} + f_{D2} + f_{D3})$$

where

F_y = minimum yield strength of the tension flange

f_{D1} = stress caused by the permanent load before the concrete attains 75% of its 28-day strength applied to the steel section alone, calculated in Table E20.3-16.

f_{D2}, f_{D3} = stresses obtained from applying the remaining permanent loads to the long-term composite section, calculated in Table E20.3-16.

$$f_{AD} = 50 - (10.13 + 0.63 + 0.78) = 38.46\text{ ksi}$$

which corresponds to an additional moment:

$$M_{AD} = f_{AD} \times S_{ST}$$

where

S_{ST} = section modulus for the short-term composite section, where $n = 8$

$$M_{AD} = (38.46)(1584) = 60{,}920\text{ k in.} = 5077\text{ k ft}$$

$$M_y = M_{D1} + M_{D2} + M_{D3} + M_{AD}$$

where M_{D1}, M_{D2}, and M_{D3} are factored moments from Table E20.3-16 at location 104:

$$M_y = 924 + 76.44 + 95.1 + 5077 = 6172\text{ k ft}$$

$$M_n = 1.3R_hM_y = 1.3\,(1.0)\,(6172) = 8024\text{ k ft}$$

$$M_n \geq M_p = 7420\text{ k ft}$$

Therefore

$$M_n = 7420\text{ k ft}$$

$$M_r = \phi_f M_n = 1.0\,(7420) = 7420\text{ k ft}$$

$$M_r > M_u = 3910\text{ k ft}$$

at location 104 from Table E20.3-1, therefore, the section provides adequate flexural strength.

d. Positive Flexure Ductility The next step is to check the ductility requirement for compact composite sections in positive flexure [A6.10.5.2.2b]. The purpose of this requirement is to make sure that the tension flange of the steel section will reach strain hardening before the concrete in the slab crushes. This article only applies if the moment due to the factored loads results in a flange stress that exceeds the yield strength of the flange. If the stress due to the moments does not exceed the yield strength, then the section is considered adequate. The reason being that there will not be enough strain in the steel at or below the yield strength for crushing of the concrete to occur in the slab. From Table E20.3-16 it is shown that the flange stress in the tension flange is 32.7 ksi, which is less than the yield strength of the flange, 50 ksi, therefore the section is adequate. The previous check illustrated was unnecessary.

Compute the noncomposite strength considering compression flange slenderness and lateral torsion buckling. The bracing is assumed to be placed at one-quarter points and the middle span is checked. This span has the largest stresses and unbraced lengths for positive flexure. Again the trial section has 0.625-in. x 12-in. top flange, 0.4375-in. x 60-in. web, and 1-in. x 16-in. bottom flange. For positive moment during construction, check the nominal yield stress:

$$f_{bu} + f_l \leq \phi_f R_h F_{yc} = 50\text{ ksi}$$

$$f_{bu} \leq \phi_f F_{nc}$$

The flange local buckling stress is next checked [A6.10.8.2.2]:

$$\lambda_f = \frac{b_{fc}}{2t_{fc}} = \frac{12}{2\,(0.625)} = 9.6$$

$$\lambda_{fp} = 0.38\sqrt{\frac{E}{F_{cy}}} = 9.1$$

$$\lambda_f \geq \lambda_{fp}$$

Therefore, the compression flange is slightly noncompact:

$$\begin{aligned} F_{yr} &= \min\left(0.7F_{yc}, F_{yw}\right) \geq 0.5F_{yc} \\ &= \min\,(35, 50) \geq 25\,\text{ksi} \\ &= 35\ \text{ksi} \end{aligned}$$

Slenderness must be checked:

$$\lambda_{fp} = 0.56\sqrt{\frac{E}{F_{cy}}} = 13.5$$

And the noncomposite flange local buckling resistance is

$$\begin{aligned} F_{nc(FLB)} &= \left[1 - \left(1 - \frac{F_{yr}}{R_h F_{yc}}\right)\left(\frac{\lambda_f - \lambda_{pf}}{\lambda_{rf} - \lambda_{pf}}\right)\right] \\ &\quad \times R_b R_h F_{yc} \\ &= \left[1 - \left(1 - \frac{35}{50}\right)\left(\frac{9.6 - 9.1}{13.5 - 9.1}\right)\right] \\ &\quad \times (1.0)\,(1.0)\,(50) \\ &= (0.97)\,(50) = 48.2\ \text{ksi} \end{aligned}$$

The lateral torsional buckling stress is next checked, beginning with the computation of the radius of gyration of the compression flange.

$$\begin{aligned} r_t &= \frac{b_{fc}}{\sqrt{12\left[1 + \frac{1}{3}\left(\frac{D_c}{b_{fc}}\right)\left(\frac{t_w}{t_{tf}}\right)\right]}} \\ &= \frac{16}{\sqrt{12\left[1 + \frac{1}{3}\left(\frac{32}{16}\right)\left(\frac{0.4375}{1.25}\right)\right]}} = 4.16\ \text{in.} \end{aligned}$$

The anchor points for LTB are

$$L_p = 1.0r_t\sqrt{\frac{E}{F_{yc}}}$$

$$L_p = 1.0\,(4.16)\sqrt{\frac{29\,000}{50}} = 100\ \text{in.}$$

$$L_r = \pi L_p = 315\ \text{in.}$$

The unbraced length is assumed to be $L_b = 24$ ft $= 288$ in, which is greater than the inelastic limit, therefore, the elastic LTB is applicable.

$$\begin{aligned} F_{cr} &= \frac{C_b R_b \pi^2 E}{\left(L_b/r_t\right)^2} \\ &= \frac{1.0\,(1.0)\,\left(\pi^2\right)(29000)}{(288/4.16)^2} = 59.7\ \text{ksi} \end{aligned}$$

The flange local and lateral torsional buckling stresses are compared and the minimum controls, therefore,

$$\begin{aligned} F_{nc} &= \min\left[F_{nc(FLB)}, F_{nc(LTB)}\right] \\ &= \min\,[48.2, 59.7] = 48.2\ \text{ksi} \end{aligned}$$

Later this is used to compare to the stresses that occur during construction.

3. *Transition Points* Transition points from the sections in positive moment regions to the sections in negative moment regions shall be located at the dead-load inflection points. These locations are chosen because composite action is not considered to be developed. The permanent-load inflection points were determined when the effective lengths were calculated previously. Going from left to the right of the bridge, inflection points exist at the following locations:

$x = 75.6$ ft   125.9 ft   192.1 ft   244.4 ft

The web size is a constant throughout the bridge. Only the flange size changes at the inflection points.

4. *Constructibility* [A6.10.3]

   a. General [A2.5.3] [A6.10.3.2.3] The resistance of the girders during construction is checked. Note that the unbraced length is important for lateral torsion buckling under the load of wet concrete. From previous computations, the $F_{nc(LTB)} = 50$ ksi for the negative moment region and 48.3 ksi in the positive moment region.

   Assuming small construction live loads, the unfactored and factored stress under 1.25(DC) are 12.82 ksi and 14.05 ksi for negative and positive moment regions, respectively. See Table E20.3-9. Significant capacity exists for additional live loads. The bracing at one-quarter points is adequate for construction. Later wind loads are checked.

   Also during construction buckling is not permitted in the web. Therefore, bend buckling must be checked per [A6.10.3.2.1].

$$f_{bu} \leq \phi_f F_{crw}$$

where

$$F_{crw} = \frac{0.9Ek}{\left(D/t_w\right)^2}$$

and

$$k = \frac{9}{\left(D_c/D\right)^2} = \frac{9}{(32/60)^2} = 31.6$$

$$F_{crw} = \frac{0.9\,(29000)\,(31.6)}{(60/0.4375)^2} = 43.9 \text{ ksi}$$

The factored stress during placement of the concrete deck, assuming live loads are small, is

$$f_{bu} = 14.05 \leq 43.9 \text{ ksi} \quad \text{OK}$$

Similarly for negative moment the factored stress is 91.7, which is less than 43.9 ksi as well. See Table E20.3-9.

b. Flexure [A6.10.8.2] Lateral support for compression flange is not available when fresh concrete is being placed [A6.10.3.2.1 and A6.10.8.2].

c. Shear [A6.10.3.3] The shear resistance is computed as the buckling resistance under construction loads in addition to nominal yield. For the latter,

$$12.82 \leq 50 \text{ ksi} \quad \text{OK}$$
$$14.05 \leq 50 \text{ ksi} \quad \text{OK}$$

The longitudinal deck stress could also be checked, however, minimum reinforcement is provided in this area and this check is not necessary. Finally, to ensure that the web does not buckle,

$$\phi V_n \leq \phi V_{cr}$$

Critical locations are checked below. The buckling to plastic shear ratios, C, are computed in the section on strength I shear located later in this example. See Table E20.3-1.

Location	Demand, V_u, kN	Buckling to Yield Ratio, C	Plastic Shear Resistance, V_p, kN	$\phi V_n = \phi CV_p$, kN	Check
100	1.25(39.8) = 49.8	0.295	761	224	OK
110	1.25(65.2) = 81.5	0.485	761	369	OK
200	1.25(63) = 78.8	0.442	761	336	OK

The shear resistance is sufficient to support construction dead load with a significant additional capacity.

d. Deck Placement [A6.10.3.4 and A6.10.3.5] Because the bridge length is relatively short, concrete placement can be achieved in one day. This article is not applicable.

5. *Service Limit State* [A6.5.2] [A6.10.4]

a. Elastic Deformations [A6.10.4.1]

(1) Optional Live-Load Deflection [A2.5.2.6.2] The deflection control is optional and depends upon the owner's specification regarding the limits. Here, the deflection limit of span/800 is computed as

$$\begin{aligned}\Delta_{\text{live load limit}} &= L/800 \\ &= 120\,(12)\,/800 \\ &= 1.8 \text{ in}\end{aligned}$$

The distribution factor for deflection is based upon uniform distribution,

$$\begin{aligned}mg_{\text{deflection}} &= 0.85\left(\frac{3 \text{ lanes}}{6 \text{ girders}}\right) \\ &= 0.43\end{aligned}$$

The live load used is the maximum of the deflection due to the

- ☐ Design truck
- ☐ Deflection resulting from 25% of design truck together with the design lane load

Use the stage three moment of inertia of 83, 461 in.[4] for the entire girder length. Modeling the bridge with BT-Beam as a prismatic beam, the design truck creates a deflection at the 204 of 0.425×10^6 ft (for $EI = 1$) and 0.45 in. with the gross M and EI properties. Using a distribution factor of 0.50 lanes per girder gives 0.23 in. Using the deflection of 25% of the design truck and design lane gives $0.25(0.23) + 0.28 = 0.33$ in. Both are far below the limit.

(2) Optional Criteria for Span-to-Depth Ratio [A2.5.2.6.3] Previously in this example, the optional span-to-depth ratio of $0.033L$ for the noncomposite steel and $0.040L$ for the total section is used to size the section. This check is shown again for completeness with respect to the service I deflection check and is computed as

$$\begin{aligned}0.033L &= 0.033\,(120)\,(12) \\ &= 47.5 \text{ in.} \leq 61.625 \text{ in.} \quad \text{OK} \\ 0.040L &= 0.040\,(120)\,(12) \\ &= 57.6 \text{ in.} \leq 70.625 \text{ in.} \quad \text{OK}\end{aligned}$$

b. Permanent Deformations [A6.10.4.2]

(1) Flexure [A6.10.4.2.2] [Appendix B—optional] [A6.10.1.9] [A6.10.1.10.1] The service limit stresses are outlined in [A6.10.4] and computed in Table E20.3-8. For the trial negative moment section [A6.10.4.2.2-3]:

$$f_f + \frac{f_l}{2} \leq 0.8R_h F_{yf} = 40 \text{ ksi}$$

$$f_f = 1.0\,(10.25) + 1.0\,(0.68) + 1.0\,(1.02) + 1.3\,(14.5)$$

$$f_f = 30.8 \leq 40 \text{ ksi} \quad \text{OK}$$

Similarly the positive moment section is checked using [A6.10.4.2.2] [Table E20.3-14]:

$$f_f + \frac{f_l}{2} \leq 0.95 R_h F_{\text{yf}} = 47.5 \text{ ksi}$$

$$f_f = 1.0\,(11.24) + 1.0\,(0.20) + 1.0\,(0.30) + 1.3\,(1.41)$$

$$f_f = 13.6 \leq 48.5 \text{ ksi} \quad \text{OK}$$

6. *Fatigue and Fracture Limit State* [A6.5.3] [A6.10.5]
   a. Fatigue [A6.10.5.1] [A6.6.1] Allowable fatigue stress range depends on load cycles and connection details. Fracture depends on material grade and temperature.

      Stress Cycles Assuming a rural interstate highway with 20,000 vehicles per lane per day,

$$\text{Fraction of trucks in traffic} = 0.20 \quad \text{(Table 4.4)} \quad \text{[Table C3.6.1.4.2-1]}$$

$$\begin{aligned} \text{ADTT} &= 0.20 \times \text{ADT} \\ &= 0.20\,(20000)\,(2 \text{ lanes}) \\ &= 8000 \text{ trucks/day} \\ p &= 0.85 \text{ (Table 4.3) [Table A3.6.1.4.2-1]} \\ \text{ADTT}_{\text{SL}} &= p \times \text{ADTT} = 0.85\,(8000) \\ &= 6800 \text{ trucks/day} \end{aligned}$$

The critical fatigue detail for a flange in tension is the location where the transverse stiffener is fillet welded to the flange. Using detail 4.1 in Table 18.3 where the category is C′. The 6800 far exceeds the 745 trucks per day for finite life. Therefore, the infinite-life resistance is used.

$$(\Delta F)_{\text{TH}} = 12 \text{ ksi}$$

The Maximum Stress Range [C6.6.1.2.5] The maximum stress range is computed using fatigue I limit state:

$$U = 1.5\,(\text{LL} + \text{IM})$$

Dynamic load allowance for fatigue is IM = 15%. $M_{\text{LL+IM}}$ is maximum in the exterior girder, no multiple presence (live-load range only):

From Tables E20.3-8 and E20.3-14, the fatigue live-load stresses are

$$f_{\text{LL+IM}} = 5.95 \quad \text{and} \quad 1.41 \text{ ksi}$$

for positive and negative moments, respectively. The opposite sense action is not included herein as it is not readily available; however, it will be significantly smaller than the primary actions/stresses and these stresses are much less than the 12-ksi limit. Fatigue limit state is satisfied for welds near the stiffeners.

   b. Fracture [A6.10.5.2] [A6.6.2] The steel specified meets fracture requirements for this non-fracture-critical system.
   c. Special Fatigue Requirements for Webs [A6.10.5.3] The shear force due the fatigue truck is determined with the use of Figure E20.1-6 (exterior girder governs and no multiple presence) [A3.6.1.1.2]. Here dead load is included with the live load. Finally, the web should not buckle under routine (fatigue I) loads. From Tables E20.3-1 and E20.3-4, the load effects are

$$\begin{aligned} V_u &= 1.0\,(65.2) + 1.0\,(17.4) + 1.0\,(6.58) \\ &\quad + 1.5\,(0.75)\left(\tfrac{1}{1.2}\right)(62.4) \\ &= 369 \text{ kips} \end{aligned}$$

At the interior support (110), the resistance is

$$\phi V_{\text{cr}} = 714 \text{ kN}$$

Therefore, the resistance is sufficient to avoid web buckling under dead load with fatigue live load.

7. *Strength Limit State* [A6.5.4] [A6.10.6]
   a. Composite Sections in Positive Flexure [A6.10.6.2.2] [A6.10.7] For positive moment regions, the non-compact negative moment region, the resistance is limited to

$$\phi M_n = \phi_f 1.3 R_h M_y$$

where ϕM_n is computed as 7420 k ft previously. The factored strength I flexural moment at 104 is

$$M_u = 3910 \leq 7420 \text{ k ft} \quad \text{OK}$$

   b. Composite Sections in Negative Flexure [A6.10.6.2.3] [A6.10.8] [Appendix A—optional] [Appendix B—optional] [D6.4—optional] The negative moment must satisfy flange local buckling and lateral torsion buckling. The resistance computed previously is 48.2 ksi, which was controlled by FLB. The load effects from Table E20.3-10 is 38.5 ksi. Therefore, the section is fine. In the resistance computation the moment gradient term was conservatively taken as $C_b = 1.0$, so a refinement should indicate that the section is sufficient.

8. *Shear Design* In general, the factored shear resistance of a girder, V_r, is taken as follows:

$$V_r = \phi_v V_n$$

where

V_n = nominal shear resistance for stiffened web

ϕ_v = resistance factor for shear = 1.0

   a. Stiffened Web Interior web panels of homogeneous girders without longitudinal stiffeners and with a transverse stiffener spacing not exceeding $3D$ are considered stiffened:

$$3D = 3\,(60) = 180 \text{ in.}$$

The nominal resistance of stiffened webs is given in AASHTO [A6.10.9.1].

(1) Handling Requirements For web panels without longitudinal stiffeners, transverse stiffeners are required if [A6.10.2.1]

$$\frac{D}{t_w} \le 150$$
$$\frac{60}{0.4375} = 137$$

Therefore, transverse stiffeners are not required for handling; however, this example demonstrates the design of transverse stiffeners as if they were required.

Maximum spacing of the transverse stiffeners is

$$d_0 \le D\left[\frac{260}{(D/t_w)}\right]^2$$
$$= 60\left(\frac{260}{137}\right)^2 = 215 \text{ in.}$$

(2) Homogeneous Sections The requirements for homogeneous sections are in AASHTO [A6.10.7.3.3]. The purpose of this section is to determine the maximum spacing of the stiffeners while maintaining adequate shear strength within the panel. Three separate sections must be examined:

1. End panels
2. Interior panels for the composite section in the positive moment region
3. Interior panels for the noncomposite section in the negative moment region

From analysis, the interior girders receive the largest shear force values (Table E20.3-1).

b. End Panels Tension field action in end panels is not permitted. The nominal shear resistance of an end panel is confined to either the shear yield or shear buckling force.

$$V_n = CV_p$$

for which

$$V_p = 0.58F_{yw}Dt_w$$
$$= 0.58\,(50)\,(60)\,(0.4375)$$
$$= 761 \text{ kips}$$

where

C = ratio of shear buckling stress to shear yield strength
k = shear buckling coefficient

The ratio, C, is determined [A6.10.9] as follows: If

$$\frac{D}{t_w} < 1.12\sqrt{\frac{Ek}{F_{yw}}}$$

then

$$C = 1.0$$

If

$$1.12\sqrt{\frac{Ek}{F_{yw}}} \le \frac{D}{t_w} \le 1.40\sqrt{\frac{Ek}{F_{yw}}}$$

then

$$C = \frac{1.12}{\frac{D}{t_w}}\sqrt{\frac{Ek}{F_{yw}}}$$

If

$$\frac{D}{t_w} > 1.40\sqrt{\frac{Ek}{F_{yw}}}$$

then

$$C = \frac{1.57}{(D/t_w)^2}\left(\frac{Ek}{F_{yw}}\right) \le 0.8$$

for which

$$k = 5 + \frac{5}{(d_0/D)^2}$$

For the end panels V_u equals 220 kips at location 100, taken from Table E20.3-3. Assume

$$\frac{D}{t_w} > 1.40\sqrt{\frac{Ek}{F_{yw}}}$$

therefore,

$$C = \frac{1.57}{(D/t_w)^2}\left(\frac{Ek}{F_{yw}}\right)$$
$$V_n = \frac{1.57}{(D/t_w)^2}\left(\frac{Ek}{F_{yw}}\right)V_p$$

Solving for k in the equation above

$$k_{\min} = \frac{V_nF_{yw}}{1.57V_pE}\left(\frac{D}{t_w}\right)^2$$

use

$$V_n = \frac{V_u}{\phi} = \frac{200}{1.0} = 220 \text{ kips}$$
$$k_{\min} = \frac{(220)\,(50)}{1.57\,(761)\,(29000)}\left(\frac{60}{0.4375}\right)^2$$
$$= 5.97$$
$$k = 5 + \frac{5}{(d_0/D)^2}$$

therefore, maximum $d_0 = 228$ in.

Checking assumption on D/t_w for $d_0 = 60$ in:

$$k = 5 + \frac{5}{(60/60)^2} = 10$$

$$1.40\sqrt{\frac{Ek}{F_{yw}}} = 1.40\sqrt{\frac{(29000)\ 10}{50}} = 107$$

$$\frac{D}{t_w} = \frac{60}{0.4375} = 137 > 107 \text{ assumption OK}$$

Place stiffeners 60 in apart ($d_0 = D$):

$$C = 0.48$$

$$\phi V_n = \phi C V_p = 369 \text{ kips} \geq 220 \text{ kips} \quad \text{OK}$$

c. Interior Panels of Compact Sections For this design, this section applies to the positive moment region only. The region considered shall be the effective lengths for the uniform unit load. These effective lengths were determined previously. First, consider the 100-ft span. The nominal shear resistance from [A6.10.9] is

$$V_n = V_p \left[C + \frac{0.87\,(1 - C)}{\sqrt{1 + (d_0/D)^2}} \right]$$

$$V_p = 0.58 F_{yw} D t_w = 761 \text{ kips}$$

where

V_n = nominal shear resistance (kips)
V_p = plastic shear force (kN) = 761 kips
ϕ_f = resistance factor for flexure = 1.0
D = web depth (in.)
d_0 = stiffener spacing (in.)
C = ratio of shear buckling stress to shear yield strength

Calculate the minimum spacing for the panels for the compact section using the maximum shear values. For the 100-ft span the compact section exists for the first 75.64 ft of the bridge. For design, the length of compact section is taken as 75 ft. Because the first interior stiffener is 60 in. from the end of the bridge and the second is 60 in. from the first and stiffeners are spaced equally across the remaining compact length.

The second stiffener is 120 in. from the end of the bridge, which corresponds to location 101. From Table E20.3-3,

$$V_u = 179 \text{ kips at 10 ft from the ends of the bridge}$$

Assume

$$\frac{D}{t_w} > 1.40\sqrt{\frac{Ek}{F_{yw}}}$$

therefore

$$C = \frac{1.57}{(D/t_w)^2}\left(\frac{Ek}{F_{yw}}\right)$$

Try six equal spacings of 130 in. over a length of 6(130) = 780 in = 65 ft:

$$k = 5 + \frac{5}{(130/60)^2} = 6.07$$

$$1.40\sqrt{\frac{(29000)\ (6.07)}{50}} = 83 < \frac{D}{t_w}$$
$$= 150 \quad \text{assumption OK}$$

so that

$$C = \frac{1.57}{(60/0.4375)^2}\left[\frac{(29000)\ (6.07)}{50}\right] = 0.295$$

$$V_n = 1.0\,(761)\left[0.295 + \frac{0.87\,(1 - 0.295)}{\sqrt{1 + (130/60)^2}}\right]$$
$$= 420 \text{ kips}$$

$$\phi_v V_n = 1.0\,(420) = 420 \text{ kips} > V_u$$
$$= 179 \text{ kips} \quad \text{OK}$$

Space transverse intermediate stiffeners at 130 in. ($d_0 = 2.16D$) from $x = 120$ in. = 10 ft to $x = 900$ in. = 75 ft along the 100-ft span.

Consider the 120-ft span for which a compact section exists from points x = 125.9 ft to x = 192.1 ft along the bridge. Therefore space transverse intermediate stiffeners equally along the 70-ft distance, therefore

$$V_n = V_p \left[C + \frac{0.87\,(1 - C)}{\sqrt{1 + (d_0/D)^2}} \right]$$

At x = 125.9 ft, using values from Table E20.3-3,

$$V_u = -\frac{170 - 139}{136 - 124}(125.9 - 124) + 179$$
$$= 174 \text{ kips}$$

Assume

$$\frac{D}{t_w} > 1.40\sqrt{\frac{Ek}{F_{yw}}}$$

therefore

$$C = \frac{1.57}{(D/t_w)^2}\left(\frac{Ek}{F_{yw}}\right)$$

Try eight stiffener spacings of 130 in. over a length of 8(105) = 840 in. = 70 ft ($d_0 = 1.70D$):

$$k = 5 + \frac{5}{(105/60)^2} = 6.6$$

$$1.40\sqrt{\frac{(29000)\ (6.6)}{50}} = 87 < \frac{D}{t_w}$$
$$= 137 \quad \text{assumption OK}$$

$$C = \frac{1.57}{(60/0.4375)^2}\left(\frac{(29000)(6.6)}{50}\right) = 0.320$$

$$V_n = (761)\left[0.320 + \frac{0.87(1-0.320)}{\sqrt{1+(105/60)^2}}\right]$$
$$= 467 \text{ kips}$$

$$\phi_v V_n = 1.0(467) = 467 \text{ kips} > V_u$$
$$= 174 \text{ kips} \quad \text{OK}$$

Space transverse intermediate stiffeners at 105 in. from $x = 125$ to 195 ft.

d. Interior Panels of Noncompact Sections This section applies to the negative moment region only. The region considered shall be the effective lengths for the uniform unit load. These effective lengths were determined previously. The noncompact section exists between $x = 75$ ft to $x = 125$ ft. Therefore, there is 25 ft on the 100-ft span and 25 ft on the 120-ft span from the bearing stiffener over the support to the assumed inflection point. The nominal shear resistance is taken as

$$V_n = V_p\left[C + \frac{0.87(1-C)}{\sqrt{1+(d_0/D)^2}}\right]$$

$$V_p = 0.58F_{yw}Dt_w = 0.58(50)(60)(0.4375)$$
$$= 761 \text{ kips}$$

where C is the ratio of shear buckling stress to the shear yield strength.

The stiffener spacing along the 100-ft span is determined first. A spacing of 60 in ($d_0 = 1.0D$) is used for all stiffeners. The spacing is checked for adequacy below. The maximum shear in the panel from $x = 75$ ft to $x = 100$ ft is $V_u = 286$ kips from Table E20.3-3.

Assume

$$\frac{D}{t_w} > 1.40\sqrt{\frac{Dk}{F_{yw}}}$$

Therefore,

$$C = \frac{1.57}{(D/t_w)^2}\left(\frac{Ek}{F_{yw}}\right)$$

$$k = 5 + \frac{5}{(60/60)^2} = 10$$

$$1.40\sqrt{\frac{(29000)(0.4375)}{50}} = 107 < \frac{D}{t_w}$$
$$= 137 \quad \text{assumption OK}$$

$$C = \frac{1.57}{(60/0.4375)^2}\left[\frac{(29000)(0.4375)}{50}\right]$$
$$= 0.485$$

$$V_n = (761)\left[0.485 + \frac{0.87(1-0.485)}{\sqrt{1+(60/60)^2}}\right]$$
$$= 610 \text{ kips}$$

$$\phi_v V_n = 1.0(610) = 610 \text{ kips} > V_u$$
$$= 286 \text{ kips} \quad \text{OK}$$

Space transverse intermediate stiffeners at 60 in. from $x = 75$ ft to $x = 100$ ft along the 100-ft span.

Determine stiffener spacing along the 120-ft span. Space stiffeners equally along the 25-ft length from $x = 100$–125 ft. The maximum shear in this panel is at $x = 100$ ft, $V_u = 284$ kips from Table E20.3-3.

Assume

$$\frac{D}{t_w} > 1.40\sqrt{\frac{Ek}{F_{yw}}}$$

Therefore,

$$C = \frac{1.57}{(D/t_w)^2}\left(\frac{Ek}{F_{yw}}\right)$$

Try four stiffener spacings of 75 in. over a length of 4(75 in.) = 300 in. ($d_0 = 1.25D$):

$$k = 5 + \frac{5}{(75/60)^2} = 8.2$$

$$1.40\sqrt{\frac{(29000)(8.2)}{50}} = 96.5 < \frac{D}{t_w} = 137 \quad \text{OK}$$

$$C = \frac{1.57}{(60/0.4375)^2}\left[\frac{(29000)(8.2)}{50}\right] = 0.442$$

$$V_n = (761)\left[0.422 + \frac{0.87(1-0.422)}{\sqrt{1+(75/60)^2}}\right]$$
$$= 576 \text{ kips}$$

$$\phi_v V_n = 1.0(576) = 576 \text{ kips} > V_u$$
$$= 284 \text{ kips} \quad \text{OK}$$

Space transverse intermediate stiffeners at 75 in. from $x = 101$–125 ft along the 120-ft span.

9. *Transverse Intermediate Stiffener Design* The LRFD specifications for stiffener design are located in AASHTO [A6.10.11]. Transverse intermediate stiffeners are composed of plates welded to either one or both sides of the web depending on the additional shear resistance the web needs. Transverse intermediate stiffeners used as connecting elements for diaphragms must extend the full depth of the web. If the stiffeners are *not* to be used as connecting elements, they must be welded against the compression flange but may not be welded to the tension

flange. The allowable distance between the end of the stiffener and the tension flange is between $4t_w$ and $6t_w$. Therefore, either cut or cope the transverse stiffeners $4t_w$ or 1.75 in. from the tension flange.

For this design, M270 Grade 50W steel is used for the stiffeners. In locations where diaphragms are to be used, a stiffener is used on each side of the web as a connecting element. For the other locations a single plate will be welded to one side of the web only. The stiffeners are designed as columns made up of either one or two plates and a centrally located strip of web.

For web in which the slenderness [A6.10.11.1.1]

$$\frac{D}{t_w} \leq 2.5\sqrt{\frac{E}{F_{yw}}}$$

$$\frac{D}{t_w} = \frac{60}{0.4375} \leq 2.5\sqrt{\frac{29000}{50}} = 60.2$$

$$137 \geq 60.2$$

Only [A6.10.11.1.2] must be checked. However, this is not the case for our slender web and [A6.10.11.1.2, A6.10.11.1.3, and A6.10.11.1.4].

a. Single-Plate Transverse Stiffeners Single-plate transverse intermediate stiffenes are used at locations where there are no connecting elements. They shall be designed based on the maximum shear for the positive and negative moment regions. This use of maximum shear is a conservative approach. The fact that the stiffeners will have more than the required strength in some areas is negligible because the amount of steel saved by changing them would be small. The stiffener size chosen is $\frac{7}{8}$ in. × 5.5 in. for both regions. The following requirements demonstrate the adequacy of this section. ($\frac{7}{8}$ in. = 0.875 in.).

b. Projecting Width The projecting width requirement is checked to prevent local buckling of the transverse stiffeners. The width of each projecting stiffener must meet the following requirements [A6.10.11.1.2]:

$$2 + \frac{d}{30} \leq b_t$$

and

$$16.0t_p \geq b_t \geq 0.25b_f$$

where

d = steel section depth (in.)
t_p = thickness of projecting element (in.)
F_{ys} = minimum yield strength of stiffener (ksi)
b_f = full width of steel flange (in.)

For the positive moment regions:

$$d = 60 + 0.625 + 1.0 = 61.625 \text{ in.}$$

$$t_p = 0.875 \text{ in.}$$
$$F_{ys} = 50 \text{ ksi}$$
$$b_f = 12 \text{ in. (compression flange)}$$
$$2 + \frac{61.625}{30} = 4.1 \text{ in.} \leq b_t = 12 \text{ in.}$$

and

$$16.0\,(0.875) = 14 \text{ in.} \geq (b_t = 5.5 \text{ in.}) \geq 0.25\,(4) = 3 \text{ in.} \quad \text{OK}$$

For the negative moment regions:

$$d = 60 + 1.25 + 1.25 = 62.5 \text{ in.}$$
$$t_p = 0.875 \text{ in.}$$
$$F_{ys} = 50 \text{ ksi}$$
$$b_f = 16 \text{ in.}$$
$$2 + \frac{62.5}{30} = 4.1 \text{ in.} \leq b_t = 5.5 \text{ in.}$$

and

$$16\,(0.875) = 14 \text{ in.} \geq (b_t = 5.5 \text{ in.}) \geq 0.25\,(16) = 4 \text{ in.} \quad \text{OK}$$

c. Moment of Inertia The moment of inertia of all transverse stiffeners must meet the following requirements [A6.10.11.1.3]:

$$I_{t1} \geq \min\left[d_0, D\right] t_w^3 J$$

for which

$$J = 2.5\left(\frac{D_p}{d_0}\right)^2 - 2.0 \geq 0.5$$

where

I_{t1} = moment of inertia of transverse stiffener taken about the edge in contact with the web for single stiffeners and about the midthickness of the web for stiffener pairs (in.^{4})
d_0 = transverse stiffener spacing (in.)
D_p = web depth for webs without longitudinal stiffeners (in.)

The transverse stiffener moment of inertia must also be greater than

$$I_{t2} \geq \frac{D^4}{40}\rho_t^{1.3}\left[\frac{F_{yw}}{E}\right]^{1.5}$$

where

$$\rho_t = \min\left[\frac{F_{yw}}{F_{crs}}, 1.0\right]$$

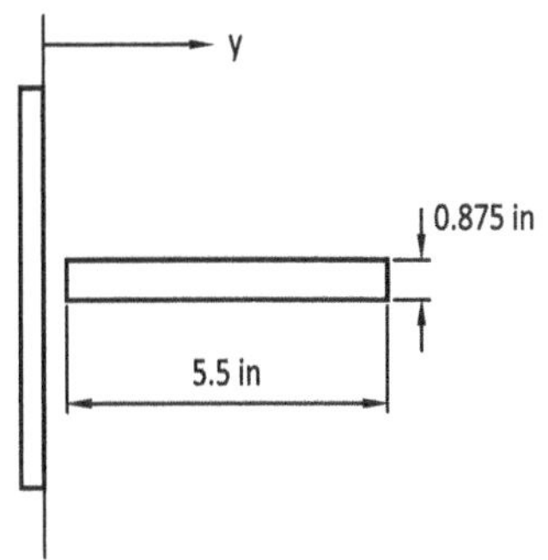

Fig. E20.3-38 Single-plate transverse intermediate stiffener.

and F_{crs} is the local buckling stress for the stiffeners,

$$F_{crs} = \frac{0.31E}{(b_t/t_p)^2} \le F_{ys}$$

For the positive moment regions use d_0 equal to 130 in. and for the negative moment regions use d_0 equal to 75 in.

The moment of inertia for a single stiffener, 0.875 in. × 5.5 in., is shown in Figure E20.3-38:

$$\begin{aligned} I &= I_0 + Ad^2 \\ &= \frac{(0.875)\,(5.5)^3}{12} + (0.875)\,(5.5)\,(2.75)^2 \\ &= 48.5 \text{ in.}^4 \end{aligned}$$

For positive moment regions:

$$\begin{aligned} J &= 2.5\left(\frac{60}{130}\right)^2 - 2.0 \\ &= -1.47 < 0.5 \end{aligned}$$

Therefore, use

$$\begin{aligned} &J = 0.5 \\ &I_t = 48.5 \ge I_{t1} = \min(60{,}130)\,(0.4375)^3\,(0.5) \\ &\quad = 2.5 \text{ in.}^4 \\ &I_t \ge I_{t1} \quad \text{OK} \end{aligned}$$

The first moment of inertia criterion is met. For the second criterion, the critical buckling stress is high, so F_{crs} is controlled by yielding and ρ_t is unity:

$$\begin{aligned} F_{crs} &= \frac{0.31E}{(b_t/t_p)^2} = \frac{0.31\,(29000)}{(5.5/0.4375)^2} = 227 \le F_{ys} \\ &= 50 \text{ ksi} \end{aligned}$$

$$\rho_t = \min\left[\frac{F_{yw}}{F_{crs}}, 1.0\right] = 1.0$$

$$I_{t2} \ge \frac{60^4}{40}(1)^{1.3}\left[\frac{50}{29000}\right]^{1.5} = 23.2 \text{ in.}^2$$

$$I_t = 48.5 \ge I_{t2} = 23.2 \quad \text{OK}$$

For negative moment regions:

$$\begin{aligned} J &= 2.5\left(\frac{60}{75}\right)^2 - 2.0 \\ &= -0.40 < 0.5 \end{aligned}$$

Therefore, use

$$\begin{aligned} &J = 0.5 \\ &I_{t1} = 5.4 \ge (75)\,(0.4375)^3\,(0.5) \\ &\quad = 3.14 \text{ in.}^4 \\ &I \ge I_{t1} \quad \text{OK} \end{aligned}$$

From the previous computation

$$I_t = 48.5 \ge I_{t2} = 23.2 \quad \text{OK}$$

Therefore, both criteria are met.

10. *Double-Plate Transverse Stiffener Design* Double-plate transverse intermediate stiffeners are used at locations where connecting elements such as diaphragms are used. For this design, they shall be based on the maximum shear for the positive and negative moment regions, respectively. This approach is conservative. The fact that the stiffeners have more than the required strength in some areas is negligible because the amount of steel saved by changing them would be small. A pair of stiffeners, 0.5 in. × 4 in. is chosen for both regions. The following requirements demonstrate the adequacy.

    a. Projecting Width The projecting width requirement is checked to prevent local buckling of the transverse stiffeners. The width of each projecting stiffener must meet the following requirements [A6.10.11.1.2]:

$$2 + \frac{d}{30} \le b_t$$

and

$$16.0t_p \ge b_t \ge 0.25b_f$$

For the positive moment regions

$$\begin{aligned} d &= 60 + 0.625 + 1.0 \\ &= 61.625 \text{ in.} \\ t_p &= 0.5 \text{ in.} \\ F_{ys} &= 50 \text{ ksi} \\ b_f &= 12 \text{ in.} \end{aligned}$$

$$2 + \frac{61.625}{30} = 4.1 \text{ in.} \approx b_t = 4 \text{ in.}$$

and

$$\begin{aligned} 16.0\,(0.5) &= 8 \text{ in.} \ge b_t = 4 \text{ in.} \ge 0.25\,(12) \\ &= 3 \text{ in.} \quad \text{OK} \end{aligned}$$

For the negative moment regions

$$d = 60 + 0.625 + 1.0 = 61.625 \text{ in.}$$
$$t_p = 0.5 \text{ in.}$$
$$F_{ys} = 50 \text{ ksi}$$
$$b_f = 16 \text{ in.}$$
$$2 + \frac{61.625}{30} = 4.1 \text{ in.} \approx b_t = 4 \text{ in.}$$

and

$$16.0\,(0.5) = 8 \text{ in.} \geq b_t = 4 \text{ in.} \geq 0.25\,(12) = 3 \text{ in.} \quad \text{OK}$$

b. Moment of Inertia The moment of inertia for a pair of stiffeners, 0.5 in. × 4 in., taken about the middle of the web, is shown in Figure E20.3-39.

$$I = I_0 + Ad^2 = \frac{2\,(0.5)\,(4)^3}{12} + 2\,(0.5)\,(4)\,(2.22)^2 = 25.0 \text{ in.}^4$$

which meets both inertia requirements.

11. Bearing Stiffeners

a. Bearing Stiffeners at Abutments The requirements for bearing stiffeners are taken from the LRFD Specifications [A6.10.11.2]. Bearing stiffeners shall be placed on the webs of plate girders at all bearing locations and at locations of concentrated loads.

The purpose of bearing stiffeners is to transmit the full bearing force from the factored loads. They consist of one or more plates welded to each side of the web and extend the full length of the web. It is also desirable to extend them to the outer edges of the flanges. At the abutments the bearing stiffeners chosen consist of one 0.875-in. × 6-in. plate on each side of the web. The following requirements demonstrate the adequacy of this section.

(1) Projecting Width To prevent local buckling of the bearing stiffener plates, the width of each projecting element has to satisfy the following [A6.10.11.2.2]:

$$b_t \leq 0.48 t_p \sqrt{\frac{E}{F_{ys}}}$$
$$= 6 \text{ in.} \leq 0.48\,(0.875)\sqrt{\frac{29000}{50}}$$
$$= 10.1 \text{ in.} \quad \text{OK}$$

(2) Bearing Resistance To get the bearing stiffener tight against the flanges, a portion in the corner must be clipped. This clipping of the stiffener is so the fillet welding of the flange and web plates can be done. By clipping the stiffener, the bearing area of the stiffener is reduced (see Fig. E20.3-40). When determining the bearing resistance, this reduced bearing area must be used. The factored bearing resistance is calculated below [A6.10.11.2.3]:

$$B_r = \phi_b\,(1.4)\,A_{pn} F_{ys}$$

where

A_{pn} = contact area of stiffener on the flange
$= 2(6 - 1.5)(0.875) = 7.88 \text{ in.}^2$

$B_r = 1.0(1.4)(7.88)(50) = 551$ kips
$= 551 \text{ kips} > R_u = 220$ kips OK

The value for R_u, which is equal to the reaction at the abutments, is taken from Table E20.3-3 at location 100 (interior girder controls).

(3) Axial Resistance of Bearing Stiffeners The axial resistance of bearing stiffeners is determined from AASHTO [A6.10.11.2.4]. The factored axial resistance P_r for components in compression is taken as [A6.9.2.1]

$$P_r = \phi_c P_n$$

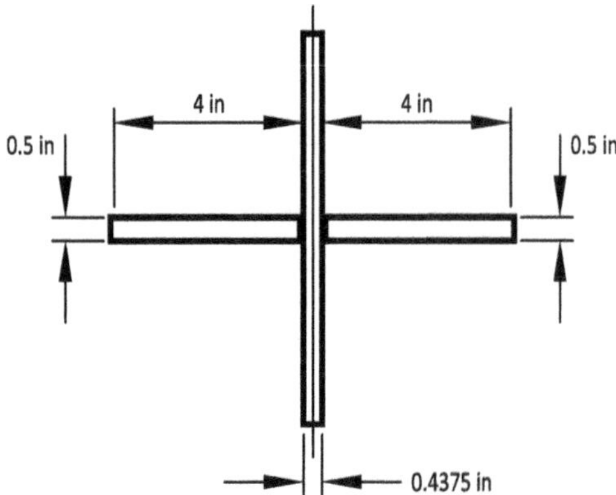

Fig. E20.3-39 Double-plate transverse intermediate stiffener.

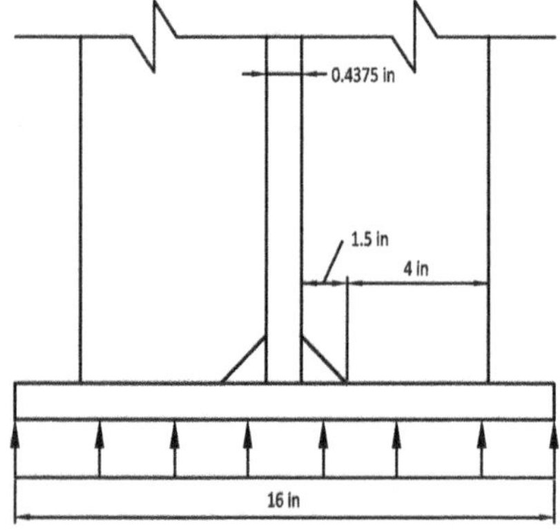

Fig. E20.3-40 Bearing stiffener at abutment.

where

ϕ_c = resistance factor for compression = 0.9

P_n = nominal compressive resistance (kN)

To calculate the nominal compressive resistance, the section properties are to be determined. The radius of gyration is computed about the center of the web and the effective length is considered to be $0.75D$, where D is the web depth. The reason the effective length is reduced is because of the end restraint provided by the flanges against column buckling.

(4) Effective Section For stiffeners welded to the web (Fig. E20.3-41), the effective column section consists of the 0.875-in. × 6-in. stiffeners [A6.10.11.2.4b].

The radius of gyration, r_s, is computed from the values in Table E20.3-17.

$$I = I_0 + Ad^2$$
$$= 108.8 + 31.5 = 140.3\,\text{in.}^4$$
$$r_s = \sqrt{\frac{I}{A}} = \sqrt{\frac{140.3}{10.5}} = 3.65\ \text{in.}$$

(5) Slenderness The limiting width-to-thickness ratio for axial compression must be checked [A6.9.4.2]. The limiting value is as follows:

$$\frac{b}{t} \le k\sqrt{\frac{E}{F_y}}$$

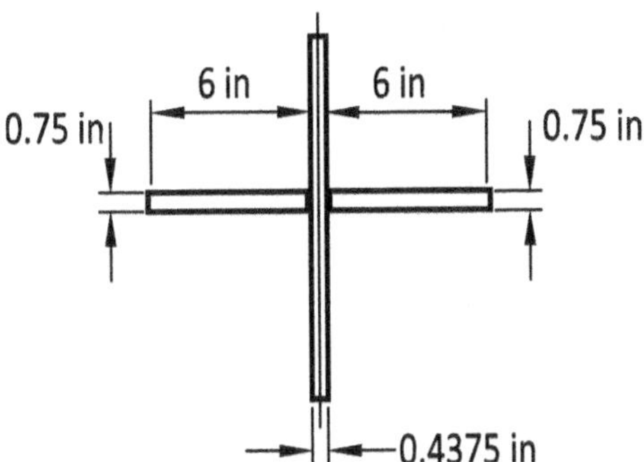

Fig. E20.3-41 Section of bearing stiffener at abutment.

Table E20.3-17 Effective Section for Bearing Stiffeners over the Abutments

Part	A	y	Ay	Ay^2	I_0
Stiffener	5.25	3.22	16.90	54.4	15.75
Stiffener	5.25	−3.22	−16.90	54.4	15.75
Total	10.50			108.8	31.50

where

k = plate buckling coefficient from Table 19.2 = 0.45

b = width of plate as specified in Table 19.2 = 6 in.

t = plate thickness = 0.875 in.

$$\frac{b}{t} = \frac{6}{0.875} = 6.86 \le 0.45\sqrt{\frac{29000}{50}} = 10.8 \quad \text{OK}$$

(6) Nominal Compressive Resistance The nominal compressive resistance is taken from AASHTO [A6.9.4.1] because the stiffeners are noncomposite members. The value of P_n is determined as follows:

$$P_o = QF_yA_g = 1.0\,(50)\,(10.5) = 525\ \text{kips}$$
$$P_e = \frac{\pi^2 EA_g}{\left(\frac{KL}{r_s}\right)^2} = \frac{\pi^2\,(29000)\,(10.5)}{\left(\frac{0.75\,(60)}{3.65}\right)^2}$$
$$= 19770\ \text{kips}$$
$$P_n = 0.658^{P_o/P_e} F_{ys} A_g$$
$$= 0.658^{525/19{,}770}\,(50)\,(10.5) = 519\ \text{kips}$$
$$P_r = \phi P_n = 0.9\,(519) = 467\ \text{kips}$$
$$P_r \ge V_u = 197\ \text{kips} \quad \text{OK}$$

where

A_s = gross cross-sectional area (in.^{2})

K = effective length factor = 0.75

L = unbraced length (in.)

r_s = radius of gyration about the plane of buckling (in.)

Therefore, stiffeners are adequate.

For the bearing stiffeners at abutments, use a pair of plates, 0.875 in. × 6 in.

b. Bearing Stiffeners at the Interior Supports The requirements that were specified for the bearing stiffeners at the abutments apply also to this section. At the interior supports the bearing stiffeners consist of two 0.875-in. × 6-in. plates on each side of the web. The two plates are spaced 6 in. apart to allow for welding. The following requirements demonstrate the adequacy of this section.

(1) Projecting Width To prevent local buckling of the bearing stiffener plates, the width of each projecting element has to satisfy the following [A6.10.11.2.2]:

$$b_t \le 0.48t_p\sqrt{\frac{E}{F_{ys}}}$$
$$= 6\ \text{in.} \le 0.48\,(0.875)\sqrt{\frac{29000}{50}}$$
$$= 10.1\ \text{in.} \quad \text{OK}$$

(2) Bearing Resistance The factored bearing resistance is calculated below (see Fig. E20.3-40) [A6.10.11.2.3]:

$$B_r = \phi_b (1.4) A_{pn} F_{ys}$$
$$A_{pn} = 4(6 - 1.5)(0.875) = 15.75 \text{ in.}^2$$
$$B_r = (1.0)(1.4)(15.75)(50)$$
$$= 1100 \text{kips} > R_u = 441 \text{ kips} \quad \text{OK}$$

The value for R_u, which is equal to the reaction at either interior support (Fig. E20.3-42) is determine from BT-Beam and is controlled by the design truck train loading. Alternatively the two shears may be conservatively added (V_{110} and V_{200}) from Table E20.3-3 (interior girder controls). This is a very conservative approximation because maximum values for shear due to the design truck are used.

(3) Axial Resistance of Bearing Stiffeners For components in compression, P_r is taken as [A6.9.2.1]

$$P_r = \phi_c P_n$$

The effective unbraced length is $0.75D = 0.75(60) = 45$ in.

(4) Effective Section The effective section criteria is found in AASHTO [A6.10.11.2.4b]. For stiffeners welded to the web, the effective column section consists of the stiffeners plus a centrally located strip of web extending $9t_w$ to each side of the stiffeners as shown in Figure E20.3-43. The spacing of the stiffeners is 6 in.

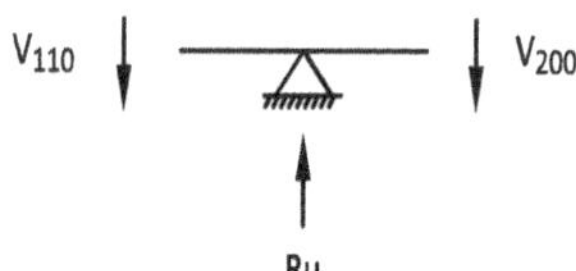

Fig. E20.3-42 Reaction at interior support.

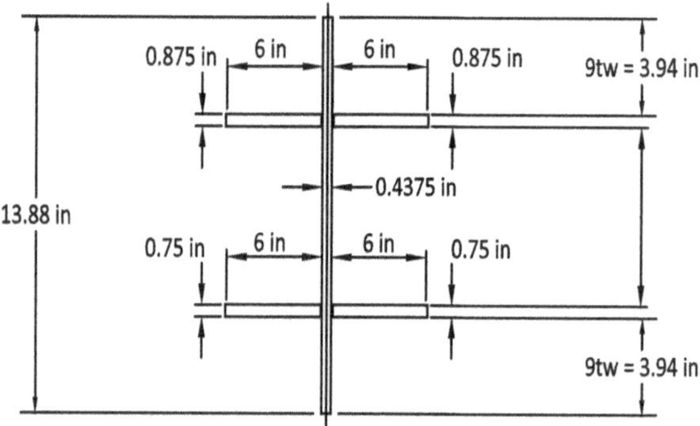

Fig. E20.3-43 Section of bearing stiffener at interior support.

Table E20.3-18 Effective Section for Bearing Stiffeners Over the Interior Supports

Part	A	y	Ay	Ay^2	I_0
Web	6.07	0	0	0	0.10
Stiffener	5.25	3.32	16.9	54.4	15.75
Stiffener	5.25	3.32	16.9	54.4	15.75
Stiffener	5.25	−3.32	−16.9	54.4	15.75
Stiffener	5.25	−3.32	−16.9	54.4	15.75
Total	20.07			217.6	63.0

The radius of gyration r_s is computed from the values in Table E20.3-18.

$$I = I_0 + Ad^2 = 63.0 + 217.6 = 280.6 \text{ in.}^4$$
$$r_s = \sqrt{\frac{I}{A}} = \sqrt{\frac{280.6}{20.07}} = 3.74 \text{ in.}$$

(5) Slenderness The limiting width-to-thickness ratio for axial compression must be checked [A6.9.4.2]. The limiting value is as follows:

$$\frac{b}{t} \le k\sqrt{\frac{E}{F_y}}$$
$$\frac{b}{t} = \frac{6}{0.875} = 6.8 < 0.45\sqrt{\frac{29000}{50}} = 10.8 \quad \text{OK}$$

(6) Nominal Compressive Resistance The nominal compressive resistance is taken from AASHTO [A6.9.4.1] because the stiffeners are noncomposite members. The value of P_n is determined as follows:

$$P_o = QF_yA_g = 1.0(50)(27.0) = 1350 \text{ kips}$$
$$P_e = \frac{\pi^2 E A_g}{\left(\frac{KL}{r_s}\right)^2} = \frac{\pi^2 (29000)(27.0)}{\left(\frac{0.75(60)}{3.74}\right)^2}$$
$$= 53{,}360 \text{ kips}$$
$$P_n = 0.658^{P_o/P_e} F_{ys} A_g$$
$$= 0.658^{1350/53{,}360}(50)(27.0) = 1335 \text{ kips}$$
$$P_r = \phi P_n = 0.9(1335) = 1200 \text{ kips}$$
$$P_r \ge R_u = 441 \text{ kips} \quad \text{OK}$$

For the bearing stiffeners at the supports, use two pairs of stiffener plates, 0.875 in. × 6 in. A summary of the stiffener design is shown in Figure E20.3-44.

12. *Shear Connectors*

a. General Design of shear connectors is specified in AASHTO [A6.10.10.4]. Stud shear connectors are to be provided at the interface between the concrete slab and the steel section. The purpose

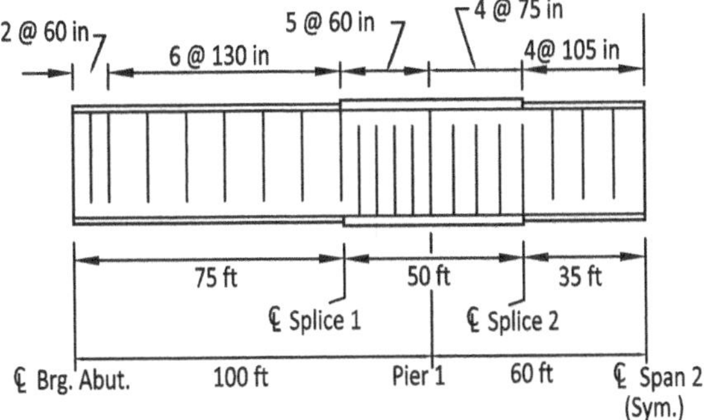

Fig. E20.3-44 Summary of stiffener design.

of the connectors is to resist the interface shear. In continuous composite bridges, shear connectors are recommended throughout the length of the bridge including negative moment regions. Before designing, the designer must consider some general information including types of shear connectors, pitch, transverse spacing, cover, and penetration.

(1) Types of Shear Connectors The two primary types of connectors used are the stud and channel shear connectors. The connectors should be chosen so that the entire surface of the connector is in contact with the concrete so that it may resist horizontal or vertical movements between the concrete and the steel section. For this design, stud shear connectors are used to provide a composite section. The ratio of the height-to-stud diameter is to be greater than 4.0. [A6.10.10.1.1]. Consider 0.75-in. diameter studs, 4-in. high, for this design:

$$\frac{4}{0.75} = 5.33 > 4 \quad \text{OK}$$

(2) Transverse Spacing Transverse spacing of the shear connectors is discussed in AASHTO [A6.10.10.1.3]. Shear connectors are placed transversely along the top flange of the steel section. The center-to-center spacing of the connectors cannot be closer than four stud diameters, or 3 in. The clear distance between the edge of the top flange and the edge of the nearest connector must be at least 1 in.

(3) Cover and Penetration Cover and penetration requirements are in AASHTO [A6.10.10.1.4]. Shear connectors should penetrate at least 2 in. into the concrete deck. Also, the clear cover over the tops of the connectors should be at least 2 in. Consider a height of 4 in. for the shear studs.

b. Fatigue Resistance Consider the fatigue resistance of shear connectors in composite sections [A6.10.10.1.2]. When exact data are not provided the ADTT is determined using a fraction of the average daily traffic volume. The average daily traffic includes cars and trucks. Using the recommendations of AASHTO [C3.6.1.4.2], the ADT can be considered to be 20,000 vehicles per lane per day. The ADTT is determined by applying the appropriate fraction from Table 8.4 to the ADT. By assuming urban interstate traffic, the fraction of trucks is 15%.

$$(\text{ADTT})_{\text{SL}} = 0.85\ (0.15)\ (20000)\ (2\ \text{lanes})$$
$$= 5100\ \text{trucks/day}$$

The number of trucks significantly exceeds the 960 trucks per day for the finite-life fatigue design. Therefore, the infinite fatigue-life criterion is used. The stud resistance is

$$Z_r\ (\text{infinite life}) = 5.5d^2 = 5.5(0.75)^2 = 3.1\ \text{kips}$$

The pitch of the shear connectors is specified in AASHTO [A6.10.10.1.2]. The pitch is to be determined to satisfy the fatigue limit state. Furthermore, the resulting number of shear connectors must not be less than the number required for the strength limit state. The minimum center-to-center pitch of the shear connectors is determined as follows:

$$p = \frac{nZ_rI}{V_{\text{sr}}Q}$$

where

p = pitch of shear connectors along the longitudinal axis (in.)
n = number of shear connectors in a cross section
I = moment of inertia of the short-term composite section (in.^{4})
Q = first moment of the transformed area about the neutral axis of the short-term composite section (in.^{3})
V_{sr} = shear force range under LL + IM determined for the fatigue limit state
Z_r = shear fatigue resistance of an individual shear connector
d_s = shear stud diameter (in)

and

$$6d_s = 4\ \text{in.} \leq p \leq 24\ \text{in.}$$

For the short-term composite section, the moment of inertia is 83,461 in.^{4}. The first moment of the transformed area about the neutral axis for the short-term composite section is determined from Figure E20.3-45:

$$Q = Ay = (8)\ (12)\ (17.93 - 4) = 1337\ \text{in.}^3$$

For this design three shear connectors are used in a cross section as shown in Figure E20.3-46:

$$\text{Stud spacing} = 4\ \text{in.} > 4\ (0.75) = 3\ \text{in.}$$
$$\text{Clear distance} = 1.625\ \text{in.} \geq 1\ \text{in.}$$

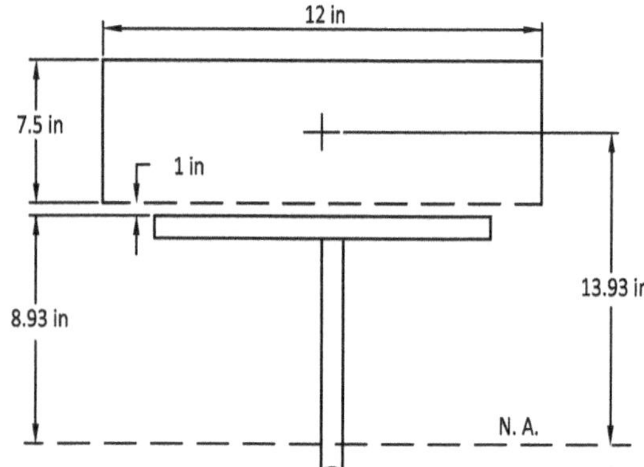

Fig. E20.3-45 Composite section properties.

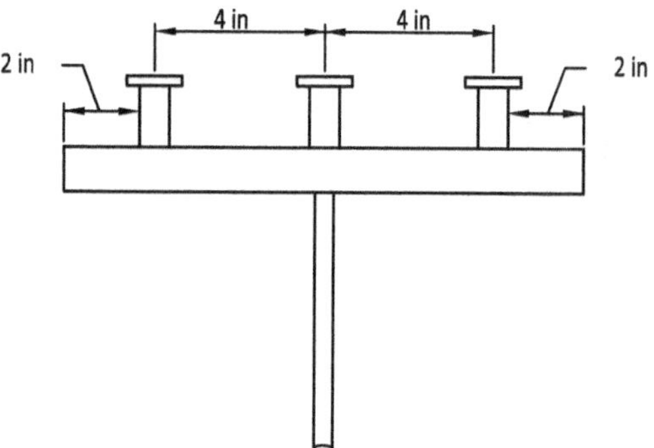

Fig. E20.3-46 Group of three shear connectors.

Therefore, the transverse spacing requirements [A6.10.10.1.3] are satisfied. The required shear connector pitch is computed at the tenth points along the spans using the shear range for the fatigue truck. The shear range V_{sr} is the maximum difference in shear at a specific point. It is computed by finding the difference in the positive and negative shears at that point due to the fatigue truck, multiplied by the dynamic load allowance for fatigue (1.15), the maximum distribution factor for one design lane loaded without multiple presence (0.75/1.2 for the exterior girder), and by the load factor for the fatigue I limit state (1.5). The shear range and the pitch at the tenth points are tabulated in Table E20.3-19. The values in the pitch column are the maximum allowable spacing at a particular location. The required spacing is plotted on Figure E20.3-47. The spacing to be used is determined from this graph. An example calculation of the pitch is performed below, for the shear range at the end of the bridge.

Table E20.3-19 Shear Range for Fatigue Loading and Maximum Shear Connector Spacing

Location		Unfactored Max. Pos. Shear (kips)	Unfactored Max. Neg. Shear (kips)	Shear Range (kips) (Includes distribution, IM, and $\gamma_{\text{fatigue I}} = 1.5$)	Pitch (in.)
Sta	(ft)				
100	0	56.04	−6.67	67.4	8.6
101	10	47.52	−6.67	56.0	10.4
102	20	39.29	−8.51	49.5	11.7
103	30	31.47	−13.33	46.4	12.5
104	40	24.16	−19.42	45.1	12.9
105	50	17.47	−27.45	46.5	12.5
106	60	11.64	−35.55	48.9	11.9
107	70	7.6	−43.2	52.6	11.0
108	80	4.42	−50.3	56.1	24[a]
109	90	1.61	−56.75	60.4	24[a]
110	100	1.61	−62.43	66.4	24[a]
200	100	62.33	−6.28	71.0	24[a]
201	112	55.57	−6.28	64.0	24[a]
202	124	48.01	−6.28	56.2	24
203	136	39.97	−9.78	51.5	11.3
204	148	31.77	−16.21	49.7	11.7
205	160	23.75	−23.75	49.2	11.8

[a]The maximum shear stud spacing is 24 in. This spacing is used in negative moment regions, assumed to be between the dead-load inflection points at 75 and 124 ft.

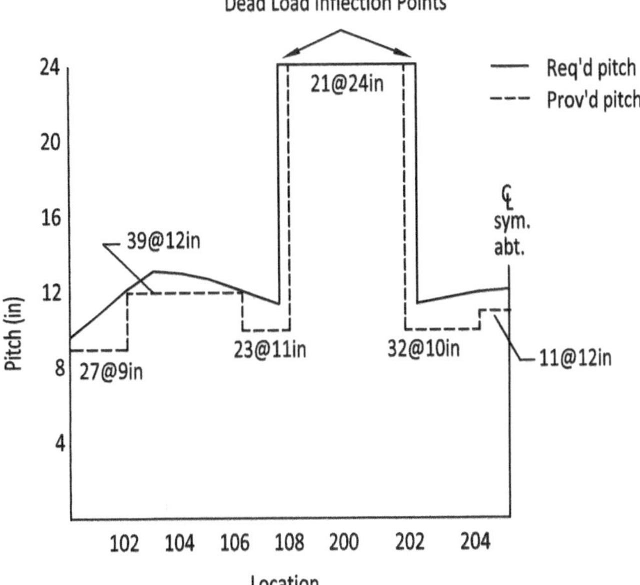

Fig. E20.3-47 Summary of shear connector spacings.

The shear is computed using fatigue limit state I and an example computation for the end of the bridge is

$$V_{\text{sr fatigue I}} = (56.04 + 6.67)\,(1.15)\left(\frac{0.75}{1.2}\right)$$
$$= 67.6 \text{ kips}$$
$$p = \frac{3\,(3.1)\,(83461)}{(67.6)\,(1337)}$$
$$= 8.6 \text{ in.}$$

c. Strength Limit State The strength limit state for shear connectors is taken from AASHTO [A6.10.7.4.4]. The factored shear resistance of an individual shear connector is as follows:

$$Q_r = \phi_{\text{sc}} Q_n$$

where

Q_n = nominal resistance of a shear connector (kips)
ϕ_{sc} = resistance factor for shear connectors = 0.85

$$Q_n = 0.5 A_{\text{sc}} \sqrt{f_c' E_c} \le A_{\text{sc}} F_u$$

where

A_{sc} = shear connector cross-sectional area (in.^{2})
E_c = modulus of elasticity of concrete (ksi)
F_u = minimum tensile strength of a stud shear connector

$$F_u = 58 \text{ ksi [A6.4.4]}$$
$$A_{\text{sc}} = \frac{\pi}{4}(0.75)^2 = 0.44 \text{ in.}^2$$
$$E_c = 1820\sqrt{f_c'} = 1820\sqrt{4} = 3600 \text{ ksi}$$
$$Q_n = 0.5\,(0.44)\,\sqrt{30\,(3600)} = 26.4 \text{ kips}$$
$$A_{\text{sc}} F_u = (0.44)\,(58)$$
$$= 125.5 \text{ kips} > 26.4 \text{ kips}$$

Use $Q_n = 25.4$ kips

Therefore

$$Q_r = (0.85)\,(25.5) = 21.7 \text{ kips}$$

The number of shear connectors required depends on the section. Between sections of maximum positive moment and points of 0 moment to either side, the number of shear connectors required is as follows:

$$n = \frac{V_h}{Q_r}$$

for which the nominal horizontal shear force is the lesser of the following:

$$V_h = 0.85 f_c' b t_s$$

or

$$V_h = F_{\text{yw}} D t_w + F_{\text{yt}} b_t t_t + F_{\text{yc}} b_c t_c$$

where

V_h = nominal horizontal shear force
b = effective slab width = 96 in.
D = web depth = 60 in.

t_s = slab thickness = 8 in.
b_c = width of compression flange = 12 in.
b_t = width of tension flange = 16 in.
t_c = thickness of compression flange = 0.625 in.
t_t = thickness of tension flange = 1.0 in.
F_y = minimum yield strengths of the respective sections = 50 ksi

$$V_h = \min \begin{cases} 0.85\,(4)\,(96)\,(8) = 2611 \text{ kips} \\ [50\,(60)\,(0.4375) + 50\,(16)\,(1) \\ +50\,(12)\,(0.625)] = 2488 \text{ kips} \end{cases}$$

$V_h = 2488$ kips

Therefore, use a nominal horizontal shear force V_h of 10,178 kN and the required number of shear connectors for this region is calculated below:

$$n = \frac{2488}{21.7} = 115$$

Therefore a minimum of 115 shear connectors are required between points of maximum positive moment and points of zero moment. From examination of Figure E20.3-47, the number of shear connectors required by the fatigue limit state exceeds the amount required from the strength limit state.

For composite sections that are continuous, the horizontal shear force between the centerline of a support and points of zero moment is determined by the reinforcement in the slab. The following calculation determines the horizontal shear force:

$$V_h = A_r F_{\text{yr}}$$

where

A_r = total area of longitudinal reinforcement in the effective width over the interior support (in.^{2}) = 1.80 + 2.17 = 3.97 in.^{2}

F_{yr} = minimum yield strength of the longitudinal reinforcement = 60 ksi

$V_h = (3.91)(60) = 238$ kips

Therefore the number of shear connectors required in this region is

$$n = \frac{238}{21.7} = 11$$

Eleven studs are required between the interior pier and the points of zero moment. The 24-in. maximum allowable spacing provides considerably more than this number of connectors. Therefore, use the shear connector spacing specified on Figure E20.3-47.

J. Dimension and Detail Requirements

1. *Diaphragms and Cross Frames* [A6.7.4] In this section, intermediate and end cross frames are designed. The framing plan is shown on Figure E20.3-48. Cross frames serve three primary purposes:

i. Lateral support of the compression flange during placement of the deck
ii. Transfer of wind load on the exterior girder to all girders
iii. Lateral distribution of wheel load

The requirements for cross-frame design are located in AASHTO [A6.7.4]. The end cross frames must transmit all the lateral forces to the bearings. All the cross frames must satisfy acceptable slenderness requirements.

a. Cross-Frame Spacing The cross frame spacings were conservatively estimated and checked for lateral torsional bucking. The assumed initial spacing were 25 and 30 ft for the end and middle spans, respectively. Here frames are aligned with the stiffeners keeping the spacing less than that initially assumed. See Figure E20.3-48 where the maximum unbraced bottom flange length is 21 ft 8 in. = 21.67 ft.

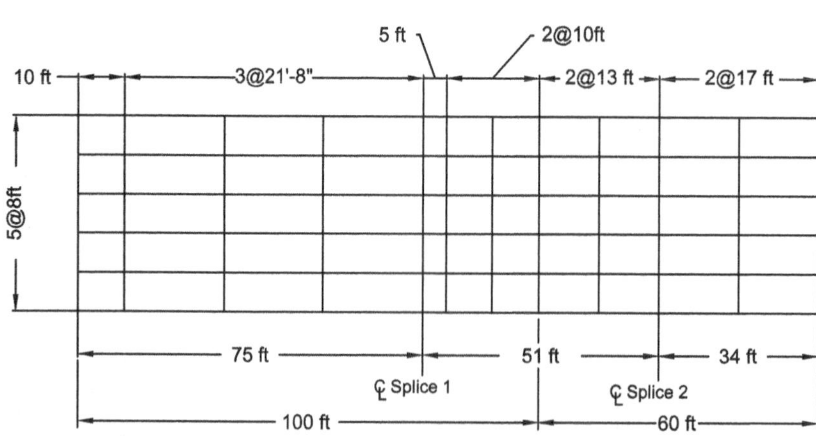

Fig. E20.3-48 Cross-frame locations.

b. Wind Load The wind load acts primarily on the exterior girders. In bridges with composite decks, the wind force acting on the upper half of the girder, deck, barrier, and vehicle is assumed to be transmitted directly to the deck. These forces are transferred to the supports through the deck acting as a horizontal diaphragm. The wind force acting on the lower half of the girder is transmitted directly to the bottom flange. For this design the wind force, W, is applied to the bottom flange only because the top flange acts compositely with the deck. The wind force is calculated as

$$W = \frac{\gamma p_B d}{2}$$

where

W = wind force per unit length applied to the flange
p_B = base horizontal wind pressure (ksf)
p_B = 0.050 ksf [A3.8.1.2]
d = depth of the member (ft)
γ = load factor for the particular group loading combination from Table 5.1 [Table A3.4.1-1], for this case strength III applies = 1.4

Consider the negative moment region first since it provides a larger value for d. The calculated wind load is conservative for the positive moment region.

$$d = 60 + 2\,(1.25) = 62.5 \text{ in.} = 5.10 \text{ ft}$$

therefore

$$W = \frac{1.4\,(0.050)\,(5.1)}{2} = 0.18 \text{ k/ft}$$

The assumed load path taken by these forces is as follows:

i. The forces in the bottom flange are transmitted to points where cross frames exist.
ii. The cross frames transfer the forces into the deck.
iii. The forces acting on the top half of the girder, deck, barriers, and vehicles are transmitted directly into the deck.
iv. The deck acts as a diaphragm transmitting the forces to the supports.
This load path is very conservative.
For this load path the maximum moment in the flange due to the wind load is as follows:

$$M_w = \frac{W L_b^2}{10}$$

where L_b = bracing point spacing

Using the maximum unbraced length for the positive moment region of 21.67 ft is conservative. Therefore the maximum lateral moment in the flange of the exterior girder due to the factored loading is

$$M_w = \frac{(0.18)\,(21.67)^2}{10} = 8.5 \text{ k ft}$$

The section modulus for the flange is

$$S_f = \tfrac{1}{6}\,(1.25)\,(16)^2 = 53.3 \text{ in.}^3$$

and the maximum bending stress in the flange is

$$f_l = \frac{M_w}{S_f} = \frac{8.5\,(12)}{53.5} = 1.9 \text{ ksi}$$

which is a very small stress and any interaction with gravity loads can be neglected.

The maximum horizontal wind force applied to each brace point is also determined using maximum spacing. Therefore, the values are conservative in most sections of the bridge.

$$P_w = W L_b = (0.18)\,(21.67) = 3.9 \text{ kips}$$

The cross frames must be designed to transfer all the lateral forces to the bearings. Figure E20.3-49 illustrates the transmittal of forces. As stated before, the forces acting on the deck, barrier, and upper half of the girder, F_1, are directly transmitted into the deck. These forces are transferred to all of the girders. The forces acting on the bottom half of the girder, F_2, are transferred to the bottom flange.

The wind force, W, was previously calculated for the bottom flange to be 0.180 k/ft. This force is referred to as F_2 in Figure E20.3-49. F_1 is calculated as

$$F_L = 1.4\,(0.050)\,(8.71) - 0.180 = 0.43 \text{ k/ft}$$

c. Intermediate Cross Frames The intermediate cross frames are designed using X-bracing along with a strut across the bottom flanges as shown in Figure E20.3-50. Single angles, Grade 50W steel,

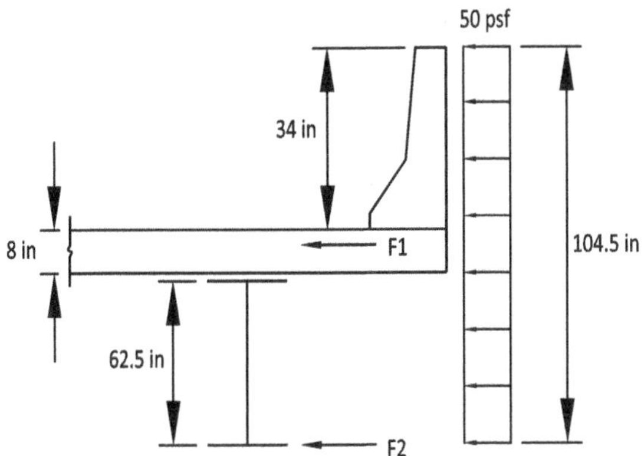

Fig. E20.3-49 Wind load acting on bridge exterior.

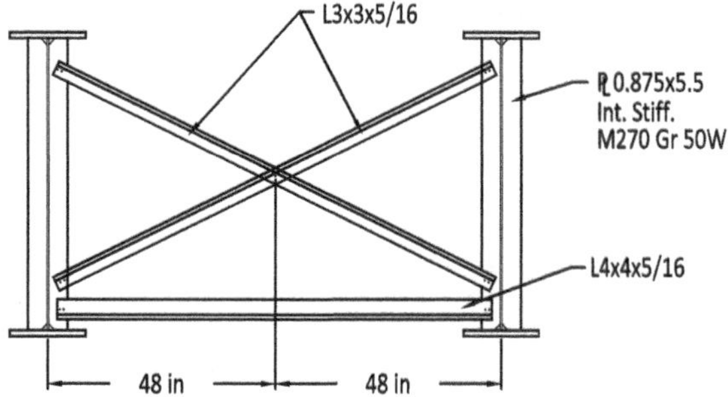

Fig. E20.3-50 Typical intermediate cross frame.

are used for the braces. For the cross braces acting in tension, $3 \times 3 \times \frac{5}{16}$ angles are used ($A_g = 1.78\,\text{in.}^2$ and $r_z = 0.583\,\text{in.}$). For the compression strut, a $4 \times 4 \times \frac{5}{16}$ angle is used ($A_g = 2.40\,\text{in.}^2$ and $r_z = 0.781\,\text{in.}$). These sections are considered as practical minimums. Section properties are given in AISC (2005).

The maximum force on the bottom flange at the brace point is

$$P_{\text{wb}} = 0.180\,(21.67) = 3.9\text{ kips}$$

In order to find forces acting in the cross brace and the compression strut, the section is treated like a truss with tension diagonals only (counters) and solved using statics. From this analysis it is determined that the cross braces should be designed for a tensile force of 4.5 kips, using 30° as the member orientation. The strut across the bottom flanges should be designed for a compressive force of 3.9 kips.

Check the $3 \times 3 \times \frac{5}{16}$ cross brace for tensile resistance [A6.8.2]:

$$P_r = \phi_y P_{\text{ny}} = \phi_y F_y A_g = (0.95)\,(50)\,(1.78) = 85\text{ kips}$$

The tensile force in the cross brace is only 4.5 kips. Therefore, the cross brace has adequate strength.

Check the limiting slenderness ratio of the cross brace for tension members [A6.8.4]. For bracing members the limiting slenderness ratio is

$$\frac{L}{r} \le 240$$

where

L = unbraced length of the cross brace (in.)
r = minimum radius of gyration of the cross brace (in.)

$$\frac{L}{r} = \frac{96/\cos(30)}{0.583} = 190 \le 240$$

Therefore use $3 \times 3 \times \frac{5}{16}$ angles for the intermediate cross braces.

Check the $4 \times 4 \times \frac{5}{16}$ strut for compressive resistance [A6.9.2.1].

Example computations for the compression element are provided in Chapter 19 and are not repeated here. From AISC (2005):

$$P_r = \phi_c P_n = 49.6\text{ kips}$$

For $KL = (0.75)(8) = 7$ ft

Therefore the $4 \times 4 \times \frac{5}{16}$ strut is adequate for the intermediate cross frames.

d. Cross Frames over Supports The cross frames over the supports are designed using X-bracing along with a strut across the bottom flanges as shown in Figure E20.3-51. Single angles, M270 Grade 50W steel, are used for the braces. The sections used for the intermediate cross frames are also used for these sections. For the cross braces acting in tension, $3 \times 3 \times \frac{5}{16}$ angles are used. For compression strut, a $4 \times 4 \times \frac{5}{16}$ angle is used.

The intermediate cross frames, through their tension diagonals, transfer the wind load between supports into the deck diaphragm. At the supports, the cross frames transfer the total tributary wind load in the deck diaphragm down to the bearings.

The force taken from the deck diaphragm by each cross frame at the supports is approximately

$$P_{\text{frame}} = \frac{(F_1 + F_2)\,L_{\text{ave}}}{5\text{ frames}} = \frac{(0.43 + 0.180)\,110}{5} = 12.8\text{ kips}$$

The force on the bottom flange of each girder that must be transmitted through the bearings to the support is approximately

$$P_{\text{girder}} = \frac{(F_1 + F_2)\,L_{\text{ave}}}{6\text{ girders}} = \frac{(0.43 + 0.180)\,110}{6} = 10.6\text{ kips}$$

In order to find the forces acting in the cross brace and the compression strut, the section is treated like a truss with counters and solved using statics. From

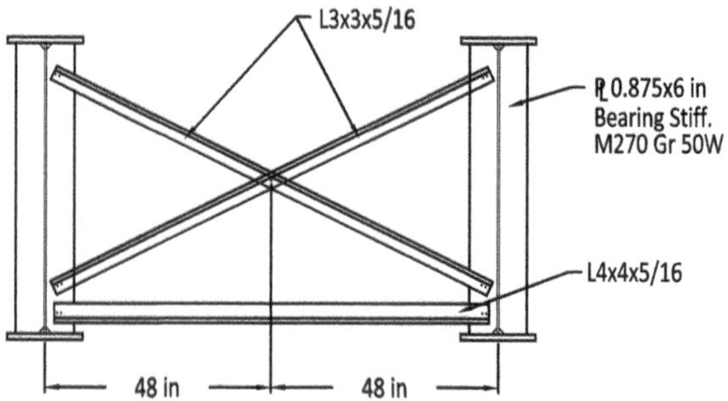

Fig. E20.3-51 Typical cross frame over supports.

this analysis, it is determined that the cross braces need to be designed for a tensile force of 12.2 kips. The strut across the bottom flanges needs to be designed for a compressive force of 12.8 kips. Because the forces above are less than the capacities of the members of the intermediate cross frames, and the same members are used, the chosen members are adequate. Therefore, use $3 \times 3 \times \frac{5}{16}$ angles for the cross bracing, and a $4 \times 4 \times \frac{5}{16}$ strut for the cross frames over the supports.

e. Cross Frames over Abutments The cross frames over the abutments are designed using an inverted V-bracing (K-bracing) along with a diaphragm across the top flange and a strut across the bottom flange as shown in Figure E20.3-52. Single angles, M 270 Grade 50W steel, are used for the braces. For the cross braces acting in tension, $3 \times 3 \times \frac{5}{16}$ angles are used. For the compression diaphragm across the top flange a W12 × 40 is used to provide additional stiffness at the discontinuous end of the bridge. For the strut across the bottom flange, a $4 \times 4 \times \frac{5}{16}$ angle is used.

Because the tributary wind load length for the abutments is 50 ft, the forces taken by the cross frames and girders can be determined from the values at the support as

$$P_{\text{frame}} = \tfrac{50}{110}(12.8) = 5.82 \text{ kips}$$
$$P_{\text{girder}} = \tfrac{50}{110}(10.6) = 4.82 \text{ kips}$$

In order to find the forces acting in the cross brace and the compression diaphragm, the section is treated like a truss with counters and solved using statics. From this analysis it is determined that the cross bracing carries 5.57 kips and the strut across the bottom flanges carries 5.82 kips. Both of these loads are less than the capacities of the members. The diaphragm across the top flanges should be designed for a compressive force of 5.82 kips.

Check the W12 × 40 diaphragm for compressive resistance [A6.9.2.1]. From AISC (2005) for KL = 8 ft:

$$P_r = \phi_c P_n = 439 \text{ kips}$$

This is much more than required, therefore use W12 × 40 diaphragm for the cross frames over the abutments.

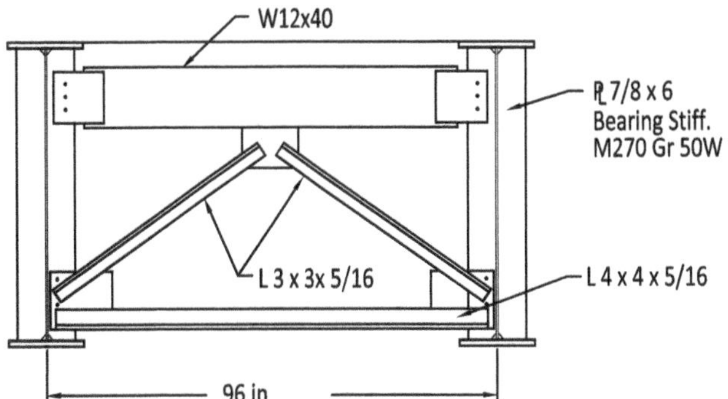

Fig. E20.3-52 Typical cross frame at abutments.

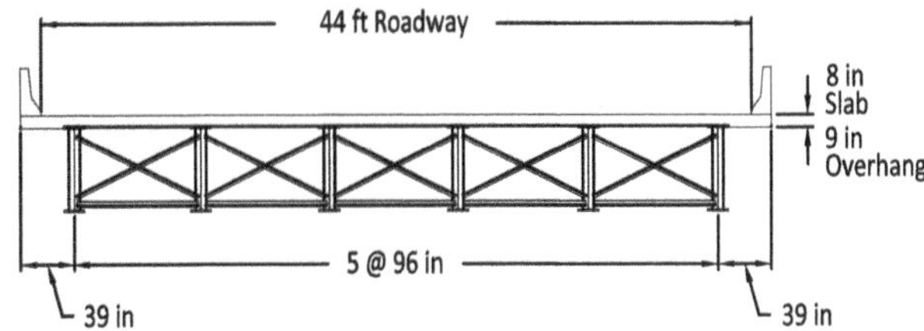

Fig. E20.3-53 Cross section of plate girder bridge showing girders and cross frames.

K. ***Design Sketch*** The design of the steel plate girder bridge is shown in Figure E20.3-53. Because the many details cannot be provided in a single drawing, the reader is referred to the figures already provided. For the cross section of the plate girder and slab, refer back to Figures E20.3-18 and E20.3-36. For stiffener spacing, refer to Figure E20.3-44. For shear stud pitch, refer back to Figure E20.3-47. For cross-frame locations, refer to Figure E20.3-48, and for cross-frame design, refer to Figures E20.3-50–E20.3-53.

The engineer must also design welds, splices, and bolted connections. These topics were not covered in this example because of lack of space.

REFERENCES

AASHTO (2010). *LRFD Bridge Design Specifications*, 5th ed., American Association of State Highway and Transportation Officials, Washington, DC.

AISC (2005). *LRFD Manual of Steel Construction*, 13 ed., American Institute of Steel Construction, Chicago.

APPENDIX A

Influence Functions for Deck Analysis

Throughout the book, several examples require the analysis of the deck for uniform and concentrated (line) loads. To facilitate analysis, influence functions were developed for a deck with five interior bays and two cantilevers. The widths (S) of the interior bays are assumed to be the same and the cantilevers are assumed to be of length (L). The required ordinates and areas are given in Table A.1. The notes at the bottom of the table describe its use. Examples are given in Chapters 9 and 14 to illustrate analysis using this table.

Table A.1 Influence Functions for Deck Analysis[a]

Location		M_{200}	M_{204}	M_{205}	M_{300}	R_{200}[b]
C	100	−1.0000	−0.4920	−0.3650	0.2700	1 + 1.270L/S
A	101	−0.9000	−0.4428	−0.3285	0.2430	1 + 1.143L/S
N	102	−0.8000	−0.3936	−0.2920	0.2160	1 + 1.016L/S
T	103	−0.7000	−0.3444	−0.2555	0.1890	1 + 0.889L/S
I	104	−0.6000	−0.2952	−0.2190	0.1620	1 + 0.762L/S
L	105	−0.5000	−0.2460	−0.1825	0.1350	1 + 0.635L/S
E	106	−0.4000	−0.1968	−0.1460	0.1080	1 + 0.508L/S
V	107	−0.3000	−0.1476	−0.1095	0.0810	1 + 0.381L/S
E	108	−0.2000	−0.0984	−0.0730	0.0540	1 + 0.254L/S
R	109	−0.1000	−0.0492	−0.0365	0.0270	1 + 0.127L/S
110 or 200		0.0000	0.0000	0.0000	0.0000	1.0000
201		0.0000	0.0494	0.0367	−0.0265	0.8735
202		0.0000	0.0994	0.0743	−0.0514	0.7486
203		0.0000	0.1508	0.1134	−0.0731	0.6269
204		0.0000	0.2040	0.1150	−0.0900	0.5100
205		0.0000	0.1598	0.1998	−0.1004	0.3996
206		0.0000	0.1189	0.1486	−0.1029	0.2971
207		0.0000	0.0818	0.1022	−0.0954	0.2044
208		0.0000	0.0491	0.0614	−0.0771	0.1229
209		0.0000	0.0217	0.0271	−0.0458	0.0542
210 or 300		0.0000	0.0000	0.0000	0.0000	0.0000

(*continued overleaf*)

Table A.1 ***(Continued)***

Location	M_{200}	M_{204}	M_{205}	M_{300}	$R_{200}{}^{b}$
301	0.0000	−0.0155	−0.0194	−0.0387	0.0387
302	0.0000	−0.0254	−0.0317	−0.0634	−0.0634
303	0.0000	−0.0305	−0.0381	−0.0761	−0.0761
304	0.0000	−0.0315	−0.0394	−0.0789	−0.0789
305	0.0000	−0.0295	−0.0368	−0.0737	−0.0737
306	0.0000	−0.0250	−0.0313	−0.0626	−0.0626
307	0.0000	−0.0191	−0.0238	−0.0476	−0.0476
308	0.0000	−0.0123	−0.0154	−0.0309	−0.0309
309	0.0000	−0.0057	−0.0072	−0.0143	−0.0143
310 or 400	0.0000	0.0000	0.0000	0.0000	0.0000
401	0.0000	0.0042	0.0052	0.0104	0.0104
402	0.0000	0.0069	0.0086	0.0171	0.0171
403	0.0000	0.0083	0.0103	0.0206	0.0206
404	0.0000	0.0086	0.0107	0.0214	0.0214
405	0.0000	0.0080	0.0100	0.0201	0.0201
406	0.0000	0.0069	0.0086	0.0171	0.0171
407	0.0000	0.0053	0.0066	0.0131	0.0131
408	0.0000	0.0034	0.0043	0.0086	0.0086
409	0.0000	0.0016	0.0020	0.0040	0.0040
410 or 500	0.0000	0.0000	0.0000	0.0000	0.0000
501	0.0000	−0.0012	−0.0015	−0.0031	−0.0031
502	0.0000	−0.0021	−0.0026	−0.0051	−0.0051
503	0.0000	−0.0026	−0.0032	−0.0064	−0.0064
504	0.0000	−0.0027	−0.0034	−0.0069	−0.0069
505	0.0000	−0.0027	−0.0033	−0.0067	−0.0067
506	0.0000	−0.0024	−0.0030	−0.0060	−0.0060
507	0.0000	−0.0020	−0.0024	−0.0049	−0.0049
508	0.0000	−0.0014	−0.0017	−0.0034	−0.0034
509	0.0000	−0.0007	−0.0009	−0.0018	−0.0018
510	0.0000	0.0000	0.0000	0.0000	0.0000
Area + (w/o cantilever)[c]	0.0000	0.0986	0.0982	0.0134	0.4464
Area − (w/o cantilever)[c]	0.0000	−0.0214	−0.0268	−0.1205	−0.0536
Area Net (w/o cantilever)[c]	0.0000	0.0772	0.0714	−0.1071	0.3928
Area + (cantilever)[d]	0.0000	0.0000	0.0000	0.1350	1.0 + 0.635L/S
Area − (cantilever)[d]	−0.5000	−0.2460	−0.1825	0.0000	0.0000
Area Net − (cantilever)[d]	−0.5000	−0.2460	−0.1825	0.1350	1.0 + 0.635L/S

[a] Multiply coefficients by the span length where the load is applied, that is, L on cantilever and S in the other spans.

[b] Do not multiply by the cantilever span length; use formulas or values given.

[c] Multiply moment area coefficient by S^2, reaction area coefficient by S.

[d] Multiply moment area coefficient by L^2, reaction area coefficient by L.

APPENDIX B

Transverse Deck Moments Per AASHTO Appendix A4

Table B.1 may be used in determining the design moments. This table is the same as AASHTO A4-1. The following assumptions, limitations and observations were used (AASHTO 2010):

- Moments are calculated using the equivalent strip method.
- Concrete slabs are supported on parallel girders.
- Multiple presence factors and the dynamic load allowance are included.
- See Article 4.6.2.1.6 for the distance between the centers of the girders to the location of the design sections for negative moments.
- Interpolation between the listed values may be used for distances other than those listed in Table B.1.
- Moments are applicable for decks supported on at least three girders and having a width of not less than 14.0 ft between the centerlines of the exterior girders.
- Moments represent the upper bound for interior regions and for any specific girder spacing were taken as the maximum value calculated using different number of girders.
- For each spacing and number of girders, the following two cases of overhang width were considered:

   a. Minimum total overhang width of 21.0 in. measured from the center of the exterior girder
   b. Maximum total overhang width equal to the smaller of 0.625 times the girder spacing and 6.0 ft

- A railing system width of 21.0 in. was used to determine the clear overhang width.
- Moments do not apply to the deck overhangs and the adjacent regions of the deck that need to be designed taking into account the provisions of Article A13.4.1 (vehicle crash for extreme limit state).
- Two 25-kip axles of the tandem, placed at 4.0 ft from each other, produced maximum effects under each of the tires approximately equal to the effect of the 32-kip truck axle.
- Tandem produces a larger total moment, but this moment is spread over a larger width. It was concluded that repeating calculations with a different strip width for the tandem would not result in a significant difference.

Table B.1 Maximum Live-Load Moments per Unit Width, kip-ft/ft (including multiple presence and dynamic load allowance)

S		M^+	M^-						
			Distance from CL of Girder to Design Section for Negative Moment						
ft.	in.		0 in.	3 in.	6 in.	9 in.	12 in.	18 in.	24 in.
4′	−0″	4.68	2.68	2.07	1.74	1.6	1.5	1.34	1.25
4′	−3″	4.66	2.73	2.25	1.95	1.74	1.57	1.33	1.2
4′	−6″	4.63	3	2.58	2.19	1.9	1.65	1.32	1.18
4′	−9″	4.64	3.38	2.9	2.43	2.07	1.74	1.29	1.2
5′	−0″	4.65	3.74	3.2	2.66	2.24	1.83	1.26	1.12
5′	−3″	4.67	4.06	3.47	2.89	2.41	1.95	1.28	0.98
5′	−6″	4.71	4.36	3.73	3.11	2.58	2.07	1.3	0.99
5′	−9″	4.77	4.63	3.97	3.31	2.73	2.19	1.32	1.02
6′	−0″	4.83	4.88	4.19	3.5	2.88	2.31	1.39	1.07
6′	−3″	4.91	5.1	4.39	3.68	3.02	2.42	1.45	1.13
6′	−6″	5	5.31	4.57	3.84	3.15	2.53	1.5	1.2
6′	−9″	5.1	5.5	4.74	3.99	3.27	2.64	1.58	1.28

(continued overleaf)

Table B.1 (*Continued*)

S		M^+	M^-						
			Distance from CL of Girder to Design Section for Negative Moment						
ft.	in.		0 in.	3 in.	6 in.	9 in.	12 in.	18 in.	24 in.
7′	−0″	5.21	5.98	5.17	4.36	3.56	2.84	1.63	1.37
7′	−3″	5.32	6.13	5.31	4.49	3.68	2.96	1.65	1.51
7′	−6″	5.44	6.26	5.43	4.61	3.78	3.15	1.88	1.72
7′	−9″	5.56	6.38	5.54	4.71	3.88	3.3	2.21	1.94
8′	−0″	5.69	6.48	5.65	4.81	3.98	3.43	2.49	2.16
8′	−3″	5.83	6.58	5.74	4.9	4.06	3.53	2.74	2.37
8′	−6″	5.99	6.66	5.82	4.98	4.14	3.61	2.96	2.58
8′	−9″	6.14	6.74	5.9	5.06	4.22	3.67	3.15	2.79
9′	−0″	6.29	6.81	5.97	5.13	4.28	3.71	3.31	3
9′	−3″	6.44	6.87	6.03	5.19	4.4	3.82	3.47	3.2
9′	−6″	6.59	7.15	6.31	5.46	4.66	4.04	3.68	3.39
9′	−9″	6.74	7.51	6.65	5.8	4.94	4.21	3.89	3.58
10′	−0″	6.89	7.85	6.99	6.13	5.26	4.41	4.09	3.77
10′	−3″	7.03	8.19	7.32	6.45	5.58	4.71	4.29	3.96
10′	−6″	7.17	8.52	7.64	6.77	5.89	5.02	4.48	4.15
10′	−9″	7.32	8.83	7.95	7.08	6.2	5.32	4.68	4.34
11′	−0″	7.46	9.14	8.26	7.38	6.5	5.62	4.86	4.52
11′	−3″	7.6	9.44	8.55	7.67	6.79	5.91	5.04	4.7
11′	−6″	7.74	9.72	8.84	7.96	7.07	6.19	5.22	4.87
11′	−9″	7.88	10.01	9.12	8.24	7.36	6.47	5.4	5.05
12′	−0″	8.01	10.28	9.4	8.51	7.63	6.74	5.56	5.21
12′	−3″	8.15	10.55	9.67	8.78	7.9	7.02	5.75	5.38
12′	−6″	8.28	10.81	9.93	9.04	8.16	7.28	5.97	5.54
12′	−9″	8.41	11.06	10.18	9.3	8.42	7.54	6.18	5.7
13′	−0″	8.54	11.31	10.43	9.55	8.67	7.79	6.38	5.86
13′	−3″	8.66	11.55	10.67	9.8	8.92	8.04	6.59	6.01
13′	−6″	8.78	11.79	10.91	10.03	9.16	8.28	6.79	6.16
13′	−9″	8.9	12.02	11.14	10.27	9.4	8.52	6.99	6.3
14′	−0″	9.02	12.24	11.37	10.5	9.63	8.76	7.18	6.45
14′	−3″	9.14	12.46	11.59	10.72	9.85	8.99	7.38	6.58
14′	−6″	9.25	12.67	11.81	10.94	10.08	9.21	7.57	6.72
14′	−9″	9.36	12.88	12.02	11.16	10.3	9.44	7.76	6.86
15′	−0″	9.47	13.09	12.23	11.37	10.51	9.65	7.94	7.02

APPENDIX C

Metal Reinforcement Information

Table C.1 Standard U.S. reinforcing bars

Designation Number	Nominal Diameter (in.)	Nominal Area (in.^{2})	Unit Weight (lb/ft)
2	0.250	0.05	0.167
3	0.375	0.11	0.376
4	0.500	0.20	0.668
5	0.625	0.31	1.043
6	0.750	0.44	1.502
7	0.875	0.60	2.044
8	1.000	0.79	2.670
9	1.128	1.00	3.400
10	1.270	1.27	4.303
11	1.410	1.56	5.313
14	1.693	2.25	7.650
18	2.257	4.00	13.600

Table C.2 Standard U.S. prestressing tendons

Tendon Type	Grade f_{pu} ksi	Nominal Diameter (in.)	Nominal Area (in.^{2})	Weight (lb/ft)
Seven-wire strand	270	0.375	0.085	0.29
	270	0.500	0.153	0.52
	270	0.600	0.216	0.74
Prestressing wire		0.192	0.0289	0.098
	250	0.196	0.0302	0.100
	240	0.250	0.0491	0.170
	235	0.276	0.0598	0.200
Prestressing bars (plain)	160	0.750	0.442	1.50
	160	0.875	0.601	2.04
	160	1.000	0.785	2.67
	160	1.125	0.994	3.38
	160	1.250	1.227	4.17
Prestressing bars (deformed)	157	0.625	0.28	0.98
	150	1.000	0.85	3.01
	150	1.250	1.25	4.39
	150	1.375	1.58	5.56

Table C.3 Cross-sectional area (in.^{2}) of combinations of U.S. bars of the same size

Number of Bars	Bar Number										
	3	4	5	6	7	8	9	10	11	14	18
1	0.11	0.20	0.31	0.44	0.60	0.79	1.00	1.27	1.56	2.25	4.00
2	0.22	0.39	0.61	0.88	1.20	1.57	2.00	2.54	3.12	4.50	8.00
3	0.33	0.58	0.91	1.32	1.80	2.35	3.00	3.79	4.68	6.75	12.00
4	0.44	0.78	1.23	1.77	2.41	3.14	4.00	5.06	6.25	9.00	16.00
5	0.55	0.98	1.53	2.21	3.01	3.93	5.00	6.33	7.81	11.25	20.00
6	0.66	1.18	1.84	2.65	3.61	4.71	6.00	7.59	9.37	13.50	24.00
7	0.77	1.37	2.15	3.09	4.21	5.50	7.00	8.86	10.94	15.75	28.00
8	0.88	1.57	2.45	3.53	4.81	6.28	8.00	10.12	12.48	18.00	32.00
9	0.99	1.77	2.76	3.98	5.41	7.07	9.00	11.39	14.06	20.25	36.00
10	1.10	1.96	3.07	4.42	6.01	7.85	10.00	12.66	15.62	22.50	40.00
11	1.21	2.16	3.37	4.84	6.61	8.64	11.00	13.92	17.19	24.75	44.00
12	1.32	2.36	3.68	5.30	7.22	9.43	12.00	15.19	18.75	27.00	48.00

Table C.4 Cross-sectional area per foot width (in.^{2}/ft) of U.S. bars of the same size

Bar Spacing (in.)	Bar Number								
	3	4	5	6	7	8	9	10	11
3.0	0.44	0.78	1.23	1.77	2.40	3.14	4.00	5.06	6.25
3.5	0.38	0.67	1.05	1.51	2.06	2.69	3.43	4.34	5.36
4.0	0.33	0.59	0.92	1.32	1.80	2.36	3.00	3.80	4.68
4.5	0.29	0.52	0.82	1.18	1.60	2.09	2.67	3.37	4.17
5.0	0.26	0.47	0.74	1.06	1.44	1.88	2.40	3.04	3.75
5.5	0.24	0.43	0.67	0.96	1.31	1.71	2.18	2.76	3.41
6.0	0.22	0.39	0.61	0.88	1.20	1.57	2.00	2.53	3.12
6.5	0.20	0.36	0.57	0.82	1.11	1.45	1.85	2.34	2.89
7.0	0.19	0.34	0.53	0.76	1.03	1.35	1.71	2.17	2.68
7.5	0.18	0.31	0.49	0.71	0.96	1.26	1.60	2.02	2.50
8.0	0.17	0.29	0.46	0.66	0.90	1.18	1.50	1.89	2.34
9.0	0.15	0.26	0.41	0.59	0.80	1.05	1.33	1.69	2.08
10.0	0.13	0.24	0.37	0.53	0.72	0.94	1.20	1.52	1.87
12.0	0.11	0.20	0.31	0.44	0.60	0.79	1.00	1.27	1.56
15.0	0.09	0.16	0.25	0.35	0.48	0.63	0.80	1.02	1.25
18.0	0.07	0.13	0.21	0.29	0.40	0.53	0.67	0.85	1.04

APPENDIX D

Refined Estimate of Time-Dependent Losses

This appendix is based upon [A5.9.5.4] and is detailed and of interest to those practicing prestressed concrete design or studying these systems. AASHTO (2010) is used as the basis for this appendix.

When members have unusual dimensions, level of prestressing, construction staging, or concrete constituent materials, a refined method of analysis or computer time-step methods shall be used. For nonsegmental prestressed members, estimates of losses due to each time-dependent source, such as creep, shrinkage, or relaxation, can lead to a better estimate of total losses compared with the values obtained by the approximate methods.

For segmental construction and posttensioned spliced precast girders, other than during preliminary design, prestress losses shall be determined by the time-step method, including consideration of the time-dependent construction stages and schedule shown in the contract documents. For components with combined pretensioning and posttensioning, and where posttensioning is applied in more than one stage, the effects of subsequent prestressing on the creep loss for previous prestressing shall be considered [A5.9.5.4.1].

In the refined analysis, the long-term loss Δf_{pLT} is the sum of the individual losses due to creep, shrinkage, and relaxation that occur separately before and after placement of deck concrete. This relationship is expressed as

$$\Delta f_{pLT} = \left(\Delta f_{pSR} + \Delta f_{pCR} + \Delta f_{pR1}\right)_{id} + \left(\Delta f_{pSD} + \Delta f_{pCD} + \Delta f_{pR2} - \Delta f_{pSS}\right)_{df} \quad \text{(D.1)}$$

where

Δf_{pSR} = prestress loss due to shrinkage of girder concrete between transfer and deck placement (ksi)

Δf_{pCR} = prestress loss due to creep of girder concrete between transfer and deck placement (ksi)

Δf_{pR1} = prestress loss due to relaxation of prestressing strands between transfer and deck placement (ksi)

Δf_{pR2} = prestress loss due to relaxation of prestressing strands in composite section after deck placement (ksi)

Δf_{pSD} = prestress loss due to shrinkage of girder concrete after deck placement (ksi)

Δf_{pCD} = prestress loss due to creep of girder concrete after deck placement (ksi)

Δf_{pSS} = prestress gain due to shrinkage of deck composite section (ksi)

$(\Delta f_{pSR} + \Delta f_{pCR} + \Delta f_{pR1})_{id}$ = sum of time-dependent prestress losses between transfer and deck placement (ksi)

$(\Delta f_{pSD} + \Delta f_{pCD} + \Delta f_{pR2} - \Delta f_{pSS})_{df}$ = sum of time-dependent prestress losses after deck placement (ksi)

Each of these individual time-dependent losses can be identified on the stress vs. time schematic of Figure 14.13.

The estimates of the individual losses are based on research published in Tadros et al. (2003). The approach additionally accounts for interaction between the precast and the cast-in-place concrete components of a composite member and for variability of creep and shrinkage properties of concrete by linking the loss formulas to the creep and shrinkage prediction formulas of Eqs. 13.24–13.26 [C5.9.5.4.1].

Shrinkage Loss of Girder Concrete Between Transfer and Deck Placement [A5.9.5.4.2a] Shrinkage of concrete is a time-dependent loss that is influenced by the curing method used, the volume-to-surface ratio V/S of the member, the water content of the concrete mix, the strength of the concrete at transfer, and the ambient relative humidity H. The total long-time shrinkage strain can range from 0.4×10^{-3} in./in. to 0.8×10^{-3} in./in. over the life of a member with about 90% occurring in the first year (see Fig. 13.11).

The shortening of the concrete due to shrinkage strain converts to a tensile prestress loss in the tendons when multiplied by E_p, so that the prestress loss due to shrinkage of the girder concrete between time of transfer to deck placement Δf_{pSR} shall be determined as [A5.9.5.4.2a]

$$\Delta f_{pSR} = \varepsilon_{bid} E_p K_{id} \quad \text{(D.2)}$$

in which

$$K_{id} = \frac{1}{1 + \frac{E_p}{E_{ci}} \frac{A_{ps}}{A_g} \left(1 + \frac{A_g e_{pg}^2}{I_g}\right) \left[1 + 0.7\psi_b \left(t_f, t_i\right)\right]} \quad \text{(D.3)}$$

where

ε_{bid} = concrete shrinkage strain of girder between the time of transfer and deck placement per Eq. 14.7

K_{id} = transformed section coefficient that accounts for time-dependent interaction between concrete and bonded steel in the section being considered for time period between transfer and deck placement

e_{pg} = eccentricity of strands with respect to centroid of girder (in.)

$\psi_b(t_f, t_i)$ = girder creep coefficient at final time due to loading introduced at transfer per Eq. 13.26

t_f = final age (days)

t_i = age at transfer (days)

The term K_{id} includes an "age-adjusted" effective modulus of elasticity to *transform the* section. By using the age-adjusted effective modulus of elasticity, elastic and creep strains can be combined and treated as if they were elastic deformations. The ratio of creep strain at time t_f to the elastic strain caused by the load applied at time of transfer is the girder creep coefficient $\psi_b(t_f, t_i)$. For constant sustained stress f_{ci}, the elastic-plus-creep strain is equal to

$$\left[1 + \psi_b\left(t_f, t_i\right)\right]\frac{f_{ci}}{E_{ci}} = \frac{f_{ci}}{E'_{ci}}$$

where E'_{ci} is an effective modulus of elasticity used to calculate the combined elastic-plus-creep strain for constant stress given by

$$E'_{ci} = \frac{E_{ci}}{1 + \psi_b\left(t_f, t_i\right)}$$

If the concrete stress varies with time, the elastic-plus-creep strain becomes

$$\left[1 + \chi\psi_b\left(t_f, t_i\right)\right]\frac{f_{ci}}{E_{ci}} = \frac{f_{ci}}{E''_{ci}}$$

where E''_{ci} is the age-adjusted effective modulus of elasticity of concrete for variable stress-inducing effects, such as prestress loss, defined as

$$E''_{ci} = \frac{E_{ci}}{1 + \chi\psi_b\left(t_f, t_i\right)}$$

and χ is the aging coefficient that accounts for concrete stress variability with time that ranges between 0.6 and 0.8 for precast prestressed concrete members and taken as 0.7 (Dilger, 1982). Because shrinkage is stress independent, the total concrete strain is

$$\varepsilon_{ci} = \varepsilon_{sh} + \frac{f_{ci}}{E''_{ci}}$$

Equating the change in strain in the prestressing steel $\Delta\varepsilon_p$ and the change in strain in concrete at the centroid of the prestressing steel $\Delta\varepsilon_c$ between the time of transfer and deck placement due to a change in the prestress force ΔP_p gives

$$\Delta\varepsilon_p = \Delta\varepsilon_c$$

$$\frac{\Delta P_p}{A_{ps}E_p} = \varepsilon_{bid} - \left(\frac{\Delta P_p}{E''_{ci}A_g} + \frac{\Delta P_p\, e_{pg}^2}{E''_{ci}\, I_g}\right)$$

Multiplication of the above equation by E_p and combination of terms gives

$$\frac{\Delta P_p}{A_{ps}}\left[1 + \frac{E_p A_{ps}}{E''_{ci}A_g}\left(1 + \frac{A_g e_{pg}^2}{I_g}\right)\right] = \varepsilon_{bid}E_p$$

Substituting the definition of E''_{ci} and the value of 0.7 for χ, the prestress loss due to shrinkage of girder concrete between the time of transfer and deck placement becomes [A5.9.5.4.2a]

$$\Delta f_{pSR} = \frac{\Delta P_p}{A_{ps}}$$

$$= \frac{\varepsilon_{bid}E_p}{1 + \frac{E_p}{E_{ci}}\frac{A_{ps}}{A_g}\left(1 + \frac{A_g e_{pg}^2}{I_g}\right)\left[1 + 0.7\psi_b\left(t_f, t_i\right)\right]}$$

$$= \varepsilon_{bid}E_p K_{id}$$

Creep Loss of Girder Concrete Between Transfer and Deck Placement [A5.9.5.4.2b] Creep of concrete is a time-dependent phenomenon in which deformation increases under constant stress due primarily to the viscous flow of the hydrated cement paste. Creep depends on the age of the concrete, the type of cement, the stiffness of the aggregate, the proportions of the concrete mixture, and the method of curing. The additional long-time concrete strains due to creep can be more than double the initial strain e_{ci} at the time load is applied.

The prestress loss due to creep of girder concrete between time of transfer and deck placement Δf_{pCR} shall be determined by [A5.9.5.4.2b]

$$\Delta f_{pCR} = \frac{E_p}{E_{ci}} f_{cgp}\psi_b\left(t_d, t_i\right)K_{id} \tag{D.4}$$

where

$\psi_b(t_d, t_i)$ = girder creep coefficient at time of deck placement due to loading introduced at transfer per Eq. 13.26

t_d = age at deck placement

At any time, the creep strain in the concrete can be related to the initial elastic strain by the creep coefficient. The concrete creep strain at the centroid of the prestressing steel for the time period between transfer and deck placement is

$$\varepsilon_{pc} = \frac{f_{cgp}}{E_{ci}}\psi_b\left(t_d, t_i\right)$$

Equating the change in strain in the prestressing steel $\Delta\varepsilon_p$ and the change in strain in concrete at the centroid of the prestressing steel $\Delta\varepsilon_{pc}$ between the time of transfer and deck placement due to a change in the prestress force ΔP_p gives

$$\Delta\varepsilon_p = \Delta\varepsilon_{pc}$$

$$\frac{\Delta P_p}{A_{ps}E_p} = \frac{f_{cgp}}{E_{ci}}\psi_b(t_d,t_i) - \left(\frac{\Delta P_p}{E''_{ci}A_g} + \frac{\Delta P_p}{E''_{ci}}\frac{e^2_{pg}}{I_g}\right)$$

Substituting the definition of E''_{ci} and the value of 0.7 for χ gives

$$\frac{\Delta P_p}{A_{ps}E_p} = \frac{f_{cgp}}{E_{ci}}\psi_b(t_d,t_i) - \left(\frac{\Delta P_p}{E_{ci}A_g} + \frac{\Delta P_p}{E_{ci}}\frac{e^2_{pg}}{I_g}\right) \times \left[1+0.7\psi_b(t_f,t_i)\right]$$

Multiplication of the above equation by E_p and combination of terms gives

$$\frac{\Delta P_p}{A_{ps}}\left\{1+\frac{E_pA_{ps}}{E_{ci}A_g}\left(1+\frac{A_g e^2_{pg}}{I_g}\right)\left[1+0.7\psi_b(t_f,t_i)\right]\right\} = \frac{E_p}{E_{ci}}f_{cgp}\psi_b(t_d,t_i)$$

so that the prestress loss due to creep of girder concrete between the time of transfer and deck placement becomes [A5.9.5.4.2b]

$$\Delta f_{pCR} = \frac{\Delta P_p}{A_{ps}} = \frac{E_p}{E_{ci}}f_{cgp}\psi_b(t_d,t_i)K_{id}$$

Relaxation Loss of Prestressing Strands Between Transfer and Deck Placement [A5.9.5.4.2c] Relaxation of the prestressing tendons is a time-dependent loss of prestress that occurs when the tendon is held at constant strain. The total relaxation loss Δf_{pR} is separated into two components

$$\Delta f_{pR} = \Delta f_{pR1} + \Delta f_{pR2} \tag{D.5}$$

where Δf_{pR1} is the relaxation loss between time of transfer of the prestressing force and deck placement and Δf_{pR2} is the relaxation loss after deck placement.

The prestress loss due to relaxation of prestressing strands between time of transfer and deck placement Δf_{pR1} shall be determined as [A5.9.5.4.2c]

$$\Delta f_{pR1} = \frac{f_{pt}}{K_L}\left(\frac{f_{pt}}{f_{py}} - 0.55\right) \tag{D.6}$$

where

f_{pt} = stress in prestressing strands immediately after transfer, taken not less than $0.55f_{py}$ in Eq. D.6

K_L = 30 for low-relaxation strands and 7 for other prestressing steel, unless more accurate manufacturer's data are available

Equation D.6 is appropriate for normal temperature ranges only. Relaxation losses increase with increasing temperatures [C5.9.5.4.2c].

If a strand is stressed and then held at a constant strain, the stress decreases with time. The decrease in stress is called intrinsic (part of its essential nature) relaxation loss. Strands commonly used in practice are low-relaxation strands. As a result, the relaxation prestress loss is relatively small: of the order of 1.8–4.0 ksi (Tadros et al., 2003) and the relaxation loss Δf_{pR1} may be assumed as 1.2 ksi for low-relaxation strands.

Tests by Magura et al. (1964) showed that the intrinsic relaxation varied in approximately linear manner with the log of the time t under stress. Based on their tests, Magura et al. (1964) recommended the following expression for the intrinsic relaxation of stress-relieved strands:

$$L_i = \frac{f_{pt}}{10}\log t\left(\frac{f_{pt}}{f_{py}} - 0.55\right)$$

where f_{pt} is the stress in prestressing strands immediately after transfer and t is the time under load in hours. This expression has become the standard of practice in many references. A modified version of the above equation is obtained by the substitution of t (days) $= t_d - t_i$ and the constant K'_L to give

$$L_i = \frac{f_{pt}}{K'_L}\frac{\log 24t_d}{\log 24t_i}\left(\frac{f_{pt}}{f_{py}} - 0.55\right)$$

where K'_L is 45 for low-relaxation strands and 10 for stress-relieved strands.

The relaxation loss from time of transfer to deck placement was further refined by Tadros et al. (2003) using the intrinsic relaxation loss L_i, the reduction factor φ_i due to creep and shrinkage of concrete, and the factor K_{id} to give

$$\Delta f_{pR1} = \phi_i L_i K_{id}$$

where

$$\phi_i = 1 - \frac{3(\Delta f_{pSH} + \Delta f_{pCR})}{f_{pt}}$$

which results in [C5.9.5.4.2c]

$$\Delta f_{pR1} = \left[1 - \frac{3(\Delta f_{pSH} + \Delta f_{pCR})}{f_{pt}}\right] \times \left[\frac{f_{pt}}{K'_L}\frac{\log 24t_d}{\log 24t_i}\left(\frac{f_{pt}}{f_{py}} - 0.55\right)\right]K_{id}$$

Equation D.6 is an approximation of the above formula with the following typical values assumed: t_i = 0.75 day, t_d = 120 days, $\phi_i = 0.67$, and $K_{id} = 0.8$.

Shrinkage Loss of Girder Concrete in the Composite Section After Deck Placement [A5.9.5.4.3a] The prestress loss due to shrinkage of girder concrete between time of deck placement and final time Δf_{pSD} shall be determined as [A5.9.5.4.3a]

$$\Delta f_{pSD} = \varepsilon_{bdf} E_p K_{df} \tag{D.7}$$

in which

$$K_{df} = \frac{1}{1 + \frac{E_p}{E_{ci}} \frac{A_{ps}}{A_c} \left(1 + \frac{A_c e_{pc}^2}{I_c}\right) \left[1 + 0.7\psi_b\left(t_f, t_i\right)\right]} \tag{D.8}$$

where

ε_{bdf} = shrinkage strain of girder between time of deck placement and final time per Eq. 13.24

K_{df} = transformed section coefficient that accounts for time-dependent interaction between concrete and bonded steel in the section being considered for time period between deck placement and final time

e_{pc} = eccentricity of strands with respect to centroid of composite section (in.)

A_c = area of section calculated using the net composite concrete section properties of the girder and the deck and the deck-to-girder modular ratio (in.^{2})

I_c = moment of inertia of section calculated using the net composite concrete section properties of the girder and the deck and the deck-to-girder modular ratio at service (in.^{4})

Equating the change in strain in the prestressing steel $\Delta\varepsilon_p$ and the change in strain in concrete at the centroid of the prestressing steel $\Delta\varepsilon_c$ between the time of deck placement and final time due to a change in the prestress force ΔP_p gives

$$\Delta\varepsilon_p = \Delta\varepsilon_c$$

$$\frac{\Delta P_p}{A_{ps} E_p} = \varepsilon_{bdf} - \left(\frac{\Delta P_p}{E''_{ci} A_c} + \frac{\Delta P_p}{E''_{ci}} \frac{e_{pc}^2}{I_c}\right)$$

Multiplication of the above equation by E_p and combination of terms gives

$$\frac{\Delta P_p}{A_{ps}} \left[1 + \frac{E_p A_{ps}}{E''_{ci} A_c} \left(1 + \frac{A_c e_{pc}^2}{I_c}\right)\right] = \varepsilon_{bdf} E_p$$

Substituting the definition of E''_{ci} and the value of 0.7 for χ, the prestress loss due to shrinkage of girder concrete between the time of deck placement and final time becomes [A5.9.5.4.3a]

$$\begin{aligned}\Delta f_{pSD} &= \frac{\Delta P_p}{A_{ps}} \\ &= \frac{\varepsilon_{bdf} E_p}{1 + \frac{E_p}{E_{ci}} \frac{A_{ps}}{A_c} \left(1 + \frac{A_c e_{pc}^2}{I_c}\right) \left[1 + 0.7\psi_b\left(t_f, t_i\right)\right]} \\ &= \varepsilon_{bdf} E_p K_{df}\end{aligned}$$

Creep Loss of Girder Concrete in the Composite Section After Deck Placement [A5.9.5.4.3b] The prestress loss due to creep of girder concrete between time of deck placement and final time Δf_{pCD} shall be determined as [A5.9.5.4.3b]

$$\begin{aligned}\Delta f_{pCD} = {}& \frac{E_p}{E_{ci}} f_{cgp} \left[\psi_b\left(t_f, t_i\right) - \psi_b\left(t_d, t_i\right)\right] K_{df} \\ & + \frac{E_p}{E_{ci}} \Delta f_{cd} \psi_b\left(t_f, t_d\right) K_{df} \geq 0.0\end{aligned} \tag{D.9}$$

where

Δf_{cd} = change in concrete stress at centroid of prestressing strands due to long-term losses between transfer and deck placement, combined with deck weight and superimposed loads (ksi)

$\psi_b(t_f, t_d)$ = girder creep coefficient at final time due to loading at deck placement per Eq. 13.26

The "≥0.0" in Eq. D.9 is needed because a negative value could result in some cases of partial prestressing, but Δf_{pCD} should not be taken as less than 0.0 [C5.9.5.4.3b].

The prestress loss due to the creep of girder concrete in the composite section is caused by two sources: (1) the initial prestressing force and the girder self-weight and (2) the deck self-weight and superimposed dead loads. The creep strain in the concrete at the centroid of the prestressing steel due to the first set of forces can be related to the difference in the elastic strains at final time and at time of deck placement by using the appropriate creep coefficients, that is,

$$\varepsilon_{pc1} = \frac{f_{cgp}}{E_{ci}} \left[\psi_b\left(t_f, t_i\right) - \psi_b\left(t_d, t_i\right)\right]$$

Equating the change in strain in the prestressing steel $\Delta\varepsilon_p$ and the change in strain in concrete at the centroid of the prestressing steel $\Delta\varepsilon_{pc1}$ between the time of deck placement and final time due to a change in the prestress force ΔP_p gives

$$\Delta\varepsilon_p = \Delta\varepsilon_{pc1}$$

$$\frac{\Delta P_p}{A_{ps} E_p} = \frac{f_{cgp}}{E_{ci}} \left[\psi_b\left(t_f, t_i\right) - \psi_b\left(t_d, t_i\right)\right] - \left(\frac{\Delta P_p}{E''_{ci} A_c} + \frac{\Delta P_p}{E''_{ci}} \frac{e_{pc}^2}{I_c}\right)$$

Substituting the definition of E''_{ci} and the value of 0.7 for χ gives

$$\frac{\Delta P_p}{A_{ps} E_p} = \frac{f_{cgp}}{E_{ci}} \left[\psi_b\left(t_f, t_i\right) - \psi_b\left(t_d, t_i\right)\right] - \left(\frac{\Delta P_p}{E_{ci} A_c} + \frac{\Delta P_p}{E_{ci}} \frac{e_{pc}^2}{I_c}\right) \left[1 + 0.7\psi_b\left(t_f, t_i\right)\right]$$

Multiplication of the above equation by E_p and combination of terms gives

$$\frac{\Delta P_p}{A_{ps}}\left\{1+\frac{E_p A_{ps}}{E_{ci}A_c}\left(1+\frac{A_c e_{pc}^2}{I_c}\right)\left[1+0.7\psi_b\left(t_f,t_i\right)\right]\right\} = \frac{E_p}{E_{ci}} f_{cgp}\left[\psi_b\left(t_f,t_i\right)-\psi_b\left(t_d,t_i\right)\right]$$

so that the prestress loss due to creep of girder concrete in the composite section between the time of deck placement and final time caused by initial prestressing and girder self-weight becomes

$$\Delta f_{pCR1} = \frac{\Delta P_p}{A_{ps}} = \frac{E_p}{E_{ci}} f_{cgp}\left[\psi_b\left(t_f,t_i\right)-\psi_b\left(t_d,t_i\right)\right]K_{df}$$

The change in creep strain in the concrete at the centroid of the prestressing steel due to the second set of forces can be related to the change in the elastic strains due to long-term losses between transfer and deck placement combined with deck weight on noncomposite section, and superimposed weight on composite section, that is,

$$\Delta\varepsilon_p = \Delta\varepsilon_{pc2}$$

$$\frac{\Delta P_p}{A_{ps}E_p} = \frac{\Delta f_{cd}}{E_{ci}}\psi_b\left(t_f,t_d\right)-\left(\frac{\Delta P_p}{E''_{ci}A_c}+\frac{\Delta P_p}{E''_{ci}}\frac{e_{pc}^2}{I_c}\right)$$

Substitution of the definition of E''_{ci} and the value of 0.7 for χ gives

$$\frac{\Delta P_p}{A_{ps}E_p} = \frac{\Delta f_{cd}}{E_{ci}}\psi_b\left(t_f,t_d\right) - \left(\frac{\Delta P_p}{E_{ci}A_c}+\frac{\Delta P_p}{E_{ci}}\frac{e_{pc}^2}{I_c}\right)\left[1+0.7\psi_b\left(t_f,t_i\right)\right]$$

Multiplication of the above equation by E_p and combination of terms gives

$$\frac{\Delta P_p}{A_{ps}}\left\{1+\frac{E_p A_{ps}}{E_{ci}A_c}\left(1+\frac{A_c e_{pc}^2}{I_c}\right)\left[1+0.7\psi_b\left(t_f,t_i\right)\right]\right\} = \frac{E_p}{E_{ci}}\Delta f_{cd}\psi_b\left(t_f,t_d\right)$$

so that the prestress loss due to creep of girder concrete in the composite section between the time of deck placement and final time caused by deck weight and superimposed dead loads becomes

$$\Delta f_{pCR2} = \frac{\Delta P_p}{A_{ps}} = \frac{E_p}{E_{ci}}\Delta f_{cd}\psi_b\left(t_f,t_d\right)K_{df}$$

The combined prestress loss due to creep of girder concrete in the composite section between the time of deck placement and final time is [A5.9.5.4.3b]

$$\Delta f_{pCR} = \Delta f_{pCD1}+\Delta f_{pCD2} = \frac{E_p}{E_{ci}} f_{cgp}\left[\psi_b\left(t_f,t_i\right)-\psi_b\left(t_d,t_i\right)\right]K_{df} + \frac{E_p}{E_{ci}}\Delta f_{cd}\psi_b\left(t_f,t_d\right)K_{df}$$

Relaxation Loss of Prestressing Strands After Deck Placement [A5.9.5.4.3c] The prestress loss due to relaxation of prestressing strands in composite section between time of deck placement and final time Δf_{pR2} shall be determined as [A5.9.5.4.3c]

$$\Delta f_{pR2} = \Delta f_{pR1} \tag{D.10}$$

Research indicates that about one-half of the losses due to relaxation occur before deck placement; therefore, the losses after deck placement are equal to the prior losses [C5.9.5.4.3c].

Shrinkage Gain of Deck Concrete in Composite Section After Deck Placement [A5.9.5.4.3d] The prestress gain due to shrinkage of deck composite section Δf_{pSS} shall be determined as [A5.9.5.4.3d]

$$\Delta f_{pSS} = \frac{E_p}{E_{ci}}\Delta f_{cdf}K_{df}\left[1+0.7\psi_b\left(t_f,t_d\right)\right] \tag{D.11}$$

in which

$$\Delta f_{cdf} = \frac{\varepsilon_{ddf}A_d E_{cd}}{\left[1+0.7\psi_d\left(t_f,t_d\right)\right]}\left(\frac{1}{A_c}+\frac{e_{pc}e_d}{I_c}\right) \tag{D.12}$$

where

Δf_{cdf} = change in concrete stress at centroid of prestressing strands due to shrinkage of deck concrete (ksi)
e_{ddf} = shrinkage strain of deck concrete between placement and final time per Eq. 13.24
A_d = area of deck concrete (in.^{2})
E_{cd} = modulus of elasticity of deck concrete (ksi)
e_d = eccentricity of deck with respect to the transformed net composite section, taken negative in common construction (in.)
$\psi_d(t_f, t_d)$ = creep coefficient of deck concrete at final time due to loading introduced shortly after deck placement (i.e., overlays, barriers, etc.) per Eq. 13.26

Deck shrinkage above the centroid of the composite section commonly creates prestress gain in the prestressing steel located below the centroid because the deck concrete shrinks more and creeps less than the precast girder concrete.

The shrinkage strain of deck concrete between time of deck placement and final time ε_{ddf} is determined by Eq. 13.24.

The shrinkage strain is related to an elastic-plus-creep stress through the age-adjusted effective modulus of the deck concrete, which gives

$$f_{\text{ddf}} = \varepsilon_{\text{ddf}} E''_{\text{cd}} = \frac{\varepsilon_{\text{ddf}} E_{\text{ci}}}{1 + \chi \psi_d (t_f, t_d)}$$

Multiplication of the stress by the area of deck concrete A_d gives a horizontal force P_{sd} in the deck due to shrinkage of deck concrete of

$$P_{\text{sd}} = \frac{\varepsilon_{\text{ddf}} A_d E_{\text{cd}}}{1 + \chi \psi_d (t_f, t_d)}$$

The change in the concrete stress at the centroid of the prestressing strands due to shrinkage of the deck concrete becomes

$$\Delta f_{\text{cdf}} = \frac{P_{\text{sd}}}{A_c} + \frac{P_{\text{sd}} e_d}{I_c} e_{\text{pc}}$$

Substitution of P_{sd} and the value of 0.7 for χ gives

$$\Delta f_{\text{cdf}} = \frac{\varepsilon_{\text{ddf}} A_d E_{\text{cd}}}{1 + 0.7\psi_d (t_f, t_d)} \left(\frac{1}{A_c} + \frac{e_{\text{pc}} e_d}{I_c} \right)$$

and through an age-adjusted effective modulus of the girder concrete, the stress produces a change in the concrete strain at the centroid of the prestressing steel of

$$\Delta \varepsilon_{\text{cdf}} = \frac{\Delta f_{\text{cdf}}}{E''_{c2}} = \frac{\Delta f_{\text{cdf}}}{E_c} \left[1 + \chi \psi_b (t_f, t_d) \right]$$

Equating the change in strain in the prestressing steel $\Delta \varepsilon_p$ and the change in strain in concrete at the centroid of the prestressing steel $\Delta \varepsilon_c$ between the time of deck placement and final time due to a change in the prestress force ΔP_p gives

$$\Delta \varepsilon_p = \Delta \varepsilon_c$$

$$\frac{\Delta P_p}{A_{\text{ps}} E_p} = \frac{\Delta f_{\text{cdf}}}{E_c} \left[1 + \chi \psi_b (t_f, t_d) \right] - \left(\frac{\Delta P_p}{E''_{\text{ci}} A_c} + \frac{\Delta P_p}{E''_{\text{ci}}} \frac{e_{\text{pc}}^2}{I_c} \right)$$

Substitution of the definition of E''_{ci} and the value of 0.7 for χ gives

$$\frac{\Delta P_p}{A_{\text{ps}} E_p} = \frac{\Delta f_{\text{cdf}}}{E_c} \left[1 + 0.7\psi_b (t_f, t_d) \right] - \left(\frac{\Delta P_p}{E_{\text{ci}} A_c} + \frac{\Delta P_p}{E_{\text{ci}}} \frac{e_{\text{pc}}^2}{I_c} \right) \left[1 + 0.7\psi_b (t_f, t_i) \right]$$

Multiplication of the above equation by E_p and combination of terms gives

$$\frac{\Delta P_p}{A_{\text{ps}}} \left\{ 1 + \frac{E_p A_{\text{ps}}}{E_{\text{ci}} A_c} \left(1 + \frac{A_c e_{\text{pc}}^2}{I_c} \right) \left[1 + 0.7\psi_b (t_f, t_i) \right] \right\} = \frac{E_p}{E_c} \Delta f_{\text{cdf}} \left[1 + 0.7\psi_b (t_f, t_d) \right]$$

so that the prestress gain due to shrinkage of the deck concrete in the composite section between the time of deck placement and final time becomes [A5.9.5.4.3d]

$$\Delta f_{\text{pSS}} = \frac{\Delta P_p}{A_{\text{ps}}} = \frac{E_p}{E_c} \Delta f_{\text{cdf}} K_{\text{df}} \left[1 + 0.7\psi_b (t_f, t_d) \right]$$

REFERENCES

AASHTO (2010). *LRFD Bridge Design Specifications*, 5th ed., American Association of State Highway and Transportation Officials, Washington, DC.

Dilger, W. H. (1982). "Methods of Structural Creep Analysis," in Z. P. Bazant and F. H. Wittman (Eds.), *Creep and Shrinkage in Concrete Structures*, Wiley, New York.

Magura, D. D., M. A. Sozen, and C. P. Siess (1964). "A Study of Stress Relaxation in Prestressing Reinforcement," *PCI Journal*, Vol. 9, No. 2, Mar.–Apr., pp. 13–57.

Tadros, M. K., N. Al-Omaishi, S. P. Seguirant, and J. G. Gallt (2003). *Prestress Losses in Pretensioned High-Strength Concrete Bridge Girders, NCHRP Report 496*, Transportation Research Board, Washington, DC.

APPENDIX E

NCHRP 12-33 PROJECT TEAM

John M. Kulicki, Principal Investigator
Dennis R. Mertz, Co-Principal Investigator
Scott A. Sabol, Program Officer (1992–1993)
Ian M. Friedland, Senior Program Officer (1988–1993)

Code Coordinating Committee
John M. Kulicki, Chair
John J. Ahlskog
Richard M. Barker
Robert C. Cassano
Paul F. Csagoly
James M. Duncan
Theodore V. Galambos
Andrzej S. Nowak
Frank D. Sears

Editorial Committee*
John M. Kulicki, Chair
Paul F. Csagoly
Dennis R. Mertz
Frank D. Sears

***With appreciation to Modjeski & Masters staff members:**
Diane M. Long
Scott R. Eshenaur
Chad M. Clancy
Robert P. Barrett
Donald T. Price
Nancy E. Kauhl
Malden B. Whipple
Raymond H. Rowand
Charles H. Johnson

NCHRP Panel
Veldo M. Goins, Chair
Roger Dorton
Steven J. Fenves
Richard S. Fountain
C. Stewart Gloyd
Stanley Gordon
Geerhard Haaijer
Clellon L. Loveall
Basile Rabbat
James E. Roberts
Arunprakash M. Shirole
James T. P. Yao
Luis Ybanez

TASK GROUPS

General Design Features
Frank D. Sears, Chair
Stanley R. Davis
Ivan M. Viest

Loads and Load Factors
Paul F. Csagoly, Chair
Peter G. Buckland
Eugene Buth
James Cooper
C. Allin Cornell
James H. Gates
Michael A. Knott
Fred Moses
Andrzej S. Nowak
Robert Scanlan

Analysis and Evaluation
Paul F. Csagoly, Chair
Peter Buckland
Ian G. Buckle
Roy A. Imbsen
Jay A. Puckett
Wallace W. Sanders
Frieder Seible
William H. Walker

Concrete Structures
Robert C. Cassano, Chair
John H. Clark
Michael P. Collins
Paul F. Csagoly
David P. Gustafson
Antonie E. Naaman
Paul Zia
Don W. Alden

Steel Structures
Frank D. Sears, Chair
John Barsom
Karl Frank
Wei Hsiong
William McGuire
Dennis R. Mertz
Roy L. Mion
Charles G. Schilling
Ivan M. Viest
Michael A. Grubb

Wood Structures
Andrzej S. Nowak, Chair
Baidar Bakht
R. Michael Caldwell
Donald J. Flemming
Hota V. S. Gangarao
Joseph F. Murphy
Michael A. Ritter
Raymond Taylor
Thomas G. Williamson

Bridge Railings
Ralph W. Bishop, Chair
Eugene Buth
James H. Hatton, Jr.
Teddy J. Hirsch
Robert A. Pege

Joints, Bearings and Accessories
Charles W. Purkiss, Chair
Ian G. Buckle
John J. Panak
David Pope
Charles W. Roeder
John F. Stanton

Earthquake Provisions Advisory Group
Ian Buckle, Chair
Robert Cassano
James Cooper
James Gates
Roy Imbsen
Geoffrey Martin

Aluminum Structures
Frank D. Sears, Chair
Teoman Pekoz

Foundations
J. Michael Duncan, Co-Chair
Richard M. Barker, Co-Chair

Deck Systems
Paul F. Csagoly, Chair
Barrington deVere Batchelor
Daniel H. Copeland
Gene R. Gilmore
Richard E. Klingner
Roman Wolchuk

Buried Structures
James Withiam, Chair
Edward P. Voytko

Walls, Piers and Abutments
J. Michael Duncan, Co-Chair
Richard M. Barker, Co-Chair
James Withiam

Calibration
Andrzej S. Nowak, Chair
C. Allin Cornell
Dan M. Frangopol
Theodore V. Galambos
Roger Green
Fred Moses
Kamal B. Rojiani

APPENDIX F

Live-Load Distribution—Rigid Method

The live-load distribution factor method for moment to the exterior girders is provided in [*A* 4.6.2.2.d]. Unlike the other equations and methods for other actions and locations of analysis, this article specifies the so-called *rigid method*. Here the bridge cross section is assumed to be rigid as illustrated in Figure F.1. It is possible under this assumption that the distribution factor could be greater than that based upon lever rule or formulas.

This appendix demonstrates the rigid method computation without undue distraction from the other more commonly used methods. The AASHTO requirements for the rigid method are outlined in [C4.6.2.2.2d-1], which is summarized below. The procedure parallels that commonly used for load distribution to piles topped with a rigid cap, to bolts in a group connected with a stiff plate, or shear walls joined by a rigid floor diaphragm in a building:

$$R = \left(\frac{N_L}{N_b}\right) + \frac{X_{\text{ext}} \sum\limits_{\text{trucks}} e_t}{\sum\limits_{N_b} x_i^2}$$

where:

R = reaction to the exterior girder
N_b = number of beams/girders in the bridge cross section
N_L = number of lanes loaded
x_i = location of beam i in the cross section
e_t = location of truck/lane in the cross section
X_{ext} = location of the exterior girder of interest

Consider one loaded lane positioned to the left side of the bridge:

$$\begin{aligned} g_{\text{moment}}^{\text{SE}} &= \left(\frac{N_L}{N_b}\right) + \frac{X_{\text{ext}} \sum\limits_{\text{trucks}} e_t}{\sum\limits_{N_b} x_i^2} \\ &= \left(\frac{1}{6}\right) + \frac{20\,(17)}{2\,(4^2 + 12^2 + 20^2)} \\ &= 0.470 \end{aligned}$$

The multiple presence factor is $m = 1.2$, therefore

$$mg_{\text{moment}}^{\text{SE}} = 1.2\,(0.470) = 0.564$$

Note the formula value from Example 6.2 is 0.55 lanes/girder. Thus, the rigid method controls, slightly over the formula in this case.

Next consider two trucks positioned to the left side of the bridge. The rigid method distribution factor is

$$\begin{aligned} g_{\text{moment}}^{\text{ME}} &= \left(\frac{N_L}{N_b}\right) + \frac{X_{\text{ext}} \sum\limits_{\text{trucks}} e_t}{\sum\limits_{N_b} x_i^2} \\ &= \left(\frac{2}{6}\right) + \frac{20\,(17+9)}{2\,(4^2 + 12^2 + 20^2)} \\ &= 0.780 \end{aligned}$$

The multiple presence factor for two lanes loaded is $m = 1.0$, therefore

$$mg_{\text{moment}}^{\text{SE}} = 1.0\,(0.780) = 0.780$$

The formula method of Example 6.2 for multiple lanes loaded is 0.71. The addition of a third truck combined with the multiple presence factor of $m = 0.85$ will not control over the two-lane case. The rigid method controls slightly over the formula in this case.

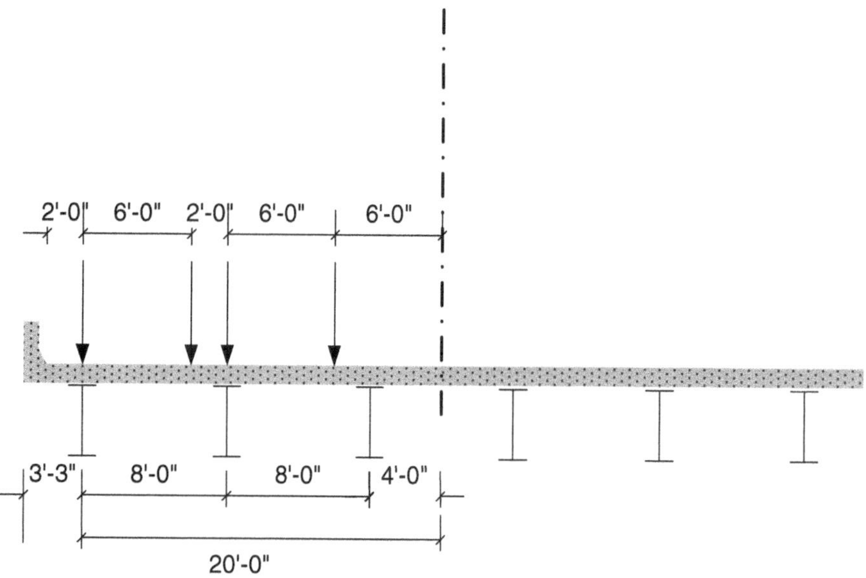

Fig. F.1 (a) Slab–girder bridge cross section (same as Example 6.2).

INDEX